S0-EMA-207

Whitworth College Library
0 0068 00118354 7

OTHER BLAINE ETHRIDGE BOOKS

Bibliography of Latin American Folklore: Tales, Myths, Festivals, Customs, Arts, Music, Magic. By Ralph Steele Boggs. Blaine Ethridge reprint of: New York--H. W. Wilson Co., 1940, 109 pp., author index.

Spanish Personal Names: Principles Governing Their Formation and Use. By Charles F. Gosnell. Blaine Ethridge reprint of: New York--H. W. Wilson Co., 1938, 112 pp., bibliography, indexes, short appendixes on alphabeting and on Portuguese names.

Reference Index to Twelve Thousand Spanish American Authors: A Guide to Bibliographies and Bio-bibliographies. By Raymond L. Grismer. Blaine Ethridge reprint of: New York--H. W. Wilson Co., 1939, 150 pp.

Diccionario de anonimos y seudonimos hispanoamericanos. By Jose Toribio Medina. Blaine Ethridge reprint of: Buenos Aires--Impr. de la Universidad, 1925, 2 volumes in 1 (as originally published). Blaine Ethridge has also reprinted the two supplements to this important work, *Errores y omisiones del Diccionario de anonimos y seudonimos hispanoamericanos de Jose Toribio Medina* and *Nueva Epanortosis al Diccionario de anonimos y seudonimos de J. T. Medina* (1928, 1929, both edited by Ricardo Victorica from notes by Medina).

Bibliography of Selected Statistical Sources of the American Nations. A Guide to the Principal Statistical Materials of the 22 American Nations. Prepared by the Inter American Statistical Institute. Blaine Ethridge reprint of: Washington--Pan American Union, 1947, xvi + 689 pp. (Updated by quarterly supplements in *Estadistica.*)

Biblioteca Andina: Essays on the Lives and Works of the Chroniclers, Or the Writers of the Sixteenth and Seventeenth Centuries who Treated of the Pre-Hispanic History and Culture of the Andean Countries. By Philip Ainsworth Means. Blaine Ethridge reprint of: New Haven--Connecticut Academy of Sciences/Yale University Press, 1928, 257 pp.

Acclimatization in the Andes. By Carlos Monge. New preface for this edition by Paul T. Baker, International Coordinator, International Biological Programme High Altitude Studies. Blaine Ethridge reprint of: Baltimore--The Johns Hopkins Press, 1948, xix + 130 pp.

History of Ancient Mexico: Anthropological, Mythological, and Social. By Bernardino de Sahagun. Translated by Fanny R. Bandelier from the Spanish version of Carlos Maria de Bustamante. Blaine Ethridge reprint of: Nashville--Fisk University Press, 1932, 315 pp., bio-bibliography.

General Censuses and Vital Statistics in the Americas. An Annotated Bibliography of the Historical Censuses and Current Vital Statistics of the 21 American Republics.... Compiled by U. S. Library of Congress. Blaine Ethridge reprint of: Washington--G.P.O., 1943, ix + 151 pp.

WRITE FOR ANNOTATED CATALOG

BLAINE ETHRIDGE--BOOKS
13977 Penrod Street, Detroit, Michigan 48223

Bibliography of Selected Statistical Sources of the American Nations

Bibliografía de Fuentes Estadísticas Escogidas de las Naciones Americanas

A guide to the principal statistical materials of the 22 American nations, including data, analyses, methodology, and laws and organization of statistical agencies.

Una guía de los principales materiales estadísticos de las 22 naciones americanas, incluyendo datos, análisis, metodología y leyes y organización de los organismos de estadística.

First edition

Primera edición

INTER AMERICAN STATISTICAL INSTITUTE, WASHINGTON, 1947

Republished by Blaine Ethridge--Books, Detroit, 1974

Dedicated to the

FIRST SESSION OF THE INTER AMERICAN STATISTICAL INSTITUTE

and the other conjoined sessions
comprising the

INTERNATIONAL STATISTICAL CONFERENCES:

United Nations, World Statistical Congress;
International Statistical Institute, Twenty-Fifth Session;
Econometric Society, Thirty-Ninth Session;
International Union for the Scientific Investigation of Population Problems;
International Association for Research in Income and Wealth (Organizing Meeting)

Washington, D. C., September 6-18, 1947

It should be noted that the material in this volume is supplemented by quarterly bibliography in *Estadistica: Journal of the American Statistical Institute*

This book was reproduced from a copy in the Sociology and Economics Department of the Detroit Public Library

Library of Congress Catalog Card Number 73-81472
International Standard Book Number 0-87917-030-1

CONTENTS—CONTENIDO

ACKNOWLEDGMENT

This publication was compiled by the Inter American Statistical Institute with the aid of statisticians and officials in charge of governmental statistical services in the 22 American nations. Preparation of the volume was under the immediate direction of Phyllis C. Andersen, head of the Bibliography Unit, during the full 3-year course of the work. Active collaborators were:

RECONOCIMIENTO

Esta publicación fué compilada por el Instituto Interamericano de Estadística con la ayuda de los estadísticos y funcionarios a cargo de servicios estadísticos gubernamentales de las 22 naciones americanas. La preparación de este volumen estuvo bajo la inmediata dirección de Phyllis C. Andersen, jefe de la Sección Bibliográfica, durante el transcurso completo de los tres años empleados en este trabajo. Fueron colaboradores activos:

Argentina

Dr. Julio E. Alizon García, Jefe, Oficina de Investigaciones Económicas, Banco Central de la República (special collaborator on economic statistics—colaborador especial en estadísticas económicas).

Ten. Col. Carlos A. Cattáneo, Director General de Estadística del Ejército Argentino, y "staff."

Prof. Robert Guye, Delegado Técnico del Instituto Interamericano de Estadística en Sur América (special collaborator on statistical standards and on statistics of labor and living conditions—colaborador especial en estandards estadísticos y en estadísticas de las condiciones de vida y de trabajo).

Bolivia

Dr. Jorge Pando Gutierrez, Director General, Dirección General de Estadística, y "staff."

Brasil

Sr. Eduardo J. Gonçalves, Instituto Brasileiro de Geografia e Estatística.

Sr. Germano G. Jardim, Estatístico do Ministério da Educação e Saúde e Assistente Técnico do Secretário Geral do Instituto Brasileiro de Geografia e Estatística (special collaborator on statistics of education—colaborador especial na estatística educacional e cultural).

Prof. Octavio A. de Moraes, Consultor Assistente, Gabinete Técnico da Comissão Censitária Nacional, e Serviço de Estatística Econômica e Financeira (special collaborator on foreign trade statistics—colaborador especial na estatística do comércio exterior).

Prof. Milton da Silva Rodrigues, Professor de Estatística, Faculdade de Filosofia, Ciências e Letras, Universidade de São Paulo (special collaborator on statistical teaching and textbooks—colaborador especial nos métodos de ensino da estatística).

Dr. M. A. Teixeira de Freitas, Presidente, Instituto Inter-Americano de Estatística, e Secretário Geral, Instituto Brasileiro de Geografia e Estatística, e "staff."

Canada

Mr. Herbert Marshall, Dominion Statistician, Dominion Bureau of Statistics, and staff.

Dr. Sedley A. Cudmore (deceased), Dominion Bureau of Statistics.

Dr. Robert M. Woodbury, Chief Statistician, International Labour Office (special collaborator on labor statistics—colaborador especial en estadísticas de trabajo).

Chile

Dr. Manuel de Viado, Jefe, Sección Estadística, Dirección General de Beneficencia y Asistencia Social.

Sr. Rafael A. Gumucio Vives, entonces Director General, Dirección General de Estadística, y "staff."

Sr. Santiago Woscoboinik, Corporación de Fomento de la Producción (special collaborator on industrial and foreign trade statistics—colaborador especial en estadísticas de industria y comercio exterior).

Colombia

Dr. Hernán Montoya (special collaborator on agricultural statistics—colaborador especial en estadísticas agrícolas).

Ing. Luis B. Ortiz C., entonces Director General, Dirección Nacional del Censo Industrial.
Sr. Eduardo Santos Rubio, Director General, Dirección Nacional de Estadística, y "staff."

Costa Rica

Sr. Ramón J. Rivera B., Director General, Dirección General de Estadística, y "staff."

Cuba

Sr. Roger Queralt Z., Director General, Dirección General de Estadística, y "staff."

Ecuador

Sr. Oswaldo Castro, Director General, Dirección General de Estadística y Censos, y "staff."
Sr. Luis Lopez Muñoz, Subdirector General, Dirección General de Estadística y Censos.
Dr. Carlos Procaccia, entonces Asesor Técnico, Dirección General de Estadística y Censos.

El Salvador

Gen. Ing. Pedro Hernández Arteaga, entonces Director General, Dirección General de Estadística, y "staff."

Guatemala

Sr. Raúl Sierra Franco, Director General, Dirección General de Estadística, y "staff."

Haiti

Mr. Marcel Fombrun, Departement Fiscal, Banque Nationale de la République d'Haïti.
Mr. René Victor, Secrétaire Général, Société d'Investigations Statistiques et Démographiques, on behalf of the Société.

Honduras

Sr. Vincente Palma, Director General, Dirección General de Estadística, y "staff."

México

Dr. Josué Sáenz, entonces Director General, Dirección General de Estadística, y "staff."

Nicaragua

Lic. Carlos Rivas O., Director General, Dirección General de Estadística, y "staff."

Panamá

Dr. Sidney Wilcox, entonces Director General, Dirección General de Estadística y Censo, y "staff."

Paraguay

Dr. Juan Kimelman, Profesor de Estadística, Universidad Nacional de Paraguay.
Dr. Carlos Alberto Soler, Director General, Dirección General de Estadística, y "staff."

Perú

Dr. Ricardo Luna Vegas, Jefe, Oficina Central de Estadística, Dirección Nacional de Estadística (special collaborator on population statistics and census methods—colaborador especial en estadísticas de población y métodos de censos).
Sr. Enrique L. Marquina P., Director, Dirección Nacional de Estadística, y "staff."
Sr. Carlos A. Uriarte, Jefe, Departamento de Estadística Demográfica, Dirección Nacional de Estadística.

República Dominicana

Sr. Vicente Tolentino Rojas, Director General, Dirección General de Estadística, y "staff."

United States

Mr. Francisco de Abrisqueta, Editor of the Journal, Inter American Statistical Institute (special collaborator on financial statistics—colaborador especial en estadísticas financieras).
Mr. James C. Capt, Director, Bureau of the Census, and staff.

Mr. Thomas F. Corcoran, Consultant, Office of the Coordinator of International Statistics, Bureau of the Census (special collaborator on economic statistics—colaborador especial en estadísticas económicas).
Dr. Hugh S. Cumming, then Director, Pan American Sanitary Bureau, and staff (special collaboration on health statistics—colaboración especial en estadísticas de sanidad).
Mr. W. Edwards Deming, Advisor on Sampling, Bureau of the Budget (special collaborator on sampling methods—colaborador especial en métodos de muestreo).
Mr. J. Edward Ely, Chief Statistician, Division of Foreign Trade Statistics, Bureau of the Census, and staff (special collaboration on foreign trade statistics—colaboración especial en estadísticas de comercio exterior).
Dr. Thomas N. E. Greville, National Office of Vital Statistics (special collaborator on life tables—colaborador especial en tablas de vida).
Mr. Newton B. Knox, Economist, Office of the Coordinator of International Statistics, Bureau of the Census (special collaborator on mining statistics—colaborador especial en estadísticas de minería).
Mr. Paul L. Koenig, Chief, Division of Agricultural Statistics, Bureau of Agricultural Economics.
Mr. L. M. LaMotte, International Business Machines Corporation, and staff (special collaboration on methods of data collection and treatment—colaboración especial en métodos de recopilación de datos y su elaboración).
Mr. Hans Muller, Statistician, National Office of Vital Statistics (special collaborator on demographic statistics—colaborador especial en estadísticas de demografía).
Mr. Eugene Owen, Bureau of Labor Statistics (special collaborator on labor and cost-of-living statistics—colaborador especial en estadísticas de trabajo y costo de la vida).
Miss Nora Powell, National Office of Vital Statistics (special collaborator on vital statistics—colaborador especial en estadísticas vitales).
Dr. Stuart A. Rice, Assistant Director of the Budget, in Charge of Statistical Standards, and staff.
Dr. Christopher Roberts, Acting Chief, International Trade Unit, Bureau of Foreign and Domestic Commerce, and staff.
Mr. L. S. Rowe (deceased), Director General, Pan American Union, and staff.
Mr. John H. Smith, Chief Statistician, Bureau of Labor Statistics (special collaborator on statistical science—colaborador especial en ciencia estadística).
Mrs. Irene Taeuber, Director, Census Library Project, Library of Congress and Bureau of the Census (special collaborator on demography—colaborador especial en demografía).
Mr. Morris B. Ullman, Division of Research and Statistics, Bureau of the Census.
Mr. J. T. Wilson, Manager, World Trade Division, International Business Machines Corporation (special collaborator on methods of data collection and treatment—colaboración especial en métodos de recopilación de datos y su elaboración).

Uruguay

Sr. Loreto M. Domínguez (special collaborator on national income—colaborador especial en renta nacional).
Dr. Eduardo Fonticelli, Director General, Dirección General de Estadística, y "staff."
Sr. Enrique Grassi Clerici, Jefe, División de Estadística del Comercio Exterior, Dirección General de Aduanas.

Venezuela

Prof. Manuel Felipe Recao, Director General, Dirección General de Estadística, y "staff."
Prof. R. de Shelly Hernández, Profesor de Estadística, Facultad de Ciencias Económicas y Sociales, Universidad Central de Venezuela.

Final review and correction of translations from one language to another were made as follows: Spanish-English and English-Spanish, by Mr. Francisco de Abrisqueta, of Colombia, now editor of the IASI quarterly journal, **Estadística**; Portuguese-English and English-Portuguese, by Mr. Germano Jardim, of the Brazilian Institute of Geography and Statistics; French-English and English-French, by Mr. O. A. Lemieux, of the Dominion Bureau of Statistics of Canada; final editorial review by Mr. Hans Muller, of the U. S. National Office of Vital Statistics.

Staff members of the Bibliography Unit at the time this volume went to press were the following: Mrs. Phyllis C. Andersen, head; Mrs. Paula Armstrong; Miss Juanita Eugere; Miss Anna Girson; Mrs. María Navas; Miss Carmen Osorno; Mrs. Mary G. Woods.

The Instituto Brasileiro de Geografia e Estatística made a substantial financial contribution toward the preparation of this volume.

La revisión y corrección final de las traducciones de un idioma a otro fueron efectuadas como sigue: Español-inglés e inglés-español por el señor Francisco de Abrisqueta, de Colombia, actualmente editor de **Estadística**, la revista trimestral del IASI; portugués-inglés e inglés-portugués, por el señor Germano Jardim, del Instituto Brasileño de Geografía y Estadística; francés-inglés e inglés-francés, por el señor O. A. Lemieux, de la Dirección General de Estadística de Canadá; la revisión editorial final, a cargo del señor Hans Muller, de la Oficina Nacional de Estadísticas Vitales de los Estados Unidos.

El personal de la Sección Bibliográfica, en la fecha de impresión de este volumen, era el siguiente: Señora Phyllis C. Andersen, jefe; señora Paula Armstrong; señorita Juanita Eugere; señorita Anna Girson; señora María Navas; señorita Carmen Osorno; y señora Mary G. Woods.

El Instituto Brasileño de Geografía y Estadística aportó una importante contribución financiera para la preparación de este volumen.

PREFACE

Statisticians and statistical agencies in all the 22 American nations have collaborated in this first edition of the Bibliography, which supersedes the preliminary working edition issued in May 1945 and circulated throughout the hemisphere for criticism and review. The present edition is the result of joint endeavor not only in content and selection of entries, but also in the formulation of purposes, criteria of selection, and organization of material.

The primary purpose of the volume is to present a list of selected references to the most important statistical sources relating to the 22 American nations, and to the body of statistical science methodology which is international in application, for consultation and use by research workers and statisticians.

Another purpose, which has developed from the interests of the collaborators, is to show the basic materials needed for collections within the statistical organizations of each nation for work in international statistics. For this purpose, publications which are considered to be minimum requirements of such a collection are indicated by asterisks.

A third important aim of the Bibliography is to present a scheme for the arrangement of statistical materials, whether publications or data. The Subject Classification Scheme, and the alphabetical index to it, are given separately as appendixes, in English and in Spanish.

Comments on the publications are given in two languages (English and one of the other three languages of the hemisphere) for maximum usefulness in the Western Hemisphere.

Approximately 2500 titles are included in this Bibliography, of which roughly 60 per cent are books and monographs, 15 per cent annuals, 15 per cent periodicals, and 10 per cent articles from periodicals.

PREFACIO

Los estadísticos y los organismos estadísticos de las 22 naciones americanas han colaborado en esta primera edición de la Bibliografía, que viene a reemplazar a la edición preliminar publicada en mayo de 1945, la cual fué distribuida en el hemisferio para su crítica y revisión. La presente edición es el resultado de una tarea conjunta, que se refiere no sólo al contenido y selección de las referencias, sino también a la formulación de los fines perseguidos, los criterios aplicados en la selección y al ordenamiento del material.

El propósito fundamental de este volumen es presentar una lista de referencias selectas de las fuentes estadísticas más importantes de los 22 países americanos y de la ciencia metodológica estadística que es de aplicación internacional, para la consulta de los investigadores y estadísticos.

Otro propósito, nacido del interés demostrado por los técnicos colaboradores, es el dar a conocer el material básico necesario para que los organismos estadísticos de cada nación dispongan de una fuente para los trabajos estadísticos de carácter internacional. Con este fin, todas aquellas publicaciones consideradas como material mínimo indispensable han sido señaladas con asteriscos.

Un tercer e importante objetivo de la Bibliografía es el de presentar un esquema para la clasificación del material estadístico, ya se refiera éste a publicaciones o a datos. El Esquema de Clasificación por Materias y su índice alfabético se anexan separadamente como apéndices, en inglés y español.

Los comentarios sobre las publicaciones se presentan en dos idiomas (inglés y una de las otras tres lenguas del hemisferio), con objeto de proporcionar el máximo aprovechamiento en el Hemisferio Occidental.

Esta Bibliografía comprende aproximadamente 2.500 títulos, de los cuales alrededor de un 60 por ciento corresponde a libros y monografías, un 15 por ciento a publicaciones anuales, otro 15 por ciento a otras publicaciones periódicas y un 10 por ciento a artículos de publicaciones periódicas.

EXPLANATORY NOTES

Plan of the bibliography

This Bibliography consists of: (1) A Main List, in which the publications are fully identified; (2) an Alphabetical Index; (3) a Subject Index; and (4) an Appendix, in English and in Spanish, showing the Subject Classification Scheme according to which the material of the volume has been classified.

Main List. The references in the Main List are arranged by country, according to the country covered in the publication rather than the country issuing it. An "International" section is given first, for publications containing international comparisons, or standards for international use, or information about international organizations dealing with statistics; and a "Non-Geographic" section last, for publications of a general methodological nature which are not about the data, methods, nor practices of any one or several countries. The countries are arranged alphabetically according to the English alphabet (hence Chile precedes Colombia instead of appearing at the end of the C's), but the names are in the language of the country (hence "República Dominicana" rather than "Dominican Republic," and "United States" rather than "Estados Unidos").

Within each country the publications are arranged by subject according to a decimal classification scheme (see appendix 1, Subject Classification Scheme for Statistical Materials). In the Main List in this volume, only the nine major headings of the Subject Classification Scheme are shown within each country, i.e., GENERAL; CENSUS; STATISTICAL SCIENCE; GEOGRAPHY; DEMOGRAPHY; ECONOMICS; HEALTH AND CULTURAL STATISTICS; GOVERNMENT AND POLITICAL STATISTICS; and LABOR AND LIVING CONDITIONS.

Articles are listed following the periodical in which they appear, in chronological order with the latest shown first.

Each reference includes: The title, subtitle (if any), author (if any), place and date of publication, and issuing agency (if any), just as

NOTAS EXPLICATIVAS

Plan de la bibliografía

Esta Bibliografía está compuesta de: (1) Una Lista Principal, en la cual las publicaciones son plenamente identificadas; (2) un Indice Alfabético; (3) un Indice de Materias; y (4) un Apéndice, en inglés y español, que contiene el Esquema de Clasificación de Materias, de acuerdo con el cual se clasificó el material de este volumen.

Lista principal. Las referencias de la Lista Principal están ordenadas por países, de acuerdo con el país al que se refiere la publicación, en vez del país editor de la publicación. Se presenta primeramente una sección "International," relativa a las publicaciones que contienen comparaciones internacionales o estandards para uso internacional, o que proporcionan información sobre las organizaciones internacionales relacionadas con la estadística; y al final una sección "Non-Geographic," para las publicaciones sobre metodología general, que no se refieren a datos estadísticos, métodos o prácticas respecto de ningún país en particular o grupo de países. Los países están ordenados alfabéticamente de acuerdo con el alfabeto inglés (por lo tanto, Chile precede a Colombia, en lugar de figurar al final de la C), pero sus nombres se expresan en el correspondiente idioma del país (en consecuencia, se escribe "República Dominicana" en lugar de "Dominican Republic," y "United States" en vez de "Estados Unidos").

Dentro de cada país, las publicaciones están ordenadas por materia, según un esquema de clasificación decimal (véase apéndice 2, Esquema de Clasificación por Materias Estadísticas). En la Lista Principal, se incluyen dentro de cada país solamente los nueve renglones mayores del Esquema por Clasificación por Materias, a saber, GENERAL; CENSOS; CIENCIA ESTADISTICA; GEOGRAFIA; DEMOGRAFIA; ECONOMIA; ESTADISTICAS DE SALUBRIDAD Y CULTURA; ESTADISTICAS GUBERNAMENTALES Y POLITICAS; y CONDICIONES DE TRABAJO Y DE VIDA.

Los artículos aparecen a continuación de la publicación periódica en la que aparecieron, en orden cronológico, figurando en primer término el de fecha más reciente.

Cada referencia incluye: El título, subtítulo (si lo hay), autor (si lo hay), lugar y fecha de publicación, y organismo editor (si lo hay), tal como aparecen en la portada de la publicación; además, el número de páginas o el número de volúmenes.

they appear on the title page of the publication; also number of pages or number of volumes.

A special feature of the Main List is the indication, by an asterisk (*) to the left of the title, of those publications considered essential for inclusion in a **basic international statistical library.**

Alphabetical index. The Alphabetical Index includes the titles, issuing agencies, authors, and subjects, in alphabetical order according to the English alphabet. Under issuing agencies and authors, the titles of their publications are given. In this index, under each subject a reference is given to a decimal subject number, indicating that publications on this subject may be found under that number in the Subject Index; and under that number within each country in the Main List. Certain special classes of publications are listed by title under the following headings: ANNUALS; BIBLIOGRAPHIES; DICTIONARIES AND GLOSSARIES; DIRECTORIES; LAWS, STATUTES, ETC.; NOMENCLATURES; ORGANIZATION CHARTS (of statistical organizations); PERIODICALS; SERIES.

Subject index. The Subject Index is the reverse of the Main List in arrangement. In it, publications are listed by **subject,** and, within each subject, by **country.** However, the Subject Index shows the publications on any given subject much more fully than the 24 sections (international, 22 countries, and non-geographic) in the Main List. In the Main List, a publication appears only once, under its primary subject, whereas in the Subject Index a publication may appear under several different subjects. Also, articles in periodicals are shown in the Main List following the periodical, hence are under the general subject of the periodical (this was done because of the impossibility of separating articles from their periodicals on the shelves in a library); whereas in the Subject Index articles are listed under the subjects to which they apply.

A publication is listed under the most specific subject which it covers, i.e., the finest subdivision in the Subject Index. For example, a book which is about telephones and telegraphs is listed under "ECONOMICS, Transport and

Una característica especial de la Lista Principal es la indicación por medio de un asterisco (*) colocado a la izquierda del título, de aquellas publicaciones consideradas esenciales para su inclusión en una **biblioteca estadística internacional básica.**

Indice alfabético. El Indice Alfabético incluye los títulos, organismos editores, autores, y materias, en orden alfabético según el alfabeto inglés. Los títulos de las publicaciones se dan, además, bajo los nombres del respectivo organismo editor y del autor. En este índice, bajo cada materia, se da una referencia del número decimal de la materia, el cual indica que bajo tal número, dicha materia puede ser encontrada tanto en el Indice de Materias, como dentro de cada país en la Lista Principal. Ciertas clases especiales de publicaciones están enumeradas por título bajo los siguientes encabezamientos: ANNUALS (publicaciones anuales); BIBLIOGRAPHIES (bibliografías); DICTIONARIES AND GLOSSARIES (diccionarios y glosarios); DIRECTORIES (directorios); LAWS, STATUTES, ETC. (leyes, estatutos, etc.); NOMENCLATURES (nomenclaturas); ORGANIZATION CHARTS (esquemas de organización de organismos estadísticos); PERIODICALS (publicaciones periódicas); SERIES (publicaciones en series).

Indice de materias. El Indice de Materias es el reverso de la Lista Principal, en cuanto a su ordenamiento. En él, las publicaciones están clasificadas por **materias,** y, dentro de cada materia, por **países.** Sin embargo, el Indice de Materias da a conocer las publicaciones sobre una determinada materia de manera mucho más completa que las 24 secciones (internacional, 22 países y no-geográfica) de la Lista Principal. En esta última, la publicación sólo aparece una vez, bajo la materia principal correspondiente, mientras que en el Indice de Materias, élla puede aparecer bajo varias diferentes materias. Asimismo, en la Lista Principal figuran los artículos de publicaciones periódicas a continuación de la respectiva publicación, encontrándose, por lo tanto, bajo la materia general de la publicación (se procedió así, debido a la imposibilidad de separar en los estantes de una biblioteca los artículos de sus respectivas publicaciones); mientras que en el Indice de Materias los artículos están incluídos bajo las materias a que específicamente se refieren.

Una publicación se clasifica de acuerdo con la materia más específica de que trata, es decir, según la subdivisión menor del Indice de Materias. Por ejemplo, un libro acerca de teléfonos y telégrafos se incluye bajo "ECONOMIA, Transporte y comunicaciones, Comuni-

communication, Communication, Telephone and telegraph"; one which covers not only telephones and telegraphs but also other phases of transport and communication is listed under "ECONOMICS, Transport and communication, General"; one which covers the entire economic life of a country, including in all probability information on telephones and telegraphs in the country, is listed under "ECONOMICS, General, General." This means that the reader wishing to find **all** the information about a specific subject such as telephones and telegraphs must look:

(a) First under the smallest applicable subheading, for publications which are about that subject only;

(b) Then under the larger general heading above it, for publications which cover that subject and also closely related subjects;

(c) Then under the main heading for books which cover that subject and also many other generally related subjects;

(d) Then, in addition, the user must look under the largest "general" section, "GENERAL," for compendiums and general reference books which cover all fields of statistics. The user is particularly directed to 1400-1499, "GENERAL, Data compilations" for statistical yearbooks and general periodical statistical publications.

Appendix 1, subject classification scheme and alphabetical index. The Subject Classification Scheme for Statistical Materials, shown separately at the end of the volume, is the basis of the arrangement of references in the volume. It has been developed in the Inter American Statistical Institute to answer the need for classification of statistical publications, data, forms, and other materials.

Indexing identification

Each publication has a call number composed of the country abbreviation, the decimal subject number, and a "Cutter" number (composed of the initial letter of the title, and a distinguishing number). The call number is used for indexing identification. For example, the identifying call number for the **Pan American Yearbook** is "Int-**1410**.P5," and for the **Canada Year Book,** "Can-**1410**.C7" (1410 being the

caciones, Teléfonos y telégrafos;" otro que trate, no sólo de teléfonos y telégrafos, sino también de otros aspectos del transporte y las comunicaciones, aparecerá bajo "ECONOMIA, Transporte y comunicaciones, General"; uno que cubra completamente la vida económica de un país, que con toda probabilidad incluye informaciones sobre los teléfonos y telégrafos del país, se clasificará bajo "ECONOMIA, General, General." Esto significa que el lector que desee encontrar **toda** la información acerca de una materia específica tal como teléfonos y telégrafos, deberá buscar:

(a) Primero, bajo la menor de las subdivisiones, referente a las publicaciones sobre dicha materia, solamente;

(b) A continuación, bajo el grupo general más amplio que siga, para las publicaciones que cubren aquel tema y materias estrechamente relacionadas con él;

(c) Luego, bajo el grupo principal, referente a los libros acerca de esa materia y de muchas otras generalmente relacionadas con el tema;

(d) Por último, el interesado deberá buscar en la sección "general" más amplia, es decir, en la sección "GENERAL," que incluye los compendios y libros de referencia general, los cuales cubren todos los campos estadísticos. En particular, el consultante deberá examinar el grupo 1400-1499, GENERAL, Compilaciones de datos," para los anuarios estadísticos y publicaciones periódicas de estadística general.

Apéndice 2, esquema de clasificación por materias, e índice alfabético. El Esquema de Clasificación por Materias Estadísticas, que se incluye separadamente al final de este volumen, es la base de la clasificación de las referencias de esta obra. Este esquema ha sido proyectado por el Instituto Interamericano de Estadística para responder a la necesidad de clasificar las publicaciones estadísticas, datos, formularios y otros materiales.

Identificación en los índices

Cada publicación tiene un número de identificación, compuesto de la abreviatura del nombre del país, del número decimal de la materia y de un número "Cutter" (compuesto de la letra inicial del título y de un número distintivo). El número de identificación se emplea para encontrar dentro de la Lista Principal las publicaciones incorporadas en los índices alfabético y de materias. Por ejemplo, el número de identificación para encontrar el **Pan American Yearbook** es "Int-**1410**.P5," y para el **Canada Year Book,** "Can-**1410**.C7"

decimal subject number for annual compilations of general data).

Criteria of selection

A high degree of selectivity has been exercised in choosing the references for inclusion in this volume, because: (a) Any attempt to obtain broad coverage of statistical subject matter necessarily calls for selectivity, to hold the undertaking within the realm of the possible; (b) increase in the number of listings would have made the volume decreasingly less useful as a handy reference source—i.e., a small, compact volume containing basic source references is of greater usefulness than one which is bulkier and quantitatively more comprehensive; (c) the limited funds at the disposal of the Institute would have precluded any other course, even had there been the desire to proceed differently.

The following criteria of selection have, in general, been used:

1. Publications must be of statistical interest; i.e., must have a direct bearing on the interests being served or investigated by statisticians, statistical organizations, or others professionally concerned with the use of statistical data.
2. Publications must relate to the American nations.
3. Generalized and compact data, methodological and reference sources, and authoritative analytical studies are given preference over other material.
4. National and international publications are given preference over local (state or municipal); in general, local material is included only in the absence of national material on the same subject.
5. Current material is given preference over older material; i.e., age has been a factor in selection. There are, of course, important exceptions, such as older volumes containing a panoramic picture covering a long period of years, methodological treatises the value of which has not decreased with age, etc.
6. "Methods and procedures" material. Many of the references included in this volume relate to collections of forms, instructions, and other fragmentary material of great value in

(siendo 1410 el número decimal de la materia para compilaciones anuales de datos generales).

Criterio de selección

Se ha aplicado un alto grado de selección para escoger las referencias que se incluyen en este volumen, porque: (a) Cualquier intento para lograr cubrir un amplio campo de materias estadísticas, exige necesariamente una selección, a fin de que la obra pueda encuadrarse dentro de lo realmente factible; (b) un aumento en el número de inclusiones habría significado una disminución del grado de utilidad del volumen como fuente manuable de referencias —es decir, un volumen pequeño y compacto, que contenga referencias de fuentes básicas, es de mucho más utilidad que uno más abultado y cuantitativamente más extenso; (c) los limitados fondos de que dispone el Instituto habrían impedido seguir otro camino, aun si el deseo hubiese sido de proceder de forma diferente.

En general, se ha seguido el siguiente criterio de selección:

1. Las publicaciones deben ser de interés estadístico, esto es, deben tener una relación directa con las materias primordiales que atienden o investigan los estadísticos, los organismos estadísticos, u otros que profesionalmente están interesados en el uso de los datos estadísticos.
2. Las publicaciones deben referirse a las naciones americanas.
3. Se da preferencia sobre cualquier otro material a los datos estadísticos generalizados y compactos, fuentes de metodología y de referencia y a estudios analíticos de reconocida autoridad.
4. Se da preferencia a obras de carácter nacional e internacional sobre las de carácter meramente local (estatales o municipales); en general, se incluye material de carácter local, solamente en ausencia de material nacional sobre el mismo tema.
5. Se da preferencia a publicaciones de actualidad sobre el material antiguo; es decir, un factor en la selección ha sido la antiguedad del material. Hay, naturalmente, importantes excepciones, tales como volúmenes antiguos que contienen un cuadro panorámico cubriendo un largo período de años, tratados de metodología cuyo valor científico no ha decrecido con el tiempo, etc.
6. Material de "métodos y procedimientos." Muchas de las referencias incluídas en este volumen bibliográfico se refieren a juegos de formularios, instrucciones y otro material fragmentario, de gran valor para los estudios sobre

studies of the methods and procedures used in the collection and processing of national statistics. This material has been sent to the Inter American Statistical Institute over a period of time by statistical agencies throughout the hemisphere, for use in connection with surveys conducted by the Institute in various subject fields.

Sources from which the references were drawn

The titles in this Bibliography represent a selection of about one-third of the total references contained in the IASI "Statistical Source File." This is a reference file of titles of statistical publications, rather than a catalog of volumes in the possession of the Institute. Actually, only a relatively small percentage of the publications listed in the file are on the shelves of the Institute. The file was started from a depository set of catalog cards for statistical publications which the U. S. Library of Congress presented to the Institute in 1942. It has been added to and maintained through a number of different channels, primarily the following:

1. Publications of a statistical nature coming to the Institute from the 22 American nations.
2. References and bibliographies in statistical materials coming under the review of Institute staff members and associates.
3. Unpublished statistical materials coming to the Institute from its collaborators and correspondents, including bibliographies compiled for special studies.
4. References to statistical materials gleaned from the current accessions of the Pan American Union Library, U. S. Library of Congress, and other libraries.
5. The collaboration obtained in the present publication did much to strengthen the file itself, since many more references were suggested than could possibly be included in the present volume.

Examination of publications

Most of the publications listed in the Bibliography were examined by members of the Bibliography Unit of the Institute, and the references prepared from information given on the title page, and from an examination of the contents. However, in some cases references

metodología y procedimientos empleados en la recolección y elaboración de las estadísticas nacionales. Este material ha sido remitido al Instituto Interamericano de Estadística durante cierto período de tiempo por los organismos estadísticos del hemisferio para su uso en los estudios dirigidos por el Instituto en diferentes especialidades.

Fuentes de las que se extrajo las referencias

Los títulos contenidos en esta Bibliografía representan una selección de aproximadamente un tercio del total de las referencias contenidas en el "Archivo de Fuentes Estadísticas" del IASI. Este es un archivo de referencias de títulos de publicaciones estadísticas, más bien que una colección de volúmenes en poder del Instituto. En realidad, sólo un porcentaje relativamente pequeño de las publicaciones catalogadas se encuentran en los estantes del Instituto. El archivo fué iniciado a base de un juego de tarjetas de catalogación de publicaciones estadísticas, que la Biblioteca del Congreso de los Estados Unidos le proporcionó al Instituto en el año 1942. Desde entonces, el archivo ha sido aumentado y mantenido a través de diferentes conductos, principalmente los siguientes:

1. Publicaciones de carácter estadístico que llegan al Instituto desde las 22 naciones americanas.
2. Referencias y bibliografías contenidas en materiales estadísticos que pasan por la revisión del personal del Instituto y de personas vinculadas a él.
3. Materiales estadísticos no publicados que recibe el Instituto de sus colaboradores y corresponsales, incluyendo bibliografías compiladas para estudios especiales.
4. Referencias de materiales estadísticos escogidas de entre las publicaciones que llegan a la Biblioteca de la Unión Panamericana, Biblioteca del Congreso de los Estados Unidos y a otras bibliotecas.
5. La colaboración que se obtuvo para la presente publicación, resultó, además, de gran provecho para el mejoramiento del archivo en sí mismo, puesto que fueron muchas las referencias que se sugirieron y que no fué posible incluir en este volumen.

Examen de las publicaciones

La mayor parte de las publicaciones contenidas en la Bibliografía fueron examinadas por el personal de la Sección Bibliográfica del Instituto, y las referencias preparadas a base de las informaciones dadas en las portadas y del examen del contenido de las obras. Sin em-

sent the Institute by collaborators in other countries proved to be unavailable in Washington, and in other cases (especially with respect to United States publications) time pressures precluded a check of the information given in some of the older entries. For every publication which was not examined by the Bibliography Unit, the reference includes the word "Unexamined" or a translation of it ("No examinado," "Não examinado," "Pas examiné").

Annotations

Nature of comments. The comments are, in general, brief and **factual** statements about the contents of the publications. Occasionally, however, pertinent history is included, and some comments which are quoted from other sources do contain **opinions** as to value or use. In many cases the comments were prepared from the table of contents, with some deletion of items and with some addition of parenthetical explanations.

Language of comments. All comments are given first in English and then in one other language, depending on the language of the publication. The following practice has been followed: Publications on Canada and Haiti, and publications of those international organizations which use English and French as their official languages (including League of Nations, International Institute of Agriculture, and International Statistical Institute) are annotated in English and French; publications on Brazil are annotated in English and Portuguese; all other publications are annotated in English and Spanish.

Information which is shown in the title or subtitle of a publication is given as the first sentence of the comment in the other language. If this information furnishes sufficient explanation of the contents or purpose of the publication, there is no other annotation.

Abbreviations used

For countries (including "International" and "Non-Geographic"):

International	Int
Argentina	Arg
Bolivia	Bol
Brazil	Bras
Canada	Can
Chile	Chile
Colombia	Col

bargo, en algunos casos resultó que las referencias enviadas al Instituto por colaboradores de otros países no se encontraban estar disponibles en Wáshington, y, en otros casos (especialmente con respecto a publicaciones de los Estados Unidos) la presión del factor tiempo impidió la verificación de la información dada en algunas de las catalogaciones más antiguas. Con respecto a cada una de las publicaciones que no pudieron ser examinadas por el personal de la Sección Bibliográfica, la referencia contiene la palabra "Unexamined" o su traducción ("No examinado," "Não examinado," "Pas examiné").

Anotaciones

Naturaleza de los comentarios. Los comentarios son, en general, explicaciones breves y **concretas acerca del contenido** de las publicaciones. Sin embargo, ocasionalmente se incluyen algunos antecedentes de la obra y se citan algunos comentarios de otras fuentes que contienen **opiniones** sobre su valor y utilidad. En muchos casos los comentarios fueron preparados con base en el índice del contenido del volumen, suprimiéndose algunos ítems y agregándose algunas explicaciones entre paréntesis.

Idioma de los comentarios. Todos los comentarios se presentan primeramente en inglés y luego en una de las otras lenguas, de acuerdo con el idioma de la publicación. Se hân seguido al respecto las siguientes prácticas: Las publicaciones del Canadá y Haití y de aquellas organizaciones internacionales que empleen el inglés y francés como idiomas oficiales (incluyendo la Sociedad de las Naciones, Instituto Internacional de Agricultura e Instituto Internacional de Estadística) se comentan en inglés y francés; las de Brasil en inglés y portugués; toda otra publicación es comentada en inglés y español.

La información de los títulos y sub-títulos de las publicaciones aparecen como primera frase en el comentario en el otro idioma. Si esta información proporciona una explicación suficiente sobre el contenido o el objeto de la publicación, no se agregan más anotaciones.

Abreviaturas usadas

Para los países (incluyendo "Internacional" y "No-Geográfico"):

Internacional	Int
Argentina	Arg
Bolivia	Bol
Brasil	Bras
Canadá	Can
Colombia	Col
Costa Rica	CR

Costa Rica	CR
Cuba	Cuba
Dominican Republic	RepDom
Ecuador	Ec
El Salvador	ElSal
Guatemala	Guat
Haiti	Haiti
Honduras	Hond
Mexico	Méx
Nicaragua	Nic
Panama	Pan
Paraguay	Par
Peru	Perú
United States	USA
Alaska	USA-Alsk
Puerto Rico	USA-PR
Uruguay	Ur
Venezuela	Ven
Other countries	
France	Fr
Great Britain	GrBr
West Indies	WI
Non-Geographic	NonG

For publications from which annotations are quoted:

General Censuses and Vital Statistics in the Americas	Gen.Cen.V.S.A.
Handbook of Latin American Studies	HBLAS
Population Index	Pop.Ind.
Statistical Activities of the American Nations	Stat.Act.

For imprints:

Compañía	Cía.
Company	Co.
Gráfico	Gráf.
Imprenta	Imp.
Litografía	Lito.
Talleres	Tall.
Serviço Gráfico do Instituto Brasileiro de Geografia e Estatística	Serviço Gráf.do IBGE
United States Government Printing Office	U.S.Govt.Print. Off.

For organizations:

Inter American Statistical Institute	IASI

Time to which the publication relates

The cut-off date established for this volume was December 31, 1946. Thus, publications carrying a later imprint date are not included. Many carrying an earlier imprint date are not included, of course, because not received in time.

Cuba	Cuba
Chile	Chile
Ecuador	Ec
El Salvador	ElSal
Estados Unidos	USA
Alaska	USA-Alsk
Puerto Rico	USA-PR
Guatemala	Guat
Haití	Haiti
Honduras	Hond
México	Méx
Nicaragua	Nic
Panamá	Pan
Paraguay	Par
Perú	Perú
República Dominicana	RepDom
Uruguay	Ur
Venezuela	Ven
Otros países	
Francia	Fr
Gran Bretaña	GrBr
Antillas	WI
No-Geográfico	NonG

Para las publicaciones de las que se han citado anotaciones:

General Censuses and Vital Statistics in the Americas	Gen.Cen.V.S.A.
Handbook of Latin American Studies	HBLAS
Population Index	Pop.Ind.
Statistical Activities of the American Nations	Stat. Act.

Para pies de imprentas:

Compañía	Cía.
Company	Co.
Gráfico	Gráf.
Imprenta	Imp.
Litografía	Lito.
Talleres	Tall.
Serviço Gráfico do Instituto Brasileiro de Geografia e Estatística	Serviço Gráf.do IBGE
United States Government Printing Office.	U.S.Govt.Print.Off.

Para organizaciones:

Instituto Interamericano de Estadística.	IASI

Fecha a la que se refiere esta publicación

La fecha límite establecida para este volumen fué el 31 de diciembre de 1946. En consecuencia, las publicaciones impresas con posterioridad a tal fecha no están incluídas. De otro lado, muchas publicaciones editadas antes de esa fecha no se incluyeron por no haberse recibido a tiempo.

MAIN LIST—LISTA PRINCIPAL

(By country and numerically by subject)

(Por país y en orden numérico por materia)

INTERNATIONAL—INTERNACIONAL

1000-1999 GENERAL

Int-**1200**.B3 **Bibliographic Series of the Pan American Union.** Starting date? Irregular. Washington, D.C.

Serie bibliográfica de la Unión Panamericana.

Int-**1200**.B5 1942 **A Bibliography of Latin American Bibliographies.** By Cecil Knight Jones. Second edition revised and enlarged by the author with the assistance of James A. Granier. Washington, D.C., U.S. Govt. Print. Off., 1942. U.S. Library of Congress. Hispanic Foundation. 311 p. Latin American series, no. 2.

1922 edition—edición de 1922: **Hispanic American Bibliographies.**

"Publications Relating to Latin America Issued by the Library of Congress," p. 309-311.

Bibliografía de bibliografías latinoamericanas. "Publicaciones Relativas a la América Latina Publicadas por la Biblioteca del Congreso" (traducción del título), p. 309-311.

Int-**1200**.E5 **Executive Agreement Series.** Starting date? Washington, D.C., U.S. Department of State.

No. 103, 108, 112, 123, 134, 162, 171, 176, 194, 210, 230, 242, 243, 297, and 301 deal with the exchange of official publications between the United States and Peru, Mexico, Chile, Cuba, Argentina, Nicaragua, Brazil, Honduras, Haiti, El Salvador, Bolivia, Panama, Dominican Republic, and Paraguay, and include lists of publications exchanged.

Serie de acuerdos ejecutivos. No. 103, 108, 112, 123, 134, 162, 171, 176, 194, 210, 230 242, 243, 297, y 301 tratan del intercambio de publicaciones oficiales entre los Estados Unidos y el Perú, México, Chile, Cuba, Argentina, Nicaragua, Brasil, Honduras, Haiti, El Salvador, Bolivia, Panamá, República Dominicana y Paraguay, e incluyen listas de las publicaciones canjeadas.

Int-**1200**.H4 **Handbook of Latin American Studies.** A selective guide to the material published . . . on anthropology, archives, art, economics, education, folklore, geography, government, history, international relations, labor and social welfare, language and literature, law, libraries, music, and philosophy. Edited for the Joint Committee on Latin American Studies of the National Research Council, the American Council of Learned Societies, and the Social Science Research Council, by Miron Burgin. 1934- . Annual. Cambridge, Massachusetts.

Manual de estudios latinoamericanos. Guía selectiva de materiales publicados sobre antropología, archivos, bellas artes, economía, educación, folklore, geografía, gobernación, historia, relaciones internacionales, trabajo y seguridad social, idiomas y literatura, derecho, bibliotecas, música, y filosofía.

1937, Phelps "Sources of Current Economic Information of Latin America." By D. M. Phelps. **Handbook of Latin American Studies,** p. 573-594.

A list and description of economic publications on Latin America, grouped under the following headings: Government publications; organizations created by governments for operation or control of industries; bank publications; private, non-institutional reviews; Latin American centers abroad; chamber of commerce and trade association publications; university; miscellaneous.

Fuentes de información económica actual sobre la América Latina. Lista y descripción de publicaciones económicas sobre la América Latina, agrupadas según los siguientes rubros: Publicaciones gubernamentales; organismos instituidos por los gobiernos para el manejo o control de establecimientos industriales; publicaciones bancarias; revistas particulares no institucionales; centros latinoamericanos en el extranjero; publicaciones de cámaras de comercio y de asociaciones comerciales; universidad; miscelánea.

Int-**1200**.I4 1943 **Index of Latin American Publications.** Washington, D.C., 1943. American Republics Unit, Bureau of Foreign and Domestic Commerce. American Republics Section, International Economics and Statistics Unit, Bureau of Foreign and Domestic Commerce. 150 p. Hectographed.

Indice de publicaciones de la América Latina.

Int-**1200.**I6 **Inter-American Bibliographical Review.** Quarterly publication of the Inter-American Bibliographical and Library Association. 1941- . Washington, D.C.

A professional publication containing articles on technical aspects of library work, dedicated to promoting eventual uniformity of standards and methods of library science throughout the nations of the Western Hemisphere; devotes a section to reviewing Latin American publications, and publications and articles dealing with various topics relating to Latin America.

Revista bibliográfica interamericana, publicación trimestral de la Asociación Interamericana Bibliográfica y Bibliotecaria. Una publicación profesional que contiene exposiciones sobre los aspectos técnicos del trabajo bibliotecario, dedicada a fomentar la eventual uniformidad de los estandards y métodos de la ciencia bibliotecaria en todas las naciones del Hemisferio Occidental; una parte de la revista se dedica al examen crítico de publicaciones latinoamericanas, así como de publicaciones y artículos que tratan de varios tópicos relacionados con la América Latina.

Int-**1200.**L3 1944 **Latin American Periodicals Currently Received in the Library of Congress and in the Library of the Department of Agriculture.** Charmion Selby, Editor. Washington, D.C., U.S. Govt. Print. Off., 1944. U.S. Library of Congress. Hispanic Foundation. 249 p. Latin American series, no. 8.

A selected and annotated list.

Publicaciones periódicas latinoamericanas, que se reciben actualmente en la Biblioteca del Congreso y en la Biblioteca del Departamento de Agricultura. Una lista escogida y anotada.

Int-**1200.**L5 **Latin American Series of the Library of Congress.** Starting date? Irregular. Washington, D.C. Library of Congress. Hispanic Foundation.

Series sobre la América Latina publicadas por la Biblioteca del Congreso.

Int-**1200.**P2 **The Pan American Book Shelf.** 1938- . Monthly. Washington, D.C. , Pan American Union. Columbus Memorial Library.

Books received in the Columbus Memorial Library.

Los libros recibidos en la Biblioteca Memorial Colón.

Int-**1200.**P5 **Publications Issued by the League of Nations.** 1935- . Annual. Geneva.

Began with **General Catalogue** in 1935, continued as yearly supplements.

Publications de la Société des Nations. Commencé avec le catalogue général en 1935 et continué comme supplément annuel.

Int-**1200.**S6 1815-1931 **Serial Publications of Foreign Governments 1815-1931.** Compiled and edited by Winifred Gregory, under the auspices of the American Council of Learned Societies, American Library Association, National Research Council. New York The H. W. Wilson Company, 1932. 720 p.

Publicaciones en series de gobiernos extranjeros, 1815-1931.

Int-**1200.**U5 1943 **Ulrich's Periodicals Directory. Directorio de Publicaciones Periódicas Ulrich.** A classified guide to a selected list of current periodicals. Guía clasificada y seleccionada de publicaciones de la actualidad. Edited by Carolyn F. Ulrich. Inter-American Edition. Edición Interamericana. New York, R. R. Bowker Company, 1943. 314 p.

Int-**1300.**A4 1946 **Acta de Sesión No. 19 Celebrada el 1° de Agosto de 1946 (Aprobada el 5 de Septiembre de 1946).** Unión Panamericana. Consejo Interamericano Económico y Social. 30 p. (aprox.) Mimeografiado.

Act of session no. 19 of the Inter-American Economic and Social Council of the Pan American Union, held August 1, 1946.

1946, Woscoboinik Anexo "B." "Memorándum de las Actividades del IASI en el Campo de Estadística de Comercio Exterior." Trabajos efectuados y planeados en los problemas de clasificación de las mercancías sobre bases uniformes, y de estandardización de definiciones y prácticas empleadas en el tratamiento de las estadísticas de comercio exterior. Preparado por Santiago Woscoboinik.

Memorandum on activities of Inter American Statistical Institute in field of foreign trade statistics: Work done and proposed on problems of classification of commodities on uniform bases, and of standardization of definitions and practices used in treatment of foreign trade statistics.

*Int-**1300.**A5 1940.3 **Proceedings of the Eighth American Scientific Congress.** Held in Washington, May 10-18, 1940, under the auspices of the government of the United States of America. Washington, D.C., Department of State, 1941-43. Eighth American Scientific Congress. 12 vol.: Vol. I, **Organization and Activities**; vol. II, **Anthropological Sciences**; vol. III, **Biological Sciences**; vol. IV, **Geological Sciences**; vol. V, **Agriculture and Conservation**; vol. VI,

Public Health and Medicine; vol. VII, **Physical and Chemical Sciences**; vol. VIII, **Statistics**; vol. IX, **History and Geography**; vol. X, **International Law, Public Law, and Jurisprudence**; vol. XI, **Economics and Sociology**; vol. XII, **Education.**

Also published in Spanish.

Publicado también en español.

1940.3, Alanís Patiño, a "Bases para el Mejoramiento de las Estadísticas Escolares en los Países de América." Por Emilio Alanís Patiño. Vol. VIII, p. 259-262.

Bases for improvement of educational statistics in the American countries.

1940.3, Alarcón Mendizábal "Cuestiones Elementales en los Censos Agrícolas Aplicables en los Países de Latino-América." Por Adolfo Alarcón Mendizábal. Vol. XIII, p. 145-151.

Fundamental questions applicable in agricultural censuses in Latin American countries.

1940.3, Arca Parró, b "Problemas y Soluciones para el Censo Demográfico Peruano de 1940." Por Alberto Arca Parró. Vol. VIII, p. 17-27.

Problems and solutions of Peruvian demographic census of 1940.

1940.3, Basabe Castellanos "Organización de la Estadística Económica en el Uruguay." Por Carlos Basabe Castellanos. Vol. VIII, p. 161-168.

Organization of economic statistics in Uruguay.

1940.3, Benavides "La Sociedad Mexicana de Geografía y Estadística." Por Francisco de A. Benavides. Vol. VIII, p. 343-344.

The Mexican Society of Geography and Statistics.

1940.3, Bergström Lourenço "A População Escolar e a Taxa de Analfabetos nas Estatísticas Educacionais Americanas." Por Manoel Bergström Lourenço. Vol. VIII, p. 315-319.

The school-age population and rate of illiteracy in inter-American statistics of education.

1940.3, Brannon "Desarrollo Histórico de la Estadística en El Salvador." Por Max P. Brannon. Vol. VIII, p. 263-278.

History of the development of statistics in El Salvador.

1940.3, Bunge, a "Métodos para Determinar el Grado de Mortalidad Infantil." Por Alejandro Ernesto Bunge. Vol. VIII, p. 29-42.

Methods for determining infant death rates.

1940.3, Bunge, b "La Población Argentina, Su Desarrollo, Sus Características y Sus Tendencias." Por Alejandro Ernesto Bunge. Vol. VIII, p. 109-111 (resumen). 79 p. typed manuscript in IASI.

The development, characteristics, and trends of the Argentine population.

1940.3, Cox B. "Estudio del Comercio Exterior de Chile en Artículos Alimenticios y su Significado." Por Ricardo Cox B. Vol. VIII, p. 169-175.

A study of foreign commerce of Chile in articles of food, and its significance.

1940.3, Fiedler "Need and Plan for a Statistical Program in Furthering Conservation of Inter-American Fisheries." By Reginald H. Fiedler. Vol. VIII, p. 177-186.

Programa estadístico en el fomento de la pesca de los países americanos: necesidad de su adopción y plan correspondiente.

1940.3, García-Mata "Un Método Práctico para Suavizar a Mano las Curvas Estadísticas." Por Carlos García-Mata. Vol. VIII, p. 279-283.

A practical method for smoothing statistical curves by hand.

1940.3, González H. "El Sistema de Coordinación en los Servicios Estadísticos Mexicanos." Por Gonzálo González H. Vol. VIII, p. 285-295.

The system of coordination of Mexican statistical services.

1940.3, Guthardt "Panorama Demográfico de la Población de Colombia." Por Emilio Carlos Guthard. Vol. VIII, p. 46 (title only). 69 page typed manuscript on microfilm in IASI.

Demographic panorama of the population in Colombia. Includes chapters on: Relative density; geographic distribution; increase; distribution by sex; distribution by age; distribution by civil status.

Incluye capítulos sobre: Densidad relativa; distribución espacial; crecimiento; distribución por sexo; distribución por edades: distribución por estado civil.

1940.3, Kingston "The Elasticity of Substitution of Brazilian Coffee." By Jorge Kingston. Vol. VIII, p. 189-200.

1940.3, León "Población y Salubridad Pública en el Nuevo Mundo." Por Alberto P. León. Vol. VIII, p. 111 (title only).
Population and public health in the New World.

1940.3, Lotka "The Place of the Intrinsic Rate of Natural Increase in Population Analysis." By Alfred J. Lotka. Vol. VIII, p. 297-313.

1940.3, Loyo González "Bases Mínimas para la Uniformidad de los Censos Nacionales de Población en el Continente Americano." Por Gilberto Loyo González. Vol. VIII, p. 49-51.
Bases for uniformity in national censuses of population in the Americas.

1940.3, Mardones Restat "Las Poblaciones y la Nutrición en el Nuevo Mundo." Por Jorge Mardones Restat. Vol. VIII, p. 113-120.
Deals with population and nutrition of the New World.

1940.3, México "Mortalidad en el Medio Rural de la República Mexicana." Obsequio de la Delegación Mexicana a los asistentes a la Sección VII del VIII Congreso Científico Americano. Trabajo redactado en 1938. México, mayo de 1940. Vol. VIII, p. 45 (title only). 27 page mimeographed article in IASI.
Mortality in rural zones of Mexico.

1940.3, Mortara "The Calculation of Life Tables for Populations Lacking Reliable Birth and Death Statistics, with Application to Brazil." By Giorgio Mortara. Vol. VIII, p. 321-334.
O cálculo de tábuas de sobrevivência para populações sem estatísticas fidedignas de nascimentos e óbitos, con aplicação ao Brasil.

1940.3, Parra Gómez "Perfeccionamiento del Registro Civil como Fuente de Información Estadística." Por Roberto Parra Gómez. Vol. VIII, p. 71-82.
Improvement of the civil register as source of statistical information.

1940.3, Rebuelto "La Sociedad Argentina Estadística." Por Emilio Rebuelto. Vol. VIII, p. 351-352.
Discusses the Argentine Statistical Society.

1940.3, Roa Reyes "Metodología Estadística Aplicada a las Ciencias Sociales y Biología Social." Por Jorge Roa Reyes y José Chelala-Aguilera. Vol. VIII, p. 335-336.
Statistical methodology applied to social and biological sciences.

1940.3, Ruíz Carranza "Estimaciones Estadísticas de los Pagos Internacionales." Por Eduardo Ruíz Carranza. Vol. VIII, p. 203-211.
Statistical estimates of international payments.

1940.3, Sálas S. "Desarrollo de Normas Comunes en la Estadística Educacional Americana." Por Irma Sálas S. Vol. VIII, p. 337-340.
Development of uniform standards in American statistics of education.

1940.3, Schiaffino "Los Problemas de la Bioestadística en la República Oriental del Uruguay." Por Rafael Schiaffino. Vol. VIII, p. 83-92.
Problems of biostatistics in Uruguay.

1940.3, Simón "El Concepto de Industria Nacional y la Protección del Estado." Por Raúl Simón. Vol. VIII, p. 213-239.
The concept of national industry and protection by the State.

1940.3, Stephan "The American Statistical Association." By Frederick Franklin Stephan. Vol. VIII, p. 353-364.
La Asociación Americana de Estadística.

1940.3, Suárez Rivadeneira "Esquema del Plan de Organización Puesto al Servicio del Levantamiento de los Censos de Edificios y Población Llevados a Cabo en Colombia en 1938." Por Antonio Suárez Rivadeneira. Vol. VIII, p. 93-94 (resume only). 142 page typed manuscript on microfilm in IASI.
Outline of plan of organization used in connection with housing and population censuses in Colombia in 1938. Contains a chart showing organization of census,

budget and time schedule for carrying through the census, etc., and chapters on: technical basis; stage of preparation; stage of execution; stage of concentration and publication; census legislation; budget.

Contiene un cuadro demostrativo de la organización del censo, presupuesto y horario para llevarlo a cabo, etc., y capítulos sobre: Bases técnicas; etapa de preparación; etapa de ejecución; etapa de concentración y publicación; legislación censal; presupuesto.

1940.3, Toscano "El Problema del Muestreo en Estadística." Por Ricardo Toscano. Vol. VIII, p. 341-342 (resume only). 45 page typed manuscript in IASI.

The problem of statistical sampling.

1940.3, Vergara "Los Censos de Población en Chile." Por Roberto Vergara. Vol. VIII, p. 95-108.

Population censuses in Chile.

1940.3, Villafuerte Flores "El Censo Comercial en México." Por Carlos Villafuerte Flores. Vol. VIII, p. 241-247.

Census of trade in Mexico.

1940.3, Zertuche "Conceptos Básicos de los Censos Industriales Recomendables para los Países Latino-Americanos." Por Albino Zertuche. Vol. VIII, p. 249-257.

Basic concepts of industrial censuses recommended for Latin American countries.

Int-**1300.**C8 **Congress and Conference Series.** 1929- . Irregular. Washington, D.C. Pan American Union.

Published also in Spanish and Portuguese—publicado también en español y portugués (Int-**1300.**S6; Int-**1300.**S5).

Int-**1300.**S5 **Serie de Congressos e Conferências.** 1929- . Irregular. Washington, D.C. União Panamericana.

Published also in English and Spanish—publicado também em inglês e espanhol (Int-**1300.** C8; Int-**1300.**S6).

Int-**1300.**S6 **Serie sobre Congresos y Conferencias.** 1929- . Irregular. Washington, D.C. Unión Panamericana.

Published also in English and Portuguese—publicado también en inglés y portugués (Int-**1300.**C8; Int-**1300.**S5).

*Int-**1400.**B5 1945 **Basic Data on the Other American Republics.** Washington, D.C., U.S. Govt. Print. Off., 1945. Office of the Coordinator of Inter-American Affairs. 172 p.

Covers almost in outline form: Political background, organization, major dates, administration, land, people, education, transportation facilities, level of living, labor, agriculture, manufacturing, trade, foreign exchange, investment, finance, current economic problems, economic relations with the United States, and contribution to war effort of each Latin American country. Gives table of basic facts at end.

Datos básicos sobre las otras repúblicas americanas. Abarcan casi en forma esquemática: Medio político, organización, fechas más importantes, administración, terreno, población, educación; conveniencias de transportación, nivel de vida, trabajo, agricultura, manufactura, comercio, cambio internacional, inversiones, finanzas, problemas económicos de actualidad, relaciones económicas con los Estados Unidos, y las contribuciones a la guerra efectuadas por cada uno de los países latinoamericanos. Al final se presenta una tabla de informaciones básicas.

*Int-**1410.**A4 **Annuaire Statistique de la Société des Nations. Statistical Year-Book of the League of Nations.** 1926- . Genève. Société des Nations. Service d'Études Économiques. League of Nations. Economic Intelligence Service.

1942-44 **Annuaire Statistique de la Société des Nations, 1942/44. Statistical Year-Book of the League of Nations, 1942/44.** Dix-septième année. Seventeenth issue. Lausanne, Suísse, Imp. Réunies, 1945. Société des Nations. Service d'Études Économiques. League of Nations. Economic Intelligence Service. 315 p.

Includes: Population and vital statistics; economic and financial statistics: Unemployment, employment; production; trade; prices; currency and banking statistics; public finance; national income.

Comprend: Statistiques démographiques; statistiques économiques et financières: Chômage, emploi; production; commerce; prix; statistiques monétaires et bancaires; finances publiques; revenu national.

*Int-**1410.**A7 1942 **Anuario Estadístico Interamericano, 1942. Inter-American Statistical Year-Book, 1942. Anuario Estatístico Interamericano, 1942. Annuaire Statistique Interaméricain, 1942.** Raul C. Migone, Director. Buenos Aires, El Ateneo; New York, The MacMillan Co.; Rio de Janeiro, Freitas Bastos &

Cía, 1942. Comisión Argentina de Altos Estudios Internacionales. Argentine Commission of High International Studies. Comissão Argentina de Altos Estudos Internacionais. Commission Argentine de Hautes Études Internationales. 1066 p. •

Includes: Population; production; industries; transport and communications; commerce; social questions; public finance; public education; army, navy and air corps; public health; international cooperation.

Incluye: Población; producción; industria; transporte y comunicaciones; comercio; cuestiones sociales; finanzas públicas; instrucción pública; ejército, armada y aviación; sanidad; cooperación internacional.

Int-**1410**.P5 **The Pan American Yearbook.** 1945- . New York. Pan American Associates.

Published also in Spanish—publicado también en español.

Anuario panamericano.

Int-**1410**.W5 **The West Indies Year Book, Including Also the Bermudas, the Bahamas, British Guiana and British Honduras, Etc.** 1928- . Annual. London, Canadian Gazette, Limited; Montreal and New York, Thomas Skinner of Canada, Limited.

Anuario de las Indias Occidentales Británicas, incluyendo Bermudas, Bahamas, Guayana Británica y Honduras Británica.

*Int-**1420**.M5 **Bulletin Mensuel de Statistique. Monthly Bulletin of Statistics.** 1919- . Genève. Nations Unies. United Nations.

From 1919 through July 1946 published by the League of Nations—de 1919 jusqu'à Août 1946 publié par la Société des Nations.

Valuable source for statistical data relative to most of the important nations of the world. Each issue contains data on some special subject in addition to regular series on: Production, stocks and exports; trade; price movement; finance; labor.

Une source importante de renseignements statistiques relatifs à la plupart des nations importantes du monde. Chaque édition contient des données sur quelque sujet spécial en plus des séries régulières sur: Production, stocks et exportations; commerce; mouvement des prix; finance; travail.

Int-**1420**.R5 **Revue de l'Institut International de Statistique: Supplément Mensuel.** Depuis? Pas publié depuis 1941? La Haye.

Monthly supplement to journal of International Statistical Institute.

Int-**1900**.C5 1943 **The Caribbean Islands and the War.** A record of progress in facing stern realities. Prepared by the United States Section of the Anglo-American Caribbean Commission. Washington, D. C., U.S. Govt. Print. Off., 1943. U.S. Department of State. 85 p. Publication, no. 2023.

Report on activities of government agencies in meeting wartime problems, and survey of non-military adjustments in civilian activities, government controls, and inter-relations of the Caribbean area. Includes discussions and statistics of food production, price control, wages, unemployment, cost of living, industrial expansion, and economic progress.

Las islas del Caribe y la guerra. Informe sobre las actividades de los organismos gubernamentales frente a los problemas de tiempo de guerra, y reseña de los ajustes no militares en las actividades de la población civil, controles gubernamentales, y relaciones mútuas en la región del Caribe. Incluye discusiones y estadísticas de la producción de alimentos, control de precios, salarios, desempleo, costo de la vida, expansión industrial, y desarrollo económico.

Int-**1900**.G5 1941 **Good Neighbors, Argentina, Brazil and Chile and Seventeen Other Countries.** By Hubert Herring. New Haven, Connecticut, Yale University Press, 1941. 381 p.

The economic, sociological, and political background of the Latin American nations. Contains much information on living conditions, including cost of living, and working conditions and regulations, including wages, economic distribution of population, and a bibliography of about 200 references.

Los buenos vecinos, Argentina, Brasil, Chile y otros diecisiete países. El medio económico, sociológico y político de las naciones latinoamericanas. Contiene muchas informaciones sobre condiciones de vida, incluyendo el costo de la vida, y sobre condiciones y reglamentos del trabajo, incluyendo salarios, distribución económica de la población, y una bibliografía de alrededor de 200 referencias.

Int-**1900**.L4 1943 **Latin America.** Based on the latest edition of the Encyclopedia Americana. New York, 1943. Americana Corporation. Latin American Division.

A general survey of Latin America presenting reference material on its history, resources, industries, cultural and political relations, and other aspects of the area. Includes information

on immigration, industrialization, labor legislation, social welfare, and social and political leaders.

La América Latina. Reseña general de la América Latina, exponiendo informaciones sobre su historia, recursos, industrias, relaciones culturales y políticas, y otros aspectos de la región. Incluye informaciones sobre inmigración, industrialización, legislación del trabajo, bienestar social, y jefes sociales y políticos.

Int-**1900**.L5 1942 **Latin America.** By Preston E. James. New York, The Odyssey Press, Inc., 1942. 908 p.

"A basic text or 'human geography' for use in courses in regional or economic geography of Latin America. . . Study of population is a controlling theme. . . Repeated attention is given to the relation between population distribution and such factors as the character of the land, political areas, racial patterns, and opportunities for economic development. . . " (Source: Preface.)

La América Latina. "Un texto básico o 'geografía humana' para uso en los cursos de geografía regional o económica de la América Latina . . . El estudio de la población representa uno de los temas más importantes. . . Repetidamente se concede atención a la relación entre la distribución de la población y otros factores, tales como las características del terreno, divisiones políticas, orígenes racial, y oportunidades de fomento económico. . . " (Traducción del comentario.)

Int-**1900**.L7 1945 **Latin America in the Future World.** By George Soule, David Efron and Norman T. Ness. New York, Farrar & Rinehart, Inc., 1945. 372 p. Unexamined.

"The writers examine Latin America's basic economy and changes during the present war, and recommend policies and specific measures for the post-war period. Subjects given special emphasis include living levels, nutrition, health, and housing. There is a chapter on the social and political status of labor and one on labor and social security legislation." (Source: **Monthly Labor Review,** vol. 60, no. 4, April 1945, p. 925.)

La América Latina en el mundo del porvenir. "Los autores analizan la economía fundamental de la América Latina y sus cambios durante la actual guerra, y recomiendan políticas y medidas específicas para el período de la postguerra. Los tópicos a que conceden atención especial incluyen niveles de vida, nutrición, salubridad, y viviendas. Se encuentra también en la obra, un capítulo sobre la situación social y política del trabajo, así como uno sobre la legislación del trabajo y sobre seguridad social." (Traducción del comentario.)

Int-**1910**.I5 **Inter-American Affairs.** An annual survey. Edited by Arthur P. Whitaker. 1941- . New York, Columbia University Press.

Asuntos interamericanos: Un estudio anual.

1944 **Inter-American Affairs, 1944.** An annual survey, no. 4. Edited by Arthur P. Whitaker. New York, Columbia University Press, 1945. 284 p.

Includes statistical tables on inter-American trade, balances of international payments, movements of capital, and a list of inter-American conferences held during 1944.

Incluye cuadros estadísticos sobre comercio interamericano, balanzas de pago internacional, movimientos de capital y una lista de las conferencias interamericanas celebradas durante el año 1944.

1943 **Inter-American Affairs, 1943.** An annual survey, no. 3. Edited by Arthur P. Whitaker. New York, Columbia University Press, 1944. 277 p.

Includes maps and charts showing voting population of the American Republics, new Brazilian federal territories created in 1943, commercial airlines in Latin America, Latin American roads, Latin American railroads; also, statistical tables showing United States Government expenditures and cash-loan transactions in and related to the twenty other American Republics for fiscal years 1941-43, net United States balance of trade with Latin American republics, etc.

Incluye mapas y cuadros que muestran la población electoral de los países americanos, los nuevos territorios federales del Brasil creados en 1943, las líneas aéreas comerciales, carreteras y líneas ferroviarias de la América Latina; también, incluye cuadros estadísticos referentes a los gastos del gobierno estadounidense y sus operaciones de préstamo en numerario relacionados con las otras 20 repúblicas americanas para los años fiscales 1941-43; extracto del balance del comercio de Estados Unidos con las repúblicas latino-americanas, etc.

1942 **Inter-American Affairs, 1942.** An annual survey, no. 2. Edited by Arthur P. Whitaker. New York, Columbia University Press, 1943. 252 p.

Includes statistical tables on area and population of countries of Western Hemisphere, production and income for the most important American countries, finance and investments, and international trade.

Incluye cuadros estadísticos sobre área y población de los países del Hemisferio Occidental, producción y renta de los países americanos más importantes, finanzas e inversiones y comercio internacional.

Int-**1910**.P5 **Political Handbook of the World.** Edited by Walter H. Mallory. 1928- . Annual. New York. Council on Foreign Relations, Inc.

Provides in compact form essential political information regarding composition of governments, programs of political parties and their leaders, and political affiliations and editors of leading newspapers and periodicals.

Guía política del mundo. Expone en forma breve, informaciones políticas, tales como composición de los distintos gobiernos, programas de partidos políticos y sus líderes, y afiliaciones políticas de los más importantes periódicos y revistas así como los nombres de sus redactores.

Int-**1910**.S5 **The Statesman's Year-Book.** Statistical and historical annual of the world. Revised after official returns. Edited by M. Epstein. 1864- . Annual. London, Macmillan and Company, Limited.

Title varies slightly—el título varía un poco.

Concise information about countries of the world, including form of government, population, religion, finance, industry, commerce, money and credit, diplomatic representatives, etc.

Anuario diplomático. Informaciones breves sobre los países del mundo, incluyendo la forma de gobierno, población, religión, finanzas, comercio, moneda y crédito, representación diplomática, etc.

Int-**1910**.W5 **The World Almanac and Book of Facts.** 1886- . Annual. New York.

A handbook in wide use throughout the United States; an annual "Book of Facts" originated by the **New York World** and continued by its successor newspaper, the **New York World-Telegram.** Includes current statistics, political, social, religious, educational, etc., recent federal laws, chronological list of important events, and much miscellaneous information. Statistics of New York City are especially complete.

Almanaque mundial y libro de datos. Un guía extensamente usado en los Estados Unidos; comenzó su publicación anual como "Libro de Datos" por el periódico **New York World,** y se continúa actualmente por los sucesores de éste, el **New York World-Telegram.** Incluye estadísticas de la actualidad sobre política, el medio social, religión, educación, etc., últimas leyes del poder federal, lista cronológica de acontecimientos importantes, muchas informaciones misceláneas. Las estadísticas sobre la ciudad de Nueva York son especialmente completas.

Int-**1930**.H4 1945 **Handbook of International Organizations in the Americas.** Prepared by Ruth D. Masters. Washington, D.C., 1945. Carnegie Endowment for International Peace. Division of International Law. 453 p.

Contains synopses of 109 international organizations with headquarters in the Western Hemisphere. Lists nearly all organizations having an international purpose, except bodies having purely national membership, purely commercial organizations, and two-country cultural or social clubs and societies; organizations originally located in Europe, but moved temporarily to the Western Hemisphere during the war, have also been excluded. Gives detailed information for each organization regarding its history, purpose and functions, membership and administration, languages, publications, and work done; quotes most important provisions of treaties establishing organizations, as well as constitutions and by-laws of each body.

Manual de las organizaciones internacionales en las Américas. Contiene sinopsis de 109 organizaciones internacionales con sede en el Hemisferio Occidental. Enumera casi todas las organizaciones que tienen propósitos internacionales, exceptuando aquellas que están formadas por personas de un mismo país, las puramente comerciales y los clubs y sociedades culturales o sociales formados por dos países. Las organizaciones originariamente con sede en Europa que motivo de la guerra fueron temporalmente instaladas en América, también han sido excluídas. Nos proporciona una amplia información sobre cada organización en lo que atañe a su historia, propósitos y funciones, personal que la forma, administración, idiomas, publicaciones y trabajo efectuado; cita las disposiciones más importantes de los tratados bases de las organizaciones, así como las actas de constitución y reglamentos.

Int-**1930**.H5 1942 **Handbook of Scientific and Technical Societies and Institutions of the United States and Canada.** United States section compiled for the National Research Council of the United States by Callie Hull and Mildred Paddock; Canadian section compiled for the National Research Council of Canada by S. J. Cook and P. A. Howard Fourth edition. Washington, D.C., 1942. National Research Council. National Academy of Sciences. 389 p. Bulletin of the National Research Council, no. 106.

Cover title—título de la cubierta: **Scientific and Technical Societies and Institutions of the United States and Canada.**

Includes notes on lists of societies and associations issued by other organizations, p. 343.

Manual de sociedades e instituciones científicas y técnicas de los Estados Unidos y Canadá. Incluye notas sobre listas de sociedades y asociaciones publicadas por otras organizaciones, p. 343.

Int-**1950**.A5 **American Nation Series.** 1927- . Irregular. Washington, D.C. Pan American Union.

Consists of a separate booklet for each country descriptive of geographical characteristics, constitution and government, resources, products and industries, transportation, educational facilities, and other phases of the country.

Serie sobre las naciones americanas editada por la Unión Panamericana. Consisten en un folleto de cada país por separado, describiendo características geográficas, constitución y gobierno, fuentes de riqueza, producción e industrias, transportes, medios de instrucción y otras aspectos del país.

Int-**1950**.C5 1939 **Commercial Travelers' Guide to Latin America.** Washington, D.C., U.S. Govt. Print. Off., 1939. U.S. Department of Commerce. Bureau of Foreign and Domestic Commerce. 3 vol.: Part I, **West Coast of South America**; part II, **East Coast of South America**; part III, **Mexico, Central America, and Caribbean Countries.**

Int-**1950**.N5 1943 **The New World Guides to the Latin American Republics.** Editor-in-chief: Earl Parker Hanson. New York, Duell, Sloan and Pearce, 1943. U.S. Coordinator of Inter-American Affairs. 2 vol.: Vol. I, **Mexico and Central America**; vol. II, **South America.**

Guía para las repúblicas latino-americanas; publicado por el organismo Coordinador de Asuntos Interamericanos de Estados Unidos.

2000-2999 CENSUS—CENSOS

Int-**2000**.C5 1950.Ib **Censo de las Américas de 1950: Antecedentes y Justificación del Proyecto.** Memorandum elevado al Instituto Interamericano de Estadística. Trabajo del estadígrafo Dr. Pedro Mendive, aprobado por la Comisión Ejecutiva. Montevideo, 1946. Consejo Interamericano de Comercio y Producción. 21 p.

1950 Census of the Americas: Antecedents and justification of the project. Memorandum transmitted to Inter American Statistical Institute by Inter American Council of Commerce and Production. A critical analysis of Ricardo Luna Vegas' study, **Métodos de los Censos de Población de las Naciones Americanas,** with suggestions regarding projected 1950 Census of the Americas.

Examen analítico del estudio **Métodos de los Censos de Población de las Naciones Americanas** del Dr. Ricardo Luna Vegas, con sugerencias en cuanto al proyectado Censo de las Américas en 1950.

Int-**2000**.P5 1881 **Projet d'un Recensement du Monde.** Étude de statistique internationale. Par József Korosi. Paris, Guillaumin et Cie., 1881. Unexamined.

Project of a world census.

*Int-**2100**.M5 1945 **Métodos de los Censos de Población de las Naciones Americanas.** Estudio preliminar para el proyectado censo continental de 1950. Por Ricardo Luna Vegas. Washington, D.C., March 1945. Inter American Statistical Institute. 127 p. Project series, no. 1.

Reprint of study published in **Estadística,** vol. III, no. 9, March 1945—reimpresión del estudio publicado en **Estadística,** vol. III, no. 9, marzo 1945.

Methods of population censuses in the American nations. A detailed study prepared for the Inter American Statistical Institute giving background material essential for development of minimum standards for proposed hemispheral census of the Americas in 1950.

Un estudio detallado hecho por el Instituto Interamericano de Estadística, que proporciona importante material básico para el desarrollo de estandards mínimos para el planeado censo de 1950 en el hemisferio americano.

*Int-**2110**.W5 1940 **Programme of the World Agricultural Census of 1940.** Rome, Print. Off. Ditto C. Colombo, 1938. International Institute of Agriculture. 50 p.

Contains extract from international convention relating to economic statistics, and resolution adopted by XIIIth general assembly of International Institute of Agriculture (Rome, October 1936), including "Resolutions on the World Agricultural Census," p. 50.

Programme du recensement agricole mondiale de 1940. Comprend un extrait de la convention internationale concernant les statistiques économiques et la résolution adoptée par la XIIIème assemblée générale de l'Institut International d'Agriculture (Rome, octobre 1936), y compris

"Résolutions sur le recensement agricole mondiale."

Int-**2110**.W5 1940.2 **The World Agricultural Census of 1940: Proposed Standard Form.** Rome, Print. Off. of the Chamber of the Deputies, Charles Colombo, 1936. International Institute of Agriculture. 32 p.

Includes explanatory notes.

Le recensement agricole mondiale de 1940: Projet de formulaire-type. Comprend des notes explicatives.

Int-**2110**.W5 1940.3 **World Agricultural Census of 1940: Questionnaires.** For the communication of census data to the International Institute of Agriculture. Rome, Villa Umberto I, 1940. International Institute of Agriculture. 58 p.

Distributed to the governments.

Recensement mondial de l'agriculture, 1940: Questionnaires pour la communication des données de recensement à l'Institut International d'Agriculture, distribués aux gouvernements.

Int-**2210**.W5 1930 **The First World Agricultural Census (1930).** Rome, Villa Umberto I, 1939. International Institute of Agriculture. 5 vol.: Vol. I, "Standard Form Proposed for the World Agricultural Census of 1930 and Documents Relating to Its Preparation"; "A Methodological Study of the Questions Contained in the Census Forms"; "Notes on Tabulation of the Census Results"; vol. II, Austria, Belgium, Czechoslovakia, Denmark, Estonia, Finland, France; vol. III, Germany, Greece, Irish Free State, Italy, Latvia, Lithuania, Netherlands, Norway, Spain, Sweden, Switzerland, and the United Kingdom (England and Wales, Scotland, and Northern Ireland); vol. IV, Canada, United States, the Outlying Territories and Possessions of the United States, Mexico, Argentina, Chile, Peru, Uruguay; vol. V, India, Japan, Algeria, Egypt, French West Africa, Kenya, Mauritius, Mozambique, Union of South Africa, Commonwealth of Australia, New Zealand.

Le premier recensement mondial de l'agriculture (1930).

3000-3999 STATISTICAL SCIENCE—CIENCIA ESTADISTICA

*Int-**3000**.D5 1944 **Directory of Statistical Personnel in the American Nations. Directorio del Personal Estadístico en las Naciones Americanas. Nominata do Pessoal Estatístico das Nações Americanas.** With biographical information—Con informaciones biográficas—Com informações biográficas. Second edition—Segunda edición—Segunda edição. Washington, D.C., May 1944. Inter American Statistical Institute. Instituto Interamericano de Estadística. Instituto Inter-Americano de Estatística. 101 p.

"This directory is a second and more complete edition of corresponding material which appeared in 1941 in the volume **Statistical Activities of the American Nations, 1940.**" (Source: Preface.)

"Esta es la segunda edición, más completa, del directorio que apareció en 1941, en el volumen **Statistical Activities of the American Nations, 1940.**" (Fuente: Prefacio.)

Int-**3000**.I5 **Intercambio Estadístico.** Un servicio por microfilm a través de las Américas por el Instituto Interamericano de Estadística. 1944-45. Irregular. Washington, D.C.

Statistical interchange: A service on microfilm across the Americas by the Inter American Statistical Institute.

May-June '45 "Source Materials: IASI Survey of Census Methods and Procedures: Nicaragua, Panama, Paraguay, Peru, Republica Dominicana, Venezuela." Vol. I, rollo 8, mayo-junio 1945.

The sixth and last reel of a series showing forms, schedules, instruction manuals, etc., used in each of the American nations in taking population censuses.

Materiales de información: Investigaciones del Instituto Interamericano de Estadística, sobre métodos y procedimientos para censos: Nicaragua, Panamá, Paraguay, Perú, República Dominicana, Venezuela. El sexto y último rollo de una serie que contiene los formularios, cuadros, manuales de instrucción, etc., usados por cada uno de los países americanos al censar la población.

Apr. '45 "Source Materials: IASI Survey of Census Methods and Procedures: Chile, Colombia, Cuba, El Salvador, Guatemala, Haiti, Honduras, Mexico." Vol. I, rollo 7, abril 1945.

The fifth reel of a series showing forms, schedules, instruction manuals, etc., used in each of the American nations in taking population censuses.

Materiales de información: Investigaciones del Instituto Interamericano de Estadística, sobre métodos y procedimientos para censos: Chile, Colombia, Cuba, El Salvador, Guatemala, Haiti, Honduras, México. El quinto rollo de una serie que contiene los formularios, cuadros, manuales de instrucción, etc., usados en cada uno de los países americanos para censar la población.

Nov. '44-Mar. '45 "Source Materials: IASI Survey of Census Methods and Procedures: Argentina, Bolivia, Brazil, Canada." Vol. 1, rollo 6, noviembre 1944-marzo 1945.

The fourth reel of a series showing forms, schedules, instruction manuals, etc., used in each of the American nations in taking the population census.

Materiales de información: Investigaciones del Instituto Interamericano de Estadística, sobre métodos y procedimientos para censos: Argentina, Bolivia, Brazil, Canadá. El cuarto rollo de una serie que contiene los formularios, cuadros, manuales de instrucción, etc., usados en cada uno de los países para censar la población.

Aug.-Oct. '44 "Source Materials: IASI Survey of Census Methods and Procedures: United States: Sixteenth Census, 1940." Vol. I, rollos 3-5, agosto-octubre, 1944.

The first three reels of a series showing forms, schedules, instruction manuals, etc., used in each of the American nations in taking population censuses.

Materiales de información: Investigaciones del Instituto Interamericano de Estadística, sobre métodos y procedimientos para censos; Estados Unidos: decimosexto censo, 1940. Los tres primeros rollos de una serie que contiene los formularios, cuadros, manuales de instrucción, etc., usados en cada uno de los países americanos para censar la población.

Int-**3000**.P5 **Project Series of the Inter American Statistical Institute. Serie de Proyectos del Instituto Interamericano de Estadística.** 1945- . Irregular. Washington, D.C.

Int-**3000**.R4 1934 **Répertoire International des Institutions Statistiques.** La Haye, 1934. Institut International de Statistique. 139 p.

International directory of statistical institutions published by the International Statistical Institute.

*Int-**3000**.R6 **Report to the Council on the Work of the . . . Session. Rapport au Conseil sur les Travaux de la . . . Session.** 1931-39. Annual. Geneva. League of Nations. Committee of Statistical Experts. Société des Nations. Comité d'Experts Statisticiens.

1939, Eng **Report to the Council on the Work of the Eighth Session.** Held in Geneva from April 22nd to 27th, 1939. Geneva, April 27, 1939. League of Nations. Committee of Statistical Experts. 10 p. Series of League of Nations publications, II, Economic and financial, 1939. II.A.5.

Published also in French—publié aussi en français (Int-**3000**.R6, 1939, Fr).

Includes—comprend: Appendix III, Indices showing variations in quantum and prices of international trade. (Other appendixes published separately as **Studies and Reports on Statistical Methods.**)

1939, Fr. **Rapport au Conseil sur les Travaux de la Huitième Session.** Tenue à Genève du 22 au 27 avril 1939. Genève, 27 avril 1939. Société des Nations. Comité d'Experts Statisticiens. 10 p. Série de publications de la Société des Nations, II Questions économiques et financières, 1939.II.A.5.

Published also in English—publié aussi en anglais (Int-**3000**.R6, 1939, Eng).

1938, Eng **Report to the Council on the Work of the Seventh Session.** Held in Geneva from July 4th to 9th, 1938. Geneva, July 9, 1938. League of Nations. Committee of Statistical Experts. 28 p. Series of League of Nations publications, II, Economic and financial, 1938. II.A.7.

Published also in French—publié aussi en français (Int-**3000**.R6, 1938, Fr).

Includes—comprend: Appendix II, Housing statistics; appendix IV, Indices of industrial production. (Other appendixes published separately as **Studies and Reports on Statistical Methods.**)

1938, Fr. **Rapport au Conseil sur les Travaux de la Septième Session.** Tenue à Genève du 4 au 9 juillet 1938. Genève, 9 juillet 1938. Société des Nations. Comité d'Experts Statisticiens. 28 p. Série de publications de la Société des Nations, II, Questions économiques et financières, 1938.II.A.7.

Published also in English—publié aussi en anglais (Int-**3000**.R6, 1938, Eng).

1937, Eng **Report to the Council on the Work of the Sixth Session.** Held in Geneva from April 19th to 24th,

1937. Geneva, April 29, 1937. League of Nations. Committee of Statistical Experts. 21 p. Series of League of Nations publications, II, Economic and financial, 1937. II.A.5.

Published also in French—publié aussi en français (Int-**3000**.R6, 1937, Fr).

Includes—comprend: Appendix I, Statistics of the gainfully-occupied population (discussion of classification); appendix II, Draft minimum nomenclature of industries; appendix III, Specimen tables (gainfully-occupied and not gainfully-occupied population; classification of the gainfully-occupied population).

1937, Fr. **Rapport au Conseil sur les Travaux de la Sixième Session.** Tenue à Genève du 19 au 24 avril 1937. Société des Nations. Comité d'Experts Statisticiens. 21 p. Série de publications de la Société des Nations, II, Questions économiques et financières, 1937.II.A.5.

Published also in English—publié aussi en anglais (Int-**3000**.R6, 1937, Eng).

1936, Eng **Report to the Council on the Work of the Fifth Session.** Held in Geneva from October 12th to 17th, 1936. Geneva, October 30, 1936. League of Nations. Committee of Statistical Experts. 21 p. Series of League of Nations publications, II, Economic and financial, 1936. II.A.23.

Published also in French—publié aussi en français (Int-**3000**.R6, 1936, Fr).

Includes—comprend: Appendix I, Recommendations on mining and metallurgical statistics; appendix II, International tourist statistics.

1936, Fr. **Rapport au Conseil sur les Travaux de la Cinquième Session.** Tenue à Genève du 12 au 17 octobre 1936. Genève, 30 octobre 1936. Société des Nations. Comité d'Experts Statisticiens. 13 p. Série de publications de la Société des Nations, II, Questions économiques et financières, 1936.II.A.23.

Published also in English—publié aussi en anglais (Int-**3000**.R6, 1936, Eng).

1935, Eng **Report to the Council on the Work of the Fourth Session.** Held in Geneva from June 3rd to 6th, 1935. Geneva, July 31, 1935. League of Nations. Committee of Statistical Experts. 44 p. Series of League of Nations publications, II, Economic and financial, 1935.II.A.10.

Published also in French—publié aussi en français (Int-**3000**.R6, 1935, Fr).

Includes—comprend: Appendix I, Minimum list of commodities for international trade statistics; appendix II, Classification of commodities by stage of production and use; appendix III, Statement regarding the principles which have guided the Committee of Experts in the preparation of the minimum list.

1935, Fr. **Rapport au Conseil sur les Travaux de la Quatrième Session.** Tenue à Genève du 3 au 6 juin 1935. Genève, 31 juillet 1935. Société des Nations. Comité d'Experts Statisticiens. 44 p. Série de publications de la Société des Nations, II, Questions économiques et financières, 1935.II.A.10.

Published also in English—publié aussi en anglais (Int-**3000**.R6, 1935, Eng).

1934, Eng **Report to the Council on the Work of the Third Session.** Held in London from April 12th to 14th, 1934. Geneva, May 11, 1934. League of Nations. Committee of Statistical Experts. 32 p. Series of League of Nations publications, II, Economic and financial, 1934. II.A.10.

Published also in French—publié aussi en français (Int-**3000**.R6, 1934, Fr).

Includes—comprend: Appendix I, Statement of the Sub-Committee of Experts for the Unification of Customs Tariff Nomenclature concerning the relations between the customs nomenclature and the minimum list of commodities; appendix II, Minimum list of commodities for international trade statistics; appendix III, statement regarding the principles which have guided the Committee of Statistical Experts in the preparation of the minimum list.

1934, Fr. **Rapport au Counseil sur les Travaux de la Troisième Session.** Tenue à Londres du 12 au 14 avril 1934. Genève, 11 mai 1934. Société des Nations. Comité d'Experts Statisticiens. 32 p. Série de publications de la Société des Nations, II, Questions économiques et financières, 1934.II.A.10.

Published also in English—publié aussi en anglais (Int-**3000**.R6, 1934, Eng).

1933, Eng **Report to the Council on the Work of the Second Session.** Held at Geneva from December 7th to 12th, 1933. Geneva, December 12, 1933. League of Nations. Committee of Statistical Experts. 18 p. Series of League of Nations publications, II, Economic and financial, 1933. II.A.31.

Published also in French—publié aussi en français (Int-**3000**.R6, 1933, Fr).

Includes—comprend: Annex, Summary of the reports received from the various countries regarding the results of the experiment in the recording of commodities by countries of provenance and destination (details of the experiment, results, and conclusions to be drawn).

1933, Fr. **Rapport au Conseil sur les Travaux de la Deuxième Session.** Tenue à Genève du 7 au 12 décembre 1933. Genève, 12 décembre 1933. Société des Nations. Comité d'Experts Statisticiens. 18 p. Série de publications de la Société des Nations, II, Questions économiques et financières, 1933.II.A.31.

Published also in English—publié aussi en anglais (Int-**3000**.R6, 1933, Eng).

1931, Eng **Report.** Geneva, March 27, 1931. League of Nations. Committee of Statistical Experts. 7 p. Series of League of Nations publications, II, Economic and financial, 1931.II.A.9.

Published also in French—publié aussi en français (Int-**3000**.R6, 1931, Fr).

"The Committee of Statistical Experts appointed on January 22nd last held its first meeting at Geneva on March 23rd, 1931." (Source: p. 1.)

Includes—comprend: Appendix I, Minimum list of countries (statistical territories to be shown in the statistics of external trade by countries); appendix II, List of commodities for the purpose of the experiment in recording external trade statistics by countries of origin, consignment, purchase, etc.

1931, Fr. **Rapport du Comité d'Experts Statisticiens.** Genève, 27 mars 1931. Société des Nations. Comité d'Experts Statisticiens. 7 p. Série de publications de la Société des Nations, II, Questions économiques et financières, 1939. II.A.9.

Published also in English—publié aussi en anglais (Int-**3000**.R6, 1931, Eng).

"Le Comité d'experts statisticiens, institué le 22 janvier dernier, a tenu sa première session à Genève le 23 mars 1931." (Source: p. 1.)

***Int-3000.S5 1940** **Statistical Activities of the American Nations, 1940.** A compendium of the statistical services and activities in 22 nations of the Western Hemisphere, together with information concerning statistical personnel in these nations. Edited under the direction of the Temporary Organizing Committee of the Inter American Statistical Institute by Elizabeth Phelps. Washington, D.C., 1941. Inter American Statistical Institute. 842 p.

Includes—incluye: "Las Actividades Estadísticas de la Argentina," por Carlos E. Dieulefait; "Las Actividades Estadísticas de Bolivia," por Jorge Pando Gutierrez; "As Atividades Estatísticas do Brasil," por M. A. Teixeira de Freitas; "The Statistical Activities of Canada," by S. A. Cudmore; "Las Actividades Estadísticas de Chile," por Roberto Vergara; "Las Actividades Estadísticas de Colombia," por Paul Hermberg; "Las Actividades Estadísticas de Costa Rica," por Sergio Carballo R.; "Las Actividades Estadísticas de Cuba," por Hugo Vivó y Escoto, Manuel F. de Vera, y Alberto Andino; "Las Actividades Estadísticas del Ecuador," por Augusto Aguirre Hernández; "Las Actividades Estadísticas de Guatemala," por Guillermo Schwartz; "The Statistical Activities of Haiti," by S. de La Rue; "Las Actividades Estadísticas de Honduras," por Miguel A. Cruz Zambrano; "Las Actividades Estadísticas de México," por Juan de Dios Bojórquez; "Las Actividades Estadísticas de Nicaragua," por Irving A. Lindberg; "Las Actividades Estadísticas de Panamá," por Hans J. Müller; "Las Actividades Estadísticas del Paraguay," por Alfonso B. Campos; "Las Actividades Estadísticas del Perú," por Friedrich Kürbs; "Las Actividades Estadísticas de la República Dominicana," por Vicente Tolentino R.; "Las Actividades Estadísticas de El Salvador," por Max P. Brannon; "Statistical Activities of the United States," prepared by the Central Statistical Board; "Las Actividades Estadísticas del Uruguay," por Eduardo J. Fonticelli; "Las Actividades Estadísticas de Venezuela," por José A. Vandellós; "Biographical Directory of Statistical Personnel"; reprint of "Notes on the Statistical Sources of Latin America" from **The Economic Literature of Latin America.**

"The object of this volume is to present a current account of the statistical services and activities of the American nations, together with a partial biographical list of the principal statistical personnel in these nations, other than the United States."(Source: Foreword.) The major part of the volume consists of sections on each of the 22 participating nations. Each section begins with a summary of basic data, statistical educational facilities, statistical library facilities, statistical societies, non-official or semi-official agencies and their serial publications, system of official statistics, national population censuses, principal government agencies which compile statistics, and principal official statistical serial publications. This is followed by an article by a statistician of the country in question, usually in the language of the country.

Actividades estadísticas de las naciones americanas, 1940. "El objeto de este volumen es ofrecer una relación al día de los servicios y actividades estadísticas de las naciones americanas, junto con una lista biográfica parcial del personal de estadística más importante en esas naciones, a excepción de los Estados Unidos de América." (Fuente: Preámbulo.) La mayor parte del volumen consiste de sendas secciones de las 22 naciones participantes. Cada sección comienza con un resumen de datos fundamentales, medios de educación estadística, organismos oficiales o semi-oficiales y sus publicaciones periódicas, sistemas de estadísticas oficiales, censo de población, principales organismos gubernativos que compilan estadísticas y principales publicaciones estadísticas periódicas oficiales. Esto va seguido de un artículo preparado por un estadístico del país que se trate, y casi todos en su respectivo idioma.

Int-3010.B5 **Bibliography Series, IASI.** 1943- . Irregular. Washington, D.C. Inter American Statistical Institute.

Bibliographies in special subjects prepared by the Inter American Statistical Institute in answer to requests for information.

Bibliografías de materias especiales que han sido preparadas por el Instituto Interamericano de Estadística respondiendo a pedidos de información.

Cen. 2, May 5, '45 **Census Methods and Procedures.** Bibliography prepared for Luna Vegas' monograph, "Censo de Población de las Naciones Americanas," published in **Estadística,** vol. II, no. 9, March 1945. Washington, D.C., May 5, 1945. Inter American Statistical Institute. 15 p. Bibliography series, no. Cen. 2.

Métodos y procedimientos para los censos. Bibliografía preparada para la monografía del Dr. Luna Vegas, intitulada: "Censo de Población de las Naciones Americanas," que ha sido publicada en **Estadística,** vol. II, no. 9, marzo 1945.

Dem. 2, Feb. 7,'45 **Vital Statistics Registration Methods in the Western Hemisphere.** Bibliography prepared in connection with IASI study on Vital Statistics Methods in 22 American Nations. Washington, D.C., February 7, 1945. Inter American Statistical Institute. 14 p. Typewritten. Bibliography series, no. Dem. 2.

Métodos usados en el Hemisferio Occidental para el registro de la estadística vital: Bibliografía preparada por el Instituto Interamericano de Estadística para el estudio sobre métodos de estadística vital en 22 naciones americanas.

Econ. 4a. Apr. 2, '45 **Bibliography of Selected References on Agricultural Statistics in the Western Hemisphere.** Drawn from the Statistical Source Files of the Inter American Statistical Institute. Washington, D.C., April 2, 1945. Inter American Statistical Institute. 23 p. Bibliography series, no. Econ. 4a.

Bibliografía de referencias seleccionadas sobre estadísticas agrícolas en el Hemisferio Occidental. Extraído del archivo de fuentes estadísticas del Instituto Interamericano de Estadística.

Econ. 6a, June 1, '45 **Organization of Agriculture Statistics.** Bibliography drawn from the Statistical Source Files of the Inter American Statistical Institute. Washington, D.C., June 1, 1945. Inter American Statistical Institute. 6 p. Bibliography series, no. Econ. 6a.

Organización de estadísticas agrícolas: Bibliografía extraída del archivo de fuentes estadísticas del Instituto Interamericano de Estadística.

Econ. 8a, Oct. 1, '45 **Foreign Trade Classification in the American Nations.** Prepared as bibliography of IASI Convertibility Index for Foreign Trade Statistical Classifications of the American Nations, Oct. 1, 1945. Washington, D.C., October 1, 1945. Inter American Statistical Institute. 7 p.

Clasificación del comercio exterior de las naciones americanas. Bibliografía preparada por el IASI como complemento de la publicación: Indice de Convertibilidad para la Clasificación Estadística del Comercio Exterior de las Naciones Americanas.

Econ. 10, Apr. 15, '46 **National Income Bibliography.** Prepared for Sr. Loreto M. Dominguez in connection with IASI project on national income. Washington, D.C., April 15, 1946. Inter American Statistical Institute. 7 p.

Typewritten. Bibliography series, no. Econ. 10.

Gov. 2, Mar. 22, '46 **Justice and Crime Statistics.** Bibliography. Washington, D.C., March 22, 1946. Inter American Statistical Institute. 4 p. Typewritten. Bibliography series, no. Gov. 2.

Estadística judicial y criminológica; Bibliografía.

Lab. 1, Jan. 30, '46 **Labor and Living Conditions in the American Nations.** Bibliography on labor conditions, employment and wages, cost of living, family living, food consumption and nutrition, and housing conditions. Prepared in connection with IASI contemplated project on Statistics of Labor and Living Conditions in the American Nations. Washington, D.C., January 30, 1946. Inter American Statistical Institute. 43 p. Typewritten. Bibliography series, no. Lab. 1.

Condiciones del trabajo y de vida en las naciones americanas: Bibliografía.

Stat. 7, Aug. 10, '45 **International Statistical Standards.** Measures to obtain international statistical comparability. Washington, D.C., August 1945. Inter American Statistical Institute. 14 p. Typewritten. Bibliography series, no. Stat. 7.

Bibliografía del Instituto Interamericano de Estadística sobre estandards internacionales de estadística.

Int-**3010**.E5 1934 **The Economist's Handbook.** A manual of statistical sources. By Gerlof Verwey and D. C. Renooij. Amsterdam, Amstel 20-22, 1934. 460 p.

Includes: Part I, **Subjects** (a) alphabetical index of subjects classified, (b) classification of subjects with indication of sources on each; part II, **Sources** (a) classification of sources according to country, giving detailed data on name, price, and character of source, for (1) general, and (2) separately for each country; and (b) alphabetical index of all sources classified. In the section classifying sources by countries, publications are identified as 10-year, 5-year, annual, semi-annual, quarterly, monthly, irregular, 1-time, etc. In the alphabetical index by title, incomplete titles are given, with dashes to indicate omissions.

Manual de fuentes estadísticas. Incluye: Parte I, **Materias** (a) índice alfabético de la clasificación por materia, (b) clasificación por materia con indicación de las fuentes de información de cada una; parte II, **Fuentes de información** (a) clasificación de las fuentes de información por países, dando datos detallados de nombre, precio y tipo, (1) en general, y (2) por separado para cada país; y (b) índice alfabético de las fuentes de información clasificadas. En la sección de las fuentes de información clasificadas por países, la publicación se identifica por decenal, quinquenal, anual, semestral, trimestral, mensual, irregular, única, etc. En el índice alfabético por títulos, los títulos se dan incompletos con guiones indicando omisiones.

*Int-**3020**.B5 **Bulletin de l'Institut International de Statistique.** 1886- . Irrégulier. Rome, etc.

Bulletin of the International Statistical Institute. Contains proceedings and papers of each session of the Institute.

Comprend les rapports et communications de chaque session de l'Institut.

1887-1938 "Tables des Matières Contenues dans les Tomes I à XXX." Tome XXX, Supplément. La Haye, 1941. 75 p.

Table of contents for volumes I to XXX.

1938, Castrilli "Coordination Internationale des Statistiques Universitaires." Origine sociale des étudiants. Par Vincenzo Castrilli. Tome XXX, 3ème livraison (Session de Prague, 1938), p. 267-275.

International coordination of statistics of universities: Social background of the students.

1938, Dore "Le Recensement Agricole Mondial de 1940." Par Valentino Dore. Tome XXX, 3ème livraison (Session de Prague, 1938), p. 163-167.

The world agricultural census of 1940.

1938, Marshall "The Canadian Census of Distribution." By H. Marshall. Tome XXX, 3ème livraison (Session de Prague, 1938), p. 168-176.

Le recensement de la distribution au Canada.

1938, Molinari "Deuxième Rapport sur les Statistiques de la Distribution." Par A. Molinari. Tome XXX, 3ème livraison (Session de Prague, 1938), p. 94-131.

Second report on statistics of distribution.

1938, Molinari, a "Statistiques du Tourisme International." Rapport de la Commission de la Statistique des Migrations. Par A. Molinari. Tome

XXX, 3ème livraison (Session de Prague, 1938), p. 1-12.

International statistics of tourism: Report of Commission on Statistics of Migration.

1938, Rice "The Organization of Statistical Services in the United States and Great Britain." By Stuart Arthur Rice, and Richard O. Lang. Tome XXX, 3ème livraison (Session de Prague, 1938), p. 224-242.

L'organisation des services statistiques aux États-Unis d'Amérique et en Grande-Bretagne.

1938, Smutny "Méthode-Frontière dans la Statistique du Tourisme." Par P. Smutny. Tome XXX, 3ème livraison (Session de Prague, 1938), p. 87-93.

Border method in statistics of tourism.

1936, Julin "Premier Rapport sur l'Organisation des Services Statistiques." Par Armand Julin. Tome XXIX, 2ème livraison (Session d'Athènes, 1936), p. 175-222.

First report on organization of statistical services.

1936, Szturm de Sztrem "Observations au Sujet du Calcul des Indices des Prix et du 'Quantum' des Importations et Exportations." Par Edouard Szturm de Sztrem. Tome XXIX, 2ème livraison (Session d'Athènes, 1936), p. 142-147.

Observations on calculation of indexes of prices and of "quantity" of imports and exports.

1934 "Statuts de l'Institut International de Statistique." Texte adopté à Londres (1934). Tome XXVIII, 1ère livraison (Session de Londres, 1934), p. 165-170.

Statutes of International Statistical Institute, adopted in London, 1934.

1925, Jensen "Report on the Representative Method in Statistics." By Adolph Jensen. Tome XXII, 1ère livraison (Session de Rome, 1925), p. 359-380.

Includes resolutions regarding study of the application of sampling in statistics adopted by the International Statistical Institute at its 9th session in Berlin, 1903, and those proposed at its 16th session in Rome, 1925.

Rapport sur la méthode représentative dans la statistique. Comprend des résolutions relatives à l'étude de l'application de la méthode représentative dans la statistique adoptées par l'Institut International de Statistique à sa IXème session à Berlin, 1903, et celles qui furent proposées à sa XVIème session à Rome, 1925.

1925, Jensen "The Representative Method in Practice." By Adolph Jensen. Tome XXII, 1ère livraison (Session de Rome, 1925), p. 381-439.

Contains a bibliography.

La méthode représentative dans la pratique. Comprend une bibliographie.

1923, Rew "Statistics of the Fisheries." By R. H. Rew. Tome XXI, 2e partie. (Session de Bruxelles, 1923), p. 260-268.

Statistique des pêcheries.

1923, Ricci "Statistique de la Production Agricole." Par Umberto Ricci. Tome XXI, 2e partie (Session de Bruxelles, 1923), p. 168-259.

Statistics of agricultural production.

1913, March "La Statistique des Etats de Culture." Par Lucien March. Tome XX, 2ème livraison (Session de Vienne, 1913), p. 649-662.

Statistics of crop conditions.

1913, Ricci "Statistique Internationale des Superficies et des Productions Agricoles." Par Umberto Ricci. Tome XX, 2ème livraison (Session de Vienne, 1913), p. 675-763.

International statistics of area and agricultural production.

1911, Ricci "Statistique Internationale des Etats des Cultures." Par Umberto Ricci. Tome XIX, 3ème livraison (Session de La Haye, 1911), p. 1-92.

International statistics of conditions of crops.

1909, Rew "International Fishery Statistics." By R. H. Rew. Tome XVIII, 1ère livraison (Session de Paris, 1909), p. 624-637.

Statistique internationale des pêcheries.

1909, Rew, a "Statistique Internationale des Pêcheries." Par R. H. Rew. Tome XVIII, 1ère livraison (Session de Paris, 1909), p. 397-405.

International fishery statistics.

Int-**3020**.C5 1919 **Conference on International Co-operation in Statistics, August 14th and 15th, 1919.** London, Harrison and Sons, 1919. League of Nations. 41 p. Unexamined.

"The Economic and Finance section of the provisional organization of the League of Nations invited members of the Institute of Agriculture at Rome, and of the International Statistical Institute at the Hague, and of the bureau founded in connection with that institute, together with other leading statisticians, to a conference at 117, Piccadilly on August 14 and 15, 1919." (Source: Text.)

Conférence sur la coopération internationale dans la statistique, le 14 et 15 août, 1919. "La Section Économique et Financière de l'organisation provisoire de la Société des Nations a invité les membres de l'Institut d'Agriculture à Rome, de l'Institut International de Statistique à La Haye, et du bureau organisé en relation avec cet institut, ainsi que d'autres statisticiens importants, à une conférence à 117, Piccadilly, le 14 et 15 août, 1919." (Traduction du commentaire.)

Int-**3030**.C5 1884-1934 **50 Années de l'Institut International de Statistique.** Par Friedrich Zahn. Munich, 1934. Institut International de Statistique. 181 p.

50 years of the International Statistical Institute.

Int-**3030**.I6 1943 **The Inter American Statistical Institute. El Instituto Interamericano de Estadística. O Instituto Inter-Americano de Estatística.** Its origin, organization, and objectives. Su origen, organización y finalidades. Sua origem, organização e finalidades. Second edition—segunda edición—segunda edição. Washington, D.C., 1943. Inter American Statistical Institute. 48 p.

Descriptive brochure of Inter American Statistical Institute, containing statutes, membership roster, committees, and other information.

Folleto descriptivo acerca del Instituto Interamericano de Estadística, incluyendo estatutos, lista de miembros componentes, comisiones y otras informaciones.

Int-**3040**.B5 **Boletín del Instituto de Investigaciones Sociales y Económicas. Bulletin of the Institute of Social and Economic Research. Bulletin de l'Institut des Recherches Sociales et Economiques. Boletim do Instituto de Investigações Sociais e Econômicas.** 1944- . Trimestral. Panamá. Universidad Interamericana. Inter-American University. Université Interamericaine. Universidade Inter-americana.

Contains articles, announcements, and survey of recent publications in fields of social science and economics.

Contiene artículos, prospectos y un estudio sobre las nuevas publicaciones en el campo de las ciencias sociales y de la economía.

Feb. '45, a "Directorio de Especialistas en Estudios Sociales y Económicos en Latinoamérica, Parte I." Vol. II, no. 3, febrero 1945, p. 317-397.

Directory of specialists in social and economic studies in Latin America, part I.

Feb. '45, b "Directorio de Instituciones de Estudios e Investigaciones Sociales y Económicos en Latinoamérica, Parte I." Vol. II, no. 3, febrero 1945, p. 398-487.

Directory of institutions of social and economic research in Latin America, part I.

Int-**3040**.E3 **Econometrica.** Journal of the Econometric Society, an international society for the advancement of economic theory in its relation to statistics and mathematics. 1933- . Quarterly. Chicago, Illinois, University of Chicago. Econometric Society.

Publicación de la Sociedad Econométrica, una sociedad internacional dedicada al progreso de la teoría económica en su relación con la estadística y las matemáticas.

July '46, Rice "The United Nations Statistical Commission." By Stuart A. Rice. Vol. 14, no. 3, July 1946, p. 242-250.

La Comisión de Estadística de las Naciones Unidas.

*Int-**3040**.E5 **Estadística.** Journal of the Inter American Statistical Institute. 1943- . Quarterly. Mexico.

Contains articles on statistical methodology, studies of the members of the Inter American Statistical Institute, and reports on activities of the organization.

Revista trimestral del Instituto Interamericano de Estadística. Contiene artículos sobre metodología estadística, estudios de los miembros del Instituto, e informes sobre las actividades del organismo.

Sept. '46, Coghlan "El Censo Escolar Nacional de 1943." Por Eduardo A. Coghlan. Vol. IV, no. 15, septiembre 1946, p. 341-352.

The 1943 school census of Argentina. Discussion of scope, organization, demarcation of territory, questionnaire utilized, characteristics and attributes registered, etc. Includes individual schedule and family schedule.

El censo escolar de 1943 de Argentina. Discusión sobre extensión, organización, demarcación del territorio, cuestionarios utilizados, características y atributos registrados, etc. Incluye "Cuestionario individual" y "Cuestionario familiar."

Sept. '46, Knox "Estadística de Metales y Minerales." Por Newton B. Knox. Vol. IV, no. 15, septiembre 1946, p. 335-340.
On statistics of metals and minerals.

Sept. '46, Montoya "Metodología en Estadísticas Agrícolas de las Américas: Parte III, El Servicio de Estadística Agrícola en los Estados Unidos; Part IV, Conclusiones." Por Hernán Montoya. Vol. IV, no. 15, septiembre 1946, p. 360-406.
Methods of agricultural statistics in the Americas: Part III, Agricultural statistics in the United States; Part IV, Conclusions. Third and final article in a series; the first one was published in **Estadística**, March 1946, the second in June 1946. Includes forms.
Artículo tercero de una serie; el primero se publicó en **Estadística** en marzo 1946, el segundo en junio 1946. Incluye formularios.

Sept. '46, Rice "ISI, IASI, and UN." By Stuart A. Rice. Vol. IV, no. 15, septiembre 1946, p. 323-334.
In English and Spanish—en inglés y español.
Relations of the International Statistical Institute and the Inter American Statistical Institute to the United Nations.
Relaciones del Instituto Internacional de Estadística y del Instituto Interamericano de Estadística con las Naciones Unidas.

Sept. '46, Ullman "Registros, Formas y Otra Documentación de una Investigación Estadística." Por Morris B. Ullman. Vol. IV, no. 15, septiembre 1946, p. 353-359.
Translation of—traducción de: "Records of a Statistical Survey," **American Archivist**, January 1942.

June '46, Domínguez "Renta Nacional, Su Significado y Medición, Parte III." Por Loreto M. Domínguez. Vol IV, no. 14, junio 1946, p. 233-251.
National income, its significance and measurement. Last article in a series of three, the first of which was published in **Estadística**, December 1945, and the second in March 1946.
La parte III finaliza esta serie de artículos. La parte I se publicó en el número de diciembre 1945 de **Estadística**, y la parte II en el número de marzo 1946.

June '46, Dunn "Statistics in the World of Tomorrow."—"La Estadística en el Mundo del Mañana." By Halbert L. Dunn. Vol. IV, no. 14, junio 1946, p. 163-171.

June '46, Montoya "Metodología en Estadísticas Agrícolas de las Américas: Parte II, Sistemas Estadísticos Agrícolas en Algunos Países Latinoamericanos." Por Hernán Montoya. Vol. IV, No. 14, junio 1946, p. 172-233.
Methods of agricultural statistics in the Americas. Part II, Systems of agricultural statistics in some Latin American countries. Second article in a series, the first of which was published in **Estadística**, vol. IV, no. 13, March 1946.
Artículo segundo de una serie, de la cual el primero fué publicado en **Estadística**, vol. IV, no. 13, marzo 1946.

Mar. '46, Domínguez 'Renta Nacional, 'Su Significado y Medición, Parte II." Por Loreto M. Domínguez. Vol. IV, no. 13, marzo 1946, p. 79-95.
National income, its significance and measurement, part II.

Mar. '46, Keyfitz "Calculation of Life Tables for Population of Countries and Regions." By Nathan Keyfitz. Vol. IV, no. 13, marzo 1946, p. 70-78.
Cálculo de tablas de mortalidad para la población de países y de regiones.

Mar. '46, Mahaney "Machine Tabulation." By John J. Mahaney. Vol. IV, no. 13, marzo 1946, p. 101-110.
Tabulación mecánica.

Mar. '46, Montoya "Metodología en Estadísticas Agrícolas de las Américas." Estudio preliminar basado en un grupo seleccionado de productos agrícolas en un número limitado de países. Incluye una exposición general de métodos de las muestras y de métodos de los censos agropecuarios. Por Hernán Montoya. Introducción y parte I. Vol. IV, no. 13, marzo 1946, p. 3-69.

Methods of agricultural statistics in the Americas. Preliminary study based on a selected group of agricultural products, in a limited number of countries. Includes a general exposition of sampling methods and methods of agricultural and livestock censuses. Introduction and part I.

Mar. '46, Roig Ocampos "Prueba de Registro de Nacimientos en Paraguay." Por Jorge Roig Ocampos y Charles G. Bennett. Vol. IV, no. 13, marzo 1946, p. 111-116.
Birth registration test in Paraguay.

Dec. '45, Domínguez "Renta Nacional, Su Significado y Medición." Parte I. Por Loreto M. Domínguez. Vol. III, no. 12, diciembre 1945, p. 483-511.
National income, its significance and measurement. Part I.

Dec. '45, Greenway "Attributes of Wage-Earner Families in Relation to Size." By Harold F. Greenway. Vol. III, no. 12, diciembre 1945, p. 546-553.
Características de las familias de asalariados en relación a su tamaño.

Dec. '45, Michalup "La Composición Familiar de Asalariados en el Distrito Federal de Venezuela." Ajustamiento de los datos de observación. Por Eric Michalup. Vol. III, no. 12, diciembre 1945, p. 554-570.
Composition of wage-earner families in the Federal District of Venezuela. Adjustment of collected data.

Sept. '45, a "Gross and Net Reproduction Rates, Canada and Provinces, 1920-1942." By the Dominion Bureau of Statistics. Vol. III, no. 11, septiembre 1945, p. 350-353.

Prepared by S. A. Cudmore and Enid Charles.

Tasas brutas y netas de reproducción, Canadá y provincias, 1920-1942.

Sept. '45, Domínguez "Precios de Exportación en la República Argentina, 1863-1943." Por Loreto M. Domínguez. Vol. III, no. 11, septiembre 1945, p. 368-380.
Prices of exports in Argentina, 1863-1943.

Sept. '45, Ely "Cambios en la Clasificación de Estadísticas de Exportación de los Estados Unidos." Muchos entraron en vigor en enero de 1945. Por J. Edward Ely y Frederick W. Harrison. Vol. III, no. 11, septiembre 1945, p. 395-401.
Translation of article published in—traducción del artículo publicado en: **Journal of International Economy,** January 6, 1945, p. 13.
Changes in classification of export statistics of the United States.

Sept. '45, Moriyama "A Method for Indexing Statistical Tabulations." By I. M. Moriyama. Vol. III, no. 11, septiembre 1945, p. 391-394.
Un método de compilar índices de tabulaciones estadísticas.

Sept. '45, Rodríguez y Rodríguez "La Oficina de Estadística del Instituto Mexicano del Seguro Social." Por Jesús Rodríguez y Rodríguez. Vol. III, no. 11, septiembre 1945, p. 385-390.
Incluye un esquema estructural.
The Statistical Section of the Mexican Institute of Social Security. Includes organization chart.

Sept. '45, Uriarte "Breves Apuntes sobre los Registros de Nacimientos y Defunciones en los Estados Unidos." Vol. III, no. 11, septiembre 1945, p. 354-367.

Brief sketch on registration of births and deaths in the United States.

Dec. '44, Hansen "A New Sample of the Population." By Morris H. Hansen and William N. Hurwitz. Vol. II, no. 8, diciembre 1944, p. 483-497.

A description of the sample designed to obtain current estimates of characteristics of the labor force and to serve related purposes, including the sampling principle, grouping primary sampling units into strata, and sampling within selected primary sampling units.

Una descripción de la muestra delineada para obtener estimaciones corrientes de las características del potencial obrero y para propósitos similares, incluyendo el fundamento de la muestra, la agrupación en estratos de las unidades primarias de muestra y la investigación sobre las mismas.

Dec. '44, Luna Vegas "La Estadística Peruana en el

Sexenio 1938-44." Por Ricardo Luna Vegas. Vol. II, no. 8, diciembre 1944, p. 498-505.

Peruvian statistics during the 6-year period, 1938-44. This history of Peruvian statistics includes a brief discussion of the census law of 1938, census organization, and planning for execution of the census of 1940; development of the National Bureau of Statistics and founding of the Peruvian Institute of Statistics; inventory of economic potential of the nation, and organic law of registers of civil status.

Esta historia de estadística peruana incluye una breve discusión sobre la ley de censos del año 1938, organización y proyecto de ejecución del censo de 1940; desarrollo de la Dirección Nacional de Estadística y creación del Instituto Peruano de Estadística; Inventario del Potencial Económico de la Nación, y Ley Orgánica de los Registros del Estado Civil.

June '44, a "Organización del Servicio de Estadística Nacional en el Perú: Decreto y Esquema." Vol. II, no. 6, junio 1944, p. 258-259.

Organization of the National Bureau of Statistics in Peru: Text of decree of January 1, 1944, and organization chart.

Texto del decreto del 1° de enero de 1944 y esquema de organización.

June '44, b "Final Reports of the Sixteenth Census of the United States, 1940." (Source: U.S. Census Bureau.) **Estadística,** vol. II, no. 6, junio 1944, p. 305-308.

A list of the volumes containing final published results of the 1940 census of the United States.

Una lista de los volúmenes que contienen los resultados finales publicados sobre el censo de 1940 de los Estados Unidos.

Mar. '44 "National Statistical Societies in the Western Hemisphere."—"Sociedades Nacionales de Estadística en el Hemisferio Occidental." Vol. II, no. 5, marzo 1944, p. 122-134.

Information about national statistical societies of Argentina, Brazil, Colombia, Mexico, Peru, United States, and Uruguay.

Información sobre las sociedades nacionales de estadística de Argentina, Brasil, Colombia, México, Perú, los Estados Unidos, y Uruguay.

Mar. '44, Alanís Patiño "Enseñanza de la Estadística en México." Por E. Alanís Patiño. Vol. II, no. 5, marzo 1944, p. 51-63.

Information about various courses of statistics given in Mexico.

Información acerca de los varios cursos de estadística, dictados en México.

Sept. '44, Taeuber "The Development of Population Predictions in Europe and the Americas." By Irene B. Taeuber. Vol. II, no. 7, septiembre 1944, p. 323-346.

Partial Spanish translation appears in **Revista de Economía Argentina,** año XXVII, tomo XLIV, no. 326, agosto 1945, p. 415-419.

Evolución de la técnica para predecir la población, en Europa y las Américas. Parte de este artículo se publicó en español en la **Revista de Economía Argentina,** año XXVII, tomo XLIV, no. 326, agosto 1945, p. 415-419.

Mar. '43, Guthardt "Orientación, Organización y Resultados de un Curso Oficial de Estadística por Correspondencia." Por Emilio C. Guthardt. Vol. I, no. 1, marzo 1943, p. 132-142.

Orientation, organization, and results of the official correspondence course in statistics in Colombia.

Texto:

*Int-**3040**.R5 **Revue de l'Institut International de Statistique.** 1933- . Trimestrielle. La Haye. Institut International de Statistique.

From 1920 to 1932 published as **Bulletin Mensuel de l'Office Permanent de l'Institut International de Statistique,** containing only data, no articles—de 1920 à 1932 publié comme **Bulletin Mensuel de l'Office Permanent de l'Institut International de Statistique,** ne contenant que de données, pas d'articles.

Review of the International Statistical Institute. Text in English, French, German, Italian, or Spanish, often with summary in English or French. Table of contents: Scientific studies of statistical nature; session of Prague; communications on statistical methodology, legislation, organization, and administration; international statistical bibliography; statistical bureaux, services, and institutions of the different countries of the world; statistical societies in the different countries of the world; statistical tables; news for the members.

Texte en anglais, français, allemand, italien ou espagnol, souvent avec sommaire en anglais ou français. Table des matières: Etudes scientifiques d'ordre statistique; session de Prague; communications sur la méthodologie, la législation, l'organisation et l'administration statistiques; bibliographie statistique internationale; bureaux, services et institutions statistiques des divers pays du monde; sociétés de statistique dans les divers pays du monde; tableaux statistiques; avis aux membres.

1941, 3-4, Dunn "A National Test for Completeness of Birth Registration in the United States of America." By Halbert L. Dunn. 10e année, livraisons 1/2, 1942, p. 75-80.

Includes an outline of birth and death registration procedure in the United States, and birth registration forms.

Une épreuve de l'intégralité de l'enregistrement des naissances dans les Etats-Unis. Comprend une esquisse des méthodes de l'enregistrement des naissances et décès dans les Etats-Unis, et formulaires de l'enregistrement des naissances.

1941, 3-4, Huber "Le Comité d'Experts Statisticiens du Bureau International du Travail (1933-1939)." Par Michel Huber. 9 année, livraisons 3/4, 1941, p. 104-112.

After explaining how the Committee of Statistical Experts of the International Labour Office was established in 1933, the writer describes the activities of the Committee and the subjects dealt with at its meetings, the results which have been obtained, and the difficulties of the principal problems submitted to the Committee.

A la suite d'une explication de l'établissement du Comité d'Experts Statisticiens en 1933, l'auteur décrit les activités du Comité, et les questions dont il s'agit dans ses réunions, ainsi que les résultats obtenus, et les difficultés des problèmes principaux soumis au Comité.

1941, 1-2, Mortara "Quelques Possibilités de Reconstruction du Mouvement d'une Population à l'Aide des Données des Recensements Démographiques." Par Giorgio Mortara. 9 année, livraisons 1/2, 1941, p. 21-35.

Shows procedure by which, in the case of a population lacking birth and death statistics, it becomes possible to determine approximately the numbers of births and deaths which occurred in the decennial interval between two censuses, provided these indicate the age distribution of the population.

Explique la méthode qui, dans le cas d'une population manquant de statistiques de naissances et de décès, rend possible la détermination approximative des chiffres de naissances et de décès qui se sont produits dans l'intervalle décennal entre deux recensements, pourvu que ces recensements indiquent la répartition de la population d'après l'âge.

1939, 2-3, Godfrey "Inquiry Concerning Statistics of Wool Stocks in the Principal Wool-Producing and Wool-Consuming Countries of the World." By Ernest H. Godfrey. 7 année, livraisons 2/3, 1939, p. 145-163.

A resumé of responses of 14 countries (including Argentina, Canada, and the United States of the Western Hemisphere) to a questionnaire sent to 16 countries is included.

Enquête statistique sur les stocks de laine dans les principaux pays producteurs et consommateurs de laine du monde. Comprend un résumé des réponses de 14 pays (y compris l'Argentine, le Canada et les Etats-Unis, de l'hémisphère occidental) à un questionnaire envoyé à 16 pays.

1939, 2-3, Huber "Le Comité d'Experts Statisticiens de la Société des Nations (1931-1939)." Par Michel Huber. 7 année, livraisons 2/3, 1939, p. 117-137.

After explaining how the Committee of Statistical Experts of the League of Nations was established in 1931, the author describes activities of the Committee and results of its projects.

A la suite d'une explication de l'établissement du Comité d'Experts Statisticiens en 1931, l'auteur décrit les activités du Comité, et les résultats de ses projets.

1938.1, Huber "Rapport sur les Travaux Préparatoires à la Vème Révision Décennale de la Nomenclature Internationale des Causes de Décès." Par Michel Huber. 6 année, livraison 1, 1938, p. 103-128.

Report on preparatory work for 5th decennial revision of International List of Causes of Death. Includes proposed nomenclature in detail (200 numbers).

Comprend projet de nomenclature détaillée des causes de décès (200 numéros).

1938.2, Bunle "Rapport de la Commission pour la Définition de la 'Population Rurale'." Par Henri Bunle. 6 année, livraison 2, 1938, p. 229-234.

INTERNATIONAL
3040

Report of Commission on Definition of "Rural Population."

1938.2, Flux "The Statistical Measurement of Changes in National Productive Equipment." By Alfred W. Flux. 6 année, livraison 2, 1938, p. 235-240.

La mesure statistique des changements de l'outillage national.

1938.2, Idenburg "Directives pour l'Elaboration des Statistiques de la Radiodiffusion dans les Divers Pays." Rapport au nom de la commission mixte entre l'Institut International de Statistique et l'Union Internationale de Radiodiffusion. Par Ph. J. Idenburg. 6 année, livraison 2, 1938, p. 241-250.

Directives for elaboration of statistics of radio broadcasting in different countries. Report for mixed commission of International Statistical Institute and International Union of Radio Broadcasting.

Includes—comprend: Introduction; statistique des postes émetteurs; statistique des programmes; statistique du personnel; statistique des auditeurs et des abonnés.

1938.2, Nyström "Commission on Building and Housing Statistics: Preliminary Report." By Bertil Nyström. Année 6, livraison 2, 1938, p. 251-263.

Rapport préliminaire de la Commission de la Statistique des Nouvelles Constructions et de l'Habitation.

1938.2, Zanten "Les Bases de Comparaison du Nombre des Accidents de la Circulation Routière." Par J. H. van Zanten. 6 année, 2 livraison, 1938, p. 264-287.

Bases of comparison of number of traffic accidents.

Includes—comprends: Le chiffre de la population; le nombre des autos; le nombre des chauffeurs; la longueur (la superficie) des route; résultats des recensements de la circulation; la consommation d'essence comme moyen de déterminer l'intensité de la circulation; combinaison de connées; conclusions.

1938.3, Huber "Conference Internationale pour la Cinquième Revision Decennale des Nomenclatures Internationales de Causes de Décès, 1938." Par M. Huber. Année 6, livraison 3, 1938, p. 428-454.

International conference on fifth decennial revision of International List of Causes of Death.

Includes—comprend: Convention; recommandations; nomenclatures internationales de causes de décès.

1938. 3, Nixon "On the Statistics Available Concerning the Occupied Population of the World and Its Distribution." By J. W. Nixon. 6 année, livraison 3, 1938, p. 377-393.

Includes tables.

Rapport sur les données statistiques actuellement disponibles sur le nombre et la distribution de la population active du monde. Comprend des tableaux.

1937.2, Bunle "Sondages Démographiques dans les Pays Sans Etat Civil Régulier." Par Henri Bunle. 5 année, livraison 2, juillet 1937, p. 132-146.

Demographic investigations in countries not having regular civil registration. Includes model card adopted.

Comprend modèle de fiche adoptée.

1937.4, Bunle "La Population Rurale." Sur l'adoption d'une définition susceptible d'être internationalement adoptée. Par Henri Bunle. 5 année, livraison 4, 1937, (janvier 1938), p. 347-357.

Rural population—on the adoption of a definition suitable for being internationally adopted.

Includes—comprend: Définitions adoptées en divers pays; examen des définitions précédentes; d'une définition internationale; suggestions pour une définition-type internationale; résumé et conclusions.

1936.1, Huber "La Statistique Internationale des Forces Motrices" Rapport par Michel Huber. 4 année, livraison 1, avril 1936, p. 62-70.

International statistics of motor power.

Includes—comprend: Rappel de la question posée; doubles emplois dans le calcul de la puissance installée; analyse des réponses reçues; conclusions; projet de recommandations.

1936.1, Nixon "Committee on Family Budgets: Preliminary Report." By J. W. Nixon. 4 année, livraison 1, avril 1936, p. 45-61.

Rapport préliminaire de la Commission sur les Budgets Familiaux.

Includes—comprend: Definition and scope of the various groups of expenditures;

receipts in kind; instalment buying; repayment of loans, etc.; conclusion.

1936.1, Nyström "Observations on the Possibility of Improving the International Comparability of Building and Housing Statistics." Paper by Bertil Nyström. 4 année, livraison 1, avril 1936, p. 71-85.

Quelques observations sur la possibilité d'avancer les comparaisons internationales des statistiques du bâtiment et du logement.

1936.2, Dore "La Statistique Forestière Internationale." Rapport sur les travaux de la commission mixte constituée par l'Institut International de Statistique et l'Institut International d'Agriculture. Par Valentino Dore. 4 année, livraison 2, juillet 1936, p. 206-211.

International forestry statistics. Report on work of mixed commission constituted by International Statistical Institute and International Institute of Agriculture.

Includes—comprend: Constitution et travaux de la Commission; buts et directives générales des travaux; propositions de la Com mission.

1936.2, Idenburg "Statistique de la Radiodiffusion." Communication du Dr. Philip J. Idenburg. 4 année, livraison 2, juillet 1936, p. 266-273.

Statistics of radio broadcasting.

Includes—comprend: Introduction; l'émission—organisation de la radiodiffusion, postes émetteurs, programmes, financement de la radiodiffusion; la réception—les auditeurs, appareils récepteurs; conclusion.

1936.2, Molinari "Observations et Propositions pour une Statistique Internationale du Tourisme." Note de A. Molinari. Année 4, livraison 2, juillet 1936, p. 243-259.

Observations and proposals for international tourist statistics.

Includes—comprend: L'importance du problème; conditions pour l'établissement d'une statistique internationale uniforme; migrations et tourisme; les travaux de la Société des Nations; observations et propositions; conclusions.

1936.2, Schäfer "Directives pour l'Elaboration des Statistiques Criminelles dans les Divers Pays." Rapport au nom de la commission mixte constituée par l'Institute International de Statistique et la Commission Internationale Pénale et Pénitentiaire. Par E. Schäfer. 4 année, livraison 2, juillet 1936, p. 195-205.

Directives for elaboration of statistics of crime in different countries. Report for mixed commission constituted by International Statistical Institute and International Penal and Penitentiary Commission.

Includes—comprend: Introduction générale aux publications statistiques criminelles officielles; plan modèle; annexe A, Plan pour une introduction générale a la statistique criminelle; annexe B, Schéma-type des cadres statistiques.

1936.4, Zanten "L'Uniformité dans la Statistique des Accidents de la Circulation." Note du J. H. van Zanten. 4 année, livraison 2, juillet 1936, p. 260-265.

Uniformity in statistics of traffic accidents.

1936.4, Godfrey "Progress in the Collection of Annual Agricultural Statistics and of Crop Reports during Growth in the Dominion of Canada." By Ernest H. Godfrey. 4 année, livraison 4, 1936 (janvier 1937), p. 518-534.

Rapport sur le progrès réalisé au Canada, dans la réunion des statistiques agricoles annuelles, et des renseignements sur les cultures durant la periode de végétation.

1934.2, Ptoukha "Méthodes pour Calculer les Taux de Mortalité Infantile selon les Mois de l'Année." Par Michel Ptoukha. 2 année, livraison 2, juillet 1934, p. 178-189.

Methods for calculating infant death-rate by months of the year.

1934.3, Gini "On a Method for Calculating the Infantile Death-Rate According to the Month of Death." By Corrado Gini. 2 année, livraison 3, octobre 1934, p. 292-300.

Discours sur une méthode pour calculer les taux de mortalité infantile selon le mois du décès.

1933.1, Nixon "Index - Numbers of Wages: A Survey." By J. W. Nixon. 1 année, livraison 1, 1933, p. 39-47.

Un aperçu des nombres-indices des salaires.

Includes—comprend: Meaning of the term, material available, purposes, representativeness of index numbers of wages; type of wage data; index numbers of wages as a measure of labour cost of production; wages as a proportion of the national income; conclusions; table I, "Series Showing the Movement of Hourly, Daily, or Weekly Wages" (in certain countries); table II, "Series Showing the Movements of Annual Wages Derived from Census of Production (or Manufacturing)" (in certain countries).

1933.2, Bowley "Methods of Investigating the Economic and Social Conditions of a Great City." By Arthur L. Bowley. 1 année, livraison 2, juillet 1933, p. 7-18.

Quelques méthodes d'enquête sur les situations économiques et sociales d'une grande ville.

Includes—comprend: Introduction; definition of poverty; the street survey—poverty maps, and classification of the population; the house sample—methods, results.

1933.2. Huber "La Statistique Internationale des Forces Motrices." Communication de Michel Huber. 1 année, livraison 2, juillet 1933, p. 40-45.

International statistics of motor power.

Includes—comprend: Recommandations antérieures; examen sommaire de la question; doubles emplois dans le compte de la puissance installée; conclusions.

1933.3, Craig "Report of the Commission for the Study of Statistics of Consumption and of Carryover of Commodities." By J. I. Craig. 1 année, livraison 3, octobre 1933, p. 47-58.

Rapport de la Commission pour l'Etude de la Statistique de Consommation et des Reports de Denrées.

1933.3, Dore "La Statistique Forestière Internationale et ses Problèmes." Par Valentino Dore. 1 année, livraison 3, octobre 1933, p. 59-65.

International forestry statistics and its problems.

*Int-3200.D5 1946 **Banking Statistics, Recommendations on Scope and Principles of Classification; Report of the Sub-Committee on Banking Statistics.** Princeton, New Jersey, March 1946. League of Nations. Committee of Statistical Experts. Sub-Committee on Banking Statistics. 56 p. Mimeographed.

Cover title—titre de la couverture: **Draft Report.**

Contents: Scope and uses of banking statistics (existing banking statistics, and information banking statistics should supply on structure and characteristics of the banking system and data on assets and liabilities, earnings, expenses and allocation of profits, volume of operations and significant asset-liability relationships, interest rates, etc.); classification of credit institutions in banking statistics (credit institutions included, principles of classification, classification by type of functions performed, by legal status, by size, by type of supervising authority, distinction between foreign and domestic banks, division by type of area of operation); detailed suggestions for the development of banking statistics (statistical statements of assets and liabilities, of earnings, expenses and allocation of profits, data on volume of operations); annex, basic classification and supplementary sub-classifications of assets and liabilities in bank balance-sheet statistics.

Statistiques bancaires, recommandations sur la portée et les principes de classification; rapport du Sous-Comité des Statistiques Bancaires. Publié par le Sous-Comité des Statistiques Bancaires du Comité d'Experts Statisticiens, Société des Nations. Matières: Portée et usages des statistiques bancaires (statistiques bancaires actuelles, et renseignements que les statistiques bancaires devraient fournir concernant la structure et les caractéristiques du régime bancaire et données sur l'actif et le passif, les profits, les dépenses et l'allocation des bénéfices, le volume des opérations et les relations importantes entre l'actif et le passif, les taux d'intérêts, etc.); classification des institutions de crédit dans les statistiques bancaires (y compris les institutions de crédit, les principes de classification, la classification selon le genre de fonction, selon l'état légal, la valeur de l'actif, le genre d'autorité de direction, distinction entre les banques étrangères et domestiques, division selon le genre de la région d'activité); suggestions détaillées pour le dévelopment des statistiques bancaires (rapports statistiques de l'actif et du passif, des profits, des dépenses, et de la répartition des bénéfices, données sur le volume des opérations); annexe, classification de base et sous-classifications supplémentaires de l'actif et du passif dans les statistiques des bilans des banques.

*Int-3200.H5 1939 **Housing Statistics.** A minimum programme of housing statistics drawn up by the Committee of Statistical Experts. Geneva, 1939. League of Nations. 14 p. Studies and reports on statistical methods, no. 5.

Published also in French—publié aussi en français (Int-3200.S5, 1939).

*Int-**3200**.I3 1939 **Indices de la Production Industrielle.** Rapport établi par le Comité d'Experts Statisticiens. Genève, 1939. Société des Nations. 16 p. Etudes et rapports sur les méthodes statistiques, no.6.

Published also in English—publié aussi en anglais (Int-**3200**.I5).

*Int-**3200**.I5 1939 **Indices of Industrial Production.** Report drawn up by the Committee of Statistical Experts. Geneva, 1939. League of Nations. 16 p. Studies and reports on statistical methods, no. 6.

Published also in French—publié aussi en français: (Int-**3200**.I3).

*Int-**3200**.S3 1938 **Statistics Relating to Capital Formation.** A note on methods by the Committee of Statistical Experts. Geneva, 1938. League of Nations. 22 p. Studies and reports on statistical methods, no. 4.

Published also in French—publié aussi en français (Int-**3200**.S7, 1938).

*Int-**3200**.S5 1939 **Statistiques de l'Habitation.** Programme minimum de statistiques de l'habitation, établi par le Comité d'Experts Statisticiens. Genève, 1939. Société des Nations. 14 p. Etudes et rapports sur les méthodes statistiques, no. 5.

Published also in English—publié aussi en anglais (Int-**3200**.H5, 1939).

*Int-**3200**.S6 1938 **Statistiques du Bois.** Programme minimum de statistiques du bois, élaboré par le Comité d'Experts Statisticiens. Genève, 1938. Sociétés des Nations. 18 p. Etudes et rapports sur les méthodes statistiques, no. 3.

Published also in English—publié aussi en anglais (Int-**3200**.T5, 1938).

*Int-**3200**.S7 1938 **Statistiques Relatives à la Formation des Capitaux.** Note sur les méthodes par le Comité d'Experts Statisticiens. Genève, 1938. Société des Nations. 24 p. Etudes et rapports sur les méthodes statistiques, no. 4.

Published also in English—publié aussi en anglais (Int-**3200**.S3, 1938).

Int-**3200**.S8 **Studies and Reports on Statistical Methods. Etudes et Rapports sur les Méthodes Statistiques.** 1938- . Irregular. Geneva. League of Nations. Committee of Statistical Experts. Société des Nations. Comité d'Experts Statisticiens.

Series of reports containing the statements and recommendations of the Committee of Statistical Experts.

Série de rapports contenant les exposés et les recommandations du Comité d'Experts Statisticiens.

*Int-**3200**.T5 1938 **Timber Statistics.** A minimum programme of timber statistics drawn up by the Committee of Statistical Experts. Geneva, 1938. League of Nations. 17 p. Studies and reports on statistical methods, no. 3.

Published also in French—publié aussi en français (Int-**3200**.S6, 1938.

*Int-**3205**.B5 1945 **Balance of Payments Statistics: Note by the Secretariat.** Princeton, New Jersey, May 1945. League of Nations. Committee of Statistical Experts. Sub-Committee on Balance of Payments Statistics. 40 p. Mimeographed.

Statistiques sur la balance des paiements: Note par le Secretariat. Etabli par le Sous-Comité des Statistiques sur la Balance des Paiements du Comité des Experts Statisticiens, Société des Nations.

Int-**3205**.B5 1945, Annexes I, II **Annex I, League of Nations Estimated Balance of International Payments for the Year . . . (8-page form for country to fill); and Annex II, Explanatory Notes to Scheme of International Balance of Payments (11 pages).** Princeton, New Jersey, 1945. League of Nations. Committee of Statistical Experts. Sub-Committee Meetings on Banking Statistics. 19 p. Mimeographed.

Annexe I, Balance des paiements internationaux estimée par la Société des Nations pour l'année . . . (formule contenant huit pages pour être remplie par le pays); et Annexe II, notes explicatives sur le projet de la balance internationale des paiements (onze pages).

Int-**3205**.B5 1945, Annex A **Note on Balance of Payments Statistics: Annex A.** Princeton, New Jersey, 1945. League of Nations. Committee of Statistical Experts. Sub-Committee on Balance of Payments Statistics. 22 p. Mimeographed. No. C.E.S./152, C.E.S./s/c B.P.6.

Note sur la statistique de la balance des paiements: Annex A.

*Int-**3210**.C4 1913.1 **Convention Respecting the Compilation of International Commercial Statistics.** Signed at Brussels, December 31, 1913. London, H.M. Stationery Off., 1924. Great Britain. Foreign Office. 43 p. Treaty series, 1924, no. 15. Unexamined.

In English and French—en anglais et français.

Convention relative à la compilation des statistiques internationales du commerce, signée à Bruxelles, 31 décembre, 1913.

*Int-3210.C4 1913.2 **Algunos Documentos Relativos a la Adopción de un Sistema de Estadística Internacional, Convención de Bruselas, 1913.** Se publica por acuerdo del Consejo Central Ejecutivo. Wáshington, Imp. Nacional, 1917. Conférence Internationale de Statistique Commerciale. 14 p.

Some documents relative to adoption of system of international statistics, convention of Brussels, 1913. Contents: Foreword; convention relative to establishment of international statistics of commerce; standard nomenclature of commodities; regulations of the International Bureau of Trade Statistics.

Contenido: Advertencia; convención relativa al establecimiento de una estadística comercial internacional; nomenclatura común de mercancías; reglamento de la Oficina Internacional de Estadísticas Comerciales.

Int-3210.C4 1913.3 **Documents et Procès-Verbaux de la Conférence Internationale de Statistique Commerciale, Bruxelles, 1913.** Bruxelles, Etablissements Généraux d'Imprimerie, 1914. Conférence Internationale de Statistique Commerciale. 350 p.

Documents and proceedings of International Conference of Commercial Statistics, Brussels, 1913. Includes: Preliminary documents; list of delegates and proceedings of sessions; explanatory notes concerning Brussels nomenclature of commodities; convention about establishment of international commercial statistics; annexes, protocol.

Comprend: Documents préliminaires; liste des délégués et procès-verbaux des séances; notes explicatives concernant la Nomenclature Commune des Marchandises; convention concernant l'établissement d'une statistique commerciale internationale; annexes, protocole.

*Int-3210.C6 1945 **Convertibility Index for Foreign Trade Statistical Classifications of the American Nations: Basic Classification Scheme Showing Detailed Export and Import Commodity Description by Classes of the Minimum List of Commodities for International Trade Statistics** This volume is the forerunner of a series of country indexes for converting the detailed foreign trade statistics of each American Nation into a common classification pattern for purposes of international comparison. Because practically all types of articles traded between the American nations are represented in the foreign trade of the United States, the approximately 55,000 inclusion terms for this volume have been drawn principally from the detailed commodity descriptions of the U.S. export and import statistical classification lists. Preliminary edition. Washington, D.C., October 1945. Inter American Statistical Institute. 1130 p.

Indice de convertibilidad para las clasificaciones de la estadística del comercio exterior de las naciones americanas: Esquema Básico de Clasificación indicando una descripción en detalle de los artículos de importación y exportación, de acuerdo con las clases de la Lista Mínima de las Mercancías para Estadísticas de Comercio Internacional. Este volumen es el precursor de una serie de índices nacionales para la conversión de la estadística detallada del comercio exterior de cada uno de los países americanos a un sistema común de clasificación que sirva de modelo para comparabilidad internacional. Por cuanto prácticamente todos los tipos de artículos comerciales entre los países americanos están representados en el comercio exterior de Estados Unidos, se han tomado los 55.000 términos que aproximadamente contiene el volumen, en su mayoría, de las descripciones detalladas de las mercancías que aparecen en las listas de clasificación estadística de exportaciones e importaciones de los Estados Unidos.

*Int-3210.D4 1937 **Draft Customs Nomenclature.** Edition revised in accordance with the observations forwarded by governments. Last previous edition was published in 1931. Geneva, 1937. League of Nations. Economic Committee. Sub-Committee of Experts for the Unification of Customs Tariff Nomenclature. Publications of the League of Nations, II, Economic and financial, 1937,II.B.51. 2 vol.

Vol. I contains classified nomenclature and alphabetical list of items; vol. II contains revised definitions and explanations.

Projet de nomenclature douanière. Edition revisée conformément aux observations communiquées par les gouvernements. Dernière édition antérieure publiée en 1931. Le vol. I contient la nomenclature classifiée et la liste alphabétique des rubriques; le vol. II contient les définitions et les explications revisées.

Int-3210.D6 1929 **Draft Framework for a Customs Tariff Nomenclature and Draft Allocation of Goods to the Various Chapters of the Framework with Explanatory Notes.** Geneva, Imp. Kundig, 1929. League of Nations. Economic Committee. Sub-Committee of Experts for the Unification of Customs Tariff Nomenclature. 96 p. Publications of the League of Nations, II, Economic and financial, 1928,II,37.

Plan de nomenclature des tarifs douaniers et projet de répartition des marchandises entre les divers chapitres du plan, avec notes explicatives. Préparé par le Sous-comité d'Experts pour l'Unification de la Nomenclature des Tarifs Douaniers, du Comité Economique de la Société des Nations.

*Int-**3210.**L5 1938 **Liste Minimum de Marchandises pour les Statistiques du Commerce International.** Edition révisée préparée par le Comité d'Experts Statisticiens. Genève, 1938. Société des Nations. 62 p. Etudes et rapports sur les méthodes statistiques, no. 2.

Published also in English—publié aussi en anglais (Int-**3210.**M5).

Une système de classification pour les statistiques du commerce extérieur. La première Liste Minimum a été preparée en 1935, et se fondait sur le projet de nomenclature douanière publié en 1931 (puisque dans beaucoup de pays les statistiques du commerce reposent dans la nomenclature douanière.

*Int-**3210.**M5 1938 **Minimum List of Commodities for International Trade Statistics.** Revised edition prepared by the Committee of Statistical Experts. Geneva, 1938. League of Nations. 62 p. Studies and reports on statistical methods, no. 2.

Published also in French—publié aussi en français (Int-**3210.**L5, 1938).

Classification scheme for foreign trade statistics. The first Minimum List was prepared in 1935, and was based on the Draft Customs Nomenclature published in 1931 (since in many countries trade statistics are based upon customs classification).

Int-**3210.**U5 1930 **Unification of Customs Nomenclature.** Report submitted to the Economic Committee by the Sub-Committee of Experts on the occasion of the eleventh assembly. Geneva, Imp.J.de G.,1930. League of Nations. 4 p. Series of League of Nations publications II, Economic and financial, 1930.II.32.

L'unification de la nomenclature douanière. Rapport soumis au Comité Economique par le Sous-Comité d'Experts dans l'occasion de la onzième assemblée.

Int-**3220.**S5 1923 **Systems of Classification of Industries and Occupations.** Report prepared for the International Conference of Labour Statisticians (29 October to 2 November 1923). Geneva, A. Kundig, December 1923. International Labour Office. 79 p. Studies and reports, series N, Statistics, no. 1. Unexamined.

Sistemas de clasificación de industrias y ocupaciones. Informe preparado para la Conferencia Internacional de Estadísticos del Trabajo (octubre 29 a noviembre 2, 1923), por la Oficina Internacional de Trabajo.

Int-**3230.**C7 1944 **Classification of Terms and Comparability of Titles through Five Revisions of the International List of Causes of Death.** Prepared under the supervision of Halbert L. Dunn. Washington, D.C., 1944. U.S. Department of Commerce. Bureau of the Census. 110 p.

Clasificación de términos y comparabilidad de títulos a través de cinco revisiones de la Lista Internacional de Causas de la Muerte.

Int-**3230.**M5 1939 **Manual of the International List of Causes of Death as Adopted for Use in the United States, Based on the Fifth Decennial Revision . . . ; Manual of Joint Causes of Death, Fourth Edition, 1939.** Washington, D.C., U.S. Govt. Print. Off., 1940. U.S. Department of Commerce. Bureau of the Census. 452 p.

Int-**3230.**N5 1929 **Nomenclaturas Internacionales de las Enfermedades (Causas de Defunción—Causas de Incapacidad del Trabajo) para Establecer las Estadísticas Nosológicas (Clasificación "Bertillon").** Cuarta sesión reunida en París, el 16 de octubre de 1929. Traducidas y publicadas por la Sección Estadística de la Sociedad de Beneficencia Pública de Lima, y revisadas por la Sección de Demografía de la Dirección de Salubridad Pública. Lima, Imp. "Tomás Valle," 1934. Sociedad de Beneficencia Pública de Lima. 164 p.

International nomenclatures of diseases (causes of death—causes of incapacity for work) to establish nosological statistics.

*Int-**3230.**N8 1938 **Nomenclatures Internationales des Causes de Décès, 1938.** (Classification Bertillon). Cinquième révision décennale effectuée par la Conférence Internationale de Paris du 3 au 7 octobre 1938. La Haye, Imp.Trio S.A., 1940. Publié avec le concours du Ministère des Affaires Etrangères de la République Française par l'Institut International de Statistique. 306 p.

Fifth revision of the international nomenclature of the causes of death, agreed upon at the international conference in Paris, 1938.

*Int-**3230.**P5 1946 **Proposed Statistical Classification of Diseases, Injuries, and Causes of Death.** Prepared by the Subcommittee on Classification of Diseases, Injuries, and Causes of Death and adopted by the U.S. Committee on Joint Causes of Death, 1946. Washington, D.C., 1946. 122 p. Part I: **Introduction and List of Categories;** part II: **Tabular List of Inclusions (Tentative Edition).**

Contents: Part I, Introduction, including general principles, historical review—early history, adoption of International List of Causes of Death, Fifth International Conference, and previous classifications of diseases for morbidity statistics — United States Committee on Joint Causes of Death, principles underlying proposed morbidity and mortality list, nature of proposed classification, description of proposed classification according to sections, supplementary classifications for special admissions, and for livebirths and stillbirths; list of 100 2-digit categories; list of categories; part II presents a tabular list of inclusions for each of these categories.

Ensayo de clasificación estadística de enfermedades, lesiones y causas de muerte, preparado por el Subcomité de Clasificación de Enfermedades, Lesiones y Causas de Muerte y adoptado por el Comité Estadounidense sobre Causas Conjuntas de Muerte, 1946. Contiene: Parte I, Introducción, incluyendo principios generales, reseña histórica—comienzo, adopción de la Lista Internacional de Causas de la Muerte, Quinta Conferencia Internacional y previas clasificaciones de las enfermedades para estadísticas sobre morbidad—Comité Estadounidense sobre Causas Conjuntas de Muerte, principios fundamentales de la propuesta lista de morbidad y mortalidad, naturaleza de la propuesta clasificación, descripción de la clasificación propuesta por secciones, clasificaciones suplementarias para inclusiones especiales, y para nacidos vivos y nacidos muertos; lista de cien categorías de 2 dígitos; lista de categorías; parte II presenta una lista tabular de inclusiones para cada una de esas categorías.

Int-**3230**.P6 1946 **Proyecto de Informe sobre Standardización de las Estadísticas de Atención Médica.** Preparado por Manuel de Viado. Montreal, Oficina Internacional del Trabajo, noviembre 1946. Comité Interamericano de Seguridad Social. 19 p. Mimeografiado.

Report on standardization of statistics of medical care, prepared by the Inter-American Committee on Social Security and published by the International Labour Office.

Int-**3230**.P7 1946 **Proyecto de Informe sobre Standardización de las Estadísticas de Morbilidad.** Preparado por Dr. Laura Bodmer. Montreal, Oficina Internacional del Trabajo, septiembre 1946. Comité Interamericano de Seguridad Social. 85 p. y 10 cuadros. Mimeografiado.

Report on standardization of statistics of morbidity, prepared by the Inter-American Committee on Social Security and published by the International Labour Office.

Int-**3240**.E5 1938 **Estadísticas de la Población Activa.** Definiciones y clasificaciones recomendadas por el Comité de Expertos Estadígrafos. Ginebra, 1938. Liga de Naciones. 32 p. Copia a máquina. Estudios e informes sobre los métodos estadísticos, no. 1.

Published originally in English and French—publicado originalmente en inglés y francés (Int-**3240**.S4; Int-**3240**.S6).

Translation into Spanish, from the French, by the National Bureau of Statistics, Division of Census, Section of Statistical Investigation, of Peru.

Incluye: **Nomenclatura Mínima de Industrias.** Traducción del francés por la Dirección Nacional de Estadística, Departamento de Censos, Sección de Investigación Estadística del Perú.

*Int-**3240**.S4 1938 **Statistics of the Gainfully-Occupied Population.** Definitions and classifications recommended by the Committee of Statistical Experts. Geneva, 1938. League of Nations. 32 p. Studies and reports on statistical methods, no. 1.

Published also in French—publié aussi en français (Int-**3240**.S6).

Includes: **Minimum Nomenclature of Industries,** p. 21-32.

*Int-**3240**.S6 1938 **Statistiques de la Population Active.** Définitions et classifications recommandées par le Comité d'Experts Statisticiens. Genève, 1938. Société des Nations. 32 p. Etudes et rapports sur les méthodes statistiques, no. 1.

Published also in English—publié aussi en anglais (Int-**3240**.S4).

Comprend: **Nomenclature Minimum des Industries.**

Int-**3300** 1944-46 **Statistical Training in the American Nations: Methods and Materials.** Background material acquired for the IASI Survey on Statistical Training Methods and Materials in the American Nations, 1944-46. Washington, D.C., 1946. Inter American Statistical Institute. 4 rolls of microfilm: Roll I, **Form Letters and Questionnaires; Argentina; Bolivia; Brasil (part 1)**; roll II, **Brasil (part 2); Canada; Chile; Colombia; Cuba; Ecuador; Guatemala; Haiti; Honduras; México; Paraguay; Perú; República Dominicana**; roll III, **United States (part 1)**; roll IV, **United States (part 2); Uruguay; Venezuela.**

Includes: Correspondence; filled questionnaires; programs of courses in statistics.

Material básico adquirido para la Encuesta del Instituto Interamericano de Estadística sobre los Métodos y Materiales en la Enseñanza Estadística en las Naciones Americanas, 1944-46. Incluye: Correspondencia; cuestionarios llenados; programas de cursos de estadística.

1944-45, questionnaires "Questionnaires and Letters Sent Out in Connection with IASI Survey on Statistical Training Methods and Materials in the American Nations." Roll I.

Includes: Letter I, to directors general of statistics, with questionnaire F-1, "Summary of Statistical Personnel"; questionnaire F-2, "Opinion of Minimum Requirements for Various Types of Statistical Personnel"; questionnaire F-3, "Biographical Information about Statisticians"; questionnaire F-4, "Possible Titles for Teaching Monographs"; letter 2, to professors of statistics, and letter 3, to statisticians, with questionnaire F-5, "Statistical Courses in Centers of Higher Learning in the Americas." All letters and questionnaires are in English, Spanish, and Portuguese, and are accompanied by a list of informants.

Cuestionarios y cartas enviadas en relación con la Encuesta del Instituto Interamericano de Estadística sobre los Métodos y Materiales en la Enseñanza Estadística en las Naciones Americanas, 1944-46. En inglés, español y portugués. Incluye una lista de los informantes.

4000-4999 GEOGRAPHY—GEOGRAFIA

Int-**4000**.M4 1945 **Index to Map of Hispanic America, 1:1,000,000.** Editor in Chief: Earl Parker Hanson, assisted by Natalie Raymond. Washington, D.C., U.S. Govt. Print. Off., 1945. American Geographical Society. 923 p. Map of Hispanic America publication, no. 5.

Indice del mapa de la América Hispana, 1:1,000.000, de la Sociedad Geográfica Americana.

Int-**4000**.M5 1942 **Map of the Americas, 1:5,000,000.** Compiled and drawn by the American Geographical Society of New York. New York, 1942. American Geographical Society of New York. 3 colored maps: Mexico, Central America, and the West Indies; South America—north; South America—south.

Issued also with title and notes in Spanish and in Portuguese—editado también con título y notas en español y portugués.

A 77-page index to the maps is also published by the Society.

Mapas en tres colores, uno de América Central y dos de Sud América, escala: 1:5,000,000. La Sociedad ha publicado también un índice de los mapas en 77 páginas.

5000-5999 DEMOGRAPHY—DEMOGRAFIA

*Int-**5000**.A5 1929-36 **Aperçu de la Démographie des Divers Pays du Monde, 1929-1936.** La Haye, 1939. Office Permanent de l'Institut International de Statistique. 433 p.

Demographic survey of the various countries of the world, 1929-36. Statistical tables show data on: Area and population of the geographic divisions of the world; populations of countries; rate of population increase; fecundity rates; marriages; divorces; births and birth rates; deaths and death rates; causes of death (according to the international nomenclature of 1920, and the international nomenclature of 1929).

Présente des tableaux statistiques sur: Superficie et population des parties du monde; population des pays; taux d'accroissement de la population; taux de fécondité; mariages; divorces; naissances, et taux de natalité; décès, et taux de mortalité; causes de décès (d'après la nomenclature internationale de 1920, et la nomenclature internationale de 1929).

Int-**5000**.B5 1939-44 **Books on Population, 1939-Mid-1944.** Washington, D.C., 1944. Population Association of America. 20 p.

Photostated from typewritten copy. List compiled by Irene B. Taeuber, in answer to the request of the American Library Association for a "list of books which could be recommended as a fitting and proper representation of the research and scholarly production in the field of population since the beginning of 1939."

Tratados sobre población, 1939-44. Copia fotostática de escritura a máquina. Lista compilada por Irene B. Taeuber, en respuesta a la solicitud de la Asociación Americana de Bibliotecarios de una lista de libros que se pueda recomendar como demostración de los trabajos de estudio e investigación que se han realizado sobre población, desde el comienzo de 1939.

Int-**5000**.C5 1937 **Congrès International de la Population, Paris, 1937.** Paris, Hermann et Cie, 1938. Congrès International de la Population. 8 vol.: Tome I, **Théorie Générale de la Population**; tome II, **Démographie Historique**; tome III, **Démographie Statistique, Etudes d'Ensemble**; tome IV, **Démographie Statistique, Etudes Spéciales, Etat de la Population, Migrations**; tome V, **Démographie Statistique, Etudes Spéciales, Natalité, Nuptialité, Mortalité**; tome VI, **Démographie de la France d'Outremer**; tome VII, **Facteurs et Conséquences de l'Evolution Démographique**; tome VIII, **Problèmes Qualitatifs de la Population.**

A compendium of papers read at the International Population Congress held in Paris in 1937 under auspices of the International Population Union. The papers are classified by subject into eight volumes, and are in French, English, German, or Italian. Those not in French have French abstracts.

Un abrégé des travaux lus au Congrès International de la Population tenue à Paris en 1937 sous les auspices de l'Union Internationale de la Population. Les travaux sont classés selon le sujet dans huit tomes, et sont écrits en français, en anglais, en allemand, ou en italien. Ceux qui ne sont pas en français ont des résumés français.

Int-**5000**.G5 1943 **General Censuses and Vital Statistics in the Americas.** An annotated bibliography of the historical censuses and current vital statistics of the 21 American republics, the American sections of the British Commonwealth of Nations, the American colonies of Denmark, France, and the Netherlands, and the American territories and possessions of the United States. Prepared under the supervision of Irene B. Taeuber, Chief, Census Library Project. Washington, D.C., U.S. Govt. Print. Off., 1943. United States Department of Commerce, Bureau of the Census; United States Library of Congress, Reference Department. 151 p.

Issued by the Census Library Project, a joint project of the Bureau of the Census of the U.S. Department of Commerce and the Reference Department of the U.S. Library of Congress. Contains list of titles of the census reports, summary of the statistical history of each country, and brief description of contents of the more important reports. It covers all of the general censuses that have ever been taken, insofar as information could be obtained. While the main effort has been to provide a complete list of censuses of population, censuses of agriculture, industry and commerce have been included. Many reports presenting population estimates are included. Reports of the censuses of provinces and cities are listed as well as national censuses. Reference is also made in many cases to yearbooks or other periodic publications which contain statistical material. There is a separate section for each country which lists in detail the current publications in the field of vital statistics.

Censos generales y estadística del movimiento demográfico en las Américas. Bibliografía comentada, editada por el Census Library Project, un proyecto conjunto de la Oficina del Censo del Departamento de Comercio de Estados Unidos y el Departamento de Referencias de la Biblioteca del Congreso. Contiene una lista de títulos de los informes censales y también presenta un resumen de la historia estadística de cada país y una breve descripción del contenido de los informes mas importantes. Cubre todos los censos generales que alguna vez se hayan tomado, siempre que la información ha sido asequible. Mientras que el principal empeño se ha puesto en proveer una lista completa de censos de población, se han incluído censos de agricultura, industria y comercio. Se han incluído también muchos informes que presentan estimaciones sobre población. Además de censos nacionales se incluyeron censos tomados en provincias y ciudades. En muchos casos se hacen referencias a anuarios y otras publicaciones periódicas que contienen material estadístico. Hay una sección por separado para cada país, que detalla las publicaciones que tratan sobre el movimiento demográfico.

Int-**5000**.I5 1943.1 Eng **Final Act of the First Inter American Demographic Congress.** Passed by the First Inter American Demographic Congress, held at Mexico City, October 12-21, 1943. 81 p. (est.) Hectographed.

Acta final del Primer Congreso Demográfico Interamericano, celebrado en México, D.F., del 12 al 21 de octubre de 1943.

Int-**5000**.M3 1946 **Papers Presented by Special Guests of the Population Association of America at Its Meetings October 25-26, 1946.** Hotel New Yorker, New York City. New York, 1946. 100 p. (approx.) Articles numbered separately. Mimeographed.

Includes—incluye: "Some Aspects of the Statistical Work of the Royal Commission on Population," by David V. Glass; "Mortality in Mexico before and during World War II," by Alberto P. León; "Recent Demographic Changes in the Netherlands," by Johannes B. D. Derksen; "Outstanding Population Problems in Sweden," by Carl-Erik Quensel; "The Future Population of Canada," by O. A. Lemieux; "The Problem of the French Population," by Alfred Sauvy; "Demographic Impact of the War in Poland," by Stefan Szulc; "Immigration and Population Growth in Brazil," by Germano Jardim; "On a Fundamental Demographic Problem of Peru: The Population Distribution by Localities," by Ricardo Luna Vegas; "Natality in Venezuela," by R. de Shelly Hernandez; "Problems of Current Demographic Data in India," by P. C. Mahalanobis; "A Summary of the Problems of Current Demographic Data in China," by Franklin C. H. Lee, "Current Problems of Demographic Data in the Philippines," by Trinidad J. Jaramillo.

Int-**5000**.P5 **Population Index.** 1935- . Quarterly. Washington, D.C. School of Public Affairs, Princeton University, and Population Association of America, Inc.

A selected and annoted bibliography of books,

articles in periodicals, and government publications grouped as to subject matter under the following headings: General; formal demography; mortality; fertility; marriage, divorce, and the family; migration; regional studies; characteristics; policy; method; references and compendia. Also contains notes and short analyses on demographic subjects, and statistics of population growth, vital rates, etc.

Indice de población. Una bibliografía selecta y comentada de libros, artículos de periódicos y publicaciones oficiales, agrupados por materias, bajo los siguientes encabezamientos: General; demografía oficial; mortalidad; fecundidad; nupcialidad, divorcio y la familia; migración; estudios regionales; características; política; método; referencias y compendio. Contiene también notas y análisis cortos sobre asuntos demográficos y estadísticas sobre el crecimiento de población, tasas vitales, etc.

Jan. '44 "Future Populations." Vol. 10, no. 1, January 1944, p. 3-14.
Poblaciones futuras.

July '43, a "Birth Rates." Vol. 9, no. 3, July 1943, p. 215-216.
Birth rates for all countries for 1921-42.
Tasas de nacimiento en todos los países, 1921-42.

July '43, b "Complete Expectation of Life at Various Ages in Selected Countries." Vol. 9, no. 3, July 1943, p. 221-223.
Completa expectación de vida a distintas edades en determinados países.

July '43, c "Death Rates." Vol. 9, no. 3, July 1943, p. 216-218.
Death rates for all countries for 1921-42.
Tasas de mortalidad en todos los países, 1921-42.

July '43, d "Infant Mortality Rates." Vol. 9, no. 3, July 1943, p. 218-220.
Infant mortality rates for all countries for 1921-42.
Tasas de mortalidad infantil en todos los países, 1921-42.

July '43, e "Marriage Rates." Vol. 9, no. 3, July 1943, p. 220-221.
Marriage rates for all countries for 1921-42.
Tasa de nupcialidad en todos los países, 1921-42.

July '43, f "Population Growth." Vol. 9, no. 3, July 1943, p. 212-214.
Shows population for each country according to last census and last official estimate.
Crecimiento de población. Muestra la población de cada país de acuerdo a los últimos censos y las mas recientes estimaciones oficiales.

Int-**5000**.S8 **Summary of Biostatistics Series.** 1943- . Washington, D.C. U.S. Department of Commerce. Bureau of the Census.

Studies of 20 Latin American countries made by the International Unit of the Vital Statistics Division, Bureau of the Census, in cooperation with the Office of the Coordinator of Inter A-merican Affairs.

Series de resúmenes bioestadísticos. Estudios de 20 países latino-americanos, realizados por la Sección Internacional de la División de Estadística, en colaboración con la Oficina Coordinadora de Asuntos Inter-Americanos.

Int-**5100**.E5 1943 **The Economic Pattern of World Population.** By John Bell Condliffe. Washington, D.C., 1943. National Planning Association. 54 p. Planning pamphlets, no. 18. Unexamined

"The growth and regional distribution of world population are interpreted in terms of the industrial revolution and its diffusion. The present correlation between population increase and meager resources will present increasingly acute problems unless international investment and trade are reorganized to provide equipment for the still growing populations. Zones of low, moderate and heavy density are contrasted, and the proliferation of cities and the diffusion of urban ways of life assessed as the most outstanding characteristics of the demographic development of the nineteenth and twentieth centuries." (Source: **Books on Population, 1939-Mid-1944.**)

La estructura económica de la población mundial. "El crecimiento y distribución regional de la población mundial son interpretados en términos de la revolución industrial y su difusión. La presente correlación entre el aumento de población y los escasos recursos presentará cada vez problemas mas serios, a no ser que el comercio y las inversiones internacionales se reorganicen de manera que provean de lo necesario a la aún creciente población. Zonas de densidad baja, mediana y alta son contrastadas, y la expansión de ciudades y la difusión de las costumbres urbanas son calificadas como las características más sobresalientes del desarrollo demográfico de

los siglos diez y nueve y veinte." (Traducción del comentario.)

Int-5100.P4 **Population.** 1935-39? Quarterly. London. International Union for the Scientific Investigation of Population Problems.

Int-5100.P8 1945 **Populations of the Other American Republics by Major Civil Divisions and by Cities of 5,000 or More Inhabitants.** Washington, D.C., U.S. Govt. Print. Off., 1945. Office of Inter American Affairs. 58 p.

Cover title—título de la cubierta: **Handbook of Latin American Population Data.**

Población de las otras repúblicas americanas, por grandes divisiones administrativas, municipios y por ciudades de 5000 o más habitantes. Publicado por la Oficina de Asuntos Interamericanos.

*Int-5100.S5 1928-34 **Statistique Internationale des Grandes Villes: Territoire et Population des Grandes Villes, 1928-1934.** La Haye, Imp. Société Anonyme "Drukkerij Trio," 1939. Office Permanent de l'Institut International de Statistique. 300 p.

Statistics of population for big cities, 1928-34; includes data on: Geographic position; territory and land ownership; estimated population on July 1st of years covered; population by sex and by age according to the latest censuses; gainfully-occupied population by sex and by profession; births; deaths; marriages and divorces; number of cases of certain contagious diseases, and deaths therefrom; causes of death; mortality by causes (rate per 100,000 inhabitants). From the Western Hemisphere, only Argentina, Brazil, Canada, Ecuador, and the United States are represented in the group of countries for which data are shown.

Présente des tableaux statistiques sur: Position géographique; territoire et propriété foncière; population calculée ou évaluée, au 1er juillet des années observées; population par sexe et par âge selon le dernier recensement; population active par sexe et par profession; naissances; décès; mariages contractés, dissolutions de mariage et excédent des mariages contractés; nombre de cas de certaines maladies contagieuses à declaration obligatoire, et nombre de décès; causes de décès; mortalité par causes (taux par 100,000 habitants). De l'hémisphère occidental, seuls l'Argentine, le Brésil, le Canada, l'Equateur, et les Etats-Unis d'Amérique sont representés dans le groupe des pays dont les chiffres sont donnés.

INTERNATIONAL
5100

Int-5120.W5 1936 **World Statistics of Aliens.** A comparative study of census returns, 1910-1920-1930. Geneva, 1936. International Labour Office. 251 p. Studies and reports, series O, Migration, no. 6.

Estadísticas mundiales sobre extranjeros. Un estudio comparativo de datos censales, 1910-1920-1930, hecho por la Oficina Internacional del Trabajo.

Int-5200.S5 **Statistical Handbook Series: The Official Vital Statistics of . . . 1925- .** Geneva. League of Nations. Health Organisation.

Includes—comprend: No. 1, Netherlands; no. 2, Belgium; no. 3, England and Wales; no. 4, Spain; no. 5, Austria; no. 6, Scandinavian countries and the Baltic Republic; no. 7, Portugal; no. 8, Czechoslovakia; no. 9, French Republic; no. 10, Hungary; no. 11, Ireland; no. 12, Switzerland; no. 13, Scotland; no. 14, Canada.

Review of practice and procedures in the collection and publication of statistics on population, births, deaths and notifiable diseases, including methods of registration and the current published reports.

Série de manuels statistiques: Statistiques vitales officielles. Revue des pratiques et procédures dans la collection et publication des statistiques sur population, naissances, décès, et maladies déclarables, y compris méthodes de l'enregistrement et les rapports courament publiés.

Int-5300.I6 1931 **International Migrations.** Edited by Walter F. Willcox. New York, 1931. National Bureau of Economic Research. Publication, no. 18. Unexamined.

Includes "Increase in the Population of the Earth and of the Continents since 1650," chapter 1, vol. II.

Migraciones internacionales.

*Int-5300.S3 1932 **Statistics of Migration.** Definitions—methods—classifications. Geneva, 1932. International Labour Office. 152 p. Studies and reports, series N, Statistics, no. 18. Unexamined.

Estadísticas sobre migración. Definiciones, métodos y clasificaciones. Informe de la Oficina Internacional del Trabajo.

Int-5300.S5 **Studies and Reports, Series O, Migration.** 1925- . Geneva. International Labour Office.

Series de estudios e informes sobre migración, hechos por la Oficina Internacional del Trabajo.

Int-5310.I4 1941.1 Eng **Final Act of the Second Inter-American Travel Congress.** Mexico, D.F., September 15-24, 1941. Washington, D.C., Travel Division, Pan American Union, 1942. Second Inter-American Travel Congress. 52 p. Congress and conference series of the Pan American Union, no. 41.

Published also in Spanish—publicado también en español (Int-**5310.**I4, 1941.1 Sp).

Int-**5310.**I4 1941.1 Sp **Acta Final del Segundo Congreso Interamericano de Turismo.** México, D.F., septiembre 15-24, 1941. Wáshington, D.C., Oficina de Turismo, Unión Panamericana, 1942. Segundo Congreso Interamericano de Turismo. 46 p. Serie sobre congresos y conferencias de la Unión Panamericana, no. 38.

Published also in English—publicado también en inglés (Int-**5310.**I4, 1941.1 Eng).

*Int-**5310.**S5 1929-1934 **Statistique Internationale des Grandes Villes: Statistique du Tourisme dans les Grandes Villes, 1929-1934.** La Haye, Imp. Société Anonyme "Drukkerij Trio," 1938. Office Permanent de l'Institut International de Statistique. 89 p.

Statistics of tourism for large cities, 1929-34. Contains statistical data on hotels, number of travelers by nationality, registration of tourists, etc. From the Western Hemisphere, only Brazil, Canada and Haiti are represented in the group of countries for which data is shown.

Contient des tableaux statistiques sur: L'hôtellerie, le nombre des voyageurs par nationalité, l'enregistrement des touristes, etc. De l'hémisphère occidental seuls le Brésil, le Canada, et Haïti figurent dans le groupe des pays dont les chiffres sont donnés.

6000-6999 ECONOMICS—ECONOMIA

Int-**6000.**A5 1943.B **Supplement to American Republics Bibliography; A List of Some of the Current Latin American Periodicals Relating to Agriculture, Commerce, Economics, Finances, Industry, Production in the Library of the Pan American Union.** Washington, D.C., 1943. U.S. Department of Commerce. Bureau of Foreign and Domestic Commerce. 19 p. Hectographed.

Suplemento de la bibliografía de las repúblicas americanas; una lista de algunos de los periódicos en circulación en la América Latina relativos a agricultura, comercio, economía, finanzas, industria, producción, en la biblioteca de la Unión Panamericana.

Int-**6000.**C4 1943 **Catalogue of Selected Publications on Economic and Financial Subjects.** A guide to the documents of value in connection with the formulation of postwar economic policies. Princeton, New Jersey, Princeton University Press, 1943. League of Nations. 69 p.

Catálogo de publicaciones escogidas de la Sociedad de las Naciones, sobre tópicos económicos y financieros—una guía de los documentos de valor relacionados con la formulación de políticas económicas de postguerra.

*Int-**6000.**C7 **Commercial Pan America.** A monthly review of commerce and finance. 1932- . Washington, D.C. Pan American Union.

Published also in Spanish—publicado también en español (Int-**6000.**P4).

Each issue is devoted to one study.

July-Aug. '46 "National Economy of Argentina." Vol. XV, no. 7-8, July-August 1946. 98 p.

Number 2 of a series of studies appearing on the national economies of the American republics. Contains: Introduction (general economic situation); agriculture; stock-raising; mining; fuel and power; industry; domestic business (cost of living, wholesale prices, stock exchange, construction, check clearings, bankruptcies); foreign trade; finance (balance of payments, public finance, private finance, national income, etc.); transportation and communications; bibliography.

Economía nacional de la Argentina.

May-June '46 "National Economy of Mexico." Vol. XV, no. 5-6, May-June 1946. 70 p.

Number 1 of a series of studies appearing on the national economies of the American republics. Contains: Introduction (general economic situation); general index of economic activity; agriculture; mineral, metallurgical, and petroleum production; industry; power production; construction and transportation; domestic economy (assets and liabilities of banks, general and consolidated balance sheet of the Bank of Mexico, indexes of purchasing power of monetary units, etc.); cost of living and national income; foreign trade; balance of payments; public finance, international cooperation and commercial policy; conclusion; bibliography.

Economía nacional de México.

Dec. '45, Ysita "Victory Minerals of Mexico." By Eugene Ysita. Vol. XIV, no. 12, December 1945. 43 p.

Published also in Spanish—publicado también en español.

Contains detailed analysis of Mexican mining and metallurgical production, with statistical tables and charts showing annual production from 1935 through 1944, and prewar and wartime comparative production by percentages and volume.

Minerales mexicanos de la victoria. Contiene un análisis detallado sobre minería y producción metalúrgica mejicana, con tablas y cuadros estadísticos indicando producción anual desde 1935 hasta 1944, y producción comparativa anterior y posterior a la guerra por medio de porcentajes y volumen.

Nov.-Dec. '44, Zier "Ecuador, a Statistical Abstract." By Julian G. Zier. Vol. XIII, no. 11-12, November-December 1944. 23 p.

Data on: Area and population; public education; agriculture; labor and wages; foreign trade; finance.

Ecuador, resumen estadístico. Datos sobre: Superficie y población; educación pública; agricultura; trabajo y salarios; comercio exterior; finanzas.

July-Aug. '44 "Mexico's Balance of Payments." Vol. XIII, no. 7-8, July-August 1944, p. 6.

Balanza de pagos de México.

July-Aug. '44, Ysita "Growth of Industry and Aviation in Mexico, 1929-1942." By Eugene Ysita. Vol. XIII, no. 7-8, July-August 1944. 2 parts.

Data and analysis on: Industrial achievements; consumption of domestic and foreign basic products; industrial production; manpower and wages; industrial accidents; Mexico's place in the world of metals; the great resources of Mexico; post-war industrial prospects and problems; investment and profits; Mexico's balance of payments; war-time financial situation; growth of aviation in Mexico, 1929-42.

Crecimiento industrial y de la aviación en México, 1929-42. Datos y análisis sobre: Realizaciones industriales, consumo de productos de primera necesidad procedentes del país y del extranjero; producción industrial; mano de obra y salarios; accidentes industriales; posición de México en el mundo de los metales; las grandes fuentes de riqueza de México; problemas y perspectivas industriales de post-guerra; inversiones y beneficios; balanza de pagos de México; situación financiera durante la guerra; crecimiento de la aviación en México, 1929-42.

Int-6000.E1 1920-35 **Economic and Financial Publications, 1920-1935.** Geneva, Imp. Réunies, Chambéry, 1935. League of Nations. Publications Department. 80 p.

List of economic and financial publications issued by the League of Nations from 1920 to 1935.

Liste des publications économiques et financières publiées par la Société des Nations de 1920 à 1935.

Int-6000.E2 **Economic Controls and Commercial Policy in the American Republics.** 1943- . Irregular. Washington, D.C. U.S. Tariff Commission.

A series of reports by country.

Controles económicos y política comercial en las repúblicas americanas. Serie de informes, por países.

Int-6000.E6 1936 **The Economic Literature of Latin America.** A tentative bibliography. Cambridge, Massachusetts, Harvard University Press, 1936. Harvard University. Bureau for Economic Research in Latin America. 2 vol.

Statistical interest in the volume centers in appendix section which contains notes on statistical sources in each of the twenty Latin American republics over a period of years, including historical data on the national censuses.

La literatura económica de la América Latina. Un ensayo de bibliografía. El interés estadístico en el volumen radica en el apéndice del mismo, el cual contiene notas sobre fuentes de información estadística en cada una de las veinte repúblicas latino-americanas abarcando un período de años e incluyendo datos históricos sobre censos nacionales.

Int-6000.E7 1944 **Economic Problems of Latin America.** Edited by Seymour E. Harris. New York, McGraw-Hill Book Co., 1944. 465 p.

Contents: Introduction, some major issues; economic problems of the Latin American republics; war and postwar agricultural problems; central banking and monetary management; fiscal policy and budget; price stabilization programs; exchanges and prices; inter-American trade policy; Argentina; Bolivia; Brazil's economy in the war and after; Chile; Colombia, with particular reference to price control; Cuba, sugar and currency; Haiti; Mexico, with special reference to its international economic relations; Paraguay, with special reference to price control; Venezuela.

Problemas económicos de la América Latina. Contiene: Introducción, algunos problemas principales; problemas económicos de las repúblicas latinoamericanas; problemas agrícolas durante la guerra y la postguerra; bancos centrales y régimen monetario; política fiscal y presupuesto; programas para la estabilización de los precios; cambios y precios; política comercial interamericana; Argentina; Bolivia; la economía del Brasil durante la guerra y después; Chile; Colombia, con especial referencia al control de precios; Cuba, azúcar y circulación monetaria; Haití; Méjico, con especial referencia a sus relaciones económicas internacionales; Paraguay, con especial referencia al control de precios; Venezuela.

Int-**6000**.E9 1960 **The Economics of 1960.** By Colin Clark. London, Macmillan and Co., Limited, 1942. 118 p.

Gives comparative data on total population and working population, mortality rates, reproduction rates, anticipated age and sex composition of the population in 1960; productivity; the supply and demand of primary products, including income and consumption; availability of capital; world equilibrium in production and consumption; historical study of the terms of trade and related phenomena; problems of 1941-60. American countries included are Argentina, Canada, Chile, United States and Uruguay.

La economía de 1960. Proporciona datos comparativos sobre población total y población activa, tasa de mortalidad, tasa de reproducción, estimación de la composición por edades y sexo de la población en 1960; productividad; la oferta y demanda de los artículos de primera necesidad, incluyendo datos sobre ingreso y consumo; disponibilidad de capital; equilibrio mundial de producción y consumo; estudio histórico de los términos del comercio y de fenómenos relacionados; problemas de 1941-60. Los países americanos incluídos son: Argentina, Canadá, Chile, Estados Unidos y Uruguay.

Int-**6000**.I2 1945 **Industrialization and Foreign Trade.** Geneva, 1945. League of Nations. Economic, Financial and Transit Department. 171 p. Series of League of Nations publications, II, Economic and financial; 1945. II.A.10.

Published also in French—publié aussi en français.

Analyzes the influence of industrialization of relatively undeveloped areas on the foreign trade of more industrially advanced countries. Traces growth of industry and trade in the world during the last seventy years, and examines in detail the statistical evidence of effects of industrial growth an the past on the trade of other countries. An annex, with full explanatory notes, shows indices of manufacturing, 1870-1938, and statistics of international trade, 1871-1938, presented in table by country and product, with yearly world totals.

Industrialisation et commerce étranger. Une analyse des influences de l'industrialisation des régions relativement peu développées sur le commerce étranger des pays plus industrialisés; rétrospective du progrès de l'industrie et du commerce dans le monde au cours des soixante-dix dernières années; étude détaillé de la preuve statistique des effets des progrès industriels du passé sur le commerce des autres pays. Une annexe, avec notes explicatives complètes, donne les indices des manufactures, 1870-1938, et les statistiques du commerce international, 1871-1938, présentées en tableaux par pays et produit, avec totaux annuels mondiaux. Publié aussi en français.

Int-**6000**.I3 **Informaciones Económicas del Banco de México.** 1943- . Irregular. México. Banco de México.

Economic information issued by the Bank of Mexico. Series of monographs.

Serie de monografías.

Int-**6000**.I45 1942, Eng **Final Act of Inter-American Conference of Systems of Economic and Financial Control, Washington, June 30-July 10, 1942.** Washington, D.C., 1942. Pan American Union. Congress and conference series, no. 39.

Published also in Portuguese and Spanish—publicado también en portugués y español (Int-**6000**.I45, 1942, Port; Int-**6000**.I45, 1942, Sp).

Int-**6000**.I45 1942, Port **Ata Final da Conferência Interamericana sôbre Medidas de Contrôle Econômico-Financeiro, Washington, 30 de Junho a 10 de Julho de 1942.** Washington, D.C., 1942. União Panamericana. 23 p. Série de congressos e conferências, no. 25.

Published also in English and Spanish—publicado também em inglês e espanhol (Int-**6000**.I45, 1942, Eng; Int-**6000**.I45, 1942, Sp).

Int-**6000**.I45 1942, Sp **Acta Final de la Conferencia Interamericana sobre Sistemas de Control Económico y Financiero, Wáshington, 30 de Junio a 10 de Julio de 1942.** Wáshington, D.C. 1942. Unión Panamericana. 25 p. Serie sobre congresos y conferencias, no. 40.

Published also in English and Portuguese—publicado también en inglés y portugués (Int-**6000**.I45, 1942, Eng; Int-**6000**.I45, 1942, Port).

*Int-**6000**.I7 1931-36 **International Abstract of Economic Statistics, 1931-1936.** The Hague, 1938. Permanent Office of the International Statistical Institute. 249 p.

Second edition of the **International Abstract of Economic Statistics.** The statistical presentation is carried out in the same manner as in the original volume, covering finance, prices and wages, trade and output, transport, employment and unemployment, with the data extended to cover the year 1936, inclusive. The economic indices are given for thirty-one countries in this edition (as compared to fifteen countries covered in the first edition), of which only three—Brazil, Canada, and the United States—represent the Western Hemisphere.

Published also in French—publié aussi en français (Int-**6000**.R4, 1931-36).

Int-**6000**.I7 1919-30 **International Abstract of Economic Statistics, 1919-1930.** Bruxelles, Imp. de l'Académie Royale de Belgique et de la Chambre des Représentants, 1934. International Conference of Economic Services, London, 1934. 218 p.

Published also in French—publié aussi en français (Int-**6000**.R4).

A compilation of classified indices of economic activity for fifteen important countries (Austria, Belgium, Canada, Czechoslovakia, France, Germany, Great Britain, Hungary, Japan, Netherlands, New Zealand, Poland, Spain, Sweden, United States), in the form of a monthly record for the period 1919 to 1930, inclusive. A table of annual statistics for each country summarizes general movements for the same period. Explanatory notes of the indices are given where necessary, and a calendar records outstanding events or disturbing factors for each country. The indices cover: Finance, prices and wages, trade and output, transport, and employment.

*Int-**6000**.I75 1928.A **International Conference on Economic Statistics, Geneva, 1928: Preparatory Documents.** Chambery, Imp. Réunies, 1928. League of Nations. 101 p. Publications of the League of Nations, II, Economic and financial, 1928, II, 6.

Discusses methods used in various countries in compiling statistics of foreign trade. Includes recommendations and opinions by the International Statistical Institute at its 15th session in Brussels, 1923, recommending use of Brussels nomenclature as basis for international trade statistics. Also discusses statistical methodology as applied to censuses of industrial production, mining and metallurgy, indexes of industrial activity, and fishery industry.

Conférence internationale concernant les statistique économiques, Genève, 1928: Documents préparatoires. Etude des méthodes employées dans les divers pays pour la compilation des statistiques du commerce extérieur. Y compris recommandations et avis de l'Institut International de Statistique à sa XVème session à Bruxelles, 1923, recommandant l'usage de la nomenclature de Bruxelles comme une base pour les statistiques du commerce international. Aussi étude de la méthodologie statistique telle qu'elle est appliquée aux recensements de la production industrielle, des mines et de la métallurgie, aux indices de l'activité industrielle et à l'industrie des pêcheries.

*Int-**6000**.I75 1928.B **Conférence Internationale Concernant les Statistiques Economiques (26 Novembre au 14 Décembre): 1, Convention Internationale Concernant les Statistiques Economiques; 2, Protocole de la Convention Internationale; 3, Acte Final de la Conférence. International Conference Relating to Economic Statistics (November 26th to December 14th, 1928): 1, International Convention Relating to Economic Statistics; 2, Protocol to the International Convention; 3, Final Act of the Conference.** Genève, Imp.de la Tribune, 1929. Société des Nations. League of Nations. 80 p. Publications de la Société des Nations, II, Questions économiques et financières, 1928.II.52.

Contains the provisions of the International Convention Relating to Economic Statistics including those concerning uniformity of international trade concepts and practices, for purposes of foreign trade statistical comparability. Includes: Annex I, **External trade statistics:** Part I, Methods of compilation and coverage, units of measure, calendar year instead of fiscal year; part II, List of countries—territories to which the statistics apply; part III (origin and destination of imports and exports), Imports—countries of origin or production, countries of consignment or provenance, countries of purchase; Export—countries of consumption; countries of consignment or destination; countries of sale; Annex II, **Statistics of fisheries;** other annexes and reports.

Comprend les dispositions de la Convention Internationale Concernant les Statistiques Economiques y compris les dispositions concernant l'uniformité des concepts et pratiques du commerce international pour la comparabilité des statistiques du commerce extérieur. Comprend: Annexe I, **Statistiques du commerce extérieur:** Partie I, Méthodes de compilation et portée, unités de mesure, l'année civile au lieu de l'année fiscale; partie II, Liste des pays—territoires statistiques; partie III (l'origine et destination des importations et des exportations), Importations—les pays d'origine ou de production; les pays de consignation ou de provenance; les pays d'achat; Exportations—les pays de consommation; les pays de consignation ou de destination; les pays de vente; Annexe II, **Statistiques des pêcheries;** autres annexes et rapports.

*Int-**6000**.I75 1928.C **Proceedings of the International Conference Relating to Economics Statistics.** Geneva, November 26th to December 14th, 1928. Geneva, Imp. Vitte, 1929. League of Nations. 361 p. Series of League of Nations publications, II, Economic and financial, 1929.II.21.

Contents: Introduction; draft convention relating to economic statistics; minutes of the plenary meetings and the meetings of the committees; documents; instruments of the conference (including **International Convention Relating to Economic Statistics**); final act. In the **Draft Convention Relating to Economic Statistics,** chapter II on "Statistical Methods" recommends distinction between "special trade" and "transit trade" to clarify compilation of trade statistics and standardize concept of export and re-export in foreign trade.

Comptes rendus de la Conférence Internationale Concernant les Statistiques Economiques, Genève, 26 novembre à 14 décembre, 1929. Matière: Introduction; projet de convention concernant les statistiques économiques; procès-verbaux des séances plénières et des séances des commissions; documents; instruments de la conférence (y compris **Convention Internationale Concernant les Statistiques Economiques**; acte finale. Au chapitre II sur les "Méthodes Statistiques" du projet de convention internationale concernant les statistiques économiques il est recommandé de faire une distinction entre "commerce spécial" et "commerce de transit" afin de clarifier la compilation des statistiques du commerce extérieur et afin d'effectuer l'uniformités dans le concept de l'exportation et de la réexportation dans le commerce extérieur.

*Int-**6000**.P4 **Panamérica Comercial.** Revista mensual de comercio y finanzas. 1932- . Wáshington. Unión Panamericana.

Published also in English—publicado también en inglés (Int-**6000**.C7).

Cada uno de los ejemplares se dedica a un estudio especial.

Sept.-Oct. '46 "Economía Nacional de Colombia." Vol. XV, no. 9-10, septiembre-octubre 1946. 342 p.

National economy of Colombia.

El tercero de una serie de estudios que se presentan sobre las economías nacionales de las repúblicas americanas. Contiene: Una introducción en la que pasa revista a la situación económica general del país; e información por separado sobre todos los aspectos de la vida económica tales como: Agricultura; producción minera; industrias manufactureras; edificaciones; transportes; comunicaciones; comercio exterior; balanza de pagos; actividad general de los negocios y finanzas del país; costo de la vida. Incluye además una bibliografía.

July-Aug. '46 "Economía Nacional de la Argentina." Vol. XV, no. 7-8, julio-agosto 1946. 256 p.

National economy of Argentina.

El segundo de una serie de estudios que se presentan sobre las economías nacionales de las repúblicas americanas. Contiene: Una introducción en la que pasa revista a la situación económica general del país; e información por separado sobre todos los aspectos de la vida económica: Agricultura; ganadería; minería; combustibles y energía; industria; situación económica interna (costo de la vida, mercado bursátil, edificación, compensaciones bancarias, quebrantos, etc.); comercio exterior; finanzas (balance de pagos, finanzas públicas, finanzas privadas, renta nacional, etc.); transportes y comunicaciones. Incluye además una bibliografía.

May-June '46 "Economía Nacional de México." Vol. XV, no. 5-6, mayo-junio 1946. 147 p.

National economy of Mexico.

El primero de una serie de estudios que se presentan sobre las economías nacionales de las repúblicas americanas. Contiene: Una introducción en la que pasa revista a la situación económica general del país; y secciones por separado sobre: Indices generales de actividad económica; agricultura; producción minerometalúrgica y de productos petrolíferos; industria; producción de energía; construcción y transportes; economía interna (recursos y obligaciones bancarias, balance general consolidado de la Central y Sucursales del Banco de México S.A., índices financieros, etc.); costo de la vida e ingreso nacional; comercio exterior; balanza de pagos; finanzas públicas; cooperación internacional y política comercial; conclusión; y una bibliografía.

Jan.-Feb. '45 "El Comercio Internacional de Chile, 1904-1944." Vol. XIV, no. 1-2, enero-febrero 1945.

International trade of Chile. Includes data on total trade, exports, and imports.

Incluye datos sobre intercambio total, exportaciones, e importaciones.

*Int-**6000**.R4 1931-36 **Recueil International de Statistiques Economiques, 1913-1936.** La Haye, 1938. Office Permanent de l'Institut International de Statistique. 249 p.

Also published in English—publié aussi en anglais (Int-**6000.I7**, 1931-36).

Recueil d'indices de l'activité économique de trente et un pays (y compris Brésil, Canada et les Etats-Unis, de l'hémisphère occidental), sous forme d'un rapport mensuel pour la période 1931 à 1936, inclusive. Un tableau des statistiques annuelles pour chaque pays résume les mouvements généraux relatifs à la période observée. Des notes explicatives des indices ont été données autant que celles ont paru nécessaires, et un calendrier mentionne les événements les plus importants, surtout ceux qui ont apporté les plus grands perturbations. Les indices comprennent: Finances, prix, commerce, production, transports et chômage.

Int-**6000.R4** 1919-30 **Recueil International de Statistiques Economiques, 1919-1930.** Bruxelles, Imp. de l'Académie Royale de Belgique et de la Chambre des Représentants, 1934. Office Permanent de l'Institut International de Statistique, La Haye, 1934. 241 p.

Published also in English—publié aussi en anglais (Int-**6000.I7**).

Receuil d'indices de l'activité économique de seize pays importante (Allemagne, Autriche, Belgique, Canada, Espagne, Etats-Unis, France, Grande-Bretagne, Hongrie, Italie, Japon, Nouvelle-Zélande, Pays-Bas, Pologne, Suède, Tchécoslovaquie), sous la forme de données mensuelles pour la période de 1919 à 1930, inclusive. Un tableau de statistiques annuelles pour chaque pays, résume les mouvements généraux relatifs à la période observée. Des notes explicatives sur les indices ont été données toutes les fois que cela était nécessaire, et, pour chaque pays, un calendrier mentionne les événements les plus importants, surtout ceux qui ont apporté le plus grand trouble. Les indices comprennent: Finances, prix et salaires, commerce et production, transports, marché du travail. L'édition anglaise ne comprende que quinze pays, puisque sur l'Italie aucunes données ont été présentées là dedans.

Int-**6000.R7** 1944 **Reports presented to the Conference of Commissions of Inter-American Development, (Held in New York, May 9-18, 1944), by the Costa Rican, Salvadorean, Guatemalan, Honduran and Nicaraguan Commissions of Inter-American Development.** On: Costa Rica—1, A brief description of Costa Rican economy; El Salvador—1, Present economic position of El Salvador; Guatemala—1, Economic conditions of Guatemala; Honduras—1, Observations on economic problems of Honduras; Nicaragua—1, Economic development and investments; 2, International trade and transportation. (Translations from the Spanish originals.) Washington, D.C., 1945. Inter-American Development Commission. 134 p.

Cover title—título de cubierta: **Central America: Reports Presented to the Conference of Commissions of Inter-American Development.**

Informes presentados a la Conferencia de Comisiones Interamericanas de Fomento (celebrada en Nueva York del 9 al 18 de mayo de 1944), por las Comisiones Interamericanas de Fomento de Costa Rica, El Salvador, Guatemala, Honduras, y Nicaragua, sobre: Costa Rica (breve reseña de la economía costarricense); El Salvador (situación actual de la economía salvadoreña); Guatemala (condiciones económicas de Guatemala); Honduras (algunas observaciones acerca de los problemas económicos de Honduras); Nicaragua (desarrollo económico e inversiones; comercio internacional y transportes).

Int-**6000.S4** 1943 **Selected Bibliography in English on Latin American Resources, Economic Development, and Trade.** September, 1943. U.S. Department of Commerce. Bureau of Foreign and Domestic Commerce. 10 p. Hectographed.

Bibliografía escogida, en inglés, sobre fuentes de riqueza, desarrollo económico y comercio de Latino-América.

Int-**6000.S6** **Studies and Reports, Series B, Economic Conditions.** Starting date? Montreal. International Labour Office.

Estudios e informes de la Oficina Internacional del Trabajo, serie B, condiciones económicas.

Int-**6000.W4** 1944 **World Economic Development.** Effects on advanced industrial countries. By Eugene Staley. Montreal, 1944. International Labour Office. 218 p. Studies and reports, series B, Economic conditions, no. 36.

Contents: Introduction and summary; effects arising out of international investment for developmental purposes (activation of economics, post-war readjustments, income distribution and the transfer problem in repayment, etc.); longer-range effects resulting from shifts in production, consumption and trade (economic development and trade prospects, measures to encourage industrial adaptation within each country, international arrangements to ease transition adjustment, etc.); some broader implications of economic development in new areas (population pressures, political power and cultural influence). The foreign trade section contains valuable statistics and charts.

Evolución económica mundial, consecuencias

en los países altamente industriales. Contenido: Introducción y resumen; resultados de las inversiones internacionales para fines de desarrollo (fomento de la economía, reajuste de postguerra, distribución de la renta y el problema de la transferencia de pagos, etc.); efectos mediatos de cambios en producción, consumo y comercio (desarrollo económico y perspectivas del comercio, medidas para estimular la adaptación industrial dentro de cada país, acuerdos internacionales para facilitar ajustes de transición); algunas consecuencias mas amplias del desenvolvimiento económico en nuevas áreas (presión demográfica, potencia política, e influencia cultural). La sección del comercio exterior contiene estadísticas y cuadros importantes

*Int-6000.W5 **World Economic Survey.** 1932- . Annual. Geneva. League of Nations.

Contains: Economic and financial repercussions of recent political events; international monetary relations; production, stocks, industrial activity; international trade and transport; and all the information available on national income in different countries.

Enquête économique mondiale. Comprend: Répercussions économiques et financières de récents événements politiques; relations monétaires internationales; production, stocks, activité industrielle; commerce et transport international; et tous les renseignements disponibles sur le revenu national des différents pays.

1942-44 **World Economic Survey, 1942-44.** Eleventh year. Princeton, New Jersey, 1945.II.A.4. League of Nations. Economic, Financial and Transit Department. 299 p. Publication, no. 1945. II.A.4.

A review of the world economic situation. Describes the mobilization of resources for war production at its peak and examines the economic effects of the war. A general picture is presented of conditions in different countries and detailed studies are given on: Production, consumption, finance, prices, trade and transport throughout the world.

Une revue de la situation économique mondiale. Décrit la mobilisation des ressources pour la production de guerre à son maximum et examine les effets économiques de la guerre. Une vue générale est presentée des conditions dans les divers pays et des études en détail sont données sur: Production, consommation, finance, prix, commerce et transports dans le monde entier.

*Int-6000.W7 **World Production and Prices.** Starting date? Annual. Geneva. League of Nations. Economic Intelligence Service.

Production mondiale et prix, publié annuellement par le Service d'Etudes Economiques, Société des Nations.

1938-39 **World Production and Prices, 1938/39.** Geneva, 1939. League of Nations. Economic Intelligence Service. 113 p.

Includes: Survey of recent changes in structure of world production; general trends of world production, 1920-38; production, employment and unemployment; industrial production by principal countries; production, stocks, and commodities by groups of commodities and industries; production and trade; price movements; industrial profits and farm income.

Comprend: Etude des variations récentes dans la structure de la production mondiale; tendances générales de production mondiale, 1920-38; production, emploiement et chômage; production industrielle par pays principaux; production, stocks, et marchandises par groupes de marchandise et par industries; production et commerce; mouvements des prix; bénéfices industrielles et revenue agricole.

Int-6000.W8 1933 **World Prosperity as Sought Through the Economic Work of the League of Nations.** By Wallace McClure. New York, The Macmillan Company, 1933. 613 p.

Contents: Introduction, the society of nations; background, instrumentalities, and progress of the economic work of the League of Nations; development of world economy through the economic work of the League of Nations: Production and distribution of wealth; international trade; finance; economics of war and peace. This book discusses the work done by the League of Nations in the sphere of world economics up to 1933, and refers to the various conventions and agreements concluded under the auspices of the League in the League's efforts to unify and standardize the practices and customs of international trade.

Prospérité mondiale comme but des travaux économiques de la Société des Nations. Matière: Introduction, la Société des Nations; l'arrière-plan, les instruments, et le progrès des travaux économiques de la Société des Nations; développement de l'économie mondiale par les travaux économiques de la Société des Nations; production et distribution de la richesse; commerce international; finance; l'économique de la guerre et de la paix. Ce livre étudie les travaux effectués par la Société des Nations concernant l'économie mondiale jusqu'à 1933, et mentionne les diverses conventions et accords conclus sous les auspices de la Société des Nations dans ses efforts pour unifier et uniformiset les pratiques et coutumes du commerce international.

Int-**6010**.E5 1924 **Economic Barometers.** Report submitted to the Economic Committee of the League of Nations. Geneva, 1924. International Labour Office. 56 p. Studies and reports, series N, Statistics, no. 5. Unexamined.

Barómetros económicos. Informe de la Oficina Internacional del Trabajo.

Int-**6010**.P5 1939-45 **Price Trends and Price Control in the American Republics, 1939-1945.** Washington, D.C., December 1945. U.S. Department of Labor. Bureau of Labor Statistics. 78 p. Mimeographed. "Preliminary Report—Subject to Revision."

Shows changes in cost of living and price indexes, and discusses price control and rationing, presenting data based on government reports and material supplied by U.S. Consular officials, for the following countries: Argentina, Bolivia, Brazil, Chile, Colombia, Costa Rica, Cuba, Ecuador, Mexico, Paraguay, Peru, United States of America, Uruguay, Venezuela. Includes two summary tables showing "Indexes of Changes in Cost of Living and Wholesale Prices from August 1939," and "Indexes of Food Prices at Retail (August 1939)," in these countries.

Tendencias y control de los precios en las repúblicas americanas. Presenta datos obtenidos de fuentes gubernamentales y de informes sometidos por oficiales consulares de los Estados unidos, exponiendo los cambios en los índices de precios y del costo de vida, y analizando los programas de racionamiento y control de precios, en los países siguientes: Argentina, Bolivia, Brasil, Chile, Colombia, Costa Rica, Cuba, Ecuador, México, Paraguay, Perú, Estados Unidos, Uruguay, Venezuela. Incluye dos cuadros que resumen, con base agosto de 1939, los índices de cambios en el costo de la vida y en los precios al por mayor, y los índices de precios de productos alimenticios al por menor, en los países citados.

*Int-**6100**.N3 1946 **National Income and Related Subjects for the United Nations.** Material prepared by the Technical Working Group. Washington, D.C., 1946. United Nations. Committee on Contributions. 250 p. (approx.) Hectographed.

Prepared in part by the United States Bureau of the Budget and Bureau of the Census. Includes: Summary tables on national income of United Nations, per capita income of United Nations, and comparisons of old and revised estimates on per capita income and on national income, population, and exchange rates; supplemental data on number and proportion of workers engaged in agriculture and related industries for United Nations (table), external assets and liabilities of United Nations (table and notes), notes on Western Hemisphere exchange rates, foreign exchange rates (tables), balance of international payments (notes), and balance of trade (notes and tables); country tables and notes on national income, per capita income, population, and alternative estimates, production, price, and other economic data, and balance of international payments for each country in the Americas, Europe, and the Near East, and for selected countries in Africa and the Far East.

Rentas nacionales y tópicos relacionados, para las Naciones Unidas. Estudio preparado, en parte, por la Dirección de Presupuestos y la Dirección de Censos de los Estados Unidos. Incluye: Cuadros que resumen las rentas nacionales de las Naciones Unidas, ingresos por persona en las Naciones Unidas, y comparaciones de estimaciones viejas y revisadas de los ingresos por persona y de rentas nacionales, población, y tipos de cambio; datos suplementarios sobre número y proporción de obreros empleados en la agricultura e industrias relacionadas de las Naciones Unidas (cuadro), activos y pasivos externos de las Naciones Unidas (cuadro y anotaciones), notas sobre tipos de cambio en el Hemisferio Occidental, tipos de cambio de divisas (cuadros), balanza de pagos internacionales (notas), y balanza comercial (notas y cuadros); cuadros y notas, por países, sobre rentas nacionales, ingresos por persona, población, y estimaciones alternativas, producción, precios, y otros datos económicos, y balanzas de pagos internacionales para cada país de las Américas, Europa y del Cercano Oriente, y para algunos países de Africa y del Lejano Oriente.

*Int-**6100**.N5 1945 **National Income Estimates of Latin America.** By Loreto M. Dominguez. Washington, D.C.? November, 1945. Conference on Research in Income and Wealth. 116 p. Multilithed.

"... Presents national income estimates for the nations of Latin America except four of the smaller republics—Guatemala, Haiti, Costa Rica, and Nicaragua ... In preparing estimates, studies by investigators within the countries have been reviewed and appraised. When no previous estimates had been prepared, special estimates appear here for the first time ... The

data . . . have been drawn from material with which the author is now working in connection with a program initiated by the Inter American Statistical Institute . . ." (Source: Foreword.)

Estimaciones de rentas nacionales de la América Latina. ". . . Expone estimaciones de rentas nacionales de las naciones latinoamericanas, con excepción de . . . Guatemala, Haiti, Costa Rica, y Nicaragua . . . En la preparación de las estimaciones, fueron examinados y avaluados varios estudios realizados por investigadores de los respectivos países. En los casos en que estimaciones anteriores no habían sido preparadas, aparecen aquí, por primera vez, estimaciones especiales. Los datos . . . han sido recogidos de materiales, con los cuales el autor está actualmente trabajando, a propósito de un programa principiado por el Instituto Interamericano de Estadística . . ." (Fuente: Prefacio; traducción del comentario.)

*Int-**6100**.R5 1946 **Report of the Sub-Committee on National Income Statistics.** Princeton, New Jersey, April 1946. League of Nations. Committee of Statistical Experts. 21 p. Mimeographed.

Contents—comprend: Preface; "Measurement of National Income and the Construction of Social Accounts," report by the Sub-Committee (including: Methods of approach, usefulness of an over-all system of social accounts, tabular framework, basis for international comparison, advantages of the recommended system of accounts); appendix "Definition and Measurement of the National Income and Related Totals," memorandum submitted by Richard Stone (including: Introduction; social accounts; a working system of social accounting; content of a system of accounts; national income and other aggregates of transactions; methods for extending the system in detail; some economic problems involved in defining and interpreting accounting systems; statistical sources and practical problems of measurement; advantages of a social accounting system).

Rapport du Sous-Comité des Statistiques sur le Revenu National, publié par le Comité d'Experts Statisticiens de la Société des Nations.

Int-**6100**.W5 1946 **Working Paper on the Status of National Income Statistics.** Prepared by Hildegarde Kneeland and Michael Sapir, Division of Statistical Standards, United States Bureau of the Budget and submitted to the Statistical Commission of the United Nations by Stuart A. Rice. With an annex, Tentative Statement on National Income Estimates for Sixty-Three Countries, prepared by Ernest M. Doblin, National Income Division, United States Department of Commerce and submitted to the Statistical Commission by Stuart A. Rice. Washington, D.C., 1946. 57 p. Mimeographed.

Contents: Summary of estimates available; concepts and methods used in measuring national income; international income comparison; applications and uses of national income statistics; recent development and recommendations; annex: Explanatory notes; derivation of estimates for individual countries; table I, national income of the United Nations; table II, per capita income of the United Nations; table III, national income of selected non-member countries; table IV, per capita income of selected non-member countries.

Documentación sobre la situación de la estadística de la renta nacional, preparada por Hildegarde Kneeland y Michael Sapir, División de Estandards Estadísticos de la Oficina del Presupuesto de Estados Unidos y sometida por Stuart A. Rice a la Comisión Estadística de las Naciones Unidas con un anexo: Informe preliminar sobre la estimación de la renta nacional para 63 países, preparado por Ernest M. Doblin, División de Renta Nacional del Departamento de Comercio de los Estados Unidos y sometido por Stuart A. Rice a la Comisión Estadística. Contiene: Resumen de estimaciones disponibles; conceptos y métodos usados para medición de la renta nacional; comparación de la renta internacional; aplicaciones y usos de las estadísticas de la renta nacional; evolución y recomendaciones recientes; anexo: Notas explicativas; derivación de estimaciones para países individuales; tabla I, renta nacional de las Naciones Unidas; tabla II, renta por habitante de las Naciones Unidas; tabla III, renta nacional de algunos países no miembros de las Naciones Unidas; tabla IV, renta por habitante de algunos países no miembros.

Int-**6200**.A45 **Agricultural, Pastoral, and Forest Industries in the American Republics.** 1944- . Irregular. Washington, D.C. United States Tariff Commission.

A series of reports by country.

Las industrias agrícolas, pastorales y forestales de las repúblicas americanas. Serie de informes, por países.

Int-**6200**.A5 1927 **Agricultural Problems in Their International Aspect: Documentation for the International Economic Conference, Geneva, 1927.** Rome, Print. Off. of the International Institute of Agriculture, 1927. International Institute of Agriculture. 662 p.

Contains statistical data on agricultural economics for various countries, including: Land ownership; rural population; marketing of farm

products; farmers' cooperatives; international agricultural credit.

Les problèmes agricoles dans leur aspect international: Documentation pour la Conférence Economique Internationale, Genève, 1927. Comprend des données statistiques sur l'économie agricole dans divers pays, y compris: Propriété de la terre; population rurale; vente des produits de la ferme; coopératives agricoles; crédit international agricole.

Int-6200.A6 1945 **Agricultural Production and Trade by Countries.** Pre-war summary by commodities. Washington, D.C., 1945. U.S. Department of Agriculture. Office of Foreign Agricultural Relations. 134 p.

A comprehensive exposition of international agricultural trade and production statistics.

Exposición comprensiva de estadísticas sobre la producción y el comercio agrícola internacional.

Int-6200.A8 **American Agricultural Series.** 1942- . Washington, D.C. Pan American Union. Division of Agricultural Cooperation.

Series de publicaciones de la Unión Panamericana sobre agricultura en las Américas.

Int-6200.B5 1914 **Les Bases Théoriques de la Statistique Agricole Internationale.** Par Umberto Ricci. Rome, Imp. de l'Institut International d'Agriculture, 1914. Institut International d'Agriculture. Service de la Statistique Générale. 314 p.

The theoretical bases of international agricultural statistics. First part, "International Statistics of Areas and Agricultural Productions" (article published in the **Bulletin of the International Statistical Institute,** volume XX, second part (Vienna Session, 1913), p. 675-763); second part, "International Statistics of Levels of Cultures" (article published in the **Bulletin of the International Statistical Institute,** volume XIX, third part (Session of the Hague, 1911), p. 1-92).

Première partie, "Statistique Internationale des Superficies et des Productions Agricoles" (article publié au **Bulletin de l'Institut International de Statistique,** tome XX, 2ème livraison (Session de Vienne, 1913), p. 675-763); deuxième partie, "Statistique Internationale des Etats des Cultures" (article publié au **Bulletin de l'Institut International de Statistique,** tome XIX, 3ème livraison (Session de la Haye, 1911), p. 1-92).

Int-6200.C5 1945.2 **Report of the First Session of the Conference, Held at the City of Quebec, Canada, October 16 to November 1, 1945.** Containing the reports of Commission A (policy and program) and Commission B (organization and administration), with supplementary data relating to resolutions and recommendations, the budget, rules of procedure, financial regulations, and the constitution. Washington, D.C., 1946. Food and Agriculture Organization of the United Nations. 89 p.

Includes reports of the committees on nutrition and food management, agriculture, forestry and forest products, fisheries, marketing, statistics, and an outline of the administrative structure of the FAO together with its constitution.

Informe de la primera sesión de la conferencia de la Organización sobre Alimentación y Agricultura de las Naciones Unidas (FAO), celebrada en Quebec, Canadá, 16 oct. a 1 nov. 1945. Incluye informes de los comités sobre alimentación y administración de productos alimenticios, agricultura, silvicultura y productos forestales, pesquería, mercados, estadísticas, y el esquema de la estructura administrativa de la FAO así como la constitución de la misma.

Int-6200.C5 1943 **Final Act of the United Nations Conference on Food and Agriculture, Hot Springs, Virginia, United States of America, 18th May-3rd June, 1943.** London, H.M. Stationery Off., 1943. Great Britain. Foreign Office. 40 p. Miscellaneous, no. 3 (1943). Unexamined.

Acta final de la conferencia de las Naciones Unidas sobre alimentos y agricultura, Hot Springs, Virginia, Estados Unidos, 18 de mayo al 3 de junio de 1943.

Int-6200.D3 1942 **Decennial Index of Publications Issued by the International Institute of Agriculture, 1930-1939.** Rome, 1942. International Institute of Agriculture. 55 p. Unexamined.

Index décennal des publications publiées par l'Institut International d'Agriculture, 1930-39.

Int-6200.D6 1944 **Documents Relating to the Food and Agriculture Organization of the United Nations, 1st August-14th December, 1944.** Presented by the Secretary of State for Foreign Affairs to Parliament by command of His Majesty. London, H.M. Stationery Off., 1945. United Nations Interim Commission on Food and Agriculture. 42 p. British Foreign Office miscellaneous, no. 4 (1945).

Includes: Constitution of the FAO; instrument of acceptance by the United Kingdom; first report to the governments of the United Nations; list of members, by countries.

Documentos relativos a la Organización de las Naciones Unidas para la Agricultura y Alimentación, agosto 1°-diciembre 14, 1944. Incluye: Constitución de la Organización; instrumento de aceptación por el Reino Unido; primer informe a los gobiernos de las Naciones Unidas; lista de miembros por países.

Int-6200.I4 1945.1 Eng **Final Act of the Third Inter-American Conference on Agriculture.** Held at Caracas, Venezuela, July 24 to August 7, 1945. Washington, D.C., 1945. Pan American Union. 113 p. Congress and conference series, no. 49.

Acta final de la Tercera Conferencia Interamericana de Agricultura.

Int-6200.I4 1945.2 **Handbook for the Use of the Delegates of the Third Inter-American Conference on Agriculture.** Caracas, Venezuela, July 24, 1945. Washington, D.C. Pan American Union. 4 vol.: Vol. 1, part one, **First and Second Sections of the Agenda: Money and Agriculture, International Basic Agricultural Commodities;** vol. 1, part two, **Second Section of the Agenda: Oil Bearing Seeds and Vegetable Oils, Animal Fats and Wool;** vol. 2, part one, **Third Section of the Agenda: Foodstuffs and Raw Materials;** vol. 2, part two, **Third Section of the Agenda: Animal Industry, Economic Entomology, Industrial Utilization of Agricultural Products, and Food and Nutrition;** vol. 3, **Fourth, Fifth and Sixth Sections of the Agenda: Markets and Transportation, Agricultural Migration in the Postwar Years, Agricultural Statistics;** vol. 4, **International Organizations Dealing with Agriculture, Farm Cost Analysis.**

Manual para uso de los delegados a la Tercera Conferencia Interamericana de Agricultura.

Int-6200.I4 1942.1 Eng **Final Act of the Second Inter-American Conference on Agriculture.** Mexico D.F., Mexico, July 6-16, 1942. Washington, D.C., Pan American Union, 1942. 59 p. Congress and conference series, no. 43.

Published also in Spanish—publicado también en español (Int-6200.I4, 1942.1 Sp).

Int-6200.I4 1942.1 Sp **Acta Final de la Segunda Conferencia Interamericana de Agricultura.** Celebrada en México, D.F., México, del 6 al 16 de julio de 1942. Wáshington, D.C., 1942. Unión Panamericana. 60 p. Serie sobre congresos y conferencias, no. 43.

Published also in English—publicado también en inglés (Int-6200.I4, 1942.1 Eng).

Int-6200.I4 1930.1 **Final Act of the Inter-American Conference on Agriculture. Acta Final de la Conferencia Interamericana de Agricultura. Acta Final da Conferência Interamericana de Agricultura. Acte Final de la Conférence Interaméricaine sur l'Agriculture.** Washington, September 8-20, 1930. Washington, D.C., U.S. Govt. Print. Off., 1930. Inter-American Conference on Agriculture. 209 p.

Int-6200.I64 1931 **The International Institute of Agriculture.** An historical and critical analysis of its organization, activities, and policies of administration. By Asher Hobson. Berkeley, California, University of California Press, 1931. International Institute of Agriculture. 356 p. University of California publications in international relations, vol. II, 1931.

Includes, in chapter IV, a discussion of the original idea of an international "clearinghouse" for agricultural statistical data and the difficulties encountered.

L'Institut International d'Agriculture. Une analyse historique et critique de son organisation, ses activités et ses plans d'administration. Comprend, au chapitre IV, une étude de l'idée première d'une "chambre de compensation internationale" pour les données statistiques agricoles et les difficultés éprouvées.

Int-6200.I65 1924 **The International Institute of Agriculture.** Its organization, activity, and results. Rome, Print. Off. of the International Institute of Agriculture, 1924. 51 p.

L'Institut International d'Agriculture: Son organisation, ses activités, ses résultats.

Int-6200.I8 **International Review of Agriculture.** 1927- . Monthly. Rome. International Institute of Agriculture.

Consists of four separate reviews—se compose de quatre revues distintes. (1) **Monthly Bulletin of Agricultural Economics and Sociology** (Int-6200.M4); (2) **Monthly Crop Report and Agricultural Statistics** (Int-6220.M5); (3) **Monthly Bulletin of Agricultural Science and Practice;** (4) **International Bulletin of Plant Protection.**

Revue internationale de l'agriculture.

*Int-6200.I9 **International Yearbook of Agricultural Statistics. Annuaire International de Statistique Agricole.** 1910—1940-41. Rome. International Institute of Agriculture. Institut International d'Agriculture.

A standard reference work relating to agricultural production, prices of agricultural products, and trade in these products.

Un ouvrage classique à consulter contenant renseignements sur la production agricole, les prix des produits agricoles et le commerce de ces produits.

Int-6200.L6 1939 **List of Publications Issued by the International Institute of Agriculture.** November 1939. Rome, Ville Umberto I, 1939. International Institute of Agriculture. 16 p. Supplement for March 1940 folded in.

Liste des publications publiées par l'Institut International d'Agriculture.

Int-**6200.**L7 1946 **Publications Editées par l'Institut International d'Agriculture: Supplément à la Liste de Janvier 1943.** Janvier 1943-Juin 1946. Rome, 1946. Institut International d'Agriculture. 2 p.

Supplement to the January 1943 list of publications issued by the International Institute of Agriculture.

Int-**6200.**L7 1943 **Liste des Publications Editées par l'Institut International d'Agriculture.** Janvier 1943. Rome, Stabilimenti Tipo. Carlo Colombo, 1943. Institut International d'Agriculture. 16 p.

List of publications issued by International Institute of Agriculture.

Int-**6200.**M2 1944 **Methods and Problems of International Statistics with Special Reference to Food and Agriculture: A List of References.** Compiled by Annie M. Hannay. Washington, D.C., December 1944. United Nations Interim Commission on Food and Agriculture. 26 p. Mimeographed.

Una lista de fuentes de referencia sobre métodos y problemas de la estadística internacional, con especial referencia a agricultura y alimentos, publicada por la Comisión Interina de Agricultura y Alimentos de las Naciones Unidas.

Int-**6200.**M4 **Monthly Bulletin of Agricultural Economics and Sociology.** 1910- . Rome. International Institute of Agriculture.

Published also in French, German, Italian, and Spanish. Title has varied in the past, and it has also appeared at irregular intervals. Since 1938 each issue has contained a section devoted to governmental measures affecting agricultural prices.

Bulletin mensuel d'économie et de sociologie agricoles, de l'Institut International d'Agriculture. Publié aussi en français, en allemand, en italien et en espagnol. Le titre a varié dans le passé, et l'ouvrage a paru à des intervalles irréguliers. Depuis 1938, chaque édition contient une section consacrée aux mesures d'Etat concernant les prix agricoles.

Int-**6200.**O6 1910-13 **L'Organisation des Services de Statistique Agricole dans les Divers Pays.** Rome, Impr.de la Chambre des Députés, 1910-13. Institut International d'Agriculture. 2 vol.

Organization of services of agricultural statistics in various countries. Includes forms for each country.

Comprend des formules pour chaque pays.

*Int-**6200.**S5 1946 **Standing Advisory Committee on Statistics: First Report to the Director-General.** Copenhagen, 1946. Food and Agriculture Organization of the United Nations. 14 p. Mimeographed.

Contents: Merging statistical work of International Institute of Agriculture into that of FAO; 1950 world census of agriculture; statistical publications of FAO; current reports; aiding governments to improve their statistical services; world food survey; relations with Inter-American Statistical Institute; relations with statistical work of Economic and Social Council of United Nations.

Comisión Consultiva Permanente de Estadística: Primer informe al Director General de la Organización de las Naciones Unidas para la Agricultura y la Alimentación. Contenido: Fusión del trabajo estadístico del Instituto Internacional de Agricultura y de la FAO; censo mundial agrícola de 1950; publicaciones estadísticas de la FAO; informes corrientes; ayuda a los gobiernos para el perfeccionamiento de sus servicios estadísticos; investigación mundial sobre alimentos; relaciones con el Instituto Interamericano de Estadística; relaciones con el trabajo estadístico del Consejo Económico y Social de las Naciones Unidas.

Int-**6200.**W4 1940-1945 **The Work of the International Institute of Agriculture During the War (1940-1945).** Rome, Villa Umberto I, 1945. International Institute of Agriculture. 106 p.

Includes a general statement of the vice-president on member states, permanent committee, staff, legal and diplomatic situation of the Institute, and work accomplished, and a report by the Secretary General on the work of the administration and on the work of the General Secretariat, the Section of Agricultural Legislation, the Bureau of General Statistics, the Bureau of Agricultural Science, the Bureau of Economic and Social Studies, and the library.

Le travail de l'Institut International d'Agriculture durant la guerre (1940-45). Comprend une déclaration générale du vice-président sur les Etats membres, le comité permanent, le personnel, la situation légale et diplomatique de l'Institut et le travail accompli, de même qu'un rapport du Secrétaire Général sur le travail d'administration et sur celui du Secrétariat Général, de la Section de la Législation Agricole, du Bureau de la Statistique Générale, du Bureau Scientifique Agricole, du Bureau des Etudes Economiques et sociales et de la bibliothèque.

Int-**6200**.W6 **The World Agricultural Situation.** 1929-30—1940-41? Annual. Rome. International Institute of Agriculture. Unexamined.

From 1929-30 to 1932-33, entitled—de 1929-30 à 1932-33 intitulé: **The Agricultural Situation.**

An economic comentary on the **International Yearbook of Agricultural Statistics,** indicating the main tendencies of the economic situation in agriculture, analyzing the position of certain branches of agricultural activity, giving some account of the action taken by the governments in the different countries, and presenting economic conditions affecting the farming classes.

La situation agricole mondiale. Un commentaire économique sur l'**Annuaire International de Statistique Agricole,** indiquant les tendances principales de la situation économique de l'agriculture, analysant la position de certaines domaines de l'activité agricole, et donnant des rapports sur l'action des gouvernements dans les différents pays, et présentant la situation économique de la classe agricole.

Int-**6210**.A4 1934-38 **Agricultural Commodities and Raw Materials: Production and Consumption in the Different Parts of the World, 1934-1938.** Rome, Fausto Failli, 1944. International Institute of Agriculture. 229 p.

Published also in French—publié aussi en français: **Denrées et Matières Premières Agricoles: Production et Consommation dans les Différentes Parties du Monde.**

Contents: General survey of distribution of agricultural production and consumption of agricultural products among different parts of world, 1934-38 (distribution of agricultural production; distribution of consumption of agricultural products; relations between production and consumption, total area and population); production, trade, consumption and prices of the several agricultural commodities; intercontinental trade in some important agricultural commodities, 1935-38.

Matière: Aperçu général de la distribution de la production agricole et de la consommation des produits agricoles parmis les différentes parties du monde, 1934-38 (distribution de la production agricole; distribution de la consommation des produits agricoles; rapport entre production et consommation, superficie totale et population); production, commerce, consommation et prix de diverses denrées agricoles; commerce intercontinental dans quelques importantes denrées agricoles, 1935-38.

Int-**6210**.C4 **Coffee Statistics.** 1940- . Frequency? New York. Pan American Coffee Bureau.

Statistical picture of the situation in the main coffee producing nations of the Western Hemisphere. Contains charts, graphs, special studies on production, consumption, imports, exports, etc.

Estadísticas del café. Panorama estadístico de la situación de los principales países productores de café del Hemisferio Occidental. Contiene cuadros, gráficos, estudios especiales sobre producción, consumo, importación y exportación.

*Int-**6210**.G5 1924-38 **Les Grands Produits Agricoles—Compendium International de Statistiques, 1924-1938.** Rome, Villa Umberto, 1944. Institut International d'Agriculture. 579 p.

". . . A vast statistical documentation on production, commerce, and consumption of agricultural products of interest to the world economy. The data, collected regularly in the **International Yearbook of Agricultural Statistics,** are brought together, elaborated and so presented as to facilitate the work of all who desire to ascertain, at a glance, the total volume of the production, consumption, exportation and importation of agricultural staples; the extent to which the leading countries and the several continents participate in these phenomena, and the changes which have occurred in their regard during the last three five-year periods preceding the war." (Translation of comment.)

". . . Une vaste documentation statistique sur la production, le commerce et la consommation des produits agricoles intéressant l'économie mondiale. . . Les données, recueillies régulièrement dans. . . l'**Annuaire International de Statistique Agricole,** ont été concentrées, élaborées et présentées de manière à faciliter la tâche de tous ceux qui désirent pouvoir se renseigner d'un coup d'oeil sur le montant total de la production et de la consommation, des exportations et des importations des produits envisagés, sur la mesure dans laquelle les principaux pays et les différents continents participent à ces phénomènes et sur les changements qui ont eu lieu à leur égard pendant les trois dernières périodes quinquennales précédant la guerre actuelle." (Source: Préface.)

Int-**6210**.P5 1945, Eng **Fourth Pan American Coffee Conference, Mexico City, September 3-14, 1945.** Washington, D.C., 1945. Pan American Union. 39 p. Congress and conference series, no. 50.

Cuarta Conferencia Panamericana del Café.

Int-**6210**.W5 **Wheat Studies.** 1925-1944. Irregular. Stanford University. Food Research Institute.

Twenty volumes. Periodical surveys and annual review of the world wheat situation and studies of special interest. No longer published.

Estudios sobre trigo. 20 volúmenes. Estudios periódicos y síntesis anual sobre la situación mundial de trigo e investigaciones de especial interés. Ha cesado su publicación.

Int-**6210**.W6 1945 **World Grain Review and Outlook, 1945.** By Helen C. Farnsworth and Vladimir P. Timoshenko. With a statistical appendix by Rosamond H. Peirce. Stanford, California, Stanford University Press, 1945. Stanford University. Food Research Institute. 319 p.

Contents: Food grains in 1943-44 (fifteen years perspective; outstanding features of the crop year; review by countries—United States, Canada, Australia, Argentina, British Isles, Continental Europe, Soviet Russia, India, other important areas); feed grains in 1943-44 (outstanding features of the crop year; review by countries—United States, Canada, Argentina); grain outlook for 1944-45 and later years (world developments; outlook by countries—United States, Canada, Australia, Argentina, Continental Europe, USSR, British Isles, and other important countries, grain policies and world trade); statistical appendix (tables, explanatory notes, classification).

Revista y perspectivas mundiales de los granos, 1945. Contiene: Granos alimenticios en 1943-44 (perspectivas de quince años; características sobresalientes de la cosecha del año; análisis por países—Estados Unidos, Canadá, Australia, Argentina, Islas Británicas, Europa continental, Rusia Soviética, India, otras zonas importantes); granos forrajeros en 1943-44 (características sobresalientes de la cosecha del año; revista por países—Estados Unidos, Canadá, Argentina); perspectivas de los granos para 1944-45 y años sucesivos (evolución mundial, perspectivas por países—Estados Unidos, Canadá, Australia, Argentina, Europa continental, USSR, Islas Británicas y otros países importantes, política y comercio mundial de granos); apéndice estadístico (tablas, notas explicativas, clasificación).

Int-**6220**.I5 1910 **International Crop-Reporting Service.** Report presented by Dr. Traugott Müller, delegate of the German Empire, adopted by the General Assembly December, 1909, concerning the establishment of an international crop-reporting service for the staples of agriculture for 1910-1911, etc. Washington, D.C., U.S. Govt. Print. Off., 1910. International Institute of Agriculture. 28 p.

Rapport presenté par le Dr. Traugott Müller, délégué de l'Empire allemand, et adopté par l'Assemblée Générale, décembre, 1909, concernant l'établissement d'un service international de renseignements sur les récoltes des principaux produits agricoles pour les années de 1910-11, etc.

*Int-**6220**.M5 **Monthly Crop Report and Agricultural Statistics.** 1910-43—1945- . Rome. International Institute of Agriculture.

Issued also in French, German, Italian, and Spanish—publié aussi en français, allemand, italien, et espagnol. Title varies—le titre varie: 1910-18, **Bulletin of Agricultural Statistics;** 1919-38, **International Crop Report.**

Estimates on agricultural production, current information on agriculture, and general surveys and studies of a statistical character.

Rapport mensuel sur les récoltes et statistique agricole, publié par l'Institut International d'Agriculture. Calcul de la production agricole, renseignements courants sur l'agriculture et enquêtes et études générales d'un caractère statistique.

Int-**6230**.A5 1933 **L'Aviculture dans le Monde.** Rome, 1933. Institut International d'Agriculture. 3 vol. Pas examiné.

Aviculture in the world.

Int-**6240**.F5 1943 **The Fisheries and Fishery Resources of the Caribbean Area.** With recommendation for their expansion and development. A report by the Caribbean Fishery Mission. Washington, D.C., 1943. Issued jointly by the U.S. Department of the Interior, Fish and Wild-Life Service, and the U.S. Coordinator of Inter-American Affairs. Unexamined.

Un informe sobre la pesca y sus recursos en el Caribe, con recomendaciones para su expansión y fomento.

Int-**6250**.C5 1926 **Actes du Ier Congrès International de Sylviculture.** Rome, 29 avril-5 mai 1926. Rome, 1926. Institut International d'Agriculture. 5 vol. Pas examiné.

Acts of the first International Congress of Forestry, Rome, April 29-May 5, 1926.

Int-**6250**.F5 1925 **Forests and Forestry—Statistical and Other Information for Certain Countries.** Rome, Printing Office of the Chamber of the Deputies, 1925. International Institute of Agriculture. Bureau of General Statistics. 425 p.

English translation of the monograph published by the International Institute of Agriculture in French in October 1924, under the title **Les Forêts.** Study on the organization of forestry statistics and nature of statistical material available in various countries. Includes United States and Canada.

Les forêts—renseignements statistiques concernant différents pays. Traduction anglaise de la monographie publiée en français par l'Institut International d'Agriculture en octobre 1924, sous le titre **Les Forêts.** Etude sur l'organisation des statistiques forestières et la nature des données statistiques disponibles dans les divers pays. Y compris les Etats-Unis et le Canada.

*Int-**6250**.I5 **International Yearbook of Forestry Statistics.** 1932- . Rome. International Institute of Agriculture.

Annuaire international de statistiques forestières.

1933-35 **International Yearbook of Forestry Statistics, 1933-35.** Rome, Printing Office "Ditta Carlo Colombo," 1936?-42. International Institute of Agriculture. Vol. I, **Europe and the U. S. S. R.**; vol. II, **America**; vol. III, **Africa**; Supplement to vol. III.

Tables in French and English—tableaux en français et en anglais.

Contents of vol. II: Summary tables: Forest area; trade by countries; general summary of trade by countries; detailed tables: Forest statistics, North and Central America, and South America; statistics of trade in wood, North and Central America, and South America; information on some wood species.

Matière du volume II: Tableaux d'ensemble: Superficie des forêts; commerce par pays; résumé général du commerce par pays; tableaux de détail: Statistiques des forêts, Amérique Septentrionale et Centrale et Amérique Méridionale; statistiques sur le commerce du bois, Amérique Septentrionale et Centrale et Amérique Méridionale; information sur quelques essences de bois.

Int-**6300**.C8 1935 **Copper Resources of the World.** Washington, D.C., 1935. Sixteenth International Geological Congress. 2 vol. Unexamined.

Includes bibliographies.

Las fuentes de cobre en el mundo. Incluye bibliografías.

Int-**6300**.F5 **Foreign Minerals Quarterly.** Starting date? Washington, D.C. U.S. Department of the Interior. Bureau of Mines.

Publicación trimestral sobre los minerales en el exterior.

Int-**6400**.I3 1946 "Industrialization in Latin America: An Outline of Its Development." By J. Rafael Oreamuno. Washington, D.C., 1946? Inter-American Development Commission. 10 p.

". . . The BULLETIN of the Pan American Union published a series of papers on inter-American relations, 1920-1945, of which this was the second. It is now being reproduced by the Inter-American Development Commission." (Source: Foreward.)

Industrialización en Latino-America: Un bosquejo de su desarrollo. "El Boletín de la Unión Panamericana publicó una serie de artículos sobre relaciones interamericanas 1920-1945, de los cuales este fué el segundo. Ha sido reproducido ahora, por la Comisión de Fomento Interamericano." (Traducción del comentario.)

Int-**6400**.I5 1945 **Industry in Latin America.** By George Wythe. New York, Columbia University Press, 1945. 371 p.

A survey of the industrial aspects of the different Latin American national economies. Contents: Background and problems; industry as the way out; nature and extent of industrialization; problems of raw materials and fuel; capital and the spirit of enterprise; the machine learns to speak Spanish and Portuguese; limitation of the market; government and industry; the twenty republics, Argentina, Uruguay, Paraguay, Brazil, the west coast of South America, the north coast of South America, Mexico, Central America, the Antillean republics; today and tomorrow; conclusion.

La industria en la América Latina. Un estudio de los aspectos industriales de las distintas economías nacionales de Latino-América. Contiene: Bases y problemas; la industria como medio de solución; naturaleza y alcance de la industrialización; problemas de materias primas y combustibles; capital y el espíritu de empresa; la máquina aprende a hablar español y portugués; la limitación del mercado; gobierno e industria; las veinte repúblicas, Argentina, Uruguay, Paraguay, Brasil, la costa oeste de América del Sud, la costa norte de América del Sud, México, América Central, las repúblicas de las Antillas; el presente y el mañana; conclusión.

Int-**6400**.M5 **Mining and Manufacturing Industries in the American Republics.** 1944- . Irregular. Washington, D.C. United States Tariff Commission.

A series of reports by country.

Las industrias manufactureras y mineras de las repúblicas americanas. Serie de informes, por países.

*Int-**6400**.S5 1934 **Statistique Internationale des Grandes Villes: Statistique de**

l'Electricité, du Gaz et de l'Eau dans les Grandes Villes, 1934. La Haye, Imp. Société Anonyme "Drukkerij Trio," 1939. Office Permenent de l'Institut International de Statistique. 97 p.

Statistics of electricity, gas and water for large cities, 1934; includes data on capacity, consumption, and rates. From the Western Hemisphere, only Argentina, Brazil, Canada and Haiti are represented in the group of countries for which data is shown.

Présente des données sur la puissance, la consommation, et le prix. De l'hémisphère occidentale, seulement l'Argentine, le Brésil, le Canada, et Haïti sont représentés dans le groupe de pays dont des chiffres sont données.

Int-**6400.**T5 1944 **The T. V. A.: Lessons for International Application.** By Herman Finer. Montreal, 1944. International Labour Office. 289 p. Studies and reports, series B, Economic conditions, no. 37.

La T. V. A.: Lecciones para aplicación internacional.

Int-**6610.**R5 1932 **Report on the Unification of Transport Statistics.** Draft international convention on transport statistics together with regulations relating respectively to maritime navigation, railways and inland navigation. Geneva, 1932. League of Nations. Committee on Unification of Transport Statistics. Publications, no. 1931.viii.21.

Rapport sur l'unification de la statistique des transports. Projet de convention internationale sur la statistique des transports y compris les règlements relatifs à la navigation maritime, aux chemins de fer et tramways et à la navigation intérieure.

Int-**6610.**T5 1942 **Transportation in Latin America.** Analysis of statistics. By Ludwig M. Homberger. Washington, D.C., 1942. American University. School of Social Sciences and Public Affairs. 13 p. Loose-leaf. Unexamined.

Transportes en la América Latina: Análisis de estadísticas.

Int-**6611.**I5 1944 **International Civil Aviation Conference, Chicago, Illinois, November 1 to December 7, 1944: Final Act and Related Documents.** Washington, D.C., U.S. Govt. Print. Off., 1945. U.S. Department of State. 284 p. Conference series, no. 64.

Conferencia Internacional sobre Aviación Civil, Chicago, Illinois, del 1° de noviembre al 7 de diciembre de 1944: Acta final y documentos relacionados con ella.

INTERNATIONAL
6400

Int-**6614.**B5 **Boletín de la Asociación Permanente.** 1916- . Bimensual. Buenos Aires. Asociación Permanente del Congreso Sudamericano de Ferrocarriles.

Formerly called—anteriormente intitulado: **Boletín del Congreso Sudamericano de Ferrocarriles.**

Bulletin of the Permanent Association of the South American Congress of Railroads. Covers the field of railway transportation in Latin America; publishes statistical information on railway construction and operation, and current developments.

Abarca el campo del transporte ferroviario en la América Latina; publica informaciones estadísticas sobre construcción, funcionamiento y desarrollo actual de los ferrocarriles.

Int-**6614.**I5 1939 **International Comparisons of Railway Ton-Mile Revenues.** Washington, D.C., 1939. U.S. Interstate Commerce Commission. Bureau of Transport Economics and Statistics. 16 p. Mimeographed. Unexamined.

Comparación internacional del beneficio por tonelada-milla de los ferrocarriles.

Int-**6615.**I5 1940, Eng **Final Act of the Inter-American Maritime Conference, Washington, November 25-December 2, 1940.** Washington, D.C., 1940. Pan American Union. 16 p. Congress and conference series, no. 33.

Published also in Spanish—publicado también en español (Int-**6615.**I5, 1940, Sp).

Int-**6615.**I5 1940, Sp **Acta Final de la Conferencia Marítima Interamericana, Unión Panamericana, Wáshington, del 25 de Noviembre al 2 de Diciembre de 1940.** Wáshington, D.C., 1941. Unión Panamericana. 15 p. Serie sobre congresos y conferencias, no. 33.

Published also in English—publicado también en inglés (Int-**6615.**I5, 1940, Eng).

Int-**6700.**B5 **Boletín Informativo del Consejo Interamericano de Comercio y Producción.** Serie A: Crónica y orientaciones. Fecha de iniciación? Mensual. Montevideo. Consejo Interamericano de Comercio y Producción.

Published through December 1944 as—publicado hasta el fin de diciembre 1944, como: **Boletín Informativo del Consejo Permanente de Asociaciones Americanas de Comercio y Producción.**

Bulletin of the Inter-American Council of Commerce and Production, Series A, Chronicle and orientations.

Int-**6700.**C4 **Circulares Documentales de la Comisión Ejecutiva del Consejo**

Interamericano de Comercio y Producción. 1943- . Montevideo.

Documentary circulars of the Executive Commission of the Inter-American Council of Commerce and Production.

Series of monographs.

Serie de monografías.

Int-**6700.C6** 1945 **Consejo Interamericano de Comercio y Producción: Recopilación de sus Acuerdos y Recomendaciones.** Montevideo, Impresora Uruguaya, 1945. Consejo Interamericano de Comercio y Producción. 2 vol.: Primer cuaderno, **Acta de Constitución y Normas de Funcionamiento;** segundo cuaderno, **Programas de Habilitación, Producción Agraria, Forestal y Pesquera.**

Collection of resolutions and recommendations of the Inter-American Council of Commerce and Production.

Int-**6700.I6** 1944.2 **Policies Adopted by the Second Plenary Meeting of the Permanent Council of American Associations of Commerce and Production.** Economic proposals for the Western Hemisphere. Washington, D.C., 1944? United States Committee to the Permanent Council of American Associations of Commerce and Production. 38 p.

Contains recommendations that the Inter American Statistical Institute establish "indices and schedules of equivalent statistical classifications for the American nations with special reference to their export trade," and that the national statistics of each American nation be made to agree with "recognized inter-American standards" for which purposes relations with the activities of the IASI are to be fostered.

Políticas adoptadas por la segunda sesión plenaria del Consejo Permanente de las Asociaciones Americanas de Comercio y Producción. Propuestas económicas para el Hemisferio Occidental. Contiene la recomendación de que el Instituto Interamericano de Estadística establezca índices y escalas de clasificaciones estadísticas equivalentes para las naciones americanas, con especial referencia a su comercio de exportación, que la estadística nacional de cada país sea llevada de acuerdo con los estandards interamericanos reconocidos, y que para este fin sean fomentadas las propuestas relaciones con las actividades del IASI.

Int-**6700.I6** 1942.1 **Resolutions of the First Plenary Meeting of the Permanent Council of American Associations of Commerce and Production.** Economic questions of interest to the Americas. Washington, D.C., 1942? United States Committee to the Permanent Council of American Associations of Commerce and Production. 37 p.

Contains recommendation that the IASI adapt, for inter-American purposes, the resolutions of the Geneva Conference of 1928 regarding the preparation of uniform trade statistics, with particular regard to classification, valuation, quantities, countries of origin of imports and countries of destination of exports.

Resoluciones tomadas por la primera sesión plenaria del Consejo Permanente de las Asociaciones Americanas de Comercio y Producción. Asuntos económicos de interés para las Américas. Contiene la recomendación de que el IASI adapte para propósitos interamericanos, las resoluciones tomadas por la Conferencia de Ginebra de 1928 relativas a la preparación de estadísticas comerciales uniformes, con especial atención a clasificación, valuación, cantidades, países de origen de importaciones y países de destino de exportaciones.

Int-**6710.S4** **Studies and Reports, Series H, Cooperation.** Starting date? Geneva. International Labour Office.

Estudios e informes de la Oficina Internacional de Trabajo, series H, cooperación.

Int-**6720.I6** 1944 **International Currency Experience, Lessons of the Inter-War Period.** New York, Columbia University Press, 1944. League of Nations. 249 p. Unexamined.

"Treats successively over-all monetary demand-supply conditions, exchange standards, central bank policy and practice, exchange fluctuations, stabilization funds, exchange control, and underlying non-monetary conditions. The statistical appendix contains material in compact form concerning world gold supply, and central bank assets and reserves on a country basis." (Source: **Journal of the American Statistical Association,** vol. 40, no. 229, March 1945, p. 129-130.)

L'expérience monétaire internationale, enseignements de la période écoulée entre les deux guerres. Traite successivement de la situation générale de l'offre et la demande monétaires, des étalons du change, des systèmes et des pratiques de la banque centrale, des fluctuations du change, des fonds de stabilisation, du contrôle du change, et des conditions fondamentales non-monétaires. L'appendice statistique comprend un précis relatif à la réserve mondiale d'or, et à l'actif et aux réserves de la banque centrale selon le pays.

Int-**6720.I8** 1944 **International Monetary Fund and International Bank for Reconstruction and Development.** Articles of agreement, United Nations Monetary and Financial Conference, Bretton Woods, New

Hampshire, July 1 to 22, 1944. Washington, D.C., U.S. Govt. Print. Off., 1944. U.S. Department of the Treasury. 88 p.

Fondo Monetario Internacional y Banco Internacional de Reconstrucción y Fomento. Acuerdos de la Conferencia Monetaria y Financiera de las Naciones Unidas, Bretton Woods, New Hampshire, julio 1° al 22, 1944.

*Int-**6720**.M6 1942-44 **Money and Banking, 1942-44.** Geneva, distributed in the United States by Columbia University Press, 1945. League of Nations. Economic, Financial and Transit Department. 224 p. League of Nations Publication 1945, II.A.I.

Contents: International summaries of essential data on currency, banking, and interest rates; central and commercial banks; balance sheet position and profit-and-loss accounts; synopsis of tables. A compendium of the banking and monetary statistics of the world. Fifty-four central bank and forty-five commercial bank systems are covered in annual and quarterly tables, extending from 1937 to September 1944. The statistics are accompanied by brief notes explaining the data given and summarizing recent changes in banking legislation.

Monnaie et banques, 1942-44. Matière: Résumés internationaux des données importantes sur la monnaie, les banques et les taux d'intérêts; banques centrales et commerciales; situation du bilan et comptes de profits et pertes; synopsis des tableaux. Un recueil de statistique bancaire et monétaire du monde. Les régimes de cinquante-quatre banques centrales et de quarante-cinq banques commerciales sont couverts dans les tableaux annuels et trimestriels, traitant la période de 1937 jusqu'à septembre, 1944. Les statistiques sont accompagnées de notes brèves expliquant les données presentées et donnant un résumé des changements récents dans la législation bancaire.

Int-**6720**.R5 1946 **Memoria, Primera Reunión de Técnicos sobre Problemas de Banca Central del Continente Americano.** Celebrada en la ciudad de México del 15 al 30 de agosto de 1946, a invitación y bajo los auspicios del Banco de México ,S.A. México, 1946. 534 p.

Report of the first meeting of experts on problems of central banks in the Western hemisphere. Contents: Preamble; invitation; agenda; directory of participants; officers; inaugural session; committee 1, regulation of money and credit; committee 2, problems of balance of payments and foreign exchange; committee 3, cooperation among the departments of economic studies; closed session; permanent committee.

Contiene: Preámbulo; invitación; agenda; directorio de participantes; funcionarios; sesión inaugural; comité 1, regulación monetaria y del crédito; comité 2, problemas de balanza de pagos y cambio exterior; comité 3, cooperación entre los departamentos de estudios económicos; sesión de clausura; comité permanente.

Int-**6720**.U6 1944 **United Nations Monetary and Financial Conference, Bretton Woods, New Hampshire, July 1 to July 22, 1944: Final Act and Related Documents.** Washington, D.C., U.S. Govt. Print. Off., 1944. U.S. Department of State. 122 p. Conference series, no. 55.

Conferencia Monetaria y Financiera de las Naciones Unidas, Bretton Woods, New Hampshire, julio 1° al 22 de 1944: Acta final y documentos relativos.

Int-**6721**.I5 1945 **Inversiones Internacionales en América Latina.** Problemas y perspectivas. Por Javier Márquez. México, 1945. Banco de México. 189 p. Informaciones económicas, no. 3.

International investments in Latin America: Problems and perspectives.

Int-**6900**.A5 **Annuaire Statistique du Commerce des Armes et des Munitions. Statistical Year-Book of the Trade in Arms and Ammunition.** 1924—1939-1940. Genève. Société des Nations. League of Nations.

1939-40 issue in English only.

1938 **Annuaire Statistique du Commerce des Armes et des Munitions, 1938. Statistical Year-Book of the Trade in Arms and Ammunition, 1938.** Genève, 1938. Société des Nations. League of Nations. 340 p.

Includes: (1) Summary trade tables, a series of statistical tables for each of the countries, showing the exports and imports of arms and ammunition, according to either the class of goods or the countries of destination or origin; (2) general summary, showing all the countries together and furnishing data concerning the values of exports and imports of arms, ammunition and implements of war expressed in former U.S. gold dollars; (3) detailed trade tables, taken from the official and public documents of the countries, on which information the two preceding parts are based.

Comprend: (1) Tableux sommaires du commerce, une série de tableaux statistiques qui indiquent les exportations et les importations d'armes et de munitions, pour chaque pays, soit par catégories de marchandises, soit par pays de destination ou de provenance; (2) récapitulation générale, qui mentionnent, par ordre alphabétique, tous les pays, et qui donnent des renseignments sur la valeur des exportations et des importations d'armes, munitions et matériels de guerre, exprimées en anciens dollars-or des Etats-Unis; (3) tableaux détaillés du commerce, tirés des documents officiels et publics de chacun des pays, sur lesquels sont basées les deux premières parties ci-dessus mentionnées.

Int-**6900**.A6 1946 **Anteproyecto de Carta para la Organización Internacional de Comercio de las Naciones Unidas.** Ampliación de las propuestas de los Estadas Unidos para la expansión del comercio internacional y del empleo, preparadas por los técnicos del gobierno de los Estados Unidos y presentadas como base para la discusión pública. Montevideo, "Impresora Uruguaya" S.A., 1946. Consejo Interamericano de Comercio y Producción. 74 p. Cuadernos comerciales del Consejo, no. 3.

Suggested charter for an international trade organization of the United Nations. Amplification of the United States' proposals for the expansion of trade and employment; prepared by the United States government technicians and presented as a basis for public discussion.

Int-**6900**.B5 1944.1 **Bibliography of the Latest Available Source Material on Foreign Trade Statistics.** Washington, D.C., 1944? U.S. Bureau of Foreign and Domestic Commerce. Division of International Economy. International Economics and Statistics Unit. 70 p. (approx.) Hectographed.

Bibliografía de las últimas fuentes de informaciones disponibles, sobre estadísticas del comercio exterior.

Int-**6900**.B5 1944.2 **Supplement-A to Bibliography of the Latest Available Source Material on Foreign Trade Statistics.** Washington, D.C., 1944? U.S. Bureau of Foreign and Domestic Commerce. Division of International Economy. International Economics and Statistics Units. 32 p. (est.) Hectographed.

Suplemento A a la bibliografía de las últimas fuentes de información disponibles, sobre estadísticas del comercio exterior.

Int-**6900**.B5 1943.1 **Bibliography of the Latest Available Source Material on Foreign Trade Statistics Consulted by the International Economics and Statistics Unit.** Washington, D.C., 1943. U.S. Bureau of Foreign and Domestic Commerce. Division of International Economy. International Economics and Statistics Unit. 149 p. Hectographed.

Bibliografía de las últimas fuentes de informaciones disponibles sobre estadísticas del comercio exterior.

Int-**6900**.B5 1943.2 **List of Publications References in Field of Foreign Trade Statistics for the American Republics.** Revised August 1, 1943. Washington, D.C., 1943. U.S. Bureau of Foreign and Domestic Commerce. Division of International Economy. International Economics and Statistics Unit. 90 p.

Lista de referencias sobre publicaciones concernientes a las estadísticas del comercio exterior de las repúblicas americanas. Revisada el 1° de agosto de 1943.

Int-**6900**.C3 1943 **Comercio Exterior de la América Latina.** Informe sobre el comercio de la América Latina con referencia especial a su comercio con los Estados Unidos. Wáshington, D.C., U.S. Govt. Print. Off., 1943. Comisión Arancelaria de los Estados Unidos. Informe no. 146. 4 vol.: Parte I, **Comercio de la América Latina con el Mundo y con los Estados Unidos**; parte II, **Política y Relaciones Comerciales de Cada Uno de los Países Latinoamericanos**: Tomo I, **Las Repúblicas Sudamericanas**; tomo 2, **México y las Repúblicas de la América Central y las Indias Occidentales**; parte III, **Principales Productos de Exportación de la América Latina.**

Published also in English—publicado también en inglés (Int-**6900**.F6).

Int-**6900**.C5 1943 **Commercial Policies and Trade Relations of European Possessions in the Caribbean Area.** A report on recent developments in the trade of the European possessions in the Caribbean area, with special reference to trade with the United States. Washington, D.C., U.S. Govt. Print. Off., 1943. United States Tariff Commission. 324 p. Report no. 151, second series.

Políticas y relaciones comerciales de las posesiones europeas en el Caribe: Un informe de la reciente evolución del comercio de las posesiones europeas en el Caribe, con especial referencia al comercio con los Estados Unidos.

Int-**6900**.C7 **Commodities of Commerce Series.** Starting date? Irregular. Washington, D.C. Pan American Union.

Serie de estudios sobre artículos comerciales.

Int-**6900**.F4 1944-45 **Foreign Commerce Series under the Fellowship Program of 1944-45.** Background material on foreign trade classification in the various American nations. Articles, prepared by interns under this program, are hectographed. Washington, D.C. U.S. Department of Commerce. Bureau of Foreign and Domestic Commerce.

Series del comercio exterior conforme al programa de becas de 1944-45, del Departamento de Comercio de los Estados Unidos. Material básico sobre clasificación del comercio exterior en varias naciones americanas. Artículos preparados por quienes intervienen en este programa.

Int-**6900**.F6 1942 **The Foreign Trade of Latin America.** A report on the trade of Latin America with special reference to trade with the United States under the provisions of title III, part II, section 332 of the Tariff Act of 1930. Revised. Washington, D.C., U.S. Govt. Print. Office, 1942. United States Tariff Commission. Report 146, second series. 4 vol. Part I, **Trade of Latin America with the World and with the United States**; part II, **Commercial Policies and Trade Relations of Individual Latin American Countries**: Vol. 1, **The South American Republics**; vol. 2, **Mexico and the Republics of Central America and the West Indies**; part III, **Selected Latin American Export Commodities.**

El comercio exterior de la América Latina. Un informe sobre el comercio de la América Latina, con especial referencia al comercio con los Estados Unidos.

Int-**6900**.F8 **Foreign Trade Series.** 1926- Washington, D.C. Pan American Union.

Individual trade reports on all Latin American republics, and general surveys of Latin American trade. Statistics of imports and exports by countries of origin or destination.

Series sobre comercio exterior. Informes individuales sobre todas las repúblicas latinoamericanas, y un estudio general del comercio latino-americano. Estadísticas de importación y exportación, por países de origen y destino.

Int-**6900**.I7 1944 **International Trade in Wood.** Rome, 1944. International Institute of Agriculture. Unexamined.

"The compilation of this volume, which was published in August, 1944, was undertaken under agreements with the International Forestry Centre. The Bureau of Statistics has assembled in it for all the countries of the world, the data on imports and exports of wood and wood pulp, during the period 1925-1939. These data, drawn from the customs statistics of the several countries, have been classified and presented as far as possible on uniform lines. The first part of the volume deals with the volume and value of the trade of each country. The second part consists of a series of summarised tables showing for each country, for each continent, and for the world as a whole the total value of the imports and exports of the products considered." (Source: **The Work of the International Institute of Agriculture during the War (1940-1945)**, p. 74-75.)

Commerce international du bois. "La compilation de ce volume, qui fut publié en août, 1940, a été entreprise en vertu d'accords avec l'International Forestry Centre. Le Bureau de la Statistique y a réuni les données sur les importations et les exportations de bois et de pâte de bois pour tous les pays du monde, pendant la période de 1925-39. Ces données, prises des statistiques douanières de divers pays, ont été classées et presentées autant que possible d'une manière uniforme. La première partie de l'ouvrage traite du volume et la valeur du commerce de chaque pays. La deuxième comprend une série de tableaux abrégés présentant pour chaque pays, pour chaque continent et pour le monde entier la valeur totale des importations et des exportations des produits en question." (Traduction du commentaire.)

Int-**6900**.L4 1941 **Latin America As a Source of Strategic and Other Essential Materials.** A report on strategic and other essential materials, and their production and trade, with special reference to the Latin American countries and to the United States. Washington, D.C., U.S. Govt. Print. Off., 1941. United States Tariff Commission. 397 p. Report no. 144, second series.

La América Latina como fuente de material estratégico y otros esenciales: Un informe sobre materiales estratégicos y otros esenciales, su producción y comercio, con especial referencia a los países latino-americanos y a Estados Unidos.

Int-**6900**.M7 **Memorandum on International Trade and Balances of Payments.** 1910-1923—1931-1932. Annual. Geneva. League of Nations.

Title varies slightly—titre varie un peu. From 1933 on, published as three separate annuals under the titles—depuis 1933, publié en trois annuels séparés sous les titres: **Review of World Trade; Balance of Payments; International Trade Statistics.**

Int-**6900**.N5 1942 **The Network of World Trade.** A companion volume to **Europe's Trade.** Princeton, New Jersey, Princeton University Press, 1942. League of Nations. Economic Intelligence Service. 172 p.

Published also in French—publié aussi en français: **Le Réseau du Commerce Mondial.**

A comprehensive exposition on the essential unity of world trade; includes: Introduction (scope of statistics presented); magnitude and composition of world trade; direction of trade;

trade of "Anglo-American Group"; system of multilateral trade; statistical tables covering the period 1928-1938, showing world trade in merchandise, world trade in gold and silver, and world trade by countries of provenance and destination.

Le réseau du commerce mondial, étude faisant suite au **Commerce de l'Europe.** Un exposé complet de l'unité essentielle du commerce mondial; y compris: Introduction (portée des statistiques presentées); ampleur et composition du commerce mondial; direction du commerce; commerce du "Groupe anglo-américain"; les systèmes de commerce entres plus de deux pays; tableaux statistiques couvrant la période 1928-38, exposant le commerce mondial des marchandises, le commerce mondial de l'or et de l'argent, et le commerce mondial par pays de provenance et de destination.

Int-**6900**.N5 1942.A **Addendum to Annex III of "The Network of World Trade" (pp. 106-171).** Originally excluded for economy of space. By Mr. Chapman. League of Nations. 33 p.

Prepared for the Inter American Statistical Institute by the author.

Addition à l'Annexe III de **Le Réseau du Commerce Mondial,** preparée pour l'Institut Interaméricain de Statistique par l'auteur.

Int-**6900**.R4 **Rapport sur le . . . Exercice.** 1891-92—1937? Annuel. Bruxelles. Bureau International pour la Publication des Tarifs Douaniers. Pas examiné.

Published also in German, English, Spanish, and Italian—publié aussi en allemand, anglais, espagnol, et italien.

Annual report of the International Bureau for the Publication of Customs Tariffs. It is understood that as late as 1939 the Bureau collected and translated into five languages the customs tariff of every country in the world.

Int-**6900**.R5 **Recent Developments in the Foreign Trade of the American Republics.** 1944- . Irregular. Washington, D.C. United States Tariff Commission.

A series of reports by country.

Recientes acontecimientos del comercio exterior de las repúblicas americanas. Serie de informes, por cada país.

Int-**6900**.R6 **Reports of the United States Tariff Commission.** Starting date? Irregular. Washington, D.C.

Informes de la Comisión de Tarifas de los Estados Unidos.

*Int-**6900**.R8 **Review of World Trade.** 1932- . Annual. Geneva. League of Nations. Economic Intelligence Service.

Before 1932, published as part of the annual **Memorandum on International Trade and Balances of Payments.**

Revue annuelle du commerce mondial, publiée par la Société des Nations. Avant 1932 publiée comme partie du **Memorandum on International Trade and Balances of Payments** (Mémoire annuel sur le commerce international et sur la balance des paiements).

Int-**6900**.S2 1946 **Some Technical Aspects of Foreign Trade Statistics with Special Reference to Valuation.** By Nicholas Michael Petruzzelli. Washington, D.C., Murray & Heister, (distributed by Sheiry Press), 1946. 252 p.

Contents: Uses and importance of foreign trade statistics; impediments in comparative foreign trade statistics; major foreign trade statistical problems other than valuation; analysis of valuation practices employed in recording foreign trade; foreign trade valuation objective and problems; problems involved in reducing national values to a common basis; conclusions and proposals; appendix I, definitions of exports and imports and bases of values in foreign trade statistics of specified countries, 1928-39; appendix II, valuation of exports and imports for statistical purposes in specified countries, 1927-39; appendix III, local units of weight and measure in specified countries with United States and metric equivalents; appendix IV, improvement, warehousing, and direct transit and transshipment trade of specified countries in various years; appendix V, ships' stores including bunker fuel in exports of specified countries; appendix VI, international convention relating to economic statistics, Geneva, December 14, 1928, annex I, external trade statistics; bibliography.

Algunos aspectos técnicos de la estadística del comercio exterior, con especial referencia a valuación. Contenido: Usos e importancia de la estadística del comercio exterior; impedimentos en la comparación de las estadísticas del comercio exterior; principales problemas estadísticos del comercio exterior distintos a valuación; análisis de las prácticas de valuación empleadas en el registro del comercio exterior; objeto y problemas de la valuación en el comercio exterior; problemas relativos a la reducción de valores nacionales a una base común; conclusiones y propuestas; apéndice I, definiciones de importaciones y exportaciones y bases de valores en las estadísticas del comercio exterior de ciertos países 1928-39; apéndice II, valuación de exportaciones e importaciones para fines estadísticos en ciertos países, 1927-39; apéndice III, unidades locales de peso y medida en ciertos países, con equivalencias estadounidenses y métricas; apéndice IV, mejoramiento, depósitos y comercio de tránsito y reembarco de ciertos países en varios años; apéndice V, abastecimientos para barcos incluyendo combustible, en las exportaciones de ciertos países; apéndice VI, convención internacional relativa a estadísticas económicas, Ginebra, diciembre 14, 1928, anexo I, estadísticas del comercio exterior; bibliografía.

*Int-**6900**.S6 **Statistiques du Commerce International. International Trade Statistics.** 1931-32— . Annuel. Genève. Société des Nations. Service d'Etudes Economiques. League of Nations. Economic Intelligence Service.

Before 1931-32 published as part of—publié avant 1931-32 comme section de: **Memorandum on International Trade and Balances of Payments.**

1938 **Statistiques du Commerce International, 1938. International Trade Statistics, 1938.** Genève, 1939. Société des Nations. Service d'Etudes Economiques. League of Nations. Economic Intelligence Service. 345 p. Série de publications de la Société des Nations, II, Questions économiques et financières, 1939.II.A.21.

Contains statistics of foreign trade of sixty-seven countries for 1936-38.

Contient la statistique du commerce international de soixante-sept pays, pour 1936-38.

1937 **Statistiques du Commerce International, 1937. International Trade Statistics, 1937.** Genève, 1938. Société des Nations. Service d'Etudes Economiques. League of Nations. Economic Intelligence Service. 449 p.

This particular edition, of which relatively few copies are available in the U.S., is important because it contains special historical material relative to Draft Customs Nomenclature and Minimum List. Contains: (1) Recapitulation of subdivisions of Draft Customs Nomenclature and Minimum List, by sections, chapters, items, and net total; also items split and sub-items split, p. 18-19; (2) tariff numbers, p. 29-55; (3) tariff classification numbers keyed to Minimum List numbers, p. 56-58. Summaries are based on 1928 edition of Minimum List.

Cette édition particulière, dont il n'existe aux Etats Unis que relativement peu d'exemplaires, est importante en ce qu'elle contient de la matière historique speciale relative au Projet de Nomenclature Douanière et à la Liste Minimum. Contient: (1) Récapitulation des divisions du Projet de Nomenclature Douanière et de la Liste Minimum par sections, chapitres, positions et le total net; aussi les positions divisées et les sous-positions divisées, p. 18-19; (2) concordance entre les positions et sous-positions de la Liste Minimum et les chiffres douaniers, p. 29-55; concordance entre les chiffres de la classification douanière et ceux de la Liste Minimum, p. 56-58. Les résumés se basent sur l'édition de 1928 de la Liste Minimum.

Int-**6900**.W4 1943 **World Trade in Agricultural Products.** By Henry C. Taylor and Anne D. Taylor. New York, Macmillan Co., 1943. 286 p.

A concise yet comprehensive survey of facts and figures concerning world trade in agricultural commodities. Discusses national policies, imperial preferences and international agreements that have influenced extent, character and direction of this trade. Each major commodity is treated in a separate chapter which is accompanied by charts, maps, and statistical tables which supply data on world imports and exports by countries. Commodities so treated are: Cotton, wool, silk, rubber, tobacco, coffee, tea, sugar, wheat, rice, feed grains, meat and live animals, fats and oils. The survey of each commodity extends down to the outbreak of the war.

Comercio mundial de productos agrícolas. Un estudio conciso pero comprensivo de hechos y cifras concernientes al comercio mundial de productos agrícolas. Examina políticas nacionales, privilegios imperialistas y tratados internacionales que han influído en la extensión, carácter y dirección de este comercio. Cada producto principal está tratado en un capítulo separado, llevando cuadros, mapas y tablas estadísticas con datos sobre importación y exportación por países. Los productos así tratados son: Algodón, lana, seda, caucho, tabaco, café, té, azúcar, trigo, arroz, granos forrajeros, carne y animales vivos, grasas y aceites. El estudio de cada producto se extiende hasta el comienzo de la guerra.

Int-**6900**.W5 1940 **World Trade in Agricultural Products.** Its growth; its crisis; and the new trade policies. By L. B. Bacon and H. C. Schloemer. Rome, Printing Office Carlo Colombo, 1940. International Institute of Agriculture. 1102 p.

Part I, analysis of world trade by commodities, deals with market developments on a world-wide scale for the most important commodities, representative of the various classes of agricultural products. Part II, analysis of world trade by

countries, deals with national policies affecting external trade in farm products in fifteen countries.

Commerce mondial des produits agricoles. Son développement; sa crise; et les politiques nouvelles du commerce. Partie I, analyse du commerce mondial par denrées, traite des développements du marché mondial pour les denrées les plus importantes, représentant les diverses classes des produits agricoles. Partie II, analyse du commerce mondial par pays, traite des politiques nationales affectant le commerce extérieur des produits de ferme dans quinze pays.

*Int-**6910**.B5 **Balance of Payments.** Annual. 1931-32— . Geneva. League of Nations. Economic Intelligence Service.

Before 1931-32, published as part of the annual —avant 1931-32 publié comme section de l'annuaire: **Memorandum on International Trade and Balances of Payment.**

Balance des paiements.

1938 **Balance of Payments, 1938.** Geneva, 1939. League of Nations. Economic Intelligence Service. 148 p. Series of League of Nations publications, II, Economic and financial, 1939.II.A.20.

Includes balance of payments statements for individual countries (only Argentina, Canada, and the United States of the countries of the Western Hemisphere are included). The standard scheme recommended by the League of Nations for compilation of balances of payments is given on p. 8-9.

Comprend des exposés relatifs à la balance des paiements pour pays individuels (y compris seuls l'Argentine, le Canada, et les Etats-Unis de l'hémisphère occidental). Le plan type pour la compilation des balances des paiements recommandé par la Société des Nations est donné sur les pages 8-9.

7000-7999 HEALTH AND CULTURAL STATISTICS—ESTADISTICAS DE SALUBRIDAD Y CULTURA

*Int-**7100**.B6 **Bulletin of the Health Organisation of the League of Nations.** 1932- . Bimonthly. Geneva. League of Nations. Health Organisation.

From 1932 to 1936 issued quarterly and entitled **Quarterly Bulletin of the Health Organisation.** Contains the technical reports prepared for the Health Organisation by its commissions and its individual experts.

Bulletin bimensuel de la Division de l'Hygiène de la Société des Nations. Publié trimestriellement de 1932 à 1936 et intitulé **Quarterly Bulletin of the Health Organisation.** Contient des rapports techniques préparés pour la Division de l'Hygiène par ses commissions et ses experts.

1943-44, Biraud "Glossary of Communicable Diseases." By Yves Biraud. Vol. X, no. 3, 1943-44. Unexamined.

This glossary contains the names of the communicable diseases in 24 languages and follows the classification of the Detailed International List of 1938.

Glossaire des maladies transmissibles. Comprend les noms des maladies transmissibles en vingt-quatre langues, d'apres la Liste Internationale Détaillée de 1938.

Dec. '38 "International List of Causes of Death Adopted by the Fifth International Conference for Revision, Paris, October 3rd-7th, 1938." Vol. VII, no. 6, December 1938, p. 944-987.

Liste internationale des causes de décès adoptée par la Cinquième Conference Internationale pour la Révision, Paris, 3-7 octobre 1938.

Int-**7100**.C5 1943 **Classified List of Medical and Public Health Journals of Latin America.** Compiled by Jeannette M. de la Garza. Washington, D.C., 1943. Pan American Sanitary Bureau. 68 p. Publication no. 199. Unexamined.

Lista clasificada de revistas latino-americanas de medicina y sanidad pública.

*Int-**7100**.I6 **International Health Year-Book.** 1924- . Annual. Geneva. League of Nations. Health Organisation.

Annuaire international de l'hygiène, publié par l'Organisation de l'Hygiène de la Société des Nations.

Int-**7100**.R5 1940 **Revistas de Medicina y Sanidad de la América.** Wáshington, D.C., agosto 1940. Oficina Sanitaria Panamericana. 51 p. Publicación no. 152. No examinado.

Journals of medicine and health in America. Bibliography.

Int-**7101**.I5 1945 **International Health Organisation.** London, 1945. United Nations Information Organization. Reference Division. 8 p. Information paper, no. 5.

Contains, in summary form, the plan for establishment of an international health organization under auspices of the United Nations, and brief recapitulation of work done by other international health organizations, such as International Public Health Office, Health Organisa-

tion of the League of Nations, Pan-American Sanitary Bureau, International Labour Organisation, and others.

Organización internacional de sanidad. Contiene en forma sumaria, el plan para el establecimiento de una organización internacional de sanidad bajo los auspicios de las Naciones Unidas, y una breve recapitulación del trabajo realizado por otras organizaciones internacionales de sanidad, como ser: Oficina Internacional de Salud Pública, Organización Sanitaria de la Liga de las Naciones, Oficina Sanitaria Panamericana, Organización Internacional del Trabajo, etc.

*Int-**7110**.B5 **Boletín de la Oficina Sanitaria Panamericana.** 1922- . Mensual. Washington, D.C.

Articles published in Spanish, Portuguese, French and English—Artículos publicados en español, portugués, francés e inglés.

Monthly bulletin of the Pan American Sanitary Bureau.

Sept. '46, Navarro "La Sanidad en Bolivia." Por Hubert E. Navarro. Vol. 25, no. 9, septiembre 1946, p. 778-784.

Public health in Bolivia.

Aug. '46, Thomen "La Sanidad de la República Dominicana." Por L. F. Thomen. Vol. 25, no. 8, agosto 1946, p. 673-678.

Public health in the Dominican Republic.

July '46, a "La Sanidad en Nicaragua." Año 25, no. 7, julio 1946, p. 577-587.

Public health in Nicaragua.

June '46, Barros Barreto "A Saúde Pública no Brasil." Panorama das atividades empreendidas pelo Departamento Nacional de Saúde. Pelo Dr. João Barros Barreto. Año 25, no. 6, junio 1946, p. 481-511.

Public health in Brazil. Panorama of activities of National Department of Health.

Mar. '46, Avila Camacho "La Sanidad en México." Por Manuel Avila Camacho. Año 25, no. 3, marzo 1946, p. 193-195.

Public health in Mexico.

Feb. '46, Carvallo "La Sanidad en el Perú." Por Constantino J. Carvallo. Año 25, no. 2, febrero 1946, p. 97-108.

Public health in Peru.

Jan. '46, Herrera "La Sanidad en Guatemala." Por Julio Roberto Herrera. Año 25, no. 1, enero 1946, p. 1-4.

Public health in Guatemala.

May '45 "Indices Sanitarios para las Principales Ciudades de Chile." Año 24, no. 5, mayo 1945, p. 428-434.

Health indexes for principal cities of Chile.

May '45, Collins "La Organización de los Servicios de Estadística Vital y Sanitaria en el Gobierno Federal de los Estados Unidos." Por Selwyn D. Collins. Vol. 24, no. 5, mayo 1945, p. 423-427.

Organization of vital and sanitary statistics in the federal government of the United States.

Mar. '45, Barros Barreto "A Saúde Pública no Brasil." Por João de Barros Barreto. Vol. 24, no. 3, marzo 1945, p. 193-210.

Public health in Brazil.

Feb. '45, Cabello González "La Estadística Biodemográfica y Epidemiológica en Chile." Por Octavio Cabello González. Vol. 24, no. 2, febrero 1945, p. 100-118.

Vital and epidemiological statistics in Chile.

May '44, Granillo "El Servicio de Bioestadística en México." Por Ricardo Granillo. Año 23, no. 5, mayo 1944, p. 419-426.

The Biostatistics Service in Mexico.

Aug. '43 "Indices Sanitarios Panamericanos." Año 22, no. 8, agosto 1943, p. 682-685.

Pan American health indexes.

Aug. '43, Linder "Clasificación y Análisis de los Datos de Estadística Vital." Por Forrest Edward Linder. Año 22, no. 8, agosto 1943, p. 687-690.

Continued in the following issues of the **Boletín**—continúa en las siguientes ediciones del **Boletín**: Año 22, no. 9, septiembre 1943, p. 802-810; año 23, no. 3, marzo 1944, p. 224-239; año 23, no. 8, agosto 1944, p. 700-704; año 24, no. 7, julio 1945, p. 611-618.

Classification and analysis of vital statistics data.

June '42 "Técnicas y Procedimientos para las Encuestas o Censos Dietéticos." Año 21, no. 6, junio 1942, p. 564-569.

Techniques and procedures for dietetic surveys or censuses. Report of conference held in Atlantic City October 18, 1941, under sponsorship of Rockefeller Foundation, for determining standards, techniques, and procedures for effecting studies and censuses on nutrition.

Informe de la conferencia realizada en Atlantic City el 18 de octubre de 1941, bajo los auspicios de la Rockefeller Foundation, con el propósito de determinar estandards, técnicas y procedimientos para la realización de estudios y censos sobre nutrición.

June-July '41 "Clave Diagnóstica para la Clasificación en Columna (Tabulación) de las Causas de Morbidad." Junio-julio 1941. 18 p. Reimpresión. Publicación de la Oficina Sanitaria Panamericana, no. 164.

Diagnostic key for columnar classification (tabulation) of causes of death.

Jan. '40 "Demografía de las Repúblicas Americanas." Año 19, no. 1, enero 1940, p. 15-29.

Demography of the American republics.

June-Aug. '39 "Nomenclatura Internacional de las Causas de Muerte." Junio, julio y agosto, 1939. 25 p. Reimpresión. Publicación de la Oficina Sanitaria Panamericana, no. 137.

International nomenclature of causes of death.

Aug. '36 "Nomenclatura para Hospitales Dedicados a Enfermedades Mentales." Año 15, no. 8, agosto 1936, p. 752-779.

Nomenclature for hospitals for mental diseases. Translation of a manual prepared by American Association of Psychiatry and Committee of Mental Hygiene.

Traducción de un manual preparado por la Asociación Americana de Psiquiatría y el Comité de Higiene Mental.

Int-7110.E4 **Epidemiological Information Bulletin.** 1945- . Semi-monthly. Washington, D.C. United Nations Relief and Rehabilitation Administration.

Boletín informativo de la UNRRA sobre epidemiología.

Int-7110.E8 **Estadística Epidemiológica para los Países de América. Epidemiological Statistics for the American Republics.** 1943?- . Anual. Washington, D.C. Oficina Sanitaria Panamericana. Pan American Sanitary Bureau.

In Spanish and English—en español e inglés.

Int-7110.I4 1943 **Informe Biodemográfico y Epidemiológico de las Américas. Biostatistical and Epidemiological Report on the Americas.** Por Rafael H. Martínez. Wáshington, D.C., Feb. 1943. Oficina Sanitaria Panamericana. Pan American Sanitary Bureau. 84 p. Publicación, no. 195.

Contains comparative statistical data on principal biostatistical rates, pestilential diseases, principal causes of death. Material is arranged alphabetically by countries. In general data are given for the years 1936 through 1941, but in some cases the series extend back to 1930.

Contiene datos estadísticos comparativos sobre las principales tasas bioestadísticas, enfermedades pestilenciales, principales causas de muerte. Ordenado alfabéticamente por países. En general los datos corresponden a los años 1936 a 1941, pero en ciertos casos las series remontan a 1930.

*Int-7110.I6 **Informe Epidemiológico Mensual. Monthly Epidemiological Report.** 1944- . Washington, D.C. Oficina Sanitaria Panamericana. Pan American Sanitary Bureau.

Contains information on incidence and mortality for specified diseases, and other vital statistics formerly included in **Boletín de la Oficina Sanitaria Panamericana.** Prior to 1946 called **Monthly Report, Biostatistics, Epidemiology.**

Contiene información sobre frecuencia y mortalidad por enfermedades determinadas, y otras bioestadísticas anteriormente incluídas en el **Boletín de la Oficina Sanitaria Panamericana.** Antes de 1946, su nombre era **Informe Mensual, Bioestadística, Epidemiología.**

June '44 "Inhabitants per Physician." —"Habitantes por Médico." No. 4, junio 1944.

Map of United States and Latin America, showing number of inhabitants per physician by country, folded in at back of no. 4.

Mapa de los Estados Unidos y de la América Latina, indicando el número de habitantes por médico en cada país, plegado al fín del no. 4.

Int-7110.P4 1944 **Quinta Conferencia Panamericana de Directores Nacionales de Sanidad, Abril 22-29, 1944. Fifth Pan American Conference of National Directors of Health, April 22-29, 1944.** Tema general: La sanidad durante la guerra y la post-guerra. General subject: Wartime and postwar health programs. Washington, D.C., 1944. Oficina Sanitaria Panamericana. Pan American Sanitary Bureau. no. 16 p.

Int-**7110**.P6 1942 **Public Health in the Americas.** Washington, D.C., 1942. Pan American Sanitary Bureau. 70 p. Publication, no. 177.

Published also in Spanish—publicada también en español (Int-**7110**.S5).

"In honor of the second Pan American health day (December 2, 1941) national health authorities and other prominent sanitarians of the Americas were invited to prepare (articles) for publication in the **Boletín de la Oficina Sanitaria Panamericana** . . . It was decided to reprint the translations in a separate booklet." (Source: p. viii.)

Int-**7110**.P8 **Publications of the Pan American Sanitary Bureau. Publicaciones de la Oficina Sanitaria Panamericana.** Starting date? Irregular. Washington, D.C. Pan American Sanitary Bureau.

Series of monographs.

Serie de monografías.

Int-**7110**.R4 **Rapport Epidémiologique. Epidemiological Report.** Supplément statistique au **Relevé Epidémiologique Hebdomadaire.** Statistical supplement to the **Weekly Epidemiological Record.** Depuis?-1940. Starting date?-1940. Mensuel. Monthly. Genève. Société des Nations. Section d'Hygiène. League of Nations. Health Section.

Publication suspended May 1940—publication terminée mai 1940.

Contains current figures on the various contagious diseases, for all countries and for the principal cities of the world.

Comprend chiffres courants sur certaines maladies contagieuses, pour tous les pays et pour les grandes villes principales du monde.

*Int-**7110**.R5 **Rapport Epidémiologique Annuel: Statistiques Corrigées des Maladies à Declaration Obligatoire. Annual Epidemiological Report: Corrected Statistics of Notifiable Diseases.** Depuis? Genève. Société des Nations. Organisation d'Hygiène. League of Nations. Health Organisation.

1938 **Rapport Epidémiologique Annuel: Statistiques Corrigées des Maladies à Declaration Obligatoire pour l'Année 1938.** Publiées par la Section d'Hygiène. **Annual Epidemiological Report: Corrected Statistics of Notifiable Diseases for the Year 1938.** Published by the Health Section. Genève, 1941. Société des Nations. Organisation d'Hygiène. League of Nations. Health Organisation. 114 p.

". . . Contains statistics of notifiable diseases in various countries for the year 1938, together with demographic data and figures for mortality from contagious diseases in the great towns of the world. . . . The recent corrected figures have been supplemented by the publication of retrospective tables, giving for a number of countries the rates of births, general mortality and infantile mortality. A table gives, for thirty-five countries, the number of deaths and the death rate for each item of the International Abridged List of Causes of Death. A special table gives for the same countries the death rate from all causes, specific for age and sex, for several recent years and also the corresponding standardised general mortality rates." (Source: Introduction.)

". . . Contient pour l'année 1938 des statistiques des maladies soumises à la déclaration obligatoire dans les divers pays, et des données sur la démographie et la mortalité par maladies contagieuses dans les grandes villes du monde. . . . Les données récentes corrigées ont été complétées par la publication de tableaux rétrospectifs, donnant pour une série de pays les taux de natalité, de mortalité générale, de mortalité infantile. Un tableau donne pour trent-cinq pays le nombre de décès et le taux de mortalité pour chacune des rubriques de la nomenclature internationale abrégée des causes de décès. Un tableau spécial donne pour les mêmes pays les taux de mortalité par toutes causes à chaque âge et pour chaque sexe pour quelques années récentes et les taux de mortalité générale standardisée correspondants." (Source: Préface.)

Int-**7110**.R6 **Relevé Epidémiologique Hebdomadaire. Weekly Epidemiological Record.** Extraits des renseignements recueillis par la Section d'Hygiène du Secrétariat de la Société des Nations. Extracts from information received by the Health Section of the Secretariat of the League of Nations. Depuis? Starting date? Genève.

Contains current figures on the cases and deaths of contagious diseases reported in certain countries in Europe, Africa, and America, including figures on birth, death, and infant mortality rates in certain countries and large towns.

Comprend des chiffres courants sur les cas et décès de maladies contagieuses déclarés dans divers pays d'Europe, d'Afrique, et d'Amérique, y compris chiffres sur les taux de natalité, de mortalité, et de mortalité infantile dans divers pays et grandes villes.

Int-**7110**.S5 1942 **La Sanidad en las Américas.** Wáshington, D.C., 1942. Oficina Sanitaria Panamericana. 95 p. Publicación no. 190. No examinado.

Published also in English—publicada también en inglés (Int-**7110**.P6).

"En honor del segundo Día Panamericano de la Salud (2 de diciembre de 1941), invitose a las autoridades nacionales de sanidad y a otros higienistas prominentes de las Américas a preparar trabajos para publicación en el **Boletín de la Oficina Sanitaria Panamericana** . . . Resultaron tan interesantes los informes recibidos, que decidiose publicarlos en forma de folleto."

Int-**7150**.G5 1935 **A Geography of Diseases.** A preliminary survey of the incidence and distribution of tropical and certain other diseases. By Earl Baldwin McKinley. Washington, D.C., The George Washington University Press, 1935. National Research Council. Division of Medical Sciences. 495 p. Supplement to **The American Journal of Tropical Medicine,** vol. 15, no. 5, September 1935.

Una geografía de enfermedades. Estudio preliminar sobre frecuencia y distribución de enfermedades tropicales y de ciertas otras enfermedades.

Int-**7160**.M5 1929 **Methods of Compiling Statistics of Railway Accidents.** Geneva, 1929. International Labour Office. 82 p. Studies and reports, series N, Statistics, no. 15. Unexamined.

Métodos de compilación de estadísticas de accidentes ferroviarios. Informe de la Oficina Internacional del Trabajo.

Int-**7200**.A5 1941 **Analfabetismo y Cultura Popular en América.** Por Roberto Moreno y García. México, D.F., Editorial Atlante, S.A., 1941. 112 p.

Illiteracy and popular culture in America. Contents: General features in America; popular education in the world; popular education in America; legal and social restrictions against illiterates; analysis by countries; illiteracy increases in America; direction of schools; educational budget in America; Indian problem; measures against illiteracy; conclusion; bibliographical note.

Contenido: Rasgos generales de América; el movimiento por la educación popular en el mundo; la educación popular en América; restricciones legales y sociales a los analfabetos; análisis por países; el analfabetismo crece en América; la dirección escolar; los presupuestos de educación en América; el problema indígena; las medidas contra el analfabetismo; conclusión; nota bibliográfica.

Int-**7200**.A7 **Annual Report of the Director of the Institute of International Education.** Starting date? New York. Institute of International Education.

Contains: Information on Latin American and other foreign students in the United States; statistics on foreign students studying in United States colleges and universities on scholarships and fellowships granted under the auspices of the Institute; list of publications of the Institute.

Informe anual del director del Instituto de Educación Internacional en Nueva York. Contiene: Información sobre estudiantes de la América Latina y de otros países extranjeros en los Estados Unidos; estadísticas sobre estudiantes extranjeros en colegios y universidades en los Estados Unidos con becas otorgadas bajo los auspicios del Instituto; lista de publicaciones del Instituto.

Int-**7200**.C6 1943.1, Eng **Final Act of the First Conference of Ministers and Directors of Education of the American Republics, Held in Panama, September 27 to October 4, 1943.** Washington, D.C., 1943. Pan American Union. 57 p. Congress and conference series, no. 45.

Published also in Spanish and Portuguese—publicado también en español y portugués (Int-**7200**.C6, 1943.1, Sp; Bras-**7200**.R5, Aug. '44).

Int-**7200**.C6 1943.1, Sp **Acta Final de la Primera Conferencia de Ministros y Directores de Educación de las Repúblicas Americanas Celebrada en Panamá del 27 de Septiembre al 4 de Octubre de 1943.** Wáshington, D.C., 1943. Unión Panamericana. 57 p. Serie sobre congresos y conferencias, no. 45.

Published also in English and Portuguese—publicado también en inglés y portugués (Int-**7200**.C6, 1943.1, Eng; Bras-**7200**.R5, Aug. '44).

Int-**7200**.E6 1942 **Educational Trends in Latin America.** Number 3. July 1942. By Francisco S. Céspedes. Washington, D.C., July 1942. Pan American Union. Division of Intellectual Cooperation. 21 p. Number 3.

Summary of educational conditions in the various Latin American countries limited to most urgent problems in field of elementary education and to solutions sought by Latin American educators. Contains list of bulletins, monographs, and books consulted, p. 19-21.

Tendencia de la educación en Latino-América. Resumen de las condiciones de la educación en los distintos países latino-americanos, limitado a problemas más apremiantes en el campo de la educación elemental y a soluciones previstas por los educadores latino-americanos. Contiene una lista de boletines, monografías y libros que han sido consultados, p. 19-21.

Int-**7210.**H5 **Higher Education in Latin America.** 1944- . Washington, D.C. Pan American Union. Division of Intellectual Cooperation.

Series of studies on higher education in the Latin American countries.

Serie de estudios de la Unión Panamericana acerca de la educación superior en los países latinoamericanos.

Int-**7300.**I5 1946.2 Eng **Inter-American Conference of Experts on Copyright, Pan American Union, June 1-22, 1946: Proceedings.** Washington, D.C., 1946. Pan American Union. 180 p. Congress and conference series, no. 51.

Published also in Spanish—publicado también en español (Int-**7300.**I5, 1946.2, Sp).

Int-**7300.**I5 1946.2, Sp **Conferencia Interamericana de Expertos para la Protección de los Derechos de Autor, 1-22 de Junio de 1946: Actas y Documentos.** Washington, D.C., 1946. Unión Panamericana. 185 p. Serie sobre congresos y conferencias, no. 50.

Published also in English—publicado también en inglés (Int-**7300.**I5, 1946.2, Eng).

Int-**7300.**J5 1944 **Journals Dealing with the Natural, Physical and Mathematical Sciences Published in Latin America.** A tentative directory. By Katherine Lenore Morgan. Washington, D.C., 1944. Pan American Union. Division of Intellectual Cooperation. 62 p.

Revistas tratando de ciencias naturales, físicas y matemáticas, publicadas en la América Latina. Ensayo de directorio.

8000-8999 GOVERNMENT AND POLITICAL STATISTICS—ESTADISTICAS GUBERNAMENTALES Y POLITICAS

Int-**8000.**C5 1941.1 **Acta Final del Segundo Congreso Inter-Americano de Municipios, Santiago de Chile, 15 al 21 de Septiembre de 1941.** Santiago?, 1941. Congreso Inter-Americano de Municipios. 10 p.

Final act of the Second Inter-American Congress of Municipalities, Santiago, Chile, September 15-21, 1941.

Int-**8200.**B4 **Boletim da União Panamericana.** 1911- . Mensal. Washington, D.C. União Panamericana.

Other bulletins published in English and Spanish—outros boletins publicados em inglês e espanhol (Int-**8200.**B7; Int-**8200.**B5).

Int-**8200.**B5 **Boletín de la Unión Panamericana.** 1911- . Mensual. Washington, D.C. Unión Panamericana.

Other bulletins published in English and Portuguese—Otros boletines publicados en inglés y portugués (Int-**8200.**B7; Int-**8200.**B4).

Int-**8200.**B7 **Bulletin of the Pan American Union.** 1893- . Monthly. Washington, D.C. Pan American Union.

Other bulletins published in Spanish and Portuguese—Otros boletines publicados en español y portugués (Int-**8200.**B5; Int-**8200.**B4).

Int-**8200.**C6 1944.1, Eng. **Conference of Commissions of Inter - American Development, May 9 to 18, 1944: Final Act.** Washington, D.C., 1944. Inter American Development Commission 45 p.

Published also in Spanish—publicado también en español (Int-**8200.**C6, 1944.1 Sp).

Int-**8200.**C6 1944.1 Sp **Acta Final de la Conferencia de Comisiones de Fomento Interamericano, Nueva York, del 9 al 18 de Mayo de 1944.** Washington, D.C., 1944. Inter-American Development Commission. 48 p.

Published also in English—publicado también en inglés (Int-**8200.**C6, 1944.1 Eng).

Int-**8200.**E5 1945 **The Economic and Financial Organization of the League of Nations.** A survey of twenty-five years' experience. By Martin Hill. Washington, D.C., 1945. Carnegie Endowment for International Peace. Division of International Law. 153 p. Studies in the administration of international law and organization. "Confidential: For restricted distribution only."

La organización económica y financiera de la sociedad de las Naciones.

Int-**8200.**H5 1938 **Handbook of International Organisations.** Geneva, 1938. League of Nations. 490 p.

Published also in French, German, Italian, and Spanish—publié aussi en français, allemand, italien et espagnol.

Manuel des organismes internationaux, publié par la Société des Nations.

Int-**8200.**I26 **Informe Anual del Director General de la Unión Panamericana.** Fecha de iniciación? Washington, D.C.

Published also in English—publicado también en inglés.

Annual report of the Director General of the Pan American Union.

Int-**8200.**I3 1945.2, Eng **Inter-American Conference on Problems of War and Peace,**

Mexico City, February 21-March 8, 1945. Report submitted to the Governing Board of the Pan American Union by the Director General. Washington, D.C., 1945. Pan American Union. 115 p. Congress and conference series, no. 47.

Published also in Spanish and Portuguese—publicado también en español y portugués (Int-**8200**.I3, 1945.2, Sp; Int **8200**.I3, 1945.2, Port).

Int-**8200**.I3 1945.2, Port **Conferência Interamericana sôbre Problemas da Guerra e da Paz, México, D.F., México, 21 de Fevereiro a 8 de Março de 1945.** Relatório apresentado ao Conselho Diretor da União Panamericana pelo Diretor Geral. Washington, D.C., 1945. União Panamericana. 124 p. Série sôbre congressos e conferências, no. 29.

Published also in English and Spanish—publicado também em inglês e espanhol (Int-**8200**.I3, 1945.2, Eng; Int-**8200**.I3, 1945.2, Sp).

Int-**8200**.I3 1945.2, Sp **Conferencia Interamericana sobre Problemas de la Guerra y de la Paz, México, D.F., México, Febrero 21-Marzo 8, 1945.** Informe sobre los resultados de la conferencia presentado al Consejo Directivo de la Unión Panamericana. Washington, D.C., 1945. Unión Panamericana. 100 p. Serie sobre congresos y conferencias, no. 47.

Published also in English and Portuguese—publicado también en inglés y portugués (Int-**8200**.I3, 1945.2, Eng; Int-**8200**.I3, 1945.2, Port).

Int-**8200**.I45 1945, Eng **Inter-American Economic and Social Council: Report on the Organization of the Council Approved by the Governing Board of the Pan American Union at the Session of August 29, 1945.** Washington, D.C., 1945. Pan American Union. 6 p. Mimeographed.

Published also in Spanish—publicado también en español (Int-**8200**.I45, 1945, Sp).

Int-**8200**.I45 1945, Sp **Consejo Interamericano Económico y Social: Informe sobre la Organización del Consejo Aprobado por el Consejo Directivo de la Unión Panamericana en la Sesión del 29 de Agosto de 1945.** Washington, D.C., 1945. Unión Panamericana. 7 p. Mimeografiado.

Published also in English—publicado también en inglés (Int-**8200**.I45, 1945, Eng).

Int-**8200**.I48 **Inter-American Series.** Starting date? Washington, D.C. U.S. Department of State.

Serie de publicaciones del Departmento de Relaciones Exteriores de los Estados Unidos sobre asuntos interamericanos.

Int-**8200**.I49 1946 **International Agencies in which the United States Participates.** Washington, D.C., U.S. Govt. Print. Off., 1946. U.S. Department of State. 322 p. Publication, 2699.

Includes for each organization: Origin and development; membership; purposes, powers, and functions; structure; finances; United States relations; relations with other international organizations; present status; basic texts and publications.

Organizaciones internacionales en las cuales participan los Estados Unidos. Incluye para cada organización los siguientes datos: Origen y desarrollo; componentes; propósitos, poderes y funciones; estructura; finanzas; relaciones con Estados Unidos; relaciones con otras organizaciones internacionales; estado actual; documentos básicos y publicaciones.

Int-**8200**.I5 1945 **The International Administration of an International Secretariat.** By Chester Purves. London, New York, etc., 1945. Royal Institute of International Affairs. 78 p.

A discussion of post-war problems, with notes based on the experience of the League of Nations.

La administración internacional de un secretariado internacional. Una discusión sobre problemas de postguerra, con notas basadas sobre experiencias de la Sociedad de las Naciones.

Int-**8200**.I55 **International Conciliation.** 1907- . Monthly except July and August. New York. Carnegie Endowment for International Peace. Division of Intercourse and Education.

Revista mensual sobre conciliación internacional.

Nov. '46 "Suggested Charter for an International Trade Organization of the United Nations." Presented by the United States as a basis for discussion by the preparatory committee for the International Conference on Trade and Employment. No. 425, November 1946, p. 487-543.

Anteproyecto de carta para la organización internacional del comercio de las Naciones Unidas, presentado por Estados Unidos como base de discusión para el comité preparatorio de la Conferencia Internacional sobre Comercio y Ocupación.

Sept. '45 "The United Nations Charter." With explanatory notes of its development at San Francisco by the executive officers of the four commissions in the conference, Malcolm W. Davis, Huntington Gilchrist, Grayson Kirk, Norman J. Padel.

ford. No. 413, September 1945, section I, p. 441-538.

Text of the charter is given, p. 480-531, with organization chart of the United Nations, p. 507.

Carta constitucional de las Naciones Unidas, y notas explicativas acerca de su desarrollo en San Francisco, por los oficiales ejecutivos de las cuatro comisiones de la conferencia. Se incluye el texto de la carta, p. 480-531, así como el esquema de organización de las Naciones Unidas, p. 507.

Int-8200.I6 1933-1940, Eng **The International Conferences of American States First Supplement, 1933-1940.** Conventions, recommendations, resolutions and motions adopted by the seventh and eighth International Conferences of the American States, The Inter-American Conference for the Maintenance of Peace, and the first and second meetings of the Ministers of Foreign Affairs of the American Republics for consultation, at Panama and Habana, together with documents relating to the organization of the conferences and information concerning technical Pan American conferences, commissions and other bodies. Washington, D.C., 1940. Carnegie Endowment for International Peace 558 p.

Published also in Spanish—publicado también en español.

Int-8200.I6 1889-1936, Sp **Conferencias Internacionales Americanas, 1889-1936.** Recopilación de los tratados, convenciones, resoluciones y mociones adoptadas por las siete primeras Conferencias Internacionales Americanas, la Conferencia Internacional Americana de Consolidación y Arbitraje y la Conferencia Interamericana de Consolidación de la Paz; con varios documentos relativos a la organización de las referidas conferencias. Prefacio por Leo S. Rowe . . . introducción por James Brown Scott. Washington, D.C., 1938. Dotación Carnegie para la Paz Internacional. 746 p.

Published also in English—publicado también en inglés.

Int-8200.I6 1889-1928, Eng **The International Conferences of American States, 1889-1928.** A collection of the conventions, recommendations, resolutions, reports, and motions adopted by the first six International Conferences of the American States, and documents relating to the organization of the conferences. Edited by James Brown Scott. New York, Oxford University Press, 1931. Carnegie Endowment for International Peace. 551 p.

Published also in Spanish—publicado también en español.

INTERNATIONAL
8200

Int-8200.J3 **Journal of the Economic and Social Council. Journal du Conseil Economique et Social.** 1946- . Frequency? New York. United Nations. Economic and Social Council. Nations Unies. Conseil Economique et Social.

Revista del Consejo Económico y Social de las Naciones Unidas.

May 31, '46 "Report of the Statistical Commission to the Economic and Social Council."—"Rapport de la Commission de Statistique au Conseil Economique et Social." (Document E/39.) First session, 1 to 15 May 1946. Première session, ler au 15 mai 1946. No. 17, Friday, 31 May 1946, p. 223-239.

Contents: Summary of recommendations; composition and terms of reference of Statistical Commission; statistical relationships of United Nations with specialized agencies; statistical organization and functions of United Nations Secretariat; statistical activities of League of Nations; relationships of United Nations with quasi-governmental and non-governmental statistical organizations; relationships of United Nations with regional statistical organizations; Sub-Commission on Statistical Sampling.

Informe de la Comisión Estadística al Consejo Económico y Social de las Naciones Unidas. Incluye: Resumen de recomendaciones; composición y términos de referencia de la Comisión Estadística; relaciones estadísticas de las Naciones Unidas con los organismos especializados; organización y funciones estadísticas de la Secretaría de las Naciones Unidas; actividades estadísticas de la Sociedad de las Naciones; relaciones de las Naciones Unidas con organizaciones estadísticas de carácter casi-gubernamental y no-gubernamental; relaciones de las Naciones Unidas con organizaciones estadísticas regionales; Subcomisión sobre Muestras Estadísticas.

Int-8200.M3 1945 **Minutes of the Special Meeting of the Governing Board of the Pan American Union Held on August 29, 1945.** Approved at the meeting held on September 26, 1945. Washington, D.C., 1945. Pan American Union. 35 p. (approx.)

Includes, in appendix, a report on the organization of the Inter-American Economic and Social Council, with organization chart.

Minutas de la sesión especial del Consejo Directivo de la Unión Panamericana del 29 de agosto de 1945. El apéndice incluye un informe sobre la organización del Consejo Interamericano Económico y Social, con esquema de organización.

Int-8200.P45 1946 **Preliminary Cumulative List of Documents in Unrestricted Series Issued by the General Assembly, Security Council, Atomic Energy Commission, Economic and Social Council and its Commissions, International Health Conference, to 31 July, 1946.** New York?, 1946. 83 p.

Lista preliminar cumulativa de documentos en series públicas editadas por la Asamblea General de las Naciones Unidas, el Consejo de Seguridad, la Comisión de Energía Atómica, el Consejo Económico y Social y sus comisiones, y la Conferencia Internacional de Salud, hasta el 31 de julio, 1946.

Int-8200.R3 **Report of the Anglo-American Caribbean Commission to the Governments of the United States and Great Britain.** 1942-43?— . Annual. Washington, D.C. Anglo-American Caribbean Commission.

Informe anual de la Comisión Anglo-americana del Caribe a los gobiernos de Estados Unidos y Gran Bretaña.

Int-8200.R7 **Report on the Work of the League.** Submitted by the Acting Secretary-General. Starting date? Annual. Geneva. League of Nations.

Rapport annuel sur les travaux de la Société des Nations, présenté par le Secrétaire Général Suppléant.

1939-1945 **Report on the Work of the League During the War.** Submitted to the Assembly by the Acting Secretary-General. Geneva, 1945. League of Nations. 167 p. Unexamined.

"The first chapters of the report describe the non-political activities which were carried on in Geneva, in America and in London during the war. Such activities covered various fields: Economic, financial, communications and transit, health, control of the drug traffic, social questions, legal questions, etc. Other chapters deal with internal questions, such as budget and staff. The problems connected with a possible transfer to the United Nations are also mentioned." (Source: **Monthly Bulletin of Statistics,** vol. XXVII, no. 1, January, 1946, back cover.)

Also published in French—publié aussi en français.

Rapport sur le travail de la Société des Nations durant la guerre. Soumis à l'assemblée par le secrétaire général suppléant. "Les premiers chapitres du rapport décrivent l'activité non politique exercée à Genève, en Amérique et à Londres durant la guerre. Cette activité embrasse divers domaines: économique, financier, communications et transports, hygiène, répression du trafic des stupéfiants, questions sociales, questions légales etc. Les autres chapitres traitent de questions internes, telles que le budget et le personnel. Il y est aussi fait mention des problèmes inhérants à un transfert possible aux Nations Unies. (Traduction du commentaire.)

Int-8210.C6 1946 **Charter of the United Nations and Statute of the International Court of Justice. Charte des Nations Unies et Statut de la Cour Internationale de Justice.** New York, 1946. 125 p.

Int-8300.I5 1942 **Final Act of the Inter-American Conference of Police and Judicial Authorities, Buenos Aires, Argentina, May 27-June 9, 1942.** Washington, D.C., 1943. Pan American Union. 38 p. Congress and conference series, no. 38.

Int-8400.S5 1942 **A Selected List of Latin American Periodicals Containing Laws and Legal Information.** Compiled in the Columbus Memorial Library of the Pan American Union. Washington, D.C., November, 1942. Pan American Union. 27 p. Bibliographic series, no. 28.

Lista escogida de publicaciones periódicas latinoamericanas, las cuales exponen leyes y decretos gubernamentales así como otras informaciones legales.

9000-9999 LABOR AND LIVING CONDITIONS—CONDICIONES DEL TRABAJO Y DE VIDA

*Int-9000.A5 **Annuaire des Statistiques du Travail. Anuario de Estadísticas del Trabajo. Year Book of Labour Statistics.** 1935-36— . Montreal. Bureau International du Travail. Oficina Internacional del Trabajo. International Labour Office.

1943-44 **Annuaire des Statistiques du Travail, 1943-44. Anuario de Estadísticas del Trabajo, 1943-44. Year Book of Labour Statistics, 1943-44.** Huitième édition. Octava edición. Eighth issue. Montreal, Inland Press Ltd., 1945. Bureau International du Travail. Oficina Internacional del Trabajo. International Labour Office. 265 p.

Includes: Total and gainfully occupied population; employment and unemployment; hours of work; wages; cost of living and retail prices; family living studies; migration; industrial accidents; industrial disputes.

Comprend: Population totale et population active; emploi et chômage; heures de travail; salaires; coût de la vie et prix de détail; enquêtes sur les conditions de vie des familles; migrations; accidents du travail; conflits du travail.

Int-**9000**.B5 1944 **Bibliography on Labor and Social Welfare in Latin America.** Compiled by Sylvia Pollack Bernstein. Second edition. Washington, D.C., May 1944. Pan American Union. Division of Labor and Social Information. 76 p.

A selected list of materials in English published from 1930 through December 1943, classified by subjects and by country, to which is appended a list of the periodicals consulted in preparation of this bibliography.

Bibliografía sobre trabajo y bienestar social en la América Latina. Una lista selecta de material publicado en inglés desde 1930 hasta diciembre de 1943, clasificada por materias y por países; adjunto lleva una lista de periódicos consultados en la preparación de esta bibliografía.

Int-**9000**.C8 **Cuadernos Sociales del Consejo.** Fecha de iniciación? Irregular. Montevideo. Consejo Interamericano de Comercio y Producción.

Social pamphlets of the Inter American Council of Commerce and Production.

Series of monographs.

Serie de monografías.

*Int-**9000**.E5 1943 **La Estandardización Internacional de las Estadísticas del Trabajo.** Una reseña de la obra de la Oficina Internacional del Trabajo y de varias conferencias internacionales de estadística. Montreal, 1943. Oficina Internacional del Trabajo. 179 p. Estudios y documentos, serie N, Estadísticas, no. 25.

Published also in English—publicado también en inglés (Int-**9000**.I7, 1943).

1943, a "La Clasificación de Industrias y Profesiones en las Estadísticas del Trabajo." **La Estandardización Internacional de las Estadísticas del Trabajo** p. 69-71.

Classification of industries and professions in labor statistics. Includes "Minimum Nomenclature of Industries by the Committee of Statistical Experts of the League of Nations," a translation into Spanish by the International Labour Office.

Incluye "Nomenclatura mínima de las industrias recomendada por el Comité de Expertos Estadígrafos de la Sociedad de las Naciones," traducción por la Oficina Internacional del Trabajo.

Int-**9000**.E7 **Estudios y Documentos, Serie N, Estadísticas.** Fecha de iniciación? Irregular. Montreal. Oficina Internacional del Trabajo.

Published also in English—publicado también en inglés (Int-**9000**.S5).

Monografías cuya extensión varía entre 40 y 200 páginas, tratando principalmente sobre métodos de compilación de estadísticas sobre salarios, horas de trabajo, disputas industriales, acuerdos colectivos, vivienda, accidentes de la industria del carbón mineral, accidentes ferroviarios, enfermedades profesionales y mortalidad, migración, comparación internacional del costo de la vida y estandardización internacional de estadísticas del trabajo.

Int-**9000**.I2 1945 **Index to Publications and Articles on Latin America Issued by the United States Bureau of Labor Statistics, 1902-43.** Prepared by Eugene D. Owen. Washington, D.C., 1945. Pan American Union. 39 p. Mimeographed. Bibliographic series, no. 31.

Indice de las publicaciones y artículos sobre Latino-América editados por la Oficina de Estadísticas del Trabajo de los Estados Unidos, 1902-43.

*Int-**9000**.I5 **International Labour Review.** 1921- . Monthly. Montreal. International Labour Office.

Published also in Spanish—publicado también en español (Int-**9000**.R6).

Contains current statistics on cost of living, general level of wages, and general level of employment and unemployment and hours of work, each of the three sets of figures appearing once each quarter. Publishes once a year results of an inquiry into wages, hours of work, and retail prices of food and fuel in certain countries. Also contains from time to time articles on statistical methods, results of family living studies, nutrition surveys, indexes of international comparisons of food costs and other statistical topics.

*Int-**9000**.I7 1943 **The International Standardisation of Labour Statistics.** A review of the statistical work of the International Labour Office and of various international statistical conferences. Montreal, 1943. International Labour Office. 169 p. Studies and reports, series N, Statistics, no. 25 (revision of no. 19.)

Published also in Spanish—publicado también en español (Int-**9000**.E5, 1943).

Int-9000.L6 1941-42 **Labor Trends and Social Welfare in Latin America, 1941 and 1942.** By Ernesto Galarza. Washington, D.C. July 1943. Pan American Union. Division of Labor and Social Information. 153 p.

A summary of labor and social welfare developments in Latin America with a survey of activities by country with discussion, statistics, legislation, and reports. Subjects covered are: Activity among labor organizations; governmental measures to regulate prices, rents, and consumption; cost of living indexes and trends; wage and salary situation; migration of workers; colonization and land settlement programs; cooperatives; public housing programs; food distribution and nutrition; maternity and child care; social security programs; effects of war conditions on the various countries.

Un resumen sobre evolución del trabajo y del bienestar social en Latino-América, 1941 y 1942, con un estudio de actividades por países, completado con discusiones, estadísticas, legislación e informes. Los asuntos incluidos son: Actividades de las organizaciones del trabajo; medidas gubernamentales para regulación de precios, alquileres y consumo; índices del costo de la vida y su curso; la situación de los sueldos y salarios; migración de trabajadores; programas de colonización y establecimiento sobre la tierra; cooperativas; programas de la vivienda popular; distribución de alimentos y nutrición; maternidad y protección a la infancia; programas de previsión social; efectos de las condiciones de guerra en los distintos países.

*Int-9000.R6 **Revista Internacional del Trabajo.** 1914- . Mensual. Montreal. Oficina Internacional del Trabajo.

Published also in English—publicada también en inglés (Int-**9000**.I5).

Contiene estadísticas corrientes sobre costo de la vida, el nivel general de los salarios y el nivel general de la ocupación y desocupación, y horas de trabajo; cada una de las tres series de cifras aparecen trimestralmente. Publica una vez al año un estudio sobre los salarios, horas de trabajo y precios al por menor de comestibles y combustibles en varios países. Contiene también de vez en cuando, artículos sobre métodos estadísticos, resultados de estudios sobre vida de la familia, nutrición, índices de comparaciones internacionales del costo de los alimentos y otros temas estadísticos.

Int-9000.S5 **Studies and Reports, Series N, Statistics.** 1923- . Irregular. Montreal. International Labour Office.

Published also in Spanish—publicado también en español (Int-**9000**.E7).

Monographs varying in length from 40 to 200 pages, dealing mainly with methods of compiling statistics of wages, hours of labour, industrial disputes, collective agreements, housing, coal-mining accidents, railway accidents, occupational morbidity and mortality, migration, international comparisons of cost of living and international standardization of labour statistics.

Int-9100.N5 1946 **Notas para una Encuesta Continental sobre Productividad Obrera.** Por Alberto Kurlat. Montevideo, 1946. Consejo Interamericano de Comercio y Producción. 53 p. Cuadernos sociales del Consejo, no. 3.

Notes for a continental survey of labor productivity.

Int-9200.E4 1945 **Employment of United States Citizens in Latin America.** Washington, D.C., 1945. Pan American Union. Division of Labor and Social Information. 18 p. Mimeographed. Unexamined.

"Contains a summary, by country, of legislation restricting employment of aliens in the Latin American Republics; discussion of immigration policies and technical and vocational education; data on wages and living standards in selected countries; information on requirements for employment; an analysis of employment possibilities in 11 fields; and other pertinent data for Latin America; bibliography." (Source: **Monthly Labor Review,** vol. 61, no. 5, November 1945, p. 1058.)

Empleo de ciudadanos estadounidenses en Latino-América. "Contiene un resumen por países, sobre legislación restringiendo el empleo de extranjeros en las repúblicas latino-americanas; un examen de las políticas de immigración y de la educación técnica y profesional; datos sobre salarios y estandards de vida en ciertos países; información sobre requisitos para empleo; un análisis sobre posibilidades de empleo en 11 ramas de ocupación; otros datos pertinentes a Latino-América; bibliografía." (Traducción del comentario.)

Int-9200.E7 1931-41 **Estimates of the Working Population of Certain Countries in 1931 and 1941.** By Arthur Lyon Bowley. Submitted to the Preparatory Committee for the International Economic Conference, Geneva, 1926. Lausanne, Imp. Réunies S.A., 1925. League of Nations. Economic and Financial Section. 19 p. Series of League of Nations publications, II, economic and financial series, 1926. II.67. Unexamined.

Evaluation de la population active de divers pays en 1931 et 1941, par Arthur Lyon Bowley. Présenté au Comité Préparatoire pour la Conférence Internationale Économique, Genève,

1926. Publié par la Section Économique et Financière de la Société des Nations.

*Int-**9200**.M5 1925 **Methods of Statistics of Unemployment.** Report prepared for the International Conference of Labour Statisticians (April 1925). Geneva, A. Livron, 1925. International Labour Office. 66 p. Studies and reports, series N, Statistics, no. 7. Unexamined.

Métodos de la estadística de desocupación. Informe de la Oficina Internacional del Trabajo.

Int-**9200**.M5 1922 **Methods of Compiling Statistics of Unemployment.** Replies of the governments. October, 1922. Geneva, A. Kundig, 1922. International Labour Office. 101 p. Studies and reports, series C (Unemployment), no. 7. Unexamined.

Métodos de compilación de estadísticas sobre desocupación: Respuesta de los gobiernos. Oficina Internacional del Trabajo.

*Int-**9200**.M7 1923 **Methods of Statistics of Wages and Hours of Labour.** Report prepared for the International Conference of Labour Statisticians, 1923. Geneva, A. Kundig, 1923? International Labour Office. 69 p. Studies and reports, series N, Statistics, no. 2. Unexamined.

Métodos de la estadística de salarios y horas de trabajo. Informe de la Oficina Internacional del Trabajo.

Int-**9200**.S7 **Studies and Reports, Series C, Employment and Unemployment.** 1920- . Geneva. • International Labour Office.

Series of monographs—serie de monografías. Estudios y documentos, serie C, Ocupación y desocupación, publicados por la Oficina Internacional del Trabajo.

Int-**9200**.S8 **Studies and Reports, Series I, Employment of Women and Children.** 1939- . Irregular. Montreal. International Labour Office.

Estudios y documentos, serie I, Empleo de mujeres y niños. Publicaciones de la Oficina Internacional del Trabajo.

Int-**9200**.W5 1946 **The War and Women's Employment.** The experience of the United Kingdom and the United States. Montreal, Published by P. S. King & Staples Ltd., London, 1946. International Labour Office. 287 p. Studies and reports, series I, Employment of women and children, new series no. 1.

La guerra y el empleo de mujeres: La experiencia del Reino Unido y de los Estados Unidos. Informe de la Oficina Internacional del Trabajo.

INTERNATIONAL
9200

*Int-**9400**.I3 1938 **Industrial Accident Statistics.** London, P. S. King & Son, Ltd., 1938. International Labour Office. 129 p. Studies and reports, series N, Statistics, no. 22.

Estadísticas de accidentes industriales Informe de la Oficina Internacional del Trabajo.

Int-**9400**.M5 1923 **Methods of Statistics of Industrial Accidents.** Report prepared for the International Conference of Labour Statisticians, 1923. Geneva, A. Kundig, 1923? International Labour Office. 63 p. Studies and reports, series N, Statistics, no. 3. Unexamined.

Métodos de estadística de accidentes industriales. Informe de la Oficina Internacional del Trabajo.

*Int-**9400**.S5 1930 **Statistical Methods for Measuring Occupational Morbidity and Mortality.** Geneva, 1930. International Labour Office. 208 p. Studies and reports, series N, Statistics, no. 16.

Métodos estadísticos para medir la mortalidad y la morbilidad profesional.

Int-**9500**.C4 1932 **A Contribution to the Study of International Comparisons of Costs of Living.** Geneva, 1932. International Labour Office. 246 p. Studies and reports, series N, Statistics, no. 17.

Una contribución al estudio de la comparación internacional del costo de la vida, por la Oficina Internacional del Trabajo.

Int-**9500**.C8 1943 **El Costo de la Vida Obrera en América.** Por Ana Mekler. Washington, D.C., 1943. Unión Panamericana. Oficina de Información Obrera y Social. 139 p.

A compilation of cost of living indexes for countries of the Western Hemisphere. Includes a bibliography.

*Int-**9500**.I5 1934 **International Comparisons of Cost of Living.** Geneva, 1934. International Labour Office. 146 p. Studies and reports, series N, Statistics, no. 20. Unexamined.

Comparación internacional del costo de la vida. Monografía por la Oficina Internacional del Trabajo.

Int-**9500**.M5 1925 **Methods of Compiling Cost-of-Living Index Numbers.** Report prepared for the second International Conference of Labour Statisticians. Geneva, 1925. International Labour Office. 64 p. Studies and reports, series N, Statistics, no. 6.

Métodos de compilación de números índices del costo de la vida. Informe de la Oficina Internacional del Trabajo.

Int-**9600**.C5 1943 **Consumo y Subconsumo.** Alimentos, vestidos y vivienda popular. Texto concordado con las resoluciones adoptadas por la Comisión Ejecutiva en la sesión de Rio de Janeiro el 16 de octubre de 1943. Montevideo, 1943. Consejo Permanente de Asociaciones Americanas de Comercio y Producción. 28 p. Circulares documentales de la Comisión Ejecutiva, no. 1.

Consumption and under-consumption of food, clothing, and housing. Documentary circular of the Permanent Council of American Associations of Commerce and Production.

*Int-**9700**.M4 1941 **Méthodes d'Enquête sur les Conditions de Vie des Familles.** Revenu, dépenses, consommation. Par Robert Morse Woodbury. Genève, B.I.T., 1941. Bureau International du Travail. 170 p. Études et documents, série N, Statistique, no. 23. Pas examiné.

Published also in English and Spanish—publié aussi en anglais et espagnol (Int-**9700**.M6; Int-**9700**.M7).

Int-**9700**.M5 1926 **Methods of Conducting Family Budget Enquiries.** Geneva, 1926. International Labour Office. 100 p. Studies and reports, series, N, Statistics, no. 9. Unexamined.

Métodos para la realización de encuestas sobre presupuestos familiares. Informe de la Oficina Internacional del Trabajo.

*Int-**9700**.M6 1942 **Methods of Family Living Studies.** Income, expenditure, consumption. By Robert Morse Woodbury. Montreal, 1942. International Labour Office. 144 p. Studies and reports, series N, Statistics, no. 23. Unexamined.

Published also in Spanish and French—publicado también en español y francés (Int-**9700**.M7; Int-**9700**.M4).

*Int-**9700**.M7 1942 **Métodos de Encuestas sobre las Condiciones de Vida de las Familias.** Ingresos, gastos, consumo. Por Robert Morse Woodbury. Montreal, 1942. Oficina Internacional del Trabajo. 168 p. Estudios y documentos, Serie N, Estadísticas, no. 23.

Published also in English and French—publicado también en inglés y francés (Int-**9700**.M6; Int-**9700**.M4).

Int-**9700**.S5 1935 **Studies of Family Living in the United States and Other Countries.** An analysis of material and methods. By Faith M. Williams and Carle C. Zimmerman. Washington, D.C., 1935. U.S. Department of Agriculture. 617 p. Miscellaneous publication, no. 223.

Estudios sobre condiciones de vida de las familias, en los Estados Unidos y en otros países: Un análisis del material y de los métodos.

*Int-**9800**.E5 1942 **Encuestas sobre el Consumo de Alimentos y la Nutrición en las Américas.** Resultados—métodos. Informe presentado por la Oficina Internacional del Trabajo a la Undécima Reunión de la Conferencia Sanitaria Pan-Americana a realizarse en Rio de Janeiro del 7 al 18 de septiembre de 1942. Por Robert Morse Woodbury. Buenos Aires, 1942. Oficina Internacional del Trabajo. 95 p.

Published also in English—publicado también en inglés (Int-**9800**.F4).

Incluye una lista de referencias a los estudios sobre las condiciones de vida de las familias en los países americanos, los cuales contienen encuestas sobre el consumo de alimentos.

*Int-**9800**.F4 1942 **Food Consumption and Dietary Surveys in the Americas.** Results—methods. Report presented by the International Labour Office to the Eleventh Pan American Sanitary Conference held in Rio de Janeiro, 7-18 September 1942. By Robert Morse Woodbury. Montreal, 1942. International Labour Office. 64 p.

Published also in Spanish—publicado también en español (Int-**9800**.E5).

Includes a list of references to family living studies in American countries which include food consumption surveys, p. 62-63.

*Int-**9800**.I5 1942 **International Comparisons of Food Costs.** Montreal, 1942. International Labour Office. 83 p. Studies and reports, series N, Statistics, no. 24.

Appendix I contains the texts of resolutions adopted on this subject by the Conference, the Governing Body, and the Committee of Statistical Experts, and appendix II contains a series of mathematical tables.

Comparación internacional del costo de los alimentos, por la Oficina Internacional del Trabajo. Apéndice I contiene el texto de las resoluciones adoptadas sobre esta materia, por la Conferencia, el Consejo Directivo y el Comité de Expertos Estadígrafos; el apéndice II contiene una serie de tablas matemáticas.

Int-**9800**.S5 1942 **El Subconsumo de Alimentos en América del Sur.** Por Emilio Llorens. Buenos Aires, Editorial Sudamericana, 1942. Biblioteca de Orientación Económica. 262 p.

Under-consumption of food in South America.

Includes tables and graphs.
Incluye tablas y gráficas.

*Int-**9800**.W4 1946 **World Food Situation, 1946.** Washington, D.C., 1946. U.S. Department of Agriculture. Office of Foreign Agricultural Relations. 181 p.

A comprehensive survey presenting data on world production and consumption of food.

Estudio comprensivo, exponiendo datos sobre producción y consumo mundial de alimentos.

Int-**9900**.L5 1943 **Low-Cost Housing in Latin America.** By Robert C. Jones. Washington, D.C., 1943. Pan American Union. Division of Labor and Social Information. 20 p. Mimeographed.

A brief introduction to developments in public housing in Latin America, dealing exclusively with low-cost housing programs. Contains references to housing censuses taken, surveys made, and boards created in this field, in each Latin American nation.

Una breve introducción al desarrollo de la vivienda popular en Latino-América, tratando exclusivamente sobre programas de vivienda de bajo costo. Contiene referencias a los censos de viviendas, estudios realizados y organismos creados con ese fin en cada una de las repúblicas americanas.

*Int-**9900**.M5 1928 **Methods of Compiling Housing Statistics.** Geneva, 1928. International Labour Office. 119 p. Studies and reports, series N, Statistics, no. 13. Unexamined.

Includes bibliography.

Métodos de compilación de estadísticas de la vivienda. Informe de la Oficina Internacional del Trabajo. Incluye una bibliografía.

*Int-**9900**.S5 1928-34 **Statistique Internationale des Grandes Villes: Statistique du Logement dans les Grandes Villes, 1928-1934.** La Haye, Imp. Société Anonyme "Drukkerij Trio," 1940. Office Permanent de l'Institut International de Statistique. 170 p.

Housing statistics for large cities, 1928-1934. Includes statistical tables on: Number of inhabited and vacant units; distribution, size, average number of dwellers, average number of rooms; increase and decrease in number of units; new units and financing of their construction. From the Western Hemisphere, only Argentina, Canada and the United States are represented in the group of countries for which data is shown.

Contient des tableaux statistiques sur: Nombre des logements habités et vacants; distribution, grandeur, nombre moyen d'habitants, nombre moyen de pièces; augmentation et diminution du nombre des logements; nouveaux logements et le financement de leur construction. De l'hémisphère occidentale, seulement l'Argentine, le Canada, et les Etats-Unis d'Amérique figurent dans le groupe des pays dont des données sont présentées.

Int-**9950**.H5 1942 **Housing and Hospital Projects of Latin American Social Security Systems.** Washington, D.C., May 1942. Pan American Sanitary Bureau. 15 p. Publication, no. 175.

Los proyectos sobre vivienda y hospitales de los sistemas de previsión social de Latino-América.

Int-**9951**.P5 1942, Eng **Final Act of the Eighth Pan American Child Congress, Washington, D.C., May 2-9, 1942.** Washington, D.C., 1942. Pan American Union. 31 p. Congress and conference series, no. 37.

Published also in Portuguese and Spanish—publicado también en portugués y español (Int-**9951**.P5, 1942, Port; Int-**9951**.P5, 1942, Sp.)

Int-**9951**.P5 1942, Port **Ata Final do Oitavo Congresso Panamericano da Criança, Washington, 2-9 de Maio de 1942.** Washington, D.C., 1942. União Panamericana. 28 p. Série de congressos e conferências, no 24.

Published also in English and Spanish—publicado também em inglês e espanhol (Int-**9951**.P5, 1942, Eng; Int-**9951**.P5, 1942, Sp).

Int-**9951**.P5 1942, Sp **Acta Final del Octavo Congreso Panamericano del Niño, Wáshington del 2 al 9 de Mayo de 1942.** Wáshington, D. C. 1942. Unión Panamericana. 33 p. Serie sobre congresos y conferencias, no. 37.

Published also in English and Portuguese—publicado también en inglés y portugués (Int-**9951**.P5, 1942, Eng; Int-**9951**.P5, 1942, Port).

ARGENTINA

1000-1999 GENERAL

Arg-**1200**.B6 **Boletín Bibliográfico Argentino.** 1937- . Irregular. Buenos Aires. Ministerio de Justicia e Instrucción Pública. Comisión Nacional de Cooperación Intelectual.

Bibliographical bulletin of Argentina, issued by the National Commission of Intellectual Cooperation.

Arg-**1200**.G5 1945 **A Guide to the Official Publications of the Other American Republics: I, Argentina.** James B. Childs, general editor. Washington, D.C., U.S. Govt.Print. Off., 1945. The Library of Congress. 124 p. Latin American series, no. 9.

Una guía de las publicaciones oficiales de las otras repúblicas americanas: I, Argentina.

*Arg-**1410**.A5 **Anuario Geográfico Argentino.** 1941- . Anual. Buenos Aires. Comité Nacional de Geografía.

A general statistical compilation which constitutes the statistical yearbook of Argentina.

Una compilación estadística general que constituye el anuario estadístico de Argentina.

1942 **Anuario Geográfico Argentino, Suplemento 1942.** Buenos Aires, 1943. Comité Nacional de Geografía. 342 p. No examinado.

1941 **Anuario Geográfico Argentino.** Buenos Aires, 1941. Comité Nacional de Geografía. 651 p.

Includes: Historical synthesis; the Argentine state; territory; climatology, hydrology, seismology; population; production; industry; commerce; finance; economics; public culture; social security, social assistance, and labor; public health; the federal capital, provinces, and territories; appendix: The National Committee of Geography.

Incluye: Síntesis histórica; el Estado argentino; el territorio argentino; climatología, hidrología, sismicidad; población; producción; industria; comercio; finanzas; instrumental económico; cultura pública; previsión social, asistencia social y trabajo; higiene pública; capital federal, provincias y territorios; apéndice: El Comité Nacional de Geografía.

Arg-**1900**.I5 **Informaciones Argentinas.** 1938- . Mensual. Buenos Aires. Ministerio de Relaciones Exteriores y Culto. Dirección de Información al Exterior.

Published also in English—publicado también en inglés: **Argentine News.**

Contains many news items and timely statistics used as bases for contributed articles.

Contiene muchas informaciones y estadísticas actuales usadas como base de los artículos de colaboradores.

Arg-**1900**.N5 1940 **Una Nueva Argentina.** Por Alejandro Ernesto Bunge. Buenos Aires, Editorial Guillermo Kraft Ltda., 1940. 513 p.

A new Argentina. In three parts: Problems of population; economy and economic policy; social policy. "Contains a considerable amount of statistical information and will thus be of value as a work of reference as well as for its interesting discussions of current problems." (Source: **International Labour Review,** October 1942, p. 519.)

En tres partes: Los problemas de la población; economía y política económica; política social. "Contiene una cantidad considerable de informaciones estadísticas y por ello será de valor como obra de referencia, así como también por sus interesantes discusiones sobre problemas corrientes." (Traducción del comentario.)

2000-2999 CENSUS-CENSOS

Arg-**2100**.C5 1945 **Argentine Fourth General Census Postponed.** Prepared by Clarence E. Birgfeld. Buenos Aires, 1945. American Embassy, Buenos Aires, Argentina. 5 p. Hectographed.

Includes attached English translation of Decree No. 30.512/45, authorizing postponement of census which was to have been held before December 1, 1945. Also contains description of items on forms which had been proposed for use in census.

Incluye adjunto una traducción al inglés del Decreto No. 30.512/45, por el cual se postergó el Censo General, que se iba a levantar antes del 1° de diciembre de 1945. Contiene también, la descripción de los ítems especificados en los formularios que se habían propuesto para uso de este censo.

Arg-**2110**.L5 1936 **Ley No. 12.343, Diciembre 29 de 1936.** Buenos Aires, 1936. 6 p. Copia máquina.

Includes—incluye: **Decreto No. 100.277, Reglamentario de la Ley No. 12.343.**

"The Executive Branch of the government will take a general agricultural census of the country." (Translation of text.)

"El Poder Ejecutivo procederá a levantar en

todo el territorio de la República un censo general agropecuario." (Fuente: Texto.)

Arg-**2130**.C5 1943.1 **Instrucciones Complementarias a las "Normas para la Organización y el Levantamiento del Censo Escolar de la Nación."** Argentina, 1943. Ministerio de Justicia e Instrucción Pública. Consejo Nacional de Educación. Censo Escolar de la Nación. 19 p.

Complementary instructions to "Standards for Organization and Taking of the School Census of the Nation" (translation of title).

Arg-**2130**.C5 1943.2 **Normas para la Organización y Levantamiento del Censo Escolar de la Nación en la Capital Federal.** Argentina, 1943. Ministerio de Justicia e Instrucción Pública. Consejo Nacional de Educación. Censo Escolar de la Nación. 14 p.

Standards for organization and taking of national school census in federal capital.

Arg-**2130**.C5 1943.3 **Normas para la Organización y Levantamiento del Censo Escolar de la Nación en las Provincias.** Argentina, 1943. Ministerio de Justicia e Instrucción Pública. Consejo Nacional de Educación. Censo Escolar de la Nación. 14 p.

Standards for organization and taking of national school census in the provinces.

Arg-**2130**.C5 1943.4 **Normas para la Organización y Levantamiento del Censo Escolar de la Nación en los Territorios.** Argentina, 1943. Ministerio de Justicia e Instrucción Pública. Consejo Nacional de Educación. Censo Escolâr de la Nación. 14 p.

Standards for organization and taking of national school census in the territories of Argentina.

Arg-**2130**.C5 1943.5 **Censo Escolar de la Nación, Ley No. 12.723: Instrucciones para los Censistas.** Disposiciones de la ley No. 12.-723, relacionadas con la labor de los censistas. Ministerio de Justicia e Instrucción Pública. Consejo Nacional de Educación. Censo Escolar de la Nación. 4 p. Schedule.

National school census of the nation: Instructions for enumerators.

Arg-**2130**.C5 1943.6 **Censo Escolar de la Nación, Ley No. 12.723: Cuestionario Familiar.** Para familias que tengan hijos de 0 a 21 años cumplidos y pupilos de 0 a 5 años cumplidos. Ministerio de Justicia e Instrucción Pública. Consejo Nacional de Educación. Censo Escolar de la Nación. 1 p.

Family questionnaire for national school census.

ARGENTINA
2130

Arg-**2130**.C5 1943.7 **Censo Escolar de la Nación Ley No. 12.723: Cuestionario Individual.** Para varones y mujeres de 6 a 21 años cumplidos inclusive. Ministerio de Justicia e Instrucción Pública. Consejo Nacional de Educación. Censo Escolar de la Nación. 1 p.

National school census of the nation: Individual questionnaire.

Arg-**2130**.C5 1943.8 **Censo Escolar de la Nación, Ley No. 12.723: Cuestionario de Convivencia.** Para varones y mujeres de 0 a 5 años cumplidos inclusive, que estén a cargo exclusivo de asilos, orfelinatos. Ministerio de Justicia e Instrucción Pública. Consejo Nacional de Educación. Censo Escolar de la Nación. 2 p.

National school census of the nation: Questionnaire on inhabitants, from 0 to 5 years of age, of orphanages and asylums.

*Arg-**2200**.C4 1936 **Cuarto Censo General, 1936: Población 22-X-1936.** Buenos Aires, Tall. Gráf. de Guillermo Kraft, Ltda., 1938-40. Buenos Aires, Comisión Técnica Encargada de Realizar el Cuarto Censo General. 4 vol.: Vol 1, **Informe Preliminar**; vol. II, **Masculinidad, Lugar de Nacimiento, Alfabetismo**; vol. III, **Estado Civil, País de Matrimonio, Religión**; vol. IV, **Fecundidad, Familias.**

Fourth general census of Buenos Aires, 1936: Population.

*Arg-**2200**.C5 1914 **Tercer Censo Nacional, Levantado el 1 de Junio de 1914.** Buenos Aires, Tall. Gráf. de L. J. Rosso y Cía, 1916-19. **Comisión Nacional del Censo.** 10 vol.: Tomo I, **Antecedentes y Comentarios**; tomo II-IV, **Población**; tomo V, **Explotaciones Agropecuarias**; tomo VI, **Censo Ganadero**; tomo VII, **Censo de las Industrias**; tomo VIII, **Censo del Comercio, Fortuna Nacional, Diversas Estadísticas**; tomo IX, **Instrucción Pública, Bienes del Estado**; tomo X, **Valores Mobiliarios y Estadísticas Diversas.**

Third national census, taken June 1, 1914. Covers population, agriculture, industry, commerce and education.

Arg-**2200**.C5 1895 **Segundo Censo de la República Argentina, Mayo 10 de 1895.** Buenos Aires, Tall. Tip. de la Penitenciaría Nacional, 1898. Comisión Directiva del Censo. 3 vol.: Tomo I, capítulos I-II, **Territorio**; tomo II, capítulo III, **Población**; tomo III, capítulos IV-XV, **Censos Complementarios.**

Second census of the Republic of Argentina, May 10, 1895.

Arg-**2200**.C5 1869 **Primer Censo de la República Argentina, Verificado en los**

Días 15, 16 y 17 de Setiembre de 1869. Buenos Aires, Imp. del Porvenir, 1872. Superintendente del Censo. 746 p.

First census of Argentina, September 15-17, 1869.

Arg-**2210**.C4 1935-36 **Censo Algodonero de la República Argentina, Año 1935-36.** Buenos Aires, Imp. Lopez, 1936. Ministerio de Agricultura. Junta Nacional del Algodón. 309 p. (Publicaciones de la Junta), no. 16.

Cotton census of Argentina, 1935-36. Includes for each province and territory: Area cultivated in 1935-36; number and average size of cotton-growing holding; laborers; type of holding; nationality of the farmers.

Incluye, por cada provincia y territorio: El área cultivada en el año 1935-36; número y dimensión media de las explotaciones algodoneras; braceros; régimen de explotación de la tierra; nacionalidad de los agricultores.

Arg-**2210**.C5 1930 **Censo Ganadero Nacional, Existencia al 1.° de Julio de 1930.** Levantado por la Comisión Nacional designada por decreto del P. E. de fecha 23 de diciembre de 1929. Ley n.° 11.563. Buenos Aires, Tall. Gráf. del Ministerio de Agricultura de la Nación, 1932. Ministerio de Agricultura de la Nación. 794 p.

National livestock census, stocks as of July 1, 1930. Includes: Cattle; sheep; swine; horses; goats; donkeys and mules; poultry; domesticated wild animals; totals of each species shown by districts or departments; economic maps showing totals of main species by districts or departments; organization of census work; handling of schedules.

Incluye: Vacunos; lanares; porcinos; yeguarizos; caprinos; asnales y mulares; animales de corral; animales silvestres en domesticidad; totales de cada especie distribuidos por partidos o departamentos; mapas económicos con los totales de las principales especies por partidos o departamentos; organización de los trabajos censales; recepción de fichas.

*Arg-**2210**.C6 1937 **Censo Nacional Agropecuario, Año 1937.** Ley n.° 12.343. Buenos Aires, Guillermo Kraft, Ltda., 1939-40. Ministerio de Agricultura. Comisión Nacional del Censo Agropecuario. 4 vol.: Pte. I, **Agricultura;** pte. II, **Ganadería;** pte. III, **Economía Rural;** pte. IV, **Economía Rural** (II Parte), **Industrias Derivadas, Varios.**

National census of agriculture and livestock, 1937. Part I, **Agriculture,** includes: Graphs; general tables; grains; industrial crops; feed crops; truck and vegetable crops; various other crops; gardens and floriculture; fruit products; forest products; appendix I, Classification of wheat crops by type; appendix II, floriculture. Part II, **Livestock,** includes: Graphs; general tables; cattle; sheep; swine; horses; goats; donkeys and mules; classification of livestocks by species and by type of farms on which they were counted; livestock owners classified according to the number of heads declared; consumption of cattle on the farm or ranch, and production of wool. Part III, **Rural Economy,** includes: Economic and legal system regulating farm and livestock tenure; area and location of holdings; housing and facilities. Part IV, **Rural Economy (Part II); Derivative Industries; Various,** includes: (a) Rural economy (part II): Classification of farmers, farm families, and laborers who work on the farms; duties of the farm personnel; farm machinery, equipment and tools, wagons and other vehicles; (b) census of sugar cane plantations; (c) industries developed from farm and livestock production.

Pte. I, **Agricultura,** incluye: Gráficos; cuadros generales; granos; cultivos industriales; plantas forrajeras; cultivos de hortalizas y legumbres; cultivos varios; jardines y floricultura; frutales; forestales; apéndice I, Clasificación de los cultivos de trigo por variedades; apéndice II, floricultura. Pte. II, **Ganadería,** incluye: Gráficos; cuadros generales; vacunos; lanares; porcinos; caballares; caprinos; asnales y mulares; clasificación de la existencia ganadera por especies y explotaciones en que fueron censadas; poseedores de ganado clasificados de acuerdo al número de cabezas que han denunciado; consumo de ganado en los establecimientos y producción de lana. Pte. III, **Economía Rural,** incluye: Régimen económico y jurídico de las explotaciones agropecuarias; superficie y ubicación de las explotaciones; vivienda e instalaciones. Pte. IV, **Economía Rural (II Parte); Industrias Derivadas; Varios,** incluye: (a) Economía rural (II parte): Clasificaciones del productor, de su familia y de las personas que trabajan en las explotaciones; oficios del personal que trabaja en las explotaciones; máquinas agrícolas, utiles y herramientas, rodados y vehículos; (b) censo de las plantaciones de caña de azúcar; (c) industrias derivadas de la producción agropecuaria.

Arg-**2210**.C7 1938-39 **Primer Censo Nacional de Floricultura, Ley 12.343, Años 1938-39.** Buenos Aires, Dirección de Propaganda y Publicaciones, 1940. Ministerio de Agricultura. Dirección de Economía Rural y Estadística. 40 p. Unexamined.

First national census of floriculture, law 12.343, years 1938-39.

*Arg-**2230**.C5 1943.1 **El Analfabetismo en la Argentina.** Estudio comparativo desde 1869 a 1943. Informe de la Dirección del Censo Escolar de la Nación. Buenos Aires, Tall. Gráf.del Consejo N. de Educación, 1944. Consejo Nacional de Educación. 25 p.

Includes: Determination of illiteracy from point of view of statistics; illiteracy from 1869 to 1943; evolution of illiteracy by provinces; illiteracy by age groups in 1943; school age population and school attendance; comparative illiteracy.

Incluye: Determinación del analfabetismo desde el punto de vista estadístico; el analfabetismo desde 1869 a 1943; evolución del analfabetismo por provincias; analfabetismo por grupos de edades en 1943; la población en edad escolar y su concurrencia a la escuela; analfabetismo comparado.

*Arg-2230.C5 1943.2 **La Distribución por Zonas de la Población Argentina y su Relación con los Hechos Culturales, Económicos y Sociales.** Informe del Director del Censo Escolar de la Nación. Buenos Aires, 1945. Consejo Nacional de Educación. Censo Escolar de la Nación. 23 p.

Distribution of Argentine population by zones and its relation to cultural, economic, and social facts. Report of director of school census.

*Arg-2230.C5 1943.3 **La Vivienda Popular.** Informe de los resultados que arrojó el censo escolar de la nación, en 1943. Buenos Aires, Tall.Gráf.del Consejo Nacional de Educación, 1945. Consejo Nacional de Educación. Censo Escolar de la Nación. 12 p.

Low-cost housing: Report on housing conditions, based on results of the national school census of 1943.

*Arg-2250.E5 **Estadística Industrial.** Resultados del relevamiento practicado al 31 de diciembre de 19 . 1935- . Bienal. Buenos Aires. Ministerio del Interior. Consejo Nacional de Estadística y Censos. Dirección General de Estadística y Censos de la Nación.

Title varies—título varía: 1935, **Censo Industrial de 1935.**

Industrial statistics. Results of the biennial census of industry.

Resultados del censo bienal de industrias.

3000-3999 STATISTICAL SCIENCE—CIENCIA ESTADISTICA

Arg-3030.L5 1894 **Ley de Estadística y Decreto Reglamentario de la Misma, Ley 3180.** Buenos Aires, 1924. 8 p. Copia máquina.

Law of statistics, no. 3180, passed November 9, 1894, establishing the General Bureau of Statistics, and listing its various duties and prerogatives. Includes decree of December 12, 1894, regulating Law 3180.

Promulgada el 9 de noviembre de 1894, estableciendo la Dirección General de Estadística de la Nación y enumerando sus deberes y prerrogativas. Incluye el Decreto del 12 de diciembre de 1894, que es el reglamentario de la Ley 3180.

Arg-3030.U5 1944 **Ultimas Disposiciones del Poder Ejecutivo Nacional Sobre Estadística y Censos.** Buenos Aires, Imprenta del Congreso Nacional, Octubre de 1944. Ministerio del Interior. Consejo Nacional de Estadística y Censos. 14 p.

Latest national executive regulations on statistics and censuses. Contents: Law no. 10.783/943, Fourth general census of the nation and decennial censuses; Law no. 13.940/944, Creation of the National Board of Statistics and Censuses; Law no. 24.883/944, Fourth general census of the nation (modifications to law no. 10.783/943.)

Contenido: Decreto Ley No. 10.783/943, IV Censo General de la Nación y Censos Decenales; Decreto Ley No. 13.940/944: Creación del Consejo Nacional de Estadística y Censos; Decreto Ley No. 24.883/944: IV Censo General de la Nación (Modificaciones al decreto ley No. 10.783/943).

Arg-3040.A5 **Anales de la Sociedad Científica Argentina.** 1876- . Mensual. Buenos Aires. Sociedad Científica Argentina.

Annals of the Argentine Scientific Society.

Oct. '45, Schneider "Sobre la Interpolación y Reducción de Series Climatológicas." Tomo CXL, entrega IV, octubre 1945, p. 257-302.

On the interpolation and reduction of climatological series. Includes: Mathematical concept of interpolation; cases in which it is applied to climatological statistics; general principles upon which methods of interpolation are based; the condition of homogeneity; methods of interpolation for pluviometric and other data of the elements; limitations of the application of interpolation.

Incluye: Concepto matemático de la interpolación; casos en que se aplica en la climatología estadística; principios generales en que se basan los métodos de interpolación; la condición de homogeneidad; los métodos de interpolación para datos pluviométricos y otros elementos; los límites de la aplicación de la interpolación.

Apr. '45, Dieulefait "Sobre las Ecuaciones Diferenciales Ordinarias a Coeficientes Constantes y el Cálculo Operacional." Por C. E. Dieulefait. Tomo CXXXIX, entrega IV, abril 1945, p. 147-151.

Ordinary differential equations with constant coefficients and operational calculus.

Arg-**3040**.B5 **Boletín Censal y Estadístico.** 1945- . Irregular. Buenos Aires. Ministerio del Interior. Consejo Nacional de Estadística y Censos.

Census and statistical bulletin.

Arg-**3040**.R3 **Revista de Ciencias Económicas.** Publicación de la Facultad de Ciencias Económicas, Centro de Estudiantes y Colegio de Graduados. 1913- . Mensual. Buenos Aires. Universidad de Buenos Aires. Facultad de Ciencias Económicas.

Review of economic sciences published by School of Economics of University of Buenos Aires. In addition to leading articles, this journal publishes current notes under headings: National economic information; national financial information; world economic and financial information; social information; bibliographic information.

Además de los artículos de fondo, esta revista publica notas bajo los siguientes títulos: Información económica nacional; información financiera nacional; información económico-financiera mundial; información social; información bibliográfica.

*Arg-**3040**.R5 **Revista de Economía Argentina.** Publicación mensual del Instituto Alejandro E. Bunge de Investigaciones Económicas y Sociales. 1918- . Buenos Aires. Founded by Alejandro E. Bunge.

Argentine Economic Review. Contains analytical articles as well as statistical data on all phases of Argentine economy—agricultural and industrial production, consumption of various products, prices and index numbers, labor, commerce, finance, population movements.

Contiene artículos analíticos como así también datos estadísticos sobre todas las fases de la economía argentina—producción agrícola e industrial, consumo de diversos productos, precios y números índices, trabajo, comercio, finanzas, movimiento de la población.

July '46, a "Bosquejo de una Economía Argentina para 1955: Segunda Parte (continuación)—Estudios Especiales." Año XXVIII, tomo XLV, no. 337, julio 1946, p. 215-223.

Draft of an Argentine economy for 1955: Second part (continued)—special studies. Includes: Energy.

Incluye: Energía.

June '46, a "Bosquejo de una Economía Argentina para 1955: Segunda Parte (continuación)—Estudios Especiales." Año XXVII, tomo XLV, no. 336, junio 1946, p. 183-189.

Draft of an Argentine economy for 1955: Second part (continued)—special studies. Includes: Primary necessities: Food, clothing, housing.

Incluye: Necesidades primarias: Alimentación, vestido y vivienda.

May '46, a "Bosquejo de una Economía Argentina para 1955: Segunda Parte (continuación) — Estudios Especiales." Año XXVII, tomo XLV, no. 335, mayo1946, p. 143-153.

Draft of an Argentine economy for 1955: Second part(continued)—special studies. Includes: Possibilities of industrial development.

Incluye: Posibilidades de desarrollo industrial.

Apr. '46 "Bosquejo de una Economía Argentina para 1955: Segunda Parte—Estudios Especiales." Año XXVII, tomo XLV, no. 334, abril 1946, p. 105-110.

Draft of an Argentine economy for 1955: Second part—special studies. Includes: Possibilities of development of agriculture and livestock.

Incluye: Posibilidades de desarrollo agropecuario.

Mar. '46 "Bosquejo de una Economía Argentina para 1955: Primera Parte—Plan General (continuación)." Año XXVII, tomo XLV, no. 333, marzo 1946, p. 71-76.

Draft of an Argentine economy for 1955: First part—general plan (continued). Includes: Possible increase of the national income, by sectors.

Incluye: Posible crecimiento de la renta nacional, por sectores.

Feb. '46 "Renta Nacional." Año XXVII, tomo XLV, no. 332, febrero 1946, p. 28-33.

Tables and charts of the Central Bank of Argentina, showing national income and its component parts from 1935 to 1945.

Tablas y cuadros del Banco Central de la República Argentina, referentes a la renta nacional y a las partes que la componen, correspondientes al período 1935-1945.

Feb. '46, a "Bosquejo de una Economía Argentina para 1955: Primera Parte—Plan General." Año XXVII, tomo XLV, no. 332, febrero 1946, p. 34-35.

Draft of an Argentine economy for 1955: First part—general plan. Includes: Purposes of the study; some aspects of the present economy (population and occupation, national wealth, production, national income, balance of payments); comparison with other countries (income distribution by sectors of the population, improvements needed); possible increase in the national income, by sectors (agriculture, industry, commerce and transportation, private and State services).

Incluye: Propósitos del estudio; algunos aspectos de la realidad económica actual (población y ocupación, patrimonio nacional, producción, renta nacional, balance de pagos); comparación con otros países (la distribución de las rentas por sectores de la población, mejoras a alcanzar); posible crecimiento de la renta nacional, por sectores (actividades agropecuarias, industria, comercio y transportes, servicios privados y del Estado).

Dec. '45, Coghlan "La Denatalidad en la Argentina." Aspectos del problema; causas y remedios. A propósito del reciente informe de la Comisión Nacional de la Denatalidad. 3a parte. Por Eduardo A. Coghlan y César H. Belaunde. Año XXVII, tomo XLIV, no. 330, diciembre 1945, p. 577-581.

Decline in birth rate in Argentina. Aspects of the problem, causes and remedies. 3rd part.

Nov. '45, Coghlan "La Denatalidad en la Argentina." Aspectos del problema; causas y remedios. A propósito del reciente informe de la Comisión Nacional de la Denatalidad. 2a parte. Por Eduardo A. Coghlan y César H. Belaunde. Año XXVII, tomo XLIV, no. 329, noviembre 1945, p. 529-535.

Decline in birth rate in Argentina. Aspects of the problem; causes and remedies. 2nd part.

Oct. '45, Coghlan, a "La Agrícola Argentina, País de Población Urbana." Por Eduardo A. Coghlan y Carlos Correa Avila. Año XXVII, tomo XLIV, no. 328, octubre 1945, p. 480-483.

Argentina, agricultural country with urban population.

ARGENTINA
3040

Oct. '45, Coghlan, b "La Denatalidad en la Argentina." Aspectos del problema; causas y remedios. A propósito del reciente informe de la Comisión Nacional de la Denatalidad. 1a. parte. Por Eduardo A. Coghlan y César H. Belaunde. Año XXVII, tomo XLIV, no. 328, octubre 1945, p. 494-496.

Decline in birth rate in Argentina. Aspects of the problem; causes and remedies. 1st part.

Sept. '45, Correa Avila "La Fecundidad y la Natalidad en el Campo y en las Ciudades." Por Carlos Correa Avila. Año XXVII, tomo XLIV, no. 327, septiembre 1945, p. 430-435.

Fecundity and birth rate in the country and the cities.

Aug. '45, Taeuber "Las Estimaciones sobre la Población Futura de América." Por Irene Barnes Taeuber. Año XXVII, tomo XLIV, no. 326, agosto 1945, p. 415-419.

Partial translation of—traducción parcial de: "The Development of Population Predictions in Europe and the Americas," **Estadística,** vol. II, no. 7, septiembre 1944, p. 323-346.

Estimates of the future population of America.

July '45, a "A Propósito del IV Censo General." Año XXVII, tomo XLIV, no. 325. julio 1945, p. 363-367.

The fourth general census (continuation of article in June 1945 issue). Contents: Approximate estimates, brief description of the units of the census, simultaneity and methods of census taking; census takers; instructions.

Continuación del artículo de la edición de junio 1945. Incluye: Cálculos aproximativos, descripción sumaria y previa de las unidades a censar; simultaneidad y métodos de levantamientos; de los censistas; instrucciones.

June '45, a "A Propósito del Cuarto Censo General." Año XXVII, tomo XLIV, no. 324, junio 1945, p. 319-321.

Census techniques for preparation and taking of fourth general census, and preparation of questionnaires.

Incluye: Técnica censual de su preparación y levantamiento; preparación de los cuestionarios.

June '45, b "Area Sembrada y Volumen Físico de la Producción: Años 1900 a 1944." Año XXVII, tomo XLIV, no. 324, junio 1945, p. 211-228.

Cultivated area and physical volume of production, 1900-44.

June '45, Correa Avila "El Analfabetismo en la Argentina." Por Carlos Correa Avila. Año XXVII, tomo XLIV, no. 324, junio 1945, p. 283-287.

Illiteracy in Argentina.

June '45, Daus "Distribución de la Población Argentina por Regiones Geográficas." Por Federico A. Daus y Roberto García Gache. Año XXVII, tomo XLIV, no. 324, junio 1945, p. 294-296.

Distribution of Argentine population by geographic regions.

June '45, Di Tella "La Población Ocupada y las Rentas Individuales en la Argentina." Por Torcuato Di Tella. Año XXVII, tomo XLIV, no. 324, junio 1945, p. 256-267.

Employed population and individual income in Argentina. "With the method expounded by Professor Colin Clark in his book 'Conditions of Economic Progress,' Torcuato Di Tella has prepared for Argentina a study on the evolution of the working population and its individual incomes, in the three main groups of activities (primary, secondary and tertiary), as well as on the phenomena of the constant afflux of those occupied in the first group of activities to the second and third." (Translation of comment.)

"Con el método expuesto por el profesor Colin Clark en su obra 'Conditions of Economic Progress,' el Ing. Torcuato Di Tella ha preparado para la República Argentina un trabajo sobre la evolución de la población ocupada y las rentas individuales de ésta en los tres grandes grupos de actividades (primarias, secundarias y terciarias), como así también sobre el fenómeno del constante afluir de los que están ocupados en la primera de las actividades nombradas a la segunda y tercera." (Fuente: Nota de la Redacción.)

May '45 "La Interpretación de los Indices del Costo de la Vida de la Familia Obrera." Año XXVII, tomo XLIV, no. 323, mayo 1945, p. 200-202.

Interpretation of indexes of the cost of living of the working class family.

Apr. '45 "Algunas Sugerencias a Propósito del Proyectado IV Censo General." Año XXVII, no. 322, abril 1945, p. 142-144.

Suggestions concerning projected fourth general census.

Dec. '44, a "Estimación de la Renta Nacional Correspondiente al Año 1941." Año XXVI, tomo XLIII, no. 318, diciembre 1944, p. 406-412.

National income estimate for Argentina for 1941, with total estimate and estimates by sectors, from a study by the Department of Finance.

Contiene un cálculo total y cálculos por sectores, obtenido de un estudio del Ministerio de Hacienda.

Feb. '43, Casiello "Protección a la Familia Argentina." Por Francisco Casiello. Febrero 1943, p. 46-53. No examinado.

Protection for the Argentine family. A study of the general aspect of family allowances, outline of a proposed scheme of family allowances for Argentina, the most recent available statistics of the operation of family allowances in eight countries throughout the world, and the names and beginning dates of schemes of family allowances already in operation in Argentina.

Un estudio sobre el aspecto general de ayuda a la familia, bosquejo de un proyecto propuesto para ayuda a la familia en la Argentina, las mas recientes estadísticas aprovechables sobre ayuda a la familia en ocho países del mundo, nombres y fechas de iniciación de las organizaciones de ayuda a la familia actualmente en funcionamiento en la Argentina.

*Arg-**3040**.R7 **Revista de Economía y Estadística.** 1939- . Trimestral. Córdoba. Universidad Nacional de Córdoba. Escuela de Ciencias Económicas.

Review of economics and statistics of the School of Economic Sciences of the National University of Cordoba. Devoted primarily to theoretical problems. Articles on questions of theoretical and applied economics, and occasional articles on specific Argentine problems. The permanent section "Crónica económica" is a review of the most important economic events in the world and Argentina.

Dedicada principalmente a problemas teóricos. Artículos sobre cuestiones teóricas y aplicadas de la economía, y de vez en cuando artículos sobre problemas específicos argentinos. La sección permanente "Crónica económica" consta de un examen de los acontecimientos económicos mas importantes en el mundo y en la Argentina.

Jan.-June '46, a "Precio Medio de la Habitación Ofrecida en la Ciudad de Buenos Aires (1943-1945)." Por el Instituto de Estadística. Informe a cargo de Jacobo Bleger. Año VIII, no. 1-2, lo. y 2o. trimestre 1946, p. 37-70.

Average cost of housing in Buenos Aires, 1943-45.

Oct.-Dec. '44 "Precio Medio de la Habitación Ofrecida en la Ciudad de Buenos Aires, 1943-44." Año VI, no. 4, octubre-diciembre 1944, p. 285-308.

Average price of housing in Buenos Aires, 1943-44.

*Arg-**3040**.R8 **Revista de la Facultad de Ciencias Económicas, Comerciales y Políticas.** 1926- . Trimestral. Rosario. Universidad Nacional del Litoral. Facultad de Ciencias Económicas, Comerciales y Políticas.

Journal of the School of Economic, Commercial, and Political Sciences of the National University of Litoral. Series 1 and 2 consist of 1 volume each, 3 numbers to each volume; series 3 closes with volume X, no. 3, September-December 1941; series 4 begins with volume 1, no. 1, January-April 1942. Includes articles on statistics.

Las series 1 y 2 consisten de un tomo cada una, con tres números de volumen; serie tres termina con el tomo X, no. 3, septiembre-diciembre 1941; la serie 4 comienza con el tomo 1, no. 1, enero-abril 1942. Incluye artículos sobre estadística.

Jan.-Apr. '45, Cortés "Sobre los Servicios Estadísticos y Censales en la República Argentina." Por Fernando Cortés. Serie 4a, tomo IV, no. 1, enero-abril 1945, p. 241-265.

Statistical services and censuses in Argentina. Includes: Necessity of periodic censuses, bases; creation of a permanent agency for statistics and censuses, its advantages; census tecnique: Fixing date for taking census, publicity, editing questionnaires, obtaining data; the Constitution and the censuses.

Incluye: Necesidad de la elaboración periódica de censos, fundamentos; creación de un organismo estable de estadística y censos, sus ventajas; técnica censal: Fijación de la fecha de empadronamiento, propaganda, redacción de los cuestionarios, obtención de los datos; nuestra Constitución y los censos.

Jan.-Apr. '45, Santone "Sobre Censo General: Observaciones y Sugerencias." Por Adolfo Santone. Serie 4a, tomo IV, no. 1, enero-abril 1945, p. 287-305.

Observations and suggestions on the general census. Contents: Preliminary instruction; permanent interest of the subject; the census as a necessity of the nation; specific characteristics of modern censuses; organization and early preparation for a census; importance and popularization of census results; centralization or decentralization of direction; financial problems of the census; permanent census; the census as a public charge; Argentina and census statistics; concluding aspects of the subject.

Contenido: Propedéutica; permanente actualidad del tema; los censos, una necesidad del Estado; características específicas de los modernos censos; organización y previa preparación del censo; trascendencia y popularización de los resultados del censo; centralización o no centralización de las direcciones; problemas financieros del censo; censo permanente; del censo como carga pública; la República Argentina y la estadística censal; aspectos conclusiones del tema.

Arg-**3040**.R9 **Revista de la Facultad de Ciencias Físicomatemáticas.** Publicaciones serie segunda. 1939- . Irregular. La Plata. Universidad Nacional de La Plata. Facultad de Ciencias Físicomatemáticas.

Journal of the School of Physicomathematical Sciences of the National University of La Plata.

Arg-**3200**.C5 1937.A **Clasificador de Bienes.** Reglamentación. Buenos Aires, Tall. Gráf.del Ministerio de Obras Públicas, 1937. Ministerio de Obras Públicas. Comisión del Censo Nacional de Bienes del Estado. 282 p. No examinado.

Classification of properties. Regulations.

Arg-**3200**.C5 1937.B **Indice General Alfabético por Materia de los Bienes Nacionales del Estado.** Buenos Aires, Tall.Gráf.del Ministerio de Obras Públicas, 1937. Ministerio de Obras Públicas. Comisión del Censo Nacional de Bienes del Estado. 61 p. No examinado.

General alphabetical index of national properties.

Arg-3300 1944-46 **Enseñanza Estadística en Argentina: Métodos y Materiales.** Material básico adquirido para la Encuesta del Instituto Interamericano de Estadística sobre los Métodos y Materiales en la Enseñanza Estadística en las Naciones Americanas, 1944-46.

Statistical training in Argentina: Methods and materials. Background material acquired for the IASI Survey on Statistical Training Methods and Materials in the American Nations, 1944-46. Includes: Filled questionnaires; outlines of statistical courses.

Incluye: Cuestionarios llenados; programas de cursos de estadística.

See also—véase también: Int-**3000**.S5.

Arg-**3320**.E5 **Estadística: Programa Analítico.** En tres partes. Pte. I: Estadística metodológica; pte. II: Estadística demográfica; pte. III: Estadística social. Buenos Aires. Instituto Alejandro E. Bunge de Investigaciones Económicas y Sociales. 5 p. Copia máquina.

Analytical program in statistics, Alejandro E. Bunge Institute of Economic and Social Research.

Arg-**3320**.P3 1943 **Programa de Biometría.** Curso de actuarios. Buenos Aires, Porter Hermanos, 1943. Universidad de Buenos Aires. Facultad de Ciencias Económicas. 11 p.

Program in biometrics, University of Buenos Aires.

Arg-**3320**.P5 1944 **Programa de Estadística.** 2° año. Buenos Aires, La Vanguardia, 1944. Universidad de Buenos Aires. Facultad de Ciencias Económicas. 9 p.

Program in statistics, University of Buenos Aires.

Arg-**3320**.P6 1944 **Programa de Matemática Actuarial.** Complemento. Curso de actuarios. Buenos Aires, La Vanguardia, 1944. Universidad de Buenos Aires. Facultad de Ciencias Económicas. 6 p.

Program in actuarial mathematics, University of Buenos Aires.

Arg-**3320**.P7 1944 **Programa de Matemática Financiera y Actuarial.** Buenos Aires, La Vanguardia, 1944. Universidad de Buenos Aires, Facultad de Ciencias Económicas. 18 p.

Program in financial and actuarial mathematics, University of Buenos Aires.

Arg-**3320**.P8 1944 **Programa de Matemáticas.** Buenos Aires, La Vanguardia, 1944. Universidad de Buenos Aires. Facultad de Ciencias Económicas. 19 p.

Program in mathematics, University of Buenos Aires.

4000-4999 GEOGRAPHY—GEOGRAFIA

Arg-**4200**.R5 1943? **Registro de Observaciones Pluviométricas.** Buenos Aires?, 1943? Ministerio de Agricultura. Dirección de Meteorología, Geofísica e Hidrología. División Hidrología. Sección Pluviométrica. 1 vol.

Registry of pluviometric observations. Contains forms.

Contiene formularios.

5000-5999 DEMOGRAPHY—DEMOGRAFIA

Arg-**5000**.A5 1945 **Argentina: Summary of Biostatistics.** Maps and charts, population, natality and mortality statistics. Washington, D.C., 1945. Prepared by U.S. Department of Commerce, Bureau of the Census, in cooperation with Office of the Coordinator of Inter-American Affairs. 152 p.

Resumen bioestadístico de Argentina. Incluye mapas y cuadros y datos sobre población, natalidad y mortalidad.

Arg-**5000**.B5 1536-1936 **Buenos Aires, Puerto del Río de la Plata, Capital de la Argentina; Estudio Crítico de Su Población, 1536-1936.** Por Nicolás Besio Moreno. Buenos Aires, Tall. Gráf.Tuduri, 1939. 500 p.

Buenos Aires, harbor of the Plata river, capital of Argentina; a critical study of its population, 1536-1936.

Arg-**5000**.E5 **Estadística Demográfica.** Territorios nacionales y bases navales. 1939- . Frecuencia? Buenos Aires. Argentina. Ministerio de Justicia e Instrucción Pública. Inspección General de Justicia.

Demographic statistics of the national territories and naval bases.

*Arg-**5000**.P2 **Informe, Serie D, Demografía.** 1926- . Irregular. Buenos Aires. Ministerio del Interior. Consejo Nacional de Estadística y Censos. Dirección General de Estadística y Censos de la Nación.

Series of reports on demography, including the annual publication, **La Población y el Movimiento Demográfico de la República Argentina.**

Serie de informes sobre demografía, incluyendo la publicación anual **La Población y el Movimiento Demográfico de la República Argentina.**

1944, No. 12 **La Población y el Movimiento Demográfico de la República Argentina en los Años 1943 y 1942.** Buenos Aires, Guillermo Kraft, Ltda., 1944. Ministerio del Interior. Consejo

Nacional de Estadística y Censos. Dirección General de Estadística y Censos de la Nación. 75 p. Informe, no. 98, serie D, no. 12, Demografía.

Population and vital statistics in Argentina in 1943 and 1942. Includes population of Argentina on 31st of December in 1943 and 1942, monthly figures for 1943 and annual figures for 1942 and 1943 on births, marriages, and deaths, with a separate section on the Argentine territories.

Incluye la población de la República Argentina al 31 de diciembre de 1943 y 1942, y cifras mensuales de 1943 y cifras anuales de 1942 y 1943 sobre nacimientos, matrimonios y defunciones, con una sección por separado sobre los territorios nacionales.

1943, No. 10 **Clasificación Estadística de las Causas de las Defunciones.** (Nomenclatura internacional de 1938.) Desarrollo de los rubros de la nomenclatura detallada e índice alfabético de las causas de muerte. Buenos Aires, Guillermo Kraft, Ltda., 1943. Ministerio de Hacienda. Dirección General de Estadística y Censos de la Nación. 184 p. Informe, no. 93, serie D, Demografía, no. 10.

Statistical classification of causes of death (international nomenclature of 1938). Contains: Text of convention and recommendations approved at international conference for the fifth decennial revision of the International List of Causes of Death, held in Paris, October 3-7, 1938; development of rubrics of detailed nomenclature; alphabetical index of causes of death, and of diseases, keyed to corresponding numbers of detailed, intermediate, and abbreviated International Lists of Causes of Death, respectively.

Contiene: Texto de la convención y recomendaciones aprobadas por la conferencia internacional para la quinta revisión decenal de las Nomenclaturas Internacionales de Causas de Muerte, reunida en París del 3 al 7 de octubre de 1938; desarrollo de los rubros de la nomenclatura detallada; índice alfabético de las causas de muerte y enfermedades, con indicación del número correspondiente en las nomenclaturas detallada, intermedia y abreviada.

1942, No. 9 **El Movimiento Demográfico en los Territorios Nacionales de la República Argentina en los Años 1933 a 1941.** Buenos Aires, Guillermo Kraft, Ltda., 1942. Ministerio de Hacienda. Dirección General de Estadística de la Nación. 165 p. Informe, no. 91, serie D, no. 9, Demografía.

Vital statistics in Argentine territories from 1933 to 1941.

1940, No. 6 **La Población y el Movimiento Demográfico de la República Argentina en los Años 1939 y 1938, y Síntesis de Años Anteriores.** Buenos Aires, Guillermo Kraft, Ltda., 1940. Ministerio de Hacienda, Dirección General de Estadística de la Nación. 160 p. Informe, no. 78, serie D, no. 6, Demografía.

Population and vital statistics in Argentina in 1939 and 1938, and summary of previous years.

Arg-**5100**.E5 1946 **Ensayo sobre la Evolución de la Población de las Américas** (**Estimación en el Lapso 1938-1973**)**: II, Argentina.** Por Jorge Justo Boero Brian y Francisco Molina Serrano. Montevideo, 1946. 83 y xvi p.

Essay on the evolution of the population of the Americas (estimated for 1938-73): II, Argentina.

Arg-**5100**.P5 **La Población de la República Argentina.** Cifras provisionales. 1931- . Anual. Buenos Aires. Ministerio del Interior. Consejo Nacional de Estadística y Censos. Dirección General de Estadística y Censos de la Nación.

Annual provisional figures on population of Argentina.

Arg-**5200** **Estadística Vital y Registro en Argentina: Métodos y Procedimientos.** Material básico adquirido para la encuesta que el Instituto Interamericano de Estadística realiza sobre los métodos de registro y estadística vital en las naciones americanas, incluyendo el estudio de 1945.

Vital statistics and registration in Argentina: Methods and procedures. Background material acquired for the study conducted by the Inter American Statistical Institute in vital statistics and registration methods in the American nations, including the survey of 1945. Includes: Instructions; forms; punch cards and codes; nomenclatures; organization charts.

Incluye: Instrucciones; formularios; tarjetas de perforación y códigos; nomenclaturas; esquemas de organización.

Arg-**5200**.A5 **Anuario Demográfico.** Natalidad, nupcialidad y mortalidad. 1911- . Buenos Aires. Ministerio del Interior. Departamento Nacional de Higiene. Sección Demografía Sanitaria y Geografía Médica.

Demographic yearbook. Birth, marriage, and death rates. Published intermittently since 1911.

Arg-**5220**.T5 1936 **Tabla de Mortalidad de la República Argentina.** Construída con los datos del 3er censo nacional. Por Jorge M. Comas y Pedro Goldenberg. Buenos Aires, 1936. Universidad de Buenos Aires. Facultad de Ciencias Económicas. Instituto de Biometría. 14 p. Monografías de los alumnos del Instituto, no. 3.

Mortality table for Argentina, constructed with the data from the 3rd national census.

6000-6999 ECONOMICS—ECONOMIA

Arg-**6000**.A6 1944 **Argentina Económica e Industrial.** Publicado bajo la dirección de Pedro J. Cristia. Colaboradores: Elías Días Molano, Eduardo de la Fuente, Samuel Gorban, Fausto Hernández y Julio Martínez. Rosario, Argentina, Molachino y Scarabino, 1944. 239 p. No examinado.

Economic and industrial Argentina.

Arg-**6000**.B8 **Business Conditions in Argentina.** 1915- . Quarterly. Tornquist, Ernesto & Cía., Ltd.

Reviews current business situation; reports on foreign trade, railways, insurance, banks, and public finance. Contains statistical tables and charts on most of the important types of economic movements, including estimates of the country's balance of payments.

Condiciones del comercio en la Argentina. Revista trimestral que analiza la situación general del comercio; informa sobre comercio exterior, ferrocarriles, seguro, bancos y hacienda pública. Contiene cuadros y diagramas estadísticos sobre las manifestaciones económicas mas importantes, incluyendo estimaciones del balance de pagos del país.

Arg-**6000**.E5 **Economía y Finanzas de Mendoza.** 1939- . Mendoza, Argentina. Ministerio de Economía, Obras Públicas y Riego de Mendoza. Instituto Técnico de Investigaciones y Orientación Económica de la Producción. No examinado.

Contains statistical series on demography, agricultural production, commerce, and finance of the province of Mendoza.

Contiene series estadísticas sobre demografía, producción agrícola, comercio y finanzas de la provincia de Mendoza.

Arg-**6000**.E6 1945 **Economic Controls and Commercial Policy in Argentina.** One of a series of reports on economic controls and commercial policy in the American republics. Washington, D.C., 1945. United States Tariff Commission. 46 p.

Folleto sobre Argentina publicado en la serie de informes de la Comisión de Aduanas de los Estados Unidos, sobre controles económicos y política comercial de las repúblicas americanas.

Arg-**6000**.I2 1946 **Indices Básicos de la Economía Nacional.** Buenos Aires, julio 1946. Presidencia de la Nación. Secretaría Técnica. 33 p.

Basic indexes of the Argentine economy. Graphs showing: National budget 1939-45 and its relation to national income and customs; banking operations; exchange transactions; salaries in industry, cost of living, etc.; public debt; production, consumption, and prices.

Contiene gráficos analíticos de ciertos aspectos de la economía nacional como ser: Presupuesto nacional correspondiente al período 1939-45 y su relación con renta y recaudación nacionales; operaciones bancarias, transacciones bursátiles, remuneración del trabajo en la industria, costo de la vida, etc.; deuda pública; producción, consumo y precios.

Arg-**6000**.I4 **Informaciones sobre las Actividades Económicas en Nuestro País: Informe.** 1946?- . Irregular. Buenos Aires. Banco Central de la República Argentina. Departamento de Investigaciones Económicas.

Information on economic activities of the country: Report of the Department of Economic Research of the Central Bank.

Arg-**6000**.R5 1944 **Reports Presented to the Conference of Commissions of Inter-American Development (Held in New York, May 9-18, 1944) by the Argentine Commission of Inter-American Development.** On: 1, Argentine industry; 2, present state of Argentine agriculture; 3, hydroelectric power in Argentina. (Translations from Spanish originals.) Washington, D.C., 1945? Inter-American Development Commission. 195 p.

Cover title—título de la cubierta: **Argentina: Reports Presented to the Conference of Commissions of Inter-American Development.**

Informes presentados a la Conferencia de Comisiones de Fomento Interamericano (realizada en N. York, del 9 al 18 de mayo de 1944), por la Comisión Argentina de Fomento Interamericano, sobre: 1, Industria argentina; 2, el estado actual de la agricultura argentina; 3, la fuerza hidroeléctrica en la Argentina.

Arg-**6000**.S4 **Situation in Argentina.** 1922- Monthly. Boston, Massachusetts. Boston First National Bank.

Statistical information on banking and business movements, exchange quotations, review of general economic conditions.

La situación en Argentina. Informaciones estadísticas sobre movimientos de bancos y comercio, cotizaciones del cambio, examen de las condiciones económicas generales. Mensual.

*Arg-**6000**.S8 **Boletín Estadístico.** 1937- . Mensual. Buenos Aires. Banco Central de la República Argentina. Departamento de Investigaciones Económicas.

Monthly statistical bulletin of the Central Bank. From 1937 to 1946, published as **Suplemento Estadístico de la Revista Económica;** before 1937, published as part of the **Revista Económica** (Arg-**6720**.R6). Contains current statistical series on: Statistics of banking and money; foreign exchange, and stocks; domestic transactions and activities of ports; wholesale prices in Buenos Aires; foreign trade; national revenues; statistics of agricultural and livestock products; industrial statistics; condition of the banks.

Publicado de 1939 a 1946 bajo el título de **Suplemento Estadístico de la Revista Económica;** antes de 1937, se publicó como una parte de la **Revista Económica** (Arg-**6720**.R6). Contiene series estadísticas de la actualidad sobre: Estadísticas de los bancos y del mercado monetario; cambios y títulos; negocios internos y movimiento portuario; precios mayoristas en Buenos Aires; comercio exterior; recaudaciones nacionales; estadísticas de productos agropecuarios; estadística industrial; estado de los bancos.

Arg-**6000**.V5 **Veritas.** La publicación Argentina económico - financiera mas importante, difundida y completa. 1931- . Mensual. Buenos Aires.

Contains economic studies on Argentina and other Latin American countries, and statistics of Argentine economic development, including two current statistical series, which do not appear elsewhere in the country, on failures and on corporations. Monthly.

Contiene estudios económicos sobre Argentina y otros países latino-americanos, y estadísticas sobre el desarrollo económico argentino, incluye dos series estadísticas permanentes que no aparecen en ninguna otra publicación del país, que son: "Quebrantos Comerciales y Civiles" y "Sociedades Anónimas."

Arg-**6010**.P5 **Precios.** Buenos Aires, 1940.
1940 Ministerio del Interior. Departamento Nacional del Trabajo. División de Estadística. 100 p. (estimadas). Mimeografiado.

Prices. Contents: Consumption of articles of prime necessity; markets and free markets; groceries; provinces and territories; raw materials, manufactured and partly-finished products.

Contiene: Artículos de consumo de indispensable necesidad; mercados y ferias francas; artículos de almacén; provincias y territorios; materias primas, productos elaborados y semielaborados.

*Arg-**6100**.R4 **La Renta Nacional de la República Argentina.** 1946. Buenos Aires, Platt-Establecimientos Gráficos, 1946.
1935-45
Banco Central de la República Argentina. Departamento de Investigaciones Económicas. 45 p.

A study of the Central Bank on national income during the period 1935-45. Contains analytic tables with explanations, on: Value of national income; physical volume of national income; index of the volume of goods at the disposition of the population; comparison between expenditures of the state and value of income that stays in the country. Also tables of the different sectors of national income, as: Agriculture, cattle raising, mining, industry, construction, transportation, communication and other public services, trade, financial enterprises, State services and personal services. In the last chapter there is an explanation of the procedures used to determine the income of each sector.

Contiene un estudio de la renta nacional en el período 1935-45. Incluye cuadros analíticos con comentarios aclaratorios sobre: Valor de la renta nacional; volumen físico de la renta nacional; índice del volumen de bienes a disposición de la población del país; comparación entre los gastos del Estado y el valor de la renta que queda en el país. Contiene también cuadros de los distintos sectores de renta, a saber: Agricultura, ganadería, minería, industria, construcción, transportes, comunicaciones y otros servicios públicos, comercio, empresas financieras, servicios del estado y personales. En el capítulo final explica el procedimiento utilizado para determinar la renta de cada uno de los sectores.

Arg-**6200** **Estadísticas Agrícolas en Argentina: Métodos y Procedimientos.** Material básico adquirido para el estudio que el Instituto Interamericano de Estadística realiza sobre los métodos de estadísticas agrícolas en las naciones americanas, incluyendo la encuesta de 1945-46.

Agricultural statistics in Argentina: Methods and procedures. Background material acquired for Inter American Statistical Institute study on methods in agricultural statistics in the American nations, including the Survey of Agricultural Production Statistics in the American Nations in 1945-46. Consists of: Answer to questionnaire; laws; forms; reports.

Consiste de: Respuesta al cuestionario; leyes; formularios; informes.

Arg-**6200**.A3 **Almanaque del Ministerio de Agricultura.** 1925- . Anual. Buenos Aires. Ministerio de Agricultura. Dirección de Propaganda y Publicaciones. No examinado.

Almanac of the Department of Agriculture. "Reprint of agricultural material appearing elsewhere." (Source: Stat. Act.)

Reproducción de material agrícola ya anteriormente publicado. (Traducción del comentario.)

Arg-**6200**.A4 **Anales de la Sociedad Rural Argentina.** Revista pastoril y agrícola. 1865- . Mensual. Buenos Aires.

Monthly journal of the Argentine Rural Society. Contains analytical articles and statistical data on agriculture and derivative industries, and on export trade.

Contiene artículos analíticos y datos estadísticos sobre agricultura e industrias derivadas, y sobre el comercio de exportación.

Arg-**6200**.A5 1928 **Anuario de la Sociedad Rural Argentina: Estadísticas Económicas y Agrarias.** Buenos Aires, Imp. Gotelli, 1928. Sociedad Rural Argentina. 367 p.

Yearbook of the Argentine Rural Society: Economic and agricultural statistics. Only one issue published.

Se publicó solamente una edición.

*Arg-**6200**.B5 **Boletín Estadístico del Ministerio de Agricultura.** 1940- Mensual. Buenos Aires. Ministerio de Agricultura. Dirección de Economía Rural y Estadística. Dirección de Estadística.

Title varies—el título varía: 1940-1944, **Boletín Mensual de Estadística Agropecuaria;** 1944-1945, **Estadística del Ministerio de Agricultura.**

Statistical bulletin of the Department of Agriculture. Annual summary in the December issue. Contains a variety of statistical information on agricultural and livestock production, exports, consumption, and immigration.

Resumen anual en la edición de diciembre. Contiene una variedad de informaciones estadísticas sobre producción agrícola y ganadera, exportación, consumo e inmigración.

Arg-**6200**.M5 **Memoria.** 1898- . Anual. Buenos Aires. Ministerio de Agricultura.

Annual report of the Department of Agriculture. Includes data on mines and geology, immigration and tourism, meteorology and hydrology, and national parks, as well as material on agricultural commodities, production, exports, etc.

Incluye datos sobre minas y geología, inmigración y turismo, meteorología e hidrología, y parques nacionales, como así también informaciones sobre productos agrícolas, producción, exportación, etc.

Arg-**6200**.P5 1923 **Programa Estadístico y Económico para el Ministerio de Agricultura de la República Argentina.** Por León M. Estabrook. Buenos Aires, Tall.Gráf.del Ministerio de Agricultura de la Nación, 1923. Ministerio de Agricultura. Sección Propaganda e Informes. 170 p.

Statistical and economic program for the Department of Agriculture of Argentina. Includes two outlines and a general plan of organization, in the appendix.

Incluye dos esquemas y un plan general de organización, en el apéndice.

Arg-**6210**.A4 **Boletín Mensual.** 1935- Buenos Aires. Secretaría de Industria y Comercio. Dirección de Algodón.

Monthly bulletin of the Bureau of Cotton of the Department of Industry and Commerce. Previously published by National Cotton Board of Department of Agriculture under title **Algodón.** Each issue has a statistical summary covering planting, production, prices, ginning, local consumption, and exports; a review of current prices and production conditions; notes on the cotton situation in Argentina and abroad; and short articles on technical production problems, economic problems, and legislation.

Publicado anteriormente por la Junta Nacional del Algodón del Ministerio de Agricultura bajo el título **Algodón.** Cada edición tiene un resumen estadístico sobre plantación, producción, precios, desmotación, consumo local, y exportación; un resumen de precios corrientes y condiciones de producción; notas sobre la situación algodonera argentina y mundial; y artículos cortos sobre problemas técnicos de producción, problemas económicos, y legislación.

Arg-**6210**.B3 **Boletín Frutas y Hortalizas.** 1936- . Mensual. Buenos Aires. Ministerio de Agricultura. Dirección de Frutas y Hortalizas. No examinado.

Fruits and vegetables bulletin. Contains statistics on the production, prices and commercialization of fruits and vegetables.

Contiene estadísticas sobre la producción, precios y comercialización de las frutas y hortalizas.

Arg-**6210**.B5 **Boletín Informativo de la Comisión Nacional de Granos y Elevadores.** 1936- . Bimensual. Buenos Aires. Ministerio de Agricultura. Comisión Nacional de Granos y Elevadores.

Bulletin of the National Comission of Grains and Elevators. Contains information on all phases of grain production and movement in

Argentina and in other countries, particularly those countries which import from Argentina.

Contiene información sobre todos los aspectos de la producción y movimiento de granos en la Argentina y otros países, particularmente aquellos países que importan de la Argentina.

Arg-**6210**.B7 **Boletín Informativo.** 1935- . Mensual. Buenos Aires. Ministerio de Agricultura. Dirección de Vitivinicultura.

From 1935 to 1943 published by—desde 1935 a 1943 publicado por: Junta Reguladora de Vinos.

Monthly bulletin of the Bureau of Winegrowing Production, sales for consumption, and prices of wine.

Producción, salidas al consumo y precios de vino.

Arg-**6210**.I5 **La Industria Azucarera.** Organo del Centro Azucarero Argentino. 1904- . Mensual. Buenos Aires.

The sugar industry in Argentina. Statistics of production and imports and exports of sugar. Monthly.

Estadísticas de producción y de importación y exportación de azúcar.

*Arg-**6230**.A5 **Anuario de Estadística Agro-Pecuaria.** Fecha de iniciación? Buenos Aires. Ministerio de Agricultura. Dirección de Economía Rural y Estadística.

Yearbook of statistics of livestock and agriculture.

Arg-**6230**.C5 **Compras de Ganado Bovino, Ovino y Porcino Efectuadas por los Frigoríficos en las Estancias.** 1928- Semanal. Buenos Aires. Ministerio de Agricultura. Junta Nacional de Carnes.

Purchases from ranches of cattle, sheep, and swine, made by the packing houses. Weekly.

Arg-**6300**.B5 **Boletín de Informaciones Petroleras.** 1924- . Mensual. Buenos Aires. Secretaría de Industria y Comercio. Dirección General de Yacimientos Petrolíferos Fiscales.

Monthly bulletin of the General Bureau of Petroleum. Contains statistical information on production, trade, and consumption of petroleum.

Contiene informaciones estadísticas sobre producción, comercio y consumo de petróleo.

*Arg-**6300**.E5 **Estadística Minera de la Nación.** 1919- . Anual. Buenos Aires. Ministerio de Agricultura. Dirección General de Minas, Geología e Hidrología.

Yearbook of mineral statistics.

ARGENTINA
6210

1940 **Estadística Minera de la Nación, Año 1940.** Datos económico-industriales, sobre precios, usos, zonas de producción, importación, exportación, etc., de los minerales. Buenos Aires, Tall. Cartográfico de la Dirección de Minas y Geología, 1941. Ministerio de Agricultura de la Nación. Dirección de Minas y Geología. 121 p.

Statistics on prices, uses, zones of production, imports, exports, etc., of minerals.

Arg-**6400** **Estadísticas Industriales en Argentina: Métodos y Procedimientos.** Material básico adquirido para el estudio que el Instituto Interamericano de Estadística realiza sobre los métodos de estadísticas industriales en las naciones americanas, incluyendo la encuesta de 1945-46.

Industrial statistics in Argentina: Methods and procedures. Background material acquired for the study conducted by the Inter American Statistical Institute in methods of industrial statistics in the American nations, including the survey of 1945-46. Consists of: Schedules and other forms; punch cards.

Consiste de: Cédulas y otros formularios; tarjetas de perforación.

See also—véase también: **Censo Industrial de 1935.**

Arg-**6400**.A3 **La Actividad Industrial durante los Primeros . . . Meses de . . ., Según los Números Indices Correspondientes a los Salarios Pagados, a la Ocupación de Obreros y a las Horas-Obrero de Trabajo en la Industria.** Fecha de iniciación? Mensual. Buenos Aires. Ministerio del Interior. Consejo Nacional de Estadística y Censos. Dirección General de Estadística y Censos de la Nación. Mimeografiado.

Cumulative monthly data on industrial activity, according to index numbers of salaries paid, occupati n cf workers, and man-hours worked in industry.

*Arg-**6400**.E5 **Informe, Serie I, Industria.** 1923- . Irregular. Buenos Aires. Ministerio del Interior. Consejo Nacional de Estadística y Censos. Dirección General de Estadística y Censos de la Nación.

Series of reports on industry, including the annual publication **Estadística Industrial de la República Argentina.**

Serie de informes sobre industria, incluyendo la publicación anual **Estadística Industrial de la República Argentina.**

1942, No. 5 **Estadística Industrial de la República Argentina Correspondiente al Año 1940.** Resultados del relevamiento practicado al 31 de diciembre de 1940. Buenos Aires, Guillermo Kraft, Ltda., 1942. Ministerio de Hacienda. Dirección General de Estadística de la Nación. 138 p. Informe, no. 88, serie I, no. 5, Industria.

Industrial statistics of Argentina, 1940. Includes an introduction on procedure adopted in compiling the data and discussion of results, and general statistics for the country, in addition to statistics by industrial groups.

Incluye una introducción sobre el procedimiento adoptado para la compilación de los datos y discusión sobre los resultados obtenidos, y estadísticas generales de la república, así como también estadísticas por rubros de industria.

Arg-**6400**.M7 1945 **Mining and Manufacturing Industries in Argentina.** One of a series of reports on mining and manufacturing industries in the American republics. Washington, D.C., 1945. United States Tariff Commission. 74 p.

Informe sobre Argentina, publicado en la serie de informes de la Comisión de Aduanas de los Estados Unidos, sobre las industrias mineras y manufactureras en las repúblicas americanas.

Arg-**6400**.R7 **Revista de la Unión Industrial Argentina.** 1887- . Mensual. Buenos Aires.

Previously published under the title—publicada anteriormente bajo el título: **Argentina Fabril.**

Monthly journal of Argentine Industrial Union.

*Arg-**6600**.M5 **Memoria.** Presentada al honorable congreso. 1891- . Anual. Buenos Aires. Ministerio de Obras Públicas.

Annual report of the Department of Public Works. Statistical tables and graphs on: Transportation (railroads, navigation and ports, etc.); sanitation; architecture; irrigation; register of public property; etc.

Cuadros y gráficos estadísticos sobre: Transportes (ferrocarriles, navegación y puertos, etc.); obras sanitarias; arquitectura; irrigación; registro de bienes del Estado; etc.

1941 **Memoria, Año 1941.** Presentada al honorable congreso. Buenos Aires, Tall.Gráf.del Ministerio de Obras Públicas, 1942. Ministerio de Obras Públicas. 2 vol.

Contenido: **Tomo I:** Ferrocarriles; ferrocarriles del Estado; vialidad; Comisión Nacional de Coordinación de Transportes; navegación y puertos; caja nacional de jubilaciones y pensiones de empleados ferroviarios; índice detallado del tomo I. **Tomo II:** Obras sanitarias; arquitectura; irrigación; estudios y obras del Riachuelo; contabilidad y contralor de trabajos públicos; registro de bienes del Estado; índice detallado del tomo II. Incluye cuadros y gráficos estadísticos.

Arg-**6614**.M5 **Memoria de los Ferrocarriles en Explotación.** 1891- . Anual. Buenos Aires. Ministerio de Obras Públicas. Dirección General de Ferrocarriles. No examinado.

Annual report on railroads.

Arg-**6615**.A5 **Anuario del Movimiento de los Puertos de la República Argentina.** 1938- . Buenos Aires. Ministerio de Obras Públicas. Dirección General de Navegación y Puertos.

Formerly issued as—publicado anteriormente bajo el título de: **Anuario de la Dirección General de Navegación y Puertos.**

In three parts—en tres partes: Parte I, **Puertos Fluviales;** parte II, **Puertos de Buenos Aires y La Plata;** parte III, **Puertos Marítimos.**

Annual port statistics.

*Arg-**6620**.M5 **Memoria de la Dirección General de Correos y Telecomunicaciones.** 1858- . Anual. Buenos Aires. Ministerio del Interior. Dirección General de Correos y Telecomunicaciones. No examinado.

Report of the General Bureau of Posts and Telecommunications. "Contains statistics of postal and telegraph activities." (Source: Stat. Act.)

Contiene estadísticas sobre actividades de correos y telégrafos. (Traducción del comentario.)

Arg-**6700**.B2 **Boletín de la Cámara de Comercio Argentino - Paraguaya.** 1940- . Mensual. Buenos Aires.

Monthly bulletin of the Argentino-Paraguayan Chamber of Commerce.

Contains articles, notes, and some data on commerce, finance, etc.

Contiene artículos, notas, y algunos datos sobre comercio, finanzas, etc.

Arg-**6700**.B4 **Boletín Oficial de la Bolsa de Comercio de Buenos Aires.** Fecha de iniciación? Semanal. Buenos Aires. Bolsa de Comercio de Buenos Aires. No examinado.

Official weekly bulletin of the Buenos Aires exchange. Gives information on securities and exchange quotations, prices of cereals, movement in livestock markets, reports and balances of companies, general information, etc.

"Trae informaciones sobre cotizaciones de títulos, cambios bancarios, precios de los cereales, entrada y salida de ganado en los mercados, memorias y balances de sociedades, informaciones generales, etc." (Source: **Stat.Act.**)

Arg-**6700**.B5 **Boletín Oficial de la Bolsa de Comercio de Rosario.** 1911- . Bimensual. Rosario, Argentina.

Official bulletin of the Commercial Exchange of Rosario. Contains data on production, markets for cereals and oils, statistics of failures, movement of imports and exports in the port of Rosario, etc. Bimonthly.

Contiene datos sobre producción, situación de los mercados de cereales y oleaginosos, estadística de los quebrantos, movimiento de importación y exportación en el puerto de Rosario, etc.

Arg-**6700**.M5 **Memoria de la Bolsa de Comercio de Buenos Aires.** 1881- . Anual. Buenos Aires. No examinado.

Annual report of the Buenos Aires Exchange.

Arg-**6700**.M7 **Monitor de Sociedades Anónimas.** 1903- . Mensual. Buenos Aires.

Monitor of corporations. Contains information on agriculture, commerce, industry, economics, and finance. Monthly.

Contiene información sobre agricultura, comercio, industrias, economía y finanzas.

Arg-**6720**.A5 **Annual Report.** 1935- . Buenos Aires. Banco Central de la República Argentina.

Contains sections on: Econonic activity and monetary developments: balance of payments, exchange operations and supplies of essential materials; securities; other activities of the bank; the year's work and miscellaneous items connected with the bank; balance sheet and auditor's report.

Contiene capítulos sobre: Actividades económicas y evolución monetaria; balance de pagos, operaciones de cambio y abastecimiento de los materiales esenciales; garantías; otras actividades del Banco; la labor efectuada durante el año y diversos artículos relacionados con el Banco; balance general y memoria del auditor.

Arg-**6720**.A7 **Annual Report of the National Mortgage Bank.** 1922- . Buenos Aires. National Mortgage Bank.

Informe anual del Banco Hipotecario Nacional.

Arg-**6720**.B5 1891-1941 **El Banco de la Nación Argentina en Su Cincuentenario, 1891-1941.** Buenos Aires, 1941. Banco de la Nación Argentina. 473 p.

The Bank of the Argentine Nation on its fiftieth anniversary, 1891-1941. The first third of this volume traces the monetary and banking history of the country to 1891, and the remainder of the book deals with operations of the Bank. Includes provincial maps showing location of branches, statistics on economic developments in Argentina and operations of the Bank, text of much monetary and banking legislation, and significant passages from presidential messages, reports of the Bank and of ministers of finance, and parliamentary debates relating to the Bank.

El primer tercio del volumen describe la historia monetaria y bancaria del país hasta 1891, y en la otra parte trata acerca de las operaciones del Banco. Incluye mapas provinciales señalando lugares donde existen sucursales del Banco, estadísticas sobre el desenvolvimiento económico del país y operaciones del Banco, texto de gran parte de la legislación monetaria y bancaria, disposiciones importantes en mensajes presidenciales, informes del Banco y de los ministros de hacienda y debates del Congreso relacionados con el Banco.

Arg-**6720**.R4 **Revista del Banco de la Nación Argentina.** 1937- . Trimestral. Buenos Aires. Banco de la Nación Argentina.

Journal of the Bank of Argentina. Contains statistical informations on quotations, mortgage loans, farm credit, etc.

Contiene informaciones estadísticas sobre cotizaciones, préstamos hipotecarios, crédito agrario, etc.

Arg-**6720**.R6 **Revista Económica.** 1928-38. Irregular. Buenos Aires. Banco Central de la República Argentina. Oficina de Investigaciones Económicas.

Suspended publication with vol. 7, nos. 9-12, April 1935, resumed publication in 1937 with series II, vol. 1, no. 1, ceased publication with series II, vol. 1, no. 3, 1938. The **Suplemento Estadístico** (now **Boletín Estadístico** (Arg-**6000**. S8)) was published by the bank as a separate publication. Contained statistics on banking, the general economic situation, and agricultural production.

Fué suspendida la publicación con el vol. 7, nos. 9-12, abril 1935; en 1937 se reanudó la publicación con la serie II, vol. 1, no. 1, cesando la misma serie con el vol. 1, no. 3, 1938. El **Suplemento Estadístico** (ahora: **Boletín Estadístico**) fué publicado por el Banco como una publicación separada. Contenía estadísticas bancarias, situación económica general y producción agrícola.

Arg-6900 **Estadísticas del Comercio Exterior en Argentina: Métodos y Procedimientos.** Material básico adquirido para la encuesta que el Instituto Interamericano de Estadística realiza sobre los métodos de estadísticas del comercio exterior en las naciones americanas, incluyendo el estudio de 1946-47.

Foreign trade statistics in Argentina: Methods and procedures. Background material acquired for the study conducted by the Inter American Statistical Institute in foreign trade statistics in the American nations, including the survey of 1946-47. Includes: Filled questionnaire and explanatory notes; nomenclature of countries.

Incluye: Cuestionario llenado y notas explicativas; nomenclatura de países.

*Arg-6900.A5 **Anuario del Comercio Exterior de la República Argentina.** 1915- . Buenos Aires. Ministerio del Interior. Consejo Nacional de Estadística y Censos. Dirección General de Estadística y Censos de la Nación.

Yearbook of foreign trade in Argentina. The most complete compilation in Argentina on the nation's foreign commerce. Import classification is based on Argentine tariff; export classification is unrelated to import.

De las compilaciones sobre el comercio exterior de la Argentina que se efectúan en el mismo país, es ésta la más completa. La clasificación de importaciones se basa en el arancel de aduanas de la Argentina; la clasificación de exportaciones no se relaciona en ningún respecto con la de las importaciones.

1943 **Anuario del Comercio Exterior de la República Argentina Correspondiente a 1943, y Datos Retrospectivos desde 1910 hasta 1943.** Buenos Aires, Peuser, S.A., 1944. Ministerio del Interior. Consejo Nacional de Estadística y Censos. Dirección General de Estadística y Censos de la Nación. 573 p.

Yearbook of foreign trade of Argentina for 1943, with retrospective data from 1910 to 1943. Includes: Retrospective data beginning in 1910; the foreign trade of Argentina in 1943 and 1942; summary by countries of origin and destination; free and duty imports by articles and countries of origin; exports by articles and countries of destination; imports and exports by countries of origin and destination for the principal products.

Incluye: Datos retrospectivos desde 1910; el comercio exterior argentino en 1943 y 1942; resumen por procedencias y destinos; importación sujeta y libre de derechos, por artículos y procedencias; exportación por artículos y destinos; las importaciones y exportaciones por países de procedencia y destino, según principales productos.

Arg-6900.B5 **Boletín de la Dirección General de Aduanas.** 1938- . Mensual. Buenos Aires. Ministerio de Hacienda. Dirección General de Aduanas.

Monthly bulletin of the General Bureau of Customs. Contains statistical information on official exchange quotations and customs collections by receivers.

Contiene informaciones estadísticas sobre cotizaciones oficiales del cambio y recaudaciones aduaneras por receptorías.

Arg-6900.C5 **El Comercio Exterior Argentino.** Boletín. 1887- . Semestral. Buenos Aires. Ministerio del Interior. Consejo Nacional de Estadística y Censos. Dirección General de Estadística y Censos de la Nación.

Foreign trade of Argentina. Published as two bulletins a year with slight variation in content and title. **El Comercio Argentino en . . . y Estadísticas Económicas Retrospectivas** is the annual summary publication which contains data for the year on foreign trade, navigation, public finance, and demography. **El Comercio Exterior Argentino en los Primeros Semestres de . . .** gives figures on foreign trade for the first six months of each year but omits sections on public finance and demography.

Publicado anualmente en dos ediciones, que se distinguen por ciertas variaciones del contenido y del título. **El Comercio Exterior Argentino en . . . y Estadísticas Económicas Retrospectivas** representa la publicación anual que resume los datos sobre comercio exterior, movimiento marítimo, finanzas públicas, y demografía para el año entero. **El Comercio Exterior Argentino en los Primeros Semestres de . . .** presenta cifras sobre el comercio exterior para los primeros seis meses de cada año, sin incluir los datos sobre finanzas públicas y demografía.

Jan.-June '45 **El Comercio Exterior Argentino en los Primeros Semestres de 1945 y 1944.** Boletín, n.° 236. Buenos Aires, Guillermo Kraft, Ltda., 1945. Ministerio del Interior. Consejo Nacional de Estadística y Censos. Dirección General de Estadística y Censos de la Nación. 149 p.

Foreign trade of Argentina in the first half years of 1945 and 1944. Includes: General data on foreign commerce, as well as data for preceding years; customs and port revenues; countries of origin and destination; imports; exports.

Incluye: Cifras generales y datos retrospectivos del comercio exterior; recaudaciones aduaneras y portuarias; procedencias y destinos; importación; exportación.

1944 **El Comercio Exterior Argentino en 1944 y 1943, y Estadísticas Económicas Retrospectivas.** Boletín, n.° 235. Buenos Aires, Guillermo Kraft, Ltda., 1945. Ministerio del Interior. Consejo Nacional de Estadística y Censos. Dirección General de Estadística y Censos de la Nación. 255 p.

Foreign trade of Argentina in 1944 and 1943, and retrospective economic statistics. Includes: Retrospective economic statistics on foreign trade, navigation, economy and finance, and demography; customs and port revenues; general figures on foreign commerce; countries of origin and destination; points of entry and embarkation; imports during 1944 and 1943; exports during 1944 and 1943.

Incluye: Estadísticas económicas retrospectivas sobre comercio exterior, navegación, economía y finanzas, y demografía; recaudaciones aduaneras y portuarias; cifras generales del comercio exterior; procedencias y destinos; puntos de entrada y de embarque; importación en los años 1944 y 1943; exportación en los años 1944 y 1943.

Arg-**6900**.C6 **El Comercio Exterior Argentino.** Informe, Serie C. 1923- . Trimestral. Buenos Aires. Ministerio del Interior. Consejo Nacional de Estadística y Censos. Dirección General de Estadística y Censos de la Nación.

Foreign commerce of Argentina. Summarizes very recent data on foreign commerce. Quarterly.

Síntesis de datos actuales sobre el comercio exterior.

Arg-**6900**.C8 **El Comercio Exterior Argentino en . . . y Su Comparación con el del Mismo Período del Año Anterior: Boletín Mensual.** Cifras provisionales. Fecha de iniciación? Buenos Aires. Ministerio del Interior. Consejo Nacional de Estadística y Censos. Dirección General de Estadística y Censos de la Nación.

Provisional cumulative monthly figures on foreign trade of Argentina with comparison with corresponding period of the previous year.

Arg-**6900**.E5 **Las Exportaciones en los . . . Comparadas con las del Mismo Período del Año Anterior.** Cifras provisionales. Fecha de iniciación? Mensual. Buenos Aires. Ministerio del Interior. Consejo Nacional de Estadística y Censos. Dirección General de Estadística y Censos de la Nación.

Provisional export figures of Argentina, compared with those of the same period of the preceding year.

Arg-**6900**.T5 1939 **Tarifa de Avalúos y Arancel de Importación.** Reimpresa en el año 1939 y coordinada con las modificaciones introducidas por las leyes números 11.281, 11.588 y otras sancionadas hasta la fecha. Edición oficial. Buenos Aires, 1939. Ministerio de Hacienda. Dirección General de Aduanas. 687 p. Customs code of Argentina.

Customs tariff. Reprint of 1939, with modifications introduced by laws number 11.281, 11.588 and others passed up to that date. Official edition.

7000-7999 HEALTH AND CULTURAL STATISTICS—ESTADISTICAS DE SALUBRIDAD Y CULTURA

*Arg-**7110**.B5 **Boletín Sanitario.** 1937- Irregular. Buenos Aires. Ministerio del Interior. Dirección Nacional de Salud Pública.

Sanitary bulletin of the National Bureau of Public Health.

Arg-**7200**.A4 1939 **El Analfabetismo en la República Argentina.** Interpretación de su estadística. Editada por el Instituto Social. Por Ernesto Nelson. Santa Fé, Imp.de la Universidad Nacional del Litoral, 1939. Biblioteca pedagógica, no. 4.

Illiteracy in the Argentine Republic: Statistical interpretation.

Arg-**7200**.A6 1938 **El Analfabetismo y las Funciones del Consejo Nacional de Educación.** Por Angel Acuna. Buenos Aires, Imp.de la Universidad, 1938.

Illiteracy and the functions of the National Board of Education.

Arg-**7200**.D5 1940 **Deserción Escolar y Analfabetismo.** Sus aspectos y cómputos, especialmente en la provincia de Buenos Aires. Por José M. Lunazzi. La Plata, Olivieri & Dominguez, 1940. Universidad Nacional de la Plata. Facultad de Humanidades y Ciencias de la Educación. 139 p.

Truancy and illiteracy, their aspects and computation, especially in the province of Buenos Aires.

Arg-**7200**.M5 **Memoria del Ministerio de Justicia e Instrucción Pública.**

1864- . Anual. Buenos Aires. No examinado.

Report of the Department of Justice and Public Education.

*Arg-7200.R5 **Recopilación Estadística.** 1919- . Anual. Buenos Aires. Ministerio de Justicia e Instrucción Pública. Dirección de Estadística y Personal.

Annual statistical compilation of the Department of Justice and Public Education.

1941-42 **Recopilación Estadística, Años 1941-1942.** Buenos Aires, Tall. Gráf.de la Penitenciaría Nacional de Bs.As., 1943. Ministerio de Justicia e Instrucción Pública. Dirección de Estadística y Personal. 822 p.

Includes: **Official establishments:** National colleges; normal schools; commercial schools; industrial and trade schools; women's professional schools; various institutions; quinquennium 1937-41; **incorporated institutions:** Secondary teaching; normal teaching; commercial teaching; professional teaching; industrial training; fine arts training; secondary, normal and special teaching; primary instruction; statistics; **national universities.**

Incluye: **Establecimientos oficiales:** Colegios nacionales; escuelas normales; escuelas de comercio; escuelas industriales y de artes y oficios; escuelas profesionales de mujeres; institutos varios; quinquenio 1937-41; **institutos incorporados:** Enseñanza secundaria; enseñanza normal; enseñanza comercial; enseñanza profesional; enseñanza industrial; enseñanza de bellas artes; instrucción secundaria, normal y especial; instrucción primaria; estadística; **universidades nacionales.**

8000-8999 GOVERNMENT AND POLITICAL STATISTICS—ESTADISTICAS GUBERNAMENTALES Y POLITICAS

Arg-8800.M4 **Memoria de la Contaduría General de la Nación.** 1854- . Anual. Buenos Aires. Ministerio de Hacienda. Contaduría General de la Nación.

Annual report of the Comptroller. Contains statistics on finance, expenditures, resources, and debt of the nation, with retrospective figures.

Contiene las estadísticas financieras, gastos, recursos y deuda de la nación, con cifras retrospectivas.

*Arg-8800.M6 **Memoria del Departamento de Hacienda.** 1854- . Anual. Buenos Aires. Ministerio de Hacienda.

Annual report of the Department of Public Finance, including reports of the agencies which compose it. Contains statistical data on many aspects of the Argentine economy, particularly on federal income and expenditures and national budget.

Memoria anual del Departamento de Hacienda, incluyendo las memorias de los organismos que lo componen. Contiene datos estadísticos sobre numerosos aspectos de la economía argentina, particularmente sobre renta federal, egresos y presupuesto nacional.

9000-9999 LABOR AND LIVING CONDITIONS —CONDICIONES DE TRABAJO Y DE VIDA

Arg-9000.B5 **Boletín Informativo del Departamento Nacional del Trabajo.** 1907- . Mensual. Buenos Aires. Ministerio del Interior. Departamento Nacional del Trabajo. No examinado.

Former titles—títulos anteriores: **Crónica Mensual del Departamento Nacional del Trabajo; Boletín Mensual del Departamento Nacional del Trabajo; Boletín Informativo.**

Information bulletin of the National Department of Labor. Statistics on labor accidents, employment, living costs, etc.

Estadísticas sobre accidentes de trabajo, ocupación, costo de la vida, etc.

Arg-9000.C5 1943 **Condiciones de Vida de la Familia Obrera.** La regulación colectiva del trabajo. La Plata, Ministerio de Gobierno, 1943. Buenos Aires. Departamento del Trabajo. 192 p. Investigaciones sociales del gobierno de la provincia de Buenos Aires.

Living conditions of the working class family. Includes: Studies of family budgets and cost of living index numbers; Professor Escudero's opinion; prices in the localities studied—their fluctuations in the city of La Plata—index numbers; working-class housing; the conventional collective regulation of working conditions in the Province of Buenos Aires—the statistical problem; salaries agreed upon (a different agreement for each locality cited)—(numerical tables); appendix.

Incluye: Encuestas sobre presupuestos familiares y números índices del costo de la vida; la opinión del profesor Escudero; los precios en las localidades encuestadas—sus oscilaciones en la ciudad de La Plata—números índices; la vivienda obrera; la regulación convencional colectiva de las condiciones de trabajo en la provincia de Buenos Aires—el problema estadístico; los salarios convenidos (a cada localidad citada corresponde un convenio distinto)—(cuadros numéricos); apéndice.

*Arg-9000.I5 **Investigaciones Sociales.** 1936- . Anual. Buenos Aires. Ministerio del Interior. Departamento Nacional del Trabajo. División de Estadística.

Yearbook of social statistics. Includes cost of living indexes, prices of food and other articles of prime necessity, average hourly wage rates by occupation, number of hours worked weekly, indexes of real wages in Buenos Aires, labor organizations, strikes, industrial accidents, wages of port workers in Buenos Aires, and wage indexes of packing house employees.

Incluye índices del costo de la vida, precios de los alimentos y otros artículos de primera necesidad, promedios de remuneración por hora de trabajo en distintas ocupaciones, cantidad de horas de trabajo por semana, índice de salarios efectivos en la ciudad de Buenos Aires, organizaciones del trabajo, huelgas, accidentes del trabajo, salarios en el puerto de Buenos Aires e índices de los salarios de los empleados de los frigoríficos.

Arg-9000.O5 1945 **Ordenamiento Económico-Social.** Buenos Aires, Guillermo Kraft Ltda., 1945. Vicepresidencia de la Nación. Consejo Nacional de Postguerra. 129 p.

Presents statistics on labor, including wages, hours, employment and unemployment, labor organizations, strikes, social security contributions and benefits, wholesale and retail prices and cost of living; the figures shown cover, for the most part, the period up to the end of 1943. The scheme for economic reconstruction and reorganization, planned by the government for the post-war period, is also given.

Expone datos acerca del trabajo, incluyendo salarios, horas, empleo y desempleo, organizaciones de obreros, huelgas, contribuciones al seguro social y beneficios, precios al por menor y al por mayor, y el costo de la vida; los datos se refieren, en lo mayor, al período hasta el fin de 1943. Se presenta, también, el plan de reorganización y reconstrucción económica, proyectado por el gobierno en el período de la post-guerra.

Arg-9000.O7 1941 **Organización Sindical.** Asociaciones obreras y patronales. Buenos Aires, 1941. Ministerio del Interior. Departamento Nacional del Trabajo. División de Estadística. 52 p. Serie B, Estadísticas y censos, no. 9.

Collective occupational organization. Associations of workers and of employers. Results of second national census of occupational associations of workers and employers in Argentina, through June 30, 1941, showing names of associations, location, field of activity, affiliation to larger bodies, and number of members.

Los resultados del segundo censo nacional de asociaciones gremiales de obreros y de patrones en la Argentina, hasta el 30 de junio de 1941, indicando nombres de las asociaciones, situación, campo de actividad, afiliación a organizaciones mas grandes y número de miembros.

Arg-9000.S5 **Serie B, Estadísticas y Censos, del Departamento Nacional del Trabajo.** Fecha de iniciación? Buenos Aires. Ministerio del Interior. Departamento Nacional del Trabajo. División de Estadística.

Series B, statistics and census, of the National Department of Labor.

Arg-9200.A5 1943 **Adaptación de los Salarios a las Fluctuaciones del Costo de la Vida.** Problemas que suscita, normas de aplicación práctica. Buenos Aires, 1943. Ministerio del Interior. Departamento Nacional del Trabajo. División de Estadística. 86 p. Mimeografiado. Unexamined.

Adaptation of salaries to fluctuations in the cost of living—problems which arise, norms for practical application. An account, with supporting statistics, showing how the government of Argentina is attacking the problem of relating wages to the cost of living; giving text of the legislation of 1943 providing for such relation, current wages for various industries, and the method used to compute the cost of living index.

Un informe, con estadísticas que lo apoyan, demostrando cómo el gobierno argentino combate el problema de relacionar los salarios al costo de la vida, transcribiendo la legislación pertinente dictada en 1943, salarios actuales en varias industrias y método usado para computar el índice del costo de la vida.

Arg-9200.D5 1940 **La Desocupación en la Argentina, 1940.** Buenos Aires, 1940. Ministerio del Interior. Departamento Nacional del Trabajo. División de Estadística. 57 p.

Unemployment in Argentina, 1940. An analysis of the causes of the state of employment in Argentina, reasons for unemployment, steps taken by the government and private employers to meet the situation, and recommendations, together with statistics of unemployment reported to the federal government from the provinces up to September 14, 1940.

Un análisis de las causas del estado de la ocupación en la Argentina, motivos de la desocupación, medidas tomadas por el gobierno y por empresarios particulares para combatir la situación, y recomendaciones, conjuntamente con estadísticas sobre desocupación transmitidas por las provincias al gobierno federal hasta el 14 de setiembre de 1940.

Arg-9300.E5 1940 **Estadística de las Huelgas, 1940.** Buenos Aires, 1941. Ministerio del Interior. Departamento Nacional del Trabajo. División de Estadística. 21 p. Serie B, Estadísticas y censos, no. 10.

Statistics of strikes, 1940. Number and character of strikes in Argentina in 1940, activities involved, duration, causes, methods used to settle them, and results, accompanied by brief explanations of the tables.

Indica al número y carácter de las huelgas en la Argentina durante el año 1940, las actividades afectadas, duración, causas, métodos usados para concertar las partes y resultados, acompañado por breves explicaciones de los cuadros.

Arg-9500.C5 **Costo de la Vida.** Fecha de iniciación? Mensual. Buenos Aires. Vice Presidencia de la Nación. Consejo Nacional de Postguerra. Dirección de Estadística Social.

Monthly 2-page release on cost of living. Tables on: Over-all index number, and particular index numbers of the factors used in arriving at the over-all figure; fluctuations of the cost of living and of foodstuffs; articles of prime necessity.

Publicación mensual de dos páginas, sobre el costo de la vida. Contiene cuadros de: Número índice general y números índices especiales de los elementos que lo integran; fluctuaciones del costo de la vida y alimentación; artículos de primera necesidad.

Arg-9500.I5 1945 **El Indice del Costo de la Vida.** Tablas básicas de salarios. Salario anual. Buenos Aires, 1945. Dirección de Estadística Social. 33 p. Mimeografiado.

Cost of living index. Includes: Price variations of articles and services of prime necessity; basic salary tables; adjustment to fluctuations in cost of living; professional classes; annual salaries.

Incluye: Variación de precios de artículos y servicios de primera necesidad; tablas básicas de salarios; adaptación a las fluctuaciones del costo de la vida; categorías profesionales; salario anual.

*Arg-9700.N5 1945 **Nivel de Vida de la Familia Obrera.** Evolución durante la segunda guerra mundial, 1939-1945. Buenos Aires, 1945. Dirección de Estadística Social. 168 p. Mimeografiado.

Standard of living of working-class families—developments during the second world war, 1939-1945. Includes: Articles of prime necessity; price details for articles and services of prime necessity; normal consumption; standard of living, and changes in the cost of living for the city of Buenos Aires, for the provinces, and for the territories. Price data applying to the city of Buenos Aires are given for the period from 1914 to August 1945, and for the rest of the country from 1939 to 1945.

Incluye: Artículos de primera necesidad; detalles de precios de los artículos y servicios de primera necesidad; consumo normal; el nivel de vida, y cambios en el costo de la vida en Buenos Aires, y en las provincias y territorios. Los datos sobre precios en Buenos Aires abarcan el período de 1914 hasta agosto de 1945, mientras que los datos de provincias y territorios se refieren al período de 1939 a 1945.

Arg-9800.A5 **La Alimentación de la Familia en Buenos Aires.** Boletín del Instituto Nacional de la Nutrición. 1937- . Bimestral. Buenos Aires. Ministerio del Interior. Dirección Nacional de Salud Pública. Instituto Nacional de la Nutrición.

Food and nutrition of families in Buenos Aires. Bimonthly.

Arg-9800.C5 **El Costo de la Alimentación.** Boletín del Instituto Nacional de la Nutrición. 1944- . Trimestral. Buenos Aires. Ministerio del Interior. Dirección Nacional de Salud Pública. Instituto Nacional de la Nutrición.

Indexes of the cost of food. Quarterly.

Arg-9800.M5 **Memoria Anual del Instituto Nacional de la Nutrición.** 1935- . Buenos Aires. Ministerio del Interior. Dirección Nacional de Salud Pública. Instituto Nacional de la Nutrición.

Before 1937 issued as—antes de 1937 publicado bajo el título: **Memoria del Instituto Municipal de la Nutrición.**

Annual report of the National Institute of Nutrition. Contains statistics on the work of the Institute in medicine, nutrition, education, research, etc.

Contiene estadísticas sobre la labor del Instituto, en medicina, nutrición, instrucción, investigaciones, etc.

Arg-9950.M5 **Memoria de la Caja Nacional de Jubilaciones y Pensiones Civiles.** 1905- . Anual. Buenos Aires.

Report of the National Fund for Retirement and Civil Pensions. Includes statistical information on railway workers retirements in Argentina.

Contiene informaciones estadísticas sobre las jubilaciones y jubilados ferroviarios del país.

Arg-9952.C5 1941 **Conceptos y Principios sobre Jubilaciones.** Por José Barral Souto. Buenos Aires, Imp.de la Universidad, 1941. Universidad de Buenos Aires. Facultad de Ciencias Económicas. Instituto de Biometría. Problemas del ambiente, no. 4. No examinado.

Concepts and principles of retirement funds.

BOLIVIA

1000-1999 GENERAL

Bol-**1200.**G5 1945 **A Guide to the Official Publications of the Other American Republics: II, Bolivia.** James B. Childs, general editor. Washington, D.C., U.S. Govt. Print. Off., 1945. The Library of Congress. 66 p. Latin American series, no. 10.

Una guía de las publicaciones oficiales de las otras repúblicas americanas: II, Bolivia.

*Bol-**1420.**R5 **Revista Mensual.** 1945- . La Paz. Ministerio de Hacienda y Estadística. Dirección General de Estadística.

Monthly review of the General Bureau of Statistics. Contains statistical series on finance, industrial production, petroleum, livestock, export, demography, transport, and communications.

Contiene series estadísticas sobre finanzas, producción industrial, petróleo, ganado, exportaciones, demografía, transportes y comunicaciones.

2000-2999 CENSUS—CENSOS

Bol-**2100.**C4 1900 **Censo de 1900: Métodos y Procedimientos.** Materiales empleados en el levantamiento del censo de población, del 10 de septiembre. Adquirido con motivo de la encuesta que el IASI realiza, sobre Métodos y Procedimientos de los Censos, 1944. De la Dirección General de Estadística.

Census of 1900: Methods and procedures. Materials used in taking population census, on September 10. Acquired in connection with IASI Survey of Census Methods and Procedures, 1944. Contents: Correspondence; laws and decrees.

Contiene: Correspondencia; leyes y decretos.

Bol-**2100.**P5 1942 **Plan para el Levantamiento del Censo de Población de la Ciudad de La Paz, 1942.** La Paz, agosto 1942. Ministerio de Hacienda. Dirección General de Estadística. 6 p. Copia máquina.

Plan for taking population census of La Paz, 1942.

*Bol-**2200.**C4 1942 **Censo Demográfico de la Ciudad de La Paz, 15 de Octubre de 1942.** La Paz, Imp.Universo, 1943. H. Alcaldia Municipal de La Paz. Dirección General de Estadística. 139 p.

Census of population of the city of La Paz taken October 15, 1942. Includes data on density, age, civil status, place of birth, race, language, occupation, profession, literacy, and religion.

Incluye datos sobre densidad, edad, estado civil, lugar de nacimiento, raza, idioma, ocupación, profesión, alfabetización, y religión.

Bol-**2200.**C5 1900 **Censo General de la Población de la República de Bolivia según el Empadronamiento de lo de Septiembre de 1900.** La Paz, J. M. Gamarra, 1902-04. Oficina Nacional de Inmigración, Estadística y Propaganda Geográfica. 2 vol.: Tomo I, **Resultados Generales;** tomo II, **Resultados Definitivos.**

General census of population of Bolivia, September 1, 1900. "(Tomo I) A general introduction gives the total and provincial population figures from the census of 1831, 1835, 1845, 1854, and 1882, and describes the organization and procedures for the census of 1900. Results of the 1900 census are then presented, giving for each province the population by urban-rural residence and by sex, for localities . . . (Tomo II) The first section, "Reseña geográfica y estadística de Bolivia," is followed by data on total population, age and sex, race, elementary education, marital status, religion, legal domicile, nationality, physical impairments, and occupations." (Source: GCVSA).

"(Tomo I) Una introducción general de las cifras de la población total y por provincias de los censos de 1831, 1835, 1845, 1854 y 1882, y describe la organización y procedimiento del censo de 1900. Los resultados del censo de 1900 son presentados a continuación, dando la población de cada provincia por localidades, con clasificación de residencia urbana y rural y de sexo. (Tomo II) A la primera parte, que es "Reseña geográfica y estadística de Bolivia," siguen datos sobre la población total, edad y sexo, raza, educación elemental, estado civil, religión, domicilio legal, nacionalidad, incapacidades físicas y ocupaciones." (Traducción del comentario.)

*Bol-**2210.**C5 1944-45, Tarija **Censo Experimental Agropecuario, Departamento de Tarija, 1944-1945.** La Paz, 1946. Ministerio de Agricultura, Ganadería y Colonización. Dirección General de Economía Rural. Sección Estadística. 211 p. Mimeografiado.

Experimental census of agriculture and livestock of the Department of Tarija, 1944-45. Includes: Rural population and elementary education, agricultural production, production of fruit, amounts of livestock and poultry, and production of agricultural and livestock products.

Incluye: Instrucción primaria y población rural, producción agrícola, producción frutícola, existencia ganadera, existencia avícola, y producción de derivados de agricultura y ganadería.

*Bol-**2210**.C5 1942, Chuquisaca **Censo Experimental Agropecuario del Departamento de Chuquisaca, 1942.** La Paz, Editorial Argote, 1944. Ministerio de Agricultura, Ganadería y Colonización. Dirección General de Economía Rural y Estadística Agropecuaria. 298 p.

Experimental census of agriculture in the department of Chuquisaca, 1942.

3000-3999 STATISTICAL SCIENCE—CIENCA ESTADISTICA

Bol-**3030**.R5 1946 **Reorganización del Servicio de Estadística en Bolivia.** La Paz?, 1946. Dirección General de Estadística. 3 p. Copia máquina.

Reorganization of the statistical service in Bolivia. 1946 organization chart of the General Bureau of Statistics is attached.

El vigente esquema de organización de la Dirección General de Estadística, 1946, adjunto.

Bol-**3300** 1944-46 **Enseñanza Estadística en Bolivia: Métodos y Materiales.** Material básico adquirido para la Encuesta del Instituto Interamericano de Estadística sobre los Métodos y Materiales en la Enseñanza Estadística en las Naciones Americanas, 1944-46. In IASI on microfilm (Int-**3300**.S5).

Statistical training in Bolivia: Methods and materials. Background material acquired for the IASI Survey on Statistical Training Methods and Materials in the American Nations, 1944-46. Includes—incluye: From Jorge Pando Gutierrez, a letter and filled questionnaires F-1, "Summary of Statistical Personnel," F-2, "Opinion of Minimum Requirements for Various Types of Statistical Personnel," and F-4, "Possible Titles for Teaching Monographs."

5000-5999 DEMOGRAPHY—DEMOGRAFIA

Bol-**5000**.B5 1945 **Bolivia: Summary of Biostatistics.** Maps and charts, population, natality and mortality statistics. Washington, D.C., February 1945. Prepared by U.S. Department of Commerce, Bureau of the Census, in cooperation with Office of the Coordinator of Inter-American Affairs. 69 p.

Resumen bioestadístico de Bolivia. Incluye mapas y cuadros y datos sobre población, natalidad y mortalidad.

*Bol-**5000**.D5 **Demografía.** 1937- . Anual. La Paz. Dirección General de Estadística.

Yearbook on demography.

1940 **Demografía, 1940.** La Paz, Edit. Argote, 1942? Ministerio de Hacienda. Dirección General de Estadística. 191 p.

Includes births, deaths, infant mortality, marriages, immigration and emigration.

Incluye natalidad, mortalidad general, mortalidad infantil, nupcialidad, inmigración, emigración.

Bol-**5200** **Estadística Vital y Registro en Bolivia: Métodos y Procedimientos.** Material básido adquirido para la encuesta que el Instituto Interamericano de Estadística realiza sobre los métodos de registro y estadística vital en las naciones americanas, incluyendo el estudio de 1945.

Vital statistics and registration in Bolivia: Methods and procedures. Background material acquired for the study conducted by the Inter American Statistical Institute in vital statistics and registration methods in the American nations, including the survey of 1945. Includes: Answer to questionnaire; laws and decrees; forms; organization charts.

Incluye: Respuesta al cuestionario; leyes y decretos; formularios; esquemas de organización.

Bol-**5250**.R5 1945 **Registro Civil de Bolivia.** Leyes y decretos relativos al matrimonio civil y divorcio absoluto. Disposiciones últimas relativas al registro civil. Edición oficial. La Paz, 1944. Ministerio de Justicia. 142 p.

Civil register of Bolivia.

6000-6999 ECONOMICS—ECONOMIA

Bol-**6000**.E5 1945 **Economic Controls and Commercial Policy in Bolivia.** One of a series of reports on economic controls and commercial policy in the American republics. Washington, D.C., 1945. United States Tariff Commission. 22 p.

Informe tratando de Bolivia, publicado en la serie de informes de la Comisión de Aduanas de los Estados Unidos, sobre controles económicos y política comercial de las repúblicas americanas.

*Bol-**6000**.M5 **Minería, Transportes y Comunicaciones.** Fecha de iniciación? Anual. La Paz. Dirección General de Estadística.

Title varies—título varía: 1940-1939, **Transportes, Balances Mineros.**

Yearbook of mining, transportation, and communication statistics.

1940-42 **Minería 1940-1941-1942; Transportes y Comunicaciones 1941-1942.** La Paz, Lito.e Imp.Unidas, 1946. Ministerio de Hacienda. Dirección General de Estadística. 129 p.

Mining, 1940-1941-1942; transportation and communications, 1941-1942. Statistics on mining, petroleum, transportation, and communication.

Estadísticas sobre minería, petróleo, transportes y comunicaciones.

Bol-**6000**.R5 1944 **Reports Presented to the Conference of Commissions of Inter-American Development (Held in New York, May 9-18, 1944) by the Bolivian Commission of Inter-American Development.** On the economic resources of Bolivia. (Translation from the Spanish original.) Washington, D.C., 1945? Inter-American Development Commission. 81 p.

Cover title—título de la cubierta: **Bolivia: Reports Presented to the Conference of Commissions of Inter-American Development.**

Informes presentados a la Conferencia de Comisiones de Fomento Interamericano (realizada en N. York, del 9 al 18 de mayo de 1944), por la Comisión Boliviana de Fomento Interamericano, sobre los recursos económicos de Bolivia.

Bol-**6020**.E5 1944 **Estadística de Consumo de Artículos de Procedencia Nacional en la Ciudad de Tarija, 1944.** Tarija, 1945. Ministerio de Agricultura, Ganadería y Colonización. Dirección General de Economía Rural. Sección Estudios Económicos. 165 p. Mimeografiado.

Statistics of consumption of articles of national origin in the city of Tarija, 1944.

*Bol-**6230**.E5 **Estadística Agropecuaria.** 1938- . Anual. La Paz. Dirección General de Estadística.

Yearbook of statistics of livestock.

1939-41 **Estadística Agropecuaria, 1939-1941.** La Paz, Edit. "Fenix," 1942. Ministerio de Hacienda. Dirección General de Estadística. 117 p.

Includes: Summary of the number of estates and communal holdings; slaughter of livestock, by department and province; consumption of meat; arable, pasture and rubber land registered from 1908 through 1941 in the national register of public lands; meteorological data.

Incluye: Resumen del número de propiedades y comunidades; derribe de ganado, por departamento y provincia; consumo de carne; tierras de cultivo, pastoreo y gomeras, inscritas en el Registro Nacional de Tierras Baldías, desde el año 1908 al 31 de diciembre de 1941; datos meteorológicos.

Bol-**6300**.B4 **Boletín de la Dirección General de Minas y Petróleo.** Fecha de iniciación? Semestral. La Paz. Ministerio de Economía Nacional. Dirección General de Minas y Petróleo.

Bulletin of the General Bureau of Mines and Petroleum. Contains statistics of mineral production and export.

Contiene estadísticas de la producción y exportación de minerales.

Bol-**6300**.C5 **Carta Informativa.** 1946?- Quincenal. La Paz. Banco Minero de Bolivia. Departamento de Estudios Económicos y Estadística. Mimeografiado.

Semimonthly bulletin of the Mining Bank of Bolivia. Information on mineral production,consumption, and exports. Supersedes the **Boletín del Banco Minero de Bolivia** which was suspended during the war.

Información sobre producción, consumo y exportación de minerales. Substituye al **Boletín del Banco Minero de Bolivia,** cuya publicación fué suspendida durante la guerra.

Bol-**6300**.M5 **Memoria Anual.** 1937- . La Paz. Banco Minero de Bolivia.

Annual report of the Mining Bank of Bolivia.

Bol-**6300**.T5 1941 **Tasas e Impuestos sobre la Industria Minera en Bolivia.** La Paz, Imp.Artística, Scs.de A.H.Ofero, 1941. Banco Minero de Bolivia. Sección Estadística y Estudios Económicos. 188 p. No examinado.

Taxes and duties on the mineral industry in Bolivia. Contains: Introduction; texts of the tax laws; financial results of the taxes and duties on the mining industry; statistics of mining production (1921-1940) and graphs.

Contiene: Introducción; textos de las leyes impositivas; resultado financiero de las tasas e impuestos sobre la industria minera; estadística de la producción minera (1921-1940) y gráficos.

Bol-**6400**. **Estadísticas Industriales en Bolivia: Métodos y Procedimientos.** Material básico adquirido para el estudio que el Instituto Interamericano de Estadística realiza sobre los métodos de estadísticas industriales en las naciones americanas, incluyendo la encuesta de 1945-46.

Industrial statistics in Bolivia: Methods and procedures. Background material acquired for the study conducted by the Inter American Statistical Institute on methods of industrial statistics in the American nations, including the survey of 1945-46. Consists of: Instructions; forms; punch cards.

Consiste de: Instrucciones; formularios; tarjetas de perforación.

*Bol-**6400**.I5 **Industria.** 1936- . Anual. La Paz. Dirección General de Estadística.

Title varies—título varía: 1936, **Extracto Estadístico de Bolivia—Sección Industria.**

Yearbook of industrial statistics.

1940 **Industria Fabril y Manufacturera, 1940.** La Paz, Edit. "Fenix," 1941? Ministerio de Hacienda. Dirección General de Estadística. 300 p. (aprox.) Páginas numeradas por secciones.

Manufacturing industry, 1940. Data on production, sales, capital, wages and salaries paid, personnel employed, and raw materials used in industry in Bolivia.

Datos sobre producción, ventas, capitales, sueldos y salarios pagados, personal ocupado y materias primas empleadas en la industria en Bolivia.

1938-39 **Industria Manufacturera, Fabril y Extractiva, 1938-1939.** La Paz, 1941? Ministerio de Hacienda. Dirección General de Estadística. 355 p.

Manufacturing and mining industries, 1938-39. Data on production, sales, personnel employed and salaries and wages paid, and capital and raw material used in industry in Bolivia, as well as chapters on mining balances and production of petroleum.

Datos sobre producción, ventas, personal ocupado, sueldos y salarios pagados, capitales y materias primas empleadas en la industria en Bolivia, así como capítulos sobre los balances mineros y la producción del petróleo.

Bol-**6400**.M5 1945 **Mining and Manufacturing Industries in Bolivia.** One of a series of reports on mining and manufacturing industries in the American republics. Washington, D.C., 1945. United States Tariff Commission. 28 p.

Informe tratando de Bolivia, publicado en la serie de informes de la Comisión de Aduanas de los Estados Unidos, sobre las industrias mineras y manufactureras en las repúblicas americanas.

Bol-**6400**.S5 1941 **Síntesis de Estadística Industrial, 1941.** Ministerio de Hacienda y Estadística. Dirección General de Estadística. 20 p. Mimeografiado.

Synthesis of industrial statistics, 1941.

*Bol-**6720**.B5 **Boletín del Banco Central de Bolivia.** A cargo del Departamento de Estadística y Estudios Económicos. 1929- . Trimestral. La Paz.

Quarterly bulletin of Central Bank of Bolivia, containing current statistical series on banking and finance for Bolivia—monetary circulation, banking reserves, exchange rates, etc.—and also series of more general character such as cost of living and prices.

Contiene series estadísticas sobre bancas y finanzas de Bolivia—circulación de moneda, reservas de bancos, cambio internacional, etc.,—y también series de carácter más general, como costo de la vida y precios.

Bol-**6720**.E4 **Estadística Bancaria.** Fecha de iniciación? Mensual. La Paz. Superintendencia de Bancos. Sección Estadística.

Banking statistics. Monthly.

*Bol-**6720**.E6 **Finanzas.** Fecha de iniciación? Anual. La Paz. Dirección General de Estadística.

Title varies—título varía: 1936, **Extracto Estadístico de Bolivia—Sección Finanzas;** 1939-40; **Finanzas;** 1941, **Estadística Financiera;** 1942-43— , **Finanzas.**

Yearbook of financial statistics.

1942-43 **Finanzas, 1942-43.** La Paz, Edit. "Fenix," 1944? Ministerio de Hacienda. Dirección General de Estadística. 175 p. (aprox.) Páginas numeradas por secciones.

Includes: Banking and credit institutions; sale of foreign money; securities; international exchange; prices and cost of living; public debt, social security; budget; distribution of salaries of government personnel.

Incluye: Economía bancaria; entidades de crédito; venta de divisas; valores; cambio internacional; precios y costo de la vida; deuda pública; protección social; presupuestos; distribución de haberes del personal de la administración pública.

*Bol-**6720**.M5 **Memoria Anual.** 1929- . La Paz. Banco Central de Bolivia.

Annual report of the Central Bank of Bolivia.

1945 **17a. Memoria Anual Correspondiente a la Gestión del Año 1945.** Presentada al Señor Ministro de Hacienda y Estadística. La Paz, Artística Otero y Calderón, 1946? Banco Central de Bolivia. 96 y LX.

Report on activities of Central Bank and of other banks of the country, and statistical appendix which includes data on trade balance of Bolivia from 1929 to 1944, foreign trade by categories, exports and imports by countries of origin and destination and by principal products, and public debt.

Memoria sobre las actividades del Banco Central y de los otros bancos del país, y un apéndice estadístico que incluye datos sobre el balance comercial de Bolivia 1929-1944, comercio exterior por categorías, exportaciones e importaciones por países de origen y destino y por principales productos, y la deuda pública.

Bol-**6900**.A5 1940 **Arancel Aduanero de Importaciones y Vocabulario para su Aplicación.** Edición oficial. La Paz, Escuela Tip. Salesiana, 1940. 275 p.

Customs tariff of Bolivia, based on Brussels nomenclature.

Basado en la nomenclatura de Bruselas.

*Bol-**6900**.C5 **Comercio Exterior.** 1910- Anual. La Paz. Dirección General de Estadística.

Yearbook of foreign trade statistics.

1943 **Comercio Exterior, Anuario Correspondiente al Año 1943.** Importación, exportación. La Paz, Edit. "Fenix," 1944. Ministerio de Hacienda. Dirección General de Estadística. 450 p. (aprox.) Páginas numeradas por secciones.

Yearbook of foreign trade statistics, 1943. Includes: **Imports:** General tables; general summary data arranged according to the Brussels Nomenclature; details by country, department, customs station and route; summary by country of origin and by department of destination according to the Brussels Nomenclature; **exports:** General tables; summary data according to the Brussels Nomenclature; principal exports by country, routes, customs stations, departments; summary by country of destination; monthly exports of minerals and principal products.

Incluye: **Importación:** Cuadros demostrativos; resumen general conforme a la Nomenclatura de Bruselas; detalle por rubros —países, departamentos, aduanas y vías; resumen por países de origen y por departamentos de destino, conforme a la Nomenclatura de Bruselas; **exportación:** Cuadros demostrativos; resumen de acuerdo a la Nomenclatura de Bruselas; principales exportaciones por rubros—países, vías, aduanas y departamentos; resumen por países de destino; exportación mensual de minerales y principales productos.

*Bol-**6910**.B5 1938-40 **Balanza Internacional de Pagos de Bolivia.** (Calculada en miles de $ o/a.) 1938-1940. La Paz, Edit. "Fenix," 1942. Ministerio de Hacienda y Estadística. Dirección General de Estadística. 25 p.

International balance of payments of Bolivia, 1938-40.

7000-7999 HEALTH AND CULTURAL STATISTICS—ESTADISTICAS DE SALUBRIDAD Y CULTURA

*Bol-**7110**.S5 1942 **Síntesis de Informes y Estadísticas de Salubridad, Año 1942.** La Paz. Ministerio de Trabajo, Salubridad y Previsión Social. Despacho de Salubridad. Sección Biodemografía. 100 p. (aprox.). Páginas numeradas por secciones.

Summary of reports and statistics of health for 1942.

Bol-**7200**.E5 1943 **El Estado Actual de la Educación en Bolivia.** Informe a la Misión Magruder. Por Vicente Donoso Torres. La Paz, Edit. del Estado, 1943. Consejo Nacional de Educación. 15 p. y 9 gráficos.

A report on educational conditions in Bolivia, in particular of the working class, with recommendations for educational reforms, using as a basis the facts brought out in the report of the Magruder Mission invited by the Bolivian Government to study living and working conditions of laborers, especially of the miners, in Bolivia.

Informe sobre las condiciones de educación en Bolivia, aplicado especialmente a la clase obrera, con recomendaciones para el mejoramiento del sistema de educación, usando como base los datos citados en el informe de la Misión Magruder, que vino a Bolivia invitada por el gobierno del país para estudiar las condiciones de vida y de trabajo del obrero y especialmente del minero.

9000-9999 LABOR AND LIVING CONDITIONS—CONDICIONES DE TRABAJO Y DE VIDA

Bol-**9000**.B5 **Boletín del Despacho de Previsión Social.** Fecha de iniciación? Frecuencia? La Paz. Ministerio del Trabajo, Salubridad y Previsión Social. No examinado.

Bulletin of the Office of Social Security.

June '40 "Costo de Vida." Junio 1940, p. 120-125. No examinado.

A critical analysis of the method used in computing the cost of living index by the General Bureau of Statistics, with a tentative index based on a questionnaire prepared by the Department of Labor, Health, and Social Security, with retail prices and indexes for September 1939 and March 1940.

Un análisis-crítica de los métodos usados

en el cómputo del índice del costo de la vida por la Dirección General de Estadística, con un ensayo de índice basado en el cuestionario preparado por el Ministerio del Trabajo, Salubridad y Previsión Social, con precios del comercio al por menor e índices correspondientes a septiembre 1939 y marzo 1940.

Bol-9000.M5 1943 **Memorándum: Elementos Generales sobre las Condiciones de Trabajo, de Legislación y de Otros Problemas Sociales de la República de Bolivia.** Por Victor Andrade. La Paz, Editorial Trabajo, 1943. Páginas numeradas por secciones. No examinado.

Report on labor and social conditions (employment, wages, cost of living, housing, etc.). Detailed tables present wage statistics for various enterprises, and social security operations including workmen's compensation, through 1942.

Empleo, salarios, costo de la vida, vivienda, etc. Cuadros detallados presentan estadísticas sobre salarios de varias empresas y operaciones del seguro social, incluyendo, indemnizaciones de obreros; hasta el año 1942 inclusive.

Bol-9000.P5 1941 **El Problema Social en Bolivia—Condiciones de Vida y de Trabajo.** Por Roberto Capriles Rico y Gastón Arduz Eguía. La Paz, Editorial Fénix, 1941. 172 p.

The social problem in Bolivia—labor and living conditions. Summary of a study of conditions of miners, agricultural laborers, and city workers in Bolivia, made in 1940 by two officials of the Bolivian government under the direction of the International Labour Office. Some statistics of employment and wages in specified regions in the years 1937 and 1939 are included.

Resumen de un estudio sobre las condiciones de vida y de trabajo de los mineros, agricultores y obreros de la ciudad, hecho en 1940 por dos funcionarios del gobierno boliviano bajo la dirección de la Oficina Internacional del Trabajo. Se incluyen algunas estadísticas sobre ocupación y salarios, de regiones determinadas, correspondientes a los años 1937 y 1939.

Bol-9950.M5 **Memoria de la Caja de Seguro y Ahorro Obrero.** Fecha de iniciación? Anual. La Paz. Caja de Seguro y Ahorro Obrero.

Annual report of the Workers' Insurance and Saving Fund.

Bol-9950.P5 **Protección Social.** Revista mensual de la Caja de Seguro y Ahorro Obrero. 1938- . La Paz. Caja de Seguro y Ahorro Obrero.

Social security. Monthly journal of the Workers' Insurance and Saving Fund.

1000-1999 GENERAL

*Bras-**1410**.A5 **Anuário Estatístico do Brasil.** 1936- . Rio de Janeiro. Instituto Brasileiro de Geografia e Estatística. Conselho Nacional de Estatística.

One issue covering 1908-12 published; not published 1913-35.

Key statistical yearbook of the Brazilian government. Each state also issues an annual abstract.

Principal anuário estatístico do govêrno brasileiro. Cada estado também pùblica uma sinópse anual.

1941-45 **Anuário Estatístico do Brasil, Ano VI-1941/1945.** Rio de Janeiro, Serviço Gráf. do IBGE, 1946. Instituto Brasileiro de Geografia e Estatística. Conselho Nacional de Estatística. 536 p.

Statistical yearbook of Brasil. Includes: **Physical situation:** Characteristics of territory; climatology; territorial division; **demographic situation:** State of population; movement of population; **economic situation:** Extractive production; agricultural production; industrial production; means of transportation; communication; real estate; industrial property; metal and fiduciary coin (currency); securities; financial activity; pawnshops and loan houses; commerce; salaries; consumption; bankruptcy and disputed titles; disasters and accidents; **social situation:** Urban improvements; medical-sanitary aid; aid for disabled; social welfare; labor; **cultural situation:** Education; other aspects of intellectual and artistic culture; religion; delinquency and suicides; **political and administrative situation:** Public administration; public finance; public security; repression; justice; political representation.

Inclue: **Situação física:** Caracterização do território; climatologia; divisão territorial; situação demográfica: Estado da população; movimento da população **situação econômica;** Produção extrativa; produção agrícola; produção industrial; meios de transporte; vias de comunicação; propriedade imobiliária; propriedade industrial; moeda metálica e fiduciária; títulos mobiliários; movimento bancário; casas de penhôres e montes de socorro; comércio; salários; consumo; falências e títulos protestados; sinistros e acidentes; **situação social:** Melhoramentos urbanos; assistência médico-sanitaria; assistência a desvalidos; previdência e assistência social; trabalho; **situação cultural:** Educação; outros aspectos da cultura intelectual e artística; cultos; delinquência e suicídios; **situação administrativa e política:** Administração pública; finanças públicas; segurança pública; repressão; justiça; representação política.

Bras-**1410**.B3 **Brasil.** Recursos, possibilidades. Data de início? Anual. Rio de Janeiro. Ministério das Relações Exteriores. Divisão Econômica e Comercial.

Published also in English—publicado também em inglês (Bras-**1410**.B5).

O sub-título varia. Abrange os campos da demografia, agricultura, indústria, comércio, finanças, e outros. Inclue aspectos gerais da vida social e cultural.

Bras-**1410**.B5 **Brazil.** An economic, social, and geographic survey. Starting date? Annual. Rio de Janeiro. Ministry of Foreign Affairs.

Published also in Portuguese—publicado também em português (Bras-**1410**.B3).

Subtitle varies. Subject fields include demography, agriculture, industry, trade, finance, and others. Both social and cultural aspects are presented from a general point of view.

Bras-**1420**.B5 **Boletim do Ministério do Trabalho, Indústria e Comércio.** 1934- . Mensal. Rio de Janeiro. Serviço de Estatística da Previdência e Trabalho. Ministério do Trabalho, Indústria e Comércio.

Bulletin of the Department of Labor, Industry and Commerce. Deals with economic problems affecting the worker, including index numbers on cost of living, price of foodstuffs, and labor legislation and statistics.

Trata dos problemas econômicos que se relacionam com as classes trabalhadoras, incluindo números índices do custo de vida, preços de gêneros alimentícios, e legislação e estatísticas do trabalho.

Sept. '44 "Anexo a Resolução No. 243: Nomenclatura Brasileira para a Classificação de Indústrias." Ano XI, no. 121, setembro 1944, p. 156-164.

Brazilian nomenclature for classification of industries.

June '44, Costa Miranda "A Estimativa de Renda Geral do Brasil." Por Osvaldo Gomes da

Costa Miranda. Ano X, no. 118, junho 1944, p. 209-215.

Estimate of the general income of Brasil.

Aug. '43, Marinho de Andrade "Censo Sindical Carioca em 1941." Por José Marinho de Andrade. Ano IX, no. 108, agôsto 1943, p. 99-115.

Results of the census of trade unions and syndicates in Rio de Janeiro in 1941.

Resultados do censo das organizações trabalhistas.

Dec. '42 "Censo Sindical do Brasil em 1939." Ano IX, no. 100, dezembro 1942, p. 85-107.

The census of trade unions and syndicates in 1939.

*Bras-**1420**.B7 **Boletim Estatístico do Instituto Brasileiro de Geografia e Estatística.** 1943- . Trimestral. Rio de Janeiro. Instituto Brasileiro de Geografia e Estatística. Conselho Nacional de Estatística.

Statistical bulletin of the Brazilian Institute of Geography and Statistics. Devoted to monthly statistics almost exlusively on economic surveys.

Dedicado quase exlusivamente a estatísticas mensais referentes a inquéritos econômicos.

Bras-**1900**.B4 1943 **Brazil in the Making.** By José Jobim. New York, The Macmillan Co., 1943. 318 p. Unexamined.

Includes data on: Land, climate and man; agriculture and animal production; mining industry; principal manufacturing industries; Brazil at war.

O Brasil progressivo. Inclue dados sôbre: A terra, o clima e o homem; agricultura e produção animal; a indústria mineira; as principais indústrias manufatureiras; o Brasil na guerra.

2000-2999 CENSUS—CENSOS

Bras-**2100**.R5 1940.1 **Recenseamento de 1940: Métodos e Processos.** Materials used in taking population census, on September 1. Acquired in connection with IASI Survey of Census Methods and Procedures, 1944. From Instituto Brasileiro de Geografia e Estatística, Serviço Nacional de Recenseamento.

Census of 1940: Methods and procedures. Contents: Laws, decrees, or other authorization; schedules; other forms; instructions; punch cards; codes; exposition and reports.

Documentários obtidos em relação a inquéritos a cargo do Instituto Inter-Americano de Estatistíca. Inclue: Leis, decretos e regulamentação diversa; questionários e outros formulários; instruções; cartões de perfuração e códigos; exposições e relatórios.

See also—ver também: **Coleção de Decretos-Leis sôbre o Recenseamento Geral da República em 1940; Plano de Apuração do Censo Demográfico.**

Bras-**2100**.R5 1940.3 **Coleção de Decretos - Leis sôbre o Recenseamento Geral da República em 1940.** Rio de Janeiro, Serviço Gráf. do IBGE, 1939. 47 p.

Collection of laws on the general census of Brazil in 1940.

Bras-**2100**.R5 1940.6 **Plano de Apuração do Censo Demográfico.** Rio de Janeiro. Instituto Brasileiro de Geografia e Estatística. Serviço Nacional de Recenseamento. 75 p. (aprox.)

Plan of tabulation for the demographic census. Contents: General classifications; absent residents and temporarily present persons; nativity and nationality; language spoken at home; education; activities; occupations; unions; social security; deaf, dumb, and blind; marital status; real estate ownership; private insurance; fertility; dwellings; samples of census schedules.

Conteúdo: Clasificações gerais; ausentes residentes e presentes temporários; naturalidade e nacionalidade; língua falada no lar; instrução; atividades; ocupações; sindicatos; previdência social; surdos-mudos e cegos; cônjuges e consortes; propriedade imobiliária; seguros particulares; fecundidades; domicílios; tipos de boletins de recenseamento.

Bras-**2100**.R5 1940.7 **A Glimpse into the Coming Fifth Census of Brazil (September 1st-1940).** Presented at the Eighth American Scientific Congress, Washington, D.C., May 10-18, 1940. Rio de Janeiro, Print.Off. of the BIGS, 1940. Brazilian Institute of Geography and Statistics. National Census Commission. 17 p.

Includes a resumé of all censuses to date in Brazil. An envelope of schedules and instructions accompanies the pamphlet. Contents of envelope: General schedule (family schedule) for the population census; individual schedule for the population census; list for collective domiciles; directions for filling the family schedule; note-book of the enumerator for the population census; general schedule for the agricultural census; general schedule for the industrial census; general schedule for the business census.

Um golpe de vista sôbre o próximo quinto censo do Brasil (1° de setembro de 1940). Inclue um resumo de todos os censos realizados no

Brasil. Acompanha o folheto um envelope contendo formulários e instruções, a saber: Boletim de família do censo demográfico; boletim individual do censo demográfico; lista de domicílio coletivo; instruções para o preenchimento do boletim de família; caderneta do agente recenseador do censo demográfico; boletim geral do censo agrícola; boletim geral do censo industrial; boletim geral do censo comercial.

Bras-**2110**.R5 1940 **Recenseamento Geral de 1940: Finalidades do Censo Agrícola.** Rio de Janeiro, 1939. Instituto Brasileiro de Geografia e Estatística. Serviço Nacional de Recenseamento. 14 p.

General census of 1940: Objectives of the Agricultural Census.

Bras-**2200**.A5 1940 **Análises de Resultados do Censo Demográfico.** Rio de Janeiro, 1944-45. Instituto Brasileiro de Geografia e Estatística. Serviço Nacional de Recenseamento. Gabinete Técnico. 6 vol.

A collection of 240 statistical studies about the population of Brazil, with tables, graphs and annotations, based on the data obtained in the general census of 1940. The studies are presented in chronological order, beginning with those made in 1942 through the latest ones executed in 1945. The aspects covered include: Enumerated and estimated population; education, literacy and illiteracy; racial characteristics, with special emphasis on the Japanese and their descendants; age, fecundity and prolificacy of the population; distribution of population; language, religion and other characteristics. Some of the studies in this 6-volume collection have been published previously by the IBGE.

Coletânea de duzentos e quarenta estudos estatísticos acêrca da população do Brasil, com tabelas, exposições gráficas e anotaçôes, apresentando dados recolhidos no recenseamento geral de 1940. Os estudos apresentam-se em ordem cronológica, dos primeiros realizados em 1942 aos últimos efetuados em 1945, e incluem os aspectos seguintes: População recenseada e população estimada; educação e alfabetização; características de raça, com consideração especial dos japonêses e seus descendentes; idade, fecundidade e prolificidade, assim como distribuição, lingua, religião e outras características da população do Brasil. Alguns dos estudos desta colecção de seis volumes, foram publicados anteriormente pelo IBGE.

Bras-**2200**.R4 1940.3 **Recenseamento Geral do Brasil (1.° de Setembro de 1940).** Rio de Janeiro, Serviço Gráf.do IBGE, 1943- . Instituto Brasileiro de Geografia e Estatística. Comissão Censitária Nacional. Série Nacional, vol. I, **Introdução,** tomo I, **A Cultura Brasileira: Introdução ao Estudo da Cultura no Brasil,** por Fernando de Azevedo. 535 p.

Will be published also in English—será publicado também en inglês (Bras-**2200**.R4, 1940.3 Eng).

Sumário: Introdução; os fatores da cultura (o país e a raça, o trabalho humano, as formações urbanas, a evolução social e política, psicologia do povo brasileiro); a cultura (instituições e crenças religiosas, a vida intelectual, as profissões liberais, a vida literária, a cultura científica, a cultura artística); a transmissão da cultura (o sentido da educação colonial, as origens das instituições escolares, a descentralização e a dualidade de sistemas, a renovação e a unificação do sistema educativo, o ensino geral e os ensinos especiais).

Bras-**2200**.R4 1940.3, Eng **Brazilian Culture.** An introduction to the study of culture in Brazil. By Fernando de Azevedo. Translated by William Rex Crawford. New York, Macmillan Co., 1947?

Translation of the introductory first volume of the 1940 general census of Brazil, to be published in 1947. Tradução do primeiro volume introdutório do recenseamento geral do Brasil em 1940 (Bras-**2200**.R4, 1940.3), a ser publicada em 1947.

Contents: The factors of culture (land and race, the work of man, the development of urban life, social and political evolution, the psychology of the Brazilian people); culture (religious institutions and beliefs, the intellectual life, the liberal professions, literary life, science, art); the transmission of culture (the significance of colonial education, the origins of scholastic institutions, decentralization and the dual system, the reform and unification of the educational system, general education and various types of special education).

*Bras-**2200**.R4 1940.4 **Recenseamento Geral do Brasil, 1.° de Setembro de 1940: Sinopse do Censo Demográfico.** Dados Gerais. Rio de Janeiro, Serviço Gráf. do IBGE, 1946. 47 p. Instituto Brasileiro de Geografia e Estatística. Comissão Censitária Nacional.

Includes: Population of Brazil according to the general censuses, by physiographical regions and federal units, and their areas and population densities, on September 1, 1940; population of Brazil and of the federal units on September 1, 1940, by sex and age groups, according to principal characteristics—color, marital status, nationality, education, religion, and principal occupation—with related complementary information.

Inclue: População de fato do Brasil nas datas dos recenseamentos gerais, segúndo as regiões fisiográficas e Unidades da Fedcração, com indicação das respectivas superfícies e densidades demográficas em 1.° de setembro de 1940; população de fato do Brasil e das Unidades da Federação em 1.° de setembro de 1940, por sexo e grupos de idades, segundo os principais caracteres individuais—côr, estado conjugal, nacionalidade, instrução, religião e ramo da atividade principal—com indicações complementares correlatas.

Bras-2200.R4 1920 **Recenseamento do Brasil Realizado em 1° de Setembro de 1920.** Rio de Janeiro, Typ.da Estatística, 1922-30. Ministério da Agricultura, Industria e Comercio. Diretoria Geral de Estatística. Serviço de Estatística da Produção. 5 vol.: Vol. I, **Introducção:** Aspecto physico do Brasil; geologia, flora e fauna; evolução do povo brasileiro; resumo historico dos inqueritos demographicos: **Annexos:** Decretos, instrucções, e modelos das cadernetas e dos questionarios para a execução do recenseamento; vol. II, parte I, **População do Rio de Janeiro:** Histórico da cidade e dos inqueritos censitarios; crescimento, densidade, e distribuição da população segundo o sexo, o estado civil, a nacionalidade, a idade, o gráo de instrucção, os defeitos physicos e as profissões; parte 2, **Agricultura e Indústrias; Districto Federal;** parte 3, **Estatística Predíal e Domiciliaria da Cidade do Rio de Janeiro, Districto Federal;** vol. III, **Agricultura;** vol IV, parte I, **População:** População do Brasil por estados, municipios e districtos, segundo o sexo, o estado civil e a nacionalidade; parte 2, **População:** População do Brasil por estados e municipios, segundo o sexo, a idade e a nacionalidade; parte 3, **População:** População do Brasil por estados e municipios, segundo os defeitos physicos, por idade, sexo, e nacionalidade; parte 4, **População:** População do Brasil por estados, municipios e districtos, segundo o gráo de instrucção, por idade, sexo e nacionalidade; parte 5, **População:** População do Brasil por estados e municípios segundo o sexo, a nacionalidade, a idade e as profissões; parte 6, **Estatística Predíal e Domiciliaria do Brasil;** vol. V, parte I, **Indústria;** parte 2, **Salarios;** parte 3, **Estatísticas Complementares do Censo Econômico.** Não examinado.

Census of Brazil, taken Sept. 1, 1920.

3000-3999 STATISTICAL SCIENCE—CIENCIA ESTADISTICA

Bras-3020.C5 1936 **Convenção Nacional de Estatística.** Rio de Janeiro. Diretoria de Estatística de Produção, Secção de Publicidade, 1936. Convenção Nacional de Estatística. 23 p.

National Statistical Convention. Background material for study of the organization of the Brazilian Institute of Geography and Statistics. "Text of the agreement entered into at the National Convention of Statistics by the Federal Government and all the state governments of Brazil to effect, under decree no. 24,609, of July 6, 1934, the establishment and regulation of the National Council of Statistics and to provide the necessary measures for the integration of the federated statistical organs within the National Institute of Statistics." (Translation of comment.)

Documentação para o estudo da organização do Instituto Brasileiro de Geografia e Estatística. "Instrumento da Convenção Nacional de Estatística, que entre si fazem o Govêrno Federal e a unanimidade dos Govêrnos das Unidades Políticas da República dos Estados Unidos do Brasil para, nos termos do Decreto n. 24.609, de 6 de julho de 1934, aprovar as bases da constituição e regulamentação do Conselho Nacional de Estatística e assentar as medidas necessárias à integração do quadro federativo do Instituto Nacional de Estatística." (Fonte: Texto.)

Bras-3020.R4 1945 **Resolução Especial e Conjunta do Conselho Nacional de Estatística, Conselho Nacional de Geografia e Comissão Censitária Nacional, no. IX Aniversário do Instituto em 29 de Maio de 1945.** Rio de Janeiro, Serviço Gráf.do IBGE, 1945? Instituto Brasileiro de Geografia e Estatística. 28 p.

Special and joint resolution of the National Council of Statistics, National Council of Geography, and the National Census Commission. Records the accomplishments of the previous nine years of the IBGE's life as well as the objectives toward which its activities are developing. The last paragraph mentions the 1950 Census of the Americas, to be coordinated through the over-all guidance of the IASI, and that the Brazilian census in that year should also be taken under the minimum standards as may be agreed upon by the 22 American nations.

Resolução especial e conjunta do Conselho Nacional de Estatística, Conselho Nacional de Geografia e Comissão Censitária Nacional. Registra as realizações durante os nove anos decorridos na existência do IBGE e aponta os objetivos na direção dos quais se desenvolvem suas atividades. O último parágrafo menciona o Censo de 1950 das Américas, a ser coordenado sob a orientação geral do IASI, e que o Censo Brasileiro, naquele ano, deverá ser levado a efeito também segundo os padrões mínimos que forem assentados pelas 22 nacões americanas.

Bras-3030.A5 1941 **A Administração Pública e a Estatística.** O papel e a missão de um orgão central de estatística no quadro das realizações do Governo . . . Por Germano G. Jardim. Rio de Janiero, D.I.P., 1941. 169 p.

Public administration and statistics. The role and accomplishments of a central office of statistics within the governmental sphere.

Bras-3030.C5 1944 **Convênios Nacionais de Estatística Municipal.** Rio de Janeiro, Serviço Gráf.do IBGE, 1944. Instituto Brasileiro de Geografia e Estatística (IBGE). Conselho Nacional de Estatística. 139 p.

Contains texts of municipal covenants and provisions of the law imposing a stamp-tax to finance the program of local statistical agencies in Brazil.

Contém os textos dos convênios municipais e os dispositivos legais sôbre o sêlo imposto para o financiamento do programma de agências municipais de estatística no Brasil.

Bras-3030.I5 1940 **Instituto Brasileiro de Geografia e Estatística.** Lembrança da Exposição Nacional dos Mapas Municipais, inaugurada na Capital Federal no dia 29 de maio de 1940, em comemoração ao 4° aniversário da instalação do Instituto. Rio de Janeiro, Serviço Gráf.do IBGE. Instituto Brasileiro de Geografia e Estatística. 12 p.

Contains charts of the organization and structure of the Brazilian Institute of Geography and Statistics and an organization chart of the general census of the Republic of September 1, 1940.

Inclue: Cartograma das Cidades do Brasil; Esquema Fundamental do Instituto Brasileiro de Geografia e Estatística; duas expressivas mensagens do Conselho Nacional de Estatística; esquema estrutural do Instituto Brasileiro de Geografia e Estatística; recenseamento geral da República em 1.° de Setembro de 1940: Esquema Geral da Operação; um compromisso do governo; um desafio aos brasileiros.

Bras-3030.R4 **Relatório do Instituto Brasileiro de Geografia e Estatística.** Data de início? Anual. Instituto Brasileiro de Geografia e Estatística.

Annual report of the Brazilian Institute of Geography and Statistics. Report of the president on the program and activities of the Institute.

Relatório do presidente sôbre o programa e actividades do Instituto.

Bras-3030.R6 **Resoluções da Junta Executiva Central do Conselho Nacional de Estatística.** 1937- . Rio de Janeiro. Instituto Brasileiro de Geografia e Estatística.

Resolutions of the Central Executive Board of the National Council of Statistics.

Bras-3030.R8 1944 **Rumos da Organização Estatística Brasileira. Trends of Statistical Organization in Brazil.** Coletânea de documentos sôbre a origem, evolução e atividades dos serviços estatísticos brasileiros. Collection of documentary accounts on the origin, development and activities of Brazilian statistical services. Por Germano G. Jardim. Rio de Janeiro, Serviço Gráf. do IBGE, 1944. 171 p.

Contents: Preliminary observations; fifth anniversary of the Institute's installation; general assemblies of the National Councils of Geography and Statistics; statistics and national organization; appendix: General system of the Brazilian statistical and geographical services (structural scheme); National Council of Statistics—general alignment of the Brazilian statistical fields (fundamental scheme); National Council of Geography—services performed and under way (schematic outline); National Census Commission—general census of the Republic, September 1st, 1940 (general scheme of the operation).

Conteúdo: Observações preliminares; quinto aniversário da instalação do Instituto; assembléias gerais dos Conselhos Nacionais de Geografia e Estatística; a estatística e a organização nacional; apêndice: Sistema geral dos serviços estatístico-geográficos brasileiros (esquema estrutural); Conselho Nacional de Estatística—ordenação geral dos assuntos da estatística brasileira (esquema fundamental); Conselho Nacional de Geografia—empreendimentos realizados e em realização (quadro esquemático); Comissão Censitária Nacional—recenseamento geral da república em 1.° de setembro de 1940 (esquema geral da operação).

Bras-3030.S7 1940 **Statute, Basic Principles, Social Leadership of the Brazilian Statistical System.** Three characteristic documents presented to the perusal of the Eighth American Scientific Congress. Rio de Janeiro, Print.Off. of the BIGS, 1940. Brazilian Institute of Geography and Statistics. 56 p.

Contents: National Convention of Statistics: Inter-governmental pact, August 11, 1936; fundamentals of the present scope of Brazilian statistics; Resolution No. 135, August 21, 1939, of the General Assembly of the National Council of Statistics (two charts appended: A. statistical organization; B. statistical fields); suggestions concerning the Brazilian policy of social assistance: Resolution No. 127, August 19, 1939, of the General Assembly of the National Council of Statistics.

Estatutos, princípios fundamentais, liderança social do sistema estatístico brasileiro. Sumário: Convenção Nacional de Estatística: Pacto intergovernamental de 11 de agosto de 1936; bases do âmbito atual da estatística brasileira: Resolução n. 135, de 21 de agôsto de 1939, da Assembléia Geral do Conselho Nacional de Estatística (dois esquemas anexos: A) organização estatística; B) assuntos estatísticos); sugestões concernentes à política brasileira de assistência social: Resolução n. 127, de 19 de agosto de 1939, da Assembléia Geral do Conselho Nacional de Estatística.

*Bras-**3040**.R5 **Revista Brasileira de Estatística:** Órgão oficial do Conselho Nacional de Estatística e da Sociedade Brasileira de Estatística. 1940- . Trimestral. Rio de Janeiro. Instituto Brasileiro de Geografia e Estatística.

Brazilian journal of statistics. Official organ of the National Council of Statistics and the Brazilian Society of Statistics. Contains technical papers, news notes, bibliographical material.

Contém artigos técnicos, noticiário, material bibliográfico.

Apr.-June '46, Teixeira de Freitas "O Ensino Primário Brasileiro no Decênio 1932-1941." Excerpto do **Relatório** de 1945. Por M. A. Teixeira de Freitas. Ano VII, no. 26, abril-junho 1946. 62 p. Separata.

Elementary education in Brazil, 1932-41.

Apr.-June '46, Dieulefait "Curso de Estatística: Capítulo II, Séries Estadísticas, Análise das Séries Econômicas e Teoria da População." Por C. E. Dieulefait. Ano VI, no. 22, abril-junho, 1945, p. 147-192.

Course in statistics: Chapter II, Statistical series, analysis of economic series, and population theory.

Jan.-Mar. '46, Dunn "A Estatística no Mundo de Amanhã." Por Halbert L. Dunn. Ano VII, no. 25, janeiro-março 1946, p. 3-8.

Statistics in the world of tomorrow.

Jan.-Mar. '46, Gualberto "Sôbre o Consumo de Alguns Gêneros Alimentícios." Por Virgílio Gualberto. Ano VII, no. 25, janeiro-março 1946. 12 p. Separata.

About the consumption of certain foods.

Oct.-Dec. '45 "Estudos sôbre a Mortalidade no Distrito Federal e no Município de São Paulo." Ano VI, no. 24, outubro-dezembro 1945, p. 583-776.

Studies on mortality in the Federal District and the municipality of São Paulo.

Contents—sumário: I, "A Mortalidade, Segundo Grupos de Idade, por sexo, no Período 1939-41, em Comparação com 1920-21"; II, "Construção e Ajustamento de Tábuas de Sobrevivência, Conforme a Mortalidade do Período 1939-41"; III, "Tábuas de Sobrevivência não Ajustadas e Ajustadas, Conforme a Mortalidade do Período 1939-41"; IV, "Comparações entre as Tábuas de Sobrevivência do Distrito Federal e do Município de São Paulo para os Períodos 1939-41 e 1920-21"; V, "Elaborações Complementares das Tábuas de Sobrevivência de 1939-41"; VI, "Tábuas de Sobrevivência Ajustadas para o Distrito Federal, de 1939-41, Retificadas Conforme a Correção da Mortalidade no Primeiro Ano de Idade"; VII, "Ajustamento das Tábuas de Sobrevivência de 1939-41, Segundo a Fórmula de Gompertz-Makeham"; VIII, "Tábuas de Sobrevivência de 1939-41, Ajustadas Segundo a Fórmula de Gompertz-Makeham"; IX, "A Mortalidade, Segundo Grupos de Causas de Óbito, em Geral e em Relação ao Sexo e à Idade, no Período 1939-41"; X, "Discriminação dos Óbitos Constantes das Tábuas de Sobrevivência de 1939-41, Segundo Grupos de Causas"; XI, "Comparação entre a Discriminação dos Óbitos Constantes das Tábuas de Sobrevivência de 1939-41, Segundo Grupos de Causas, para o Distrito Federal e o Município de São Paulo"; XII, "Ensaio de Retificação das Taxas de Mortalidade, Segundo Grupos de Causas, no Primeiro Ano de Idade, no Distrito Federal"; XIII, "A Marcha da Taxa de Mortalidade Geral no Distrito Federal e no Município de São Paulo, nos Anos de 1920 a 1943"; XIV, "A Mortalidade do Distrito Federal e do Município de São Paulo, no Quadro Internacional."

July-Sept. '45, Carvalho "Pela Livre Circulação de Mercadorias." Plano de simplificação de documentos e formalidades em tôrno de uma guia nacional de circulação. Por Afrânio de Carvalho. Ano VI, no. 23, julho-setembro 1945, p. 430-482.

Free circulation of merchandise. Contents: Bureaucracy versus circulation; unification of permit forms; free direct and indirect transit of goods; reduction of documents to the common denominator of the permit form; issuance of permits and their simultaneous application; statistical and fiscal use of the permit form; simplification of formalities in shipments; conclusion.

Sumário: Burocracia versus circulação; unificação das guias de circulação; liberdade de trânsito direto e indireto; redução de documentos ao denominador comum da guia; tiragem da guia e sua movimentação simultânea; utilização estatística e fiscal da guia; simplificação de formalidades dos embarques; conclusão.

Jan.-Mar. '44 & July-Sept. '44, Mortara "As Tábuas de Sobrevivência e Suas Aplicações na Demografia." Por Giorgio Mortara. Ano V, no. 17, janeiro-março 1944, p. 67-100 e ano V, no. 19, julho-setembro 1944, p. 317-358. Separata.

Life tables and their applications in demography. Following an introduction on two types of life tables and their significance, the study is developed in three chapters: The first describes the elements of the life table; the second, co-ordination of life table with reproduction table, for determination of reproduction rate; the third analyzes application of life table to study of hypothetical populations as characterized by mortality and reproduction rates observed on real generations or populations. Each chapter is divided into two sections, one devoted to theoretical discussion and the other to practical examples and applications. The study concludes with a brief review of modern methods for analysis of the natural movement of population.

Após uma introdução sôbre dois tipos de tábua de sobrevivência e sua significação, o estudo desenvolve-se em três capítulos: o primeiro descreve os elementos da tábua de sobrevivência; o segundo a coordenação entre a tábua de sobrevivência e a tábua de reprodução, para a determinação do coeficiente de reprodução; o terceiro analisa a aplicação da tábua de sobrevivência ao estudo de populações supostas, caracterizadas por taxas de mortalidade e de reprodução observadas em gerações ou populações reais. Cada capítulo está dividido em duas secções, uma dedicada á exposição teórica, e a outra a exemplos práticos e aplicações. O estudo termina por uma breve conclusão sôbre os processos modernos para o estudo de movimento natural da população.

Apr.-June '44, Silva Rodrigues "Vocabulário Brasileiro de Estatística." Por Mílton da Silva Rodrigues. Ano V, no. 18, abril-junho 1944, p. 187-297.

Brazilian vocabulary of statistics. A detailed vocabulary of 283 pages is followed by two appendixes: I, Comparative glossary in Portuguese, French, English, German, Spanish, and Italian; II, Bilingual glossaries of each language with Portuguese.

A um compreensivo vocabulário de 283 páginas seguem-se dois apêndices: I, Glossário comparado em português, francês, inglês, alemão, espanhol, e italiano; II, Glossários bilingues de cada língua com o português.

Jan.-Mar. '44, Dieulefait "Curso de Estatística: Capítulo I, Introdução, Valores Indicativos de Distribuicões Estatísticas, Elementos de Interpolação e de Ajustamento pelo Processo dos Mínimos Quadrados," Por Carlos E. Dieulefait. Tradução de Constança Alvin Pessoa. Ano V, no. 17, janeiro-março 1944, p. 3-48.

Course in statistics: Chapter I, Introduction, indicative values of statistical distributions, elements of interpolation and adjustment by method of least squares.

Oct.-Dec. '43, Morais "'Números Índices', Suas Aplicações na Estatística Brasileira e Sugestões Atinentes à Sua Padronização Continental." Por O. Alexander de Morais. Ano IV, no. 16, outubro-dezembro 1943, p. 641-652. Separata.

Index numbers: Their application in Brazilian statistics and suggestions concerning continental standardization.

Oct.-Dec. '43, Mortara, 2 "Estudos de Demografia Interamericana: III, Tábua de Mortalidade e de Sobrevivência para o México (1929-33); IV, Tábua de Mortalidade e de Sobrevivência para a Cidade de Lima (1933-35)." Por Giorgio Mortara. Ano IV, no. 16, outubro-dezembro 1943, p. 687-695. Separata.

Studies in inter-American demography: III, Mortality and life table for Mexico, 1929-33; mortality and life table for the city of Lima, 1933-35.

Oct.-Dec. '43, Teixeira de Freitas "O Problema do Município no Brasil Atual." Por M. A. Teixeira de Freitas. Ano

IV, no. 16, outubro-dezembro 1943, p. 697-714.

The problem of the municipality in Brazil of today. Study of economic, social and political conditions and needs of municipal units of Brazil.

Estudo sôbre as condições atuais e necessidades econômicas, sociais e políticas das unidades municipais do Brasil.

July-Sept. '43, Mortara, 1 "Estudos de Demografia Interamericana: I, Tábua de Mortalidade e de Sobrevivência para a Colômbia (1939-41); II, Ajustamento de Tábua de Sobrevivência para a Colômbia (1939-41), segundo a Fórmula de Gompertz-Makeham." Por Giorgio Mortara. Ano IV, no. 15, julho-setembro 1943, p. 531-541. Separata.

Studies in inter-American demography: I, Mortality and life table for Colombia, 1939-41; II, Adjustment of the life table for Colombia, 1939-41, according to the Gompertz-Makeham formula.

Apr.-June '43, a Organization and activities of the statistical offices of the central government in Brazil; series of articles. **Revista Brasileira de Estatística,** ano IV, no. 14, abril-junho 1943, p. 196-248.

Organização e atividades dos serviços de estatística do govêrno central do Brasil; série de artigos.

Apr.-June '43, b Organization and activities of the statistical offices of the individual states; series of articles. Ano IV, no. 14, abril-junho 1943, p. 279-424.

Organização e atividades dos serviços estaduais de estatística; série de artigos.

Jan.-Mar. '42, Freitas Filho "O Clínico e a Bio-Estatística." Por Lincoln de Freitas Filho. Ano III, no. 9, janeiro-março 1942, p. 153-174.

The physician and vital statistics. Includes registration forms, and detailed international nomenclature of causes of death, p. 167-172; intermediate international nomenclature of causes of death, p. 172-173; abbreviated international nomenclature of causes of death, p. 173-174.

Inclue modêlos de registros, e **Nomenclatura Internacional Detalhada de Causas de Morte,** p. 167-172; **Nomenclatura Internacional Intermediaria de Causas de Morte,** p. 172-173; **Nomenclatura Internacional Abreviada de Causas de Morte,** p. 173-174.

Bras-**3200**.N5 1939 **Normas de Apresentação da Estatística Brasileira.** Resolução no. 75, de 18 de julho de 1938, modificada pela de número 158, de 22 de julho de 1939. Apêndice, decreto-lei no. 292, de 23 de fevereiro de 1938, que regula o uso da ortografia nacional. Formulário ortográfico. Rio de Janeiro, Serviço Gráf. do IBGE, 1939. Instituto Brasileiro de Geografia e Estatística. Conselho Nacional de Estatística. 47 p.

Rules for the presentation of Brazilian statistics.

Bras-**3300** 1944-46 **Aperfeiçoamento Estatístico no Brasil: Métodos e Recursos.** Background material acquired for the IASI Survey on Statistical Training Methods and Materials in the American Nations, 1944-46.

Statistical training in Brazil: Methods and materials. Includes: Correspondence, filled questionnaires, programs of statistical courses, and requirements for admission of statistical personnel in Brazil.

Documentários obtidos em relação a inquéritos a cargo do Instituto Inter-Americano de Estatística. Inclue: Correspondência, questionários preenchidos, programas de cursos estatísticos, e exigências para admissão de pessoal estatístico no Brasil.

See also—ver também: **Anexo aos Questionários F-1, 2 e 3 do IASI** . . . (Bras-**3300**.A5, 1945).

Bras-**3300**.A5 1945 **Anexo aos Questionários F-1, 2 e 3 do I.A.S.I. para Estabelecimento de uma Futura Definição das Exigências Mínimas para Admissão do Pessoal Estatístico.** Rio de Janeiro, 1945. Cópia máquina. Instituto Brasileiro de Geografia e Estatística. 10 vol.: Vol. 1°, **Estrutura de Serviços de Estatística;** vol. 2°, **Efetivos, Categorias, Remuneração e Quadros de Pessoal Estatístico;** vol. 3°, **Atribuções Inherentes aos Cargos e Funções;** vol. 4°, **Formas e Requisitos para Admissão de Pessoal;** vol. 5°, **Requisitos para Promoção e Substituição de Funcionários;** vol. 6°, **Aperfeiçoamento Técnico do Pessoal Estatístico;** vol. 7°, **Programas para Concurso;** vol. 8°, **Regimentos de Concursos e Exigências para Admissão de Pessoal;** vol. 9°, **Programas de Estatística;** vol. 10°, **Bibliografia Estatística Legislativa e Metodológica do Brasil.**

Annex to the IASI questionnaires F-1, 2 and 3 for the establishment of a future definition of minimum requirements for admission of statistical personnel. 10 vol.: 1st, Structure of statistical services; 2nd, number, categories, remuneration of statistical staffs; 3rd, duties inherent in positions and functions; 4th, methods and requirements for admission of personnel; 5th, requirements for promotion and substitution of employees; 6th, technical training of statistical personnel; 7th, programs for competitions; 8th, rules for competitions and requirements for admission of personnel; 9th, statistical programs; 10th, legal and methodological statistical bibliography of Brazil.

Bras-**3320**.P5 1944 **Programa de Estatística Geral e Aplicada.** (VI cadeira). Rio de Janeiro, 1944. Ministério de Educação e Saúde. Universidade do Brasil. Faculdade Nacional de Filosofia. 6 p.

Outline of course in general and applied statistics in the University of Brazil.

4000-4999 GEOGRAPHY — GEOGRAFIA

Bras-**4000**.B5 **Boletim Geográfico do Conselho Nacional de Geografia.** Informações, notícias, bibliografia, legislação. 1943- . Mensal. Rio de Janeiro. Instituto Brasileiro de Geografia e Estatística. Conselho Nacional de Geografia.

First three numbers have title—os três primeiros números tinham o título: **Boletim do Conselho Nacional de Geografia.**

Geographical bulletin of the National Council of Statistics. Information, news, bibliography, legislation.

Bras-**4000**.R5 **Revista Brasileira de Geografia.** Órgão do Conselho Nacional de Geografia. 1939- . Trimestral. Rio de Janeiro. Instituto Brasileiro de Geografia e Estatística. Conselho Nacional de Geografia.

Brazilian journal of geography. Organ of the National Council of Geography. Includes articles on physical and biological geography, with summaries in Spanish, French, German, Italian, English and Esperanto; reports of the Instituto and associated organizations, and bibliographical reviews.

Inclue: Artigos sôbre geografia física e biológica, com resumos em espanhol, francês, alemão, italiano, inglês e esperanto; relatórios do Instituto e órgãos filiados, e registros bibliográficos.

Bras-**4100**.D6 1945 **Divisão Territorial do Brasil.** Quadro da divisão administrativa regional decretada, nos termos da lei n.° 311, de 2 de março de 1938, para vigorar de 1944 a 1948. Rio de Janeiro, Serviço Gráf. do IBGE, 1945. Instituto Brasileiro de Geografia e Estatística. 224 p.

Territorial division of Brazil.

BRASIL
3320

5000-5999 DEMOGRAPHY—DEMOGRAFIA

Bras-**5000**.B5 1945 **Brazil: Summary of Biostatistics.** Maps and charts, population, natality and mortality statistics. Washington, D.C., January 1945. Prepared by U. S. Department of Commerce, Bureau of the Census, in cooperation with Office of the Coordinator of Inter-American Affairs. 130 p.

Resumo bio-estatístico do Brasil. Inclue mapas, quadros e dados sôbre população, natalidade e mortalidade.

Bras-**5000**.E5 **Estudos Brasileiros de Demografia.** 1946- . Irregular. Rio de Janeiro, São Paulo, Porto Alegre. Fundação Getúlio Vargas.

Brazilian studies on demography. Circulated in the form of monographs and issued irregularly. Their purpose is to develop scientific research on population phenomena of Brazil, and at the same time: To contribute to improvement of collection, tabulation, and analysis of data relating to these phenomena in the country; to collaborate in extensive research in the international field; and to promote projects of a general, theoretical or methodological nature.

Circulam na forma de monografias, sem periodicidade fixa. Seu objetivo é promover o desenvolvimento das pesquisas científicas sôbre os fenômenos da população no Brasil, visando ao mesmo tempo: Contribuir para o aperfeiçoamento da coleta, elaboração e análise dos dados referentes a êsses fenômenos no âmbito nacional; colaborar em investigações extensivas ao campo internacional; e estimular os trabalhos de caráter geral, teóricos ou metodológicos.

Bras-**5200** **Estatística Vital e Registro Civil no Brasil: Normas e Métodos.** Background material acquired for the study conducted by the Inter American Statistical Institute in vital statistics and registration methods in the American nations, including the survey of 1945.

Vital statistics and registration in Brazil: Methods and procedures.

Documentários obtidos em relação a inquéritos a cargo do Instituto Inter-Americano de Estatística.

See also—ver também: **Vital Statistics and Registration Methods** (Bras-**5200**.V5, 1946).

*Bras-**5200**.A5 **Anuário de Bioestatística.** 1927-28— . Rio de Janeiro. Ministério de Educação e Saúde. Departamento Nacional de Saúde. Serviço Federal de Bioestatística.

Yearbook of vital statistics.

Bras-5200.B5 **Boletim Mensal do Serviço Federal de Bio-Estatística.** 1941- . Rio de Janeiro. Ministério da Educação e Saúde. Departamento Nacional de Saúde. Serviço Federal de Bio-Estatística.

Monthly bulletin of the Federal Service of Vital Statistics.

Bras-5200.R5 **Resumo da Bio-Estatística de Cidades Brasileiras.** Data de início? Semanal. Rio de Janeiro. Ministério da Educação e Saúde. Departamento Nacional de Saúde. Serviço Federal de Bio-Estatística.

Summary of vital statistics of Brazilian cities. Uncorrected vital statistics for a group of Brazilian cities, based on the civil register.

Apresenta estatística vital, não revista, para um grupo de cidades brasileiras, com base no registro civil.

Bras-5200.V5 1946 **Vital Statistics and Registration Methods.** Re letter of May 5, 1945, from the Inter American Statistical Institute. By Eduardo J. Gonçalves. Rio de Janeiro, 1946. Instituto Brasileiro de Geografia e Estatística. 3 vol.

Includes: Laws, decrees, and resolutions; instructions; forms; nomenclatures; bibliography.

Estatística vital e registro civil no Brasil. Inclue: Leis, decretos, e resoluções; instruções; formulários; nomenclaturas; bibliografias.

Bras-5212.E5 1945 **Estudos sôbre a Mortalidade nas Grandes Cidades Brasileiras.** Rio de Janeiro, 1945. Instituto Brasileiro de Geografia e Estatística. Serviço Nacional de Recenseamento. Gabinete Técnico. 164 p.

Studies on mortality in large Brazilian cities, no. 5A, 9-15, 27-37.

Inclue no. 5A, 9-15, 27-37.

Bras-5220.T3 1946 **Tábuas Brasileiras de Mortalidade e Sobrevivência.** Escolhidas e coordenadas por Giorgio Mortara. Fundação Getúlio Vargas. 132 p. Estudos brasileiros de demografia, ano I, vol. I, fevereiro 1946, monografia, no. 1.

Study on life tables of Brazil. Includes: General introductory concepts about life tables; information on sources and explanation of the methods of construction of the Brazilian life tables; life tables for Brazil; some comparisons between various Brazilian life tables and those of other countries.

Inclue: Noções introdutivas sôbre as tábuas de mortalidade e sobrevivência; informações sôbre as fontes, e esclarecimentos sôbre os processos de construção das tábuas de mortalidade e sobrevivência brasileiras; tábuas de mortalidade e sobrevivência brasileiras; algumas comparações entre as diversas tábuas brasileiras e entre estas e tábuas estrangeiras.

Bras-5220.T52 1946 **Tábuas de Mortalidade e Sobrevivência Brasileiras: Distrito Federal e Município de São Paulo.** Edição especial em homenagem à diretoria do Instituto Interamericano de Estatística, ao ensejo de sua primeira reunião no Brasil (7 a 11—I—1946). Rio de Janeiro, Serviço Gráf. do IBGE, 1946. Instituto Brasileiro de Geografia e Estatística. 194 p.

Mortality and life tables for the Federal District and the municipality of São Paulo. Special edition in honor of the directorate of the Inter American Statistical Institute on occasion of its first meeting in Brazil, January 1946. Includes comparative data for 1920-21 and 1939-41.

Inclue dados comparativos de 1920-21 e 1939-41.

Bras-5220.T53 1920-21 **Tábuas de Sobrevivência, Conforme a Mortalidade do Período 1920-21, para o Distrito Federal e o Município de São Paulo.** Rio de Janeiro, 1944. Instituto Brasileiro de Geografia e Estatística. Serviço Nacional de Recenseamento. Gabinete Técnico. 62 p.

Life tables, according to mortality in the period 1920-21, for the Federal District and the municipality of São Paulo. Includes: Studies of mortality in the large Brazilian cities, no. 1-4, 6A, 7A, 8A.

Inclue: Estudos sôbre a mortalidade nas grandes cidades brasileiras, no. 1-4, 6A, 7A, 8A.

Bras-5220.T54 1939-41 **Tábuas de Sobrevivência, Conforme a Mortalidade do Período 1939-41, para o Distrito Federal e o Município de São Paulo.** Rio de Janeiro, 1944. Instituto Brasileiro de Geografia e Estatística. Serviço Nacional de Recenseamento. Gabinete Técnico. 138 p.

Life tables, according to mortality in the period 1939-41, for the Federal District and the municipality of São Paulo. Includes: Mortality and life tables, based on 1940 census, for the two Brazilian capital cities, and international comparisons of mortality.

Inclue: Tábuas de mortalidade e sobrevivência, baseadas no censo de 1940, para as duas capitais brasileiras, e comparações internacionais da mortalidade.

*Bras-**5300**.E5. **Estatística Geral Migratoria.** Data de início? Anual. Rio de Janeiro. Ministério do Trabalho, Indústria e Comércio. Departamento Nacional de Imigração.

Annual statistics of migration.

Bras-**5300**.R5 **Revista de Imigração e Colonização.** Órgão oficial do Conselho de Imigração e Colonização. 1940- . Trimestral. Rio de Janeiro. Conselho de Imigração e Colonização.

Quarterly journal of immigration and colonization. Articles, legislation, news notes, as well as a section of statistical series on population movements.

Artigos, legislação, noticiário, e uma secção de séries estatísticas sôbre movimentos da população.

6000-6999 ECONOMICS — ECONOMIA

Bras-**6000**.B65 1946 **Brazil.** By Preston E. James. New York, Odyssey Press, 1946. 262 p. Unexamined.

" . . . A reprint of the sections on Brazil from the author's **Latin America** with facts brought up to date and new material added . . . Forty-four legible maps and many excellent photographs supplement the text. The main theme of the volume is the pattern and expansion of settlement as related to the physical characteristics of the land, the types and stages of economic pursuits." (Source: **The Annals of the American Academy of Political and Social Science,** vol. 247, September 1946, p. 195.)

" . . . Uma separata das sécções referentes ao Brasil do livro **America Latina** pelo mesmo autor, com os fatos atualizados e acréscimos de material novo adicional . . . Quarenta e quatro mapas e diversas excelentes fotografias suplementam o texto. O tema principal do volume é o padrão e a expansão do povoamento em relação ás caraterísticas físicas da terra e dos tipos e fáses de alcance econômico." (Tradução do comentário.)

Bras-**6000**.E3 1945 **Economic Controls and Commercial Policy in Brazil.** One of a series of reports on economic controls and commercial policy in the American republics. Washington, D.C., 1945. United States Tariff Commission. 49 p.

Publicado na série de relatórios da Comissão Aduaneira dos Estados Unidos, sôbre controles econômicos e política comercial das repúblicas americanas.

BRASIL
5300

*Bras-**6000**.E6 **Estatísticas Econômicas. Brasil.** 1935- . Mensal. Rio de Janeiro. Ministério da Fazenda. Tesouro Nacional. Serviço de Estatística Econômica e Financeira.

Monthly statistical data on foreign trade, mineral and metal production, cost of living, and banking.

Quadros estatísticos sôbre comércio exterior, produção mineral e metalúrgica, custo da vida, e bancos.

Jan.-Dec. '45 **Estatísticas Econômicas, Brasil, Janeiro a Dezembro 1935-1945.** Rio de Janeiro, Imp. Nacional, 1946. Ministério da Fazenda. Tesouro Nacional. Serviço de Estatística Econômica e Financeira. 44 p.

Contains summary tables of statistical data for the period 1935-45, and monthly figures for 1943-45 on: Foreign commerce, imports and exports by classes of articles; principal exported and imported articles; foreign commerce according to principal American countries; coastwise shipping, by classes of articles; average value, by ton, of merchandise of foreign commerce and coastwise shipping; mineral and metallurgical production; cost of living in Rio de Janeiro; principal items of assets and liabilities of national and foreign banks and the Bank of Brazil; average rates of interest collected by the Bank of Brazil; monetary potential; gold purchases by the federal government; securities transacted in the principal stock exchanges; exchange fluctuations in the Rio de Janeiro market; quotations on securities of the Federal Government in the London and New York market; quotations of bonds, stocks and other federal securities on Rio de Janeiro stock exchange.

Contém quadros sumários das estatísticas referentes ao período 1935-45, e dados mensais de 1943-45 sôbre: Comércio exterior, importação e exportação por classe de mercadorias; principais mercadorias de importação e exportação; comércio exterior, segundo os principais países americanos; comércio de cabotagem, por classe de mercadorias; valor médio, por tonelada, das mercadorias do comércio exterior e de cabotagem; produção mineral e metalúrgica; custo da vida no Rio de Janeiro; principais contas do ativa e passivo dos bancos nacionais e estrangeiros e do Banco do Brasil; taxas médias de juros cobrados pelo Banco do Brasil; potencial monetário; compras de ouro pelo govêrno federal; títulos negociados nas principais bôlsas; curso do cambio na praça do Rio de Janeiro; cotações do govêrno federal nos mercados de Londres e Nova Iórque; cotações de apólices e outros títulos federais na bolsa do Rio de Janeiro.

Bras-**6000.**I3 Sept. '44 **Informações Relativas ao Distrito Federal, em Setembro de 1944.** Rio de Janeiro, 1944. Instituto Brasileiro de Geografia e Estatística. 7 p. Mimeografado. Inquéritos econômicos para a defesa nacional, comunicado no. 57, de 20 de novembro de 1944.

Economic information relating to the Federal District, September 1944.

Bras-**6000.**I5 **Inquéritos Econômicos para a Defesa Nacional.** Data de início? Rio de Janeiro. Instituto Brasileiro de Geografia e Estatística. Conselho Nacional de Estatística.

Series of economic studies for national defense.

Bras-**6000.**Q5 **Quadros Estatísticos.** Resumo anual de estatísticas econômicas. Data de início? Rio de Janeiro. Ministério da Fazenda. Tesouro Nacional. Diretoria Estatística Econômica e Financeira.

Yearbook of statistical tables on economic and financial aspects of Brazil, including industrial, agricultural and mining production; domestic and foreign trade; transportation and communication; population and immigration.

Anuário de quadros estatísticos de aspectos econômicos e financeiros do Brasil, incluindo produção industrial, agrícola e mineira; comércio interno e externo; transportes e comunicações; população e imigração.

Bras-**6000.**R5 1944 **Reports Presented to the Conference of Commissions of Inter-American Development (Held in New York on May 9-18, 1944) by the Brazilian Commission of Inter-American Development.** On: 1, Hydroelectric energy; 2, transportation; 3, uniformity of technical standards in the Americas. (Translations from the Portuguese originals.) Washington D.C., 1945? Inter-American Development Commission. 30 p.

Cover title—título da capa: **Brazil: Reports Presented to the Conference of Commissions of Inter-American Development.**

Relatórios apresentados á Conferência de Comissões de Desenvolvimento Interamericano (reunida em Nova Iórque 9-18 de maio de 1944) pela Comissão Brasileira de Desenvolvimento Interamericano, sôbre: 1, Energia hidroelétrica; 2, transportes; 3, uniformidade de regras técnicas nas Américas.

Bras-**6000.**S7 Apr. '44 **Situação e Movimento dos Estabelecimentos Comerciais e Industriais em Abril de 1944.** Rio de Janeiro, 1944. Instituto Brasileiro de Geografia e Estatística. Conselho Nacional de Estatística. 14 p. Mimeografado. Inquéritos econômicos para a defesa nacional, comunicado no. 58, de 5 de dezembro de 1944.

Situation and movement of commercial and industrial establishments, April 1944.

Bras-**6010.**D5 **Dados e Índices Econômico-Financeiros Relativos ao Distrito Federal e á Cidade de São Paulo.** 1941- . Mensal. Rio de Janeiro. Ministério da Fazenda. Tesouro Nacional. Serviço de Estatística Econômica e Financeira.

Monthly economic and financial data and indexes for the Federal District and the city of São Paulo. Combines some statistical series—construction, consumption, production, etc.—which reflect the position of the national economy, through the means of indexes referring to the two largest cities of the country.

Reúne algumas séries estatísticas—construção, consumo, produção—capazes de refletir a posição de economia nacional, posta em foco mediante os índices relativos ás duas maiores praças do país.

Bras-**6020.**E5 1939 **A Estatística da Produção no Estado de Minas Gerais.** Por Joaquim Ribeiro Costa. Rio de Janeiro, 1939. Instituto Brasileiro de Geografia e Estatística. Conselho Nacional de Estatística. 184 p.

Statistics of production of the State of Minas Gerais. Last half of publication shows model questionnaires.

Metade da publicação consiste de modêlos de questionários.

Bras-**6100.**R5 1944 **Renda Nacional do Brasil: Colecção de Estudos Diversos.** Reunidos e remetidos ao Instituto Interamericano de Estatística pelo Instituto Brasileiro de Geografia e Estatística, 1944.

Studies on national income of Brazil.

Inclue—includes: "A Estimativa da Renda Geral do Brasil," por Osvaldo Gomes da Costa Miranda do **Boletim do Ministério do Trabalho, Indústria e Comércio,** ano X, no. 118, junho 1944, p. 209-215; "Notas sôbre o Cálculo da Renda Nacional," por Luiz Dodsworth Martins; estudo "Alguns Aspectos da Política Econômica Mais Conveniente ao Brasil no Período de Após-Guerra: Geografia e Política Industrial," por Roberto C. Simonsen.

Bras-**6200** **Estatísticas Agrícolas no Brasil: Métodos e Processos.** Background material acquired for Inter American Statistical Institute study on methods in agricultural statistics in the American nations, including the Survey of Agricultural Production Statistics in the American Nations in 1945-46.

Agricultural statistics in Brazil: Methods and procedures. Consists of: Correspondence; instructions; forms; reports.

Documentários obtidos em relação a inquéritos a cargo do Instituto Inter-Americano de Estatística. Consiste de: Correspondência; instruções; formulários; relatórios.

Includes—inclue: "Relatório ao Presidente do Bureau Panamericano do Café" (inclue: Exportação do café; importação do café; necessidade da uniformização do sistema de exprimir volumes de café; necessidade da uniformização do criterio para a divisão, no tempo, da produção cafeeira; censo da produção cafeeira em todo o hemisferio ocidental), e "Notas sôbre as Estatísticas da Produção de Café na América" do Departamento Nacional do Café.

Bras-**6200**.A2 1946 **Agricultural, Pastoral, and Forest Industries in Brazil.** One of a series of reports on agricultural, pastoral, and forest industries in the American republics. Washington, D.C., U.S.Govt.Print.Off., 1946. United States Tariff Commission. 78 p.

Publicado na série de relatórios da Comissão Aduaneira dos Estados Unidos, sôbre as indústrias agro-pecuárias e florestais nas repúblicas americanas.

Bras-**6200**.M5 **Mensário de Estatística da Produção.** Data de início? Mensal. Rio de Janeiro. Brasil. Ministério da Agricultura. Serviço de Estatística da Produção. Não examinado.

Monthly agricultural production statistics. "Statistical information and commentary on the Brazilian agricultural life." (Source: Stat. Act.)

"Informações estatísticas e comentários sôbre a vida agrícola brasileira." (Fonte: Stat.Act.)

Bras-**6210**.A4 **Anuário Açucareiro.** Data de início? Rio de Janeiro. Instituto do Açucar e do Alcool. Não examinado.

Sugar yearbook.

Bras-**6210**.A5 **Anuário Algodoeiro.** Data de início? Rio de Janeiro. Bolsa de Mercadorias de São Paulo. Não examinado.

Cotton yearbook. "Cotton production in Brazil; cotton textiles. Exports and imports of raw cotton and cotton manufactures. Statistical tables and graphs." (Source: HBLAS.)

"A produção algodeira no Brasil; tecidos de algodão. Exportação e importação de algodão e produtos de algodão. Tabelas e gráficos estatísticos." (Fonte: HBLAS.)

Bras-**6210**.A6 **Anuário Estatístico do Café.** 1934-35— . Departamento Nacional do Café. Secção de Estatística.

Statistical yearbook of coffee. Includes: Abstract of exports of national products, 1821-1939, emphasizing the coffee position; average prices by months; variations in value of exported coffee according to exchange transactions during 1937-1939, and many other data regarding a variety of features of the coffee economy, in 3992 tables.

Inclue: Uma sinopse da exportação dos produtos nacionais no período de 1821 a 1939, realçando a posição do café; a descriminação do preço medio levantado mensalmente; as variações do valor do café exportado nas transações cambiais no trienio 1937-39 e muitos outros dados referentes aos mais variados aspectos da economia cafeeira, em 3992 tabelas.

Bras-**6210**.B6 **Boletim do Instituto do Açucar e do Álcool.** Data de início? Mensal. Rio de Janeiro. Instituto do Açucar e do Álcool. Secção de Estatística. Não examinado.

Monthly bulletin of the Institute of Sugar and Alcohol.

Bras-**6210**.B8 **Brasil Açucareiro.** Data de início? Mensal. Rio de Janeiro? Instituto do Açucar e do Álcool. Não examinado.

Monthly review of statistics of the sugar industry.

Uma revista mensal de estatística da indústria açucareira.

Bras-**6210**.R5 **Revista do Departamento Nacional do Café.** 1933? - . Mensal. Rio de Janeiro.

Monthly review of the National Department of Coffee, with statistical tables.

Revista do comércio do café, com quadros estatísticos.

Bras-**6400** **Estatísticas Industriais no Brasil: Métodos e Processos.** Background material acquired for the study conducted by the Inter American Statistical Institute in methods of industrial statistics in the American nations, including the survey of 1945-46.

Industrial statistics in Brazil: Methods and procedures. Includes: Registro industrial em 1946 (questionnaire).

Documentários obtidos em relação a inquéritos a cargo do Instituto Inter-Americano de Estatística. Inclue: Registro industrial em 1946 (questionário).

Bras-**6400.**E7 1944 **Os Estoques de Matérias e Produtos Têxteis do Distrito Federal em 30 de Junho de 1943 e em 30 de Junho de 1944.** Rio de Janeiro, 1944. Instituto Brasileiro de Geografia e Estatística. Conselho Nacional de Estatística. 9 p. Mimeografado. Inquéritos econômicos para a defesa nacional, comunicado no. 54, de 7 de outubro de 1944.

Stocks of textile materials and products in Federal District, June 30, 1943 and June 30, 1944.

Bras-**6400.**I5 1946 **Inquérito Continental Sôbre Fomento e Coordenação de Indústrias.** Resposta referente ao Brasil, organizada pelo Departamento de Economia Industrial (FIESP), sob a direção do Eng. Dr. Roberto C. Simonsen, Presidente da Federação das Indústrias do Estado de São Paulo e Vice-Presidente da Secção Brasileira do Conselho Interamericano. Montevidéu, 1946. Conselho Interamericano de Comércio e Produção. 507 p.

Continental inquiry on the development and coordination of industries. Reply referring to Brazil, organized by the Department of Industrial Economy (FIESP). Contents: Production and exports of manufactured products (general evolution of the transforming industries, 1938-42, with tables showing, by type of industry, the number of establishments and of employees, value of production, capital-value of installations (1941), value of securities, and application of capital—average capital invested per company; exports of transformed or manufactured products by categories and destination 1938-42 with tables on exports and imports, summary by categories, principal products, imports and exports of consumers' goods and of producers' goods according to durability, total imports and exports of Brazil); financing of industries; machinery and equipment for industry; personnel and its preparation; taxation policies and tax laws; special legislation for industrial development and control; coordination of industries; statistical information on transforming industries, 1938-42, by specific groups; annexes and tables on various economic features. Spanish summary of contents of each chapter.

Sumário: Produção e exportação de produtos fabrís (evolução geral das indústrias transformadoras, 1938-42, com quadros, segundo as indústrias, sôbre número de estabelecimentos e de empregados, valor da produção, capital—valor das instalações (1941), valor dos títulos, aplicação do capital; média do capital investido, por companhia; exportação de produtos transformados ou manufaturados por rubricas e destino 1938-42 com quadros sôbre exportação e importação, resumo por categorias, principais produtos, importação e exportação de bens de consumo e de bens de produção segundo a durabilidade, importação e exportação total do Brasil); financiamento de indústrias; maquinas e equipamentos para indústria; pessoal e sua preparação; política tributária e leis impositivas; legislação especial de fomento e contrôle industrial; coordenação de indústrias; informações estatísticas de indústrias transformadoras, 1938-42, por grupos específicos; anexos e quadros sôbre varios aspectos econômicos. Sumário em espanhol do conteúdo de cada capítulo.

Bras-**6400.**M5 1945 **Mining and Manufacturing Industries in Brazil.** One of a series of reports on mining and manufacturing industries in the American republics. Washington, D.C., 1945. United States Tariff Commission. 75 p.

Publicado na série de relatórios da Comissão Aduaneira dos Estados Unidos, sôbre as indústrias mineiras e manufatureiras das repúblicas americanas.

Bras-**6614.**E5 **Estatística das Estradas de Ferro do Brasil.** 1923- . Anual. Rio de Janeiro. Ministério da Viação e Obras Públicas. Departamento Nacional de Estradas de Ferro. Não examinado.

Statistics of railways in Brazil.

Bras-**6615.**M6 **Movimento Marítimo e Fluvial do Brasil.** Bienal. Rio de Janeiro. Ministério da Fazenda. Tesouro Nacional. Serviço de Estatística Econômica e Financeira.

Maritime and river navigation in Brazil. Includes statistical data on the number of incoming and outgoing vessels in each port according to nationality, tonnage, distribution of the shippers in the various ports and the nationality of the shippers.

Inclue dados estatísticos sôbre o número de entradas e de saídas de embarcações em cada pôrto, segundo bandeira, tonelagem, distribuição das empresas armadoras nos diversos pôrtos, e a nacionalidade das mesmas.

Bras-**6720.**B5 **Boletim Estatístico. Statistical Bulletin.** Data de início? Anual. Rio de Janeiro. Banco do Brasil. Departamento de Estatística e Estudos Econômicos. Bank of Brazil, Department of Statistics and Economic Research.

Formerly issued quarterly—Anteriormente em edição trimestral.
Yearly statistical bulletin of the Bank of Brazil. Contains statistics of economic activities of the country which include population, immigration, production, foreign trade, and shipping, as well as financial and banking statistics for the country and the Bank.
Contém estatísticas das atividades econômicas do país, incluindo população, imigração, produção, comércio exterior, e navegação, assim como estatísticas financeiras e bancárias relativas ao país e ao Banco.

Bras-**6720.**M4 **Movimento Bancário do Brasil.** Data de início? Mensal. Rio de Janeiro. Ministério da Fazenda. Tesouro Nacional. Serviço de Estatística Econômica e Financeira.
Monthly banking statistics.

Bras-**6720.**M5 **Movimento Bancário do Brasil.** Data de início? Anual. Rio de Janeiro. Ministério da Fazenda. Tesouro Nacional. Serviço de Estatística Econômica e Financeira.
Yearly banking statistics. ". . . Intends to complete, by means of demonstrative tables, the information published in the monthly bulletins on the movement of transactions conducted by the credit establishments operating in the country. The task of revising and coordinating these data falls also within the scope of the annual bulletin." (Translation of comment.)
". . . Visa completar, por meios de quadros demonstrativos, as informações contidas nos boletins mensais sôbre o movimento das operações realizadas pelos estabelecimentos de crédito que funcionam no país. Estenderam-se ao boletim anual os trabalhos de revisão e padronização." (Fonte: **Movimento Bancário do Brasil, 1942-1943,** "Introdução.")

Bras-**6720.**R4 **Relatório do Banco do Brasil.** Data de início? Anual. Rio de Janeiro. Banco do Brasil.
Report of the Bank of Brazil. "The annual report of the Banco do Brasil is a most important source of information of an economic and financial character." (Source: HBLAS.)
"O relatório anual do Banco do Brasil é a fonte mais importante de informações de natureza econômica e financeira." (Fonte: HBLAS.)

Bras-**6720.**R6 **Resumen de la Memoria del Banco do Brasil.** Data de início? Anual? Rio de Janeiro. Banco do Brasil.
Summary of report of the Bank of Brazil. In Spanish. Contains summary text and tables on the economic and financial position of Brazil and on the activities of the Bank.
Contém textos resumidos e tabelas sôbre a situação econômica e financeira do Brasil e sôbre as atividades do banco.

Bras-**6900.**B6 **Boletim Estatístico da Diretoria das Rendas Aduaneiras.** 1939- . Anual. Rio de Janeiro. Ministério da Fazenda. Tesouro Nacional. Diretoria das Rendas Aduaneiras. Não examinado.
Statistical bulletin of the Bureau of Customs Collections.

Bras-**6900.**B8 **Brazilian Customs Tariff. Tarifa das Alfândegas do Brasil.** 1944 Authorised by the Ministry of Finance. In English and Portuguese. Rio de Janeiro, 1944? British Chamber of Commerce in Brazil, Inc. 287 p.
Decree law 2878 of December 18, 1940, tariff law, and loose supplements through 1943.
Autorizada pelo Ministério da Fazenda. Em inglês e português. Decreto-lei 2878 de 18 de Dezembro de 1940, pauta alfandegária e suplementos avulsos até 1943.

*Bras-**6900.**C3 **Comércio Exterior do Brasil.** Data de início? Anual. Rio de Janeiro. Ministério da Fazenda. Tesouro Nacional. Serviço de Estatística Econômica e Financeira.
Subtitle varies—subtítulo varia: 1928-1932, 1937-1938, Importação, exportação; 1939-1940, Importação e exportação de mercadorias por países. 1928-1932, 1937-1938 published also in English and French—publicado também em inglês e francês.
Statistical data on Brazil's exports and imports. Classification of goods, by countries of origin and destination.

1939-40 **Comércio Exterior do Brasil: Importação e Exportação de Mercadorias por Países, 1939-1940.** Rio de Janeiro, Imp. Nacional, 1941. Ministério da Fazenda. Tesouro Nacional. Serviço de Estatística Econômica e Financeira. 342 p.
Statistical data on imported and exported goods, by countries of origin and destination.

Bras-**6900.**C4 **Comércio Exterior do Brasil, Intercambio por Países.** 1937-39— Anual. Rio de Janeiro. Ministério das Relações Exteriores.
Foreign commerce of Brazil, trade by countries. Annual.

Bras-**6900.**C5 **Comércio Exterior do Brasil: Resumo Mensal.** Data de início? Mensal. Rio de Janeiro. Ministério da Fazenda. Tesouro Nacional. Serviço de Estatística Econômica e Financeira.

Foreign trade of Brazil: Monthly summary. Each issue contains data for the year to date, compared with corresponding periods of previous years. Statistics of foreign trade classified by commodities and countries. A volume containing data on an annual basis is also issued, though irregularly. For example, the volume appearing in 1940 covered the years 1933-37 and the one in 1936 covered the years 1928-32.

Cada número contém dados para o ano a que se refere, em comparação com correspondentes períodos anteriores. Estatísticas do comércio exterior classificadas por mercadorias e países. É publicado também um volume de dados anuais, com irregularidade, porém. Por exemplo, o volume aparecido em 1940, abrangendo os anos de 1933-37, e outro, saído em 1936, relativo aos anos de 1928-32.

Bras-**6900**.C6 **Comércio Exterior do Brasil: Resumo por Mercadorias.** 1900- . Anual. Rio de Janeiro. Ministério da Fazenda. Tesouro Nacional. Serviço de Estatística Econômica e Financeira.

Foreign trade of Brazil: Annual summary by commodities.

7000-7999 HEALTH AND CULTURAL STATISTICS—ESTADISTICAS DE SALUBRIDAD Y CULTURA

Bras-**7000**.E5 1942 **Educação e Saúde.** Comunicados do órgão central de estatística do Ministério da Educação e Saúde. Rio de Janeiro, Serviço Gráf.do IBGE, 1942. Instituto Brasileiro de Geografia e Estatística. Conselho Nacional de Estatística. 511 p.

Education and health: Communications of the central office of statistics of the Department of Education and Health. In this volume the IBGE presents the press releases which were distributed by the former General Office of Information, Statistics and Publicity, of the Department of Education and Health, during the decennium 1931-40.

IBGE divulga, neste volume, os comunicados distribuídos à imprensa no decênio 1931-40 pela antiga Diretoria Geral de Informações, Estatística e Divulgação, do Ministério da Educação e Saúde.

Bras-**7100**.A5 **Arquivos de Higiene.** 1927- . Irregular. Rio de Janeiro. Ministério da Educação e Saúde. Departamento Nacional de Saúde.

Archives of Hygiene. Irregular. Contains articles on hygiene and related sciences, and on administration of public health.

Contém artigos sôbre higiene e ciências correlatas e sôbre a administração da saúde pública.

Dec. '44, Freitas Filho "O Problema da Seleção da Causa Primária do Óbito no Brasil, II." Por Lincoln de Freitas Filho. Ano 14, no. 3, dezembro 1944, p. 7-107.

The problem of selecting the primary cause of death in Brazil, part II.

Dec. '43, Scorzelli "Mortinatalidade nas Capitais Brasileiras, 1932-1941." Por Achilles Scorzelli. Ano 13, no. 3, dezembro 1943, p. 7-25.

Stillbirths in Brazilian capital cities, 1932-41.

Aug. '43, Freitas Filho "O Problema da Seleção da Causa Primária do Óbito no Brasil, I." Por Lincoln de Freitas Filho. Ano 13, no. 2, agosto 1943, p. 7-64.

The problem of selecting the primary cause of death in Brazil, part I.

Bras-**7101**.D5 1943 **O Departamento Nacional de Saúde em 1943.** Por João de Barros Barreto. Rio de Janeiro, 1944. Ministério da Educação e Saúde. Departamento Nacional de Saúde. 574 p. Não examinado.

The National Health Department in 1943.

Bras-**7200**.B5 **Boletim do Instituto Nacional de Estudos Pedagógicos.** 1939- . Rio de Janeiro. Ministério da Educação e Saúde. Instituto Nacional de Estudos Pedagógicos.

Bulletin of the National Institute of Pedagogical Studies. Serial publication on administration, research and statistics of education in Brazil.

Publicação em séries relativas à administração, pesquisas e estatísticas da educação no Brasil.

Bras-**7200**.E3 1942 **O Ensino no Brasil no Quinquênio 1936-1940.** Rio de Janeiro, Serviço Gráf.do IBGE, 1942. Ministério da Educação e Saúde. Instituto de Estudos Pedagógicos. 88 p. Boletím, no. 25.

Education in Brazil, 1936-40.

Bras-**7200**.E4 1939 **O Ensino no Brasil no Quinquênio 1932-1936.** Rio de Janeiro, Serviço Gráf.do Ministério da Educação e Saúde, 1939. Ministério da Educação e Saúde.

Instituto Nacional de Estudos Pedagógicos. 83 p. Boletim, no. 1.
Education in Brazil during the quinquennium 1932-1936.

Bras-7200.E6 1929 **Estatística Intellectual do Brasil, 1929.** Rio de Janeiro, Typ.de Dep.Nacional de Estatística, 1931-32. Instituto Brasileiro de Geografia e Estatística. 2 vol. Não examinado.
Intellectual statistics of Brazil, 1929. Includes statistics for 1929, and lists of educational establishments, libraries, learned institutions and societies, periodicals, publishers, and theatres in Brazil in 1930.
Inclue estatísticas relativas a 1929 e enumera os estabelecimentos de educação, bibliotecas instituições e sociedades culturais, periódicos, editores, e teatros do Brasil em 1930.

Bras-7200.O5 1946 **Outline of Education in Brazil.** Prepared by the "National Institute of Pedagogic Studies" (Instituto Nacional de Estudos Pedagógicos), Ministério da Educação, Rio de Janeiro. New York, 1946. Brazilian Government Trade Bureau. 90 p. (Booklets of the Brazilian Government Trade Bureau), no. 6, January 1946.
Esquema geral da educação brasileira; preparado pelo Instituto Nacional de Estudos Pedagógicos.

Bras-7200.R5 **Revista Brasileira de Estudos Pedagógicos.** Publicada pelo Instituto Nacional de Estudos Pedagógicos, Ministério da Educação. 1944- . Mensal. Rio de Janeiro.
Monthly journal of pedagogical studies. Publishes results of the studies and research projects of the Brazilian Ministry of Education; edited under the supervision of the National Institute for Pedagogical Studies, with the object of presenting and discussing general questions of pedagogy, and in particular, the problems of education in Brazil. Publishes the results of investigations carried out by the various agencies of the national ministry and the different state departments of education; bibliography section lists national and foreign works in education.
Orgão dos estudos e pesquísas do Ministério da Educação; publica-se sob a responsabilidade do Instituto Nacional de Estudos Pedagógicos, e tem por fim, expor e discutir questões gerais da pedagogia e, de modo especial, os problemas da vida educacional brasileira. Publica os resultados de trabalhos realizados pelos diferentes orgãos do Ministério e dos Departamentos Estaduais de Educação; mantém seção bibliográfica, dedicada aos estudos pedagógicos nacionais e estrangeiros.

May '45, Jardim "A Coleta da Estatística Educacional (III)." Por Germano Jardim. Vol. IV, no. 11, maio 1945, p. 229-250.
The collection of educational statistics (III). Includes: School registration methods; samples of forms.
Inclue: Escrituração escolar; modelos de formulários.

Nov. '44, Jardim "A Coleta da Estatística Educacional." Por Germano Jardim. Vol. II, no. 5, novembro 1944, p. 221-230.
The collection of educational statistics. Includes: Description of federal and regional systems of collecting educational statistics.
Inclue: Descrição do mecanismo colector federal e regional da estatística educacional.

Sept. '44, Jardim "A Coleta da Estatística Educacional." Por Germano Jardim. Vol. I, no. 3, setembro 1944, p. 361-370.
The collection of the educational statistics. Includes: Covenant of 1931; scope of statistical inquiries; classification of education.
Inclue: O convênio de 1931; a compreensão dos inquéritos estatísticos; classificação do ensino.

Aug. '44 "I Conferência de Ministros e Diretores de Educação das Repúblicas Americanas." Vol. I, no. 2, agôsto 1944, p. 226-280.
First Conference of Ministers and Directors of Education of the American Republics. The complete text of the final act of the conference. Includes: Final act; resolutions, recommendations, agreements and conventions.
O texto completo da ata final dos trabalhos da Conferência. Inclue: Ata final; resoluções, recomendações, acôrdos e convenções.

8000-8999 GOVERNMENT AND POLITICAL STATISTICS— ESTADISTICAS GUBERNAMENTALES Y POLITICAS

Bras-8000.R4 **Relatório do Departamento Administrativo do Serviço Público. 1941?-** . Anual. Rio de Janeiro.
Report of the Administrative Department of Public Service. Covers organization of public service administration in Brazil, with statistical tables analyzing employment and personnel

changes by agency; job classification; "in-service" training programs for public employees, etc.

Relatorio do Departamento Administrativo do Serviço Público. Trata da organização dos serviços públicos da administração brasileira com tabelas estatísticas sôbre os quadros e movimento do funcionalismo por repartições; classificação dos cargos, programas de aperfeiçoamento dos funcionários, etc.

9000-9999 LABOR AND LIVING CONDITIONS —CONDICIONES DE TRABAJO Y DE VIDA

Bras-9200.S5 1940 **Salário Mínimo.** Rio de Janeiro, 1940- . Ministério do Trabalho, Indústria e Comércio. Serviço de Estatística da Previdência e Trabalho. Vol.: Vol. I **Legislação, Estatística e Doutrina.**

Includes: Legislation of 1930 to 1940, concerning the establishment of a minimum wage in Brazil; results of the survey to investigate living conditions and to show the lowest types of remuneration in effect in the country; summary of the survey on minimum wage; statistical graphs and tables.

Inclue: Legislação de 1930 a 1940, relativa à instituição do salário mínimo no Brasil; resultados do inquérito para investigar as condições de vida e indicar os salários mais baixos efetivamente pagos no país; resumo do inquérito sôbre o salário mínimo; gráficos e quadros estatísticos.

Bras-9800.V5 1945 **A Velocidade de Circulação dos Estoques de Produtos Alimentícios nos Mercados do Distrito Federal e de São Paulo, em 1944.** Rio de Janeiro, 1945. Instituto Brasileiro de Geografia e Estatística. Conselho Nacional de Estatística. 6 p. Mimeografado. Inquéritos econômicos para a defesa nacional, comunicado no. 86.

The speed of circulation of food stocks in the markets of the Federal District and São Paulo, 1944.

Bras-9950.R5 **Relatório Anual do Serviço de Estatística da Previdência e Trabalho.** Data de início? Rio de Janeiro. Ministério do Trabalho, Indústria e Comércio. Serviço de Estatística da Previdência e Trabalho.

Annual report of the Bureau of Social Insurance and Labor Statistics. Includes organization charts, laws, and statistics of cost-of-living, income, salaries, etc.

Inclue esquemas da organização, legislação, e estatísticas do custo de vida, renda, salários, etc.

1000-1999 GENERAL

Can-**1100.M2** 1942 **Manual of Instructions: Balance Sheets, Revenues and Expenditures and Other Accounting Statements of Municipal Corporations.** Ottawa, E. Cloutier, 1942. Department of Trade and Commerce. Dominion Bureau of Statistics. Finance Statistics Branch. Loose-leaf.

Includes forms.

Manuel d'instructions: Bilans, revenus et dépenses et autres états comptables des corporations municipales. Comprend formules.

Can-**1400.B5** **Bulletins No. F.** Starting date? Ottawa. Department of Trade and Commerce. Dominion Bureau of Statistics.

Series of monographs.

Séries de monographies.

*Can-**1410.A5** **Annuaire du Canada.** Répertoire statistique officiel des resources, de l'histoire, des institutions et de la situation économique et sociologique du Dominion. Depuis? Ottawa. Ministère du Commerce. Bureau Fédéral de la Statistique.

Published also in Fnglish under the title **The Canada Year Book** (Can-**1410.C7**).

Publié aussi en anglais sous titre **The Canada Year Book** (Can-**1410.C7**).

Can-**1410.C5** **Canada, the Official Handbook of Present Conditions and Recent Progress.** Starting date? Annual. Ottawa. Department of Trade and Commerce. Dominion Bureau of Statistics.

Published also in French, Spanish, and Portuguese—publié aussi en français, espagnol, et portugais.

Usually referred to as **The Canadian Official Handbook.** A comprehensive survey of the current Canadian situation, in brief and attractive form. Contains data on population, vital statistics, agriculture, production, transportation and communication, labor, employment and unemployment, foreign and domestic commerce, prices, public finance, banking, and education.

Canada, le manuel officiel des conditions présentes et de progrès récents. Généralement cité comme **Le Manuel Officiel du Canada.** Une étude complète de la situation actuelle au Canada, dans une forme concise et attrayante. Contient des données sur la population, les statistiques démographiques, l'agriculture, la production, les transports et les communications, le travail—emploi et chômage, le commerce extérieur et intérieur, les prix, les finances publiques, les banques, et l'éducation.

*Can-**1410.C7** **The Canada Year Book.** The official statistical annual of resources, history, institutions, and social and economic conditions of the Dominion. 1905- . Ottawa. Department of Trade and Commerce. Dominion Bureau of Statistics.

Issued also in French—publié aussi en français (Can-**1410.A5**).

1945, a "Publications of the Dominion Bureau of Statistics." Including reports, bulletins, press releases, etc. **The Canada Year Book, 1945.** 16 p. Reprint.

Publications du Bureau Fédéral de la Statistique.

2000-2999 CENSUS—CENSOS

Can-**2100.C5** 1941.1 **Eighth Census of Canada, 1941: Methods and Procedures.** Materials used in taking population census. Acquired in connection with IASI Survey of Census Methods and Procedures, 1944. Department of Trade and Commerce. Dominion Bureau of Statistics.

Consists of: Correspondence; laws, decrees, or other authorization; publicity and propaganda; schedules; other forms; instructions; punch cards; codes.

Huitième recensement du Canada, 1941: Méthodes et procédés. Matériel employé pour le recensement de la population. Acquis en relation avec l'Étude des Méthodes et Procédés de Recensement, efectuée par l'Institut Interaméricain de Statistique. Comprend: Correspondance; lois, décrets, ou autres autorisations; publicité et propagande; questionnaires; autres formules; instructions; cartes à poinçonner; codes.

Can-**2100.C5** 1941.2 **Eighth Census of Canada, 1941: Instructions to Commissioners and Enumerators. Huitième Recensement du Canada, 1941: Instructions aux Commissaires et Enumérateurs.** Ottawa, Edmond Cloutier, 1941. Dominion Bureau of Statistics. Bureau Fédéral de la Statistique. 250 p.

Can-**2100.C5** 1941.3 **Eighth Census of Canada, 1941: Administrative Report of the Dominion Statistician. Huitième Recensement du Canada, 1941: Rapport Administratif du Statisticien du Dominion.** Ottawa, Edmond Cloutier, 1945. Dominion Bureau of Statistics. Bureau Fédéral de la Statistique. 147 p.

Includes census legislation in Canada, scope of the census and relation to other statistics, organization for taking the 1941 census, schedules, compilation and tabulation, new census machines, sampling in the census, preliminary reports of the results, plan and distribution of the census reports, and summary of the results. Contains appendixes, which include sections of the Statistical Act, enumeration of the armed forces in the census, assignment of not stated cases, table of rates and allowances to enumerators, bulletins, forms and schedules.

Comprend lois de recensement au Canada, portée du recensement et sa relation avec les autres statistiques, organisation du recensement de 1941, questionnaires, compilation et mise en tableaux, nouvelles machines de recensement, échantillonnage dans le recensement, rapports préliminaires des résultats, plan et distribution des rapports sur le recensement et sommaire des résultats. Y compris appendices, contenant articles de la loi de la statistique, l'énumération des forces armées au recensement, l'attribution des cas non spécifiés, tableaux des taux et allocations aux énumérateurs, bulletins, formules, et questionnaires.

Can-**2100**.C5 1931 **Seventh Census of 1931: Administrative Report of the Dominion Statistician.** (Reprint from Volume I of the Report of the Census of Canada, 1931.) Ottawa, J. O. Patenaude, 1934. Department of Trade and Commerce. Dominion Bureau of Statistics. 100 p.

Septième recensement du Canada, 1931: Rapport administratif du Statisticien du Dominion. (Réimpression du volume I du Rapport du Recensement du Canada, 1931.)

*Can-**2200**.C5 1941.2 **Eighth Census of Canada, 1941. Huitième Recensement du Canada, 1941.** Ottawa, 1944- . Dominion Bureau of Statistics. Bureau Fédéral de la Statistique. vol.: vol. II, **Population by Local Subdivisions—Population par Subdivisions Locales**; vol. X, **Merchandising and Service Establishments,** part I, **Retail Merchandise Trade—Établissements de Commerce et de Service,** partie I, **Commerce de Merchandises en Détail.**

Can-**2200**.C5 1941.2a **Eighth Census of Canada, 1941: Population.** Ottawa, 1942-43. Department of Trade and Commerce. Dominion Bureau of Statistics.

Bulletins for Canada and each province, presenting final population figures of the 1941 census by countries and census divisions, conjugal condition, age, racial origin, religion, birthplace, immigration and citizenship, school attendance and years of schooling, language, occupations, housing, and agriculture.

Huitième recensement du Canada, 1941: Population. Bulletins pour le Canada et chaque province, présentant les chiffres définitifs de la population du recensement de 1941 par comtés et divisions du recensement, état conjugal, âge, origine raciale, religion, lieu de naissance, immigration et citoyenneté, fréquentation scolaire et années à l'école, langue, occupations, logement, et agriculture.

Can-**2200**.C5 1931.2 **Seventh Census of Canada, 1931. Septième Recensement du Canada, 1931.** Ottawa, J. O. Patenaude, 1933-42. Dominion Bureau of Statistics. Bureau Fédéral de la Statistique. 13 vol.: Vol. I, **Population—Summary**; vol. II, **Population by Areas—Population** (par divisions géographiques); vol. III, **Ages of the People—Ages de la Population;** vol. IV, **Cross-Classification—Classifications Recoupées**; vol. V, **Earnings of Wage-Earners, Dwellings, Households, Families, Blind and Deaf-Mutes—Gains des Employés à Gages, Habitations, Ménages, Familles, Aveugles et Sourds-Muets**; vol. VI, **Unemployment—Chômage**; vol. VII, **Occupations and Industries**; vol. VIII, **Agriculture**; vol. IX, **Institutions**; vol. X, **Merchandising and Service Establishments—Part I, Retail Merchandise Trade—Commerce et Services, Partie I**; vol. XI, **Merchandising and Service Establishments—Part II, Retail Services—Wholesale Trade—Retail Chains—Hotels—Commerce et Services, Partie II**; vol. XII, **Monographs: The Canadian Family; Fertility of the Population of Canada; Housing in Canada; Illiteracy and School Attendance; The Age Distribution of the Canadian People; Canadian Life Tables**; vol. XIII, **Monographs: Unemployment; Dependency of Youth; Rural and Urban Composition of the Canadian Population; Racial Origins and Nativity of the Canadian People.**

Can-**2200**.C6 1944 **Census of Indians in Canada, 1944. Recensement des Indiens du Canada, 1944.** Ottawa, Edmond Cloutier, 1945. Department of Mines and Resources. Ministère des Mines et des Ressources. Indian Affairs Branch. Division des Affaires Indiennes. 32 p.

Census of Indians, arranged under provinces, territories, districts, and Indian agencies.

Recensement des Indiens, établi par provinces, territoires, districts, et agences.

Can-**2210**.C5 1941.2a **Eighth Census of Canada, 1941: Census of Agriculture: Nova Scotia. Huitième Recensement du Ca-**

nada, 1941: Recensement Agricole: Nouvelle-Écosse. Ottawa, Edmond Cloutier, 1945. Dominion Bureau of Statistics. Bureau Fédéral de la Statistique. 188 p.

"The present bulletin, one of a series which will deal with each province in turn, gives for Nova Scotia, the final results of the Census of Agriculture, taken as of June 2, 1941." (Source: Preface.)

"Le présent bulletin, l'un d'une série qui traitera de chacune des provinces à tour de rôle, donne les résultats définitifs du recensement de l'Agriculture fait en Nouvelle-Écosse le 2 juin 1941." (Source: Préface.)

3000-3999 STATISTICAL SCIENCE—CIENCIA ESTADISTICA

Can-**3030.**D5 1935 **The Dominion Bureau of Statistics.** Its origin, constitution and organization. Ottawa, J. O. Patenaude, 1935. Department of Trade and Commerce. Dominion Bureau of Statistics. 88 p.

Includes list of publications of the Bureau, and a copy of the law establishing the Bureau.

Le Bureau Fédéral de la Statistique. Comprend une liste des publications du Bureau, et une copie de la loi établissant le Bureau.

*Can-**3040.**C5 **Canadian Journal of Economics and Political Science.** 1935- . Quarterly. Canadian Political Science Association.

Includes articles on statistical methodology.

Le journal canadien d'économie et de science politiques. Comprend des articles sur les méthodes statistiques.

Aug. '45 "A Bibliography of Current Publications on Canadian Economics." Vol. II, no. 3, August 1945, p. 486-498.

Une bibliographie des publications courantes sur l'économie canadienne.

Aug. '45, Keyfitz "The Sampling Approach to Economic Data." By Nathan Keyfitz. Vol. II, no. 3, August 1945, p. 467-477.

L'emploi de la méthode d'échantillonnage comme moyen de traiter les données économiques.

May '45 "Recent Studies on National Income." Vol. II, no. 2, May 1945, p. 270-280.

Second of two parts.

Études récentes sur le revenu national. La deuxième de deux parties.

Feb. '45 "Recent Studies on National Income." Vol. II, no. 1, February 1945, p. 115-129.

First of two parts.

Études récentes sur le revenu national. La première de deux parties.

Aug. '43, Hope "Agriculture's Share of the National Income." By E. C. Hope. Vol. 9, no. 3, August 1943. p. 384-393.

La part de l'agriculture dans le revenu national.

Can-**3220.**C5 1941 **Classification of Industries: Eighth Census of Canada, 1941.** Ottawa, Edmond Cloutier, 1941. Department of Trade and Commerce. Dominion Bureau of Statistics. 57 p.

Classification des industries: Huitième recensement du Canada.

Can-**3240.**C5 1941 **Classification of Occupations: Eighth Census of Canada, 1941.** Ottawa, Edmond Cloutier, 1941. Department of Trade and Commerce. Dominion Bureau of Statistics. 328 p.

Classification des occupations: Huitième recensement du Canada, 1941.

Can-**3300** 1944-46 **Statistical Training in Canada: Methods and Materials.** Background material acquired for the IASI Survey on Statistical Training Methods and Materials in the American Nations, 1944-46.

Includes: Filled questionnaires on statistical courses in Canada.

Enseignement statistique au Canada: Méthodes et matériaux. Matériel fondamental recueilli pour le Relevé de l'Institut Interaméricain de Statistique sur les Méthodes et Matériaux Employés dans l'Enseignement Statistique dans les Nations Américaines. Comprend: Questionnaires remplis concernant les cours de statistique au Canada.

5000-5999 DEMOGRAPHY—DEMOGRAFIA

Can-**5100.**E7 1946 **Ensayo sobre la Evolución de la Población de las Américas: I, Canadá.** Por Jorge Justo Boero Brian-y-Francisco Molina Serrano. Montevideo, 1946. 40 p. (aprox.)

Essay on the evolution of the population of Americas: I, Canada.

Can-**5100.**F5 1946 **The Future Population of Canada.** Ottawa, 1946. Department of Trade and Commerce. Dominion Bureau of Statistics. 61 p. Bulletin, no. F-4.

Population trends in Canada, with statistical

tables and explanations showing estimates of the population up to 1971.

La population future du Canada. Tendances de la population au Canada; tableaux statistiques et explications donnant la population estimative jusqu'à 1971.

Can-**5120**.A5 1943 **The American-Born in Canada, a Statistical Interpretation.** The relations of Canada and the United States, a series of studies prepared under the direction of the Carnegie Endowment for International Peace. By Robert H. Coats, and M. G. MacLean. Toronto, Ryerson Press; New Haven, Yale University Press, 1943. 176 p.

"Part I, 'The distribution of the American-born and its significance', describes the absolute and relative distribution of the American-born and their occupations, and discusses the causal factors involved in producing the observed patterns. There is a digression on the Canadian-born in the United States and an addendum on the day-to-day movements across the border. Part II, 'Statistical analysis of the American-born in Canada', is an analysis of official Canadian statistics from the decennial censuses of 1871-1931 and the quinquennial censuses of the Prairie Provinces, 1906-36. Subjects covered include distribution by area; age, sex, and rural-urban composition; marital status, family composition and vital statistics; racial origin; language and mother-tongue; citizenship; school attendance and illiteracy; religion; and occupations and unemployment." (Source: **Books on Population, 1939-Mid 1944.**)

Les personnes de naissance américaine au Canada, une interprétation statistique. Les relations entre le Canada et les États-Unis, une série d'études preparée sous la direction du Carnegie Endowment for International Peace. "Partie I 'La répartition des personnes de naissance américaine et sa signification', décrit la répartition absolue et relative des personnes de naissance américaine et leurs professions, et étudie les facteurs causaux qui entrent en jeu dans la production des exemples observés. Il y a une digression concernant les personnes de naissance canadienne aux États-Unis et une addition concernant le mouvement frontalier quotidien. Partie II, 'Une analyse statistique des personnes de naissance américaine au Canada', présente une analyse des statistiques officielles du Canada tirées des recensements décennaux de 1871-1931 et des recensements quinquennaux des provinces de Prairies, 1906-36. Les sujets traités comprennent: Distribution par région; âge, sexe, et composition rurale et urbaine; état conjugal, composition de la famille et statistiques démographiques; origine raciale; langues et langues maternelles; citoyenneté; fréquentation scolaire et analphabétisme; religion; et professions et chômage." (Traduction du commentaire.)

Can-**5200** **Vital Statistics and Registration in Canada: Methods and Procedures.** Background material for the study conducted by the Inter American Statistical Institute in vital statistics and registration methods in the American nations, including the survey of 1945. Includes: Answer to questionnaire; instructions; punch cards and codes.

Statistiques vitales et enregistrement au Canada: Méthodes et procédés. Matériel fondamental recueilli pour l'étude effectuée par l'Institut Interaméricain de Statistique sur les méthodes d'enregistrement et les statistiques vitales dans les nations américaines, y compris l'enquête de 1945. Comprend: Réponse au questionnaire; instructions; cartes poinçonnées et codes.

Can-**5200**.P4 **Preliminary Annual Report, Vital Statistics of Canada. Rapport Prêliminaire Annuel, Statistiques Vitales du Canada.** (Exclusive of Yukon and the Northwest Territories.) (Ne comprend pas le Yukon et les Territoires du Nord-Ouest.) Starting date? Ottawa. Dominion Bureau of Statistics. Vital Statistics Branch. Bureau Fédéral de la Statistique. Branche des Statistique Vitales.

1944,a **Preliminary Annual Report, Vital Statistics of Canada, 1944, Tentative Figures. Rapport Préliminaire Annuel, Statistiques Vitales du Canada, 1944, Chiffres Provisoires.** (Exclusive of Yukon and the Northwest Territories.) With rates computed on estimated population of 11,958,000 for the nine provinces. (Ne comprend pas le Yukon et les Territoires du Nord-Ouest.) Avec taux calculés d'après une population estimative de 11,958,000 pour les neuf provinces. Ottawa, 1946. Dominion Bureau of Statistics. Vital Statistics Branch. Bureau Fédéral de la Statistique. Branche des Statistiques Vitales. 35 p. Multilithed.

1944,b **Preliminary Annual Report, Vital Statistics of Canada, 1944, Final Figures. Rapport Préliminaire Annuel, Statistiques Vitales du Canada, 1944, Chiffres Définitifs.** (Exclusive of Yukon and the Northwest Territories.) With rates computed on estimated population of 11,958,000 for the nine provinces. (Ne comprend pas le Yukon et les Territoires du Nord-Ouest.) Avec taux calculés d'après une population estimative de 11,958,000 pour les neuf provinces. Ottawa, 1946. Dominion Bureau of Statistics. Vital Statistics Branch. Bureau Fédéral de la Statistique. Branche des Statistique Vitales. 35 p. Multilithed.

Can-**5200**.P5 **Preliminary Report on Births, Deaths and Marriages. Rapport Préliminaire sur les Naissances, les Décès et les Mariages.** Starting date? Quarterly. Ottawa. Dominion Bureau of Statistics. Vital Statistics Branch. Bureau Fédéral de la Statistique. Branche des Statistiques Vitales. Multilithed.

Summary of births, deaths and marriages by provinces; infant mortality, neo-natal mortality, maternal mortality by certain specified causes; at all ages by specified causes, by provinces.

Résumé trimestrel des naissances, décès et mariages, par province; mortalité infantile, mortalité des nouveau-nés, mortalité puerpérale par certaines causes spécifiées; décès à tous âges par causes spécifiées, par province.

Can-**5200**.R5 **Registration of Births, Deaths and Marriages. Enregistrement des Naissances, Décès et Mariages.** Starting date? Monthly. Ottawa. Dominion Bureau of Statistics. Vital Statistics Branch. Bureau Fédéral de la Statistique. Branche des Statistiques Vitales.

Single mimeographed sheets. The number of registrations of births, deaths and marriages during the month in the cities, towns and villages having a population of 10,000 and over.

Feuillets miméographiés d'une seule page. Le nombre de naissances, décès et mariages enregistrés pendant l'espace d'un mois dans les cités, villes et villages d'une population de 10,000 et plus.

*Can-**5200**.V5 **Vital Statistics. Statistiques Vitales.** Annual report. Rapport annuel. 1921- . Ottawa. Dominion Bureau of Statistics. Bureau Fédéral de la Statistique.

Births, infant mortality, general mortality, causes of death, marriages.

Naissances, mortalité infantile, mortalité générale, causes de décès, mariages.

Can-**5212**.S5 1945 **A Study in Maternal, Infant and Neo-Natal Mortality in Canada. Étude sur la Mortalité Puerpérale, Infantile et Neonatale au Canada.** Prepared by the Dominion Bureau of Statistics in collaboration with the Department of National Health and Welfare. Préparée par le Bureau Fédéral de la Statistique en collaboration avec le Ministère de la Santé Nationale et du Bien-Etre Social. Ottawa, 1945. 49 p.

A reprint of the original study published in 1942, with the addition of statistics for the years 1941-43 to the text and tables.

Une réimpression de l'étude originale publiée en 1942, avec des statistiques supplémentaires pour les années 1941-43 dans le texte et dans les tableaux.

Can-**5220**.C5 **Canadian Life Tables.** Starting date? Decennial. Ottawa. Department of Trade and Commerce. Dominion Bureau of Statistics. Social Analysis and Economic Research Branch.

Tables de longévité canadienne.

1871-1931 **Canadian Abridged Life Tables, 1871, 1881, 1921, 1931.** Ottawa, 1939. Department of Trade and Commerce. Dominion Bureau of Statistics. Unexamined.

Tables de longévité canadienne abrégées, 1871, 1881, 1921, 1931.

Can-**5240**.V5 1942 **The Vital Statistics Instruction Manual for Canada. Manuel d'Instructions sur les Statistiques Vitales.** Department of Trade and Commerce. Dominion Bureau of Statistics. Division of Vital Statistics. Series of 1942. Multilithed.

Contents: Introduction to the manual; standard vital statistics codes for place of birth, racial origin, citizenship (nationality), religious denominations; geographical codes for the provinces.

Matière: Introduction au Manuel; code regulier des statistiques vitales du lieu de naissance, de l'origine raciale, citoyenneté (nationalité), des dénominations religieuses; codes géographiques pour les provinces.

Can-**5250**.H2 1937 **Handbook on Death Registration and Certification.** Containing "International List of Causes of Death." Ottawa, J. O. Patenaude, 1937. Department of Trade and Commerce. Dominion Bureau of Statistics. 44 p.

Manuel sur la certification et l'enregistrement des décès, comprenant "Liste internationale des causes de décès."

Can-**5250**.N2 1940.A **National Registration, August 1940.** Preliminary statistical tabulation. Ottawa, 1941. Published by Department of National War Services and Dominion Bureau of Statistics. 44 p. Unexamined.

L'enregistrement national, août 1940. Calcules statistiques provisoires.

Can-**5250**.N2 1940.B **Specialized Occupations, National Registration, 1940. Occupations Spécialisées, l'Enregistrement National, 1940.** Ottawa, 1941. Department of

National War Services and Dominion Bureau of Statistics. Ministère des Services Nationaux de Guerre et Bureau Fédéral de la Statistique. 73 p. Unexamined.

Can-**5250**.N2 1940.C **National Registration Regulations, 1940.** Issued by the Chief Registrar for Canada. Ottawa, J. O. Patenaude, 1940. 10 p. Unexamined.

Published also in French—publié aussi en français.

Règlements concernant l'inscription nationale, 1940. Publié par le registraire en chef du Canada.

Can-**5250**.R5 1943 **Report on Completeness of Birth Registration in Canada.** Prepared by John T. Marshall, Chief, Division of Vital Statistics. Ottawa, 1943. Dominion Bureau of Statistics. 44 p. Mimeographed. Confidential.

Contains exhibits.

Rapport sur l'integralité de l'enregistrement des naissances au Canada. Comprend des pièces à l'appui.

6000-6999 ECONOMICS — ECONOMIA

Can-**6000**.B5 **Business Summary of the Bank of Montreal.** Monthly. Montreal.

A review of business conditions throughout Canada, Newfoundland, and the United States.

Résumé des affaires de la Banque de Montréal. Une revue mensuelle de la situation commerciale dans tout le Canada, Terre-Neuve et les États-Unis.

Can-**6000**.C5 1945 **Canada.** Foreign commerce yearbook country series. Washington, D.C., U.S. Govt. Print. Off., 1945. U.S. Department of Commerce. Bureau of Foreign and Domestic Commerce. International Reference Service, vol. 2, no. 36. Unexamined.

Contains information about area, population, climate, agriculture, commerce and industry, mining and manufacturing, communication and transportation, power and light, national economic indexes, national income, average daily wage rates, foreign trade, etc.

Le Canada. Séries par pays de l'annuaire du commerce extérieur des États-Unis. Comprend des renseignements sur la superficie, la population, le climat, l'agriculture, le commerce et l'industrie, les mines et les manufactures, les transports et les communications, l'énergie et l'éclairage, les indices économiques nationales, le revenu national, le moyenne quotidienne des salaires, le commerce extérieur, etc.

*Can-**6000**.M5 **Monthly Review of Business Statistics.** 1926- . Ottawa. Department of Trade and Commerce. Dominion Bureau of Statistics. Business Statistics Branch.

A statistical summary, with tables, charts, and text, of current economic conditions in Canada, including production, construction, foreign and domestic commerce, transportation, employment, prices, and finance.

Revue mensuelle de la situation économique. Précis statistique, avec tableaux, graphiques et texte, de la situation économique courante au Canada, comprenant production, construction, commerce extérieur et intérieur, transports, emploi, prix, et finance.

Can-**6010**.P3 1943 **The Price Control and Subsidy Program in Canada.** By Jules Backman. Washington, D.C., 1943. Brooking Institution. 68 p. Pamphlet, no. 52. Unexamined.

Le programme de régie des prix et de subventions au Canada.

Can-**6010**.P5 **Prices and Price Indexes.** Starting date? Monthly. Ottawa. Department of Trade and Commerce. Dominion Bureau of Statistics. Business Statistics Branch.

Has a semi-annual supplement **World Price Movement,** containing wholesale price and cost of living.

Prix et indices des prix. Avec un supplément semestriel **World Price Movement,** comprenant les prix de gros et le coût de la vie.

Can-**6010**.P6 **Prices and Price Indexes.** Wholesale prices, cost of living, security prices. 1923- . Annual. Ottawa. Department of Trade and Commerce. Dominion Bureau of Statistics. Prices Branch.

Each issue gives comparative statistics from 1913. Publication suspended temporarily in 1941; data for 1941-43 consolidated in the 1913-43 issue. Before 1943, subtitle read: Commodities, securities, foreign exchange.

Prix et indices des prix. Prix de gros, coût de la vie, prix des valeurs mobilières. De 1923 jusqu'à présent, chaque édition donne des chiffres comparatifs à comptes de 1913. La publication a été suspendue temporairement en 1941; les données de 1941-43 réunies dans l'édition de 1913-43. Avant 1943, le sous-titre était: Produits, valeurs mobilières, change étranger.

1913-43 **Prices and Price Indexes, 1913-1943.** Wholesale prices, cost of living, security prices. Ottawa, Edmond Cloutier, 1945. Dominion Bureau of Statistics. Prices Branch. 87 p.

Contains a historical summary of all Bureau price data available, but deals chiefly with the years 1940-43. Includes tables and charts on: Index numbers of wholesale prices; wholesale price movement in Canada; cost of living in Canada, 1941-43; security price indexes; world price movements.

Contient une rétrospective de toutes les données sur prix disponibles au Bureau, mais traite principalement des chiffres des années 1940-43. Comprend tableaux et graphiques sur: Nombres-indices des prix de gros; mouvement des prix de gros au Canada; coût de la vie au Canada, 1941-43; indices des prix des valeurs mobilières; mouvement des prix mondiaux.

Can-**6020**.S5 **Survey of Production in Canada.** Starting date? Annual. Ottawa. Department of Trade and Commerce. Dominion Bureau of Statistics. Business Statistics Branch.

Un relevé de la production au Canada. Annuel.

1938-43 **Survey of Production in Canada, 1938-1943.** Ottawa, 1946. Canada. Department of Trade and Commerce. Dominion Bureau of Statistics. Business Statistics Branch. 39 p. Multilithed.

Value of Canadian production, gross and net value of production, percentage analysis of production, percentage increase in the volume of production 1938-1943, production adjusted for price change, value of Canadian production by groups, value of manufacturing production by groups, processing industries, provincial distribution of production, and forest products.

Valeur de la production canadienne, valeur brute et nette de la production, analyse proportionnelle de la production; pourcentage d'augmentation du volume de la production, 1938-43, production adjustée selon les variations des prix, valeur de la production canadienne par groupes, valeur de la production manufacturière par groupes, industries de transformation, distribution provinciale de la production, et produits forestiers.

Can-**6100**.N4 1926-36 **National Income, Canada, 1926-36.** By Donald MacGregor. Ottawa, 1939.

Revenu national, Canada, 1926-36.

Can-**6100**.N5 1919-38 **National Income of Canada, 1919-1938.** Prepared by Sydney B. Smith. Ottawa, 1941. Department of Trade and Commerce. Dominion Bureau of Statistics. Part I: **General Analysis, Scope of Inquiry and Method of Approach.**

Revenu national du Canada, 1919-38.

*Can-**6200**.Q5 **Quarterly Bulletin of Agricultural Statistics.** 1908- . Ottawa. Dominion Bureau of Statistics. Agricultural Statistics Branch.

Before 1940 issued as—avant 1940 publié comme: **Monthly Bulletin of Agricultural Statistics.**

Presents currently the outstanding agricultural records and statistics relating to acreage, production, quantity and value of crops, livestock, farm labor, wages, value of farm lands, prices, etc., together with numerous special articles on subjects of interest to Canadian agriculture.

Bulletin trimestriel de la statistique agricole. Donne couramment les principaux rapports et la statistique concernant l'acréage, la production, la quantité et la valeur des récoltes, le bétail, les employés dans l'agriculture, les gages, la valeur des terres agricoles, prix, etc., et divers articles d'intérêt particulier pour l'agriculture canadienne.

Can-**6210**.R5 **Report of the Grain Trade of Canada.** Starting date? Annual. Ottawa. Department of Trade and Commerce. Dominion Bureau of Statistics. Agricultural Statistics Branch.

Contains a summary of the production and distribution of the principal grain crops.

Rapport sur le commerce du grain du Canada. Comprend un sommaire de la production et de la distribution des principales récoltes de grain.

Can-**6230**.L5 **Live Stock and Animal Products Statistics. Statistique de Bétail et des Produits Animaux.** 1920- . Annual. Ottawa. Dominion Bureau of Statistics. Agricultural Branch. Bureau Fédéral de la Statistique. Branche de l'Agriculture.

*Can-**6240**.F5 **Fisheries Statistics of Canada. Statistique des Pêcheries.** (Prepared in collaboration with Dominion and Provincial Fisheries Departments.) (En collaboration avec les ministères des Pêcheries du Gouvernement fédéral et des Provinces.) Starting date? Annual. Ottawa. Department of Trade and Commerce. Dominion Bureau of Statistics. Fisheries and Animal Products Branch. Ministère du Commerce. Bureau Fédéral de la Statistique. Branche des Pêcheries et des Produits Animaux.

1944 **Fisheries Statistics of Canada, 1944. Statistiques des Pêcheries,**

1944. (Prepared in collaboration with Dominion and Provincial Fisheries Departments.) (En collaboration avec les ministères des Pêcheries du Gouvernement fédéral et des Provinces.) Ottawa, Edmond Cloutier, 1946. Department of Trade and Commerce. Dominion Bureau of Statistics. Fisheries and Animal Products Branch. Ministère du Commerce. Bureau Fédéral de la Statistique. Branche des Pêcheries et des Produits Animaux. 284 p.

Contents: **Summary tables:** Capital equipment and employment; review by provinces; fisheries bounty; historical review; foreign trade; **detailed tables:** Fish caught and marketed, by provinces, 1944; capital equipment and employment in primary operations, by provinces, 1944; capital equipment and employment in fish processing establishments, by provinces, 1944; classification of fish processing establishments by value of production; vessels fishing for principal kinds of sea fish, 1944; miscellaneous.

Matière: **Tableaux sommaires:** Outillage-capital et employés; revue par province; primes de pêche; revue rétrospective; commerce extérieur; **tableaux détaillés:** Poisson pris et vendu, par province, 1944; outillage-capital et employés dans les opérations primaires, par province, 1944; outillage-capital et employés dans les conserveries et saurisseries, par province, 1944; classification des conserveries et saurisseries, selon la valeur de la production; vaisseaux faisant la pêche des principales espèces de poisson de mer, 1944; divers.

*Can-**6300**.A5 **Annual Report on the Mineral Production of Canada.** 1886- . Ottawa. Canada. Department of Trade and Commerce. Dominion Bureau of Statistics. Mining, Metallurgical and Chemical Branch.

Final data on the production of Canada's mines, together with details of capital employed in the industry, salaries and wages paid, number of employees, amounts expended on fuel and power, power producing equipment installed, and process supplies purchased.

Rapport annuel de la production minérale du Canada. Chiffres définitifs sur la production des mines du Canada, avec détails sur le capital engagé dans cette industrie, les salaires et gages, le nombre d'employés, les sommes dépensées en combustible et énergie, les installations pour produire l'énergie, le matériel de conditionnement acheté.

Can-**6300**.C5 **Coal Statistics for Canada.** Starting date? Annual. Ottawa. Department of Trade and Commerce. Dominion Bureau of Statistics. Mining, Metallurgical, and Chemical Branch.

Annuaire statistique du charbon pour le Canada.

1944 **Coal Statistics for Canada for the Calendar Year 1944.** Ottawa, Edmond Cloutier, 1946. Department of Trade and Commerce. Dominion Bureau of Statistics. Mining, Metallurgical and Chemical Branch. 96 p.

Contains tables of data for the Dominion and each province on: Output; tonnage lost; disposition; shipments; exports; sizes of coal; imports; consumption; average calorific value of fuels; fuel consumption by steam railways; prices; average value per ton raised, and labor and fuel costs per ton; employees, salaries and wages, output per man-day by districts; fuel and electricity; power.

Comprend des tableaux pour le Dominion et pour chaque province sur: Production; perte de tonnage; écoulement; expéditions; exportations; grosseur du charbon; importations; consommation; valeur moyenne calorifique des combustibles; consommation de combustible par les chemins de fer à vapeur; prix; valeur moyenne par tonne produite, et les dépenses pour la main-d'oeuvre et le combustible par tonne; employés, salaires et gages, production par homme-jour par district; combustible et électricité; force motrice.

Can-**6400** **Industrial Statistics in Canada: Methods and Procedures.** Background material acquired for the study conducted by the Inter American Statistical Institute in methods of industrial statistics in the American nations, including the survey of 1945-46.

Consists of: Instructions; schedules; forms; publications.

Statistiques industrielles au Canada: Méthodes et procédés. Matériel fondamental recueilli pour l'étude efectuée par l'Institut Interaméricain de Statistique concernant les méthodes de statistiques industrielles dans les nations américaines, y compris l'enquête de 1945-46.

Comprend: Instructions; questionnaires; formules; publications.

Can-**6400**.C6 **Central Electric Stations in Canada.** (Prepared in collaboration with the Dominion Water and Power Bureau, Department of Mines and Resources.) 1917- . Annual. Ottawa. Department of

Trade and Commerce. Dominion Bureau of Statistics. Transportation and Public Utilities Branch.

Published also in French—publié aussi en français: **Usines Électriques Centrales du Canada.**

Can-**6400.**I7 **Iron and Steel and Their Products in Canada.** 1919- . Biennial. Ottawa. Department of Trade and Commerce. Dominion Bureau of Statistics. Mining, Metallurgical and Chemical Branch.

Fer et acier et leurs produits au Canada.

Can-**6400.**M5 1939-43 **Manufactures of the Non-Ferrous Metals in Canada, 1939-1943.** Ottawa, Edmond Cloutier, 1946. Department of Trade and Commerce. Dominion Bureau of Statistics. Mining, Metallurgical and Chemical Branch. 124 p.

Presents statistics on production, capital, employes, etc., for the manufacturing plants in Canada which use non-ferrous metals as their principal materials. Includes chapters on: Aluminum products (other than smelter output); brass and copper products; white metal alloys, including solders, babbitt, etc.; jewelry and silverware; electrical apparatus and supplies; non-ferrous metal smelting and refining; miscellaneous non-ferrous metal products.

Manufactures des métaux non ferreux au Canada, 1939-1943. Donne la statistique de la production, du capital, du personnel, etc., dans les établissements canadiens qui font usage des métaux non ferreux comme principales matières premières. Comprend des chapitres sur: Les produits de l'aluminum (autre que la production des smelters); les produits du laiton et du cuivre; les alliages de métal blanc, y compris les soudures, et les garnitures de métal blanc, etc.; la bijouterie et l'argenterie les appareils et les fournitures électriques; la réduction et l'affinage des métaux non ferreux; les divers produits de métaux non ferreux.

*Can-**6400.**M6 **The Manufacturing Industries of Canada.** Summary report. Starting date? Annual. Ottawa. Department of Trade and Commerce. Dominion Bureau of Statistics. General Manufactures Branch.

Les industries manufacturières du Canada. Annuel.

1943 **The Manufacturing Industries of Canada, 1943.** Summary report. Ottawa, Edmond Cloutier, 1946. Canada. Department of Trade and Commerce. Dominion Bureau of Statistics. General Manufactures Branch. 134 p.

Growth of manufacturing in Canada, value and volume of manufactured products, production by industrial groups and individual industries, capital employed, employment in manufactures, wages and salaries, size of establishments, power and fuel, and provincial and local distribution of manufacturing production.

Rapport sommaire. Progrès des industries manufacturières au Canada, valeur et volume de la fabrication de produits ouvrés, production par groupes industriels et par industrie, capital immobilisé, emploi dans les manufactures, salaires et gages, importance des établissements, force motrice et combustible, et distribution provinciale et locale de la production manufacturière.

Can-**6500.**H5 **Housing Statistics. Statistiques du Logement.** Dwelling units and type of buildings reported by municipalities and other areas for . . . months ending . . . Unités de logement et genres de maisons declarés par les municipalités et d'autres régions pour les . . . mois terminés . . . Starting date? Irregular. Ottawa. Department of Trade and Commerce. Dominion Bureau of Statistics. Public Finance Statistics Branch.

Can-**6500.**R5 **Reports on the Construction Industry in Canada.** Starting date? Annual. Ottawa. Department of Trade and Commerce. Dominion Bureau of Statistics. Construction Branch.

Rapport sur l'industrie de la construction au Canada. Annuel.

Can-**6611.**C3 **Civil Aviation.** Operating statistics. 1941- . Annual. Ottawa. Department of Trade and Commerce. Dominion Bureau of Statistics. Transportation and Public Utilities Branch. Unexamined.

Aviation civile. Statistique d'opération. Annuel.

Can-**6611.**C5 **Civil Aviation.** Operating statistics. 1941- . Monthly. Ottawa. Department of Trade and Commerce. Dominion Bureau of Statistics. Transportation and Public Utilities Branch. Unexamined.

Aviation civile. Statistique d'opération. Mensuel.

Can-**6612.**H5 **The Highway and the Motor Vehicle in Canada.** 1928- . Annual. Ottawa. Department of Trade and Commerce. Dominion Bureau of Statistics. Transportation and Public Utilities Branch.

Includes statistics on registrations of motor vehicles, highway expenditures, revenues from

motor vehicle registrations, accidents, gasoline sales and tax, etc.

Grandes routes et véhicules à moteur. Comprend statistique sur les véhicules à moteur enregistrés, les dépenses pour les grandes routes, les revenus provenant des enregistrements, les accidents, les ventes et les taxes sur l'essence, etc.

Can-**6612**.M5 **Motor Carriers.** Starting date? Annual. Ottawa. Department of Trade and Commerce. Dominion Bureau of Statistics. Transportation and Public Utilities Branch. Unexamined.

Voiturage motorisé.

Can-**6614**.E5 **Electric Railways of Canada.** Starting date? Annual. Ottawa. Department of Trade and Commerce. Dominion Bureau of Statistics. Transportation and Public Utilities Branch.

Les tramways électriques du Canada. Annuel.

Can-**6614**.S5 **Statistics of Steam Railways of Canada.** Starting date? Annual. Ottawa. Department of Trade and Commerce. Dominion Bureau of Statistics. Transportation and Public Utilities Branch.

Statistique des chemins de fer à vapeur du Canada. Annuel.

Can-**6614**.S8 **Summary of Monthly Railway Traffic Reports.** Starting date? Ottawa. Department of Trade and Commerce. Dominion Bureau of Statistics. Transportation and Public Utilities Branch. Unexamined.

Résumé des rapports mensuels sur le trafic des chemins de fer.

Can-**6615**.C5 **Canal Statistics.** Prepared in collaboration with the Department of Transport. 1919- . Annual. Ottawa. Department of Trade and Commerce. Dominion Bureau of Statistics. Transportation and Public Utilities Branch.

Published also in French—publié aussi en français: **Statistiques des Canaux.**

Statistiques des canaux. Dressée en collaboration avec le Ministère des Transports.

1945 **Canal Statistics for the Year Ended December 31, 1945.** Prepared in collaboration with the Department of Transport. Ottawa, 1946. Department of Trade and Commerce. Dominion Bureau of Statistics. Transportation and Public Utilities Branch. 41 p.

Contains statistics of traffic on Canadian canals, including the St. Lawrence-Great Lakes traffic, an analysis of Welland Ship Canal traffic, freight rates by water for grains, season of navigation, description of canals of Canada, total traffic through each canal, number and net registered tonnage of vessels using Canadian canals, total freight using Canadian canals, and tables on each individual canal.

Comprend statistique du trafic des canaux canadiens, y compris le trafic du S.-Laurent et des Grands Lacs, une analyse du trafic sur le canal maritime Welland, taux du fret par eau pour les grains, saison de navigation, description des canaux du Canada, trafic global de chaque canal, nombre et tonnage net enregistré des navires passant par les canaux canadiens, fret total transporté par les canaux canadiens et tableaux sur chaque canal.

Can-**6622**.T5 **Telegraph Statistics. Statistiques des Télégraphes.** 1911/12- . Annual. Ottawa. Department of Trade and Commerce. Dominion Bureau of Statistics. Transportation and Public Utilities Branch. Bureau Fédéral de la Statistique. Branche des Transports et des Utilités Publiques.

Includes statistics on revenues, operating expenses, income, employees, telegrams sent and received, etc.

Comprend statistique sur recettes, dépenses d'exploitation, revenu, employés, télégrammes envoyés et reçus, etc.

Can-**6622**.T6 **Telephone Statistics. Statistiques du Téléphone. 1910/11-** . Annual. Ottawa. Department of Trade and Commerce. Dominion Bureau of Statistics. Transportation and Public Utilities Branch.

Contains statistics on number of telephones, telephone systems, revenue, capital, expenditures, employees, wages and salaries, etc.

Comprend statistique sur le nombre d'appareils, les réseaux téléphoniques, les recettes, le capital immobilisé, les dépenses, les employés, les salaires et gages, etc.

Can-**6710**.C5 1929-41 **Census of Merchandising and Service Establishments.** Summary of monthly indexes of retail sales in Canada, 1929-1941. Revised base. Average, 1935-1939 = 100. Ottawa, 1942. Department of Trade and Commerce. Dominion Bureau of Statistics. 26 p. Unexamined.

Recensement des établissements de commerce et de service. Resumé des indices mensuels des ventes de détail au Canada, 1929-41.

Can-**6710**.M5 **Monthly Index of Retail Sales.** 1929- . Ottawa. Department of Trade and Commerce. Dominion Bureau of Statistics. Business Statistics Branch. Mimeographed.

Indice mensuel des ventes de détail.

Can-**6710**.M6 **Monthly Index of Wholesale Sales.** Starting date? Ottawa. Department of Trade and Commerce. Dominion Bureau of Statistics. Business Statistics Branch. Mimeographed.

Indice mensuel des ventes de gros.

Can-**6720**.A5 **Annual Report of Cheques Cashed Against Individual Accounts and Equation of Exchange.** 1924- . Ottawa. Department of Trade and Commerce. Dominion Bureau of Statistics.

Title varies—titre varie: 1924-34, **Annual Report of Bank Debits to Individual Accounts or Amount of Cheques Passing through the Banks at Clearing House Centres in Canada;** 1935-36, **Annual Report of 1, Bank Debits to Individual Accounts; 2, Notes on the Equation of Exchange in the Post-War Period.**

Rapport annuel des chèques débités aux comptes des particuliers et sur l'égalisation du change.

Can-**6720**.C5 **Commercial Failures in Canada.** 19 - . Quarterly. Ottawa. Department of Trade and Commerce. Dominion Bureau of Statistics. General Statistics Branch.

December issue in the cumulative volume for the year—l'édition de décembre est le volume cumulatif pour l'année.

Faillites commerciales au Canada.

Can-**6900** **Foreign Trade Statistics in Canada: Methods and Procedures.** Background material acquired for the study conducted by the Inter American Statistical Institute in methods of foreign trade statistics in the American nations, including the survey of 1946-47.

Consists of: Filled questionnaire and explanatory notes; forms.

Statistiques du commerce extérieur au Canada: Méthodes et procédés. Matériel fondamental recueilli pour l'étude effectuée par l'Institut Interaméricain de Statistique sur les méthodes de statistiques du commerce extérieur dans les nations américaines, y compris l'enquête de 1946-47. Comprend: Questionnaire rempli et notes explicatives; formules.

Can-**6900**.C7 1944 **Customs Tariff and Amendments with Index to January 20, 1944.** Department of Trade and Commerce. Dominion Bureau of Statistics.

Tarif douanier et modifications avec un index jusqu'au 20 janvier, 1944.

*Can-**6900**.T5 **Trade of Canada. Commerce du Canada.** Starting date? Annual. Ottawa. Department of Trade and Commerce. Dominion Bureau of Statistics. Ministère du Commerce. Bureau Fédéral de la Statistique.

Previously published as—intitulé autrefois: **Annual Report of the Trade of Canada.**

Gives figures for each article in detail, analyzing current trade trends from numerous angles, and giving a historical review of total trade by countries and class of commodities.

Donne les chiffres pour chaque article en détail, analysant les tendances courantes du commerce, à plusieurs points de vue et donnant une revue historique du total du commerce pour chaque classe de marchandise et pour chaque pays. Annuel.

1945 **Trade of Canada, Year Ended December 31, 1945. Commerce du Canada, Année Terminée le 31 Décembre, 1945.** Ottawa, Edmond Cloutier, 1946. Department of Trade and Commerce. Dominion Bureau of Statistics. Ministère du Commerce. Bureau Fédéral de la Statistique. 3 vol.: Vol. I, **Summary and Analytical Tables—Tableaux Sommaires et Analytiques;** vol. II, **Exports—Exportations;** vol. III, **Imports—Importations.**

Can-**6900**.T6 **Trade of Canada: Exports of Canadian Produce.** Starting date? Monthly. Ottawa? Department of Trade and Commerce. Dominion Bureau of Statistics. External Trade Branch.

Each monthly issue contains cumulative figures for the year—chaque édition mensuelle contient les chiffres cumulatifs pour l'année.

Contains detailed and summary tables on the exports of Canadian produce, by countries and by articles, and summary tables on exports of foreign produce.

Commerce du Canada: Exportations de produits canadiens. Contient tableaux détaillés et sommaires sur les exportations de produits canadiens, par pays et par articles, et tableaux sommaires sur les exportations de produits étrangers.

Can-**6900**.T7 **Trade of Canada: Imports Entered for Consumption.** Starting date? Monthly. Ottawa? Department of Trade and Commerce. Dominion Bureau of Statistics. External Trade Branch.

Each monthly issue contains the cumulative figures for the year—chaque édition mensuelle contient les chiffres cumulatifs pour l'année.

Contains detailed and summary tables on imports into Canada for consumption, by countries and by articles.

Commerce du Canada: Importations pour consommation. Contient tableaux détaillés et sommaires sur les importations entrant au Canada pour consommation, par pays et par articles.

Can-6910.C5 **The Canadian Balance of International Payments.** Revised statements and preliminary statement. Starting date? Annual. Ottawa. Department of Trade and Commerce. Dominion Bureau of Statistics. International Payments Branch.

La balance Canadienne des paiements internationaux.

7000-7999 HEALTH AND CULTURAL STATISTICS—ESTADISTICAS DE SALUBRIDAD Y CULTURA

Can-7110.C5 **The Canadian Journal of Public Health.** Official journal of the Canadian Public Health Association. 1910- Monthly, Toronto. Canadian Public Health Association.

Le journal canadien de la santé publique. Journal officiel de l'Association Canadienne de la Santé Publique.

Sept. '46 "Report of the Committee on Form and Content of Annual Vital Statistics Reports." By the Vital Statistics Section, Canadian Public Health Association. Vol. 37, no. 9, September 1946, p. 359-368.

Contains the committee's recommendations and two appendixes: Appendix I includes a revised list of tentative summary tables on: Population; births, deaths, and marriages; stillbirths; marriages; infant mortality; maternal mortality; cause of death; mortality from notifiable diseases; accidental deaths; Appendix II presents a table prepared by the Vital Statistics Branch of the Dominion Bureau of Statistics showing the tables published in their annual report for 1942.

Rapport du Comité sur la Forme et le Contenu des Rapports Annuels des Statistiques Démographiques. Comprend les recommandations du comité et deux appendices: Appendice I, comprend une liste révisée de tableaux provisoires sommaires sur: Population; naissances, décès, mariages; mort-nés; mariages; mortalité infantile; mortalité puerpérale; cause de décès; mortalité de maladies déclarables; morts accidentelles; Appendice II, présente un tableau preparé par la Branche des Statistiques Vitales du Bureau Fédéral de la Statistique donnant les tableaux publiés dans le rapport annuel de 1942.

June '46, Dunn "International Statistical Needs of Public Health." By Halbert L. Dunn. Vol. 37, no. 6, June 1946, p. 227-233.

Besoins de statistiques internationales pour l'hygiène publique.

Feb. '46, Cameron "State of Health of the People of Canada in 1944." By G. D. W. Cameron and J. T. Marshall. Vol. 37, no. 2, February 1946, p. 57-64.

L'état de santé de la population canadienne en 1944.

Can-7110.R5 1945 **Report of the National Health Survey.** Conducted by the Canadian Medical Procurement and Assignment Board. Ottawa, King's Printer, 1945. Canadian Medical Procurement and Assignment Board. 336 p. Unexamined.

" . . . Source book of information . . . with charts, diagrams, and statistical tables . . . inventory of health services . . . available in Canada when the goverment's reform programme gets under way . . . Conditions of health and distribution of illness . . . only incidentally dealt with . . . " (Source: **The Canadian Journal of Economics and Political Science,** vol. 12, no. 1, February 1946, p. 98-99.)

Rapport de l'enquête sur la santé nationale. " . . . Manuel de renseignements . . . graphiques, diagrammes et tableaux statistiques . . . inventaire des services de santé . . . sera distribué au Canada lorsque le programme de réformes de l'État sera inauguré . . . Situation de la santé et distribution de la maladie . . . étudiées incidemment seulement . . . " (Traduction du commentaire.)

Can-7130.A4 **Annual Report of Mental Institutions. Rapport Annuel sur les Institutions pour Maladies Mentales.** 1932- . Ottawa. Department of Trade and Commerce. Dominion Bureau of Statistics. Institutional Statistics Branch. Ministère du Commerce. Bureau Fédéral de la Statistique. Branche de la Statistique des Institutions.

Can-7130.A6 **Annual Report of Tuberculosis Institutions. Rapport Annuel sur les Institutions pour Tuberculeux.** 1943?- Ottawa. Department of Trade and Commerce.

Dominion Bureau of Statistics. Institutional Statistics Branch. Ministère du Commerce. Bureau Fédéral de la Statistique. Branche de la Statistique des Institutions.

1944 **Annual Report of Tuberculosis Institutions for the Year 1944.** Ottawa, Edmond Cloutier, 1946. Dominion Bureau of Statistics. Institutional Statistics Branch. 95 p.

Contains data on tuberculosis institutions, including admissions, discharges, and deaths of patients. Includes a section on tuberculosis clinics.

Contient des renseignements sur les institutions pour tuberculeux, y compris les admissions, les sorties et les décès de patients. Comprend une section sur les cliniques antituberculeuses.

*Can-7200.B5 **Biennial Survey of Education in Canada.** 1921- . Ottawa. Department of Trade and Commerce. Dominion Bureau of Statistics. Education Statistics Branch.

Title varies—le titre varie.

Relevé biennal de l'education au Canada.

Can-7200.L5 1942-44 **Libraries in Canada, 1942-44.** Being part III of the **Biennial Survey of Education in Canada, 1942-44.** Ottawa, Edmond Cloutier, 1946. Department of Trade and Commerce. Dominion Bureau of Statistics. Education Statistics Branch. 54 p.

Includes summary and analysis of data on public libraries, university and college libraries, government, business and technical society libraries, and professional progress, and also a section with the libraries listed individually.

Bibliothèques au Canada, 1942-44, partie III du relevé biennal de l'éducation au Canada. Comprend un sommaire et analyse des données sur les bibliothèques publiques, les bibliothèques des universités et des collèges, les bibliothèques du gouvernement, les bibliothèques des sociétés commerciales et professionelles, et sur les progrès professionels, en plus d'une section où les bibliothèques sont cataloguées individuellement.

Can-7210.E5 1944 **Educational Institutions in Canada.** Prepared by John E. Robbins. Ottawa, 1944. Department of Trade and Commerce. Dominion Bureau of Statistics. Education Statistics Branch. 93 p. Unexamined.

Maisons d'enseignement au Canada.

8000-8999 GOVERNMENT AND POLITICAL STATISTICS—ESTADISTICAS GUBERNAMENTALES Y POLITICAS

Can-8300.A5 **Annual Report of Statistics of Criminal and Other Offences. Rapport Annuel sur la Statistique de la Criminalité.** 1876- . Ottawa. Department of Trade and Commerce. Dominion Bureau of Statistics. Judicial Statistics Branch. Ministère du Commerce. Bureau Fédéral de la Statistique. Section de la Statistique Judiciaire.

1945 **Seventieth Annual Report of Statistics of Criminal and Other Offences for the Year Ended September 30, 1945. Soixante-Dixième Rapport Annuel sur la Statistique de la Criminalité pour l'Année Expirée le 30 Septembre 1945.** Offences by adults—juvenile delinquency—court proceedings—police statistics—prison statistics—pardons and commutations—appeals. Délits des adultes—délits juvéniles— jugements et leurs formes—statistiques de la police—statistiques des établissements pénitentiaires—grâces et commutations—appels. Ottawa, Edmond Cloutier, 1946. Dominion Bureau of Statistics. Judicial Statistics Branch. Bureau Fédéral de la Statistique. Section de la Statistique Judiciaire. 200 p.

Can-8320.J5 1945 **Juvenile Delinquents, for the Year Ended September 30, 1945. Jeunes Délinquants pour l'Année Expirée le 30 Septembre 1945.** Ottawa, Edmond Cloutier, 1946. Dominion Bureau of Statistics. Judicial Statistics Branch. Bureau Fédéral de la Statistique. Section de la Statistique Judiciaire. 45 p.

Reprinted from the **Annual Report of Criminal Statistics, 1945.**

Réimprimé du rapport annuel sur la **Statistique de la Criminalité, 1945.**

9000-9999 LABOR AND LIVING CONDITIONS—CONDICIONES DE TRABAJO Y DE VIDA

Can-9200.E5 **The Employment Situation . . . Together With Payrolls.** (As reported by employers having 15 or more employees.) 19 - . Monthly. Ottawa. Department of Trade and Commerce. Dominion Bureau of Statistics.

Includes tables and charts and index numbers of employment by economic areas and industries.

La situation de l'emploi . . . et bordereaux de paye, tels que déclarés par les patrons employant 15 personnes ou plus. Comprend tableaux et cartes et nombres-indices de l'emploi par région économique et par industrie.

Can-**9200**.L5 **Labour Force Bulletin.** 1946- Quarterly? Ottawa. Dominion Bureau of Statistics. Labour Force Survey.

Shows the results of the quarterly surveys of the civilian labor force in Canada.

Bulletin de la main-d'oeuvre. Donne les résultats des relevés trimestriels de la main-d'ceuvre civile au Canada.

Can-**9200**.M5 **Man Hours and Hourly Earnings.** Starting date? Monthly. Ottawa. Department of Trade and Commerce. Dominion Bureau of Statistics. Unexamined.

Heures-ouvrier et gains par heure.

Can-**9200**.O5 1891-1931 **Occupational Trends in Canada, 1891-1931.** Ottawa, 1939. Department of Trade and Commerce. Dominion Bureau of Statistics. 29 p.

Tendances dans les professions au Canada, 1891-1931.

Can-**9500**.C4 **Canadian Farm Family Living Costs.** 1941- . Semiannual. Ottawa. Department of Trade and Commerce. Dominion Bureau of Statistics. Business Statistics Branch. Unexamined.

Coût de la vie de la famille agricole au Canada.

Can-**9500**.C6 1913-42 **Cost of Living Index Numbers for Canada, 1913-1942.** Ottawa, 1943. Department of Trade and Commerce. Dominion Bureau of Statistics. 15 p. Unexamined.

Nombres-indices du coût de la vie au Canada, 1913-42.

Can-**9950**.R5 1943 **Report on Social Security for Canada.** Prepared by Leonard Charles Marsh. Ottawa, E. Cloutier, 1943. Advisory Committee on Reconstruction. 145 p.

An analysis in Spanish of this report appeared in the **Revista Internacional del Trabajo,** vol. XXVII, no. 5, mayo 1943.

Rapport sur la sécurité sociale au Canada. Une analyse de ce rapport fut publiée en espagnol dans la **Revista Internacional del Trabajo,** tome XXVII, n° 5, mai 1943.

Can-**9952**.A5 **Annual Report on Current Benefit Years Under the Unemployment Insurance Act.** Starting date? Ottawa. Department of Trade and Commerce. Dominion Bureau of Statistics. Unexamined.

Rapport annuel sur les années de prestation courantes en vertu de la loi d'assurance-chômage.

Can-**9952**.S5 **Statistical Report on the Operation of the Unemployment Insurance Act.** 1942- . Monthly. Ottawa. Department of Trade and Commerce. Dominion Bureau of Statistics. Unexamined.

Rapport statistique sur l'application de la loi d'assurance-chômage. Mensuel.

CHILE

1000-1999 GENERAL

Chile-**1400.**S5 1933 **Sinopsis Geográfico-Estadística de la República de Chile, 1933.** Santiago, Soc. Imp. y Lit. Universo, 1933. Ministerio de Economía y Comercio. Dirección General de Estadística.

Synopsis of geographical and statistical data of Chile, 1933, including tables and charts.

Incluye cuadros y diagramas.

*Chile-**1420.**E5 **Estadística Chilena.** 1928- Mensual. Santiago. Ministerio de Economía y Comercio. Dirección General de Estadística.

Monthly summary of national statistics, including: Demography; agriculture and livestock; mining; industry; building; prices; transportation; foreign trade; money and capital; public finance; labor; social security; justice and police; etc. Contains a bibliography. December issue gives synopsis of year's activities and includes comparative data for preceding years. Since December 1941 issue, results of 1940 census of population have appeared in this publication:

Resumen mensual de la estadística nacional, incluyendo: Demografía; agricultura y ganadería; minería; industria; edificación; precios; transportes; comercio exterior; dinero y capitales; hacienda pública; trabajo; previsión y asistencia social; policía y justicia; etc. Contiene una bibliografía. La edición de diciembre contiene sinopsis de las actividades desarrolladas durante el año incluyendo datos retrospectivos para comparación. Desde la edición de diciembre de 1941, en esta publicación han aparecido los resultados del censo de población de 1940.

Apr. '46 "Estudios sobre Mortalidad por Tuberculosis, Gripe y Sífilis." Año XIX, no. 4, abril 1946, p. 149-151.

Studies on mortality from tuberculosis, grippe, and syphilis. Data for 1928-43.

Datos para 1928-43.

Mar. '46 "Distribución Abecedaria de Apellidos." Año XIX, no. 3, marzo 1946, p. 144.

A table showing frequency distribution of names beginning with each letter of the alphabet, for use in setting up archives, card files, etc.

Un cuadro que indica la frecuencia abecedaria de apellidos empezando con las distintas letras del alfabeto, para uso en el establecimiento de archivos de apellidos de fichas y tarjetas, etc.

Nov. '45,a "Censo Continental de 1950." Año XVIII, no. 11, noviembre 1945, p. 461-462.

Comments on the projected 1950 census of the Americas.

Mar.-Apr. '45 "Pesca: Producción, Valor y Distribución de la Producción Pesquera." Año XVIII, no. 3-4, marzo-abril 1945, p. 41.

Fish: Production, value, and distribution of fishery production.

Nov. '44 "Número Total de Empleados Públicos." Año XVII, no. 11, noviembre 1944, p. 786-787.

Total number of public employees. Includes: Total number and basic cost of public employees, by departments, 1938 and 1942-44.

Incluye: Número y costo base total de los empleados públicos, por ministerios, 1938 y 1942-44.

June '44,a "Existencia de Anuarios, Censos y Otras Publicaciones Editadas por la Dirección General de Estadística." Año XVII, no. 6, junio 1944, inside back cover—cara interna de la tapa posterior.

List of annuals, censuses, and other publications issued by the General Bureau of Statistics.

*Dec. '41 "XI Censo General de Población." Año XIV, no. 12, diciembre 1941.

XI general census of population, 1940: National population summaries by province, department, commune, and district. Other issues of **Estadística Chilena** which give results from the 1940 population census are: June 1942 (**number of dwellings, by category, district, locality, and sex; population by province, department, commune, and district, classified by sex**); July 1942 (number of dwellings, by category, commune, district, and locality: Prov. of Antofagasta); Aug. 1942, p. 356-362 (number of dwellings . . .: Prov. of Atacama); Sept. 1942, p. 408-420 (number of dwellings . . .: Prov. of Coquimbo); Oct. 1942, p. 466-474 (number of dwellings . . .: Prov. of Aconcagua; urban and rural population, with distinction of sex by provinces, departments, communes, and districts: Prov. of Tarapaca, Antofagasta, and Atacama); Nov. 1942, p. 518-522 (number of dwellings . . .: Prov. of Valparaiso); Dec. 1942, p. 653-675 (**population, by provinces, in 1940 and in previous censuses; rate of increase of the population, by provinces, in 1940 and in previous censuses;** number of dwellings . . .: Prov. of Santiago and O'Higgins; urban and rural population . . .: Prov. of Coquimbo, Aconcagua, and Valparaiso); Jan.-Feb. 1943, p. 45-53 (number of dwellings . . .: Prov. of Colchagua and Curico); April 1943, p. 152-177 (number of dwellings . . .: Prov. of Talca and Maule; urban and rural population . . .: Prov. of Santiago, O'Higgins, Colchagua, Curico, Talca, and Maule); May 1943, p. 227-236 (prov. of Linares: Number of dwellings . . .; urban and rural population . . .); June 1943, p. 301-318 (prov. of Nuble: Number of dwellings . . .; urban and rural population . . .); July 1943, p. 363-390 (prov. of Concepcion and Arauco: Number of dwellings . . .; urban and rural population . . .); Aug. 1943, p. 431-449 (prov. of Bio-Bio: Number of dwellings . . .; urban and rural population . . .); Sept. 1943, p. 509-545 (prov. of Malleco and Cautin: Number of dwellings . . .; urban and rural population . . .); Oct. 1943, p. 597-607 (number of dwellings . . .: Prov. of Valdivia and Osorno); Dec. 1943, p. 771-791 (number of dwellings . . .: Prov. of Llanquihue, Chiloe, Aysen, and Magallanes; urban and rural population . . .: Prov. of Valdivia, Osorno, Llanquihue Chiloe, Aysen, and Magallanes); Jan.-Feb. 1944, p. 52-55 (**population of cities with more than 5,000 inhabitants in 1940 and in previous censuses; towns of 1,000 to 5,000 inhabitants in 1940 and in previous censuses**); May 1945, p. 75 (**working population of the country**); June 1945, p. 118-137 (**gainfully-occupied men and women by work and status; foreigners, by province and sex;** prov. of Tarapaca and Antofagasta: Religion; age, civil status, and education; activities by sex, status, and commune; occupation by sex and commune); July 1945, p. 194-207 (prov. of Atacamba y Coquimbo: Religion; age . . .; activities . . .; occupation . . .; Aug. 1945, p. 260-282 (prov. of Aconcagua and Valparaiso: Religion; age . . .; activities . . .; occupation . . .); Oct. 1945, p. 381-418 (prov. of Santiago and O'Higgins: Religion; age . . .; activities . . .; occupation . . .); Nov. 1945, p. 471-488 (prov. of Colchagua y Curico: Religion; age . . .; activities . . .; occupation . . .); Dec. 1945, p. 567-588 (**education of the population by sex from 1854 to 1940 and percentage of literates; illiteracy of those over 15 years of age for each 100 inhabitants; population of the country, by age, sex, and province; Chile—age, civil status, and education—total men and total women;** prov. of Talca and Maule: Religion; age . . .; activities . . .; occupation . . .); Jan.-Feb. 1946, p. 15-36 (prov. of Linares and Nuble: Religion; age . . .; activities . . .; occupation . . .; **education of the population of the country according to the population censuses of 1930 and 1940, by age and sex (literacy and illiteracy)**); March 1946, p. 92-111 (**education of the population of the country, by provinces, population of school age without education, and illiterate population of post-school age;** prov. of Concepcion and Arauco: Religion; age . . .; activities . . .; occupation . . .); April 1946, p. 157-174 (prov. of Bio-Bio and Malleco: Religion; age . . .; activities . . .; occupations . . .); June 1946, p. 326-360 (prov. of Cautin, Valdivia, Osorno, and Llanquihue: Religion; age . . .; activities . . .; occupations . . .); July, 1946, p. 416-439 (prov. of Chiloe, Aysen, and Magallanes: Religion; age . . .; activities . . .; occupations . . .; **country total—religion**); Sept. 1946, p. 546-564 (**population of country, by occupation and sex; detail of activities of population of country**).

Resúmenes de la población nacional por provincias, departamentos, comunas y distritos. Otras ediciones de **Estadística Chilena** que dan los resultados del censo de población de 1940, son: Junio 1942, p. 217-236 (**número de viviendas por categoría, distritos, localidades, y sexos; habitantes por provincias, departamentos, comunas y distritos, clasificados por sexo**); julio 1942, p. 296-300 (número de viviendas por categoría, comunas, distritos y localidades: Prov. de Antofagasta); ag. 1942, p. 356-362 (número de viviendas . . .: Prov. de Atacama); sept. 1942, p. 408-420 (número de viviendas . . .: Prov. de Coquimbo); oct. 1942, p. 466-474 (número de viviendas . . .: Prov. de Aconcagua; población urbana y rural, con distinción del sexo por provincias, departamentos, comunas y distritos: Prov. de Tarapacá, Antofagasta y Atacama); nov. 1942, p. 518-522 (número de viviendas . . .: Prov. de Valparaíso); dic. 1942, p. 653-675 (**población, por provincias, en 1940 y en censos anteriores; intensidad del desarrollo de la población, por provincias, en 1940 y en censos anteriores;** número de viviendas . . .: Prov. de Santiago y O'Higgins;

población urbana y rural . . .: Prov. de Coquimbo, Aconcagua y Valparaíso); en.-feb. 1943, p. 45-53 (número de viviendas . . .: Prov. de Colchagua y Curicó); abril 1943, p. 152-177 (número de viviendas . . .: Prov. de Talca y Maule; población urbana y rural . . .: Prov. de Santiago, O'Higgins, Colchagua, Curicó, Talca y Maule); mayo 1943, p. 227-236 (Prov. de Linares: Número de viviendas . . .; población urbana y rural . . .); junio 1943, p. 301-318 (Prov. de Nuble: Número de viviendas . . .; población urbana y rural . . .); julio 1943, p. 363-390. (Prov. deConcepción y Arauco: Número de viviendas; población urbana y rural...); ag. 1943, p. 431-449 (Prov. de Bio-Bio: Número de viviendas . . .: población urbana y rural . . .); sept. 1943, p. 509-545 (Prov. de Malleco y Cautín: Número de viviendas . . .: población urbana y rural . . .); oct. 1943, p. 597-607 (número de viviendas . . .: Prov. de Valdivia y Osorno); dic. 1943, p. 771-791 (número de viviendas . . .: Prov. de Llanquihue, Chiloé, Aysen y Magallanes; población urbana y rural . . .: Prov. de Valdivia, Osorno, Llanquihue, Chiloé, Aysen y Magallanes); en.-feb. 1944, p. 52-54 (**población de las ciudades con más de 5,000 habitantes en 1940 y censos anteriores; pueblos de 1.000 a 5.000 habitantes según el censo de 1940 y en censos anteriores**); mayo 1945, p. 75 (**población activa del país**); junio 1945, p. 118-137 (**hombres y mujeres activos según trabajo y situación; extranjeros, por provincia y sexo;** prov. de Tarapacá y Antofagasta: Religiones; edad, estado civil e instrucción; actividades por sexo, situación y comunas; ocupaciones por sexo y comunas); julio 1945, p. 194-207 (prov. de Atacamba y Coquimbo: Religiones; edad . . .; actividades . . .; ocupaciones . . .); ag. 1945, p. 260-282 (prov. de Aconcagua y Valparaíso: Religiones; edad . . .; actividades . . .; ocupaciones . . .); oct. 1945, p. 381-418 (prov. de Santiago y O'Higgins: Religiones; edad . . .; actividades . . .; ocupaciones . . .); nov. 1945, p. 471-488 (prov. de Colchagua y Curicó: Religiones; edad . . .; actividades . . .; ocupaciones . . .); dic. 1945, p. 567-588 (**instrucción de los habitantes por sexos, desde 1854 hasta 1940 y porcentaje de alfabetos; analfabetismo de los mayores de 15 años por cien habitantes; población del país por edades, sexo y provincias; Chile—edad, estado civil e instrucción—total hombres y total mujeres;** prov. de Talca e Maule (religiones; edad . . .; actividades . . .; ocupaciones . . .); en.-feb. 1946, p. 15-36 (prov. de Linares y Nuble: Religiones; edad . . .; actividades . . .; ocupaciones . . .; **instrucción de los habitantes de la República, según censos de población de 1930 y 1940, por edades y sexo (alfabetismo y analfabetismo)**); marzo 1946, p. 92-111 (**instrucción de los habitantes de la República, por provincias, población de edad escolar sin instrucción y población analfabeta de edad post-escolar;** prov. de Concepción y Arauco: Religiones; edad . . .; actividades . . .; ocupaciones . . .); abril 1946, p. 157-174 (prov. de Bio-Bio y Malleco: Religiones; edad . . .; actividades . . .; ocupaciones . . .); junio 1946, p. 326-360 (prov. de Cautín, Valdivia, Osorno y Llanquihue: Religiones; edad . . .; actividades . . .; ocupaciones . . .); julio 1946, p. 416-439 (prov. de Chiloé, Aysen y Magallanes: Religiones; edad . . .; actividades . . .; ocupaciones . . .; **total de país—religiones**); sept. 1946, p. 546-564 (**habitantes del país, según ocupación y sexo; detalle de las actividades de la población del país**).

Chile-**1900**.E5 1943 **La Economía Nacional y el Problema de las Subsistencias en Chile.** Por Elías Nehgme Rodríguez. Santiago, 1943. Universidad de Chile. 2 vol. No examinado.

The national economy and the problem of subsistence in Chile. ". . . A summary of the economic and social situation in Chile, with an abundance of statistics . . . The introduction deals with the bases of the problem of subsistence: Population and natural resources; data related to the problem of food, clothing, and lodgings; and the consequences. Part I presents the fundamental concepts of the problem of subsistence, and part II the social and economic causes that affect the problem, both general causes such as concentration of landownership and imperialism, and specific causes—money, production, commerce, consumption, income, and State intervention. The third part treats the fundamentals relative to a plan to lower the cost of living, with general conclusions. There is also a general bibliography." (Translation of comment.)

" . . . Resumen de la situación económica y social de Chile con abundancia de estadísticas . . . La introducción trata de los antecedentes del problema de subsistencia: Población y recursos naturales; los datos relacionados con el problema de alimentación, vestido y alojamiento; y las consecuencias. La parte I expresa los conceptos fundamentales del problema de subsistencia, y la parte II las causas económicas y sociales que afectan el problema, asimismo las causas generales, el latifundismo y el imperialismo, y causas específicas—dinero, producción, comercio, consumo, ingresos e intervención del Estado. La tercera parte se dedica a los fundamentos relativos a un plan para diminuir el costo de la vida, con conclusiones generales. También hay una bibliografía general." (Fuente: **Revista Internacional del Trabajo,** vol. XXX, no. 1, julio 1944.)

2000-2999 CENSUS — CENSOS

Chile-**2100**.C5 1940.1 **Censo de 1940: Métodos y Procedimientos.** Material usado en el levantamiento de los censos de población y de viviendas. Adquirido por el IASI con motivo de su estudio sobre Métodos y Procedimientos de los Censos, 1944. De la Dirección General de Estadística.

Census of 1940: Methods and procedures. Material used in taking population and housing census. Acquired in connection with IASI Survey of Census Methods and Procedures, 1944. Includes: Instructions.

Incluye: Instrucciones.

Chile-**2100**.C5 1940.2 **Instrucciones para la División del Territorio Comunal en Zonas de Empadronamiento: XI Censo de Población.** Santiago, Imp. y Lito. Universo, S.A., 1940. Ministerio de Economía y Comercio. Dirección General de Estadística. Comisión Directiva del Censo. 26 p. Folleto 2.

Instructions for the division of communal territory into census zones, for 1940 population census.

Chile-**2100**.C5 1940.3 **Libreta de Empadronador: XI Censo de Población.** Santiago, Imp.y Lito.Universo, S.A., 1940. Ministerio de Economía y Comercio. Dirección General de Estadística. 32 p. Comisión Directiva del Censo. Folleto 3.

Enumerator's booklet for the 1940 census of population.

Chile-**2100**.C5 1940.4 **Reglamento del XI Censo de Población.** Dtos. 136 y 483. Santiago, Imp.y Lito.Universo, S.A., 1940. Ministerio de Economía y Comercio. Dirección General de Estadística. Comisión Directiva del Censo. 14 p. Folleto no. 1.

Regulation for the 1940 population census: Decrees no. 136 and 483.

Chile-**2150**.C7 1943 **El Censo Económico General de 1943.** Antecedentes sobre su organización. Santiago, Imp.y Lit.Universo, S.A., 1943. Ministerio de Economía y Comercio. Dirección General de Estadística. 50 p. (aprox.)

General economic census of 1943: Antecedents concerning its organization. Includes: Decree no. 214, of April 9, 1943, authorizing census; instructions for municipalities; instructions for enumerators; housing schedule; agriculture and livestock questionnaire; industrial questionnaire (includes mining, commerce, and health and educational institutions); special form for employees, tenants, and copartners of farms; description of the census zone (for rural zones only); control list for enumerators; communal list.

Incluye: Decreto supremo no. 214, de 9 de abril de 1943, que ordena la realización del censo económico general; instrucciones para las municipalidades; instrucciones para las empadronadores; cédula de vivienda; cuestionario agrícola y ganadero; cuestionario industrial (incluso minería, comercio y establecimientos sanitarios y educacionales); formulario especial para empleados, inquilinos y medieros; descripción de la zona de empadronamiento (solamente para las zonas de empadronamiento rurales); lista de control para empadronadores; lista comunal.

Chile-**2200**.C5 1930 **Resultados del X Censo de la Población Efectuado el 27 de Noviembre de 1930 y Estadísticas Comparativas con Censos Anteriores.** Santiago, Imp.Universo, 1931-35. Dirección General de Estadística. 3 vol.: Vol. I, **Estadísticas Comparativas con Censos Anteriores; Población Urbana y Rural, según Sexo, por Provincias, Departamentos y Comunas;** vol. II, **Edad, Estado Civil, Nacionalidad, Religión e Instrucción;** vol. III, **Ocupaciones.**

Results of the tenth census of population, November 27, 1930.

*Chile-**2210**.A5 1935-36 **Agricultura 1935-36 Censo.** Santiago, Imp.Universo, 1938. Dirección General de Estadística. 704 p.

Census of agriculture, 1935-36. Includes schedules, forms, and instructions, p. 674-704.

Incluye cédulas, formularios e instrucciones, p. 674-704.

*Chile-**2230**.C5 1933 **Censo de Educación, Año 1933.** Santiago, Imp. Universo, 1934. Dirección General de Estadística. 84 p.

Census of education, 1943. Condition of the premises, number and age of students, marital status, nationality, and training of teachers in primary, secondary, and special schools and universities, and data on public libraries, museums of natural history, and the press.

Condición de los locales, número y edad de los

alumnos, y estado civil, nacionalidad y preparación del profesorado en escuelas primarias, secundarias, especiales y en universidades, y datos sobre bibliotecas públicas, museos de historia natural, y la prensa.

*Chile-**2250**.C5 1937 **Censo Industrial y Comercial, Año 1937.** Santiago, Imp.y Lito.Universo, S.A., 1939. Dirección General de Estadística. 241 p.

Census of industry and commerce, 1937. Includes number of establishments, employes, laborers, monthly salaries and weekly wages, in industry and commerce. The preface explains the methods employed and definitions used in taking the census.

Incluye el número de establecimientos, empleados, obreros, sueldos mensuales y salarios semanales en la industria y el comercio. El prefacio explica los métodos seguidos y definiciones usadas en la realización del censo.

3000-3999 STATISTICAL SCIENCE—CIENCIA ESTADISTICA

Chile-**3030**.L5 1932 **Dirección General de Estadística: Ley Orgánica.** Santiago, Imp.Dir.Gral.de Estadística, 1932? 12 p.

Organic law of the General Bureau of Statistics.

*Chile-**3040**.E5 **Economía.** Revista de la Facultad de Economía y Comercio. 1939- . Mensual. Santiago. Universidad de Chile. Facultad de Economía y Comercio.

Monthly review of the School of Economics and Commerce of the University of Chile. Includes articles on statistics.

Incluye artículos sobre la estadística.

Aug. '45, Levine Bawden "Observaciones Preliminares sobre la Disponibilidad de Bienes de Consumo y la Creación de Bienes de Capital." Por Flavián Levine Bawden y Pedro Irañeta Lepe. Año VI, nos. 15-16, agosto 1945, p. 72-84.

Preliminary observations on disposition of consumer goods and creation of capital goods.

Apr. '45, Levine Bawden "La Población Chilena: Segunda Parte, El Trabajo de la Población Chilena." Por Flavián Levine Bawden y Juan Crocco Ferrari. Año VI, no. 14, abril 1945, p. 17-85.

The Chilean population: Second part, work of the population. Contents: Labor supply; determination of salaries; causes and effects of geographic distribution and concentration of activities; unemployment, causes and remedies; statistical sources; methodology of classification and estimation; gainfully-occupied population of Chile; concentration of population in industrial centers; problem of unemployment in Chile; specialized workers; absenteeism; salaries; conclusions of the second part.

Contiene: La oferta de trabajo; la determinación de los salarios; causas y efectos de la distribución y concentración geográfica de la actividad; la desocupación, causas y remedios; las fuentes estadísticas; metodología de las clasificaciones y estimaciones; la población activa chilena; la concentración de la población en los centros industriales; el problema de la desocupación en Chile; la mano de obra especializada; el ausentismo obrero; los salarios en Chile; conclusiones de la segunda parte.

July '44, Levine Bawden, a "Indices de Producción Física." Por Flavián Levine Bawden. Año V, no. 10-11, julio 1944, p. 69-98.

Indexes of physical production.

July '44, Levine Bawden, b "La Población Chilena: Primera Parte, Las Características Demográficas." Por Flavián Levine Bawden y Juan Crocco Ferrari. Año V, no. 10-11, julio 1944, p. 31-68.

The Chilean population: First part, demographic characteristics. Contents: Historical observations; censuses; increase of the population; net rate of reproduction of the Chilean population; causes of variations in rate of increase; conclusions of the first part.

Contiene: Observaciones históricas; los censos; el crecimiento de la población; la tasa neta de reproducción de la población chilena; las causas de las variaciones de la tasa de crecimiento; conclusiones de la primera parte.

Chile-**3210**.L5 1943 **Lista en Castellano e Inglés de los Artículos Importables desde los Estados Unidos de Norte América (con Números del "Schedule B" y Unidades).** Santiago? 1943. Consejo Nacional de Comercio Exterior. No examinado.

List in Spanish and English of articles that can be imported from the United States, with "Schedule B" numbers, and units.

Chile-**3230**.N5 1929 **Nomenclatura de las Enfermedades.** (Estadística de morbilidad.) (Estadística de las causas de muerte.) Acordada por la Comisión internacional encargada de la revisión de las nomenclaturas internacionales de las enfermedades . . . Santiago, Balcells & Co., 1930. Dirección General de Estadística. 304 p.

Nomenclature of diseases for statistics of morbidity and causes of death.

Chile-**3300** 1944-46 **Enseñanza Estadística en Chile: Métodos y Materiales.** Material básico adquirido para la Encuesta del Instituto Interamericano de Estadística sobre los Métodos y Materiales en la Enseñanza Estadística en las Naciones Americanas, 1944-46.

Statistical training in Chile: Methods and materials. Background material acquired for the IASI Survey on Statistical Training Methods and Materials in the American Nations, 1944-46. Includes: Letters and information on statistical courses in Chile.

Incluye: Cartas e informes sobre cursos de estadística en Chile.

5000-5999 DEMOGRAPHY—DEMOGRAFIA

Chile-**5000**.C5 1943 **Chile: Demographic Data.** Analytical discussion, graphic presentation, population, natality and mortality statistics. Washington, D.C., July 1943. Prepared by U.S. Department of Commerce, Bureau of the Census, in cooperation with Office of the Coordinator of Inter-American Affairs. 100 p.

Resumen bioestadístico de Chile. Incluye mapas y cuadros y datos sobre población, natalidad y mortalidad.

*Chile-**5000**.D5 **Demografía y Asistencia Social.** Fecha de iniciación? Anual. Santiago. Ministerio de Economía y Comercio. Dirección General de Estadística.

Yearbook of demographic statistics and statistics of social assistance.

1942 **Demografía y Asistencia Social, Año 1942.** Santiago, Imp.y Lito. Universo, S.A., 1945. Dirección General de Estadística. 152 p.

Includes: **Demography:** Territory and population; marriage rate; birth rate; stillbirth rate; general mortality; infant mortality; **social assistance:** Hospitals; scientific institutions and laboratories; public assistance; nurseries; asylums; insane asylums; almshouses; Legal Medical Institute; clinics and dispensaries; hygiene; health of the armed forces.

Incluye: **Demografía:** Territorio y población; nupcialidad; natalidad; mortinatalidad; mortalidad general; mortalidad infantil; **asistencia social:** Servicio hospitalario; institutos científicos y laboratorios; asistencias públicas; gotas de leche; asilos; manicomios; hospicios; Instituto Médico Legal; policlínicos y dispensarios; higiene; sanidad de las fuerzas armadas.

Chile-**5200** **Estadística Vital y Registro en Chile: Métodos y Procedimientos.** Material básico adquirido para la encuesta que el Instituto Interamericano de Estadística realiza sobre los métodos de registro y estadística vital en las naciones americanas, incluyendo el estudio de 1945.

Vital statistics and registration in Chile: Methods and procedures. Background material acquired for the study conducted by the Inter American Statistical Institute in vital statistics and registration methods in the American nations, including survey of 1945. Includes: Answer to questionnaire; decrees; forms; punch cards and codes.

Incluye: Respuesta al cuestionario; decretos; formularios; tarjetas perforadas y códigos.

Chile-**5250**.M5 1945 **Manual Práctico del Oficial Civil.** Recopilación con los últimos reglamentos y disposiciones que se relacionan con el Servicio del Registro Civil. Por Manfredo Urzúa Acevedo. Rancágua, Chile, La Tribuna, 1945? 77 p.

Manual of the civil register.

6000-6999 ECONOMICS—ECONOMIA

Chile-**6000**.C3 1945 **Chile.** Foreign commerce yearbook country series. Prepared in the Division of International Economy from data compiled by the American Republics unit, and assembled for publication by the International Trade Unit. Washington, D.C., August 1945. U.S. Department of Commerce. Bureau of Foreign and Domestic Commerce. Division of International Economy. 85 p. (approx.)

Contains information about area, population, climate, agriculture, mining and manufacturing, commerce and industry, communication and transportation, power and light, national economic indices, average wage rates, foreign trade, etc.

Chile. Anuario del comercio exterior de los Estados Unidos, series por países. Contiene información sobre superficie, población, clima, agricultura, minería y elaboración, comercio e industria, transportes y comunicaciones, fuerza y luz, índices de la economía nacional, promedio de salarios, comercio exterior, etc.

Chile-**6000**.C34 1944 **Chile a Través de Sus Zonas Geográfico-Económicas.** Por Santiago Chicharro Valdosera. Santiago, 1944. 236 p. Seminario de ciencias económicas de la Facultad de Ciencias Jurídicas y Sociales de la Universidad de Chile, ser. E-I. No examinado.

Chile through the geographic-economic zones of the country. "This work will be of immense value to students interested in examining Chilean government statistics. It deserves a place beside the governmental statistical volumes that were the only sources used in its preparation. . . . A standard outline of discussion is followed: Surface, coasts and islands, climate, means of communication, demography, production, and commerce." (Source: **Geographical Review,** vol. XXXVI, no. 4, October 1946.)

"Esta obra es de inmenso valor para los estudiantes interesados en examinar las estadísticas del gobierno chileno. Merece un puesto al lado de los volúmenes estadísticos oficiales que fueron las únicas fuentes de información usadas en su preparación. . . . El autor sigue un plan tradicional de exposición: Superficie, costas e islas, clima, medios de comunicación, demografía, producción y comercio." (Traducción del comentario.)

Chile-**6000**.C35 1945 **Chile: An Economy in Transition.** By P. T. Ellsworth. New York, The Macmillan Company, 1945. 183 p.

An examination of the changes which have occurred in the Chilean economy during the period 1929-42, with special consideration of the country's increasing industrialization, seen in the light of the government's program of adjustment in the international balance of payments. Includes: The impact of the depression; recovery, readjustment, and recuperation; monetary developments; foreign trade controls, government intervention in economic activity; the resumption of inflation, 1939-42; prospects for economic development; appendix of statistical tables.

La economía chilena en transición. Análisis de los cambios que han ocurrido en la economía de Chile durante el período de 1929 a 1942, con consideración especial de la progresiva industrialización del país, a la luz del programa gubernamental de ajuste de la balanza de pagos internacionales. Incluye: Los efectos de la depresión, recuperación, reajuste, y reestablecimiento; desarrollo monetario; control del comercio exterior; intervención del Estado en las actividades económicas; reaparición de la inflación, 1939-42; expectativas del desarrollo económico; apéndice de cuadros estadísticos.

Chile-**6000**.C4 1939-43 **Cinco Años de Labor, 1939-1943.** Santiago, 1944? Corporación de Fomento de la Producción. 365 p. No examinado.

Report by the Corporation of Development of Production on five years' work, 1939-43.

*Chile-**6000**.E6 **Economía y Finanzas.** Observador internacional. 1937- . Mensual. Santiago.

Monthly journal containing editorials, articles and statistics on commerce, economics, finance, and mining.

Contiene artículos de fondo, artículos y estadísticas sobre comercio, economía, hacienda y minería.

Chile-**6000**.E65 1945 **Economic Controls and Commercial Policy in Chile.** One of a series of reports on economic controls and commercial policy in the American republics. Washington, D.C., 1945. United States Tariff Commission. 25 p.

Folleto sobre Chile, en la serie de informes de la Comisión de Aduanas de los Estados Unidos, sobre controles económicos y política comercial de las repúblicas americanas.

Chile-**6000**.R5 1944 **Reports Presented to the Conference of Commissions of Inter-American Development (Held in New York on May 9-18, 1944) by the Chilean Commission of Inter-American Development.** On: 1, Chilean fuels; 2, electric power development in Chile. (Translations from Spanish originals.) Washington, D.C., 1945? Inter-American Development Commission. 20 p.

Cover title—título de la cubierta: **Chile: Reports Presented to the Conference of Commissions of the Inter-American Development.**

Informes presentados a la Conferencia de Comisiones de Fomento Interamericano (realizada en N. York, del 9 al 18 de mayo de 1944), por la Comisión Chilena de Fomento Interamericano, sobre: 1, Combustibles chilenos; 2, el desarrollo de la fuerza eléctrica en Chile.

Chile-**6100**.D5 1935 **Determinación de la Entrada Nacional de Chile.** Por Raúl Simón. Santiago, Imp. Nascimiento, 1935. No examinado.

Determination of the national income of Chile.

Chile-**6100**.N5 1934-42 **The National Income of Chile.** 1934-1942. Preliminary draft. Prepared by Loreto M. Domínguez. Washington, D.C., November 30, 1944. Inter American Statistical Institute. 132 p.

La renta nacional de Chile, 1934-42.

*Chile-**6100**.R5 1940-45 **Renta Nacional, 1940-1945.** Santiago, Imp. Universitaria,

1946. Corporación de Fomento de la Producción. 2 vol.

National Income, 1940-45. The first part of Volume I deals with the principal theoretical problems of national income: Meaning, classification, concept according to several authors, methodological problems of valuation, basic schemes of statistical presentation, etc. In the second part of this first volume there is an analysis of general economic problems in relation to national income—economic cycle, average level of living and its capitalization, tendency to import. The third part presents: Extent of the study, principles and basic ideas that have guided the investigation, inherent difficulties of the absence of complete statistics, methodology and classification adopted and their analysis, etc.; a summary in outline form, but complete, of all the most important conclusions that it is possible to deduce from the estimate of the national income of the country, and tables of the essentials of the composition, distribution and classification of the national income of Chile; discussion and analysis of former studies of national income made in the country. In Volume II there is a complete and detailed analysis of the national income estimate of Chile, including the incomes of agriculture, fishing, mining, industry, building, transportation, public services, trade, finance, government.

La primera parte del volumen I trata los problemas teóricos fundamentales sobre renta nacional: Su significado, su clasificación, su concepto según diversos autores, los problemas metodológicos de su valoración, los esquemas básicos de su presentación estadística, etc. En la segunda parte de este primer volumen se analizan problemas de orden económico general, relacionados con la renta nacional—el ciclo económico, el nivel medio de la vida y su capitalización, la propensión a importar. En la tercera parte se presentan: Alcance del estudio, principios e ideas básicas que han guiado la investigación, dificultades inherentes por razón de la falta de estadísticas completas, metodología y clasificación adoptadas y su análisis, etc.; un resumen, en forma esquemática, pero completa, de todas las conclusiones más importantes que es posible deducir del cálculo de la renta nacional del país, y cuadros de la composición, distribución y clasificación esenciales de la renta nacional chilena; discusión y análisis de estudios anteriores de la renta nacional efectuados en el país. En el volumen II se presenta el análisis completo y detallado del cálculo de la renta nacional de Chile, incluyendo las rentas de la agricultura, pesca, minería, industria, construcción, transportes, servicios de utilidad pública, comercio, finanzas, gobierno.

Chile-**6200** **Estadísticas Agrícolas en Chile: Métodos y Procedimientos.** Material básico adquirido para el estudio que el Instituto Interamericano de Estadística realiza sobre los métodos de estadísticas agrícolas en las naciones americanas, incluyendo la encuesta de 1945-46.

Agricultural statistics in Chile: Methods and procedures. Background material acquired for Inter American Statistical Institute study on methods in agricultural statistics in the American nations, including the Survey of Agricultural Production Statistics in the American Nations in 1945-46. Consists of: Correspondence; forms; laws.

Consiste de: Correspondencia; formularios; leyes.

See also—véase también: **Ley Orgánica de la Dirección General de Estadística** (Chile-**3030**.L5, 1932).

*Chile-**6200**.A5 **Agricultura e Industrias Agropecuarias.** 1879- . Anual. Santiago. Ministerio de Economía y Comercio. Dirección General de Estadística.

Yearbook of statistics of agriculture and agricultural industry.

1943-44 **Agricultura e Industrias Agropecuarias, Año Agrícola 1943-1944.** Santiago, Dirección General de Estadística. 59 p.

Includes: Cereals and small-farm produce; cattle raising; values of production; cattle markets and industries; aviculture, vineyards and wine production; tobacco; mills; sawmills; transportation of agricultural produce; prices; geography; meteorology.

Incluye: Cereales y chácaras; ganadería; valores de la producción; ferias e industrias ganaderas; avicultura; viñas y producción vitivinícola; tabacos; molinos; aserraderos; transportes agrícolas; precios; geografía; meteorología.

Chile-**6200**.A7 1945 **Agricultural, Pastoral, and Forest Industries in Chile.** One of a series of reports on agricultural, pastoral, and forest industries in the American republics. Washington, D.C., 1945. United States Tariff Commission. 46 p.

Folleto sobre Chile, publicado en la serie de informes de la Comisión de Aduanas en los Estados Unidos sobre las industrias agrícolas, campestres y forestales en las repúblicas americanas.

Chile-**6200**.B5 **Boletín de la Sección Estadística de la Caja de Colonización Agrícola.** Fecha de iniciación? Trimestral. Santiago. Caja de Colonización Agrícola. Sección Estadística.

Quarterly bulletin of the Statistical Section of the Board of Agricultural Colonization.

Chile-**6200**.M5 **Memoria de la Caja de Crédito Agrario.** Fecha de iniciación? Anual. Santiago. Caja de Crédito Agrario.

Annual report of the Farm Credit Fund.

*Chile-**6300**.M5 **Minería.** Fecha de iniciación? Anual. Santiago. Ministerio de Economía y Comercio. Dirección General de Estadística.

Yearbook of mining statistics. 1939 issue, which included statistics of industry, entitled: **Minería e Industria.**

Edición 1939, que incluye estadísticas industriales, bajo el título: **Minería e Industria.**

Chile-**6400** **Estadísticas Industriales en Chile: Métodos y Procedimientos.** Material básico adquirido para el estudio que el Instituto Interamericano de Estadística realiza sobre los métodos de estadísticas industriales en las naciones americanas, incluyendo la encuesta de 1945-46.

Industrial statistics in Chile: Methods and procedures. Background material acquired for the study conducted by the Inter American Statistical Institute on methods of industrial statistics in the American nations, including the survey of 1945-46. Consists of: Correspondence; publications.

Consiste de: Correspondencia; publicaciones.

See also—véase también: **Censo Industrial y Comercial, Año 1937.**

*Chile-**6400**.I5 **Industrias.** Fecha de iniciación? Anual. Santiago. Ministerio de Economía y Comercio. Dirección General de Estadística.

Yearbook of industrial statistics. 1939 edition combined with **Minería** (Chile-**6300**.M5). Edición 1939, combinada con **Minería** (Chile-**6300**.M5).

1943 **Industrias, Año 1943.** Santiago, Dirección General de Prisiones—Imp., 1946. Dirección General de Estadística. 67 p.

Includes: **General statistics; detailed statistics:** Textile industry; leather industry; canned goods; chemistry products; iron and steel industry; oil refineries; industrial and edible fats; fish; tobacco; building; **index numbers:** Production in important branches of industry.

Incluye: **Estadística general; estadistica detallada:** Industria textil; industria de cuero; conservas; productos químicos; industria siderúrgica; fábricas de aceites; grasas, industriales y comestibles; pesca; tabacos; edificación; **números índices:** Producción en ramos industriales importantes.

Chile-**6400**.M5 1945 **Mining and Manufacturing Industries in Chile.** One of a series of reports on mining and manufacturing industries in the American republics. Washington, D.C., 1945. United States Tariff Commission. 75 p.

Folleto sobre Chile, publicado en la serie de informes de la Comisión de Aduanas de los Estados Unidos, sobre las industrias mineras y manufactureras en las repúblicas americanas.

*Chile-**6700**.C5 **Comercio Interior y Comunicaciones.** Fecha de iniciación? Anual. Santiago. Ministerio de Economía y Comercio. Dirección General de Estadística.

Title varies—título varía: 1941, **Anuario Comercio Interior y Comunicaciones.**

Yearbook of statistics of domestic commerce and communications.

1943 **Comercio Interior y Comunicaciones, Año 1943.** Santiago, Imp. y Lito.Universo, S.A., 1946. Dirección General de Estadística. 168 p.

Includes: **Domestic commerce:** Prices; **communications:** Migration; maritime movement; maritime passengers; maritime accidents; officials of the merchant marine; national merchant marine; air travel; railways; vehicles; roads, bridges and rivèrs; mail service and telegraphs of the State; telegraphs and cables; telephones; wireless; **coastal shipping.**

Incluye: **Comercio interior:** Precios; **comunicaciones:** Movimiento migratorio; movimiento marítimo; pasajeros por vía marítima; accidentes marítimos; oficiales de la marina mercante; marina mercante nacional; navegación aérea; ferrocarriles; vehículos; caminos, puentes y vías fluviales; correos y telégrafos del Estado; telégrafos y cables; teléfonos; radiotelefonía; **comercio de cabotaje.**

*Chile-**6720**.B5 **Boletín Mensual del Banco Central de Chile.** 1928-Santiago.

Monthly bulletin of the Central Bank of Chile.

Chile-**6720.E5** **Estadística Bancaria.** Fecha de iniciación? Anual. Santiago. Ministerio de Hacienda. Superintendencia de Bancos.

Annual statistics on banking.

*Chile-**6720.F5** **Finanzas, Bancos y Cajas Sociales.** Fecha de iniciación? Anual. Santiago. Ministerio de Economía y Comercio. Dirección General de Estadística.

Yearbook of statistics of finance, banking and social security.

*Chile-**6720.M5** **Memoria Anual Presentada a la Superintendencia de Bancos.** 1925- . Santiago. Banco Central de Chile.

Annual report of the Central Bank of Chile. Includes a survey of economic conditions in Chile, summary of the texts of decrees and laws of the year relating to the Central Bank, and balance sheets of the Bank.

Incluye un estudio sobre las condiciones económicas de Chile, resumen de los textos de las leyes y decretos referentes al Banco Central dictados durante el año, y el balance del Banco.

Chile-**6900** **Estadísticas del Comercio Exterior en Chile: Métodos y Procedimientos.** Material básico adquirido para la encuesta que el Intituto Interamericano de Estadística realiza sobre los métodos de estadísticas del comercio exterior en las naciones americanas, incluyendo el estudio de 1946-47.

Foreign trade statistics in Chile: Methods and procedures. Background material acquired for the study conducted by the Inter American Statistical Institute in foreign trade statistics in the American nations, including the survey of 1946-47. Includes: Filled questionnaire; nomenclature of countries.

Incluye: Cuestionario llenado; nomenclatura de países.

Chile-**6900.A5** **1944 Arancel Aduanero.** 7.a
1944 edición oficial. Ley n.° 4321, publicada en el **Diario Oficial** n.° 15010 de 27 de febrero de 1928, modificada por el decreto con fuerza de ley n.° 296 de 20 de mayo de 1931, por los decretos leyes n.os 547 y 548 de 5 de septiembre de 1932 y diversas otras leyes. Valparaíso, Imp.y Lito.Universo, 1944. 297 p.

1944 customs tariff.

*Chile-**6900.C5** **Comercio Exterior.** 1886- . Anual. Santiago. Ministerio de Economía y Comercio. Dirección General de Estadística.

Title varies—título varía: 1942, **Anuario de Comercio Exterior.**

Yearbook of statistics of foreign trade.

1944 **Comercio Exterior, Año 1944.** Santiago, Imp.y Lito.Universo, S.A., 1946. Dirección General de Estadística. 197 p.

Includes: Imports; exports; re-exports; goods in transit; loading of ships' stores; loading of foreign ships' stores (export); tonnage; custom receipts.

Incluye: Importación; exportación; reexportación; comercio en tránsito; reembarque a rancho; embarque a rancho de naves extranjeras (exportación); tonelaje movilizado; rentas aduaneras.

Chile-**6900.R5** **Recent Developments in the**
1945 **Foreign Trade of Chile.** One of a series of reports on recent developments in the foreign trade of the American republics. Washington, D.C., 1945. United States Tariff Commission. 80 p.

Folleto sobre Chile, publicado en la serie de informes de la Comisión de Aduanas de los Estados Unidos, sobre desarrollos recientes en el comercio exterior de las repúblicas americanas.

*Chile-**6910.B5** **Balanza de Pagos de Chile.** 1942- . Anual. Santiago. Banco Central de Chile.

Annual balance of payments of Chile, prepared by the Central Bank.

1944 **Balanza de Pagos de Chile, Año 1944.** Estudio realizado por la Sección de Estadística e Investigaciones Económicas del Banco Central de Chile. Santiago, Imp. Universitaria, 1946. Banco Central de Chile. 118 p.

Contents: Methodological considerations; contents of information presented in balance of payments (exports, imports, services, government and private transactions, movement of capital, gold and cash), interpretation of the balance of payments.

Contiene: Consideraciones metodológicas; contenido de las informaciones presentadas en la balanza de pagos (exportaciones, importaciones, servicio de mercaderías, transacciones del gobierno, transacciones privadas, movimiento de capitales, oro y numerario); interpretación de la balanza de pagos.

7000-7999 HEALTH AND CULTURAL STATISTICS--ESTADISTICAS DE SALUBRIDAD Y CULTURA

Chile-**7100**.R5 1939 **La Realidad Médico-Social Chilena.** Síntesis. Santiago, Imp.Lathrop, 1939. Ministerio de Salubridad, Previsión y Asistencia Social. 216 p. No examinado.

Medical-social conditions in Chile. "Part I (geographic and demographic antecedents) summarizes trends in population increase, age composition, fertility and mortality for various periods through 1938, in some cases from 1848. Part II (living conditions of working class) is concerned with economic conditions and the analysis of mortality for specific diseases in the light of those conditions." (Source: Gen.Cen. V.S.A.)

"Parte I, 'Algunos antecedentes geográficos y demográficos,' resume el curso del aumento de población, composición por edades, fertilidad y mortalidad a través de varios períodos hasta 1938 y desde 1848 en algunos casos. Parte II, 'Condiciones de vida de las clases trabajadoras,' trata de las condiciones económicas y el análisis de mortalidad por enfermedades específicas, en vista de esas condiciones." (Traducción del comentario.)

*Chile-**7100**.R7 **Revista Chilena de Higiene y Medicina Preventiva.** 1937- . Trimestral. Santiago. Ministerio de Salubridad, Previsión y Asistencia Social.

Chilean review of hygiene and preventive medicine.

Mar. '44, Cabello González "Población y Natalidad." Por Octavio Cabello González. Vol. VI, no. 4, marzo 1944, p. 221-258. No examinado.

Population and natality.

Jan. '43, Moroder "Mortalidad Infantil en Chile." Por Juan Moroder y Francisca Lopez. Vol. V, enero 1943, p. 279-301. Reimpresión.

Infant mortality in Chile.

Jan. '42, Mardones Restat "Sobre el Cálculo de la Duración Media de la Vida en Chile." Por Jorge Mardones Restat. No. 13, enero 1942, p. 3-10. No examinado.

On the calculation of the average length of life in Chile.

Chile-**7110**.M5 **Memoria del Servicio Nacional de Salubridad.** Fecha de iniciación? Anual. Santiago. Ministerio de Salubridad, Previsión y Asistencia Social.

Annual report of the National Health Service.

*Chile-**7110**.M55 **Anexo Estadístico a la Memoria del Servicio Nacional de Salubridad.** Fecha de iniciación? Anual. Santiago. Ministerio de Salubridad, Previsión y Asistencia Social.

Statistical annex to the annual report of the National Health Service.

Chile-**7110**.P5 **Previsión Social.** 1934- . Bimestral. Santiago. Ministerio de Salubridad, Previsión y Asistencia Social.

Social security. Bimonthly.

*Chile-**7110**.S5 **Sinopsis Estadística de los Servicios de Beneficencia y Asistencia Social de Chile.** 1944- . Anual. Santiago. Servicios de Beneficencia y Asistencia Social. Dirección General. Sección Estadística.

Statistical synopsis of social welfare and assistance in Chile.

1945 **Sinopsis Estadística de los Servicios de Beneficencia y Asistencia Social de Chile en 1945.** Santiago, Imp. "Roma," 1946. Servicios de Beneficencia y Asistencia Social. Departamento de Estadística. 91 p.

Contents: Means of hospital care, pregnancies, births, and abortions; natality; morbidity; tuberculosis; venereal diseases; infant care; collective care; mortality; duration of hospitalization.

Contenido: Medios de atención hospitalaria; embarazos, partos y abortos; natalidad; morbilidad; tuberculosis; venéreas; atención infantil; atenciones colectivas; mortalidad; permanencia hospitalaria.

Chile-**7130**.M5 1946 **Movimiento de los Principales Hospitales durante los Meses de Enero a Abril de 1946.** Santiago de Chile, junio 1946. Servicios de Beneficencia y Asistencia Social. Dirección General. Departamento de Estadística. 10 p. Mimeografiado. Publicación, no. 1.

Statistics of the principal hospitals for January-April 1946.

Chile-**7130**.P4 1946 **La Primera Zona Hospitalaria: Tarapaca.** Santiago de Chile, junio 1946. Servicios de Beneficencia y Asistencia Social. Dirección General. Departamento de Estadística. 15 p. Mimeografiado. Publicación, no. 2.

The first hospital zone: Tarapaca.

Chile-7130.P5 **Publicaciones de la Dirección General de Servicios de Beneficencia y Asistencia Social.** 1946- . Irregular. Santiago de Chile. Servicios de Beneficencia y Asistencia Social. Dirección General. Departamento de Estadística.

Series of monographs issued by the General Bureau of Charities and Social Assistance.

Serie de monografías.

Chile-7130.S5 1946 **La Segunda Zona Hospitalaria: Antofagasta.** Santiago de Chile, julio 1946. Servicios de Beneficencia y Asistencia Social. Dirección General. Departamento de Estadística. 15 p. Mimeografiado. Publicación, no. 3.

The second hospital zone: Antofagasta.

Chile-7130.S6 1946 **Un Semestre de Labor en los Sanatorios.** Santiago de Chile, agosto 1946. Servicios de Beneficencia y Asistencia Social. Dirección General. Departamento de Estadística. 15 p. Publicación, no. 4.

A semester's work in the sanatoriums.

Chile-7200.E5 1945 **Education in Chile.** By Cameron D. Ebaugh. Washington, D.C., U.S.Govt.Print.Off., 1945. U.S. Federal Security Agency. U.S. Office of Education. 123 p. Bulletin, 1945, no. 10.

Discussion of the Chilean educational system, covering elementary, secondary, vocational and higher education, including Chilean universities, and giving background data on evolution of education in Chile.

Un estudio sobre el sistema de educación en Chile, comprendiendo escuelas elementales, secundarias, vocacionales y profesionales, incluyendo las universidades chilenas, y citando datos básicos acerca del desarrollo del sistema educacional en Chile.

Chile-7210.U5 1944 **The Universities of Chile.** By Katherine Lenore Morgan. Washington, D.C., 1944. Pan American Union. Divisions of Intellectual Cooperation. 157 p. Higher education in Latin America, vol. 2.

Las universidades de Chile.

8000-8999 GOVERNMENT AND POLITICAL STATISTICS—ESTADISTICAS GUBERNAMENTALES Y POLITICAS

*Chile-8000.P5 **Política, Administración, Justicia y Educación.** Fecha de iniciación? Anual. Santiago. Ministerio de Economía y Comercio. Dirección General de Estadística.

Yearbook of statistics on government, administration, justice, and education.

1943 **Política, Administración, Justicia y Educación, Año 1943.** Santiago, Dirección General de Prisiones—Imp., 1945. Dirección General de Estadística. 99 p.

Includes: **Politics and administration:** Electoral population of the Republic; administrative divisions; diplomatic and consular corps; citizenship papers; **justice:** Administration and justice; police; fire departments; penitentiaries and prisons; **education:** Primary education; secondary education; special education; university education; public libraries; museums of natural history; the press.

Incluye: **Política y administración:** Población electoral de la República; división administrativa; cuerpo diplomático y consular; cartas de nacionalización; **justicia:** Administración y justicia; policías; cuerpo de bomberos; penitenciarías, presidios y cárceles; **educación:** Educación primaria; educación secundaria; educación especial; educación universitaria; bibliotecas públicas; museos de historia natural; prensa.

Chile-8800.M5 **Memoria de la Contraloría General de la República y Balance General de la Hacienda Pública.** Fecha de iniciación? Anual. Santiago. Ministerio de Hacienda. Contraloría General de la República.

Report of the Office of the Comptroller General of the Republic and general balance of the Treasury.

9000-9999 LABOR AND LIVING CONDITIONS—CONDICIONES DE TRABAJO Y DE VIDA

*Chile-9000.V5 1945 **Veinte Años de Legislación Social.** Santiago, Imp. y Lito.Universo, S.A., 1945. Dirección General de Estadística. 255 p.

Twenty years of social legislation. Includes 200 pages of data and comments of the General Bureau of Labor on wages in industry and agriculture, employment, unemployment, cost of living, determination of the minimum living wage, family allotments, labor conflicts, absenteeism, trade unions in 1943, industrial accidents, labor courts, and classification of workers by occupation, as well as information on social security and health and medicine.

Incluye 200 páginas de datos y explicaciones de la Dirección General del Trabajo sobre salario industrial, salario agrícola, ocupación, desocupación, costo de la vida, determinación del salario mínimo vital, asignación familiar, conflictos colectivos, ausentismo, corporaciones sindicales año 1943, accidentes del trabajo, tribunales del trabajo, y clasificación de obreros por profesiones, como también información sobre previsión social y sanidad y medicina.

Chile-9600.E5 1943 **Encuesta Continental sobre el Consumo de Productos de Alimentación y Vestido y sobre la Vivienda Popular: Respuestatipo Referente a la República de Chile.** Por Bruno Leuschner. Montevideo, 1943. Consejo Interamericano de Comercio y Producción. 155 p. Anexo a la circular documental de la Comisión Ejecutiva, no. 1, sobre **Consumo y Subconsumo.** No examinado.

Continental survey on consumption of food products, clothing, and low-cost housing: Reply from Chile.

Chile-9950.E5 **Estadísticas de la Caja de Seguro Obligatorio.** 1941- Anual. Santiago. Caja de Seguro Obligatorio.

Annual statistics of the Compulsory Workers' Insurance Fund.

Chile-9950.S5 1942 **La Seguridad Social en Chile.** Por Julio Bustos A. Santiago, 1942. Ministerio de Salubridad, Previsión y Asistencia Social. Departamento de Previsión Social. No examinado.

Social security in Chile.

COLOMBIA

1000-1999 GENERAL

Col-**1400**.C5 1944 **Colombia en Cifras.** Edición extraordinaria de **El Mes Financiero y Económico.** Bogotá, Editorial Antena, 1944. El Mes Financiero y Económico. 805 p. No examinado.

Colombia in figures.

Col-**1400**.S5 **Síntesis Estadística de Colombia.** 1941- . Irregular. Bogotá. Contraloría General de la República. Dirección Nacional de Estadística.

Statistical summary of Colombia.

1939-43 **Síntesis Estadística de Colombia, 1939-1943.** Anexo al **Anuario General de Estadística** de 1943. Bogotá, Imp.Nacional, 1944. Contraloría General de la República. Dirección Nacional de Estadística. 214 p.

Contents: Territory and meteorology; population; extractive industries; foreign trade; finance; property transactions; transportation and communication; economic indexes; corporations; urban building; labor; prices; consumption, public assistance; education; periodic publications; shows; lotteries; justice; fiscal statistics; political-administrative statistics; manufactures; electric plants; cooperatives; income tax returns, inheritance, and utility surpluses.

Contenido; Territorio y meteorología; población; industrias extractivas; comercio exterior; finanzas; propiedad raíz; transportes y comunicaciones; índices del movimiento económico; sociedades anónimas; edificaciones urbanas; trabajo; precios; consumos; asistencia pública; educación; publicaciones periódicas; espectáculos; loterías; justicia; fiscal; político-administrativa; industria manufacturera; plantas eléctricas; cooperativas; liquidaciones del impuesto sobre la renta, patrimonio, y exceso de utilidades.

*Col-**1410**.A5 **Anuario General de Estadística.** 1905, 1915- . Bogotá. Contraloría General de la República. Dirección Nacional de Estadística.

Statistical yearbook of Colombia.

1943 **Anuario General de Estadística, Colombia, 1943.** Bogotá, Imp. Nacional, 1945. Contraloría General de la República. Dirección Nacional de Estadística. 550 p.

Includes: Geography; meteorology; population; social assistance and public health; industrial production; transportation and communication; foreign trade; finance; prices and consumption; labor; education; statistics of justice; fiscal statistics.

Incluye: Geografía; meteorología; población; asistencia social e higiene; producción industrial; transportes y comunicaciones; comercio exterior; finanzas; precios y consumo; trabajo; educación; estadística de justicia; estadística fiscal.

*Col-**1420**.A5 **Anales de Economía y Estadística.** Revista de la Contraloría General de la República. 1938- . Mensual. Bogotá.

Subtitle varies—subtítulo varía.

Annals of economy and statistics. Principal object is to disseminate information on the studies of a technical character undertaken by the Technical Council on Statistics (formerly Center of Studies and Coordination). Contains data on production, prices, cost of living, public finance, and other phases of the economy.

Su principal fin es divulgar información sobre los estudios técnicos a cargo del Consejo Técnico de Estadística (anteriormente Centro de Estudios y Coordinación). Contiene datos sobre producción, precios, costo de la vida, finanza pública y otras fases de la economía.

Oct. '46 "Las Condiciones Económico-Sociales y el Costo de la Vida de la Clase Media en Bogotá." Segunda época, año II, suplemento a los no. 19 y 20 (julio y agosto), octubre 1946. 99 p.

Economic and social conditions and cost of living of the middle class in Bogotá. Contents: Economic and social analysis of the family (demographic characteristics, units of consumption, family composition, length of matrimonial life, age, profession, culture, education, syndicalism, income, expenses, etc.); nutrition of the middle class (food, housing, consumption of fuel, etc.); the typical family and normal consumption (selection of the economically and demographically typical family, normal consumption, basic and total cost); construction of middle class cost-of-living index.

Contiene: Análisis económico-social de la familia (características demográficas, unidades de consumo, composición de la familia, duración de la vida matrimonial, edades, profesiones, cultura, instrucción, sindicalización, rentas, gastos, etc.); régimen alimenticio de la clase media (alimentación, vivienda, consumo de combustibles y varios); la familia típica y el consumo normal (elección de la familia económicamente típica y la demográficamente típica, consumo normal, costo básico y total); formación del índice del costo de la vida de la clase media.

June '46 "Las Condiciones Económico-Sociales y el Costo de la Vida de la Clase Obrera en la Ciudad de Honda." Segunda época, año II, suplemento a los no. 15 y 16 (marzo-abril), junio 1946. 95 p.

The economic and social conditions and the cost of living of the working classes in the city of Honda. Supplemented fully by an appendix of statistical tables, and presenting many charts and graphs, this study includes chapters on: Demographic characteristics of working class families; occupation, extent of education, and unionization; income; expenditures (including family budgets and distribution of outlay); nutrition; housing, facilities and home furnishings; clothing, fuel, and miscellaneous consumption items; the typical family and its normal consumption.

Completado extensamente por un apéndice de cuadros estadísticos, y presentando muchos otros cuadros y gráficos, este estudio incluye capítulos sobre: Características demográficas de las familias obreras; ocupaciones, grado de educación, y sindicalización; ingresos y rentas; gastos (incluyendo presupuestos familiares y distribución de gastos); régimen alimenticio; viviendas, servicios y mobiliarios; vestido, combustibles, y consumos diversos; la familia típica y el consumo normal.

May-June '46, Abrisqueta "El Indice de Precios del Comercio en General." Por Francisco de Abrisqueta. Segunda época, año II, no. 17-18, mayo-junio 1946, p. 57-88.

The general commodity price index. Contents: Review of the antecedents of the price index in Colombia; object of the index of commodity prices in general; geographic scope; branches of commerce (types of classification in countries similar to Colombia); classification schemes with basic distribution according to the economic nature of the products and with redistribution according to their commercial nature; articles, numbers, denominations, classes, units of measure, containers; classification of products by economic nature, by commercial nature, and by economic destination; use of existing information and sources for compilation of prices; frequency of compilation; forms, registers, and inspectors; control and criticism of the prices; previous analysis of the prices; average of the prices; classification of the indexes and formulas for calculation; simple and weighted indexes; weighting; base and retroactive investigation; sensitive prices and indexes in gold; bibliography.

Contiene: Reseña de los antecedentes del índice de precios en Colombia; objeto de los índices de precios del comercio en general; extensión geográfica; campos comerciales (tipos de clasificación en países semejantes a Colombia); esquemas sobre la distribución básica según la naturaleza económica de los productos y sobre la redistribución según la naturaleza comercial de los mismos; artículos, números, denominaciones, clases, unidades de medida y envases; clasificación de los productos según naturaleza económica, según naturaleza comercial, y por su destino económico; aprovechamiento de las informaciones existentes y fuentes de recolección de los precios; periodicidad de la recolección; formularios, registros e inspectores; control y crítica de los precios; análisis previo de los precios; promedio de los precios; clasificación de los índices y fórmulas de cálculo; índices simples y ponderados; ponderación; base e investigación retroactiva; precios reagibles e índices en oro; bibliografía.

Mar-Apr. '46, Abrisqueta, 1 "Algunas Instituciones Económicas Ecuatorianas: El Banco Central del Ecuador y Su Oficina de Investigaciones Económicas." Por Francisco de Abrisqueta. Segunda época, año 2, no. 15-16, marzo-abril 1946, p. 64-67.

Description of the Office of Economic Research of the Central Bank of Ecuador.

Mar.-Apr. '46, Abrisqueta 2 "La Estadística Nacional Ecuatoriana." Por Francisco de Abrisqueta. Segunda epoca, año 2, no. 15-16, marzo-abril 1946, p. 68-69.

National statistics of Ecuador.

Nov.-Dec. '45, Dunn "La Estadística y el Mundo de Mañana." Por Halbert L. Dunn. Segunda época, año I, no. 11-12, noviembre-diciembre 1945, p. 73-75.

Statistics and the world of tomorrow.

Nov. '45, Abrisqueta "Reglamentación sobre el Cálculo de los Medios de Pagos en Circulación y de las Exigibilidades Monetarias en Colombia." Por Francisco de Abrisqueta. Segunda época, año I, suplemento al no. 7 y 8 (julio y agosto), noviembre 1945. 20 p.

Regulation of the estimation of mediums of payment in circulation and monetary requirements in Colombia.

Sept.-Oct. '45, Abrisqueta "La Estadística Aplicada a la Economía y a los Negocios." Por Francisco de Abrisqueta. Segunda época, año I, no. 9-10, septiembre-octubre 1945, p. 65-67.

Statistics applied to economy and business.

Sept.-Oct.'45, Arévalo Correal "La Clasificación Estadística del Comercio Exterior de Colombia." Sus problemas y algunas posibles soluciones. Por Benjamín Arévalo Correal. Segunda época, año I, no. 9-10, septiembre-octubre 1945, p. 68-73.

Statistical classification of foreign trade of Colombia. Problems and solutions.

July-Aug. '45, Abrisqueta "El Problema de los Arrendamientos de Viviendas en Bogotá." Por Francisco de Abrisqueta. Segunda época, año I, no. 7-8, julio-agosto 1945, p. 99-103.

Problem of rents in Bogota.

May '45, Abrisqueta "El Desarrollo de Bogotá y las Investigaciones del Costo de la Vida Obrera." Un análisis sobre la recolección de precio para el cálculo de los índices del costo de la vida obrera en Bogotá. Por Francisco de Abrisqueta. Segunda época, año I, no. 5, mayo 1945, p. 41-83.

Development of Bogota and research on workers' cost of living. Analysis of collection of prices for computation of cost of living indexes for workers in Bogota.

Apr. '45, Abrisqueta "Factores de Examen de la Coyuntura Económica Colombiana." Por Francisco de Abrisqueta. Segunda época, año I, no. 4, abril 1945, p. 27-31.

Factors for analysis of economic system of Colombia. Includes a classification of economic statistics and a chart showing the structure of economics.

Incluye una clasificación de las estadísticas económicas y un cuadro demostrativo de la estructura económica.

Mar. '45, Pineda de Castro "La Investigación del Costo de la Vida y las Condiciones Social-Económicas de los Obreros del Ferrocarril." Por Alvaro Pineda de Castro. Segunda época, año I, no. 3, marzo 1945, p. 31-45.

Study of cost of living and economic and social conditions of railroad workers.

Feb. '45, Abrisqueta "Estadísticas de la Producción." Por Francisco de Abrisqueta. Segunda época, año I, no. 2, febrero 1945, p. 20-27.

Production statistics.

Feb. '45, Montoya Canal "Estadística de la Delincuencia en Colombia." Por Aníbal Montoya Canal. Segunda época, año I, no. 2, febrero 1945, p. 37-53.

Statistics of delinquency in Colombia.

Feb. '45, Ortiz C. "El Censo Industrial de Colombia de 1945." Sus características principales. Por Luis B. Ortiz C. Segunda época, año I, no. 2, febrero 1945, p. 70-72.

Principal characteristics of the 1945 industrial census of Colombia.

Oct. 20, '43, Montoya "Plan General para el Levantamiento del Primer Censo Agropecuario." Por Hernán Montoya. Tomo VI, no. 19-20, octubre 20, 1943, p. 17-51.

Detailed description of plans for first agricultural census, including reproduction of schedules and definitions.

Descripción detallada de los planes para el censo, incluyendo una reproducción de las cédulas y definiciones.

July '43, Mortara "Tabla de Mortalidad y Supervivencia para Colombia." Por Jorge Mortara. Tomo VI, no. 13, julio 5, 1943 p. 18-25.

Mortality and life table for Colombia.

Mar. 20, '43, Rodríguez "Tabla de Mortalidad de Colombia." Por Jorge Rodríguez. Tomo VI, no. 6, marzo 20, 1943, p. 23-29.

Mortality table of Colombia.

Aug. 5, '42, Guthardt "Un Curso Oficial de Estadística." Por Emilio C. Guthardt. Tomo V, no. 14, agosto 5, 1942, p. 27-30.

An official course in statistics.

May 5, '42, a "Normas para la Elaboración de los Cuadros Estadísticos." Recomendadas por el Instituto Internacional de Estadística en su XXIII ava sesión, Atenas, 1939. Tomo V, no. 8, mayo 5, 1942, p. 64-65.

Standards for construction of statistical tables, recommended by the International Statistical Institute.

May 5, '42, b "Programa de Estudios del Curso General de Estadística de la Contraloría General de la República." Tomo V, no. 8, mayo 5, 1942, p. 66-67.

Outline of course of study in statistics given in the office of the Comptroller General of Colombia.

May 5, '42, Rodríguez "Analfabetismo Urbano y Rural." Por Jorge Rodríguez. Tomo V, no. 8, mayo 5, 1942, p. 53.

Urban and rural illiteracy.

Feb. 20, '42, Bermúdez V. "Diez Años de Educación y Cultura en Colombia." Por Luis A. Bermúdez V. Tomo V, no. 3, febrero 20, 1942, p. 47-82.

Ten years of education and culture in Colombia.

Feb. 5, '42 "El Curso de Estadística de la Contraloría General." Por el Ministro de Educación Nacional. Tomo V, no. 2, febrero 5, 1942, p. 3.

Includes resolution no. 1589 of December 11, 1941, which authorizes the courses of statistics given in the department of the Comptroller General.

Contiene Resolución Número 1589 de 1941 (diciembre 11), por la cual se autorizan los cursos de estadística, dados en la Contraloría General.

Apr. 25, '40, Higuita "Estudio Histórico Analítico de la Población Colombiana en 170 Años." Por Juan de D. Higuita. Tomo III, no. 2, abril 25, 1940, suplemento, p. 1-113.

Analytical historical study of the Colombian population for 170 years, presented to the Eighth American Scientific Congress in Washington, May 10-18, 1940.

Presentado al VIII Congreso Científico Americano reunido en Wáshington del 10 al 18 de mayo de 1940.

Aug. '39 "Informe que la Contraloría General de la República Rinde al Sr. Ministro de Gobierno y a las Honorables Cámaras sobre el Levantamiento del Censo Civil de 1938." Tomo II, no. 4, agosto 1939, p. 1-45.

Report presented by the Comptroller General to the Secretary of Government and to Congress on the general census of 1938. Includes general description, cost, and summary results.

Incluye descripción general, costo, y resumen de resultados.

July '39, Ortiz C. "Ensayo sobre las Condiciones de la Vida Rural en el Municipio de Moniquirá-Boyacá." Por Luis B. Ortiz C. Tomo II, no. 3, julio 1939, p. 27-65.

Essay on rural living conditions in municipality of Moniquira-Boyaca. Includes questionnaires and forms.

Incluye cuestionarios y formularios.

July '39, Ricaurte Montoya "Estudio sobre Estadística Agrícola." Por Alberto Ricaurte Montoya. Tomo II, no. 3, julio 1939, p. 1-26.

Study on agricultural statistics. Includes: Statistics of agriculture in certain countries; statistics of agricultural production; the census of agriculture in certain countries (Chile, Mexico, Uruguay, the United States, and Canada are the countries in the Western Hemisphere that are covered); program of the International Institute of Agriculture; conclusions.

Incluye: La estadística agrícola en algunos países; la estadística agrícola de producción; el censo agrícola en algunos países (de los países del Hemisferio Occidental quedan incluidos Chile, México, Uruguay, los Estado Unidos y el Canadá); programa del Instituto Internacional de Agricultura; conclusiones.

1938, Hermberg "El Costo de la Vida de la Clase Obrera en Bogotá." Por Paul Hermberg. Tomo I, no. 1, 1938, p. 1-104.

Cost of living of the working class in Bogota.

Col-**1900**.C5 **Colombia.** Organo de la Contraloría General de la República. 1944- . Mensual. Bogotá. Contraloría General de la República.

Monthly journal of the Comptroller General. Covers all phases of economic, social, cultural, and political life of the country. Contains sections on geography, geology, climatology, minerals, agriculture and livestock, transportation and communication, public finance, etc.

Abarca todas las fases de la vida económica, social, cultural y política del país. Contiene secciones sobre geografía, geología, climatología, minerales, agricultura y ganadería, transportes y comunicaciones, hacienda pública, etc.

2000-2999 CENSUS—CENSOS

Col-**2100**.C5 1950 "Censo Continental: Una Sugerencia del Instituto Interamericano de Estadística." Por Eduardo Santos Rubio y Bernardo Aguilera Camacho. Bogotá, 1945. Contraloría General de la República. Dirección Nacional de Estadística. 6 p. Copia máquina.

Comment on projected inter-American census of 1950, with the conclusion that Colombia postpone her census of 1948, to conforme with date proposed by the Inter American Statistical Institute. Includes: Frequency of censuses in Colombia from 1825 to 1938; cost and budget; legal regulation.

Comentario sobre el proyectado censo continental de 1950, con la conclusión de que Colombia efectue la postergación de su censo de 1948 a 1950, para estar conforme con la fecha propuesta por el Intituto Interamericano de Estadística. Incluye: Periodicidad de los censos en Colombia, desde 1825 a 1938; costo y presupuesto; régimen legal.

Col-**2100**.C5 1938 **Decretos y Bases de Organización para la Ejecución de los Censos, 1938.** Bogotá, Editorial "El Gráfico," 1937? Contraloría General de la República. Sección de Censos Nacionales. 58 p.

Legislation and organizational basis for the censuses of 1938. Bound with **Instrucciones Generales para la Ejecución de los Censos de Edificios y Población,** which includes organization chart. Census schedules bound in.

Encuadernado con las **Instrucciones Generales para la Ejecución de los Censos de Edificios y Población,** que incluye esquema estructural. Las cédulas del censo están encuadernadas en el mismo tomo.

Col-**2150**.C5 1945, a **Censo Industrial de Colombia, 1945: Métodos y Procedimientos.** Material adquirido con motivo del estudio que el Instituto Interamericano de Estadística realiza sobre estadísticas industriales de las naciones americanas, 1945. Contraloría General de la República. Dirección Nacional del Censo Industrial.

Material on industrial census of Colombia, 1945, acquired in connection with IASI Survey of Industrial Statistics in the American Nations. Consists of: Correspondence; instructions; forms; laws; posters.

Consiste de: Correspondencia, instrucciones; formularios; leyes; carteles.

Col-**2150**.C5 1945, b **Primer Censo Industrial de Colombia, 1945: Cartilla de Instrucciones.** Bogotá, Imp. Nacional, 1945. Contraloría General de la República. Dirección Nacional del Censo Industrial. 103 p. Cuaderno, no. 43.

First industrial census of Colombia, 1945: Instructions.

Col-**2150**.C5 1945, c **Primer Censo Industrial de Colombia, 1945: Boleta General.** Bogotá, Imp.Nacional, 1945. Contraloría General de la República. Dirección Nacional del Censo Industrial. 13 p. Cédula.

First industrial census of Colombia, 1945: General schedule.

*Col-**2200**.C5 1938 **Censo General de Población, 5 de Julio de 1938.** Ordenado por la ley 67 de 1917. Bogotá, Imp.Nacional, 1942. Contraloría General de la República. Estadística Nacional. Dirección General de los Censos. 16 vol.: Tomo I, **Departamento de Antioquia;** tomo II, **Departamento de Atlántico;** tomo III, **Departamento de Bolívar;** tomo IV, **Departamento de Boyacá;** tomo V, **Departamento de Caldas;** tomo VI, **Departamento de Cauca;** tomo VII, **Departamento de Cundinamarca;** tomo VIII, **Departamento de Huila;** tomo IX, **Departamento del Magdalena;** tomo X, **Departamento de Nariño;** tomo XI, **Departamento Norte de Santander;** tomo XII, **Departamento de Santander;** tomo XIII, **Departamento de Tolima;** tomo XIV, **Departamento del Valle de Cauca;** tomo XV, **Intendencias y Comisarías;** tomo XVI, **Resumen General del País.**

General census of population, July 5, 1938. Each provincial volume includes: **Census of population:** Density; increase; urban-rural population; age, sex, and marital status; education, by age and sex; nationality; gainfully occupied population and occupation; **census of buildings:** Number of buildings with and without plumbing, electricity, and sewers; classification of buildings according to whether occupied by owners or tenants, with distinction of Colombian or foreign proprietors; use to which the buildings are put. The final summary volume contains data for the entire country, with comparison of the total population with that of previous censuses.

Cada volumen provincial incluye: **Censo de**

población: Densidad; crecimiento; población urbana y rural; edad, sexo y estado civil; instrucción, por edad y sexo; nacionalidad; población activa y ocupación; **censo de edificios:** Número de edificios con acueducto, luz eléctrica, alcantarillado, o sin estos servicios; clasificación de los edificios según estén ocupados por dueños o inquilinos, y con discriminación de propietarios nacionales o extranjeros; uso a que están destinados los edificios. El volumen del resumen general da datos del país entero, con comparación de la población total con la de los censos previos.

*Col-**2220**.C5 1938 **Primer Censo Nacional de Edificios, Efectuado el 20 de Abril de 1938.** Bogotá, Imp.Nacional, 1939. Contraloría General de la República. Estadística Nacional. Dirección General de los Censos. 393 p.

First national census of buildings, taken April 20, 1938. Includes: Number of buildings with and without plumbing, electricity, and sewers; classification of buildings according to whether occupied by owners or by tenants, with distinction of Colombian or foreign proprietors; use to which the buildings are put.

Incluye: Edificios con acueducto, luz eléctrica, alcantarillado, o sin estos servicios; clasificación de los edificios según estén ocupados por dueños o inquilinos, y con discriminación de propietarios nacionales o extranjeros; uso a que están destinados los edificios.

3000-3999 STATISTICAL SCIENCE—CIENCIA ESTADISTICA

Col-**3030**.E3 1938 **La Estadística Nacional.** Su organización, sus problemas. Por Carlos Lleras Restrepo. Bogotá, Imp.Nacional, 1938. Contraloría General de la República. 415 p. Colección de estudios administrativos.

A reference handbook on the organization of national statistics in Colombia.

Col-**3030**.E5 1945 **Estatuto Orgánico de la Estadística Nacional.** Bogotá, Imp. Nacional, 1945. Contraloría General de la República. Dirección General de Estadística. 95 y 363 p. Cuaderno, no. 39.

Organic law of national statistics. Contains laws, procedures, and forms regulating official statistical activities of Colombia. Includes: Resolution of October 25, 1944, cancelling plans for agricultural census; organic resolutions and agreements of national statistics (law 82 of December 23, 1935, and regulatory resolutions no. 632 and 633 of October 31, 1944, and no. 656 of January 25, 1945); first industrial census of Colombia, 1945—legislation (decree no. 85 of January 19, 1945, and regulatory resolution no. 657 of January 25, 1945, with organization chart of the General Bureau of Census); industrial classification (branches of economic activity) according to the recommendations of the Committee of Statistical Experts of the League of Nations; schedules for the industrial census; basic forms for statistical information (demography and health; industry, labor, and transportation; foreign trade; economic and social research and studies; fiscal and administrative; cultural, including school register; justice; mechanical tabulation); nomenclature of raw materials used and products obtained in industries of transformation.

Contiene leyes, procedimientos y formularios reglamentando las actividades estadísticas oficiales de Colombia. Incluye: Resolución del 25 de octubre de 1944 anulando los planes para un censo agropecuario; resoluciones y acuerdos orgánicos de la estadística nacional (Ley 82 del 23 de diciembre de 1935, resoluciones reglamentarias no. 632 y 633 del 31 de octubre de 1944, no. 656 del 25 de enero de 1945); primer censo industrial de Colombia, 1945—disposiciones legales (Decreto No. 85 del 19 de enero de 1945, resolución reglamentaria No. 657 del 25 de enero de 1945, con el esquema estructural de la Dirección General del Censo); clasificación industrial (ramas de actividad económica) según las recomendaciones del Comité de Expertos Estadísticos de la Sociedad de las Naciones; cédulas para el censo industrial; formularios básicos de información estadística (demografía y sanidad; industrias, trabajo y transportes; comercio exterior; investigaciones y estudios económico-sociales; fiscal y administrativa; cultural, incluyendo el registro escolar; justicia; tabulación mecánica); nomenclatura de las materias primas utilizadas y productos obtenidos en las industrias de transformación.

1945, No. 82 "Ley 82 de 1935 (Diciembre 23)." Sobre estadística. **Estatuto Orgánico de la Estadística Nacional,** p. 3-4.

Law which places the central direction of official statistical activities of Colombia under the Comptroller General.

Ley que deposita en la Contraloría General de la República la dirección de las actividades estadísticas oficiales de Colombia.

1945, No. 75 "Decreto Número 75 de 1939 (Enero 14)." **Estatuto Orgánico de la Estadística Nacional,** p. 197.

Law which imposes upon the head of each educational institution in Colombia the responsibility for collecting and submitting reliable statistical data on education.

Ley que obliga a los dirigentes de cada una de las instituciones educacionales a coleccionar y transmitir datos estadísticos fidedignos sobre educación.

Col-**3030.**I5 1945 **Instituto Nacional de Estadística de la Universidad de Colombia.** Bogotá, 1945. Universidad Nacional de Colombia. Instituto Nacional de Estadística. 16 p. Copia máquina.

National Statistical Institute of the National University of Colombia.

Includes—incluye: Estatutos; informe que rinde la Comisión Organizadora a la Asamblea General, con motivo de la inauguración del Instituto, el día 1° de junio de 1945; palabras pronunciadas por el Dr. Antonio García, director del Instituto de Ciencias Económicas y Catedrático de Economía de la Universidad Nacional; miembros activos del Instituto . . .; candidatos para la directiva del Instituto y temas de trabajo; acuerdo número 85 de 1945 (mayo 15), acta número 19, por el cual se incorpora el Instituto de Estadística en el de Ciencias Económicas de la Universidad.

Col-**3030.**05 1946 "Organización de Servicios de Estadística: Resolución No. 64 de Enero de 1946." Bogotá, 1947. Dirección Nacional de Estadística. Copia máquina.

Organization of statistical services: Resolution no. 64 of January, 1946. Summary of new regulations covering statistical services in Colombia. Covers division of General Bureau of Statistics into three large sections: I, Demographic and social (groups: demography, public welfare, education, judicial and penitentiary); II, economic and fiscal (labor, industry, transportation, finance, general economy); III, foreign trade; all statistical information released inside and outside the country is centralized in the Secretariat of the General Bureau of Statistics.

Sumario de los nuevos reglamentos de los servicios estadísticos en Colombia. Divide la Dirección Nacional de Estadística en tres grandes secciones: Sección I—Demográfica y sociales (grupos: demografía, asistencia pública, educación, judicial y carcelaria); Sección II—Económicas y fiscales (estadísticas de trabajo, industrias, transporte, fisco, economía general); Sección III—Comercio exterior; Secretaría de la Dirección Nacional de estadística, en la cual están centralizadas las informaciones estadísticas que se suministran al interior y al extranjero.

Col-**3040.**E5 **Estadística.** Organo de la Sociedad Colombiana de Estudios Estadísticos. 1943- . Irregular. Bogotá. Sociedad Colombiana de Estudios Estadísticos.

Title varies—título varía: 1943-194?, **Revista de la Sociedad Colombiana de Estudios Estadísticos.**

Journal of the Colombian Society of Statistical Studies. Articles, legislation, news of the Society.

Artículos, legislación, noticias de la Sociedad.

Col-**3210.**C5 1936 **Codificación Estadística del Comercio Exterior.** Bogotá, 1936. Contraloría General de la República. Dirección Nacional de Estadística. 298 p. Copia máquina.

Statistical classification of foreign trade.

Col-**3230.**N5 1942 **Nomenclatura de Morbilidad y Mortalidad.** Aprobada por resolución número 456 de 1942. Por Ignacio Moreno Pérez. Bogotá, Imp.Nacional, 1943. Contraloría General de la República. Dirección Nacional de Estadística. 94 p.

Morbidity and mortality nomenclature.

Col-**3300** 1944-46 **Enseñanza Estadística en Colombia: Métodos y Materiales.** Material básico adquirido para la Encuesta del Instituto Interamericano de Estadística sobre los Métodos y Materiales en la Enseñanza Estadística en las Naciones Americanas, 1944-46.

Statistical training in Colombia: Methods and materials. Background material acquired for the IASI Survey on Statistical Training Methods and Materials in the American Nations, 1944-46. Includes: Letters, filled questionnaires, and programs of statistical courses in Colombia.

Incluye: Cartas, cuestionarios llenados y programas de cursos de estadística en Colombia.

4000-4999 GEOGRAPHY—GEOGRAFIA

Col-**4000.**M5 1946 **Mapa Físico de Colombia.** Bogotá, Lito. Colombia, 1946. Banco de la República. Sección de Investigaciones Económicas. 1 p. Boletín gráfico, no. 4, junio de 1946.

A physical map of Colombia, showing heights of mountains, lengths of rivers, area of the country and of lakes, etc.

Un mapa físico de Colombia, conteniendo la altura de las montañas, largo de los ríos, extensión del país y de los lagos, etc.

Col-**4200.**A5 **Anuario Meteorológico.** Fecha de iniciación? Bogotá. Ministerio de la Economía Nacional. Departamento de Irrigación. Sección de Meteorología y Aforos.

Meteorological yearbook.

Col-4200.M5 1945 **Meteorología.** Bogotá, Imp.del Banco de la República, 1945. Banco de la República. Sección de Investigaciones Económicas. 2 p. Boletín gráfico, no. 2, septiembre de 1945.

A meteorological map of Colombia showing rainfall and temperatures, with statistical tables and explanations.

Un mapa meteorológico de Colombia, exponiendo precipitación pluvial y temperaturas, con cuadros estadísticos y explicaciones.

5000-5999 DEMOGRAPHY—DEMOGRAFIA

Col-5000.C5 1944 **Colombia: Summary of Biostatistics.** Maps and charts, population, natality and mortality statistics. Washington, D.C., December 1944. Prepared by U.S. Department of Commerce, Bureau of the Census, in cooperation with Office of the Coordinator of Inter-American Affairs. 138 p.

Resumen bioestadístico de Colombia. Incluye mapas y cuadros y datos sobre población, natalidad y mortalidad.

Col-5110.P5 1946 **Población de Colombia por Ramas de Actividad Económica (Según Censo del 5 de Julio de 1938).** Bogotá, julio 1946. Banco de la República. Sección de Investigaciones Económicas. Boletín gráfico, no. 5.

Population of Colombia by economic activity; taken from the census of July 5, 1938.

Col-5200 **Estadística Vital y Registro en Colombia: Métodos y Procedimientos.** Material básico adquirido para la encuesta que el Instituto Interamericano de Estadística realiza sobre los métodos de registro y estadística vital en las naciones americanas, incluyendo el estudio de 1945.

Vital statistics and registration in Colombia: Methods and procedures. Background material acquired for the study conducted by the Inter American Statistical Institute in vital statistics and registration methods in the American nations, including study of 1945. Includes: Answer to questionnaire; laws, decrees; forms.

Incluye: Respuesta al cuestionario; leyes, decretos; formularios.

Col-5250.R5 1940 **Registro del Estado Civil de las Personas.** Instrucción a los notarios, alcaldes, corregidores o inspectores de policía y cónsules de Colombia en el exterior sobre el modo de llevarlo. Bogotá, Imp.Nacional, 1940. Ministerio de Gobierno. Departamento de Justicia. 115 p.

Civil status register. Instructions for notaries, mayors, sheriffs, police inspectors and Colombian consuls abroad on the way to keep the register. Legal dispositions on registration, p. 13-32.

"Disposiciones legales sobre el registro civil," p. 13-32.

Col-5250.R7 1934 **Registro del Estado Civil: Decreto 540 de 1934 y Disposiciones Vigentes sobre la Materia.** Bogotá, Imp. Nacional, 1934. 66 p. No examinado.

Register of civil status: Decree 540 of 1934 and legislation in force on the subject.

6000-6999 ECONOMICS—ECONOMIA

Col-6000.B5 **Boletín Gráfico.** 1945- . Irregular. Bogotá. Banco de la República. Sección de Investigaciones Económicas.

Graphic presentations of economic and geographic subjects, published by the Bank of the Republic.

Exposiciones gráficas de aspectos económicos y geográficos.

*Col-6000.C5 **Comercio e Industrias.** Organo del Departamento de Comercio e Industrias. 1930- . Mensual. Bogotá. Ministerio de la Economía Nacional. Departamento de Comercio e Industrias.

Title varies—título varía: 1930-33, **Boletín de Comercio e Industria;** 1934-35, **Boletín de Comercio;** 1935-37, **Comercio;** 1938-40, **Boletín de Comercio e Industrias.**

Monthly organ of the Bureau of Commerce and Industry. Contains information and statistical data on: Production; domestic and foreign trade; prices in Colombia and price quotations on world markets; transportation; general economic conditions.

Contiene información y datos estadísticos sobre: Producción; comercio interior y exterior; precios en Colombia y cotizaciones de precios en distintos mercados del mundo; transportes; condiciones económicas generales.

Col-6000.E5 1945 **Economic Controls and Commercial Policy in Colombia.** One of a series of reports on economic controls and commercial policy in the American republics. Washington, D.C., 1945. United States Tariff Commission. 25 p.

Folleto sobre Colombia, en la serie de informes de la Comisión de Aduanas de los Estados Unidos, sobre controles económicos y política comercial de las repúblicas americanas.

Col-6000.M5 **Memoria del Ministerio de la Economía Nacional.** Fecha de iniciación? Anual. Bogotá. Ministerio de la Economía Nacional.

Previous to 1942 called—anteriormente a 1942 intitulado: **Informe del Ministerio de la Economía Nacional.**

Annual report of the Department of National Economy. Contains statistical data on agricultural and livestock production, commerce and industry, mining and other branches of the national economy.

Contiene datos estadísticos sobre producción agrícola y ganadera, comercio e industria, minería y otras ramas de la economía nacional.

*Col-**6000**.M7 **El Mes Financiero y Económico.** La revista para comerciantes, industriales y hombres de negocios. 1937- . Mensual. Bogotá.

Monthly review of economics and finance. Includes a regular statistical section.

Incluye una sección estadística regular.

Col-**6000**.R5 1944 **Report Presented to the Conference of Commissions of Inter-American Development (Held in New York, May 9-18, 1944) by the Colombian Commission of Inter-American Development.** On an outline of the present industrial situation in Colombia. (Translation from the Spanish original.) Washington, D.C., 1945? Inter-American Development Commission. 20 p.

Cover title—título de la cubierta: **Colombia: Report Presented to the Conference of Commissions of Inter-American Development.**

Informe presentado a la Conferencia de Comisiones de Fomento Interamericano (realizada en N. York, del 9 al 18 de mayo de 1944), por la Comisión Colombiana de Fomento Interamericano, sobre una reseña de la actual situación industrial de Colombia.

Col-**6200** **Estadísticas Agrícolas en Colombia: Métodos y Procedimientos.** Material básico adquirido para el estudio que el Instituto Interamericano de Estadística realiza sobre los métodos de estadísticas agrícolas en las naciones americanas, incluyendo la encuesta de 1945-46.

Agricultural statistics in Colombia: Methods and procedures. Background material acquired for Inter American Statistical Institute study on methods in agricultural statistics in the American nations, including the Survey of Agricultural Production Statistics in the American Nations in 1945-46. Consists of: Correspondence; reports.

Consiste de: Correspondencia; informes.

Col-**6200**.A4 1945 **Agricultural, Pastoral, and Forest Industries in Colombia.** One of a series of reports on agricultural, pastoral, and forest industries in the American republics. Washington, D.C., 1945. United States Tariff Commission. 44 p.

Publicado en la serie de informes de la Comisión de Aduanas en los Estados Unidos, sobre las industrias agrícolas, campestres y forestales en las repúblicas americanas.

Col-**6200**.T5 **Tierras y Aguas.** Organo de los Departamentos de Tierras y Aguas. 1938- . Mensual. Bogotá. Ministerio de la Economía Nacional.

Monthly bulletin of the Bureaus of Land and Water.

Col-**6210**.B5 **Boletín de Estadística de la Federación Nacional de Cafeteros.** 1932- . Irregular. Bogotá.

Statistical bulletin of the National Federation of Coffee Growers. Data on coffee production and trade.

Datos sobre producción y comercio del café.

Apr. '44 **Boletín de Estadística, Organo de la Federación Nacional de Cafeteros de Colombia.** Año XI, vol. V, no. 25, abril 1944. Federación Nacional de Cafeteros de Colombia.

Fairly complete data on national, state, and municipal production of coffee for Colombia, an estimate of number of coffee trees in the country based on current production, and other information of general interest.

Datos bastante completos sobre la producción nacional, departamental y municipal de café colombiano, un cálculo sobre el número de árboles de café en este país basado en la producción actual, y otras informaciones de interés general.

Dec. '40 "Convenio Interamericano del Café, Noviembre 28 de 1940." Año VIII, vol. V. no. 21, diciembre 1940, p. 260-270.

Text of the Interamerican Coffee Agreement regulating coffee trade between Latin American countries and United States on basis of export quotas assigned to each coffee-producing country.

Convenio que reglamenta el comercio del café entre los países latino-americanos y Estados Unidos, a base de cuotas de exportación para cada país productor de café.

Aug. '39, a "Acuerdo de la Oficina Panamericana del Café en Nueva York, Acerca de la Reglamentación de la Estadística Cafetera Mundial." Año VIII, vol. V, no. 19, agosto 1939, p. 84-88.

Text and discussion of agreement of Pan American Coffee Bureau in New York, regarding co-ordination of international coffee statistics by the Bureau and setting up of necessary statistical organization within the Bureau to implement the agreement.

Texto y discusión de los acuerdos de la Oficina Panamericana del Café en Nueva York, acerca de la coordinación de la estadística internacional del café por esa Oficina, y el establecimiento de los órganos estadísticos necesarios dentro de la misma Oficina a fín de llevar a la práctica los acuerdos.

Aug. '39, b "El Censo Cafetero 1939-40." Año VIII, vol. V, no. 19, agosto 1939, p. 79-83.

A general editorial discussion of the proposed coffee census of 1939-40. Includes reproduction of official questionnaire.

Editorial con discusión general sobre el propuesto censo cafetero. Incluye una reproducción del cuestionario oficial.

Col-**6400.**M5 1945 **Mining and Manufacturing Industries in Colombia.** One of a series of reports on mining and manufacturing industries in the American republics. Washington, D.C., 1945. United States Tariff Commission. 45 p.

Folleto sobre Colombia, de la serie de informes de la Comisión de Aduanas de los Estados Unidos, sobre las industrias mineras y manufactureras en las repúblicas americanas.

Col-**6614.**R5 **Revista del Consejo Administrativo de los Ferrocarriles Nacionales.** 1931- . Irregular. Bogotá.

Review of the Administrative Council of National Railroads. Contains detailed information on the railroads of the country. Last number of each year is the annual report on national railroads, entitled **Informe—Contabilidad y Estadística.**

Contiene información detallada sobre los ferrocarriles del país. El ultimo número de cada año es el informe anual sobre ferrocarriles nacionales, intitulado: **Informe—Contabilidad y Estadística.**

Col-**6720.**A5 **Annual Report Presented by the Manager to the Board of Directors.** 1922-23— . Bogotá. Banco de la República. English edition.

Published also in Spanish—publicado también en español.

Informe anual presentado por el gerente del Banco de la República al Directorio.

Col-**6720.**B5 **Boletín de Información Mensual.** Control de cambios, importaciones y exportaciones. 1945- . Bogotá. Banco de la República.

Monthly bulletin of information on exchange control, imports, and exports, published by the Bank of the Republic.

*Col-**6720.**B7 **Boletín de la Superintendencia Bancaria.** 1924- . Irregular. Bogotá.

Bulletin of the Superintendent of Banks. Gives detailed information concerning all banks of the country.

Contiene información detallada acerca de todos los bancos del país.

Col-**6720.**C8 1928-39 **Cuadros Estadísticos, 1928-39.** Bogotá, Editorial Minerva, S.A. Superintendencia Bancaria. 6 p.

Statistical tables of the Superintendent of Banks covering 1928-39.

Col-**6720.**I7 **Informe y Datos Estadísticos Presentados por el Superintendente Bancario al Señor Ministro de Hacienda y Crédito Público.** Fecha de iniciación? Anual. Bogotá.

Annual report and statistical data of the Superintendent of Banks.

*Col-**6720.**R5 **Revista del Banco de la República.** 1927- . Mensual. Bogotá. Banco de la República.

Monthly bulletin of the Bank of the Republic. Approximately two-thirds of the publication is devoted to statistical data in both table and chart form. Data are presented on government finance, foreign trade, foreign exchange, stock market quotations, cost of living, banking operations, commodity prices, and other economic movements.

Aproximadamente dos tercios de la publicación está dedicada a datos estadísticos en forma de tablas y cuadros. Presenta datos sobre finanzas del gobierno, comercio exterior, cambio exterior, cotizaciones de bolsa, costo de la vida, operaciones bancarias, precios de productos y otras actividades económicas.

Col-**6900** **Estadísticas del Comercio Exterior en Colombia: Métodos y Procedimientos.** Material básico adquirido para el estudio que el Instituto Interamericano de Estadística realiza sobre los métodos de estadísticas del comercio exterior en las naciones americanas, incluyendo la encuesta de 1946-47.

Foreign trade statistics in Colombia: Methods and procedures. Background material acquired for the study conducted by the Inter American Statistical Institute in methods of foreign trade statistics in the American nations, including the survey of 1946-47. Includes: Filled questionnaire and explanatory notes; codes.

Incluye: Cuestionario llenado y notas explicativas; códigos.

*Col-**6900**.A5 **Anuario de Comercio Exterior.** Fecha de iniciación? Bogotá. Contraloría General de la República. Dirección Nacional de Estadística.

Title varies—título varía: **Comercio Exterior de Colombia.**

Yearbook of foreign trade statistics.

1945 **Anuario de Comercio Exterior, 1945.** Bogotá, Imp. Nacional, 1946. Contraloría General de la República. Dirección Nacional de Estadística. 876 y XLIV p.

Contents: Alphabetical indexes of articles imported and exported, 1945; quotations of the principal foreign currency in Bogota, 1936-45; imports and exports by years; imports by country of origin and purchase, 1943-45; exports by country of sale, 1943-45; detail of imports by articles, 1945; detail of exports by articles, 1945; tariff exemptions; international maritime movement.

Contenido: Indices alfabéticos de los artículos importados y exportados en el año de 1945; cotizaciones de las principales monedas extranjeras en Bogotá, 1936-45; importación y exportación por años; importación por países de origen y de compra, 1943-45; exportación por países de venta, 1943-45; detalle de la importación por artículos, 1945; detalle de la exportación por artículos, 1945; exenciones aduaneras; movimiento marítimo internacional.

Col-**6900**.C5 **Comercio Exterior de Colombia.** Fecha de iniciación? Semestral. Bogotá. Contraloría General de la República. Dirección Nacional de Estadística.

Foreign trade of Colombia. Semiannual.

Col-**6900**.T5 1944 **Tarifa de Aduanas de 1944.** Compilada y revisada por Domingo V. de la Espriella. Bogotá, Editorial Minerva, 1944.

Tariff of Colombia. Includes law no. 62 of 1931 and legislative decrees and acts of the Supreme Court of Customs which modify and augment it.

Ley 62 de 1931 y decretos de carácter legislativo, actas del Tribunal Supremo de Aduanas que la modifican y adicionan.

7000-7999 HEALTH AND CULTURAL STATISTICS—ESTADISTICAS DE SALUBRIDAD Y CULTURA

Col-**7150**.E5 1936 **Estadísticas Demográficas y Nosológicas, Año de 1936.** Publicación dirigida por Miguel Angel Escobar, director de la Sección de Demografía de la Dirección Nacional de Estadística. Editada por cuenta del Ministerio de Trabajo, Higiene y Previsión Social. Bogotá, Imp. Nacional, 1938. Contraloría General de la República. Dirección Nacional de Estadística. 63 p.

Demographic and nosological statistics, 1936.

Col-**7200**.E5 1946 **Education in Colombia.** By John H. Furbay. Washington, D.C., U.S.Govt.Print.Off., 1946. U. S. Federal Security Agency. U. S. Office of Education. 111 p. Bulletin, 1946, no. 6.

Study of the Colombian educational system, including background material on country and people and the evolution of education in Colombia, covering kindergarten and elementary education, secondary education, vocational education, training and status of teachers, the national university, other institutions of higher learning, and agencies of public and pupil welfare.

Estudio del sistema de educación de Colombia, incluyendo datos básicos generales sobre el país, la población y la evolución de la educación en Colombia. Abarca jardines infantiles y educación elemental, educación secundaria, educación profesional, preparación y estado de maestros, la universidad nacional, otras instituciones de estudios superiores, y organizaciones para el beneficio del público y de los estudiantes.

Col-**7210**.U5 1945 **The Universities of Colombia.** By Katherine Lenore Morgan. Washington, D.C., 1945. Pan American Union. Division of Intellectual Cooperation. 180 p. Higher education in Latin America, vol. 3.

Las universidades de Colombia.

Col-**7300**.G5 1945 **Geología.** Bogotá, 1945. Banco de la República. Sección de Investigaciones Económicas. 2 p. Boletín gráfico, no. 3, diciembre de 1945.

A geological map of Colombia with explanation of data, prepared by the National Geological Service.

Un mapa geológico de Colombia con explicación de datos, preparado por el Servicio Geológico Nacional.

8000-8999 GOVERNMENT AND POLITICAL STATISTICS—ESTADISTICAS GUBERNAMENTALES Y POLITICAS

*Col-**8800.**E5 **Estadística Fiscal y Administrativa.** Anexo al **Informe Financiero** del Contralor General. Fecha de iniciación? Anual. Bogotá. Contraloría General de la República. Dirección Nacional de Estadística.

Yearbook of fiscal and administrative statistics. Annex to the financial report of the Comptroller General.

1944 **Estadística Fiscal y Administrativa, Colombia—1944.** Anexo al **Informe Financiero** del Contralor General. Bogotá, 1946. Contraloría General de la República. Dirección Nacional de Estadística. 299 p.

National, departmental, territorial, municipal, and administrative fiscal statistics of income, expenditures, and debt.

Estadística fiscal nacional, departamental, territorial, municipal, y administrativa de ingresos, gastos, y deuda.

Col-**8800.**I3 **Información Fiscal de Colombia.** 1936- . Mensual. Bogotá. Contraloría de la República.

Title varies—título varía: 1936-Jan.'39, **Información Económica y Estadística de Colombia.**

Monthly statistical statement of government finances.

Col-**8800.**I5 **Informe Financiero del Contralor General.** 1923- . Anual. Bogotá. Contraloría General de la República.

Annual report on the financial situation in the country by the Comptroller General.

1944 **Informe Financiero del Contralor General de la República de Colombia . . . Correspondiente al Año Fiscal de 1944.** Bogotá, Imp.Nacional, 1945. 220 p.

Includes the report of the Comptroller General and data on: The financial balance; relation of income to expenditures; expenditures for previous fiscal years; national public debt; national railroads.

Incluye el informe del Contralor General y datos sobre: Balance financiero; relaciones de rentas y gastos; gastos por pagar de vigencias anteriores; deuda pública nacional; ferrocarriles nacionales.

Col-**8800.**R5 **Revista de Hacienda.** 1939-1941. Frecuencia? Bogotá. Ministerio de Hacienda y Crédito Público.

Superseded—substituía a: **Boletín de Hacienda.**

Journal of finance.

Nov. '39, Hermberg "El Esquema de la Balanza de Pagos." Por Paul Hermberg. Año I, no. 2, noviembre 1939, p. 93-113. No examinado.

Outline of balance of payments. Includes notes on plans and practices adapted to the situation of Colombia.

Incluye notas sobre planes y prácticas adaptadas a la situación de Colombia.

1000-1999 GENERAL

CR-**1400.**R5 **Resúmenes Estadísticos, Años**
1883-1910 **1883 a 1910.** Comercio, agricultura, industria. San José, Imp.Nacional, 1912. Secretaría de Hacienda y Comercio. Dirección General de Estadística. 135 p. No examinado.

Statistical summaries, 1883-1910. Includes: Commercial census, 1907, p. 102-104; agricultural census, 1905 y 1910, p. 107-113; industrial census, 1907, p. 126-133.

Incluye: Censo comercial, 1907, p. 102-104; censo agrícola, 1905 y 1910, p. 107-113; censo industrial, 1907, p. 126-133.

*CR-**1410.**A5 **Anuario Estadístico.** 1897- . San José. Secretaría de Hacienda y Comercio. Dirección General de Estadística.

Statistical yearbook of Costa Rica.

1942 **Anuario Estadístico, Año 1942.** Tomo cuadragésimosexto. San José, Imp. Nacional, 1945. Dirección General de Estadística. 293 p.

Includes summary data on coastal trade, trade balance, livestock, tourism, and railroad traffic, in addition to detailed data on exports, imports, and demography.

Además de datos detallados sobre exportación, importación, y demografía, incluye datos sumarios sobre cabotaje, balanza comercial, ganadería, turismo, y tráfico de las empresas ferroviarias.

CR-**1410.**I5 **Informe de la Dirección General de Estadística.** Fecha de iniciación? Anual. San José. Secretaría de Hacienda y Comercio. Dirección General de Estadística.

Yearly report of the General Bureau of Statistics. The contents of this publication are also included in **Memoria de la Secretaría de Hacienda y Comercio** (CR-**8800.**M5).

El contenido de esta publicación se incluye también en la **Memoria de la Secretaría de Hacienda y Comercio** (CR-**8800.**M5).

1945 **Informe de la Dirección General de Estadística, Año 1945.** San José, Imp.Nacional, 1946. 91 p.

Contains statistical tables and notes on the commercial situation of the country, imports, exports, industry, coastal trade, tourism, and demography.

Contiene cuadros estadísticos y notas sobre la situación comercial del país, importaciones, exportaciones, industrias, movimiento de cabotaje, turismo, y demografía.

2000-2999 CENSUS—CENSOS

CR-**2200.**C5 **Censo de Personas sin Traba-**
1932 **jo, 1932.** San José, Imp.Nacional, 1933. Secretaría de Hacienda y Comercio. Dirección General de Estadística. 35 p.

Census of unemployment, 1932. Unemployed persons are given by age, sex, duration of employment and customary occupation. Persons not in labor force are not recorded.

Personas desocupadas con clasificación de edad, sexo, duración en el empleo y ocupación habitual. No se incluyen las personas que no pertenecen al potencial obrero.

CR-**2200.**P5 **Población de la República de**
1927 **Costa Rica según el Censo General de Población Levantado al 11 de Mayo de 1927.** Por provincias, cantones y distritos. San José, María V. de Lines, Librería Española, Imp., 1927. Secretaría de Hacienda y Comercio. Oficina Nacional del Censo. 20 p.

Population of Costa Rica according to 1927 census. Total populations are given for provinces, cantons, districts, and capital cities of the provinces.

Presenta población total por provincias, cantones, distritos y capitales de provincias.

CR-**2230.**A5 **Alfabetismo y Analfabetismo en**
1927 **Costa Rica según el Censo General de Población, 11 de Mayo de 1927.** San José, Imp., Librería y Encuadernación Alsina, 1928. Secretaría de Hacienda y Comercio. Oficina Nacional de Censo. 65 p.

Literacy and illiteracy in Costa Rica according to the 1927 census of population. "This survey of the literacy of the population nine years of age and over includes classifications of the province, canton, and district populations by sex and age (under 9, 9 or over). Comparative statistics are presented for 1864 and 1892. Historical series are included on school enrollment." (Source: Gen.Cen.V.S.A.)

"Este estudio del alfabetismo de la población de nueve y más años de edad, incluye clasificación de la población por provincias, cantones y distritos y por sexo y edad (menores de 9, y 9 y más años). Contiene estadísticas comparativas para los años 1864 y 1892. Se incluyen series históricas sobre matriculación en escuelas." (Traducción del comentario.)

CR-**2250**.C5 1915 **Censo Comercial, 1915.** San José, Imp.Nacional, 1917. Secretaría de Hacienda y Comercio. Dirección General de Estadística. 210 p.

Commercial census, 1915.

5000-5999 DEMOGRAPHY—DEMOGRAFIA

*CR-**5000**.A5 **Apéndice al Informe de la Dirección General de Estadística.** Fecha de iniciación? Anual. San José. Secretaría de Hacienda y Comercio. Dirección General de Estadística.

A yearly appendix to the report of the General Bureau of Statistics, devoted principally to demographic statistics, presenting data by provinces, as well as national totals.

Un apéndice anual dedicado principalmente a estadísticas demográficas, presentando datos por provincias, como así también los totales nacionales.

CR-**5000**.C5 1944 **Costa Rica: Summary of Biostatistics.** Maps and charts, population, natality and mortality statistics. Washington, D.C., February 1944. Prepared by U. S. Department of Commerce, Bureau of the Census, in cooperation with Office of the Coordinator of Inter-American Affairs. 92 p.

Resumen bioestadístico de Costa Rica. Incluye mapas y cuadros y datos sobre población, natalidad y mortalidad.

CR-**5200**.A5 1940 **Aspectos Biodemográficos de la Población de Costa Rica.** Informe correspondiente al año 1940 presentado por el Dr. Pablo Luros. Extracto de la **Memoria Anual de la Secretaría . . .** San José, Imp.Nacional, 1942. Secretaría de Salubridad Pública y Protección Social. 183 p.

Biodemographic aspects of the population of Costa Rica. Contains data on births, deaths, fecundity, causes of death classified according to international nomenclature and grouped by diseases, diseases by age, diseases by province, etc.

Contiene datos sobre natalidad, mortalidad, fecundidad, causas de muerte clasificadas de acuerdo a la nomenclatura internacional y agrupadas por enfermedades, enfermedades por edad, por provincias, etc.

6000-6999 ECONOMICS — ECONOMIA

CR-**6000**.E5 1945 **Economic Controls and Commercial Policy in Costa Rica.** One of a series of reports on economic controls and commercial policy in the American republics. Washington, D.C., 1945. United States Tariff Commission. 24 p.

Folleto sobre Costa Rica, en la serie de informes de la Comisión de Aduanas de los Estados Unidos, sobre controles económicos y política comercial de las repúblicas americanas.

CR-**6400**.M5 1945 **Mining and Manufacturing Industries in Costa Rica.** One of a series of reports on mining and manufacturing industries in the American republics. Washington, D.C., 1945. United States Tariff Commission. 11 p.

Folleto sobre Costa Rica, en la serie de informes de la Comisión de Aduanas de los Estados Unidos, sobre las industrias mineras y manufactureras en las repúblicas americanas.

CR-**6720**.B6 **Boletín Estadístico.** Fecha de iniciación? Frecuencia? San José. Banco Nacional de Costa Rica.

Statistical bulletin of the National Bank of Costa Rica. Includes data on: The National Bank; cost of living index; wholesale prices index; exchange rates; index of currency in circulation; commercial banks; mortgage banks; National Insurance Bank; public finance.

Incluye datos sobre: El Banco Nacional; índice del costo de la vida; índice de precios al por mayor; tipos de cambio; índice del medio circulante; bancos comerciales; bancos hipotecarios; Banco Nacional de Seguros; hacienda pública.

*CR-**6720**.R5 **Revista del Banco Nacional de Costa Rica.** Fecha de iniciación? Mensual. San José.

Monthly journal of the National Bank of Costa Rica. Contains statistical summaries on a variety of topics, as well as information on current commercial and monetary affairs of Costa Rica.

Contiene resúmenes estadísticos sobre variedad de tópicos, como también información sobre asuntos comerciales y monetarios corrientes de Costa Rica.

CR-**6720**.R6 **Revista del Banco Nacional de Seguros.** 1944- . Irregular. San José.

Journal of the National Insurance Bank.

Dec. '45 "Memoria Anual de 1944, del Banco Nacional de Seguros." Año I, no. 4, diciembre de 1945, p. 3-71 (publicación entera).

Annual report of the National Insurance Bank of Costa Rica for the year 1944.

CR-**6900** **Estadísticas del Comercio Exterior en Costa Rica: Métodos y Procedimientos.** Material básico adquirido para el estudio que el Instituto Interamericano de Estadística realiza sobre los métodos de estadísticas del comercio exterior en las naciones americanas, incluyendo la encuesta de 1946-47.

Foreign trade statistics in Costa Rica: Methods and procedures. Background material acquired for the study conducted by the Inter American Statistical Institute in methods of foreign trade statistics in the American nations, including the survey of 1946-47. Consists of: Filled questionnaire and explanatory notes.

Consiste de: Questionario llenado y notas explicativas.

CR-6900.A5 1933 **Arancel de Aduanas de la República de Costa Rica.** 2a. edición. San José, Imp.Nacional, 1933. 288 p.

Customs tariff of Costa Rica. Includes export duties and index. Corrections and modifications appear in the **Diario Oficial.**

Incluye derechos de aduanas e índice de exportación. Las correcciones y modificaciones aparecen en el **Diario Oficial.**

*CR-6900.B5 **Boletín de Exportación.** 1941- . Anual. San José. Secretaría de Hacienda y Comercio. Dirección General de Estadística.

Yearly export statistics of Costa Rica. Incorporates **Boletín de Exportación de Café** (bulletin of exportation of coffee), which was published until 1941 as a separate publication. Annual tabular report of the agricultural and industrial exports of the country. Besides data referring to the present year, statistics of previous years are published which contain a detailed study of coffee, the principal product of the country.

Incorpora **Boletín de Exportación del Café,** el cual se publicó hasta 1941 como una publicación distinta. Informe tabular anual de la exportación agrícola e industrial del país. A más del dato que se refiere al presente año, se publican estadísticas de años anteriores que contienen un estudio detallado de café, el principal producto del país.

*CR-6900.I5 **Importación por Artículos.** Volumen, precio y procedencia (complemento al **Informe**). Fecha de iniciación? Anual. San José. Secretaría de Hacienda y Comercio. Dirección General de Estadística.

Annual statistics of imports, by volume, price and origin. Supplement to the annual report of the General Bureau of Statistics.

7000-7999 HEALTH AND CULTURAL STATISTICS—ESTADISTICAS DE SALUBRIDAD Y CULTURA.

CR-7110.M5 **Memoria de la Secretaría de Salubridad Pública y Protección Social.** Fecha de iniciación? Anual. San José. Secretaría de Salubridad Pública y Protección Social.

Annual report of the Department of Public Health and Social Protection.

*CR-7110.S5 **Salud.** 1936- . Anual. San José. Secretaría de Salubridad Pública y Protección Social. Departamento de Estadística Vital.

Health: Annual publication of the Bureau of Vital Statistics.

CR-7200.E5 1946 **Education in Costa Rica.** By John H. Furbay. Washington, D.C., U.S. Govt.Print.Off., 1946. U. S. Federal Security Agency. U. S. Office of Education. 62 p. Bulletin, 1946, no. 4.

Study of the Costa Rican educational system, including background material on country and people and the evolution of education in Costa Rica, covering kindergarten and elementary education, secondary education, training and status of teachers, higher education, and agencies of public and pupil welfare.

Estudio del sistema de educación de Costa Rica, incluyendo datos básicos generales sobre el país, la población y la evolución de la educación en Costa Rica. Abarca jardines infantiles y educación elemental, educación secundaria, preparación y situación profesional de los maestros, educación superior y organizaciones para el beneficio del público y de los estudiantes.

8000-8999 GOVERNMENT AND POLITICAL STATISTICS — ESTADISTICAS GUBERNAMENTALES Y POLITICAS

*CR-8800.M5 **Memoria de la Secretaría de Hacienda y Comercio.** Fecha de iniciación? Anual. San José. Secretaría de Hacienda y Comercio.

Annual report of the Department of Finance and Commerce. Contains statistical data from the various bureaus of this ministry.

Contiene datos estadísticos de varios departamentos de este ministerio.

9000-9999 LABOR AND LIVING CONDITIONS — CONDICIONES DE TRABAJO Y DE VIDA

CR-9952.C5 1944 **La Conmutación de Pensiones Futuras como Reparación Civil en Caso de Muerte.** Estudio jurídico-actuarial. Por Harry Zurcher A. y J. Walter Dittel. San José, 1944. 36 p. Hectographed.

Contains a mortality table for Costa Rica and various tables based thereon.

Contiene una tabla de mortalidad para Costa Rica y varias tablas basadas en la misma.

1000-1999 GENERAL

Cuba-**1200.**G5 1945 **A Guide to the Official Publications of the Other American Republics: VII, Cuba.** James B. Childs, general editor. Washington, D.C., U.S. Govt. Print. Off., 1945. The Library of Congress. 40 p. Latin American series, no. 11.

Una guía de las publicaciones oficiales de las otras repúblicas americanas publicada por la Biblioteca del Congreso de los Estados Unidos: VII, Cuba.

Cuba-**1400.**C5 1944 **Condiciones Económicas y Sociales de la República de Cuba.** Por Carlos M. Raggi Ageo. La Habana, Edit. Lex, 1944. Ministerio de Trabajo. Oficina de Estudios del Plan de Seguridad Social. 215 p. Publicaciones de la revista trabajo, vol. I.

Compact survey of the Cuban national economy, using figures of the 1919 census to compare with data obtained in 1943; discusses population, economic conditions, budgets, cost of living, wages, unemployment, labor relations, and estimate of the national income.

Breve estudio analítico de la economía nacional de Cuba, exponiendo cifras del censo de 1919 en comparación con datos obtenidos en 1943; trata acerca de población, economía, presupuestos, costo de la vida, salarios, desempleo, relaciones de trabajo, y estimación de la renta nacional.

Cuba-**1410.**A5 **Anuario Estadístico de la República de Cuba.** 1914. Habana. No examinado.

Statistical yearbook of Cuba. Only one issue ever published. Composed principally of data from government agencies or from official publications.

Se ha publicado una sola edición. Formado principalmente con datos facilitados por las oficinas de gobierno o contenidos en publicaciones oficiales.

*Cuba-**1420.**B5 **Boletín Mensual de Estadísticas.** 1945- . Habana. Ministerio de Hacienda. Dirección General de Estadística.

Monthly bulletin of statistics. Contains a summary of the economic situation of the country, and tables on: Prices; money; banking; stock quotations; national revenues; foreign trade; transportation; wages; building.

Contiene un sumario de la situación económica del país, y tablas sobre: Precios de las mercancías; estadísticas monetarias; estadísticas bancarias; mercado de valores; ingresos del estado; comercio exterior; transporte; salarios; edificaciones.

Apr. '45, Queralt "El Censo Ganadero de 1945." Por Roger A. Queralt. Vol. I, no. 4, abril 1945, p. 97.

One-page comment on completion of livestock census, by the Director General of Statistics, followed by a table of provisional results.

Un comentario de una página, sobre terminación del censo, por el Director General de Estadística, seguido de un cuadro de resultados provisionales.

2000-2999 CENSUS — CENSOS

Cuba-**2100.**C5 1943.1 **Censo de 1943: Métodos y Procedimientos.** Material usado en el levantamiento del censo de población y del censo electoral, del 25 de julio. Adquirido con motivo del estudio que el Instituto Interamericano de Estadística realiza sobre métodos y procedimientos de los censos, 1944. De la Dirección General del Censo.

Materials used in taking population and electoral censuses, on July 25. Acquired in connection with Inter American Statistical Institute Survey of Census Methods and Procedures, 1944. Consists of: Laws, decrees, or other authorization; schedules; other forms; instructions; punch cards; codes.

Consiste de: Leyes, decretos u otras disposiciones; cédulas; otros formularios; instrucciones; tarjetas perforadas; claves.

Cuba-**2100.**C5 1943.2 **Censo de 1943.** Ley del censo; instrucciones generales a los enumeradores; instrucciones para la identificación dactiloscópica; instrucciones para llenar el modelo número 10 (informe de población). Habana, Imp. Cultura, S.A., 1943. Dirección General del Censo. 74 p.

Census of 1943: Census law; general instructions for enumerators; instructions for fingerprinting; instructions for filling form no. 10 (report on population).

*Cuba-**2200.**C5 1943.2 **Censo de 1943.** Habana, P. Fernández y Cía., 1945. Junta Nacional del Censo. Dirección General del Censo de 1943. 1373 p.

Cover title—título de la cubierta: **Censo del Año 1943.** Half-title—título en la anteportada: **Informe General del Censo de 1943.**

Census of 1943. Includes general description of Cuba, history, economic resources, political

and administrative organization, organization of the central government, arts and letters, tourism and sports, former censuses of Cuba, analysis of the present census, and statistical tables on population, industry, and occupation.

Incluye una descripción general de Cuba, historia, recursos económicos, organización política y administrativa, organización del gobierno central, arte y letras, turismo y deportes, censos anteriores de Cuba, análisis del presente censo, y tablas estadísticas de población, industrias y ocupaciones.

Cuba-**2200.**C5 1931-38 **Censo de 1931.** Recopilación de datos estadísticos referentes a la población de la República de Cuba, desde el 21 de septiembre de 1931 (día del último censo decenal) al 31 de diciembre de 1938. Habana, 1939. Dirección General del Censo. 1 p. Diagrama.

Chart showing summary population figures of 1931 census, and annual estimates to 1938.

3000-3999 STATISTICAL SCIENCE—CIENCIA ESTADISTICA

Cuba-**3030.**C5 1945 **Consejo Nacional de Estadística.** Decreto-ley no. 23 de junio 27 de 1935; ley orgánica del Ministerio de Hacienda; decreto presidencial no. 734 de marzo 20 de 1944. Habana, 1945. 16 p.

National Council of Statistics: Decree-law no. 23, June 27, 1935; organic law of the Department of Finance; presidential decree no. 734, March 20, 1944.

Cuba-**3030.**G5 1935 **Gaceta Oficial, Edición Extraordinaria No. 115, de 28 de Junio de 1935, Creando la Dirección General de Estadística.** Habana, marzo 20, 1944. 6 p. Copia máquina.

Law creating the General Bureau of Statistics.

Cuba-**3030.**O5 1945? **Organización de las Secciones de la Dirección General de Estadística.** Indicando las distintas funciones encomendadas a cada una de ellas, y la importancia de las mismas como auxiliar directo de la administración pública. Habana, 1945? Ministerio de Hacienda. Dirección General de Estadística. 4 p. Copia máquina.

Organization of the sections of the General Bureau of Statistics, showing the functions of each. Charts stapled in showing, on front, organization of the National Council of Statistics, and on back, legislation on the co-ordination of national statistics.

Cuadros adjuntos muestran, en el frente, la organización del Consejo Nacional de Estadística y en el reverso "Legislación de la Coordinación de las Estadísticas Nacionales."

Cuba-**3030.**R5 1945 **Reorganización de la Dirección General de Estadística de la República de Cuba.** Traslado de sus oficinas. Ampliación. Por C. Ramos. Habana, mayo 1945. Ministerio de Hacienda. Dirección General de Estadística. 4 p. Copia a máquina.

Reorganization of the General Bureau of Statistics.

Cuba-**3300** 1944-46 **Enseñanza Estadística en Cuba: Métodos y Materiales.** Material básico adquirido para la encuesta del Instituto Interamericano de Estadística sobre los Métodos y Materiales en la Enseñanza Estadística en las Naciones Americanas, 1944-46.

Statistical training in Cuba: Methods and materials. Background material acquired for the IASI Survey on Statistical Training Methods and Materials in the American Nations, 1944-46. Includes: Letters, filled questionnaire, and program of statistical courses.

Incluye: cartas, cuestionario llenado, y un programa de cursos de estadística.

Cuba-**3320.**P5 1944 **Programa de Estadística Social y Metodología Estadística.** Habana, González y Compañía, 1944. Universidad de Habana. Facultad de Ciencias Sociales y Derecho Público. 11 p.

Outline of course in social statistics and statistical methodology.

5000-5999 DEMOGRAPHY—DEMOGRAFIA

Cuba-**5000.**C5 1945 **Cuba: Summary of Biostatistics.** Maps and charts, population, natality and mortality, statistics. Washington, D.C., 1945. Prepared by U. S. Department of Commerce, Bureau of the Census, in cooperation with Office of the Coordinator of Inter-American Affairs. 113 p.

Resumen bioestadístico de Cuba. Incluye mapas y cuadros y datos sobre población, natalidad y mortalidad.

Cuba-**5200** **Estadística Vital y Registro en Cuba: Métodos y Procedimientos.** Material básico adquirido para la encuesta que el Instituto Interamericano de Estadística realiza sobre los métodos de registro y estadística vital en las naciones americanas, incluyendo el estudio de 1945.

Vital statistics and registration in Cuba: Methods and procedures. Background material acquired for the study conducted by the Inter American Statistical Institute in vital statistics and registration methods in the American nations, including the survey of 1945. Consists of: Answer to questionnaire; laws and decrees; forms; organization chart; publications.

Consiste de: Respuesta al cuestionario; leyes y decretos; formularios; esquema estructural; publicaciones.

See also—véase también: **Consejo Nacional de Estadística** (decreto ley no. 23 de junio de 1935); **Boletín Oficial de la Secretaría de Hacienda.**

Cuba-**5200**.M5 **Movimiento de Población.** 1903- . Anual. Habana. Ministerio de Hacienda. Dirección General de Estadística.

Population movement: Annual population estimates and birth and death data.

Estimaciones de población y datos sobre nacimientos y mortalidad.

Cuba-**5220**.T5 1920 **Tablas de Vida de la Ciudad de la Habana.** Por Vega Lamar. Habana, Imp."La Propagandista," 1920. 24 p.

"Life tables for the city of Havana based on population estimated for period 1904-14 from censuses of 1900 and 1907 and deaths occurring in 1904-14." (Translation of comment.)

"Tablas de la vida de la ciudad de la Habana basadas en la población calculada para el decenio de 1904-14 de los censos de 1900 y 1907 y las defunciones ocurridas en los 10 años 1904-14." (Fuente: Texto.)

6000-6999 ECONOMICS — ECONOMIA

Cuba-**6000**.C5 **Cuba Económica y Financiera.** 1926- . Mensual. Habana.

Monthly economic and financial review, with news and data on all phases of Cuba's economic life.

Noticias y datos sobre todos los aspectos de la vida económica de Cuba.

Cuba-**6000**.E5 1946 **Economic Controls and Commercial Policy in Cuba.** One of a series of reports on economic controls and commercial policy in the American republics. Washington, D.C., 1946. United States Tariff Commission. 49 p.

Folleto sobre Cuba, en la serie de informes de la Comisión de Aduanas de los Estados Unidos, sobre controles económicos y política comercial de las repúblicas americanas.

*Cuba-**6010**.B5 **Boletín de la Oficina de Números Indices.** Fecha de iniciación? Mensual. Habana. Ministerio de Agricultura. Dirección de Industrias. Negociado de Economía Agrícola e Industrial. Oficina de Números Indices.

Monthly bulletin of the Office of Index Numbers.

Cuba-**6010**.I5 1945 **Informes sobre la Estabilización de Precios, el Control de la Inflación y Solicitudes de Abastecimientos Adicionales para Cuba en el Año de 1945.** La Habana, Imp. "Cuba Intelectual," 1945. Ministerio de Comercio. Oficina de Regulación de Precios y Abastecimiento. 69 p.

Reports on stabilization of prices, control of inflation, and applications for additional supplies, for Cuba, 1945. Contents: Report on fight against inflation in Cuba, principally in regard to food; report of assessors of Commission on Sale of Sugar Crop of 1945, on supply of population of Cuba; memorandum of Commission on supply of beans, potatoes, fertilizer, raw materials and machinery for manufacturing gums and resins; statistical charts, tables, and comparisons.

Incluye: Informe sobre los resultados de la lucha contra la inflación en Cuba, principalmente en relación con los productos alimenticios; informe de los asesores de la Comisión para la Venta de la Zafra Azucarera de 1945, sobre el abastecimiento de la población de Cuba; memorándum a la Comisión para la Venta de la Zafra Azucarera de 1945, sobre el abastecimiento de frijoles, papas, fertilizantes, materias primas y maquinaria para la fabricación de gomas y resina; cuadros, tablas y comparaciones estadísticas.

Cuba-**6010**.N5 1938 **Los Números Indices en Cuba.** Por Hugo Vivó. Habana, junio 1938. Secretaría de Agricultura. Dirección de Industrias. Negociado de Economía Agrícola e Industrial. Oficina de Números Indices. 43 p.

Index numbers in Cuba.

Cuba-**6200**.B5 **Boletín Agrícola para el Campesino Cubano.** Fecha de iniciación? Mensual. Habana. Ministerio de Agricultura. No examinado.

Monthly agricultural bulletin. Contains statistics of agricultural production and consumption, and the results of special studies conducted by various sections of the Department of Agriculture.

Contiene estadísticas sobre producción y consumo de productos agrícolas, y resultados de los estudios especiales hechos en distintas secciones del Ministerio.

Cuba-**6210**.A5 **Anuario Azucarero de Cuba. Cuba Sugar Year Book.** 1937- . Habana, Cuba Económica y Financiera.

In English and Spanish—en inglés y español.

Contains material on cultivation, mills, sugar, molasses, refineries, labor, communications, etc.

Contiene datos sobre cultivo, ingenios, azúcar, melaza, refinerías, trabajo, comunicaciones, etc.

Cuba-**6210**.B5 **Boletín del Café.** Fecha de iniciación? Mensual. Habana. Ministerio de Agricultura. Instituto Cubano de Estabilización del Café. No examinado.

Monthly bulletin on coffee.

Cuba-**6210**.C5 **Compilación Estadística sobre Café.** Fecha de iniciación? Anual. Habana. Ministerio de Agricultura. Dirección de Industrias. Negociado de Estadística.

Official annual report on the coffee crop of Cuba.

Cuba-**6210**.Z5 **Zafra.** Fecha de iniciación? Anual. Habana. Ministerio de Agricultura. Dirección de Industrias. Negociado de Asuntos Azucareros.

Official annual report on the sugar crop of Cuba.

Cuba-**6300**.B5 **Boletín de Minas.** Fecha de iniciación? Anual. Habana. Ministerio de Agricultura. Dirección de Montes, Minas y Aguas.

Annual containing statistics of mineral resources and production.

Contiene estadísticas sobre recursos y fuentes minerales.

Cuba-**6610**.M5 **Memoria Oficial de la Comisión Nacional de Transportes.** Fecha de iniciación? Anual. Habana. No examinado.

Annual report of the National Commission of Transportation.

Cuba-**6700**.D5 1941 **Directorio Oficial de Exportación, Importación, Producción y Turismo, 1941.** Habana? Cámara de Comercio de la República de Cuba. No examinado.

Directory of exports, imports, production, and tourism, 1944, by the Chamber of Commerce. Contains a national income estimate for 1938.

Incluye una estimación de la renta nacional para 1938.

Cuba-**6900**.A5 1945 **Arancel y Repertorio de Aduanas de la República de Cuba.** Por Rafael Pérez Lobo. Habana, "Cultura, S.A.," 1945. 2 vol. No revisado.

Customs tariff of Cuba. "Contains . . . all the legislation relating to the subject, trade treaties, and Supreme Court pronouncements on the application of the tariff to 1945, arranged by items, in the following parts: Customs tariff; legal dispositions; jurisprudence; official repertoire; indexes." (Translation of comment.)

"Contiene . . . toda la legislación complementaria sobre la materia, Tratados de Comercio y sentencias del Tribunal Supremo sobre aplicación del Arancel dictadas hasta 1945, ordenadas por Partidas, distribuídos en las siguientes partes: Arancel de Aduanas, Disposiciones Legales, Jurisprudencia, Repertorio Oficial e Indices." (Fuente: **Cuba Económica y Financiera,** vol. XX, no. 234, septiembre 1945).

*Cuba-**6900**.C5 **Comercio Exterior.** Fecha de iniciación? Anual. La Habana. Ministerio de Hacienda. Dirección General de Estadística.

Yearbook of foreign commerce.

1935-44 **Comercio Exterior, 1935-39 a 1944.** La Habana, Imp. P. Fernández y Cía., S.en C., 1946. Ministerio de Hacienda. Dirección General de Estadística. 630 p.

Foreign commerce of Cuba from 1935-39 to 1944. Includes: Graph showing imports and exports, 1902-44; comparative summary of Cuba's foreign commerce 1899-1944; trade balance between Cuba and various countries during 1940-44; and average for 1935-39; comparative summaries by countries, months, customs stations, classes of articles, and by classes of merchandise and countries, of the total value of the imported and exported merchandise during the calendar years 1940-44, and averages for 1935-39; comparison between the quantities and values of imported and exported articles, shown by groups and countries, and by items and countries.

Incluye: Un cuadro gráfico de la exportación e importación durante los años 1902-44; resumen comparativo del comercio exterior de Cuba durante los años de 1899-1944; saldos de la balanza comercial de la República con los distintos países, 1940-44, y promedio de 1935-39; resumenes comparativos por países, meses, aduanas, clases de artículos, y por clases de mercancías y países, del valor total de las mercancías importadas y exportadas durante 1940-44, y promedio de 1935-39; estados comparativos de cantidad y valor de las mercancías importadas y de las mercancías exportadas por grupos y países, y por partidas y países.

Cuba-**6900**.I5 **Importación y Exportación de la República de Cuba.** 1923-

Mensual. Habana. Mimeografiado. Ministerio de Hacienda. Dirección General de Estadística.

Monthly bulletin on imports and exports. Summaries of foreign trade for the current and previous periods, classified by countries.

Resúmenes del comercio exterior del período actual y de los anteriores, clasificado por países.

7000-7999 HEALTH AND CULTURAL STATISTICS—ESTADISTICAS DE SALUBRIDAD Y CULTURA

Cuba-**7110**.B5 **Boletín Semanal . . . Epidemiológico Sanitario.** Fecha de iniciación? Semanal. Habana. Instituto Finlay. Sección de Publicaciones Científicas, Biblioteca y Museo.

Weekly bulletin on epidemiology.

*Cuba-**7110**.S5 **Salubridad y Asistencia Social.** 1909- . Mensual. Habana. Ministerio de Salubridad y Asistencia Social.

Monthly bulletin on health and social assistance. Contains health and demographic data such as monthly list of causes of death in Havana classified according to the International List.

Contiene estadísticas de la salud y datos demográficos, tales como una lista mensual de las causas de muerte en Habana clasificada de acuerdo a la Lista Internacional.

Cuba-**7200**.E5 1943 **Education in Cuba.** By Severin K. Turosienski. Washington, D.C., U.S. Govt. Print Off., 1943. Federal Security Agency. U.S. Office of Education. 90 p. Bulletin, 1943, no. 1.

A discussion of the Cuban educational system, covering kindergarten and elementary, secondary, vocational, and professional schools, including the University of Havana, and giving background data on area, population, government, and the development of educational institutions in Cuba.

Un estudio sobre el sistema de educación en Cuba, comprendiendo jardines infantiles, escuelas elementales, secundarias, vocacionales y profesionales, incluyendo la Universidad de la Habana, y dando datos básicos sobre área, población, política, y el desarrollo de las instituciones de educación en Cuba.

8000-8999 GOVERNMENT AND POLITICAL STATISTICS—ESTADISTICAS GUBERNAMENTALES Y POLITICAS

Cuba-**8800**.B5 **Boletín Oficial del Ministerio de Hacienda.** 1909- . Mensual? Habana.

Official bulletin of the Department of Finance.

ECUADOR

1000-1999 GENERAL

*Ec-**1400**.E5 **Ecuador en Cifras, 1938 a 1942.**
1938-42 Quito, Imp.del Ministerio de Hacienda, 1944. Dirección Nacional de Estadística. 515 p.

Ecuador in figures, 1938 to 1942. The basic compilation of statistics of Ecuador. Includes: General data; position of Ecuador on the continent; meteorologic conditions; demographic conditions; cultural conditions; judicial, penal and police conditions; welfare, health, and labor; economic and social conditions; agricultural, industrial, and mining production; roads, communications, and transports; fiscal conditions; credit and money situation; foreign trade.

Compilación básica de estadísticas del Ecuador. Incluye: Datos generales; situación del Ecuador en el continente; situación meteorológica; situación demográfica; situación cultural; situación judicial, penal y policial; previsión, salubridad y trabajo; situación económico-social; producción agrícola, industrial y minera; vialidad, comunicaciones y transportes; situación fiscal; situación monetaria y crediticia; comercio exterior.

Ec-**1420**.E5 **Estadística y Censos.** Información mensual del gobierno del Ecuador. Actividades nacionales. Marzo 1939-octubre 1939. Quito. Ministerio de Hacienda. Dirección General de Estadística.

Statistics and censuses, bulletin of the General Bureau of Statistics, published monthly from March to October 1939. Summarized all current statistical data collected in Ecuador at the time, including population estimates.

Resumen de todos los datos estadísticos de la actualidad que en ese tiempo se recogieron en el Ecuador, incluyendo los cálculos de la población.

*Ec-**1420**.T5 **El Trimestre Estadístico del Ecuador.** Revista trimestral de la Dirección General de Estadística y Censos, adscrita al Ministerio de Economía. 1945- . Quito.

Quarterly review of the General Bureau of Statistics of Ecuador. Includes studies and reports of the Bureau, signed articles, bibliographical notes, and current statistical series on meteorology, vital statistics, public health, production, education, transportation, prices, consumption, foreign commerce, fiscal statistics, and banking and money.

Incluye estudios e informes de la Dirección General, artículos por autores individuales, notas bibliográficas, y datos estadísticos recientes sobre meteorología, movimiento demográfico, estadísticas sanitarias, producción, educación, transportes, precios, consumos, comercio exterior, estadísticas fiscales y bancos y moneda.

July-Dec. '45, a "Comentarios y Sugestiones Respecto a las Bases Mínimas para el Censo de las Américas de 1950." Por la Dirección General de Estadística y Censos del Ecuador. Año I, no. 3-4, julio-diciembre 1945, p. 1-23.

Comments and suggestions regarding minimum bases for the 1950 Census of the Americas.

July-Dec. '45, b "Indices del Costo de la Vivienda Popular en Quito de 1938 a 1945." Año I, no. 3-4, julio-diciembre 1945, p. 24-32.

Indexes of cost of low-cost housing in Quito from 1938 to 1945. "Since the stock of the issue of 'Cuaderno de Estadística No. 1': 'Indices del Costo de la Vivienda Popular en Quito, de 1938 a 1944', published in February of 1945, has rapidly become exhausted, we think it timely to reproduce the substantial part of its contents on occasion of the publication of the results of this investigation for the year 1945." (Source: Note at head of article; translation.)

"Por haberse agotado rápidamente la edición del 'Cuaderno de Estadística No. 1': 'Indices del Costo de la Vivienda Popular en Quito, de 1938 a 1944,' publicado en febrero de 1945, creemos oportuno reproducir la parte substancial de su contenido, con ocasión de publicarse los resultados de esta investigación para el año de 1945." (Fuente: Nota bajo el encabezamiento del artículo.)

July-Dec. '45, c "Modelos para Registrar las Cuentas Internacionales." Por el Comité de Expertos Estadísticos de la Sociedad de las Naciones, Sub-Comité sobre Estadísticas de Balanzas de Pagos. Año I, no. 3-4, julio-diciembre 1945, p. 33-44.

Model forms for registering international accounts, by the Subcommittee on Balance of Payments Statistics of the Committee of Statistical Experts of the League of Nations.

Apr.-June '45, Procaccia "El Censo Continental de las Américas y el Primer Censo Demográfico

Nacional." Por Carlos Procaccia. Año I, no. 2, abril-junio 1945, p. 7-17.

Includes population estimates for Ecuador for Jan. 1, 1944 and Jan. 1, 1945.

Incluye estimaciones de población para Ecuador, enero 1, 1944 y enero 1, 1945.

Apr.-June '45, Romero "Síntesis de Estadística Educacional." Por Oscar A. Romero. Año I, no. 2, abril-junio 1945, p. 18-30.

Synthesis of educational statistics. Includes: Pre-school and primary instruction; rural normal schools; professional schools; special schools; complementary schools; secondary schools; universities; budgets for education.

Incluye: Enseñanza pre-escolar y primaria; normales rurales; escuelas profesionales; escuelas especiales; escuelas complementarias; colegios secundarios; universidades; presupuestos de educación.

May '45 "Ley de Estadística (de 9 de Agosto de 1944), Decreto No. 760." Año I, no. 1, mayo 1945, p. 24-26.

Law of statistics of August 9, 1944, decree no. 760, creating the General Bureau of Statistics and Censuses and outlining its functions.

Crea la Dirección General de Estadística y Censos y detalla sus funciones.

May '45, Lopez Muñoz "Labores de la Dirección General de Estadística y Censos." Por Luis Lopez Muñoz. Año I, no. 1, mayo 1945, p. 3-6.

Work of the General Bureau of Statistics and Censuses.

May '45, Procaccia "Las Estadísticas del Movimiento Demográfico en el Ecuador, y su Perfeccionamiento." Por Carlos Procaccia. Año I, no. I, mayo 1945, p. 7-23.

Statistics of demographic movement in Ecuador, and their improvement. Includes birth and death registration forms.

Incluye formularios de registro de nacimientos y defunciones.

3000-3999 STATISTICAL SCIENCE—CIENCIA ESTADISTICA

Ec-**3000**.C5 **Cuadernos de Estadística.** 1945- . Irregular. Quito. Ministerio de Economía. Dirección General de Estadística y Censos.

Statistical studies.

Ec-**3210**.C5 1945 **Clasificación del Arancel de Aduanas del Ecuador según la Lista Mínimum de Mercaderías para las Estadísticas del Comercio Internacional.** Por Carlos Procaccia. Quito, Imp.del Ministerio del Tesoro, 1945. Ministerio de Economía. Dirección General de Estadística y Censos. 96 p. Cuadernos de estadística, no. 2.

Classification of the customs tariff of Ecuador according to the Minimum List of Commodities for International Trade Statistics. Contains: Classification of sections and chapters of the Minimum List; classification of the paragraphs and rubrics of the customs tariff of Ecuador, according to sections, chapters, items, and subitems of the Minimum List; complete translation into Spanish of the 1938 revision of the original edition of the Minimum List, with the additions and corrections of 1939; classification according to stage of production and to use of items of the Minimum List and of paragraphs and rubrics of the customs tariff of Ecuador; complete list of the paragraphs and rubrics of the customs tariff of Ecuador, indicating the corresponding items and sub-items of the Minimum List; division for statistical purposes of the paragraphs (numbers) and rubrics (letters) of the customs tariff of Ecuador, whose contents must be classified under more than one item or sub-item of the Minimum List.

Contiene: Clasificación de las secciones y capítulos de la Lista Mínimum; clasificación de los párrafos e incisos del Arancel de Aduanas del Ecuador, según las secciones, capítulos, posiciones y subposiciones de la Lista Mínimum; traducción integral al castellano de la Lista Mínimum, de la edición original revisada de 1938, con los aumentos y correcciones de 1939; clasificación según el grado de preparación y el uso, de las posiciones de la Lista Mínimum y de los párrafos e incisos del Arancel de Aduanas del Ecuador; elenco completo de los párrafos e incisos del Arancel de Aduanas del Ecuador, con indicación de las posiciones y subposiciones correspondientes de la Lista Mínimum; fraccionamiento para fines estadísticos de los párrafos (números) e incisos (letras) del Arancel de Aduanas del Ecuador, cuyo contenido debe clasificarse bajo más de una posición o subposición de la Lista Mínimum.

Ec-**3300** 1944-46 **Enseñanza Estadística en Ecuador: Métodos y Materiales.** Material básico adquirido para la Encuesta del Instituto Interamericano de Estadística sobre los Métodos y Materiales en la Enseñanza Estadística en las Naciones Americanas, 1944-46.

Statistical training in Ecuador: Methods and

materials. Background material acquired for the IASI Survey on Statistical Training Methods and Materials in the American Nations. Includes: Correspondence and filled questionnaires on statistical courses in Ecuador.

Incluye: Correspondencia y cuestionarios llenados sobre cursos de estadística en Ecuador.

5000-5999 DEMOGRAPHY—DEMOGRAFIA

Ec-**5000**.E5 1944 **Ecuador: Summary of Biostatistics.** Maps and charts, population, natality and mortality statistics. Washington, D.C., June 1944. Prepared by U. S. Department of Commerce, Bureau of the Census, in cooperation with Office of the Coordinator of Inter-American Affairs. 61 p.

Resumen bioestadístico de Ecuador. Incluye mapas y cuadros y datos sobre población, natalidad y mortalidad.

Ec-**5100**.P5 1944 "Población de la República del Ecuador, al 1° de Enero de 1944." Por provincias, cantones y ciudades; urbana y rural. **Registro Oficial,** no. 49, julio 29, p. 332-334.

Population of Ecuador for January 1, 1944, by provinces, cantons, and cities; urban and rural.

Ec-**5100**.P5 1942 **La Población del Ecuador.** Como contribución al estudio de la población del Ecuador, lo publica el Ministerio de Previsión Social. Por Luis Telmo Paz y Miño. Quito, Tall.Gráf.de Educación, 1942. 51 p. Reimprimido de **Ecuador,** no. 3, septiembre 1936.

The population of Ecuador. Contains figures from all earlier estimates, and gives official estimates for 1935, on the basis of the complete and partial censuses taken since 1777.

Contiene cifras de las mas antiguas estimaciones, y presenta las estimaciones oficiales para 1935, sobre la base de los censos totales y parciales tomados desde 1777.

Ec-**5110**.D5 1780-1935 **La Distribución Geográfica de la Población del Ecuador.** Por Luis Telmo Paz y Miño. Quito, Imp. de la Universidad Central, 1938. 32 p.

The geographic distribution of the population of Ecuador. Study of population movement in Ecuador from 1780 to 1935 based on data obtained from censuses of 1780 and 1824. Contains graphs and tables showing percentage of growth and density of population by geographic regions.

Un estudio del movimiento de población de Ecuador durante el período 1780-1935, basado en los datos obtenidos de los censos de 1780 y 1824. Contiene gráficos y cuadros demostrativos del porcentaje de crecimiento y de la densidad de la población por regiones geográficas.

Ec-**5200**.B5 1943 **Bioestadística en el Cosmos Ecuatoriano.** Por Juan José Samaniego. Quito, Tall.Gráf.de Educación, 1943. Ministerio de Previsión Social. 23 p. Publicaciones del Ministerio de Previsión Social.

Vital statistics in Ecuador.

Ec-**5250**.L5 1927 **Ley de Registro Civil 25/29 de Octubre de 1.900 y Reglamento a la Misma 19 de Julio de 1.927.** Quito, 1927. 13 y 12 p. Copia máquina.

Civil register law of October 25-29, 1900, and its regulatory statute of July 19, 1927.

6000-6999 ECONOMICS — ECONOMIA

Ec-**6000**.E5 1945 **Economic Controls and Commercial Policy in Ecuador.** One of a series of reports on controls and commercial policy in the American republics. Washington, D.C., 1945. United States Tariff Commission. 25 p.

Folleto sobre Ecuador, de la serie de informes de la Comisión de Aduanas de los Estados Unidos, sobre controles económicos y política comercial de las repúblicas americanas.

Ec-**6000**.I5 **Información Económica.** Organo del Ministerio de Economía. 1944-45. Mensual. Quito.

No longer published—ya no se publica. Only 6 numbers published, from October 1944 to March 1945—sólo 6 números publicados, de octubre de 1944 a marzo de 1945.

Economic information. " . . . Object is to maintain contact between the country's commerce, agriculture, and industry—factors of economic activity—and the work carried on by the Department of Economy." (Translation of comment.)

" . . . Tiene por objeto mantener en contacto al comercio, agricultura e industria nacionales—factores de actividad económica—con las labores que desarrolla el Ministerio de Economía." (Fuente: Editorial, año I, no. 1, octubre 1944, p. 2.)

Ec-**6000**.I5A **Datos Estadísticos.** Suplemento a **Información Económica.** 1944-45. Mensual. Quito. Ministerio de Economía. Dirección General de Estadística y Censos.

Supplement to **Información Económica,** issued by the General Bureau of Statistics and Census, containing general statistical data. Only five issues published, from the end of 1944 through the first months of 1945.

Suplemento de la **Información Económica** publicada por la Dirección General de Estadística y Censos, y exponiendo datos estadísticos generales. Sólo han aparecido cinco ediciones, desde el fin de 1944 hasta los primeros meses de 1945.

Ec-**6000**.M5 **Memoria del Ministerio de Economía.** 1945-46— . Anual. Quito.

Annual report of the Department of Economy.

Ec-**6000**.R5 1944 **Report Presented to the Conference of Commissions of Inter-American Development by the Ecuadorian Commission of Inter-American Development.** On Ecuador and its natural resources. (Translation from the Spanish original.) Washington, D.C., 1945? Inter-American Development Commission. 89 p.

Cover title—título de la cubierta: **Ecuador: Report Presented to the Conference of Commissions of Inter-American Development.**

Informe presentado a la Conferencia de Comisiones de Fomento Interamericano, por la Comisión Ecuatoriana de Fomento Interamericano, sobre Ecuador y sus recursos naturales.

Ec-**6300**.M5 **La Minería y el Petróleo en el Ecuador: Anuario.** 1939-42. Quito. Ministerio de Agricultura, Industrias, Minas y Turismo. Dirección General de Minería y Petróleos.

No longer published—se suspendió su publicación.

Yearbook on mining and petroleum in Ecuador.

Ec-**6400**.M5 1945 **Mining and Manufacturing Industries in Ecuador.** One of a series of reports on mining and manufacturing industries in the American republics. Washington, D.C., 1945. United States Tariff Commission. 19 p.

Folleto sobre Ecuador, de la serie de informes de la Comisión de Aduanas de los Estados Unidos, sobre las industrias mineras y manufactureras en las repúblicas americanas.

Ec-**6720**.B3 **Boletín de la Superintendencia de Bancos.** 1938- . Trimestral. Quito. Ministerio de Economía. Superintendencia de Bancos.

Quarterly bulletin of the Superintendent of Banking. Includes a general report on banking institutions, laws and regulations pertaining to banking, and statistics on banks and insurance.

Incluye: Informe general sobre las instituciones bancarias, leyes y decretos relativos a bancos, y estadística bancaria y de seguros.

Ec-**6720**.B4 **Boletín del Banco Hipotecario del Ecuador.** 1937-43. Trimestral. Quito.

No longer published—ya no se publica.

Bulletin of the Mortgage Bank of Ecuador. Especial attention is given to agricultural statistics.

Presenta especialmente datos estadísticos sobre la agricultura.

*Ec-**6720**.B5 **Banco Central del Ecuador, Boletín.** 1928- . Mensual. Quito.

Title varies—título varía: 1928-1944: **Boletín Mensual del Banco Central del Ecuador.**

Monthly bulletin of the Central Bank of Ecuador, devoted to economics, money, and banking in Ecuador. Includes signed articles, reports of the bank, and statistical data including money in circulation, exchange quotations, value of imports and exports.

Dedicado a la economía, moneda y actividades bancarias del Ecuador. Incluye artículos por autores individuales, informes del banco, y datos estadísticos sobre la circulación de moneda, cotización de cambio, valor de importación y exportación.

Feb.-Mar. '46 "Memoria Anual que el Consejo de Administración del Banco Central del Ecuador Presenta a la Asamblea Ordinaria de Accionistas por el Ejercicio Económico de 1945." Año xix, no. 223-224, febrero-marzo 1946, p. 5-52.

Annual report of the Central Bank of Ecuador for 1945.

Apr.-May '43, Riofrio Villagomez "Algunos Datos sobre la Evolución Financiera y Económica en el Ecuador." Por E. Riofrio Villagomez. Año xvi, no. 189-190, abril-mayo 1943, p. 21-41.

Some data on the financial and economic development in Ecuador. Includes a national income estimate.

Incluye una estimación de la renta nacional.

Mar. '42, Laso "Breves Consideraciones sobre la Economía Ecuatoriana." Por Luis Eduardo Laso. Marzo 1942, p. 14. No examinado.

Brief considerations on the economy of Ecuador. Includes apparently the first national income estimate for Ecuador, which indicates only the number of families in the higher and lower income brackets.

Incluye una estimación, tal vez la primera, de la renta nacional del Ecuador, indicando solamente el número de las familias correspondientes a las altas y bajas categorías de ingresos.

Ec-**6720**.E5 1946 **Estadísticas y Otras Informaciones para el Fondo Monetario Internacional.** Quito, noviembre 1946. Banco Central del Ecuador. 25 p. (aprox.)

Statistics and other information for the International Monetary Fund. Contents: Summary of existing regulations for control of foreign exchange and foreign trade in Ecuador; information about exchange, payment, and financial agreements of Ecuador with other countries; legal procedure for buying gold obtained in the country; and a statistical section (dollar quotation, 1944-46; reserves and stock of gold, and foreign exchange; composition of the total reserve, 1937-46; total and legal reserves and the liabilities of the reserve, 1946; gold production, 1938-45; gold export, 1937-45; balance of exchange 1942-46; import, export and balance, 1937-46; basic composition of the exports, 1937-46; price indexes of articles of prime necessity in Quito, 1939-46).

Contiene: Síntesis de disposiciones vigentes en el Ecuador sobre control de cambios extranjeros y del comercio exterior; información concerniente a acuerdos de cambio, de pagos y de finanzas del Ecuador con otros países; régimen legal de compras internas del oro extraído del país; y una sección de estadísticas (cotización del dólar 1944-46; reservas y otras existencias de oro y cambio extranjero; composición de la reserva total 1937-46; reserva legal y total y exigibilidades sujetas a la reserva, 1946; producción de oro, 1938-45; exportaciones de oro, 1937-45; balanza de cambios, 1942-46; importaciones, exportaciones y saldos, 1937-46; composición básica de las exportaciones, 1937-46; índices de precios de artículos de primera necesidad en Quito, 1939-46.)

Ec-**6720**.I5 **Informe del Superintendente de Bancos al Señor Ministro de Economía.** Fecha de iniciación? Anual. Quito. Ministerio de Economía. Superintendencia de Bancos.

Annual report of the Superintendent of Banks.

Ec-**6900** **Estadísticas del Comercio Exterior en Ecuador: Métodos y Procedimientos.** Material básico adquirido para la encuesta que el Instituto Interamericano de Estadística realiza sobre los métodos de estadísticas del comercio exterior en las naciones americanas.

Foreign trade statistics in Ecuador: Methods and procedures. Background material acquired for the study conducted by the Inter American Statistical Institute in foreign trade statistics in the American nations. Includes: Filled questionnaire; definitions, instructions, and conversion tables for measures, from the tariff law of 1943; geographic code.

Incluye: Cuestionario llenado; definiciones, instrucciones y tablas para conversión de medidas de la ley arancelaria de aduanas de 1943; clave de orden geográfico.

Ec-**6900**.A4 **Aduanas del Ecuador.** Organo informativo de la Dirección General de Aduanas del Ecuador. 1946- . Semestral. Guayaquil, Ecuador. Ministerio del Tesoro. Dirección General de Aduanas.

Customs of Ecuador: Quarterly journal of the General Bureau of Customs. Contains statistical data concerning customs activities such as: Trade fluctuations, trade movements, dollar quotation, international trade, treaties, value of imported and exported goods, etc.

Contiene información estadística de todo lo que respecta al movimiento aduanero, como ser: Fluctuaciones del comercio, movimiento del comercio, cotización del dólar, comercio internacional, tratados, valores de las mercaderías importadas y exportadas, etc.

*Ec-**6900**.A5 **Anuario de Comercio Exterior.** 1889-91, 1905, 1913-27, 1943-44— . Quito. Ministerio de Economía. Dirección General de Estadística y Censos.

Yearbook of foreign trade. Has been issued by various agencies at different times, and under various titles; 1925-27; **Comercio Exterior de la República del Ecuador.**

Ha sido publicado en varias épocas por distintos organismos y bajo diferentes títulos; 1925-27: **Comercio Exterior de la República del Ecuador.**

Ec-**6900**.E3 1942 **Ecuador.** Washington, D. C., 1942. Pan American Union. 6 p. Foreign trade series, no. 195.

Ec-**6900**.E5 1940-42 **Estadísticas de las Exportaciones del Ecuador, 1940-1942.** (Con series históricas anteriores.) Sistematizadas por el Dr. Hanns Heiman G. Washington, D.C., 1943-45. Ministerio de Economía. Dirección General de Estadística y Censos. 200 p. (aprox.)

Contents—contenido: Parte I, **Su Desarrollo en los Ultimos Decenios hasta el Año 1940;** parte II, **Su Desarrollo en los Años de 1940 a 1942.**

"The present compilation of data about Ecuadorean exports brings together and systematizes the statistics of exports elaborated by the administrative organism of the National Customs stations, and is based on the detailed classified annual data of the General Bureau of Customs, and on the systematization of the same data, accomplished previously by the chief of statistics

. . . and by the director general of statistics . . ." (Source: Preliminary note; translation.)

"La presente compilación de datos sobre las exportaciones del Ecuador, reune y sistematiza las estadísticas de exportaciones elaboradas por los organismos administrativos de las Aduanas Nacionales, y se basa sobre los detalles anuales clasificados de la Dirección General de Aduanas, y sobre las sistematizaciones de los mismos datos, efectuadas, en épocas anteriores, por el jefe de estadística . . . y por el director general de estadística . . ." (Fuente: Nota preliminar.)

Ec-**6900**.L5 1943 **Ley Arancelaria de Aduanas, 1943.** Quito, Imp. Ministerio de Hacienda, 1943? Ministerio de Hacienda. 239 p.

Customs code of Ecuador. Law of March 30, 1938. Classification is the same as that of Paraguay.

Según la Ley del 30 de marzo de 1938. La clasificación es la misma que la del Paraguay.

7000-7999 HEALTH AND CULTURAL STATISTICS—ESTADISTICAS DE SALUBRIDAD Y CULTURA

Ec-**7110**.I5 **Informe que el Señor Director General de Sanidad Presenta ante el Señor Ministro del Ramo acerca de las Labores Desarrolladas.** 1910-24, 1926- . Anual. Guayaquil. Ministerio de Previsión Social y Trabajo. Dirección General de Sanidad.

Annual report of the General Bureau of Public Health.

1941-42 **Informe que el Señor Director General de Sanidad Presenta ante el Señor Ministro del Ramo, acerca de las Labores Desarrolladas entre los Meses de Abril a Diciembre de 1941 y de Enero a Abril de 1942.** Guayaquil, Ecuador, 1942. 56 p. y apéndices.

Report of the Director General of Public Health for 1941-42. Includes the report of the National Institute of Hygiene, and a statistical appendix which includes statistics of the sanitary departments and demographic information.

Incluye "Memoria del Instituto Nacional de Higiene," p. 21-33, y un apéndice estadístico "Estadística Sanitaria" que incluye estadísticas de los departamentos sanitarios e información demográfica.

Ec-**7200**.E5 1942 **Estadística y Escalafón.** Quito, Tall.Gráf. de Educación, 1942. Ministerio de Educación Pública. 486 p. y anexos.

Statistics and roster. Includes tables with all kinds of data and references on kindergarten, primary, specialized, professional, rural, secondary, and higher education in the country.

Incluye cuadros numéricos, con toda clase de datos y referencias, de las enseñanzas preescolar, primaria, especial, profesional, rural, secundaria y superior de la república.

8000-8999 GOVERNMENT AND POLITICAL STATISTICS—ESTADISTICAS GUBERNAMENTALES Y POLITICAS

Ec-**8800**.B5 **Boletín del Ministerio del Tesoro.** Organo de información trimestral. 1945- . Quito.

Quarterly bulletin of the Department of the Treasury, containing official information of the Department and its branches, studies in economics and finance, and fiscal statistics.

Contiene informaciones oficiales del ministerio y de sus organismos, así como reseñas económicas y financieras, y estadísticas fiscales.

Ec-**8800**.I5 **Informe que el Ministro de Hacienda y Crédito Público Presenta a la Nación.** 1833-1943. Anual. Quito. Ministerio de Hacienda y Crédito Público.

No longer published—se suspendió su publicación.

Report of the Minister of Finance and Public Credit. Covers public finance, banking, and foreign trade.

Incluye finanzas públicas, bancos y comercio exterior.

Ec-**8800**.R5 **Revista de Hacienda.** 1928-43. Bimensual. Quito. Ministerio de Hacienda y Crédito Público.

No longer published. Until 1940, published monthly as **Boletín de Hacienda.**

Se suspendió su publicación. Publicada hasta 1940 mensualmente, con el título **Boletín de Hacienda.**

Public finance review. Articles and statistics on foreign trade, public finance, banking, and other economic subjects.

Artículos y estadísticas sobre comercio exterior, finanzas públicas, bancos, y otros tópicos económicos.

9000-9999 LABOR AND LIVING CONDICIONS—CONDICIONES DE TRABAJO Y DE VIDA

Ec-**9000**.D5 **Documentos Anexos al Informe que el Ministro de Previsión Social y Trabajo Presenta a la Nación.** 1941?- . Anual. Quito. Ministerio de Previsión Social, Trabajo y Asistencia Pública.

Statistical appendix to the annual report of the Minister of Social Security and Labor, with data on public health, social assistance, social security, and labor.

Apéndice estadístico del informe del ministerio, con datos sobre salud pública, asistencia y previsión social, y trabajo.

Ec-**9000**.I5 **Informe a la Nación (del Ministro de Previsión Social y Trabajo).** Fecha de iniciación? Anual. Quito. Ministerio de Previsión Social, Trabajo y Asistencia Pública.

Title varies—título varía: 19 -1939-40: **Memoria (del Ministro de Previsión Social).**

Report of the Minister of Social Security and Labor. Covers social security, labor, and public health.

Incluye previsión social, trabajo y salud pública.

Ec-**9800**.I5 1945 **Indices del Costo de la Vivienda Popular en Quito, de 1938 a 1944.** Quito, Imp.del Ministerio del Tesoro, 1945. Ministerio de Economía. Dirección General de Estadística y Censos. 28 p. Cuadernos de estadística, no. 1.

Indexes of low-cost housing in Quito, 1938-44. Contains a detailed methodological explanation of procedure adopted for the collection and selection of the data and the calculation of the indexes.

Contiene una detallada explicación metodológica del procedimiento adoptado para la recolección y selección de los datos y el cálculo de los índices.

EL SALVADOR

1000-1999 GENERAL

*ElSal-**1410**.A5 **Anuario Estadístico.** 1911-1923, 1927- . San Salvador. Ministerio de Economía. Dirección General de Estadística.
Statistical yearbook of El Salvador.

1944 **Anuario Estadístico Correspondiente al Año de 1944.** Editado por la Dirección General de Estadística. San Salvador, Imp.Nacional, 1946. 3 vol. Publicaciones del Ministerio de Economía.

Contents: Volume I: Climatology; demographic percentages and indexes; urban and rural population; vital statistics (marriage, divorce, birth, death, and retrospective figures 1940-44); migration; education (primary and secondary); police statistics (seizures and captures, arrests, crimes and offenses) and judicial statistics (crimes against the public estate and criminal branch); volume II: Services of the interior (mail, telephone, telegraph, etc.); transportation; prices of principal articles of food and principal products of the country; foreign trade (general exports and exports of principal products and imports); finance (money rates, balances of the banks, national budget, movement of funds, etc.); volume III: Statistics of agriculture and livestock; mineral statistics; industrial statistics (production of salt, of milk and its derivatives, of cement, etc., and of the cotton fabric industry).

Contenido: Tomo I: Climatología; porcentajes e índices demográficos; población urbana y rural; movimiento de población (matrimonios, divorcios, nacimientos, defunciones y cifras retrospectivas 1940-1944); migración; instrucción pública (enseñanza primaria y secundaria); estadísticas policial (comisos y capturas, arrestados, delitos y faltas) y judicial (delitos contra la hacienda pública y ramo criminal); tomo II: Servicios del interior (correo, teléfono, telégrafo, etc.); movimiento de transportes; precios de los principales artículos alimenticios y principales productos del país; comercio exterior (exportación general, y por separado de los principales productos, e importación); finanzas (cotizaciones de moneda, balances de bancos, presupuesto de la nación, movimiento de fondos, etc.); tomo III: Estadística agrícola-ganadera; estadística minera; estadística industrial (producción de sal, de leche y sus derivadas, de cemento, etc., e industria de tejidos de algodón).

*ElSal-**1420**.B5 **Boletín Estadístico de la Dirección General de Estadística.** Fecha de iniciación? Mensual. San Salvador. Ministerio de Economía. Dirección General de Estadística.
Statistical bulletin of the General Bureau of Statistics. Issued in sections, covering such branches as demography, agricultural production, foreign trade. Contains pictorial charts as well as statistical tables. Includes current statistical series. Also issues a semi-annual issue "Extraordinario."
Editado por secciones que abarcan distintas ramas, como ser demografía, producción agrícola, comercio exterior. Contiene cuadros gráficos y cuadros estadísticos. Incluye series estadísticas corrientes. También aparece semestralmente una edición que se llama "Extraordinario."

ElSal-**1900**.P5 1939 **Prontuario—Geográfico, Comercial, Estadístico, y Servicios Administrativos de El Salvador.** Tercera edición. Por José Tomás Calderon. San Salvador, 1939.
Compendium of geography, commerce, statistics, and administrative services of El Salvador. A compendium of general information, in the nature of an almanac.
Un compendio de informaciones generales en forma de almanaque.

ElSal-**1950**.G5 **La Guía Turística de El Salvador.** 1937- . Anual. San Salvador. Junta Nacional de Turismo.
Tourist guide of El Salvador.

2000-2999 CENSUS—CENSOS

ElSal-**2100**.C5 1930.1 **Censo de 1930: Métodos y Procedimientos.** Material usado en el levantamiento del censo de población del 1 de mayo. Adquirido con motivo de la encuesta que el Instituto Interamericano de Estadística realizó sobre Métodos y Procedimientos de los Censos en las Naciones Americanas, 1944. Del Ministerio de Economía, Dirección General de Estadística.

Census of 1930: Methods and procedures. Materials used in taking population and electoral censuses, on May 1. Acquired for Inter American Statistical Institute Survey of Census Methods and Procedures in the American Nations, 1944. Consists of: Correspondence; laws, decrees, and resolutions; schedules; other forms; instructions.

Consiste de: Correspondencia; leyes, decretos y resoluciones; cédulas; otros formularios; instrucciones.

See also—véase también: **Población de la República de El Salvador,** Censo de 1° de mayo de 1930; **Adiciones al Reglamento del Censo de Población Decretadas el 14 de Mayo de 1930,** Reglamento y plan ejecutivo; **Instrucciones al Personal del Censo de la República.**

ElSal-**2100.**C5 1930.2 **Adiciones al Reglamento del Censo de Población.** Decretadas el 14 de mayo de 1930 y publicadas en el **Diario Oficial** número 108 del mismo mes y año. San Salvador, Tip. "La Unión," 1930. Ministerio de Fomento. Dirección General del Censo. 15 p.

Additions to legislation on population census, May 14, 1930.

ElSal-**2100.**C5 1930.3 **Censo de Población de 1930.** San Salvador, Tip. "La Unión" S.S., 1929. Ministerio de Fomento. Dirección General del Censo. 44 p.

Population census of 1930. Contents: Statistical law (decree of May 2, 1926); law of the population census of 1930; executive plan of the census of May 10, 1930.

Contenido: Ley de estadística (decreto de 2 de mayo de 1926); reglamento del censo de población de 1930; plan ejecutivo del censo de 1930.

ElSal-**2100.**C5 1930.4 **Instrucciones al Personal del Censo de la República.** San Salvador, Tip. "La Unión." Ministerio de Fomento. Dirección General del Censo. 18 p.

Instructions to personnel of the census.

*ElSal-**2200.**C5 1930 **Población de la República de El Salvador.** Censo del 1° de mayo de 1930. San Salvador, Tall. Nacional de Grabados, 1942. Ministerio de Hacienda. Dirección General de Estadística. 520 p.

Population of El Salvador. Census of May 1, 1930.

ElSal-**2210.**C5 1940 **Primer Censo Nacional del Café, 1938-39.** San Salvador, Tall.Gráf.Cisneros, 1940. Asociación Cafetalera de El Salvador. 20 p.

First national census of coffee. Contents: History of the plans for this coffee census, which was carried out between November 1938 and the last days of March 1939; decree of October 21, 1938, authorizing the census; schedules, forms, and questionnaires used, with annotations and explanations; results of the census; list of owners of coffee-producing properties; list of proprietors of coffee trade-marks; map of El Salvador showing coffee-producing areas, etc.

Contiene: Historia de los planes para este censo del café que se ha llevado a cabo entre noviembre 1938 y los últimos días de marzo de 1939; decreto del 21 de octubre de 1938 autorizando el censo; cédulas, formularios y cuestionarios usados, con anotaciones y explicaciones; resultados del censo; lista de poseedores de propiedades productoras de café; lista de propietarios de marcas autorizadas de café; mapa de El Salvador indicando áreas de producción de café.

3000-3999 STATISTICAL SCIENCE—CIENCIA ESTADISTICA

ElSal-**3030.**I5 1935 **La Investigación Estadística, Origen, Desarrollo y Estado Actual Comparada con la de Varios Países Americanos y del Caribe.** Un plan de reorganización de los servicios. Por Max P. Brannon. San Salvador, 1935. Ministerio de Hacienda, Crédito Público, Industria y Comercio. No examinado.

The origin, development, and present status of statistical research in El Salvador compared with that of various American and Caribbean countries. A plan of reorganization of the services.

ElSal-**3030.**L5 1940 **Ley Orgánica del Servicio Estadístico.** San Salvador, Imp.Nacional, 1940. 14 p. Publicaciones del Ministerio de Hacienda.

Basic law of the statistical service.

ElSal-**3200.**C5 1943 **Claves y Nomenclaturas Usadas por la Dirección General de Estadística.** San Salvador, Imp. Nacional, 1943. Ministerio de Hacienda y Crédito Público. Dirección General de Estadística. 269 p.

Codes and nomenclatures used by the General Bureau of Statistics.

4000-4999 GEOGRAPHY—GEOGRAFIA

ElSal-**4000.**D5 1945 **Diccionario Geográfico de la República de El Salvador.** Editado por la Dirección General de Estadística. Año de 1945. San Salvador, Imp. Nacional, 1945. 246 p. Publicaciones del Ministerio de Economía.

At head of cover title—a la cabeza del título de la cubierta: Antonio Cardona Lazo.

Geographic dictionary of El Salvador.

5000-5999 DEMOGRAPHY—DEMOGRAFIA

ElSal-**5000.**E5 1944 **El Salvador: Summary of Biostatistics.** Maps and charts, population, natality and mortality statistics. Washington, D.C., March 1944. Prepared by

U. S. Department of Commerce, Bureau of the Census, in cooperation with Office of the Coordinator of Inter-American Affairs. 111 p.

Resumen bioestadístico de El Salvador. Incluye mapas y cuadros y datos sobre población, natalidad y mortalidad.

ElSal-**5100**.P5 1942 **La Población de El Salvador.** Estudio acerca de su desenvolvimiento desde la época prehispánica hasta nuestros días. Por Rodolfo Barón Castro. Madrid, Consejo Superior de Investigaciones Científicas, Instituto Gonzalo Fernández de Oviedo, 1942. 644 p.

The population of El Salvador: Study relating to its development from the pre-Hispanic period up to the present time.

ElSal-**5200** **Estadística Vital y Registro en El Salvador: Métodos y Procedimientos.** Material básico adquirido para la encuesta que el Instituto Interamericano de Estadística realiza sobre los métodos de registro y estadística vital en las naciones americanas, incluyendo el estudio de 1945.

Vital statistics and registration in El Salvador: Methods and procedures. Background material acquired for the study conducted by the Inter American Statistical Institute in vital statistics and registration methods in the American nations, including the survey of 1945. Includes: Answer to questionnaire; laws, decrees; forms; punch cards and codes.

Incluye: Respuesta al cuestionario; leyes y decretos; formularios; tarjetas perforadas y códigos.

ElSal-**5212**.D5 **Defunciones en la República de El Salvador.** Fecha de iniciación? Anual. San Salvador. Ministerio del Interior. Dirección General de Sanidad.

Deaths in El Salvador. Annual. Tables of provisional figures classified according to the Intermediate International Nomenclature, 1938. Prior to 1943 the 1929 form was used.

Cuadros de cifras provisionales clasificadas de acuerdo a la Nomenclatura Internacional Intermediaria, 1938. Anteriormente a 1943, fué usada la forma de 1929.

6000-6999 ECONOMICS—ECONOMIA

ElSal-**6000**.M5 **Memoria de Economía.** Fecha de iniciación? Anual. San Salvador. Ministerio de Economía. No examinado.

Annual report of the Department of Economy.

ElSal-**6200**.E5 1943 **Estadística Forestal y Agraria, Año 1943.** San Salvador, marzo 1943. Ministerio de Hacienda. Dirección General de Estadística.

Forestry and rural statistics, 1943. "Investigation of the extent of land occupied by forests, pastures, fruit trees, etc." (Translation of comment.)

"Investigación de la extensión de los terrenos ocupados por bosques, pastos, frutales, etc." (Fuente: Nota Explicativa.)

ElSal-**6210**.C5 **El Café.** Información económica sobre los mercados de producción y de consumo del grano. Fecha de iniciación? Mensual. San Salvador. Ministerio de Economía. No examinado.

Economic information on the markets of production and consumption of coffee. Monthly.

ElSal-**6210**.C6 **El Café de El Salvador.** Revista de la Asociación Cafetalera de El Salvador. 1930- . Mensual. San Salvador? Asociación Cafetalera de El Salvador.

The coffee of El Salvador. Monthly journal of the Association of Coffee-Growers of El Salvador.

ElSal-**6210**.M5 1942-43— **Memoria de la Compañía Salvadoreña de Café, S.A.** . Anual. San Salvador.

Annual report of the Salvadorian Coffee Company. Shows statistical data on the coffee industry of El Salvador.

Expone datos estadísticos acerca de la industria de café de El Salvador.

ElSal-**6600**.B5 **Boletín.** Fecha de iniciación? Mensual. San Salvador. Ministerio del Interior. No examinado.

Bulletin of the Department of Interior. Contains principally statistics of railroads and highways.

Contiene principalmente estadísticas sobre ferrocarriles y carreteras.

ElSal-**6720**.M5 **Memoria del Banco Central de Reserva de El Salvador.** 1924- . Anual. San Salvador.

Annual report of the Central Reserve Bank of El Salvador. Contains data on the banking situation of the country and the progress of the Bank.

Contiene datos sobre la situación bancaria del país y de la marcha del Banco.

*ElSal-**6720**.R5 **Revista Mensual del Banco Central de Reserva de El Salvador.** 1934- . San Salvador.

Monthly review of the Central Reserve Bank of El Salvador. Contains information on the activities of the bank and those of other banks, movement of deposits, currency circulation, exchange quotations, collection of customs duties, etc.

Contiene información sobre las actividades de este Banco y de los otros, movimiento de depósitos, circulación monetaria, cotizaciones de cambio, cobro de derechos de aduana, etc.

ElSal-**6900** **Estadísticas del Comercio Exterior en El Salvador: Métodos y Procedimientos.** Material básico adquirido para el estudio que el Instituto Interamericano de Estadística realiza sobre los métodos de estadísticas del comercio exterior en las naciones americanas, incluyendo la encuesta de 1946-47.

Foreign trade statistics in El Salvador: Methods and procedures. Background material acquired for the study conducted by the Inter American Statistical Institute in methods of foreign trade statistics in the American nations, including the survey of 1946-47. Consists of: Filled questionnaire and explanatory notes; country code.

Consiste de: Cuestionario llenado y notas explicativas; clave de países.

ElSal-**6900**.I5 1943 **Intercambio Comercial de El Salvador en el Año 1943.** Resumen comparativo que establece la balanza comercial. San Salvador, junio 1944. Ministerio de Hacienda, Crédito Público, Industria y Comercio. Dirección General de Estadística. 2 p. Copia máquina.

Trade balance of El Salvador, 1943.

ElSal-**6900**.T5 1941 **Tarifa de Aforos.** Décima edición, 1941. San Salvador, Tall.Nacional de Grabados, 1941. Ministerio de Hacienda, Crédito Público, Industria y Comercio. 557 p.

Valuation schedule. Decree No. 38, July 24, 1941. Modified by laws appearing in the **Diario Oficial.** Customs code of El Salvador. Classification closely follows Brussels nomenclature.

Modificado por las leyes que aparecen en el **Diario Oficial.** Decreto No. 38 del 24 de julio de 1941. Código de aduanas de El Salvador. La clasificación sigue a la nomenclatura de Bruselas.

7000-7999 HEALTH AND CULTURAL STATISTICS—ESTADISTICAS DE SALUBRIDAD Y CULTURA

ElSal-**7000**.I5 **Informe de las Labores Desarrolladas por el Poder Ejecutivo en los Ramos de Cultura y Asistencia Social.** Rendido ante la honorable Asamblea Nacional Legislativa. 1945- . Anual. San Salvador. Ministerio de Cultura y Asistencia Social.

Supersedes—sustituye a: **Memoria de Cultura Popular** (ElSal-**7000**.M5) y **Memoria de Asistencia Social y Beneficencia** (ElSal-**9951**.M5).

Yearly report of the Department of Education and Social Welfare of El Salvador.

1945 **Informe de las Labores Desarrolladas por el Poder Ejecutivo en los Ramos de Cultura y Asistencia Social durante el Año de 1945.** Rendido ante la honorable Asamblea Nacional Legislativa el 26 de marzo de 1946. San Salvador, Imp. Nacional, 1946? 252 p.

Cover title—título de la cubierta: **Labor del Supremo Gobierno en los Ramos de Cultura y Asistencia Social durante el Año de 1945.**

Contains summaries of the department's activities in education, cultural institutions, welfare societies, and hospitals, with tables on statistics of education. Also includes resumes of the most important decrees and resolutions affecting this field.

Contiene resúmenes de las actividades del ministerio relativo a educación, instituciones culturales, sociedades de asistencia social y hospitales, con tablas de estadística de educación. También se incluyen resúmenes de los decretos y acuerdos más importantes en el ramo.

ElSal-**7000**.M5 **Memoria de Cultura Popular.** Fecha de iniciación? Anual. San Salvador. Ministerio de Cultura Popular.

Since 1945 superseded by—desde 1945 sustituido por: **Informe de los Labores . . . en los Ramos de Cultura Asistencia Social** (ElSal-**7000**.I5).

Annual report of the Department of Education of El Salvador.

9000-9999 LABOR AND LIVING CONDITIONS—CONDICIONES DE TRABAJO Y DE VIDA

ElSal-**9951**.M5 **Memoria de Asistencia Social y Beneficencia.** Fecha de iniciación? Anual. San Salvador. Ministerio del Interior. Dirección General de Sanidad. No examinado.

Since 1945 superseded by—desde 1945 sustituido por: **Informe de los Labores . . . en los Ramos de Cultura y Asistencia Social** (ElSal-**7000**.I5).

Yearly report of the General Bureau of Health of El Salvador on social welfare.

GUATEMALA

1000-1999 GENERAL

Guat-**1410**.E5 **Estadísticas Gráficas.** Guatemala, 1939. Secretaría de Hacienda y Crédito Público. No examinado.

Graphic statistics. A publication of statistical graphs only, on: Demography; education; railroads; motor vehicles; production, imports and exports of commodities; exchange, etc.

Una publicación de gráficos estadísticos unicamente, sobre: Demografía, educación, ferrocarriles; vehículos a motor; producción, importación y exportación de productos; cambios, etc.

*Guat-**1410**.M5 **Memoria de las Labores del Poder Ejecutivo en el Ramo de Hacienda y Crédito Público.** Fecha de iniciación? Anual. Guatemala, C.A. Ministerio de Hacienda y Crédito Público.

Title varies slightly—el título varía un poco: 193 -1940: **Memoria de las Labores del Ejecutivo en el Ramo de Hacienda y Crédito Público.**

Annual report of the Department of Finance and Public Credit. Constitutes the statistical yearbook of Guatemala.

Representa el anuario estadístico de Guatemala.

1942 **Memoria de las Labores del Poder Ejecutivo en el Ramo de Hacienda y Crédito Público durante el Año Administrativo de 1942.** Guatemala, C.A., Tipo. Nacional, 1943. 736 p.

Includes: **Expositive part:** Income; expenditures; public credit; money, banks, insurance, securities and taxes of corporations; laws and statutes; administration; changes in the general budget; **appendixes:** Graphic tables; tables of the General Bureau of Accounts, Department of Money and Banking, General Bureau of Customs; General Bureau of Revenues, General Bureau of Statistics (including demography, nosography, public welfare, consumption, current prices, communications, real estate, public finance, commercial quotations, electric power, construction of houses, migration).

Incluye: **Parte expositiva:** Ingresos; egresos; crédito público; moneda, bancos, seguros, fianzas e impuestos a cargo de sociedades accionadas; leyes y reglamentos; labor administrativa; alteraciones en el presupuesto general de gastos; **anexos:** Estadísticas gráficas; cuadros de la Dirección General de Cuentas, del Departamento Monetario y Bancario, de la Dirección General de Aduanas, de la Dirección General de Rentas, de la Dirección General de Estadística (incluyendo demografía, nosografía, asistencia pública, consumos, precios corrientes, comunicaciones, propiedad inmueble, hacienda, cotizaciones, energía eléctrica, construcciones de casas, movimiento migratorio).

*Guat-**1420**.B5 **Boletín de la Dirección General de Estadística.** Publicación de la Dirección General de Estadística, dependencia del Ministerio de Economía y Trabajo. 1946- . Bimensual. Guatemala.

Bimonthly bulletin of the General Bureau of Statistics. Includes data on demography, balance of payments, budget, etc., as well as monthly statistical series on production, consumption, index numbers of prices and salaries, transportation and communication, fiscal statistics, foreign trade, and monetary statistics.

Incluye datos sobre demografía, balance de pagos, presupuesto, etc., y también series estadísticas mensuales sobre producción, consumo, números índices de precios y salarios, transportes y comunicaciones, estadísticas fiscales, comercio exterior, y estadísticas monetarias.

Nov. '46 "Censo Industrial." No. 4, noviembre 1946, p. 1-2.

On the importance of an industrial census to Guatemala. Includes decree authorizing the taking of an industrial census by the General Bureau of Statistics in collaboration with the General Bureau of Commerce and Industry.

La importancia que tiene un censo industrial en Guatemala y el decreto por el cual se autoriza a levantar un censo industrial, estando a cargo de la Dirección General de Estadística en colaboración de la Dirección General de Comercio e Industria.

July '46, a "Empresas Eléctricas de Servicio Público." Resultado de la encuesta practicada sobre las empresas eléctricas de servicio público. No. 2, julio 1946, p. 10-20.

Results of a study on public utility electricity enterprises. Includes data on plants, service, consumption, personnel employed, and rates.

Incluye datos sobre plantas eléctricas, servicio, consumo, personal empleado, y tarifas.

July '46, b "Resultados del Censo Escolar Levantado el 23 de Enero de 1946: Municipio de Guatemala." No. 2, julio 1946, p. 45-46.

Results of the school census taken January 23, 1946, in the municipality of Guatemala.

May '46, a "Censo Escolar." No. 1, mayo 1946, p. 58.

Contains a table of the number of children of school age in the Republic, from the results of the school census of January 23, 1946.

Contiene un cuadro de "Número de Niños de Edad Escolar en la República," tomado del censo escolar del 23 de enero de 1946.

May '46, b "Balanza Comercial, 1935 a 1945." No. 1, mayo 1946, p. 56.

Trade balance of Guatemala from 1935 to 1945.

Guat-**1950.**G5 1942 **Guía Kilométrica de las 23 Rutas Nacionales de la República de Guatemala.** Guatemala, Tipo. Nacional, 1942. Ministerio de Comunicaciones y Obras Públicas. Dirección General de Caminos. 185 p. No examinado.

Kilometric guide of the 23 national routes of Guatemala.

Guat-**1950.**N5 1937 **Nomenclator de Guatemala.** Diccionario geográfico y guía de comunicaciones. Por Emilio Descamps. Guatemala, Tall.Tip. "San Antonio," 1937. 598 p. No examinado.

Gazetteer of Guatemala. Geographic dictionary and communications guide.

2000-2999 CENSUS—CENSOS

Guat-**2100.**C5 1940.1 **Censo de 1940: Métodos y Procedimientos.** Material usado en el levantamiento del censo de población del 7 de abril. Adquirido con motivo de la encuesta que el Instituto Interamericano de Estadística realizó sobre Métodos y Procedimientos de los Censos en las Naciones Americanas, 1944. Del Ministerio de Economía y Trabajo, Dirección General de Estadística.

Census of 1940: Methods and procedures. Materials used in taking population census, on April 7. Acquired in connection with Inter American Statistical Institute Survey on Census Methods and Procedures in the American Nations, 1944. Consists of: Laws, decrees, or other dispositions; publicity and propaganda; schedules; other forms; instructions; punch cards.

Se compone de: Leyes, decretos u otras disposiciones; publicidad y propaganda; cédulas; otros formularios; instrucciones; tarjetas de perforación.

See also—véase también: **Reglamento para el Censo General de Población del Año 1940** (Guat-**2100.**C5, 1940.2).

Guat-**2100.**C5 1940.2 **Reglamento para el Censo General de Población del Año 1940.** Guatemala, Tipo.Nacional, 1939. 23 p.

Regulations for the general census of population of 1940.

Guat-**2130.**C5 1946.1 **Censo Escolar de 1946: Métodos y Procedimientos.** Material usado en el levantamiento del censo. Adquirido con motivo de la encuesta que el Instituto Interamericano de Estadística realiza sobre Métodos y Procedimientos de los Censos en las Naciones Americanas, 1944. Del Ministerio de Economía y Trabajo, Dirección General de Estadística.

School census of 1946: Methods and procedures. Material used in taking the census. Acquired in connection with Inter American Statistical Institute Survey of Census Methods and Procedures in the American Nations. Consists of: Laws; instructions; forms; schedules; publicity.

Consiste de: Leyes; instrucciones; formularios; cédulas; propaganda.

See also—véase también: **Instrucciones Generales para los Jefes de Empadronadores y Personal Auxiliar; Acuerdo Disponiendo el Levantamiento del Censo Escolar y Reglamento General del Mismo.**

Guat-**2130.**C5 1946.2 **Acuerdo Disponiendo el Levantamiento del Censo Escolar y Reglamento General del Mismo.** Acuerdos gubernativos de 10 de octubre y de 7 de noviembre de 1945. Guatemala, Tipo.Nacional, 1945. Ministerio de Economía y Trabajo. Dirección General de Estadística. 31 p.

Decree directing the taking of the school census, and general regulations for the census. Executive decrees of October 10 and November 7, 1945.

Guat-**2130.**C5 1946.3 **Instrucciones Generales para los Jefes de Empadronadores y Personal Auxiliar.** Guatemala, Tipo.Nacional, 1946. Ministerio de Economía y Trabajo. Dirección General de Estadística. Censo Escolar. 56 p.

General instructions for the supervisors of enumerators and auxiliary personnel of the school census.

*Guat-**2200.**C5 1940 **Quinto Censo General de Población, Levantado el 7 de Abril de 1940.** Guatemala, Tipo.Nacional, 1942. Secretaría de Hacienda y Crédito Público. Dirección General de Estadística. 885 p.

Fifth general census of population, April 7, 1940. Comparisons are made between censuses of 1880, 1893, 1921, and 1940 by regions and departments. For 1940 the following tabulations are included: Total population by regions and departments; urban and rural population; ages, race, education, and school age population by sex; civil status; physical handicaps; nationality by sex; profession, occupation, and trade by sex. The majority of the tabulations are given in detail for both departments and municipalities.

Se presentan comparaciones de los datos censales de 1880, 1893, 1921 y 1940, por regiones y departamentos. Para 1940 se incluye: Población total por regiones y departamentos; población urbana y rural; edades, raza, instrucción y edad escolar por sexos; condición civil; impedimento físico; nacionalidad por sexos; profesión, ocupación y oficio, por sexos. La mayoría de los datos tabulados se presentan detalladamente por departamentos así como por municipalidades.

Guat-**2200.**C5 1921 **Censo de la Población de la República, Levantado el 28 de Agosto de 1921.** Guatemala, 1924-26. Ministerio de Fomento. Dirección General de Estadística. 3 vol.: Parte I, **Población Clasificada por Municipios, Departamentos y Zonas con Distinción de Población Urbana y Rural, Instrucción, Raza, Sexo y Edades;** parte II, **División Política y Administrativa, Estado Civil, Nacionalidad y Ocupaciones;** complemento de la parte II, **Estado Civil, Nacionalidad, Ocupaciones.**

Census of population of Guatemala, August 28, 1921. In 3 volumes: Part I includes population by municipalities, departments, and zones, with distinction by urban and rural population, education, race, sex, and age; part II includes political and administrative division, civil status, nationality, and occupations; supplement to part II includes civil status, nationality, occupations.

3000-3999 STATISTICAL SCIENCE—CIENCIA ESTADISTICA

Guat-**3000.**S5 **Serie de Estudios Monográficos.** 1946- . Irregular. Guatemala. Ministerio de Economía y Trabajo. Dirección General de Estadística.

Monograph series of the General Bureau of Statistics.

Guat-**3030.**A5 1945 **Acuerda: . . . Dictar las Disposiciones del Caso a Efecto de que la Dirección General de Estadística Llene su Cometido.** Guatemala, 1945. 2 p. Mimeografiado.

Legal disposition to effect the General Bureau of Statistics' fulfilment of its functions. Present organization of the General Bureau of Statistics of Guatemala, 1 p. attached. Supplement to basic statistical law, decree no. 1820.

"Actual Organización de la Dirección General de Estadística de Guatemala," 1 p. adherida. Suplemento a la ley estadística fundamental, Decreto No. 1820.

Guat-**3030.**A6 1945 **Acuerdo Gubernativo de 2 de Mayo de 1945 por el Cual Fué Creada la Oficina Permanente del Censo.** Guatemala, 1945. Dirección General de Estadística. Oficina Permanente del Censo. 1 p. Copia máquina.

Executive decree of May 2, 1945, establishing the permanent census office.

Guat-**3030.**L5 1936 **Ley de Estadística: Decreto Número 1820.** Guatemala, Tip.Nacional, 1936. 27 p.

Law of statistics.

Guat-**3030.**S5 1945 **Statistics in Guatemala (A Brief Survey).** By Thomas F. Corcoran. Washington, D.C., December 1945. Inter-American Development Commission. 24 p. Multilithed.

Translated into Spanish—traducido al español (Guat-**6000.**R5, May-Aug. '46, Corcoran).

Report on official activities and procedures in statistics in Guatemala, including population statistics, the school census, vital statistics, agricultural statistics, foreign trade statistics, industrial and commercial statistics, tourist statistics, and cost-of-living data.

Guat-**3030.**U5 1946 **Ultimas Disposiciones Oficiales en Relación con la Estadística.** Guatemala, 1946. 2 p. Mimeografiado.

Recent official regulations relating to statistics.

Guat-**3210.**I5 1943 **Instrucciones a los Importadores de Artículos y Materiales en Tránsito o Procedentes de los Estados Unidos de Norte América o del Canadá.** Guatemala, 1943. Ministerio de Economía. Sección de Coordinación Económica Financiera. 109 p.

Instructions to importers of articles and materials in transit or coming from the United States or Canada. Translation into Spanish of United States Schedule B for commodities imported into Guatemala.

Traducción al español del Cuadro B de Estados Unidos para artículos importados por Guatemala.

Guat-**3300** 1944-46 **Enseñanza Estadística en Guatemala: Métodos y Materiales.** Material básico adquirido para la Encuesta del Instituto Interamericano de Estadística sobre los Métodos y Materiales en la Enseñanza Estadística en las Naciones Americanas, 1944-46.

Statistical training in Guatemala: Methods and materials. Background material acquired for the IASI Survey on Statistical Training Methods and Materials in the American Nations, 1944-46. Includes: Letter and filled questionnaire on statistical courses in Guatemala.

Incluye carta y cuestionario llenado sobre cursos de estadística en Guatemala.

5000-5999 DEMOGRAPHY—DEMOGRAFIA

Guat-**5000.**G5 1944 **Guatemala: Summary of Biostatistics.** Maps and charts, population, natality and mortality statistics. Washington, D.C., March 1944. Prepared by U. S. Department of Commerce, Bureau of the Census, in cooperation with Office of the Coordinator of Inter-American Affairs. 98 p.

Resumen bioestadístico de Guatemala. Incluye mapas y cuadros y datos sobre población, natalidad y mortalidad.

Guat-**5200** **Estadística Vital y Registro en Guatemala: Métodos y Procedimientos.** Material básico adquirido para la encuesta que el Instituto Interamericano de Estadística realiza sobre los métodos de registro y estadística vital en las naciones americanas, incluyendo el estudio de 1945.

Vital statistics and registration in Guatemala: Methods and procedures. Background material acquired for the study conducted by the Inter American Statistical Institute in vital statistics and registration methods in the American nations, including the survey of 1945. Includes: Laws; forms; codes.

Incluye: Leyes; formularios; claves.

6000-6999 ECONOMICS—ECONOMIA

Guat-**6000.**I5 1944 **Informe Presentado por la Delegación de Guatemala ante la Conferencia de Comisiones Nacionales de Fomento Interamericano sobre la Estructura y Condiciones Económicas de Su País.** Guatemala, Tipo.Nacional, 1944. 55 p.

Report presented by the delegation of Guatemala to the Conference of Commissions of Inter-American Development on the economic structure and conditions of their country. Contents: Characterization of essential features of Guatemalan economy; utilization of natural wealth; cost of living; circulation of currency and banking statistics; organizations and forms of credit; national debt; industrial aspect; commercial aspect; various taxes and their amounts in the last years; tariff agreements and import restrictions; transportation services; tourism; summary.

Contenido: Caracterización de los rasgos esenciales de la economía guatemalteca; aprovechamiento de las riquezas naturales del país; costo de la vida; circulación monetaria y estadísticas bancarias; organismos y formas de crédito; deudas del estado; aspecto industrial; aspecto comercial; diferentes impuestos y su monto en los últimos años; convenios aduaneros y restricciones a la importación; servicios de transporte; turismo; y síntesis.

Guat-**6000.**M5 **Memoria del Ministerio de Comunicaciones y Obras Públicas.** Fecha de iniciación? Anual. Guatemala. Ministerio de Comunicaciones y Obras Públicas. No examinado.

Annual report of the Department of Communications and Public Works. Deals with public postal service, telecommunication, mining, industry, commerce, electrical power production, civil aeronautics, and railways.

Trata sobre servicio público postal, telecomunicaciones, minería, industria, comercio, producción de energía eléctrica, aeronáutica civil y ferrocarriles.

*Guat-**6000.**R5 **Revista de Economía.** Publicación bimensual del Ministerio de Economía y Trabajo. 1946- Guatemala.

Economic review published bimonthly by the Department of Economics and Labor. Regularly includes articles, opinions of the Economic Council, legislation, data on the financial status of the banks of Guatemala, and a bibliography of the last books acquired by the library of the Department.

Incluye secciones regulares de estudios, reproducciones y colaboraciones, dictámenes del Consejo de Economía, legislación, datos sobre los estados financieros de los bancos de Guatemala, y una bibliografía de los últimos libros adquiridos por la biblioteca del Ministerio.

May-Aug. '46, Corcoran "La Estadística en Guatemala: Una Breve Reseña." Por Thomas F. Corcoran. Año I, no. 3-4, mayo-agosto 1946, p. 76-107.

Translation of—traducción de: **Statistics in Guatemala (a Brief Survey)** (Guat-**3030.**S5.)

Reseña de actividades y prácticas oficiales

de estadística en Guatemala, incluyendo estadísticas de población, el censo escolar, estadísticas vitales, estadísticas agrícolas, estadísticas del comercio exterior, estadísticas industriales y comerciales, estadísticas de turismo, y datos sobre el costo de la vida.

Guat-**6200**.M5 **Memoria de las Labores del Poder Ejecutivo en el Ramo de Agricultura.** Fecha de iniciación? Anual. Guatemala. Ministerio de Agricultura y Minería.

Cover title for 1941-42—título de la cubierta de la edición de 1941-42: **Memoria de la Secretaría de Agricultura.**

Annual report of the Department of Agriculture. Contains diverse statistical data in the field of agriculture, presented in tables and charts, as well as production and trade figures for sugar, coffee, tobacco, lumber, rubber, cotton, bananas, rice, wheat, beans, corn, etc.

Contiene varios datos estadísticos sobre la agricultura, presentados en tablas y cuadros, así como cifras de la producción y del comercio de azúcar, café, tabaco, madera, caucho, algodón, bananos, arroz, trigo, frijoles, maíz, etc.

Guat-**6200**.R4 **Revista Agrícola.** Organo de divulgación del Ministerio de Agricultura. 1924- . Semestral. Guatemala. Ministerio de Agricultura. Dirección de Enseñanza y Divulgación Agrícola.

Formerly called—anteriormente intitulado: **Boletín de Agricultura y Caminos de Guatemala.** Until January, 1946, published monthly—hasta enero de 1946, publicado mensualmente. Formerly issued by—anteriormente editado por: Secretaría de Agricultura y Minería (Jan.-Feb. '45—enero-febrero '45); Secretaría de Economía y Agricultura (Dec. '44—diciembre '44); Secretaría de Agricultura (before Dec. '44—antes de diciembre '44).

Agricultural review. Organ of information of the Department of Agriculture. Contains various informative articles on agriculture, including livestock, agronomy, soil conservation, etc., occasionally including some statistics on crop production.

Contiene diversos artículos informativos sobre agricultura, incluyendo ganadería, agronomía, conservación de suelos, etc.; de vez en cuando incluye estadísticas sobre producción agrícola.

Guat-**6210**.R5 **Revista Cafetalera de Guatemala.** Organo de publicidad de la Oficina Central del Café, y su junta asesora. 1944?- . Mensual? Guatemala. Oficina Central del Café.

Coffee journal of Guatemala. Publicity organ of the Central Coffee Bureau and its advisory board.

Guat-**6300**.R5 1946 **Report of the Utilization of the Mineral Resources of the Republic of Guatemala.** By Blandford C. Burgess. Washington, D.C, March 1946. Inter-American Development Commission. 107 p.

Informe sobre la utilización de las riquezas minerales de Guatemala.

Guat-**6400**.I5 1946 **Industrial Report on the Republic of Guatemala.** By N. K. Ovalle. Washington, D.C., March 1946. Inter-American Development Commission. 163 p. Mimeographed.

Report of an industrial engineer, designated by the Inter-American Development Commission to survey the possibilities of industrial development in Guatemala through the utilization of the natural resources of the country. Covers: Livestock, crops (e.g., corn, black beans, etc.), sugar, alcohol, cotton, wool, wheat, paints, varnishes and lacquers, soaps, vegetable oils, cement, vegetable fibers, paper factory, rubber, fuel, acid factory, meteorological observations, electric power, roads and transportation, industrial institute, shoes and tanneries. Includes statistical tables showing data on production of the items discussed covering, generally, the period 1941-45, estimates of costs involved in further development of the various industries, and other statistical data, including population figures given in detailed tables by departments.

Informe del ingeniero industrial, encargado por la Comisión de Fomento Interamericano de realizar un estudio acerca de las posibilidades del desarrollo industrial de Guatemala a través del aprovechamiento de las riquezas naturales del país. Incluye: Ganado, siembras (de maíz, frijoles, etc.), azúcar, alcohol, algodón, lana, trigo, pinturas, barnices y lacas, jabones, aceites vegetales, cemento, fibras vegetales, fábrica de papel, caucho, combustible, fábrica de ácidos, observaciones meteorológicas, energía eléctrica, carreteras y transportación, instituto industrial, zapatos y curtidurías. Presenta cuadros estadísticos exponiendo datos acerca de la producción de los mencionados artículos, que abarcan en lo general el período 1941-45, estimaciones del costo del desarrollo más amplio de distintas industrias, y otros datos estadísticos, incluyendo cifras de población expuestas en cuadros detallados por departamentos.

Guat-**6600**.V5 1941 **Vías de Comunicación . . .** Carreteras; puentes; canales. Guatemala, Tipo.Nacional, 1941. Partido Liberal Progresista. 415 p. Publicaciones del Partido Liberal Progresista.

Transportation routes . . . Highways; bridges; canals. Contents: Seven thousand kilometers of highways in eight years (including number and type of equipment used in construction; bridges constructed during 1940; number of bridges constructed in eight years; principal highways; extension and connections of the road system of the country; construction, improvement and maintenance of highways; regulation of road signs and signals; bridges constructed over the different highways, 1931-39; map of the transportation system of the Republic); the Chiquimulilla Canal; album of the canal.

Contenido: Siete mil kilometros de carreteras en ocho años (incluyendo el número y tipo de los equipos usados en construcción; los puentes construídos durante el año de 1940; la cantidad de puentes construídos en ocho años; las principales carreteras; extensión y conexiones de la red vial del país; construcción, mejoramiento y mantenimiento de carreteras; reglamento de indicadores y señales para caminos; puentes construídos en las diferentes carreteras, 1931-39; mapa de vialidad de la República); el canal de Chiquimulilla; album del canal.

Guat-**6620**.G5 **Gaceta de Comunicaciones.** Organo oficial del Servicio de Comunicaciones. 1939- . Mensual. Guatemala. No examinado.

Journal of communications. Monthly organ of the Bureau of Communications. Succeeds the **Gaceta Postal** of the General Bureau of Posts.

Substituye a **Gaceta Postal** de la Dirección General de Correos.

Guat-**6900** **Estadísticas del Comercio Exterior en Guatemala: Métodos y Procedimientos.** Material básico adquirido para el estudio que el Instituto Interamericano de Estadística realiza sobre los métodos de estadísticas del comercio exterior en las naciones americanas, incluyendo la encuesta de 1946-47.

Foreign trade statistics in Guatemala: Methods and procedures. Background material acquired for the study conducted by the Inter American Statistical Institute in methods of foreign trade statistics in the American nations, including the survey of 1946-47. Consists of: Filled questionnaire and explanatory notes.

Consiste de: Cuestionario llenado y notas explicativas.

Guat-**6900**.C5 **Cuadros de la Dirección General de Aduanas.** Fecha de iniciación? Anual. Guatemala. Ministerio de Hacienda y Crédito Público. Dirección General de Aduanas.

Tables of the General Bureau of Customs. Covers export and import movements. Merchandise classified according to Brussels system of terminology.

Abarca el movimiento de exportación e importación. Mercaderías clasificadas de acuerdo al sistema de terminología de Bruselas.

Guat-**6900**.E5 1935 **Decreto Gubernativo Número 1765, Diciembre 7, 1935: Arancel de Aduanas de la República de Guatemala.** No. 516. Guatemala, Tipo.Nacional. 522 p.

Governmental decree number 1765, December 7, 1935: Customs tariff of Guatemala. Number 516. Includes: Decree number 1765 (with export duties and index); decree number 2064, customs code; decree number 2748, duty on imports of medicinal products, and toilet articles.

Incluye: Decreto número 1765 (con derechos de exportación e índice); Decreto número 2064, código de aduanas; Decreto número 2748, impuestos sobre la importación de productos medicinales y artículos de tocador.

7000-7999 HEALTH AND CULTURAL STATISTICS—ESTADISTICAS DE SALUBRIDAD Y CULTURA

Guat-**7110**.B5 **Boletín Sanitario de Guatemala.** Organo de la Dirección General de Sanidad Pública. 1930- . Anual. Guatemala. Ministerio de Salud Pública y Asistencia Social. Dirección General de Sanidad Pública.

Health bulletin of Guatemala. Annual publication of the General Bureau of Public Health.

Guat-**7110**.M5 1938 **A Medical Survey of the Republic of Guatemala.** By George Cheever Shattuck. Washington, D.C., Standard Engraving Co., July 1938. Carnegie Institution of Washington. 253 p. Publication, no. 499.

Tables, graphs.

Un estudio médico sobre la República de Guatemala. Cuadros y gráficos.

Guat-**7110**.R5 **Revista de Sanidad.** Fecha de iniciación? Trimestral. Ministerio de Salud Pública y Asistencia Social. Dirección General de Sanidad Pública. No examinado.

Quarterly journal of health. Contains demographic data.

Contiene datos demográficos.

Guat-**7130**.M5 **Memoria de las Labores Realizadas por el Hospital General y Sus Dependencias.** Fecha de iniciación? Anual. Guatemala. Ministerio de Salud Pública y Asistencia Social. Hospital General.

Annual report of the work of the General Hospital and its dependencies. Includes statistical tables on: Movement of patients; hospital accounts; patients treated; operations completed; etc.

Incluye cuadros estadísticos sobre: Movimiento de enfermos; cuentas de los hospitales; enfermos asistidos; operaciones efectuadas; etc.

Guat-**7200.**E3 1946 **La Edad de los Escolares y Su Valor como Determinante en el Fenómeno Educativo.** Un índice de su evolución. Guatemala, Tipo.Nacional, 1946. Ministerio de Educación Pública. Departamento de Estadística y Escalafón. 35 p. Publicaciones del Ministerio de Educación Pública, no. 2.

The ages of students and its value as a determinant in educational phenomena. Presents statistical tables on the students in primary schools of the capital, and analyses the phenomena that affect education.

Presenta cuadros estadísticos sobre los alumnos en las escuelas primarias de la capital y analiza los fenómenos que influyen en la educación.

Guat-**7200.**E5 1946 **Estudio sobre la Enseñanza Parvularia.** Guatemala, Unión Tip.—Castañeda, Avila & Cía., 1946. Ministerio de Educación Pública. Departamento de Estadística y Escalafón. 28 p. Publicaciones del Ministerio de Educación Pública, no. 1.

A study of pre-primary education in Guatemala. Contains statistical data on the number of students in public and in private schools, urban and rural, by sex, and the number of teachers by age, sex, preparation, salaries, etc.

Contiene datos estadísticos sobre la concurrencia de alumnos a escuelas oficiales y particulares, urbanas y rurales, clasificando por sexo el alumnado. Tiene también la clasificación del personal encargado de la enseñanza según edad, sexo, preparación, sueldos que perciben, etc.

Guat-**7200.**M5 **Memoria de las Labores del Poder Ejecutivo en el Ramo de Educación Pública.** 1884- . Anual. Guatemala. Ministerio de Educación Pública.

Cover title—título de la cubierta: **Memoria de la Secretaría de Educación Pública.**

Report of the work of the Department of Public Education.

1944 **Memoria de las Labores del Poder Ejecutivo en el Ramo de Educación Pública durante el Año Administrativo de 1943.** Presentada a la Asamblea Legislativa en sus sesiones ordinarias de 1944. Guatemala, Tipo. Nacional, 1944. 395 p.

Contents: Administrative service; rural education; kindergarten education; urban primary education; night instruction; secondary education; normal education; commercial and language education; arts, crafts, and agriculture; fine arts education; university education; cultural extension; Guatemalan Institute of Summer Courses; museums and botanical gardens; libraries; physical education; manual work; student hygiene; congresses and conferences; general inspection of education; general summary.

Contenido: Servicio administrativo; enseñanza rural; educación de párvulos; enseñanza primaria urbana; enseñanza nocturna; enseñanza secundaria; enseñanza normal; enseñanza comercial y de idiomas; artes y oficios y agricultura; enseñanza de bellas artes; enseñanza universitaria; extensión cultural; Instituto Guatemalteco de Cursos de Verano; museos y jardín botánico; bibliotecas; educación física; trabajo manual; higiene escolar; congresos y conferencias; inspección general de educación; resumen general.

Guat-**7200.**P5 **Publicaciones del Ministerio de Educación Pública.** 1946- . Irregular. Guatemala. Ministerio de Educación Pública. Departamento de Estadística y Escalafón.

Series of monographs of the Department of Public Education.

Serie de monografías.

Guat-**7210.**C5 1946 **Cuatro Estudios Estadístico-Educativos de la Escuela Guatemalteca.** Guatemala, Imp.Hispania, julio 1946. Departamento de Estadística y Escalafón. 43 p. Publicaciones del Ministerio de Educación Pública, no. 4.

Four statistical studies of education in Guatemala, on night schools, physical education, singing instruction, and the National Academy of Languages.

Incluye estudios sobre escuelas nocturnas, educación física, enseñanza de canto y Academia Nacional de Idiomas.

Guat-**7210.**E5 1946 **Estudio sobre el Costo de la Enseñanza Primaria.** Guatemala, Imp.Hispania, agosto 1946. Departamento de Estadística y Escalafón. 36 p. Publicaciones del Ministerio de Educación Pública, no. 5.

Study on the cost of elementary education.

Guat-**7310**.C5 1946 **Características Antropométricas de los Niños de Guatemala.** Guatemala, Imp.Hispania, junio 1946. Departamento de Estadística y Escalafón. 22 p. Publicaciones del Ministerio de Educación Pública, no. 3.

Anthropometric characteristics of children in Guatemala.

9000-9999 LABOR AND LIVING CONDITIONS—CONDICIONES DE TRABAJO Y DE VIDA

Guat-**9200**.A5 1946 **Análisis de los Sueldos Pagados por el Estado.** Guatemala, junio 1946. Dirección General de Estadística. 21 p. Mimeografiado. Serie de estudios monográficos, no. 1.

Analysis of government salaries in Guatemala. Notes, tables, and graphs showing distribution and statistical relations of public funds paid out in salaries by the various government departments, to the total income and expenditures of the government, the total number of government employes, etc.

Notas, cuadros y gráficos, exponiendo distribución y relaciones estadísticas entre los fondos públicos utilizados para sueldos en los varios organismos del gobierno, y los ingresos y gastos totales del Estado, el número de empleados gubernamentales, etc.

HAITI

1000-1999 GENERAL

Haiti-**1200**.B5 1941 **Bibliographie Générale et Méthodique d'Haïti.** Par Ulrick Duvivier. Port-au-Prince, Imp.de l'État, 1941. 2 vol.

General and systematic bibliography of Haiti.

Haiti-**1400**.M5 1932 **Monograph, Republic of Haiti, Compiled 1932.** Revisions regularly made from date of compilation. U. S. Marine Corps. Division of Operations and Training. Intelligence Section. 914 p.

"For official use only. The object of this book is to provide . . . information upon the Republic of Haiti."

Monographie, la République d'Haïti, compilée en 1932. Révisée régulièrement depuis la date de compilation. "A l'usage officiel seulement. L'objet de ce livre est de fournir des renseignements . . . sur la République d'Haïti." (Traduction du commentaire.)

Haiti-**1410**.B5 **Bulletin de Statistique.** 1942-43— . Annuel. Port-au-Prince. Administration Communale de Port-au-Prince.

Annual statistical bulletin of the commune of Port-au-Prince.

1943-44 **Bulletin de Statistique, No. 2, Exercice 1943-1944.** Port-au-Prince, Imp.de l'Etat, 1945. Administration Communale de Port-au-Prince. 155 p.

1942-43 **Bulletin de Statistique, No. 1, Exercice 1942-1943.** Port-au-Prince, Imp.de l'Etat, 1944. Administration Communale de Port-au-Prince. 114 p.

*Haiti-**1420**.B5 **Bulletin Mensuel du Département Fiscal.** Depuis? Port-au-Prince. Banque Nationale de la République d'Haïti. Département Fiscal.

Published also in English—publié aussi en anglais (Haiti-**1420**.M5).

Antérieurement publié sous le titre **Bulletin Mensuel.** Compris dans le sous-titre jusqu'à juin 1945: Publié par le Département Fiscal. Services collaborant: Direction Générale des Travaux Publics; Service National d'Hygiène et d'Assistance Publique; Service National de la Production Agricole.

Comprend des données sur les importations et les exportations classées par les pays d'origine et de destination, par mois, par produits; il y a aussi des données sur les douanes et sur d'autres aspects de la finance publique.

*Haiti-**1420**.M5 **Monthly Bulletin of the Fiscal Department.** 1923- . Port-au-Prince. National Bank of the Republic of Haiti. Fiscal Department.

Published also in French—publié aussi en français (Haiti-**1420**.B5). English edition contains some parts in French. Formerly issued by the Office of the Fiscal Representative. Previously published under the title **Monthly Bulletin.** Subtitle until June 1945 included: Published by the Fiscal Department. Contributing services: National Public Health Service; Agricultural Service; Public Works Administration.

Includes data on imports and exports, by countries of origin and destination, by months, by commodities; also data on customs and other aspects of public finance.

Haiti-**1900**.C5 **Cahiers d'Haïti.** 1943- . Mensuel. Port-au-Prince.

Journal of Haiti.

Dec. '43, Young "Les Statistiques Vitales." Par Chester W. Young. No. 5, décembre 1943, p. 32-35.

Vital statistics.

Haiti-**1900**.H5 1941 **The Haitian People.** By James G. Leyburn. New Haven, Connecticut, Yale University Press, 1941. 342 p.

La nation haïtienne.

*Haiti-**1900**.M5 Sept.13, '19 "Résultats du Recensement Général des Communes de la République d'Haïti (Septembre 1918—Août 1919)." Effectué sous l'Administration de Monsieur B. Dartinguenave, Secrétaire d'Etat au Département de l'Intérieur. **Le Moniteur,** 74ème année, no. 63, 13 septembre 1919, p. 470-471.

Results of the general census of communes of Haiti, September 1918-August 1919.

2000-2999 CENSUS-CENSOS

Haiti-**2100**.R5 1918-19 **Recensement Général des Communes de la République d'Haïti, Septembre 1918—Août 1919: Méthodes et Procédés et Résultats.** Matériel employé pour la recensement de la population. Acquis en relation avec l'Étude des Méthodes et Procédés de Recensement dans les nations américaines, efectuée par l'Institut Interaméricain de Sta

tistique, 1944. Envoyé par l'Ambassade américaine.

General census of the Communes of Haiti, September 1918—August 1919: Methods and procedures and results. Materials used in taking population census. Acquired in connection with the Inter American Statistical Institute's Survey of Census Methods and Procedures in the American Nations, 1944. Sent by the American Embassy. Consists of: Correspondence; publicity and propaganda; reports.

Comprend: Correspondance; publicité et propagande; rapports.

Haiti-**2100**.S5 1942 **Sampling Plans for a Census of Haiti.** Prepared by the Subcommittee on Sampling Techniques. December, 1942. Washington, D.C., 1942. U. S. Department of Commerce. Bureau of the Census. 11 p. Mimeographed.

A memorandum from the Subcommittee on Sampling Techniques to the Committee on Sample Censuses.

Plans d'échantillonnage statistique pour un recensement d'Haïti, par le Bureau du Recensement des États-Unis.

3000-3999 STATISTICAL SCIENCE—CIENCIA ESTADISTICA

Haiti-**3300** 1944-46 **Enseignement Statistique en Haïti: Méthodes et Materiaux.** Matériel fondamental recueilli pour le Relevé de l'Institut Interaméricain de Statistique sur les Méthodes et Materiaux Employés dans l'Enseignement Statistique dans les Nations Américaines, 1944-46.

Statistical training in Haiti: Methods and materials. Background material acquired for the Inter American Statistical Institute Survey on Statistical Training Methods and Materials in the American Nations, 1944-46. Includes: Correspondence and filled questionnaire on statistical personnel and statistical courses in Haiti.

Comprend: Correspondance et questionnaire rempli concernant le personnel statistique et les cours de statistique en Haïti.

4000-4999 GEOGRAPHY—GEOGRAFIA

Haiti-**4000**.G5 1924 **Geology of the Republic of Haiti.** By Wendell P. Woodring, John S. Brown, and Wilbur S. Burbank of the U. S. Geological Survey. Baltimore, Maryland, The Lord Baltimore Press, 1923. Department of Public Works. 631 p.

Includes: Introduction; geography; geology; geomorphology; mineral resources; water resources; bibliography.

Géologie de la République d'Haïti. Comprend: Introduction; géographie; géologie; géomorphologie; ressources minérales; ressources d'eau; bibliographie.

Haiti-**4200**.B5 **Bulletin Annuel de l'Observatoire Météorologique du Petit Séminaire Collège St. Martial.** 1910- . Port-au-Prince. Pas examiné.

Annual bulletin of the meteorological observatory of the Petit Séminaire Collège St. Martial; contains meteorological data from observations carried out in Port-au-Prince, including: Barometric pressures, temperature, rains, relative humidity, vapor pressures, wind velocity and direction, cloudiness, sunshine, cloud currents, data from rainfall measuring stations throughout the country, and seismological observations.

Contient les observations météorologiques faites à Port-au-Prince, comprenant: pressions barométriques, température, pluies, humidité relative, tensions de la vapeur d'eau, direction et vitesse du vent, nébulosité, présence du soleil, direction des nuages, données des stations pluviométriques à travers le pays, et observations sismologiques.

5000-5999 DEMOGRAPHY—DEMOGRAFIA

Haiti-**5000**.H5 1945 **Haiti: Summary of Biostatistics.** Maps and charts, population, natality and mortality statistics. Washington, D.C., June 1945. Prepared by U. S. Department of Commerce, Bureau of the Census, in cooperation with Office of the Coordinator of Inter-American Affairs. 72 p.

Résumé biostatistique d'Haïti. Comprend des cartes et des graphiques, ainsi que des données sur la population, la natalité et la mortalité.

Haiti-**5000**.R5 1944 **Recensement et Démographie.** Par René Victor. Introduction de Chester W. Young. Port-au-Prince, Imp.de l'Etat, 1944. 131 p.

Census and demography.

Haiti-**5200** **Statistiques Vitales et Enregistrement en Haïti: Méthodes et Procédés.** Matériel fondamental recueilli pour l'étude effectuée par l'Institut Interaméricain de Statistique sur les méthodes d'enregistrement et les statistiques vitales dans les nations américaines, y compris l'enquête de 1945.

Vital statistics and registration in Haiti: Methods and procedures. Background material for the study conducted by the Inter American Statistical Institute in vital statistics and registration methods in the American nations, including the survey of 1945. Includes: Answer to questionnaire; laws.

Comprend: Réponse au questionnaire; lois.

Haiti-**5203**.L5 Dec. 22, '22 **Loi sur le Service de l'État Civil.** Donné au Palais Législatif, à Port au Prince, le 22 décembre 1922, an 119e de l'indépendance. Port-au-Prince, 1945. 4 p. Copie dactylographiée.

Law on the Service of Civil Status.

6000-6999 ECONOMICS—ECONOMIA

*Haiti-**6000**.A3 **Annual Report of the Fiscal Department.** Starting date? Port-au-Prince. Banque Nationale de la République d'Haïti.

Also published in French—publié aussi en Français (Haiti-**6000**.R5).

The most important annual publication dealing with economic and financial statistics in Haiti. Contains information about the economic life of the country, including statistics.

1944-45 **Annual Report of the Fiscal Department for the Fiscal Year October 1944-September 1945.** Submitted to the Secretary of State for Finance, Commerce and National Economy by the Board of Directors of the Banque Nationale de la République d'Haïti. Port-au-Prince, Imp. de l'Etat, 1946? Banque Nationale de la République d'Haïti. 112 p.

Includes: Government revenues and expenditures, budget and public debt, banking and currency and position of the treasury, foreign trade values, imports and exports, air and maritime traffic, and building activity; appendixes of statistical tables showing detailed and comparative data. Also published in French.

Comprend: Revenus et dépenses de l'État, budget et dette nationale, banques et monnaie et situation du trésor, valeur du commerce extérieur, importations et exportations, transport aérien et maritime, et activités de construction; appendices de tableaux statistiques qui exposent des données comparatives et en détail. Publié aussi en français.

Haiti-**6000**.A5 1944 **Aspects de l'Économie et des Finances d'Haïti.** Problèmes de temps actuals; problèmes d'après guerre. Par Louis R. E. Gation. Port-au-Prince, Imp. du Collège Vertières, 1944. 353 p.

Aspects of the economy and finances of Haiti: Problems of the present time; post-war problems. Contents: Contribution of the banana; the possibilities of production; vital statistics of Haiti; the war and the present situation; the planning of the economy of Haiti after independence; financial policy of Haiti; national currency and the devaluation of the American dollar; public and private finances; Haiti and contemporary realizations; conclusions; bibliography.

Matière: Contribution de la figue-banane; les possibilités de la production; statistiques vitales d'Haïti; la guerre et la situation actuelle; planification de l'économie haïtienne après l'indépendance; politique financière d'Haïti; la monnaie nationale et la dévaluation du dollar Américan; finances publiques et finances privées; Haïti et les réalisations contemporaines; conclusions; bibliographie.

Haiti-**6000**.E5 1946 **Economic Controls and Commercial Policy in Haiti.** One of a series of reports on economic controls and commercial policy in the American republics. Washington, D.C., 1946. United States Tariff Commission. 26 p.

Rapport sur Haïti faisant partie de la série qui traite des règlements économiques et de la politique commerciale dans les républiques américaines. Publié par la Commission du Tarif des États-Unis.

*Haiti-**6000**.R5 **Rapport Annuel de Département Fiscal.** 1916-17- . Port-au-Prince. Banque Nationale de la République d'Haïti.

Published also in English—publié aussi en anglais (Haiti-**6000**.A3).

Title has varied—la titre a changé: 1916-17—1922-23, **Haitians Customs Receivership;** 1923-24—1932-33, **Rapport Annuel du Conseiller Financier—Receveur Général;** 1933-34—1939-40, **Rapport Annuel de Représentant Fiscal.**

La publication annuelle la plus importante traitant de la statistique économique et financière d'Haïti. Contient des renseignements sur la vie économique du pays, y compris des statistiques.

1944-45 **Rapport Annuel du Département Fiscal, 1944-1945.** Port-au-Prince, 1946? Banque Nationale de la République d'Haiti. Pas examiné.

Now being printed.

Publication sous presse.

1943-44 **Rapport Annuel du Département Fiscal pour l'Exercice Octobre 1943-Septembre 1944.** Soumis au Secrétaire d'Etat des Finances, du Commerce et de l'Economie Nationale par le Conseil d'Administration de la Banque Nationale de la République d'Haïti. Port-au-Prince, Imp.de l'Etat, 1945? Banque Nationale de la République d'Haïti. 119 p.

Annual report of the Fiscal Department of the Banque Nationale de la République d'Haïti, for the fiscal year October 1943 to

September 1944. Includes: Internal revenue, government expenditures, treasury position, public debt, national budget, banking and currency, imports and exports (broken down by countries and principal products, volume and value), with appendices of statistical tables showing comparative data of previous years. Also published in English.

Inclut: Revenus de l'État, dépenses de l'État, situation du Trésor, dette publique, budget national, banques et monnaie, importations et exportations (classées par pays et denrées principales, volume et valeur), avec appendices de tableaux statistiques exposant des données comparatives des années antérieures. Publié aussi en anglais.

Haiti-**6200**.B5 **Bulletins du Service National de la Production Agricole et de l'Enseignement Rural.** Depuis? Irrégulier. Port-au-Prince. Secrétairerie d'État de l'Agriculture et du Travail. Service National de la Production Agricole et de l'Enseignement Rural.

Bulletins of the Agricultural Production and Rural Education Service.

*Haiti-**6200**.R5 **Rapport Annuel du Service National de la Production Agricole et de l'Enseignement Rural.** 1924-25— . Port-au-Prince. Secrétairerie d'État de l'Agriculture et du Travail. Service National de la Production Agricole et de l'Enseignement Rural.

Annual report of the Agricultural Production and Rural Education Service.

Haiti-**6400**.M5 1945 **Mining and Manufacturing Industries in Haiti.** One of a series of reports on mining and manufacturing industries in the American republics. Washington, D.C. 1945. United States Tariff Commission. 9 p.

Rapport sur Haïti, faisant partie de la série qui traite des industries minière et manufacturière dans les républiques américaines. Publié par la Commission du Tarif des États-Unis.

Haiti-**6900** **Statistique du Commerce Extérieur d'Haïti: Méthodes et Procédés.** Matériel fondamental recueilli pour l'étude effectuée par l'Institut Interaméricain de Statistique sur les méthodes de statistiques du commerce extérieur dans les nations américaines, y compris l'enquête de 1946-47.

Foreign trade statistics in Haiti: Methods and procedures. Background material acquired for the study conducted by the Inter American Statistical Institute in methods of foreign trade statistics in the American nations, including the survey of 1946-47. Consists of: Filled questionnaire and explanatory notes.

Comprend: Questionnaire rempli et notes explicatives.

Haiti-**6900**.H5 1935 **Handbook and Trade Directory of Haiti.** The revised tariff of Haiti as of June 1, 1935. New York, 1935. Haiti Consulate in New York. 231 p.

Includes "Law of August 9, 1926," tariff law, p. 94-226. Classification of tariff similar to that of Cuba, Nicaragua, and Dominican Republic.

Manuel et directoire du commerce d'Haïti. Le tarif revisé d'Haïti jusqu'au 1 juin, 1935. Comprend "Loi du 9 août, 1926," loi du tarif, p. 94-226. La classification du tarif est similaire à celle de Cuba, Nicaragua, et de la République Dominicaine.

Haiti-**6900**.Q5 1942 **Quelques Données Statistiques sur le Commerce d'Haïti.** Port-au-Prince, Imp.de l'État, 1942. Département de l'Agriculture. Service National de la Production Agricole et de l'Enseignement Rural. Bureau d'Économie Rurale et de Statistiques Agricoles. 32 p. Bulletin du Service National de la Production Agricole et de l'Enseignement Rural, no. 23.

Some statistical data on the trade of Haiti.

7000-7999 HEALTH AND CULTURAL STATISTICS—ESTADISTICAS DE SALUBRIDAD Y CULTURA

Haiti-**7110**.B5 **Bulletin du Service National d'Hygiène et d'Assistance Publique.** 1934- . Trimestriel. Port-au-Prince. Secrétairerie d'État de l'Intérieur. Service National d'Hygiène et d'Assistance Publique.

Quarterly bulletin of the National Service of Public Health and Public Assistance.

*Haiti-**7110**.R5 **Rapport Annuel du Directeur Général.** 1917-18— . Port-au-Prince. Secrétairerie d'État de l'Intérieur. Service National d'Hygiène et d'Assistance Publique.

Published also in English—publié aussi en anglais.

Annual report of the Director General of the National Service of Public Health and Public Assistance.

Haiti-**7200**.R5 1943-44 **Les Résultats de la Troisième Année de Réforme de l'Enseignement Urbain, 1943-44.** Port-au-Prince, Imp.de l'État, 1945. Département de l'Instruction Publique. Direction Générale de l'Enseignement Urbain. 95 p. Pas examiné.

The results of the third year of reform of urban education, 1943-44.

Haiti-**7200.**T5 1942-43. **Tables et Graphiques Statistiques de l'Enseignement Urbain, 1942-1943.** Port-au-Prince, Imp.de l'État, 1945? Département de l'Instruction Publique. Direction Générale de l'Enseignement Urbain. 62 p.

Statistical tables and graphs on urban education, 1942-43. "This brochure contains, in tabular and graphic form, the statistics of primary, occupational, and secondary schools, both public and private, under the direction of the General Bureau of Urban Education; they show school attendance and the number of schools and teachers." (Translation of comment.)

"Cette brochure contient, sous forme de tables et graphiques, les statistiques des écoles Primaires, Profesionnelles et Secondaires publiques et privées relevant de la Direction Générale de l'Enseignement Urbain; elles indiquent la fréquentation scolaire, le nombre des écoles et des instituteurs." (Source: Introduction.)

Haiti-**7200.**T6 1942 **Tables Statistiques de la Fréquentation des Écoles de l'Enseignement Rural, 1939-1940 et 1940-1941.** Port-au-Prince, Imp.de l'État, 1942. Service National de la Production Agricole et de l'Enseignement Rural. Section des Statistiques et Recherches. Enseignement Rural. 41 p. Bulletin du Service National de la Production Agricole et de l'Enseignement Rural, no. 18.

Statistical tables of attendance in the schools of rural education, 1939-40 and 1940-41.

8000-8999 GOVERNMENT AND POLITICAL STATISTICS—ESTADISTICAS GUBERNAMENTALES Y POLITICAS

Haiti-**8000.**R5 **Rapport Annuel du Sous-Secrétaire d'État et Ingénieur en Chêf.** 1921-22- . Port-au-Prince. Secrétairerie d'État des Travaux Publics. Direction Générale des Travaux Publics.

Title varies—titre varie.

Annual report of the Under Secretary of Public Works and Chief Engineer.

Haiti-**8300.**R5 **Rapport Annuel.** Depuis? Port-au-Prince. Garde d'Haiti.

Published also in English—publié aussi en anglais.

Annual report of the army and police force of Haiti.

Haiti-**8800.**R3 **Rapport Annuel de l'Administration Générale des Contributions.** 1924-25— . Port-au-Prince.

Annual report of the Internal Revenue Service.

HONDURAS

2000-2999 CENSUS—CENSOS

Hond-**2100.C**5 1945.1 **Censo General de Población Levantado el 24 de Junio de 1945: Métodos y Procedimientos.** Material usado en el levantamiento del censo. Adquirido con motivo de la encuesta que el Instituto Interamericano de Estadística realiza sobre Métodos y Procedimientos de los Censos en las Naciones Americanas. De la Secretaría de Gobernación, Justicia, Sanidad y Beneficencia, Dirección General de Estadística.

Census of population, June 24, 1945: Methods and procedures. Acquired in connection with Inter American Statistical Institute study on Census Methods and Procedures in the American nations. Consists of: Laws; instructions; forms. Se compone de: Leyes; instrucciones; formularios.

See also—véase también: Hond-**2100.C**5, 1945.2; Hond-**2100.C**5, 1945.3.

Hond-**2100.C**5 1945.2 **Instrucciones para la Preparación y Práctica del Censo de Población de 1945.** Tegucigalpa, Tall.Tipo. Nacionales, 1945. 14 p.

Instructions for preparing and taking the population census of 1945.

Hond-**2100.C**5 1945.3 **Reglamento del Censo de Población de 1945.** Tegucigalpa, Tall.Tipo.Nacionales, 1945. Secretaría de Gobernación, Justicia, Sanidad y Beneficencia. 8 p.

Law on population census of 1945.

Hond-**2100.C**5 1940.1 **Censo de 1940: Métodos y Procedimientos.** Material usado en el levantamiento de los censos electoral y de población del 30 de junio. Adquirido con motivo de la encuesta que el Instituto Interamericano de Estadística realizó sobre Métodos y Procedimientos de los Censos en las Naciones Americanas, 1944. De la Secretaría de Gobernación, Justicia, Sanidad y Beneficencia, Dirección General de Estadística.

Census of 1940: Methods and procedures. Materials used in taking population and electoral censuses, on June 30. Acquired in connection with Inter American Statistical Institute Survey on Census Methods and Procedures in the American Nations, 1944. Consists of: Correspondence; laws; instructions; schedules; other forms; publications.

Se compone de: Correspondencia; leyes; instrucciones; cédulas; otros formularios; publicaciones.

See also—véase también: **Reglamento para el Levantamiento del Censo de Población de 1940** (Hond-**2100.C**5, 1940.3); **Instrucciones para la Preparación y Práctica del Censo de 1940** (Hond-**2100.C**5, 1940.2); **Resumen del Censo General de Población, Levantado el 30 de Junio de 1940** (Hond-**2200.C**5, 1940).

Hond-**2100.C**5 1940.2 **Instrucciones para la Preparación y Práctica del Censo de 1940.** Tegucigalpa, D.C., Tall.Tipo. Nacionales. 11 p.

Instructions for preparing and taking population census of 1940.

Hond-**2100.C**5 1940.3 **Reglamento para el Levantamiento del Censo de Población de 1940.** Tegucigalpa, D.C., Tall.Tipo. Nacionales, 1939. 8 p.

Law for taking population census of 1940.

*Hond-**2200.C**5 1940 **Resumen del Censo General de Población Levantado el 30 de Junio de 1940.** Tegucigalpa, D.C., Tall. Tipo.Nacionales, 1942. Dirección General de Estadística, dependiente de la Secretaría de Gobernación. 214 p.

Summary of general population census taken June 30, 1940. Statistics for major and minor civil divisions, age and sex, urban and rural distributions, civil status, race, nationality, occupation, religion, education and literacy.

Presenta estadísticas por divisiones políticas mayores y menores, edad y sexo, distribución urbana y rural, estado civil, raza, nacionalidad, ocupación, religión, educación y alfabetismo.

3000-3999 STATISTICAL SCIENCE—CIENCIA ESTADISTICA

Hond-**3030.R**5 1944 **Reglamento de la Estadística Nacional de la República de Honduras.** Tegucigalpa, Tall.Tipo.Nacionales, 1944. 27 p.

Regulation of national statistics in Honduras.

Hond-**3300** 1944-46 **Enseñanza Estadística en Honduras: Métodos y Materiales.** Material básico adquirido para la Encuesta del Instituto Interamericano de Estadística sobre los Métodos y Materiales en la Enseñanza Estadística en las Naciones Americanas, 1944-46.

Statistical training in Honduras: Methods and materials. Background material acquired for the IASI Survey on Statistical Training Methods and Materials in the American Nations, 1944-46. Includes: Correspondence and filled questionnaires on statistical personnel and statistical courses in Honduras.

Incluye: Correspondencia y cuestionarios llenados sobre personal estadístico y cursos de estadística en Honduras.

5000-5999 DEMOGRAPHY—DEMOGRAFIA

Hond-5000.H5 1944 **Honduras: Summary of Biostatistics.** Maps and charts, population, natality and mortality statistics. Washington, D.C., July 1944. Prepared by U. S. Department of Commerce, Bureau of the Census, in cooperation with Office of the Coordinator of Inter-American Affairs. 119 p.

Resumen bioestadístico de Honduras. Incluye mapas y cuadros y datos sobre población, natalidad y mortalidad.

Hond-5200 **Estadística Vital y Registro en Honduras: Métodos y Procedimientos.** Material básico adquirido para la encuesta que el Instituto Interamericano de Estadística realiza sobre los métodos de registro y estadística vital en las naciones americanas, incluyendo el estudio de 1945.

Vital statistics and registration in Honduras: Methods and procedures. Background material acquired for the study conducted by the Inter American Statistical Institute in vital statistics and registration methods in the American nations, including the survey of 1945. Includes: Answer to questionnaire; laws; forms; organization chart; codes.

Incluye: Respuesta al cuestionario; leyes; formularios; esquema de organización; códigos.

Hond-5200.M5 1937-38 **Movimiento de Población, Nacimientos, Defunciones y Matrimonios Habidos en la República durante el Año Económico de 1937 a 1938.** Tegucigalpa, enero 1940. Secretaría de Hacienda, Crédito Público y Comercio. Dirección General de Estadística. 20 p. Copia máquina. Enclosure no. 3, despatch no. 934, American Legation.

Vital statistics, births, deaths and marriages in Honduras, during fiscal year 1937-38. Includes data for 1933-37.

Incluye datos de los años 1933-37.

6000-6999 ECONOMICS—ECONOMIA

Hond-6000.I5 **Informe del Secretario de Agricultura, Fomento y Obras Públicas.** 1909-10— Anual. Tegucigalpa, D.C.

Annual report of the Secretary of Agriculture, Development and Public Works. Contains diverse statistical data, presented in charts and tables, on agricultural production and trade, railroads, mining, and postal, air and telecommunications.

Contiene datos estadísticos diversos, presentados en cuadros y diagramas, sobre comercio y producción agrícola, ferrocarriles, minería y comunicaciones (aéreas, postales y telecomunicaciones).

Hond-6200.C5 1945 **Cuestionario de la Comisión Nacional de Alimentación y Agricultura.** Tegucigalpa, mayo 1945. Secretaría de Fomento, Agricultura y Trabajo. Comisión Nacional de Alimentación y Agricultura. 15 p. Con 2 páginas de "Nota" explicativa.

Questionnaire of the National Commission of Food and Agriculture, designed to obtain statistical data on socio-economic life and activities of the country, national resources, and manpower, such data to furnish background material for study of the problem of increasing agricultural production and development of agricultural industries in Honduras.

Un cuestionario dedicado a obtener datos estadísticos sobre la vida económica-social y actividades del país, sobre sus recursos económicos y sobre su potencial obrero, datos tales que suministren material básico para el estudio del problema del incremento de la producción agrícola y el desarrollo de industrias agrícolas en Honduras.

Hond-6200.H5 **Honduras Agrícola.** Organo de la Comisión Nacional de Alimentación y Agricultura. 1944- . Mensual. Tegucigalpa. Secretaría de Agricultura, Fomento y Obras Públicas. Comisión Nacional de Alimentación y Agricultura.

Monthly journal of the National Commission of Food and Agriculture.

Hond-6900.A5 1934 **Arancel de Aduanas, 1934.** Tegucigalpa, Imp.Calderón, 1934? Ministerio de Hacienda, Crédito Público y Comercio. 114 p.

Customs tariff of Honduras, decree of April 11, 1934. Includes export section, and customs regulations.

Decreto de 11 de abril, 1934. Incluye sección de exportación y reglamentos de aduana.

7000-7999 HEALTH AND CULTURAL STATISTICS — ESTADISTICAS DE SALUBRIDAD Y CULTURA

Hond-7101.R5 1925 **Reglamento Interior de la Dirección General de Sanidad.** Tegucigalpa, Tipografía Nacional, 1925. 26 p.

Internal regulations of the General Bureau of Health.

Hond-**7110**.B5 **Boletín Sanitario de la Dirección General de Sanidad.** Fecha de iniciación? Trimestral. Tegucigalpa. Secretaría de Gobernación, Justicia, Sanidad y Beneficencia. Dirección General de Sanidad.

Quarterly sanitary bulletin of the General Bureau of Health.

8000-8999 GOVERNMENT AND POLITICAL STATISTICS — ESTADISTICAS GUBERNAMENTALES Y POLITICAS

Hond-**8000**.I5 **Informe de los Actos Realizados por el Poder Ejecutivo en los Ramos de Gobernación, Justicia, Sanidad y Beneficencia.** 1938-39— . Anual. Tegucigalpa.

Yearly report of the Department of Government, Justice, Health and Welfare. Includes criminal, judicial, and vital statistics.

Incluye estadísticas criminales, judiciales y demográficas.

*Hond-**8800**.I5 **Informe de Hacienda, Crédito Público y Comercio.** Presentado al Soberano Congreso Nacional. Fecha de iniciación? Anual. Tegucigalpa. Secretaría de Hacienda, Crédito Público y Comercio.

Published formerly under the title—publicado anteriormente bajo el título: **Memoria del Ramo de Hacienda, Crédito Público y Comercio.**

Annual report of the Department of Finance, Public Credit and Commerce. Contains principally statistics of foreign trade and public finance.

Contiene principalmente estadísticas de comercio exterior y hacienda pública.

1943-44 **Informe de Hacienda, Crédito Público y Comercio.** Presentado al Soberano Congreso Nacional . . . 1943 a 1944. Tegucigalpa, Tall.Tipo-Litográficos "Ariston," 1945? Secretaría de Hacienda, Crédito Público y Comercio. 900 p. (aprox.)

Includes: Section of government revenue and budget; public credit branch; section of accounts centralization; trade statistics and maritime movement; emergency measures; section of finance offices; detail of general imports . . . conforming to the customs tariff nomenclature, and detail of general exports, conforming to the customs tariff nomenclature.

Incluye: Sección de rentas y presupuesto; ramo de crédito público; sección de centralización de cuentas; estadística comercial y movimiento marítimo; medidas de emergencia; sección de oficinas de hacienda; "Detalle de la Importación General . . . Conforme a la Nomenclatura del Arancel de Aduanas" y "Detalle de la Exportación General, Conforme a la Nomenclatura del Arancel de Aduanas."

1000-1999 GENERAL

Méx-1200.B5 **Boletín Bibliográfico Mexicano.** 1939- . Mensual. México, D.F. Instituto Panamericano de Bibliografía y Documentación.

Monthly bibliographical bulletin. A summary of books and pamphlets published in Mexico.

Un resumen de libros y folletos publicados en México.

Méx-1200.P5 1933-42 **Publicaciones Oficiales, 1933-1942.** México, 1943. Secretaría de la Economía Nacional. 62 p.

Official publications, 1933-1942. Library of the Book Fair and National Press Exhibition. "The compilation and editing of the bibliographical cards were in the hands of Antonio Magaña Esquivel." (Translation of comment.)

Biblioteca de la Feria del Libro y Exposición Nacional de Periodismo. "La compilación y la redacción de las fichas bibliográficas estuvieron a cargo de Antonio Magaña Esquivel." (Fuente: Texto.)

Méx-1400.M5 1938 **México en Cifras.** 1938. México, D.F., Tall.Gráf. de la Nación, 1939. Secretaría de la Economía Nacional. Dirección General de Estadística. 70 p.

Mexico in figures. A statistical atlas. First edition published in 1934.

Un atlas estadístico. Primera edición publicada en 1934.

*Méx-1410.A5 **Anuario Estadístico de los Estados Unidos Mexicanos.** 1925- . México, D.F. Secretaría de la Economía Nacional. Dirección General de Estadística.

Statistical yearbook of the United States of Mexico. Covers population, commerce, industry, labor, agriculture, education, finance, etc.

Abarca población, comercio, industria, trabajo, agricultura, educación, finanzas, etc.

1941 **Anuario Estadístico de los Estados Unidos Mexicanos, 1941.** México, 1943. Secretaría de la Economía Nacional. Dirección General de Estadística. 1071 p.

Contents: Territory and meteorology; population (estimates, distribution, sex and age, culture, origin, religion, housing, occupation, etc.); vital statistics; education (schools, libraries, art, sports, etc.); social assistance; judicial statistics (delinquents, sentences, suicides, etc.); labor (labor groups, labor conflicts, strikes, lockouts, industrial accidents and disease, wages, unemployment, etc.); agriculture and fisheries (usable land and amount used, principal crops, production and yield, livestock, health of livestock, forest products, rural credit societies, irrigation, fish products, etc.); industries (production of gold, silver, coal and petroleum, cotton thread and fabric, sugar, alcoholic beverages, etc.); communication and transportation; foreign trade; domestic commerce (average prices, consumption, etc.); finance; wages and cost of living.

Contiene información estadística sobre: Territorio y meteorología; población del país (estimaciones, distribución, sexo y edad, cultura, origen, religión, vivienda, ocupación, etc.); movimiento de la población; educación (escuelas, bibliotecas, arte, deportes, etc.); asistencia social; judicial (delincuentes, sentenciados, suicidios, etc.); trabajo (agrupaciones obreras, conflictos del trabajo, huelgas, paros, víctimas de accidentes y de enfermedades profesionales del trabajo, salarios, desocupación, etc.); agricultura y pesca (tierra aprovechable y aprovechada, principales cultivos, su producción y rendimiento, ganadería, sanidad de los ganados, productos forestales, sociedades de crédito ejidal, riego, productos de la pesca, etc.); industrias (producción de oro, plata, carbón mineral y petróleo, hilados y tejidos de lana y algodón, azúcar, bebidas alcohólicas, etc.); comunicaciones y transportes; comercio exterior; comercio interior (precios medios, consumo, etc.); finanzas; salarios y costo de la vida.

Méx-1410.C5 **Compendio Estadístico.** 1941- . Anual? México, D.F. Secretaría de la Economía Nacional. Dirección General de Estadística.

Statistical abstract. Simplified abstract prepared particularly for use by professors, students, business men, and workers. Contains statistical data on population, education, commerce, transportation and communication, finance, and other national activities taken principally from the results of the most recent censuses effectuated in the respective fields.

Compendio simplificado preparado especialmente para uso de profesores, estudiantes, comerciantes y trabajadores. Contiene datos estadísticos sobre población, educación, comercio, transportes y comunicaciones, finanzas y otras actividades nacionales, tomados principalmente de los resultados de los censos mas recientes en las respectivas materias.

*Méx-**1420.R5** **Revista de Estadística.** 1938- . Mensual. México, D.F. Secretaría de la Economía Nacional. Dirección General de Estadística.

Formerly published under the title—anteriormente publicado bajo el título: **Revista de Economía y Estadística.**

Monthly statistical review. Has sections on finance, agricultural production, index numbers, foreign commerce, social statistics, etc. Contains current statistical series.

Contiene secciones sobre finanzas, producción agrícola, números índices, comercio exterior, estadísticas sociales, etc. Contiene series estadísticas corrientes.

2000-2999 CENSUS—CENSOS

Méx-**2100.**C5 1940.1 **Censo de 1940: Métodos y Procedimientos.** Material usado en el levantamiento del censo de población del 6 de marzo. Adquirido con motivo de la encuesta que el Instituto Interamericano de Estadística realizó sobre Métodos y Procedimientos en las Naciones Americanas, 1944. De la Secretaría de la Economía Nacional, Dirección General de Estadística.

Census of 1940: Methods and procedures. Materials used in taking population census, on March 6. Acquired in connection with Inter American Statistical Institute Survey of Census Methods and Procedures in the American Nations, 1944. Consists of: Instructions; decrees; reports; forms; schedules; maps; punch cards.

Consiste de: Instrucciones; decretos; informes; formularios; cédulas; mapas; tarjetas de perforación.

See also—véase también: **Decreto y Bases de Organización de los Siete Censos Nacionales y de los Padrones Estadísticos que se Levantarán en 1939-40; Instrucciones para la Perforación de Tarjetas del 6° Censo de Población de 1940.**

Méx-**2100.**C5 1940.2 **Decreto y Bases de Organización de los Siete Censos Nacionales y de los Padrones Estadísticos Que Se Levantarán en 1939-40.** Chapultepec, D.F., Tall.Gráf. de la Nación, 1939. Secretaría de la Economía Nacional. Dirección General de Estadística. Oficina Central de los Censos. 32 p.

Law and basis for the organization of the seven national censuses and the statistical registers to be effected in 1939-40.

Méx-**2100.**C5 1940.3 **Instrucciones para la Perforación de Tarjetas del 6° Censo de Población de 1940.** México, D.F., 1940. Secretaría de la Economía Nacional. Dirección General de Estadística. 15 p.

Instructions for punching cards of the sixth population census of 1940.

Méx-**2110.**C5 1940.1 **Segundo Censo Agrícola-Ganadero, Marzo de 1940: Métodos y Procedimientos.** Material usado en el levantamiento del censo. Adquirido con motivo de la encuesta que el Instituto Interamericano de Estadística realizó sobre Métodos y Procedimientos de los Censos en las Naciones Americanas, 1944. De la Secretaría de la Economía Nacional, Dirección General de Estadística.

Second census of agriculture and livestock, March 1940: Methods and procedures. Materials used in taking the census. Acquired in connection with Inter American Statistical Institute Survey of Census Methods and Procedures in the American Nations, 1944. Consists of: Laws, decrees, and other authorization; schedules; other forms; instructions; punch cards; codes; exposition and reports; cost.

Se compone de: Leyes, decretos, u otras disposiciones; cédulas; otros formularios; instrucciones; tarjetas de perforación; códigos; exposición e informes; costo.

Méx-**2110.**C7 1940.1 **Segundo Censo Ejidal de los Estados Unidos Mexicanos, 6 de Marzo de 1940: Métodos y Procedimientos.** Material usado en el levantamiento del censo. Adquirido con motivo de la encuesta que el Instituto Interamericano de Estadística realizó sobre Métodos y Procedimientos de los Censos en las Naciones Americanas, 1944. De la Secretaría de la Economía Nacional, Dirección General de Estadística.

Second census of communal land, March 6, 1940: Methods and procedures. Material used in taking the census. Acquired in connection with Inter American Statistical Institute Survey of Census Methods and Procedures in the American Nations, 1944. Consists of: Laws, decrees, and other authorization; schedules; other forms; instructions; punch cards and codes; exposition and reports; cost.

Se compone de: Leyes, decretos, u otras disposiciones; cédulas; otros formularios; instrucciones; tarjetas de perforación y códigos; exposición e informes; costo.

Méx-**2120.**C5 1939.1 **Segundo Censo de Edificios de los Estados Unidos Mexicanos, 20 de Octubre de 1939: Métodos y Procedimientos.** Material usado en el levantamiento del censo. Adquirido con motivo de la encuesta que el Instituto Interamericano de Estadística realizó sobre Métodos y Procedimientos de los Censos en las Naciones Americanas, 1944. De la Secretaría de la Economía Nacional. Dirección General de Estadística.

Second census of buildings, October 20, 1939: Methods and procedures. Materials used in taking the census. Acquired in connection with Inter American Statistical Institute Survey of Census Methods and Procedures in the American Nations, 1944. Consists of: Instructions; forms; codes; punch cards.

Se compone de: Instrucciones; formularios; códigos; tarjetas de perforación.

See also—véase también: **Segundo Censo de Edificios de los Estados Unidos Mexicanos: Resumen General.**

Méx-**2150.**C3 1940.1 **Primer Censo Comercial, Marzo de 1940: Métodos y Procedimientos.** Material usado en el levantamiento del censo. Adquirido con motivo de la encuesta que el Instituto Interamericano de Estadística realizó sobre Métodos y Procedimientos de los Censos en las Naciones Americanas, 1944. De la Secretaría de la Economía Nacional, Dirección General de Estadística.

First commercial census, March 1940: Methods and procedures. Material used in taking the census. Acquired in connection with Inter American Statistical Institute Survey of Census Methods and Procedures in the American Nations, 1944. Consists of: Laws, decrees, and other authorization; schedules; other forms; instructions; punch cards and codes; exposition and reports; cost.

Se compone de: Leyes, decretos u otras disposiciones; cédulas; otros formularios; instrucciones; tarjetas de perforación y códigos; exposición e informes; costo.

Méx-**2150.**C5 1945.1 **Los Censos Cuarto Industrial, Segundo Comercial y Segundo de Transportes, 1945: Métodos y Procedimientos.** Material usado en el levantamiento de los censos. Adquirido con motivo del estudio que el Instituto Interamericano de Estadística realiza sobre Métodos y Procedimientos de los Censos en las Naciones Americanas. De la Secretaría de la Economía Nacional, Dirección General de Estadística.

Fourth industrial census, second commercial census, and second census of transportation, 1945: Methods and procedures. Materials used in taking the censuses. Acquired in connection with Inter American Statistical Institute study on Census Methods and Procedures in the American Nations. Consists of: Correspondence; laws, decrees, or other authorization; schedule; instructions; exposition.

Se compone de: Correspondencia; leyes, decretos u otras disposiciones; cédula; instrucciones; exposición.

Méx-**2150.**C5 1940.1 **Tercer Censo Industrial de los Estados Unidos Mexicanos, 1940: Métodos y Procedimientos.** Material usado en el levantamiento del censo. Adquirido con motivo de la encuesta que el Instituto Interamericano de Estadística realizó sobre Métodos y Procedimientos de los Censos en las Naciones Americanas, 1944. De la Secretaría de la Economía Nacional, Dirección General de Estadística.

Third industrial census, 1940: Methods and procedures. Materials used in taking the census. Acquired in connection with the Inter American Statistical Institute Survey of Census Methods and Procedures in the American Nations, 1944. Consists of: Laws, decrees, and other authorization; schedules; other forms; instructions; punch cards and codes; exposition and reports; cost.

Se compone de: Leyes, decretos y otras disposiciones; cédulas; otros formularios; instrucciones; tarjetas de perforación y códigos; exposición e informes; costo.

Méx-**2180.**C5 1940.1 **Primer Censo de Transportes, Febrero y Marzo de 1940: Métodos y Procedimientos.** Material usado en el levantamiento del censo. Adquirido con motivo de la encuesta que el Instituto Interamericano de Estadística realizó sobre Métodos y Procedimientos de los Censos en las Naciones Americanas, 1944. De la Secretaría de la Economía Nacional, Dirección General de Estadística.

First census of transportation, February and March 1940: Methods and procedures. Materials used in taking census. Acquired in connection with Inter American Statistical Institute Survey of Census Methods and Procedures in the American Nations, 1944. Consists of: Laws, decrees, and other authorization; schedules; other forms; instructions; punch cards and codes; exposition and reports; cost.

Se compone de: Leyes, decretos y otras disposiciones; cédulas; otros formularios; instrucciones; tarjetas de perforación y códigos; exposición e informes; costo.

*Méx-**2200.**C5 1940.1 **Estados Unidos Mexicanos, Sexto Censo de Población, 1940.** México, 1943- . Secretaría de la Economía Nacional. Dirección General de Estadística. Un volumen para cada estado, y uno de resumen general.

Sixth population census of Mexico, 1940. A volume for each state and a general summary volume.

Méx-2200.C5 1940.2 **Sexto Censo General de Población de los Estados Unidos Mexicanos, 6 de Marzo de 1940: Población Municipal.** México, D.F., Tall.Gráf.de la Nación, 1942. Secretaría de la Economía Nacional. Dirección General de Estadistica. 56 p.

Sixth general census of population of the United States of Mexico, March 6, 1940: Municipal population.

Méx-2200.C5 1930.1 **Memoria de los Censos Generales de Población, Agrícola, Ganadero e Industrial de 1930.** México, D.F., Tall.Gráf.de la Nación, 1932. Departamento de la Estadística Nacional. Dirección Nacional de Estadística. 212 p.

Report of the population, agricultural, livestock, and industrial censuses of 1930.

Méx-2200.C5 1930.2 **Quinto Censo de Población, 15 de Mayo de 1930.** Resumen general. México, D.F., Tall.Gráf.de la Nación, 1934. Secretaría de la Economía Nacional. Dirección General de Estadística. 269 p.

Fifth population census, May 15, 1930. General summary.

*Méx-2210.C7 1940 **Segundo Censo Ejidal de los Estados Unidos Mexicanos, 6 de Marzo de 1940.** México, D.F., 1942. Secretaría de la Economía Nacional. Dirección General de Estadística. Un volumen por separado para cada estado.

Second census of communal lands of the United States of Mexico, March 6, 1940. Separate volumes for each state.

Méx-2210.C7 1935 **Primer Censo Ejidal, 1935.** México, D.F., Tall.Gráf.de la Nación, 1937. Secretaría de la Economía Nacional. Dirección General de Estadística. 2 vol.: Vol. I: **Resumen General;** vol. II: partes por separado para varios estados.

First census of communal lands, 1935. In two volumes: Vol. I: **General Summary;** vol. II: separate parts for various states.

*Méx-2220.C5 1939.1 **Segundo Censo de Edificios de los Estados Unidos Mexicanos: Resumen General.** México, D.F., 1943. Secretaría de la Economía Nacional. Dirección General de Estadística. 254 p.

Second census of buildings of the United States of Mexico: General summary.

Méx-2220.C5 1939.2 **Segundo Censo de Edificios, 20 de Octubre de 1939: Datos Definitivos.** México, D.F., Tall.Gráf.de la Nación, 1941-43. Secretaría de la Economía Nacional. Dirección General de Estadística. 32 vol.

Second census of buildings, October 20, 1939: Final data. Data are presented by provinces and municipalities, covering type, material and location of housing, occupancy, number of pieces of furniture, sewing machines and radios, and access to water. There is a volume for each state, and a summary volume. Population data accompany the housing information.

Los datos son presentados por provincias y municipalidades, abarcando tipo, clase de material y ubicación de la vivienda, ocupantes, cantidad de muebles, máquinas de coser y radios, y el abastecimiento de agua. Se compone de un volumen para cada estado y un volumen sumario. También son incluidos datos sobre la población relacionados con la vivienda.

*Méx-2250.C5 1940 **Tercer Censo Industrial de los Estados Unidos Mexicanos, 1940.** México, D.F., 1934- . Secretaría de la Economía Nacional. Dirección General de Estadística. Un folleto por separado para cada industria.

Third industrial census of Mexico, 1940. A separate pamphlet for each industry.

Méx-2250.C5 1940.3 **Características Principales de la Industria: Censo de 1940.** México, 1945. Secretaría de la Economía Nacional. Dirección General de Estadística. 3 p. Copia máquina.

Summary data from the 1940 census of industry on number of establishments, investments (buildings, machinery, etc.), raw material consumed, production, personnel employed, wages and salaries paid.

Datos sumarios del censo de industria de 1940 sobre número de establecimientos, inversiones (edificios, maquinaria, etc.), materias primas consumidas, personal ocupado, sueldos y salarios pagados.

Méx-2250.C5 1935 **Segundo Censo Industrial de 1935.** México, D.F., 1941. Secretaría de la Economía Nacional. Dirección General de Estadística. Resultados del censo publicados en tres partes: Parte I: **Resumen General** (en dos partes); parte II: un volumen por separado para cada estado; parte III: varios volúmenes que tratan sobre materias primas.

Second industrial census of 1935. Results of the Census published in three parts: Part I: **General Summary** (in two parts); part II: a separate volume for each state; part III: various volumes dealing with raw materials.

Méx-2250.C5 1930 **Primer Censo Industrial de 1930.** México, D.F., 1935.

Secretaría de la Economía Nacional. Dirección General de Estadística. No revisado. 2 vol.
First industrial census, 1930.

3000-3999 STATISTICAL SCIENCE—CIENCIA ESTADISTICA

Méx-3010.B5 1942 **Bibliografía Mexicana de Estadística.** México, D.F., Tall. Gráf.de la Nación, 1942. Secretaría de la Economía Nacional. Dirección General de Estadística. 2 vol.

A comprehensive guide to statistical publications on Mexico. Both official and non-official sources are given, and works of foreign as well as national origin are included. Brief discussion of methods. Includes sections on population, social, economic, administrative, and geographic statistics. Both geographic and author indexes.

Una guía comprensiva de las publicaciones estadísticas referentes a México. Incluye fuentes de información oficiales y no oficiales y trabajos tanto hechos en el extranjero como en el interior. Discusión breve sobre métodos. Incluye secciones sobre estadísticas social, económica, administrativa y geográfica. Indice geográfico y de autores.

Méx-3030.I5 1941 **Informes sobre las Principales Estadísticas Mexicanas.** México, D.F., 1941. Secretaría de la Economía Nacional. Dirección General de Estadística. 174 p.

Series of reports on the development and present status of the statistical activities of the Mexican government, including population censuses, housing censuses, population movements, morbidity and health, etc.

Series de informes sobre la evolución y estado actual de las actividades estadísticas del gobierno mexicano, incluyendo censos de población, censos de viviendas, movimiento demográfico de la población, morbilidad y salud, etc.

Méx-3030.L5 1941 **Ley Federal de Estadística y su Reglamento.** México, Tall. Gráf.de la Penitenciaría del D.F., 1941. 85 p.
Federal law on statistics and its regulation.

*Méx-3040.B5 **Boletín de la Sociedad Mexicana de Geografía y Estadística.** 1839- . Irregular. México, D.F.

Bulletin of the Mexican Society of Geography and Statistics. Includes articles on statistical questions.

Incluye artículos sobre asuntos estadísticos.

Jan.-Feb. '46, León y Alvaro Aldama C. "Problemas de Población en la América del Caribe y Central." Por Alberto P. León y Alvaro Aldama C. Tomo LXI, no. 1, enero-febrero 1946, p. 3-10 y 7 páginas de tablas y gráficas.
Population problems of Central America and the Caribbean area.

Mar.-Apr.'45, Fernández y Fernández "Indices Generales de la Producción Agrícola." Por Ramón Fernández y Fernández. Tomo LX, no. 2, marzo-abril 1945, p. 239-255.
Also published in—también publicado en el **Boletín Mensual de la Dirección de Economía Rural,** no. 226, marzo 1945, p. 120-129.
General indexes of agricultural production.

Méx-3220.C5 1945 **Catálogo de Industrias, 1945.** México, D.F.?, 1945. Dirección General de Estadística. Oficina del Censo Industrial. 20 p. (aprox.) Mimeografiado.
List of industries, 1945.

Méx-3220.T5 1940 **Tercer Censo Industrial: Catálogo General de las Industrias de Transformación y Extractivas, 1940.** México, 1940? Secretaría de la Economía Nacional. Dirección General de Estadística. 25 p.

Third industrial census: General list of manufacturing and extractive industries, 1940. Includes: Lists of transforming and extractive industries based on the national nomenclature of occupations.

Incluye: Catálogos de las industrias de transformación y de las industrias extractivas con base en la nomenclatura nacional de ocupaciones; catálogo de las industrias extractivas, con base en la nomenclatura nacional de ocupaciones.

Méx-3230.S5 1931 **Sinonimias Populares Mexicanas de las Enfermedades.** México, D.F., 1931. Secretaría de la Economía Nacional. Dirección General de Estadística. 23 p.
Popular Mexican synonyms of diseases.

Méx-3240.N5 1940 **Nomenclatura Nacional de Ocupaciones, 1940.** México, D.F., 1941. Secretaría de la Economía Nacional. Dirección General de Estadística. 1063 p.

National nomenclature of occupations, 1940. Manual of categories employed in the 1940 census.

Manual de las categorías empleadas en el censo de 1940.

Méx-3300 1944-46 **Enseñanza Estadística en México: Métodos y Materiales.** Material básico adquirido para la Encuesta del Instituto

Interamericano de Estadística sobre los Métodos y Materiales en la Enseñanza Estadística en las Naciones Americanas, 1944-46.

Statistical training in Mexico: Methods and materials. Background material acquired for the IASI Survey on Statistical Training Methods and Materials in the American Nations, 1944-46. Includes: Correspondence; filled questionnaires on statistical personnel and statistical courses in Mexico; programs of courses.

Incluye: Correspondencia; cuestionarios llenados sobre personal estadístico y cursos de estadística en México; programas de cursos.

Méx-**3320**.C5 1945 **Cursos de Invierno de 1945.** México, D.F., 1945. Universidad Nacional Autónoma de México. Facultad de Derecho y Ciencias Sociales. Escuela Nacional de Economía. 21 p.

Winter courses at the National University of Mexico, 1945.

5000-5999 DEMOGRAPHY—DEMOGRAFIA

Méx-**5000**.I5 1942 **Instructivo para la Elaboración de la Estadística Demográfica en las Oficinas Sanitarias de los Estados y Territorios Federales.** México, D.F., mayo 1942. Dirección General de Epidemiología. Oficina de Demografía y Estadística Sanitaria. 18 p. Multigrafiado.

Instructions for the elaboration of demographic statistics in the Health Offices of the states and federal territories. Includes tables, graphs, and charts.

Incluye cuadros, gráficos y diagramas.

Méx-**5000**.M5 1943 **Mexico: Demographic Data.** Analytical discussion, graphic presentation, population, natality and mortality statistics. Washington, D.C., July 1943. Prepared by the U. S. Department of Commerce, Bureau of the Census, in cooperation with Office of the Coordinator of Inter-American Affairs. 100 p.

Resumen bioestadístico de México. Incluye mapas y cuadros y datos sobre población, natalidad y mortalidad.

Méx-**5100**.P5 1921-38 **Población del País Calculada al 30 de Junio de Cada Año del Período 1921-1938.** México, D.F., 1938. Secretaría de la Economía Nacional. Dirección General de Estadística. Oficina de Estadística Social. 1 p. Mimeografiado.

Population of the country calculated for the 30th of June of each year of the period 1921-38.

Méx-**5200** **Estadística Vital y Registro en México: Métodos y Procedimientos.** Material básico adquirido para la encuesta que el Instituto Interamericano de Estadística realiza sobre los métodos de registro y estadística vital en las naciones americanas, incluyendo el estudio de 1945.

Vital statistics and registration in Mexico: Methods and procedures. Background material acquired for the study conducted by the Inter American Statistical Institute in vital statistics and registration methods in the American nations, including the survey of 1945. Includes: Answer to questionnaire; laws, decrees; forms; punch cards and codes; instructions; reports.

Incluye: Respuesta al cuestionario; leyes, decretos; formularios; tarjetas de perforación y códigos; instrucciones; informes.

Méx-**5212**.M5 1922-39 **Mortalidad en México.** México, Publicaciones S.E.N., 1940? Secretaría de Economía Nacional. Dirección General de Estadística. 119 p.

Mortality in Mexico. Graphs and statistical tables. Data for 1922-39.

Gráficas y cuadros estadísticos. Datos para 1922-39.

Méx-**5300**.D5 1923-41 **Datos Estadísticos del Departamento de Terrenos Nacionales.** México, 1939. Secretaría de Agricultura y Fomento. Dirección General de Población Rural, Tierras Nacionales y Colonización. 29 p.

Statistical data of the Bureau of Public Land. Covers the years 1923-39, with supplements for 1940 and 1941.

Abarca los años 1923-39, con suplementos para 1940 y 1941.

Méx-**5310**.I5 1942 **Intercambio Turístico Continental con la República Mexicana.** Informe sobre las estadísticas mexicanas de turismo. Adopción de la tarjeta única para turistas y plan mínimo de estadísticas continentales de turismo. Ensayo o aplicación de este plan con datos de la República Mexicana. Conclusiones. Por Roberto Parra Gómez. México, septiembre 1942. Congreso Inter-Americano de Turismo. 23 p. Mimeografiado.

Continental exchange of tourists with the Mexican Republic. Report on Mexican statistics of tourism. Adoption of the single tourist card and minimum plan for continental statistics of tourism. Trial or application of this plan with data of the Mexican Republic.

6000-6999 ECONOMICS—ECONOMIA

Méx-**6000**.E4 1946 **Economic Controls and Commercial Policy in Mexico.** One of a series of reports on economic controls and commercial policy in the American republics.

Washington, D.C., 1946. United States Tariff Commission. 46 p.

Folleto sobre México, de la serie de informes de la Comisión de Aduanas de los Estados Unidos, sobre controles económicos y política comercial de las repúblicas americanas.

Méx-6000.I5 **Investigación Económica.** Revista de la Escuela Nacional de Economía. 1941- . Trimestral. México, D.F. Universidad Nacional Autónoma de México. Escuela Nacional de Economía.

Economic survey: Quarterly journal of the National School of Economics. Articles on economic theory, with application to Mexican economic history and problems.

Artículos sobre teoría económica, con aplicación a la historia y los problemas económicos de México.

Méx-6000.M5 **Memoria de la Secretaría de la Economía Nacional.** Fecha de iniciación? Anual. México, D.F.

Annual report of the Department of National Economy. Contains some statistical data on petroleum, mining, and other industries, and on agricultural production.

Contiene algunos datos estadísticos sobre petróleo, minería y otras industrias, y sobre producción agrícola.

Méx-6000.M7 1942 **The Mexican Economy.** Preliminary edition. Washington, D.C., September 1942. U. S. Board of Economic Warfare. Blockade and Supply Branch. American Hemisphere Division. 231 p. Mimeographed. Unexamined.

"The purpose of this preliminary compilation is to provide officials of the Board of Economic Warfare with basic data of the economy of Mexico." (Source: Text.)

"El propósito de este tratado es, proveer a los miembros del Consejo de Guerra Económica de datos fundamentales sobre la economía mexicana." (Traducción del comentario.)

Méx-6000.P5 **Planificación Económica.** 1943- . Mensual. México, D.F. Secretaría de la Economía Nacional. Comisión de Coordinación Económica Nacional.

Economic planning.

Méx-6000.R4 1944 **Reports Presented to the Conference of Commissions of Inter-American Development (Held in New York, May 9-18, 1944) by the Mexican Commission of Inter-American Development.** On: 1, Notes on past and present industry in Mexico; 2, Mexican industry—a descriptive survey and its recent evolution; 3, the commerce of the Americas during the war. (Translations from Spanish original.) Washington, D.C., 1945? Inter-American Development Commission. 63 p.

Cover title—título de la cubierta: **Mexico: Reports Presented to the Conference of Commissions of Inter-American Development.**

Informes presentados a la Conferencia de Comisiones de Fomento Interamericano (realizada en N. York, del 9 al 18 de mayo de 1944), por la Comisión Mexicana de Fomento Interamericano, sobre: 1, Datos sobre la industria del pasado y presente en Méjico; 2, industria mejicana—un estudio descriptivo y su reciente evolución; 3, el comercio de las Américas durante la guerra.

Méx-6000.R5 **Revista de Economía.** 1937- Mensual. México, D.F., Imp. Aldina, Robredo & Rosell S.R.L.

Monthly journal of economics. Contains tables with price quotations. The data are taken from other sources, which are shown after each table. Covers agriculture, industry, commerce, communications, money and credit, public finance, etc.

Contiene cuadros con cotización de precios. Los datos han sido tomados de otras fuentes de información que figuran después de cada cuadro. Abarca agricultura, industria, comercio, comunicaciones, moneda y crédito, hacienda pública, etc.

*Méx-6000.T5 **El Trimestre Económico.** 1934- . Trimestral. México, D.F., Fondo de Cultura Económica.

Quarterly economic review. Articles on general economic theory and on specific economic problems of Mexico. Contains bibliography.

Artículos sobre teoría económica en general y sobre problemas económicos específicos de México. Contiene una bibliografía.

Oct.-Dec. '46, Ortiz Mena "La Balanza de Pagos y el Ingreso Nacional." Por Raúl Ortiz Mena. Vol. XIII, no. 3, octubre-diciembre 1946, p. 451-462.

Balance of payments and national income.

Jan.-Mar. '46, Alanís Patiño "Observaciones Sobre Algunas Estadísticas Agrícolas." Por E. Alanís Patiño y E. Vargas Torres. Vol. XII, no. 4, enero-marzo 1946, p. 589-625.

Observations on some agricultural statistics. Includes: Introduction; agricultural statistics from 1893 to 1907; agricultural statistics from 1925 to 1945; agricultural statistics and public opinion; agricultural and industrial statistics; agricultural and transportation statistics; agricultural and commercial statistics; agricultural and family consumption statistics; continuous and censal statistics; partial and total agricultural statistics; agricultural and climatological statistics; forecasts and preliminary and definitive statistics; important recommendations; résumé; propositions.

Incluye: Introducción; estadísticas agrícolas de 1893 a 1907; estadísticas agrícolas de 1925 a 1945; las estadísticas agrícolas y la opinión pública; estadísticas agrícolas e industriales; estadísticas agrícolas y de transportes; estadísticas agrícolas y comerciales; estadísticas agrícolas y de consumos familiares; estadísticas agrícolas continuas y censales; estadísticas agrícolas parciales y totales; estadísticas agrícolas y climatológicas; pronósticos, estadísticas preliminares y definitivas; advertencias importantes; resumen; proposiciones.

Oct. Dec. '45, Luna Olmedo "Factores que Influyen en la Balanza de Pagos de México." Por Agustín Luna Olmedo. Vol. XII, no. 3, octubre-diciembre 1945, p. 371-403.

Factors influencing the balance of payments of Mexico. Includes: General remarks; public expenditures and circulating media; circulation and prices; prices and monetary reserve; monetary reserve and exchange rate; fundamental concepts of balance and its fluctuations; influence of the international value of the peso; external balance and national economy; balance of payments and economic fluctuations; economic policy and exchange policy; two statistical tables.

Incluye: Generalidades; los gastos públicos y el medio circulante; el medio circulante y los precios; los precios y la reserva monetaria; la reserva monetaria y el tipo de cambio; los conceptos fundamentales de la balanza y sus fluctuaciones; influencia del valor internacional del peso; el saldo exterior y la economía nacional; la balanza de pagos y las fluctuaciones económicas; política económica y política de cambios; dos cuadros estadísticos.

Oct.-Dec. '45, Ortiz Mena "Producto Nacional e Ingreso Nacional." Por Raúl Ortiz Mena. Vol. XII, no. 3, octubre-diciembre 1945, p. 433-453. Reimpresión.

National product and national income. Includes: Definition and method of calculation; methods of valuation.

Incluye: Definición y método de cálculo; métodos de valoración.

Méx-6200 **Estadísticas Agrícolas en México: Métodos y Procedimientos.** Material básico adquirido para el estudio que el Instituto Interamericano de Estadística realiza sobre los métodos de estadísticas agrícolas en las naciones americanas, incluyendo la encuesta de 1945-46.

Agricultural statistics in Mexico: Methods and procedures. Background material acquired for Inter American Statistical Institute study on methods in agricultural statistics in the American nations, including the Survey of Agricultural Production Statistics in the American Nations in 1945-46. Consists of: Answers to questionnaires; instructions; forms; reports.

Consiste de: Respuestas a los cuestionarios; instrucciones; formularios; informes.

See also—véase también: **Boletín Mensual de la Dirección de Economía Rural,** enero 1943; **Instrucciones Generales para los Corresponsales Estadísticos.**

*Méx-6200.B5 **Boletín Mensual de la Dirección General de Economía Rural.** 1926- . México, D.F. Secretaría de Agricultura y Fomento. Dirección General de Economía Rural.

Monthly bulletin of the General Bureau of Rural Economy. Contains detailed statistics of area, production, yield, average price, value of crops, consumption, wholesale prices in Mexico City, foreign and domestic trade. Each issue contains also a statistical-economic study of an individual product.

Contiene estadísticas detalladas sobre área, producción, rendimiento, precio promedio, valor de la producción de la cosecha, consumo, precios al por mayor en la ciudad de México, comercio interior y exterior. Cada edición contiene también un estudio estadístico-económico de un determinado producto.

Nov. '46, González H. "El Desarrollo y Estado Actual de las Estadísticas Agrícolas en México." Por Gonzalo González H. No. 246, noviembre 1946, p. 1089-1100.

The development and the present status of agricultural statistics in Mexico.

Mar. '45 "Indices Generales de la Producción Agrícola." Colaboración del Ing. Ramón Fernández y Fernández. No. 226, marzo 1945, p. 120-129.

General indexes of agricultural production. Includes sections on: Antecedents; current indexes; weighting procedure; conclusion; movement of indexes during the period 1939-44.

Incluye secciones: Antecedentes; índices que se elaboran actualmente; el procedimiento de ponderación; conclusión; el movimiento de índices de 1939 a 1944.

Jan. '45 "Indices Económicos." Números índices que se calculan en México y que tienen relación con la economía agrícola del país. No. 224, enero 1945, p. 1242-1256.

Economic index numbers calculated in Mexico and which are related to the agricultural economy of the country.

Jan. '43 "Historia y Evolución de las Estadísticas Agrícolas Continuas, durante el Período 1925-42." No. 200, enero 1943, p. 1095-1096.

History and evolution of continuous agricultural statistics, during the period 1925-42.

Méx-**6200**.M5 **Memoria de la Secretaría de Agricultura y Fomento.** Fecha de iniciación? Anual. México, D.F. No examinado.

Annual report of the Department of Agriculture and Development. Reviews industrial and agricultural programs. Contains statistics of crops produced, areas planted, agricultural exports and imports; also detailed exposition of irrigation and irrigation projects.

Examina los programas industrial y agrícola. Contiene estadísticas de la producción de cosechas, áreas sembradas, importaciones y exportaciones agrícolas; también una exposición sobre irrigación y proyectos de irrigación.

Méx-**6200**.M8 **México Agrario.** Revista sociológica. 1939- . Trimestral. México, D.F.

Rural Mexico. Quarterly sociological journal. Discussion and analysis of the social aspects of Mexican agriculture. Contains editorials and news relating to the activities of the Mexican Rural Society, as well as essays and articles on current questions of social and agrarian policies.

Discusión y análisis de los aspectos sociales de la agricultura mexicana. Contiene artículos de fondo y noticias relativas a las actividades de la Sociedad Agronómica Mexicana, como así también ensayos y artículos sobre cuestiones de política social y agraria.

Méx-**6200**.P5 1945 **Programa e Instructivo para el Servicio de Estadística Agrícola en las Agencias Generales.** México, D.F., 1945. Secretaría de Agricultura y Fomento. Dirección de Economía Rural. 50 p. (aprox.)

Program and instructions for the agricultural statistics service in the rural agencies of the Department of Agriculture and Development. Includes questionnaires and forms.

Incluye cuestionarios y formularios.

Méx-**6200**.R8 1946 **Report on the Crop and Livestock Statistics and Agricultural Census Work in Mexico.** By Joseph A. Becker and Irvin Holmes. Washington, D.C., June 1946. United States Department of Agriculture. Office of Foreign Agricultural Relations. 26 p. Hectographed.

Includes a section on the 1940 Mexican census of agriculture (background, questionnaires, maps, enumeration methods, processing and publication, appraisal, recommendations), and one on crop and livestock statistics in Mexico (organization, questionnaires, processing of returns, publications, appraisal, personnel training, recommendations).

Informe sobre la estadística de la cosecha y ganadería y la obra del censo agrícola de Méjico; publicado por la Oficina de Relaciones Agrícolas Exteriores del Departamento de Agricultura de Estados Unidos. Incluye una sección sobre el censo agrícola mejicano de 1940 (base, cuestionarios, mapas, enumeración, métodos, procedimiento y publicación, valuación, recomendaciones) y otra sección sobre estadísticas de cosecha y ganadería en Méjico (organización, cuestionario, el manejo de los datos obtenidos, publicación, valuación, preparación de personal, recomendaciones).

Méx-**6220**.I5 1942 **Instrucciones Generales para los Corresponsales Estadísticos.** México, abril 20, 1942. Secretaría de Agricultura y Fomento. Dirección General de Economía Rural. 30 p. (aprox.) Mimeografiado.

General instructions for statistical correspondents of the Department of Agriculture. Includes forms and instructions for crop estimating.

Incluye formularios e instrucciones para estimaciones de cosechas.

Méx-**6300**.M5 1944 **Minería y Riqueza Minera de México.** Por Jenaro González Reina. México, D.F., Gráfica Panamericana, 1944. Banco de México, S.A. 211 p. Monografías industriales, no. 2.

Mining and mineral wealth in Mexico. Production, value, exports of gold, silver, zinc, copper, mercury, etc.

Producción, valor, exportación de oro, plata, cinc, cobre, mercurio, etc.

Méx-**6400** **Estadísticas Industriales en México: Métodos y Procedimientos.** Material básico adquirido para el estudio que el Instituto Interamericano de Estadística realiza sobre los métodos de estadísticas industriales en las naciones americanas, incluyendo la encuesta de 1945-46.

Industrial statistics in Mexico: Methods and procedures. Background material acquired for the study conducted by the Inter American Statistical Institute in methods of industrial statistics in the American nations, including the survey of 1945-46. Consists of: Decrees and resolutions; instructions; forms; reports; publications.

Consiste de: Decretos y resoluciones; instrucciones; formularios; informes; publicaciones.

See also—véase también: **Tercer Censo Industrial: Catálogo General de las Industrias de Transformación y Extractivas, 1940.**

Méx-**6400.**I5 1943 **Industrial Development in Mexico.** Prepared by the American Republics Unit of the Bureau of Foreign and Domestic Commerce. Washington, D.C., November 1943. U. S. Department of Commerce. Bureau of Foreign and Domestic Commerce. 42 p.

At head of title—a la cabeza del título: Inquiry reference service.

Contents: Introduction; historical sketch of industrial development; factors conditioning industrial development; status of industrial development; developments in industrialization as a result of the war; postwar problems arising out of industrialization.

Desarrollo industrial de México. Contenido: Introducción; reseña histórica de la evolución industrial; factores que influyen en el desenvolvimiento industrial; estado de la explotación industrial; evoluciones en la industrialización como resultado de la guerra; problemas de postguerra que se presentan en la industrialización.

Méx-**6400.**M3 1946 **Mining and Manufacturing Industries in Mexico.** One of a series of reports on mining and manufacturing industries in the American republics. Washington, D.C., 1946. United States Tariff Commission. 103 p.

Folleto sobre México, de la serie de informes de la Comisión de Aduanas de los Estados Unidos, sobre las industrias mineras y manufactureras en las repúblicas americanas.

Méx-**6400.**M5 **Monografías Industriales del Banco de México, S.A.** 1943?- . Irregular. México, D.F.

Industrial monographs of the Bank of Mexico.

Méx-**6720.**B5 **Banco de México, S.A.: Asamblea General Ordinaria de Accionistas.** Fecha de iniciación? Anual? México, D.F. Banco de México, S.A.

Meeting of stockholders of the Bank of Mexico.

1946 **Banco de México, S.A.: Vigésimacuarta Asamblea General Ordinaria de Accionistas.** México, D.F., 1946. Banco de México, S.A. 133 p.

Contains reports on international operations, production, transports, prices, banks and credit, public finance, and annexes of statistical tables and charts, including foreign trade data, index of food prices in Mexico City (p. 79), indexes of wholesale prices in the United States and Mexico (p. 80-81), index of wholesale prices in Mexico City (p. 84-85), and other data. The statistics shown cover the year 1945, and are compared with figures of preceding years.

Contiene informes acerca de operaciones internacionales, producción, transportes, precios, bancos y crédito, finanzas públicas, y anexos con cuadros y gráficas estadísticas, incluyendo comercio exterior, índice del costo de la alimentación en la Ciudad de México (p. 79), índices de precios al mayoreo en México y Estados Unidos de Norteamérica (p. 80-81), índice de precios al mayoreo en la Cuidad de México (p. 84-85), y otros datos. Las estadísticas expuestas abarcan el año 1945, y se comparan con cifras de años anteriores.

Méx-**6720.**R5 **Revista del Banco de la República.** Fecha de iniciación? Mensual. México, D.F. No examinado.

Monthly journal of the Bank of the Republic. "The Banco de la Republica has for some years published a monthly Revista containing data on government finance, foreign trade, commodity prices, cost of living, and other economic phases." (Source: Stat. Act.)

"El Banco de la República ha publicado mensualmente durante algunos años una Revista que contiene datos sobre hacienda pública, comercio exterior, precios de productos, costo de la vida y otras fases económicas." (Traducción del comentario.)

Méx-**6900** **Estadísticas del Comercio Exterior en México: Métodos y Procedimientos.** Material básico adquirido para el estudio que el Instituto Interamericano de Estadística realiza sobre los métodos de estadísticas del comercio exterior en las naciones americanas, incluyendo el estudio de 1946-47.

Foreign trade statistics in Mexico: Methods and procedures. Background material acquired for the study conducted by the Inter American Statistical Institute in methods of foreign trade statistics in the American nations, including the survey of 1946-47. Consists of: Filled questionnaire; country nomenclature.

Consiste de: Cuestionario llenado; clave de países.

*Méx-6900.A5 **Anuario Estadístico del Comercio Exterior de los Estados Unidos Mexicanos.** 1933- . México, D.F. Secretaría de la Economía Nacional. Dirección General de Estadística.

Statistical yearbook of foreign commerce of Mexico.

Méx-6900.B5 **Boletín de Aduanas.** Fecha de iniciación? Mensual. México, D.F. Secretaría de Hacienda y Crédito Público. Departamento de Biblioteca y Archivos Económicos.

Bulletin of customhouses.

Méx-6900.R5 **Revista del Comercio Exterior.** 1938- . Bimestral. México, D.F. Secretaría de Relaciones Exteriores. Dirección General de Comercio Exterior.

Foreign commerce review. Bimonthly. Contains statistics on foreign trade of Mexico and other countries, wholesale prices of Mexican commodities in other countries, etc.

Contiene estadísticas sobre el comercio exterior de México y otros países, precios al por mayor de mercaderías mexicanas en otros países, etc.

Méx-6900.T5 1937 **Tarifas de los Derechos de Importación y Exportación de los Estados Unidos Mexicanos.** 2a edición de 1o de Diciembre de 1937 con reformas al día. México, Información Aduanera de México. 965 p.

Tariff of import and export duties in Mexico.

Méx-6900.T6 **Tarifa del Impuesto de Exportación.** Con la nueva nomenclatura y lista de aforos para el cobro del 12% sobre los productos que se exporten. Hojas substituibles. México, Información Aduanera de México. 1200 p. (aprox.)

Export tariff. With the new nomenclature and valuation schedule for the assessment of the 12% duty on exported products. Loose leaf edition.

7000-7999 HEALTH AND CULTURAL STATISTICS—ESTADISTICAS DE SALUBRIDAD Y CULTURA

Méx-7110.B5 1943 **Bibliografía Que el Departamento de Salubridad Pública Presenta en la Feria del Libro y Exposición Nacional de Periodismo.** México, 1943. Departamento de Salubridad Pública. Dirección General de Educación Higiénica. 63 p.

Bibliography presented by the Department of Public Health at the Book Fair and National Press Exhibition. Contents: Preface, three stages of public health; analytic bibliographical critique; bibliographical and hemerographical summary.

Contenidos: Prólogo, tres etapas de la salubridad pública; crítica analítica bibliográfica; resumen bibliográfico y hemerográfico.

Méx-7110.B6 **Boletín de la Oficina de Demografía y Estadística Sanitaria.** Vol. I, no. 1, junio 1942. México, D.F. Secretaría de Salubridad y Asistencia Pública. Departamento de Estadística y Demografía.

Only one number of this bulletin appeared—solamente un número de este boletín ha aparecido.

Bulletin of the Office of Demography and Health Statistics. Includes rate comparisons of Mexico with other countries of the world, for selected causes of death.

Incluye comparaciones de las tasas de determinadas causas de muerte, de México con otros países del mundo.

Méx-7110.B7 **Salubridad y Asistencia.** Organo de la Secretaría de Salubridad y Asistencia. 19 - . Trimestral. México, D.F.

Title varies—título varía: 19 -1938, **Salubridad;** 1938-41, **Boletín de Salubridad e Higiene;** 1942-45? **Boletín del Departamento de Salubridad Pública.**

Quarterly journal of the Department of Health and Assistance.

Méx-7110.M5 **Memoria del Departamento de Salubridad y Asistencia Pública.** Fecha de iniciación? Anual. México, D.F.

Annual report of the Department of Health and Public Assistance. Detailed account of health work for the year, and the health program.

Rendición detallada de la labor realizada durante el año y programa de sanidad.

Méx-7110.R5 **Revista del Instituto de Salubridad y Enfermedades Tropicales.** 1940- . Trimestral. México, D.F., Oficina Técnica de Educación y Propaganda, Departamento de Salubridad Pública. No examinado.

Review of the Institute of Health and Tropic Diseases.

Bustamante "Tablas de Vida de los Habitantes de los Estados Unidos Mexicanos." Por Miguel E. Bustamante y Alvaro Aldama Contreras. No examinado.

Life tables of the inhabitants of the United States of Mexico. Appears as a series in various issues of the **Revista.**

Aparece como una serie en varias ediciones de la **Revista.**

June '41 "Esperanza de Vida en Veinte Estados de la República Mexicana." Vol. 2, no. 1, junio 1941, p. 5-18. No examinado.

Life expectancy in twenty states of the Mexican Republic.

Nov. '40, Bustamante "Distribución de la Población de la República Mexicana por Edades y por Sexos, 1921 a 1936." Por Miguel E. Bustamante. Vol. 1, no. 1, noviembre 1940, p. 39-52. No examinado.

Distribution of population of Mexico by age and sex, 1921-36. A statistical analysis of changes in age groups.

Un análisis estadístico de las variaciones en los grupos de edad.

8000-8999 GOVERNMENT AND POLITICAL STATISTICS—ESTADISTICAS GUBERNAMENTALES Y POLITICAS

Méx-8000.M5 **Memoria de la Secretaría de Gobernación.** Fecha de iniciación? Anual. México. Secretaría de Gobernación.

Annual report of the Department of Government.

Méx-8300.T5 1939 **Tendencia y Rítmo de la Criminalidad en México, D.F.** Por Alfonso Quiroz, José Gómez Robleda y Benjámín Agüelles. México, D.A.P.P., septiembre 1939. Instituto de Investigaciones Estadísticas. 136 p. Publicaciones, tomo I, no. 1.

Tendency and rhythm of criminality in the Federal District of Mexico.

Méx-8800.B5 1943 **Bibliografía de la Secretaría de Hacienda y Crédito Público.** México, D.F., 1943. Secretaría de Hacienda y Crédito Público. 226 p.

Bibliography of the Department of Finance and Public Credit. Library of the Book Fair and National Press Exhibition, 1943. "The bibliographical research and compilation for the period from 1821 to 1899 were in the hands of Mr. Román Beltrán, while Miss Luz García Núñez undertook the part corresponding to the period 1900-1942." (Translation of comment.)

Biblioteca de la Feria del Libro y Exposición Nacional del Periodismo, 1943. "La investigación y formulación bibliográficas del período comprendido entre 1821 y 1899 estuvieron a cargo del Sr. Román Beltrán, la Srta. Luz García Núñez elaboró la parte correspondiente de 1900 a 1942." (Fuente: Texto.)

9000-9999 LABOR AND LIVING CONDITIONS—CONDICIONES DE TRABAJO Y DE VIDA

*Méx-9000.A5 **Anuario de Estadísticas del Trabajo.** 1941- . México, D.F. Secretaría del Trabajo y Previsión Social.

Yearbook of labor statistics.

Méx-9000.M5 **Memoria de Labores.** Fecha de iniciación? Anual. México, D.F. Secretaría del Trabajo y Previsión Social.

Report on the work of the Department of Labor and Social Security. Covers work of women and children, social security activities, labor arbitration, labor organizations, strikes.

Abarca trabajo de mujeres y niños, actividades del seguro social, mediación del trabajo, organizaciones obreras, huelgas.

Méx-9000.R5 **Revista Mexicana de Sociología.** 1939- . Cada cuatro meses. México, D.F. Universidad Nacional de México. Instituto de Investigaciones Sociales.

Mexican review of sociology.

Méx.9000.T5 **Trabajo y Previsión Social.** 1937- . Mensual. México, D.F. Secretaría del Trabajo y Previsión Social.

Title varies—título varía: 1937-40, **Revista del Trabajo.**

Labor and social welfare: Monthly review. Contains occasional statistics on industrial accidents, industrial conflicts, minimum wages, etc.

Contiene estadísticas irregulares sobre accidentes industriales, conflictos industriales, salarios mínimos, etc.

Méx-9200.E5 1942 **Estadística de Salarios.** Por Pedro Merla. México, D.F., Tall.Gráf.de la Nación, 1942. Secretaría del Trabajo y Previsión Social. 27 p.

Statistics of salaries.

Méx-9500.E5 1935 **Un Estudio del Costo de la Vida.** Por Federico Bach. México, D.F., 1935. 40 p. No examinado.

Study of the cost of living.

Méx-9600.E5 "Las Estadísticas sobre los Presupuestos Familiares en los Grupos Indígenas." México, 1941. Por Emilio Alanís Patiño. Mimeografiado. No examinado.

Statistics on family budgets in the Indian groups. Paper presented to the First Inter-American Indian Congress, summarizing the objectives and difficulties of Indian family budget studies in Mexico and presenting schedules for the first Mexican study on this subject.

Documentos presentados al Primer Congreso Indígena Inter-Americano, resumiendo los objetos y dificultades del estudio del presupuesto familiar indígena en México, y presentando cuadros del primer estudio mexicano sobre este particular.

MEXICO
9600

NICARAGUA

1000-1999 GENERAL

*Nic-**1410**.A5 **Anuario Estadístico de la República de Nicaragua.** Fecha de iniciación? Managua. Ministerio de Hacienda y Crédito Público. Dirección General de Estadística.

Statistical yearbook of Nicaragua.

1943 **Anuario Estadístico de la República de Nicaragua, 1943.** Managua, Tall.Nacionales, 1945. Dirección General de Estadística. 149 p.

Contents: Meteorology; area and population density; demography; migration; health and social assistance; judicial statistics; school statistics; shows; libraries and museums; fire department; economic and financial section.

Contiene: Meteorología; superficie y densidad de población; demografía; movimiento migratorio; sanidad y asistencia social; estadística judicial; estadística escolar; espectáculos públicos; bibliotecas y museos; cuerpo de bomberos; sección económica y financiera.

Nic-**1410**.G5 **Gráficos Estadísticos de Nicaragua.** Fecha de iniciación? Anual? Managua. Ministerio de Hacienda y Crédito Público. Dirección General de Estadística.

Statistical graphs of Nicaragua.

Nic-**1420**.B5 **Boletín Mensual de Estadística.** Organo de la Dirección General de Estadística de Nicaragua. 1944- . Managua. Ministerio de Hacienda y Crédito Público. Dirección General de Estadística.

Title varies—título varía: Abril, 1945-mayo 1944, **Estadística.**

Monthly bulletin of statistics. Contains data compiled by the General Bureau of Statistics in the fields of demography, economics and administration. This bulletin is intended to supplement the publication **Economía y Finanzas** by the same Bureau.

Contiene datos compilados por la Dirección General, referentes a demografía, economía y administración. Este boletín tiene el propósito de suplementar a **Economía y Finanzas,** publicación de la misma Dirección.

Apr.-June, '46, a "Monografía de la República de Nicaragua." Año 2, tomo 2, no. 24-26, abril-junio 1946, p. 7-23.

Includes a section on geography with details on hydrography, mountains, climate, area, situation, etc.; a brief historical résumé; a description of the sources of wealth, races, religion, government, and political divisions. The principal section contains general information on demography, culture, education, communication, agriculture, industry, commerce, finance and banking, all in the form of statistical tables.

Contiene una sección geográfica con detalles de hidrografía, orografía, clima, area, situación, etc.; una breve reseña historica; una descripción de fuentes de riqueza, razas, religión, gobierno y división política. La parte principal la constituye la información general que contiene sobre demografía, cultura, instrucción, comunicaciones, agricultura, industria, comercio, finanzas y banca; todas esas informaciones están dadas en forma de cuadros.

Apr.-June, '46, b "Proceso Evolutivo de la Población de Nicaragua Durante los Ultimos 25 Años." Año 2, tomo 2, no. 24-26, abril-junio 1946, p. 26.

Evolution of the population of Nicaragua in the last twenty-five years.

Sept.-Oct. '45 "Población de la República de Nicaragua, Calculada hasta el 31 de Diciembre de 1944." Distribuida por grupos de edades y por departamentos. Cálculos basados en los resultados del censo de población de 1940. Año 2, no. 17-18, septiembre-octubre 1945, p. 7.

July, '45 "Estadística Vital, Año 1944." Año 2, tomo 2, no. 15, julio 15, 1945, p. 12-20.

Vital statistics, 1944. Includes demographic indexes, marriages, births and mortality.

Incluye índices demográficos, matrimonios, nacimientos y mortalidad.

June 15, '45, a "Balanza de Comercio, Año 1944." Año 2, no. 14, junio 15, 1945, p. 17-18.

Trade balance, 1944.

June 15, '45, b "Inmigrantes, Según Nacionalidad, Edades y Sexo, Año de 1944." Año 2, no. 14, junio 15, 1945, p. 27.

Immigrants in 1944, by nationality, age and sex. Statistical table.
Cuadro estadístico.

June 15, '45, c "Movimiento Migratorio, Año 1944." Año 2, no. 14, junio 15, 1945, p. 26.
Migration, 1944. Statistical table.
Cuadro estadístico.

June 15, '45, d "Movimiento Migratorio Registrado en la República Según Procedencia y Destino, Año 1944." Año 1, no. 14, junio 15, 1945, p. 26.
Migration registered in the Republic, by origin and destination, 1944. Statistical table.
Cuadro estadístico.

May 15, '45 "Métodos Usados en los Registros del Estado Civil de las Personas y Estadística Vital." Año 2, no. 13, mayo 15, 1945, p. 8-9, 19.
Methods used in civil status registration and vital statistics.

Apr. 15, '45, a "Actividades de la Población de Nicaragua." Año 2, no. 12, abril 15, 1945, p. 7-15.
1940 population census results for occupation. Includes tables on: Economically active population; economically inactive population; population distributed by departments according to profession and occupation.
Resultados del censo de población de 1940 sobre ocupación. Incluye tablas: Población economicamente activa; población economicamente inactiva; población distribuida por departamentos según profesión, ocupación u oficio.

Sept. 15, '44 "Movimiento Migratorio, Año 1943." Año 1, no. 5, septiembre 15, 1944, p. 21.
Migration, 1943. Statistical table.
Cuadro estadístico.

July 15, '44 "Ley Orgánica de la Estadística Nacional." Año 1, no. 3, julio 15, 1944. 17 p. Reimpresión.
Organic law of national statistics.

June 15, '44, a "Información y Estudio sobre los Aspectos Estadísticos de la Población Extranjera en Nicaragua—1937-1942." Año 1, no. 2, junio 15, 1944, p. 7-9, 18.
Information and study on the statistical aspects of the alien population in Nicaragua, 1937-42.

June 15, '44, b "Ley sobre Estadística de Producción Agrícola." Año I, no. 2, junio 15, 1944, p. 17.
Law on statistics of agricultural production, of August 4, 1941.
Ley del 4 de agosto de 1941.

June 15, '44, c "Primera Ley Republicana que Centraliza la Estadística." Año I, no. 2, junio 15, 1944, p. 17.
First national law centralizing statistics, of March 2, 1881.
Ley promulgada el 2 de marzo de 1881.

June 15, '44, Castro Silva "Registro Nacional de Identidad y del Estado Civil." Por J. M. Castro Silva. Año 1, no. 2, junio 15, 1944, p. 6, 20.
National register of identity and civil status.

July-Dec. '40 "Cifras Preliminares de Población por Provincias y Capitales de Provincias, según el Censo de 1940." No. 20-25, julio-diciembre 1940, p. 57-B.
Preliminary figures of population by provinces and provincial capitals, according to the 1940 census.

NICARAGUA
1900

Nic-**1900.**G5 **La Gaceta.** Diario Oficial. 1896- . Managua.
Official publication containing laws, decrees, and decisions of the Federal government.
Leyes, decretos y resoluciones del gobierno federal.

June 12, '44 "Ley No. 21 de 12 de Junio de 1944, Obligación de Todos los Agricultores de la República . . . " Año XLVIII, no. 120, junio 12, 1944, p. 1034.
Law no. 21 of June 12, 1944, on agricultural statistics.
Ley de estadística agrícola.

Apr. 25, '44 "Ley No. 91 de 25 de Abril de 1944, Créase la Oficina de Estadística Distritorial." Año XLVIII, no. 83, abril 25, 1944, p. 726.
Law no. 91 of April 25, 1944, creating the district statistical office.

Nic-**1950.**N5 **Nicaragua.** Guía general ilustrada. 1940-41— . Frecuencia? Managua. No examinado.
Nicaragua: General illustrated guide. "(First issue) contains . . . data on mining, industrial

and commercial establishments, a list of stock-raisers, and a summary of a land valuation census taken in 1935." (Source: HBLAS.)

"(Primera edición) contiene . . . datos sobre minería, establecimientos comerciales e industriales, una lista de criadores de ganado, y un resumen del censo de valuación de tierras levantado en 1935. (Traducción del comentario.)

2000-2999 CENSUS—CENSOS

Nic-2100.C5 1940.1 **Censo de 1940: Métodos y Procedimientos.** Material usado en el levantamiento del censo de población del 23 de mayo. Adquirido con motivo de la encuesta que el Instituto Interamericano de Estadística realizó sobre Métodos y Procedimientos de los Censos en las Naciones Americanas, 1944. Del Ministerio de Hacienda y Crédito Público, Dirección General de Estadística.

Census of 1940: Methods and procedures. Materials used in taking population census on May 23. Acquired for the Inter American Statistical Institute Survey of Census Methods and Procedures in the American Nations, 1944. Consists of: Instructions; decrees; schedules: forms.

Consiste de: Instrucciones; decretos; cédulas; formularios.

See also—véase también: **Primer Censo General de Población, 1940: Instructivo para Empadronadores: Interpretación de la Boleta de Población.**

Nic-2100.C5 1940.2 **Primer Censo General de Población, 1940: Instructivo para Empadronadores; Interpretación de la Boleta de Población.** Managua, D.C. Tall.Nacionales. Ministerio de Hacienda y Crédito Público. Dirección General de Estadística. 17 p.

First general census of population, 1940: Instructions for enumerators: Interpretation of the population schedule.

Nic-2120.C5 1940 **Primer Censo de Edificios: 18 de Enero de 1940: Cédula.** Managua, 1940. Ministerio de Hacienda y Crédito Público. Dirección General de Estadística. 1p. Forma 1—C.E.

First census of buildings, January 18, 1940: Schedule.

Nic-2150.C3 1940 **Primer Censo Comercial de Nicaragua, 1940: Instructivo.** Interpretación de la Boleta. Managua, Tall. Nacionales, 1940? Ministerio de Hacienda y Crédito Público. Dirección General de Estadística. 37 p.

First industrial census of Nicaragua, 1940: Instructions. Interpretation of the census schedule.

Nic-2150.C5 1940 **Primer Censo Industrial de Nicaragua, 1940: Instructivo para Proporcionar los Datos a los Empadronadores.** Managua, Tall.Nacionales, 1940? Ministerio de Hacienda y Crédito Público. Dirección General de Estadística. 41 p.

First industrial census of Nicaragua, 1940: Instructions for furnishing the data to the enumerators.

Nic-2200.C5 1940.2 **Población de Nicaragua.** Censo de 1940. Managua, 1940. Ministerio de Hacienda y Crédito Público. Dirección General de Estadística. 16 p. Copia máquina.

Population of Nicaragua. Census figures for May 1940 by urban and rural population of municipalities, and estimates for December 1940 and December 1941.

Cifras del censo correspondientes a mayo de 1940, por población urbana y rural de los municipios, y estimaciones para diciembre de 1940 y diciembre de 1941

Nic-2200.C5 1940.3 **Población de Nicaragua, por Sexos, Urbana y Rural (Censo de 1940).** Managua, 1942? Ministerio de Hacienda y Crédito Público. Dirección General de Estadística. 16 p. Copia máquina.

Population of Nicaragua, by sex, urban and rural (1940 census).

Nic-2200.C5 1940.4 **Población según Edades y Sexo (Censo 1940).** Managua, 1942? Ministerio de Hacienda y Crédito Público. Dirección General de Estadística. 16 p. Copia máquina.

Population of Nicaragua, by age and sex, (1940 census).

Nic-2200.C5 1920 **Censo General de 1920.** Managua, Tipo. Nacional, 1920. Ministerio de Hacienda y Crédito Público. Dirección General de Estadística. 327 p. No examinado.

General census of 1920.

*Nic-2220.C5 1940 **Censo de Edificios, 18 de Enero de 1940: Resumen por Departamentos y Municipios.** Urbano y rural. Managua, 1940? Ministerio de Hacienda y Crédito Público. Dirección General de Estadística. 16 p. Copia máquina.

Census of buildings, January 18, 1940: Summary by departments and municipalities, urban and rural.

3000-3999 STATISTICAL SCIENCE—CIENCIA ESTADISTICA

Nic-3000.E5 **Estudios Especiales de la Dirección General de Estadística.**

Fecha de iniciación? Managua. Ministerio de Hacienda y Crédito Público. Dirección General de Estadística.

Special studies of the General Bureau of Statistics. Series of monographs.

Serie de monografías.

Nic-**3030**.L5 1941 **Ley Orgánica de la Estadística Nacional.** Managua, Tall. Nacionales, 1941. 19 p.

Appears also in—aparece también en: **Estadística,** año I, no. 3, julio 15, 1944, p. 22-25 (Nic-**1420**.E5).

Organic law of national statistics.

5000-5999 DEMOGRAPHY—DEMOGRAFIA

*Nic-**5000**.I5 1933 **Indices Demográficos de Nicaragua, 1933-1942.** Managua. Ministerio de Hacienda y Crédito Público. Dirección General de Estadística. 14 p. Mimeografiado. Estudios especiales de la Dirección General de Estadística.

Demographic indexes of Nicaragua, 1933-42.

Nic-**5000**.N5 1945 **Nicaragua: Summary of Biostatistics.** Maps and charts, population, natality and mortality statistics. Washington, D.C., May 1945. Prepared by U. S. Department of Commerce, Bureau of the Census, in cooperation with Office of the Coordinator of Inter-American Affairs. 89 p.

Resumen bioestadístico de Nicaragua. Incluye mapas y cuadros, y datos sobre población, natalidad y mortalidad.

Nic-**5200** **Estadística Vital y Registro en Nicaragua: Métodos y Procedimientos.** Material básico adquirido para la encuesta que el Instituto Interamericano de Estadística realiza sobre los métodos de registro y estadística vital en las naciones americanas, incluyendo el estudio de 1945.

Vital statistics and registration in Nicaragua: Methods and procedures. Background material acquired for the study conducted by the Inter American Statistical Institute in vital statistics and registration methods in the American nations, including the survey of 1945. Includes: Answer to questionnaire; laws; forms; organization chart.

Incluye: Respuesta al cuestionario; leyes; formularios; esquema de organización.

Nic-**5212**.S5 1938-40 **Sección Demográfica: Mortalidad Infantil, 1938-1939-1940.** Informe especial. Managua, agosto 1941. Ministerio de Hacienda y Crédito Público. Dirección General de Estadística. 3 p.

Special report on infant mortality, 1938-40.

6000-6999 ECONOMICS—ECONOMIA

Nic-**6000**.E5 **Economía y Finanzas.** Organo de la Superintendencia de Bancos y de la Dirección General de Estadística. 1943- . Bimestral. Managua. Superintendencia de Bancos, y Ministerio de Hacienda y Crédito Público. Dirección General de Estadística.

Economics and finance. Quarterly publication of the Superintendent of Banks and of the General Bureau of Statistics.

Apr. 15, '44 "Población Indígena en Nicaragua." 15 abril 1944, p. 10.
The Indian population in Nicaragua.

Nic-**6200**.I5 1945 **Informe sobre Labores Agrícolas Presentado por la Delegación de Nicaragua a la Tercera Conferencia Interamericana de Agricultura, en Caracas, Estados Unidos de Venezuela.** Managua, D.N., Guardian & Co. Ltd., 1945. Ministerio de Agricultura y Trabajo. 23 p.

Report on agricultural activities, presented by the Nicaraguan delegation to the Third Inter-American Conference on Agriculture in Caracas, Venezuela.

Nic-**6200**.M5 **Memoria del Ministerio de Agricultura y Trabajo.** Fecha de iniciación? Anual. Managua. Ministerio de Agricultura y Trabajo. No examinado.

Report of the Department of Agriculture and Labor. Contains some statistical data.

Contiene algunos datos estadísticos.

Nic-**6230**.I5 1945 **Informe sobre la Industria Pecuaria Presentada por la Delegación de Nicaragua a la 3a Conferencia Interamericana de Agricultura, en Caracas, EE. UU. de Venezuela.** Managua, D.N., Guardian & Co., Ltd., 1945. Ministerio de Agricultura y Trabajo. 10 p.

Report on the cattle industry, presented by the Nicaraguan delegation to the Third Inter-American Conference on Agriculture in Caracas, Venezuela.

Nic-**6300**.E5 **Estadística Minera.** Fecha de iniciación? Anual? Managua. Ministerio de Hacienda y Crédito Público. Dirección General de Estadística. Sección Industrial.

Mining statistics.

Nic-**6400**.M5 1945 **Mining and Manufacturing Industries in Nicaragua.** One of a series of reports on mining and manufacturing industries in the American republics. Washington, D.C., 1945. United States Tariff Commission. 11 p.

Folleto sobre Nicaragua, de la serie de informes de la Comisión de Aduanas de los Estados Unidos, sobre las industrias mineras y manufactureras en las repúblicas americanas.

Nic-**6720.**F5 1941-44 **Fondo de Nivelación de Cambios, 1941-1944.** Managua, 1945? Ministerio de Hacienda y Crédito Público. Dirección General de Estadística. 6 p. Mimeografiado. Estudios especiales de la Dirección General de Estadística.

Exchange stabilization fund, 1941-44.

Nic-**6720.**I5 **Informe y Balances del Banco Hipotecario de Nicaragua.** Fecha de iniciación? Anual. Managua, D.N. Banco Hipotecario de Nicaragua.

Published semi-annually before 1941—publicado semestralmente antes de 1941.

Annual report of the Mortgage Bank of Nicaragua. Presents financial, commercial and agricultural trade data.

Presenta datos financieros, comerciales y agrícolas.

Nic-**6720.**R5 **Revista del Banco Nacional de Nicaragua.** Fecha de iniciación? Bimestral. Managua. Banco Nacional de Nicaragua. No examinado.

Journal of the National Bank of Nicaragua. A bi-monthly review of the banking and financial situation. Contains statistical data.

Un análisis bimestral de la situación bancaria y financiera. Contiene datos estadísticos.

Nic-**6720.**S5 1945 **Situación Monetaria, Año 1945.** Managua, mayo 1945. Ministerio de Hacienda y Crédito Público. Dirección General de Estadística. 7 p. Mimeografiado.

Monetary situation, 1945.

Nic-**6900.**L5 1942 **Ley Arancelaria de 1918 de la República de Nicaragua y Sus Reformas hasta el 30 de Junio de 1942.** Managua, Imp.Tall.Robelo, 1942. 358 p.

The customs code of 1918 of the Republic of Nicaragua, and its modifications through June 30, 1942. Modified by laws appearing in the **Gaceta Oficial.** Classification very similar to that of Cuba, Haiti, and Dominican Republic.

Ley arancelaria de Nicaragua. Modificada por leyes que aparecen en la **Gaceta Oficial.** Clasificación muy similar a las de Cuba, Haití y República Dominicana.

Nic-**6900.**M5 **Memoria del Recaudador General de Aduanas y Alta Comisión.** Fecha de iniciación? Anual. Managua. Ministerio de Hacienda y Crédito Público. Administración de Aduanas y Alta Comisión.

Annual report of the Collector of Customs and the High Commission. Commercial and financial customs statistics.

Estadísticas comerciales y financieras de aduanas.

7000-7999 HEALTH AND CULTURAL STATISTICS — ESTADISTICAS DE SALUBRIDAD Y CULTURA

Nic-**7110.**B5 **Boletín Sanitario.** Organo de la Dirección General de Sanidad. 1941-2 . Mensual. Managua. Ministerio de Higiene y Beneficencia Pública. Dirección General de Sanidad.

Health bulletin. Monthly publication of the General Bureau of Health.

Nic-**7110.**I5 **Informe de la Dirección General de Sanidad.** Fecha de iniciación? Anual. Managua. Ministerio de Higiene y Beneficencia Pública. Dirección General de Sanidad.

Annual report of the General Bureau of Health.

Nic-**7130.**M5 1943-44 **Movimiento Registrado en los Hospitales de la República Durante los Años de 1944 y 1943.** Managua, D.N., Tall.Nacionales de Imp. y Encuadernación, 1945. Ministerio de Hacienda y Crédito Público. Dirección General de Estadística. 102 p. Publicaciones especiales de la Dirección General de Estadística.

Registration in the hospitals in Nicaragua, 1943-44. Contains detailed analysis of morbidity in hospitals.

Contiene un análisis detallado de la morbilidad hospitalaria.

8000-8999 GOVERNMENT AND POLITICAL STATISTICS—ESTADISTICAS GUBERNAMENTALES Y POLITICAS

Nic-**8000.**M5 **Memoria de Fomento y Obras Públicas.** Fecha de iniciación? Anual. Managua. Ministerio de Fomento y Obras Públicas.

Annual report of the Department of Development and Public Work. Contains some statistical data.

Contiene algunos datos estadísticos.

Nic-**8800.**D7 1943 **Deuda Pública—Rentas—Bancos—Comercio Exterior, 1943.** Managua, 1945? Ministerio de Hacienda y Crédito Público. Dirección General de Estadística. 60 p. Mimeografiado. Estudios especiales de la Dirección General de Estadística.

Public debt, income, banks, foreign trade, 1943.

Nic-**8800**.M5 **Memoria de Hacienda y Crédito Público.** Fecha de iniciación? Anual. Managua. Ministerio de Hacienda y Crédito Público.

Annual report of the Department of Finance and Public Credit. Contains some statistical tables and charts.

Contiene algunos cuadros y diagramas estadísticos.

Nic-**8800**.M7 **Movimiento Gráfico de los Presupuestos Nacionales, 1931/32 a 1943/44.** Managua. Ministerio de Hacienda y Crédito Público. Dirección General de Estadística.

Graphs of the national budgets, 1931-32 to 1943-44.

Nic-**8800**.P5 **Presupuesto General de Egresos e Ingresos.** Fecha de iniciación? Anual. Managua, D.N. Ministerio de Hacienda y Crédito Público. Sección de Auditoria y Presupuesto.

Yearly national budget of income and expenses for fiscal year, July 1st through June 30th.

Presupuesto del estado para el año fiscal, 1° de julio al 30 de junio.

PANAMA

1000-1999 GENERAL

*Pan-1420.E5 **Estadística Panamena.** Publicación mensual de la Contraloría General. 1941-2 . Panamá. Contraloría de la República. Dirección de Estadística y Censo.

Monthly bulletin of statistics of the Bureau of Statistics and Census.

2000-2999 CENSUS—CENSOS

Pan-2100.C5 1940.1 **Censo de 1940: Métodos y Procedimientos.** Material usado en el levantamiento de los censos electoral y de población del 8 de septiembre. Adquirido con motivo de la encuesta que el Instituto Interamericano de Estadística realizó sobre Métodos y Procedimientos de los Censos en las Naciones Americanas, 1944. De la Contraloría General de la República, Dirección General de Estadística, Oficina del Censo.

Census of 1940: Methods and procedures. Materials used in taking population and electoral censuses, on September 8. Acquired in connection with Inter American Statistical Institute Survey on Census Methods and Procedures in the American Nations, 1944. Consists of: Correspondence; instructions; punch cards and codes; exposition and reports; maps; publications.

Se compone de: Correspondencia: instrucciones; tarjetas de perforación y códigos; exposición e informes; mapas; publicaciones.

See also—véase también: **Cuarto Censo Decenal de Población, 1940: Instrucciones a los Empadronadores e Inspectores; Censo de Población, 1940:** vol. X, **Compendio General.**

Pan-2100.C5 1940.2 **Cuarto Censo Decenal de Población, 1940: Instrucciones a los Empadronadores e Inspectores.** Panamá, Imp.Nacional, 1940. Secretaría de Trabajo, Comercio e Industrias. Oficina del Censo. 35 p.

Fourth decennial population census, 1940: Instructions for the enumerators and inspectors.

Pan-2110 1945 **Censo Agro-Pecuario de las Provincias de Herrera y Los Santos, Diciembre 1945: Métodos y Procedimientos.** Conjunto de los materiales empleados por la Sección de Economía Agrícola del Ministerio de Agricultura, Comercio e Industrias de la República de Panamá, en el levantamiento del censo.

Collection of forms and instructions used in taking the agricultural census of the provinces of Herrera and Los Santos in Panama, December 1945.

Pan-2110.C5 1845;1 **Censo Agro-Pecuario de las Provincias de Herrera y Los Santos, Diciembre 1945: Instrucciones a los Empadronadores.** Panamá, Imp.Nacional, 1945. Ministerio de Agricultura, Comercio e Industrias. Sección de Economía Agrícola. 46 p.

Agricultural and livestock census of the provinces of Herrera and Los Santos, December 1945: Instructions to the enumerators. Schedule folded in.

Lleva un formulario adherido.

Pan-2110.C5 1943.1 **Censo Agro-Pecuario, Diciembre de 1943, Distrito de Penonomé: Métodos y Procedimientos.** Material usado en el levantamiento del censo. Adquirido con motivo del estudio que el Instituto Interamericano de Estadística realiza sobre Métodos y Procedimientos de los Censos en las Naciones Americanas. Del Ministerio de Agricultura y Comercio, Sección de Economía Agrícola.

Agricultural census of the District of Penonomé, December 1943: Methods and procedures. Material used in taking the census. Acquired in connection with Inter American Statistical Institute study of Census Methods and Procedures in the American nations. Consists of: Instructions: schedules.

Se compone de: Instrucciones; cédulas.

*Pan-2200.C5 1940.1 **Censo de Población, 1940.** Panamá, Imp.Nacional, 1943. Contraloría General de la República. Oficina del Censo. 10 vol.: Vol. I, **Provincia del Darién;** vol. II, **Provincia de Panamá;** vol. III, **Provincia de Colón;** vol. IV, **Provincia de Chiriquí;** vol. V, **Provincia de Coclé;** vol. VI, **Provincia de Veraguas;** vol. VII y VIII, **Provincia de los Santos y Herrera;** vol. IX, **Provincia de Bocas del Toro;** vol. X, **Compendio General.**

Population census, 1940. A volume for each province and a summary volume.

*Pan-2210.C5 1943.1 **Censo Agro-Pecuario del Distrito de Penonomé, Diciembre 1943.** Por Juan Rivera Z. y Ofelia Hooper, bajo la direccón de Thomas F. Corcoran. Panamá, The Star and Herald Co., 1945. Ministerio de Agricultura y Comercio. Sección de

Economía Agrícola, y Contraloría General de la República. Dirección General de Estadística. 99 p.

Census of agriculture and livestock in the District of Penonome, December 1943.

*Pan-**2210**.C5 1942 **Censo Agro-Pecuario, 1942.** Por provincias y distritos. Panamá, Cía.Editora Nacional, 1943. Ministerio de Agricultura y Comercio. Sección de Economía Agrícola. 57 p.

Census of agriculture and livestock, 1942, by provinces and districts.

5000-5999 DEMOGRAPHY—DEMOGRAFIA

Pan-**5000**.P5 1945 **Panama: Summary of Biostatistics.** Maps and charts, population, natality and mortality statistics. Washington, D. C., June 1945. Prepared by U. S. Department of Commerce, Bureau of the Census, in cooperation with Office of the Coordinator of Inter-American Affairs. 81 p.

Resumen bioestadístico de Panamá. Incluye mapas y cuadros y datos sobre población, natalidad y mortalidad.

Pan-**5200** **Estadística Vital y Registro en Panamá: Métodos y Procedimientos.** Material básico adquirido para la encuesta que el Instituto Interamericano de Estadística realiza sobre los métodos de registro y estadística vital en las naciones americanas, incluyendo el estudío de 1945.

Vital statistics and registration in Panama: Methods and procedures. Background material acquired for the study conducted by the Inter American Statistical Institute in vital statistics and registration methods in the American nations, including the survey of 1945. Includes: Answer to questionnaire; laws and decrees; instructions; forms; punch cards and codes; exposition.

Incluye: Respuesta al cuestionario; leyes y decretos; instrucciones; formularios; tarj tas perforadas y códigos; exposición.

6000-6999 ECONOMICS—ECONOMIA

Pan-**6000**.E5 1943 **Examen Económico de la República de Panamá para el Año de 1943.** Por Thomas F. Corcoran. Panamá, Estrella de Panamá, septiembre 1944. Contraloría General de la República. Dirección General de Estadística. 19 p.

Economic study of the Republic of Panama for 1943.

*Pan-**6000**.R5 **Revista de Agricultura y Comercio.** Organo oficial del Ministerio de Agricultura y Comercio. 1941- Mensual. Panamá.

Succeeds—sucede a: **Boletín de Trabajo, Comercio e Industrias.**

Official monthly publication of the Department of Agriculture and Commerce. Carries varied current statistical series.

Contiene series estadísticas corrientes.

Apr. '39 "Análisis de los Resultados del Primer Censo Oficial de Empresas Industriales y Comerciales Radicadas en Panamá, Ejecutado en el Año de 1938." **Boletín de Trabajo, Comercio e Industrias,** no. 13, abril, 1939, p. 2-26. No examinado.

Analysis of the results of the first official census of industrial and commercial enterprises in Panama, 1938. "The analysis of the 1938 census, which covered the entire Republic, includes the geographical distribution of industry by type. There are data on citizenship status of proprietors by various characteristics of their business and nationality by industry." (Source: Gen. Cen.V.S.A.)

"El análisis del censo de 1938 que abarca toda la República, incluye la distribución geográfica de la industria, por tipo. Contiene datos sobre ciudadanía de los propietarios por varias características de sus comercios, y nacionalidad por industrias." (Traducción del comentario.)

Pan-**6400**.M5 1945 **Mining and Manufacturing Industries in Panama.** One of a series of reports on mining and manufacturing industries in the American republics. Washington, D. C., 1945. United States Tariff Commission. 13 p.

Folleto sobre Panamá, de la serie de informes de la Comisión de Aduanas de los Estados Unidos, sobre las industrias mineras y manufactureras en las repúblicas americanas.

Pan-**6900** **Estadísticas del Comercio Exterior en Panamá: Métodos y Procedimientos.** Material básico adquirido para el estudio que el Instituto Interamericano de Estadística realiza sobre los métodos de estadísticas del comercio exterior en las naciones americanas, incluyendo la encuesta de 1946-47.

Foreign trade statistics in Panama: Methods and procedures. Background material acquired for the study conducted by the Inter American Statistical Institute in methods of foreign trade statistics in the American nations, including the survey of 1946-47. Consists of: Filled questionnaire and explanatory notes; country code; form.

Consiste de: Cuestionario llenado y notas explicativas; clave de países; formulario.

Pan-6900.A5 1938 **Arancel de Importación de la República de Panamá.** (Ley 69 de 1934.) Edición oficial. Panamá, Imp. Nacional, 1938. 118 y XXVIII p.

Customs tariff for imports of Panama. Classification based on Brussels nomenclature.

Clasificación basada en la nomenclatura de Bruselas.

7000-7999 HEALTH AND CULTURAL STATISTICS — ESTADISTICAS DE SALUBRIDAD Y CULTURA

Pan-7110.B5 **Bio-Estadística—Informes Especiales.** 1944- . Semanal. Panamá. Ministerio de Salubridad y Obras Públicas. Departamento de Salubridad. División de Bio-Estadística y Educación Sanitaria.

Vital statistics—special reports. Weekly.

Pan-7110.E5 **Estadísticas Epidemiológicas.** 1944- . Semanal? Panamá. Ministerio de Salubridad y Obras Públicas. Departamento de Salubridad. División de Bio-Estadística y Educación Sanitaria.

Epidemiological statistics.

Pan-7110.H5 **Higiene y Sanidad.** Organo del Departamento de Salubridad. Sección de Salubridad. 1932?- . Trimestral. Panamá. Ministerio de Salubridad y Obras Públicas. Departamento de Salubridad. Sección de Salubridad.

Hygiene and health. Organ of the Department of Health, Section of Public Health.

Pan-7110.M5 **Memoria del Ministerio de Salubridad y Obras Públicas.** Fecha de iniciación? Anual. Panamá.

Report of the Department of Health and Public Works. Includes the following reports: Section of Hygiene and Welfare; Section of Roads; Section of Design and Construction; Section of Aqueducts and Sewers; Administrative Section and National Railroad of Chiriqui. The annex to the 1938 edition contains a compilation of resolutions and decrees.

Incluye memorias de: Sección de Higiene y Beneficencia; Sección de Caminos; Sección de Diseños y Construcciones; Sección de Acueductos y Alcantarillados; Sección Administrativa y Ferrocarril Nacional de Chiriquí. El anexo a la edición de 1938 contiene una compilación de resoluciones y decretos.

*Pan-7210.E5 **Estadística Cultural.** Publicación anual. 1943-44— . Anual. Panamá. Ministerio de Educación. Departamento de Estadística y Archivos.

Annual compilation of educational statistics of the Minister of Education.

1945-46 **Estadística Cultural (Publicación Anual) Número 3, Año Escolar 1945-1946.** Panamá, Imp. Nacional, 1946. Ministerio de Educación. Departamento de Estadística y Archivos.

Contains data presented in tables and charts on the failures, promotions attendance, cost of teaching, scale of salaries, matriculation, teachers, etc. in the following grades: Kindergarten (public and private); primary (public and private); secondary (public and private); university (public).

Contiene datos presentados en cuadros y tablas sobre los aplazados, los promovidos, la asistencia, el costo de la enseñanza, la escala de sueldos, la matrícula, los profesores, etc. en las siguientes: Enseñanza pre-escolar (oficial y privada); enseñanza primaria (oficial y privada); enseñanza post-primaria (oficial y privada); enseñanza universitaria (oficial).

8000-8999 GOVERNMENT AND POLITICAL STATISTICS — ESTADISTICAS GUBERNAMENTALES Y POLITICAS

Pan-8800.I5 **Informe del Contralor General de la República.** Fecha de iniciación? Anual. Panamá. Contraloría General de la República.

Annual report of the Comptroller General of the Republic.

9000-9999 LABOR AND LIVING CONDITIONS—CONDICIONES DEL TRABAJO Y DE VIDA

Pan-9950.B5 **Boletín de Previsión Social.** 1946- . Trimestral. Panamá. Ministerio de Trabajo, Previsión Social y Salud Pública.

Bulletin of social welfare. Each issue contains an organization chart of the Department of Labor, Social Welfare and Public Health.

Cada edición contiene un esquema de organización del Ministerio de Trabajo, Previsión Social y Salud Pública.

PARAGUAY

1000-1999 GENERAL

Par-**1200**.S5 1943 **Strategic Index of the Americas: Preliminary Bibliography of Paraguay.** Washington, D. C., 1943. U. S. Office for Emergency Management. Coordinator of Inter-American Affairs. Research Division. 32 p.

Titles grouped under headings: General and miscellaneous; geography; economics; social conditions; immigration and colonization; Indians.

Indice estratégico de las Américas: Bibliografía preliminar del Paraguay. Títulos agrupados bajo los siguientes encabezamientos: Generales y varios; geografía; economía; condiciones sociales; inmigración y colonización; indios.

Par-**1400**.P5 1939 **Paraguay, Datos y Cifras Estadísticas.** Población, producción, exportación, industrias, vialidad, comercio, instrucción pública. Asunción, 1939. Ministerio de Industria y Comercio. 30 p.

Paraguay, statistical data. Population, production, exports, industries, roads, business, public education.

*Par-**1410**.A5 **Anuario Estadístico de la República del Paraguay.** 1888- . Asunción. Ministerio de Hacienda. Dirección General de Estadística.

Title varies—título varía: 19 -1939, **Memoria de la Dirección General de Estadística.**

Statistical yearbook of Paraguay.

1942 **Anuario Estadístico de la República del Paraguay, 1942.** Asunción, Imp.Nacional, 1944. Ministerio de Hacienda. Dirección General de Estadística. 265 p.

Includes: Climatology; demography; economics and finance (foreign trade, agriculture, livestock, industry, money, banks and real estate, prices of articles of prime necessity and computation of the cost of living, public finance); communications; public health; public education; judicial, police and prison statistics; statistics of municipalities.

Incluye: Climatología; demografía; economía y finanzas (comercio exterior, agricultura, ganadería, industria, emisión, bancos y registro de la propiedad, precios de los artículos de primera necesidad y cálculo del costo de la vida, finanzas públicas); comunicaciones; salúd pública; instrucción pública; movimiento judicial, policial y carcelario; estadísticas de municipalidades.

1937, a "Censo de Población." **Memoria de la Dirección General de Estadística . . . 1937,** p. 7-13.

Exposition on the census of population taken in 1936. Includes a memorandum from the provisional president of the Republic listing the deficiencies and obstacles in the preparation and execution of the work, p. 10-11.

Exposición sobre el censo de población de 1936. Incluye un memorándum del presidente provisional de la República sobre los obstáculos que se presentaron en la preparación y la realización del trabajo, p. 10-11.

1936, a "Censo de Población." **Memoria de la Dirección General de Estadística . . . 1936,** p. 7-14.

Exposition and report on the census of population of 1936, including legislation authorizing the census.

Exposición e informe sobre el censo de población de 1936, incluyendo los decretos autorizando dicho censo.

1936, b "Ley de Estadística." **Memoria de la Dirección General de Estadística . . . 1936,** p. 14-17.

Law regulating the General Bureau of Statistics of October 4, 1935.

Reglamento de la Dirección General de Estadística. Incluye "Ley N. 1.492 (Octubre 4 de 1935): Dirección de Estadística."

*Par-**1410**.B5 **Boletín Semestral de la Dirección General de Estadística.** 1915- . Asunción. Ministerio de Hacienda. Dirección General de Estadística.

Prior to 1936 issued quarterly—antes de 1936 publicábase trimestralmente.

Semi-annual bulletin of the General Bureau of Statistics. Contains data on foreign trade, population, communications, health, climate, and other subjects.

Contiene datos sobre comercio exterior, población, comunicaciones, sanidad, clima y otros.

2000-2999 CENSUS—CENSOS

Par-**2100**.C5 1936.1 **Censo de 1936: Métodos y Procedimientos.** Material usado en el levantamiento del censo de población. Adquirido con motivo de la encuesta que el Instituto Interamericano de Estadística realizó sobre Métodos y Procedimientos de los Censos en las Naciones Americanas, 1944. Del Ministerio de Hacienda, Dirección General de Estadística.

Census of 1935: Methods and procedures. Materials used in taking population census. Acquired in connection with Inter American Statistical Institute Survey of Census Methods and Procedures in the American Nations, 1944. Consists of: Correspondence; laws, decrees, or other authorization; instructions; schedules; forms; reports.

Consiste de: Correspondencia; leyes, decretos u otras disposiciones; instrucciones; cédulas; otros formularios; informes.

See also—véase también: **Libreta de Familia** (registro del estado civil de las personas) (Par-**5250**.L5, 1944).

Par-**2100**.C5 1936.2 **Censo de Población, Decreto No. 2120 de 18 de Junio de 1936.** Que declara carga pública para todos los ciudadanos del país los trabajos del censo de población. Asunción, junio 1936. 3 p. Mimeografiado.

Census of population. Decree of June 1936, declaring the work of the population census a public duty for all citizens.

Par-**2100**.C5 1936.3 **Decreto No. 60.287 de 21 de Noviembre de 1935.** Que faculta a la Dirección General de Estadística a proceder al levantamiento del censo de población de la República. Asunción, noviembre 1935. 1 p. Mimeografiado.

Decree of November 1935, authorizing the General Bureau of Statistics to take the population census of Paraguay.

Par-**2100**.C5 1936.4 **Ley No. 1.509 de 1° de Noviembre de 1935.** Por la que se autoriza al poder ejecutivo a invertir hasta 10.000.000.—de pesos de curso legal en la formación del censo de la República. Asunción, noviembre 1935. 1 p. Mimeografiado.

Law of November 1935, authorizing the Executive to spend up to 10,000,000 pesos of legal currency for taking a census of Paraguay.

Par-**2110**.C5 1944.5 **Censo Familiar: Cédula.** Departamento de Caraguatay, Partido de Piribebuy. Asunción?, La Colmena, S. A., 1944? 1 p.

Schedule of a family census taken in the district of Piribebuy in 1945, in connection with a nutrition survey made by the Food Supply Division of the Office of Inter-American Affairs.

Cédula del censo familiar levantado en el Partido de Piribebuy en 1945, en conexión con un estudio de alimentación realizado por la División de Provisión de Alimentos, de la Oficina de Asuntos Interamericanos.

3000-3999 STATISTICAL SCIENCE—CIENCIA ESTADISTICA

Par-**3030**.D5 1942 **Decreto Ley 11.126 que Reorganiza y Coordina los Servicios Estadísticos de la República.** Asunción, Imp. Nacional, 1942.

Law on reorganization and coordination of the statistical services in the Republic, 1942.

Par-**3300** 1944-46 **Enseñanza Estadística en Paraguay: Métodos y Materiales.** Material básico adquirido para la Encuesta del Instituto Interamericano de Estadística sobre los Métodos y Materiales en la Enseñanza Estadística en las Naciones Americanas, 1944-46.

Statistical training in Paraguay: Methods and materials. Background material acquired for the IASI Survey on Statistical Training Methods and Materials in the American Nations, 1944-46. Includes: Letter and filled questionnaire on statistical courses in Paraguay.

Incluye: Carta y cuestionario llenado sobre cursos de estadística en Paraguay.

5000-5999 DEMOGRAPHY—DEMOGRAFIA

Par-**5000**.P5 1944 **Paraguay: Summary of Biostatistics.** Maps and charts, population, natality and mortality statistics, Washington, D. C., October 1944. Prepared by U. S. Department of Commerce, Bureau of the Census, in cooperation with Office of the Coordinator of Inter-American Affairs. 66 p.

Resumen bioestadístico de Paraguay. Incluye mapas y cuadros y datos sobre población, natalidad y mortalidad.

Par-**5200** **Estadística Vital y Registro en Paraguay: Métodos y Procedimientos.** Material básico adquirido para la encuesta que el Instituto Interamericano de Estadística realiza sobre los métodos de registro y estadística vital en las naciones americanas, incluyendo el estudio de 1945.

Vital statistics and registration in Paraguay: Methods and procedures. Background material acquired for the study conducted by the Inter American Statistical Institute in vital statistics and registration methods in the American nations, including the survey of 1945. Includes: Answer to questionnaire; laws and decrees; instructions; forms; punch cards and codes.

Incluye: Respuesta al cuestionario; leyes y decretos; instrucciones; formularios; tarjetas de perforación y códigos.

Par-**5250**.L3 1940 **Leyes del Registro del Estado Civil y Decretos Reglamentarios.** Asunción, Imp.Nacional, 1940. 92 p.

Laws on the register of civil status, and regulatory decrees.

Par-**5250**.L5 1944 **Registro del Estado Civil de las Personas: Libreta de Familia.** Asunción, Imp.Nacional, 1944. 24 p.

Booklet for the family, in which are registered marriages, births, deaths, and other acts listed in register of civil status. Includes most important legal dispositions of laws of register of civil status.

"La 'Libreta de Familia' estará destinada a registrar los matrimonios, nacimientos, defunciones y demás actos que se hayan inscrito en los Registros del Estado Civil . . . Contendrá impresas las disposiciones más importantes de las Leyes . . . del Registro del Estado Civil . . ." (Fuente: p. 4.)

6000-6999 ECONOMICS—ECONOMIA

Par-**6000**.A5 1938-39 **Anuario del Ministerio de Economía, 1938-39.** Asunción. 546 p.

Only one issue has appeared—solamente una edición ha sido publicada.

Yearbook of the Department of Economy. Includes: Demography; cost of living index; banking movements; foreign trade; education; etc.

Incluye: Demografía; índice numérico del costo de vida; movimiento bancario; comercio exterior; instrucción pública; etc.

Par-**6000**.B5 **Boletín del Ministerio de Agricultura, Comercio e Industrias.** 1940-43. Trimestral. Asunción. Ministerio de Agricultura.

No longer published—cesado su publicación. Superseded by—substituido por: **Paraguay Industrial y Comercial** and **Revista de Agricultura y Ganadería.** Title varied—título varió: June-junio '43, **Agricultura, Comercio e Industrias.**

Bulletin of the Department of Agriculture, Commerce and Industries. Contained statistics of agricultural and livestock production, foreign trade, etc.

Contenía estadísticas sobre producción agrícola y ganadera, comercio exterior, etc.

Par-**6000**.E5 1945 **Economic Controls and Commercial Policy in Paraguay.** One of a series of reports on economic controls and commercial policy in the American republics. Washington, D.C., 1945. United States Tariff Commission. 23 p.

Folleto sobre Paraguay, de la serie de informes de la Comisión de Aduanas de los Estados Unidos, sobre controles económicos y política comercial de las repúblicas americanas.

*Par-**6000**.P5 **Paraguay Industrial y Comercial.** Organo oficial del Ministerio de Industria y Comercio. 1944- Mensual? Asunción.

Industrial and commercial Paraguay. Official publication of the Department of Industry and Commerce.

May '45 "Comercio Exterior del Paraguay, Año 1944." Año I, no. 9, mayo 1945, p. 1-3.

Foreign commerce of Paraguay, 1944.

Par-**6000**.R5 1944 **Report Presented to the Conference of Commissions of Inter-American Development (Held in New York, May 9-18, 1944) by the Paraguayan Commission of Inter-American Development.** On economic problems of Paraguay. Washington, D. C., 1946? Inter-American Development Commission. 20 p.

Cover title—título de la cubierta: **Paraguay: Report Presented to the Conference of Commissions of Inter-American Development.**

Informe presentado a la Conferencia de Comisiones de Fomento Interamericano (realizada en N. York, del 9 al 18 de mayo de 1944) por la Comisión Paraguaya de Fomento Interamericano sobre problemas económicos del Paraguay.

Par-**6010**.P5 1943 **Paraguay, a Study of Price Control, Cost-of-Living and Rationing.** Washington, D.C., March 1943. U. S. Office of Price Administration. Foreign Information Branch. 43 p. Mimeographed.

Paraguay, un estudio sobre control de precios, costo de la vida y racionamiento, por la Oficina de Administración de Precios de los Estados Unidos.

*Par-**6200**.R5 **Revista de Agricultura y Ganadería.** 1944- . Mensual. Asunción. Ministerio de Agricultura.

Journal of agriculture and livestock.

Par-**6200**.T5 1945 **Technical Mission to Paraguay: Reports on the Forests, the Forest Industries, and Trade in Forest Products, Hides and Leather Industries, Edible Vegetable Oils Industries of Paraguay.** Washington, D.C., 1945. Inter-American Development Commission. 166 p.

Includes tables on production, exports, prices, etc., of the industries studied.

Delegación técnica al Paraguay de la Comisión de Fomento Interamericano: Informes sobre bosques, industria forestal y comercio de productos forestales, industria del cuero y de las pieles, industria del aceite comestible vegetal en Paraguay. Incluye cuadros sobre producción, exportación, precios, etc. de las industrias estudiadas.

Par-**6400** **Estadísticas Industriales en Paraguay: Métodos y Procedimientos.** Material básico adquirido para el estudio que el Instituto Interamericano de Estadística realiza sobre los métodos de estadísticas industriales en las naciones americanas, incluyendo la encuesta de 1945-46.

Industrial statistics in Paraguay: Methods and procedures. Background material acquired for the study conducted by the Inter American Statistical Institute in methods of industrial statistics in the American nations, including the survey of 1945-46. Consists of: Correspondence; publications.

Consiste de: Correspondencia; publicaciones.

Par-**6400**.C5 **Corporación Paraguaya de Alcoholes.** 1943?- . Anual. Asunción. Corporación Paraguaya de Alcoholes.

Annual report of the Paraguayan Alcohol Corporation.

Memoria de la Corporación Paraguaya de Alcoholes.

1943 **Corporación Paraguaya de Alcoholes.** Asunción, Tall. Gráf.Cándido Zamphirópolos, 1944. Corporación Paraguaya de Alcoholes. 100 p.

Report of the Paraguayan Alcohol Corporation for the year 1943. Includes data on production and deliveries of alcohol and cane in 1943, and quotas for production for 1944.

Memoria del ejercicio 1943 de la Corporación Paraguaya de Alcoholes. Incluye datos sobre producción y entregas de alcoholes y caña en 1943, y cupos de producción para el año 1944.

Par-**6400**.M5 **Mining and Manufacturing**
1945 **Industries in Paraguay.** One of a series of reports on mining and manufacturing industries in the American republics. Washington, D.C., 1945. United States Tariff Commission. 15 p.

Folleto sobre Paraguay, de la serie de informes de la Comisión de Aduanas de los Estados Unidos, sobre las industrias mineras y manufactureras en las repúblicas americanas.

*Par-**6720**.M5 **Memoria.** Fecha de iniciación? Anual. Asunción. Banco del Paraguay.

Annual report of the Bank of Paraguay.

1945 **Memoria Correspondiente al Ejercicio 1945.** Asunción, La Colmena S. A., 1946? Banco del Paraguay. 110 p.

Economic and financial development of the country during 1945, and balance of the bank for this period.

Desenvolvimiento económico-financiero del país durante el período, y balance del Banco correspondiente al mismo.

Par-**6900**.C5 **Consolidación de los Adi-**
1940 **cionales ad-Valorem de Importación, Decreto N. 105, Vigente desde el 15 de Marzo de 1940; Lista de los Cambios Efectuados en los Derechos de la Tarifa y Arancel de Aduanas, Ley N. 667; Nueva Tarifa de Exportación, Decreto N. 383 Vigente desde el 25 de Marzo de 1940.** Asunción, Imp. Nacional, 1940. 16 p.

Consolidation of import surtaxes, decree no. 105, effective March 15, 1940; list of tariff changes, law 667; new export tariff, decree no. 383, effective March 25, 1940.

Par-**6900**.I5 **Indice General por Orden Alfa-**
1924 **bético de la Tarifa y Arancel de Aduanas** (**Ley N. 667 de 27 de Septiembre de 1924**). Asunción, Imp.Nacional, 1925. 16 p.

General alphabetic index of the customs tariff schedule (law no. 667 of Sept. 27, 1924).

Par-**6900**.L5 **Ley N. 667, de Tarifa y Arancel**
1924 **de Aduanas, Septiembre 27 de 1924.** Asunción, Imp.Nacional, 1925. 73 p.

Customs tariff, law no. 667 of September 27, 1924. Customs code of Paraguay; includes export section which was later superseded. Modifications appear in **Gaceta Oficial.** Classification like that of Ecuador tariff.

Leyes aduaneras del Paraguay; incluye sección de exportación que mas tarde fue substituida. Las modificaciones aparecen en la **Gaceta Oficial.** La clasificación es similar a la del Ecuador.

7000-7999 HEALTH AND CULTURAL STATISTICS—ESTADISTICAS DE SALUBRIDAD Y CULTURA

*Par-**7110**.B5 **Boletín del Ministerio de Salud Pública y Previsión**

Social. 1940- . Semestral. Asunción.
Bulletin of the Department of Public Health and Social Welfare.

Par-7110.M5 **Memoria de la Sección Estadística.** Fecha de iniciación? Anual. Asunción. Ministerio de Salud Pública.
Title varies—título varía: 1938, **Memoria de la Sección Estadística del Ministerio de Salud Pública.**
Annual report of the Statistical Section of the Department of Public Health. Includes tables on natality, mortality, and marriages, and summary of the work carried on during the year by the different divisions of the department.
Incluye cuadros sobre natalidad, mortalidad y nupcialidad, y resumen de los trabajos efectuados por las distintas dependencias del ministerio durante el año.

Par-7200.R5 **Revista de Educación.** 1938- . Frecuencia? Asunción. Ministerio de Educación. No examinado.
Review of education, published by the Department of Education.

8000-8999 GOVERNMENT AND POLITICAL STATISTICS—ESTADISTICAS DE SALUBRIDAD Y CULTURA

Par-8000.M5 **M. O. P.** 1942- . Irregular. Asunción. Ministerio de Obras Públicas y Comunicaciones.
Bulletin of the Department of Public Works and Communications.

Par-8800.B3 **Boletín de la Dirección General de Impuestos Internos.** 1941- . Trimestral. Asunción. Dirección General de Estadística.
Report of the General Bureau of Internal Revenue, including data, laws, etc. of the divisions.
Informe de la Dirección General de Impuestos Internos, incluyendo datos, leyes, etc. de las secciones.

Par-8800.B5 **Boletín del Tesoro.** 1915- . Trimestral? Asunción. Contaduría General y Dirección del Tesoro.
Bulletin of the Treasury. Data on public resources, expenditures, income, budget, public debt, etc.
Datos sobre recursos, gastos, renta, presupuesto, deuda pública, etc.

PERU

1000-1999 GENERAL

*Perú-**1410**.E5 **Anuario Estadístico del Perú.** 19- . Anual. Lima. Ministerio de Hacienda y Comercio. Dirección Nacional de Estadística.

Title varies—título varía: 19 -1943: **Extracto Estadístico del Perú.**

Statistical yearbook of Peru.

1944-45, a **Anuario Estadístico del Perú, 1944-1945: Fascículo: Geografía—Demografía—Salud Pública y Asistencia Social.** Lima, Imp.Americana, 1946. Ministerio de Hacienda y Comercio. Dirección Nacional de Estadística. XXXVI y 146 p.

Statistical yearbook of Perú, 1944-45: Fascicle: Geography, demography, public health and social assistance. "The present 'Statistical Yearbook of Peru' contains some substantial innovations in comparison with the 'Statistical Abstract of Peru', of which it is the continuation." (Translation of comment.)

"El presente 'Anuario Estadístico del Perú' contiene algunas innovaciones sustanciales respecto del 'Extracto Estadístico del Perú', del cual es continuación." (Fuente: Introducción.)

1943 **Extracto Estadístico del Perú, 1943.** Lima, Imp.Americana, 1945? Ministerio de Hacienda y Comercio. Dirección Nacional de Estadística. CXLIV y 734 p.

Contents: Territory; population; vital statistics; migration; social assistance (health, hospital assistance, and food); agriculture, livestock, forestry, fishery; mining; manufacturing industries; transportation and communication; foreign commerce; domestic commerce; money, banking, and insurance; labor; finance; administration; justice, crime, and police; education and culture; religion.

Contiene: Territorio; población; movimiento de la población; migraciones; asistencia social (salubridad y asistencia hospitalaria y alimentación); agricultura, ganadería, silvicultura y pesca; minería; industrias de transformación; transportes y comunicaciones; comercio exterior; comercio interior; moneda, bancos y seguros; trabajo; finanzas; administración; justicia, criminalidad y hechos de policía; educación y cultura; culto.

1942, a "Decreto Supremo de 20 de Agosto de 1943 Que Ordena Formar el Inventario del Potencial Económico de la Nación." **Extracto Estadístico del Perú, 1942,** p. 592-593. (Anexo no. 9.)

Decree of August 20, 1943, instructing the National Bureau of Statistics to undertake a special investigation of the economic potential of Peru, including raw materials of agriculture, cattle, mining, industry and any other class.

Ley que ordena a la Dirección Nacional de Estadística a realizar una investigación especial acerca del potencial económico del Perú, inclusive materias primas, de carácter agrícola, ganadero, minero, industrial y de otro orden.

1942, b "Decreto Supremo de 1° de Enero de 1944, por el Que se Organiza el Servicio de Estadística Nacional y Crea el Consejo Superior de Estadística." **Extracto Estadístico del Perú, 1942,** p. 588-590. (Anexo no. 6.)

Decree of January 1, 1944, organizing, national statistical service and creating Superior Council of Statistics.

1942, c "Encuesta sobre Presupuestos Familiares Obreros, en 1940, Correspondiente a Lima (Cercado)." **Extracto Estadístico del Perú, 1942,** p. 547-553.

Survey on working class family budgets in the metropolitan area of Lima in 1940. Tables on the nutrition of a group of 75 working class families; last statistical tables of the survey undertaken by the school of Economy and Finance of the University of San Marcos. Includes: Weekly expenses of the families, by income and food products; distribution of the expenses; average expenses.

Cuadros relativos a la alimentación del grupo de 75 familias obreras; últimos cuadros estadísticos de la encuesta realizada por el Seminario de Economía y Finanzas de la Universidad Mayor de San Marcos. Incluye: Gastos semanales de las familias, por ingresos y productos alimenticios; distribución de los gastos; promedio de gastos.

1942, d "Ley No. 7567, de 5 de Agosto de 1932, que Transforma en Nacional la Dirección de Estadística y

Establece las Normas para el Funcionamiento del Servicio de Estadística Nacional." **Extracto Estadístico del Perú, 1942,** p. 581-585. (Anexo no. 1.)

Law no. 7567 of August 5, 1932, transforming the Bureau of Statistics into a National Bureau and establishing rules for the functioning of the national statistical service.

1942, e "Producción de Energía." **Extracto Estadístico del Perú, 1942,** p. 232-234.

Production of energy. Includes tables on: Water concessions for power purposes, 1923-42; water concessions for power purposes, 1940-42, by departments; number and generating capacity of electric plants, 1940, by departments; generating capacity of electric plants, 1940, by departments and class of use.

Incluye (cuadros): Concesiones de agua para fuerza motriz, 1923-1942; concesiones de agua para fuerza motriz, 1940-1942, por departamentos; centrales eléctricas, número y capacidad generadora, en 1940, por departamentos; centrales eléctricas, capacidad generadora, en 1940, por departamentos y clase de explotación.

1942, f "Resolución Suprema que Aprueba el Diagrama de Organización del Servicio de Estadística Nacional." **Extracto Estadístico del Perú,** 1942, p. 590-591. (Anexo no. 7.)

Resolution approving the organization chart of the national statistical service. Regulation establishing the departments, sections, groups and offices of the national statistical service.

Regulación que establece los departamentos, secciones, grupos y oficinas del servicio de estadística nacional.

1940, a "Encuesta sobre Presupuestos Familiares Obreros, en 1940, Correspondiente a Lima (Cercado)." **Extracto Estadístico del Perú, 1940,** p. 720-727. No examinado.

Survey of working class family budgets in Lima, 1940. "First statistical tables of . . . the investigation undertaken by the seminar of economics and finance of the University of San Marcos." (Translation of comment.)

"Los primeros cuadros estadísticos de . . . la encuesta realizada por el Seminario de Economía y Finanzas de la Universidad Mayor de San Marcos . . ." (Fuente: **Extracto Estadístico del Perú, 1942,** p. 547.)

1933 **Extracto Estadístico y Censo Electoral de la República.** Lima, Tall. de Linotipía, 1933. Dirección Nacional de Estadística. Servicio de Estadística Electoral. 232 p.

Electoral population of the Republic by districts, provinces, and sections, classified by race, grade of instruction, civil status, profession, age, and height.

Población electora de la República por distritos, provincias, y departamentos, clasificada por razas, grado de instrucción, estado civil, profesiones, edades, y estaturas.

*Perú-**1420**.B4 **Boletín de la Dirección Nacional de Estadística.** 1928- . Mensual. Lima. Ministerio de Hacienda y Comercio. Dirección Nacional de Estadística.

Monthly bulletin of the General Bureau of Statistics. National statistics in summary form on population, commerce, agriculture, mining, money and banking, and other fields.

Presenta en forma sumaria estadísticas sobre población, comercio, agricultura, minería, moneda y operaciones bancarias, etc.

2000-2999 CENSUS—CENSOS

*Perú-**2200**.C5 1940.1 **Censo Nacional de Población y Ocupación, 1940.** Lima, Imp. Torres-Aguirre, 1943- . Ministerio de Hacienda y Comercio. Dirección Nacional de Estadística. vol.: Vol. I, **Resúmenes Generales.**

Census of population and occupation, 1940. Volume I, general summary, contains: Exposition and comments; graphs; **statistical tables:** Population and territory; age; sex; civil status; fecundity; race; language; education; religion; economically active and inactive population; place of birth; defects and infirmities; families; housing; foreigners; **annexes:** Law no. 8695, July 1, 1938, authorizing the census law no. 7567, August 5, 1932, which makes the Bureau of Statistics national and establishes rules for its functioning; regulation of law no. 8695, and other decrees, acts, and resolutions; organization chart of agencies charged with preparation and execution of census of 1940; forms and schedules; instructions for enumerators; occupational code; cost of the census.

Vol. I contiene: Exposición y comentarios; gráficos; **cuadros estadísticos:** Población y territorio; edad; sexo; estado civil; fecundidad de las madres; raza; idiomas; instrucción; religión; población económicamente activa e inactiva; lugar de nacimiento; defectos y enfermedades; familias; viviendas; extranjeros; **anexos:** Ley No. 8695, de 1o. de julio de 1938, que manda levantar el censo nacional de población; Ley No. 7567, de 5 de agosto de 1932, que transforma en nacional la Dirección de Estadística y establece las normas para el funcionamiento del servicio de estadística nacional; reglamento de la Ley No. 8695, y otros decretos, actas y resoluciones; diagrama de los organismos encargados de la preparación y ejecución del censo nacional de 1940; formularios y cédulas; instrucciones al empadronador; clave de ocupaciones; costo del censo.

Perú-**2200**.C5 1940.2 **Censo Nacional de 1940: Resultados Generales.** Primer informe oficial. Lima, 1941. Ministerio de Hacienda y Comercio. Dirección Nacional de Estadística. Departamento de Censos. 68 p.

National census of 1940: General results. First official report. Contents: Census decrees and plans, organization and execution of the census, plans of analysis and publication; general results; estimates of omissions; summary of population increase; analytical study of the composition of the population; age, sex and race for departments; the census and the control of national population statistics.

Contenido: Decretos y planes para el censo, organización y ejecución del censo, planes de análisis y publicación; resultados generales; estimaciones de las omisiones; resumen del aumento de población; estudio analítico de la composición de la población; edad, sexo y raza por departamentos; el censo y el control de la estadística nacional de población.

Perú-**2200**.C5 1940.4 **Estado de la Instrucción en el Perú según el Censo Nacional de 1940.** Informe especial. Lima, Imp.Torres Aguirre, 1942. Ministerio de Hacienda y Comercio. Dirección Nacional de Estadística. 67 p.

State of education in Peru according to the national census of 1940. The school age population is given by regions.

La población de edad escolar está dada por regiones.

3000-3999 STATISTICAL SCIENCE—CIENCIA ESTADISTICA

*Perú-**3040**.E5 **Estadística Peruana.** Publicación trimestral del Instituto Peruano de Estadística. 1945- . Lima.

Peruvian statistics. Quarterly publication of the Peruvian Statistical Institute. Articles and notes on statistical theory and on statistics in Peru.

Artículos y notas sobre teoría estadística y sobre estadística en el Perú.

July '46, Barreto "Naturaleza y Alcances de la Llamada Balanza Internacional de Pagos." Por Emilio G. Barreto. Año II, no. 4, julio 1946, p. 62-66.

Fundamental character and scope of the international balance of payments.

July '46, García Frías "Ensayo de Tabla de Mortalidad para los Habitantes de Lima, Ciudad Capital." Por Roque García Frías. Año II, no. 4, julio 1946, p. 85-102.

Mortality tables for the inhabitants of Lima.

July '46, Kürbs "Síntesis Estadística de los Problemas Económicos de los Estados Unidos de América en la Guerra y Postguerra". Por Friedrich Kürbs. Año II, no. 4, julio 1946, p. 9-22.

A statistical synthesis of the economic problems of the United States in war and post war.

July '46, Luna Vegas "Apuntes sobre la Estadística en los Estados Unidos." Por Ricardo Luna Vegas. Año II, no. 4, julio 1946. p. 23-26.

Notes on statistics in the United States.

July '46, Rose Jibaja "Aplicabilidad en la Agricultura Peruana del Método Estadístico de los 'Ejemplares-Tipo' (1)." Por Jorge Rose Jibaja. Año II, no. 4, julio 1946, p. 27-49.

The applicability of the statistical method of sampling in Peruvian agriculture.

July '46, Téllez Sarzola "Las Enfermedades Infecto-Contagiosas, Casos por Departamentos en 1945 y en la República en 1939-1945." Por Oscar Téllez Sarzola. Año II, no. 4, julio 1946, p. 56-61.

Cases of contagious diseases reported, by departments for 1945, and for the entire country from 1939 to 1945.

July '46, Velarde B. "Estadística Económica Militar." Por César A. Velarde B. Año II, no. 4, julio 1946, p. 50-55.

Economic military statistics. Article deals principally with industrial military statistics.

El artículo se refiere principalmente a la estadística industrial militar.

Nov. '45, Arca Parró Perú. "El Medio Geográfico y la Población del Perú." Conclusión. Por

Alberto Arca Parró. Año I, no. 3, noviembre 1945, p. 7-36.

The geographic setting and the population of Peru. Includes: The geographical regions and the political-administrative division; the geographical regions and the national population process.

Incluye: Las regiones geográficas y la división político-administrativa; las regiones geográficas y el proceso nacional de la población.

Nov. '45, García Frías "Determinación Estadística Matemática de la Población Electoral del Perú en Junio de 1945." Por Roque García Frías. Año I, no. 3, noviembre 1945, p. 37-54.

Statistical-Mathematical determination of the voting population of Peru in July 1945.

Nov. '45, Ricci "Estadística de las Superficies." Por Umberto Ricci. Año I, no. 3, noviembre 1945, p. 55-67.

Statistics of areas. First part of the Spanish translation by Pascual Saco Lanfranco, published in Lima in 1920, of **Les Bases Théoriques de la Statisque Agricole Internationale,** by Umberto Ricci (Int-**6200**.B5, 1914).

Primera parte de la traducción castellana de Pascual Saco Lanfranco, publicada en Lima en 1920, de **Les Bases Théoriques de la Statisque Agrícole Internationale,** por Umberto Ricci (Int-**6200**.B5, 1914).

Jan. '45, a "El Inventario del Potencial Económico de la Nación." Encuesta agro-pecuaria. Vol. I, no. 1, enero, 1945, p. 52-59.

The inventory of the national economic potential survey of agriculture and livestock.

Jan. '45, b "Legislación Estadística." Vol. I, no. 1, enero 1945, p. 93-114.

Statistical legislation. Includes: Law no. 7567, of August 5, 1932, transforming the Bureau of Statistics into a national bureau and establishing rules for the functioning of the national statistical service; resolution of December 7, 1938, directing municipal councils in the republic to submit monthly statistical data; decree of Feb. 9, 1940, incorporating, for statistical purposes, the neighboring districts into the capital of the republic; decree of Aug. 20, 1943, directing an inventory of the national economic potential; memorandum of the National Bureau of Statistics on the inventory of the national economic potential; decree of January 1, 1944, in accordance with law no. 7567, organizing the national statistical service and creating the Statistical High Council; resolution of January 1, 1944, approving the organization chart of the national statistical service; note of the Secretary of the Treasury, President of the Statistical High Council, to the other secretaries, about organization of the national statistical service; explanation by the National Bureau of Statistics of the change in the base of the wholesale price index and the cost of living index in Lima; recommendations by the III regional agronomic conference focussing on the multiple problems of agriculture and cattle raising (August 14-18, 1944); statutes of Peruvian Statistical Institute.

Contiene: Ley No. 7567, de 5 de agosto de 1932, que transforma en Nacional la Dirección de Estadística y establece las normas para el funcionamiento del servicio de estadística nacional; resolución suprema de 7 de diciembre de 1938, por la que se ordena a los concejos municipales de la República, la remisión mensual de los datos estadísticos; decreto supremo de 9 de febrero de 1940, por el que se incorpora, para fines estadísticos, los distritos aledaños, a la ciudad capital de la república; decreto supremo de 20 de agosto de 1943 que ordena formar el inventario del potencial económico de la nación; memorandum de la Dirección Nacional de Estadística, sobre el inventario del potencial económico de la nación; decreto supremo de 1° de enero de 1944, conforme a la Ley N° 7567, por el que se organiza el servicio de estadística nacional y crea el consejo superior de estadística; resolución suprema de 1° de enero de 1944, por la que se aprueba el diagrama de organización del servicio de estadística nacional; nota del Sr. Ministro de Hacienda, Presidente del Consejo Superior de Estadística, a los demás Ministros de Estado, sobre la organización del servicio de estadística nacional; exposición de la Dirección Nacional de Estadística, sobre el cambio de "Base" de los números indicadores de precios al por mayor y del costo de la vida en Lima; recomendaciones de la III convención agronómica regional enfocando los multiples problemas de la agricultura y ganadería (14-18 de agosto de 1944); estatutos del Instituto Peruano de Estadística.

Perú-**3040**.R5 **Revista de la Facultad de Ciencias Económicas.** 1929?- . Semi-anual? Lima. Universidad Mayor de San Marcos. Facultad de Ciencias Económicas.

Journal of the College of Economic Science of the University of San Marcos. Includes statistical material.

Incluye material estadístico.

1944, Palacios "Encuesta sobre Presupuestos Familiares Obreros Realizada en la Ciudad de Lima, en 1940." Por Leoncio Miguel Palacios. No. 29 y 30, abril y agosto, 1944. 213 p. Reimpresión.

Survey of working class family budgets carried out in the city of Lima in 1940.

Perú-**3300** **Enseñanza Estadística en Perú:**
1944-46 **Métodos y Materiales.** Material básico adquirido para la Encuesta del Instituto Interamericano de Estadística sobre los Métodos y Materiales en la Enseñanza Estadística en las Naciones Americanas, 1944-46.

Statistical training in Peru: Methods and materials. Background material acquired for the IASI Survey on Statistical Training Methods and Materials in the American Nations, 1944-46. Includes: Correspondence and filled questionnaires on statistical personnel and statistical courses in Peru.

Incluye: Correspondencia y cuestionarios llenados sobre personal estadístico y cursos de estadística en el Perú.

4000-4999 GEOGRAPHY — GEOGRAFIA

Perú-**4000**.D5 **Demarcación Política del**
1821-1946 **Perú: Recopilación de Leyes y Decretos (1821-1946).** Por Justino M. Tarazona S. Lima, 1946. Ministerio de Hacienda y Comercio. Dirección Nacional de Estadística. 512 p.

Political demarcation of Peru: Collection of laws and decrees, 1821-1946. Contains: A report by the author on the procedure followed in the work; prologue by Dr. Ricardo Luna Vegas on the antecedents and object of the volume; bibliography consulted; and, in chronological order, all the laws, decrees, regulations and resolutions that have been promulgated with respect to the political demarcation of the country (laws on: Creation of provinces, departments, districts, and municipalities; suffrage; political authorities; geographic nomenclature; etc.)

Contiene: Un informe del autor acerca del procedimiento seguido para esta obra; prólogo del Dr. Ricardo Luna Vegas referente a los antecedentes de esta obra y su objeto; bibliografía consultada; y por orden cronológico todas las leyes, decretos, reglamentos y disposiciones que con respecto a la demarcación política del país se han dictado (leyes sobre: Creación de provincias, departamentos distritos y municipios; sufragio; autoridades políticas; nomenclatura geográfica; etc.).

Perú-**4200**.B4 **Boletín Anual Meteorológico.** Fecha de iniciación? Lima. Ministerio de Fomento y Obras Públicas. Dirección de Comunicaciones y Meteorología de Aeronáutica. Departamento de Meteorología.

Annual meteorological bulletin.

5000-5999 DEMOGRAPHY — DEMOGRAFIA

Perú-**5000**.P5 **Peru: Summary of Biostatis-**
1944 **tics.** Maps and charts, population, natality and mortality statistics. Washington, D.C., September 1944. Prepared by U. S. Department of Commerce, Bureau of the Census, in cooperation with Office of the Coordinator of Inter-American Affairs. 79 p.

Resumen bioestadístico de Perú. Incluye mapas y cuadros y datos sobre población, natalidad y mortalidad.

Perú-**5200** **Estadística Vital y Registro en Perú: Métodos y Procedimientos.** Material básico adquirido para la encuesta que el Instituto Interamericano de Estadística realiza sobre los métodos de registro y estadística vital en las naciones americanas, incluyendo el estudio de 1945.

Vital statistics and registration in Peru: Methods and procedures. Background material acquired for the study conducted by the Inter American Statistical Institute in vital statistics and registraton methods in the American nations, including the survey of 1945. Includes: Laws, decrees and resolutions; forms; punch cards and codes; instructions.

Incluye: Leyes, decretos y resoluciones; formularios; tarjetas perforadas y códigos; instrucciones.

6000-6999 ECONOMICS — ECONOMIA

Perú-**6000**.E5 **Economic Controls and Com-**
1945 **mercial Policy in Perú.** One of a series of reports on economic controls and commercial policy in the American republics. Washington, D.C., 1945. United States Tariff Commission. 25 p.

Folleto sobre Perú, de la serie de informes de la Comisión de Aduanas de los Estados Unidos, sobre controles económicos y política comercial de las repúblicas americanas.

Perú-**6000**.G5 **La Gaceta Económica y Financiera.** Fecha de iniciación?

Frecuencia? Lima? No examinado.

Economic and financial journal. Articles on economics and social subjects, including statistical material.

Artículos sobre asuntos económicos y sociales, incluyendo material estadístico.

Perú-6000.I5 **Informaciones Comerciales, Económicas y Financieras del Perú.** Fecha de iniciación? Bimestral. Lima. Ministerio de Relaciones Exteriores y Culto. Departamento Comercial.

Commercial, economical and financial information of Peru. Statistical information on economic conditions, as well as news, legislation, etc. Bimonthly.

Información estadística sobre las condiciones económicas como también noticias, legislación, etc.

Perú-6000.M5 **Memoria del Ministerio de Fomento y Obras Públicas.** Fecha de iniciación? Anual. Lima.

Annual report of the Department of Public Work and Development. Contains some statistical data.

Contiene algunos datos estadísticos.

Perú-6000.N8 **La Nueva Economía.** Organo mensual del Bureau Industrial del Peru. 1935- . Lima.

The new economy: Monthly organ of the Industrial Bureau of Peru. Includes current statistical material.

Incluye material estadístico corriente.

Perú-6000.P5 1945 **Peru.** Foreign commerce yearbook country series. Washington, D.C., U.S.Govt.Print.Off., 1945. U. S. Department of Commerce. Bureau of Foreign and Domestic Commerce. International Reference Service, vol. 2, no. 45. Unexamined.

Contains information about area, population, climate, agriculture, commerce and industry, mining and manufacturing, communication and transportation, power and light, national economic indexes, national income, average wages, foreign trade, etc.

Perú. Anuario del comercio exterior de los Estados Unidos, series por países. Contiene información sobre área, población, clima, agricultura, comercio e industria, minería y fabricación, comunicaciones y transportes, fuerza y luz, índices de economía nacional, renta nacional, promedio de salarios, comercio exterior, etc.

Perú-6000.R3 1944 Inter-American **Reports Presented to the Conference of Commissions of Inter-American Development (Held in New York, May 9-18, 1944) by the Peruvian Commission of Inter-American Development.** On: 1, The present utilization of Peru's resources; 2, diversification of agriculture. Washington, D.C., 1945? Inter-American Development Commission. 14 p.

Cover title—título de la cubierta: **Peru: Reports Presented to the Conference of Commissions of Inter-American Development.**

Informes presentados a la Conferencia de Comisiónes de Fomento Interamericano (realizada en N. York, del 9 al 18 de mayo de 1944), por la Comisión Peruana de Fomento Interamericano, sobre: 1, La utilización actual de las riquezas peruanas; 2, diversificación de la agricultúra.

Perú-6000.R5 **Revista de Economía y Finanzas.** Fecha de iniciación? Mensual. Lima.

Journal of economics and finance. Includes notes, economic news, legislation, as well as articles on economic life of Peru. Monthly.

Incluye notas, noticias económicas, legislación, como también artículos sobre la vida económica del Perú.

Perú-6000.R7 **Revista de Hacienda.** 1940- . Trimestral. Lima. Ministerio de Hacienda y Comercio.

Journal of public finance. Contains some statistical material showing economic position and foreign trade of Peru, international balance of payments, etc. Each issue contains one or two special studies based on some economic or social aspect of Peruvian life.

Contiene algunos datos estadísticos sobre la situación económica y comercio exterior del Perú, balance internacional de pagos, etc. Cada edición contiene uno o dos estudios especiales basados en alguno de los aspectos sociales o económicos de la vida en el Perú.

Perú-6200 **Estadísticas Agrícolas en Perú: Métodos y Procedimientos.** Material básico adquirido para el estudio que el Instituto Interamericano de Estadística realiza sobre los métodos de estadísticas agrícolas en las naciones americanas, incluyendo la encuesta de 1945-46.

Agricultural statistics in Peru: Methods and procedures. Background material acquired for Inter American Statistical Institute study on methods in agricultural statistics in the American nations, including the Survey of Agricultural Production Statistics in the American Nations in 1945-46. Consists of: Answers to questionnaires; reports.

Consiste de: Respuestas a los cuestionarios; informes.

*Perú-**6200**.B5 **Boletín de la Dirección de Agricultura.** 1931- . Mensual. Lima. Ministerio de Agricultura. Dirección de Agricultura.

Title varies—título varía: 1931- ?; **Boletín de la Sección Técnica de Estadística.**

Bulletin of the Bureau of Agriculture. Monthly publication dealing principally with production of cotton, sugar, rice, and wheat.

Trata principalmente sobre la producción de algodón, azúcar, arroz y trigo.

Perú-**6200**.M5 **Memoria de la Sociedad Nacional Agraria.** Fecha de iniciación? Anual. Lima.

Annual report of the National Agrarian Society. Contains statistical information on major crops of Peru.

Contiene información estadística sobre las principales cosechas del Perú.

1943-1944 **Memoria que la Junta Directiva de la Sociedad Nacional Agraria Presenta a la Asamblea General Ordinaria, Año 1943-44.** Lima, Librería e Imp.Gil, S.A., 1944. Sociedad Nacional Agraria. 125 p.

Report of the Directing Board of the National Agricultural Society. Includes brief reports about the various agricultural committees of the country, and statistics on the outstanding crops for 1943.

Incluye breves memorias acerca de las varias comisiones agrícolas del país y estadística sobre las cosechas salientes de 1943.

Perú-**6200**.S5 1945, Eng. **Statistical Program for the Ministry of Agriculture of Peru.** By H. H. Schutz. Washington, D.C., 1945. U. S. Executive Office of the President. Office for Emergency Management. Office of the Coordinator of Inter-American Affairs. 56 p. Mimeographed.

Published also in Spanish—publicado también en español (Perú-**6200**.S5, 1945, Sp).

Reproduced and distributed by U. S. Department of Agriculture, Bureau of Agricultural Economics. Includes questionnaires and forms.

Perú-**6200**.S5 1945, Sp. **Programa Estadístico para el Ministerio de Agricultura del Perú.** Por H. H. Schutz. Washington, D.C., 1945. U. S. Executive Office of the President. Office for Emergency Management. Office of the Coordinator of Inter-American Affairs. 57 p. Mimeografiado.

Published also in English—publicado también en inglés (Perú-**6200**.S5, 1945, Eng).

Distribuido y reproducido por U. S. Department of Agriculture, Bureau of Agricultural Economics. Incluye cuestionarios y formularios.

Perú-**6200**.V5 **La Vida Agrícola.** Revista de agricultura y ganadería. 1924- . Mensual. Lima.

A review of the agricultural and livestock situation in Peru, containing various statistical data in these fields.

Una revisión de la situación agrícola y ganadera del Perú. Contiene datos estadísticos diversos al respecto.

Perú-**6210**.A5 **Algodón.** Boletín de la Cámara Algodonera del Perú. 1940- . Mensual. Lima.

Cotton. Bulletin of the Chamber of Cotton of Peru. Official information on cotton in Peru. Each issue has as a rule one article of national or international interest, and a number of statistical tables.

Información oficial sobre algodón en el Perú. Cada edición contiene, por lo general, un artículo de interés nacional o internacional y algunos cuadros estadísticos.

Perú-**6210**.E5 **Estadística de la Producción del Algodón en el Perú.** Fecha de iniciación? Anual. Lima. Ministerio de Agricultura. Dirección de Agricultura. Departamento de Estadística Agropecuaria.

Annual statistics of cotton production in Peru.

1943 **Estadística de la Producción de Algodón en el Perú Correspondiente al Año 1943.** Publicación oficial. Lima, Imp.El Universal, 1943. Ministerio de Agricultura. Dirección General de Agricultura. 40 p.

Includes: Statistics of production of cotton: Area; production of fibre; consumption of cotton by factories; exports of cotton derivatives; imports; general summary 1916-43.

Incluye: Estadística de la producción de algodón: Superficie; producción de fibra; consumo de algodón por fábricas; exportación de los derivados del algodón; importación; resumen general 1916-43.

Perú-**6210**.E6 **Estadística de la Producción de Arroz.** Fecha de iniciación? Anual. Lima. Ministerio de Agricultura. Dirección de Agricultura. Departamento de Estadística Agropecuaria.

Annual statistics of rice production.

Perú-**6210**.E7 **Estadística de la Producción de Caña de Azúcar y Azúcar de Caña en el Perú.** Fecha de iniciación? Anual. Lima. Ministerio de Agricultura. Dirección de Agricultura. Departamento de Estadística Agropecuaria.

Annual statistics of production of sugar cane and cane sugar in Peru.

Perú-**6210**.E8 **Estadística de la Producción del Trigo en el Perú.** Fecha de iniciación? Anual. Lima. Ministerio de Agricultura. Dirección de Agricultura. Departamento de Estadística Agropecuaria.

Annual statistics of wheat production in Peru.

Perú-**6210**.M5 **Memoria Anual de la Cámara Algodonera del Perú.** 1940?- Lima.

Annual report of the Chamber of Cotton of Peru.

1945 **Memoria Anual Correspondiente al Año 1945.** Lima, Imp.H.C. Rozas, 1946. Cámara Algodonera del Perú. 121 p.

Includes data on area sown, size of the crop, sales, exports, cotton textile industry, etc. Also includes statistical tables and graphs on production, sales, prices, exports, etc., and also the production of flax.

Incluye datos sobre área sembrada, volumen de la cosecha, venta, exportaciones, industria textil derivada, etc. Incluye además cuadros y gráficos estadísticos referentes a producción, ventas, precios, exportaciones, etc., y también sobre la producción de lino.

Perú-**6220**.B4 **Aguas e Irrigación.** 1933?- . Trimestral. Lima. Ministerio de Fomento y Obras Públicas. Dirección de Aguas e Irrigación.

Waters and irrigation. Lists and descriptions of irrigation projects in Peru; news of similar activities abroad; legislation. Quarterly.

Lista y descripciones de los planes de irrigación en el Perú; noticias de las actividades similares en el exterior; legislación.

Perú-**6220**.B6 **Boletín de la Compañía Administradora del Guano.** 1925- . Mensual. Lima.

Monthly bulletin of the Guano Administration Company.

Perú-**6220**.M5 **Memoria de la Compañía Administradora del Guano.** Fecha de iniciación? Anual. Lima. No examinado.

Annual report of the Guano Administration Company.

Perú-**6230**.E5 1944 **Estudio de la Situación de la Industria Lechera en el Departamento de Lambayeque.** Informe preparado por el Departamento de Costos de Producción, División Económica del Servicio Cooperativo Inter-Americano de Producción de Alimentos. Lima, Ministerio de Agricultura, 1944. Ministerio de Agricultura. Servicio Cooperativo Inter-Americano de Producción de Alimentos. 110 p. Mimeografiado.

Study of the situation of the milk industry in the Department of Lambayeque. Report prepared by the Department of Production Costs, Economic Division of the Inter-American Cooperative Food Production Service.

*Perú-**6300**.A5 **Anuario de la Industria Minera en el Perú.** 1903- . Lima. Ministerio de Fomento y Obras Públicas. Dirección de Minas y Petróleo.

Title varies—título varía: 1903-1922, **Estadística Minera**; 1924-1931, **La Industria Minera-Metales, Petróleo.**

Yearbook of the mining industry.

1944 **Anuario de la Industria Minera en el Perú, 1944.** Boletín oficial. Lima, Imp.Americana, 1945. Ministerio de Fomento. Dirección de Minas y Petróleo. 296 p.

Yearbook of the mining industry in Peru. Production, prices, exports and imports of metals and non-metals.

Producción, precios, exportación e importación de metales y productos nometálicos.

Perú-**6300**.B7 **Boletín Oficial de la Dirección de Minas y Petróleo.** 1923- . Irregular. Lima. Ministerio de Fomento y Obras Públicas. Dirección de Minas y Petróleo.

Official Bulletin of the Bureau of Mines and Petroleum. An official mining journal containing articles and statistics of interest to the mining industry of Peru.

Un boletín minero oficial que contiene artículos y estadísticas de interés para la industria minera del Perú.

Perú-**6300**.B5 **Boletín del Cuerpo de Ingenieros de Minas.** Fecha de iniciación? Anual. Lima. Ministerio de Fomento y Obras Públicas. Cuerpo de Ingenieros de Minas.

Annual bulletin of the Board of Mining Engineers. Report on the mining production of the country. Lists publications of the Ministerio on the inside covers.

Informe sobre la producción minera del país. En la parte interior de las tapas lleva una lista de las publicaciones del Ministerio.

Perú-6400.B5 Perú. **Boletín de la Dirección de Industrias del Ministerio de Fomento y Obras Públicas.** Fecha de iniciación? Frecuencia? Lima.

Most of the articles in this magazine contain data.

La mayoría de los artículos en este boletín contienen datos.

Aug. '46 "Estadística Industrial." No. 3, agosto 1946. 151 p.

Industrial statistics. Contents: Distribution of the industries included in the census in 1944 (map); textile industries (spinning mills, cotton fabric, wool fabric, artificial silk fabric, knitted fabric, passementerie); leather industries (tanneries); apparel industries (footwear); ceramic industries (glass and crystal); paper and pasteboard industries (manufacture of paper and pasteboard); food industries (soda water, beer); rubber; graph showing the principal characteristics of the industries included in the survey. "This offering will be augmented in the next publication with figures for other industries. . . . In this way we will arrive, in the near future, at having surveyed all the national industry . . ." (Translation of comment.)

Contenido: Distribución de las varias industrias censadas en 1944 (mapa); industrias textiles (hilanderías, tejidos de algodón, tejidos de lana, tejidos de seda artificial, tejidos de punto, pasamanería); industrias de los cueros (curtiembres); industrias del vestuario (calzado); industria cerámica (vidrios y cristales); industrias del papel y cartón (fabricación de papel y cartón); industrias de la alimentación (aguas gaseosas, cerveza); caucho o jebe; gráfico de las principales características de las industrias censadas. "Este aporte se verá aumentado en la próxima publicación con las cifras de nuevas industrias. . . De esta manera llegaremos, en un futuro muy próximo, a tener censada toda la gran industria nacional. . . ." (Fuente: Prefacio.)

Sept. '45, Arbulú Casanova "Cifras Estadísticas de la Industria Textil." Por Héctor Arbulú Casanova. No. 2, septiembre 1945, p. 96-163.

Statistical data on the textile industry of Peru.

Perú-6400.E5 1942 **Estadística de los Servicios Eléctricos del Perú.** Lima, Imp.Torres Aguirre, 1942. Ministerio de Fomento y Obras Públicas. Servicio Técnico de Electricidad. 173 p.

Statistics of electric power services in Peru.

Perú-6400.M5 1945 **Mining and Manufacturing Industries in Perú.** One of a series of reports on mining and manufacturing industries in the American republics. Washington, D.C., 1945. United States Tariff Commission. 31 p.

Folleto sobre Perú, de la serie de informes de la Comisión de Aduanas de los Estados Unidos, sobre las industrias mineras y manufactureras en las repúblicas americanas.

Perú-6610.B5 **Boletín de la Dirección de Caminos y Ferrocarriles.** 1938- . Trimestral. Lima. Ministerio de Fomento y Obras Públicas. Dirección de Caminos y Ferrocarriles.

Quarterly bulletin of the Bureau of Highways and Railways.

Perú-6610.M5 **Memoria de la Dirección General de Tránsito.** Fecha de iniciación? Anual. Lima. Ministerio de Gobierno y Policía. Dirección General de Tránsito. No examinado.

Annual report of the General Transit Bureau on traffic and vehicles.

Informe anual sobre tránsito y vehículos.

Perú-6611.B5 **Boletín de la Dirección de Aeronáutica.** 1936- . Mensual. Lima. Ministerio de Fomento y Obras Públicas. Dirección General de Aeronáutica. No examinado.

Formerly called—anteriormente publicado bajo el título: **Boletín de la Dirección de Aviación Comercial y Civil.** Bulletin of the General Bureau of Aeronautics. Legislation and statistics.

*Perú-6720.B5 **Boletín Mensual del Banco Central de Reserva del Perú.** 1930- . Lima.

Monthly bulletin of the Central Reserve Bank of Peru. Money and banking statistics and other information. Each issue has a short summary of the Peruvian economic situation; a statistical section on banking, exchange rates, foreign trade, prices, living costs, security prices, and bills protested; a balance sheet of the bank, and a brief comment on the special laws governing the operations of the bank.

Estadísticas sobre moneda y operaciones bancarias y otras informaciones. Cada edición tiene un breve sumario sobre la situación económica del país; una sección estadística sobre operaciones bancarias, cotizaciones de cambio, comercio exterior, precios, costo de la vida, precios de títulos, y documentos protestados; un balance del Banco y un breve comentario sobre las leyes especiales que rigen las operaciones del Banco.

Aug. '45 Barreto "La Renta Nacional del Perú." Por Emilio G. Barreto. Año XV, no. 165, agosto 1945, p. 5-8.

National income of Peru. Includes a summary table of the national income for 1942.

Incluye un cuadro sumario de la renta nacional en el año 1942.

July '45 "Resumen de la Balanza de Operaciones Internacionales del Perú." Año XV, no. 164, julio 1945, p. 8.

Summary of the balance of international operations of Peru.

Feb. '45, Barreto "La Balanza Internacional de Pagos del Perú, Años 1939-1943." Por Emilio G. Barreto. Año XV, no. 159, febrero 1945, p. 7-12.

Balance of international payments of Peru, 1939-43.

July '45, Barreto "Naturaleza y Alcances de la Llamada Balanza Internacional de Pagos." Por Emilio G. Barreto. Año XV, no. 164, julio 1945, p. 5-7.

Nature and scope of the balance of international payments.

Perú-**6720**.M4 **Memoria de la Superintendencia de Bancos y Estadística Bancaria de Seguros y Capitalización.** Fecha de iniciación? Anual. Lima.

Annual report of the Superintendent of Banks and statistics of banking, insurance, and capitalization.

Perú-**6720**.M6 **Memoria del Banco Agrícola del Perú.** 1932- . Anual. Lima.

Report of the Farm Bank of Peru.

Perú-**6720**.M7 **Memoria del Banco Central de Reserva del Perú.** Fecha de iniciación? Anual. Lima.

Report of the Central Reserve Bank of Peru.

1945 **Memoria del Banco Central de Reserva del Perú, 1945.** Lima?, Imp.Torres Aguirre, S.A., 1945? Banco Central de Reserva del Peru. 79 p.

Contains: General financial statement of the Central Reserve Bank of Perú for 1945; brief reports on the capital and reserve, cash balance, credit transactions, gold purchases, investments, Chamber of Compensation, utilities, amortization of the public debt, fractional money, National House of Currency, appraisal, capital of the Industrial Bank, Mining Bank, liquidation of the German Transatlantic Bank, credit of the Export-Import Bank, directory; and tables, including cost-of-living, foreign commerce, etc., as well as financial statistics.

Contiene: Memoria financiera general de 1945; breves informes sobre el capital y reservas, encaje, operaciones de crédito, compras de oro, inversiones, Cámara de Compensación, utilidades, amortización de la deuda pública, moneda fraccionaria, Casa Nacional de Moneda, especies valoradas, el capital del Banco Industrial, Banco Minero, liquidación del Banco Alemán Transatlántico, crédito del Export-Import Bank, directorio; y cuadros, incluyendo costo de vida, comercio exterior, etc., así como estadística financiera.

Perú-**6720**.M8 **Memoria del Banco Industrial del Perú.** 1937- . Anual. Lima.

Report of the Industrial Bank of Peru.

Perú-**6720**.S5 **Situación de las Empresas Bancarias del Perú.** Fecha de iniciación? Anual. Lima. Superintendencia de Bancos. In IR.

Situation of banking establishments in Peru.

Perú-**6900** **Estadísticas del Comercio Exterior en Perú: Métodos y Procedimientos.** Material básico adquirido para el estudio que el Instituto Interamericano de Estadística realiza sobre los métodos de estadísticas del comercio exterior en las naciones americanas, incluyendo la encuesta de 1946-47.

Foreign trade statistics in Peru: Methods and procedures. Background material acquired for the study conducted by the Inter American Statistical Institute in methods of foreign trade statistics in the American nations, including the survey of 1946-47. Consists of: Filled questionnaire.

Consiste de: Cuestionario llenado.

*Perú-**6900**.A5 **Anuario del Comercio Exterior del Perú.** Fecha de iniciación? Callao. Ministerio de Hacienda y Comercio. Superintendencia General de Aduanas. Departamento de Estadística General de Aduanas.

Yearbook of foreign trade of Peru. Detailed imports and exports, classified by countries of origin and destination, by customs duties, weight, and value; also coastal trade. This information is also published in condensed form in the **Anuario Estadístico del Perú.**

Detalla importaciones y exportaciones, clasificadas por países de origen y destino, por derechos de aduana, peso y valor; también comercio de cabotaje. Esta información está publicada también en forma condensada en el **Anuario Estadístico del Perú.**

Perú-**6900**.B5 **Boletín de Aduanas.** 1923- . Mensual. Callao: Ministerio de Hacienda y Comercio. Superintendencia General de Aduanas. Departamento de Estadística General de Aduanas. No examinado.

Monthly bulletin of customs. Tariff information and import statistics.

Información arancelaria y estadísticas de importación.

Perú-**6900**.P5 **Peru—Year Book of Foreign Trade Exporters, Importers, and Merchants.** Starting date? Lima. Compañía Peruana de Negocios Internacionales.

Contains data on the geography of Peru; population; agricultural, mining and industrial production; foreign and domestic trade; banking, insurance, and public finance; legislation. Lists Peruvian Consular Service in American countries; representative societies and associations of commerce, industry, and agriculture; banking institutions; insurance companies; exporters and importers.

Perú—anuario de exportadores, importadores y comerciantes. Contiene datos sobre geografía del Perú; población; producción agrícola, minera e industrial; comercio interior y exterior; operaciones bancarias, seguro y hacienda pública; legislación. Enumera los servicios consulares del Perú en países americanos; sociedades y asociaciones representativas del comercio, la industria, y la agricultura; instituciones bancarias; compañías de seguro; exportadores e importadores.

Perú-**6900**.T5 1938 **Tarifa de Derechos de Importación.** Edición corregida. Incluye nuevas partidas, asimilaciones del cuerpo consultivo de aranceles de aduana, Convenios internacionales, Impuestos adicionales, Anotaciones, Concordancias. Callao, Editora Peruana, 1938. Ministerio de Hacienda y Comercio. Superintendencia General de Aduanas. 259 p. Ley 8044 de 1935.

Import tariff, law 8044 of 1935. Corrected edition. Includes new items, resolutions of the consulting board on tariff schedules, international agreements, additional taxes, notes, agreements.

7000-7999 HEALTH AND CULTURAL STATISTICS—ESTADISTICAS DE SALUBRIDAD Y CULTURA

Perú-**7110**.B4 **Boletín de la Dirección General de Salubridad.** Fecha de iniciación? Anual. Lima. Ministerio de Salud Pública y Asistencia Social. Dirección General de Salubridad.

Annual bulletin of the General Bureau of Health. Devoted to public health questions. General articles accompanied by statistics.

Dedicado a cuestiones de salud pública. Artículos generales acompañados de estadísticas.

Perú-**7110**.B6 **Boletín del Departamento de Bioestadística.** Fecha de iniciación? Frecuencia? Lima. No examinado.

Bulletin of the Department of Vital Statistics.

*Perú-**7110**.M5 **Memoria que Presenta al Congreso Nacional el Ministro de Salud Pública y Asistencia Social.** Fecha de iniciación? Anual. Lima.

Annual report of the Department of Public Health and Social Assistance. Contains data on demography, health, Indian affairs, nutrition, social security, etc.

Contiene datos sobre demografía, sanidad, asuntos indígenas, nutrición, previsión social, etc.

Perú-**7130**.D5 **Demografía y Estadística de Hospitales y de Otros Establecimientos.** Fecha de iniciación? Anual. Lima. Sociedad de Beneficencia Pública de Lima. Sección de Estadística.

Demography and statistics of hospitals and other institutions.

Perú-**7160**.B5 **Boletín del Cuerpo de Investigación y Vigilancia.** Fecha de iniciación? Mensual. Lima. Ministerio de Gobierno y Policía. Dirección General de Tránsito. No examinado.

Bulletin of the Board of Investigation and Vigilance of the General Transit Bureau. Monthly report on traffic accidents.

Informe mensual sobre accidentes de tránsito.

Perú-**7200**.E5 1946 **Education in Peru.** By Cameron D. Ebaugh. Washington, D.C., U.S.Govt.Print. Off., 1946. U. S. Federal Security Agency. Office of Education. 91 p. Bulletin, 1946, no. 3.

Study of the Peruvian educational system, covering elementary, secondary, vocational, and higher education, including the Peruvian universities, and giving background data on the evolution of education in Peru.

Un estudio sobre el sistema de educación en Perú, comprendiendo escuelas elementales, secundarias, vocacionales y profesionales, incluyendo las universidades de Perú, y citando datos básicos acerca del desarrollo del sistema educacional en Perú.

9000-9999 LABOR AND LIVING CONDITIONS—CONDICIONES DE TRABAJO Y DE VIDA

*Perú-9000.B5 **Boletín Estadístico de la Dirección de Trabajo y de la Dirección de Previsión Social.** 1936- . Anual. Lima. Ministerio de Salud Pública y Asistencia Social.

Annual statistical bulletin of the Bureau of Labor and the Bureau of Social Security.

Perú-9000.E5 1939 **Estadística del Trabajo.** Investigación preliminar realizada por la Caja de Seguro Social. Lima, 1939. Caja Nacional de Seguro Social. Biblioteca. 35 p. (Publicación), no. 7.

Labor statistics. Preliminary study undertaken by the Social Security Agency.

Perú-9000.I5 **Informaciones Sociales.** Organo de la Caja Nacional de Seguro Social. 1939- . Trimestral. Lima.

Previously published monthly—anteriormente publicado mensualmente.

Social information. Quarterly publication of the National Social Security Agency. Basic source of information on social security in Peru.

Fuente básica de información sobre seguro social en Perú.

Perú-9800.F5 1943 **The Food Situation in Southern Peru.** Report prepared by the Economic Division, Servicio Cooperativo Interamericano de Producción de Alimentos. Lima, Ministry of Agriculture, October 1943. 107 p.

La situación alimenticia en el sud del Perú.

Perú-9800.S4 1945 **La Situación Alimenticia en el Perú, 1943-1944.** Preparado por Luis Rose Ugarte. Lima, Editora Médica Peruana S.A., 1945. Ministerio de Agricultura. Servicio Cooperativo Inter-Americano de Producción de Alimentos. 225 p.

The food situation in Peru, 1943-1944. Includes: Prologue and introduction on nature, means, process, and importance of the study; chapters on area and population (geographic-economic zones, hydrography, climate, economic characteristics of the population, etc.); land and production (area cultivated, classes of cultivation, distribution of cultivated land by products, problems of ownership of the land, rents, etc.), consumption, prices, imports and exports, wages; and a section of maps and tables completing the data of the preceding chapters.

Contiene: Prólogo e introducción comentando la naturaleza del estudio, medios, proceso e importancia; capítulos sobre la superficie y población (zonas geo-económicas, hidrografía, clima, características económicas de la población, etc.); tierra y producción (área cultivada, clases de cultivos, distribución de las tierras cultivadas por productos, problemas de la propiedad de la tierra, arrendamientos, etc.), consumo, precios, importación y exportación, jornales; y una sección de mapas y cuadros completando los datos de los capítulos precedentes.

Perú-9800.S5 1944 **Situación Alimenticia en la Ciudad de Iquitos.** Informe, preparado por Luis Rose Ugarte. Lima, julio 1944. Ministerio de Agricultura. Servicio Cooperativo Inter-Americano de Producción de Alimentos. 64 p. Mimeografiado.

Food situation in the city of Iquitos. Includes tables and maps.

Incluye tablas y mapas.

Perú-9800.S7 1944 **Situación Alimenticia en los Departamentos del Norte.** Informe preparado por Luis Rose Ugarte. Traducción del original inglés por Irma Landazuri S. Lima, 1944. Ministerio de Agricultura. Servicio Cooperativo Inter-Americano de Producción de Alimentos. 146 p. Mimeografiado. 2 vol.

The food situation in the northern departments. Report prepared by Luis Rose Ugarte, translated from the English original by Irma Landazuri S. Includes tables and maps.

Incluye tablas y mapas.

Perú-9950.M5 **Memoria Anual de la Caja Nacional de Seguro Social.** Fecha de iniciación? Lima. No examinado.

Annual report of the National Social Security Agency.

REPUBLICA DOMINICANA

1000-1999 GENERAL

*RepDom-**1410**.A5 **Anuario Estadístico de la República Dominicana.** 1936- . Ciudad de Trujillo. Dirección General de Estadística.

Statistical yearbook of the Dominican Republic. Each edition is in two volumes. The 1937 issue includes a section, p. 84-111, on the history of population estimates from 1514 to 1929 and the following data from the 1935 census: Density by province, sex and citizenship status, urban and rural residence, nationality, marital conditions, race or color, literacy, religion, labor force, and physical defects. The 1938 issue repeats the history of population estimates and the census data for 1935 published in the 1937 issue, and presents additional data on population by sex in urban centers, marital status, nationality, race or color, religion, literacy, and labor force; a graph showing estimated or enumerated total population by five year periods from 1900 to 1938 is included. The 1939 issue gives the **housing census** of 1935 on p. 76-117 and the **agricultural census** of 1935 on p. 178-300 of the first volume and the **livestock census** of 1935 on p. 14-176 of the second volume.

Cada edición consiste de dos volúmenes. La edición de 1937 incluye una sección, p. 84-111, sobre la historia de las estimaciones de la población desde 1514 a 1929 y los datos siguientes tomados del censo de 1935: Densidad por provincias, sexo y ciudadanía, residencia urbana y rural, nacionalidad, estado civil, raza o color, alfabetismo, religión, potencial obrero y defectos físicos. La edición de 1938 repite la historia de las estimaciones de población y los datos del censo de 1935 publicados en la edición de 1937, y trae también datos adicionales sobre población por sexo en centros urbanos, estado civil, nacionalidad, raza o color, religión, alfabetismo y potencial obrero; incluye un gráfico demostrando la población estimada o censada por períodos quinquenales desde 1900 a 1938. La edición de 1939 presenta el **censo de habitaciones** de 1935, p. 76-117, y el **censo agrícola** de 1935, p. 178-300 del primer volumen; el **censo pecuario** de 1935, p. 14-176 del segundo volumen.

RepDom-**1950**.R5 1944 **República Dominicana, Album Estadístico Gráfico.** Ciudad Trujillo, Tall.Lito.de Ferrua y Hnos.,1944. Dirección General de Estadística. 40 p.

Album of statistical graphs, on the Dominican Republic, issued by the General Bureau of Statistics.

2000-2999 CENSUS — CENSOS

RepDom-**2100**.C5 1935.1 **Censo de 1935: Métodos y Procedimientos.** Material usado en el levantamiento del censo de población del 13 de mayo. Adquirido con motivo de la encuesta que el Instituto Interamericano de Estadística realizó sobre Métodos y Materiales de los Censos en las Naciones Americanas, 1944. De la Dirección General de Estadística, Sección del Censo.

Census of 1935: Methods and procedures. Materials used in taking population census, on May 13. Acquired in connection with Inter American Statistical Institute Survey on Census Methods and Procedures in the American Nations, 1944. Consists of: Correspondence; laws, decrees, or other authorization; instructions; reports; forms.

Se compone de: Correspondencia; leyes, decretos u otras disposiciones; instrucciones; informes; formularios.

See also—véase también: **Ley del Congreso Nacional que Dispone las Condiciones en que Deben Desarrollarse las Operaciones del Censo; Instrucciones Generales sobre las Funciones de un Enumerador; Instrucciones a los Enumeradores del Censo de Población.**

RepDom-**2100**.C5 1935.2 **Censo de 1935: Ley del Congreso Nacional que Dispone las Condiciones en que Deben Desarrollarse las Operaciones del Censo.** Santo Domingo, Imp.La Opinión, 1935? 10 p.

Law on census operations of 1935.

RepDom-**2100**.C5 1935.3 **Censo de 1935: Instrucciones a los Enumeradores del Censo de Población.** Sto.Domingo, Sindicato, 1935? Dirección General de Estadística. Sección del Censo. 16 p.

Instructions to enumerators for the 1935 census of population.

RepDom-**2100**.C5 1935.4 **Censo de 1935: Instrucciones Generales sobre las Funciones de un Enumerador.** Santo Domingo, Tip.Listin Diario, 1935? Dirección General de Estadística. Sección del Censo. 8 p.

RepDom-**2100**.C6 1945 **Censo Especial Urbano de Ciudad Trujillo, 1945.** Levantado por la Dirección General de Estadística. Ciudad Trujillo?, Pol Hermanos, 1945.

Consejo Administrativo del Distrito de Santo Domingo. 2 p.

Consists of two schedules—consta de dos cédulas: **Formulario No. 1: Censo de Edificios; Formulario No. 2: Censo de Población.**

Special urban census of Cuidad Trujillo, 1945.

RepDom-**2100**.C7 1944 **Censo Especial Urbano San Cristobal, 1944: Instrucciones a los Enumeradores.** Dirección General de Estadística. 13 p.

Instructions to enumerators for special urban census of San Cristóbal, 1944.

RepDom-**2110**.C5 1940.1 **Censo Agro-Pecuario de 1940: Métodos y Procedimientos.** Material usado en el levantamiento del censo, en enero. Adquirido con motivo de la encuesta que el Instituto Interamericano de Estadística realizó sobre Métodos y Procedimientos de los Censos en las Naciones Americanas, 1944. De la Dirección General de Estadística.

Census of agriculture and livestock, 1940: Methods and procedures. Material used in taking census. Acquired in connection with Inter American Statistical Institute Survey of Census Methods and Procedures in the American Nations, 1944. Consists of: Laws, decrees, or other authorization; publicity and propaganda; schedules; other forms; instructions; exposition and reports.

Se compone de: Leyes, decretos, u otras disposiciones; publicidad y propaganda; cédulas; otros formularios; instrucciones; exposición e informes.

*RepDom-**2200**.C5 1935-46 **Población de la República Dominicana, Según las Sucesivas Modificaciones Territoriales a Partir del 13 de Mayo de 1935, Día en que Se Levantó el Segundo Censo Nacional de Población, Hasta el 1 de Enero de 1946.** Con un índice alfabético de comunes, distritos municipales y secciones municipales existentes al 31 de dic., 1945, y unas notas sobre el método utilizado para la determinación de la población probable. Ciudad Trujillo, Sección de Publicaciones, 1946. Dirección General de Estadística. 354 p.

Population of the Dominican Republic according to successive territorial modifications from May 13, 1935, day in which the Second National Census of Population was taken, until January 1, 1946; with an alphabetical index of communities, municipal districts and municipal sections existing on December 31, 1945, and some notes on the method used in determining the probable population. The first, second, and third parts give the population by municipal sections of the communities of the Republic, according to the census of 1935. The fourth part gives the probable population as of January 1 and July 1 of each year, from 1900 to 1950 for the whole Republic, and from January 1, 1935 until July 1, 1944 for the provinces and communities in accordance with the successive modifications.

La primera, segunda, y tercera parte presentan la población por secciones municipales de las comunes de la República, según el censo de 1935. La cuarta parte presenta la población probable al 1ro. de enero y 1ro. de julio de cada año, desde 1900 a 1950 para toda la República y desde el 1ro. de enero de 1935 hasta el 1ro. de julio de 1944 para las provincias y comunes de acuerdo con las sucesivas modificaciones.

RepDom-**2200**.C5 1935.1 **Población de la República Dominicana, Distribuida por Nacionalidades: Cifras del Censo Nacional de 1935.** Ciudad Trujillo, 1937. Dirección General de Estadística Nacional. Sección del Censo. 19 p. Mimeografiado.

Population of the Dominican Republic, by nationalities: Figures from the 1935 national census.

RepDom-**2200**.C5 1935.2 **Población de la República Censada el 13 de Mayo de 1935, Clasificada por Grupos de Edades, Zona y Sexo.** Dirección General de Estadística. 1 p. Copia máquina.

Population of the Republic enumerated on March 13, 1935, classified by age groups, zone and sex.

RepDom-**2200**.C5 1935.3 **Censo de Población Urbana y Rural, por Provincias y Comunas.** Ciudad Trujillo, Sección de Publicaciones de la Dirección General de Estadística, 1939. Dirección General de Estadística. Sección del Censo. 4 p. Mimeografiado.

Census of urban and rural population, by provinces and communes. Data from the 1935 census.

Cifras del censo de 1935.

RepDom-**2200**.C5 1920 **Población de la República, según el Censo Levantado el 24 de Diciembre de 1920, Clasificada por Sexo y Grupos de Edades.** Dirección General de Estadística. 1 p. Copia máquina.

Population of Dominican Republic according to 1920 census, classified by sex and age groups. 1-page, typewritten, prepared by the General Bureau of Statistics.

RepDom-**2210**.C5 1943.1 **Primer Censo Cafetalero Nacional, 1943.** Ciudad

Trujillo, Tall.Tipo."Rincon," 1944. Comisión de Defensa del Café y del Cacao. 172 p.

First census of coffee plantations of the Dominican Republic, 1943. Includes: Principal national results and detail by provinces, counties and districts; classification by provinces according to the area dedicated to coffee; detailed results by districts.

Incluye: Principales resultados nacionales y detalle por provincias, comunes y distritos municipales; clasificaciones por provincias según la extensión dedicada al café; resultados generales detallados por secciones municipales.

3000-3999 STATISTICAL SCIENCE —CIENCIA ESTÁDISTICA

RepDom-**3000.**M5 **Monografías Estadísticas de la Dirección General de Estadística, 1941-** . Cuidad de Trujillo. Dirección General de Estadística.

Series of monographs prepared by the Section of Special Studies of the General Bureau of Statistics.

Series de monografías preparadas por la Sección de Estudios Especiales.

RepDom-**3030.**C5 1944? **Cuadros de Materias y Programas Oficiales de la Comisión Nacional del Servicio Civil.** (Para aspirantes a cargos en la Administración Pública). Para las diferentes categorías de los servicios en el ramo de estadística (para todas las oficinas dependientes de la Dirección General de Estadística, de acuerdo con la Ley Número 43 (ley del servicio civil), publicada en la **Gaceta Oficial** número 5775, de fecha 22 de julio de 1942, reformada por la ley número 89, publicada en la **Gaceta Oficial** número 5807, de fecha 3 de octubre de 1942. Cuidad Trujillo, R.D., 1944? Comisión Nacional del Servicio Civil. 28 p.

Official programs and subjects of the National Civil Service Commission for the different categories of services in the field of statistics.

RepDom-**3030.**T5 1945 **Trujillo y la Estadística.** Por Nicolas Rizik H. Cuidad Trujillo, Editora Montalvo, 1945. 135 p.

Trujillo and the statistical service. Includes: Origin of statistics, and its developments in the Dominican Republic; some aspects of the national economy (agriculture, finance, trade); social welfare (public health, health and social welfare, education).

Contiene: Origen de la estadística y su evolución en la República Dominicana; algunos aspectos de la economía nacional (agricultura, finanzas, comercio); asistencia social (salubridad, sanidad y asistencia social, instruccion pública).

RepDom-**3300** 1944-46 **Enseñanza Estadística en la República Dominicana: Métodos y Materiales.** Material básico adquirido para la Encuesta del Instituto Interamericano de Estadística sobre los Métodos y Materiales en la Enseñanza Estadística en las Naciones Americanas 1944-46. In IASI on microfilm (Int-**3300.**S5).

Statistical training in the Dominican Republic: Methods and materials. Background material acquired for the IASI Survey on Statistical Training Methods and Materials in the American Nations, 1944-46. Includes: Correspondence, filled questionnaires and training programs for statistical personnel in the Dominican Republic.

Incluye: Correspondencia, cuestionarios llenados y programas de enseñanza para el personal estadístico en la República Dominicana.

RepDom-**3320.**P5 1940 **Programa de un Curso de Estadística Metodológica (Introducción a la Estadística Matemática) Seguido de un Curso de Matemáticas Generales Necesarias para su Estudio.** A cargo de José V. Montesino Samperio. Ciudad Trujillo, 1940. Dirección General de Estadística Nacional. 22 p.

Program of a course in statistical methods (introduction to mathematical statistics) followed by a course in general mathematics required for the study of statistical methods.

5000-5999 DEMOGRAPHY—DEMOGRAFIA

RepDom-**5000.**D5 1945 **Dominican Republic: Summary of Biostatistics.** Maps and charts, population, natality and mortality statistics. Washington, D.C., 1945. Prepared by U. S. Department of Commerce, Bureau of the Census, in cooperation with Office of the Coordinator of Inter-American Affairs. 93 p.

Resumen bioestadístico de la República Dominicana. Incluye mapas y cuadros y datos sobre población, natalidad y mortalidad.

RepDom-**5200.**C5 1946 **Commentary, Explanations and Sample Copies of the Forms and Manuals Used in Vital Records and Vital Statistics Activities in the Dominican Republic, November 1946.** By Chester W. Young. Ciudad Trujillo, November 1946. U. S. Public Health Service. Pan American Sanitary Bureau. 74 p. Typewritten.

Includes forms and instruction manuals used by the Central Office of Civil Status of the Department of Justice, the Demographic Section of the General Bureau of Statistics, and the Statistical Section of the Department of Health and Public Assistance.

Explicaciones y muestras de formularios utilizados para documentación y estadística vital en la República Dominicana, noviembre de 1946. Incluye formularios y manuales instructivos usados por la Oficina Central del Estado Civil del Departamento de Justicia, por la Sección Demográfica de la Dirección General de Estadística y por la Sección de Estadística de la Secretaría de Estado de Sanidad y Asistencia Pública.

RepDom-**5212**.S5 **Suicidios y Tentativas de Suicidios Verificados en la República.** 1939- . Anual. Ciudad de Trujillo. Dirección General de Estadística. Sección Sociográfica. Mimeografiado.

Annual data on suicides and attempted suicides.

RepDom-**5300**.R5 1942 **Refugee Settlement in the Dominican Republic.** A survey conducted under the auspices of the Brookings Institution and directed by Dana G. Munro. Washington, D.C., 1942. Brookings Institution. 410 p.

"The first part of the report presents a general survey of the economics of the refugee settlement, with emphasis on the problems in the already overpopulated Caribbean. The second part consists of an economic survey of the Dominican Republic, while the third part deals with the history, program, achievements, and the problems of the Sosua colony." (Source: **Books on Population, 1939-Mid-1944**.)

Radicación de refugiados en la República Dominicana. Un estudio conducido bajo los auspicios de la Institución Brooking y dirigido por Dana G. Munro. "La primera parte del informe presenta un estudio general sobre la economía de la colonización de refugiados, con especial referencia a los problemas de las ya excesivamente pobladas Antillas. La segunda parte consiste de un estudio económico sobre la República Dominicana, mientras que la tercera trata sobre la historia, programa, logro y problemas de la colonia Sosua." (Traducción del comentario.)

6000-6999 ECONOMICS—ECONOMIA

RepDom-**6000**.D5 1946 **Dominican Republic.** Foreign commerce yearbook country series. Washington, D.C., U. S. Govt.Print.Off., 1946. U. S. Department of Commerce. Bureau of Foreign and Domestic Commerce. International Reference Service, vol. 3. Unexamined.

Contains information about area, population, climate, agriculture, mining and manufacturing, commerce and industry, communication and transportation, power and light, national economic indices, average wages, foreign trade, etc.

República Dominicana. Anuario del comercio exterior de los Estados Unidos, series por países. Contiene información sobre superficie, población, clima, agricultura, minería y manufacturas, comercio e industria, transportes y comunicaciones, fuerza y luz, índices de la economía nacional, promedio de salarios, comercio exterior, etc.

RepDom-**6000**.E5 1946 **Economic Controls and Commercial Policy in the Dominican Republic.** One of a series of reports on economic controls and commercial policy in the American republics. Washington, D.C., 1946. United States Tariff Commission. 27 p.

Folleto sobre la República Dominicana, de la serie de informes de la Comisión de Aduanas de los Estados Unidos, sobre controles económicos y política comercial de las repúblicas americanas.

RepDom-**6000**.R5 1944 **Report Presented to the Conference of Commissions of Inter-American Development (Held in New York, May 9-18, 1944) by the Dominican Commission of Inter-American Development.** On the economic development of the Dominican government. (Translation from the Spanish original.) Washington, D.C., 1946? Inter-American Development Commission. 52 p.

Cover title—título de la cubierta: **Dominican Republic: Report Presented to the Conference of Commissions of Inter-American Development.**

Informe presentado a la Conferencia de Comisiones de Fomento Interamericano (realizada en N.York, del 9 al 18 mayo de 1944) por la Comisión Dominicana de Fomento Interamericano sobre el fomento económico del gobierno dominicano.

RepDom-**6230**.S5 **Sacrificio de Ganado.** 1938- . Anual. Ciudad Trujillo. Dirección General de Estadística. Sección de Producción y Economía. Mimeografiado.

Livestock slaughter. Data on livestock killed, and estimates on the consumption of meat in each commune. Annual.

Datos sobre ganado sacrificado y estimaciones del consumo de carne en cada municipio.

RepDom-**6400**.R5 1938 **Resumen General de la Producción de la Pequeña Industria Rural, Año 1938.** Ciudad Trujillo. Dirección General de Estadística. Mimeografiado. No examinado.

General summary of production of small rural industry, 1938.

RepDom-**6600**.M5 **Movimiento Postal, Telefónica y Radiotelegráfico. Carreteras de la República Dominicana.** 1936- . Anual. Ciudad Trujillo. Dirección General de Estadística. Sección de Gobierno y Administración. Mimeografiado.

Detailed and summary figures on posts, telephone, telegraph, and roads of the Dominican Republic. Annual.

Cifras detalladas y sumarias sobre movimiento postal, telefónico, telegráfico y de las carreteras del país.

RepDom-**6720**.E5 **Estadística Bancaria.** 1936- Semestral. Ciudad Trujillo. Dirección General de Estadística. Sección de Producción y Economía. Mimeografiado.

Previously published as—previamente publicado bajo el título: **Boletín de Estadística Bancaria.**

Banking statistics. Data on the operation and condition of all the banks in the country. Semiannual.

Datos sobre funcionamiento y situación de todos los bancos del país.

1944 **Estadística Bancaria, 1944.** Vol. IX, no. 2. Ciudad Trujillo, Sección de Publicaciones, 1945. Dirección General de Estadística. Sección de Producción y Economía. 41 p. Mimeografiado.

Includes: Deposits in commercial banks; deposits and withdrawals; credits issued by commercial banks; cash; main banking operations; banking operations in the main cities of the country; main banking indexes; general balance of the Reserve Bank of the Republic; New York quotations of Latin American currencies; banking series.

Incluye: Depósitos de los bancos comerciales; cargos en cuentas de depósitos; colocaciones por los bancos comerciales; caja; principales operaciones bancarias; movimiento bancario en las principales ciudades de la República; principales índices bancarios; balance general del Banco de Reservas de la República; cambio sobre Nueva York de monedas latinoamericanas; series bancarias.

RepDom-**6721**.R5 **Registro Público.** Estadística de operaciones registradas. 1939- . Anual. Ciudad Trujillo. Dirección General de Estadística. Sección de Producción y Economía. Mimeografiado.

Detailed and summary data for communes and provinces on the operation of the Public Register: Rents, mortgages, etc. Annual.

Datos detallados y resumidos de comunas y provincias, sobre el funcionamiento del Registro Público: Rentas, hipotecas, etc.

RepDom-**6730**.E5 **Estadística de los Negocios de Seguros.** 1936- . Anual. Ciudad Trujillo. Dirección General de Estadística. Sección de Producción y Economía. Mimeografiado.

Statistics of insurance transactions.

1942-44 **Estadıstica de los Negocios de Seguros, 1942-1943-1944.** Ciudad Trujillo, 1946. Dirección General de Estadística. Sección de Producción y Economía. 71 p.

Statistical tables and comparative charts on: Insurance activities during 1941-44 insurance companies authorized by the executive branch of the government and guarantees deposited by them; class of insurance and origin of the company (national or foreign). The principal data are: Number of policies, value insured, net premiums and damages paid and pending. Life insurance in separate tables.

Consta de cuadros estadísticos y esquemas comparativos sobre: Operaciones de seguros durante los años 1941-44, compañías de seguros autorizadas por el Poder Ejecutivo y fianzas depositadas por las mismas; clase de seguro y procedencia de las compañías (nacionales y extranjeras). Los datos principales son: Número de pólizas, valor asegurado, primas netas y siniestros pagados y pendientes. Los seguros sobre la vida están en cuadros por separado.

RepDom-**6900** **Estadísticas del Comercio Exterior en la República Dominicana: Métodos y Procedimientos.** Material básico adquirido para el estudio que el Instituto Interamericano de Estadística realiza sobre los métodos de estadísticas del comercio exterior en las naciones americanas, incluyendo la encuesta de 1946-47.

Foreign trade statistics in the Dominican Republic: Methods and procedures. Background material acquired for the study conducted by the Inter American Statistical Institute in methods of foreign trade statistics in the American nations, including the survey of 1946-47.

Consists of: Filled questionnaire and explanatory notes; codes.

Consiste de: Cuestionario llenado y notas explicativas; códigos.

RepDom-**6900**.E3 **Estudio Estadístico de Al-**
1941 **gunos Aspectos del Co-**

mercio Exterior de la República Dominicana, 1920-1939. Ciudad Trujillo, Sección de Publicaciones, 1941. Dirección General de Estadística. 125 p. Monografías estadísticas, no. 1.

Statistical study of some aspects of the foreign trade of the Dominican Republic, 1920-39.

RepDom-**6900**.E5 1920 **Executive Order #332: . . . Tariff Law on Imports and Exports . . . Effective on and after January 1, 1920.** San Juan, Bureau of Supplies, Printing, and Transportation, 1920. 212 p.

Published also in Spanish—publicado también en español (RepDom-**6900**.05, 1920).

Only five changes made in this tariff from 1920 to 1945. Classification very similar to that of Cuba, Haiti, and Nicaragua.

*RepDom-**6900**.E7 **Exportación de la República Dominicana.** 1931- . Mensual. Ciudad Trujillo. Dirección General de Estadística. Sección de Producción y Economía.

Exports of the Dominican Republic. Issued monthly, with a separate volume of annual summary, through 1944; 1945 has each month and annual summary, no. 1-13, in one volume. Each number has comparative data for the corresponding period of the preceding year.

Publicado mensualmente, con un volumen por separado conteniendo el resumen anual, hasta 1944 inclusive; los números mensuales y el anual de 1945 (no.1-13) están en un solo volumen. Cada número tiene datos comparativos del período respectivo del año anterior.

1945 **Exportación de la República Dominicana, 1945.** Vol. XIV, nos. 1-13. Ciudad Trujillo, Sección de Publicaciones, 1946. Dirección General de Estadística. Sección de Producción y Economía. 294 p.

Contents: Exports by article and country, by country and article, and by customs station and article, for 1944 and 1945.

Contiene: Exportación por artículos y países, países y artículos y por aduanas y artículos, 1944 y 1945.

*RepDom-**6900**.I5 **Importación de la República Dominicana.** 1939- . Mensual. Ciudad Trujillo. Dirección General de Estadística. Sección de Producción y Economía.

Imports of the Dominican Republic. Issued monthly, with a separate volume of annual summary, through 1943; 1944 and 1945 have all the months and annual summary, no. 1-13, in one volume. Each number has comparative data for the corresponding period of the preceding year.

Publicado mensualmente y un volumen por separado con el resumen anual hasta 1943 inclusive; para 1944 y 1945 están todos los meses y el resumen anual, no. 1-13, en un solo volumen. Cada número tiene datos comparativos del período correspondiente del año inmediato anterior.

1945 **Importación de la República Dominicana, 1945.** Vol. VI, no. 1-13. Ciudad Trujillo, Sección de Publicaciones, 1946. Dirección General de Estadística. Sección de Producción y Economía. 497 p.

Contents: Imports by article and country; value of imports by country; value of imports by customs stations; weight by kilos and value imported.

Contiene: Importación por artículos y países; valor de las importaciones por països; valor de las importaciones por aduanas; kilos de carga y valor importado.

RepDom-**6900**.M5 1926-37 **Maderas de Exportación.** Ciudad Trujillo. Dirección General de Estadística. Mimeografiado. No examinado.

Data on exports of wood, 1926-37.

Datos sobre exportación de maderas, 1926-37.

RepDom-**6900**.O5 1920 "Orden Ejecutiva No. 332 (de Septiembre 25, 1919) Que Promulga la Ley sobre Aranceles de Importación y Exportación que Estará en Vigor desde el Día 1° de Enero de 1920." **Gaceta Oficial,** año XXXV, no. 3051, octubre 1, 1919, p. 1-151.

Published also in English—publicado también en inglés (RepDom-**6900**.E5, 1920).

Solamente cinco correcciones se han hecho en esta Ley de Aranceles durante los años 1920 a 1945. Clasificación muy similar a las de Cuba, Haití y Nicaragua.

RepDom-**6900**.R5 1937 **Resumen General de la Producción de Algunos Artículos de Consumo y Exportación de la República, durante el Año 1937.** Ciudad Trujillo. Mimeografiado. No examinado.

Summary of the production of some articles for consumption and export in the Republic during the year 1937.

7000-7999 HEALTH AND CULTURAL STATISTICS—ESTADISTICAS DE SALUBRIDAD Y CULTURA

RepDom-**7100**.C5 1946 **Commentary Explanations and Sample Copies of the Forms Used in Medical and Public Health Statistics Activities (Not Including Vital Statis-**

tics) in the Dominican Republic, November 1946. By Chester W. Young. Ciudad Trujillo, November 1946. U. S. Public Health Service. Pan American Sanitary Bureau. 40 p. Typewritten.

Includes forms and instruction manuals used by the Statistical Section of the Department of Health and Public Assistance and the Sociographical Section of the General Bureau of Statistics.

Explicaciones y muestras de formularios utilizadas en estadísticas médicas y sanitarias en la República Dominicana noviembre de 1946 (no incluyendo estadística vital). Incluye formularios y manuales instructivos usados por la Sección de Estadística de la Secretaría de Estado de Sanidad y Asistencia Pública y por la Sección Sociográfica de la Dirección General de Estadística.

RepDom-**7110.**E5 **Estadística Sanitaria.** Fecha de iniciación? Anual. Ciudad Trujillo. Dirección General de Estadística. Sección Sociográfica. Mimeografiado.

Annual public health statistics.

RepDom-**7160.**A5 **Accidentes de Tránsito Terrestre por Circulación de Vehículos.** 1936- . Anual. Ciudad Trujillo. Mimeografiado. Dirección General de Estadística. Sección Sociográfica.

Traffic accidents.

8000-8999 GOVERNMENT AND POLITICAL STATISTICS—ESTADISTICAS GUBERNAMENTALES Y POLITICAS

RepDom-**8800.**F5 **Finanzas Municipales.** 1936- . Anual. Ciudad Trujillo. Dirección General de Estadística. Sección de Gobierno y Administración. Mimeografiado.

Annual statistics on municipal finance. Summaries and detailed figures on each municipality. Annual.

Presenta resúmenes de datos y cifras detalladas relativas a finanzas de cada una de las municipalidades.

UNITED STATES

1000-1999 GENERAL

USA-**1400**.F5 1938 **The Federal Chart Book.** Prepared by the staffs of the Central Statistical Board and National Resources Committee, under the direction of the Industrial Committee of the National Resources Committee. Washington, D.C., January 1938. Central Statistical Board. National Resource Committee. 143 p. Technical paper, no. 4.

Official compilation of graphs showing changes in general business and social conditions.

Compilación oficial de gráficos demostrando cambios en las condiciones sociales y del comercio.

USA-**1400**.T5 1946 **Trends in American Progress.** Facts and figures about the growth of economic life in America. Minneapolis, Minnesota, 1946? Investors Syndicate. 66 p.

A compilation of data covering the period 1854-1945, including vital statistics, employment, enrollment in schools, transportation and communication, housing, and foreign trade as well as data on money and finance, insurance, and principal agricultural and mining products. A predecessor volume was entitled: **A Half Century of Progress.**

Curso del progreso americano: Hechos y cifras relativas al crecimiento de la vida económica en Norteamérica. Una compilación de datos correspondientes al período 1894-1945, incluyendo movimiento demográfico, ocupación, matrícula escolar, transportes y comunicaciones, vivienda y comercio exterior, como también datos sobre moneda y finanzas, seguros y principales productos agrícolas y mineros. Un volumen precedente a éste se había publicado bajo el título de: **A Half Century of Progress.**

*USA-**1410**.S5 **Statistical Abstract of the United States.** 1879- . Annual. Washington, D.C. Department of Commerce. Bureau of the Census.

The official compendium of the federal government, summarizing available statistical information from governmental and non-governmental sources. The source of each table is indicated and can serve as a guide to more detailed information. A bibliography of sources of statistical data is included.

Resumen estadístico de Estados Unidos. El compendio oficial del gobierno federal que resume las informaciones estadísticas disponibles, obtenidas de fuentes oficiales y no oficiales. Se indica la fuente de información de cada cuadro de manera que sirve de guía para informaciones mas detalladas. Se incluye también una bibliografía de las fuentes de información estadística.

1946 **Statistical Abstract of the United States, 1946.** Sixty-seventh number. Washington, D.C., U.S.Govt.Print. Off., 1946. Department of Commerce, Bureau of the Census. 1039 p.

Contents: Area and population; vital statistics (including health and medical care); crime and criminals; immigration, emigration, and citizenship; education; climate; public lands; labor force (including employment, hours, and earnings); military services and veterans' affairs; social security and related programs; income and expenditures; prices; election; national government finances; state and local government finances and employment; banking and finance; communication systems (including postal service); power; public roads and motor vehicles; transportation, air and land — steam and electric railways, express companies, motor busses, and civil aeronautics; waterways, water traffic, and shipping; irrigation and drainage; farms — general statistics; farm production and related statistics; farm animals and animal products; farm crops and foodstuffs; forests and forest products; fisheries; mining and mineral products; construction and housing; manufactures; foreign commerce of the United States; commerce of territories and possessions; distribution and services.

Contiene: Superficie y población; movimiento demográfico (incluyendo sanidad y asistencia médica); crímenes y criminales; inmigración; emigración y ciudadanía; educación; clima; tierras públicas; potencial obrero (incluyendo ocupación, horas diarias de trabajo y remuneración); servicio militar, veteranos; seguro social y programas concernientes; ingresos y gastos; precios; elecciones, finanzas del gobierno nacional; finanzas y empleados de los gobiernos locales y de los estados; finanzas y operaciones bancarias; sistema de comunicaciones (incluyendo servicio postal); energía; carreteras y vehículos automotores; transportes aéreos y terrestres—ferrocarriles a vapor y eléctricos, compañías de expresos, líneas de ómnibus y aeronáutica civil; rutas marítimas, tránsito marítimo y navegación; irrigación y drenaje; estadísticas generales de granjas; producción agrícola y estadísticas relativas;

animales de granja y sus productos; cosechas de granja y productos alimenticios; bosques y productos forestales; minería y productos mineros; construcción y vivienda; fabricación; comercio exterior de los Estados Unidos; comercio de los territorios y posesiones; distribución y servicios.

USA-1910.A5 **The American Year Book.** A record of events and progress. 1910- . New York and London, D. Appleton and Co.

Publication suspended from 1920-24, inclusive—publicación suspendida desde 1920 a 1924 inclusive.

A compendium of American life, including science, humanities, arts, education, etc.

El anuario estadounidense. Crónica de acontecimientos y progreso. Un compendio sobre la vida en los Estados Unidos, incluyendo ciencias, humanidades, artes, educación, etc.

2000-2999 CENSUS—CENSOS

USA-2100.C5 1940.1 **Sixteenth Census of the United States: 1940: Methods and Procedures.** Materials used in taking population census on May 1. Acquired in connection with IASI Survey of Census Methods and Procedures in the American Nations, 1944. Department of Commerce. Bureau of the Census.

Consists of: Correspondence; laws, decrees, or other authorization; schedules; other forms; instructions; punch cards; codes; exposition and reports.

Décimosexto censo de los Estados Unidos, 1940: Métodos y procedimientos. Material usado en el levantamiento del censo del 1 de mayo. Adquirido con motivo del estudio que el Instituto Interamericano de Estadística realizó sobre Métodos y Procedimientos de los Censos en las Naciones Americanas, 1944. Consiste de: Correspondencia; leyes, decretos u otras disposiciones; cédulas; otros formularios; instrucciones; tarjetas de perforación; códigos; exposición e informes.

See also—véase también: **Enumerator's Record Book; Handbook for Employees of the Bureau of the Census; Classified Index of Occupations.**

USA-2100.C5 1940.4 **Sixteenth Census of the United States, 1940: Enumerator's Record Book.** Form P-6. Washington, D.C., U.S.Govt.Print.Off., 1940. Department of Commerce. Bureau of the Census. 45 p.

Décimosexto censo de los Estados Unidos, 1940: Registro del enumerador.

USA-2100.C5 1940.5 **Sixteenth Decennial Census of the United States: Instructions to Enumerators: Population and Agriculture, 1940.** Form PA-1. Washington, D.C. Department of Commerce. Bureau of the Census. 173 p.

Décimosexto censo decenal de los Estados Unidos: Instrucciones a los enumeradores: población y agricultura, 1940.

USA-2100.C5 1940.6 **Instructions: Preparation of Plans of Division by Enumeration Districts, 1940.** Revised. Washington, D. C., June 1939. Department of Commerce. Bureau of the Census. 25 p. Mimeographed.

Instrucciones: Preparación de planos de división por distritos de enumeración, censo de 1940.

USA-2100.C5 1940.7 **Instructions for Processing Data of the Sixteenth Decennial Census.** Washington, D.C., 1940. Department of Commerce. Bureau of the Census.

Instrucciones para el procesamiento de los datos del décimosexto censo decenal.

USA-2110.C5 1940.1 **Censo de Agricultura de los Estados Unidos.** Washington, D.C., 1945. Department of Commerce. Bureau of the Census. 6 p. Copia máquina.

Census of agriculture of the United States. Includes: Purpose of the census; history of the census; the mechanism of tabulation and procedures; preparation of schedules and questionnaire for 1940; enumeration method; office procedure.

Incluye: Propósito del censo; historia del censo; el mecanismo de tabulación y procedimientos; preparación de las boletas y cuestionario de 1940; método de enumeración; procedimiento en la oficina.

USA-2110.C5 1940.2 **Agriculture Handbook.** Sixteenth census of the United States: 1940. Description and illustrations of the uses of agriculture census statistics in education, business, research, and visual analysis; with explanations of the technique of tabulation and procedures. Washington, D.C., U.S.Govt. Print.Off., 1943. Department of Commerce. Bureau of the Census. 245 p.

Manual agrícola. Décimosexto censo de los Estados Unidos, 1940. Descripción e ilustraciones sobre los usos de las estadísticas de los censos agrícolas en educación, negocios, investigación y análisis visual; con explicaciones de la técnica de la tabulación y procedimientos.

USA-2120.C5 1940.1 **Sixteenth Decennial Census of the United States: Instructions to Enumerators, Housing, 1940.** Form

HC-1. Washington, D.C., U.S.Govt.Print. Off., 1940. Department of Commerce. Bureau of the Census. 29 p.

Décimosexto censo decenal de los Estados Unidos: Instrucciones a los enumeradores sobre vivienda, 1940.

USA-2150.C5 1939.1 **Sixteenth Census of the United States: 1940: Instructions to Enumerators for Business and Manufactures, 1939.** Washington, D.C., Department of Commerce. Bureau of the Census. 121 p.

Décimosexto censo de los Estado Unidos, 1940: Instrucciones a los enumeradores para comercios y fábricas, 1939.

USA-2200.C3 1944 **Census of Congested Production Areas: 1944.** Washington, D. C., 1944. Department of Commerce. Bureau of the Census. Series CA-1, **Final Population Figures for the Area and Its Constituent Parts;** series CA-2, **Wartime Changes in Population and Family Characteristics;** series CA-3, **Characteristics of the Population, Labor Force, Families, and Housing.**

This census was taken in cooperation with the Committee for Congested Production Areas. Reports are made for ten congested production areas: Mobile, San Diego, Charleston (S.C.), San Francisco Bay, Los Angeles, Portland-Vancouver, Hampton Roads, Puget Sound, Muskegon, and Detroit-Willow Run.

Censo de las áreas de producción congestionadas: 1944. Este censo fué levantado en cooperación con la Comisión de Areas de Producción Congestionadas. Contiene informes hechos sobre diez áreas de producción congestionadas. Mobile, San Diego, Charleston (S.C.), San Francisco Bay, Los Angeles, Portland-Vancouver, Hampton Roads, Puget Sound, Muskegon y Detroit-Willow Run.

*USA-2200.C5 1940.1 **Sixteenth Census of the United States: 1940: Population.** Washington, D.C., U.S.Govt.Print. Off., 1940-43. Department of Commerce. Bureau of the Census. 4 vol.: Vol. I, **Number of Inhabitants.** Total population for states, counties, and minor civil divisions; for urban and rural areas; for incorporated places; for metropolitan districts; for census tracts; vol. II (in 7 parts). **Characteristics of the Population.** Sex, age, race, nativity, citizenship, country of birth of foreign-born white, school attendance, education, employment status, class of worker, major occupation group, and industry group; reports by states; vol. III (in 5 parts), **The Labor Force.** Occupation, industry, employment, and income; reports by states; vol. IV (in 4 parts), **Characteristics by Age.** Marital status, relationship, education and citizenship; reports by states.

Décimosexto censo de los Estados Unidos, 1940: Población. 4 vol.: Vol. I, número de habitantes: Total de la población por estados, distritos y subdivisiones menores; por áreas rurales y urbanas; por zonas incorporadas; por distritos metropolitanos; por sectores censales; vol. II (en 7 partes), características de la población: Sexo, edad, raza, natividad, ciudadanía, país de origen de los blancos nacidos en el extranjero, asistencia escolar, educación, estado ocupacional, clase de trabajador, grupo principal de ocupación, y grupo industrial; informes por estados; vol. III (en 5 partes), el potencial obrero: Ocupación, industria, empleo y renta; informes por estados; vol. IV (en 4 partes), características por edad: Estado civil, parentesco, educación y ciudadanía; informes por estados.

USA-2200.C5 1940.2 **Sixteenth Census of the United States: 1940: Population: Characteristics of Persons Not in the Labor Force 14 Years Old or Over.** Age, sex, color, household relationship, months worked in 1939, and usual major occupation group. Washington, D.C., U.S.Govt.Print.Off., 1943. Department of Commerce. Bureau of the Census. 117 p.

Décimosexto censo de los Estados Unidos, 1940: Población: Características de las personas no comprendidas en el potencial obrero, de 14 o más años de edad. Edad, sexo, color, relación con el jefe de familia, meses trabajados en 1939 y principales grupos de ocupación habitual.

USA-2200.C5 1940.3 **Sixteenth Census of the United States: 1940: Population: Characteristics of the Nonwhite Population by Race.** Washington, D.C.,U.S.Govt.Print. Off., 1943. Department of Commerce. Bureau of the Census. 112 p.

Décimosexto censo de los Estados Unidos, 1940: Población: Características de la población no blanca, por razas.

USA-2200.C5 1940.4 **Sixteenth Census of the United States: 1940; Population: Comparative Occupation Statistics for the United States, 1870 to 1940.** A Comparison of the 1930 and the 1940 census occupation and industry classifications and statistics; a comparable series of occupation statistics, 1870 to 1930; and a social-economic grouping of the labor force, 1910 to 1940. By Alba M. Edwards. Washington, D. C., U. S. Govt. Print. Off., 1943. Department of Commerce. Bureau of the Census. 206 p.

Décimosexto censo de los Estados Unidos, 1940; población: Estadísticas comparativas de

ocupación en los Estados Unidos durante el período 1870-1940. Una comparación de las clasificaciones y estadísticas de ocupación e industrias de los censos de 1930 y 1940; una serie comparativa de estadísticas sobre ocupación, 1870 a 1930; y una agrupación económico-social del potencial obrero, 1910 a 1940.

USA-2200.C5 1940.5a **Sixteenth Census of the United States: 1940: Population: Differential Fertility, 1940 and 1910: Fertility for States and Large Cities.** Washington, D.C., U.S.Govt.Print.Off., 1943. Department of Commerce. Bureau of the Census. 281p.

Décimosexto censo de los Estados Unidos, 1940: Población: Fertilidad diferencial, 1940 y 1910: Fertilidad por estados y grandes ciudades.

USA-2200.C5 1940.5b **Sixteenth Census of the United States: 1940: Population: Differential Fertility, 1940 and 1910: Standardized Fertility Rates and Reproduction Rates.** Washington, D.C., U.S.Govt.Print.Off., 1944. Department of Commerce. Bureau of the Census. 40 p.

Décimosexto censo de los Estados Unidos, 1940: Población: Fertilidad diferencial, 1940 y 1910: Coeficientes estandardizados de fertilidad y reproducción.

USA-2200.C5 1940.5c **Sixteenth Census of the United States: 1940: Population: Differential Fertility, 1940 and 1910: Women by Number of Children Under 5 Years Old.** Washington, D. C., U. S. Govt. Print. Off., 1945. Department of Commerce. Bureau of the Census. 265 p.

Décimosexto censo de Estados Unidos, 1940; población: Fertilidad diferencial, 1940 y 1910: Mujeres por número de hijos menores de 5 años.

USA-2200.C5 1940.5d **Sixteenth Census of the United States: 1940: Population: Differential Fertility, 1940 and 1910: Women by Number of Children Ever Born.** Washington, D.C., U.S.Govt.Print.Off., 1945. Department of Commerce. Bureau of the Census. 410 p. Unexamined.

Décimosexto censo de los Estados Unidos, 1940: Población: Fertilidad diferencial, 1940 y 1910: Mujeres por número de hijos nacidos.

USA-2200.C5 1940.6 **Sixteenth Census of the United States: 1940: Population: Education, Occupation and Household Relationship of Males 18 to 44 Years Old.** Prepared by the Division of Population . . . in cooperation with the Special Services Division of the War Department. Washington, D.C., U.S.Govt. Print.Off., 1943. Department of Commerce. Bureau of the Census. 23 p.

Décimosexto censo de los Estados Unidos 1940: Poblacion: Varones de 18 a 44 años de edad: educación, ocupación y relación con el jefe de familia.

USA-2200.C5 1940.7a **Sixteenth Census of the United States: 1940: Population and Housing: Families: General Characteristics.** States, cities of 100,000 or more, and metropolitan districts of 200,000 or more. Washington, D.C., U.S.Govt.Print.Off., 1943. Department of Commerce. Bureau of the Census. 332 p.

Décimosexto censo de los Estados Unidos, 1940: Población y vivienda: Familias: Características generales. Estados, ciudades de 100.000 o más habitantes y distritos metropolitanos de 200.000 o más.

USA-2200.C5 1940.7b **Sixteenth Census of the United States: 1940: Population and Housing: Families: Characteristics of Rural-Farm Families.** Regions and divisions. Washington, D.C., U.S.Govt.Print.Off., 1943. Department of Commerce. Bureau of the Census. 82 p.

Décimosexto censo de los Estados Unidos, 1940: Población y vivienda: Familias: Características de familias rurales. Regiones y divisiones.

USA-2200.C5 1940.7c **Sixteenth Census of the United States: 1940: Population: Families: Employment Status.** Regions and cities of 1,000 or more. Washington, D. C., U.S.Govt.Print.Off., 1943. Department of Commerce. Bureau of the Census. 110 p.

Décimosexto censo de los Estados Unidos, 1940: Población: Familias: Situación de la ocupación. Regiones y ciudades de 1.000 o más habitantes.

USA-2200.C5 1940.7d **Sixteenth Census of the United States: 1940: Population: Families: Family Wage or Salary Income in 1939.** Regions and cities of 1,000,000 or more. Washington, D.C., U.S.Govt.Print.Off., 1943. Department of Commerce. Bureau of the Census. 156 p.

Décimosexto censo de los Estados Unidos, 1940: Población: Familias: Ingresos familiares por sueldos o salarios en 1939. Regiones y ciudades de 1.000.000 o más habitantes.

USA-2200.C5 1940.7e **Sixteenth Census of the United States: 1940: Population and Housing: Families: Income and Rent** Regions, cities of 1,000,000 or more, and metro-

politan districts of 1,000,000 or more. Washington, D.C., U.S.Govt.Print .Off., 1943. Department of Commerce. Bureau of the Census. 237 p.

Décimosexto censo de Estados Unidos, 1940; población y vivienda: Familias: Renta y alquiler. Regiones, ciudades de 1.000.000 o más habitantes y distritos metropolitanos de 1.000.000 o más habitantes.

USA-2200.C5 1940.7f **Sixteenth Census of the United States: 1940: Population: Families: Size of Family and Age of Head.** Regions and cities of 1,000,000 or more. Washington, D.C., U.S.Govt.Print.Off., 1944. Department of Commerce. Bureau of the Census. 125 p.

Décimosexto censo de los Estados Unidos, 1940: Población: Familias: Número de componentes de la familia y edad del jefe. Regiones y ciudades de 1.000.000 o más habitantes.

USA-2200.C5 1940.7g **Sixteenth Census of the United States: 1940: Population and Housing: Families: Tenure and Rent.** Regiones, cities of 1,000,000 or more, and metropolitan districts of 500,000 or more. Washington, D.C., U.S.Govt.Print.Off., 1943. Department of Commerce. Bureau of the Census. 141 p.

Décimosexto censo de los Estados Unidos, 1940: Población y vivienda: Familias: Tenenencia y alquileres. Regiones, ciudades de 1.000.000 o más habitantes y distritos metropolitanos de 500.000 o más habitantes.

USA-2200.C5 1940.7h **Sixteenth Census of the United States, 1940: Population: Families: Types of Families.** Regions and cities of 1,000,000 or more. Washington, D.C., U. S. Govt. Print. Off., 1943. Department of Commerce. Bureau of the Census. 221 p.

Décimosexto censo en los Estados Unidos, 1940: Población: Familias: Tipo de familias. Regiones y ciudades de 1.000.000 o más habitantes.

USA-2200.C5 1940.8 **Sixteenth Census of the United States: 1940: Population: Internal Migration 1935 to 1940.** Color and sex of migrants. Washington, D.C., U.S.Govt. Print.Off., 1943. Department of Commerce. Bureau of the Census. 490 p.

Décimosexto censo de los Estados Unidos, 1940: Población: Migración interna 1935 a 1940. Color y sexo de los migrantes.

USA-2200.C5 1940.9 **Sixteenth Census of the United States: 1940: Population: Special Report on Institutional Population, 14 Years Old and Over.** Characteristics of inmates in penal institutions, and in institutions for the delinquent, defective, and dependent. Washington, D.C., U.S.Govt.Print.Off., 1943. 361 p.

Décimosexto censo de los Estados Unidos, 1940: Población: Informe especial sobre población institucional de 14 y más años de edad.

USA-2200.C5 1940.10a **Sixteenth Census of the United States: 1940: Population: The Labor Force (Sample Statistics): Employment and Family Characteristics of Women.** Washington, D.C., U.S.Govt.Print.Off., 1943. Department of Commerce. Bureau of the Census. 212 p.

Décimosexto censo de los Estados Unidos, 1940: Población: Potencial obrero (estadísticas de muestra): Empleo y características familiares de mujeres.

USA-2200.C5 1940.10b **Sixteenth Census of the United States: 1940: Population: The Labor Force (Sample Statistics): Employment and Personal Characteristics.** Washington, D.C., U.S.Govt.Print.Off., 1943. Department of Commerce. Bureau of the Census. 177 p.

Décimosexto censo de los Estados Unidos, 1940: Población: Potencial obrero (estadísticas de muestra): Ocupación y características personales.

USA-2200.C5 1940.10c **Sixteenth Census of the United States: 1940: Population: The Labor Force (Sample Statistics): Industrial Characteristics.** Washington, D.C., U. S.Govt.Print.Off., 1943. Department of Commerce. Bureau of the Census. 174 p.

Décimosexto censo de los Estados Unidos 1940: Población: Potencial obrero (estadísticas de muestra): Características industriales.

USA-2200.C5 1940.10d **Sixteenth Census of the United States: 1940: Population: The Labor Force (Sample Statistics): Occupational Characteristics.** Washington, D.C., U.S.Govt.Print.Off., 1943. Department of Commerce. Bureau of the Census. 256 p.

Décimosexto censo de los Estados Unidos, 1940: Población: Potencial obrero (estadísticas de muestra): Características de la ocupación.

USA-2200.C5 1940.10e **Sixteenth Census of the United States, 1940: Population: The Labor Force (Sample Statistics): Usual Occupation.** Washington, D.C., U.S.Govt. Print.Off., 1943. Department of Commerce. Bureau of the Census. 63 p.

Décimosexto censo de los Estados Unidos, 1940: Población: Potencial obrero (estadísticas de muestra): Ocupación habitual.

USA-2200.C5 1940.10f **Sixteenth Census of the United States: 1940: Population: The Labor Force (Sample Statistics): Wage or Salary Income in 1939.** Washington, D.C., U.S.Govt.Print.Off., 1943. Department of Commerce. Bureau of the Census. 194 p.
Décimosexto censo de los Estados Unidos. 1940: Población: Potencial obrero (estadísticas de muestra): Sueldos y salarios en 1939.

USA-2200.C5 1940.11a **Sixteenth Census of the United States: 1940: Population: Nativity and Parentage of the White Population: General Characteristics.** Age, marital status, and education for states and large cities. Washington, D.C., U.S.Govt.Print. Off., 1943. Department of Commerce. Bureau of the Census. 279 p.
Décimosexto censo de los Estados Unidos, 1940: Población: Lugar de nacimiento y origen de la población blanca: Características generales. Edad, estado civil y educación, por estados y grandes ciudades.

USA-2200.C5 1940.11b **Sixteenth Census of the United States: 1940: Population: Nativity and Parentage of the White Population: Country of Origin of Foreign Stock.** By nativity, citizenship, age, and value or rent of home, for states and large cities. Washington, D.C., U.S.Govt.Print.Off., 1943. Department of Commerce. Bureau of the Census. 122 p.
Décimosexto censo de los Estados Unidos, 1940: Población: Lugar de nacimiento y origen de la población blanca: País de origen de los extranjeros. Por lugar de nacimiento, ciudadanía, edad, y valor o alquiler de la casa, por estados y grandes ciudades.

USA-2200.C5 1940.11c **Sixteenth Census of the United States: 1940: Population: Nativity and Parentage of the White Population: Mother Tongue.** By nativity, parentage, country of origin, and age, for states and large cities. Washington, D.C., U.S.Govt.Print. Off., 1943. Department of Commerce. Bureau of the Census. 58 p.
Décimosexto censo de los Estados Unidos: Población: Lugar de nacimiento y origen de la población blanca: Lengua materna. Por lugar de nacimiento, origen, país de origen y edad, por estados y grandes ciudades.

USA-2200.C5 1940.12 **Sixteenth Census of the United States, 1940: Areas of the United States, 1940.** Washington, D. C., U. S. Govt. Print. Off., 1942. Department of Commerce. Bureau of the Census. 465 p.
Décimosexto censo de los Estado Unidos, 1940: Superficies de los Estados Unidos, 1940.

USA-2200.C5 1940.13 **Sixteenth Census of the United States: 1940: Population: Unincorporated Communities.** United States, by states. Total population of unincorporated communities having 500 or more inhabitants for which separate figures could be compiled. Washington, D.C., U.S.Govt.Print. Off., 1943. Department of Commerce. Bureau of the Census. 32 p. Unexamined.
Décimosexto censo de los Estados Unidos, 1940: Población: Comunidades no organizadas de 500 o más habitantes, para las cuales se ha podido recopilar cifras por separado.

USA-2200.C5 1940.14 **Sixteenth Census of the United States: 1940: Population: State of Birth of the Native Population.** Washington, D.C., U.S. Govt.Print.Off., 1944. Department of Commerce. Bureau of the Census. 78 p.
Décimosexto censo de los Estados Unidos, 1940: Población: Estado de nacimiento de la población nativa.

USA-2200.C5 1940.15 **Sixteenth Census of the United States: 1940: Population: Estimates of Labor Force, Employment and Unemployment in the United States, 1940 and 1930.** Washington, D.C., U.S.Govt.Print. Off., 1944. Department of Commerce. Bureau of the Census. 18 p.
Décimosexto censo de los Estados Unidos, 1940: Población: Estimaciones sobre potencial obrero; ocupación y desocupación en los Estados Unidos, 1940 y 1930.

USA-2200.C5 1930 **Fifteenth Census of the United States, 1930: Abstract of the Fifteenth Census of the United States,** Washington, D.C., U.S.Govt.Print.Off., 1933. Department of Commerce. Bureau of the Census. 968 p.
Décimoquinto censo de los Estados Unidos, 1930: Resumen.

USA-2200.C5 1930.2 **Fifteenth Census of the United States: 1930: The Indian Population of the United States and Alaska.** Washington, D.C., U.S.Govt.Print.Off., 1937. Department of Commerce. Bureau of the Census. 238 p.
Décimoquinto censo de los Estados Unidos, 1930: La población indígena de los Estados Unidos y Alaska.

USA-2200.C7 1937 **Census of Unemployment, 1937: Final Report on Total and Partial Unemployment.** Washington, D.C., U.S.Govt.Print.Off., 1938. Census of Partial Employment, Unemployment and Occupations. 4 vol.: Vol. I, **United States Summary; Geographic Divisions; Reports by States, Alabama-Indiana**; vol. II **Reports by States, Iowa-New York;** vol. III, **Reports by States, North Carolina-Wyoming; Alaska and Hawaii**; vol. IV, **The Enumerative Check Census.**

Censo de desocupación, 1937: Informe final sobre la desocupación total y parcial.

USA-2200.C7 1937.A **Census of Unemployment, 1937: Final Report on Total and Partial Unemployment for Geographic Divisions.** Including data for registrants in each industry group. This report is a supplement to the United States summary. Washington, D.C., U.S.Govt.Print.Off., 1938. Census of Partial Employment, Unemployment, and Occupations. 89 p.

Censo de desocupación, 1937: Informe final sobre desocupación total y parcial por divisiones geográficas. Incluyendo datos de registrados en cada uno de los grupos industriales. Este informe es un suplemento al resumen de los Estados Unidos.

USA-2200.C8 1910 **Indian Population in the United States and Alaska, 1910.** Washington, D.C., U.S.Govt.Print. Off., 1915, Department of Commerce. Bureau of the Census. 285 p.

Población indígena en los Estados Unidos y Alaska, 1910.

*USA-2210.C5 1940.1 **Sixteenth Census of the United States: 1940: Agriculture.** Washington, D.C., U.S.Govt.Print. Off., 1942-43. Department of Commerce. Bureau of the Census. 3 vol.: Vol. I (in 6 parts), **First and Second Series State Reports.** Statistics for countries; farms and farm property, with related information for farms and farm operators, livestock and livestock products, and crops; vol. II (in 3 parts), **Third Series State Reports.** Statistics for counties; value of farm products, farms classified by major sources of income, farms classified by value of products; vol. III, **General Report.** Statistics by subjects.

Décimosexto censo de los Estados Unidos, 1940: Agricultura. 3 vol.: Vol. I (en 6 partes), primera y segunda serie de informes de los estados (estadísticas por distritos; granjas y propiedades granjeras con información relativa a granjas y explotadores de granjas, ganadería y productos ganaderos, y cosechas); vol. II, (en 3 partes) tercera serie de informes de los estados (estadísticas por distritos; valor de los productos de granja, granjas clasificadas por principales fuentes de ingresos, granjas clasificadas por el valor de los productos); vol. III, informe general (estadísticas por materia).

USA-2210.C5 1940.2 **Sixteenth Census of the United States: 1940: Agriculture: Cross-Line Acreage.** A special study; farms reporting and acreage by place of enumeration and by location on acreage, with relationship to all farms, by counties, with a summary for the United States; 1940 and 1935. Washington, D. C., U.S.Govt.Print.Off., 1943. Department of Commerce. Bureau of the Census. 311 p.

Décimosexto censo de los Estados Unidos, 1940 agricultura: Cantidad de acres correspondientes a fincas que se hallan ubicadas en mas de una división administrativa. Un estudio especial.

USA-2210.C5 1940.3 **Sixteenth Census of the United States: 1940: Agriculture: Value of Farm Products by Color and Tenure of Farm Operator.** Statistics, by color and tenure of farm operator, on the value of farm products sold, traded, or used by farm households. Sample data from the sixteenth census were used as a basis for obtaining these statistics. The figures are, therefore, subject to sampling errors. A special study. By Irvin Holmes. Washington, D.C., U.S.Govt.Print.Off. 1944. Department of Commerce. Bureau of the Census. 291 p.

Estudio especial. Datos expuestos según color y clase de posesión del agricultor, sobre el valor de los productos agrícolas que se vendieron, se traficaron o se usaron en los hogares agrícolas.

USA-2210.C5 1940.4 **Sixteenth Census of the United States: 1940: Drainage of Agricultural Lands.** Land in drainage enterprises, capital invested and drainage works. Statistics for counties with state and United States summaries and a synopsis of drainage laws. Washington, D.C., U.S.Govt.Print.Off., 1942. Department of Commerce. Bureau of the Census. 683 p.

Décimosexto censo de los Estados Unidos 1940: Drenaje de las tierras agrícolas.

USA-2210.C5 1940.6 **Sixteenth Census of the United States: 1940: Irrigation o Agricultural Lands.** Irrigation enterprises, areas, irrigation works, investment, indebtedness, maintenance and operation, water used, pay roll and employees, and irrigated crops, with detailed statistics for drainage basins and counties, and summaries for states and the

United States. Washington, D.C., U.S.Govt. Print. Off., 1942. Department of Commerce. Bureau of the Census. 689 p.

Décimosexto censo de los Estados Unidos: 1940: Irrigación de las tierras agrícolas.

USA-**2210**.C5 1940.15 **Sixteenth Census of the United States: 1940: Agriculture: Abandoned or Idle Farms.** Number and acreage, with classification of those reporting acreage by cause of nonoperation and by year of abandonment. Statistics for counties and a summary for the United States. A special study. Washington, D.C., U.S.Govt.Print.Off., 1943. Department of Commerce. Bureau of the Census. 201 p.

Décimosexto censo de los Estados Unidos: 1940: Agricultura: Tierras abandonadas y estériles.

USA-**2210**.C5 1940.17 **Sixteenth Census of the United States: 1940: Agriculture: Cows Milked and Dairy Products.** Number of cows milked, milk produced, disposition of dairy products, and number of cows kept mainly for milk production, classified by number of cows milked, by counties; with related data for other classes of livestock and livestock products, for the states and for the United States. Washington, D.C., U.S.Govt.Print.Off., 1942. Department of Commerce. Bureau of the Census. 556 p.

Décimosexto censo de los Estados Unidos: 1940: Agricultura: Vacas lecheras y productos de la industria lechera.

USA-**2210**.C5 1940.25 **Sixteenth Census of the United States: 1940: Agriculture: Special Cotton Report.** Farms reporting, acreage, and production and value of farm products by number of bales harvested. With statistics for counties and a summary for the United States. Washington, D. C., U. S. Govt. Print. Off., 1943. Department of Commerce. Bureau of the Census. 265 p.

Décimosexto censo de los Estados Unidos: 1940: Agricultura: Informe especial sobre algodón.

USA-**2210**.C5 1940.27 **Sixteenth Census of the United States: 1940: Agriculture: Special Poultry Report.** Statistics by geographic divisions and states for poultry of all kinds on hand and raised; by counties for chickens and chicken egg production by number of chickens on hand; and by counties for farms reporting chickens and turkeys raised by numbers raised. Washington, D.C., U.S.Govt. Print. Off., 1942. Department of Commerce. Bureau of the Census. 739 p.

Décimosexto censo de los Estados Unidos: 1940: Agricultura: Informe especial sobre aves de corral.

*USA-**2220**.C5 1940.1 **Sixteenth Census of the United States: 1940: Housing.** Washington, D.C., U.S.Govt.Print.Off., 1943. Department of Commerce. Bureau of the Census. 4 vol.: Vol. I (in 2 parts), **Data for Small Areas.** Selected housing statistics for states, counties, and minor civil divisions; for rural and urban areas; for incorporated places; and for metropolitan districts; reports by states; vol. II (in 5 parts), **General Characteristics.** Occupancy and tenure status, value of home or monthly rent, size of household and race of head, type of structure, exterior material, year built, conversion, state of repair, number of rooms, housing facilities and equipment, and mortgage status; reports by states; vol. III (in 3 parts), **Characteristics by Monthly Rent or Value.** These statistics form the basis for determining the relationship between rent or value and such characteristics as type and age of structure, state of repair, number of rooms, size of household, race of head, persons per room, housing facilities and equipment, and mortgage status; reports by states; vol. IV (in 3 parts), **Mortgages on Owner-Occupied Nonfarm Homes.** First mortgages by amount outstanding, type of payment, frequency and amount of payment, interest rate, and holder of mortgage; all mortgaged properties by value of property, estimated rental value, year built, and color of occupants . . .; supplements A and B show mortgage data for homes built in 1935-40, and homes occupied by non-white owners, respectively; reports by states.

Décimosexto censo de los Estados Unidos, 1940: Vivienda. 4 vol.: Vol. I (en 2 partes), datos para pequeñas áreas (seleccionadas estadísticas de viviendas, por estados, distritos y subdivisiones civiles menores; por áreas urbanas y rurales; por lugares incorporados; y por distritos metropolitanos; informes por estados); vol. II (en 5 partes), características generales (ocupación y estado de propriedad, valor de la casa o monto del alquiler mensual, número de ocupantes y raza del principal, tipo de construcción, material exterior, año en que fué edificado, transformación, estado de conservación, número de habitaciones, facilidades de construcción y suministro de accesorios, y situación hipotecaria; informes por estados); vol. III (en tres partes), características por alquiler mensual o valor (estas estadísticas forman la base para determinar las relaciones entre el alquiler o valor y carac-

terísticas tales como tipo y año de construcción, estado de conservación, número de habitaciones, número de ocupantes, raza del principal, personas por habitación, facilidades de construcción y suministro de accesorios, situación hipotecaria; informes por estados); vol. IV (en 3 partes), hipotecas sobre viviendas urbanas ocupadas por sus dueños (primeras hipotecas por saldos a amortizar, tipo de pago, frecuencia y monto de las cuotas, tasa de interés y acreedor hipotecario; todas las propiedades hipotecadas por valor de la propiedad, estimación del valor de alquiler, año de construcción y color de los ocupantes . . . ; los suplementos A y B contienen datos sobre hipotecas de casas construídas en 1935-40, y viviendas ocupadas por propietarios no blancos, respectivamente; informes por estados).

USA-**2220**.C5 1940.2 **Sixteenth Census of the United States: 1940: Housing: Analytical Maps . . . ; Block Statistics.** Prepared under the supervision of Dr. Leon E. Truesdell, Chief Statistician for Population, Bureau of the Census. Prepared and duplicated by New York City WPA War Services. Washington, D.C., 1943? Work Projects Administration, New York City. 96 vol. Consists of a volume showing housing characteristics by blocks for each of ninety-two cities of the United States and four boroughs of New York City. Unexamined.

Décimosexto censo de los Estados Unidos, 1940: Vivienda: mapas analíticos . . . ; estadísticas por cuadra. Consiste de un volumen sobre características de las viviendas, por cuadras, de cada una de 92 ciudades de los Estados Unidos y los cuatro distritos de la ciudad de Nueva York.

USA-**2220**.C5 1940.3 **Sixteenth Census of the United States: 1940: Housing: Characteristics by Type of Structure.** Regions, states, cities of 100,000 or more, and principal metropolitan districts. Washington, D.C., U.S.Govt.Print.Off., 1945. Department of Commerce. Bureau of the Census. 402 p. Unexamined.

Décimosexto censo de los Estados Unidos, 1940: Vivienda: Características por tipo de construcción. Regiones, estados, ciudades de 100,000 o más habitantes y principales distritos metropolitanos.

USA-**2250**.C3 **Census of Electrical Industries: Telephones and Telegraphs.** 1902-30. Quinquennial. Washington, D. C. Department of Commerce. Bureau of the Census. Unexamined.

The statistics for 1902 (issued as Bulletin 17 and also as **Special Reports**) and those for 1912 (issued as Bulletin 123) have title: **Telephones and Telegraphs, 1902 and 1912** respectively. 1907 has title, **Telegraph Systems, 1907**; statistics for 1932 are found in **Census of Electrical Industries, 1932: Telephones and Telegraphs;** 1922 in two issues; 1917 has full title, **Census of Electrical Industries, 1917: Telegraphs and Municipal Electric Fire-Alarm and Police-Patrol Signaling Systems.**

Censo de industrias eléctricas; teléfonos y telégrafos. Quinquenal.

*USA-**2250**.C5 1940 **Sixteenth Census of the United States: 1940: Census of Business, 1939.** Washington, D.C., U.S. Govt.Print.Off., 1942-43. Department of Commerce. Bureau of the Census. 5 vol.: Vol. I, **Retail Trade:** Part I, **United States Summaries and General Analyses;** part 2, **Commodity Sales and Analysis by Sales Size;** part 3, **Kinds of Business, by Area, States, Counties, and Cities;** vol. II, **Wholesale Trade;** vol. III, **Service Establishments, Places of Amusement, Hotels, Tourist Courts and Tourist Camps;** vol. IV, **Construction;** vol. V, **Distribution of Manufacturers' Sales.**

Décimosexto censo de los Estados Unidos, 1940: Censo del comercio, 1939. 5 vol.: Vol. I, comercio al por menor: Parte 1, resúmenes de Estados Unidos y análisis general; parte 2, venta de artículos y análisis por importancia de las ventas; parte 3, clases de negocios, por área, estados, distritos y ciudades; vol. II, comercio al por mayor; vol. III, establecimientos de servicios, lugares de diversión, hoteles, mansiones y campos para turistas; vol. IV, construcción; vol. V, distribución de las ventas de fabricantes.

*USA-**2250**.C7 1939.1 **Sixteenth Census of the United States: 1940: Manufactures, 1939.** Washington, D.C., U.S.Govt. Print.Off., 1942. Department of Commerce. Bureau of the Census. 4 vol.: Vol. I, **Statistics by Subjects;** vol. II (in 2 parts), **Reports by Industries;** vol. III, **Reports for States and Outlying Areas.**

Décimosexto censo de los Estados Unidos, 1940: Fabricación, 1939.

USA-**2260**.R5 1936 **Religious Bodies, 1936.** Selected statistics for the United States by denominations and geographic divisions. Washington, D.C., U.S.Govt.Print. Off., 1941. Department of Commerce. Bureau of the Census. 185 p.

These statistics, comprising 256 denominations, were published first in 78 separate bulle-

tins with title: **Census of Religious Bodies, 1936, Bulletin.**

Cuerpos religiosos, 1936. Selección de estadísticas de Estados Unidos, por denominaciones y por divisiones geográficas. Estas estadísticas que abarcan 256 denominaciones, fueron publicadas primeramente en 78 boletines por separado, bajo el título de: **Census of Religious Bodies, 1936 Bulletin.**

*USA-2270.C5 1939 **Sixteenth Census of the United States: 1940: Mineral Industries, 1939.** Washington, D. C., U.S. Govt. Print. Off., 1944. Department of Commerce. Bureau of the Census. 2 vol.: Vol. I, **General Summary and Industry Statistics**; vol. II, **State and County Statistics.**

Décimosexto censo de los Estados Unidos, 1940: Industrias minerales, 1939.

3000-3999 STATISTICAL SCIENCE — CIENCIA ESTADISTICA

USA-3000.G7 1946 **Government Statistics for Business Use.** Edited by Philip M. Hauser and William R. Leonard. New York, John Wiley & Sons, Inc.; London, Chapman & Hall, Ltd., 1946. 432 p.

A volume addressed to the business community indicating the major types of data available in the Federal government for management, production, and marketing needs.

Estadísticas oficiales para uso comercial. Un volumen dirigido al comercio con indicación de los principales tipos de datos disponibles en manos del gobierno federal, para necesidades de la administración, producción y comercialización.

Contents—contenido: Introduction (history, organization, effect of war on government statistics); "National Income and Other Business Indicators," by Milton Gilbert and Louis Paradiso; "Manufacturing," by Maxwell R. Conklin, Frank Hanna, and H. B. McCoy; "Mineral Statistics," by Y. S. Leong; "Agriculture," by Conrad Taeuber; "Wholesale, Retail, and Service Trades," by Walter F. Crowder; "International Trade and Payments," by J. Edward Ely and August Maffry; "Transportation and Other Public Utilities," by Frank L. Barton and Patricia Van Deraa; "Accounting Statistics," by William W. Cooper; "Money, Credit, and Banking," by Edward T. Crowder; "Prices," by Lester S. Kellogg; "Population," by Philip M. Hauser; "Housing and Construction," by Howard G. Brunsman; "Labor," by Charles D. Stewart; selected list of publications and authors.

USA-3030.A6 **Annual Report of the Central Statistical Board.** 1934—1937-38. Washington, D.C. Central Statistical Board.

The first report of the Central Statistical Board covers the calendar year 1934, the second the period from January 1, 1935 to June 30, 1936, and the third and fourth the fiscal years of 1936-37 and 1937-38.

Informe anual de la Junta Central de Estadística de los Estados Unidos. El primer informe de la Junta Central de Estadística abarca el año civil de 1934, el segundo ciñe el período del 1° de enero de 1935 al 30 de junio de 1936, y los informes tercero y cuarto comprenden los años fiscales de 1936-37 y 1937-38.

1937-38 **Fourth Annual Report of Central Statistical Board, July 1, 1937, to June 30, 1938.** Washington, D.C., U.S.Govt.Print.Off., 1938. Central Statistical Board. 21 p.

Report covers: Summary of chief activities; coordination in cases of overlapping statistical jurisdiction; review of statistical plans and publications; standard industrial classification; administrative developments; special technical advice and assistance; statistical clearing house; report to the President on returns made by the public to the Federal Government.

El informe abarca: Resumen de las actividades principales; coordinación de los casos de sobrepuesta jurisdicción estadística; exámen de planes y publicaciones estadísticas; la clasificación industrial standard; nuevos acontecimientos administrativos; ayuda y consejo técnico especiales; centro de intercambio estadístico; informe al Presidente sobre las declaraciones sometidas por el público al gobierno federal.

1934 **First Annual Report of the Central Statistical Board, Year Ended December 31, 1934.** Washington, D.C., U.S.Govt.Print.Off., 1935. Central Statistical Board. 50 p.

Covers the organization of statistical services in the United States and analyzes, by subject-matter fields, the statistical programs of Federal agencies.

Primer informe anual de la Junta Central de Estadística, abarcando el año civil de 1934. Presenta la organización de los servicios de estadística en los Estados Unidos, y analiza, según materia tratada, los programas estadísticos de las agencias federales.

USA-3030.B5 1929 **The Bureau of the Census: Its History, Activities and Organization.** By William Stull Holt. Washington, D.C., 1929. Brookings Institution. Institute for Government Research. 224 p. Service monographs of the United States Government, no. 53. Unexamined.

El Departamento del Censo: Su historia, actividades y organización.

USA-3030.C5 1946 **Collection of Economic Statistics in the United States.** Report prepared by the Division of Statistical Standards, Bureau of the Budget for the Subcommittee on Economic and Statistical Information of the Committee to Study Reform of the State (France). Washington, D.C.? February 1, 1946. Executive Office of the President. Bureau of the Budget. Division of Statistical Standards. 45 p. (approx.) Hectographed.

Contents: I, Statistical Organization and Methods in the United States; II, Principal Types of Economic Statistics (production, price, employment, financial, banking, foreign trade, national income and national product statistics, and statistics on the balance of international payments); Appendix I, Collection of Statistics by Federal Agencies; Appendix II, Bibliography of Statistical Publications.

Colección de estadísticas económicas en los Estados Unidos. Contenido: I, Organización y Métodos de Estadística en los Estados Unidos; II, Clases principales de estadísticas económicas (estadísticas de producción, precios, empleo, comercio exterior, renta nacional y producto nacional, balanza de pagos internacionales); Apéndice I, Recopilación de Datos Estadísticos por las Agencias del Gobierno Federal; Apéndice II, Bibliografía de Publicaciones Estadísticas.

USA-3030.F5 **Federal Statistical Directory.** A list of professional and technical personnel of federal agencies. 1935- . Annual. Washington, D.C. United States. Executive Office of the President. Bureau of the Budget. Division of Statistical Standards.

Subtitle varies—subtítulo varía: 1942, A list of administrative, research, and data-collecting personnel; 1943-1944, A list of professional and administrative personnel of Federal statistical and research agencies. Title varies—título varía: **Directory of Federal Statistical Agencies.**

Una lista de personal estadístico técnico y profesional en organismos federales. Anual.

1946 **Federal Statistical Directory.** A list of professional and technical personnel of federal agencies. Tenth edition. August 1946. Washington, D.C., 1946. Bureau of the Budget. Division of Statistical Standards. 148 p.

Contains lists of the personnel within federal agencies engaged in research and statistical activities, presented by divisional and sectional organization within the agency, with an alphabetical index of the individuals listed.

Contiene una lista de personal ocupado en trabajos de estadística y de investigación en los organismos federales, por orden de división y sección en el organismo, con un índice alfabético de las personas incluídas.

USA-3030.H5 1940 **Handbook for Employees of the Bureau of the Census.** Washington, D.C., U.S.Govt.Print.Off., 1940. Department of Commerce. Bureau of the Census. 41 p.

Contains background material of the Bureau, regulation, activities.

Manual para los empleados del Departamento del Censo. Contiene material básico del Departamento, reglamentación y actividades.

USA-3040.A5 **The Annals of the American Academy of Political and Social Science.** 1890- . Bimonthly. Philadelphia, Pennsylvania.

Anales de la Academia Estadounidense de Ciencias Políticas y Sociales.

Jan. '45, Dublin "Trends in Longevity." By Louis I. Dublin and Alfred J. Lotka. Vol. 237, January 1945, p. 123-133.

After a broad résumé of the increasing expectation of life at different ages in various countries, the social, economic, and especially medical factors responsible for increasing longevity are discussed. The prospects for further improvements are outlined, and American and British forecasts of future trends noted.

Tendencia de la longevidad. Después de un amplio sumario sobre el aumento en la expectativa de vida, a diferentes edades, en distintos países, se discuten los factores sociales, económicos y especialmente médicos del aumento en la longevidad. Se presenta un bosquejo sobre los propósitos de perfeccionamiento más amplio, y predicciones estadounidenses y británicas sobre observadas tendencias del futuro.

Jan. '45, Dunn "Demographic Status of South America." By Halbert Louis Dunn, Hope Tisdale Eldridge and Nora P. Powell. Vol. 237, January 1945, p. 22-23.

Available census and vital statistics are summarized with reference to population size, intercensal growth, age structures, racial composition, fertility, mortality by age and cause, and growth potential.

Estado demográfico de América del Sud. Los censos y las estadísticas vitales disponibles están resumidos con referencia al volumen de la población, crecimiento intercensal, estructura por edad, composición racial, fertilidad, mortalidad por edad y causa, y crecimiento potencial.

Jan. '45, Durand "The Trend Toward an Older Population." By John D. Durand. Vol. 237, January 1945, p. 142-151.

Present and probable future age distributions, classified broadly into the young (under 15), the productive (15-64), and the aged (65 and over), are described for the various regions of the world. Major emphasis is placed on the economic, social, and military implications of aging populations.

La tendencia hacia una población de más edad. La distribución por edades en el presente y la probable del futuro. Clasificación general de las edades, en jóvenes (menores de 15), activos (15-64) y viejos (de 65 y más) con descripción en distintas regiones del mundo. Presta particular atención a los cambios económicos, sociales y militares que implicaría un aumento en la edad de la población.

Jan. '45, Hauser "The Changing Population of the United States." By Philip M. Hauser and Conrad Taeuber. Vol. 237, January 1945, p. 12-21.

A broad survey of historical and contemporary size, composition, characteristics, vital statistics, and migration, together with some discussion of future trends in numbers.

Cambios en la población de los Estados Unidos. Una investigación amplia del volumen histórico y contemporáneo, composición, características, estadística vital y migración, conjuntamente con alguna discusión sobre tendencias futuras en números.

Jan. '45, León "Population Problems in Central and Caribbean America." By Alberto P. León and Alvaro Aldama C. Vol. 237, January 1945, p. 34-44.

Analysis of density, fertility, mortality, natural increase, the vital index, age distributions, population potential, and population quality, relying primarily on the official statistics of Mexico, Costa Rica, and El Salvador. Trends are fitted to vital rates and population, and the population of Mexico is projected to the year 2350.

Problemas de población en América Central y en las Antillas. Análisis de densidad, fertilidad, mortalidad, aumento natural, el índice vital, distribución por edad, potencial de la población y calidad de la población, basada principalmente sobre las estadísticas oficiales de Méjico, Costa Rica y El Salvador. Se adaptan tendencias a tasas vitales y población, y la población de Méjico está estimada hasta el año 2350.

Jan. '45, Spengler "Population and Per Capita Income." By Joseph J. Spengler. Vol. 237, January 1945, p. 182-192.

The influence of the population factor on per capita real income is discussed in relation to the following: Ration of population to productive factors; substitutability of labor; convertibility of labor; occupational composition; pattern of consumption; economics of specialization; external trading relations; age composition; and fullness of employment.

Población y renta por habitante. La influencia del factor de población sobre el ingreso real por habitante es discutida en relación a lo siguiente: Porporción productiva de la población; substituibilidad del trabajo; convertibilidad del trabajo; composición por ocupaciones; modelo de consumo; economía de la especialización; relaciones comerciales exteriores; composición por edad; y ocupación completa.

Jan. '45, Whelpton "Trends, Determinants, and Control in Human Fertility." By P. K. Whelpton and Clyde V. Kiser. Vol. 237, January 1945, p. 112-122.

After a detailed tabular presentation and brief discussion of trends in fertility in various countries of the world, there is a more detailed discussion of differences between the rates of racial, regional, socio-economic, educational, religious, and other groups. This latter discussion is based almost entirely on United States experience.

Tendencia, determinantes y control de la fertilidad humana. Después de una detallada presentación tabular y corta discusión sobre las tendencias de la fertilidad en varios países del mundo, hay una discusión mas detallada sobre diferencias entre las tasas de los grupos raciales, regionales, económico-sociales, educacionales, religiosos y otros. Esta última discusión está basada casi por completo sobre la experiencia en los Estados Unidos.

Jan. '45, Yerushalmy "Infant and Maternal Mortality in the Modern World." By Jacob Yerushalmy. Vol. 237, January 1945, p. 134-141.

For both infant and maternal deaths, there is a resume of levels and trends in various countries and a discussion of the major factors producing the observed changes.

Mortalidad infantil y materna en el mundo moderno. Hay un resumen de nivel y tendencia de la mortalidad infantil y materna en varios países, y una discusión sobre los factores principales que producen los cambios observados.

Nov. '36 "The American People: Studies in Population." Dealing with the composition, distribution and growth of the population, and its relation to resources in the United States. Vol. 188, November 1936. Entire issue devoted to the above subject.

El pueblo estadounidense: Estudios sobre población. Trata sobre composición, distribución y crecimiento de la población y su relación con los recursos, en los Estados Unidos.

USA-**3040**.B5 **Biometrics Bulletin.** 1945- . Bimonthly. Washington, D. C. American Statistical Association. Biometrics Section.

Boletín de biometría.

June '45 "Teaching Statistics at the Department of Agriculture Graduate School in Washington." Vol. I, no. 3, June 1945, p. 33-34.

Enseñanza de estadística en la Escuela de Estudios Avanzados de la Secretaría de Agricultura en Washington.

June '45, Schumacher "Statistical Method in Forestry." By F. X. Schumacher. Vol. I, no. 3, p. 29-32.

El método estadístico en la industria forestal.

Apr. '45 "Graduate Work in Statistics at Columbia University." Vol. 1, no. 2, April 1945, p. 22-23.

Labor avanzada en estadística en Columbia University.

*USA-**3040**.J5 **Journal of the American Statistical Association.** 1888-89- . Quarterly. Washington, D.C.

Revista de la Asociación Estadística Estadounidense.

Sept. '46, Latham "One Statistical World." By Earl Latham. Vol. 41, no. 235, September 1946, p. 275-292.

Description of international statistical agencies and statistical commissions in international organizations.

Descripción de organismos estadísticos internacionales y comisiones estadísticas pertenecientes a organizaciones internacionales.

Sept. '45, Deming "On Training in Sampling." By W. Edwards Deming. Vol. 40, no. 231, September 1945, p. 307-316.

Sobre la enseñanza de muestreo.

June '45, a "The Federal Government's Statistical Program for Reconversion and Postwar Adjustment: A Round Table." Vol. 40, no. 230, June 1945, p. 229-236.

El programa estadístico del gobierno federal para la reconversión y ajustes de postguerra: Una discusión de mesa redonda.

Includes—incluye: "Industry and Business Statistics," by M. Joseph Meehan, and discussion by Rufus S. Tucker; "Population," by Leon E. Truesdell; "Farm Wage Statistics for a Full Employment Program," by Conrad Taeuber; "Labor Statistics for a Full Employment Program," by Arnold Tolles.

June '45, b "Management of Field Work and Collection of Statistics: A Round Table." Vol. 40, no. 230, June 1945, p. 245-250.

Contents—contenido: "Management of Field Staffs in the Opinion Research Field," by Joseph C. Bevis; "Problems of Field Management in Army Opinion Research," by William McPeak; "Management of Field Work and Collection of Statistics of the Labor Force," by A. Ross Eckler.

Dirección del trabajo sobre el terreno y recolección de datos estadísticos: Una discusión de mesa redonda.

June '45, c "Problems of Integrating Federal Statistics: A Round Table." Vol. 40, no. 230, June 1945, p. 237-244.

Includes remarks by—incluye observaciones hechas por: Stuart A. Rice, A. Ford Hinricks, Howard R. Tolley, and Philip M. Hauser.

Los problemas de la integración de las estadísticas federales: Una discusión de mesa redonda.

June '45, Silva Rodrigues "On the Training of Statisticians." By Milton da Silva Rodrigues. Vol. 40, no. 230, June 1945, p. 172-174.

Sobre la preparación de estadígrafos.

Mar. '45 Opportunities for statisticians; a series of articles on statistical work in the federal government, in state and local governments, in teaching and research, and in industry. Vol. 40, no. 229, March 1945, p. 57-79.

Oportunidades para estadígrafos; una serie de artículos sobre trabajo estadístico en el gobierno federal, en los gobiernos de estados y locales, en la enseñanza e investigación y en la industria.

Mar. '45, Jessen "The Master Sample of Agriculture: II, Design." By R. J. Jessen. Vol. 40, no. 229, March 1945, p. 46-56.

El patron muestra de la agricultura: II, Proyecto.

Mar. '45, King "The Master Sample of Agriculture: I, Development and Use." By A. J. King, Vol. 40, no. 229, March 1945, p. 38-45.

El patrón muestra de la agricultura: I, Desarrollo y uso.

June '44, Edwards "Methods Used in Processing Data from the Physical Examination Reports of the Selective Service System." By Thomas I. Edwards and Louis P. Hellman. Vol. 39, no. 226, June 1944, p. 165-182.

Métodos usados en el procesamiento de los datos de los informes del examen físico del Sistema del Servicio Selectivo.

June '44, Kimelman "Some Problems of Statistical Education and Teaching." By Juan Kimelman. Vol. 39, June 1944, p. 229-235. Reprint.

Translation of the article appearing in **Estadística,** vol. I, no. 3, under the title **Algunos Problemas de Educación y Enseñanza Estadística.**

Algunos problemas de educación y enseñanza estadística. Traducción del artículo que aparece en **Estadística,** vol. I, no. 3.

Mar. '44, Kolesnikoff "Standard Commodity Classification." By S. Vladimir Kolesnikoff. Vol. 39, no. 225, March 1944, p. 42-52.

Clasificación estandard de los géneros.

Sept. '43, Hansen "On Some Census Aids to Sampling." By Morris H. Hansen and W. Edwards Deming. Vol. 38, no. 223, September 1943, p. 353-357.

Sobre ciertas ayudas del censo al muestreo.

Mar. '43, Eckler "Marketing and Sampling Uses of Population and Housing Data." By A. Ross Eckler and E. P. Staudt. Vol. 38, no. 221, March 1943, p. 87-92.

Usos de los datos de población y vivienda para estudios del mercado consumidor y para muestreos.

Dec. '42, Deming "Errors in Card Punching." By W. Edwards Deming, Benjamin J. Tepping and Leon Geoffrey. Vol. 37, no. 220, December 1942, p. 525-536.

Errores en la perforación de fichas.

June'42, McPherson "Mathematical Operations with Punched Cards." By J. C. McPherson. Vol. 37, no. 218, June 1942, p. 275-281.

Operaciones matemáticas con fichas perforadas.

Mar. '42, Hansen "Relative Efficiencies of Various Sampling Units in Population Inquiries." By Morris H. Hansen and William N. Hurwitz. Vol. 37, no. 217, March 1942, p. 89-94.

Eficiencias relativas de varias unidades de muestra en investigaciones sobre población.

Mar. '42, Snedecor "Recenr Development in Sampling for Agricultural Statistics." By George Waddel Snedecor and Arnold J. King. Vol. 37, no. 217, March 1942, p. 95-102.

Reciente evolución del muestreo en las estadísticas agrícolas.

Dec. '41, Berkson "A Punch Card Designed to Contain Written Data and Coding." By Joseph Berkson. Vol. 36, no. 216, December 1941, p. 535-538.

Una ficha perforada designada para contener datos escritos y datos codificados.

Sept. '41, Deming "On Sample Inspection in the Processing of Census Returns." By W. Edwards Deming and Leon Geoffrey. Vol. 36, no. 215, September 1941, p. 351-360.
Sobre inspección de muesastr en el procesamiento de los formularios llenados del censo.

Dec. 40, Stephan "The Sampling Procedure of the 1940 Population Census." By Frederick Franklin Stephan, W. Edwards Deming and Morris H. Hansen. Vol., 35 no. 212, December 1940, p. 615-630. Reprint.
El procedimiento del muestreo en el censo de población de 1940.

June '39, Kellogg "Some Problems in Teaching Elementary Statistics." By Lester S. Kellogg. Vol. 34, no. 206, June 1939, p. 299-308.
Algunos problemas en la enseñanza de estadística elemental.

Ma.. '35, Dunn "Application of Punched Card Methods to Hospital Statistics." By Halbert Louis Dunn and Leonard Townsend. Vol. 30, no. 189A, March 1935, p. 244-248.
Aplicación de los métodos de fichas perforadas a las estadísticas de hospitales.

Sept. '32, Dunn "Adaptation of New Geometric Code to Multiple Punching in Mechanical Tabulation." By Halbert Louis Dunn. Vol. 27, no. 179, September 1932, p. 279-286. Reprint.
Adaptación de nuevas claves geométricas a perforaciones múltiples en la tabulación mecánica.

Mar. '28, Blackett "Fundamental Problem in the Case Method of Instruction in Statistics." By Olin W. Blackett. Vol. 23, no. 161, March 1928, Supplement, p. 103-104.
Problema fundamental en el método de ejemplos prácticos en la enseñanza estadística.

Dec. '26 "Requirements for Statisticians and Their Training." Vol. XXI, no. 156, December 1926, p. 419-446. In 5 parts: Part I, "Statistical Teaching in American Colleges and Universities," by James W. Glover, p. 419-424; part II, "Standards and Requirements in Statistics," by Donald R. Belcher, p. 424-430; part III, "Content and Purpose of Training in Elementary and in Advanced Statistics," by Willford I. King, p. 430-435; part IV, "Mathematical Background for the Study of Statistics," by H. L. Rietz and A. R. Crathorne, p. 435-440; part V, "Statistics in Insurance Serv ce," by Edwin W. Kopf, p. 440-446.
Part I is a summary of replies to a questionnaire sent out in 1925 to American colleges and universities, by the Committee for Education and Professional Standards, requesting data in relation to statistical courses.
Requisitos para estadísticos y su entrenamiento. En 5 partes. La parte I representa el resumen de las respueasts al cuestionario solicitando datos sobre cursos en estadística, que fué remitido a las universidades y altas instituciones de educación en los Estados Unidos, por la Comisión de Normas Educacionales y Profesionales, en el año 1925.

USA-**3040**.M5 **Mathematical Reviews.** 1940- . Monthly. Lancaster, Pennsylvania- American Mathematical Society.
Abstracts of articles.
Revistas matemáticas. Resúmenes de artículos, publicados mensualmente por la Sociedad Americana de Matemática.

USA-**3040**.P5 **Psychometrika.** A journal devoted to the development of psychology as a quantitative rational science. 1936- . Quarterly. New York, Columbia University. Institute of Educational Research. Unexamined.
Includes articles on statistical methodology.
Una revista trimestral dedicada al fomento de la psicología como ciencia racional cuantitativa. Incluye artículos sobre metodología estadística.

USA-**3040**.R5 **The Review of Economic Statistics.** 1919- . Quarterly. Cambridge, Mass. Harvard University. Committee on Economic Research.
Revista trimestral de estadísticas económicas, de la Universidad Harvard.

USA-**3040**.S5 **Statistical Reporter.** Starting date? Monthly. Washington, D.C. Bureau of the Budget. Division of Statistical Standards.
Previously semimonthly. Title was **Report to Federal Statistical Agencies** before September

1944. Covers current developments relating to governmental statistics of sufficient importance to be brought to the general attention of Federal statistical agencies.

Relator estadístico. Anteriormente publicado bimensualmente. Antes de septiembre de 1944 el título era **Report to Federal Statistical Agencies.** Incluye desarrollos de la actualidad relacionados con las estadísticas gubernamentales de suficiente importancia para que se llame sobre ellos la atención general de los órganos federales de estadística.

USA-3110.L5 1945 **List of Schedules Used by the Bureau of the Census for Collecting Data, 1945.** Washington, D.C., August 1945. Department of Commerce. Bureau of the Census. 32 p. Photostated.

Lista de formularios usados por el Departamento del Censo, para la recolección de datos, 1945.

USA-3110.P5 1933 **Procedure for Selecting, Analyzing, Assembling and Perforating into Tabulating Cards the Economic and Financial Data Reported on Income Tax Returns.** Washington, D.C., 1933. Department of the Treasury. Bureau of Internal Revenue. Reproduced from typewritten copy. Part I, **Individual Income Tax Returns for 1932—Form 1040.** Unexamined.

Procedimiento para la selección, análisis, recolección y perforación en fichas tabuladoras, de los datos económicos y financieros de las declaraciones del impuesto sobre la renta.

USA-3110.Q5 1943 **Questionnaire Manual.** Washington, D.C., U.S.Govt.Print.Off., September 1943. War Production Board. Bureau of Planning and Statistics. Office of Survey Standards. 20 p.

Manual de cuestionarios del Consejo de Producción de Guerra.

USA-3110.R5 1942-43 **Report, October 9, 1942-April 13, 1943, from Advisory Committee on Government Questionnaires.** To American Retail Federation, American Trade Association Executives, Chamber of Commerce of the United States, Controller's Institute of America, National Association of Manufacturers. Washington, D.C., 1943. Advisory Committee on Government Questionnaires. 61 p. Unexamined.

Includes—incluye: "Functions, Organization and Problems of the Division of Statistical Standards of the Bureau of the Budget." p. 23-26.

Informe, octubre 9 de 1942 a abril 13 de 1943, de la Comisión Consultiva sobre Cuestionarios Gubernativos.

USA-3200.R5 1918 **Report of the Committee on the Standardization of Mining Statistics.** Compiled by Albert Hill Fay. Washington, D.C., U.S.Govt.Print.Off., 1918. Department of the Interior. Bureau of Mines. 39 p. Technical paper, no. 194. Unexamined.

Informe de la Comisión de Estandardización de las Estadísticas Mineras.

USA-3200.S5 1916 **Suggested Standard Forms for Mine Statistics with Especial Reference to Accidentes.** By Albert Hill Fay. Washington, D.C., U.S.Govt.Print.Off., 1916. Department of Interior. Bureau of Mines. 33 p. Unexamined.

Formularios estandards recomendados para la estadística minera, con especial referencia a accidentes.

USA-3200.T5 **Technical Papers of the Division of Statistical Standards.** Starting date? Irregular. Washington, D.C. Executive Office of the President. Bureau of the Budget. Division of Statistical Standards.

Documentos técnicos del Departamento de Estandards Estadísticos.

USA-3210.C5 **Custom House Guide.** Foreign traders' encyclopedia. United States custom tariff; custom posts; internal revenue code; customs, shipping and commerce regulations; reciprocal trade agreements. 1862- . Annual. New York.

Guía de aduana. Enciclopedia anual del comercio exterior. Arancel de aduana de los Estados Unidos; aduanas; código de ingresos internos; reglamentos sobre aduanas, navegación y comercio; acuerdos comerciales recíprocos.

1943, a "United States Custom Tariff, Act of 1930." Effective—June 18, 1930; revised to March 1, 1943. 1st section, alphabetical import commodities index, 30.000 commodities and duties including reciprocal trade agreement rates, 2nd section, tariff act of 1930, dutiable list, free list, special provisions, administrative provisions, index. **Custom House Guide, 1943,** p. 551-994.

Tarifa de aduana de los Estados Unidos, Ley de 1930. Efectiva—18 de junio de 1930, revisada al 1° de marzo de 1943. 1a. sección, índice alfabético de artículos de importación, 30.000 artículos y derechos incluyendo tarifa de los acuerdos recíprocos; 2a. sección, ley de arancel de 1930, lista de los artículos imponibles, lista de los artículos libres de derechos, estipulaciones especiales, estipulaciones administrativas, índice.

USA-3210.S2 1940 **Schedule A, Statistical Classification of Imports into the United States, Arranged by Tariff Schedules and Tariff Paragraphs of Act of 1930.** Effective January 1, 1939 (with authorized changes through May 16, 1940). Richmond, Virginia, 1940. U. S. Tariff Commission. 550 p. Multilithed. W.P.A. statistical project 65-2-31-57.

Import data presented in tariff classification order with corresponding Schedule A numbers.

Cuadro A, clasificación estadística de importaciones de los Estados Unidos, ordenado de acuerdo con los cuadros y parágrafos arancelarios del acta de 1930. Efectivo enero 1° de 1939 (con reformas autorizadas hasta el 16 de mayo de 1940). Datos de importación.

USA-3210.S3 1943 **Schedule A, Statistical Classification of Imports into the United States.** With rates of duty and regulations governing the preparation of monthly, quarterly, and annual statements of imports, effective January 1, 1943. Washington, D.C., U.S.Govt.Print.Off., 1943. Department of Commerce. Bureau of the Census. 235 p.

Formulario A, clasificación estadística de importaciones de los Estados Unidos, con tasas de impuestos y reglamentos que rigen sobre la preparación mensual, trimestral y anual de informes sobre importación; puesto en vigor el 1° de enero de 1943.

USA-3210.S4 1945 **Schedule B, Statistical Classification of Domestic and Foreign Commodities Exported from the United States.** And regulations governing statistical returns of exports of commodities. January 1, 1945 edition. Washington, D.C., U.S.Govt. Print.Off., 1945. Department of Commerce. Bureau of the Census. 2 vol.: Part I, **Alphabetic Index;** part II, **Numbered Classifications and Articles Included.**

Cuadro B, clasificación estadística de productos nacionales y extranjeros exportados por los Estados Unidos, y reglamentos sobre nóminas estadísticas de la exportación de productos.

USA-3210.S6 1944 **Standard Commodity Classification: Vol. II: Alphabetic Index.** Prepared by the Technical Committee on Standard Commodity Classification. Washington, D.C., U.S.Govt.Print.Off., 1944. Executive Office of the President, Bureau of the Budget; War Production Board; Department of the Treasury, Procurement Division. 766 p. Technical paper, no. 27.

Clasificación estandard de los géneros, vol. II: Indice alfabético.

USA-3210.S6 1943 **Standard Commodity Classification: Vol. I: Standard Classified List of Commodities.** Prepared by the Technical Committee on Standard Commodity Classification. Washington, D.C., U.S.Govt. Print.Off., 1943. Executive Office of the President, Bureau of the Budget; War Production Board; Department of the Treasury, Procurement Division. 652 p. Technical paper, no. 26.

Clasificación estandard de los géneros, vol. I: Lista clasificada estandard de los géneros.

USA-3220.C5 1939 **Sixteenth Census of the United States, 1940: Manufactures, 1939: Industry Classifications for the Census of Manufactures, 1939.** Prepared under the supervision of Thomas J. Fitzgerald. Washington, D.C., U.S. Govt. Print. Off., 1942. U. S. Department of Commerce. Bureau of the Census. 34 p.

Clasificación industrial empleada en el censo de manufacturas levantado en los Estados Unidos en 1939.

USA-3220.S5 1942 **Standard Industrial Classification Manual.** Prepared by Technical Committee on Industrial Classification. Washington, D.C., U.S.Govt.Print.Off., 1942. Executive Office of the President. Bureau of the Budget. 2 vol.: Vol. I, **Manufacturing Industries;** vol. II, **Non-manufacturing Industries.**

Manual de clasificación industrial estandard.

USA-3230.C5 1940 **Classification of Causes of Stillbirth.** List of causes of stillbirth, tabular list and joint causes of stillbirth, as approved by the Committee on Research and Standards of the American Public Health Association for publication and trial in the United States, Oct. 7, 1940. Department of Labor. Children's Bureau. 30 p. Mimeographed.

Lista de las causas de mortinatalidad, lista tabular y causas conjuntas de mortinatalidad, aprobadas por la Comisión de Investigación y Estandards de la Asociación Americana de Salud Pública para publicación y ensayo en los Estados Unidos, 7 de octubre de 1940.

USA-3230.M5 1944 **Manual for Coding Causes of Illness According to a Diagnosis Code for Tabulating Morbidity Statistics.** Prepared by direction of the Surgeon General in the Division of Public Health Methods, R. S. Public Health Service, with the cooperation and assistance of the Division of Vital Statistics, U. S. Bureau of the Census, the Johns Hopkins

Hospital and School of Hygiene and Public Health, and the Mayo Clinic. Washington, D.C., U.S.Govt.Print.Off., 1944. Federal Security Agency. U. S. Public Health Service. 489 p. Miscellaneous publication, no. 32.

Manual para la codificación de causas de enfermedad de acuerdo con un código de diagnósticos para tabular estadísticas de morbilidad.

USA-3240.C5 1940.1 **Sixteenth Census of the United States: Classified Index of Occupations.** Occupation classification based on the standard classification. Prepared by Alba M. Edwards. Washington, D.C., U.S.Govt.Print. Off., 1940. Department of Commerce. Bureau of the Census. 199 p.

Decimosexto censo de los Estados Unidos: Indice clasificado de ocupaciones.

USA-3240.C5 1940.2 **Sixteenth Census of the United States: Alphabetical Index of Occupations and Industries.** Occupation and industry classifications based on the respective standard classifications. Prepared by Alba M. Edwards. Washington, D.C., U.S.Govt.Print. Off., 1940. Department of Commerce. Bureau of the Census. 607 p.

Decimosexto censo de los Estados Unidos: Indice alfabético de ocupaciones e industrias.

USA-3270.F5 1944 **Foreign Weights and Measures with Factors for Conversion to United States Units.** By Caroline G. Gries. Washington, D.C., June 1944. Department of Agriculture. Office of Foreign Agricultural Relations. 23 p. Multilithed.

Includes: Common metric units used to measure area, capacity and weight with United States equivalents; common United States units of area, capacity and weight with metric equivalents; old Spanish measures of area, capacity and weights with metric and United States equivalents; common units of measurement in specified foreign countries with United States equivalents; equivalents for converting hundred weight and metric weights to United States units; miscellaneous conversion factors and equivalents; units of measure classified by principal countries and territories.

Pesos y medidas extranjeras, con coeficientes para la conversión a unidades estadounidenses. Incluye: Unidades comunes usadas para medir superficies, capacidad y peso, con sus equivalentes estadounidenses; unidades de superficie, capacidad y peso comunes en los Estados Unidos, con equivalentes métricos; antiguas medidas españolas de área, capacidad y peso, con sus equivalentes métricos y estadounidenses; unidades de medida comunes en ciertos países extranjeros con sus equivalents en los Estados Unidos; equivalents para la conversión de "hundred weight" y pesos métricos, a unidades estadounidenses; coeficientes para conversiones varias y sus equivalentes; unidades de medida, clasificadas por principales países y territorios.

USA-3300 1944-46 **Statistical Training in the United States: Methods and Materials.** Background material acquired for the Inter American Statistical Institute Survey on Statistical Training Methods and Materials in the American Nations, 1944-46.

Includes programs of courses offered by the following universities—incluye programas de cursos dictados en las siguientes universidades: American University, U. of California, U. of Chicago, Columbia U., George Washington U., Harvard U., U. of Iowa, Johns Hopkins U., U. of Michigan, U. of Minnesota, U. of North Carolina, Princeton U., U. of Wisconsin, Yale U., and United States Department of Agriculture Graduate School.

Enseñanza estadística en los Estados Unidos: Métodos y materiales. Material básico adquirido para la Encuesta del Instituto Interamericano de Estadística sobre los métodos y materiales en la enseñanza estadística en las naciones americanas, 1944-46.

USA-3300.O5 1945 "An Outline of Training Activities in the U. S. Bureau of the Census." By James Stockard. Washington, D.C., 1945. 2 p. Typewritten.

Acquired in connection with IASI Survey of Statistical Training Methods and Materials in the American Nations, 1944.

Compilado con relación a la encuesta del IASI sobre Métodos y Materiales en la Enseñanza Estadística en las Naciones Americanas, 1944.

4000-4999 GEOGRAPHY—GEOGRAFIA

USA-4000.G5 **Geographical Review.** 1916- . Quarterly. New York. American Geographical Society.

Revista geográfica.

Jan. '46 "The Map of Hispanic America on the Scale of 1:1,000,000." Vol. XXXVI, no. 1, January 1946, p. 1-28.

Includes: Relation to the international map; quantity and quality of source material; catalogue of maps of Hispanic America; contributions of official bureaus and development companies; uses of the map, in boundary disputes; uses for aerial navigation; use for strategic purposes; appendix—note on the international map.

El mapa de la América Hispana en la escala de 1:1,000,000. Incluye: Relación con el mapa internacional; cantidad y calidad de las fuentes; catálogo de mapas de la América Hispana; contribuciones de los órganos oficiales y de las compañías de fomento; usos del mapa en disputas de fronteras; usos para navegación aérea; usos para fines estratégicos; apéndice—nota acerca del mapa internacional.

Jan. '42, Arca Parró "Census of Peru' 1940." By Alberto Arca Parró. Vol. XXXII, no. 1, January 1942, p. 1-20. Reprint.

Describes background, plan of organization, scope and preliminary results of 1940 census.

Censo de Perú, 1940. Describe la base, plan de organización, objeto y resultados preliminares del censo de 1940.

Apr. '37, Platt "The Millionth Map of Hispanic America." By Raye R. Platt. Vol. XXVII, no. 2, April 1937, p. 301-308.

El mapa de Hispano-América en escala 1:1,000,000.

5000-5999 DEMOGRAPHY—DEMOGRAFIA

USA-5000.M5 **Milbank Memorial Fund Quarterly.** 1923- . New York.

Publicación trimestral del Milbank Memorial Fund.

USA-5000.N5 1939 **The Natural History of Population.** By Raymond Pearl. New York, Oxford University Press, 1939. 416 p.

"This volume gives a summary and synthesis of many of Pearl's recent studies on fertility and related population problems. Chapters I and II concern 'The biological background' and 'The biology of fertility.' Chapter III, 'Human and animal reproduction patterns', is a detailed analysis of American data on nativity, race and age differentials, changes from 1920 to 1930 in the ratio of actual to potential mothers, the proportion of first births, age patterns in fertility and the extent of adequate reproductivity, and parity patterns in relation to age." (Source: **Books on Population, 1939-Mid 1944.**)

"La historia natural de la población. Este volumen presenta un resumen y análisis de muchos de los recientes estudios realizados por Pearl sobre fertilidad y problemas de población relacionados con ella. Los capítulos I y II se refieren al 'fundamento biológico' y a la 'biología de la fertilidad.' El capítulo III, 'normas de reproducción animal y humana,' es un análisis detallado de los datos de Estados Unidos, sobre diferenciales de natividad, raza y edad, cambios desde 1920 a 1930 en la proporción de las madres actuales a las potenciales, la proporción de primeros nacimientos, la edad en relación a la fertilidad y el alcance de la reproductividad adecuada y el número de hijos tenidos en relación a la edad de la madre." (Traducción del comentario.)

USA-5000.S5 1940 **Studies in American Demography.** By Walter Francis Willcox. Ithaca, New York, Cornell University Press, 1940. 556 p.

The first part, "Studies in American Census Statistics", is a combination of selected published studies, revised, with unpublished lecture notes. The second part, "Studies in American Registration Statistics", is based on the author's **Introduction to the Vital Statistics of the United States, 1900-1930.**

Estudios sobre demografía de los Estados Unidos. La primera parte, estudios sobre estadísticas de censos en América, es una combinación de selectos estudios publicados, revisados y con notas no publicadas. La segunda parte, estudios sobre estadísticas de registro en América, está basada en la obra del mismo autor **Introduction to the Vital Statistics of the United States, 1900-1930** (USA-I.5, 1900-30).

USA-5100.E6 1940-2000 **Estimates of Future Population of the United States, 1940-2000.** Prepared by Warren S. Thompson and P. K. Whelpton, of the Scripps Foundation for Research in Population Problems, for the Committee on Population Problems of the National Resources Planning Board. Washington, D.C., U.S.Govt.Print.Off., 1943. 137 p.

"Twelve series of estimates, prepared on the basis of three alternative assumptions as to both fertility and mortality. The effect of an average net immigration of 100,000 foreign born whites was computed under three combinations of fertility and mortality assumptions. The data are presented by five year age groups at five year intervals, by sex, for native white, foreign born white, and colored. Correction tables for both war losses and deficits of births were computed in terms of units of 100,000 for whites and 10,000 for colored in order to permit correction of the estimates when the magnitude of the losses becomes known." (Source: **Books on Population, 1939-Mid-1944.**)

Estimaciones de la población futura de los Estados Unidos. "Doce series de estimaciones fueron preparadas sobre la base de tres suposiciones alternativas con respecto a fertilidad y mortalidad. El efecto de un promedio neto de la inmigración de 100,000 blancos nacidos en el extranjero fué computado bajo tres combinaciones de suposiciones sobre fertilidad y mortalidad. Los datos están presentados por grupos quinquenales de edad, a intervalos de cinco años, por sexo, blancos nativos, blancos nacidos en el extranjero y negros. Tablas de corrección para pérdidas de guerra y déficits de nacimientos fueron computadas en proporción de 100.000 blancos y 10,000 negros, a fín de permitir una corrección en las estimaciones cuando la magnitud de las pérdidas llega a ser conocida." (Traducción del comentario.)

USA-**5100**.F5 1940 **Foundations of American Population Policy.** By Frank Lorimer, Ellen Winston, and Louise K. Kiser. New York and London, Harper & Brothers, 1940. 178 p.

"This volume, prepared under the auspices of the National Economic and Planning Association, represents a systematic attempt to state possible lines of population policy, to clarify basic issues, and to reveal critical areas requiring further study. The final chapter, 'Toward a national population policy', attempts to outline the positive measures needed to adapt the American economy to changing population trends, and to achieve voluntary reproduction adequate from both the quantitative and qualitative standpoint, at the same time maintaining a democratic society." (Source: **Books on Population, 1939-Mid-1944.**)

Fundamentos de la política demográfica americana. "Este volumen . . . representa un intento sistemático para establecer posibles direcciones en la política de población, esclarecer problemas básicos y revelar materias difíciles que requieren más estudio. El capítulo final, 'Hacia una política nacional de población', trata de delinear las medidas que deben ser tomadas para adaptar la economía americana a cambios en el rumbo de la población y para lograr reproducción voluntaria adecuada desde ambos puntos de vista, cuantitativo y cualitativo, manteniendo al mismo tiempo una sociedad democrática." (Traducción del comentario.)

USA-**5100**.M5 1943 **Membership List of the Population Association of America, 1943.** Population Association of America. 12 p. Hectographed.

Lista de miembros de la Asociación Americana de Población.

USA-**5110**.C5 1790-1900 **A Century of Population Growth.** From the first census of the United States to the twelfth, 1790-1900. Washington, D.C., U.S.Govt.Print. Off., 1909. Department of Commerce. Bureau of the Census. 303 p.

Un siglo de crecimiento de la población de los Estados Unidos.

USA-**5110**.F5 1944 **Final Report of the Committee for Congested Production Areas.** December 1944. Washington, D.C., U.S.Govt.Print.Off., 1944. Executive Office of the President. Committee for Congested Production Areas. 37 p.

Describes problems of congestion which faced the nation when Committee was created in April 1943, methods employed by Committee in coping with these problems, and results accomplished. Contains flow charts of in-migration.

Informe final de la Comisión para las Zonas Productivas Congestionadas. Describe los problemas de la congestión que envolvía a la nación al tiempo que esta Comisión fué creada (abril 1943), los métodos empleados por la Comisión para afrontar estos problemas y los resultados obtenidos. Contiene esquemas demostrando el movimiento de migración interna.

USA-**5120**.C5 1850-1930 **The Canadian-Born in the United States.** An analysis of the statistics of the Canadian element in the population of the United States, 1850 to 1930. The relations of Canada and the United States, a series of studies prepared under the direction of the Carnegie Endowment for International Peace. By Leon E. Truesdell. New Haven, Yale University Press, 1943. 263 p. Unexamined.

"This study consists of a series of comprehensive chapters on special topics, each chapter including source of data, definitions, comments on outstanding relationships, and historical or other probable explanations of observed relationships. All aspects of demography for which information is available are covered." (Source: **Books on Population, 1939-Mid-1944.**)

Los canadienses-nativos en los Estados Unidos. Un análisis de las estadísticas del elemento canadiense en la población de los Estados Unidos, 1850 a 1930. "Este estudio consiste en una serie de capítulos comprensivos sobre asuntos especiales, cada capítulo incluyendo fuente de información, definiciones, comentarios sobre prominentes relaciones, y explicaciones históricas y otras probables de ciertas relaciones. Todos los aspectos de la demografía cuya información es asequible, han sido incluidos." (Traducción del comentario.)

USA-**5120**.J5 1943 **Jewish Population Studies.** Edited by Sophia M. Robison with the assistance of Joshua Starr. New York,

1943. Conference on Jewish Relations. 189 p. Jewish social studies, publications, no. 3. Unexamined.

Contents—contenido: "Methods of Gathering Data on the Jewish Population," by Sophia M. Robison; "The Jewish Population of Trenton, 1937," by Sophia M. Robison; "The Jewish Population of Passaic, 1937," by Sophia M. Robison; "The Jewish Population of Buffalo, 1938," by U. Z. Engelman; "A Comparative Study of the Jewish Communities of Norwich and New London, 1938," by Bessie B. Wessel; "A Sample Study of the Jewish Population of Pittsburgh, 1938," by Maurice Taylor; "A Study of Detroit Jewry, 1935," by H. J. Meyer; "A Study of Chicago Jewry, 1930, Based on Death Certificates," by A. J. Jaffe; "The Jewish Population of Minneapolis, 1936," by Sophia M. Robison; "A Study of San Francisco Jewry, 1938," by Samuel Moment; "Conclusion," by Sophia M. Robison.

Estudios sobre la población judía.

USA-**5120**.N4 1790-1915 **Negro Population, 1790-1915.** Prepared by John Cummings. Washington, D.C., U.S.Govt.Print.Off., 1918. Department of Commerce. Bureau of the Census. 844 p. Unexamined.

Población negra de los Estados Unidos, 1790-1915.

USA-**5120**.N5 1920-32 **Negroes in the United States, 1920-32.** Prepared by Charles E. Hall. Washington, D.C., U.S. Govt. Print. Off., 1935. Department of Commerce. Bureau of the Census. 845 p. Unexamined.

Negros en los Estados Unidos, 1920-32.

USA-**5120**.R5 1943 **Report on Alien Registration.** Submitted to the Attorney General, February 1, 1943. Washington, D.C., 1943. Department of Justice. Immigration and Naturalization Service. 10 p. Unexamined.

Informe sobre registro de extranjeros.

USA-**5200** **Vital Statistics and Registration in the United States: Methods and Procedures.** Background material acquired for the study conducted by the Inter American Statistical Institute in vital statistics and registration methods in the American nations, including the survey of 1945.

Includes: Answer to questionnaire; instructions; forms; punch cards; exposition.

Estadística vital y registro en los Estados Unidos: Métodos y procedimientos. Material básico adquirido para la encuesta que el Instituto Interamericano de Estadística realiza sobre los métodos de registro y estadística vital en las naciones americanas, incluyendo el estudio de 1945. Incluye: Respuesta la cuestionario; instrucciones; formularios; tarjetas de perforación; exposición.

USA-**5200**.I5 1900-30 **Introduction to the Vital Statistics of the United States, 1900 to 1930.** By Walter Francis Willcox. Washington, D.C., U.S.Govt.Print.Off., 1933. Department of Commerce. Bureau of the Census. 138 p.

Introducción a la estadística vital de los Estados Unidos.

USA-**5200**.P5 1944 **Plan for the Coordination of Vital Records and Vital Statistics in the United States.** As recommended by the Council on Vital Records and Vital Statistics. Washington, D.C., November 1944. Department of Commerce. Bureau of the Census. 24 p. Mimeographed.

Plan para la coordinación de los registros vitales con la estadística vital en los Estados Unidos.

USA-**5200**.V5 1944-45 **Vital Statistics Instruction Manual.** Washington, D.C., 1944-45. Department of Commerce. Bureau of the Census. Hectographed. 6 vol.: Part I, **Coding and Punching Geographic and Personal Particulars: Births, Deaths, and Stillbirths Occurring during 1945**; part I, **Supplement B, Special Instructions for Coding and Punching Motor Vehicle Accident Death Transcripts: Deaths Occurring during 1944**; part II, **Assignment of Causes of Death: Mortality Transcripts, 1945**; part III, **Instructions for Sorting and Tabulating, 1943. Birth, Death, and Stillbirth Punch Cards**; part IV, **Geographic Code, 1943 Edition (Based on 1940 Population Census)**; part V, **Current Mortality Sample: Coding, Punching and Tabulating Monthly Samples for 1945.**

Manual de instrucciones para la estadística vital, 6 vol.: Parte I, Codificación y perforación de los detalles geográficos y personales: Nacimientos, defunciones y nacidos muertos durante 1945; parte I, suplemento B, Instrucciones especiales para codificación y perforación de copias de declaraciones de muertes producidas por accidentes de vehículos automotores: muertes durante 1944; parte II, Asignación de las causas de muerte: copias de declaraciones de muerte, 1945; parte III, Instrucciones para clasificacion y tabulación de tarjetas perforadas correspondientes a nacimientos, muertes y nacimientos muertos; parte IV, Código geográfico, edición 1943 (basado en el censo de 1940); parte

V, Muestra de la mortalidad corriente: Codificación, perforación y tabulación de muestras mensuales de 1945.

*USA-5200.V6 **Vital Statistics of the United States.** 1937- . Annual. Washington, D.C. Department of Commerce. Bureau of the Census.

Estadística vital en los Estados Unidos. Anual.

USA-5200.V7 1900-40 **Vital Statistics Rates in the United States, 1900-1940.** By Forrest Edward Linder and Robert D. Grove. Washington, D.C., U.S.Govt.Print.Off., 1943. Department of Commerce. Bureau of the Census. 1051 p.

Chapters I to V translated into Portuguese by the Brazilian Institute of Geography and Statistics.

Tasas de estadística vital en los Estados Unidos, 1900-1940. Capítulos I al V traducidos al portugués por el Instituto Brasilero de Geografía y Estadística.

USA-5200.V8 **Vital Statistics—Special Reports.** 1934- Irregular. Washington, D.C. Department of Commerce. Bureau of the Census.

Estadística vital—informes especiales.

May 2, '40 "International Vital Statistics: Summary." Population, birth, death, marriage, divorce, and migration statistics for the United States and specified foreign countries including death rates for selected causes, time trends, and international comparisons. Vol. 9, no. 36, May 2, 1940, p. 345-462.

Resumen de datos bioestadísticos internacionales. Estadísticas de población, nacimientos, defunciones, matrimonios, divorcios, y movimiento migratorio, para los Estados Unidos y algunos países extranjeros, incluyendo tasas de defunciones según causas específicas, tendencias cronológicas, y comparaciones internacionales.

USA-5210.M5 **Monthly Vital Statistics Bulletin.** 1937- . Washington, D.C. Department of Commerce. Bureau of the Census.

Telegraphic report of birth and death certificates filed in 42 state offices.

Boletín mensual de estadística vital. Informe telegráfico sobre certificados de nacimientos y defunciones registrados en 42 oficinas de estados.

USA-5211.H5 **Human Fertility.** Devoted to the biological and clinical aspects of human fertility and its control. 1935- . Bimonthly. Baltimore, Maryland. Planned Parenthood Federation of America, Inc.

Fertilidad humana. Revista bimestral dedicada a los aspectos biológicos y clínicos de la fertilidad humana y de su control.

USA-5211.R5 1920 **Ratio of Children to Women, 1920.** A study in the differential rate of natural increase. By Warren Simpson Thompson. Washington, D.C., U.S. Govt.Print.Off., 1931. Department of Commerce. Bureau of the Census. 242 p. Census monograph, no. XI.

Proporción de niños a mujeres, 1920.

USA-5212.C5 **Current Mortality Analysis.** 1943- . Monthly. Washington, D.C. Department of Commerce. Bureau of the Census.

Based on the returns of a 10-percent sample of death certificates received in vital statistics offices.

Análisis de mortalidad corriente. Basado en las cifras de una muestra del 10 por ciento de los certificados recibídos en las oficinas de estadística vital.

USA-5212.D5 1934 **Death Rates by Occupation.** Based on data of the U. S. Bureau of the Census, 1930. By Jessamine S. Whitney. New York, 1934. National Tuberculosis Association. 32 p.

Tasas de mortalidad por ocupación, basado en los datos del Departamento del Censo de Estados Unidos, 1930.

USA-5212.P5 1945 **Procedures for Current Mortality Sample, 1945.** Instructions and information for special agents of the Bureau of the Census for the current mortality sample. Washington, D.C., February 1945. Department of Commerce. Bureau of the Census. 6 p.

Procedimientos para muestra de mortalidad corriente, 1945. Instrucciones e información, para agentes especiales del Departamento del Censo.

USA-5212.T5 1911-35 **Twenty-Five Years of Health Progress.** A study of the mortality experience among the industrial policy holders of the Metropolitan Life Insurance Company, 1911, 1935. By Louis I. Dublin and Alfred J. Lotka. New York, 1937. Metropolitan Life Insurance Company. 611 p.

Veinticinco años de progreso en sanidad. Un estudio sobre la experiencia de mortalidad obtenida por la Compañía Metropolitana de Seguro de Vida entre sus asegurados industriales, 1911-35.

USA-5212.W5 **Weekly Mortality Index.** 1930- . Washington, D. C., Department of Commerce. Bureau of the Census.

Telegraphic report of death certificates filed in 93 major cities.

Informe telegráfico de certificados de defunción registrados en 93 ciudades importantes. Semanal.

USA-5220.U5 1939 **United States Abridged Life Tables, 1939.** Urban and rural, by regions, color, and sex. Washington, D.C., 1943. Department of Commerce. Bureau of the Census. 18 p. Unexamined.

Abridged life tables are presented by urban and rural residence, color, and sex, for the continental United States and three major regions. Life tables values are shown separately for: (1) Cities of 100,000 or more; (2) other urban places; and (3) rural territory.

Presenta tablas abreviadas de vida por residencia urbana y rural, por color y sexo, para Estados Unidos continental y tres regiones principales. Las tablas de vida contienen cifras por separado para: (1) Ciudades de 100,000 y mas, (2) otros lugares urbanos, y (3) territorio rural.

USA-5220.U5 1930-39 **United States Abridged Life Tables, 1930-1939.** Preliminary. By geographic divisions, color and sex. Washington, D.C., 1942. Department of Commerce. Bureau of the Census. 15 p. Unexamined.

"These life tables, the first separate tables for geographic divisions covering the entire United States, are based on the mortality of 1930-1939. Life table values are also shown for the death registration states of 1920 and 1930." (Source: Gen.Cen.V.S.A.).

"Estas primeras tablas de vida para divisiones geográficas abarcando todo Estados Unidos, están basadas sobre la mortalidad de 1930-1939. Cifras de las tablas de vida se presentan también para las áreas de registro de defunciones de 1920 y 1930." (Traducción del comentario.)

USA-5220.U5 1890-1910 **United States Life Tables, 1890, 1901, 1910, and 1901-1910.** Explanatory text, mathematical theory, computations, graphs, and original statistics, also tables of United States Life annuities, life tables of foreign countries, mortality tables of life insurance companies. Prepared by James Waterman Glover. Washington, D.C., U.S.Govt.Print.Off., 1921. Department of Commerce. Bureau of the Census. 496 p.

Tablas de vida de Estados Unidos, 1890, 1901, 1910, y 1901-1910. Texto explicativo, teoría matemática, computaciones, gráficos y estadísticas originales, y tablas de renta vitalicia de Estados Unidos, tablas de vida de países extranjeros, tablas de mortalidad de compañías de seguros de vida.

USA-5250.B5 1942 **Birth Certificates, a Digest of the Laws and Regulations of the Various States.** By Earl Harrison Davis. New York, The H. W. Wilson Co., 1942. 136 p.

Certificados de nacimiento, una recopilación de leyes y reglamentos de los distintos estados de los Estados Unidos.

USA-5250.P5 1943 **Physician's Handbook on Birth and Death Registration.** Containing international list of causes of death. 9th edition, rev. Washington, D.C., U.S. Govt.Print.Off., 1943. Department of Commerce. Bureau of the Census. 94 p.

Manual del médico sobre registro de nacimiento y defunción.

USA-5300.I4 1943 **Immigration in the United States: A Selected List of Recent References.** Compiled by Anne L. Baden. Washington, D.C., 1943. Library of Congress. Division of Bibliography. 94 p. Unexamined.

Inmigración en los Estados Unidos: Una lista escogida de referencias recientemente aparecidas.

USA-5300.I6 1870-1930 **Interstate Migrations among the Native White as Indicated by Differences between State of Birth and State of Residence.** A series of maps based on the census, 1870-1930. By Charles Josiah Galpin and Theodore Bergen Manny. Washington, D.C., 1934. Department of Agriculture. Bureau of Agricultural Economics. 105 p. Unexamined.

Migración de blancos nativos entre estados, tal como la indican las diferencias entre estado de nacimiento y estado de residencia. Mapas basados en los censos, 1870-1930.

USA-5300.R4 1942 **A Research Memorandum on Internal Migration Resulting from the War Effort.** By Conrad Taeuber and Irene Barnes Taeuber. New York, May 1942. Social Science Research Council. 36 p. Mimeographed.

Prepared for the Committee on Research on Social Aspects of the War.

Un documento sobre investigación de la migración interna ocasionada por el esfuerzo bélico. Preparado para la Comisión de Estudios sobre los Aspectos Sociales de la Guerra.

USA-5300.R6 1939 **Rural Migration in the United States.** By C. E. Lively and Conrad Taeuber. Washington, D.C., U.S. Govt.Print.Off., 1939. Works Progress Administration. 192 p. Research monograph, no. 19. Unexamined.

"The Division of Research of the Works Progress Administration and the Bureau of Agricultural Economics of the Department of Agriculture collaborated in the comprehensive study of the nature and characteristics of rural migration, with particular reference to the relations of migration or the failure to migrate to problems of relief and economic opportunity." (Source: **Books on Population, 1939-Mid-1944.**)

Migración rural en los Estados Unidos. "La División de Investigaciones de la Administración del Progreso del Trabajo y la Dirección de Economía Agrícola del Ministerio de Agricultura, colaboraron en el amplio estudio de la naturaleza y características de la migración rural, con particular referencia a las relaciones de la migración o falta de migración con los problemas de subsistencia y oportunidad económica." (Traducción del comentario.)

6000-6999 ECONOMICS—ECONOMIA

USA-6000.A5 **American Economic Review.** The journal of the American Economic Association. 1911- . Quarterly. Evanston, Illinois.

Revista económica estadounidense. Publicación de la Asociación Económica Estadounidense.

Sept. '46 "Handbook: List of Members of the American Economic Association, 1946." Vol. XXXVI, no. 4, part 2, Supplement, September 1946. 143 p.

Manual de la Asociación Económica Americana, conteniendo una lista de los miembros, correspondiente al año 1946.

USA-6000.A6 **Annual Report of the National Bureau of Economic Research.** 1920-21— . New York.

Memoria anual de la Oficina Nacional de Investigación Económica.

1945-46 **Twenty-Sixth Annual Report of the National Bureau of Economic Research.** By Arthur F. Burns. New York, June 1946. National Bureau of Economic Research, Inc. 69 p.

Title on cover page—título sobre la cubierta: **Economic Research and the Keynesian Thinking in Our Times.**

Includes: Part one, "Economic Research and the Keynesian Thinking in Our Times," by Arthur F. Burns; part two, "Activities During the Year and Some Plans for the Future."

Incluye: "Investigación económica y el pensamiento de Keynes en nuestro tiempo;" "Actividades durante el año y algunos planes para el futuro."

USA-6000.D5 **Dun's Review.** 1893- . Monthly. New York. Dun and Bradstreet, Inc.

Jun '45, George "Gross National Product Projections for Full Employment: III, Measuring the Labor Force in 1950." By Edwin B. George. June 1945. 8 p. Reprint.

Last of three parts. Reviews and compares the various 1950 labor force estimates, and examines developments which will affect the size of the future labor force.

La última de tres partes de un estudio sobre proyecciones de la renta bruta nacional. Critica y compara las diversas estimaciones de la población activa y del potencial obrero en 1950, y analiza los fenómenos que afectarán su tamaño en el futuro.

May '45, George "Gross National Product Projections for Full Employment: II, Contrasting Estimates—Range and Reasons." By Edwin B. George. May 1945, 8 p. Reprint.

Second of three parts. Puts the most commonly used estimates on as nearly as comparable basis as possible, eliminating differences due to concept, base years, price level, and examines the reasons for the remaining differences. Includes tables summarizing estimates and comparisons.

La segunda de tres partes de un estudio sobre proyecciones del producto bruto nacional. Establece, sobre bases tan comparables como es posible hacerlo en la práctica, las estimaciones más comúnmente empleadas, eliminando las diferencias debidas a conceptos, años bases, nivel de precios, y analiza las razones de las diferencias que quedan. Incluye cuadros que resumen las estimaciones y comparaciones.

March '45, George "Gross National Product Projections: I, The Background and Relation to Current Issues." By Edwin B. George. March 1945. 11 p. Reprint.

First of three parts of a study on gross national product projections. This first article is an introductory explanation of methodology.

La primera de tres partes de un estudio acerca de proyecciones del producto bruto national. Este primer artículo sirve de introducción sobre la metodología.

USA-6000.I4 1939 **Industrial Change and Employment Opportunity—A Selected Bibliography.** Prepared under the supervision of Alexander Courvitch with the assistance of Carolyn Blanks, Marion Hayes, Esther Skala, and Sophie Udin. Philadelphia, Pennsylvania, 1939. Work Projects Administration. National Research Project. 254 p. Report, no. G-5.

Contents: General background; changes in industry; changes in labor supply, employment, and unemployment; attitudes of labor toward technological changes; adjustments and remedies.

Evolución industrial y oportunidad de ocupación—una bibliografía escogida. Contiene: Fundamento general; evolución de la industria; cambios en la oferta de mano de obra, y en la ocupación y desocupación; posición del trabajo frente a la evolución técnica; adaptaciones y remedios.

USA-6000.O4 **Occasional Papers of the National Bureau of Economic Research.** 1942?- . Irregular. New York.

Documentos ocasionales de la Oficina Nacional de Investigación Económica.

USA-6000.P5 **Publications of the National Bureau of Economic Research.** Starting date? Irregular. New York.

Series of monographs.

Publicaciones de la Oficina Nacional de Investigación Económica. Serie de monografías.

USA-6000.Q5 **Quarterly Journal of Economics.** 1886- Cambridge, Massachusetts, Harvard University Press. Unexamined.

Revista trimestral de economía de la Universidad Harvard.

USA-6000.R5 **Reports of the National Research Project on Reemployment Opportunities and Recent Changes in Industrial Techniques.** 1936?- . Irregular. Philadelphia, Pennsylvania. Work Projects Administration.

Series of monographs on: Types and rates of technological change; production, productivity, and employment; effects of industrial change on labor markets.

Informes sobre el Proyecto Nacional de Investigación de las Oportunidades de Reempleo y Cambios Recientes en la Técnica Industrial. Series de monografías sobre: Tipos y proporciones de los cambios tecnológicos; producción; productividad y ocupación; efectos de los cambios industriales sobre el mercado del trabajo.

USA-6000.S5 1945 **The Social Framework of the American Economy.** An introduction to economics. By J. R. Hicks and Albert Gailord Hart. New York, Oxford University Press, 1945. 261 p.

Includes: Economic facts and economic theory; production and exchange; goods and services; consumption and investment; population and its history; the economics of population; the specialization of labor; the effort of labor; capital goods and their varieties; private property in capital; national capital; social output and social income; foreign payments and the national income; the state and the national income; national income of the United States in 1939; national income in real terms—index numbers; national income and economic progress; the inequality of incomes; conclusion, further horizons; appendix: A, on the definition of production; B, on the idea of an optimum population; C, on the depreciation of capital; D, on what is meant by a favorable or adverse balance of payments; E, on the place of indirect taxes in the national income; F, on comparison between the real national incomes of different countries. An American edition based on J. R. Hicks' **The Social Framework: An Introduction to Economics,** first published in 1942.

La estructura social de la economía estadounidense. Incluye: Hechos económicos y teoría económica; producción e intercambio; bienes y servicios; consumo e inversión; la población y su historia; la economía de la población; la especialización del trabajo; el esfuerzo del trabajo; bienes de capital y sus variedades; propiedad privada del capital; capital nacional; producción social y renta social; pagos internacionales y renta nacional; el estado y la renta nacional; la renta nacional de Estados Unidos en 1939; la renta nacional en términos reales—números índices; renta nacional y progreso económico; la iniquidad de las rentas; conclusión, horizontes mas amplios; apéndice: A, sobre la definición de producción; B, sobre la idea de una población óptima; C, sobre la depreciación del capital; D, sobre lo que significa una favorable o adversa balanza de pagos; E, sobre la posición de los impuestos indirectos en la renta nacional; F, sobre la comparación internacional de las rentas nacionales de diferentes países. Una edición americana basada sobre la obra de J. R. Hicks **The**

Social Framework: An Introduction to Economics, primeramente publicada en 1942.

USA-6000.S7 **Survey of Current Business.** 1921- . Monthly. Washington, D.C. Department of Commerce. Bureau of Foreign and Domestic Commerce.

Gives monthly or quarterly figures on production, prices, sales, stocks, distribution, employment and other factors that reflect business conditions for basic industries and trade of the United States, and national income estimates. Has an annual supplement summarizing the monthly statistics for a number of years.

Investigación sobre negocios corrientes. Presenta cifras mensuales o trimestrales sobre producción, precios, ventas, existencias, distribución, ocupación, y otros factores que reflejan las condiciones de los negocios en las industrias básicas y en el comercio de los Estados Unidos, y estimaciones sobre renta nacional. Tiene un suplemento anual resumiendo las estadísticas mensuales de varios años.

USA-6010.W5 1939-41 **Wartime Prices: Part I, August 1939 to Pearl Harbor.** By John M. Blair and Melville J. Ulmer. Washington, D.C., 1944. Department of Labor. Bureau of Labor Statistics. 272 p. Bulletin no. 749. Unexamined.

Study of wholesale prices in primary markets in the United States from August 1939 to December 1941.

Precios en tiempo de guerra: Parte I, de agosto 1939 a Pearl Harbor. Estudio de los precios al por mayor en los principales mercados de los Estados Unidos desde agosto de 1939 a diciembre de 1941.

USA-6020.P5 1945 **Production: Wartime Achievements and the Reconversion Outlook.** A report to the War Production Board. By J. A. Krug. Washington, D.C., October 1945. Executive Office of the President. Office for Emergency Management. War Production Board. 113 p. WPB document, no. 334.

A summary report presenting a picture of the war effort in the United States, largely in quantitative terms. Charts and tables, in general for the years 1939-45, cover such topics as: Wartime expansion in general; manufacturing facilities; manpower; lend-lease and foreign trade; munitions production; construction activity, etc.

Producción: Logros en tiempo de guerra y perspectivas de transformación. Informe al Consejo de Producción de Guerra. Un informe sumario que presenta el panorama del esfuerzo de guerra en los Estados Unidos, generalmente en términos cuantitativos. Cuadros y gráficos, en general para los años 1939-45, referentes a los siguientes tópicos: Expansión de guerra en general; facilidades de fabricación; potencial humano; préstamos y comercio exterior; producción de municiones; construcción, etc.

USA-6020.S5 **Statistics of War Production.** 1942- . Monthly. Washington, D.C. Executive Office of the President. Office for Emergency Management. War Production Board. Confidential.

Key statistical data relating to the production sector of the war economy; a source-book for government statisticians.

Estadísticas mensuales de la producción de guerra. Importantes datos estadísticos relativos al sector de producción de la economía de guerra; un libro informativo para estadígrafos oficiales.

USA-6100.I4 1921-22 **Income in the United States: Its Amount and Distribution.** By Wesley C. Mitchell, Willford Isbell King, Frederick R. Macaulay, and O. W. Knauth. New York, 1921-22. National Bureau of Economic Research. 2 vol. Unexamined.

Includes figures on the United States national income for 1909-18.

Renta en los Estados Unidos: Su monto y distribución. Incluye cifras de la renta nacional de Estados Unidos para 1909-18.

USA-6100.I6 1938 **The Income Structure of the United States.** By Maurice Leven. Washington, D.C., 1938. Brookings Institution. 177 p. Publication, no. 74. Unexamined.

La estructura de la renta en Estados Unidos.

USA-6100.N2 1946 **National Income: A Summary of Findings.** By Simon Kuznets. New York, 1946. National Bureau of Economic Research. 160 p. Twenty-fifth anniversary series, no. 1. Unexamined.

Contents: Structure, 1919-1938 (total and per capita; distribution by industrial origin, type, size, type of use; regional and community-size differences); long term changes, 1869-1938 (total and per capita; distribution by industrial origin, type, size, type of use; fluctuations in rates of growth); changes during business cycles, 1900-1938 (magnitude of changes; national aggregates in current and constant prices; differences among industries, among types of payments, among shares of upper and lower income groups, among final use components); problems of interpretation (the distinction between net and gross; gross national product and national income; national income and welfare; consistent valuation; directions of future work).

Renta nacional: Un resumen de conclusiones.

Contiene: Estructura, 1919-1938 (total y por habitante; distribución por origen industrial, tipo, monto, clase de uso; diferencias regionales y diferencias en relación a la importancia numérica de las localidades); cambios a largo plazo, 1869-1938 (total y por habitante; distribución por origen industrial, tipo, monto, clase de uso, fluctuaciones de la tasa de incremento); cambios durante los ciclos del comercio, 1900-1938 (magnitud de los cambios; conjunto nacional de precios corrientes y constantes; diferencias entre las industrias, entre los tipos de pago, entre las categorías altas y bajas de ingreso, entre componentes según uso definitivo); problemas de la interpretación (la diferencia entre ingreso bruto e ingreso neto; producción nacional bruta y renta nacional bruta; renta nacional y bienestar; valuación consistente; dirección de las futuras investigaciones).

USA-6100.N3 1919-35 **National Income and Capital Formation. 1919-1935.** By Simon Smith Kuznets. New York, MacMillan Publishing Co., 1938. 86 p. Unexamined.

This is a later and independent edition of an earlier publication of the National Bureau of Economic Research.

Renta nacional y formación de capital, 1919-35. Es una edición reciente e independiente de una publicación anterior de la Oficina Nacional de Investigaciones Económicas.

USA-6100.N4 1941 **National Income and Its Composition, 1919-1938.** By Simon Smith Kuznets, Lillian Epstein and Elizabeth Jenks. New York, 1941. National Bureau of Economic Research. Publication, no. 40. 2 vol.

Renta nacional y su composición, 1919-38.

USA-6100.N5 1929-37 **National Income in the United States, 1929-1937.** By Robert R. Nathan. Washington, D.C., U.S.Govt.Print. Off., 1938. Department of Commerce. Bureau of Foreign and Domestic Commerce. Unexamined.

Renta nacional en los Estados Unidos, 1929-37.

USA-6100.N7 1945 **National Product in Wartime.** By Simon Smith Kuznets. New York, 1945. National Bureau of Economic Research. 156 p. Publication, no. 44.

Contents: Treatment of war output; national product, 1939-43; national product in World Wars I and II. "The author states that estimates of national product must be made in accordance with assumptions about which there is no general agreement. These differences of concepts of national product are magnified by a major war. The problems are discussed and attempts are made to apply tentative solutions in measuring changes in national product during World Wars I and II. Estimates for the period of the first World War and the years immediately following the war are based upon estimates formerly published by the National Bureau of Economic Research, with adjustments to attain closer comparability with the estimates for recent years." (Source: **Monthly Labor Review,** vol. 61, no. 3, September 1945, p. 610.)

Producto nacional en tiempo de guerra. Contenido: Tratamiento de la producción de guerra; producto nacional, 1939-43; producto nacional en las guerras mundiales I y II. "El autor establece que las estimaciones del producto nacional deben ser hechas de acuerdo con suposiciones acerca de las cuales no hay acuerdo general. Estas diferencias de concepto sobre producto nacional están aumentadas por una guerra de grandes proporciones. Se discuten los problemas, y se trata aplicar soluciones provisionales para medir los cambios en el producto nacional durante las dos guerras mundiales. Las estimaciones para el período de la primer guerra mundial y los años inmediatamente posteriores, están basadas sobre las estimaciones publicadas anteriormente por la Oficina Nacional de Investigaciones Económicas, con correcciones para obtener mas estrecha comparabilidad con las estimaciones correspondientes a los años recientes." (Traducción del comentario.)

USA-6100.O5 1940 **Our National Resources: Facts and Problems.** Washington, D.C., U.S.Govt.Print.Off., 1940. National Resources Planning Board. 45 p.

Recursos nacionales de los Estados Unidos: Realidades y problemas.

USA-6100.P5 1939 **Patterns of Resource Use.** A technical report. Washington, D.C., U.S.Govt.Print.Off., 1939. National Resources Committee. Industrial Committee. 149 p.

Modalidades de la utilización de los recursos: Un informe técnico.

USA-6100.W5 1850-1900 **Wealth and Income of the People of the United States.** By Willford Isbell King. 1915. Unexamined.

Gives rates of growth of real income in the United States, 1850-1900.

Riqueza y renta del pueblo estadounidense. Presenta tasas de crecimiento de la renta real en los Estados Unidos, 1850-1900.

USA-6200 **Agricultural Statistics in the United States: Methods and Procedures.** Background material acquired for Inter American Statistical Institute study on methods in agricultural statistics in the American nations, including the Survey of Agricultural Production Statistics in the American Nations in 1945-46.

Consists of: Schedules; other forms; instructions; questionnaires.

Estadísticas agrícolas en los Estados Unidos: Métodos y procedimientos. Material básico adquirido para el estudio que el Instituto Interamericano de Estadística realiza sobre los métodos de estadísticas agrícolas en las naciones americanas, incluyendo la encuesta de 1945-46. Consiste de: Cédulas; otros formularios; instrucciones; cuestionarios.

USA-6200.A4 **Agricultural Finance Review.** Farm credit, farm insurance, farm taxation. 1938- . Annual. Washington, D.C. United States Department of Agriculture. Bureau of Agricultural Economics.

Semiannual from 1938 to 1941. Current developments and research in the field of farm credit, including amount of outstanding farm mortgage loans held by leading agencies at end of each year; agricultural loans held by all insured commercial banks; interest rates charged on farm mortgages by credit agencies under the supervision of the Department of Agriculture; farm bankruptcy reports; and insurance companies owned and controlled by farmers, and operating on a mutual plan.

Revista anual de finanzas agrícolas de la Secretaría de Agricultura; publicado semestralmente de 1938 a 1941. Nuevos acontecimientos e investigaciones en el ramo del crédito agrícola, incluyendo suma de hipotecas agrícolas pendientes en manos de los órganos principales al fin de cada año; préstamos agrícolas en manos de todos los bancos comerciales asegurados; tipos de interés de las hipotecas agrícolas cargados por los órganos de crédito bajo la superintendencia de la Secretaría de Agricultura; informes de bancarrotas agrícolas; así como compañías de seguro en posesión de, y controladas por agricultores, y funcionando según un plan mutuo.

1945 **Agricultural Finance Review.** Farm credit, farm insurance, farm taxation. Volume 8, November 1945. Washington, D.C. United States Department of Agriculture. Bureau of Agricultural Economics. 104 p.

Articles and a statistical appendix on farm-mortgage credit, non-real-estate credit, farm cooperative credit, farm taxation, agricultural insurance, and other related data.

Artículos y apéndice estadístico sobre crédito de hipotecas agrícolas, crédito no relacionado con bienes raíces, crédito de cooperativas agrícolas, impuestos agrícolas, seguro agrícola, y otros datos relacionados.

USA-6200.A45 **Agricultural Prices.** Starting date? Monthly. Washington, D.C. United States Department of Agriculture. Bureau of Agricultural Economics.

Prices received by farmers for all groups of farm products, index numbers of prices received and paid by farmers, and parity prices.

Informe mensual de precios agrícolas; precios recibidos por los agricultores para toda clase de productos agrícolas, números índices de precios recibidos y pagados por los agricultores, y precios a la par (precios "parity").

USA-6200.A5 **Agricultural Statistics.** 1926- . Annual. Washington, D.C. United States Department of Agriculture.

A collection of the more important series of agricultural statistics; acreage, yield and production of crops, commercial crops, prices received by farmers, livestock production, market supplies and prices, and imports and exports. Prior to 1936 the information contained in these volumes was published in the statistical section of the Yearbook of the Department of Agriculture.

Estadísticas agrícolas. Colección anual de las más importantes series de estadísticas agrícolas; número de acres, cosecha y producción de granos, cosechas para el comercio, precios recibidos por los agricultores, producción de ganado, provisiones de mercados y precios, e importaciones y exportaciones. Antes del año 1936, las informaciones contenidas en estos volúmenes fueron publicadas en la sección estadística del Anuario de la Secretaría de Agricultura.

1944 **Agricultural Statistics, 1944.** Washington, D.C., U.S.Govt.Print.Off., 1944. United States Department of Agriculture. 587 p.

Includes sections on: Grains; cotton, sugar, and tobacco; oilseeds, fats, and oils; fruits, vegetables, melons, and tree nuts; hay, seeds, and minor field crops; beef cattle, hogs, sheep, horses, and mules; dairy and poultry products; farm capital and income; agricultural conservation and adjustment; miscellaneous statistics.

Incluye secciones sobre: Granos; algodón, azúcar y tabaco; semillas oleaginosas, grasas y aceites; frutas, vegetales, melones, y nueces; heno, semillas, y cosechas secundarias; ganado vacuno, cerdos, ovejas, caballos y mulas; productos de lechería y aves de corral; capital e ingreso agrícola; conservación y adaptación agrícola; estadísticas misceláneas.

USA-6200.A7 1942 **American Agriculture, 1899-1939: A Study of Output, Employment and Productivity.** By Harold Barger and Hans H. Landsberg. New York, 1942. National Bureau of Economic Research. 440 p. Publication, no. 42.

"This book presents an intensive study of the annual volume of net agricultural output in the United States, by individual items and overall of labor used in agriculture, of food consumption, and of farm productivity per acre and per man." (Source: **Journal of the American Statistical Association,** vol. 38, no. 224, December 1943.)

Agricultura estadounidense, 1899-1939: Un estudio sobre producción, empleo y productividad. "Este libro presenta un estudio intensivo del volumen anual de la producción agrícola neta en los Estados Unidos por artículos individuales y total de mano de obra utilizado en la agricultura, del consumo de alimentos, y de la productividad agrícola per acre y per hombre." (Traducción del comentario.)

USA-6200.A8 1943 **Analysis of Specified Farm Characteristics for Farms Classified by Total Value of Products.** Washington, D.C., U.S.Govt.Print.Off., 1943. Department of Commerce. Bureau of the Census. 221 p.

Análisis de ciertas características de la granja, para granjas clasificadas según el valor de sus productos.

USA-6200.C6 **Crops and Markets.** Published quarterly by the United States Department of Agriculture. 1924- . Washington, D.C.

Previous to 1927 issued weekly. Crop report and market statistics on livestock, meats, wool, dairy and poultry products, grain, feed and seed, and cotton; also data on prices received by farmers, and other special material of a marketing and economic nature.

Informe trimestral de cosecha y estadísticas de mercados, referentes a ganado, carnes, lana, productos de lechería y aves de corral, grano, forrajes y semillas, y algodón; también, datos sobre precios recibidos por los agricultores, y otras informaciones especiales de carácter económico y mercantil. Antes de 1927, se publicó semanalmente.

USA-6200.F3 1939 **The Farm-Housing Survey.** Directed by Bureau of Home Economics in cooperation with Bureau of Agricultural Engineering, Extension Service and Office of the Secretary. Washington, D.C., U.S.Govt.Print.Off., 1939. Department of Agriculture. 42 p. Miscellaneous publication, no. 323. Unexamined.

Estudio de la vivienda rural.

USA-6200.F4 **The Farm Income Situation.** Starting date? Monthly. Washington, D.C. United States Department of Agriculture. Bureau of Agricultural Economics.

Tables indicating cash income from farm marketings, index numbers of income, income by commodities, and government payments.

Informe mensual sobre la situación de las rentas agrícolas. Contiene cuadros exponiendo ingresos al contado por la venta de productos agrícolas, número índices de ingresos, ingresos según comodidades, y sumas pagadas por el gobierno.

USA-6200.F7 **The Farm Real Estate Situation.** 1925- . Annual. Washington, D.C. Department of Agriculture. Bureau of Agricultural Economics. Unexamined.

La situación de los bienes raíces rurales. Anual.

USA-6200.I4 1944 **The Impact of the War on the Financial Structure of Agriculture.** By Alvin S. Tostlebe. Washington, D.C., 1944. Bureau of Agricultural Economics. 191 p.

"Presents facts on the extent of wartime changes in the financial structure of agriculture, interprets the economic significance of these changes and explains the effect of possible postwar developments on the financial structure of agriculture. A consolidated balance sheet of all farms of the United States is developed as of January 1 for each year, 1940 to 1944. The balance sheet lists the value in dollars of tangible assets, including real estate, livestock, machinery and equipment, crops and household equipment, and lists intangible assets including currency and deposits, U. S. savings bonds, warehouse receipts, and other assets. Detailed statements of the methods used in making the estimates are found in the appendix." (Source: **Journal of the American Statistical Association,** vol. 40, no. 229, March 1945, p. 136-137.)

El efecto de la guerra sobre la estructura financiera de la agricultura. "Presenta ejemplos sobre el alcance de los cambios en la estructura financiera de la agricultura en tiempos de guerra, interpreta el significado económico de esos cambios y explica los efectos de posibles evoluciones de postguerra en la estructura financiera de la agricultura. Un balance consolidado de todas las granjas de los Estados Unidos se establece para el 1° de enero de cada año, 1940 a 1944. El balance enumera los valores en dólares de los activos materiales incluyendo bienes raíces, ganados, maquinarias y herramientas, cosechas y muebles de la vivienda, y enumera los activos inmateriales incluyendo moneda y depósitos, títulos del Estado, recibos de almacenes y otros. El apéndice expone detalladamente los métodos usados para las estimaciones." (Traducción del comentario.)

USA-**6200**.J5 **Journal of Farm Economics.** 1919- . Quarterly. Menasha, Wisconsin. American Farm Economic Association.

Includes articles on methods and development of agricultural statistics, statistical sampling in agriculture etc.

Revista trimestral de economía rural de la Asociación Americana de Economía Rural. Incluye artículos sobre métodos y evolución de las estadísticas agrícolas, muestreo estadístico en agricultura, etc.

USA-**6200**.R5 **Research Bulletin.** 1911- . Irregular. Ames, Iowa. Iowa State College of Agriculture and Mechanic Arts. Agricultural Experiment Station.

Each issue contains complete report on results of research and investigation conducted about a specific problem. Always published in cooperation with other agency or agencies which helped on the specific projects.

Boletín de investigación, del Colegio Agrícola del Estado de Iowa. Cada edición contiene un informe completo sobre los resultados del estudio e investigación realizados sobre un problema específico. Publicado siempre en cooperación con otro organismo u organismos que ayudaron en los proyectos específicos.

USA-**6200**.Y5 **Yearbook of the Department of Agriculture.** 1895- . Washington, D.C.

Prior to 1936 this yearbook contained the agricultural statistics later published in **Agricultural Statistics** (USA-**6200**.A5).

Anuario de la Secretaría de Agricultura. Anteriormente a 1936 este anuario contenía las estadísticas agrícolas que luego se publicaron en **Agricultural Statistics** (USA-**6200**.A5).

USA-**6220**.C5 1933 **The Crop and Livestock Reporting Service of the United States.** Washington, D.C., U.S.Govt.Print. Off., 1933. Department of Agriculture. Bureau of Agricultural Economics. 104 p.

El servicio de información sobre cosecha y ganadería de Estados Unidos.

USA-**6220**.C7 **Crop Production.** Starting date? Monthly. Washington, D.C. United States Department of Agriculture. Bureau of Agricultural Economics. Crop Reporting Board.

Monthly during growing season. Contains yield per acre, total production of major crops.

Informe de la producción de las cosechas; contiene producción por "acre", y producción total de los cosechas principales. Publicado mensualmente durante la temporada del crecimiento del cultivo.

USA-**6240**.F4 **Fishery Statistics of the United States.** 1939- . Annual. Washington, D.C. Department of the Interior. Fish and Wildlife Service. Statistical digest series. Unexamined.

"Supersedes the annual report of the Division of Fishing Industries entitled **Fishery Industries of the United States.**" (Source: 1939 issue, p. 1.)

Estadística anual de la pesca en los Estados Unidos. "Substituye el informe anual de la Dirección de Industrias de la Pesca, intitulado **Fishery Industries of the United States.**" (Traducción del comentario.)

USA-**6240**.S6 **Statistical Digest Series of the Fish and Wildlife Service.** 1942- . Washington, D.C. Department of the Interior. Fish and Wildlife Service.

Series which includes the annual issues of **Fishery Statistics of the United States.**

Serie de recopilaciones estadísticas del Servicio de Peces y Animales Salvajes. Serie que incluye las ediciones anuales de **Fishery Statistics of the United States.**

*USA-**6300**.M5 **Minerals Yearbook.** Starting date? Annual. Washington, D.C. Department of the Interior. Bureau of Mines.

A compendium of mineral statistics of the United States and the principal mineral-producing countries of the world.

Compendio de estadísticas de minerales de los Estados Unidos, y de los principales países de producción minera en el mundo.

1944 **Minerals Yearbook, 1944.** Washington, D.C., U.S.Govt.Print.Off., 1946. United States Department of the Interior. Bureau of Mines. 1636 p.

Contents: Review of mineral industries in 1944; statistical summary of mineral production, with tables on quantity and value of mineral production of the United States and principal producing states, and mineral products of United States, by states; statistics on metals and non-metals, including domestic production, consumption, stocks, prices, foreign trade — exports and imports, world production — review by countries, with separate chapters on mine production by states; mine safety, with figures on employment and accidents in mineral industries; foreign minerals review, by country, with emphasis on the Latin American countries.

Contiene: Síntesis de la actividad minera en 1944; resumen estadístico de la producción minera, con cuadros sobre cantidad y valor de la producción minera de los Estados Unidos y de los principales Estados productores, y producción por Estados; estadísticas sobre metales y metaloides, incluyendo producción doméstica, consumo, existencias, precios, comercio exterior (importaciones y exportaciones), síntesis de la producción mundial por países, y capítulos sobre producción minera por Estados seguridad en las minas, con cifras sobre ocupación y accidentes en las industrias mineras; síntesis de los minerales en el exterior, por países, con especial atención a los países latino-americanos.

USA-**6300**.M7 1899-1939 **The Mining Industries, 1899-1939: A Study of Output, Employment, and Productivity.** By Harold Barger and Sam H. Schurr. New York, 1944. National Bureau of Economic Research. 452 p.

Las industrias mineras, 1899-1939: Un estudio sobre producción, empleo y productividad.

USA-**6300**.T5 **Technical Papers of the Bureau of Mines.** Starting date? Washington, D.C. Department of the Interior. Bureau of Mines.

Series of monographs.

Documentos técnicos de la Dirección de Minas. Series de monografías.

USA-**6400**.I3 **Industrial Corporation Reports.** Starting date? Frequency? Washington, D.C. Federal Trade Commission.

Financial data on operations for principal manufacturing industries.

Informes de compañías industriales. Datos financieros sobre las operaciones de las principales industrias manufactureras.

1939 **Industrial Corporation Reports for 1939: Summary by Industry Groups.** Washington, D.C., 1941. Federal Trade Commission. 25 p.

Summary of principal assets and sales, investment and rates of return, dividends paid and surplus accumulation, and operating ratios or cents per dollar of sales for 1939 of the most important corporations in 76 industries.

Informes sobre compañías industriales para 1939: resumen por grupos de industrias. Resúmenes de los principales activos y ventas, inversiones y tipos de ganancias, dividendos pagados y sobrantes acumulados, así como de proporciones de operaciones, o sea centavos por cada dólar de ventas realizadas en 1939, de las más importantes compañías en 76 ramos de industria.

USA-**6400**.O4 1929-39 **Output of Manufactured Commodities, 1929-1939.** Washington, D.C., 1942. Department of Commerce. Bureau of Foreign and Domestic Commerce. 195 p. Unexamined.

Producción total de artículos manufacturados, 1929-39.*

USA-**6400**.O6 1940 **The Output of Manufacturing Industries, 1899-1937.** By Solomon Fabricant and Julius Shiskin. New York, 1940. National Bureau of Economic Research. 685 p. Publication, no. 39.

La producción de las industrias manufactureras, 1899-1937.

USA-**6400**.S5 **Statistics of Electric Utilities in the United States, Classes A and B Privately Owned Companies.** Starting date? Annual. Washington, D.C. Federal Power Commission.

Financial statistics of privately owned companies of electric utilities, consisting of balance sheets, income and earned surplus statements, capital stocks and bonds, electric operating revenues, customers and sales by classes of service, operating expenses, utility plant, and physical quantities.

Estadísticas de empresas eléctricas de servicio público en los Estados Unidos—compañías privadas clases "A" y "B". Estadísticas financieras de compañías privadas de empresas eléctricas de servicio público, constando de balances, declaraciones de ingresos y ganancias sobrantes, acciones y títulos capitales, ingresos proveniendo de las operaciones eléctricas, clientes y ventas según clases de servicio, gastos de operaciones, plantas de servicio y cantidades físicas.

USA-**6500**.C3 **Construction.** Employment, volume, prices, wages, rents. Starting date? Monthly. Washington, D.C.

United States Department of Labor. Bureau of Labor Statistics.

Includes: Retail prices of building materials in eight principal markets; employment and pay rolls on construction; estimated construction expenditures, by type of construction, by month; building construction started in urban areas grouped by size of city, geographic division, State, and urban area, and by type; Federal construction; residential rents and vacancy rates; wholesale prices of building materials and construction machinery; prevailing wage rates in selected cities.

Informe mensual sobre construcción; incluye: Precios al por menor de los materiales de construcción en ocho mercados principales; empleo y total de salarios en construcciones; gastos estimados de construcción, por clase de construcción, y por mes; construcciones de edificios empezadas en áreas urbanos, clasificadas por tamaño de la ciudad, división geográfica, estado, y área urbana, y por tipo; construcción federal; arrendamientos de habitaciones y tipos de vacantes; precios al por mayor de los materiales de construcción, y de maquinaria para construcción; tipos de salarios predominantes en ciudades escogidas.

USA-6500.C4 1944 **The Construction Industry in the United States.** Prepared by Division of Construction and Public Employment. Washington, D.C., U.S.Govt.Print.Off. 1944. United States Department of Labor. Bureau of Labor Statistics. 149 p. Bulletin, no. 786.

Includes: Trend of construction expenditures, 1915-43; employment in the construction industry, 1929-43; volume of federal construction, 1935-43; building construction (including costs, trends, valuation of urban construction, new dwelling units in non-farm areas, and indexes of building construction, 1929-43). Appendixes of statistical tables show breakdown of wages and hours in the construction industry for the period 1934-43, and volume and valuation of building construction in 2,707 cities, by city, for 1942 and 1943.

La industria de construcción y edificación en los Estados Unidos. Incluye: Tendencias de los gastos de construcción, 1915 a 1943; empleo en la industria de construcción, 1929 a 1943; volumen de construcción del gobierno federal, 1935 a 1943; edificación (incluyendo costos, tendencias, valuación de edificación urbana, construcción de nuevas viviendas en las regiones no-agrícolas, e índices de edificación, 1929 a 1943). Apéndices de cuadros estadísticos presentan detalles de salarios y horas en la industria de construcción para el período 1934-43, así como el volumen y la valuación de edificaciones en 2,707 ciudades, ciudad por ciudad, en 1942 y 1943.

USA-6610.S5 **Statements of the Bureau of Transport Economics and Statistics.** Starting date? Irregular. Washington, D.C. Interstate Commerce Commission. Bureau of Transport Economics and Statistics.

Series of special studies.

Informes del Departamento de Economía y Estadística de Transporte.

USA-6610.T5 **Transport Fact Book.** Annual summary of basic data and trends in the transit industry of the United States. Starting date? New York. American Transit Association.

Resumen anual de datos fundamentales y curso de la industria del transporte en los Estados Unidos.

USA-6611.A5 **Annual Airline Statistics: Domestic Carriers.** Starting date? Washington, D.C. Civil Aeronautics Board. Economic Bureau. Accounting and Rates Division.

Detailed data on traffic, operations, equipment, finances and employment, for domestic air carriers, by company.

Estadísticas anuales de las compañías de transporte aéreo. Presenta datos detallados acerca de tránsito, operaciones, equipo, finanzas, y empleo de las empresas de transporte aéreo en el interior de los Estados Unidos, por compañías individuales.

USA-6611.S5 1944 **Statistical Handbook of Civil Aviation.** Washington, D.C., October 1944. U. S. Department of Commerce. Civil Aeronautics Administration. 86 p. Looseleaf, mimeographed. Civil Aeronautics Administration.

All official data on the development of civil aviation in the United States available for immediate reference. Supplemented by periodic releases. Many of the statistics go back to 1926. Airports and landing fields by States, classes, types, and number by years; U. S. air transport routes, July 1, 1944; operations—operators, number of aircraft, average available seats, average speed, route mileage, miles flown, passengers transported, trips, express and freight, fuel consumed, passenger-miles flown, passenger load factor, personnel employed; air carrier accidents, showing vital statistics and causes; number and classes of airmen; production.

Manual estadístico de aviación civil; todos los

datos oficiales sobre el desarrollo de la aviación civil en los Estados Unidos, asequibles para consulta inmediata. Suplementado con publicaciones periódicas. Muchas de las estadísticas se remontan a 1926. Aeropuertos y campos de aterrizaje por Estados, clases, tipos, y número por años; vías de transporte aéreo en los Estados Unidos, el 1° de julio de 1944; operaciones—empresarios, número de aviones, promedio de asientos disponibles, promedio de velocidad, millas de vías, millas de vuelo, pasajeros transportados, viajes, expresos y fletes, combustible consumido, pasajeros-millas de vuelo, factor de la carga de pasajeros, personal empleado; accidentes en las empresas de transporte aéreo, presentando estadísticas vitales y causas; número y clases de aviadores; producción.

USA-6613.R5 1921-41 **A Review of Statistics of Oil Pipe Lines, 1921-1941.** Washington, D.C., 1942. Interstate Commerce Commission. Bureau of Transport Economics and Statistics. 68 p. Mimeographed. Statement, no. 4280. Unexamined.

Una revista de estadísticas de los oleoductos, 1921-41.

USA-6614.S6 **Statistics of Railways in the United States.** 1888- . Annual. Washington, D.C. Interstate Commerce Commission. Bureau of Transport Economics and Statistics.

Estadísticas de ferrocarriles en los Estados Unidos. Anual.

1944 **Fifty-Eighth Annual Report on the Statistics of Railways in the United States for the Year Ended December 31, 1944.** Including selected data for the Pullman Co., Railway Express Agency, Inc., electric railways, carriers by water, oil pipe lines, motor carriers, freight forwarders, and private car owners subject to the Interstate Commerce Act for the year 1944. Prepared by the Bureau of Transport Economics and Statistics. Washington, D.C., U.S.Govt.Print.Off., 1946. Interstate Commerce Commission. 578 p.

Detailed data on roadway and track, equipment, traffic, operation, employees, fuel, accidents, revenues, expenses, income and earned surplus, dividends and interest, balance sheet, investment in road and equipment, capitalization, receiverships and trusteeships, etc.

Datos detallados sobre firmes y vías ferroviarias, equipo, tránsito, operaciones, empleados, combustibles, accidentes, ingresos, gastos, beneficios y ganancias sobrantes, dividendos e intereses, balances, inversiones para caminos y equipos, capitalización, receptorías y cargos de fideicomisario, etc.

USA-6615.A5.I **Annual Report of the Chief of Engineers, U. S. Army, Part I.** Starting date? Washington, D.C. War Department. Office of the Chief of Engineers.

Report on rivers and harbors and flood control.

La parte primera de la memoria anual del jefe del Cuerpo de Ingenieros del ejército de los Estados Unidos consta del informe sobre los ríos y puertos, y sobre el control de anegaciones.

USA-6615.A5.II **Annual Report of the Chief of Engineers; Part 2 Commercial Statistics, Water-Borne Commerce of the United States.** Starting date? Washington, D.C. War Department. Office of the Chief of Engineers.

Tonnage of water-borne commerce by ports, rivers, canals and connecting channels by chief commodity.

Parte segunda de la memoria anual del jefe del Cuerpo de Ingenieros del ejército de los Estados Unidos, presentando estadísticas comerciales, expone datos del tonelaje del comercio fluvial, por puertos, ríos, canales y vías comunicantes, según mercancías principales.

USA-6620.S5 **Statistics of the Communications Industry in the United States.** Starting date? Annual. Washington, D.C. Federal Communications Commission.

Financial and operating data relative to telephone, wire-telegraph, ocean-cable, and radiotelegraph carriers and controlling companies, and standard radio broadcast stations and networks. Also employment, compensation and accidents, and number of telephones and radios in homes by State and region.

Estadísticas de la industria de comunicaciones en los Estados Unidos. Datos financieros y de operaciones sobre compañías transmisoras y compañías de mando de comunicaciones telefónicas, telegráficas, cablegráficas y radiotelegráficas, así como radioemisoras y redes regulares de transmisión. También, empleo, compensación y accidentes, y número de teléfonos y radioreceptores en hogares, por estado y región.

USA-6700.D4 **Domestic Commerce.** A bulletin of national economy. Starting date? Monthly. Washington, D.C. Department of Commerce.

Formerly publishe l as a weekly — antes publicado semanalmente.

Comercio interior. Mensual. Un boletín sobre economía nacional.

USA-6700.D6 **Domestic Commerce Series of the Bureau of Foreign and Domestic Commerce.** Starting date? Irregular. Washington, D.C. Department of Commerce. Bureau of Foreign and Domestic Commerce.

Serie sobre comercio interior de la Dirección de Comercio Exterior e Interior.

USA-6720.A3 **Annual Report of the Board of Governors of the Federal Reserve System.** 1914- . Washington, D.C. Board of Governors of the Federal Reserve System.

Condition of Federal Reserve banks; gold, foreign exchange and money in circulation; discount and money rates; member and nonmember banks; business conditions.

Memoria anual del Consejo Directivo del Sistema de la Reserva Federal. Condiciones de los bancos de la Reserva Federal; oro, divisas y moneda en circulación; descuento y tipos de cambio; bancos asociados con el sistema y bancos no asociados; situación comercial.

USA-6720.A5 **Annual Report of the Comptroller of the Currency.** 1863- . Washington, D.C. Treasury Department. Office of the Comptroller of the Currency.

Statistics on all banks, with particular emphasis on the national banking system, giving assets, liabilities, earnings, expenses and dividends. Summary tables show principal items of assets and liabilities by size of banks, by State for national banks, and for banks other than national. Data on deposits by State and by type and bank in receivership are included.

Memoria anual del Contralor del Cambio; presenta estadísticas de todos los bancos, con consideración especial del sistema bancario nacional, indicando activos, pasivos, ganancias, gastos y dividendos. Cuadros de resumen exponen los principales detalles de los activos y pasivos según importancia de los bancos, bancos nacionales por Estados, y otras clases de bancos. Se incluyen datos de depósitos, por Estado y por clase, así como por bancos sometidos a comisiones receptoras.

USA-6720.A7 **Annual Report of the Farm Credit Administration.** 1933-34— . Washington, D.C. Department of Agriculture. Farm Credit Administration.

Data on various agricultural credit institutions operating under the Farm Credit Administration. Financial Statements for production credit associations, Federal intermediate banks, banks for cooperatives, Federal land banks, regional agricultural credit corporations, and joint stock land banks.

Memoria anual de la Administración de Crédito Agrícola. Datos sobre varias instituciones de crédito agrícola funcionando bajo la dirección de la Administración de Crédito Agrícola. Declaraciones financieras de asociaciones de crédito de producción, bancos federales intermediarios, bancos para cooperativas, bancos federales hipotecarios, corporaciones regionales de crédito agrícola, y bancos hipotecarios de acciones mixtas.

*USA-6720.B5 1943 **Banking and Monetary Statistics.** Washington, D.C., National Capital Press, 1943. Board of Governors of the Federal Reserve System. 979 p.

Basic reference book containing background banking and monetary data. The statistics in most cases cover the period since 1914, when the Federal Reserve System was established. Data relates largely to the condition and operation of the Federal Reserve Banks and member banks. There are also data on the condition and operation of all banks, including State banks not members of the Federal Reserve System, and statistics of bank debits, bank earnings, bank suspensions, branch, group, and chain banking, currency, money rates, security markets, Treasury finance, production and movement of gold, and international financial developments.

Estadísticas bancarias y monetarias. Fuente básica de referencia que contiene datos fundamentales sobre bancos y moneda. En la mayoría de los casos, las estadísticas abarcan el período desde 1914, año en el cual fué establecido el Sistema de la Reserva Federal. Se relacionan los datos, en gran parte, a las condiciones y operaciones de los Bancos de la Reserva Federal y de bancos asociados con el sistema. Al mismo tiempo, se exponen también datos sobre las condiciones y operaciones de todos los bancos, incluyendo los bancos estaduales que no son asociados con el Sistema de la Reserva Federal, y estadísticas sobre débitos bancarios, ganancias bancarias, suspensiones de bancos, operaciones bancarias de sucursales así como de grupos y cadenas de bancos, moneda, tipos de cambio, mercados de valores, finanzas de la Tesorería, producción y movimiento de oro, y desarrollos financieros internacionales.

USA-6720.C5 1945 **Chart Book I, Federal Reserve Charts on Bank Credit, Money Rates, and Business.** Space for plotting through 1946. Washington, D.C., February 1945. Board of Governors of the Federal Reserve System. 56 p. Unexamined.

Libro gráfico I, gráficos del Sistema de Reserva Federal sobre créditos bancarios, cotizaciones de monedas y comercio.

USA-6720.F2 **Farm Credit Quarterly.** 1936- . Washington, D.C. United States Department of Agriculture. Farm Credit Administration.

Statements of the condition of Federal land banks, Federal Farm Mortgage Corporation, Federal intermediate credit banks, banks for cooperatives, and the production credit system.

Revista trimestral de crédito agrícola. Informes sobre las condiciones de los bancos hipotecarios federales y de la Corporación Federal de Hipotecas Agrícolas, bancos federales intermediarios de crédito, bancos para cooperativas, y el sistema de crédito de producción.

*USA-6720.F4 **Federal Reserve Bulletin.** 1915- . Monthly. Washington, D.C. Board of Governors of the Federal Reserve System.

Contains the principal available statistics relating to financial and business developments in the United States, including: Federal Reserve bank statistics, reserve, credit, discount rates, rates on industrial loans, deposits; statement of condition of Federal Reserve banks by weeks; U. S. money in circulation; reports on all banks by classes, their loans, investments, deposits; security prices and new issues, corporate earnings and dividends; U. S. Treasury finance; government corporations and credit agencies; business indexes. Also, international financial statistics relating to gold, international capital transactions of the United States, and financial developments abroad, such as: Gold reserve, production, and movements; capital movements to the United States; money rates and foreign exchange rates; central and commercial banks; price movements.

Boletín mensual de la Reserva Federal. Contiene las principales estadísticas asequibles sobre desarrollos fomentos financieros y comerciales en los Estados Unidos, incluyendo: Estadísticas bancarias de la Reserva Federal, reserva, crédito, tipos del descuento, tipos de préstamos industriales, depósitos; estado de cuentas de los bancos de la Reserva Federal en cifras semanales; circulación del cambio nacional en los Estados Unidos; informes sobre todos los bancos por clases, sus préstamos, inversiones, depósitos; cotizaciones de títulos y nuevas emisiones, ganancias corporativas y dividendos; finanzas de la Secretaría del Tesoro de los Estados Unidos; corporaciones y cajas de crédito gubernamentales; índices comerciales. También, estadísticas financieras internacionales relacionadas con el oro, transacciones internacionales de capitales de los Estados Unidos, y desarrollos financieros en el extranjero, tales como: Reserva, producción y movimientos del oro; movimientos de capitales hacia los Estados Unidos; tipos de cambios y cotizaciones de cambios internacionales; bancos centrales y comerciales; movimientos de precios.

Mar. '46 "Intern tional Financial Statistics." Vol. 32, no. 3, March 1946, p. 340-357.

Includes: Gold reserves of central banks and governments; gold production; gold movements; net capital movements to United States since January 2, 1935; central banks; money rates in foreign countries; commercial banks; foreign exchange rates; price movements (wholesale prices, retail food prices and cost of living, security prices.)

Incluye: Reservas de oro de bancos centrales y gobiernos; producción de oro; movimiento del oro; movimiento de capital neto hacia los Estados Unidos desde enero 2, 1935; bancos centrales; rédito monetario en los países extranjeros; bancos comerciales; tarifa del cambio en el extranjero; movimiento de precios (precios al por mayor, precios de alimentos al por menor y costo de vida, precios de valores).

USA-6720.R5 **Report of the Reconstruction Finance Corporation.** 1932- . Quarterly. Washington, D.C.

Cumulative report on loan and other authorizations, number of borrowers by States and classes, and statement of condition at the close of business.

Memoria de la Corporación Financiera de Reconstrucción. Informe acumulativo sobre préstamos y otras autorizaciones, número de prestatarios según Estados y clases, y declaración de condición al pasar balance.

USA-6720.S5 **Survey of American Listed Corporations: Data on Profits and Operations.** 1936- . Annual. Washington, D.C. Securities and Exchange Commission.

Abstracted and summarized data from registration statements on securities and the annual reports supplemental thereto filed under the Securities Exchange Act of 1934. Data are given by company grouped according to industry.

Examen de compañías americanas registradas: datos sobre ganancias y operaciones. Anual. Datos extractados y resumidos de las declaraciones registradas sobre valores, y de los informes anuales suplementarios sometidos de acuerdo con la Ley de Bolsas de Valores de 1934. Se exponen los datos por cada compañía, clasificada según ramo de industria.

USA-6721.A5 **Annual Report of the Securities and Exchange Commission.** 1934- . Washington, D.C. Securities and Exchange Commission.

Statistical appendix contains data on value of issues registered, grouped by purpose, method of distribution, type of security and industry.

Memoria anual de la Comisión de Valores y Bolsas. El apéndice contiene datos estadísticos sobre el valor de emisiones registradas, clasificadas según intento, método de distribución, clase de títulos e industria.

USA-6721.C3 1941 **Census of Foreign-Owned Assets in the United States.** Washington, D.C., U.S.Govt.Print.Off., 1945. Department of the Treasury. Office of the Secretary. 88 p. Unexamined.

"This report shows the results of a census of foreign-owned assets in the United States taken in 1941 by the Treasury Department's Foreign Funds Control. The data provided by the census are available not only for use in unfreezing foreign-owned assets over which the Foreign Funds Control exercised wartime supervision, but also for their application to various postwar problems of international concern. This census was taken immediately after 'freezing control' was extended in June, 1941 to Germany, Italy and the remainder of continental Europe. In order to obtain full information the Treasury circularized tens of thousands of questionnaires to business concerns and individuals." (Source: **The Department of State Bulletin,** vol. XIV, no. 350, March 17, 1946, back cover.)

"Este informe expone los resultados de un censo de bienes y capitales de propiedad extranjera, levantado en los Estados Unidos en 1941, por el Servicio de Control de Cambios Extranjeros de la Secretaría del Tesoro. La información proporcionada por el censo está disponible no solamente con relación al desbloqueo de los bienes y capitales de propiedad extranjera, los cuales fueron administrados y regulados durante la guerra por el Servicio de Control de Fondos Extranjeros, sino también en relación con la deliberación de los varios problemas de interés internacional, que surjan en el período de la postguerra. Este censo se levantó inmediatamente después de la extensión del 'control de bloqueo' en junio de 1941 a Alemania, Italia y al resto del continente europeo. Con el objeto de obtener informaciones completas, la Secretaría del Tesoro circuló decenas de miles de cuestionarios a individuos y establecimientos comerciales." (Traducción del comentario.)

USA-6721.C5 1914-43 **Corporate Cash Balances, 1914-1943.** By Friedrich A. Lutz. New York, 1945. National Bureau of Economic Research. 132 p. Unexamined.

"Eighth in a series of studies in business financing prepared under the Financial Research Programme of the National Bureau of Economic Research. It outlines an investigation of the behaviour of the cash balances of a number of manufacturing and trade corporations in the United States between the years 1914 and 1943 . . . " (Source: **Journal of the Royal Statistical Society,** vol. CVIII, parts I-II, 1945, p. 237.)

Estado de caja de sociedades, 1914-43. "Este es el octavo de una serie de estudios sobre financiación de negocios, preparado bajo el Programa de Investigación Financiera de la Oficina Nacional de Investigación Económica. Reseña una investigación del funcionamiento de los balances de caja de un número de entidades manufactureras y comerciales de los Estados Unidos correspondiente al período 1914-43." (Traducción del comentario.)

USA-6721.S5 **Statistical Bulletin.** 1942- . Monthly. Washington, D.C. Securities and Exchange Commission.

Security offerings, security trading on exchange, and related statistics on the security markets.

Boletín estadístico mensual de la Comisión de Valores y Bolsas. Ofertas de valores, comercio de valores en bolsas, y otras estadísticas relacionadas con mercados de valores.

USA-6730.A5 **Annual Report of the Federal Deposit Insurance Corporation.** Starting date? Washington, D.C. Federal Deposit Insurance Corporation.

Statistics covering financial statements and operations of the corporation, number of banks, offices and deposits by State and by type; assets and liabilities of operating banks, examiners' evaluations; earnings, expenses and dividends and disbursements to protect deposits.

Memoria anual de la Corporación Federal de Seguro de Depósitos; estadísticas de las declaraciones financieras y las transacciones de la compañía, número de bancos, oficinas y depósitos, por Estados y por tipos; activos y pasivos de los bancos en operación, avalúos de los inspectores; ganancias, gastos y dividendos, y desembolsos a favor de depósitos respaldados.

USA-**6730**.R5 **The Record of the American Institute of Actuaries.** 1909- . Semiannual. Chicago, Illinois. Unexamined.

Informe del Instituto Americano de Actuarios. Semestral.

June '43, Greville "Short Methods of Constructing Abridged Life Tables". By T. N. E. Greville. Vol. XXXII, no. 65, part 1, June 1943, p. 29-43. Reprint.

Methods for constructing abridged life tables, devised primarily by nonactuaries, the mathematical basis for some of the purely empirical formulas indicated, and techniques outlined.

Métodos breves para la construcción de tablas abreviadas de vida. Métodos para la construcción de tablas abreviadas de vida, primeramente ideadas por no-actuarios; la base matemática para algunas de las fórmulas puramente empíricas está indicada, y la técnica está resumida.

Oct. '42, Greville "Actuarial Note: 'Census' Methods of Constructing Mortality Tables and Their Relation to 'Insurance' Methods." By T. N. E. Greville. Vol. XXXI, part 2, no. 64, October 1942, p. 367-373. Reprint.

A Portuguese translation of this article appears in **Revista Brasileira de Atuária,** vol. III, no. 2, julho 1943 (Bras-**6730**.R5).

Nota actuarial: "Métodos censo" de construcción de tablas de mortalidad y sus relaciones con los métodos de seguro. Una traducción al portugués de este artículo, aparece en la **Revista Brasileira de Atuária,** vol. III, no. 2, julio de 1943 (Bras-**6730**.R5).

USA-**6730**.S5 **Statistical Bulletin of the Metropolitan Life Insurance Company.** 1920- . Monthly. New York.

Boletín estadístico de la Compañía Metropolitana de Seguro de Vida.

USA-**6900** **Foreign Trade Statistics in the United States: Methods and Procedures.** Background material acquired for the study conducted by the Inter American Statistical Institute in met ιods of foreign trade statistics in the American nations, including survey of 1946-47.

Consists of: Filled questionnaire and explanatory notes.

Estadísticas del comercio exterior en los Estados Unidos: Métodos y procedimientos. Material básico adquirido para la encuesta que el Instituto Interamericano de Estadística realiza sobre los métodos de estadísticas del comercio exterior en las naciones americanas, incluyendo el estudio de 1946-47. Consiste de: Cuestionario llenado y notas explicativas.

USA-**6900**.E5 1946 **Los Estados Unidos y el Comercio Mundial durante la Guerra 1939-1945.** Informe del Departamento de Comercio Exterior de la Cámara de Comercio de los Estados Unidos. Montevideo, "Impresora Uruguaya" S.A., 1946. Consejo Interamericano de Comercio y Producción. 70 p. Cuadernos comerciales del Consejo, no. 4.

The United States and world commerce during the war, 1939-45. Contents: Summary of exports and imports (lend-lease, value and quantity); exports (lend-lease and commercial exports and analysis by principal articles); imports (geographic distribution and analysis by principal articles); transition in 1945; exports and imports of gold and silver (table).

Contenido: Resumen de exportación e importación (préstamo y arriendo, valor y cantidad); exportación (exportaciones realizadas de acuerdo con el programa de préstamo y arriendo, exportaciones comerciales y análisis de los principales artículos); importación (distribución geográfica y análisis de los principales artículos); la transición en 1945; exportación e importación de oro y plata (cuadro).

USA-**6900**.E6 1938 **Export and Import Practice.** By F. R. Eldridge and other officials of the Bureau of Foreign and Domestic Commerce. Washington, D.C., U.S.Govt. Print.Off., 1938. Department of Commerce. Bureau of Foreign and Domestic Commerce. 306 p. Trade promotion series, no. 175.

Contents: **Export Practice:** Getting the order; The export shipment (incl. export documents); receiving payment; **Import practice:** The Import purchase; the import shipment (incl. customs procedure); selling the American market; appendixes (incl. International Convention for the Unification of Certain Rules Relating to Bills of Lading, Brussels, 1924).

Procedimientos de importación y exportación. Contenido: Gestiones para exportación: Obtención de la orden; el embarque en la exportación (incluyendo documentos de exportación); obtención del pago; gestiones para importación: La compra de lo que se va a importar; el embarque en la importación (incluye gestiones de aduana); venta en el mercado estadounidense; apéndices (incluyen la Convención Internacional para la Unificación de Ciertas Reglas Relativas a Conocimientos de Embarque, Bruselas, 1924).

*USA-6900.F4 **Foreign Commerce and Navigation of the United States.** Starting date? Annual. Washington, D.C. Department of Commerce. Bureau of the Census. Foreign Trade Division.

Before 1942, published by — anteriormente a 1942 publicado por: Bureau of Foreign and Domestic Commerce, Department of Commerce.

Contains annual statistics showing the quantity and value of exports and imports of the United States, classified by articles and cross-classified by countries of destination or origin.

Comercio exterior y navegación de los Estados Unidos. Contiene estadísticas anuales indicando cantidad y valor de las exportaciones e importaciones de Estados Unidos, clasificado por artículos y doble clasificación por países de destino u origen.

USA-6900.F7 **Foreign Commerce Weekly.** 1940- . Washington, D.C. Department of Commerce. Bureau of Foreign and Domestic Commerce.

Supersedes — substituye: **Commerce Reports of the Bureau of Foreign and Domestic Commerce.**

Includes section "New Books and Reports," and the regular departments, "News by Countries" and "News by Commodities," have much information on current developments in Latin America.

Semanario del comercio exterior. Incluye una sección de nuevos libros e informes, y las secciones permanentes de "noticias por países" y "noticias por artículos" contienen amplia información sobre desarrollos en la América Latina.

Aug. 3, '46, a "Latin America's Economy as World Conflict Ended: Part IV: South American West-Coast Republics." Prepared in American Republics Division, Office of International Trade, Department of Commerce (with collaboration of Myron Burgin). Vol. XXIV, no. 5, August 3, 1946, p. 7-10, 42-43, 48.

Posición de la economía latinoamericana al fin de la guerra mundial—parte cuarta: Repúblicas sudamericanas del litoral del oeste.

Aug. 3, '46, b "Living Costs in Brazil." By Gertrude E. Heare. Vol. XXIV, no. 5, August 3, 1946, p. 11, 41-42.

"This article is extracted from a circular by the same title which has been prepared for publication in the International Reference Service. The information is intended for Americans who plan to reside or travel in Brazil." (Source: Editor's note.)

Custo da vida no Brasil. "Éste artigo é um extracto de uma circular publicada, sob o mesmo título, pelo Serviço de Referências Internacionais. As informações destinam-se aos americanos que pretendam residir ou viajar no Brasil." (Tradução do comentario.)

July 27, '46 "Latin America's Economy as World Conflict Ended: Part III: South American North Coast and East Coast Republics." Prepared in American Republics Division, Office of International Trade, Department of Commerce (with collaboration of Myron Burgin). Vol. XXIV, no. 4, July 27, 1946. p. 5-8, 41-44.

Posición de la economía latinoamericana al fin de la guerra mundial—parte tercera: Repúblicas sudamericanas del litoral del norte y del este.

July 27, '46, Bernstein "Latin-American Price Trends." Prepared by Sylvia P. Bernstein. Vol. XXIV, no. 4, July 27, 1946, p. 9.

Indexes of general cost of living, food prices, and general wholesale prices for 1939, 1944, and 1945, and, where available, for 1946.

El curso de los precios en la América Latina. Indices del costo de la vida en general, precio de los alimentos, y precios generales al por mayor correspondientes a los años 1939, 1944 y 1945 y los disponibles de 1946.

July 13, '46 "Latin America's Economy as World Conflict Ended: Part II: Mexico, Central America, and the Caribbean Republics." Prepared in American Republics Division, Office of International Trade, Department of Commerce. Vol. XXIV, no. 2, July 13, 1946, p. 3-4, 51-58.

Posición de la economía latinoamericana al fin de la guerra mundial—parte segunda: México, Centro-América y las Repúblicas del Caribe.

July 6, '46 "Latin America's Economy as World Conflict Ended: Part I: The Region Readjusts to 'Postwar Imperatives'." Prepared in American Republics Division, Office of International Trade. Vol. XXIV, no. 1, July 6, 1946, p. 3-7.

Posición de la economía latinoamericana al fin de la guerra mundial—parte primera:

Reajuste de la región a las necesidades del período de la postguerra.

*USA-6900.F8 **Foreign Commerce Yearbook.** Starting date? Washington, D.C. Department of Commerce. Bureau of Foreign and Domestic Commerce.

Contains statistics on production, trade and market conditions for each of 63 important foreign countries, and summary comparative tables on climate, population, agriculture, mining, manufacturing, transportation, international trade and finance by countries.

Anuario del comercio exterior. Contiene estadísticas sobre producción, comercio y condiciones del mercado de cada uno de 63 países extranjeros importantes y de resúmenes comparativos sobre clima, población, agricultura, minería, fabricación, transportes, comercio internacional y finanzas por países.

USA-6900.M5 **Monthly Summary of Foreign Commerce of the United States.** Starting date? Washington, D.C. Department of Commerce. Bureau of the Census.

Give monthly figures on the quantity and value of exports and imports of the United States by articles and totals by countries.

Resumen mensual del comercio exterior de los Estados Unidos. Presenta cifras mensuales sobre cantidad y valor de las exportaciones e importaciones de los Estados Unidos, por artículos y totales por países.

USA-6900.S5 **Summary of Foreign Trade of the United States.** 1941- . Annual. Washington, D.C. Department of Commerce. Bureau of Foreign and Domestic Commerce. Division of International Economy. International Economics and Statistics Unit.

Previously issued as—anteriormente publicado como: Trade Information Bulletins, no. 791, 822, 831, and 839.

Resumen anual del comercio exterior de los Estados Unidos.

USA-6910.B5 **Balance of International Payments of the United States.** 1922- . Annual. Washington, D.C. Department of Commerce. Bureau of Foreign and Domestic Commerce.

Balanza de pagos internacionales de Estados Unidos.

USA-6910.U5 1943 **The United States in the World Economy.** The international transactions of the United States during the interwar period. By Hal B. Lary. Washington, D.C., U.S.Govt.Print.Off., 1943. Department of Commerce. Bureau of Foreign and Domestic Commerce. 216 p. Economic series, no. 23.

"This study represents the culmination of more than 20 years of research by the Department of Commerce on the Balance-of-payments and international investment position of the United States." (Source: **Estadística,** vol. I, no. 4, diciembre 1943.)

Estados Unidos en la economía mundial. Las transacciones internacionales de los Estados Unidos durante el período comprendido entre las dos guerras. "Este estudio representa la culminación de más de 20 años de investigación de la balanza de pagos y estado de las inversiones internacionales de Estados Unidos, por la Secretaría de Comercio." (Traducción del comentario.)

7000-7999 HEALTH AND CULTURAL STATISTICS—ESTADISTICAS DE SALUBRIDAD Y CULTURA

USA-7100.S5 1944 **Selected Bibliography on Medical Economics.** By Helen Hollingsworth and Margaret C. Klem. Washington, D.C., November 1944. Federal Security Agency. Social Security Board. Bureau of Research and Statistics. 33 p. Mimeographed. Bureau memorandum, no. 60.

Includes sections on: Health status; expenditures for medical care; medical facilities; public health; industrial medical care; medical care for recipients of public assistance or general relief; health insurance and prepayment hospital and medical care in the United States; health insurance in foreign countries.

Bibliografía escogida de economía médica. Incluye secciones sobre: Salubridad; gastos para atención médica; facilidades médicas; salud pública; atención médica industrial; atención médica para beneficiarios de asistencia pública o de beneficiencia general; seguros contra enfermedad y atención médica y hospitalaria con pago previo en los Estados Unidos; seguros contra enfermedad en el exterior.

USA-7110.A4 **American Journal of Public Health and the Nation's Health.** 1911- . Monthly. New York. American Public Health Association.

Revista estadounidense de salud pública y sanidad nacional.

Mar. '44, Collins "Vital and Health Statistics in the Federal Government." By Selwyn De Witt Collins. Vol. 34, no. 3, March 1944, p. 219-223.

Estadística vital y sanitaria del gobierno federal.

Feb. '44, Sibley "Problems in Population Estimation." By Eldridge Sibley. Vol. 34, no. 2, February 1944, p. 174-179. Reprint.

Discussion by—observación hecha por: Grace C. Scholz, p. 179-181.

"Methods of estimating population surveyed, with emphasis on the values and possibilities of a continuous population registration system." (Source: Pop.Ind.)

Problemas en la estimación de la población. "Reseña de métodos de estimación de la población con especial detenimiento sobre el valor y las posibilidades de un sistema contínuo de registro de la población." (Traducción del comentario).

July '36, Baehne "Tabulating Machinery for Indexing and General Tabulation in the Vital Statistics Office." By G. W. Baehne and Henry L. Porsche. Vol. 26, No. 7, July 1936, p. 681-685. Reprint.

Maquinaria de tabulación para establecer índices y para tabulación general en la oficina de estadística vital.

June '36, Berkson "A System of Codification of Medical Diagnoses for Application to Punch Cards, with a Plan for Operation." By Joseph Berkson. Vol. 26, no. 6, June 1936, p. 606-612. Reprint.

Un sistema de codificación de los diagnosticos médicos para aplicación a fichas perforadas, con un plan de operación.

USA-7110.A5 **American Public Health Association Year Book.** Starting date? Annual. New York.

Annual supplement to—suplemento anual a: **American Journal of Public Health.**

Anuario de la Asociación Estadounidense de Salud Pública.

USA-7110.A7 **Annual Report of the Surgeon General of the U. S. Public Health Service.** Starting date? Washington, D.C. Federal Security Agency. United States Public Health Service.

Informe anual del Cirujano General del Servicio de Salud Pública de Estados Unidos.

USA-7110.P5 **Public Health Reports.** 1878- Weekly. Washington, D.C. Federal Security Agency. Public Health Service.

Informes sobre salud pública. Semanal.

1933-43 "The Incidence of Illness and the Volume of Medical Services among 9,000 Canvassed Families." By Selwyn De Witt Collins. **Public Health Reports,** 1933-43. Washington, D.C., U.S. Govt.Print.Off., 1944.

A collection of 23 reprints in one volume.

La incidencia de las enfermedades y el volumen de servicios médicos entre 9,000 familias investigadas. Una colección de 23 reimpresiones en un solo volumen.

USA-7130.P5 **Patients in Mental Institutions.** 1923- . Annual. Washington, D.C. Department of Commerce. Bureau of the Census.

In 1938 the reports entitled **Mental Defectives and Epileptics in Institutions and Patients in Hospitals for Mental Disease** were combined to form the present publication. Not issued for the years 1924 and 1925.

Enfermos en instituciones de enfermedades mentales. En 1938 los informes intitulados **Mental Defectives and Epileptics in Institutions** and **Patients in Hospitals for Mental Disease** fueron combinados para formar la presente publicación. No fueron editados los correspondientes a los años 1924 y 1925.

USA-7150.M4 1940 **Manual of Procedure for the Mechanical System of Reporting Morbidity, Treatment-Progress and Control of the Venereal Diseases.** Washington, D.C., March 1940. Federal Security Agency. United States Public Health Service. 108 p.

Contains instructions and forms for the adaptation of punch card methods to this phase of public health work.

Manual de procedimiento para el sistema mecánico de información sobre morbidad, progreso de tratamiento y control de las enfermedades venéreas. Contiene instrucciones y formularios para la adaptación de los métodos de fichas perforadas a esta fase del trabajo sobre salud pública.

USA-7150.M6 **Morbidity: Instructions for Coding and Punching.** Washington, D.C. Committee on the Cost of Medical Care. 75 p. (approx.) Mimeographed.

Includes instructions for: The general case card; family card, individual history and exposure; occupation code; maternity card; surgical, hospital and clinic card; dental, eye. preventive medicine card.

Morbidad: Instrucciones para la codificación y perforación. Incluye instrucciones para: La ficha general del caso; ficha familiar, historia individual, y posibilidades de contagio o infección; orden de ocupación; ficha de maternidad; fichas de cirugía, hospital y clínica; fichas dentales, de la vista y de medicina preventiva.

USA-7150.T5 **Tuberculosis Reference Statistical Handbook.** Morbidity and mortality, New York and New Jersey; mortality large American cities; clinic reports New York city; sanatorium and hospital service metropolitan New York. Starting date? Annual. New York. New York Tuberculosis and Health Association.

Manual de consulta de estadísticas sobre tuberculosis.

USA-7150.V5 **Venereal Disease Information.** 1920- . Monthly. Washington, D.C. Federal Security Agency. Public Health Service. Unexamined.

Información sobre enfermedades venéreas.

USA-7160.A5 **Accident Facts.** 1927- . Annual. Chicago, Illinois. National Safety Council. Statistical Bureau. Public safety series.

Datos sobre accidentes ocurridos. Anual.

USA-7160.U5 1943 **Uniform Definitions of Motor Vehicle Accidents.** Prepared by Iwao M. Moriyama. Washington, D.C., U.S.Govt.Print.Off., 1943. National Conference on Uniform Traffic Accident Statistics. Committee on Definitions. 17 p.

Definiciones uniformes de accidentes automotores.

USA-7170.B5 **The Blind and Deaf-Mutes in the United States.** 1900- . Decennial. Washington, D.C. Department of Commerce. Bureau of the Census.

Los ciegos y sordomudos en los Estados Unidos.

USA-7190.M5 1943 **Medical Care and Costs in Relation to Family Income.** A statistical source book including selected data on characteristics of illness. By Helen Hollingsworth and Margaret C. Klem. Washington, D.C., March 1943. Federal Security Agency. Social Security Board. Bureau of Research and Statistics. 219 p. Bureau memorandum, no. 51.

Contains: Summary of findings; description of studies; bibliography; tables; index to tables.

Atención médica y costo en relación al ingreso familiar. Un libro estadístico de consulta, que contiene una selección de datos sobre características de enfermedades. Contiene: Resumen de conclusiones; descripción de estudios; bibliografía; cuadros; índices de cuadros.

*USA-7200.B3 **Biennial Survey of Education.** Starting date? Washington, D.C. Federal Security Agency. U. S. Office of Education.

Basic statistics on enrollment, personnel, finances, and administrative units for various types of educational institutions and libraries.

Reseña bienal de educación; estadísticas básicas sobre matriculación, personal, finanzas y unidades administrativas de varias clases de instituciones educacionales y de bibliotecas.

1940-42 **Biennial Survey of Education in the United States, 1940-42: Volume II, Chapter II, Statistical Summary of Education, 1941-42.** Washington, D.C., U.S.Govt.Print.Off., 1944. Federal Security Agency. U. S. Office of Education. 42 p.

Includes sections on: Enrollments; public high schools by size and type; private schools; high school graduates; urban and rural public schools; selected data for state school systems; trends in elementary and secondary education, 1869 to 1942; higher education; schools of nursing; teachers; schools for Negroes; school survival rates; educational attainment of the population; sources of income for education; expenditures for public and private education; value of property and endowments of educational institutions; federally aided vocational education; libraries.

Resumen estadístico 1941 a 1942. Incluye secciones sobre: Matriculación; escuelas secundarias públicas según importancia y clase; escuelas privadas; graduados de escuelas secundarias; escuelas públicas urbanas y rurales; algunos datos sobre sistemas estaduales de instrucción; progreso de la educación elemental y secundaria, 1869 a 1942; educación superior; escuelas para enfermeras; instructores; escuelas para los negros; coeficientes de sobrevivencia escolar; nivel educational de la población; fuentes de ingresos para educación; gastos para instrucción pública y privada; valor de bienes y dotaciones de instituciones educacionales; educación profesional apoyada por subsidios federales; bibliotecas.

USA-7200.J5 **Journal of Educational Psychology.** Devoted primarily to the scientific study of problems of learning and teaching. 1910- . Monthly September to May. Baltimore, Maryland, Warwick and York, Inc.

Includes articles on statistics.

Revista de psicología educacional. Dedicada principalmente a estudios científicos de los problemas de la enseñanza y aprendizaje. Incluye artículos sobre estadísticas.

USA-7200.R5 **Research Bulletin of the National Education Association.** 1923- . Quarterly. Washington, D.C. National Education Association of the United States. Research Division. Cover-title: **National Education Association Research Bulletin.**

Each issue contains a single study.

Boletín trimestral de investigaciones de la Asociación Nacional de Educación. Cada edición contiene un solo estudio.

USA-7300.A5 **American Journal of Mathematics.** 1878- . Quarterly. Baltimore, Maryland. American Mathematical Society. Unexamined.

Revista estadounidense de matemáticas.

USA-7300.Q5 **The Quarterly Review of Biology.** 1926- . Baltimore, Maryland, The Williams and Wilkins Company.

La revista trimestral de biología.

8000-8999 GOVERNMENT AND POLITICAL STATISTICS—ESTADISTICAS GUBERNAMENTALES Y POLITICAS

USA-8000.M5 **Municipal Year Book.** The authoritative resumé of activities and statistical data of American cities. 1934- . Chicago, Illinois. International City Manager Association. Unexamined.

Anuario municipal. Resumen de actividades y datos estadísticos de ciudades estadounidenses.

USA-8100.R5 **Report to the President, the Senate and the House of Representatives by the Director of War Mobilization and Reconversion.** 1945- . Quarterly. Washington, D.C. Office of War Mobilization and Reconversion.

Informe trimestral del Director de Movilización para la Guerra y de Reconversión.

USA-8200.D5 **The Department of State Bulletin.** 1939- . Weekly. Washington, D.C.

Includes information about international organizations, treaties and conventions, conferences and congresses, etc.

Boletín semanal de la Secretaría de Relaciones Exteriores. Incluye información sobre organizaciones internacionales, tratados y convenciones, conferencias y congresos, etc.

USA-8210.C5 **Conference Series of the United States Department of State.** Starting date? Washington, D.C.

Serie de conferencias de la Secretaría de Relaciones Exteriores de Estados Unidos.

USA-8300.F5 **Federal Prisons.** Starting date? Annual. Washington, D.C. United States Department of Justice. Bureau of Prisons.

Before 1941—antes de 1941: **Federal Offenders.**

Statistics on the prison population of the United States.

Informe de las prisiones federales. Estadísticas acerca de la población de las prisiones de los Estados Unidos.

USA-8300.T5 1930-39 **Ten Years of Uniform Crime Reporting, 1930-1939.** A report. Washington, D.C., 1939. Department of Justice. Federal Bureau of Investigation. Loose-leaf.

Diez años de información uniforme sobre crímenes, 1930-39: Un informe.

USA-8300.U5 **Uniform Crime Reports for the United States and Its Possessions.** 1930- . Semiannual. Washington, D.C. United States Department of Justice. Federal Bureau of Investigation.

Annual and monthly data on offenses known to the police, persons charged, persons found guilty, age and sex distribution of persons arrested, and number of police department employees. No. 1 of each year includes monthly data only; no. 2 includes annual data.

Informes uniformes sobre crímenes en los Estados Unidos y sus posesiones. Publicación semestral. Datos anuales y mensuales sobre delitos conocidos por la policía, personas acusadas, personas declaradas culpables, distribución por sexo y edad de personas arrestadas, y número de empleados de los servicios de policía. El número 1° de cada año expone sólo datos mensuales, mientras que el número 2° incluye datos anuales.

USA-8310.J5 **Judicial Criminal Statistics.** Statistics relating to the disposition of defendants in criminal cases definitely before trial courts of general criminal jurisdiction in 30 states. 1932- . Annual. Washington, D.C. Department of Commerce. Bureau of the Census.

Estadísticas judicales criminales.

USA-8320.J3 **Juvenile Court Statistics.** Starting date? Annual. Washington, D.C. U. S. Department of Labor. Children's Bureau. Division of Statistical Research. Social-Statistics Section.

Data on number and types of cases and disposition of cases in selected juvenile courts.

Estadísticas de las cortes juveniles. Datos

sobre número y clases de casos y sus disposiciones en algunas cortes juveniles escogidas.

USA-8330.P6 **Prisoners in State and Federal Prisons and Reformatories.** 1923- . Annual. Washington, D.C. Department of Commerce. Bureau of the Census.

Prisioneros en cárceles y reformatorios federales y de los estados.

USA-8500.A5 1943 **Aims, Organization, and Publications of the National Bureau of Economic Research.** New York, 1943. National Bureau of Economic Research. 12 p.

Propósitos, organización y publicaciones de la Oficina Nacional de Investigación Económica.

USA-8500.P3 **Pamphlets of the Brookings Institution.** Starting date? Washington, D.C.

Folletos de la Institución Brookings.

USA-8500.P5 **Planning Pamphlets of the National Planning Association.** Starting date? Washington, D.C.

Folletos de planeamiento de la Asociación Nacional de Planeamiento.

USA-8500.P8 1934-43 **Publications of the National Resources Planning Board, 1934-1943.** Washington, D.C., 1943. National Resources Planning Board. 4 p.

Lista de publicaciones del Consejo de Planeamiento de los Recursos Nacionales, 1934-43.

USA-8500.R3 **Report of the National Research Council.** Organized in 1916 at the request of the President of the United States by the National Academy of Sciences under its congressional charter. Starting date? Annual. Washington, D.C.

Memoria del Consejo Nacional de Investigaciones, organizado en 1916 conforme a su carta congresional por la Academia Nacional de Ciencias, a la instancia del Presidente de los Estados Unidos.

USA-8500.T5 **Technical Papers of the National Resources Planning Board.** 1941- . Washington, D.C.

Documentos técnicos del Consejo de Planeamiento de los Recursos Nacionales.

USA-8600.P5 **Public Opinion Quarterly.** 1937- . Princeton, New Jersey. Princeton University. School of Public Affairs.

Publicación trimestral sobre opinión pública.

USA-8800.A5 **Annual Report of the Secretary of the Treasury on the State of the Finances.** Starting date? Washington D.C. Treasury Department.

Informe anual del Secretario del Tesoro de los Estados Unidos sobre el estado de las finanzas.

1945 **Annual Report of the Secretary of the Treasury on the State of the Finances for the Fiscal Year Ended June 30, 1945.** Washington, D.C., U.S.Govt.Print. Off., 1946. 779 p.

Comprehensive statement of the financial condition of the U.S., including the report on operations of the Treasury Department, the reports of the bureaus and divisions, and statistical data on receipts and expenditures, war activities program, public debt, Treasury notes, trusts, government corporations, stock and circulation of money, ownership of government securities, and customs.

Exposición anual amplia de la situación financiera de los Estados Unidos, incluyendo la memoria sobre las operaciones de la Secretaría del Tesoro, los informes de las direcciones y divisiones, y datos estadísticos acerca de ingresos y gastos, programa de actividades de guerra, deuda pública, billetes de la Tesorería, trusts, corporaciones gubernamentales, depósitos y circulación de moneda, propiedad de títulos del gobierno, y aduanas.

USA-8800.B7 **The Budget of the United States Government.** Starting date? Annual. Washington, D.C. Bureau of the Budget.

Official annual publication by the Bureau of the Budget presenting requested appropriations for all agencies for the coming fiscal year, estimates of expenditures for the current fiscal year, and actual expenditures for the preceding fiscal year.

Presupuesto del gobierno de los Estados Unidos. Publicación oficial anual de la Dirección de Presupuestos exponiendo las apropiaciones solicitadas por todos los órganos gubernamentales para el año fiscal entrante, estimaciones de gastos para el año fiscal corriente, y gastos reales efectuados durante el año fiscal anterior.

USA-8800.C4 **City Finances.** 1898- . Annual. Washington, D.C. Department of Commerce. Bureau of the Census.

Previously issued as—anteriormente publicada como: **Financial Statistics of Cities.**

Finanzas municipales. Anual.

USA-8800.C6 **County Finances.** Starting date? Annual. Washington,

D.C. Department of Commerce. Bureau of the Census.

Previously issued as—previamente publicada como: **Financial Statistics of Counties.**

Finanzas de los distritos. Anual.

USA-8800.F5 **Financial Statistics of State and Local Government (Wealth, Public Debt, and Taxation).** Revenue receipts, governmental cost payments, public debt, assessed valuations, and tax levies for the government of the states, counties, cities, towns, villages, and boroughs, school districts, townships, and other civil divisions. 1880- . Decennial. Washington, D.C. Department of Commerce. Bureau of the Census.

Estadísticas financieras de los estados y de los gobiernos locales (riqueza, deuda pública e impuestos). Ingresos, gastos de gobierno, deuda pública, tasación impositiva y gravámenes de los gobiernos de estados, distritos, ciudades, pueblos, aldeas y barrios, distritos escolares, municipios y otras subdivisiones civiles.

USA-8800.R5 1945 **Twentieth Report to Congress on Lend-Lease Operations, for the Period Ended June 30, 1945.** Washington, D.C., U.S.Govt.Print.Off., 1945. The President of the United States. 63 p.

Final report by the President of the United States on the operation of lend-lease program. Separate chapters summarize lend-lease aid to the United Kingdom, the Soviet Union, China, India, Australia, New Zealand, and Continental Europe. The participation of Latin American countries in the program is shown in the statistical tables included.

Vigésimo informe al Congreso sobre operaciones de préstamo y arriendo del período extinguido el 30 de junio de 1945. Informe del presidente de los Estados Unidos, sobre el programa de préstamo y arriendo. Capítulos separados resumiendo la ayuda de préstamo y arriendo al Reino Unido, la Unión Soviética, China, India, Australia, Nueva Zelandia y Europa Continental. La participación de los países latinoamericanos en este programa está indicada en los cuadros estadísticos incluídos.

USA-8800.S5 **State Finances.** 1915- . Annual. Washington, D.C. Department of Commerce. Bureau of the Census.

Previously issued as—previamente publicado como: **Financial Statistics of States.**

Finanzas de los estados. Anual.

USA-8800.S6 **Statement of the Public Debt.** Starting date? Monthly. Washington, D.C. Treasury Department.

Final complete statement of the public debt as of the last day of the preceding month.

Memoria mensual completa y final sobre la deuda pública, según su estado en el día último del mes anterior.

USA-8800.S7 **Statistics of Income.** Starting date? Annual. Washington, D.C. U. S. Treasury Department. Bureau of Internal Revenue. Statistics of income compiled from individual tax, corporation income tax, and other tax returns filed with the Federal government.

Estadísticas de rentas, recopiladas de las declaraciones de ingresos personales, de declaraciones de ingresos de sociedades anónimas, y de otras declaraciones contributivas sometidas al gobierno federal.

USA-8800.T5 **Treasury Bulletin.** 1939- . Monthly. Washington, D.C. United States Treasury Department. Office of the Secretary.

Includes, besides articles, statistics for the current year and certain comparative data for the previous years on receipts and expenditures, internal revenue collections, public debt, capital movements between the United States and foreign countries, monetary data, and law enforcement.

Boletín de la Secretaría del Tesoro. Además de artículos, se incluyen estadísticas para el año corriente y ciertos datos comparativos de años anteriores sobre ingresos y gastos, recaudación de los impuestos internos, deuda pública, movimientos de capital entre Estados Unidos y países extranjeros, datos monetarios, ejecución de las leyes.

9000-9999 LABOR AND LIVING CONDITIONS—CONDICIONES DE TRABAJO Y DE VIDA

USA-9000.A4 **American Journal of Sociology.** 1895- . Bimonthly. Chicago, Illinois.

Includes articles on social statistics, population, family living, etc.

Revista estadounidense de sociología; bimensual. Incluye artículos sobre estadísticas sociales, población, vida de la familia, etc.

USA-9000.A5 1943 **American Society in Wartime.** By William F. Ogburn. Chicago, Illinois, University of Chicago Press, 1943. 237 p. Unexamined.

La sociedad estadounidense en tiempos de guerra.

USA-9000.A6 **American Sociological Review.** The official journal of the American Sociological Society. 1936- . Bimonthly. Minneapolis, Minnesota.

Includes articles on statistics.

Revista sociológica estadounidense. Publicación bimensual de la Sociedad Sociológica Estadounidense. Incluye artículos sobre estadísticas.

Aug. '46, Wolfbein "Demographic Factors in Labor Force Growth." By S. L. Wolfbein and A. J. Jaffe. Vol. 11, no. 4, August 1946, p. 392-396.

Paper read before the annual meeting of the American Sociological Society, Cleveland, Ohio, March 1-3, 1946.

Factores demográficos en el crecimiento del potencial obrero. Estudio leído ante la reunión anual de la Sociedad Estadounidense de Sociología, en Cleveland, Ohio, del 1° al 3° de marzo de 1946.

Dec. '44 "Representative Books in Sociology since 1938." Vol. 9, no. 6, December 1944, p. 693-695.

"The American Library Association early in 1944 asked for a list of books in Sociology recognized as . . . representation of the research . . . in this field . . .." (Source: Text.)

Libros representativos en la ciencia sociológica desde 1938. "La Asociación Estadounidense de Bibliotecas ha pedido al principio de 1944, una lista de los libros de sociología reconocidos como . . . representación de las investigaciones . . . en este campo . . . ." (Traducción del comentario.)

Aug. '44, Deming "On Errors in Surveys." By W. Edwards Deming. Vol. IX, no. 4, August 1944, p. 359-369. Reprint.

Sobre errores en las investigaciones.

Feg. '44, Hagood "What Level of Living Indexes Measure." By Margaret Jarman Hagood and Louis J. Ducoff. Vol. 9, no. 1, February 1944, p. 78-84.

Lo que miden los índices del nivel de vida.

June '43, Reuss "An Appraisal of the 1936 Religious Census." By Carl F. Reuss. Vol. 8, no. 3, June 1943, p. 342-345.

Una avaluación del censo religioso de 1936.

USA-9000.B4 **Bulletins of the Bureau of Labor Statistics.** Starting date? Irregular. Washington, D.C. Department of Labor. Bureau of Labor Statistics.

Series of special studies.

Boletines del Departamento de Estadísticas del Trabajo. Serie de estudios especiales.

USA-9000.B6 **Bulletins of the Social Science Research Council.** Starting date? Irregular. New York.

Series of special studies.

Boletines del Consejo de Investigaciones de las Ciencias Sociales. Serie de estudios especiales.

USA-9000.H5 **Handbook of Labor Statistics,** 1941 Washington, D.C., U.S.Govt. Print.Off., 1943. Department of Labor. Bureau of Labor Statistics. Bulletin, no. 694. 2 vol.: Vol. 1, **All Topics except Wages;** vol. 2, **Wages and Wage Regulations.**

Manual de estadísticas del trabajo, 1941.

USA-9000.L4 **Labor Conditions in Latin America.** Latin American series. 1939- . Irregular. Washington, D.C. Department of Labor. Bureau of Labor Statistics.

Reprints of a regular section in the **Monthly Labor Review.**

Condiciones del trabajo en la América Latina. Reimpresiones de una sección permanente de **Monthly Labor Review.**

USA-9000.L6 **Labor in the United States:** 1938 **Basic Statistics for Social Security.** By Wladimir S. Woytinsky. Washington, D.C., 1938. Social Science Research Council. Committee on Social Security. 333 p.

El trabajo en los Estados Unidos: Estadísticas básicas para el seguro social.

*USA-9000.M5 **Monthly Labor Review.** 1915- . Washington, D.C. Department of Labor. Bureau of Labor Statistics.

Contains current information on labor, construction activities, cost of living, prices for the United States, and a section devoted to labor conditions in Latin America.

Revista mensual del trabajo. Contiene información corriente sobre trabajo, actividades de construcción, costo de la vida, precios en los Estados Unidos, y una sección dedicada a las condiciones del trabajo en la América Latina.

June '46, Cárcamo C. "Work Accidents in Chile." By Luis Cárcamo C. Vol. 62, no. 6, June 1946, p. 914.

Accidents del trabajo en Chile.

May 1946 "Employment Situation in Latin America." Vol. 62, no. 5, May 1946, p. 741-752.

Covers: Employment situation during the war, and employment controls; effect of the war's end on employment, and the situation in 1945; government employment agencies; distribution of labor force, by principal groups.

Situación del empleo en la América Latina; incluye: Empleo durante la guerra, y control del empleo; efecto del fin de la guerra sobre el empleo, y la situación en 1945; agencias gubernamentales de empleo; distribución del potencial obrero, por grupos principales.

May '45, Hosea "Fact-Finding Activities of the Bureau of Labor Statistics." Prepared by Harold R. Hosea. Vol. 60, no. 5, May 1945, p. 927-953.

Las investigaciones del Departamento de Estadísticas del Trabajo.

USA-9000.S3 1943 **Selected List of the Publications of the Bureau of Labor Statistics.** Compiled by Elizabeth A. Johnson. 1943 supplement to 1940 edition. Washington, D.C., U.S.Govt.Print.Off., 1943. Department of Labor. Bureau of Labor Statistics. 45 p. Bulletin, no. 747.

Una lista escogida de publicaciones del Departamento de Estadísticas del Trabajo. Suplemento 1943 a la edición de 1940.

USA-9000.S3 1940 **A Selected List of the Publications of the Bureau of Labor Statistics.** Compiled by Elizabeth A. Johnson. 1940 edition. Washington, D.C., U.S.Govt. Print.Off., 1941. Department of Labor. Bureau of Labor Statistics. 48 p. Bulletin, no. 683.

Contains list of publications to 1940 under the Consumer Purchases Study of 1935-36.

Una lista escogida de publicaciones del Departamento de Estadísticas del Trabajo. Edición 1940. Contiene una lista de las publicaciones aparecidas hasta 1940, correspondientes al estudio de adquisiciones del consumidor de 1935-36.

USA-9000.S5 **Social Forces.** A scientific medium of social study and interpretation. 1922- . Four times a year. Baltimore, Maryland, Williams and Wilkins Co.

Poderes sociales: Un medio científico de estudios e interpretaciones sociales. Trimestral.

Mar. '43, Hagood "Statistical Methods for Delineation of Regions Applied to Data on Agriculture and Population." By Margaret Jarman Hagood. Vol. 21, no. 3, March 1943, p. 287-297.

Métodos estadísticos para delineación de regiones en relación con datos sobre agricultura y población.

USA-9000.S7 1944 **A Source List of Selected Labor Statistics.** Preliminary ed. Compiled by a committee of the Social Science Group. New York, 1944. Special Libraries Association. Social Science Group. 37 p. Unexamined.

Una lista de fuentes escogidas de estadísticas del trabajo. Edición preliminar. Compilada por una Comisión del Grupo de Ciencias Sociales, de la Asociación de Bibliotecas Especiales.

USA-9100.L5 1899-1939 **Labor Savings in American Industry, 1899-1939.** By Solomon Fabricant. New York, 1945. National Bureau of Economic Research. 52 p. Occasional paper, no. 23. Unexamined.

Economización de trabajo en la industria estadounidense.

USA-9200.C5 1899-1939 **Changes in Distribution of Manufacturing Wage Earners, 1899-1939.** A comparative study. Washington, D.C., U.S.Govt.Print.Off., 1942. Prepared by Bureau of the Census and Bureau of Agricultural Economics. 268 p. Unexamined.

Cambios en la distribución de los asalariados en la industria manufacturera, 1899-1939.

USA-9200.E4 **Employment and Pay Rolls.** Detailed report. 1915- . Monthly. Washington, D.C. Department of Labor. Bureau of Labor Statistics. Employment and Occupational Outlook Branch. Division of Employment Statistics.

Mimeographed bulletin containing current information on employment and payrolls detailed by industry, including manufacturing and other industries.

Empleo y nóminas. Boletín mimeografiado conteniendo informaciones de actualidad sobre empleo y nóminas, detalladas por industrias e incluyendo industria manufacturera y otros ramos industriales.

USA-9200.E5 1899-1939 **Employment in Manufacturing, 1899-1939, an Analysis of Its Relation to the Volume of Production.** By Solomon Fabricant. New York, 1942. National Bureau of Economic Research. 362 p. Publications, no. 41. Unexamined.

Ocupación en las fábricas, 1899-1939, un análisis de su relación con el volumen de producción.

USA-9200.F5 **Farm Labor.** Starting date? Monthly. Washington, D.C. United States Department of Agriculture. Bureau of Agricultural Economics. Mimeographed.

Statistics on farm employment and farm wage rates.

Estadísticas sobre empleo agrícola y tipos de salarios agrícolas.

USA-9200.H4 **Hours and Earnings.** 1916?- . Monthly. Washington, D.C. Department of Labor. Bureau of Labor Statistics. Employment and Occupational Outlook Branch. Division of Employment Statistics.

Mimeographed releases containing current information on average weekly earnings, average hours worked per week, and average hourly earnings detailed by industry, including manufacturing and other industries.

Horas e ingresos. Boletín mimeografiado conteniendo informaciones de actualidad sobre promedio de salarios semanales, promedio de horas trabajadas en una semana, y promedio de ingresos por hora, detalladas por industrias e incluyendo industria manufacturera y otros ramos industriales.

USA-9200.H5 1932-40 **Hours and Earnings in the United States, 1932-40.** With supplement for 1941. By Alice Olenin and Thomas F. Corcoran. Washington, D.C., U.S. Govt.Print.Off., 1942. Department of Labor. Bureau of Labor Statistics. 207 p. Bulletin, no. 697.

Horas e ingresos en los Estados Unidos, 1932-40.

USA-9200.I4 1945 **Income from Independent Professional Practice.** By Milton Friedman and Simon Kuznets. New York, 1945. National Bureau of Economic Research. 599 p. Publications of the National Bureau of Economic Research, no. 45.

Los ingresos por el ejercicio de profesiones liberales.

USA-9200.I6 1922 **Industrial Unemployment.** A statistical study of its extent and causes. By Ernest S. Bradford. Washington, D.C., U.S.Govt.Print.Off., 1922. Department of Labor. Bureau of Labor Statistics. 52 p. Bulletin, no. 310. Unexamined.

Desocupación industrial. Un estudio estadístico de sus causas y alcance.

USA-9200.L5 **The Labor Market.** Starting date? Monthly. Washington, D.C. U. S. Department of Labor. U. S. Employment Service.

Summary of trends, changes and outlook in the labor market nationally and for major industries and areas, presented in non technical style. Tables summarize placement activities by State with more detailed statistics for placements by industry, sex and color by major occupational groups.

El mercado del trabajo; resumen de tendencias, cambios y perspectivas en el mercado del trabajo del país entero y de las industrias y areas principales, expuesto de modo no técnico. Cuadros que resumen las actividades de colocación por estados, presentan estadísticas más detalladas sobre colocaciones por industrias, sexo y color, según grupos principales de ocupaciones.

USA-9200.L6 **Labor Turnover.** 1929- . Monthly. Washington, D.C. U. S. Department of Labor. Bureau of Labor Statistics. Employment and Occupational Outlook Branch.

Mimeographed releases containing current information on separation and hiring rates in manufacturing and a few non-manufacturing industries, including rates by sex.

Movimiento de la mano de obra. Boletín mensual mimeografiado conteniendo informaciones de actualidad sobre tasas de empleados cesantes y nuevos empleados en las industrias manufactureras y en algunas industrias no manufactureras, incluyendo tasas por sexo.

USA-9200.M7 **Monthly Report of Employment, Executive Branch of the Federal Government.** Starting date? Washington, D.C. United States Civil Service Commission.

Tabulated statistics of civilian employment in the Federal government, by department and by employment group.

Informe mensual sobre los empleados del ramo ejecutivo del gobierno federal. Cuadros estadísticos sobre el empleo de personal civil en el gobierno federal, según departamentos y grupos de empleo.

USA-9200.S5 **Special Salary Tabulations, 1944-45.** 1945- . Irregular. Washington, D.C. National Education Association of the United States. Research Division.

Series of special studies on salaries of personnel in education.

Serie de estudios especiales sobre los salarios del personal en la enseñanza.

USA-9200.W7 1945 **The Wages of Farm and Factory Laborers.** By Daniel J. Ahearn, Jr. New York, Columbia Univer-

sity Press, 1945. 245 p. Studies in history, economics and public law, no. 518. Unexamined.

A comparison of the wages of farm laborers and factory workers from 1914 to 1944. Contains: Extraordinary rise from 1914 to 1920 with its subsequent deflation; comparative stability following the recovery of 1922; depression-revival changes of 1929-30; includes four charts and fifty-one tables.

Salarios de los obreros ocupados en la agricultura y en fábricas. Una comparación entre los salarios de los trabajadores agrícolas y de los obreros de fábricas correspondiente a los años 1914 a 1944. Contiene: Aumento extraordinario desde 1914 a 1920 con la consiguiente deflación; estabilidad comparativa posterior al restablecimiento de 1922; cambios de depresión y restauración de 1929-30; incluye cuatro diagramas y 51 cuadros.

USA-9200.W9 **Women in Factories.** Estimated number of women wage earners employed in manufacturing industries. 1943- . Semiannual. Washington, D.C. Department of Labor. Bureau of Labor Statistics. Employment and Occupational Outlook Branch. Division of Employment Statistics.

Estimated employment of women wage earners, percent of women wage earners, number of women per 100 wage earners in manufacturing industries, classified by major manufacturing, durable goods, and non-durable goods industries.

Empleo de mujeres en las fábricas. Informe semestral. Número calculado de mujeres asalariadas en las industrias manufactureras, porcentaje de mujeres asalariadas, número de mujeres por cada cien personas asalariadas en las industrias manufactureras, clasificadas por industrias principales de manufactura, industrias de productos durables, e industrias de productos no durables.

USA-9300.C5 **Conference Board Economic Record.** 1938?- . Monthly. New York. National Industrial Conference Board. Unexamined.

Informe económico del Consejo Nacional de Conferencia Industrial. Mensual.

USA-9500.C4 **Changes in Cost of Living.** 1934- . Washington, D.C. Department of Labor. Bureau of Labor Statistics.

Issued as reprints from—editados como reimpresiones de: **Monthly Labor Review.**

Cambios en el costo de la vida.

USA-9500.R3 1945 **Report of the President's Committee on the Cost of Living.** Washington, D.C., U.S.Govt.Print.Off., 1945. Office of Economic Stabilization. 423 p.

A compilation of the reports of the President's Committee on the Cost of Living. Gives the conclusions reached by the committee regarding the Bureau of Labor Statistics cost-of-living index, including discussions on statistical methods used, meaning of the index, adequacy for certain purposes.

Una compilación de los informes del Comité Presidencial sobre Costo de la Vida. Presenta las conclusiones obtenidas por el Comité con respecto al índice del costo de la vida del Departamento de Estadística del Trabajo. Incluye también los debates sobre métodos estadísticos usados, interpretación del índice y su utilidad para ciertos fines.

USA-9600.B5 **Bulletins of the Women's Bureau.** Starting date? Washington, D.C. Department of Labor. Women's Bureau.

Boletines de la Dirección de Mujeres.

USA-9600.J5 **Jewish Social Studies, Publications.** Starting date? New York. Conference on Jewish Relations.

Series of monographs.

Serie de monografías. Publicaciones de estudios sociales judíos.

USA-9600.M3 1945 **Mexican War Workers in the United States.** The Mexico-United States manpower recruiting program and its operation. By Robert C. Jones. Washington, D.C., 1945. Pan American Union. Division of Labor and Social Information. 46 p.

Contents: Agricultural Labor Program and Railroad Labor Program.

Trabajadores mexicanos de guerra en los Estados Unidos. El programa mexicano-estadounidense del alistamiento de mano de obra y su funcionamiento. Contiene: Programas de trabajo agrícola y de trabajo ferroviario.

USA-9600.N5 1943 **The Negro's Share.** A study of income, consumption, housing and public assistance. By Richard Sterner. New York, Harper, 1943. 433 p. Unexamined.

"This study is one of several monographs on the Negro in America. The problem of the Negro's economic position in American society is considered from the dual standpoints of his participation in production and his level of living as influenced by his position as income-receiver and consumer. General statistical analyses of the Negro's flight from agriculture, trends in employment and unemployment, and the size and composition of the Negro family precede the more detailed sections on income, expenditures,

and social welfare provisions." (Source: **Books on Population, 1939-Mid-1944.**)

La participación de los negros. Un estudio sobre rentas, consumo, vivienda y asistencia pública. "Este estudio es una de las varias monografías sobre los negros en Estados Unidos. El problema de la situación económica de los negros en la sociedad estadounidense es considerado desde ambos puntos de vista de su participación en la producción y de su nivel de vida, como percibidores de rentas y como consumidores. Un análisis estadístico general del alejamiento de los negros de la agricultura, tendencia de la ocupación y desocupación y el tamaño y composición de la familia negra precede los capítulos mas importantes sobre renta, gastos y previsión social." (Traducción del comentario.)

USA-9700.C5 1935-36 **Consumer Expenditures in the United States, Estimates for 1935-36.** Washington, D.C., U.S.Govt.Print. Off., 1939. National Resources Committee. 195 p.

Contents: Average patterns of consumer spending; national consumption in 1935-36; statistical tables for reference use; sources and methods used in the study; effect of size and distribution of income on national consumption. This report is a companion volume to the study of **Consumers Incomes in the United States** published in 1938. Part of the Consumer Purchases Study.

Gastos de los consumidores en Estados Unidos, estimaciones para 1935-36. Contenido: Características promedias de los gastos del consumidor; consumo nacional en 1935-36; cuadros estadísticos para uso de referencia; fuentes de información y métodos usados en este estudio; efecto de la cuantía y distribución de la renta sobre el consumo nacional. Este informe es un volumen que acompaña al estudio de **Consumers Incomes in the United States** publicado en 1938. Parte del estudio de compras del consumidor.

USA-9700.C6 1935-36 **Consumer Incomes in the United States, Their Distribution in 1935-36.** Washington, D.C., U.S.Govt. Print.Off., 1938. National Resources Committee. 104 p.

Contents: Incomes of all consumers; incomes of various consumer groups; comparison with other income studies; sources and methods used in the study; statistical tables for reference use. Part of the Consumer Purchases Study.

Rentas del consumidor en los Estados Unidos, su distribución en 1935-36. Contenido: Rentas de todos los consumidores; rentas de distintos grupos de consumidores; comparación con otros estudios sobre rentas; fuentes de información y métodos usados en este estudio; cuadros estadísticos para uso de referencia. Parte del estudio sobre compras del consumidor.

USA-9700.C7 1935-36 **Consumer Purchases Study, 1935-36.** A nation-wide survey of family income and expenditures, in the planning and execution of which three agencies participated—the U. S. National Resources Committee, the U. S. Department of Labor, Bureau of Labor Statistics, and the U. S. Department of Agriculture, Bureau of Home Economics. Each agency issued its own series of publications on the results of the particular phase of the study under its auspices. Over-all national estimates were made by the National Resources Committee.

Estudio de las compras del consumidor, 1935-36. Una amplia investigación nacional del ingreso y de los gastos de la familia, en cuyo planeamiento y ejecución participaron tres organismos—Comisión Nacional de Recursos de Estados Unidos; Departamento de Estadísticas del Trabajo de la Secretaría del Trabajo; y el Departamento de Economía Doméstica de la Secretaría de Agricultura. Cada organismo ha editado su propia serie de publicaciones sobre los resultados de la fase particular estudiada bajo sus auspicios. La Comisión Nacional de Recursos ha hecho estimaciones generales para el país entero.

For publications included in this bibliography see—para publicaciones incluídas en esta bibliografía véase: **Family Income and Expenditures, Five Regions, Part 2, Family Expenditures, Farm Series** (USA-9700.F5I1); **Family Food Consumption and Dietary Levels, Five Regions, Farm Series** (USA-9700.F5F1); **Family Income and Expenditures, Five Regions, Part 2, Family Expenditures, Urban and Village Series** (USA-9700.F5I2); **A Selected List of Publications of the Bureau of Labor Statistics** (USA-9000.S3); **Family Income in Chicago, 1935-36** (USA-9700.F5C); **Printed Publications for Research Workers** (USA-9700.P5); **Consumer Incomes in the United States, Their Distribution in 1935-36** (USA-9700.C5); **Family Expenditures in the United States, Statistical Tables and Appendixes** (USA-9700.F5E9); **Consumer Expenditures in the United States, Estimates for 1935-1936** (USA-9700.C5).

USA-9700.F5C 1935-36 **Family Income in Chicago, 1935-36.** Study of consumer purchases, urban series. Washington, D.C., U.S. Govt.Print.Off., 1939. Bureau of Labor Statistics, Department of Labor, in cooperation with Works Progress Administration. Bulletin, no. 642. 2 vol.: Vol. I, **Family Income**; vol. II, **Family Expenditure.**

Renta de la familia en Chicago, 1935-36. Estudio de las compras del consumidor, serie urbana.

USA-9700.F5E9 1935-36 **Family Expenditures in the United States, Statistical Tables and Appendixes.** A National Resources Committee publication released by the National Resources Planning Board. Washington, D.C., U.S.Govt.Print.Off., 1941. National Resources Planning Board. 209 p.

Contents: Foreword; summary tables; methodology tables; reference tables; sources and methods used in the study; schedule forms for study of consumer purchases; index of tables. Part of the Consumer Purchases Study of 1935-36.

Gastos de la familia en Estados Unidos, cuadros estadísticos y apéndices. Una publicación de la Comisión Nacional de Recursos puesta en circulación por el Consejo Nacional de Planeamiento de Recursos. Contenido: Prefacio; cuadros sinópticos; cuadros de metodológía; cuadros de referencia; fuentes de información y métodos usados en este estudio; formularios para el estudio de las compras del consumidor; índice de los cuadros. Parte del estudio de las compras del consumidor, 1935-36.

USA-9700.F5F1 1941 **Family Food Consumption and Dietary Levels, Five Regions, Farm Series.** Consumer purchases study. By Hazel K. Stiebeling, Day Monroe, Callie M. Coons, Esther F. Phipard, and Faith Clark. Washington, D.C., 1941. U. S. Department of Agriculture, Bureau of Home Economics in cooperation with the Work Projects Administration. 393 p. Miscellaneous publication, no. 405.

Consumo familiar de alimentos y nivel de alimentación, cinco regiones, serie rural. Estudio de las compras del consumidor.

USA-9700.F5I1 1941 **Family Income and Expenditures, Five Regions, Part 2, Family Expenditures, Farm Series.** Consumer purchases study. By Day Monroe, Dorothy S. Brady, Margaret Perry, Kathryn Cronister, and Edith Dyer Rainboth. Washington, D.C., 1941. U. S. Department of Agriculture, Bureau of Home Economics in cooperation with the Work Projects Administration. 366 p. Miscellaneous publication, no. 465.

Renta y gastos de la familia, cinco regiones, parte 2, gastos de la familia, serie rural. Estudio de las compras del consumidor.

USA-9700.F5I2 1940 **Family Income and Expenditures, Five Regions, Part 2, Family Expenditures, Urban and Village Series.** Consumer purchases study. By Dorothy S. Brady, Day Monroe, Elizabeth Phelps, and Edith Dyer Rainboth. Washington, D.C., 1940. U. S. Department of Agriculture, Bureau of Home Economics in cooperation with the Work Projects Administration. 410 p. Miscellaneous publication, no. 396.

Renta y gastos de la familia, cinco regiones, parte 2, gastos de la familia, serie urbana y semiurbana. Estudio de las compras del consumidor.

USA-9700.I5 1942 **Income and Spending and Saving of City Families in Wartime.** By Alice Cable Hanson. Washington, D.C.,U.S.Govt.Print.Off., 1942. Department of Labor. Bureau of Labor Statistics. 31 p. Bulletin, no. 724. Unexamined.

Rentas, gastos y ahorros de las familias urbanas en tiempo de guerra.

USA-9700.P5 1944 **Printed Publications for the Research Worker.** Washington, D.C., November 1944. Department of Agriculture. Agricultural Research Administration. Bureau of Human Nutrition and Home Economics. 2 p. Mimeographed.

Contains list of publications under the Consumer Purchases Study of 1935-36.

Publicaciones impresas para investigadores. Contiene una lista de publicaciones del estudio de las compras de los consumidores de 1935-36.

USA-9700.R4 1943 **Rural Family Spending and Saving in Wartime.** Washington, D.C., U.S.Govt.Print.Off., 1943. Department of Agriculture. 163 p. Miscellaneous publication, no. 520.

Gastos y ahorros de la familia rural en tiempo de guerra.

USA-9700.R6 1940 **Rural Level of Living Indexes for Counties of the United States, 1940.** By Margaret Jarman Hagood. Washington, D.C., 1943. Department of Agriculture. Bureau of Agricultural Economics. 43 p. Mimeographed. Unexamined.

Rural farm, rural nonfarm, and composite rural indexes are given by county and state.

Indices rurales del nivel de vida por distritos de los Estados Unidos. Indíces rurales agrícolas, rurales no agrícolas, e índices rurales compuestos, por distritos y estados.

USA-9700.S5 1942 **Spending and Saving of the Nation's Families in Wartime.** By Alice Cable Hanson. Washington, D.C., U.S.Govt.Print.Off., 1942. Department of Labor. Bureau of Labor Statistics. 22 p. Bulletin, no. 723. Unexamined.

Gastos y ahorros de las familias nacionales en tiempo de guerra.

USA-9800.D5 1939 **Diets of Families of Employed Wage Earners and Clerical Workers in Cities.** By Hazel K. Stiebeling and Esther F. Phipard. Washington, D.C., U.S. Govt.Print.Off., 1939. Department of Agriculture. 141 p. Circular, no. 507.

Régimen alimenticio de las familias de los obreros y empleados de oficina en las ciudades.

USA-9800.W5 1945 **Wartime Food Purchases.** Washington, D.C., U.S.Govt. Print.Off., 1945. Department of Labor. Bureau of Labor Statistics. 26 p. Bulletin, no. 838. Reprint from **Monthly Labor Review,** vol. 60, no. 6, June 1945, p. 1143-1157, with additional data prepared by Leonore A. Epstein. Unexamined.

Compras de alimentos en tiempos de guerra.

USA-9900.A5 **Annual Report, U. S. National Housing Agency.** 1942- . Washington, D.C. National Housing Agency.

Informe anual de la Dirección Nacional de la Vivienda de los Estados Unidos.

USA-9900.F5 **Federal Home Loan Bank Review.** 1935- . Monthly. Washington, D.C. National Housing Agency. Federal Home Loan Bank Administration.

Data on residential construction, building costs, savings, and loan lending, mortgage recordings, foreclosures, financial statements of Federal Home Loan Banks. Annual statistical review issue in March.

Revista del Banco Federal de Préstamos Hipotecarios. Datos sobre construcción de habitaciones, costos de construcción, ahorros, y préstamos, registro de hipotecas, juicios hipotecarios (embargos), declaraciones financieras de los Bancos Federales de Préstamos Hipotecarios. Aparece en marzo de cada año una edición dedicada a la síntesis estadística anual.

USA-9900.W5 1945 **War Housing in the United States.** Prepared for use at the Conference of the United Nations in San Francisco, April 1945. Washington, D.C., 1945. National Housing Agency. 36 p. Unexamined.

La vivienda durante la guerra en los Estados Unidos. Preparado para uso en la Conferencia de las Naciones Unidas en San Francisco, abril de 1945.

USA-9950.B5 **Bureau Memorandums of the Bureau of Research and Statistics of the Social Security Board.** Starting date? Irregular. Washington, D.C. Federal Security Agency. Social Security Board. Bureau of Research and Statistics.

UNITED STATES
9800

Series of monographs.

Memorándums de oficina de la Dirección de Investigación y Estadística del Consejo de Seguro Social. Serie de monografías.

USA-9950.E5 **Employment Security Activities.** 1945- . Monthly. Washington, D.C. Federal Security Agency. Social Security Board. Bureau of Employment Security. Program Division.

Operations of State unemployment compensation agencies; claims, benefits, appeals, etc.

Funcionamiento de las agencias estatales de compensación del desempleo; reclamaciones, beneficios, apelaciones, etc.

*USA-9950.S4 **Social Security Bulletin.** 1938- . Monthly. Washington, D.C. Federal Security Agency. Social Security Board.

Articles and data on unemployment claims and benefits, public assistance, retirement, etc.

Boletín del seguro social. Artículos y datos sobre reclamaciones y beneficios de los desocupados, asistencia pública, jubilación, etc.

July, '46, Cohen "The First Two Years of Social Insurance in Mexico." By Wilbur J. Cohen. Vol. 9, no. 7, July 1946, p. 22-27.

Includes organization chart of the Mexican Social Insurance Institute, and tables on the number of employers and employees under the social insurance program, benefits as percent of basic wage and duration of benefits, income and disbursements under the program, etc.

Reseña sobre los dos primeros años del funcionamiento del seguro social en México. Incluye esquema de organización del Instituto Mexicano del Seguro Social, y cuadros sobre el número de empresarios y empleados que están comprendidos en el programa de seguro social, los beneficios como porcentaje del salario básico, así como duración de los beneficios, ingresos y desembolsos del programa, etc.

Mar. '46, Woytinsky "Postwar Economic Perspectives: IV, Aftermath of the War." By W. S. Woytinsky. Vol. 9, no. 3, March 1946, p. 11-25.

Perspectivas económicas de postguerra: IV, Consecuencias de la guerra.

Feb. '46, Woytinsky "Postwar Economic Perspectives: III, Prewar Experience—Production and

Consumption." By W. S. Woytinsky. Vol. 9, no. 2, February 1946, p. 9-16.

Perspectivas económicas de postguerra: III, Experiencia anterior a la guerra—producción y consumo.

Jan. '46, Woytinsky "Postwar Economic Perspectives: II, Prewar Experience—The Labor Force and Employment." By W. S. Woytinsky. Vol. 9, no. 1, January 1946, p. 8-16.

Perspectivas económicas de postguerra: II, Experiencia anterior a la guerra—potencial obrero y empleo.

Dec. '45 ,Woytinsky "Postwar Economic Perspectives: I, Experience After World War I." By W. S. Woytinsky. Vol. 8, no. 12, December1945, p. 18-30.

Perspectivas económicas de postguerra: I, Experiencias después de la primera guerra mundial.

USA-9950.S6 **Social Security Yearbook.** 1939- . Washington, D.C. Federal Security Agency. Social Security Board.

Annual supplement to the **Social Security Bulletin.**

Anuario de previsión social. Suplemento anual del **Social Security Bulletin.**

USA-9950.S8 1944 **Survey of Social Security Statistics.** By Herman Sturm. Washington, D.C., 1944. Federal Security Agency. Social Security Board. Bureau of Research and Statistics. 71 p. Bureau memorandum, no. 56.

"The primary purpose of this report is to furnish a guide to the scope, character, and sources of statistics made available by the operation of the three major programs of social security . . . and of statistical series maintained by the Social Security Board on related programs." (Source: p. 1.)

Reseña de las estadísticas del seguro social. "El principal propósito de este informe es el de suministrar una guía del carácter, extensión y fuentes de estadísticas, que por el funcionamiento de los tres programas principales de seguro social son aprovechables . . . y de series estadísticas mantenidas por el Consejo de Seguro Social sobre otros programas relacionados con ellos." (Traducción del comentario.)

USA-9951.P5 **Publications of the Children's Bureau.** Starting date? Washington, D.C. Department of Labor Children's Bureau.

Series of monographs.

Publicaciones de la Dirección de Niños. Serie de monografías.

USA-9952.A5 **Annual Report of the Railroad Retirement Board.** Starting date? Washington, D.C. Railroad Retirement Board.

Reports on railroad retirement and railroad unemployment insurance operations.

Memoria anual de la Junta de Jubilación de Empleados Ferroviarios. Informes anuales sobre jubilación de empleados ferroviarios y operaciones de los seguros de desempleo ferroviario.

USA-9952.M5 **The Monthly Review of the Railroad Retirement Board.** 1940- . Chicago, Illinois.

Monthly reports on railroad retirement and railroad unemployment insurance operations.

Revista mensual de la Junta de Jubilación de Empleados Ferroviarios. Informes mensuales sobre jubilación de empleados ferroviarios y operaciones de los seguros de desempleo ferroviario.

ALASKA

2000-2999 CENSUS — CENSOS

*USA Alsk-2200.C5 1940 **Sixteenth Census of the United States: 1940: Alaska: Population: Characteristics of the Population (with Limited Data on Housing).** Second series. Washington, D.C., U.S.Govt. Print.Off., 1943. Department of Commerce. Bureau of the Census. 20 p.

Décimosexto censo de los Estados Unidos, 1940: Alaska: Población: Características de la población (con algunos datos sobre vivienda).

PUERTO RICO

1000-1999 GENERAL

USA PR-1410.A5 **Annual Report of the Governor of Puerto Rico** Starting date? San Juan.

Informe anual del gobernador de Puerto Rico.

USA PR-1420.P5 **Puerto Rico Monthly Statistical Report.** 1943- . San Juan. Office of the Governor. Office of Statistics.

Informe estadístico mensual de Puerto Rico.

2000-2999 CENSUS — CENSOS

USA PR-2100.C5 1940.1 **Sixteenth Census of the United States: 1940: Methods and Procedures; Forms and Instructions Used in Taking the Census of Puerto Rico.** Acquired in connection with Inter American Statistical Institute Survey of Census Methods and Procedures, 1944. From the Department of Commerce. Bureau of the Census.

Consists of: Announcements of census; instructions to enumerators (pamphlet); codes and instructions to be used with punch cards; schedules.

Décimosexto censo de los Estados Unidos, 1940: Métodos y Procedimientos; Formularios e instrucciones usados en el levantamiento del censo de Puerto Rico. Adquirido con motivo de la encuesta que el Instituto Interamericano de Estadística realizó sobre Métodos y Procedimientos de los Censos en las Naciones Americanas. Consiste de: Propaganda; instrucciones a los empadronadores; códigos e instrucciones para uso con las tarjetas de perforación; cédulas.

USA PR-2100.C5 1940.2 **Censo Décimosexto de los Estados Unidos: 1940: Instrucciones a los Enumeradores: Puerto Rico.** Forma PR-10. Wáshington, D.C., Imp.del Gobierno, 1940. Departamento de Comercio. Oficina de Censo. 81 p.

Sixteenth census of the United States, 1940: Instructions to the enumerators, Puerto Rico.

*USA PR-2200.C5 1940.1 **Sixteenth Census of the United States: 1940: Puerto Rico, Población, Boletín No. 1: Número de los Habitantes. Puerto Rico, Population, Bulletin No. 1; Number of Inhabitants.** Washington, D.C., U.S.Govt.Print.Off., 1942. Department of Commerce. Bureau of the Census. 12 p.

*USA PR-2200.C5 1940.2 **Censo Décimosexto de los Estados Unidos: 1940: Puerto Rico, Población, Boletín Núm. 2: Características de la Población. Sixteenth Census of the United States, 1940: Puerto Rico, Population, Bulletin No. 2: Characteristics of the Population.** Washington, D.C., U.S.Govt.Print.Off., 1943. Department of Commerce. Bureau of the Census. 82 p.

*USA PR-2200.C5 1940.3 **Sixteenth Census of the United States: 1940: Puerto Rico, Población, Boletín No. 3: Ocupaciones y Otras Características por Edad. Puerto Rico, Population, Bulletin No. 3: Occupations and Other Characteristics by Age.** Washington, D.C., U.S.Govt.Print.Off., 1943. Department of Commerce. Bureau of the Census. Unexamined.

*USA PR-2210.C5 1940 **Censo Décimosexto de los Estados Unidos: 1940: Agricultura. Sixteenth Census of the United States: 1940: Agriculture: Puerto Rico.** Fincas, propiedad agrícola, animales domésticos, y cultivos. Farms, farm property, livestock, and crops. Washington, D.C., U.S.Govt.Print.Off., 1942. United States Department of Commerce. Bureau of the Census. 84 p.

Cover-title—título sobre la cubierta: **Puerto Rico: Agricultura.**

The appendix shows the schedule used in this census.

El apéndice muestra la cédula usada en este censo.

*USA PR-2200.C5 1940 **Censo Décimosexto de los Estados Unidos: 1940: Puerto Rico: Viviendas. Sixteenth Census of the United States: 1940: Puerto Rico: Housing.** Características generales. General characteristics. Washington, D.C., U.S.Govt.Print.Off., 1943. United States Department of Commerce. Bureau of the Census. 121 p.

Includes an appendix showing schedule used in this census.

Incluye un apéndice que muestra la cédula empleada en este censo.

5000-5999 DEMOGRAPHY—DEMOGRAFIA

USA PR-5110.S5 1940 **Staff Report to the Interdepartmental Committee on Puerto Rico.** By Erich W. Zimmerman. Washington, D.C., 1940. Appendix A, **Estimated Rates of Growth of the Population of Puerto Rico,** prepared by the U. S. Bureau of the Census. Unexamined.

Informe a la Comisión Interdepartamental sobre Puerto Rico: Apéndice A, Tasas estimadas del crecimiento de la población de Puerto Rico, preparado por el Departamento del Censo de los Estados Unidos.

6000-6999 ECONOMICS — ECONOMIA

USA PR-**6910**.B5 1941-43 **The Balance of External Payments of Puerto Rico, 1941-42 and 1942-43.** By Robert L. Sammons. San Juan, April 1945. University of Puerto Rico and the Office of Statistics, Office of the Governor, cooperating. 80 p. Advanced mimeographed version.

La balanza internacional de pagos de Puerto Rico, 1941-42 y 1942-43.

7000-7999 HEALTH AND CULTURAL STATISTICS—ESTADISTICAS DE SALUBRIDAD Y CULTURA

USA PR-**7110**.P5 **The Puerto Rico Journal of Public Health and Tropical Medicine.** 1925- Quarterly. San Juan.

Official bulletin of the Department of Health and the School of Tropical Medicine of the University of Puerto Rico under the auspices of Columbia University.

La revista portoriqueña de salud pública y medicina tropical. Boletín oficial del Departamento de Salubridad y de la Escuela de Medicina Tropical de la Universidad de Puerto Rico, bajo los auspicios de la Universidad de Columbia.

USA PR-**7200**.A5 1945 **El Analfabetismo en Puerto Rico . . .** Por Ismael Rodriguez Pou. San Juan, Imp.Venezuela, 1945. Universidad de Puerto Rico. Consejo Superior de Enseñanza. 130 p. Publicaciones pedagógicas, serie II, no. II. Unexamined.

Illiteracy in Puerto Rico.

URUGUAY

1000-1999 GENERAL

*Ur-1410.A5 **Anuario Estadístico de la República Oriental del Uruguay.** 1884- . Montevideo. Ministerio de Hacienda. Dirección General de Estadística.

Statistical yearbook of Uruguay. Volume I covers practically all phases of the economic life of the country, including demography, hygiene and public health, social welfare and labor, education, justice, agriculture and livestock, industry, communication and transportation, banking, finance, and public debt. Also contains a section devoted to international comparisons. Volume II related exclusively to foreign trade. Comparative data for several years are carried in both volumes.

El volumen I abarca practicamente todas las fases de la vida económica del país, incluyendo demografía, higiene y salud pública, bienestar social y trabajo, educación, justicia, agricultura y ganadería, industria, comunicaciones y transportes, banca, finanzas y deuda pública. También contiene una sección dedicada a comparaciones internacionales. El volumen II se refiere exclusivamente al comercio exterior. Datos retrospectivos de varios años se incluyen en ambos volúmenes.

1908 "Censo General de la República en 1908." *Año* 1908, tomo II, parte III, p. 755-1260.

General census, 1908.

Ur-1410.S5 **Síntesis Estadística de la República Oriental del Uruguay.** Fecha de iniciación? Anual. Montevideo. Ministerio de Hacienda. Dirección General de Estadística.

Statistical abstract of Uruguay. Statistical summaries of material appearing in **Anuario Estadístico.** Comparative data for several years.

Síntesis estadística de los mismos datos que aparecen en el **Anuario Estadístico.** Datos comparativos de varios años.

Ur-1420.B5 **Boletín Estadístico de la República Oriental del Uruguay** 1927-39. Irregular. Montevideo. Ministerio de Hacienda. Dirección General de Estadística.

Temporarily suspended — suspendido temporalmente.

Statistical bulletin of Uruguay. Principally statistical tables in fields of trade, agriculture and livestock, banking and finance.

Principalmente cuadros estadísticos sobre comercio, agricultura y ganadería, operaciones bancarias y finanzas.

2000-2999 CENSUS — CENSOS

Ur-2110.C5 1943 **Censo Ganadero 1943: Ley de Enero 27 de 1943.** Montevideo, Imp.Nacional, 1943. Ministerio de Ganadería y Agricultura. Dirección de Agronomía. 17 p.

Livestock census, 1943: Law of January 27, 1943. Contains instructions, forms, and schedules.

Contiene instrucciones, formularios y cédulas.

*Ur-2210.C5 1937 **Censo Agropecuario, Año 1937.** Montevideo, 1939-40. Ministerio de Ganadería y Agricultura. Dirección de Agronomía. Sección de Economía y Estadística Agraria. 5 vol.: Pte. I, **Stock Vacuno y Ovino;** pte. II, **Lechería, Porcinos, Equinos, Asnal y Mular, Cabrios, Avicultura, Apicultura;** pte. III, **Fruticultura;** pte. IV, a. **Población Trabajadora Agraria;** b. **Régimen de la Propiedad Agraria;** c. **Sistemas de Explotación;** d. **Relaciones de los Sistemas de Explotación, con el Régimen de la Propiedad;** pte. V.

Census of agriculture and livestock, 1937.

*Ur-2250.C5 1936 **Censo Industrial de 1936.** Montevideo, 1939. Ministerio de Industrias y Trabajo. Dirección de Estadística Económica. 58 p.

Industrial census, 1936. Includes number of establishments, workers, capital, wages, and productivity by provinces and industrial groups.

Incluye número de establecimientos, obreros, capital, salarios y productividad, por provincias y grupos industriales.

3000-3999 STATISTICAL SCIENCE — CIENCIA ESTADISTICA

Ur-3000.C5 **Cuadernos del Instituto de Estadística.** 1942- . Montevideo.

Pamphlets of the Institute of Statistics. Series of monographs.

Serie de monografías.

Ur-3030.E4 1914 **Estadística General de la República, Ley de Reorganización.** Montevideo, Imp.Artística de Juan J. Dornaliche, 1914. 16 p.

Law on reorganization of statistics in Uruguay.

*Ur-3040.R5 **Revista de la Facultad de Ciencias Económicas y de**

Administración. 1940- . Irregular. Montevideo. Universidad de la República. Facultad de Ciencias Económicas y de Administración.

Journal of the School of Economics and Administration of the University of the Republic. Includes articles on statistics.

Incluye artículos sobre estadística.

Ur-3300 1944-46 **Enseñanza Estadística en Uruguay: Métodos y Materiales.** Material básico adquirido para la Encuesta del Instituto Interamericano de Estadística sobre los Métodos y Materiales en la Enseñanza Estadística en las Naciones Americanas, 1944-46.

Statistical training in Uruguay: Methods and materials. Background material acquired for the IASI Survey on Statistical Training Methods and Materials in the American Nations, 1944-46. Includes: Filled questionnaires on statistical courses in Uruguay.

Incluye: Cuestionarios llenados sobre cursos de estadística en el Uruguay.

5000-5999 DEMOGRAPHY — DEMOGRAFIA

Ur-5000.U5 1944 **Uruguay: Summary of Biostatistics.** Maps and charts, population, natality, and mortality statistics. Washington, D.C., November 1944. Prepared by the U. S. Department of Commerce, Bureau of the Census, in cooperation with Office of the Coordinator of Inter-American Affairs. 129 p.

Resumen bioestadístico de Uruguay. Incluye mapas y cuadros y datos sobre población, natalidad y mortalidad.

Ur-5100.H5 1939 **Historia y Análisis Estadístico de la Población del Uruguay.** Por Agustín Ruano Fournier, Edmundo M. Narancio, Federico Capurro Calamet. Montevideo, Peña y Cía., Impresores, 1939. Universidad de la República. Facultad de Derecho y Ciencias Sociales de Montevideo. Biblioteca de Publicaciones Oficiales. 290 p.

History and statistical analysis of the population of Uruguay.

Ur-5200 **Estadística Vital y Registro en Uruguay: Métodos y Procedimientos.** Material básico adquirido para la encuesta que el Instituto Interamericano de Estadística realiza sobre los métodos de registro y estadística vital en las naciones americanas, incluyendo el estudio de 1945.

Vital statistics and registration in Uruguay: Methods and procedures. Background material acquired for the study conducted by the Inter American Statistical Institute in vital statistics and registration methods in the American nations, including the survey of 1945. Includes: Forms; punch cards.

Incluye: Formularios; tarjetas de perforación.

Ur-5200.M5 **El Movimiento del Estado Civil y la Mortalidad de la República Oriental del Uruguay.** Anuario de la Dirección General del Registro del Estado Civil. Fecha de iniciación? Montevideo. Ministerio de Instrucción Pública y Previsión Social. Instituto de Jubilaciones y Pensiones. Dirección General del Registro del Estado Civil.

Civil status and mortality in Uruguay: Yearbook of the General Bureau of Civil Status Registration. Contains data on births, deaths, and marriages, by localities and for the nation as a whole.

Contiene datos sobre nacimientos, muertes y matrimonios, por localidades y del total de la nación.

6000-6999 ECONOMICS — ECONOMIA

Ur-6000.B5 **Boletín de Hacienda.** Economía, finanzas, estadística. 1914- . Mensual. Montevideo. Ministerio de Hacienda. Contaduría General de la Nación.

Journal of finance, monthly publication of the Department of Finance. Economics, finance, statistics. Includes statistical series and charts on foreign commerce, production, prices, public debt, and other phases of the country's economy.

Incluye series y cuadros estadísticos sobre comercio exterior, producción, precios, deuda pública y otras fases de la economía del país.

*Ur-6000.B6 **Boletín de Hacienda, Suplemento.** Revista de índices económicos. 1944- . Mensual. Montevideo. Ministerio de Hacienda. Contaduría General de la Nación.

Monthly supplement to **Boletín de Hacienda:** Economic indexes.

Ur-6000.B7 **Boletín de la Dirección de Estadística Económica.** Fecha de iniciación? Trimestral. Montevideo. Ministerio de Industrias y Trabajo. Dirección de Estadística Económica. No revisado.

Bulletin of the Bureau of Economic Statistics. "Quarterly bulletin containing statistics of commerce, industry, and labor; and cost of living, employment, and other indexes." (Source: Stat. Act.)

"Boletín trimestral que contiene estadísticas del comercio, industria y trabajo; e índices del costo de la vida, ocupación, etc." (Traducción del comentario.)

Ur-6000.E5 1945 **Economic Controls and Commercial Policy in Uruguay.** One of a series of reports on economic controls and commercial policy in the American republics. Washington, D.C., 1945. United States Tariff Commission. 27 p.

Folleto sobre Uruguay, de la serie de informes de la Comisión de Aduanas de los Estados Unidos, sobre controles económicos y política comercial de las repúblicas americanas.

*Ur-6000.S5 **Suplemento Estadístico de la Revista Económica.** 1944- . Mensual. Montevideo. Banco de la República Oriental del Uruguay. Departamento de Investigaciones Económicas.

Monthly statistical supplement to **Revista Económica** of the Bank of the Republic.

Ur-6200 **Estadísticas Agrícolas en Uruguay: Métodos y Procedimientos.** Material básico adquirido para el estudio que el Instituto Interamericano de Estadística realiza sobre los métodos de estadísticas agrícolas en las naciones americanas, incluyendo la encuesta de 1945-46.

Agricultural statistics in Uruguay: Methods and procedures. Background material acquired for Inter American Statistical Institute study on methods in agricultural statistics in the American nations, including the Survey of Agricultural Production Statistics in the American Nations in 1945-46. Consists of: Correspondence; laws, decrees, or resolutions; instructions; forms; publications.

Consiste de: Correspondencia; leyes, decretos o resoluciones; instrucciones; formularios; publicaciones.

See also—véase también: **Censo Ganadero 1943, según la Ley de Enero de 1943; Estadística General de la República: Ley de Reorganización; Ley y Reglamentación de la Estadística Agrícola; Estadística de Cereales y Oleaginosas; Estimaciones e Informes Finales de Area Sembrada y Producción: Maíz, Arroz, Cereales . . . 1943-45; Mercados del Mundo.**

*Ur-6200.A5 **Anuario de Estadística Agrícola.** 1892-1894, 1913- . Montevideo. Ministerio de Ganadería y Agricultura. Dirección de Agronomía. Sección de Economía y Estadística Agraria.

Agricultural statistics yearbook. Contains statistics of livestock and crops.

Contiene estadísticas de cosechas y ganadería.

Ur-6200.E4 1943 **Estadística Agrícola.** Montevideo, Imp.Nacional, 1943. Ministerio de Ganadería y Agricultura. Dirección de Agronomía. Sección de Economía y Estadística Agrícola. 80 p. (aprox.)

Agricultural statistics: Questionnaire form, instructions, and decrees of February 27, 1930, and May 28, 1930.

Formulario de cuestionario, instrucciones y decretos del 27 de febrero de 1930 y del 28 de mayo de 1930.

Ur-6200.L5 1918 **Ley y Reglamentación de la Estadística Agrícola.** Montevideo, Imp.Nacional, 1918. Issued by the Ministerio de Industrias, Oficina de Estadística Agrícola. 21 p.

Law and regulations on agricultural statistics.

Ur-6200.M5 **Mercados del Mundo.** Información agroeconómica. 1937?- . Frecuencia? Montevideo. Ministerio de Ganadería y Agricultura. Dirección de Agronomía. Sección de Economía y Estadística Agraria.

World markets. Agricultural-economic information.

Ur-6210.E5 1944 **Estadística de Cereales y Oleaginosos.** Montevideo, Imp. Nacional, 1944. Ministerio de Ganadería y Agricultura. Dirección de Agronomía. Sección de Economía y Estadística Agraria. 60 p. (aprox.)

Statistics of cereals and oleaginous plants.

Ur-6210.E6 1943-45 **Estimaciones e Informes Finales de Area Sembrada y Producción: Maíz, Arroz, Cereales . . . 1943-45.** Ministerio de Ganadería y Agricultura. Dirección de Agronomía. Sección de Economía Estadística Agraria. 33 p. Mimeografiado.

Estimates and final reports on cultivated area and production of corn, rice, cereals, 1943-45.

Ur-6400 **Estadísticas Industriales en Uruguay: Métodos y Procedimientos.** Material básico adquirido para el estudio que el Instituto Interamericano de Estadística realiza sobre los métodos de estadísticas industriales en las naciones americanas, incluyendo la encuesta de 1945-46.

Industrial statistics in Uruguay: Methods and procedures. Background material acquired for the study conducted by the Inter American Statistical Institute in methods of industrial statistics in the American nations, including the survey of 1945-46. Consists of: Laws and decrees; forms.

Consiste de: Leyes y decretos; formularios.

Ur-**6400**.M5 1945 **Mining and Manufacturing Industries in Uruguay.** One of a series of reports on mining and manufacturing industries in the American republics. Washington, D.C., 1945. United States Tariff Commission. 23 p.

Folleto sobre Uruguay, de la serie de informes de la Comisión de Aduanas de los Estados Unidos, sobre las industrias mineras y manufactureras en las repúblicas americanas.

Ur-**6700**.E5 **Estadística de la Cámara Nacional de Comercio.** Fecha de iniciación? Anual. Montevideo. Cámara Nacional de Comercio. Bolsa de Comercio.

Annual statistics of the National Chamber of Commerce.

Ur-**6720**.M5 **Memoria del Directorio del Banco de la República Oriental del Uruguay.** Balances de situación y resultados e información estadística. Fecha de iniciación? Anual. Montevideo.

Annual report of the Board of Directors of the Bank of the Republic of Uruguay. Financial statements and statistical information.

1945 **Memoria y Balance General Correspondientes al Ejercicio Terminado en 31 de Diciembre de 1945.** Montevideo, Tall.Gráf.del Banco de la República Oriental del Uruguay, 1946. Banco de la República Oriental del Uruguay. 153 p.

Contains in the introduction a sketch of the world situation, the economic characteristics of 1945, the situation of the country, the means of payment, balance of foreign trade, the internal commercial activity, and the credit position. Gives a detailed balance of the bank, including retrospective figures. Includes an analysis of the activities of the bank by departments (commercial affairs, credit, etc.)

Contiene por introducción un bosquejo sobre la situación universal, características económicas del año 1945, la situación del país, sus medios de pago, las cifras del intercambio, la actividad comercial interna y la posición del crédito. A continuación detalla el balance del Banco, dando algunas cifras retrospectivas. Incluye un análisis de las actividades del Banco poı departamentos (asuntos comerciales, crédito, etc.).

Ur-**6720**.R5 **Revista del Banco de la República Oriental del Uruguay.** 1942- . Trimestral. Montevideo. Banco de la República Oriental del Uruguay. Departamento de Investigaciones Económicas.

Quarterly review of the Bank of the Republic. Information on banking and business, with statistical data.

Informaciones sobre bancos y negocios, con datos estadísticos.

Jan. '46 "Balances de Pagos del Uruguay, Años 1940-1944." Año 4, no. 16, enero 1946, p. 18.

Balance of payments of Uruguay, 1940-44.

Ur-**6900** **Estadísticas del Comercio Exterior en Uruguay: Métodos y Procedimientos.** Material básico adquirido para la encuesta que el Instituto Interamericano de Estadística realiza sobre los métodos de estadísticas del comercio exterior en las naciones americanas.

Foreign trade statistics in Uruguay: Methods and procedures. Background material acquired for the study conducted by the Inter American Statistical Institute in foreign trade statistics in the American nations. Includes: Filled questionnaire.

Incluye: Cuestionario llenado.

Ur-**6900**.B5 **Boletín Aduanero.** 1936- . Mensual. Montevideo. Ministerio de Hacienda. Dirección General de Aduanas.

Monthly customs bulletin. Contains current statistical series.

Series de estadísticas aduaneras corrientes.

Ur-**6900**.M5 **Memoria del Contralor de Exportaciones e Importaciones.** Fecha de iniciación? Anual. Montevideo. Ministerio de Hacienda. Contralor de Exportaciones e Importaciones.

Annual report of the Comptroller of Exports and Imports.

Ur-**6900**.S5 1942 **Situación y Perspectivas para la Organización del Intercambio Comercial Interamericano.** Trabajo laureado por el Banco de la República Oriental del Uruguay con el Premio "Julio Mailhos (hijo)" año 1942. Por Juan Manuel Aguirre Gonzalez. Montevideo, Tall.Gráf.del Banco de la República Oriental del Uruguay, 1943. 61 p. and tables.

Situation and prospects for inter american trade organization. Principal tables: Foreign trade of the American nations, 1930-35; strategic materials, 1938-41; principal products in the exports of each American country (percentages); principal production in the American countries.

Los cuadros importantes son: Comercio exterior de las naciones Americanas (1930 a 1935);

materiales estratégicos (1938-1941); productos principales en las exportaciones de cada uno de los países americanos (porcentajes); principales producciones en los países de América.

Ur-6900.T5 **Tarifa de Aforos de Importación.** Montevideo, Imp.Nacional, 1940-42. Ministerio de Hacienda. Dirección General de Aduanas. 21 vol. (from 15 to 165 pages each).
1940-42

Import tariff. Customs code of Uruguay. Classification based on League of Nations Draft Customs Nomenclature. Modifications appear in the **Diario Oficial.**

Legislación aduanera del Uruguay. Clasificación basada sobre el proyecto de nomenclatura aduanera de la Sociedad de las Naciones. Las modificaciones aparecen en el **Diario Oficial.**

7000-7999 HEALTH AND CULTURAL STATISTICS—ESTADISTICAS DE SALUBRIDAD Y CULTURA

Ur-7110.M5 **Memoria del Ministerio de Salud Pública.** Fecha de iniciación? Frecuencia? Montevideo. Ministerio de Salud Pública.

Report of the Department of Public Health.

Ur-7200.A5 **El Analfabetismo.** Por Julio Castro. Montevideo, Ed.Imp. Nacional, 1940. No examinado.
1940

Illiteracy.

8000-8999 GOVERNMENT AND POLITICAL STATISTICS—ESTADISTICAS GUBERNAMENTALES Y POLITICAS

Ur-8800.D5 **Deuda Pública Nacional.** Fecha de iniciación? Anual. Montevideo. Departamento de Estadigrafía e Investigaciones Financieras.

National public debt. Includes statistical tables. Annual.

Incluye cuadros estadísticos.

Ur-8800.P5 **Presupuesto General del Estado.** 1942- . Anual. Montevideo. Ministerio de Hacienda. Contaduría General de la Nación.

National budget. Annual.

9000-9999 LABOR AND LIVING CONDITIONS—CONDICIONES DE TRABAJO Y DE VIDA

Ur-9500.C5 **Costo de la Vida Obrera en Montevideo.** 1942- . Mensual. Montevideo. Ministerio de Industrias y Trabajo. Dirección General de Asuntos Económicos.

Cost of living of workers in Montevideo.

Ur-9800.B5 **Boletín Mensual del Instituto Nacional de Alimentación.** 1944- . Montevideo.

Monthly bulletin of the National Food Institute.

VENEZUELA

1000-1999 GENERAL

Ven-**1200**.A5 **Anuario Bibliográfico Venezolano.** 1942- . Caracas. Biblioteca Nacional. No examinado.

Bibliographic annual of Venezuela.

*Ven-**1410**.A5 **Anuario Estadístico de Venezuela.** 1877-1912, 1938- . Caracas. Ministerio de Fomento. Dirección General de Estadística y de Censos.

None published 1913-37 — no publicado 1913-37.

Statistical yearbook of Venezuela.

1944 **Anuario Estadístico de Venezuela, 1944.** Caracas, Editorial Grafolit, 1945. Ministerio de Fomento Dirección General de Estadística. 597 p.

Includes: Geographical and meteorological data; population and vital statistics; mail and long distance communication; communication and transportation; production and consumption; prices and economic indexes; economic and financial statistics; cultural, social and administrative statistics; foreign trade.

Incluye: Datos geográficos y meteorológicos; estado y movimiento de la población; correos y telecomunicaciones; comunicaciones y transportes; producción y consumo; precios e índices económicos; estadísticas económicas y financieras; estadísticas culturales, sociales y administrativas; comercio exterior.

Ven-**1420**.B5 **Boletín Mensual de Estadística.** 1941- . Caracas. Ministerio de Fomento. Dirección General de Estadística y de Censos.

Pubished as quarterly since 1942 — editado como publicación trimestral desde 1942.

Bulletin of statistics. Contains: Foreign trade statistics; economic index numbers and price series; production statistics; demographic and vital statistics; criminal and judicial and other administrative statistics.

Ven-**1900**.G5 **Gaceta Oficial de los Estados Unidos de Venezuela.** 1871?- . Diariamente? Caracas.

Official publication of the United States of Venezuela, containing laws, decrees, etc.

Nov. 27, '44 "Ley de Estadística y de Censos Nacionales." Año LXXIII, mes II, no. 21.572, noviembre 27, 1944, p. 1-3.

Law of statistics and national censuses.

Contiene: Estadísticas del comercio exterior; números índices económicos y otras series de precios; estadísticas de la producción; estadística vital y demográfica; estadística criminal y judicial, y otras estadísticas administrativas.

Ven-**1950**.V5 **Venezuela, 1945.** Caracas? 1945 El Mes Financiero y Económico, 1945. 659 p. Unexamined.

" . . . Contains sections on the history, races, culture, and politics of the country; statistics of population, trade, and production of various commodities; cattle raising and agriculture; oil and other industries; government agencies, cities, and education; poetry, music, folklore, art, and sports. There are separate sections on the Federal District and the several States." (Source: **Agriculture in the Americas,** vol. VI, no. 10, October 1946, p. 170.)

" . . . Contiene capítulos de historia, razas, cultura y política del país; estadísticas de población, comercio y producción diversa; cría de ganado y agricultura; petróleo y otras industrias; agencias gubernamentales, ciudades y educación; poesía, música, folklore, arte y deportes. Hay capítulos por separado del Distrito Federal y de los estados." (Traducción del comentario.)

2000-2999 CENSUS — CENSOS

Ven-**2100**.C5 **Censo de 1941: Métodos y** 1941.1 **Procedimientos.** Material usado en el levantamiento del censo de población del 7 de diciembre. Adquirido con motivo del estudio que el Instituto Interamericano de Estadística realizó sobre Métodos y Procedimientos de los Censos en las Naciones Americanas, 1944. Del Ministerio de Fomento. Dirección General de Estadística y de Censos.

Census of 1941: Methods and procedures. Materials used in taking population census, on December 7. Acquired in connection with IASI Survey of Census Methods and Procedures, 1944. Consists of: Correspondence; instructions; expositions and reports; punch cards; codes, schedules; other forms.

Consiste de: Correspondencia; instrucciones exposiciones e informes; tarjetas de perforación; claves; cédulas; otros formularios.

See also — véase también: **Instrucciones y Modelos del Séptimo Censo Nacional de**

Población; Séptimo Censo Nacional de Población Levantado el 7 de Diciembre de 1941: Tomo I, Distrito Federal y Estado Anzoategui.

Ven-2100.C5 1941.2 **Séptimo Censo Nacional de Población, 1941: Instrucciones y Modelos.** Caracas, Lit.y Tip.Casa de Especialidades, 1941. Ministerio de Fomento. Dirección General de Estadística. 70 p.

Seventh national population census, 1941: Instructions and models.

*Ven-2200.C5 1941 **Séptimo Censo Nacional de Población, Levantado el 7 de Diciembre de 1941.** Caracas, 1944- . Ministerio de Fomento. Dirección General de Estadística y de Censos. 7 tomos: Tomo I, **Distrito Federal y Estado Anzoategui** (incluye cédulas, instrucciones, formularios, etc., empleados en el censo); tomo II, **Estados Apure, Aragua, Barinas y Bolívar;** tomo III, **Estados Carabobo, Cojedes y Falcón;** tomo IV, **Estado Guarico, Lara y Mérida;** tomo V, **Estados Miranda, Monagas y Nueva Esparta;** tomo VI, **Estados Portuguesa, Sucre y Tachira;** tomo VII, **Estados Trujillo, Yaracuy, Zulia, Territorio Federal Amazonas, Territorio Federal Delta Amacuro y Dependencias Federales.**

Seventh national census of population, December 7, 1941. For each of the Venezuelan states statistical tables give data on demographic density, coefficients of intercensal and annual increase, and various comparisons between the censuses of 1936 and 1941. The rest of the information is published under the following eight headings: Age and civil status; education and school attendance; nationality and place of birth; religion; vaccination and mental and physical defects; economic activities; urban, suburban, and rural population, and localities of more than 500 inhabitants; census of housing, and sanitary data. The first volume includes schedules, instructions, forms, etc., which were used in the census.

Para cada uno de los estados venezolanos, cuadros estadísticos exponen datos sobre densidad demográfica, coeficientes de aumento intercensal y anual, y comparaciones diversas de los censos de 1936 y 1941; el resto de la información se publica según los ocho encabezamientos siguientes: Edades y estado civil; instrucción y asistencia escolar; nacionalidad y lugar de nacimiento; religión; vacunación y defectos físicos y mentales; actividades económicas; población urbana, suburbana y rural, y localidades de más de 500 habitantes; censo de casas, y datos sanitarios.

Ven-2200.C5 1936.2 **Sexto Censo Nacional, 1936.** Caracas, Tipo.Americana, 1939-40. Ministerio de Fomento. Dirección General de Estadística. 3 vol.: Vol. 1, **Distrito Federal;** vol. 2, **Estados Anzoátegui, Apure, Aragua, Bolívar, Carabobo, Cojedes, Falcón, Guárico, Lara y Mérida;** vol. 3, **Estados Miranda, Monagas, Nueva Esparta, Portuguesa, Sucre, Táchira, Trujillo, Yaracuy, Zamora y Zulia, y Territorio Federal Amazonas, Territorio Federal Delta Amacuro, Dependencias Federales y los Resúmenes Generales de Población de Todas las Entidades de la República.**

Sixth national census, 1936.

*Ven-2210.C5 1937 **Censos Agrícola y Pecuario 1937.** Caracas, 1937-40. Ministerio de Fomento. Dirección General de Estadística. 23 vol.: Un volumen separado para cada Estado.

Census of agriculture and livestock, 1937. A separate volume for each state. Covers size of farms and ranges, proprietorship and other forms of tenure, acreage cultivated, produce, value of land, buildings and tools topographical location and incumbrances.

Abarca: Extensión de los fundos; propiedad y otras formas de explotación; área cultivada; productos; valor de tierras, casas y herramientas; ubicación topográfica; hipotecas.

*Ven-2250.C5 1936 **Censos Industrial, Comercial y Empresas que Prestan Servicios, 1936.** Edición oficial. Caracas, 1938-41. Ministerio de Fomento. Dirección General de Estadística. 21 vol: Un volumen separado para cada Estado.

Census of industries, commerce and services, 1936. Each volume contains statistics by industrial, commercial and service groups. Includes number of establishments, capital, expenses of production, volume of sales, value added by manufacture, number of employees by sex and nationality, wages and salaries and profits.

Cada volumen contiene estadística por grupos de industrias, de comercios, y de impresas que prestan servicios. Incluye número de industrias, capital, costo de la producción, volumen de ventas, valor agregado por los procesos de manufactura, número de personal por sexo y nacionalidad, salarios y sueldos, y beneficios. A separate volume for each state.

3000-3999 STATISTICAL SCIENCE—CIENCIA ESTADISTICA

Ven-3030.I5 **Informe de la Dirección General de Estadística y de Censos.**

Fecha de iniciación? Anual. Caracas. Ministerio de Fomento. Dirección General de Estadística y de Censos.

Annual report of the General Bureau of Statistics and Censuses.

Extracted from — extraído de la: **Memoria del Ministerio de Fomento.**

Ven-3300 1944-46 **Enseñanza Estadística en Venezuela: Métodos y Materiales.** Material básico adquirido para la Encuesta del Instituto Interamericano de Estadística sobre los Métodos y Materiales en la Enseñanza Estadística en las Naciones Americanas, 1944-46.

Statistical training in Venezuela: Methods and materials. Background material acquired for the IASI Survey on Statistical Training Methods and Materials in the American Nations, 1944-46. Includes: Correspondence and filled questionnaires on statistical personnel and statistical courses in Venezuela.

Incluye: Correspondencia y cuestionarios llenados sobre personal estadístico y cursos de estadística en Venezuela.

4000-4999 GEOGRAPHY — GEOGRAFIA

Ven-4100.D5 1945 **Suplemento de la División Político-Territorial de la República.** Conteniendo las modificaciones efectuadas por las asambleas legislativas estadales en sus sesiones de 1945. Edición oficial. Caracas, 1945. Ministerio de Fomento. Dirección General de Estadística. 6 p.

Supplement to **División Político-Territorial de la República,** containing the modifications made by the state legislatures in 1945.

Ven-4100.D5 1944 **División Político-Territorial de la República.** Edición oficial. Caracas, Lit.y Tip.Vargas, 1944. Ministerio de Fomento. Dirección General de Estadística. 97 p.

Political-territorial divisions of Venezuela.

Ven-4100.N5 1941 **Nomenclador General de Areas y Lugares Habitados de Venezuela.** Segun el VII censo nacional de población levantado el 7 de diciembre de 1941. Caracas, Edit. Bolívar, 1944. Ministerio de Fomento. Dirección General de Estadística. 766 p.

General gazetteer of areas and inhabited places in Venezuela, according to the VII national population census taken December 7, 1941. Names of localities appear in alphabetical order in the first column; in the second column the place is classified by category — village, ranch, farm, hamlet, etc. The other three columns show the name of the municipality, district, and federal unit to which the place belongs.

Los nombres de las localidades aparecen en orden alfabético en la primera columna; en la segunda columna se clasifica el lugar según su categoría, ya bien en caserío, hato, hacienda, vecindario, etc. En las otras tres columnas se anota el nombre del municipio, distrito y entidad federal a que pertenece.

5000-5999 DEMOGRAPHY—DEMOGRAFIA

Ven-5000.V5 1944 **Venezuela: Summary of Biostatistics.** Maps and charts, population, natality and mortality statistics. Washington, D.C., August 1944. Prepared by U. S. Department of Commerce, Bureau of the Census, in cooperation with Office of the Coordinator of Inter-American Affairs. 118 p.

Resumen bioestadístico de Venezuela. Incluye mapas y cuadros y datos sobre población, natalidad y mortalidad.

Ven-5200 **Estadística Vital y Registro en Venezuela: Métodos y Procedimientos.** Material básico adquirido para la encuesta que el Instituto Interamericano de Estadística realiza sobre los métodos de registro y estadística vital en las naciones americanas, incluyendo el estudio de 1945.

Vital statistics and registration methods in Venezuela: Methods and procedures. Background material acquired for the study conducted by the Inter American Statistical Institute in vital statistics and registration methods in the American nations, including the survey of 1945. Includes: Answer to questionnaire; instructions; forms; punch cards and codes.

Incluye: Respuesta al cuestionario; instrucciones; formularios; tarjetas de perforación y códigos.

*Ven-5200.A5 **Anuario de Epidemiología y Estadística Vital.** 1938- . Caracas. Ministerio de Sanidad y Asistencia Social. Dirección de Salubridad Pública. División de Epidemiología y Estadística Vital.

Title varies — título varía: 1938-39, **Relación Anual de la Sección de Estadística Vital.**

Yearbook of epidemiological and vital statistics.

1944 **Anuario de Epidemiología y Estadística Vital, Año 1944.** Caracas, Lit.y Tipo.del Comercio, 1945. Ministerio de Sanidad y Asistencia Social.

Dirección de Salubridad Pública. División de Epidemiología y Estadística Vital. 475 p.

Includes: Epidemiology (vaccinations; morbidity); vital statistics (population; natality; stillbirths; mortality).

Incluye: Epidemiología (vacunaciones; morbilidad); estadística vital (población; natalidad; mortinatalidad; mortalidad).

6000-6999 ECONOMICS — ECONOMIA

Ven-6000.E3 1945 **Economic Controls and Commercial Policy in Venezuela.** One of a series of reports on economic controls and commercial policy in the American republics. Washington, D.C., 1945. United States Tariff Commission. 26 p.

Folleto sobre Venezuela, de la serie de informes de la Comisión de Aduanas de los Estados Unidos, sobre controles económicos y política comercial de las repúblicas americanas.

Ven-6000.G5 1943 **Guía Industrial y Comercial de Venezuela, 1943.** Caracas, Editorial Crisol, 1943-44. Ministerio de Fomento. Dirección General de Estadística. 3 vol.: Tomo I, **Distrito Federal;** tomo II, **Estados Aragua, Carabobo, Lara, Mérida, Miranda, Sucre, Táchira, y Zulia;** tomo III, **Estados Anzoátegui, Apure, Barinas, Bolívar, Cojedes, Falcón, Guárico, Monagas, Nueva Esparta, Portuguesa, Trujillo, Yaracuy, Territorio Amazonas, Territorio Delta-Amacuro.**

Industrial and commercial directory of Venezuela, 1943.

Ven-6000.G5 1943.A **Suplemento de la Guía Industrial y Comercial de Venezuela Correspondiente al Año 1943.** Estados Anzoátegui, Apure, Barinas, Bolívar, Cojedes, Falcón, Guárico, Monagas, Nueva Esparta, Portuguesa, Trujillo, Yaraguay, y territorios federales Amazonas y Delta Amacuro. Caracas, Editorial Bolívar, 1945. Ministerio de Fomento. Dirección General de Estadística. 324 p.

"... Contains data relative to the foundation and dissolution of establishments for the two halves of 1943," in the mentioned states and territories. (Translation of comment.)

"... Contiene los datos relativos a la fundación y disolución de empresas durante los dos semestres de 1943," en los estados y territorios citados. (Fuente: Introducción.)

*Ven-6000.M3 **Memoria del Ministerio de Fomento.** Fecha de iniciación? Anual. Caracas. Ministerio de Fomento.

Annual report of the Department of Development. Covers activities in fields of mining, petroleum industry, general manufacturing industries and commerce, as well as report of General Bureau of Statistics.

Incluye las actividades mineras, industria petrolera, industrias de transformación y comercio, como así también el informe de la Dirección General de Estadística.

1945 **Memoria del Ministerio de Fomento Presentada al Congreso de los Estados Unidos de Venezuela, 1945.** Caracas, Tipo.Americana, 1945. Ministerio de Fomento. 632 p.

Contains reports of the Juridical Consulting Bureau, Cabinet, Bureau of Mines, Hydrocarbon Technical Office, Bureau of Manufacturing, Bureau of Commerce, General Bureau of Statistics, Bureau of Administration and reports on the activities of the Technical Service of Mining and Geology, National Laboratory, the Labor Bank; graphs, tables, documents; statements of accounts.

Contiene informes de la Consultoria Jurídica, Dirección del Gabinete, Dirección de Minas, Oficina Técnica de Hidrocarburos, Dirección de Industria, Dirección de Comercio, Dirección General de Estadística, Dirección de Administración, e informes sobre las actividades del Servicio Técnico de Minería y Geología, Laboratorio Nacional, el Banco Obrero; gráficos, cuadros; documentos; cuenta.

Ven-6000.M5 **Memoria y Cuenta del Ministerio de Obras Públicas.** Fecha de iniciación? Anual. Caracas. Ministerio de Obras Públicas.

Annual report of the Department of Public Works.

Ven-6000.R3 1944 **Reports Presented to the Conference of Commissions of Inter-American Development (Held in New York, May 9-18, 1944) by the Venezuelan Commission of Inter-American Development.** On: 1, Present state of Venezuelan industry; 2, brief exposition of the industrial needs of Venezuela; 3, general description of the industrial possibilities of Venezuela. (Translations from the Spanish originals.) Washington, D.C., 1945? Inter-American Development Commission. 39 p.

Cover title — título de la cubierta: **Venezuela: Reports Presented to the Conference of Commissions of Inter-American Development.**

Informes presentados a la Conferencia de Comisiones de Fomento Interamericano (realizada en Nueva York, del 9 al 18 de mayo de 1944), por la Comisión Venezolana de Fomento Interamericano, sobre: 1, El estado presente de la industria venezolana; 2, breve exposición de las necesidades industriales de Venezuela; 3, descripción general de las posibilidades industriales de Venezuela.

Ven-6000.R5 **Revista de Fomento.** 1937- . Irregular. Caracas. Ministerio de Fomento.

Supersedes — substituye a: **Boletín del Ministerio de Fomento.**

Journal of the Department of Development. Review of principal index numbers and other economic series, to which the General Bureau of Statistics contributes analyses, comments, and charts.

Esta revista contiene los principales números índices y series económicas de otra índole, acompañados de comentarios, análisis y cuadros suministrados por la Dirección General de Estadística.

Ven-6000.V5 1946 **Venezuela.** Foreign commerce yearbook country series. Washington, D.C., U.S.Govt.Print.Off., 1946. Department of Commerce. Bureau of Foreign and Domestic Commerce. International Reference Service, vol. 3. Unexamined.

Contains information about area, population, climate, agriculture, commerce and industry, mining and manufacturing, communication and transportation, power and light, national economic indexes, national income, average wages, foreign trade, etc.

Contiene información sobre superficie, población, clima, agricultura, comercio e industria, minería y manufactura, comunicaciones y transportes, energía y luz, índices de economía nacional, renta nacional, promedio de salarios, comercio exterior, etc.

Venezuela; anuario del comercio exterior de los Estados Unidos, serie por países.

Ven-6200.A5 1945 **Agricultural, Pastoral, and Forest Industries in Venezuela.** One of a series of reports on agricultural, pastoral, and forest industries in the American republics. Washington, D.C., 1945. United States Tariff Commission. 27 p.

Folleto sobre Venezuela, de la serie de informes de la Comisión de Aduanas en los Estados Unidos sobre las industrias agrícolas, campestres y forestales en las repúblicas americanas.

*Ven-6200.M5 **Memoria y Cuenta del Ministerio de Agricultura y Cría.** Fecha de iniciación? Anual. Caracas. Ministerio de Agricultura y Cría.

Annual report of the Department of Agriculture and Livestock.

Ven-6210.R5 **Revista del Instituto Nacional del Café.** Fecha de iniciación? — 1943 Frecuencia? Caracas. Instituto Nacional del Café.

Journal of the National Coffee Institute. Ceased publication in 1943 when the National Coffee Institute was liquidated by presidential decree, and its functions were taken over by the Agricultural and Livestock Bank.

Cesó su publicación en 1943 cuando el Instituto Nacional del Café fué disuelto por decreto presidencial y sus funciones fueron asumidos por el Banco Agrícola y Pecuario.

Ven-6230.B5 **Boletín Mensual de la Dirección de Ganadería.** 1917- . Caracas. Ministerio de Agricultura y Cría. Dirección de Ganadería.

Monthly journal of the Bureau of Livestock.

Ven-6300.E5 **Estadística del Petróleo.** 1936-38- . Anual. Caracas. Ministerio de Fomento. Dirección General de Estadística y de Censos.

Statistics of petroleum. Annual.

Ven-6400 **Estadísticas Industriales en Venezuela: Métodos y Procedimientos.** Material básico adquirido para el estudio que el Instituto Interamericano de Estadística realiza sobre los métodos de estadísticas industriales en las naciones americanas, incluyendo la encuesta de 1945-46.

Industrial statistics in Venezuela: Methods and procedures. Background material acquired for the study conducted by the Inter American Statistical Institute in methods of industrial statistics in the American nations, including the survey of 1945-46. Consists of: Correspondence; instructions; schedules.

Consiste de: Correspondencia; instrucciones; cédulas.

Ven-6400.M5 1945 **Mining and Manufacturing Industries in Venezuela.** One of a series of reports on mining and manufacturing industries in the American republics. Washington, D.C., 1945. United States Tariff Commission. 31 p.

Folleto sobre Venezuela, de la serie de informes de la Comisión de Aduanas de los Estados Unidos, sobre las industrias mineras y manufactureras en las repúblicas americanas.

Ven-6620.I5 1944 **Indicador Nacional de Vías de Comunicación de Venezuela.** Caracas, 1944. Ministerio de Fomento. Dirección General de Estadística y de Censos. No examinado.

National index of routes of communication in Venezuela.

Ven-6720.B3 **Banco Central de Venezuela: Memoria.** Fecha de iniciación? Anual. Caracas.

Annual report of the Central Bank of Venezuela. Activities of the bank (including assets, liabilities, deposits, transactions, etc.), state of the public treasury, stock market, circulation of money, foreign exchange, wholesale price indexes.

Actividades del banco (incluyendo capitales activos y pasivos, depósitos, transacciones, etc.), situación del tesoro nacional, mercado de valores, circulación monetaria, cambio exterior, índices de precios al por mayor.

1945 **Banco Central de Venezuela: Memoria Correspondiente al Ejercicio Anual 1945.** Caracas, Lito. del Comercio, 1945? Banco Central de Venezuela. 72 p. (aprox.)

*Ven-6720.B5 **Boletín del Banco Central de Venezuela.** 1941- . Trimestral. Caracas.

Quarterly bulletin of the Central Bank of Venezuela. Contains tables of prices and various types of index numbers, articles on monetary circulation, commercial banks, exchange, foreign trade and public finance.

Incluye tablas de precios y varios tipos de números índices, artículos sobre circulación monetaria, bancos comerciales, cambio, comercio exterior y finanzas públicas.

Ven-6900 **Estadísticas del Comercio Exterior en Venezuela: Métodos y Procedimientos.** Material básico adquirido para la encuesta que el Instituto Interamericano de Estadística realiza sobre los métodos de estadísticas del comercio exterior en las naciones americanas.

Foreign trade statistics in Venezuela: Methods and procedures. Background material acquired for the study conducted by the Inter American Statistical Institute in foreign trade statistics in the American nations. Includes: Filled questionnaire, with complementary notes; country code.

Incluye: Cuestionario llenado, con notas complementarias; código de países.

Ven-6900.A5 1913-36 **Análisis del Comercio Exterior.** Caracas, 1938. Ministerio de Relaciones Exteriores.

Analysis of foreign trade. Covers movements of Venezuelan commerce for the years 1913-36, with a special analysis of selected items of export by country of destination and items of import by country of origin.

Contiene el movimiento del comercio venezolano durante los años 1913-36; con especial análisis de ciertos artículos de exportación por países de destino y artículos de importación por países de origen.

*Ven-6900.E5 **Estadística Mercantil y Marítima.** Fecha de iniciación? Anual. Caracas. Ministerio de Fomento. Dirección General de Estadística y de Censos.

Trade and maritime statistics.

Ven-6900.L5 1941 **Ley de Arancel de Aduanas.** Edición especial, contiene todas las resoluciones y reformas efectuadas desde el 23 de octubre de 1936 hasta el 15 de enero de 1941. Caracas, Librería y Editorial "Las Novedades," 1941. 270 p.

Customs tariff law, with revisions.

Ven-6900.R5 1945 **Recent Developments in the Foreign Trade of Venezuela.** One of a series of reports on recent developments in the foreign trade of the American republics. Washington, D.C., 1945. United States Tariff Commission. 83 p.

Folleto sobre Venezuela, de la serie de informes de la Comisión de Aduanas de los Estados Unidos, sobre desarrollos recientes en el comercio exterior de las repúblicas americanas.

7000-7999 HEALTH AND CULTURAL STATISTICS—ESTADISTICAS DE SALUBRIDAD Y CULTURA

Ven-7110.M5 **Memoria y Cuenta del Ministerio de Sanidad y Asistencia Social.** Fecha de iniciación? Anual. Caracas. Ministerio de Sanidad y Asistencia Social.

Annual report of the Department of Health and Public Assistance.

*Ven-7110.R5 **Revista de Sanidad y Asistencia Social.** 1936- . Bimestral. Caracas. Ministerio de Sanidad y Asistencia Social.

Volumes I-II titled—volumenes I-II intitulados: **Boletín del Ministerio de Sanidad y Asistencia Social.**

Bimonthly journal of public health and public assistance.

*Ven-7200.M5 **Memoria que el Ministro de Educación Nacional Presenta al Congreso Nacional.** Fecha de iniciación?

Anual. Caracas. Ministerio de Educación Nacional.

Annual report of the Department of National Education.

1944 **Memoria que el Ministro de Educación Nacional Presenta al Congreso Nacional en Sus Sesiones Ordinarias de 1945; Contiene la Actuación del Despacho en el Año Civil de 1944.** Caracas, Tipo.Americana, 1945. 680 p.

Report of the activities of the Department of National Education during 1944. Includes: School statistics (attendance, expenditures, etc.) for urban and rural primary, secondary and normal schools; statistics on special and technical education; adult education programs; schools of fine arts; statistics on higher education; organization of student welfare; education extension service through correspondence, radio and movies; activities of various academies and museums, and of the National Library; physical education programs; professional welfare activities; breakdown of budget expended during 1944, and comparison with figures of previous years (charts and graphs).

Incluye: Estadística escolar (asistencia, gastos, etc.) de la educación urbana y rural primaria, secundaria y normal; datos sobre educación especial y técnica; programas de educación de adultos; educación artística; estadísticas de educación superior; organización de bienestar estudiantil; extensión pedagógica por correspondencia, radio y cine; actividades de varios museos y academias, y de la Biblioteca Nacional; programas de educación física; actividades de previsión social profesional; inversión de gastos del presupuesto en 1944, con cifras comparativas de años anteriores (cuadros y gráficas).

8000-8999 GOVERNMENT AND POLITICAL STATISTICS—ESTADISTICAS GUBERNAMENTALES Y POLITICAS

Ven-8800.B5 **Boletín Informativo del Ministerio de Hacienda.** 1939- . Mensual. Caracas.

Information and data on the activities and functions of the various divisions of the Department of Finance. Monthly bulletin.

Informaciones y datos sobre las actividades y funciones de las distintas secciones del Ministerio.

Ven-8800.C5 **Cuenta General de Rentas y Gastos Públicos; Bienes Nacionales Inclusive Materias, y Cuenta del Departamento de Hacienda, Rentas y Gastos.** Fecha de iniciación? Anual. Caracas. Ministerio de Hacienda.

Annual containing a general computation of national income and expenditures and national property, by the Department of Finance.

Ven-8800.I5 **Informe de la Contraloría General de la Nación.** Fecha de iniciación? Anual. Caracas. Contraloría General de la Nación.

Report of the General Comptroller of the Nation.

*Ven-8800.M5 **Memoria de Hacienda.** Fecha de iniciación? Anual. Caracas. Ministerio de Hacienda.

Annual report of the Department of Finance. Covers national income, condition of the treasury, public debt, customs tariff and other taxes, etc.

Abarca renta nacional, estado de la tesorería, deuda pública, arancel de aduana y otros impuestos, etc.

1943 **Memoria de Hacienda Presentada al Congreso Nacional en las Sesiones de 1944; Contiene la Gestión del Despacho en el Año Civil de 1943.** Caracas, C.A.Artes Gráf., 1944. 200 p. (aprox.) Páginas numeradas por secciones.

Report of the activity of the Department of Finance for 1943. Includes reports of the Cabinet, the Bureau of Administration, the Bureau of Customhouses, the Bureau of Internal Revenue, the Bureau of the Treasury, the Bureau of Public Credit, and the Bureau of Economy and Finance. Second section, comprised of 90 documents, includes statistical tables.

Incluye informes de la Dirección de Gabinete, la Dirección General de Administración, la Dirección de Aduanas, la Dirección de la Renta Interna, la Dirección del Tesoro, la Dirección de Crédito Público, la Dirección de Economía y Finanzas. Sección segunda, abarcada de 90 documentos, incluye cuadros estadísticos.

Ven-8800.R5 **Revista de Hacienda.** Organo del Ministerio de Hacienda. Semestral. Caracas. Ministerio de Hacienda.

Semiannual journal of the Department of Finance.

Mar. '46 "Resumen de la Primera Parte de la Encuesta sobre las Condiciones de Vida de las Familias de Caracas." Año XI, no. 20, marzo 1946, p. 117-157.

Resumé of the first part of a survey on living conditions of Caracas families.

June 1944, a "Exposición de Motivos del Proyecto de Ley de Presupuesto General de Rentas y Gastos Públicos para el Año Económico 1944-1945." Año IX, no. 16, junio 1944, p. 97-108.

Exposition of purposes of projected law on budget of public income and expenditures for fiscal year 1944-45.

June 1944, b "Movimiento General de las Rentas Nacionales por Años Económicos, desde el 1° de Julio de 1908 hasta el 30 de Junio de 1943." Año IX, no. 16, junio 1944, p. 125-126. Cuadro.

Table showing general movement of national income from July 1, 1908 to June 30, 1943.

June 1944, c "Presupuestos de Ingresos y Gastos Públicos de Venezuela en los Años Económicos de 1935-1936 a 1942-1943." Año IX, no. 16, junio 1944, p. 124.

Budget of public income and expenditures of Venezuela for fiscal years 1935-36 to 1942-43.

June 1944, d "Resumen del Presupuesto General de Rentas y Gastos Públicos que Regirá en el Año Económico de 1° de Julio de 1944 a 30 de Junio de 1945." Año IX, no. 16, junio 1944, p. 109-110.

Summary of the general budget of incomes and expenditures for the fiscal year July 1, 1944-June 30, 1945.

9000-9999 LABOR AND LIVING CONDITIONS — CONDICIONES DE TRABAJO Y DE VIDA

*Ven-9000.M5 **Memoria del Ministerio del Trabajo y de Comunicaciones.** Fecha de iniciación? Anual. Caracas. Ministerio del Trabajo y de Comunicaciones.

Annual report of the Department of Labor and Comunications. Covers the major aspects of employment, labor relations, cost of living and standards for laborers, social security, and postal and other communications.

Contiene los principales aspectos de la ocupación, relaciones del trabajo, costo de la vida y estandards para obreros, previsión social, y comunicaciones (postal y otras).

Ven-9500.I5 1939 **Investigación sobre el Costo de la Vida en Caracas, los Presupuestos Familiares, 1939.** Por José Vandellós. Edición oficial. Caracas, Tipo. Garrido, 1940. Ministerio de Fomento. Dirección General de Estadística. 79 p.

Investigation of the cost of living in Caracas, and analysis of data collected on the family budget. References to similar studies in other countries.

Investigación del costo de la vida y análisis de datos recogidos sobre presupuesto familiar. Referencias a estudios similares efectuados en otros países.

FRANCE

3000-3999 STATISTICAL SCIENCE—CIENCIA ESTADISTICA

Fr-3040.J5 **Journal de la Société de Statistique de Paris.** (Reconnue établissement d'utilité publique par décret du 19 juin 1869). 1860- . Bimestriel. Paris. Société de Statistique de Paris.

Bimonthly journal of the Statistical Society of Paris.

GREAT BRITAIN

3000-3999 STATISTICAL SCIENCE— CIENCIA ESTADISTICA

GrBr-**3010**.G5 **Guide to Current Official Statistics of the United Kingdom.** 1922- . Annual. London. Permanent Consultative Committee on Official Statistics.

Guía de las estadísticas oficiales corrientes del Reino Unido.

GrBr-**3040**.B5 **Biometrika.** A journal for the statistical study of biological problems. 1913- . Quarterly. London.

Una revista trimestral dedicada al estudio estadístico de problemas biológicos.

GrBr-**3040**.J5 **Journal of the Royal Statistical Society.** Starting date? Quarterly. London.

Revista trimestral de la Real Sociedad Estadística.

Part 1, '46, Yates "A Review of Recent Statistical Developments in Sampling and Sampling Surveys." By F. Yates. Vol. CIX, part 1, 1946, p. 12-43.

Includes a discussion of the paper, p. 31-43.

Un tratado acerca de recientes evoluciones estadísticas en el campo de muestras e investigaciones por muestras. Incluye un debate sobre el artículo, p. 31-43.

7000-7999 HEALTH AND CULTURAL STATISTICS — ESTADISTICAS DE SALUBRIDAD Y CULTURA

GrBr-**7300**.A5 **Annals of Eugenics.** A journal devoted to the genetic study of human populations. 1925?- . Quarterly. London?

Anales de eugenesia. Una revista trimestral dedicada al estudio genético de las poblaciones humanas.

WEST INDIES

2000-2999 CENSUS — CENSOS

WI-2200.E5 1943 **Eighth Census of Jamaica and Its Dependencies, 1943: Population, Housing and Agriculture.** Kingston, Jamaica, B.W.I., The Government Printer, 1945. Central Bureau of Statistics—(formerly Census Office). cii and 571 p.

Final report and analysis of data obtained in the 1943 census. Includes: Administrative report; population and housing (including growth, place of birth, nationality, age, conjugal condition, racial origin, literacy, standard of education, school attendance, religious denominations, gainfully occupied and unemployed, earnings of wage earners, housing and households, blind and deaf mutes), agriculture (including classifications by groups and parishes) for Jamaica and for Turks and Caicos Islands and Cayman Islands; appendices (schedules, instructions, census regulations of 1942, circulars, personnel, descriptions and maps of census districts).

Octavo censo de Jamaica y sus dependencias, 1943: Población, vivienda y agricultura. Informe final y análisis de los datos recogidos. Incluye: Informe administrativo; población y viviendas (incluyendo crecimiento, lugar de nacimiento, nacionalidad, edad, estado civil, origen racial, alfabetismo, estandard de educación, asistencia escolar, religión, población activa y desocupada, salarios de obreros, viviendas y hogares familiares, ciegos y sordomudos), agricultura (incluyendo clasificación según grupos y parroquias) para Jamaica y para las Islas Turks e Islas Caicos así como para las Islas Cayman; apéndices (cédulas y formularios, instrucciones, reglamentos del censo de 1942, propaganda, personal, descripciones y mapas de los distritos censales).

WI-2200.E5 1943.1 **Eighth Census of Jamaica, 1943: Administrative Report.** (Part 1 of the census report.) Kingston, Jamaica, The Government Printer, 1945. Census Office, Kingston, Jamaica, B.W.I. 31 p. (approx.)

This report is included in the volume — este informe se incluye en el tomo: **Eighth Census of Jamaica and Its Dependencies, 1943: Population, Housing and Agriculture.**

Report on the methods and procedures used in taking the 1943 Census of Jamaica. Includes schedules used to collect the data on population, agriculture, livestock, housing, etc., and punched tabulation card models.

Informe administrativo sobre métodos y procedimientos del octavo censo de Jamaica, levantado en 1943. Incluye cédulas y formularios empleados en la colección de datos sobre población, agricultura, ganadería, viviendas, etc., así como modelos de las tarjetas perforadas usadas en la tabulación.

WI-2200.E5 1943.2 **Eighth Census of Jamaica, 1943: Introduction to Population and Housing.** (Part 2 of the census report). Kingston, Jamaica, The Government Printer, 1945. Census Office, Kingston, Jamaica, B.W.I. 60 p. (approx.)

This report is included in the volume — este informe se incluye en el tomo: **Eighth Census of Jamaica and Its Dependencies, 1943: Population, Housing and Agriculture.**

Data and some comparative analysis.

Introducción a los datos sobre población y viviendas, recogidos en el octavo censo de Jamaica, levantado en 1943. Datos, y algunas análisis comparativos.

6000-6999 ECONOMICS — ECONOMIA

WI-6000.D5 **Development and Welfare in the West Indies, Bulletins.** Starting date? Bridgetown, Barbados, Advocate Company, Ltd.

Desarrollo y bienestar en las Antillas.

WI-6100.N3 1942 **The National Income of Barbados, 1942.** By Frederic Benham. Bridgetown, Barbados, Advocate Co., Ltd., 1944. 24 p. Development and welfare in the West Indies, bulletin no. 9. Unexamined.

La renta nacional de Barbados, 1942.

WI-6100.N5 1942 **The National Income of Jamaica, 1942.** By Frederic Benham. Bridgetown, Barbados, Advocate Co., Ltd., 1944. 29 p. Development and welfare in the West Indies, bulletin no. 5.

La renta nacional de Jamaica, 1942.

WI-6100.N7 1942 **The National Income of St. Vincent, 1942.** By Frederic Benham. Bridgetown, Barbados, Advocate Co., Ltd., 1944. 6 p. Development and welfare in the West Indies, bulletin, no. 8.

La renta nacional de San Vicente, 1942.

NON-GEOGRAPHIC

1000-1999 GENERAL

NonG-**1600.**R5 1943 **The Repair and Preservation of Records.** By Adelaide E. Minogue. Washington, D.C., U.S. Govt.Print.Off., 1943. National Archives. 56 p. Bulletin, no. 5.

La reparación y conservación de documentos.

NonG-**1600.**S5 1937 **Summary Report of National Bureau of Standards Research on Preservation of Records.** By A. E. Kimberly and B. W. Scribner. Washington, D.C., U.S.Govt.Print.Off., 1937. Department of Commerce. National Bureau of Standards. 28 p. Miscellaneous publication, no. M144.

Informe sumario de la investigación sobre conservación de documentos, efectuada por el Departamento Nacional de Estandards.

NonG-**1600.**S7 1939 **Summary Report of Research at the National Bureau of Standards on the Stability and Preservation of Records on Photographic Film.** By B. W. Scribner. Washington, D.C., U.S.Govt. Print.Off., May 1939. Department of Commerce. National Bureau of Standards. 17 p. Miscellaneous publication, no. M162.

Informe sumario de la investigación en el Departamento Nacional de Estandards sobre la estabilidad y conservación de documentos en películas fotográficas.

NonG-**1800.**M5 1936 **Manual on Methods of Reproducing Research Materials.** A survey made for the Joint Committee on Materials for Research of the Social Science Research Council and the American Council of Learned Societies. Robert C. Binkley. Ann Arbor, Michigan, Edwards Brothers, Inc., 1936. 206 p.

Manual sobre métodos de reproducción de materiales de investigación.

Non-**1940.**E5 1937 **Encyclopaedia of the Social Sciences.** New York, The Macmillan Co., 1937. 8 vol.

Enciclopedia de las ciencias sociales.

3000-3999 STATISTICAL SCIENCE—CIENCIA ESTADISTICA

NonG-**3005.S4** 1939 **Statistical Dictionary of Terms and Symbols.** By Albert Kenneth Kurtz and Harold A. Edgerton. New York, J. Wiley and Sons, Incorporated; London, Chapman and Hall, Limited, 1939. 191 p.

"Does not purport to be an attempt to standardize terminology but rather to present the usual meanings attached to statistical terms and symbols." (Source: Preface.)

Diccionario estadístico de términos y símbolos. "No tiene por fín la estandardización de la terminología sino que presenta el significado común que se le da a los términos y símbolos estadísticos." (Traducción del comentario.)

NonG-**3010.**S5 1941 **The Second Yearbook of Research and Statistical Methodology.** Edited by Oscar Krisen Buros. Highland Park, New Jersey, The Gryphon Press, 1941. 383 p.

Consists of reviews of books quoted from leading periodicals. The first yearbook in this series edited by Oscar Krisen Buros was published under the title **Research and Statistical Methodology, Books and Reviews of 1933-38.**

El segundo anuario de investigación y metodología estadística. Consiste de revistas de libros, tomadas de periódicos importantes. El primer anuario de esta serie editada por Oscar Krisen Buros, fué publicado bajo el título **Research and Statistical Methodology, Books and Reviews of 1933-38.**

NonG-**3040.**A3 **The Annals of Mathematical Statistics.** The official journal of the Institute of Mathematical Statistics. 1930- . Quarterly. Baltimore, Maryland.

Anales de estadística matemática. Publicación del Instituto de Estadística Matemática.

Mar. '46 "Constitution of the Institute of Mathematical Statistics."
Vol. XVII, no. 1, March 1946, p. 117-121.

NonG-**3100.**T5 1935 **Teoría de la Correlación.** Por Carlos E. Dieulefait. Rosario, Argentina, Establ.Gráf.Pomponio, 1935. Universidad Nacional de Litoral. Facultad de Ciencias Económicas, Comerciales y Políticas. Instituto de Estadística. 73 p.

Theory of correlation.

NonG-**3105.**I5 1944 **An Introduction to Statistical Analysis,** By Clarence Hudson Richardson. Rev. ed. New York, Harcourt, Brace and Company, 1944. 498 p.

Una introducción al análisis estadístico.

NonG-**3105.**S4 1943 **Statistical Adjustment of Data.** By W. Edwards Deming. New York, John Wiley & Sons, Inc.; London, Chapman & Hall, Ltd., 1943. 261 p.

Contents: Some simple adjustments (the meaning of adjustment, simple illustrations of curve fitting); least squares solution of more complicated problems (propagation of error, the general problem in least squares); conditions without parameters (geometric conditions, systematic computation for geometric conditions, adjusting sample frequencies to expected marginal totals); conditions containing parameters (curve fitting in more complicated circumstances, systematic computation for fitting curves by least squares); exercises and notes (exercises in fitting various functions, four examples in curve fitting); appendix (tables for making random observations for class illustration).

Ajuste estadístico de datos. Contenido: Algunos ajustamientos simples (el significado de ajuste, ilustración sencilla sobre ajuste de curvas); solución por mínimos cuadrados de problemas mas complejos (propagación del error, el problema general de los mínimos cuadrados); condiciones sin parámetros (condiciones geométricas, el ajuste de frecuencias de muestras a totales marginales previstos); condiciones conteniendo parámetros (ajuste de curvas en circunstancias mas complejas, cómputo sistemático para el ajuste de curvas por mínimos cuadrados); ejercicios y notas (ejercicios de ajuste de varias funciones, cuatro ejemplos de ajuste de curvas); apéndice (cuadros para hacer observaciones al azar como ilustración en clase).

NonG-**3110**.A5 1936 **Accounting Control.** New York, 1936. International Business Machines Corporation. 8 p. Machine methods of accounting, no. AM-8.

Métodos Mecánicos de control de contabilidad.

NonG-**3110**.B5 1944 **Bibliography: Use of IBM Electric Accounting Machines and Test Scoring Machine in Mathematics, Statistics and Scientific Research.** September 1, 1944. New York, 1944. International Business Machines Corporation. 4 p. and annex, **Additions — 1944,** 4 p., and annex, **Supplementary Bibliography,** 6 p.

Bibliografía sobre empleo de máquinas eléctricas IBM de contabilidad y máquinas para marcar los resultados de exámenes, en matemática, estadística e investigaciones científicas.

NonG-**3110**.C5 1943 **Consumer and Opinion Research.** The questionnaire technique. By Albert B. Blankenship. New York and London, Harper & Brothers Publishers, 1943. 238 p.

Contents: Introduction (origin, development, scope of questionnaire surveys); the general technique of conducting surveys; statement of the problem; method of collecting the data (mail, telephone, and personal interviews); preliminary phrasing of the questions; preliminary assembly of the questionnaire; testing the questionnaire; securing a cross-section (the random sample; the controlled or stratified sample; checking the returns); the size of the sample; preparing and distributing materials; the process of interviewing; summarizing the results; preparing the report; measuring the adequacy of surveys; the case against the questionnaire survey.

Investigaciones sobre opinión pública y consumidores: La técnica del cuestionario. Contiene: Introducción (origen, desarrollo, alcance de las investigaciones del cuestionario); la técnica general de la ejecución de las investigaciones; exposición del problema; método para la obtención de los datos (correo, teléfono y entrevistas personales); redacción preliminar de las preguntas; preparación del cuestionario; ensayo del cuestionario; elección de una sección transversal (la muestra obtenida por elección al azar; la muestra por elección controlada o estratificada; comprobación de los datos obtenidos); la magnitud de la muestra; preparación y distribución de materiales; el proceso de entrevistar; resumen de los resultados; preparación del informe; análisis del valor práctico de las investigaciones; el caso en contra a la investigación del cuestionario.

NonG-**3110**.D4 1936 **The Design of Tabulating Cards.** New York, 1936. International Business Machines Corporation. 16 p. Machine methods of accounting, no. AM-4-1.

La proyección de las fichas tabuladoras.

NonG-**3110**.E5 1936 **The Electric Accounting Machine Card.** New York, 1936. International Business Machines Corporation. 20. p. Machine methods of accounting, no. AM-3-2.

La ficha de la máquina eléctrica de contabilidad.

NonG-**3110**.H5 1942 **How to Use Crank-Driven Calculators.** A twenty-unit course. By Albert Stern and Mary Stuart. New York, Chicago, etc., The Gregg Publishing Company, 1942. 92 p.

Un curso en el uso de máquinas calculadoras a manubrio.

NonG-**3110**.M3 **Machine Methods of Accounting.** 1936- . Irregular. New York. International Business Machines Corporation.

Series of monographs.

Métodos mecánicos de contabilidad. Serie de monografías.

NonG-**3110.**M5 1940 **Manual of Electric Punched Card Accounting Machines.** New York, 1940. International Business Machines Corporation. Electric Accounting Machines Division. Loose-leaf. Unexamined.

Manual de las máquinas eléctricas de contabilidad de fichas perforadas.

NonG-**3110.**O5 1936 **Organization and Supervision of the Tabulating Department.** New York, 1936. International Business Machines Corporation. 16 p. Machine methods of accounting, no. AM-6.

Organización y dirección del departamento de tabulación.

NonG-**3110.**P1 1935 **Practical Applications of the Punched Card Method in Colleges and Universities.** Edited by G. W. Baehne. Morningside Heights, New York, Columbia University Press, 1935. 442 p.

A compilation of contributed articles by various authors, grouped under the following headings: Development and principles of the punched card method; applications in the registrar's office; applications in the university business offices; miscellaneous administrative applications; applications in psychology and educational research; applications in medical research and in hospitals; applications in legal research; applications in agricultural research; miscellaneous research applications; applications of specific methods to the solution of statistical problems.

Aplicaciones prácticas del método de la tarjeta perforada en colegios y universidades. Una compilación de artículos contribuidos por varios autores agrupados bajo los títulos siguientes: Desarrollo y principios del método de la ficha perforada; aplicaciones en la oficina del registrador; aplicaciones en las oficinas comerciales de la universidad; aplicaciones administrativas misceláneas; aplicaciones en psicología y en la investigación pedagógica; aplicaciones en investigaciones médicas y en hospitales; aplicaciones en investigaciones legales; aplicaciones en investigaciones agrícolas; aplicaciones misceláneas en investigaciones; aplicaciones de métodos específicos a la solución de problemas estadísticos.

NonG-**3110.**P5 1942 **Principles of Punch-Card Machine Operation.** How to operate punch-card tabulating and alphabetic accounting machines. By Harry Pelle Hartkemeier. New York, Thomas Y. Crowell Company, 1942. 26 p.

Principios del manejo de las máquinas de fichas perforadas.

NonG-**3110.**P6 1943 **Principles of Operation of the Alphabetical Accounting Machine.** Type 405. Revised January 1, 1943. International Business Machines Corporation. Educational Department. 90 p.

Principios del funcionamiento de la máquina alfabética de contabilidad. Tipo 405.

NonG-**3110.**S5 1943 **Statistics: Collecting, Organizing, and Interpreting Data.** By Raleigh Schorling, John R. Clarck, and Frances G. Lankford. Yonkers, New York, World Book Co., 1943. 76 p.

A monograph designed for use as a high-school text.

Estadística: Recolección, organización e interpretación de datos. Una monografía designada para uso como texto en la escuela secundaria.

NonG-**3110.**T5 1942 **Terminología, Inglés-Español, Español-Inglés, Contiene las Voces y Expresiones mas Usuales con Respecto a las Máquinas Eléctricas de Contabilidad "International", Sus Funciones y Su Uso.** Buenos Aires, 1942. International Business Machines Corporation. 72 p.

English-Spanish and Spanish-English terminology of terms and expressions most commonly used in connection with "International" electric accounting machines, their functions and use.

NonG-**3120.**E5 1943 **Elementos Matemáticos para el Análisis de las Series de Tiempo.** Por Alfredo Fernández. Montevideo, 1943. 89 p. Cuaderno, no. 2.

Reprint from—reimpresión de: **Revista de la Facultad de Ciencias Económicas y de Administración,** no. 4, año 1943.

Elements of mathematics for analysis of time series.

NonG-**3120.**M5 1938 **The Making and Using of Index Numbers.** By Wesley C. Mitchell. Washington, D.C., U.S.Govt. Print.Off., 1938. U. S. Department of Labor. Bureau of Labor Statistics. 114 p. Bulletin, no. 656.

La construcción y uso de números índices.

NonG-**3120.**N5 1945 **Números Indices.** Naturaleza de los números índices; números índices de precios al por mayor; otras clases de números índices. La Habana, Tall. Tipo.de Editorial Lex, 1945. 61 p.

Index numbers. Nature of index numbers; index numbers of wholesale prices; other kinds of index numbers. This work is part of the new edition of **Estadística Comercial,** by Ernesto Pino, which is used as a textbook in the School of Commerce of the University of Habana.

Esta obra es parte de la nueva edición de **Estadística Comercial,** por Ernesto Pino, que es usada como libro de texto de la Escuela de Comercio de la Universidad de La Habana.

NonG-**3130.**N5 1938 **Notes on Statistical Mapping with Special Reference to the Mapping of Population Phenomena.** By John Kirtland Wright, Lloyd A. Jones, Leonard Stone, and T. W. Birch. New York, Washington, 1938. American Geographical Society; Population Association of America. 37 p.

Apuntes sobre la cartografía estadística, con especial referencia a la cartografía de los fenómenos de población.

NonG-**3140.**A6 1943 **The Advanced Theory of Statistics.** By Maurice G. Kendall. London, Charles Griffin, 1943. 457 p. Vol. I.

La teoría estadística avanzada.

NonG-**3140.**A8 1946 **Algunos Problemas de Inferencia Estadística.** Por S. C. Mazza. Montevideo, Casa Imp. "La Industrial", 1946. Universidad de la República, Uruguay. Facultad de Ciencias Económicas y de Administración. Instituto de Estadística. 23 p. Cuaderno, no. 5.

Extract from—apartado de la: **Revista del Centro de Estudiantes de Ciencias Económicas y de Administración,** vol. II, no. 3, December 1945.

Problems of statistical inference; covers mathematical aspects of sampling, testing, and estimating, as applied to statistics of population.

Trata de algunos aspectos matemáticos de los métodos de muestra, prueba y estimación, aplicados a las estadísticas de población.

NonG-**3140.**E5 1936 **Elementos de Estadística Matemática.** Por C. Charliér. Versión castellana de José González Galé. Buenos Aires, Imp.del Boletín Matemático, 1936. Universidad Nacional de Buenos Aires. Facultad de Ciencias Económicas. Instituto de Biometría. 95 p. Cuadernos de trabajo, no. 1.

Elements of mathematical statistics.

NonG-**3140.**E7 1939 **Estadística Matemática.** Por Georges Darmois. Versión castellana de los actuarios Emilia Amelia Journet y José F. Domínguez, con prólogo del Dr. José González Galé. Buenos Aires, 1939. Universidad Nacional de Buenos Aires. Facultad de Ciencias Económicas. Instituto de Biometría. 285 p. Cuaderno de trabajo, no. 8.

Mathematical statistics.

NonG-**3140.**H5 1924 **Handbook of Mathematical Statistics.** By Henry Lewis Rietz, editor-in-chief, H. C. Carver, A. R. Crathorne, W. L. Crum, James Glover, E. V. Huntington, Truman L. Kelley, Warren M. Persons, Allyn A. Young, members of the Committee on the Mathematical Analysis of Statistics of the Division of Physical Sciences of the National Research Council. Boston, New York, etc., Houghton Mifflin Company, 1924. 221 p.

A convenient source of reference for mathematical formulas.

Manual de estadística matemática. Una fuente útil de referencia sobre fórmulas matemáticas.

NonG-**3140.**I5 1938 **Interpolación y Ajustamiento de la Curva Logística Generalizada.** Por J. Barral Souto. Buenos Aires, Imp. de la Universidad, 1938. Universidad de Buenos Aires. Facultad de Ciencias Económicas. Instituto de Biometría. 16 p. Cuadernos de trabajo, no. 7.

Interpolation and adjustment of the generalized logistic curve.

NonG-**3140.**M5 1927 **Mathematical Statistics.** By Henry Lewis Rietz. Chicago, Illinois, Open Court Publishing Company, 1927. Mathematical Association of America. 181 p. The Carus mathematical monographs, no. 3.

A brief treatment of the basic mathematical theory for non-technical readers.

La estadística matemática. Un tratado breve sobre la teoría matemática básica para el público no técnico.

NonG-**3140.**R5 1943 **Random Sampling Distributions.** By Alan Edward Treoar. Minneapolis, Minnesota, Burgess Publishing Co., 1943. 94 p.

Distribuciones del muestreo al azar.

NonG-**3140.**S5 1941 **Statistical Deflation in the Analysis of Economic Series.** By John Henry Smith. Chicago, Illinois, 1941. 123 p.

La deflación estadística en el análisis de series económicas.

NonG-**3140.**S7 1940 **Suma de Productos Formados por Combinaciones con Números Dados.** Por José Barral Souto. Buenos Aires, 1940. Universidad de Buenos Aires. Facultad de Ciencias Económicas. Instituto de Biometría. 22 p. Cuaderno de trabajo, no. 9.

Sum of products formed by combinations of given numbers.

NonG-**3140**.T5 1939 **Tests of Significance, What They Mean and How to Use Them.** By John Henry Smith. Chicago, Illinois, The University of Chicago Press, 1939. 90 p. Studies in business administration, vol. X, no. 1. Unexamined.

Pruebas para determinar la significación, lo que ellas representan, y el modo de usarlas.

NonG-**3150**.R5 1927 **Random Sampling Numbers.** Arranged by Leonard Henry Caleb Tippett. Cambridge, England, University Press, 1927. University of London. University College. Department of Applied Statistics. 34 p. Tracts for computers, no. XV.

Números de muestreo al azar.

NonG-**3150**.S3 "Sample Surveys in Census Work." By Philip M. Hauser and Morris H. Hansen. Washington, D.C., 194-? United States. Department of Commerce. Bureau of the Census. 12 p. Hectographed.

Investigación por muestra en trabajos censales.

NonG-**3150**.S5 1944 **Sampling Inspection Tables: Single and Double Sampling.** By Harold F. Dodge and Harry G. Romig. New York, John Wiley & Sons, Inc.; London, Chapman & Hall, Ltd. 106 p.

"The sampling inspection tables presented in this book were developed for use in the manufacture of communication and apparatus for the Bell Telephone System. . . . Chapter I outlines some of the factors to be considered in setting up inspection plans and develops a basis for minimizing the amount of inspection. Chapter II covers double sampling, the 'average outgoing quality limit' . . . and the mathematical background of the tables. Chapter III . . . outlines the shop procedure for applying the tables." (Source: Preface.)

"Cuadros de inspección del muestreo: Muestreo simple y doble. Los cuadros de inspección del muestreo presentados en este libro fueron creados para el uso en la manufactura de medios de comunicación y aparatos de la Bell Telephone System . . . Capítulo I es un bosquejo sobre algunos de los factores a considerarse en la fijación de los planes de inspección, y desarrolla una base para reducir al mínimo la labor de inspección. Capítulo II abarca el muestreo doble, el límite de calidad media de los productos examinados . . . y la base matemática de los cuadros. Capítulo III resume el procedimiento para aplicación de los cuadros en el taller." (Traducción del comentario.)

NonG-**3150**.S7 1945 **Sampling Statistics and Applications.** By James G. Smith and Acheson J. Duncan. New York, McGraw-Hill Book Company, Inc., 1945. 696 p. **Fundamentals of the Theory of Statistics.** vol. II.

Contains: **General theory of frequency curves:** Definitions and basic concepts, probability and the probability calculus, the symmetrical binomial distribution and the normal curve, the Pearsonian system of frequency curves, the Gram-Charlier system of frequency curves, summary of the theory of frequency curves and some samples, numerical calculations for frequency curves; **elementary theory of random sampling:** A preview of sampling theory, sampling from a discrete twofold population, sampling from continuous normal populations, use of the sampling distributions, sampling fluctuations in correlation statistics; **advanced sampling problems:** Sampling from a discrete manifold population, joint sampling fluctuations in mean and standard deviation, sampling fluctuations in regression statistics, problems involving two samples, analysis of variance, the problem of nonnormality.

La estadística de muestreo y sus aplicaciones. Contiene: **Teoría general de curvas de frecuencia:** Definiciones y conceptos básicos, probabilidad y el cálculo de probabilidad, la distribución simétrica binomial y la curva normal, sistema de Pearson de curvas de frecuencia, sistema de Gram-Charlier de curvas de frecuencia, resumen de la teoría de curvas de frecuencia y algunas muestras, cálculos numéricos para curvas de frecuencia; **teoría elemental de muestreo al azar:** Un tratado preliminar sobre la teoría del muestreo, el muestreo aplicado a un universo discreto de dos características, el muestreo aplicado a una población continua normal, uso de las distribuciones de muestreo, fluctuaciones de muestreo en estadísticas de correlación; **problemas avanzadas del método de muestreo:** Muestreo aplicado a un universo discreto multiple, fluctuaciones combinadas de muestreo en la media aritmética y en la desviación standard, fluctuaciones de muestreo en estadísticas de regresión, problemas relacionados con dos muestras, análisis de variación, el problema de no-normalidad.

NonG-**3150**.S8 1942 **Sobre la Teoría de las Muestras.** Por Sigfrido C. Mazza. Montevideo, 1942. Universidad de la República. Facultad de Ciencias Económicas y de Administración. Instituto de Estadística. 19 p. Cuaderno no. 1. Reimpresión de la **Revista de la Facultad de Ciencias Económicas y de Administración,** no. 4, año 1942.

On the theory of sampling.

NonG-**3150.**T4 1939 **Tables of Random Sampling Numbers.** By Maurice George Kendall and Bernard Babington Smith. Cambridge, England, University Press, 1939. University of London. University College. Department of Applied Statistics. 60 p. Tracts for computers, no. XXIV.

Tablas de números del muestreo al azar.

NonG-**3150.**T8 1942 **The Theory of Sampling.** With special reference to the collection and interpretation of agricultural statistics. By Walter A. Hendricks. Raleigh, North Carolina, 1942. U. S. Department of Agriculture, Bureau of Agricultural Economics, and North Carolina State College of Agriculture and Engineering, Department of Experimental Statistics cooperating. 122 p. Mimeographed.

Includes: Some principles of notation, summation signs, averages, frequency distributions, measurement of variation, samples and populations, unbiased estimates, concept of probability, sample means as estimates of populations means, variance, pooled variance and the significance of the difference between two averages, analysis of variance, short cuts in computation, application of analysis of variance to sampling problems, some general principles of sampling, random sampling, stratified sampling, subsampling, fiducial limits for means from stratified samples and subsamples, sampling units and expansion factors, linear regression and correlation, multiple regression and multiple correlation, joint regression equations.

La teoría del método de la muestra, con referencia especial a la recollección e interpretación de estadísticas agrícolas. Incluye: Algunos principios de notación, signos de suma, promedios, distribución de frecuencias, medida de variación, muestras y poblaciones, estimaciones imparciales, concepto de probabilidad, medias de muestra como estimaciones de medias de población, variación, variación combinada y la significación de la diferencia entre dos promedios, análisis de variación, métodos cortos en el cálculo, aplicación del análisis de variación a problemas del método de muestra, algunos principios generales del método de muestra, obtención de muestras al azar, al método estratificado de muestra método de la sub-muestra, límites fiduciales para medias de muestras estratificadas y sub-muestras, unidades de muestreo y factores de expansión, regresión lineal y correlación, regresión múltiple y correlación múltiple, ecuaciones conjuntas de regresión.

NonG-**3160.**A5 1945 **A.S.T.M. Manual on Presentation of Data.** Including supplement A, Presenting + limits of uncertainty of an observed average; supplement b, "Control chart" method of analysis and presentation of data; and tables of squares and square roots. Sponsored by Committee E-1 on Methods of Testing. Philadelphia, 1945. American Society for Testing Materials. 73 p. Unexamined.

"This is the latest printing of the well known manual used by engineers for the past twelve years . . . The main body of this manual . . . explains the presentation and interpretation of frequency distributions . . ." (Source: **Journal of the American Statistical Association,** vol. 40, no. 231, September 1945.)

Manual del A.S.T.M. (Sociedad Americana para Exámenes de Materiales) sobre la presentación de datos. Incluye suplemento A, presentando + límites de incertidumbre de un promedio observado; suplemento B, método de diagrama de control de análisis y presentación de datos; y tablas de números cuadrados y raices cuadradas. "Esta es la impresión más reciente del manual popular usado por ingenieros durante los últimos doce años . . . La parte principal de este manual . . . explica la representación e interpretación de distribuiciones de frecuencias . . ." (Traducción del comentario.)

NonG-**3160.**C5 1942 **Control Chart Method of Controlling Quality during Production.** New York, 1942. American Standards Association. 41 p. American war standards, ZI.3-1942. Unexamined.

Método del diagrama de control en la inspección de calidad durante la producción.

NonG-**3160.**E5 1931 **Economic Control of Quality of Manufactured Product.** By Walter Andrew Shewhart. New York. D. Van Nostrand Company, Inc., 1931. 501 p.

Control económico de la calidad de productos manufacturados.

NonG-**3160.**G5 1941 **American War Standards: Guide for Quality Control and Control Chart Method of Analyzing Data.** Approved May, 1941. New York, 1941. American Standards Association 66 p. ASA 21.1-1941, 21.2-1941.

Two separate monographs in the American War Standards series, and an appendix: " 'Control Chart' Method of Analysis and Presentation of Data."

Standards de guerra estadounidenses: Guía para control cualitativo y método del diagrama de control para análisis de datos. Dos monografías por separado en las series de standards de guerra estadounidenses, y un apéndice sobre 'diagrama de control' método de análisis y presentación de datos.

NonG-**3160**.S4 1936 **Statistical Tables; Their Structure and Use.** By Helen Mary Walker and Walter N. Durost. New York, Teachers College, Columbia University, 1936. 76 p.

Cuadros estadísticos; su construcción y uso.

NonG-**3160**.S5 1939 **Statistical Method from the Viewpoint of Quality Control.** By Walter Andrew Shewhart . . . with editorial assistance of W. Edwards Deming. Washington, D.C., 1939. United States. Department of Agriculture. Graduate School. 155 p.

A series of four lectures.

Método estadístico desde el punto de vista del control cualitativo. Una serie de cuatro lecciones.

NonG-**3260**.A5 1943 **American Standard Engineering and Scientific Graphs for Publications.** Approved by American Standards Association January, 1943. New York, the American Society of Mechanical Engineers, 1943. The American Society of Mechanical Engineers. 27 p. American Standard ASA Z15,3-1943.

Includes: Design and layout; construction of original graph to be used for reproduction (including discussion of paper and ink, size, appearance, lettering); handling of finished illustrations.

Estandards americanos para gráficos de ingeniería y científicos para uso en publicaciones. Aprobados por la Asociación Americana de Estandards en enero, 1943. Incluye: Proyecto y disposición; construcción del gráfico original para reproducción (incluyendo discusión de papel y tinta, dimensiones, apariencia, inscripción); manejo de ilustraciones terminadas.

NonG-**3260**.G5 1939 **Graphic Presentation.** By Willard Cope Brinton. New York, Brinton Associates, 1939. 511 p.

Contents: Introduction (history of development of graphic methods); graphic narrative; tabulation; classification charts; geneology and genetics charts; organization charts; relationship charts; flow charts; sector charts; bar charts of various kinds; maps of various kinds; rating charts; chronology charts; progress charts; curve charts; frequency charts; correlation charts; Ogive and Lorenz charts; ratio charts; three-dimensional methods; composite charts; suggestions for making a chart; standards for time series charts; the camera and its use; lantern slides; preparation of illustrations; color and its use; methods of reproducing; methods of printing; selection of paper; binding techniques; graphic charts in advertising; quantitative cartoons; quantitative posters; displays and exhibits; dioramas; graphic charts in conference rooms.

Representación gráfica. Contiene: Introducción (historia de la evolución de los métodos gráficos); narración gráfica; tabulación; gráficos de clasificación; gráficos genealógicos y genéticos; gráficos de organización; gráficos de relación; esquemas de disposición de trabajo; diagramas seccionales; gráficos de barras de varios tipos; varios tipos de mapas; gráficos de tasas; diagramas cronológicos, diagramas de progreso; diagramas de curvas; diagramas de frecuencia; diagramas de correlación; curvas ojivas y curvas de Lorenz, gráficos de proporción; métodos tridimensionales; diagramas compuestos; sugestiones para la confección de diagramas; standards para gráficos de series cronológicas; la cámara y sus usos; diapositiva; preparación de ilustraciones; color y sus usos; métodos de reproducción; métodos de imprimir; selección de papel; técnica de encuadernación; gráficos en los anuncios; diagramas cuantitativos; afiches cuantitativos; exposiciones y exhibiciones; dioramas; diagramas en la sala de conferencia.

NonG-**3260**.G7 1940 **Graphs: How to Make and Use Them.** By Herbert Arkin and Raymond R. Colton. Revised edition. New York and London, Harper & Brothers Publishers, 1940. 236 p.

Contents: The principle of the graph; construction of the graph; graphic layout; equipment for graphic presentation; line graph; ratio chart; bar chart; area and solid diagrams; graphing relationships; map graphs; organization charts; graphs for computations; control charts; reproduction of graphs; statistical tables and reports.

Gráficos: Su confección y modo de usarlos. Contiene: Los principios del gráfico; confección del gráfico; planeamiento gráfico; equipo empleado en la representación gráfica; diagrama de lineas; gráfico de proporción; gráfico de barras; gráficos de superficie y de volumen; representación gráfica de relaciones; mapas gráficos; cuadros de organización; diagrama de computación; gráficos de control; reproducción de los gráficos; tablas e informes estadísticos.

NonG-**3260.**H4 1932 **Handbook of Statistical Monographs, Tables, and Formulas.** By Jack W. Dunlap and Albert K. Kurtz. Yonkers-on-Hudson, New York, World Book Company, 1932. 163 p.

Manual de monografías, tablas y fórmulas estadísticas.

NonG-**3260.**H6 1937 **How to Use Pictorial Statistics.** By Rudolf Modley. New York and London, Harper and Brothers, 1937. 170 p. Unexamined.

"A guide to the construction and use of pictographs." (Source: Croxton and Crowden, **Applied General Statistics,** 1943.)

Como se usan estadísticas pictográficas. "Una guía de la construcción y del uso de pictográficos." (Traducción del comentario.)

NonG-**3260.**P5 **The Preparation of Statistical Tables: A Handbook.** Washington, D.C. U. S. Department of Agriculture. Bureau of Agricultural Economics. Unexamined.

"A mimeographed set of recommendations, designed to promote uniformity of tables, prepared by the Bureau of Agricultural Economics." (Source: Croxton and Crowden, **Applied General Statistics,** 1943.)

La preparación de cuadros estadísticos—un manual. "Una colección mimeografiada de recomendaciones dedicadas al fomento de uniformidad de cuadros, preparada por la Dirección de Economía Agrícola." (Traducción del comentario.)

NonG-**3260.**S3 1944 **Standards of Presentation.** (Tabular). Published statistical reports of headquarters, ASF. Washington, D.C., May 1944. United States. Army Service Forces. Control Division. Statistics and Progress Branch. 40 p.

Contents: General considerations, format, text presentation, tabular presentation, general presentation. "This volume contains standard procedures and specifications for use in the preparing of published statistical reports at Headquarters, ASF. In addition to presenting detailed instructions concerning style and form of presentation, it is also intended to serve as a guide in planning the contents of the report in such a way as to tell most effectively the story of activities in a particular field." (Source: Foreword.)

Estandards de presentación. Contenido: Consideraciones generales, formato, presentación del texto, presentación tabular, presentación general. "Este volumen contiene procedimientos y especificaciones estandards para uso en la preparación de informes estadísticos publicados en la Oficina Central de Servicios del Ejército. Además de dar instrucciones detalladas sobre estilo y forma de presentación, intenta también servir de guía para la disposición del contenido de manera tal que se muestre de la manera mas efectiva la historia de las actividades en un campo particular." (Traducción del comentario.)

NonG-**3260.**S7 1938 **Statistical Tables for Biological, Agricultural and Medical Research.** By R. A. Fisher and F. Yates. London, Oliver and Boyd, 1938. 90 p.

Cuadros estadísticos para investigaciones biológicas, agrícolas y médicas.

NonG-**3260.**T5 1938 **Time-Series Charts.** A manual of design and construction. American Standard. Approved by American Standards Association, November 1938. New York, 1938. Prepared by Committee on Standards for Graphic Presentation. The American Society of Mechanical Engineers, sponsor body. 68 p. ASA Z15.2-1938.

Diagramas de series cronológicas. Un manual de proyecto y construcción. Aprobado por la Asociación Americana de Estandards, noviembre 1938.

NonG-**3330.**Almendras 1945 **Estadística General y Matemática.** Por Domingo Almendras. Chile, 1945. Universidad de Chile. Escuela de Economía y Comercio. 336 p.

General and mathematical statistics. Contents: Definitions, historical review, the field of statistics and its importance; divisions of statistics and compilation of data; characteristic values; and an extensive chapter on graphic presentation.

Contiene: Definiciones, reseña histórica, importancia y campo de la estadística; divisiones de la estadística y recopilación de datos; números característicos; y un extenso capítulo sobre representación gráfica.

NonG-**3330.**Arenas Paz 1935 **Manual de Estadística y de Información General.** Con cifras comparativas. I. Estadística metodológica. II. Estadística aplicada. III. Estadística e información gráfica. (450 páginas, 75 cuadros numéricos, 25 cuadros sinópticos, 73 diagramas.) Por Belisario Arenas Paz. Bogotá, Escuelas Salesianas de Tipografía y Zincograbado, 1935. 469 p.

Manual of statistics and general information with comparative figures. Includes: Statistical methodology; applied statistics; statistics and graphic information.

NonG-**3330**.Arkin 1939 **An Outline of Statistical Methods.** As applied to economics, business, education, social and physical sciences, etc. By Herbert Arkin and Raymond R. Colton. Fourth edition. New York, Barnes & Noble, Inc., 1939. 224 and 47 p. College outline series.

Includes: Statistical series; frequency distribution and its analysis—central tendency and dispersion and skewness; time series analysis—trend and seasonal and cyclical analysis; correlation; the normal curve; theory of sampling; index numbers; collection of data; statistical tables; graphic presentation; special techniques in education, psychology, and biology; tables of squares, square roots, cubes, and cube roots.

Reseña de metodología estadística. Incluye: Series estadísticas; distribución de frecuencias y su análisis—tendencia central y dispersión y asimetría; análisis de series cronológicas—análisis de tendencias, de fenómenos estacionales y ciclos; correlación; la curva normal; teoría del método de la muestra; números índices; recopilación de datos; cuadros estadísticos; representación gráfica; técnicas especiales en la enseñanza, psicología y biología; tablas de números cuadrados, raíces cuadradas, cubos, y raíces cúbicas.

NonG-**3330**.Arrús 1945 **Estadística Metodológica.** Por Oscar F. Arrús. Apuntes tomados de las clases dictadas en la Facultad de Ciencias Económicas y Comerciales de la Universidad Nacional Mayor de San Marcos. Lima, Tall.Gráf.de la Editorial Lumen S.A., 1945. 127 p.

Statistical methodology. Includes: Historical development of statistics; nature and definition of statistics; characteristic attributes of statistics; investigation or collection; statistical criticism; classification, computation or enumeration; exposition of the results; general notions on the probabilities and the law of errors; statistical series; averages; dispersion or variation; correlation and covariance.

Incluye: Desenvolvimiento histórico de la estadística; naturaleza y definición de la estadística; caracteres propios de la estadística; investigación o recolección; crítica estadística; el despojo, cómputo o recuento; exposición de los resultados; nociones generales sobre las probabilidades y la ley de los errores; las series estadísticas; promedios; dispersión o variación; correlación y covariación.

NonG-**3330**.ATC 1945 **Statistical Control Manual.** Prepared by Headquarters, Air Transport Command. Washington, D.C., July 1945. Army Air Forces. Air Transport Command. 70 p. (approx.) ATC manual, 80-15-1.

Prepared as an aid in preparation of statistical reports, analyses, and presentations. In addition to a section on statistical control specifically written for the Air Transport Command, the manual includes chapters on statistical methods, sampling methods and procedures, methods of graphic and tabular presentation, and suggested standardized charts and tables.

Manual de control estadístico. Preparado como ayuda en la preparación de informes análisis y presentaciones estadísticas. En adición a una sección sobre control estadístico escrito específicamente para el Comando de Transporte Aéreo, el manual incluye capítulos sobre métodos estadísticos, métodos y procedimientos de muestras, métodos sobre presentaciones gráficas y tabulares, y gráficos y cuadros estandards recomendados.

NonG-**3330**.Bertillon 1906 **Compendio de la Obra Intitulada: Curso Elemental de Estadística Administrativa.** Por Jacques Bertillon. Extractado y traducido por Carlos García. San Luis Potosí, México, Imp.de la Librería Universal, 1906. 105 p.

Elementary course in administrative statistics. Contents: First part, Statistics—its necessity—general methods; second part, technique of statistics; third part, the way to utilize statistical data; fourth part, elements of demography; appendix (examples of forms, statistical charts, other statistical exposition). (Translation of table of contents.)

Contiene: Primera parte, De la estadística—su necesidad—métodos generales; segunda parte, técnica de la estadística; tercera parte, del modo de utilizar los datos estadísticos; cuarta parte, elementos de demografía; anexo (ejemplos de formularios, cuadros estadísticos, otra exposición estadística).

NonG-**3330**.Boddington 1942 **Estadística y Su Aplicación al Comercio.** Por A. Lester Boddington. Traducción de la quinta edición inglesa por Pedro Segura. Reimpresión. Barcelona, Buenos Aires, etc., Editorial Labor, S.A., 1942. 249 p.

Translated from English—traducción del inglés (NonG-**3330**.Boddington, 1939).

Contenido: Compilación de datos estadísticos (homogeneidad de los datos, necesidad de obtener datos completos, estabilidad de los datos; datos representativos; procedimiento para recoger datos, etc; exactitud y aproximación; tipos y promedios); presentación y usos de los datos estadísticos (preparación de tablas o cuadros estadísticos; el método gráfico; métodos de comparación; números índices; correlación; el censo como auxiliar en la dirección científica de los negocios; investigación de los negocios; auxiliares mecánicos en el trabajo estadístico.)

NonG-**3330.**Boddington 1939 **Statistics and Their Application to Commerce.** By A. Lester Boddington. Eighth edition. London, Sir Isaac Pitman & Sons, Ltd., 1939, 360 p.

Translated into Spanish—traducido al español (NonG-**3330.**Boddington, 1942).

Contents: Compilation of statistical data (homogeneity, completeness, stability of the data; sampling; methods of collecting data, etc; accuracy and approximations; types and averages); presentation and uses of statistical data (tabulation; the graphic method; methods of comparison; index numbers; correlation; the census as an aid to scientific business; business research; mechanical aids to statistical work.)

NonG-**3330.**Bowley 1937 **Elements of Statistics.** By Arthur Lyon Bowley. 6th ed. London, P. S. King & Son, Ltd.; New York, C. Scribner's Sons, 1937. 503 p. Unexamined.

Elementos de estadística.

NonG-**3330.**Brumbaugh 1941 **Business Statistics.** By Martin Allen Brumbaugh, Lester S. Kellogg, and Irene J. Graham. Chicago, Illinois, R. D. Irwin, Inc., 1941. 913 p. Unexamined.

Estadísticas comerciales.

NonG-**3330.**Chaddock 1925 **Principles and Methods of Statistics.** By Robert Emmet Chaddock. Boston, New York, Houghton Mifflin Co., 1925. 471 p. Unexamined.

Principios y métodos de estadística.

NonG-**3330.**Colombia 1937-38 **Cartilla de Estadística.** Bogotá, 1937-38. Colombia. Contraloría General de la República. Dirección Nacional de Estadística. 5 vol. No examinado.

Statistical notes.

*NonG-**3330.**Croxton 1 1943 **Applied General Statistics.** By Frederick E. Croxton and Dudley J. Crowden. New York, Prentice-Hall, Inc., 1943. 944 p.

Spanish and Portuguese translations are in preparation—traducciones al español y portugués en curso de preparación.

Includes: Statistical data; statistical tables; graphic presentation—simple curves, the semi-logarithmic or ratio chart, other types of charts; ratios and percentages; the frequency distribution; measures of central tendency; dispersion, skewness, and kurtosis; describing a frequency distribution by a fitted curve; reliability and significance of statistical measures—arithmetic means, percentages, standard deviations, variances, and the criterion of likelihood; the problem of time series; analysis of time series—secular trend, other trend types, periodic movements, types of seasonal movements, cyclical movements; fundamentals in index number construction; index number theory and practice; simple correlation, non-linear correlation, multiple and partial correlation; correlation of time series and forecasting.

Estadística general aplicada. Incluye: Datos estadísticos; tablas estadísticas; representación gráfica—curvas sencillas; el gráfico semilogarítimico o proporcional; otros tipos de gráficos; proporciones y porcentajes; la distribución de frecuencias; medidas de tendencia central; dispersión, asimetría, y kurtosis; descripción de una distribución de frecuencias por medio de una curva ajustada; confiabilidad y significación de medidas estadísticas—medias aritméticas, porcentajes, desviación standard, variaciones, y el criterio de probabilidad; el problema de series cronológicas; análisis de series cronológicas—tendencia secular, otros tipos de tendencias, movimientos periódicos, tipos de movimientos estacionales; movimientos cíclicos, fundamentos de construcción de números índices; teoría y práctica de los números índices; correlación simple, correlación no-lineal, correlación múltiple y parcial, correlación de series cronológicas, y pronósticos.

NonG-**3330.**Croxton 2 1941 **Workbook in Applied General Statistics.** By Frederick Emory Croxton. Rev. ed. New York, Prentice-Hall, Inc., 1941. 181 p., and 58 p. of tables, tabular forms, and graphic forms.

"To accompany the author's **Applied General Statistics.**" (Source: Foreword.)

Manual de ejercicios de estadística general aplicada. "Para acompañar a **Applied General Statistics** del mismo autor." (Traducción del comentario.)

NonG-**3330.**Croxton 3 1934 **Practical Business Statistics.** By Frederick E. Croxton and Dudley J. Cowden. New York, Prentice-Hall, Inc., 1934. 529 p.

Contents: Introduction; statistical data; statistical tables; graphic presentation (curves, the semi-logarithmic chart, other types of charts); ratios and percentages; frequency distribution; measures of central tendency; dispersion and skewness; unreliability; normal curve; problem of time series; analysis of time series; construction of index numbers; some current indexes; simple correlation; types of correlation; forecasting; appendices (including selected list of readily available sources).

Estadísticas prácticas del comercio. Contenido: Introducción; datos estadísticos; cuadros estadísticos; representación gráfica (curvas, diagrama semi-logarítmico, otros tipos de diagramas); proporciones y porcentajes; distribución de frecuencias; medidas de tendencia central; dispersión y asimetría; incertidumbre; curva normal; problema de las series cronológicas; análisis de las series cronológicas; construcción de números índices; algunos índices comunes; correlación simple; tipos de correlación; pronósticos; apéndices (incluyendo una lista de fuentes de información disponibles).

NonG-**3330.**Crum 1946 **Rudimentary Mathematics for Economists and Statisticians.** By W. L. Crum and Joseph A. Schumpeter. New York, McGraw-Hill Book Co., 1946. 183 p.

Contents: Graphic Analysis — simplest case; graphic analysis—curves and equations; limits; rates and derivatives; maxima and minima; differential equations; determinants.

Matemática elemental para economistas y estadígrafos. Contenido: Análisis gráfico — el caso mas simple; análisis gráfico — curvas y ecuaciones; límites; tasas y derivadas; máxima y mínima; ecuaciones diferenciales; determinantes.

NonG-**3330.**Crum 1938 **Introduction to Economic Statistics.** By William Leonard Crum, Alson Currie Patton, and Arthur Rothwell Tebbutt. First edition. New York, London, McGraw-Hill Book Company, Inc., 1938. 423 p.

Contents: Statistical data (introduction, variables and homogeneity, sources and their use, collection of primary data, construction of general tables, summary tables, charting categorical series, charting time series, charting frequency series); general analytical methods (summary numbers, averages for frequency series, dispersion, normal law of error, skewness and moments, correlation table, coefficient of correlation, further correlation methods); analysis of time series (relatives and index numbers, weighted index numbers, secular trend, seasonal variation, cyclical fluctuations, lagging correlation and forecasting sequence); appendix (laboratory procedure, normal curve of error and other frequency curves, logarithms of numbers).

Introducción a las estadísticas económicas. Contenido: Datos estadísticos (introducción, variabilidad y homogeneidad, fuentes y su uso, recolección de datos primarios, elaboración de cuadros generales, cuadros de resúmenes, representación gráfica de series de categorías, de tiempo y de frecuencia); métodos analíticos generales (cifras sumarias, promedios de series de frecuencia, dispersión, ley normal del error, asimetría y momentos, cuadro de correlación, coeficiente de correlación, otros métodos de correlación); análisis de series de tiempo (números relativos y números índices, índices ponderados, tendencia secular, variación estacional, fluctuaciones cíclicas, correlación y secuencia de predicción); apéndice (procedimientos de laboratorio, curva normal del error y otras curvas de frecuencia, logaritmos de los números).

NonG-**3330.**Dieulefait 1938-40 **Elementos de Estadística Metodológica.** Según el curso de la materia dictado en la Facultad de Ciencias Económicas, Comerciales y Políticas de la Universidad Nacional del Litoral, por Carlos E. Dieulefait. Redactado por . . . José Blasco. Rosario, Imp. de la Universidad, 1938-40. Universidad Nacional del Litoral. Facultad de Ciencias Económicas, Comerciales y Políticas. Instituto de Estadística. (Arg.) 3 vol.

Elements of statistical methodology.

NonG-**3330.**Ezekiel 1941 **Methods of Correlation Analysis.** By Mordecai Ezekiel. Second edition. New York, John Wiley & Sons, Inc.; London, Chapman & Hall, Limited, 1941. 531 p.

A comprehensive treatment on the methods of correlation analysis.

Un tratado amplio sobre métodos de análisis de correlación.

NonG-**3330.**Fisher 1 1944 **Statistical Methods for Research Workers.** By Fisher. Ninth edition. London, Oliver and Boyd, Ltd., 1944. 350 p. Unexamined.

Métodos estadísticos para investigadores.

NonG-**3330.**Fontenelle 1933 **O Methodo Estatistico em Biologia e em Educação.** Por J. P. Fontenelle. Rio de Janeiro, J. R. de Oliveira & C., 1933. 222 p.

Statistical methods in biology and education. Includes: Mensuration and values; statistical work; graphic presentation; frequency distribution; central tendency; variability; normal frequency curve; applications of the normal curve; association; correlation.

Inclue: Mensuração e valores; o trabalho estatístico; representação gráfica; distribuição de frequência; tendencia central; variabilidade; curva normal de frequência; aplicações da curva normal; associação; correlação.

NonG-**3330.**García Pérez 1945 **Elementos de Método Estadístico.** Por Andrés García Pérez. México, Imp. Universitaria, 1945. 284 p.

Elements of statistical methodology. Includes: Definitions; study of statistical phenomena; comparison of intensities of several phenomena; study of distribution; relation between distribution and attributes of phenomena; graphic presentation.

Incluye: Definiciones; estudio de los fenómenos estadísticos; comparaciones entre las intensidades de varios fenómenos; estudio de la distribución; relaciones entre las distribuciones y las modalidades de los fenómenos; representaciones gráficas.

NonG-**3330.**Garrett 1937 **Statistics in Psychology and Education.** By Henry Edward Garrett. Second edition. New York, Longmans, Green & Co., 1937. Unexamined.

Adapted to the needs of workers in these fields who do not have an extensive mathematical background.

La estadística en la psicología y en la educación. Adaptado a las necesidades de las personas que trabajan en ese campo y que no poseen un extenso conocimiento de matemáticas.

NonG-**3330.**Gavett 1941 **Principios de Metodología Estadística.** Por George Irving Gavett. Traducción y adaptación del Ing. Miguel Gleason Alvarez. México, 1941. Universidad Nacional Autónoma de México. Escuela Nacional de Economía. No examinado.

Translation of—traducción de: **A First Course in Statistical Method** (NonG-3330.Gavett, 1937).

Principles of statistical methodology.

NonG-**3330.**Gavett 1937 **A First Course in Statistical Method.** By G. Irving Gavett. Second edition. New York and London, McGraw-Hill Book Company, Inc., 1937. 400 p.

Published also in Spanish — publicado también en español (NonG-**3330.**Gavett, 1941).

Contents: Introduction; tabulation, frequency distribution; graphical representation, frequency graphs; averages; dispersion; skewness; probable error; curve fitting; correlation, regression; multiple and partial correlation; logarithmic graphical representation; appendixes (logarithms; permutations, combinations, binomial expansion; laws of probability; derivatives and integrals; determinants; mathematical tables).

Un curso de iniciación en métodos estadísticos. Contenido: Introducción; tabulación, distribución de frecuencia; representación gráfica, diagramas de frecuencia; promedios; dispersión; asimetría; error probable; ajustamiento de curvas; correlación, regresión; correlación multiple y parcial; representación gráfica logarítmica; apéndices (logaritmos; permutaciones, combinaciones, expansión binomial; leyes de probabilidad; derivadas e integrales; determinantes; tablas matemáticas).

NonG-**3330.**Gini 2 1935 **Curso de Estadística.** Por Corrado Gini. Con un apéndice matemático por Luigi Galvani. Traducción del italiano por José A. Vandellós Barcelona, Madrid, Buenos Aires, Editorial Labor, S. A., 1935. Enciclopedia de ciencias jurídicas y sociales, sección ciencias sociales. 426 p.

Course in statistics. Includes: Brief notes about the origin and development of statistics; fundamental principles of statistical methods; collection of data; measurement of statistical phenomena; integration and logical elaboration of data; elemental notions of mathematics with application to statistics.

Incluye: Algunos apuntes sobre el origen y desarrollo de la estadística; principios fundamentales de la estadística metodológica; recopilación de datos; medición de los fenómenos estadísticos; integración y elaboración lógica de los datos; nociones elementales de matemáticas con aplicación a la estadística.

NonG-**3330.**Hagood 1941 **Statistics for Sociologists.** By Margaret Jarman Hagood. New York, Reynal and Hitchcock, Inc., 1941. 934 p. Unexamined.

"This text, designed for the first year in statistics for students concentrating in psychology, devotes its fifth part, a section of five chapters, to selected techniques for population data." (Source: **Books on Population, 1939-Mid-1944.**)

Estadísticas para sociólogos. "Este texto, designado para el primer año en estadística para estudiantes que se especializan en psicología, dedica su quinta parte (una sección de cinco capítulos) a determinadas técnicas para la obtención de datos sobre población." (Traducción del comentario.)

NonG-**3330.**Jansen de Mello 1944 **Utilização de Medidas Estatísticas em Biologia, Medicina e Saúde Pública.** Por Eder Jansen de Mello. Rio de Janeiro, Gráf. Barbero, 1944. Brasil. Ministério da Educação e Saúde. Serviço Nacional de Educação Sanitária. 152 p.

Use of statistical measurements in biology, medicine, and public health. Includes: Generalization of statistical knowledge — errors of sampling; probabilities — binomial frequency distribution; normal frequency distribution — tables of probability of deviation measured in "sigmas"; tests of significance; significance of the difference between two averages and between two frequencies in large samples; significance of the difference between two averages and between two frequencies in small samples; tests of significance for more than two frequencies — proof of agreement of the facts with theory; degree of association — coefficient of correlation and other measures and their significance; tables and nomograms of significance.

Inclue: Generalização do conhecimento estatístico — êrros de amostra; probabilidades — distribuição binomial de frequências; distribuição normal de frequências — tábuas de probabilidades de desvios medidos em "sigmas"; os testes de significância; significância da diferença entre duas médias e entre duas frequências de grandes amostras; significância da diferença entre duas médias e entre duas frequências de pequenas amostras; teste de significância para mais de duas frequências — prova de concordância dos fatos com a teoria; grau de associação — coeficiente de correlação e outras medidas e sua significância; tábuas e nomogramas de significância.

NonG-**3330.**Julin 1924? **Curso de Estadística General y Aplicada.** Por Armand Julin. Traducción y notas por Felipe S. Paz. Madrid, Imp. de Juan Pueyo, 1924? 312 p.

Course in general and applied statistics. Contains: General statistics (theoretical methodology, technical methodology); applied statistics: (a) Production statistics (industrial statistics, agricultural statistics); (b) commercial statistics (foreign trade statistics, transportation statistics); (c) price statistics.

Incluye: Estadística general (metodología teórica, metodología técnica); estadística aplicada: (a) Estadística de la producción (estadística industrial, estadística agrícola); (b) estadística de los intercambios (estadística del comercio exterior, estadística de los transportes); (c) estadística de los precios.

NonG-**3330.**Kafuri 1934 **Lições de Estatística Matemática.** Por Jorge Kafuri. Rio de Janeiro, Flores & Mano Edit., 1934. 4 vol.: Vol. 1, **Instituição e Cálculo dos** Indices Estatísticos; vol. II, **Leis de Distribuição de Frequência;** vol. III, **Problema de Correlação;** vol. IV, **Aplicações e Tabelas.**

Lessons in mathematical statistics. 4 vol.: Vol. I, Definition and calculation of statistical coefficients; vol. II, laws of frequency distributions; vol. III, the correlation problem; vol. IV, applications and tables. Volume I, the only one published, deals with observation methods and classification in frequency distributions, their characteristic properties, coefficients, and averages.

Volume I, o único publicado, trata dos métodos de observação e classificação nas distribuições de frequência, suas propriedades caracteristicas, índices e promédios.

NonG-**3330.**Kingston 1945 **A Teoria da Indução Estatística.** Por Jorge Kingston. Rio de Janeiro, Serviço Gráf. do IBGE, 1945. 119 p.

Theory of statistical induction. Includes: The problems of theoretical statistics; significance of averages and other statistics; the binomial distribution and its approximations; dispersion of Lexis — distributions of contagion; chi-square and verification of empirical laws; distribution of Student; estimation and comparison of variances; analysis of variance; verification of the interdependence of phenomena—analysis of covariance.

Inclue: Os problemas de estatística teórica; significância de médias e outras estatísticas; a distribuição binomial e suas aproximições; a dispersão lexiana—distribuições de contágio; chi - quadrado e a verifição de leis empíricas; a distribuição de Student; a estimação e comparação de variâncias; a análise da variância; verificação da interdependência entre fenômenos—análise da convariância.

NonG-**3330.**Kürbs **Curso de Estadística Expositiva o Aplicada.** Por Friederich Kürbs. Lima, Editorial Horizonte. 270 p. Mimeografiado.

Course in expositive or applied statistics. Includes: Statistics of territory and population; statistics of economics; circulation or exchange; statistics of public authority and its organization; economic statistics of business.

Incluye: La estadística del territorio y de la población; la estadística de la ciencia económica: circulación o intercambio; la estadística del poder público y su organización; la estadística económica de las empresas.

NonG-**3330.**Mateus 1926 **Principios Técnicos de Estadística.** Metodológica y aplicada. Traducidos, recopilados y arreglados en forma de texto para el uso de las escuelas y oficinas colombianas. Por George Mateus. Bogotá, Tall.de Ediciones Colombia, 1926. 213 p. Unexamined.

Technical principles of statistics, methodological and applied.

NonG-**3330.**Mather 1943 **Statistical Analysis in Biology.** By K. Mather. Foreword by R. A. Fisher. New York, Interscience Publishers, Inc., 1943. Unexamined.

"The . . . volume is by a geneticist for geneticists and other experimental biologists. It begins with the relation between sample and population . . . . Following a discussion of probability, significance and the basic random sampling distributions, he continues with tests of significance, the analysis of variance, experimental design, relations of two variables, analysis of frequency data, and concludes with estimation and information . . . " (Source: **Science,** vol. 102, no. 2641, Aug. 10, 1945, p. 161-162.)

El análisis estadístico aplicado a la biología. "El . . . volumen esta escrito por un especialista en genética para especialistas en genética y otros biólogos experimentales. Empieza con la relación entre la muestra y la población . . . Después de una discusión sobre probabilidad, significación y las distribuciones básicas de muestras al azar, el autor continúa con pruebas de significación, el análisis de variación, plan experimental, relaciones de dos variables análisis de datos de frecuencia, y concluye con estimación e información . . ." (Traducción del comentario.)

NonG-**3330.**Mazza 1 1942 **Apuntes de Estadística.** Del curso dictado en el año 1942 por el Prof. S. C. Mazza, director del Instituto. Montevideo, 1942. Universidad de la República (Uruguay). Facultad de Ciencias Económicas y de Administración. Instituto de Estadística. 75 p. (aprox.) Mimeografiado. Fasc. I, **Distribuciones de Frecuencias; Correlación.**

Statistical notes from the course dictated by S. C. Mazza at the Institute of Statistics in Montevideo during the year 1942; Part I, Frequency distribution; correlation.

NonG-**3330.**Mazza 2 1940 **Cuestiones de Matemáticas Referentes a la Estadística. Fascículo I.** Por Sigfrido C. Mazza. Rosario de S.Fé, Argentina, Tall.Gráf.Emilio Fenner, 1940. Universidad Nacional del Litoral. Facultad de Ciencias, Económicas Comerciales y Políticas. Instituto de Estadística.

Questions of mathematics referring to statistics, part I. Includes: Determinants, systems of linear equations; orthogonal polynomials; least squares.

Incluye: Determinantes; sistemas de ecuaciones lineales; polinomios ortogonales; mínimos cuadrados.

*NonG-**3330.**Mills 1940, Eng. **Statistical Methods Applied to Economics and Business.** By Frederick Cecil Mills. Revised. New York, Henry Holt and Company, 1940. 736 p.

Published also in Spanish—publicado también en español (NonG-**3330.**Mills, 1940, Sp.). Portuguese translation in preparation—traducción al portugués en curso de preparación.

A comprehensive basic textbook. Includes: Statistical methods and economic and commercial problems; graphic presentation; elaboration of statistical data—frequency distribution; study of frequency distribution—averages, measures of variability and of dispersion; index numbers of prices—nature of the index numbers; analysis of time series—measures of tendency, measures of cyclical and seasonal fluctuations; indexes of material volume; measure of relation—linear correlation, between time series, non linear correlation, the problem of estimation, multiple and partial correlation; fundamental probability and the normal curve of error; statistical induction and the problem of adequate selection of representative groups; the method of least squares and its application to certain statistical problems.

NonG-**3330.**Mills 1940, Sp. **Métodos Estadísticos Aplicados a la Economía y los Negocios.** Por FrederickCecil Mills. Versión española de Juan Ruiz Magán y Enrique Gastardi. 2. ed. Madrid, M.Aguilar, 1940. 517 p.

Published also in English—publicado también en inglés. NonG-**3330.** Mills, 1940, Eng.) Portuguese translation in preparation—traducción al portugués en curso de preparación.

Un texto básico comprensivo. Incluye: Los metodos estadísticos y los problemas económicos y comerciales; representación gráfica; elaboración de los datos estadísticos—distribución de frecuencias; estudio de la distribución de frecuencias—promedios, medidas de variabilidad y de dispersión; números índices de precios—naturaleza de los números índices; análisis de las series cronológicas—medidas de la tendencia, medidas de las fluctuaciones cíclicas y estacionales; índices de volumen material; medida de la relación—correlación lineal, entre series cronológicas, correlación no lineal, el problema de la estimación, correlación múltiple y parcial; la probabilidad elemental y la curva normal de error; la inducción estadística y el problema de la adecuada elección de grupos representativos; el método de los mínimos cuadrados y su aplicación a ciertos problemas estadísticos.

NonG-**3330.**Moncetz 1941 **Métodos Estadísticos.** Una iniciación. Por A. de Moncetz. Traducción de José Ferrell y Joaquín Ramírez Cabañas. México, Editorial América, 1941. 114 p.

Introduction to statistical methods.

NonG-**3330.**Neiswanger 1943 **Elementary Statistical Methods.** As Applied to Business and Economic Data. By William Addison Neiswanger. New York, The Macmillan Company, 1943. 740 p.

Includes: Statistical methods—importance and use; cautions in the use and interpretation of the statistical method; initiating and conducting a statistical investigation; sampling and concept of error; tabular presentation of statistical data; graphic comparisons of simple magnitudes; charting time series; tabular and graphic descriptions of frequency distributions; averages; dispersion and skewness; dispersion of statistics computed from samples; index numbers; current index numbers and their use; analysis of time series; seasonal variations; statistical concept of normal and cyclical variations; analysis of functional relationships; measurement of correlation.

Métodos estadísticos elementales, aplicados a los negocios y a los datos económicos. Incluye: Métodos estadísticos—importancia y uso; precauciones en el uso e interpretación del método estadístico; el iniciar y el dirigir de una investigación estadística; el método de la muestra y el concepto de error; representación tabular de datos estadísticos; comparaciones gráficas de magnitudes simples; representación gráfica de series cronológicas; descripción tabular y gráfica de la distribución de frecuencias; promedios; dispersión y asimetría; dispersión de datos computados de muestras; números índices; números índices de la actualidad y sus usos; análises de series cronológicas; variación estacional; el concepto estadístico de variaciones normales y cíclicas; análisis de relaciones funcionales; medida de correlación.

NonG-**3330.**Pagano 1942 **Licões de Estatística.** Por Aúthos Pagano. Edição comemorativa do 40.° aniversario da fundação da Escola de Comércio "Alvares Penteado." São Paulo, Brasil, Tipo. Rossolillo, 1942. 2 vol.: Vol. I, **Parte Geral;** vol. II, **Cálculo das Probabilidades e Questões Demográficas.**

Lessons in statistics. First volume includes: History of statistics; definitions; science or method; normalcy of the large numbers and permanency of the small numbers; the data; series or statistical distributions; tabulation of data; graphic presentation; elements of central tendency; geometric mean; harmonic mean; other measures; elements of dispersion or variability; moments; interpolation and adjustment—elementary processes; finite differences—Newton's formula of interpolation; adjustment by the method of the least squares; adjustment of statistical data with the help of non-linear equations; correlations. Second volume includes: Probability (deviation, frequency, etc.); some demographic questions (growth, tables, etc.); table of values of the normal curve.

Vol. I inclue: Historia da estatística; definições; ciência ou método; da normalidade dos grandes números e da permanência dos pequenos números; os dados; séries ou distribuições estatísticas; tabulação dos dados; representação gráfica; elementos de tendência central; média geométrica; média harmônica; outras medidas; elementos de dispersão ou de variabilidade; momentos; interpolação e ajustamento—processos elementares; diferenças finitas—fórmula de interpolação de Newton; adjustamento pelo método dos mínimos quadrados; ajustamento de dados estatísticos com o auxilio de equações não lineares; correlações. Vol. II inclue: Cálculo das probabilidades (desvio, frequência, etc.); algumas questões demográficas (crescimento, tábuas, etc.); tábua dos valores da curva normal.

NonG-**3330.**Posada Gaviria 1934 **Principios de Estadística General.** Por Vincente Posada Gaviria. Bogotá, 1934. No examinado.

Principles of general statistics.

NonG-**3330.**Procaccia 1944 **Lecciones de Estadística Metodológica.** Por Carlos Procaccia. Quito, Imp. del Ministerio del Tesoro, 1944. Ministerio de Economía. Dirección General de Estadística y Censos. 117 p. Mimeografiado.

Lessons in statistical methods. Includes: Definition of statistics; direct gathering of statistical quantities; primary statistical elaboration; graphic presentation of statistical quantities;

arithmetic methods of simplification and presentation; norms, classes, and examples of graphic presentation of statistical quantities; algebraic methods of statistical elaboration, presentation and investigation; curves of binomial frequency, and application to statistics; index numbers; systematization, comparison, critique and analysis of elaborated data.

Incluye: Definición de estadística; captación directa de las cantidades estadísticas; la elaboración estadística primaria; representación gráfica de las cantidades estadísticas; procedimientos aritméticos de simplificación y representación; normas, clases y ejemplos de representación gráfica de las cantidades estadísticas; procedimientos algébricos de elaboración, representación e investigación estadísticas; las curvas de frecuencia binomial y su aplicación a la estadística; los números índices; sistematización, comparación, crítica y análisis de los datos elaborados.

NonG-3330.Roa Reyes 1929 **Curso de Estadística.** Por Jorge Roa, Habana, Imp.de la Universidad, 1929. Universidad de la Habana. Facultad de Derecho y de Ciencias Sociales. 7 vol. en 1.: Sílabo I, **Introducción**; sílabo II, **De las Unidades**; sílabo III, **De la Tabulación**; sílabo IV, **Representación Diagramática**; sílabo V, **Representación Gráfica**; sílabo VI, **Promedios: La Media, la Mediana y el Modo**; sílabo VII, **Indices, Desviaciones, Dispersiones, Correlaciones, Comparaciones.**

Course in statistics. Syllabus I, **Introduction**; syllabus II, **the Unities**; syllabus III, **Tabulation**; syllabus IV, **Diagramatic Presentation**; syllabus V, **Graphic Presentation**; syllabus VI, **Averages: the Mean, the Median and the Mode**; syllabus **VII, Indexes, Deviations, Dispersions, Correlations, Comparisons.**

NonG-3330. Rodríguez 1946 **Lecciones de Estadística.** Por Jorge Rodríguez. Cuarta edición. Medellín, Colombia, Tall.Tipo. Bedout, 1946. 285 p.

Contents: Introduction (definition and concept, history, and utility of statistics); statistical technique (statistical investigations; preparation; collection of data; criticism of data; elaboration; relative numbers; index numbers; publication; law of large numbers; interpretation of statistics); expository statistics (divisions of statistics; territorial statistics; population statistics; economic statistics; educational statistics; criminal statistics; political statistics; financial statistics); appendix: History of statistics in Colombia; sex determination; legislation (law 82 of 1935).

Contiene: Preliminares (definición y concepto, historia y utilidad de la estadística); técnica estadística (de las investigaciones estadísticas; preparación; recolección de los datos; crítica de los datos; elaboración; números relativos; números índices; publicación; ley de los grandes números; interpretación de la estadística); estadística expositiva (división de la estadística; estadística del territorio; estadística de la población; estadística económica; estadística intelectual; estadística moral; estadística política; estadística financiera); apéndice: Historia de la estadística en Colombia; determinismo sexual; legislación (ley 82 de 1935).

NonG-3330.Schott 1934 **Estadística.** Por Sigmund Schott. Traducción de la 3a edición alemana y anotada por Manuel Sánchez Sarto. Segunda edición. Barcelona, Buenos Aires, Editorial Labor, S.A., 1934. 206 p. Colección labor de la biblioteca de iniciación cultural, sección X, economía, no. 186.

Statistics. Includes: Essence and object of statistics; the tools of statistics; gathering and elaboration of the numerical material; publication, simplification and interpretation of the results; principal branches of social statistics; chronological table showing the development of social statistics.

Incluye: Esencia y objeto de la estadística; los órganos de la estadística; obtención y elaboración del material numérico; publicación, simplificación y interpretación de los resultados; sectores principales de la estadística social; cuadro cronológico relativo al desarrollo de la estadística social.

NonG-3330.Secrist 1 1925 **An Introduction to Statistical Methods.** A textbook for college students, a manual for statisticians and business executives. By Horace Secrist. Rev. ed. New York, The Macmillan Co., 1925. 584 p.

Una introducción a métodos estadísticos.

NonG-3330.Shelly Hernández 1941 **La Estadística Aplicada a las Ciencias Biológicas.** Por R. de Shelly Hernández. Primera edición. Caracas, Lito.del Comercio, 1941. 386 p.

Statistics as applied to the biological sciences. Includes: Elementary biomathematics; collection of data; statistical tables; averages; population; absolute numbers and coefficients or rates; graphs; indexes; statistical series; probability; correlation and regression; curves, interpolation and graduation; life tables.

Incluye: Biomatemáticas elementales; recolección de datos; cuadros estadísticos; promedios;

población; cifras absolutas y coeficientes o tasas; gráficos; índices; la serie estadística; cálculo de probabilidades; correlación y regresión; curvas, interpolaciones y graduación; la tabla de vida.

NonG-**3330.**Silva Rodrigues 1 1941 **Elementos de Estatística Geral.** Por Milton Camargo da Silva Rodrigues. 3. ed., rev. e aumentada. São Paulo, Cia.Editora Nacional, 1941. 424 p.

Elements of general statistics.

NonG-**3330.**Silva Rodrigues 2 1942 **Iniciação à Estatística Econômica.** Por Milton da Silva Rodrigues. São Paulo, Rio de Janeiro, etc., Companhia Editora Nacional, 1942. 252 p. Biblioteca de estudos comerciais e econômicos, vol. 30.

Introduction to economic statistics. Includes: Graphic presentation; relative numbers; frequency distributions; measures of position; measures of dispersion and skewness; probabilities of precision; time series; growth; analysis of time series; index numbers; secular tendency and its measurement; seasonal variations and their measurements; functions and their graphic presentation.

Inclue: Representação gráfica; números relativos; as distribuições por frequência; medidas de posição; medidas de dispersão e assimetria; probabilidades de precisão; marchas; crescimento; a análise das séries cronológicas; números índices; a tendência secular e sua medida; as variações estacionais e suas medidas; funções e suas representações gráficas.

NonG-**3330.**Smith **Fundamentals of the Theory of Statistics.** By James G. Smith and Acheson J. Duncan. 1944- New York, McGraw-Hill Book Company, Inc.

Series of textbooks on statistics, each appearing as a separate volume and under a different title.

Fundamentos de la teoría de estatística. Serie de textos sobre estadística; cada uno aparece como un volumen separado, y bajo un título distinto.

NonG-**3330.**Smith 1 1944 **Elementary Statistics and Applications.** By James G. Smith and Acheson J. Duncan. New York, McGraw-Hill Book Company, Inc., 1944. 720 p. **Fundamentals of the Theory of Statistics,** vol. 1.

Contains: **Introduction:** Statistics in the arts and sciences, gathering statistics, sources of statistics, presentation of statistics, statistics—a study of variation; **analysis of frequency distributions:** Summarization and comparison, illustration of frequency-distribution analysis; **the normal frequency curve:** Probability, probability distributions, probability calculus, symmetrical binomial distribution and the normal curve, use of normal frequency curve in sampling analysis; **study of bivariates and multivariates:** Simple correlation, computation of **r** and other measures of correlation, nonlinear correlation, multiple and partial correlation, analysis of multivariate frequency distribution illustrated, normal frequency surface; **study of dynamic variability:** Index numbers, rational basis of the analysis of time series, trend analysis, orthogonal-polynomial trends, time-series analysis—seasonal variation, determination of cycle; **forecasting:** The art of forecasting with statistics.

La estadística elemental y sus aplicaciones. Contiene: **Introducción:** La estadística en las artes y ciencias, recolección de datos, fuentes de estadísticas, presentación de datos; estadística—un estudio de variación; **análisis de la distribución de frecuencias:** Resumen y comparación, ilustración del análisis de la distribución de frecuencias; **curva normal de frecuencia:** Probabilidad, distribuciones de probabilidad, cálculo de probabilidad, distribución simétrica binomial y la curva normal, uso de la curva normal de frecuencia en el análisis de muestras; **estudio de bivariables y multivariables:** Correlación simple, cálculo de **r** y otras medidas de correlación, correlación no lineal, correlación múltiple y parcial, ilustración del análisis de la distribución multivariable de frecuencias, superficie normal de frecuencia; **estudio de la variabilidad dinámica:** Números índices, basis racional del análisis de series cronológicas, análisis de tendencias, tendencias ortogonal-polinomiales, análisis de series cronológicas—variación estacional, determinación de ciclos; **pronósticos:** El arte de pronosticar con estadísticas.

NonG-**3330.**Snedecor 1946 **Statistical Methods Applied to Experiments in Agriculture and Biology.** By George W. Snedecor. Fourth edition. Ames, Iowa, The Collegiate Press, Inc., 1946. 485 p. Unexamined.

Métodos estadísticos aplicados a experimentos en agricultura y biología.

NonG-**3330.**Tippett 1943 **Statistics.** By L. H. C. Tippett. London, New York, etc., Oxford University Press, 1943. 184 p. The home university library of modern knowledge, CLVI.

Treats the raw material of statistics, arrangement and presentation of material, numerical measures, sampling, calculation of chance, statistical laws, statistical reasoning, statistics in affairs, and statistics and other sciences.

Trata de la materia prima de la estadística, disposición y presentación del material, medidas numéricas, muestras, cálculo de probabilidades, leyes estadísticas, razonamientos estadísticos, las estadísticas en los negocios, la estadística y las otras ciencias.

NonG-**3330.**Treloar 1 1943 **Correlation Analysis.** By Alan Edward Treloar. Minneapolis, Minnesota, Burgess Publishing Co., 1943. 64 p.
Análisis de correlación.

NonG-**3330.**Treloar 2 1939 **Elements of Statistical Reasoning.** By Alan Edward Treloar. New York, John Wiley and Sons, Inc., 1939. 251 p. Unexamined.
Elementos de razonamiento estadístico.

NonG-**3330.**Trucco 1944 **Análisis Estadístico Aplicado a los Trabajos de Investigación en Agricultura y Biología.** Por Sixto E. Trucco. Buenos Aires, "El Ateneo," 1944. 287 p.
Statistical analysis applied to research work in agriculture and biology. Includes: Statistical series—qualitative characters; quantitative characters; graphic presentation; measures of position; measures of intensity; measures of variability; Pearson's theory of moments; measures of dispersion; measures of skewness and of kurtosis; systematic study of a time series; systematic study of a frequency series; correlation; regression; errors; normal curve; methods of adjustment; analysis of variation.
Incluye: Series estadísticas—caracteres cualitativos; caracteres cuantitativos; representaciones gráficas; medidas de posición; medidas de intensidad; medidas de variabilidad; teoría de los momentos de Pearson; medidas de dispersión; medidas de asimetría y de kurtosis; estudio sistemático de una serie histórica; estudio sistemático de una serie de frecuencia; correlación; regresión; errores; curva normal; métodos de ajustamiento; análisis de la variación.

NonG-**3330.**Velarde B. 1945 **Curso de Estadística.** Por César A. Velarde B. Segunda edición, corregida y aumentada. Lima, Imp.del Ministerio de Guerra, 1945. 140 p.
Course in statistics. Includes: Statistical methodology (collection, graphic presentation, publication, interpretation, mathematical elaboration of data, deviations, distribution, and other statistical relationships); applied statistics (demographic, vital, economic, industrial, agricultural and livestock, financial, transport and communication statistics, military statistics, and military utilization of statistical data).
Incluye: Estadística metodológica (recopilación, representación gráfica, publicación, interpretación y elaboración matemática de los datos, desviaciones, distribución y otras relaciones estadísticas); estadística aplicada (estadística demográfica, vital, económica, industrial, financiera, agropecuaria, de transportes y comunicaciones, estadística militar, y utilización militar de los datos estadísticos.)

NonG-**3330.**Virgilii 1908 **Manual de Estatística.** Por Filippo Virgilii. Rio de Janeiro, Typ.da Directoría Geral de Estatística, 1908. 172 p.
Manual of statistics. Includes: Statistical methodology (collection of the facts, criticism of the data, elaboration of the data, presentation of the results); expository statistics (territory and climate, economic life, intellectual life, moral life, political life). Portuguese translation from the fourth Italian edition.
Inclue: Estatística metodológica (collecta dos factos, critica dos dados, elaboração dos dados, exposição dos resultados); estatística expositiva (território e clima, vida económica, vida intellectual, vida moral, vida política). Tradução portuguesa da quarta edição italiana.

NonG-**3330.**Viveiros de Castro 1938 **Pontos de Estatística.** Por Lauro Sodre' Viveiros de Castro. Rio de Janeiro, Departamento de Estatística e Publicidade, Ministério de Trabalho, Indústria e Comércio, 1938. 117 p.
Notes on statistics. Includes: Frequency distribution; tabular presentation—its characteristics; graphic presentation—utility and characteristics; statistical bar graphs, curve and sector graphs, frequency polygon and histogram; the simple and weighted arithmetic mean; mode and median; percentiles and quartiles; standard deviation; index numbers; probabilities; correlation; measures of dispersion and skewness; the geometric and harmonic mean; problems to be solved; appendix (organization of Brazilian statistics).
Inclue: Distribuição de frequência; representação tabular—seus caraterísticos; representação gráfica—utilidade e caraterísticos; gráficos estatísticos em barras, curvas e setores; histograma e poligono de frequência; média aritmética simple e ponderada; moda e mediana; percentílios e quartílios; desvio padrão; números índices; probabilidades; correlação; medidas de dispersão e assimetria; média geométrica e harmônica; problemas a resolver; apendice (organização da estatística brasileira).

NonG-**3330.**Wolfenden 1942 **The Fundamental Principles of Mathe-**

matical Statistics. With special reference to the requirements of actuaries and vital statisticians and an outline of a course in graduation. By Hugh Herbert Wolfenden. Toronto, Pub. for the Actuarial Society of America, New York, by the Macmillan Company of Canada Limited, 1942. 379 p.

Contains: The nature of the problems; the classical approach; the combination of observations; theory of random sampling; generalization of the binomial law—the "multinomial" distribution; frequency distributions and curves in general; the fitting of curves, and graduation; tests of goodness of fit; recent researches; an outline of a course in graduation, history, mathematics and interpretations, applications.

Los principios fundamentales de la estadística matemática, con referencia especial a los requisitos de actuarios y técnicos de estadística vital y esquema general de un curso de graduación. Contiene: La naturaleza de los problemas; el concepto clásico; la combinación de observaciones; teoría de obtención de muestras al azar; generalización de la ley binomial—la distribución "multinomial"; distribuciones de frecuencias y curvas en general; el ajuste de curvas, y graduación; pruebas de fineza del ajuste; investigaciones recientes; esquema de un curso de graduación, historia, matemática e interpretaciones, aplicaciones.

NonG-**3330**.Yule 1940 **An Introduction to the Theory of Statistics.** With 55 diagrams and 4 folding plates. By George Udny Yule and Kendall. 12th ed., rev. London, C. Griffin & Co., Ltd., 1940. 570 p. Unexamined.

Una introducción a la teoría de la estadística.

4000-4999 GEOGRAPHY—GEOGRAFIA

NonG-**4000**.A5 1943 **Aerial Photographs and Their Applications.** By Harold Theodore Uhr Smith. New York, London, D. Appleton-Century Company, Inc., 1943. 372 p. Unexamined.

Fotografías aéreas y sus aplicaciones.

NonG-**4000**.S5 **Surveying and Mapping.** Quarterly journal, American Congress on Surveying and Mapping. 1941- . Washington, D.C.

Issue of January 1945, vol. V, no. 1, published also in Spanish—edición de enero de 1945, vol. V, no. 1, publicada también en español.

Topografía y levantamiento de mapas.

5000-5999 DEMOGRAPHY—DEMOGRAFIA

NonG-**5000**.B5 **Biología Social.** Investigaciones comparadas de estadística social. Por Jorge Roa Reyes. Habana, 1939. Universidad de la Habana. Facultad de Ciencias Sociales y de Derecho Diplomático y Consular. 300 p.

Social biology. Contents: Introduction (statistics and the social sciences; population; the statistical objective; the statistical method); vegetative demography (birth and population; marriage; mortality; morbidity); organic demography (social organization; suicide; crime); ethnographic demography (ethnographic statistics; migration); physical demography (the physical environment); social demography (the social environment); bibliographic notes.

Contenido: Introducción (la estadística y las ciencias sociales; la población; objetivo estadístico; el método estadístico); demología vegetativa (la natalidad y la población; la nupcialidad; la mortalidad; la morbilidad); demología orgánica (organización social; el suicidio; la criminalidad); demología etnográfica (estadística etnográfica; las migraciones); demología física (el ambiente físico); demología social (el ambiente social); notas bibliográficas.

NonG-**5000**.H5 **Human Biology.** 1929- Quarterly. Baltimore, Maryland, Johns Hopkins Press.

Biología humana. Revista trimestral.

Dec. '45, Janer "Population Growth in Puerto Rico and Its Relation to Time Changes in Vital Statistics." By José L. Janer. Vol. 17, no. 4, December 1945, p. 267-313.

Includes: Analysis of population growth; time changes in the life tables for Puerto Rico, 1910 to 1940.

Incluye: Análisis del desarrollo de la población de Puerto Rico; cambios en las tablas de vida para Puerto Rico, 1910 a 1940.

Sept. '40, Pearl "A Comparative Examination of Certain Aspects of the Population of the New World." By Raymond Pearl. Vol. 12, no. 3, September 1940, p. 359-402.

Un examen comparativo de ciertos aspectos de la población del Nuevo Mundo.

NonG-**5000**.P3 1944 **Plenty of People.** By Warren Simpson Thompson. Lancaster, Pennsylvania, The Jaques Cattel Press, 1944. 246 p.

A broad description of the population growth of the world since 1800 and of trends in birth and death rates is followed by discussions of the relations of war to population growth, the probable future population of the nations and regions of the world, and the political and economic implications of differential national growth. The next major section covers distribution, including historical, contemporary and probable future problems of migration. The concluding chapters cover respectively, the social and economic effects of age changes and of the slower growth of population, the biological fitness of the people, minorities, population policies, and considerations involved in the formulation of a population policy for the United States.

Abundancia de población. Una amplia descripción del crecimiento de la población en el mundo, desde 1800, y de las tendencias de la natalidad y mortalidad; a ello siguen exposiciones sobre las relaciones de la guerra con el crecimiento de la población, la probable población futura de las naciones y regiones del mundo y el significado político y económico del diferente crecimiento de las naciones. La siguiente sección principal abarca la distribución, incluyendo problemas de migración (históricos, contemporáneos y probables del futuro). Los capítulos finales tratan respectivamente sobre los efectos sociales y económicos de los cambios en la edad y del crecimiento mas lento de la población, la idoneidad biológica de la población, minorías, políticas de población, y consideraciones relativas a la formulación de una política de población para Estados Unidos.

NonG-**5000**.S6 1924 **Studies in Human Biology.** By Raymond Pearl. Baltimore, Maryland, Williams and Wilkins Press, 1924. 653 p. Unexamined.

Includes logistic curves for United States, Australia, Belgium, Denmark, England and Wales, France, Germany, Hungary, Italy, Norway, Scotland, Serbia, Sweden, Japan, Java, and Philippine Islands.

Estudios sobre biología humana. Incluye curvas logísticas correspondientes a Estados Unidos, Australia, Bélgica, Dinamarca, Inglaterra, Gales, Francia, Alemania, Hungría, Italia, Noruega, Escocia, Servia, Suecia, Japón, Java y las Islas Filipinas.

NonG-**5100**.P2 1939 **People, the Quantity and Quality of Population.** By Henry P. Fairchild. New York, Henry Holt, 1939. 315 p. Unexamined.

"This volume, written for the general reader, proceeds from an analysis of prehuman reproduction, genetic geometry, and the interrelations of land, economic culture, and number of people to a consideration of Malthusian principles. An analysis of population increase in its historical development leads to consideration of optimum population and problems of population quality, with the emphasis placed on the basic significance of man's control of his numbers through voluntary limitation." (Source: **Books on population, 1939-Mid-1944.**)

La cantidad y calidad de la población. "Este volumen escrito para el lector en general, procede de un análisis de la reproducción prehumana, geometría genética, y de las relaciones entre suelo, cultura económica y número de habitantes a la consideración de los principios de Malthus. Un análisis del aumento de población en su desarrollo histórico orienta a una consideración de población óptima y los problemas de la calidad de la población, con especial atención al significado básico del control por el hombre de su número por medio de su limitación voluntaria." (Traducción del comentario.)

NonG-**5100**.P5 1940 **Population, a Problem for Democracy.** The Godkin lectures, 1938. By Gunnar Myrdal. Cambridge, Massachusetts, Harvard University Press, 1940. 237 p. Unexamined.

"Preliminary chapters sketch the inter-relationships of the great secular swings in population and economic theory, and stress the basic importance of population replacement in the long-time destinies of democracies. The population situation in Sweden and the ideological crisis produced by the severe decline in reproduction form the background for an analysis of the attitudes of individuals toward the population problem as they are influenced by conservative or radical affiliation, family status, and national identification. The effects of population decline on the economy of a nation are then discussed. An attempt is made to outline the broad general principles for a positive population policy in a democratic country." (Source: **Books on Population, 1939-Mid-1944.**)

Población, un problema para la democracia. "Capítulos preliminares bosquejan las relaciones entre las grandes oscilaciones seculares en las teorías económicas y demográficas, y destacan la importancia básica de la restitución de la población en los lejanos destinos de las democracias. El estado de la población en Suecia y la crisis ideológica producida por la pronunciada declinación de la reproducción, constituyen la base para un análisis de las actitudes de los individuos frente al problema de la población estando influenciados por su afiliación conservativa o radical, su estado de familia, e identificación nacional. Se expone los efectos de la disminución de la población sobre la economía de un país. Se hace un intento de resumen de los principios generales para una política positiva de población en un país democrático." (Traducción del comentario.)

NonG-5100.P63 1942 **Population Problems.** By Warren S. Thompson. 3d ed. New York and London, McGraw-Hill Book Company, Inc., 1942. 471 p.

"The third edition of this standard text embodies the most important results of recent population study and research, and also utilizes more recent census and vital statistics data." (Source: **Books on Population, 1939 Mid-1944.**)

Problemas de población. "La tercera edición de este texto estandard engloba los resultados mas importantes de los recientes estudios e investigaciones sobre población, y también utiliza los datos de los censos y estadísticas vitales mas recientes." (Traducción del comentario.)

NonG-5100.P65 1943 **Population Problems, a Cultural Interpretation.** By Paul Henry Landis. New York, Cincinnati, etc., American Book Company, 1943. 500 p. Unexamined.

"Part I, Population facts and population trends, defines the field of population within a sociological frame of reference, and outlines broad trends in growth. Part II, Cultural forces in vital process, devotes its ten chapters to the subject of general and differential fertility and mortality. Part III, Sex, age, and ethnic composition. Part IV, Sociocultural factors in the distribution of population, and Part V, Problems of migration, survey recent studies for the United States and interpret them in sociological terms. A final chapter considers a population policy for the United States." (Source: Pop. Ind., vol. 10, no. 1, January, 1944.)

Problemas de población, una interpretación cultural. "Parte I, Realidades y tendencias de la población, define el campo de la población en un marco sociológico de referencia, y resume las tendencias generales del crecimiento. Parte II, Fuerzas culturales en el proceso vital, dedica sus diez capítulos al tema de fertilidad general y diferencial, y de mortalidad. Parte III, Composición por sexo, edad y étnica, Parte IV, Realidades socioculturales en la distribución de la población, y Parte V, Problemas de migración, investigan los recientes estudios sobre Estados Unidos y los interpretan en términos sociológicos. Un capítulo final considera una política de población para los Estados Unidos." (Traducción del comentario.)

NonG-5100.T4 1937 **Teoría Matemática de la Población.** Por Carlos E. Dieulefait. Buenos Aires, Tall.Gráf.Porter Hnos., 1937. Universidad de Buenos Aires. Facultad de Ciencias Económicas. Instituto de Biometría. 12 p. Cuadernos de trabajo, no. 2.

Mathematical theory of population.

NonG-5100.T5 1943 **Teoría Matemática del Desarrollo Numérico de la Población.** Por Antonio H. Picón. Montevideo, Peña y Cía., 1943. Universidad de la República. Facultad de Ciencias Económicas y de Administración. Instituto de Estadística. 48 p. Cuaderno, no. 4.

Mathematical theory of the numerical development of population.

NonG-5110.E5 1939 **The Economics of a Declining Population.** By William B. Reddaway. New York, The Macmillan Co., 1939. 138 p. Unexamined.

"The phrase 'declining population' is used to include all the coming changes in the size and age-composition of the population of England and Wales. The first part gives a general discussion of population trends and population forecasting, and summarizes the present position in Great Britain on the basis of Enid Charles' estimates. The second part discusses the implications of these trends for various aspects of economic life." (Source: **Books on Population, 1939-Mid-1944.**)

La economía de una población en declino. "La frase 'población en declino' se usa para incluir los cambios a producirse en el volumen y composición por edad de la población de Inglaterra y Gales. La primer parte presenta una exposición general de la tendencias y pronósticos de la población, y resume el actual estado en Gran Bretaña, sobre la base de las estimaciones hechas por Enid Charles. La segunda parte es una exposición de las implicaciones de esas tendencias en distintos aspectos de la vida económica." (Traducción del comentario.)

NonG-5200.C5 1944 **Conceptos de Estadística Demográfica.** San Salvador, Tip.La Unión, Dutriz Hnos., 1944. El Salvador. Dirección General de Sanidad. 173 p. Publicaciones del Servicio Cooperativo Interamericano de Salud Pública.

Concepts of demographic statistics.

NonG-5200.M5 1942 **Manual de Estadística Vital.** Por Forrest Edward Linder. Con un prólogo del Dr. Rafael Schiaffino. Montevideo, Imp.Administración de Lotería, 1942. Uruguay. Ministerio de Salud Pública. 72 p.

Handbook of vital statistics. Includes: Importance of vital statistics; classification; global index; specific index; relation between specific index and global index; elements for the organization of vital statistics; summary of the principal factors of the interpretation of vital statistics.

Incluye: Importancia de la estadística vital;

clasificación; índices globales; índices específicos; relación entre los índices específicos y los índices globales; elementos de organización de la estadística vital; resumen de los factores principales en la interpretación de estadísticas vitales.

NonG-**5200**.V4 1923 **Vital Statistics.** An introduction to the science of demography. By George Chandler Whipple. Second edition. New York, John Wiley & Sons, Inc.; London, Chapman & Hall, Limited, 1923. 579 p.

Contents: Demography; statistical arithmetic; statistical graphics; enumeration and registration; prediction of future population; general death-rates, birth-rates, marriage-rates; specific death-rates; cause of death; analysis of death-rates; statistics of particular diseases; studies of deaths by age periods; probability; correlation; life tables; appendixes (including bibliography and model laws.)

Estadística vital. Introducción a la ciencia demográfica. Contenido: Demografía; aritmética estadística; gráficos estadísticos; enumeración y registro; predicciones sobre población futura; tasas generales de mortalidad, natalidad y nupcialidad; tasas específicas de mortalidad; causa de muerte; análisis de las tasas de mortalidad; estadísticas de enfermedades determinadas; estudios sobre mortalidad por períodos de edad; probabilidad; correlación; tablas de vida; apéndices (incluyendo bibliografía y modelos de leyes).

NonG-**5200**.V6 1943 **Un Vocabulario Bioestadístico. A Vital Statistics Vocabulary.** Términos estadísticos, matemáticos, demográficos y médicos. Inglés-español; español-inglés. Statistical, mathematical, demographic and medical terms. Washington, D.C., 1943. United States Department of Commerce. Bureau of the Census. 109 p.

NonG-**5211**.M5 1940 **Multiple Human Births.** Twins, triplets, quadruplets and quintuplets. New York, Doubleday, Doran and Company, Inc., 1941. 214 p.

Published in England under the title—publicado en Inglaterra bajo el título: **Twins and Super Twins.**

Nacimientos humanos múltiples.

NonG-**5220**.L5 1936 **Length of Life.** A study of the life table. By Louis I. Dublin and Alfred J. Lotka. New York, The Ronald Press Company, 1936. 400 p.

Duración de la vida. Un estudio de la tabla de vida.

NON-GEOGRAPHIC
5200

6000-6999 ECONOMICS—ECONOMIA

NonG-**6000**.B5 1940 **Business Cycles and Forecasting.** By Elmer C. Bratt. Revised edition. Chicago, Illinois, Business Publications, Inc., 1940. 814 p.

Contents: Seasonal variations in industry; concepts of balance; industrial growth and decadence; statistical methods for measuring and analyzing the business cycle; factors responsible for the cyclical nature of business; prominent business cycle theories; classification of business-cycle theories; opposing theoretical positions; business cycle history; the "great" depression; the 1937 downturn and since; schemes proposed for artificial control of cyclical movements; business stabilization and the desirability of moderate cyclical movements; business barometers (integral measures of general business conditions; measure of business processes and component parts of business change; credit and financial conditions); characteristic fluctuations of the business cycle; the forecasting of business cycles; problems of forecasting economic change.

Ciclos económicos y su previsión. Contiene: Variaciones estacionales en la industria; conceptos de balance; crecimiento industrial y decadencia; métodos estadísticos para la medición análisis del ciclo económico; factores causantes de la naturaleza cíclica de los negocios; teorías prominentes sobre los ciclos económicos; clasificación de las teorías sobre los ciclos económicos; teorías opuestas; historia de los ciclos económicos; "gran" depresión; la tendencia a bajar del año 1937 y el período siguiente; esquemas propuestos para el control artificial de los movimientos cíclicos; estabilización de los negocios y la conveniencia de moderados movimientos cíclicos; barómetros económicos (medidas integrales de condiciones económicas generales; medidas de procesos económicos y partes componentes de cambios económicos; crédito y condiciones financieras); fluctuaciones características del ciclo económico; la previsión de los ciclos económicos; problemas de la previsión de cambios económicos.

NonG-**6000**.B6 1928 **Business Cycles, The Problem and Its Setting.** By Wesley C. Mitchell. New York, 1928. National Bureau of Economic Research. Unexamined.

"This book should be read in its entirety by anyone seriously interested in time series analysis. Chapter III is on statistical analysis, while Chapter V contains a summary of the entire book." (Source: Croxton & Crowden, **Applied General Statistics,** 1943.)

Ciclos económicos—el problema y su solución. "Este libro debe ser leído enteramente por todos aquellos, que se interesen seriamente en el análisis de series cronológicas. El capítulo III trata del análisis estadístico, mientras que el capítulo V contiene un resumen del libro entero." (Traducción del comentario.)

NonG-**6000**.C4 1938 **Commodity Flow and Capital Formation.** By Simon Smith Kuznets. New York, 1938. National Bureau of Economic Research. Publication, no. 34. Unexamined.

Circulación de los productos y formación de capital.

NonG-**6000**.C6 1940 **The Conditions of Economic Progress.** By Colin Clark. London, Macmillan and Company, Limited, 1940. 504 p.

Contents: Summary and conclusions; purpose of economics; present level of economic welfare in different countries; unused potentialities of production; rate of growth of real income; flow of labour to tertiary production; occupational adjustments and maldistributions; productivity of primary industry; productivity of secondary industry; productivity of tertiary industry; morphology of economic growth; role of capital in economic progress; distribution of incomes between individuals; changing tendencies of consumption; terms of exchange; relations between investments and income; chart showing real income per head of employed population on 48-hour per week basis.

Las condiciones del progreso económico. Contenido: Resumen y conclusiones; propósito de la economía; nivel actual del bienestar económico en diferentes países; potencialidades de producción no aprovechadas; tasa de crecimiento de la renta real; movimiento del trabajo hasta la producción terciaria; ajustamientos y maldistribuciones en las ocupaciones; productividad de la industria primaria; productividad de la industria secundaria; productividad de la industria terciaria; morfología del crecimiento económico; papel que desempeña el capital en el progreso económico; distribución de la renta entre los individuos; tendencias a cambios en el consumo; términos de cambio; relaciones entre inversiones e ingreso; gráfico demonstrativo de la renta real por cabeza de la población ocupada, sobre la base de 48 horas por semana.

NonG-**6000**.S7 1940 **Survey of Economic Theory on Technological Change and Employment.** By Alexander Gourvitch. Philadelphia, Pennsylvania, 1940. U. S. Federal Works Agency. Work Projects Administration. National Research Project on Reemployment Opportunities and Recent Changes in Industrial Techniques. 252 p. Report, no. G-6.

Contents: Scope of the survey; historical introduction; general equilibrium theory in neoclassical economics; technological progress and cyclical fluctuations in employment; long-term trends.

Estudio de la teoría económica sobre cambios y ocupación tecnológica. Contenido: Alcance del estudio; introducción histórica; teoría general del equilibrio en la economía neoclásica; progreso tecnológico y fluctuaciones cíclicas en la ocupación; tendencias de largo término.

NonG-**6000**.T6 1944 **Theory of Games and Economic Behavior.** By John von Neumann and Oskar Morgenstern. Princeton, New Jersey, Princeton University Press, 1944. 625 p. Unexamined.

Teoría de los juegos y conducta económica.

NonG-**6100**.I5 1943 **International Trade and the National Income Multiplier.** By Fritz Machlup. Philadelphia, Pennsylvania. The Bakiston Co., 1943. 237 p.

An analysis of the interrelationships of international trade, capital movements, investment, and national income; and international economics, business cycle theory and monetary theory in general.

Comercio internacional y multiplicador de la renta nacional. Un análisis de las relaciones recíprocas del comercio internacional, movimientos de capital, inversiones e ingreso nacional; y economía internacional, teoría de los ciclos económicos y teoría monetaria en general.

NonG-**6100**.N6 1937 **National Income and Outlay.** By Colin Clark. London, Macmillan and Co., Ltd., 1937. 304 p.

Contents: Purpose and definition of national income measurements; population and income-earners; income assessable to income-tax, 1924-33; the whole national income; the distribution of the product of industry; redistribution of income through taxation; independent determination of the national income; the accumulation of capital; prices, real income and short-period determination of the national income; historical statistics of national income; productive sources of income; changes in consumption, investment, prices and cost in the trade cycle, 1929-36; the rate of economic progress; the number of incomes; number of incomes above tax exemption limit; stocks.

Renta y gasto nacional. Contenido: Propósitos y definiciones de las medidas de la renta nacional; población y percibidores de ingresos; ingresos sujetos al impuesto a la renta 1924-

33; la renta nacional completa; la distribución del producto de la industria; redistribución de la renta a través de la imposición; determinación independiente de la renta nacional; la acumulación de capital; precios, renta real y determinación a corto plazo del ingreso nacional; estadísticas históricas del ingreso nacional; fuentes productivas de rentas; cambios en el consumo, inversión, precios y costo en el ciclo comercial, 1929-36; la tasa del progreso económico; el número de ingresos; número de ingresos que sobrepasan el límite de la exención del impuesto; existencias.

NonG-**6100**.S3 **Studies in Income and Wealth.** By the Conference of Research in Income and Wealth. New York, National Bureau of Economic Research, 1937- . 8 vol.: **Vol. I:** Part one, "Concepts of National Income," by M. A. Copeland; part two, "Accounting Methodology in the Measurement of National Income," by Clark Warburton; part three, "On the Treatment of Corporate Savings in the Measurement of National Income," by Solomon Fabricant; part four, "Changing Inventory Valuations and Their Effect on Business Savings and on National Income Produced," by Simon Kuznets; part five, "Public Revenue and Public Expenditure in National Income," by Gerhard Colm; part six, "The Distinction Between 'Net' and 'Gross' in Income Taxation," by Carl Shoup; part seven, "Some Problems in Measuring Per Capita Labor Income," by Solomon Kuznets; part eight, "Income Parity for Agriculture," by O. C. Stine; **vol. II:** Part one, "On the Measurement of National Wealth," by Simon Kuznets; part two, "The Correction of Wealth and Income Estimates for Price Changes," by M. A. Copeland and E. M. Martin; part three, "National Income, Savings, and Investment," by Gottfried Haberler; part four, "Capital Gains in Income Theory and Taxation Policy," by Roy Blough and W. W. Hewett; part five, "Problems in Estimating National Income Arising from Production by Government," by G. C. Means (Laughlin Currie and R. R. Nathan concurring); part six, "Allocation of Benefits from Government Expenditures," by R. W. Nelson and Donald Jackson; **vol. III:** Part one, "American Studies of the Distribution of Wealth and Income by Size," by C. L. Merwin; part two, "Income Capitalization as a Method of Estimating the Distribution of Wealth by Size Groups," by Charles Stewart; part three, "The Use of Income Tax Data in the National Resources Committee Estimate of the Distribution of Income by Size," by Enid Baird and Selma Fine; part four, "The Volume and Components of Saving in the United States, 1933-1937," by R. W. Goldsmith with the assistance of Walter Salant; part five, "Three Estimates of the Value of the Nation's Output of Commodities and Services: A Comparison," by Clark Warburton; part six, "Some Problems Involved in Allocating Incomes by States," by R. R. Nathan; part seven, "Income and the Measurement of the Relative Capacities of the State," by P. H. Wueller; **vol. IV, Outlay and Income in the United States, 1921-1938,** by Harold Barger; **vol. V, Income Size Distributions in the United States, Part I; vol. VI:** "Income Measurement as Affected by Government Operations," by John Lindeman; "Measuring the Economic Impact of Armament Expenditures," by R. W. Goldsmith; "The Concept of Income Parity for Agriculture," by E. W. Grove; "Significance of International Transactions in National Income," by Rollin F. Bennett; "Forecasting National Income and Related Measures," by F. L. Thomsen and P. H. Bollinger; "A Statistical Study of Income Differences among Communities," by Herbert E. Klarman; "Adequacy of Estimates Available for Computing Net Capital Formation," by Wendell D. Hance; **vol. VII: Changes in Income Distribution during the great Depression,** by Horst Mendershausen; vol. VIII: "Measuring National Income in Colonial Territories," by Phyllis Deane; "Studies on Income and Wealth," by Paul Studensky. Vol. VI-VIII unexamined.

Estudios sobre renta y riqueza, de la Conferencia de Investigaciones sobre Renta y Riqueza. 8 vol.

NonG-**6100**.S5 1942 **Studies in the National Income, 1924-1938.** By Arthur Lyon Bowley. Cambridge, England, Cambridge University Press, 1942. National Institute of Economic and Social Research. 255 p. Economic and social studies, I. Unexamined.

The volume consists of four chapters: The definition of national income; some constituents of the national income; the census of production; price movements: Index of real income.

Estudios de la renta nacional en Gran Bretaña, 1924-38. El volumen consiste de cuatro capítulos: La definición de la renta nacional; algunos componentes de la renta nacional; el censo de producción; fluctuación de precios: Indice de la renta real.

NonG-**6200**.S5 1942 **Statistical Investigation of a Sample Survey for Obtaining Farm Facts.** By Raymond J. Jessen. Ames, Iowa, June 1942. Iowa State College of Agriculture and Mechanic Arts. Agricultural Experiment Station. 104 p. Research bulletin, no. 304.

Statistical Section, Rural Science Section, Agricultural Economics Subsection, in cooperation with U. S. Department of Agriculture, Bureau of Agricultural Economics, Agricultural Marketing Service, and Works Progress Administration.

Investigación estadística de una encuesta a base de muestras para la obtención de datos relativos a los establecimientos rurales.

NonG-**6250**.S5 1942 **Sampling Methods in Forestry and Range Management.** By F. X. Schumacher and R. A. Chapman. Durham, North Carolina, 1942. Duke University. School of Forestry. 213 p. Bulletin, no. 7. Unexamined.

"This book presents a treatment of statistical methods applicable . . . to a wide variety of field investigations common to many branches of science . . . The book is divided into three parts: Statistical background, direct estimates by sampling, and indirect estimates through regression, and concludes with a discussion of 'certain practical aspects of sampling.' " (Source: **Journal of the American Statistical Association,** vol. 39, no. 225, March 1944.)

Métodos de muestreo en la administración de la silvicultura y de campos de pastoreo. "Este libro presenta un tratado de métodos estadísticos aplicables . . . a amplia variedad de investigaciones en el campo, comunes a muchas ramas de la ciencia . . . El libro está dividido en tres partes: Fundamento estadístico, estimaciones directas por medio del muestreo, e indirectas por medio de regresiones, y concluye con una exposición de ciertos aspectos prácticos del muestreo." (Traducción del comentario.)

NonG-**6400**.S5 1943 **Statistical Methods in Industry.** A course of six lectures given in Sheffield by Mr. Leonard Henry Caleb Tippett, with an introductory lecture by Professor E. S. Pearson, and a preface by Sir William J. Larke. London, 1943. British Iron and Steel Federation. Iron and Steel Industrial Research Council. 74 p. Unexamined.

Métodos estadísticos en la industria. Un curso de seis clases dictadas en Sheffield por Leonard Henry Caleb Tippett, con una clase introductora dictada por el Prof. E. S. Pearson, y un prefacio de William J. Larke.

NonG-**6614**.R5 1941 **Railway Statistical Terms.** A collection of definitions of words and phrases frequently used in discussions of railway statistics. Washington, D.C., 1941. U. S. Interstate Commerce Commission. Bureau of Transport Economics and Statistics. 58 p. Mimeographed. Statement, no. 4119. Unexamined.

Términos de la estadística ferroviaria. Una colección de definiciones de palabras y frases usadas con frecuencia en exposiciones sobre estadísticas ferroviarias.

7000-7999 HEALTH AND CULTURAL STATISTICS—ESTADISTICAS DE SALUBRIDAD Y CULTURA

NonG-**7110**.P5 1942 **Public Health Statistics.** By Marguerite Franklyn Hall. New York, London, P. B. Hoeber, Inc., 1942. 408 p.

Contents: Introduction and orientation; general principles of computation and handling of mass data; tabular presentation; graphical presentation; population data; estimates of populations; standard millions; vital events; rates; mortality rates; vital statistics other than mortality; variation and trends in rates; ratios and index numbers; construction of a life table; mass data, preliminary definitions; measures of central tendency; measures of variability; measures of relationship and trend; theory of probability; point binomial; normal curve; measures of reliability; prerequisites to the application of statistical methods.

La estadística de salud pública. Incluye: Introducción y orientación; principios generales de cálculo y manejo de datos colectivos; presentación tabular; representación gráfica; datos de población; estimaciones de poblaciones; el millón estandard; acontecimientos vitales; tasas; tasas de mortalidad; estadística vital aparte de la mortalidad; variación y tendencias de tasas; proporciones y números índices; construcción de una tabla de vida; datos colectivos, definiciones preliminares; medidas de tendencia central; medidas de variabilidad; medidas de relación y tendencia; teoría de probabilidad; punto binomial; curva normal; medidas de confianza; requisitos previos exigidos para la aplicación de métodos estadísticos.

NonG-**7150**.P5 1939 **Problems of Aging, Biological and Medical Aspects.** Edited by E. V. Cowdry. Sponsored by the Josiah Macy, Jr. Foundation. Baltimore, Maryland, Williams & Wilkins, 1939. 758 p. Unexamined.

"Current knowledge concerning the problems of aging and senescence in plants, animals, and man is synthesized and interpreted in a series of analyses by 25 specialists . . . Two of the contributions concern human demography. Clark Wissler, in a discussion of 'Human cultural levels', contrasts modern vital phenomena with those of other culture groups, while Louis I. Dublin discusses 'Longevity in retrospect and

prospect'." (Source: **Books on Population, 1939-Mid-1944.**)

Problemas del envejecimiento, aspectos médicos y biológicos. "El conocimiento general relativo a los problemas del envejecimiento y senectud de las plantas y animales y del hombre está sintetizado e interpretado en una serie de análisis, por 25 especialistas . . . Dos de las colaboraciones se refieren a demografía humana. Clark Wissler, en un tratado sobre nivel de cultura humana contrasta fenómenos vitales modernos con aquéllos de otros grupos de cultura, mientras Louis I. Dublin trata sobre longevidad en retrospectiva y en perspectiva." (Traducción del comentario.)

NonG-**7180.**I5 1940 **Introduction to Medical Biometry and Statistics.** By Raymond Pearl. Third edition, revised and enlarged. Philadelphia and London, W. B. Saunders Company, 1940. 537 p.

Includes: Definitions and orientation; landmarks in the history of biostatistics; the raw data of biostatistics; tabular presentation of statistical data; original scientific records and their translation to tabular form; graphic presentation of statistical data; rates and ratios; life tables; standardized and corrected death-rates; the sampling error concept; elementary theory of probability; special theorems in probability; measurement of variation; measurement of correlation; partial correlation; simple curve fitting; the logistic curve.

Introducción a la biometría médica y estadística. Incluye: Definiciones y orientación; rasgos sobresalientes en la historia de la bioestadística; los datos crudos de bioestadística; representación tabular de datos estadísticos; registros científicos originales y sus traducciones en forma tabular; representación gráfica de datos estadísticos; tasas y proporciones; tablas de vida; tasas de mortalidad estandardizadas y corregidas; el concepto del error de muestreo; teoría elemental de probabilidad; teoremas especiales de probabilidad; medida de variación; medida de correlación; correlación parcial; ajuste de curvas simples; la curva logística.

NonG-**7180.**P5 1939 **Principles of Medical Statistics.** By A. Bradford Hill. Second edition, revised and enlarged. London, The Lancet Limited, 1939. 189 p.

Includes: Aim of statistical methods; selection; presentation of statistics; variability of observations; calculation of the standard deviation; problems of sampling (averages, proportions, differences, chi square); coefficient of correlation; calculation of the correlation coefficient; life tables and survival after treatment; common fallacies and difficulties; calculation of standardised death-rates; summary and conclusions.

Principios de la estadística médica. Incluye: Fin del método estadístico; selección; presentación de datos; variabilidad de observaciones; cálculo de la desviación estandard; problemas del método de muestra (promedios, porcentajes, diferencias, chi al cuadrado); coeficiente de correlación; cálculo del coeficiente de correlación; tablas de vida, y supervivencia después de tratamiento; fallas y dificultades comunes; cálculo de tasas estandardizadas de mortalidad; resumen y conclusiones.

NonG-**7300.**S5 1946 **Sobre la Cuantificación del Estilo Literario.** Una contribución al estudio de la unidad de autor en "La Celestina" de Fernando de Rojas. Por José V. Montesino Samperio. Caracas, Escuela Técnica Industrial Tall.de Artes Gráf., 1946. 63 p.

Published also in—publicado también en: **Revista Nacional de Cultura,** no. 55-56, 1946.

Quantitative method applied to the study of literary style.

NonG-**7320.**S5 1944 **The Statistical Study of Literary Vocabulary.** By G. Udny Yule. Cambridge, England, Cambridge University Press, 1944. 306 p. Unexamined.

El estudio estadístico del vocabulario literario.

9000-9999 LABOR AND LIVING CONDITIONS—CONDICIONES DE TRABAJO Y DE VIDA

NonG-**9000.**S5 1942 **Social Research.** By George A. Lundberg. New York, London, Longman's, Green and Co., 1942.

Includes chapters on sample methods, schedule construction, questionnaire techniques, measurement of attitudes and opinions, evaluation of institutional performance, interviewing social bookkeeping.

Investigación social. Incluye capítulos sobre métodos de muestra, proyección de cédulas, técnicas de cuestionarios, medida de actitudes y opiniones, avalúo de funcionamento de instituciones, entrevista, contabilidad social.

NonG-**9000.**T5 1942 **Teoría y Métodos de Estadística del Trabajo.** Por José Figuerola. Con un prólogo de Robert Guye. Buenos Aires, Editorial Labor, S.A., 1942. 608 p. Enciclopedia de ciencias jurídicas y sociales, sección ciencias sociales.

Theory and methods of labor statistics. Includes: Preliminary observations on basic principles, measurement of social phenomena, origin and development of labor statistics in Argentina; **the social scene:** Classification of occupations;

the gainfully occupied population; organization of labor unions; **working and living conditions of laborers:** Hiring of laborers; wages and duration of work; the cost of living; **abnormalities of work:** Unemployment; labor conflicts; occupational accidents and professional diseases; **appendixes:** Organization of labor statistics services and the training of technicians; the technique of averages; Argentine legal stipulations applying to labor statistics; international conventions; tentative bibliography; graphic presentation—rules to achieve uniformity.

Incluye: Observaciones preliminares sobre principios básicos, medición de los hechos sociales, origen y desarrollo de la estadística del trabajo en la Argentina; **el medio social:** Clasificación de actividades; la población activa; organización sindical; **condiciones de trabajo y de vida de los trabajadores:** Contratación de trabajadores; salarios y duración del trabajo; el costo de la vida; **anormalidades del trabajo:** Desocupación; conflictos del trabajo; accidentes del trabajo y enfermedades profesionales; **apéndices:** Organización del servicio estadístico del trabajo y formación de técnicos; la técnica de los promedios; disposiciones legales argentinas sobre estadística del trabajo; acuerdos internacionales; ensayo bibliográfico; representación gráfica—reglas para lograr su uniformidad.

NOTE

The Alphabetical Index is a consolidated index giving titles, issuing agencies, authors, and subjects, in alphabetical order according to the English alphabet. Under *issuing agencies and authors,* the titles of their publications are given. Under each *subject,* a reference is given to a decimal subject number, indicating that publications on this subject may be found under that number in the Subject Index, and under that number within each country in the Main List. Certain special classes of publications ("class pull-togethers") are listed by title under the following headings: ANNUALS; BIBLIOGRAPHIES; DICTIONARIES AND GLOSSARIES; DIRECTORIES; LAWS, STATUTES, ETC.; NOMENCLATURES; ORGANIZATION CHARTS (of statistical organizations); PERIODICALS; SERIES.

NOTA

El Indice Alfabético es un índice consolidado que contiene títulos, organismos editores, autores, y materias, en orden alfabético de acuerdo con el alfabeto inglés. Bajo los *organismos editores y los autores* se presentan los títulos de sus respectivas publicaciones. Bajo cada *materia* se proporciona una referencia del número decimal de la materia, que indica que las publicaciones acerca de esa materia pueden ser encontradas bajo dicho número en el Indice de Materias y, bajo ese mismo número, dentro de cada país en la Lista Principal. Ciertas clases especiales de publicaciones están enumeradas por títulos bajo los siguientes encabezamientos: ANNUALS (publicaciones anuales); BIBLIOGRAPHIES (bibliografías); DICTIONARIES AND GLOSSARIES (diccionarios y glosarios); DIRECTORIES (directorios); LAWS, STATUTES, ETC. (leyes, estatutos, etc.); NOMENCLATURES (nomenclaturas); ORGANIZATION CHARTS (esquemas de organización de organismos estadísticos); PERIODICALS (publicaciones periódicas); SERIES (publicaciones en series).

ALPHABETICAL INDEX — INDICE ALFABETICO

(Titles, authors, issuing-agency, and topics—Títulos, autores, organismos editores y tópicos)

ABASTECIMIENTOS DE MERCANCIAS Y MATERIALES:
GENERAL—*véase* 6020.
ESPECIFICOS — *véase* las respectivas clasificaciones.
ABORTION—*see* 5212.
ABORTOS—*véase* 5212.
Abrisqueta, Francisco de.
Col-*1420*.A5, May-June '46. "El Indice de Precios del Comercio en General."
Col-*1420*.A5, Mar.-Apr. '46. "Algunas Instituciones Económicas Ecuatorianas: El Banco Central del Ecuador y Su Oficina de Investigaciones Económicas."
Col-*1420*.A5, Nov. '45. "Reglamentación sobre el Cálculo de los Medios de Pagos en Circulación y de las Exigibilidades Monetarias en Colombia."
Col-*1420*.A5, July-Aug. '45. "El Problema de los Arrendamientos de Viviendas en Bogotá."
Col-*1420*.A5, May '45. "El Desarrollo de Bogotá y las Investigaciones del Costo de la Vida Obrera."
Col-*1420*.A5, Apr. '45. "Factores de Examen de la Coyuntura Económica Colombiana."
Col-*1420*.A5, Feb. '45. "Estadísticas de la Producción."
Accident Facts. USA-*7160*.A5.
ACCIDENTES:
GENERAL—*véase* 7160; 9400.
DE TRANSITO Y TRANSPORTES—*véase* 7160.
DEFUNCIONES PROVENIENTES DE —*véase* 5212; 7160; 9400.
INDUSTRIALES (incluyendo PROTECCION)—*véase* 9400.
MORBILIDAD POR—*véase* 7160.
PREVENCION DE—*véase* 9400; 7160.
SEGUROS CONTRA—*véase* 6730; 9952.
Accidentes de Tránsito Terrestre por Circulación de Vehículos. RepDom-*7160*.A5.
ACCIDENTS:
GENERAL—*see* 7160; 9400.
DEATHS—*see* 5212; 7160; 9400.
INDUSTRIAL—*see* 9400.
INSURANCE—*see* 6730; 9952.
MORBIDITY—*see* 7160.
PREVENTION—*see* 9400; 7160.
TRAFFIC—*see* 7160.
ACCIONES, TITULOS Y VALORES (COMERCIALES, FINANCIEROS, INDUSTRIALES, incluyendo PRECIOS E INDICES)—*véase* 6721.
ACCOUNTING—*see* 1100.
Accounting Control. NonG-*3110*.A5, 1936.
"Accounting Methodology in the Measurement of National Income." NonG-*6100*.S3.
"Accounting Statistics." USA-*3000*.G7, 1946.
ACERO E HIERRO—*véase* 6400.
ACREAGE OF FARMS AND RANCHES—*see* 6200.
Acta de Sesión No. 19 Celebrada el 1° de Agosto de 1946 (Aprobada el 5 de Septiembre de 1946) (Consejo Interamericano Económico y Social). Int-*1300*.A4, 1946.
Acta Final da Conferência Interamericana de Agricultura. Int-*6200*.I4, 1930.1.
Acta Final de la Conferencia de Comisiones de Fomento Interamericano, Nueva York, del 9 al 18 de Mayo de 1944. Int-*8200*.C6, 1944.1, Sp.
Acta Final de la Conferencia Interamericana de Agricultura. Int-*6200*.I4, 1930.1.
Acta Final de la Conferencia Interamericana sobre Sistemas de Control Económico y Financiero, Washington, 30 de Junio a 10 de Julio de 1942. Int-*6000*.I45, 1942, Sp.
Acta Final de la Conferencia Marítima Interamericana, Unión Panamericana, Wáshington, del 25 de Noviembre al 2 de Diciembre de 1940. Int-*6615*.I5, 1940, Sp.
Acta Final de la Primera Conferencia de Ministros y Directores de Educación de las Repúblicas Americanas Celebrada en Panamá del 27 de Septiembre al 4 de Octubre de 1943. Int-*7200*.C6, 1943.1, Sp.
Acta Final de la Segunda Conferencia Interamericana de Agricultura. Int-*6200*.I4, 1942.1, Sp.
Acta Final del Octavo Congreso Panamericano del Niño, Wáshington del 2 al 9 de Mayo de 1942. Int-*9951*.P5, 1942, Sp.
Acta Final del Segundo Congreso Inter-Americano de Municipios, Santiago de Chile, 15 al 21 de Septiembre de 1941. Int-*8000*.C5, 1941.1.
Acta Final del Segundo Congreso Interamericano de Turismo. Int-*5300*.I4, 1941.1 Sp.
Acte Final de la Conférence Interaméricaine sur l'Agriculture. Int-*6200*.I4, 1930.1.
Actes du Ier Congrès International de Sylviculture. Int-*6250*.C5, 1926.
La Actividad Industrial durante los Primeros . . . Meses de . . . , según los Números Indices Correspondientes a los Salarios Pagados, a la Ocupación de Obreros y a las Horas-Obrero

ALUMBRADO Y ENERGIA ELECTRICA —*véase* 6400.
ALUMINIO—*véase* 6400.
ALUMINUM—*see* 6400.
Alvim Pessoa, Constança.
Bras-*3040*.R5, Jan.-Mar. '44, Dieulefait. "Curso de Estatística: Capítulo I, Introdução, Valores Indicativos de Distribuicões Estatísticas, Elementos de Interpolação e de Ajustamento pelo Processo dos Mínimos Quadrados."
American Academy of Political and Social Science.
USA-*3040*.A5. *The Annals of the American Academy of Political and Social Science.*
USA-*3040*.A5, Nov. '36. "The American People: Studies in Population."
American Agricultural Series. Int-*6200*.A8.
American Agriculture, 1899-1939: A Study of Output, Employment and Productivity. USA-*6200*.A7, 1942.
The American-Born in Canada, a Statistical Interpretation. Can-*5120*.A5, 1943.
American Congress on Surveying and Mapping.
NonG-*4000*.S5. *Surveying and Mapping.*
American Council of Learned Societies.
Int-*1200*.H4. *Handbook of Latin American Studies.*
Int-*1200*.S6, 1815-1931. *Serial Publications of Foreign Governments, 1815-1931.*
NonG-*1800*.M5, 1936. *Manual on Methods of Reproducing Research Materials.*
American Economic Association.
USA-*6000*.A5. *American Economic Review.*
USA-*6000*.A5, Sept. '46. "Handbook: List of Members of the American Economic Association, 1946."
American Economic Review. USA-*6000*.A5.
American Embassy, Buenos Aires, Argentina.
Arg-*2100*.C5, 1945. *Argentine Fourth General Census Postponed.*
American Farm Economic Association.
USA-*6200*.J5. *Journal of Farm Economics.*
American Geographical Society.
Int-*4000*.M4, 1945. *Index to Map of Hispanic America, 1:1,000.000.*
USA-*4000*.G5. *Geographical Review.*
USA-*4000*.G5. Jan. '46. "The Map of Hispanic America on the Scale of 1:1,000.000."
American Geographical Society of New York.
Int-*4000*.M5, 1942. *Map of the Americas, 1:5,000,000.*
American Institute of Actuaries.
USA-*6730*.R5. *The Record of the American Institute of Actuaries.*
American Journal of Mathematics. USA-*7300*.A5.
American Journal of Public Health and the Nation's Health. USA-*7110*.A4.
American Journal of Sociology. USA-*9000*.A4.
American Library Association.
Int-*1200*.S6, 1815-1931. *Serial Publications of Foreign Governments, 1815-1931.*
NonG-*1500*.L2, 1942. *A Library Classification for Public Administration Materials.*
American Mathematical Society.
USA-*3040*.M5. *Mathematical Reviews.*
USA-*7300*.A5. *American Journal of Mathematics.*
American Nation Series. Int-*1950*.A5.
"The American People: Studies in Population." USA-*3040*.A5, Nov. '36.
American Public Health Association.
USA-*7110*.A4. *American Journal of Public Health and the Nation's Health.*
USA-*7110*.A5. *American Public Health Association Year Book.*
American Scientific Congress.
Int-*1300*.A5, 1940.3. *Proceedings of the Eighth American Scientific Congress.*
Bras-*2100*.R5, 1940.7. *A Glimpse into the Coming Fifth Census of Brazil (September 1st-1940).*
Bras-*3030*.S7, 1940. *Statute, Basic Principles, Social Leadership of the Brazilian Statistical System.*
Col-*1420*.A5, Apr. 25, '40, Higuita. "Estudio Histórico Analítico de la Población Colombiana en 170 Años."
American Scientific Congress, Mexican Delegation.
Int-*1300*.A5, 1940.3, México. "Mortalidad en el Medio Rural de la República Mexicana."
American Society for Testing Materials.
NonG-*3160*.A5, 1945. *A.S.T.M. Manual on Presentation of Data.*
American Society in Wartime. USA-*9000*.A5, 1943.
American Society of Mechanical Engineers.
NonG-*3260*.A5, 1943. *American Standard Engineering and Scientific Graphs for Publications.*
NonG-*3260*.T5, 1938. *Time-Series Charts.*
American Sociological Review. USA-*9000*.A6.

American Sociological Society.
USA-*9000*.A6. *American Sociological Review.*
USA-*9000*.A6, Dec. '44. "Representative Books in Sociology since 1938."
American Standard Engineering and Scientific Graphs for Publications. NonG-*3260*. A5, 1943.
American Standards Association.
NonG-*3160*.C5, 1942. *Control Chart Method of Controlling Quality during Production.*
NonG-*3160*.G5, 1941. *American War Standards: Guide for Quality Control and Control Chart Method of Analyzing Data.*
NonG-*3260*.A5, 1943. *American Standard Engineering and Scientific Graphs for Publications.*
NonG-*3260*.T5, 1938. *Time-Series Charts.*
"The American Statistical Association." Int-*1300*.A5, 1940.3, Stephan.
American Statistical Association.
USA-*3040*.J5. *Journal of the American Statistical Association.*
USA-*3040*.J5, June '45, a. "The Federal Government's Statistical Program for Reconversion and Postwar Adjustment: A Round Table."
USA-*3040*.J5, June '45, b. "Management of Field Work and Collection of Statistics: A Round Table."
USA-*3040*.J5, June '45, c. "Problems of Integrating Federal Statistics: A Round Table."
USA-*3040*.J5, Mar. '45. Opportunities for statisticians; a series of articles on statistical work in the federal government, in state and local governments, in teaching and research, and in industry. *Journal of the American Statistical Association.*
USA-*3040*.J5, Dec. '26. "Requirements for Statisticians and Their Training."
——. Biometrics Section.
USA-*3040*.B5. *Biometrics Bulletin.*
USA-*3040*.B5, June '45. "Teaching Statistics at the Department of Agriculture Graduate School in Washington."
USA-*3040*.B5, June '45, Schumacher. "Statistical Method in Forestry."
USA-*3040*.B5, Apr. '45. "Graduate Work in Statistics at Columbia University."
"American Studies of the Distribution of Wealth and Income by Size." NonG-*6100*.S3.
American Transit Association.
USA-*6610*.T5. *Transport Fact Book.*
American University. School of Social Sciences and Public Affairs.
Int-*6610*.T5, 1942. *Transportation in Latin America.*
American War Standards: Guide for Quality Control and Control Chart Method of Analyzing Data. NonG-*3160*.G5, 1941.
The American Year Book. USA-*1910*.A5.
Americana Corporation. Latin American Division.
Int-*1900*.L4, 1943. *Latin America.*
AMMUNITION—*see* 6400.
Anales de Economía y Estadística. Col-*1420*.A5.
Anales de la Sociedad Científica Argentina. Arg-*3040*.A5.
Anales de la Sociedad Rural Argentina. Arg-*6200*.A4.
ANALFABETISMO—*véase* 7200; 5110.
El Analfabetismo. Ur-*7200*.A5, 1940.
El Analfabetismo en la Argentina. Arg-*2230*.C5, 1943.1.
"El Analfabetismo en la Argentina." Arg-*3040*.R5, June '45, Correa Avila.
El Analfabetismo en la República Argentina. Arg-*7200*.A4, 1939.
El Analfabetismo en Puerto Rico . . . USA PR-*7200*.A5, 1945.
"Analfabetismo Urbano y Rural." Col-*1420*.A5, May 5, '42, Rodríguez.
Analfabetismo y Cultura Popular en América. Int-*7200*.A5, 1941.
El Analfabetismo y las Funciones del Consejo Nacional de Educación. Arg-*7200*.A6, 1938.
Análises de Resultados do Censo Demográfico. Bras-*2200*.A5, 1940.
ANALISIS:
DE COSTOS—*véase* 1100.
ESTADISTICA—*véase* 3105.
"Análisis de los Resultados del Primer Censo Oficial de Empresas Industriales y Comerciales Radicadas en Panamá, Ejecutado en el Año de 1938." Pan-*6000*.R5, Apr. '39.
Análisis de los Sueldos Pagados por el Estado. Guat-*9200*.A5, 1946.
Análisis del Comercio Exterior. Ven-*6900*.A5, 1913-36.
Análisis Estadístico Aplicado a los Trabajos de Investigación en Agricultura y Biología. NonG-*3330*.Trucco, 1944.
ANALYSIS:
COST—*see* 1100.
STATISTICAL—*see* 3105.
Analysis of Specified Farm Characteristics for Farms Classified by Total Value of Products. USA-*6200*.A8, 1943.
ANATOMIA—*véase* 7180.
ANATOMY—*see* 7180.

ANCIANOS, ASILOS PARA—*véase* 9951.
Andino, Alberto.
In-*3000*.S5, 1940. "Las Actividades Estadísticas de Cuba."
Andrade, Víctor.
Bol-*9000*.M5, 1943. *Memorándum: Elementos Generales sobre las Condiciones de Trabajo, de Legislación y de Otros Problemas Sociales de la República de Bolivia.*
"Anexo a Resolução No. 243: Nomenclatura Brasileira para a Classificação de Industrias." Bras-*1420*.B5, Sept. '44.
Anexo aos Questionários F 1, 2 e 3 do I.A.S.I. para Estabelecimento de uma Futura Definição das Exigências Mínimas para Admissão do Pessoal Estatístico. Bras-*3300*.A5, 1945.
Anexo Estatístico a la Memoria del Servicio Nacional de Salubridad. Chile-*7110*.M55.
Anglo-American Caribbean Commission.
Int-*1900*.C5, 1943. *The Caribbean Islands and the War.*
Int-*8200*.R3. *Report of the Anglo-American Commission to the Governments of the United States and Great Britain.*
The Anglo-South American Handbook. Int-*1900*.S5.
ANIMALES, ENFERMEDADES DE—*véase* 6230.
ANIMALS, DISEASES OF—*see* 6230.
Annals of Eugenics. GrBr-*7300*.A5.
The Annals of Mathematical Statistics. NonG-*3040*.A3.
The Annals of the American Academy of Political and Social Science. USA-*3040*.A5.
Annex I, League of Nations Estimated Balance of International Payments for the Year . . . (8-page form for country to fill); and Annex II, Explanatory Notes to Scheme of International Balance of Payments (11 pages). Int-*3205*.B5, 1945, Annexes.
Annuaire des Statistiques du Travail. Int-*9000*.A5.
Annuaire du Canada. Can-*1410*.A5.
Annuaire International de Statistique Agricole. Int-*6200*.I9.
Annuaire Statistique de la Société des Nations. Int-*1410*.A4.
Annuaire Statistique Interaméricain. Int-*1410*.A7.
Annual Airline Statistics: Domestic Carriers. USA-*6611*.A5.
Annual Epidemiological Report: Corrected Statistics of Notifiable Diseases for the Year 1938. Int-*7110*.R5, 1938.
Annual Report (Banco Central de la República Argentina). Arg-*6720*.A5.
Annual Report of Cheques Cashed Against Individual Accounts and Equation of Exchange. Can-*6720*.A5.
Annual Report of Mental Institutions. Can-*7130*.A4.
Annual Report of Statistics of Criminal and Other Offences. Can-*8300*.A5.
Annual Report of the Board of Governors of the Federal Reserve System. USA-*6720*.A3.
Annual Report of the Central Statistical Board. USA-*3030*.A6.
Annual Report of the Chief of Engineers, U. S. Army, Part I. USA-*6615*.A5.I.
Annual Report of the Chief of Engineers; Part 2, Commercial Statistics, Water-Borne Commerce of the United States. USA-*6615*.A5.II.
Annual Report of the Comptroller of the Currency. USA-*6720*.A5.
Annual Report of the Director of the Institute of International Education. Int-*7200*.A7.
Annual Report of the Farm Credit Administration. USA-*6720*.A7.
Annual Report of the Federal Deposit Insurance Corporation. USA-*6730*.A5.
Annual Report of the Fiscal Department (Banque Nationale de la République d'Haïti). Haiti-*6000*.A3.
Annual Report of the Governor of Puerto Rico. USA PR-*1410*.A5.
Annual Report of the National Bureau of Economic Research. USA-*6000*.A6.
Annual Report of the Railroad Retirement Board. USA-*9952*.A5.
Annual Report of the Secretary of the Treasury on the State of the Finances. USA-*8800*.A5.
Annual Report of the Securities and Exchange Commission. USA-*6721*.A5.
Annual Report of the Surgeon General of the U. S. Public Health Service. USA-*7110*.A7.
Annual Report of the Trade of Canada. Can-*6900*.T5.
Annual Report of Tuberculosis Institutions. Can-*7130*.A6.
Annual Report of Current Benefit Years Under the Unemployment Insurance Act. Can-*9952*.A5.
Annual Report of the Mineral Production of Canada. Can-*6300*.A5.
Annual Report on Vital Statistics of Canada. Can-*5200*.V5.
Annual Report Presented by the Manager to the Board of Directors (Banco de la República). Col-*6720*.A5.
Annual Report, U. S. National Housing Agency. USA-*9900*.A5.
ANNUALS
In specific subject fields, *see* the particular field.
GENERAL STATISTICAL—*see* 1410.

ANNUALS—cont.

Arg-*9950*.M5. *Memoria de la Caja Nacional de Jubilaciones y Pensiones Civiles.*
Bol-*5000*.D5. *Demografía.*
Bol-*6000*.M5. *Minería, Transportes y Comunicaciones.*
Bol-*6230*.E5. *Estadística Agropecuaria.*
Bol-*6300*.M5. *Memoria Anual* (Banco Minero de Bolivia).
Bol-*6400*.15. *Industria.*
Bol-*6720*.E6. *Finanzas.*
Bol-*6720*.M5. *Memoria Anual* (Banco Central de Bolivia).
Bol-*6900*.C5. *Comercio Exterior.*
Bol-*9950*.M5. *Memoria de la Caja de Seguro y Ahorro Obrero.*
Bras-*1410*.A5. *Anuário Estatístico do Brasil.*
Bras-*1410*.B3. *Brasil* (Ministério das Relações Exteriores).
Bras-*1410*.B5. *Brazil* (Ministry of Foreign Affairs).
Bras-*3030*.R4. *Relatório do Instituto Brasileiro de Geografia e Estatística.*
Bras-*5200*.A5. *Anuário de Bioestatística.*
Bras-*5300*.E5. *Estatística Geral Migratoria.*
Bras-*6000*.Q5. *Quadros Estatísticos* (Diretoria Estatística Econômica e Financeira).
Bras-*6210*.A4. *Anuário Açucareiro.*
Bras-*6210*.A5. *Anuário Algodoeiro.*
Bras-*6210*.A6. *Anuário Estatístico do Café.*
Bras-*6614*.B5. *Estatística das Estradas de Ferro do Brasil.*
Bras-*6615*.M6. *Movimento Marítimo e Fluvial do Brasil.*
Bras-*6720*.B5. *Boletim Estatístico* (Banco do Brasil). *Statistical Bulletin* (Bank of Brazil).
Bras-*6720*.R4. *Relatório do Banco do Brasil.*
Bras-*6720*.R6. *Resumen de la Memoria del Banco do Brasil.*
Bras-*6900*.B6. *Boletim Estatístico da Diretoria das Rendas Aduaneiras.*
Bras-*6900*.C3. *Comércio Exterior do Brasil.*
Bras-*6900*.C4. *Comércio Exterior do Brasil, Intercambio por Países.*
Bras-*6900*.C6. *Comércio Exterior do Brasil: Resumo por Mercadorias.*
Bras-*8000*.R4. *Relatório do Departamento Administrativo do Serviço Público.*
Bras-*9950*.R5. *Relatório Anual do Serviço de Estatística da Previdência e Trabalho.*
Can-*1410*.C5. *Canada, the Official Handbook of Present Conditions and Recent Progress.*
Can-*1410*.C7. *The Canada Year Book.*
Can-*5200*.P4. *Preliminary Annual Report, Vital Statistics of Canada. Rapport Préliminaire Annuel, Statistiques Vitales du Canada.*
Can-*5200*.V5. *Vital Statistics. Statistiques Vitales.*
Can-*5220*.*C5*. *Canadian Life Tables.*
Can-*6010*.P6. *Prices and Price Indexes.*
Can-*6020*.S5. *Survey of Production in Canada.*
Can-*6210*.R5. *Report of the Grain Trade of Canada.*
Can-*6230*.L5. *Live Stock and Animal Products Statistics. Statistique de Bétail et des Produits Animaux.*
Can-*6240*.F5. *Fisheries Statistics of Canada. Statistique des Pêcheries.*
Can-*6300*.A5. *Annual Report on the Mineral Production of Canada.*
Can-*6300*.C5. *Coal Statistics for Canada.*
Can-*6400*.C6. *Central Electric Stations in Canada.*
Can-*6400*.I7. *Iron and Steel and Their Products in Canada.*
Can-*6400*.M6. *The Manufacturing Industries of Canada.*
Can-*6500*.R5. *Report on the Construction Industry in Canada.*
Can-*6611*.C3. *Civil Aviation.*
Can-*6612*.H5. *The Highway and the Motor Vehicle in Canada.*
Can-*6612*.M5. *Motor Carriers.*
Can-*6614*.E5. *Electric Railways of Canada.*
Can-*6614*.S5. *Statistics of Steam Railways of Canada.*
Can-*6615*.C5. *Canal Statistics.*
Can-*6622*.T5. *Telegraph Statistics. Statistiques des Télégraphes.*
Can-*6622*.T6. *Telephone Statistics. Statistiques du Téléphone.*
Can-*6720*.A5. *Annual Report of Cheques Cashed Against Individual Accounts and Equation of Exchange.*
Can-*6720*.C5. *Commercial Failures in Canada.*
Can-*6900*.T5. *Trade of Canada. Commerce du Canada.*
Can-*6910*.C5. *The Canadian Balance of International Payments.*
Can-*7130*.A4. *Annual Report of Mental Institutions. Rapport Annuel sur les Institutions pour Maladies Mentales.*
Can-*7130*.A6. *Annual Report of Tuberculosis Institutions. Rapport Annuel sur les Institutions pour Tuberculeux.*
Can-*7200*.B5. *Biennial Survey of Education in Canada.*
Can-*8300*.A5. *Annual Report of Statistics of Criminal and Other Offences. Rapport*

ANNUALS—cont.

ElSal-*7000*.I5. *Informe de las Labores Desarrolladas por el Poder Ejecutivo en los Ramos de Cultura y Asistencia Social.*

Guat-*1410*.E5. *Estadísticas Gráficas* (Secretaría de Hacienda y Crédito Público).

Guat-*1410*.M5. *Memoria de las Labores del Poder Ejecutivo en el Ramo de Hacienda y Crédito Público.*

Guat-*6000*.M5. *Memoria del Ministerio de Comunicaciones y Obras Públicas.*

Guat-*6200*.M5. *Memoria de las Labores del Poder Ejecutivo en el Ramo de Agricultura.*

Guat-*6200*.R4. *Revista Agrícola.*

Guat-*6900*.C5. *Cuadros de la Dirección General de Aduanas.*

Guat-*7110*.B5. *Boletín Sanitario de Guatemala.*

Guat-*7130*.M5. *Memoria de las Labores Realizadas por el Hospital General y Sus Dependencias.*

Guat-*7200*.M5. *Memoria de las Labores del Poder Ejecutivo en el Ramo de Educación Pública.*

Haiti-*4200*.B5. *Bulletin Annuel de l'Observatoire Météorologique du Petit Séminaire Collége St. Martial.*

Haiti-*6000*.A3. *Annual Report of the Fiscal Department* (Banque Nationale de la République d'Haïti).

Haiti-*6000*.R5. *Rapport Annuel du Département Fiscal* (Banque Nationale de la République d'Haïti).

Haiti-*6200*.R5. *Rapport Annuel du Service National de la Production Agricole et de l'Enseignement Rural.*

Haiti-*7110*.R5. *Rapport Annuel du Directeur Général* (Service National d'Hygiène et d'Assistance Publique).

Haiti-*8000*.R5. *Rapport Annuel du Sous-Secrétaire d'Etat et Ingénieur en Chêf.*

Haiti-*8300*.R5. *Rapport Annuel* (Garde d'Haïti).

Haiti-*8800*.R3. *Rapport Annuel de l'Administration Générale des Contributions.*

Hond-*6000*.I5. *Informe del Secretario de Agricultura, Fomento y Obras Públicas.*

Hond-*8000*.I5. *Informe de los Actos Realizados por el Poder Ejecutivo en los Ramos de Gobernación, Justicia, Sanidad y Beneficencia.*

Hond-*8800*.I5. *Informe de Hacienda, Crédito Público y Comercio.*

Méx-*1410*.A5. *Anuario Estadístico de los Estados Unidos Mexicanos.*

Méx-*1410*.C5. *Compendio Estadístico* (Dirección General de Estadística).

Méx-*6000*.M5. *Memoria de la Secretaría de la Economía Nacional.*

Méx-*6200*.M5. *Memoria de la Secretaría de Agricultura y Fomento.*

Méx-*6900*.A5. *Anuario Estadístico del Comercio Exterior de los Estados Unidos Mexicanos.*

Méx-*7110*.M5. *Memoria del Departamento de Salubridad y Asistencia Pública.*

Méx-*8000*.M5. *Memoria de la Secretaría de Gobernación.*

Méx-*9000*.M5. *Memoria de Labores* (Secretaría del Trabajo y Previsión Social).

Nic-*1410*.A5. *Anuario Estadístico de la República de Nicaragua.*

Nic-*1410*.G5. *Gráficos Estadísticos de Nicaragua.*

Nic-*6200*.M5. *Memoria del Ministerio de Agricultura y Trabajo.*

Nic-*6300*.E5. *Estadística Minera.*

Nic-*6720*.I5. *Informe y Balances del Banco Hipotecario de Nicaragua.*

Nic-*6900*.M5. *Memoria del Recaudador General de Aduanas y Alta Comisión.*

Nic-*7110*.I5. *Informe de la Dirección General de Sanidad.*

Nic-*8000*.M5. *Memoria de Fomento y Obras Públicas.*

Nic-*8800*.M5. *Memoria de Hacienda y Crédito Público.*

Nic-*8800*.P5. *Presupuesto General de Egresos e Ingresos.*

Pan-*7110*.M5. *Memoria del Ministerio de Salubridad y Obras Públicas.*

Pan-*7210*.E5. *Estadística Cultural.*

Pan-*8800*.*I5*. *Informe del Contralor General de la República.*

Par-*1410*.A5. *Anuario Estadístico de la República del Paraguay.*

Par-*1410*.B5. *Boletín Semestral de la Dirección General de Estadística.*

Par-*6000*.A5, 1938-39. *Anuario del Ministerio de Economía, 1938-39.*

Par-*6720*.M5. *Memoria* (Banco del Paraguay).

Par-*7110*.M5. *Memoria de la Sección Estadística* (Ministerio de Salud Pública).

Perú-*1410*.E5. *Anuario Estadístico de Perú.*

Perú-*3040*.R5. *Revista de la Facultad de Ciencias Económicas.*

Perú-*4200*.B4. *Boletín Anual Meteorológico.*

Perú-*6000*.M5. *Memoria del Ministerio de Fomento y Obras Públicas.*

Perú-*6200*.M5. *Memoria de la Sociedad Nacional Agraria.*

ANNUALS—cont.

USA-*6900*.S5. *Summary of Foreign Trade of the United States.*
USA-*6910*.B5. *Balance of International Payments of the United States.*
USA-*7110*.A5. *American Public Health Association Year Book.*
USA-*7110*.A7. *Annual Report of the Surgeon General of the U. S. Public Health Service.*
USA-*7130*.P5. *Patients in Mental Institutions.*
USA-*7150*.T5. *Tuberculosis Reference Statistical Handbook.*
USA-*7160*.A5. *Accident Facts.*
USA-*7170*.B5. *The Blind and Deaf-Mutes in the United States.*
USA-*7200*.B3. *Biennial Survey of Education.*
USA-*8000*.M5. *Municipal Year Book.*
USA-*8300*.F5. *Federal Prisons.*
USA-*8300*.U5. *Uniform Crime Reports for the United States and Its Possessions.*
USA-*8310*.J5. *Judicial Criminal Statistics.*
USA-*8320*.J3. *Juvenile Court Statistics.*
USA-*8330*.P6. *Prisoners in State and Federal Prisons and Reformatories.*
USA-*8500*.R3. *Report of the National Research Council.*
USA-*8800*.A5. *Annual Report of the Secretary of the Treasury on the State of the Finances.*
USA-*8800*.B7. *The Budget of the United States Government.*
USA-*8800*.C4. *City Finances.*
USA-*8800*.C6. *County Finances.*
USA-*8800*.F5. *Financial Statistics of State and Local Government (Wealth, Public Debt, and Taxation).*
USA-*8800*.S5. *State Finances.*
USA-*8800*.S7. *Statistics of Income.*
USA-*9200*.W9. *Women in Factories.*
USA-*9950*.S6. *Social Security Yearbook.*
USA-*9952*.A5. *Annual Report of the Railroad Retirement Board.*
USA PR-*1410*.A5. *Annual Report of the Governor of Puerto Rico.*
Ur-*1410*.A5. *Anuario Estadístico de la República Oriental del Uruguay.*
Ur-*1410*.S5. *Síntesis Estadística de la República Oriental del Uruguay.*
Ur-*5200*.M5. *El Movimiento del Estado Civil y la Mortalidad de la República Oriental del Uruguay.*
Ur-*6200*.A5. *Anuario de Estadística Agrícola.*
Ur-*6700*.E5. *Estadística de la Cámara Nacional de Comercio.*
Ur-*6720*.M5. *Memoria del Directorio del Banco de la República Oriental del Uruguay.*
Ur-*6900*.M5. *Memoria del Contralor de Exportaciones e Importaciones.*
Ur-*7110*.M5. *Memoria del Ministerio de Salud Pública.*
Ur-*8800*.D5. *Deuda Pública Nacional.*
Ur-*8800*.P5. *Presupuesto General del Estado.*
Ven-*1200*.A5. *Anuario Bibliográfico Venezolano.*
Ven-*1410*.A5. *Anuario Estadístico de Venezuela.*
Ven-*3030*.I5. *Informe de la Dirección General de Estadística y de Censos.*
Ven-*5200*.A5. *Anuario de Epidemiología y Estadística Vital.*
Ven-*6000*.M3. *Memoria del Ministerio de Fomento.*
Ven-*6000*.M5. *Memoria y Cuenta del Ministerio de Obras Públicas.*
Ven-*6200*.M5. *Memoria y Cuenta del Ministerio de Agricultura y Cría.*
Ven-*6300*.E5. *Estadística del Petróleo.*
Ven-*6720*.B3. *Banco Central de Venezuela: Memoria.*
Ven-*6900*.E5. *Estadística Mercantil y Marítima.*
Ven-*7110*.M5. *Memoria y Cuenta del Ministerio de Sanidad y Asistencia Social.*
Ven-*7200*.M5. *Memoria que el Ministro de Educación Nacional Presenta al Congreso Nacional.*
Ven-*8800*.C5. *Cuenta General de Rentas y Gastos Públicos; Bienes Nacionales Inclusive Materias, y Cuenta del Departamento de Hacienda, Rentas y Gastos.*
Ven-*8800*.I5. *Informe de la Contraloría General de la Nación.*
Ven-*8800*.M5. *Memoria de Hacienda.*
Ven-*9000*.M5. *Memoria del Ministerio del Trabajo y de Comunicaciones.*
GrBr-*3010*.G5. *Guide to Current Official Statistics of the United Kingdom.*

ANNUITIES—*see* 6730.

Anteproyecto de Carta para la Organización Internacional de Comercio de las Naciones Unidas. Int-*6900*.A6, 1946.

ANTHROPOMETRY AND ANTHROPOMETRIC MEASUREMENTS—*see* 7310.

ANTROPOMETRIA Y MEDIDAS ANTROPOMETRICAS—*véase* 7310.

ANUALIDADES—*véase* 6730.

Anuário Açucareiro. Bras-*6210*.A4.

Anuário Algodoeiro. Bras-*6210*.A5.

Anuario Azucarero de Cuba. Cuba Sugar Year Book. Cuba-*6210*.A5.

Arca Parró, Alberto—cont.
Perú-*3040*.E5, Apr. '45. "El Medio Geográfico y la Población del Perú."
USA-*4000*.G5, Jan. '42. "Census of Peru, 1940."
ARCHITECTURE—*see* 7300.
ARCHIVOS:
BIBLIOGRAFICOS (EN GENERAL)—*véase* 1200.
BIBLIOGRAFICOS, DE FUENTES ESTADISTICAS—*véase* 3010.
BIOGRAFICOS (EN GENERAL)—*véase* 1930.
"Area Sembrada y Volumen Físico de la Producción: Años 1900 a 1944." Arg-*3040*.R5, June '45, b
Arenas Paz, Belisario.
NonG-*3330*, 1935. *Manual de Estadística y de Información General.*
Argentina. Banco Central de la República Argentina—*see-véase* Banco Central de la República Argentina.
——. Banco de la Nación Argentina—*see-véase* Banco de la Nación Argentina.
——. Biblioteca de Orientación Económica—*see-véase* Biblioteca de Orientación Económica, Argentina.
——. Bolsa de Comercio de Buenos Aires—*see-véase* Bolsa de Comercio de Buenos Aires.
——. Bolsa de Comercio de Rosario—*see-véase* Bolsa de Comercio de Rosario.
——. Buenos Aires (Ciudad). Comisión Técnica Encargada de Realizar el Cuarto Censo General.
Arg-*2200*.C4, 1936. *Cuarto Censo General, 1936: Población 22-X-1936.*
——. ——. Bolsa de Comercio—*see-véase* Bolsa de Comercio de Buenos Aires.
——. Buenos Aires (Prov.). Departamento del Trabajo.
Arg-*9000*.C5, 1943. *Condiciones de Vida de la Familia Obrera.*
——. Caja Nacional de Jubilaciones y Pensiones Civiles.
Arg-*9950*.M5. *Memoria de la Caja Nacional de Jubilaciones y Pensiones Civiles.*
——. Censo Escolar de la Nación—*see-véase* Argentina. Ministerio de Justicia e Instrucción Pública. Consejo Nacional de Educación. Censo Escolar de la Nación.
——. Comisión Argentina de Altos Estudios Internacionales—*see-véase* Comisión Argentina de Altos Estudios Internacionales.
——. Comisión del Censo Nacional de Bienes del Estado—*see-véase* Argentina. Ministerio de Obras Públicas. Comisión del Censo Nacional de Bienes del Estado.
——. Comisión Directiva del Censo—*see-véase* Argentina. Ministerio del Interior. Consejo Nacional de Estadística y Censos. Dirección General de Estadística y Censos de la Nación.
——. Comisión Nacional de Cooperación Intelectual—*see-véase* Argentina. Ministerio de Justicia e Instrucción Pública. Comisión Nacional de Cooperación Intelectual.
——. Comisión Nacional de Granos y Elevadores—*see-véase* Argentina. Ministerio de Agricultura. Comisión Nacional de Granos y Elevadores.
——. Comisión Nacional del Censo—*see-véase* Argentina. Ministerio del Interior. Consejo Nacional de Estadística y Censos. Dirección General de Estadística y Censos de la Nación.
——. Comisión Nacional del Censo Agropecuario—*see-véase* Argentina. Ministerio de Agricultura. Comisión Nacional del Censo Agropecuario.
——. Comité Nacional de Geografía.
Arg-*1410*.A5. *Anuario Geográfico Argentino.*
——. Consejo Nacional de Educación—*see-véase* Argentina. Ministerio de Justicia e Instrucción Pública. Consejo Nacional de Educación.
——. Consejo Nacional de Estadística y Censos—*see-véase* Argentina. Ministerio del Interior. Consejo Nacional de Estadística y Censos.
——. Consejo Nacional de Postguerra—*see-véase* Argentina. Vicepresidencia de la Nación. Consejo Nacional de Postguerra.
——. Contaduría General de la Nación—*see-véase* Argentina. Ministerio de Hacienda. Contaduría General de la Nación.
——. Departamento de Salud Pública—*see-véase* Argentina. Ministerio del Interior. Dirección Nacional de Salud Pública.
——. Departamento Nacional de Higiene—*see-véase* Argentina. Ministerio del Interior. Dirección Nacional de Salud Pública.
——. Departamento Nacional del Trabajo—*see-véase* Argentina. Ministerio del Interior. Departamento Nacional del Trabajo.
——. Dirección de Algodón—*see-véase* Argentina. Secretaría de Industria y Comercio. Dirección de Algodón.
——. Dirección de Economía Rural y Estadística—*see-véase* Argentina. Ministerio de Agricultura. Dirección de Economía Rural y Estadística.
——. Dirección de Estadística—*see-véase* Argentina. Ministerio de Agricultura. Dirección de Economía Rural y Estadística.
——. Dirección de Estadística Social—*see-véase* Argentina. Vicepresidencia de la

Argentina—cont.

Arg-*4200*.R5, 1943? *Registro de Observaciones Pluviométricas.*

——. ——. Dirección de Propaganda y Publicaciones.

Arg-*6200*.A3. *Almanaque del Ministerio de Agricultura.*

Arg-*6200*.P5, 1923. *Programa Estadístico y Económico para el Ministerio de Agricultura de la República Argentina.*

——. ——. Dirección de Vitivinicultura.

Arg-*6210*.B7. *Boletín Informativo.*

——. ——. Dirección General de Minas, Geología e Hidrología.

Arg-*6300*.E5. *Estadística Minera de la Nación.*

——. ——. Dirección General de Yacimientos Petrolíferos Fiscales—*see-véase* Argentina. Secretaría de Industria y Comercio. Dirección General de Yacimientos Petrolíferos Fiscales.

——. ——. Junta Nacional de Carnes.

Arg-*6230*.C5. *Compras de Ganado Bovino, Ovino y Porcino Efectuadas por los Frigoríficos en las Estancias.*

——. ——. Junta Nacional del Algodón—*see-véase* Argentina. Secretaría de Industria y Comercio. Dirección de Algodón.

——. ——. Junta Reguladora de Vinos—*see-véase* Argentina. Ministerio de Agricultura. Dirección de Vitivinicultura.

——. ——. Sección Propaganda e Informes—*see-véase* Argentina. Ministerio de Agricultura. Dirección de Propaganda y Publicaciones.

——. Ministerio de Hacienda.

Arg-*3040*.R5, Dec. '44, a. "Estimación de la Renta Nacional Correspondiente al Año 1941."

Arg-*8800*.M6. *Memoria del Departamento de Hacienda.*

——. ——. Contaduría General de la Nación.

Arg-*8800*.M4. *Memoria de la Contaduría...*

——. ——. Dirección General de Aduanas.

Arg-*6900*.B5. *Boletín de la Dirección...*

Arg-*6900*.T5, 1939. *Tarifa de Avalúos y Arancel de Importación.*

——. ——. Dirección General de Estadística de la Nación—*see-véase* Argentina. Ministerio del Interior. Consejo Nacional de Estadística y Censos. Dirección General de Estadística y Censos de la Nación.

——. Ministerio de Justicia e Instrucción Pública.

Arg-*7200*.M5. *Memoria del Ministerio de Justicia e Instrucción Pública.*

——. ——. Comisión Nacional de Cooperación Intelectual.

Arg-*1200*.B6. *Boletín Bibliográfico Argentino.*

——. ——. Consejo Nacional de Educación. Censo Escolar de la Nación.

Arg-*2130*.C5, 1943.1. *Instrucciones Complementarias a las "Normas para la Organización y el Levantamiento del Censo Escolar de la Nación."*

Arg-*2130*.C5, 1943.2. *Normas para la Organización y Levantamiento del Censo Escolar de la Nación en la Capital Federal.*

Arg-*2130*.C5, 1943.3. *Normas para la Organización y Levantamiento del Censo Escolar de la Nación en las Provincias.*

Arg-*2130*.C5, 1943.4. *Normas para la Organización y Levantamiento del Censo Escolar de la Nación en los Territorios.*

Arg-*2130*.C5, 1943.5. *Censo Escolar de la Nación, Ley No. 12.723: Instrucciones para los Censistas.*

Arg-*2130*.C5, 1943.6. *Censo Escolar de la Nación, Ley No. 12.723: Cuestionario Familiar.*

Arg-*2130*.C5, 1943.7. *Censo Escolar de la Nación, Ley No. 12.723: Cuestionario Individual.*

Arg-*2130*.C5, 1943.8. *Censo Escolar de la Nación, Ley No. 12.723: Cuestionario de Convivencia.*

Arg-*2230*.C5, 1943.1. *El Analfabetismo en la Argentina.*

Arg-*2230*.C5, 1943.2. *La Distribución por Zonas de la Población Argentina y Su Relación con los Hechos Culturales, Económicos y Sociales.*

Arg-*2230*.C5, 1943.3. *La Vivienda Popular.*

——. ——. Dirección General de Estadística y Personal.

Arg-*7200*.R5. *Recopilación Estadística.*

——. Ministerio de Obras Públicas.

Arg-*6600*.M5. *Memoria.*

——. ——. Comisión del Censo Nacional de Bienes del Estado.

Arg-*3200*.C5, 1937.A. *Clasificador de Bienes.*

Arg-*3200*.C5, 1937.B. *Indice General Alfabético por Materia de los Bienes Nacionales del Estado.*

——. ——. Dirección General de Ferrocarriles.

Arg-*6614*.M5. *Memoria de los Ferrocarriles en Explotación.*

——. ——. Dirección General de Navegación y Puertos.

Arg-*6615*.A5. *Anuario del Movimiento de los Puertos de la República Argentina.*

——. Ministerio de Relaciones Exteriores y Culto. Dirección de Información al Exterior.

Arg-*1900*.I5. *Informaciones Argentinas.*

Argentina—cont.

——. ——. Instituto Nacional de la Nutrición—*see-véase* Argentina. Ministerio del Interior. Dirección Nacional de Salud Pública. Instituto Nacional de la Nutrición.

——. Ministerio del Interior. Consejo Nacional de Estadística y Censos.

Arg-*3030*.U5, 1944. *Ultimas Disposiciones del Poder Ejecutivo Nacional sobre Estadística y Censos.*

Arg-*3040*.B5. *Boletín Censal y Estadístico.*

——. ——. ——. Dirección General de Estadística y Censos de la Nación.

Arg-*2200*.C5, 1914. *Tercer Censo Nacional, Levantado el 1 de Junio de 1914.*

Arg-*2200*.C5, 1895. *Segundo Censo de la República Argentina, Mayo 10 de 1895.*

Arg-*2200*.C5, 1869. *Primer Censo de la República Argentina, Verificado en los Días 15, 16 y 17 de Setiembre de 1869.*

Arg-*2250*.E5. *Estadística Industrial.*

Arg-*5000*.P2. *Informe, Serie D, Demografía.*

Arg-*5000*.P2, 1943, No. 10. *Clasificación Estadística de las Causas de las Defunciones.* (Nomenclatura internacional de 1938).

Arg-*5000*.P2, 1942, No. 9. *El Movimiento Demográfico en los Territorios Nacionales de la República Argentina en los Años 1933 a 1941.*

Arg-*5100*.P5. *La Población de la República Argentina.*

Arg-*6400*.A3. *La Actividad Industrial durante los Primeros . . . Meses de . . . , según los Números Indices Correspondientes a los Salarios Pagados, a la Ocupación de Obreros y a las Horas-Obrero de Trabajo en la Industria.*

Arg-*6400*.E5. *Informe, Serie I, Industria.*

Arg-*6900*.A5. *Anuario del Comercio Exterior de la República Argentina.*

Arg-*6900*.C5. *El Comercio Exterior Argentino.*

Arg-*6900*.C6. *El Comercio Exterior Argentino.*

Arg-*6900*.C8. *El Comercio Exterior Argentino en . . . y Su Comparación con el del Mismo Período del Año Anterior: Boletín Mensual.*

Arg-*6900*.E5. *Las Exportaciones en los . . . Comparadas con las del Mismo Período del Año Anterior.*

——. ——. Departamento de Salud Pública—*see-véase* Argentina. Ministerio del Interior. Dirección Nacional de Salud Pública.

——. ——. Departamento Nacional de Higiene—*see-véase* Argentina. Ministerio del Interior. Dirección Nacional de Salud Pública.

——. ——. Departamento Nacional del Trabajo.

Arg-*9000*.B5. *Boletín Informativo del Departamento . . .*

——. ——. ——. División de Estadística.

Arg-*6010*.P5, 1940. *Precios.*

Arg-*9000*.I5. *Investigaciones Sociales.*

Arg-*9000*.O7, 1941. *Organización Sindical.*

Arg-*9000*.S5. *Serie B, Estadísticas y Censos, del Departamento . . .*

Arg-*9200*.A5, 1943. *Adaptación de los Salarios a las Fluctuaciones del Costo de la Vida.*

Arg-*9200*.D5, 1940. *La Desocupación en la Argentina, 1940.*

Arg-*9300*.E5, 1940. *Estadísticas de las Huelgas, 1940.*

——. ——. Dirección General de Correos y Telecomunicaciones.

Arg-*6620*.M5. *Memoria de la Dirección . . .*

——. ——. Dirección Nacional de Salud Pública.

Arg-*7110*.B5. *Boletín Sanitario.*

——. ——. ——. Instituto Nacional de la Nutrición.

Arg-*9800*.A5. *La Alimentación de la Familia en Buenos Aires.*

Arg-*9800*.C5. *El Costo de la Alimentación.*

Arg-*9800*.M5. *Memoria Anual del Instituto . . .*

——. ——. ——. Sección de Demografía Sanitaria y Geografía Médica.

Arg-*5200*.A5 *Anuario Demográfico.*

——. Presidencia de la Nación. Secretaría Técnica.

Arg-*6000*.I2, 1946. *Indices Básicos de la Economía Nacional.*

——. Rosario (Ciudad). Bolsa de Comercio—*see-véase* Bolsa de Comercio de Rosario.

——. Secretaría de Industria y Comercio. Dirección de Algodón.

Arg-*2210*.C4, 1935-36. *Censo Algodonero de la República Argentina, Año 1935-36.*

Arg-*6210*.A4. *Boletín Mensual.*

——. ——. Dirección General de Yacimientos Petrolíferos Fiscales.

Arg-*6300*.B5. *Boletín de Informaciones Petroleras.*

——. Sociedad Científica Argentina—*see-véase* Sociedad Científica Argentina.

——. Sociedad Rural Argentina—*see-véase* Sociedad Rural Argentina.

——. Superintendente del Censo—*see-véase* Argentina. Ministerio del Interior. Consejo Nacional de Estadística y Censos. Dirección General de Estadística y Censos de la Nación.

Argentina—cont.
——. Universidad de Buenos Aires—*see-véase* Universidad de Buenos Aires.
——. Universidad Nacional de Córdoba—*see-véase* Universidad Nacional de Córdoba.
——. Universidad Nacional de La Plata—*see-véase* Universidad Nacional de La Plata, Argentina.
——. Universidad Nacional del Litoral—*see-véase* Universidad Nacional del Litoral.
——. Vicepresidencia de la Nación. Consejo Nacional de Postguerra.
Arg-*9000.05*, 1945. *Ordenamiento Económico-Social.*
——. ——. ——. Dirección de Estadística Social.
Arg-*9500*.C5. *Costo de la Vida.*
Arg-*9500*.I5, 1945. *El Indice del Costo de la Vida.*
Arg-*9700*.N5, 1945. *Nivel de Vida de la Familia Obrera.*
Argentina Económica e Industrial. Arg-*6000*.A6, 1944.
Argentina Fabril. Arg-*6400*.R7.
Argentina: Reports Presented to the Conference of Commissions of Inter-American Development. Arg-*6000*.R5, 1944.
Argentina: Summary of Biostatistics. Arg-*5000*.A5, 1945.
Argentine Commission of High International Studies—*see-véase* Comisión Argentina de Altos Estudios Internacionales.
Argentine Fourth General Census Postponed. Arg-*2100*.C5.
Argentine News. Arg-*1900*.I5.
Arkin, Herbert.
NonG-*3260*.G7, 1940. *Graphs: How to Make and Use Them.*
NonG-*3330*, 1939. *An Outline of Statistical Methods.*
ARQUITECTURA—*véase* 7300.
Arquivos de Higiene. Bras-*7100*.A5.
ARRENDAMIENTOS:
GENERAL (incluyendo VALORES)—*véase* 6721; 2220.
DE HABITACIONES Y VIVIENDAS—*véase* 9900; 9500; 2220.
Arrús, Oscar F.
NonG-*3330*, 1945. *Estadística Metodológica.*
ARTES—*véase* 7300-7399.
ARTS—*see* 7300-7399.
ASESINATOS—*véase* 5212; 8300.
ASILOS:
DE ASISTENCIA SOCIAL (incluyendo ASILOS PARA ANCIANOS, CIEGOS, HUERFANOS, ETC.)—*véase* 9951.
PARA INVALIDOS Y VETERANOS DE LA GUERRA (incluyendo HOSPITALES)—*véase* 7130.
ASISTENCIA SOCIAL—*véase* 9951.
Asociación Cafetalera de El Salvador.
ElSal-*2210*.C5, 1940. *Primer Censo Nacional de Café, 1938-39.*
ElSal-*6210*.C6. *El Café de El Salvador.*
Asociación Permanente del Congreso Sudamericano de Ferrocarriles.
Int-*6614*.B5. *Boletín de la Asociación Permanente.*
ASOCIACIONES—*véase* las respectivas clasificaciones.
Aspectos Biodemográficos de la Población de Costa Rica. CR-*5200*.A5, 1940.
Aspects de l'Économie et des Finances d'Haïti. Haiti-*6000*.A5, 1944.
ASSISTANCE, SOCIAL—*see* 9951.
ASSOCIATIONS—*see* the particular subject field.
A.S.T.M. Manual on Presentation of Data. NonG-*3160*.A5, 1945.
ASTRONOMIA—*véase* 7300.
ASTRONOMY—*see* 7300.
Ata Final da Conferência Interamericana sôbre Medidas de Contrôle Econômico-Financiero, Washington, 30 de Junho a 10 de Julho de 1942. Int-*6000*.I45, 1942, Port.
Ata Final do Oitavo Congresso Panamericano da Criança, Washington, 2-9 de Maio de 1942. Int-*9951*.P5, 1942, Port.
ATENCION MEDICA:
GENERAL—*véase* 7190.
COSTO DE—*véase* 7190; 9700.
ATLAS:
GENERAL—*véase* 4000.
ESTADISTICOS—*véase* 1400.
Atlas Estatístico do Brasil. Bras-*1400*.A7, 1941.
Atlas Statistique du Brésil. Bras-*1400*.A7, 1941.
ATLASES:
GENERAL—*see* 4000.
STATISTICAL—*see* 1400.
ATMOSFERA, CONDICIONES DE LA—*véase* 4200.
"Attributes of Wage-Earner Families in Relation to Size." Int-*3040*.E5, Dec. '45, Greenway.
AUTOBUSES—*véase* 6612.
AUTOMOBILE—*see* 6612.
AUTOMOBILES, MANUFACTURE OF—*see* 6400.
AUTOMOVILES—*véase* 6612.
AVENAMIENTO—*véase* DRENAJE Y RIEGO.
AVES DE CORRAL—*véase* 6230.
AVIACION:
GENERAL—*véase* 6611.
COMERCIAL—*véase* 6611.

Banco Central del Ecuador.
Ec-*6720*.B5. *Banco Central del Ecuador, Boletín.*
Ec-*6720*.B5, Feb.-Mar. '46. "Memoria Anual que el Consejo de Administración del Banco Central del Ecuador Presenta a la Asamblea Ordinaria de Accionistas por el Ejercicio Económico de 1945."
Ec-*6720*.E5, 1946. *Estadísticas y Otras Informaciones para el Fondo Monetario Internacional.*
——. Oficina de Investigaciones Económicas.
Col-*1420*.A5, Mar.-Apr. '46, Abrisqueta. "Algunas Instituciones Económicas Ecuatorianas: El Banco Central del Ecuador y Su Oficina de Investigaciones Económicas."
Banco Central del Ecuador, Boletín. Ec-*6720*.B5.
"El Banco Central del Ecuador y Su Oficina de Investigaciones Económicas." Col-*1420*.A5, Mar.-Apr. '46, Abrisqueta.
Banco de la Nación Argentina.
Arg-*6720*.B5, 1891-1941. *El Banco de la Nación Argentina en Su Cincuentenario, 1891-1941.*
Arg-*6720*.R4. *Revista del . . .*
El Banco de la Nación Argentina en Su Cincuentenario, 1891-1941.—Arg-*6720*.B5, 1891-1941.
Banco de la República, Colombia.
Col-*6720*.A5. *Annual Report Presented by the Manager to the Board of Directors.*
Col-*6720*.B5. *Boletín de Información Mensual.*
Col-*6720*.R5. *Revista del . . .*
——. Sección de Investigaciones Económicas.
Col-*4000*.M5, 1946. *Mapa Físico de Colombia.*
Col-*4200*.M5, 1945. *Meteorología.*
Col-*5110*.P5, 1946. *Población de Colombia por Ramas de Actividad Económica (Según Censo del 5 de Julio de 1938).*
Col-*6000*.B5. *Boletín Gráfico.*
Col-*7300*.G5, 1945. *Geología.*
Banco de la República, México.
Méx-*6720*.R5. *Revista del Banco de la República.*
Banco de la República Oriental del Uruguay.
Ur-*6720*.M5. *Memoria del Directorio del Banco de la República Oriental del Uruguay.*
——. Departamento de Investigaciones Económicas.
Ur-*6000*.S5. *Suplemento Estadístico de la Revista Económica.*
Ur-*6720*.R5. *Revista del . . .*
Ur-*6720*.R5, July '45. "Balance de Pagos del Uruguay: Estimación del Balance de Pagos del Uruguay Correspondiente al Año 1944."
Ur-*6720*.R5, Jan. '46. "Balances de Pagos del Uruguay, Años 1940-1944."
Banco de México, S.A.
Int-*6000*.I3. *Informaciones Económicas del Banco de México.*
Int-*6720*.R5, 1946. *Memoria, Primera Reunión de Técnicos sobre Problemas de Banca Central del Continente Americano.*
Int-*6721*.I5, 1945. *Inversiones Internacionales en América Latina.*
Méx-*6300*.M5, 1944. *Minería y Riqueza Minera de México.*
Méx-*6400*.M5. *Monografías Industriales del . . .*
Méx-*6720*.B5. *Banco de México, S.A.: Asamblea General Ordinaria de Accionistas.*
Banco de México, S.A.: Asamblea General Ordinaria de Accionistas. Méx-*6720*.B5.
Banco del Paraguay.
Par-*6720*.M5. *Memoria.*
Banco do Brasil.
Bras-*6720*.R4. *Relatório do . . .*
Bras-*6720*.R6. *Resumen de la Memoria del . . .*
——. Departamento de Estatística e Estudos Econômicos.
Bras-*6720*.B5. *Boletim Estatístico.*
Banco Hipotecario de Nicaragua.
Nic-*6720*.I5. *Informe y Balances del . . .*
Banco Hipotecario del Ecuador.
Ec-*6720*.B4. *Boletín del . . .*
Banco Industrial del Perú.
Perú-*6720*.M8. *Memoria del . . .*
Banco Internacional de Reconstrucción y Fomento—*see-véase* International Bank for Reconstruction and Development.
Banco Minero de Bolivia.
Bol-*6300*.M5. *Memoria Anual.*
——. Departamento de Estudios Económicos y Estadística.
Bol-*6300*.C5. *Carta Informativa.*
Bol-*6300*.T5, 1941. *Tasas e Impuestos sobre la Industria Minera en Bolivia.*
Banco Nacional de Costa Rica.
CR-*6720*.B6. *Boletín Estadístico.*
CR-*6720*.R5. *Revista del . . .*
Banco Nacional de Nicaragua.
Nic-*6720*.R5. *Revista del . . .*
Banco Nacional de Seguros, Costa Rica.
CR-*6720*.R6. *Revista del . . .*
CR-*6720*.R6, Dec. '45. "Memoria Anual de 1944, del . . ."
BANCOS Y OPERACIONES BANCARIAS —*véase* 6720.

Benham, Frederic—cont.
WI-*6100*.N7, 1942. *The National Income of St. Vincent, 1942.*
Bennett, Charles G.
Int-*3040*.E5, Mar. '46, Roig Ocampos. "Prueba de Registro de Nacimientos en Paraguay."
Bennett, Rollin F.
NonG-*6100*.S3. "Significance of International Transactions in National Income."
Bergström Lourenço, Manoel.
Int-*1300*.A5, 1940.3. "A População Escolar e a Taxa de Analfabetos nas Estatísticas Educacionais Americanas."
Berkson, Joseph.
USA-*3040*.J5, Dec. '41. "A Punch Card Designed to Contain Written Data and Coding."
USA-*7110*.A4, June '36. "A System of Codification of Medical Diagnoses for Application to Punch Cards, with a Plan for Operation."
Bermúdez V., Luis A.
Col-*1420*.A5, Feb. 20, '42. "Diez Años de Educación y Cultura en Colombia."
Bernstein, Sylvia Pollack.
Int-*9000*.B5, 1944. *Bibliography on Labor and Social Welfare in Latin America.*
USA-*6900*.F7, July 27, '46. "Latin-American Price Trends."
Bertillon, Jacques.
NonG-*3330*, 1906. *Compendio de la Obra Intitulada: Curso Elemental de Estadística Administrativa.*
Besio Moreno, Nicolás.
Arg-*5000*.B5, 1536-1936. *Buenos Aires, Puerto del Río de la Plata, Capital de la Argentina; Estudio Crítico de Su Población, 1536-1936.*
Betancourt, Rómulo.
Int-*7110*.B5, Apr. '46, Lairet. "La Sanidad en Venezuela."
Bevis, Joseph C.
USA-*3040*.J5, June '45, b. "Management of Field Work and Collection of Statistics: A Round Table."
Bibliografía de la Secretaría de Hacienda y Crédito Público. Méx-*8800*.B5, 1943.
Bibliografía Mexicana de Estadística. Méx-*3010*.B5, 1942.
Bibliografía que el Departamento de Salubridad Pública Presenta en la Feria del Libro y Exposición Nacional de Periodismo. Méx-*7110*.B5, 1943.
BIBLIOGRAFIAS:
LISTA DE *véase* BIBLIOGRAPHIES.
GENERAL—*véase* 1200.
ESPECIFICAS—*véase* las respectivas clasificaciones.
ESTADISTICAS—*véase* 3010.
Bibliographic Series of the Pan American Union. Int-*1200*.B3.
Bibliographie Générale et Méthodique d'Haïti. Haiti-*1200*.B5, 1941.
BIBLIOGRAPHIES
Int-*1200*.B3. *Bibliographic Series of the Pan American Union.*
Int-*1200*.B5, 1942. *A Bibliography of Latin American Bibliographies.*
Int-*1200*.E5. *Executive Agreement Series.*
Int-*1200*.H4. *Handbook of Latin American Studies.*
Int-*1200*.H4, 1942, Rohen y Gálvez. "Latin American Periodicals Dealing with Labor and Social Welfare."
Int-*1200*.H4, 1937, Phelps. "Sources of Current Economic Information of Latin America."
Int-*1200*.I4, 1943. *Index of Latin American Publications.*
Int-*1200*.I6. *Inter-American Bibliographic Review.*
Int-*1200*.I6, Summer 1941, Owen. "Sources of Information on Social and Labor Problems in Latin America."
Int-*1200*.L3, 1944. *Latin American Periodicals Currently Received in the Library of Congress and in the Library of the Department of Agriculture.*
Int-*1200*.L5. *Latin American Series of the Library of Congress.*
Int-*1200*.P2. *The Pan American Book Shelf.*
Int-*1200*.P5. *Publications Issued by the League of Nations.*
Int-*1200*.S6, 1815-1931. *Serial Publications of Foreign Governments, 1815-1931.*
Int-*1200*.U5, 1943. *Ulrich's Periodicals Directory. Directorio de Publicaciones Periódicas Ulrich.*
Int-*3000*.S5, 1940. *Statistical Activities of the American Nations, 1940.*
Int-*3010*.B5. *Bibliography Series, IASI.*
Int-*3010*.B5, Cen. 2. May 5, '45. *Census Methods and Procedures.*
Int-*3010*.B5, Dem. 2, Feb. 7, '45. *Vital Statistics Registration Methods in the Western Hemisphere.*
Int-*3010*.B5, Econ. 4a. Apr. 2, '45. *Bibliography of Selected References on Agricultural Statistics in the Western Hemisphere.*
Int-*3010*.B5, Econ. 6a, June 1, '45. *Organization of Agriculture Statistics.*
Int-*3010*.B5, Econ. 8a, Oct. 1, '45. *For-*

Boletín de Exportación. CR-*6900*.B5.
Boletín de Exportación del Café. CR-*6900*.B5.
Boletín de Hacienda. Col-*8800*.R5.
Boletín de Hacienda. Ur-*6000*.B5.
Boletín de Hacienda, Suplemento. Ur-*6000*.B6.
Boletín de Información Mensual (Banco de la República). Col-*6720*.B5.
Boletín de Informaciones Petroleras. Arg-*6300*.B5.
Boletín de la Asociación Permanente. Int-*6614*.B5.
Boletín de la Cámara de Comercio Argentino-Paraguaya. Arg-*6700*.B2.
Boletín de la Compañía Administradora del Guano. Perú-*6220*.B6.
Boletín de la Dirección de Aeronáutica. Perú-*6611*.B5.
Boletín de la Dirección de Agricultura. Perú-*6200*.B5.
Boletín de la Dirección de Aviación Comercial y Civil. Perú-*6611*.B5.
Boletín de la Dirección de Caminos y Ferrocarriles. Perú-*6610*.B5.
Boletín de la Dirección de Estadística Económica. Ur-*6000*.B7.
Boletín de la Dirección de Industrias del Ministerio de Fomento y Obras Públicas. Perú-*6400*.B5.
Boletín de la Dirección General de Aduanas. Arg-*6900*.B5.
Boletín de la Dirección General de Estadística. Guat-*1420*.B5.
Boletín de la Dirección General de Impuestos Internos. Par-*8800*.B3.
Boletín de la Dirección General de Minas y Petróleo. Bol-*6300*.B4.
Boletín de la Dirección General de Salubridad. Perú-*7110*.B4.
Boletín de la Dirección Nacional de Estadística. Perú-*1420*.B4.
Boletín de la Oficina de Demografía y Estadística Sanitaria. Méx-*7110*.B6.
Boletín de la Oficina de Números Indices. Cuba-*6010*.B5.
Boletín de la Oficina Sanitaria Panamericana. Int-*7110*.B5.
Boletín de la Sección Estadística de la Caja de Colonización Agrícola. Chile-*6200*.B5.
Boletín de la Sección Técnica de Estadística (Dirección de Agricultura). Perú-*6200*.B5.
Boletín de la Sociedad Mexicana de Geografía y Estadística. Méx-*3040*.B5.
Boletín de la Superintendencia Bancaria. Col-*6720*.B7.
Boletín de la Superintendencia de Bancos del Ecuador. Ec-*6720*.B3.
Boletín de la Unión Panamericana. Int-*8200*.B5.
Boletín de Minas. Cuba-*6300*.B5.
Boletín de Previsión Social. Pan-*9950*.B5.
Boletín de Salubridad e Higiene. Méx-*7110*.B7.
Boletín de Trabajo, Comercio e Industrias. Pan-*6000*.R5.
Boletín del Banco Central de Bolivia. Bol-*6720*.B5.
Boletín del Banco Central de Venezuela. Ven-*6720*.B5.
Boletín del Banco Central del Ecuador. Ec-*6720*.B5.
Boletín del Banco Hipotecario del Ecuador. Ec-*6720*.B4.
Boletín del Banco Minero de Bolivia. Bol-*6300*.C5.
Boletín del Café. Cuba-*6210*.B5.
Boletín del Congreso Sudamericano de Ferrocarriles. Int-*6614*.B5.
Boletín del Cuerpo de Ingenieros de Minas. Perú-*6300*.B5.
Boletín del Cuerpo de Investigación y Vigilancia (Dirección General de Tránsito). Perú-*7160*.B5.
Boletín del Departamento de Bioestadística. Perú-*7110*.B6.
Boletín del Departamento de Salubridad Pública. Méx-*7110*.B7.
Boletín del Despacho de Previsión Social. Bol-*9000*.B5.
Boletín del Instituto de Investigaciones Sociales y Económicas. Int-*3040*.B5.
Boletín del Instituto Nacional de la Nutrición. Arg-*9800*.A5.
Boletín del Ministerio de Agricultura, Comercio e Industrias. Par-*6000*.B5.
Boletín del Ministerio de Fomento. Ven-*6000*.R5.
Boletín del Ministerio de Salud Pública y Previsión Social. Par-*7110*.B5.
Boletín del Ministerio del Fomento y Obras Públicas. ElSal-*6600*.B5.
Boletín del Ministerio del Tesoro. Ec-*8800*.B5.
Boletín del Tesoro. Par-*8800*.B5.
Boletín Estadístico (Banco Central de la República Argentina. Departamento de Investigaciones Económicas). Arg-*6000*.S8.
Boletín Estadístico (Banco Nacional de Costa Rica). CR-*6720*.B6.
Boletín Estadístico de la Dirección de Trabajo y de la Dirección de Previsión Social. Perú-*9000*.B5.
Boletín Estadístico de la Dirección General de Estadística. ElSal-*1420*.B5.
Boletín Estadístico de la República Oriental del Uruguay. Ur-*1420*.B5.
Boletín Estadístico del Ministerio de Agricultura. Arg-*6200*.B5.
Boletín Frutas y Hortalizas. Arg-*6210*.B3.
Boletín Gráfico (Banco de la República. Sección

Bolivia—cont.

perimental Agropecuario, Departamento de Tarija, 1944-1945.

Bol-*2210*.C5, 1942, Chuquisaca. *Censo Experimental Agropecuario del Departamento de Chuquisaca, 1942.*

——. ——. ——. Sección Estudios Económicos.

Bol-*6020*.E5, 1944. *Estadística de Consumo de Artículos de Procedencia Nacional en la Ciudad de Tarija, 1944.*

——. Ministerio de Economía Nacional. Dirección General de Minas y Petróleo.

Bol-*6300*.B4. *Boletín de la Dirección General de Minas y Petróleo.*

——. Ministerio de Hacienda. Dirección General de Estadística—*see-véase* Bolivia. Dirección General de Estadística. Ministerio de Justicia.

Bol-*5250*.R5, 1945. *Registro Civil de Bolivia.*

——. Ministerio del Trabajo, Salubridad y Previsión Social.

Bol-*9000*.B5. *Boletín del Despacho de Previsión Social.*

——. ——. Sección Biodemografía.

Bol-*7110*.S5, 1942. *Síntesis de Informes y Estadísticas de Salubridad, Año 1942.*

——. Oficina Nacional de Inmigración, Estadística y Propaganda Geográfica—*see-véase* Bolivia. Dirección General de Estadística.

——. Superintendencia de Bancos. Sección Estadística.

Bol-*6720*.E4. *Estadística Bancaria.*

Bolivia: Reports Presented to the Conference of Commissions of Inter-American Development. Bol-*6000*.R5, 1944.

Bolivia: Summary of Biostatistics. Bol-*5000*-B5, 1945.

Bollinger, P. H.

NonG-*6100*.S3. "Forecasting National Income and Related Measures."

Bolsa de Comercio de Buenos Aires.

Arg-*6700*.B4. *Boletín Oficial de la . . .*

Arg-*6700*.M5. *Memoria de la . . .*

Bolsa de Comercio de Rosario.

Arg-*6700*.B5. *Boletín Oficial de la . . .*

Bolsa de Mercadorias de São Paulo.

Bras-*6210*.A5. *Anuário Algodoeiro.*

BOND PRICES AND INDEXES—*see* 6721.

BONOS, PRECIOS E INDICES DE—*véase* 6721.

Books on Population, 1939-Mid-1944. Int-*5000*.B5, 1939-44.

"Bosquejo de una Economía Argentina para 1955: Primera Parte—Plan General." Arg-*3040*.R5, Feb. '46, a.

"Bosquejo de una Economía Argentina para 1955: Primera Parte—Plan General (continuación)." Arg-*3040*.R5, Mar. '46.

"Bosquejo de una Economía Argentina para 1955: Segunda Parte—Estudios Especiales." Arg-*3040*.R5, Apr. '46.

"Bosquejo de una Economía Argentina para 1955: Segunda Parte—Estudios Especiales." Arg-*3040*.R5, May '46, a.

"Bosquejo de una Economía Argentina para 1955: Segunda Parte (continuación)—Estudios Especiales." Arg-*3040*.R5, June '46, a.

"Bosquejo de una Economía Argentina para 1955: Segunda Parte (continuación)—Estudios Especiales." Arg-*3040*.R5, July '46, a.

Boston First National Bank.

Arg-*6000*.S4. *Situation in Argentina.*

BOTANICA—*véase* 7300.

BOTANY—*see* 7300.

Bowley, Arthur Lyon.

Int-*3040*.R5, 1933.2. "Methods of Investigating the Economic and Social Conditions of a Great City."

Int-*9200*.E7, 1931-41. *Estimates of the Working Population of Certain Countries in 1931 and 1941.*

NonG-*3330*, 1937. *Elements of Statistics.*

NonG-*6100*.S5, 1942. *Studies in the National Income, 1924-1938.*

Bradford, Ernest S.

USA-*9200*.I6, 1922. *Industrial Unemployment.*

Brady, Dorothy S.

USA-*9700*.F511, 1941. *Family Income and Expenditures, Five Regions, Part 2, Family Expenditures, Farm Series.*

USA-*9700*.F512, 1940. *Family Income and Expenditures, Five Regions, Part 2, Family Expenditures, Urban and Village Series.*

Brannon, Max P.

Int-*1300*.A5, 1940.3. "Desarrollo Histórico de la Estadística en El Salvador."

Int-*3000*.S5, 1940. "Las Actividades Estadísticas de El Salvador."

ElSal-*3030*.15, 1935. *La Investigación Estadística, Origen, Desarrollo y Estado Actual Comparada con la de Varios Países Americanos y del Caribe.*

Brasil. Banco do Brasil—*see-véase* Banco do Brasil.

——. Centro de Estudos Econômicos—*see-véase* Centro de Estudos Econômicos.

——. Comissão Censitária Nacional—*see-véase* Brasil. Instituto Brasileiro de Geografia e Estatística. Comissão Censitária Nacional.

Brasil—cont.

Bras-*5220*.T53, 1920-21. *Tábuas de Sobrevivência, Conforme a Mortalidade do Período 1920-21, para o Distrito Federal e o Município de São Paulo.*

Bras-*5220*.T54, 1939-41. *Tábuas de Sobrevivência, Conforme a Mortalidade do Período 1939-41, para o Distrito Federal e o Município de São Paulo.*

——. ——. Conselho Nacional de Estatística.

Bras-*1410*.A5. *Anuário Estatístico do Brasil.*

Bras-*1420*.B7. *Boletim Estatístico do Instituto Brasileiro de Geografia e Estatística.*

Bras-*3030*.R6. *Resoluções da Junta Executiva Central do Conselho Nacional de Estatística.*

Bras-*3040*.R5, Apr.-June '46, a. "O Ensino Primário Brasileiro no Decênio 1932-1941."

Bras-*3040*.R5, Oct.-Dec. '45. "Estudos sôbre a Mortalidade no Distrito Federal e no Município de São Paulo."

Bras-*3200*.N5, 1939. *Normas de Apresentação da Estatística Brasileira.*

Bras-*6000*.I3, Sept. '44. *Informações Relativas ao Distrito Federal, em Setembro de 1944.*

Bras-*6000*.I5. *Inquéritos Econômico para a Defesa Nacional.*

Bras-*6000*.S7, Apr. '44. *Situação e Movimento dos Estabelecimentos Comerciais e Industriais em Abril de 1944.*

Bras-*6020*.E5, 1939. *A Estatística da Produção no Estado de Minas Gerais.*

Bras-*6400*.E7, 1944. *Os Estoques de Matérias e Produtos Têxteis do Distrito Federal em 30 de Junho de 1943 e em 30 de Junho de 1944.*

Bras-*7000*.E5, 1942. *Educação e Saúde.*

Bras-*9800*.V5, 1945. *A Velocidade de Circulação dos Estoques de Produtos Alimentícios nos Mercados do Distrito Federal e de São Paulo, em 1944.*

——. ——. Conselho Nacional de Geografia.

Bras-*4000*.B5. *Boletim Geográfico do Conselho...*

Bras-*4000*.R5. *Revista Brasileira de Geografia.*

——. ——. Serviço Nacional de Recenseamento—*see - véase* Brasil. Instituto Brasileiro de Geografia e Estatística. Comissão Censitária Nacional. Serviço Nacional de Recenseamento.

——. Instituto do Açucar e do Álcool—*see-véase* Instituto do Açucar e do Álcool, Brasil.

——. Instituto Nacional de Estudos Pedagógicos—*see-véase* Brasil. Ministério da Educação e Saúde. Instituto Nacional de Estudos Pedagógicos.

——. Ministério da Agricultura. Serviço de Estatística da Produção.

Bras-*2200*.R4, 1920. *Recenseamento do Brasil Realizado em 1° de Setembro de 1920.*

Bras-*6200*.M5. *Mensário de Estatística da Produção.*

——. Ministerio da Educação e Saúde. Departamento Nacional de Saúde.

Int-*7110*.B5, June '46, Barros Barreto. "A Saúde Pública no Brasil."

Bras-*7100*.A5. *Arquivos de Higiene.*

Bras-*7101*.D5, 1943. *O Departamento Nacional de Saúde em 1943.*

——. ——. ——. Serviço Federal de Bioestatística.

Bras-*5200*.A5. *Anuário de Bioestatística.*

Bras-*5200*.B5. *Boletim Mensal do Serviço Federal de Bio-Estatística.*

Bras-*5200*.R5. *Resumo da Bio-Estatística de Cidades Brasileiras.*

——. ——. Instituto Nacional de Estudos Pedagógicos.

Bras-*7200*.B5. *Boletim do Instituto . . .*

Bras-*7200*.E3, 1942. *O Ensino no Brasil no Quinquênio 1936-1940.*

Bras-*7200*.E4, 1939. *O Ensino no Brasil no Quinquênio 1932-1936.*

Bras-*7200*.O5, 1946. *Outline of Education in Brazil.*

Bras-*7200*.R5. *Revista Brasileira de Estudos Pedagógicos.*

——. ——. Serviço de Estatística da Educação e Saúde.

Bras-*3040*.R5, Apr.-June '46, a. "O Ensino Primário Brasileiro no Decênio 1932-1941."

——. ——. Serviço Nacional de Educação Sanitária.

NonG-*3330*.Jansen de Mello, 1944. *Utilização de Medidas Estatísticas em Biologia, Medicina e Saúde Pública.*

——. Ministério da Fazenda. Tesouro Nacional. Diretoria das Rendas Aduaneiras.

Bras-*6900*.B6. *Boletim Estatístico da Diretoria . . .*

——. Ministério da Fazenda. Tesouro Nacional. Diretoria de Estatística Econômica e Financeira—*see-véase* Brasil. Ministério da Fazenda. Tesouro Nacional. Serviço de Estatística Econômica e Financeira.

——. ——. ——. Serviço de Estatística Econômica e Financeira.

Bras-*6000*.E6. *Estatísticas Econômicas.*

British West Indies. Jamaica. Census Office —*see - véase* British West Indies. Jamaica. Central Bureau of Statistics.
British West Indies. Jamaica. Central Bureau of Statistics.
WI-*2200*.E5, 1943. *Eighth Census of Jamaica and Its Dependencies, 1943: Population, Housing and Agriculture.*
WI-*2200*.E5, 1943.1. *Eighth Census of Jamaica, 1943: Administrative Report.*
WI-*2200*.E5, 1943.2. *Eighth Census of Jamaica, 1943: Introduction to Population and Housing.*
Brookings Institution.
Can-*6010*.P3, 1943. *The Price Control and Subsidy Program in Canada.*
USA-*6100*.I6, 1938. *The Income Structure of the United States.*
USA-*8500*.P3. *Pamphlets of the . . .*
RepDom-*5300*.R5, 1942. *Refugee Settlement in the Dominican Republic.*
——. Institute for Government Research.
USA-*3030*.B5, 1929. *The Bureau of the Census: Its History, Activities and Organization.*
Brown, John S.
Haiti-*4000*.G5, 1924. *Geology of the Republic of Haiti.*
Brumbaugh, Martin Allen.
NonG-*3330*, 1941. *Business Statistics.*
Brunsman, Howard G.
USA-*3000*.G7, 1946. *Government Statistics for Business Use.*
Brussels Nomenclature. Int-*3210*.C4,1913.2.
Brussels Nomenclature. Int-*3210*.C4,1913.3.
BUDGET:
FAMILY—*see* 9700.
NATIONAL—*see* 8800.
The Budget of the United States Government. USA-*8800*.B7.
Buenos Aires (Ciudad)—*see - véase* Argentina. Buenos Aires (Ciudad).
Buenos Aires (Prov.)—*see - véase* Argentina. Buenos Aires (Prov.).
Buenos Aires, Puerto del Río de la Plata, Capital de la Argentina; Estudio Crítico de Su Población, 1536-1936. Arg-*5000*.B5, 1536-1936.
BUILDING ACTIVITY AND INDEXES—*see* 6500.
BUILDING AND LOAN ASSOCIATIONS—*see* 6720.
BUILDINGS—*see* 2220; 9900.
Bulletin Annuel de l'Observatoire Météorologique du Petit Séminaire Collège St. Martial. Haiti-*4200*.B5.
Bulletin de l'Institut des Recherches Sociales et Économiques. Int-*3040*.B5.
Bulletin de l'Institut International de Statistique. Int-*3020*.B5.
Bulletin de Statistique (Administration Communale de Port-au-Prince). Haiti-*1410*.B5.
Bulletin du Service National d'Hygiène et d'Assistance Publique. Haiti-*7110*.B5.
Bulletin Mensuel (Banque Nationale de la République d'Haïti. Département Fiscal). Haiti-*1420*.B5.
Bulletin Mensuel de l'Office Permanent de l'Institut International de Statistique. Int-*3040*.R5.
Bulletin Mensuel de Statistique (Nation Unies). Int-*1420*.M5.
Bulletin Mensuel du Département Fiscal (Banque Nationale de la République d'Haïti). Haiti-*1420*.B5.
Bulletin of Agricultural Statistics. Int-*6220*.M5.
Bulletin of the Health Organisation of the League of Nations. Int-*7100*.B6.
Bulletin of the Institute of Social and Economic Research. Int-*3040*.B5.
Bulletin of the Pan American Union. Int-*8200*.B7.
Bulletins du Service National de la Production Agricole et de l'Enseignement Rural. Haiti-*6200*.B5.
Bulletins No. F (Dominion Bureau of Statistics). Can-*1400*.B5.
Bulletins of the Bureau of Labor Statistics. USA-*9000*.B4.
Bulletins of the Social Science Research Council. USA-*9000*.B6.
Bulletins of the Women's Bureau. USA-*9000*.B5.
BULLION—*see* 6720.
Bunge, Alejandro Ernesto.
Int-*1300*.A5, 1930.3, a. "Métodos para Determinar el Grado de Mortalidad Infantil."
Int-*1300*.A5, 1940.3, b. "La Población Argentina, Su Desarrollo, Sus Características y Sus Tendencias."
Arg-*1900*.N5, 1940. *Una Nueva Argentina.*
Arg-*3040*.R5. *Revista de Economía Argentina.*
Bunle, Henri.
Int-*3040*.R5, 1938.2. "Rapport de la Commission pour la Définition de la 'Population Rurale'."
Int-*3040*.R5, 1937.2. "Sondages Démographiques dans les Pays Sans État Civil Régulier."
Int-*3040*.R5, 1937.4 "La Population Rurale."

Cameron, G. D. W.
Can-*7110*.C5, Feb. '46. "State of Health of the People of Canada in 1944."
CAMIONES—*véase* 6612.
Campos, Alfonso B.
Int-*3000*.S5, 1940. "Las Actividades Estadísticas del Paraguay."
Canada. Advisory Committee on Reconstruction.
Can-*9950*.R5, 1943. *Report on Social Security for Canada.*
——. Bank of Montreal—*see - véase* Bank of Montreal.
——. Chief Registrar.
Can-*5250*.N2, 1940.C. *National Registration Regulations, 1940.*
——. Committee on Form and Content of Annual Vital Statistics Reports—*see-véase* Committee on Form and Content of Annual Vital Statistics Reports.
——. Department of Mines and Resources. Indian Affairs Branch.
Can-*2200*.C6, 1944. *Census of Indians in Canada, 1944. Recensement des Indiens du Canada, 1944.*
——. Department of National War Services.
Can-*5250*.N2, 1940.A. *National Registration, August 1940.*
Can-*5250*.N2, 1940.B. *Specialized Occupations, National Registration, 1940. Occupations Spécialisées, l'Enregistrement National, 1940.*
——. Department of Trade and Commerce. Dominion Bureau of Statistics.
Int-*3040*.E5, Sept. '45, a. "Gross and Net Reproduction Rates, Canada and Provinces, 1920-1942."
Can-*1400*.B5. *Bulletins No. F.*
Can-*1410*.A5. *Annuaire du Canada.*
Can-1410.C5. *Canada, the Official Handbook of Present Conditions and Recent Progress.*
Can-*1410*.C7. *The Canada Year Book.*
Can-*1410*.C7, 1945, a. "Publications of the Dominion Bureau of Statistics."
Can-*2100*.C5, 1941.1. *Eighth Census of Canada, 1941: Methods and Procedures.*
Can-*2100*.C5, 1941.2. *Eighth Census of Canada, 1941: Instructions to Commissioners and Enumerators. Huitième Recensement du Canada, 1941: Instructions aux Commissaires et Énumérateurs.*
Can-*2100*.C5, 1941.3. *Eighth Census of Canada, 1941: Administrative Report of the Dominion Statistician. Huitième Recensement du Canada, 1941: Rapport Administratif du Statisticien du Dominion.*
Can-*2100*.C5, 1931. *Seventh Census of 1931: Administrative Report of the Dominion Statistician.*
Can-*2200*.C5, 1941.2. *Eighth Census of Canada, 1941. Huitième Recensement du Canada, 1941.*
Can-*2200*.C5, 1941.2a. *Eighth Census of Canada, 1941: Population.*
Can-*2200*.C5, 1931.2. *Seventh Census of Canada, 1931. Septième Recensement du Canada, 1931.*
Can-*2210*.C5, 1941.2a. *Eighth Census of Canada, 1941: Census of Agriculture: Nova Scotia. Huitième Recensement du Canada, 1941: Recensement Agricole: Nouvelle-Écosse.*
Can-*3030*.D5, 1935. *The Dominion Bureau of Statistics.*
Can-*3220*.C5, 1941. *Classification of Industries: Eighth Census of Canada, 1941.*
Can-*3240*.C5, 1941. *Classification of Occupations: Eighth Census of Canada, 1941.*
Can-*5100*.F5, 1946. *The Future Population of Canada.*
Can-*5212*.S5, 1945. *A Study in Maternal, Infant and Neo-Natal Mortality in Canada. Étude sur la Mortalité Puerpérale, Infantile et Néonatale au Canada.*
Can-*5220*.C5, 1871-1931. *Canadian Abridged Life Tables, 1871, 1881, 1921, 1931.*
Can-*5250*.H2, 1937. *Handbook on Death Registration and Certification.*
Can-*5250*.N2, 1940.A. *National Registration, August 1940.*
Can-*5250*.N2, 1940.B. *Specialized Occupations, National Registration, 1940. Occupations Spécialisées, l'Enregistrement National, 1940.*
Can-*6100*.N5, 1919-38. *National Income of Canada, 1919-1938.*
Can-*6710*.C5, 1929-41. *Census of Merchandising and Service Establishments.*
Can-*6720*.A5. *Annual Report of Cheques Cashed Against Individual Accounts and Equation of Exchange.*
Can-*6720*.C5. *Commercial Failures in Canada.*
Can-*6900*.C7, 1944. *Customs Tariff and Amendments with Index to January 20, 1944.*
Can-*9200*.E5. *The Employment Situation... Together with Payrolls.*
Can-*9200*.M5. *Man Hours and Hourly Earnings.*
Can-*9200*.O5, 1891-1931. *Occupational Trends in Canada, 1891-1931.*

Canada—cont.

Can-*6400*.C6. *Central Electric Stations in Canada.*
Can-*6611*.C3. *Civil Aviation.*
Can-*6611*.C5. *Civil Aviation.*
Can-*6612*.H5. *The Highway and the Motor Vehicle in Canada.*
Can-*6612*.M5. *Motor Carriers.*
Can-*6614*.E5. *Electric Railways of Canada.*
Can-*6614*.S5. *Statistics of Steam Railways of Canada.*
Can-*6614*.S8. *Summary of Monthly Railway Traffic Reports.*
Can-*6615*.C5. *Canal Statistics.*
Can-*6622*.T5. *Telegraph Statistics. Statistiques des Télégraphes.*
Can-*6622*.T6. *Telephone Statistics. Statistiques du Téléphone.*

——. ——. ——. Vital Statistics Branch.
Can-*5200*.P4. *Preliminary Annual Report, Vital Statistics of Canada. Rapport Prêliminaire Annuel, Statistiques Vitales du Canada.*
Can-*5200*.P5. *Preliminary Report on Births, Deaths and Marriages. Rapport Préliminaire sur les Naissances, les Décès et les Mariages.*
Can-*5200*.R5. *Registration of Births, Deaths and Marriages. Enregistrement des Naissances, Décès et Mariages.*
Can-*5200*.V5. *Vital Statistics. Statistiques Vitales.*
Can-*5240*.V5, 1942. *The Vital Statistics Instruction Manual for Canada. Manuel d'Instructions sur les Statistiques Vitales.*
Can-*5250*.R5, 1943. *Report on Completeness of Birth Registration in Canada.*

——. Dominion Bureau of Statistics—*see - véase* Canada. Department of Trade and Commerce. Dominion Bureau of Statistics.

——. Indian Affairs Branch—*see-véase* Canada. Department of Mines and Resources. Indian Affairs Branch.

——. Medical Procurement and Assignment Board.
Can-*7110*.R5. 1945. *Report of the National Health Survey.*

Canada. Can-*6000*.C5, 1945.

Canada, the Official Handbook of Present Conditions and Recent Progress. Can-*1410*.C5.

The Canada Year Book. Can-*1410*.C7.

Canadian Abridged Life Tables, 1871, 1881, 1921, 1931. Can-*5220*.C5, 1871-1931.

The Canadian Balance of International Payments. Can-*6910*.C5,

The Canadian-Born in the United States. USA-*5120*.C5, 1850-1930.

"The Canadian Census of Distribution." Int-*3020*.B5, 1938, Marshall.

Canadian Farm Family Living Costs. Can-*9500*.C4.

Canadian Journal of Economics and Political Science. Can-*3040*.C5.

The Canadian Journal of Public Health. Can-*7110*.C5.

Canadian Life Tables. Can-*5220*.C5.

The Canadian Official Handbook. Can-*1410*.C5.

Canadian Political Science Association.
Can-*3040*.C5. *Canadian Journal of Economics and Political Science.*
Can-*3040*.C5, May '45. "Recent Studies on National Income."
Can-*3040*.C5, Feb. '45. "Recent Studies on National Income."

Canadian Public Health Association.
Can-*7110*.C5. *The Canadian Journal of Public Health.*

——. Vital Statistics Section.
Can-*7110*.C5, Sept. '46. "Report of the Committee on Form and Content of Annual Vital Statistics Reports."

Canal Statistics. Can-*6615*.C5.

CANALES—*véase* 6615.

CANALS—*see* 6615.

CAÑERIAS:
CONSTRUCCION DE—*véase* 6500.
OPERACIONES DE—*véase* 6613.

CAPACIDAD CONSUMIDORA—*véase* 6000.

CAPITAL—*see-véase* 6721.

"Capital Gains in Income Theory and Taxation Policy." NonG-*6100*.S3.

CAPITAL PUNISHMENT—*see* 8300.

CAPITALES FINANCIEROS—*véase* 6721.

Capriles Rico, Roberto.
Bol-*9000*.P5, 1941. *El Problema Social en Bolivia—Condiciones de Vida y de Trabajo.*

CAR LOADINGS—*see* 6614.

Características Antropométricas de los Niños de Guatemala. Guat-*7310*.C5, 1946.

Características Principales de la Industria: Censo de 1940. Méx-*2250*.C5, 1940.3.

Carballo R., Sergio.
Int-*3000*.S5, 1940. "Las Actividades Estadísticas de Costa Rica."

CARBON—*see-véase* 6300.

Cárcamo C., Luis.
USA-*9000*.M5, June '46. "Work Accidents in Chile."

CARCELES—*véase* 8330.

Cardona Lazo, Antonio.
ElSal-*4000*.D5, 1945. *Diccionario Geográfico de la República de El Salvador.*

The Caribbean Islands and the War. Int-*1900*.C5, 1943.

Censo de 1943: Métodos y Procedimientos. Cuba-*2100*.C5, 1943.1.
Censo de 1941: Métodos y Procedimientos. Ven-*2100*.C5, 1941.1.
Censo de 1940: Métodos y Procedimientos. Chile-*2100*.C5, 1940.1.
Censo de 1940: Métodos y Procedimientos. Guat-*2100*.C5, 1940.1.
Censo de 1940: Métodos y Procedimientos. Hond-*2100*.C5, 1940.1.
Censo de 1940: Métodos y Procedimientos. Méx-*2100*.C5, 1940.1.
Censo de 1940: Métodos y Procedimientos. Nic-*2100*.C5, 1940.1.
Censo de 1940: Métodos y Procedimientos. Pan-*2100*.C5, 1940.1.
Censo de 1936: Métodos y Procedimientos. Par-*2100*.C5, 1936.1.
Censo de 1935: Instrucciones a los Enumeradores del Censo de Población. RepDom-*2100*.C5, 1935.3.
Censo de 1935: Instrucciones Generales sobre las Funciones de un Enumerador. RepDom-*2100*.C5, 1935.4.
Censo de 1935: Ley del Congreso Nacional que Dispone las Condiciones en que Deben Desarrollarse las Operaciones del Censo. RepDom-*2100*.C5, 1935.2.
Censo de 1935: Métodos y Procedimientos. RepDom-*2100*.C5, 1935.1.
Censo de 1931. Cuba-*2200*.C5, 1931-38.
Censo de 1930: Métodos y Procedimientos. ElSal-*2100*.C5, 1930.1.
Censo de 1900: Métodos y Procedimientos. Bol-*2100*.C4, 1900.
Censo de Agricultura de los Estados Unidos. USA-*2110*.C5, 1940.1.
Censo de Edificios. RepDom-*2100*.C6, 1945.
Censo de Edificios, 18 de Enero de 1940: Resumen por Departamentos y Municipios. Nic-*2220*.C5, 1940.
Censo de Educación, Año 1933. Chile-*2230*.C5, 1933.
Censo de la Población de la República, Levantado el 28 de Agosto de 1921. Guat-*2200*.C5, 1921.
"Censo de las Américas de 1950." Ec-*1420*.T5, July-Dec. '45, a.
Censo de las Américas de 1950: Antecedentes y Justificación del Proyecto. Int-*2000*.C5, 1950.1b.
Censo de Personas sin Trabajo, 1932. CR-*2200*.C5, 1932.
"Censo de Población." Par-*1410*.A5, 1937, a.
"Censo de Población." Par-*1410*.A5, 1936, a.
Censo de Población. RepDom-*2100*.C6, 1945.
Censo de Población, 1940. Pan-*2200*.C5, 1940.1.
Censo de Población de 1930. ElSal-*2100*.C5, 1930.3.
Censo de Población, Decreto N°. 2120 de 18 de Junio de 1936. Par-*2100*.C5, 1936.2.
Censo de Población Urbana y Rural, por Provincias y Comunas. RepDom-*2200*.C5, 1935.3.
Censo Décimosexto de los Estados Unidos: 1940: Agricultura. USA PR-*2210*.C5, 1940.
Censo Décimosexto de los Estados Unidos: 1940: Instrucciones a los Enumeradores: Puerto Rico. USA PR-*2100*.C5, 1940.2.
Censo Décimosexto de los Estados Unidos: 1940: Puerto Rico, Población, Boletín Núm. 2: Características de la Población. USA PR-*2200*.C5, 1940.2
Censo Décimosexto de los Estados Unidos: 1940: Puerto Rico: Viviendas. USA PR-*2220*.C5, 1940.
Censo del Año 1943. Cuba-*2200*.C5, 1943.2.
Censo Demográfico de la Ciudad de La Paz, 15 de Octubre de 1942. Bol-*2200*.C4, 1942.
El Censo Económico General de 1943. Chile-*2150*.C7, 1943.
"Censo Escolar." Guat-*1420*.B5, May '46, a.
Censo Escolar de 1946: Métodos y Procedimientos. Guat-*2130*.C5, 1946.1.
Censo Escolar de la Nación, Ley No. 12.723: Cuestionario de Convivencia. Arg-*2130*.C5, 1943.8
Censo Escolar de la Nación, Ley No. 12.723: Cuestionario Familiar. Arg-*2130*.C5, 1943.6.
Censo Escolar de la Nación, Ley No. 12.723: Cuestionario Individual. Arg-*2130*.C5.1943.7.
Censo Escolar de la Nación, Ley No. 12.723: Instrucciones para los Censistas. Arg-*2130*.C5, 1943.5.
"El Censo Escolar Nacional de 1943." Int-*3040*.E5, Sept. '46, Coghlan.
Censo Especial Urbano de Ciudad Trujillo, 1945. RepDom-*2100*.C6, 1945.
Censo Especial Urbano San Cristóbal, 1944: Instrucciones a los Enumeradores. RepDom-*2100*.C7, 1944.
Censo Experimental Agropecuario del Departamento de Chuquisaca, 1942. Bol-*2210*.C5, 1942, Chuquisaca.
Censo Experimental Agropecuario, Departamento de Tarija, 1944-1945. Bol-*2210*.C5, 1944-45, Tarija.
Censo Ganadero 1943: Ley de Enero 27 de 1943. Ur-*2110*.C5, 1943.
"El Censo Ganadero de 1945." Cuba-*1420*.B5, Apr. '45, Queralt.
Censo Ganadero Nacional, Existencia al 1.° de Julio de 1930. Arg-*2210*.C5, 1930.
Censo General de 1920. Nic-*2200*.C5, 1920.
Censo General de la Población de la República de

on Total and Partial Unemployment. USA-*2200*.C7, 1937.
Census of Unemployment, 1937: Final Report on Total and Partial Unemployment for Geographic Divisions. USA-*2200*.C7, 1937.1.
CENTER OF POPULATION—*see* 5110.
Central America: Reports Presented to the Conference of Commissions of Inter-American Development. Int-*6000*.R7, 1944.
Central Electric Stations in Canada. Can-*6400*.C6.
CENTRAL HEATING—*see* 6400; 9900.
Centro Azucarero Argentino.
Arg-*6210*.I5. *La Industria Azucarera.*
CENTROS DE POBLACION—*véase* 5110.
A Century of Population Growth. USA-*5110*.C5, 1790-1900.
Céspedes, Francisco S.
Int-*7200*.E6, 1942. *Educational Trends in Latin America.*
Chaddock, Robert Emmet.
NonG-*3330*, 1925. *Principles and Methods of Statistics.*
Changes in Cost of Living. USA-*9500*.C4.
Changes in Distribution of Manufacturing Wage Earners, 1899-1939. USA-*9200*.C5, 1899-1939.
Changes in Income Distribution during the Great Depression. NonG-*6100*.S3.
"Changing Inventory Valuations and Their Effect on Business Savings and on National Income Produced." NonG-*6100*.S3.
"The Changing Population of the United States." USA-*3040*.A5, Jan. '45, Hauser.
Chapman, J. H.
Int-*6900*.N5, 1942.A. *Addendum to Annex III of "The Network of World Trade."*
Chapman, R. A.
NonG-*6250*.S5, 1942. *Sampling Methods in Forestry and Range Management.*
CHARITY—*see* 9951.
Charles, Enid.
Int-*3040*.E5, Sept. '45, a. "Gross and Net Reproducion Rates, Canada and Provinces, 1920-1942."
Charliér, Carl Vilhelm Ludvig.
NonG-*3140*.E5, 1936. *Elementos de Estadística Matemática.*
Chart Book I, Federal Reserve Charts on Bank Credit, Money Rates, and Business. USA-*6720*.C5, 1945.
Charte des Nations Unies et Statut de la Cour Internationale de Justice. Int-*8210*.C6, 1946.
Charter of the United Nations and Statute of the International Court of Justice. Int-*8210*.C6, 1946.
CHARTS, ORGANIZATION:
OF STATISTICAL SERVICES—*see* 3030.
In specific subject fields, *see* the particular field.
Chelala-Aguilera, José.
Int-*1300*.A5, 1940.3. Roa Reyes. "Metodología Estadística Aplicada a las Ciencias Sociales y Biología Social."
CHEMICALS—*see* 6400.
CHEMISTRY—*see* 7300.
Chicharro Valdosera, Santiago.
Chile—*6000*.C34, 1944. *Chile a Través de Sus Zonas Geográfico-Económicas.*
CHILD GROWTH—*see* 7180.
CHILD WELFARE—*see* 9951.
CHILDBIRTH MORTALITY—*see* 5212.
Childs, James B.
Arg-*1200*.G5, 1945. *A Guide to the Official Publications of the Other American Republics: I, Argentina.*
Bol-*1200*.G5, 1945. *A Guide to the Official Publications of the Other American Republics: II, Bolivia.*
Cuba-*1200*.G5, 1945. *A Guide to the Official Publications of the Other American Republics: VII, Cuba.*
Chile. Banco Central de Chile—*see-véase* Banco Central de Chile.
——. Caja de Colonización Agrícola. Sección Estadística.
Chile-*6200*.B5. *Boletín de la Sección Estadística de la Caja de Colonización Agrícola.*
——. Caja de Crédito Agrario.
Chile-*6200*.M5. *Memoria de la . . .*
——. Caja de Seguro Obligatorio.
Chile-*9950*.E5. *Estadística de la . . .*
——. Consejo Nacional de Comercio Exterior.
Chile-*3210*.L5, 1943. *Lista en Castellano e Inglés de los Artículos Importables desde los Estados Unidos de Norte América (con Números del "Schedule B" y Unidades).*
——. Contraloría General de la República—*see-véase* Chile. Ministerio de Hacienda, Economía y Comercio. Contraloría General de la República.
——. Corporación de Fomento de la Producción.
Chile-*6000*.C4, 1939-43. *Cinco Años de Labor, 1939-1943.*
Chile-*6100*.R5, 1940-45. *Renta Nacional, 1940-1945.*
——. Departamento de Previsión Social—*see-véase* Chile. Ministerio de Salubridad, Previsión y Asistencia Social. Dirección General de Previsión Social.

Chile—cont.

Chile-*7130*.P5. *Publicaciones de la Dirección General de Servicios de Beneficencia y Asistencia Social.*

Chile-*7130*.S5, 1946. *La Segunda Zona Hospitalaria: Antofagasta.*

Chile-*7130*.S6, 1946. *Un Semestre de Labor en los Sanatorios.*

——. ——. Dirección General de Previsión Social.

Chile-*9950*.S5, 1942. *La Seguridad Social en Chile.*

——. Ministerio del Trabajo. Dirección General del Trabajo. Servicio de Estadística.

Chile-*9000*.V5, 1945. *Veinte Años de Legislación Social.*

——. Oficina Central de Estadística—*see-véase* Chile. Ministerio de Hacienda, Economía y Comercio. Dirección General de Estadística.

——. Servicios de Beneficencia y Asistencia Social. Dirección General. Departamento de Estadística—*see-véase* Chile. Ministerio de Salubridad, Previsión y Asistencia Social. Dirección General de Beneficencia y Asistencia Social. Sección Estadística.

——. Superintendencia de Aduanas—*see - véase* Chile. Ministerio de Hacienda, Economía y Comercio. Superintendencia de Aduanas.

——. Superintendencia de Bancos—*see - véase* Chile. Ministerio de Hacienda, Economía y Comercio. Superintendencia de Bancos.

——. Universidad de Chile—*see-véase* Universidad de Chile.

Chile. Chile-*6000*.C3, 1945.

Chile a Través de Sus Zonas Geográfico-Económicas. Chile-*6000*.C34, 1944.

Chile: An Economy in Transition. Chile-*6000*.C35, 1945.

Chile: Demographic Data. Chile-*5000*.C5, 1943.

Chile: Reports Presented to the Conference of Commissions of Inter-American Development. Chile-*6000*.R5, 1944.

CHURCHES—*see* 7400.

CIEGOS:

GENERAL—*véase* 7170.

ASILOS PARA—*véase* 9951.

CIENCIA:

GENERAL (PURA)—*véase* 7300-7399.

ESPECIFICA—*véase* las respectivas clasificaciones.

ACTUARIAL (DE SEGUROS, incluye CURSOS Y PROGRAMAS DE ENSEÑANZA)—*véase* 6730.

APLICADA—*véase* las respectivas clasificaciones.

BIBLIOGRAFICA—*véase* 7200.

DOMESTICA—*véase* 9700.

ESTADISTICA—*véase* 3000-3999.

MATEMATICA—*véase* 7300.

MEDICA—*véase* 7180.

POLITICA—*véase* 8000-8999.

SOCIAL (SOCIOLOGIA)—*véase* 9000.

"Cifras Estadísticas de la Industria Textil." Perú-*6400*.B5, Sept. '45, Arbulú Casanova.

"Cifras Preliminares de Población por Provincias y Capitales de Provincias, según el Censo de 1940." Nic-*1420*.B5, July-Dec. '40.

Cinco Años de Labor, 1939-1943. Chile-*6000*.C4, 1939-43.

50 Années de l'Institut International de Statistique. Int-*3030*.C5, 1884-1934.

CIRCULACION DE MONEDAS Y GIROS —*véase* 6720.

Circulares Documentales de la Comisión Ejecutiva del Consejo Interamericano de Comercio y Producción. Int-*6700*.C4.

CIRCULATION OFMONEY AND NOTES —*see* 6720.

CIRUGIA—*véase* 7180.

CITIZENSHIP—*see* 5110.

City Finances. USA-*8800*.C4.

CIUDADANIA—*véase* 5110.

CIVIC PLANNING—*see* 8500.

Civil Aviation. Can-*6611*.C3.

Civil Aviation. Can-*6611*.C5.

CIVIL REGISTER—*see* 5250.

CIVIL SERVICE—*see* 8000.

CIVILIAN DEFENSE—*see* 8100.

Clark, Colin.

Int-*6000*.E9, 1960. *The Economics of 1960.*

NonG-*6000*.C6, 1940. *The Conditions of Economic Progress.*

NonG-*6100*.N6, 1937. *National Income and Outlay.*

Clark, Faith.

USA-*9700*.F5F1, 1941. *Family Food Consumption and Dietary Levels, Five Regions, Farm Series.*

Clark, John R.

NonG-*3110*.S5, 1943. *Statistics: Collecting, Organizing, and Interpreting Data.*

CLASIFICACION:

ADUANERA—*véase* 3210; 6900.

DE COMERCIO EXTERIOR—*véase* 3210; 6900.

DE MERCANCIAS—*véase* 3210; 6900.

DE MORBILIDAD—*véase* 3230; 7150.

DE MORTALIDAD—*véase* 3230; 5212.

DE OCUPACIONES—*véase* 3240; 9000.

ESTADISTICA (incluye NOMENCLATURAS—*véase* 3200-3299; *véase*

Collins, Selwyn De Witt—cont.

Services among 9,000 Canvassed Families."

Colm, Gerhard.

NonG-*6100*.S3. "Public Revenue and Public Expenditure in National Income."

Colombia. Consejo Administrativo de los Ferrocarriles Nacionales.

Col-*6614*.R5. *Revista del . . .*

——. Contraloría General de la República.

Col-*1420*.A5. *Anales de Economía y Estadística.*

Col-*1420*.A5, May 5, '42, a. "Normas para la Elaboración de los Cuadros Estadísticos."

Col-*1420*.A5, Aug. '39. "Informe que la Contraloría General de la República Rinde al Sr. Ministro de Gobierno y a las Honorables Cámaras sobre el Levantamiento del Censo Civil de 1938."

Col-*1900*.C5. *Colombia.*

Col-*3030*.E3, 1938. *La Estadística Nacional.*

Col-*8800*.I3. *Información Fiscal de Colomba*

Col-*8800*.I5. *Informe Financiero del Contralor General.*

——. ——. Dirección General de los Censos.

Col-*2100*.C5, 1938. *Decretos y Bases de Organización para la Ejecución de los Censos, 1938.*

Col-*2200*.C5, 1938. *Censo General de Población, 5 de Julio de 1938.*

Col-*2220*.C5, 1938. *Primer Censo Nacional de Edificios, Efectuado el 20 de Abril de 1938.*

——. ——. Dirección Nacional de Estadística.

Col-*1400*.S5. *Síntesis Estadística de Colombia.*

Col-*1410*.A5. *Anuario General de Estadística.*

Col-*1420*.A5, Oct. '46. "Las Condiciones Económico-Sociales y el Costo de la Vida de la Clase Media en Bogotá."

Col-*1420*.A5, June '46. "Las Condiciones Económico-Sociales y el Costo de la Vida de la Clase Obrera en la Ciudad de Honda."

Col-*1420*.A5, May 5, '42, b. "Programa de Estudios del Curso General de Estadística de la Contraloría General de la República."

Col-*1420*.A5, Feb. 5, '42. "El Curso de Estadística de la Contraloría General."

Col-*2100*.C5, 1950. "Censo Continental: Una Sugerencia del Instituto Interamericano de Estadística."

Col-*3030*.E5, 1945. *Estatuto Orgánico de la Estadística Nacional.*

Col-*3030*.O5, 1946. "Organización de Servicios de Estadística: Resolución No. 64 de Enero de 194 ."

Col-*3210*.C5, 1936. *Codificación Estadística del Comercio Exterior.*

Col-*3230*.N5, 1942. *Nomenclatura de Morbilidad y Mortalidad.*

Col-*6900*.A5. *Anuario de Comercio Exterior.*

Col-*6900*.C5. *Comercio Exterior de Colombia.*

Col-*7150*.E5, 1936. *Estadísticas Demográficas y Nosológica, Año de 1936.*

Col-*8800*.E5. *Estadística Fiscal y Administrativa.*

NonG-*3330*.Colombia, 1937-38. *Cartilla de Estadística.*

——. ——. Dirección Nacional del Censo Industrial.

Col-*2150*.C5, 1945, a. *Censo Industrial de Colombia, 1945: Métodos y Procedimientos.*

Col-*2150*.C5, 1945, b. *Primer Censo Industrial de Colombia, 1945: Cartilla de Instrucciones.*

Col-*2150*.C5, 1945, c. *Primer Censo Industrial de Colombia, 1945: Boleta General.*

——. Departamento de Aguas y Meteorología—*see - véase* Colombia. Ministerio de la Economía Nacional. Departamento de Aguas y Meteorología.

——. Departamento de Comercio e Industrias—*see - véase* Colombia. Ministerio de la Economía Nacional. Departamento de Comercio e Industrias.

——. Departamento de Irrigación —*see-véase* Colombia. Ministerio de la Economía Nacional. Departamento de Aguas y Meteorología.

——. Departamento de Justicia—*see - véase* Colombia. Ministerio de Gobierno. Departamento de Justicia.

——. Dirección General de los Censos—*see - véase* Colombia. Contraloría General de la República. Dirección General de los Censos.

——. Dirección Nacional de Estadística—*see - véase* Colombia. Contraloría General de la República. Dirección Nacional de Estadística.

——. Dirección Nacional del Censo Industrial—*see - véase* Colombia. Contraloría General de la República. Dirección Nacional del Censo Industrial.

——. Federación Nacional de Cafeteros de

Comisión de Fomento Interamericano—*see-véase* Inter-American Development Commission.

"Le Comité d'Experts Statisticiens de la Société des Nations (1931-1939)." Int-*3040*.R5, 1939, 2-3, Huber.

"Le Comité d'Experts Statisticiens du Bureau International du Travail (1933-1939)." Int-*3040*.R5, 1942, 1-2, Huber.

Comité Interamericano de Seguridad Social—*see - véase* Inter-American Committee on Social Security.

Commentary Explanations and Sample Copies of the Forms and Manuals Used in Vital Records and Vital Statistics Activities in the Dominican Republic, November 1946. RepDom-*5200*.C5, 1946.

Commentary Explanations and Sample Copies of the Forms Used in Medical and Public Health Statistics Activities (Not Including Vital Statistics) in the Dominican Republic, November 1946. RepDom-*7100*.C5, 1946.

COMMERCE—*see* 6700-6799; 6900-6999.

Commerce du Canada. Can-*6900*.T5.

Commerce Exterieur du Brésil. Bras-*6900*.C3.

Commerce Reports of the Bureau of Foreign and Domestic Commerce. USA-*6900*.F7.

Commercial Failures in Canada. Can-*6720*.C5.

Commercial Pan America. Int-*6000*.C7.

Commercial Policies and Trade Relations of European Possessions in the Caribbean Area. Int-*6900*.C5, 1943.

Commercial Travelers' Guide to Latin America. Int-*1950*.C5, 1939.

Commission Argentine de Hautes Études Internationales—*see - véase* Comisión Argentina de Altos Estudios Internacionales.

"Commission on Building and Housing Statistics: Preliminary Report." Int-*3040*.R5, 1938.2, Nyström.

"Committee on Family Budgets: Preliminary Report." Int-*3040*.R5, 1936.1, Nixon.

Committee on Form and Content of Annual Vital Statistics Reports.
  Can-*7110*.C5, Sept. '46. "Report of the Committee on Form and Content of Annual Vital Statistics Reports."

Committee on the Cost of Medical Care.
  USA-*7150*.M6. *Morbidity: Instructions for Coding and Punching.*

COMMODITIES:
  AGRICULTURAL—*see* 6210.
  FOREIGN TRADE—*see* 6900.
  MANUFACTURING—*see* 6400.
  MINING—*see* 6300.

Commodities of Commerce Series. Int-*6900*.C7.

COMMODITY CLASSIFICATION—*see* 6900; 3210.

Commodity Flow and Capital Formation. NonG-*6000*.C4, 1938.

COMMODITY PRICES:
  GENERAL IN SCOPE—*see* 6010.
  SPECIFIC—*see* the particular subject field.

COMMUNICATIONS:
  GENERAL—*see* 6620-6629.
  CENSUSES OF—*see* 2280; 6620.

Compañía Peruana de Negocios Internacionales.
  Perú-*6900*.P5. *Peru—Year Book of Foreign Trade Exporters, Importers, and Merchants.*

Compañía Salvadoreña de Café, S.A.
  ElSal-*6210*.M5. *Memoria de la. . .*

COMPAÑIAS COMERCIALES:
  GENERAL—*véase* 6710.
  ESPECIFICA—*véase* las respectivas clasificaciones.

"Comparação entre a Discriminação dos Óbitos Constantes das Tábuas de Sobrevivência de 1939-41, Segundo Grupos de Causas, para o Distrito Federal e o Município de São Paulo." Bras-*3040*,R5, Oct.-Dec. '45.

"Comparaciones Internacionales de los Modos de Vivir." Int-1300.A5, 1940.3, Alanís Patiño, b.

"Comparações entre as Tábuas de Sobrevivência do Distrito Federal e do Município de São Paulo para os Períodos 1939-41 e 1920-21." Bras-*3040*.R5, Oct.-Dec. '45.

"A Comparative Examination of Certain Aspects of the Population of the New World." NonG-*5000*.H5, Sept. '40, Pearl.

Compendio de la Obra Intitulada: Curso Elemental de Estadística Administrativa. NonG-*3330*. Bertillon, 1906.

Compendio Estadístico (Dirección General de Estadística). Méx-*1410*.C5.

COMPENDIOS ESTADISTICOS—*véase* 1400-1499.

Compendium International de Statistiques, 1924-1938. Int-*6210*.G5, 1924-38.

Compilación Estadística sobre Café. Cuba-*6210*.C5.

COMPILATIONS OF DATA—*see* 1400-1499.

"Complete Expectation of Life at Various Ages in Selected Countries." Int-*5000*.P5, July '43. b.

"La Composición Familiar de Asalariados en el Distrito Federal de Venezuela." Int-*3040*.E5, Dec. '45, Michalup.

Compras de Ganado Bovino, Ovino y Porcino Efectuadas por los Frigoríficos en las Estancias. Arg-*6230*.C5.

Conference of Inter-American Development—cont.

tura y Condiciones Económicas de Su País.

Méx-*6000*.R4, 1944. *Reports Presented to the Conference of Commissions of Inter-American Development (Held in New York, May 9-18, 1944) by the Mexican Commission of Inter-American Development.*

Par-*6000*.R5, 1944. *Report Presented to the Conference of Commissions of Inter-American Development (Held in New York, May 9-18, 1944) by the Paraguayan Commission of Inter-American Development.*

Perú-*6000*.R3, 1944. *Reports Presented to the Conference of Commissions of Inter-American Development (Held in New York, May 9-18, 1944) by the Peruvian Commission of Inter-American Development.*

RepDom-*6000*.R5, 1944. *Report Presented to the Conference of Commissions of Inter-American Development (Held in New York, May 9-18, 1944) by the Dominican Commission of Inter-American Development.*

Ven-*6000*.R3, 1944. *Reports Presented to the Conference of Commissions of Inter-American Development (Held in New York, May 9-18, 1944) by the Venezuelan Commission of Inter-American Development.*

Conference of Commissions of Inter-American Development, May 9 to 18, 1944: Final Act. Int-*8200*.C6, 1944.1 Eng.

Conference of Ministers and Directors of Education of the American Republics.

Int-*7200*.C6, 1943.1, Eng. *Final Act of the First Conference of Ministers and Directors of Education of the American Republics, Held in Panama, September 27 to October 4, 1943.*

Int-*7200*.C6, 1943.1, Sp. *Acta Final de la Primera Conferencia de Ministros y Directores de Educación de las Repúblicas Americanas Celebrada en Panamá del 27 de Septiembre al 4 de Octubre de 1943.*

Bras-*7200*.R5, Aug. '44. "I Conferência de Ministros e Diretores de Educação das Repúblicas Americanas."

Conference of the Food and Agriculture Organization.

Int-*6200*.C5, 1945.2. *Report of the First Session of the Conference, Held at the City of Quebec, Canada, October 16 to November 1, 1945.*

Int-*6200*.C5, 1943. *Final Act of the United Nations Conference on Food and Agriculture, Hot Springs, Virginia, United States of America, 18th May-3rd June, 1943.*

Conference on International Co-operation in Statistics.

Int-*3020*.C5, 1919. *Conference on International Co-operation in Statistics, August 14th and 15th, 1919.*

Conference on International Co-operation in Statistics, August 14th and 15th, 1919. Int-*3020*.C5, 1919.

Conference on Jewish Relations.

USA-*5120*.J5, 1943. *Jewish Population Studies.*

USA-*9600*.J5. *Jewish Social Studies, Publications.*

Conference on Research in Income and Wealth.

Int-*6100*.N5, 1945. *National Income Estimates of Latin America.*

NonG-*6100*.S3. *Studies in Income and Wealth.*

Conference on War and Peace-*see—véase* Inter-American Conference on Problems of War and Peace.

Conference Series of the United States Department of State. USA-*8210*.C5.

CONFERENCES:

GENERAL—*see* 1300.

INTERNATIONAL—*see* 8200; *see also* the particular field.

In a specific field—*see* the particular field.

Conferencia de Comisiones de Fomento Interamericano—*see - véase* Conference of Commissions of Inter-American Development.

Conferência de Ministros e Diretores de Educação das Repúblicas Americanas—*see - véase* Conference of Ministers and Directors of Education of the American Republics.

Conferencia de Ministros y Directores de Educación de las Repúblicas Americanas—*see - véase* Conference of Ministers and Directors of Education of the American Republics.

Conferencia Interamericana de Agricultura—*see - véase* Inter-American Conference on Agriculture.

Conferencia Interamericana de Expertos para la Protección de los Derechos de Autor—*see-véase* Inter-American Conference of Experts on Copyright.

Conferencia Interamericana de Expertos para la Protección de los Derechos de Autor, 1-22 de Junio de 1946: Actas y Documentos. Int-*7300*.I5, 1946.2, Sp.

Conferência Interamericana de Peritos para a Proteção dos Direitos Autorais—*see - véase* Inter-American Conference of Experts on Copyright.

Consejo Interamericano de Comercio . . . —cont.
sôbre Fomento e Coordenação de Indústrias.
USA-*6900*.E5, 1946. *Los Estados Unidos y el Comercio Mundial durante la Guerra 1939-1945.*
——. Comisión Ejecutiva.
Int-*6700*.C4. *Circulares Documentales de la Comisión Ejecutiva del Consejo Interamericano de Comercio y Producción.*
Chile-*9600*.E5, 1943. *Encuesta Continental sobre el Consumo de Productos de Alimentación y Vestido y sobre la Vivienda Popular: Respuestatipo Referente a la República de Chile.*
Consejo Interamericano de Comercio y Producción: Recopilación de sus Acuerdos y Recomendaciones. Int-*6700*.C6, 1945.
Consejo Interamericano Económico y Social—*see-véase* Pan American Union. Inter-American Economic and Social Council.
Consejo Interamericano Económico y Social: Informe sobre la Organización del Consejo Aprobado por el Consejo Directivo de la Unión Panamericana en la Sesión del 29 de Agosto de 1945. Int-*8200*.I45, 1945, Sp.
Consejo Nacional de Educación, Bolivia—*see-véase* Bolivia. Consejo Nacional de Educación.
Consejo Nacional de Estadística. Cuba-*3030*.C5, 1945.
Consejo Permanente de Asociaciones Americanas de Comercio y Producción—*see-véase* Consejo Interamericano de Comercio y Producción.
Conselho Interamericano de Comércio e Produção—*see-véase* Consejo Interamericano de Comercio y Producción.
CONSERVACION DE REGISTROS—*véase* 1600.
Consolidación de los Adicionales ad-Valorem de Importación, Decreto N. 105, Vigente desde el 15 de Marzo de 1940; Lista de los Cambios Efectuados en los Derechos de la Tarifa y Arancel de Aduanas, Ley N. 667; Nueva Tarifa de Exportación, Decreto N. 383 Vigente desde el 25 de Marzo de 1940. Par-*6900*.C5, 1940.
"Construção e Ajustamento de Tábuas de Sobrevivência, Conforme a Mortalidade do Período 1939-41." Bras-*3040*.R5, Oct.-Dec. '45.
CONSTRUCCION:
GENERAL—*véase* 6500.
CENSOS DE—*véase* 2120; 2220; 6500.
CONSTRUCTION
GENERAL—*see* 6500.
CENSUSES OF—*see* 2120; 2220; 6500.
Construction. USA-*6500*.C3.
The Construction Industry in the United States. USA-*6500*.C4, 1944.
Consumer and Opinion Research. NonG-*3110*.C5, 1943.
Consumer Expenditures in the United States, Estimates for 1935-36. USA-*9700*.C5, 1935-36.
Consumer Incomes in the United States, Their Distribution in 1935-36. USA-*9700*.C6, 1935-36.
Consumer Purchases Study, 1935-36. USA-*9700*.C7, 1935-36.
CONSUMO:
GENERAL—*véase* 6020.
ESPECIFICO—*véase* las respectivas clasificaciones.
COOPERATIVAS DE—*véase* 6710.
DE PRODUCTOS ALIMENTICIOS (NUTRICION DE FAMILIAS)—*véase* 9800.
POR FAMILIAS (NO ALIMENTICIO)—*véase* 9700.
Consumo y Subconsumo. Int-*9600*.C5, 1943.
CONSUMPTION:
GENERAL—*see* 6020.
Of specific commodities—*see* the particular field.
FAMILY—*see* 9700.
FOOD—*see* 9800.
CONTABILIDAD (EN GENERAL, incluyendo CONTABILIDAD DE COSTOS)—*véase* 1100.
CONTRIBUCIONES (EN GENERAL Y DE TODAS CLASES, incluyendo IMPUESTOS NACIONALES, ESTATALES, MUNICIPALES Y LOCALES, SOBRE HERENCIAS, VENTAS, ETC.)—*véase* 8800.
A Contribution to the Study of International Comparisons of Costs of Living. Int-*9500*.C4, 1932.
"'Control Chart' Method of Analysis and Presentation of Data." NonG-*3160*.G5, 1941.
Control Chart Method of Analyzing Data. NonG-*3160*.G5, 1941.
Control Chart Method of Controlling Quality During Production. NonG-*3160*.C5, 1942.
Convenção Nacional de Estatística—*see-véase* Brasil. Convenção Nacional de Estatística.
Convenção Nacional de Estatística. Bras-*3020*.C5, 1936.
"Convenio Interamericano del Café, Noviembre 28 de 1940." Col-*6210*.B5, Dec. '40.
CONVENIOS:
GENERAL (incluyendo CONVENIOS POLITICOS INTERNACIONALES)—*véase* 8210.

CONVENIOS—cont.

COMERCIALES—*véase* 6900.

Convênios Nacionais de Estatística Municipal. Bras-*3030*.C5, 1944.

Convention Respecting the Compilation of International Commercial Statistics. Int-*3210*.C4, 1913.1.

CONVENTIONS—*see* 8210.

Convertibility Index for Foreign Trade Statistical Classifications of the American Nations: Basic Classification Scheme Showing Detailed Export and Import Commodity Description by Classes of the Minimum List of Commodities for International Trade Statistics. Int-*3210*.C6. 1945.

Cook, S. J.
- Int-*1930*.H5, 1942. *Handbook of Scientific and Technical Societies and Institutions of the United States and Canada.*

Coons, Callie M.
- USA-*9700*.F5F1, 1941. *Family Food Consumption and Dietary Levels, Five Regions, Farm Series.*

Cooper, William W.
- USA-*3000*.G7, 1946. *Government Statistics for Business Use.*

COOPERATIVAS:
- GENERAL—*véase* 6710.
- AGRICOLAS—*véase* 6200.
- DE CONSUMO—*véase* 6710.
- DE CREDITO—*véase* 6720.

CO-OPERATIVES:
- GENERAL—*see* 6710.
- AGRICULTURAL—*see* 6200.
- CONSUMER—*see* 6710.
- CREDIT—*see* 6720.

"Coordination Internationale des Statistiques Universitaires." Int-*3020*.B5, 1938, Castrilli.

Copeland, M. A.
- NonG-*6100*.S3. "Concepts of National Income."
- NonG-*6100*.S3. "The Correction of Wealth and Income Estimates for Price Changes."

Copper Resources of the World. Int-*6300*.C8, 1935.

COQUE—*véase* 6400.

Corcoran, Thomas F.
- Guat-*3030*.S5, 1945. *Statistics in Guatemala (A Brief Survey).*
- Guat-*6000*.R5, May-Aug. '46. "La Estadística en Guatemala: Una Breve Reseña."
- Pan-*2210*.C5, 1943.1. *Censo Agro-Pecuario del Distrito de Penonomé, Diciembre 1943.*
- Pan-*6000*.E5, 1943. *Examen Económico de la República de Panamá para el Año de 1943.*
- Pan-*6100*.R5, 1940. *Renta Nacional de la República de Panamá: National Income, 1940.*
- Par-*2110*.C5, 1944.5. *Censo Familiar: Cédula.*
- USA-*9200*.H5, 1932-40. *Hours and Earnings in the United States, 1932-40.*

Córdoba (Prov.)—*see-véase* Argentina. Córdoba (Prov.).

CORN—*see* 6210.

Corporación de Fomento de la Producción—*see-véase* Chile. Corporación de Fomento de la Producción.

Corporate Cash Balances, 1914-1943. USA-*6721*.C5, 1914-43.

CORRAL, AVES DE—*véase* 6230.

Correa Avila, Carlos.
- Arg-*3040*.R5, Oct. '45, Coghlan, b. "La Agrícola Argentina, País de Población Urbana."
- Arg-*3040*.R5, Sept. '45. "La Fecundidad y la Natalidad en el Campo y en las Ciudades."
- Arg-*3040*.R5, June '45. "El Analfabetismo en la Argentina."

CORRECCION, CASAS DE—*véase* 8330.

"The Correction of Wealth and Income Estimates for Price Changes." NonG-*6100*.S3.

Correlation Analysis. NonG-*3330*.Treloar 1, 1943.

CORREOS:
- SERVICIOS DE (ADMINISTRACION)—*véase* 8000.

Côrte Internacional de Justiça—*see-véase* United Nations. International Court of Justice.

CORTE SUPREMA—*véase* 8300.

Cortés, Fernando.
- Arg-*3040*.R8, Jan.-Apr. '45. "Sobre los Servicios Estadísticos y Censales en la República Argentina."

COSECHAS, INFORMES SOBRE—*véase* 6220.

COST:
- ACCOUNTING AND ANALYSES—*see* 1100.
- OF LIVING—*see* 9500.
- OF MEDICAL CARE—*see* 7190; 9700.
- OF PRODUCTION—*see* 6400.

Cost of Living Index Numbers for Canada. 1913-1942. Can-*9500*.C6, 1913-42.

Costa Miranda, Osvaldo Gomes da.
- Bras-*1420*.B5, June '44. "A Estimativa de Renda Geral do Brasil."

Costa Rica. Banco Nacional de Costa Rica—*see-véase* Banco Nacional de Costa Rica.

——. Banco Nacional de Seguros—*see-véase* Banco Nacional de Seguros, Costa Rica.

Cuba—cont.

——. ——. Dirección General de Estadística.
Cuba-*1420*.B5. *Boletín Mensual de Estadísticas.*
Cuba-*3030*.O5, 1945? *Organización de las Secciones de la Dirección . . .*
Cuba-*3030*.R5, 1945. *Reorganización de la Dirección General de Estadística de la República de Cuba.*
Cuba-*5200*.M5. *Movimiento de Población.*
Cuba-*6900*.C5. *Comercio Exterior.*
Cuba-*6900*.I5. *Importación y Exportación de la República de Cuba.*

——. Ministerio de Justicia. Dirección General del Censo—*see-véase* Cuba. Dirección General del Censos.

——. Ministerio de Salubridad y Asistencia Social.
Cuba-*7110*.S5. *Salubridad y Asistencia Social.*

——. Ministerio del Trabajo. Oficina de Estudios del Plan de Seguridad Social.
Cuba-*1400*.C5, 1944. *Condiciones Económicas y Sociales de la República de Cuba.*

——. Negociado de Asuntos Azucareros—*see-véase* Cuba. Ministerio de Agricultura. Dirección de Industrias. Negociado de Asuntos Azucareros.

——. Negociado de Economía Agrícola e Industrial—*see-véase* Cuba. Ministerio de Agricultura. Dirección de Industrias. Negociado de Economía Agrícola e Industrial.

——. Oficina de Números Indices—*see-véase* Cuba. Ministerio de Agricultura. Dirección de Industrias. Negociado de Economía Agrícola e Industrial. Oficina de Números Indices.

——. Oficina de Regulación de Precios y Abastecimiento—*see-véase* Cuba. Ministerio de Comercio. Oficina de Regulación de Precios y Abastecimiento.

——. Universidad de la Habana—*see-véase* Universidad de la Habana.

Cuba Económica y Financiera. Cuba-*6000*.C5.

Cuba: Summary of Biostatistics. Cuba-*5000*.C5, 1945.

Cudmore, S.A.
Int-*3000*.S5, 1940. "The Statistical Activities of Canada."
Int-*3040*.E5, Sept. '45, a. "Gross and Net Reproduction Rates, Canada and Provinces, 1920-1942."

Cuenta General de Rentas y Gastos Públicos; Bienes Nacionales Inclusive Materias, y Cuenta del Departamento de Hacienda, Rentas y Gastos. Ven-*8800*.C5.

Cuestionario de la Comisión Nacional de Alimentación y Agricultura. Hond-*6200*.C5, 1945.

CUESTIONARIOS (ESTADISTICOS)—*véase* 3100.

Cuestiones de Matemáticas Referentes a la Estadística. NonG-*3330*.Mazza 2, 1940.

"Cuestiones Elementales en los Censos Agrícolas Aplicables en los Países de Latino-América." Int-*1300*.A5, 1940.3, Alarcón Mendizábal.

CULTOS MEDICOS—*véase* 7180.

CULTS, MEDICAL—*see* 7180.

CULTURA, ESTADISTICAS DE—*véase* 7000-7999.

A Cultura Brasileira: Introdução ao Estudo da Cultura no Brasil. Bras-*2200*.R4, 1940.

CULTURAL RELATIONS PROGRAMS, INTERNATIONAL—*see* 8200.

CULTURAL STATISTICS—*see* 7000-7999.

Cummings, John.
USA-*5120*.N4, 1790-1915. *Negro Population, 1790-1915.*

Current Mortality Analysis. USA-*5212*.C5.

CURRENT STATISTICAL SERIES:
GENERAL—*see* 1420.
In special subject fields, *see* the particular field.

Currie, Laughlin.
NonG-*6100*.S3. "Problems in Estimating National Income Arising from Production by Government."

Curso de Estadística. NonG-*3330*.Gini 2, 1935.

Curso de Estadística. NonG-*3330*.Roa Reyes, 1929.

Curso de Estadística. NonG-*3330*.Velarde B., 1945.

"El Curso de Estadística de la Contraloría General." Col-*1420*.A5, Feb. 5, '42.

Curso de Estadística Expositiva o Aplicado. NonG-*3330*.Kürbs.

Curso de Estadística General y Aplicada. NonG-*3330*.Julin, 1924?

"Curso de Estatística: Capítulo I, Introdução, Valores Indicativos de Distribuições Estatísticas, Elementos de Interpolação e de Ajustamento pelo Processo dos Mínimos Quadrados." Bras-*3040*.R5, Jan.-Mar. '44, Dieulefait.

"Curso de Estatística: Capítulo II, Séries Estatísticas, Análise das Séries Econômicas e Teoria da População." Bras-*3040*.R5, Apr.-June '46, Dieulefait.

Curso Elemental de Estadística Administrativa. NonG-*3330*.Bertillon, 1906.

"Un Curso Oficial de Estadística." Col-*1420*.A5, Aug. 5, '42, Guthardt.

CURSOS DE ESTUDIOS:
ESPECIFICOS—*véase* las respectivas clasificaciones.
DE CIENCIA ACTUARIAL (DE SEGUROS)—*véase* 6730.

Income and Related Totals." Int-*6100*.R5, 1946.
DEFINITIONS, STATISTICAL—*see* 3100.
DEFORMIDADES—*véase* 7170.
DEFORMITIES—*see* 7170.
DEFUNCIONES:
GENERAL—*véase* 5212.
ACCIDENTALES—*véase* 5212; 7160; 9400.
CLASIFICACION DE—*véase* 3230; 5212.
MILITARES—*véase* 5212.
NEONATALES—*véase* 5212.
PRESUNTAS, PRACTICA SOBRE—*véase* 5250.
PROFESIONALES—*véase* 5212; 9400.
REGISTRO DE—*véase* 5250.
TASAS DE MORTALIDAD—*véase* 5212.
Defunciones en la República de El Salvador. ElSal-*5212*.D5.
DELINCUENCIA JUVENIL—*véase* 8320.
DEMANDA Y OFERTA:
GENERAL—*véase* 6000.
DE TRABAJO—*véase* 9200.
Demarcación Política del Perú: Recopilación de Leyes y Decretos (1821-1946). Perú-*4000*.D5, 1821-1946.
Deming, W. Edwards.
USA-*3040*.J5, Sept. '45. "On Training in Sampling."
USA-*3040*.J5, Sept. '43, Hansen. "On Some Census Aids to Sampling."
USA-*3040*.J5, Dec. '42. "Errors in Card Punching."
USA-*3040*.J5, Sept. '41. "On Sample Inspection in the Processing of Census Returns."
USA-*3040*.J5, Dec. '40, Stephan. "The Sampling Procedure of the 1940 Population Census."
USA-*9000*.A6, Aug. '44. "On Errors in Surveys."
NonG-*3105*.S4, 1943. *Statistical Adjustment of Data.*
DEMOGRAFIA—*véase* 5000-5999.
Demografía. Bol-*5000*.D5.
"Demografía de las Repúblicas Americanas." Int-*7110*.B5, Jan. '40.
Demografía y Asistencia Social. Chile-*5000*.D5.
Demografía y Estadística de Hospitales y de Otros Establecimientos. Perú-*7130*.D5.
"Demographic Factors in Labor Force Growth." USA-*9000*.A6, Aug. '46, Wolfbein.
"Demographic Status of South America." USA-*3040*.A5, Jan. '45, Dunn.
DEMOGRAPHY—*see* 5000-5999.
"La Denatalidad en la Argentina." Arg-*3040*.R5, Dec. '45, Coghlan.
"La Denatalidad en la Argentina." Arg-*3040*.R5, Nov. '45, Coghlan.
"La Denatalidad en la Argentina." Arg-*3040*.R5, Oct. '45, Coghlan, a.
Denrées et Matières Premières Agricoles: Production et Consommation dans les Différentes Parties du Monde. Int-*6210*.A4, 1934-38.
DENSIDAD DE LA POBLACION—*véase* 5110.
DENSITY OF POPULATION—*see* 5110.
O Departamento Nacional de Saúde em 1943. Bras-*7101*.D5, 1943.
The Department of State Bulletin. USA-*8200*.D5.
DEPENDENCY—*see* 9951.
DEPENDIENTES—*see* 9951.
DEPOSITOS BANCARIOS Y TASAS DE INTERES—*véase* 6720.
DEPOSITS AND RATES, BANK—*see* 6720.
DEPRESIONES ECONOMICAS—*véase* 6000.
DEPRESSIONS, ECONOMIC—*see* 6000.
DERECHO Y LEGISLACION:
GENERAL—*véase* 8400.
ESPECIFICO—*véase* las respectivas clasificaciones.
INTERNACIONAL—*véase* 8200-8299
DERMATOLOGIA—*véase* 7180.
DERMATOLOGY—*see* 7180.
"El Desarrollo de Bogotá y las Investigaciones del Costo de la Vida Obrera." Col.*1420*.A5, May '45, Abrisqueta.
"Desarrollo de Normas Comunes en la Estadística Educacional Americana." Int-*1300*.A5, 1940.3, Sálas S.
DESARROLLO FISICO (incluyendo DESARROLLO DEL NIÑO)—*véase* 7180.
"Desarrollo Histórico de la Estadística en El Salvador." Int-*1300*.A5, 1940.3, Brannon.
"El Desarrollo y Estado Actual de las Estadísticas Agrícolas en México." Méx-*6200*.B5, Nov. '46, González H.
Descamps, Emilio.
Guat-*1950*.N5, 1937. *Nomenclator de Guatemala.*
DESCUENTOS, TIPOS DE—*véase* 6720.
DESEMPLEO:
GENERAL—*véase* 9200.
SEGURO CONTRA EL—*véase* 9952.
Deserción Escolar y Analfabetismo. Arg-*7200*.D5, 1940.
The Design of Tabulating Cards. NonG-*3110*.D4, 1936.
DESOCUPACION—*véase* 9200.
La Desocupación en la Argentina, 1940. Arg-*9200*.D5, 1940.
DESOCUPADOS, ASISTENCIA A LOS—*véase* 9951.

DIRECTORIES—cont.
Int-*8200*.I49, 1946. *International Agencies in which the United States Participates.*
Int-*8200*.I6, 1933-1940, Eng. *The International Conferences of American States First Supplement, 1933-1940.*
Int-*8200*.I6, 1889-1936, Sp. *Conferencias Internacionales Americanas, 1889-1936.*
Bras-*7200*.E6, 1929. *Estatística Intellectual do Brasil, 1929.*
Cuba-*6700*.D5, 1941. *Directorio Oficial de Exportación, Importación, Producción y Turismo, 1941.*
ElSal-*1900*.P5, 1939. *Prontuario—Geográfico, Comercial, Estadístico, y Servicios Administrativos de El Salvador.*
ElSal-*1950*.G5. *La Guía Turística de El Salvador.*
Guat-*1950*.G5, 1942. *Guía Kilométrica de las 23 Rutas Nacionales de la República de Guatemala.*
Guat-*1950*.N5, 1937. *Nomenclator de Guatemala.*
Haiti-*6900*.H5, 1935. *Handbook and Trade Directory of Haiti.*
Nic-*1950*.N5. *Nicaragua.*
USA-*3030*.F5. *Federal Statistical Directory.*
USA-*5100*.M5, 1943. *Membership List of the Population Association of America, 1943.*
USA-*6000*.A5, Sept. '46. "Handbook: List of Members of the American Economic Association, 1946."
USA-*6610*.T5. *Transport Fact Book.*
USA-*7190*.M5, 1943. *Medical Care and Costs in Relation to Family Income.*
Ven-*6000*.G5, 1943. *Guía Industrial y Commercial de Venezuela, 1943.*
"Directorio de Especialistas en Estudios Sociales y Económicos en Latinoamérica, Part I." Int-*3040*.B5, Feb. '45, a.
"Directorio de Instituciones de Estudios e Investigaciones Sociales y Económicos en Latinoamérica, Parte I." Int-*3040*.B5, Feb. '45, b.
Directorio del Personal Estadístico en las Naciones Americanas. Int-*3000*.D5, 1944.
Directorio Oficial de Exportación, Importación, Producción y Turismo, 1941. Cuba-*6700*. D5, 1941.
DIRECTORIOS Y GUIAS:
LISTA DE—*véase* DIRECTORIES.
GENERAL—*véase* 1930.
ESPECIFICOS—*véase* las respectivas clasificaciones.
DE ORGANIZACIONES Y ORGANOS ESTADISTICOS—*véase* 3030.
Directory of Federal Statistical Agencies. USA-*3030*.F5.
Directory of Statistical Personnel in the American Nations. Int-*3000*.D5, 1944.
DISABILITY—*see* 7150; 9400; 7170.
DISCOUNT RATES—*see* 6720.
"Discriminação dos Óbitos Constantes das Tábuas de Sobrevivência de 1939-41, Segundo Grupos de Causas." Bras-*3040*.R5, Oct.-Dec. '45.
DISEASE:
GENERAL—*see* 7150.
OCCUPATIONAL—*see* 9400.
OF ANIMALS—*see* 6230.
"The Distinction Between 'Net' and 'Gross' in Income Taxation." NonG-*6100*.S3.
DISTRIBUCION:
DE FRECUENCIAS (ESTADISTICAS) —*véase* 3100-3199.
DE MERCANCIAS—*véase* MERCADOS.
"Distribución Abecedaria de Apellidos." Chile-*1420*.E5, Mar. '46.
"Distribución de la Población Argentina por Regiones Geográficas." Arg-*3040*.R5, June '45, Daus.
"Distribución de la Población de la República Mexicana por Edades y por Sexos, 1921 a 1936." Méx-*7110*.R5, Nov. '40, Bustamante.
La Distribución Geográfica de la Población del Ecuador. Ec-*5110*.D5, 1780-1935.
La Distribución por Zonas de la Población Argentina y Su Relación con lo Hechos Culturales, Económicos y Sociales. Arg-*2230*.C5, 1943.2.
Dittel, J. Walter.
CR-*9952*.C5, 1944. *La Conmutación de Pensiones Futuras como Reparación Civil en Caso de Muerte.*
Divisão Territorial do Brasil. Bras-*4100*.D6, 1945.
DIVISAS—*véase* 6720.
División Político-Territorial de la República. Ven-*4100*.D5, 1944.
DIVISIONES TERRITORIALES—*véase* 4100; 8000.
Doblin, Ernest M.
Int-*6100*.W5, 1946. "Tentative Statement on National Income Estimates for Sixty-Three Countries."
Documentos Anexos al Informe que el Ministro de Previsión Social y Trabajo Presenta a la Nación. Ec-*9000*.D5,
Documents et Procès-Verbaux de la Conférence Internationale de Statistique Commerciale, Bruxelles, 1913. Int-*3210*.C4, 1913.3.
Documents Relating to the Food and Agriculture Organization of the United Nations, 1st

Dunn, Halbert Louis—cont.
USA-*3040*.A5, Jan. '45. "Demografic Status of South America."
USA-*3040*.J5, Mar. '35. "Application of Punched Card Methods to Hospital Statistics."
USA-*3040*.J5, Sept. '32. "Adaptation of New Geometric Code to Multiple Punching in Mechanical Tabulation."
DUPLICATING PROCESSES—*see* 1800.
Durand, John Dana.
USA-*3040*.A5, Jan. '45. "The Trend Toward an Older Population."
Durost, Walter N.
NonG-*3160*.S4, 1936. *Statistical Tables; Their Structure and Use.*
DUTIES, COMMERCE—*see* 6900.
Duvivier, Ulrick.
Haiti-*1200*.B5, 1941. *Bibiographie Générale et Méthodique d'Haïti.*
DWELLINGS—*see* 9900; 6500.

EARNINGS—*see* 6721; 9700.
Ebaugh, Cameron D.
Chile-*7200*.E5, 1945. *Education in Chile.*
Perú-*7200*.E5, 1946. *Education in Peru.*
Eckler, A. Ross.
USA-*3040*.J5, June '45, b. "Management of Field Work and Collection of Statistics: A Round Table."
USA-*3040*.J5, Mar. '43. "Marketing and Sampling Uses of Population and Housing Data."
Econometric Society.
Int-*3040*.E3. *Econometrica.*
Econometrica. Int-*3040*.E3.
ECONOMIA—*véase* 6000-6999; *véase también* las respectivas clasificaciones.
ECONOMIA DOMESTICA—*véase* 9700.
Economía. Chile-*3040*.E5.
"Economía Nacional de Colombia." Int-*6000*.P4, Sept.-Oct. '46.
"Economía Nacional de la Argentina." Int-*6000*.P4, July-Aug. '46.
"Economía Nacional de México." Int-*6000*.P4, May-June '46.
La Economía Nacional y el Problema de las Subsistencias en Chile. Chile-*1900*.E5, 1943.
Economía y Finanzas. Chile-*6000*.E6.
Economía y Finanzas. Nic-*6000*.E5.
Economía y Finanzas de Mendoza. Arg-*6000*.E5.
The Economic and Financial Organization of the League of Nations. Int-*8200*.E5, 1945.
Economic and Financial Publications, 1920-1935. Int-*6000*.E1, 1920-35.
Economic Barometers. Int-*6010*.E5, 1924.
Economic Control of Quality of Manufactured Product. NonG-*3160*.E5, 1931.
Economic Controls and Commercial Policy in Argentina. Arg-*6000*.E6, 1945.
Economic Controls and Commercial Policy in Bolivia. Bol-*6000*.E5, 1945.
Economic Controls and Commercial Policy in Brazil. Bras-*6000*.E3, 1945.
Economic Controls and Commercial Policy in Chile. Chile-*6000*.E65, 1945.
Economic Controls and Commercial Policy in Colombia. Col-*6000*.E5, 1945.
Economic Controls and Commercial Policy in Costa Rica. CR-*6000*.E5, 1945.
Economic Controls and Commercial Policy in Cuba. Cuba-*6000*.E5, 1946.
Economic Controls and Commercial Policy in Ecuador. Ec-*6000*.E5, 1945.
Economic Controls and Commercial Policy in Haiti. Haiti-*6000*.E5, 1946.
Economic Controls and Commercial Policy in Mexico. Méx-*6000*.E4, 1946.
Economic Controls and Commercial Policy in Paraguay. Par-*6000*.E5, 1945.
Economic Controls and Commercial Policy in Peru. Perú-*6000*.E5, 1945.
Economic Controls and Commercial Policy in the American Republics. Int-*6000*.E2.
Economic Controls and Commercial Policy in the Dominican Republic. RepDom-*6000*.E5, 1946.
Economic Controls and Commercial Policy in Uruguay. Ur-*6000*.E5, 1945.
Economic Controls and Commercial Policy in Venezuela. Ven-*6000*.E3, 1945.
ECONOMIC GEOGRAPHY—*see* 6000.
ECONOMIC INDEXES:
GENERAL IN SCOPE—*see* 6010.
In specific subject fields, *see* the particular field.
The Economic Literature of Latin America. Int-*6000*.E6, 1936.
The Economic Pattern of World Population. Int-*5100*.E5, 1943.
ECONOMIC PLANNING—*see* 8500.
Economic Problems of Latin America. Int-*6000*.E7, 1944.
ECONOMICS—*see* 6000-6999.
The Economics of 1960. Int-*6000*.E9, 1960.
The Economics of a Declining Population. NonG-*5110*.E5, 1939.
The Economist's Handbook. Int-*3010*.E5, 1934.
Ecuador, Banco Central del Ecuador—*see-véase* Banco Central del Ecuador.
——. Banco Hipotecario del Ecuador—*see-véase* Banco Hipotecario del Ecuador.
——. Dirección General de Aduanas—*see-véase* Ecuador. Ministerio del Tesoro. Dirección General de Aduanas.
——. Dirección General de Estadística y

Ecuador—cont.

Censos—*see-véase* Ecuador. Ministerio d Economía. Dirección General de Estadística y Censos.

——. Dirección General de Industrias y Minas—*see-véase* Ecuador. Ministerio de Economía. Dirección General de Industrias y Minas.

——. Dirección General de Sanidad—*see-véase* Ecuador. Ministerio de Previsión Social y Trabajo. Dirección General de Sanidad.

——. Dirección Nacional de Estadística—*see-véase* Ecuador. Ministerio de Economía. Dirección General de Estadística y Censos.

——. Ministerio de Agricu'tura, Industrias, Minas y Turismo—*see-véase* Ecuador. Ministerio de Economía.

——. ——. Dirección General de Minería y Petróleos—*see-véase* Ecuador. Ministerio de Economía. Dirección General de Industrias y Minas.

——. Ministerio de Economía—*see also-véase también* Ecuador. Ministerio de Hacienda y Crédito Público.

Ec-*6000*.I5. *Información Económica.*

Ec-*6000*.M5. *Memoria del . . .*

——. ——. Dirección General de Estadística y Censos.

Ec-*1400*.E5, 1938-42. *Ecuador en Cifras, 1938 a 1942.*

Ec-*1420*.E5. *Estadística y Censos.*

Ec-*1420*.T5. *El Trimestre Estadístico del Ecuador.*

Ec-*1420*.T5, July-Dec. '45, a. "Comentarios y Sugestiones Respecto a las Bases Mínimas para el Censos de las Américas de 1950."

Ec-*1420*.T5, July-Dec. '45, b. "Indices del Costo de la Vivienda Popular en Quito de 1938 a 1945."

Ec-*1420*.T5, May '45. "Ley de Estadística (de 9 de Agosto de 1944), Decreto No. 760."

Ec-*3000*.C5. *Cuadernos de Estadística.*

Ec-*3210*.C5, 1945. *Clasificación del Arancel de Aduanas del Ecuador según la Lista Mínimum de Mercaderías para las Estadísticas del Comercio Internacional.*

Ec-*6000*.I5A. *Datos Estadísticos.*

Ec-*6900*.A5. *Anuario de Comercio Exterior.*

Ec-*6900*.E5, 1940-42. *Estadísticas de las Exportaciones del Ecuador, 1940-1942.*

Ec-*9800*.I5, 1945. *Indices del Costo de la Vivienda Popular en Quito, de 1938 a 1944.*

NonG-*3330*.Procaccia, 1944. *Lecciones de Estadística Metodológica.*

——. ——. Dirección General de Industrias y Minas.

Ec-*6300*.M5. *La Minería y el Petróleo en el Ecuador: Anuario.*

——. ——. Superintendencia de Bancos.

Ec-*6720*.B3. *Boletín de la Superintendencia de Bancos del Ecuador.*

Ec-*6720*.I5. *Informe del Superintendente de Bancos al Señor Ministro de Economía.*

——. Ministerio de Educación Pública.

Ec-*7200*.E5, 1942. *Estadística y Escalafón.*

——. Ministerio de Hacienda y Crédito Público—*see also-véase también* Ecuador. Ministerio de Economía *and-y* Ecuador. Ministerio del Tesoro.

Ec-*6900*.L5, 1943. *Ley Arancelaria de Aduanas, 1943.*

Ec-*8800*.I5. *Informe que el Ministro de Hacienda y Crédito Público Presenta a la Nación.*

Ec-*8800*.R5. *Revista de Hacienda.*

——. ——. Dirección Nacional de Estadística —*see-véase* Ecuador. Ministerio de Economía. Dirección General de Estadística y Censos.

——. Ministerio de Previsión Social, Trabajo y Asistencia Pública.

Ec-*5200*.B5, 1943. *Bioestadística en el Cosmos Ecuatoriano.*

Ec-*9000*.D5. *Documentos Anexos al Informe que el Ministro de Previsión Social y Trabajo Presenta a la Nación.*

Ec-*9000*.I5. *Informe a la Nación (del Ministro de Previsión Social y Trabajo).*

——. ——. Dirección General de Sanidad.

Ec-*7110*.I5. *Informe que el Señor Director General de Sanidad Presenta ante el Señor Ministro del Ramo acerca de las Labores Desarrolladas.*

——. Ministerio del Tesoro—*see also-véase también* Ecuador. Ministerio de Hacienda y Crédito Público.

Ec-*8800*.B5. *Boletín del Ministerio del Tesoro.*

——. ——. Dirección General de Aduanas.

Ec-*6900*.A4. *Aduanas del Ecuador.*

——. Superintendencia de Bancos—*see-véase* Ecuador. Ministerio de Economía. Superintendencia de Bancos.

"Ecuador, a Statistical Abstract." Int-*6000*. C7, Nov.-Dec. '44, Zier.

Ecuador en Cifras, 1938 a 1942. Ec-*1400*.E5, 1938-42.

Ecuador: Report Presented to the Conference of Commissions of Inter-American Development. Ec-*6000*.R5, 1944.

Ecuador: Summary of Biostatistics. Ec-*5000.*E5, 1944.
La Edad de los Escolares y Su Valor como Determinante en el Fenómeno Educativo. Guat-*7200.*E3, 1946.
EDADES:
ENFERMEDADES RELACIONADAS CON—*véase* 7150.
GRUPOS DE LA POBLACION POR—*véase* 5110.
Edgerton, Harold A.
NonG-*3005.*S4, 1939. *Statistical Dictionary of Terms and Symbols.*
EDIFICACIONES Y EDIFICIOS:
GENERAL—*véase* 9900; 6500.
CENSOS DE—*véase* 2120; 2220; 6500; 6721.
Educação e Saúde. Bras-*7000.*E5, 1942.
EDUCACION:
GENERAL-*véase* 7200-7299.
ESPECIFICA—*véase* las respectivas clasificaciones.
ESTADISTICA—*véase* 3300-3399.
EDUCATION:
GENERAL—*see* 7200-7299.
In specific subject fields—*see* the particular field.
STATISTICAL—*see* 3300-3399.
Education in Chile. Chile-*7200.*E5, 1945.
Education in Colombia. Col-*7200.*E5, 1946.
Education in Costa Rica. CR-*7200.*E5, 1946.
Education in Cuba. Cuba-*7200.*E5, 1943.
Education in Peru. Perú-*7200.*E5, 1946.
Educational Institutions in Canada. Can-*7210.*E5, 1944.
EDUCATIONAL PSYCHOLOGY—*see* 7200.
Educational Trends in Latin America. Int-*7200.*E6, 1942.
Edwards, Alba M.
USA-*2200.*C5, 1940.4. *Sixteenth Census of the United States: 1940; Population: Comparative Occupation Statistics for the United States, 1870 to 1940.*
USA-*3240.*C5, 1940.1. *Sixteenth Census of the United States: Classified Index of Occupations.*
USA-*3240.*C5, 1940.2. *Sixteenth Census of the United States: Alphabetical Index of Occupations and Industries.*
Edwards, Thomas I.
USA-*3040.*J5, June '44. "Methods Used in Processing Data from the Physical Examination Reports of the Selective Service System."
Efron, David.
Int-*1900.*L7, 1945. *Latin America in the Future World.*
Eguía, Gastón Arduz.
Bol-*9000.*P5, 1941. *El Problema Social en Bolivia—Condiciones de Vida y de Trabajo.*
Eighth American Scientific Congress—*see-véase* American Scientific Congress.
Eighth Census of Canada, 1941. Can-*2200.*C5, 1941.2.
Eighth Census of Canada, 1941: Administrative Report of the Dominion Statistician. Can-*2100.*C5, 1941.3.
Eighth Census of Canada, 1941: Census of Agriculture: Nova Scotia. Can-*2210.*C5, 1941.2a.
Eighth Census of Canada, 1941: Instructions to Commissioners and Enumerators. Can-*2100.*-C5, 1941.2.
Eighth Census of Canada, 1941: Methods and Procedures. Can-*2100.*C5, 1941.1.
Eighth Census of Canada, 1941: Population. Can-*2200.*C5, 1941.2a.
Eighth Census of Jamaica, 1943: Administrative Report. WI-*2200.*E5, 1943.1.
Eighth Census of Jamaica, 1943: Introduction to Population and Housing. WI-*2200.*E5, 1943.2.
Eighth Census of Jamaica and Its Dependencies, 1943: Population, Housing and Agriculture. WI-*2200.*E5, 1943.
El Salvador. Asociación Cafetalera de El Salvador—*see-véase* Asociación Cafetalera de El Salvador.
——. Banco Central de Reserva de El Salvador—*see-véase* Banco Central de Reserva de El Salvador.
——. Dirección General de Estadística—*see-véase* El Salvador. Ministerio de Economía. Dirección General de Estadística.
——. Dirección General de Sanidad—*see-véase* El Salvador. Ministerio del Interior. Dirección General de Sanidad.
——. Junta Nacional del Turismo—*see-véase* Junta Nacional del Turismo, El Salvador.
——. Leyes, Estatutos, Etc.
ElSal-*3030.*L5, 1940. *Ley Orgánica del Servicio Estadístico.*
——. Ministerio de Cultura Popular—*see-véase* El Salvador. Ministerio de Cultura y Asistencia Social.
——. Ministerio de Cultura y Asistencia Social.
ElSal-*7000.*I5. *Informe de las Labores Desarrolladas por el Poder Ejecutivo en los Ramos de Cultura y Asistencia Social.*
ElSal-*7000.*M5. *Memoria de Cultura Popular.*
——. Ministerio de Economía.
ElSal-*3030.*I5, 1935. *La Investigación Estadística, Origen, Desarrollo y Estado*

Analysis of Its Relation to the Volume of Production. USA-*9200*.E5, 1899-1939.
Employment of United States Citizens in Latin America. Int-*9200*.E4, 1945.
Employment Security Activities. USA-*9950*.E5.
"Employment Situation in Latin America." USA-*9000*.M5, May 1946.
The Employment Situation . . . Together with Payrolls. Can-*9200*.E5.
EMPRESAS:
PUBLICAS (EN GENERAL)—*véase* 8000.
PUBLICAS, ESPECIFICAS—*véase* las respectivas clasificaciones.
UTILIDADES PUBLICAS—*véase* 6400.
"Empresas Eléctricas de Servicio Público." Guat-*1420*.B5, July '46, a.
EMPRESTITOS—*véase* 6720.
ENCICLOPEDIAS—*véase* 1940.
Encuesta Continental sobre el Consumo de Productos de Alimentación y Vestido y sobre la Vivienda Popular: Respuestatipo Referente a la República de Chile. Chile-*9600*.E5, 1943.
"Encuesta sobre Presupuestos Familiares Obreros Realizada en la Ciudad de Lima, en 1940." Perú-*3040*.R5, 1944, Palacios.
"Encuesta sobre Presupuestos Familiares Obreros, en 1940, Correspondiente a Lima (Cercado)." Perú-*1410*.E5, 1942, c.
"Encuesta sobre Presupuestos Familiares Obreros, en 1940, Correspondiente a Lima Cercado)." Perú-*1410*.E5, 1940, a.
Encuestas sobre el Consumo de Alimentos y la Nutrición en las Américas. Int-*9800*.E5, 1942.
Encyclopaedia of the Social Sciences. NonG-*1940*.E5, 1937.
ENCYCLOPEDIAS—*see* 1940.
ENERGY:
ELECTRIC—*see* 6400.
HYDRAULIC—*see* 6400.
ENFERMEDADES:
GENERAL—*véase* 7150.
DE ANIMALES—*véase* 6230.
DE LA VEJEZ—*véase* 7150.
INDUSTRIALES—*véase* 9400.
PROFESIONALES—*véase* 9400.
PROVENIENTES DE OCUPACIONES—*véase* 9400.
SEGUROS CONTRA—*véase* 6730; 7190; 9952.
"Las Enfermedades Infecto-Contagiosas, Casos por Departamentos en 1945 y en la República en 1939-1945." Perú-*3040*.E5, July '46, Téllez Sarzola.
ENFERMOS MENTALES:
GENERAL—*véase* 7170; 9951.
INSTITUCIONES PARA—*véase* 7130.
Engelman, U. Z.
USA-*5120*.J5, 1943. *Jewish Population Studies.*
Engineering and Scientific Graphs for Publications. NonG-*3260*.A5, 1943.
ENGINEERING TECHNOLOGY:
GENERAL—*see* 6000.
In specific subject fields, *see* the particular field.
Enregistrement des Naissances, Décès et Mariages. Can-*5200*.R5.
"Ensaio de Retificação das Taxas de Mortalidade, Segundo Grupos de Causas, no Primeiro Ano de Idade no Distrito Federal." Bras-*3040*.R5, Oct.-Dec. '45.
"Ensayo de Tabla de Mortalidad para los Habitantes de Lima, Ciudad Capital." Perú-*3040*.E5, July '46, García Frías.
Ensayo sobre la Evolución de la Población de las Américas: I, Canadá. Can-*5100*.E7, 1946.
Ensayo sobre la Evolución de la Población de las Américas (Estimación en el Lapso 1938-1973): II, Argentina. Arg-*5100*.E5, 1946.
"Ensayo sobre las Condiciones de la Vida Rural en el Municipio de Moniquirá-Boyacá." Col-*1420*.A5, July '39, Ortiz C.
Enseignement Statistique en Haïti: Méthodes et Matériaux. Haiti-*3300*, 1944-46.
ENSEÑANZA:
GENERAL—*véase* 7200.
ESPECIFICA—*véase* las respectivas clasificaciones.
ESTADISTICA (CURSOS Y PROGRAMAS DE)—*véase* 3300-3399.
PEDAGOGICA—*véase* 7200.
"Enseñanza de la Estadística en México." Int-*3040*.E5, Mar. '44, Alanís Patiño.
Enseñanza Estadística en Argentina: Métodos y Materiales. Arg-*3300*, 1944-46.
Enseñanza Estadística en Bolivia: Métodos y Materiales. Bol-*3300*, 1944-46.
Enseñanza Estadística en Chile: Métodos y Materiales. Chile-*3300*, 1944-46.
Enseñanza Estadística en Colombia: Métodos y Materiales. Col-*3300*, 1944-46.
Enseñanza Estadística en Cuba: Métodos y Materiales. Cuba-*3300*, 1944-46.
Enseñanza Estadística en Ecuador: Métodos y Materiales. Ec-*3300*, 1944-46.
Enseñanza Estadística en Guatemala: Métodos y Materiales. Guat-*3300*, 1944-46.
Enseñanza Estadística en Honduras: Métodos y Materiales. Hond-*3300*, 1944-66.
Enseñanza Estadística en la República Dominicana: Métodos y Materiales. RepDom-*3300*, 1944-46.
Enseñanza Estadística en México: Métodos y Materiales. Méx-*3300*, 1944-46.

Estadística de las Huelgas, 1940. Arg-*9300*.E5, 1940.
"Estadística de las Superficies." Perú-*3040*.E5, Nov.'45, Ricci.
Estadística de los Negocios de Seguros. Rep Dom-*6730*.E5.
Estadística de los Servicios Eléctricos del Perú. Perú-*6400*.E5, 1942.
"Estadística de Metales y Minerales." Int-*3040*.E5, Sept.'46, Knox.
Estadística de Salarios. Méx-*9200*.E5, 1942.
Estadística del Ministerio de Agricultura. Arg-*6200*.B5.
Estadística del Petróleo. Ven-*6300*.E5.
Estadística del Trabajo. Perú-*9000*.E5, 1939.
"Estadística Económica Militar." Perú-*3040*.E5, July '46, Velarde B.
"La Estadística en el Mundo de Mañana." Int-*3040*.E5, June '46, Dunn.
"La Estadística en Guatemala: Una Breve Reseña." Guat-*6000*.R5, May-Aug. '46, Corcoran.
Estadística Epidemiológica para los Países de América. Int-*7110*.E8.
Estadística Financiera. Bol-*6720*.E6.
Estadística Fiscal y Administrativa. Col-*8800*.E5.
Estadística Forestal y Agraria, Año 1943. ElSal-*6200*.E5, 1943.
Estadística General de la República, Ley de Reorganización. Ur-*3030*.E4, 1914.
Estadística General y Matemática. NonG-*3330*.Almendras, 1945.
Estadística Industrial. Arg-*2250*.E5.
"Estadística Industrial." Perú-*6400*.B5, Aug. '46.
Estadística Industrial de la República Argentina. Arg-*6400*.E5.
Estadística Matemática. NonG-*3140*.E7, 1939.
Estadística Mercantil y Marítima. Ven-*6900*.E5.
Estadística Metodológica. NonG-*3330*.Arrús, 1945.
Estadística Minera. Nic-*6300*.E5.
Estadística Minera. Perú-*6300*.A5.
Estadística Minera de la Nación. Arg-*6300*.E5.
La Estadística Nacional. Col-*3030*.E3, 1938.
"La Estadística Nacional Ecuatoriana." Col-*1420*.A5, Mar.-Apr.'46, Abrisqueta 2.
Estadística Panameña. Pan-*1420*.E5.
Estadística Peruana. Perú-*3040*.E5.
"La Estadística Peruana en el Sexenio 1938-44." Int-*3040*.E5, Dec. '44, Luna Vegas.
Estadística: Programa Analítico. Arg-*3320*.E5.
"Estadística Sanitaria." Ec-*7110*.I5, 1941-42.
Estadística Sanitaria. RepDom-*7110*.E5.
"Estadística Vital, Año 1944." Nic-*1420*.B5, July '45.
Estadística Vital y Registro en Argentina: Métodos y Procedimientos. Arg-*5200*.
Estadística Vital y Registro en Bolivia: Métodos y Procedimientos. Bol-*5200*.
Estadística Vital y Registro en Chile: Métodos y Procedimientos. Chile-*5200*.
Estadística Vital y Registro en Colombia: Métodos y Procedimientos. Col-*5200*.
Estadística Vital y Registro en Cuba: Métodos y Procedimientos. Cuba-*5200*.
Estadística Vital y Registro en El Salvador: Métodos y Procedimientos. ElSal-*5200*.
Estadística Vital y Registro en Guatemala: Métodos y Procedimientos. Guat-*5200*.
Estadística Vital y Registro en Honduras: Métodos y Procedimientos. Hond-*5200*.
Estadística Vital y Registro en México: Métodos y Procedimientos. Méx-*5200*.
Estadística Vital y Registro en Nicaragua: Métodos y Procedimientos. Nic-*5200*.
Estadística Vital y Registro en Panamá: Métodos y Procedimientos. Pan-*5200*.
Estadística Vital y Registro en Paraguay: Métodos y Procedimientos. Par-*5200*.
Estadística Vital y Registro en Perú: Métodos y Procedimientos. Perú-*5200*.
Estadística Vital y Registro en Uruguay: Métodos y Procedimientos. Ur-*5200*.
Estadística Vital y Registro en Venezuela: Métodos y Procedimientos. Ven-*5200*.
Estadística y Censos. Ec-*1420*.E5.
"La Estadística y el Mundo de Mañana." Col-*7420*.A5, Nov.-Dec. '45, Dunn.
Estadística y Escalafón (Ministerio de Educación Pública). Ec-*7200*.E5, 1942.
Estadística y Su Aplicación al Comercio. NonG *1330*.Boddington, 1942.
Estadísticas Agrícolas en Argentina: Métodos y Procedimientos. Arg-*6200*.
Estadísticas Agrícolas en Chile: Métodos y Procedimientos. Chile-*6200*.
Estadísticas Agrícolas en Colombia: Métodos y Procedimientos. Col-*6200*.
Estadísticas Agrícolas en México: Métodos y Procedimientos. Méx-*6200*.
Estadísticas Agrícolas en Perú: Métodos y Procedimientos. Péru-*6200*.
Estadísticas Agrícolas en Uruguay: Métodos y Procedimientos. Ur-*6200*.
Estadísticas de la Caja de Seguro Obligatorio. Chile-*9950*.E5.
Estadísticas de la Población Activa. Int-*3240*.E5, 1938.
"Estadísticas de la Producción." Col-*1420*.A5, Feb. '45, Abrisqueta.
Estadísticas de las Exportaciones del Ecuador, 1940-1942. Ec-*6900*.E5, 1940-42.

"A Estimativa de Renda Geral do Brasil." Bras-*1420*.B5, June '44, Costa Miranda.
Os Estoques de Matérias e Produtos Téxteis do Distrito Federal em 30 de Junho de 1943 e em 30 de Junho de 1944. Bras-*6400*.E7, 1944.
Estudio Crítico de Su Población, 1536-1936. Arg-*5000*.B5, 1536-1936.
Estudio de la Situación de la Industria Lechera en el Departamento de Lambayeque. Perú-*6230*.E5, 1944.
"Estudio del Comercio Exterior de Chile en Artículos Alimenticios y Su Significado." Int-*1300*.A5, 1940.3, Cox B.
Un Estudio del Costo de la Vida. Méx-*9500*.E5, 1935.
Estudio Estadístico de Algunos Aspectos del Comercio Exterior de la Republica Dominicana, 1920-1939. RepDom-*6900*.E3, 1941.
"Estudio Histórico Analítico de la Población Colombiana en 170 Años." Col-*1420*.A5, Apr. 25, '40, Higuita.
Estudio sobre el Costo de la Enseñanza Primaria. Guat-*7210*.E5, 1946.
"Estudio sobre Estadística Agrícola." Col-*1420*.A5, July '39, Ricuarte Montoya.
Estudio sobre la Enseñanza Parvularia. Guat-*7200*.E5, 1946.
Estudios Especiales de la Dirección General de Estadística. Nic-*3000*.E5.
"Estudios sobre Mortalidad por Tuberculosis, Gripe y Sífilis." Chile-*1420*.E5, Apr.'46.
Estudios y Documentos, Serie N, Estadísticas. Int-*9000*.E7.
Estudos Brasileiros de Demografia. Bras-*5000*.E5.
"Estudos de Demografia Interamericana: I, Tábua de Mortalidade e de Sobrevivência para a Colômbia (1939-41); II, Ajustamento de Tábua de Sobrevivência para a Colômbia (1939-41), segundo a Fórmula de Gompertz-Makeham." Bras-*3040*.R5, July-Sept.'43, Mortara, 1.
Estudos sôbre a Mortalidade nas Grandes Cidades Brasileiras. Bras-*5212*.E5, 1945.
"Estudos sôbre a Mortalidade no Distrito Federal e no Município de São Paulo." Bras-*3040*.R5, Oct.-Dec.'45.
Étude sur la Mortalité Puerpérale, Infantile et Neonatale au Canada. Can-*5212*.S5, 1945.
Etudes et Rapports sur es Méthodes Statistiques. Int-*3200*.S8.
EUGENESIA—*véase* 7300.
EUGENICS—*see* 7300.
Evaluation of Vital Statistics Work in Alaska, August 26 to October 6, 1944. USA Alsk-*5200*.E5, 1944, Morrison.
Examen Económico de la República de Panamá para el Año de 1943. Pan-*6000*.E5, 1943.
EXAMENES FISICOS, PROGRAMAS DE—*véase* 7110.
EXCHANGE RATES—*see* 6720.
Executive Agreement Series. Int-*1200*.E5.
Executive Order #332: . . . Tariff Law on Imports and Exports . . . Effective on and after January 1, 1920. RepDom-*6900*.E5, 1920.
"Existencia de Anuarios, Censos y Otras Publicaciones Editadas por la Dirección General de Estadística." Chile-*1420*.E5, June '44, a.
EXPECTATION OF LIFE—*see* 5220.
EXPECTACION DE VIDA—*véase* 5220.
EXPENDITURES:
FAMILY—*see* 9700.
FOR DEFENSE—*see* 8800.
PUBLIC—*see* 8800.
"Explicaciones Generales sobre la Estadística del Comercio Exterior del Perú." Perú-*3040*.E5, Nov.'45, Zapata Vidaurre.
Export and Import Practice. USA-*6900*.E6, 1938.
EXPORTACION—*véase* 6900.
Exportación de la Republica Dominicana. RepDom-*6900*.E7.
Las Exportaciones en los . . . Comparadas con las del Mismo Período del Año Anterior. Arg-*6900*.E5.
EXPORTS—*see* 6900.
Exports of Canadian Produce. Can-*6900*.T6.
"Exposición de Motivos del Proyecto de Ley de Presupuesto General de Rentas y Gastos Públicos para el Año Económico 1944-1945." Ven-*8800*.R5, June 1944, a.
EXTENSION AND EDUCATION, AGRICULTURAL—*see* 6200.
EXTRACTIVES AND EXTRACTIVE INDUSTRIES—*see* 6300.
EXTRACTIVOS, INDUSTRIAS Y PRODUCTOS—*véase* 6300.
Extracto Estadístico de Bolivia—Sección Finanzas. Bol-*6720*.E6.
Extracto Estadístico de Bolivia—Sección Industria. Bol-*6400*.I5.
Extracto Estadístico del Perú. Perú-*1410*.E5.
Extracto Estadístico y Censo Electoral de la República. Perú-*1410*.E5, 1933.
EXTRANJEROS RESIDENTES:
NUMERO DE—*véase* 5120.
PROBLEMAS SOCIALES DE—*véase* 9600.
Ezekiel, Mordecai.
NonG-3330, 1941. *Methods of Correlation Analysis.*

FABRICA, MARCAS DE—*véase* 6710.
FABRICACION—*véase* 6400; 2250.

FERROCARRILES (EN GENERAL, incluyendo CARGAS DE)—*véase* 6614.
FERTILIDAD—*véase* 5211.
FERTILITY—*see* 5211.
FERTILIZANTES—*véase* 6220.
FERTILIZER—*see* 6220.
FIEBRE AFTOSA (DEL GANADO)—*véase* 6230.
Fiedler, Reginald H.
Int-*1300*.A5, 1940.3. "Need and Plan for a Statistical Program in Furthering Conservation of Inter-American Fisheries."
Fifteenth Census of the United States, 1930: Abstract of the Fifteenth Census of the United States. USA-*2200*.C5, 1930.
Fifteenth Census of the United States: 1930: The Indian Population of the United States and Alaska. USA-*2200*.C5, 1930.2.
Fifth Pan American Conference of National Directors of Health, April 22-29, 1944. Int-*7110*.P4, 1944.
Figuerola, José.
NonG-*9000*.T5, 1942. *Teoría y Métodos de Estadística del Trabajo.*
Final Act of Inter-American Conference of Systems of Economic and Financial Control, Washington, June 30-July 10, 1942. Int-*6000*.I45, 1942, Eng.
Final Act of the Conference (Relating to Economic Statistics). Int-*6000*.I75, 1928.B.
Final Act of the Eighth Pan American Child Congress, Washington, D. C., May 2-9, 1942. Int-*9951*.P5, 1942, Eng.
Final Act of the First Conference of Ministers and Directors of Education of the American Republics, Held in Panama, September 27 to October 4, 1943. Int-*7200*.C6, 1943.1, Eng.
Final Act of the First Inter American Demographic Congress. Int-*5000*.I5, 1943.1, Eng.
Final Act of the Inter-American Conference of Police and Judicial Authorities, Buenos Aires, Argentina, May 27-June 9, 1942. Int-*8300*.I5, 1942.
Final Act of the Inter-American Conference on Agriculture. Int-*6200*.I4, 1930.1.
Final Act of the Inter-American Maritime Conference, Washington, November 25-December 2, 1940. Int-*6615*.I5, 1940, Eng.
Final Act of the Second Inter-American Conference on Agriculture. Int-*6200*.I4, 1942.1, Eng.
Final Act of the Second Inter-American Travel Congress. Int-*5300*.I4, 1941.1, Eng.
Final Act of the Third Inter-American Conference on Agriculture. Int-*6200*.I4, 1945.1, Eng.
Final Act of the United Nations Conference on Food and Agriculture, Hot Springs, Virginia, United States of America, 18th May-3rd June, 1943. Int-*6200*.C5, 1943.
Final Report of the Committee for Congested Production Areas. USA-*5110*.F5, 1944.
Final Report on Total and Partial Unemployment for Geographic Divisions. USA-*2200*.C7, 1937.A.
"Final Reports of the Sixteenth Census of the United States, 1940." Int-*3040*.E5, June '44, b.
FINANCES:
GENERAL—*see* 6720-6729.
GOVERNMENT—*see* 8800.
INTERNATIONAL — *see* 6720-6729; 6910; 8200-8299.
MUNICIPAL—*see* 8800.
PUBLIC—*see* 8800.
STATE—*see* 8800.
Financial Statistics of Cities. USA-*8800*.C4.
Financial Statistics of Counties. USA-*8800*.C6.
Financial Statistics of State and Local Government (Wealth, Public Debt, and Taxation). USA-*8800*.F5.
Financial Statistics of States. USA-*8800*.S5.
FINANZAS:
GENERAL—*véase* 6720-6729.
ESTATALES—*véase* 8800.
GUBERNAMENTALES—*véase* 8800.
INTERNACIONALES—*véase* 6720-6729; 6910; 8200-8299.
MUNICIPALES—*véase* 8800.
PUBLICAS—*véase* 8800.
Finanzas. Bol-*6720*.E6.
Finanzas, Bancos y Cajas Sociales. Chile-*6720*.F5.
Finanzas Municipales. RepDom-*8800*.F5.
FINCAS—*véase* 6200.
Fine, Selma.
NonG-*6100*.S3. "The Use of Income Tax Data in the National Resources Committee Estimate of the Distribution of Income by Size."
Finer, Herman.
Int-*6400*.T5, 1944. *The T. V. A.: Lessons for International Application.*
FINGERPRINTING—*see* 5250.
First Annual Report of the Central Statistical Board, Year Ended December 31, 1934. USA-*3030*.A6, 1934.
A First Course in Statistical Method. NonG-*3330*.Gavett, 1937.
First Inter American Demographic Congress—*see* - *véase* Inter American Demographic Congress.
"The First Two Years of Social Insurance in Mexico." USA-*9950*.S4, July '46, Cohen.

Fournier, Agustin Ruano.
Ur-*5100*.H5, 1939. *Historia y Análisis Estadístico de la Población del Uruguay.*
Fourth Annual Report of the Central Statistical Board, July 1, 1937, to June 30, 1938. USA-*3030*.A6, 1937-38.
Fourth Pan American Coffee Conference, Mexico City, September 3-14, 1945. Int-*6210*.P5, 1945, Eng.
France. Société de Statistique de Paris—*see - véase* Société de Statistique de Paris.
FRECUENCIAS, DISTRIBUCION DE—*véase* 3100-3199.
FREE PORTS—*see* 6900.
Freitas Filho, Lincoln de.
Bras-*3040*.R5, Jan.-Mar. '42, "O Clínico e a Bio-Estatística."
Bras-*7100*.A5, Dec. '44. "O Problema da Seleção da Causa Primária do Óbito no Brasil, II."
Bras-*7100*.A5, Aug. '43. "O Problema da Seleção da Causa Primária do Óbito no Brasil, I."
FREQUENCY DISTRIBUTION—*see* 3100-3199.
Friedman, Milton.
USA-*9200*.I4, 1945. *Income from Independent Professional Practice.*
FUERZA HIDRAULICA—*véase* 6400.
FUERZAS AEREAS—*véase* 8110.
FUNCIONARIOS—*véase* 8000.
Fundação Getúlio Vargas.
Bras-*5000*.E5. *Estudos Brasileiros de Demografia.*
Bras-*5220*.T3, 1946. *Tábuas Brasileiras de Mortalidade e Sobrevivência.*
The Fundamental Principles of Mathematical Statistics. NonG-*3330*.Wolfenden, 1942.
"Fundamental Problem in the Case Method of Instruction in Statistics." USA-*3040*.J5, Mar. '28, Blackett.
Fundamentals of the Present Scope of Brazilian Statistics. Bras-*3030*.S7, 1940.
Fundamentals of the Theory of Statistics. NonG-*3330*.Smith.
FUNDOS (EN GENERAL, incluyendo SUPERFICIE DE)—*véase* 6200.
Furbay, John H.
Col-*7200*.E5, 1946. *Education in Colombia.*
CR-*7200*.E5, 1946. *Education in Costa Rica.*
"The Future Population of Canada." Int-*5000*.M3, 1946.
The Future Population of Canada. Can-*5100*.F5, 1946.
"Future Populations." Int-*5000*.P5, Jan. '44.
La Gaceta. Nic-*1900*.G5.
Gaceta de Comunicaciones. Guat-*6620*.G5.
La Gaceta Económica y Financiera. Perú-*6000*.G5.
Gaceta Oficial de los Estados Unidos de Venezuela. Ven-*1900*.G5.
Gaceta Oficial, Edición Extraordinaria No. 115, de 28 de Junio de 1935, Creando la Dirección General de Estadística. Cuba-*3030*.G5, 1935.
Gaceta Postal. Guat-*6620*.G5.
GAINFULLY EMPLOYED—*see* 9200; 5110.
Galarza, Ernesto.
Int-*9000*.L6, 1941-42. *Labor Trends and Social Welfare in Latin America, 1941 and 1942.*
Galpin, Charles Josiah.
USA-*5300*.I6, 1870-1930. *Interstate Migrations among the Native White as Indicated by Differences between State of Birth and State of Residence.*
Galvani, Luigi.
NonG-*3330*.Gini 2, 1935. *Curso de Estadística.*
GAME:
PROTECTION—*see* 6230.
RESERVES AND RESOURCES—*see* 6230.
GANADERIA:
GENERAL—*véase* 6230.
CENSOS DE—*véase* 2110; 2210; 6230.
GANADO—*véase* 6230.
GANANCIAS—*véase* 6721.
García, Carlos.
NonG-*3330*.Bertillon, 1906. *Compendio de la Obra Intitulada: Curso Elemental de Estadística Administrativa.*
García Frías, Roque.
Perú-*3040*.E5, July '46. "Ensayo de Tabla de Mortalidad para los Habitantes de Lima, Ciudad Capital."
Perú-*3040*.E5, Nov.'45. "Determinación Estadística Matemática de la Población Electoral del Perú en Junio de 1945."
García Gache, Roberto.
Arg-*3040*.R5, June '45, Daus. "Distribución de la Población Argentina por Regiones Geográficas."
García-Mata, Carlos.
Int-*1300*.A5, 1940.3 "Un Método Práctico para Suavizar a Mano las Curvas Estadísticas."
García Pérez, Andrés.
NonG-*3330*, 1945. *Elementos de Método Estadístico.*
GARDENING—*see* 6220.
Garrett, Henry Edward.
NonG-*3330*, 1937. *Statistics in Psychology and Education.*

Godfrey, Ernest H.—cont.
Principal Wool-Producing and Wool-Consuming Countries of the World."
Int-*3040*.R5, 1936.4. "Progress in the Collection of Annual Agricultural Statistics and of Crop Reports during Growth in the Dominion of Canada."
GOLD RESERVES—*see* 6720.
Goldenberg, Pedro.
Arg-*5220*.T5, 1936. *Tabla de Mortalidad de la República Argentina.*
Goldsmith, R. W.
NonG-*6100*.S3. "Measuring the Economic Impact of Armament Expenditures."
NonG-*6100*.S3. "The Volume and Components of Saving in the United States, 1933-1937."
Gonçalves, Eduardo J.
Bras-*5200*.V5, 1946. *Vital Statistics and Registration Methods.*
González Galé, José A.
NonG-*3140*.E5, 1936. *Elementos de Estadística Matemática.*
González H., Gonzalo.
Int-*1300*.A5, 1940.3. "El Sistema de Coordinación en los Servicios Estadísticos Mexicanos."
Méx-*6200*.B5, Nov. '46. "El Desarrollo y Estado Actual de las Estadísticas Agrícolas en México."
González Reina, Jenaro.
Méx-*6300*.M5, 1944. *Minería y Riqueza Minera de México.*
Good Neighbors, Argentina, Brazil and Chile and Seventeen Other Countries. Int-*1900*.G5, 1941.
Gourvitch, Alexander.
USA-*6000*.I4, 1939. *Industrial Change and Employment Opportunity—A Selected Bibliography.*
NonG-*6000*.S7, 1940. *Survey of Economic Theory on Technological Change and Employment.*
GOVERNMENT AND POLITICAL STATISTICS—*see* 8000-8999.
Government Statistics for Business Use. USA-*3000*.G7, 1946.
GRAFICOS (ESTADISTICOS)—*véase* 3100.
Gráficos Estadísticos de Nicaragua. Nic-*1410*.G5.
Graham, Irene J.
NonG-*3330*.Brumbaugh, 1941. *Business Statistics.*
Les Grands Produits Agricoles. Int-*6210*.G5, 1924-38.
Granier, James A.
Int-*1200*.B5, 1942. *A Bibliography of Latin American Bibliographies.*
Granillo, Ricardo.
Int-*7110*.B5, May '44. "El Servicio de Bioestadística en México."
GRANJAS—*véase* 6200.
GRAPHIC PRESENTATION—*see* 3260.
Graphic Presentation. NonG-*3260*.G5, 1939.
GRAPHS—*see* 3100.
Graphs: How to Make and Use Them. NonG-*3260*.G7, 1940.
Great Britain. Foreign Office.
Int-*3210*.C4, 1913.1. *Convention Respecting the Compilation of International Commercial Statistics.*
Int-*6200*.C5, 1943. *Final Act of the United Nations Conference on Food and Agriculture, Hot Springs, Virginia, United States of America, 18th May-3rd June, 1943.*
——. Permanent Consultative Committee on Official Statistics.
GrBr-*3010*.G5. *Guide to Current Official Statistics of the United Kingdom.*
——. Royal Institute of International Affairs—*see* - *véase* Royal Institute of International Affairs, Great Britain.
——. Royal Statistical Society—*see* - *véase* Royal Statistical Society, Great Britain.
Greenway, Harold F.
Int-*3040*.E5, Dec. '45. "Attributes of Wage-Earner Families in Relation to Size."
Gregory, Winifred.
Int-*1200*.S6, 1815-1931. *Serial Publications of Foreign Governments, 1815-1931.*
Greville, Thomas N. E.
USA-*6730*.R5, June '43. "Short Methods of Constructing Abridged Life Tables."
USA-*6730*.R5, Oct. '42. "Actuarial Note: 'Census' Methods of Constructing Mortality Tables and Their Relation to 'Insurance' Methods."
Gries, Caroline G.
USA-*3270*.F5, 1944. *Foreign Weights and Measures with Factors for Conversion to United States Units.*
"Gross and Net Reproduction Rates, Canada and Provinces, 1920-1942." Int-*3040*.E5, Sept. '45, a.
"Gross National Product Projections: I, The Background and Relation to Current Issues." USA-*6000*.D5, March '45, George.
"Gross National Product Projections for Full Employment: II, Contrasting Estimates—Range and Reasons." USA-*6000*.D5, May '45, George.
"Gross National Product Projections for Full Employment: III, Measuring the Labor Force in 1950." USA-*6000*.D5, June '45, George.

GROUP MEDICINE—*see* 7190.
Grove, E. W.
NonG-*6100*.S3. "The Concept of Income Parity for Agriculture."
Grove, Robert D.
USA-*5200*.V7, 1900-40. *Vital Statistics Rates in the United States, 1900-1940.*
GROWTH:
CURVES—*see* 3100.
OF CHILDREN—*see* 7180.
OF POPULATION—*see* 5110.
"Growth of Industry and Aviation in Mexico, 1929-1942." Int-*6000*.C7, July-Aug. '44, Ysita.
Gualberto, Virgilio.
Bras-*3040*.R5, Jan.-Mar. '46. "Sôbre o Consumo de Alguns Gêneros Alimentícios."
GUANO—*see - véase* 6220.
Guatemala. Departamento de Coordinación Económica Financiera—*see - véase* Guatemala. Ministerio de Economía y Trabajo. Departamento de Coordinación Económica Financiera.
——. Departamento de Estadística y Escalafón—*see - véase* Guatemala. Ministerio de Educación Pública. Departamento de Estadística y Escalafón.
——. Dirección de Enseñanza y Divulgación Agrícola—*see - véase* Guatemala. Ministerio de Agricultura y Minería. Dirección de Enseñanza y Divulgación Agrícola.
——. Dirección General de Aduanas—*see - véase* Guatemala. Ministerio de Hacienda y Crédito Público. Dirección General de Aduanas.
——. Dirección General de Caminos—*see - véase* Guatemala. Ministerio de Comunicaciones y Obras Públicas. Dirección General de Caminos.
——. Dirección General de Estadística—*see - véase* Guatemala. Ministerio de Economía y Trabajo. Dirección General de Estadística.
——. Dirección General de Sanidad Pública —*see - véase* Guatemala. Ministerio de Salud Pública y Asistencia Social. Dirección General de Sanidad Pública.
——. Leyes, Estatutos, Etc.
Guat-*2100*.C5, 1940.2. *Reglamento para el Censo General de Población del Año 1940.*
Guat-*3030*.A5, 1945. *Acuerda: . . . Dictar las Disposiciones del Caso a Efecto de que la Dirección General de Estadística Llene Su Cometido.*
Guat-*3030*.L5, 1936. *Ley de Estadística: Decreto Número 1820.*
Guat-*6900*.E5, 1935. *Decreto Gubernativo Número 1765, Diciembre 7, 1935: Arancel de Aduanas de la República de Guatemala.*
——. Ministerio de Agricultura y Minería.
Guat-*6200*.M5. *Memoria de las Labores del Poder Ejecutivo en el Ramo de Agricultura.*
——. ——. Dirección de Enseñanza y Divulgación Agrícola.
Guat-*6200*.R4. *Revista Agrícola.*
——. Ministerio de Comunicaciones y Obras Públicas.
Guat-*6000*.M5. *Memoria del . . .*
Guat-*6620*.G5. *Gaceta de Comunicaciones.*
——. ——. Dirección General de Caminos.
Guat-*1950*.G5, 1942. *Guía Kilométrica de las 23 Rutas Nacionales de la República de Guatemala.*
——. Ministerio de Economía y Trabajo.
Guat-*6000*.R5. *Revista de Economía.*
——. ——. Departamento de Coordinación Económica Financiera.
Guat-*3210*.I5, 1943. *Instrucciones a los Importadores de Artículos y Materiales en Tránsito o Procedentes de los Estados Unidos de Norte América o del Canadá.*
——. ——. Dirección General de Estadística.
Guat-*1420*.B5. *Boletín de la Dirección . . .*
Guat-*1420*.B5, Nov. '46. "Censo Industrial."
Guat-*1420*.B5, July '46, a. "Empresas Eléctricas de Servicio Público."
Guat-*1420*.B5, July '46, b. "Resultados del Censo Escolar Levantado el 23 de Enero de 1946: Municipio de Guatemala."
Guat-*1420*.B5, May '46, a. "Censo Escolar."
Guat-*1420*.B5, May '46, b. "Balanza Comercial, 1935 a 1945."
Guat-*2100*.C5, 1940.1. *Censo de 1940: Métodos y Procedimientos.*
Guat-*2130*.C5, 1946.1. *Censo Escolar de 1946: Métodos y Procedimientos.*
Guat-*2130*.C5, 1946.2. *Acuerdo Disponiendo el Levantamiento del Censo Escolar y Reglamento General del Mismo.*
Guat-*2130*.C5, 1946.3. *Instrucciones Generales para los Jefes de Empadronadores y Personal Auxiliar.*
Guat-*2200*.C5, 1940. *Quinto Censo General de Población, Levantado el 7 de Abril de 1940.*
Guat-*2200*.C5, 1921. *Censo de la Población de la República, Levantado el 28 de Agosto de 1921.*

Guatemala—cont.

Guat-*3000*.S5. *Serie de Estudios Monográficos.*

Guat-*9200*.A5, 1946. *Análisis de los Sueldos Pagados por el Estado.*

——. ——. ——. Oficina Permanente del Censo.

Guat-*3030*.A6, 1945. *Acuerdo Gubernativo de 2 Mayo de 1,945 por el Cual Fué Creada la Oficina Permanente del Censo.*

——. Ministerio de Educación Pública.

Guat-*7200*.M5. *Memoria de las Labores del Poder Ejecutivo en el Ramo de Educación Pública.*

——. ——. Departamento de Estadística y Escalafón.

Guat-*7200*.E3, 1946. *La Edad de los Escolares y Su Valor como Determinante en el Fenómeno Educativo.*

Guat-*7200*.E5, 1946. *Estudio sobre la Enseñanza Parvularia.*

Guat-*7200*.P5. *Publicaciones del Ministerio de Educación Pública.*

Guat-*7210*.C5, 1946. *Cuatro Estudios Estadístico-Educativos de la Escuela Guatemalteca.*

Guat-*7210*.E5, 1946. *Estudio sobre el Costo de la Enseñanza Primaria.*

Guat-*7310*.C5, 1946. *Características Antropométricas de los Niños de Guatemala.*

——. Ministerio de Fomento—*see - véase* Guatemala. Ministerio de Comunicaciones y Obras Públicas.

——. Ministerio de Hacienda y Crédito Público.

Guat-*1410*.E5. *Estadísticas Gráficas.*

Guat-*1410*.M5. *Memoria de las Labores del Poder Ejecutivo en el Ramo de Hacienda y Crédito Público.*

——. ——. Dirección General de Aduanas.

Guat-*6900*.C5. *Cuadros de la Dirección General de Aduanas.*

——. ——. Dirección General de Estadística—*see - véase* Guatemala. Ministerio de Economía y Trabajo. Dirección General de Estadística.

——. Ministerio de Salud Pública y Asistencia Social. Dirección General de Sanidad Pública.

Guat-*7110*.B5. *Boletín Sanitario de Guatemala.*

Guat-*7110*.R5. *Revista de Sanidad.*

——. ——. Hospital General.

Guat-*7130*.M5. *Memoria de las Labores Realizadas por el Hospital General y Sus Dependencias.*

——. Oficina Central del Café.

Guat-*6210*.R5. *Revista Cafetalera de Guatemala.*

——. Oficina Permanente del Censo—*see - véase* Guatemala. Ministerio de Economía y Trabajo. Dirección General de Estadística. Oficina Permanente del Censo.

——. Secretaría de Agricultura—*see - véase* Guatemala. Ministerio de Agricultura y Minería.

——. ——. Dirección General de Caminos *see - véase* Guatemala. Ministerio de Comunicaciones y Obras Públicas. Dirección General de Caminos.

——. Secretaría de Fomento—*see - véase* Guatemala. Ministerio de Comunicaciones y Obras Públicas.

——. Secretaría de Hacienda y Crédito Público—*see - véase* Guatemala. Ministerio de Hacienda y Crédito Público.

——. Secretaría de Hacienda y Crédito Público. Dirección General de Estadística —*see - véase* Guatemala. Ministerio de Economía y Trabajo. Dirección General de Estadística.

——. Servicio de Comunicaciones—*see - véase* Guatemala. Ministerio de Comunicaciones y Obras Públicas.

Guatemala: Summary of Biostatistics. Guat-*5000*.G5, 1944.

GUERRA—*véase* 8100-8199.

Guía Industrial y Comercial de Venezuela, 1943. Ven-*6000*.G5, 1943.

Guía Kilométrica de las 23 Rutas Nacionales de la República de Guatemala. Guat-*1950*.G5, 1942.

La Guía Turística de El Salvador. *ElSal-1950*.G5.

GUIAS—*véase* 1950.

GUIDE BOOKS—*see* 1950.

Guide for Quality Control and Control Chart Method of Analyzing Data. NonG-*3160*.G5, 1941.

A Guide to the Official Publications of the Other American Republics: I, Argentina. Arg-*1200*.G5, 1945.

A Guide to the Official Publications of the Other American Republics: II, Bolivia. Bol-*1200*.-G5, 1945.

A Guide to the Official Publications of the Other American Republics: VII, Cuba. Cuba-*1200*.G5, 1945.

Guthardt, Emilio Carlos.

Int-*1300*.A5, 1940.3. "Panorama Demográfico de la Población de Colombia."

Int-*3040*.E5, Mar. '43. "Orientación, Organización y Resultados de un Curso Oficial de Estadística por Correspondencia."

Haïti—cont.

——. Secrétairerie d'État des Finances. Administration Générale des Contributions.
Haiti-*8800*.R3. *Rapport Annuel de l'Administration Générale des Contributions.*

——. Secrétairerie d'État des Travaux Publics.
Haiti-*4000*.G5, 1924. *Geology of the Republic of Haiti.*

——. ——. Direction Générale des Travaux Publics.
Haiti-*8000*.R5. *Rapport Annuel du Sous-Secrétaire d'État et Ingénieur en Chêf.*

——. Service National d'Hygiène et d'Assistance Publique—*see - véase* Haïti. Secrétairerie d'État de l'Intérieur. Service National d'Hygiène et d'Assistance Publique.

——. Service National de la Production Agricole et de l'Enseignement Rural—*see - véase* Haïti. Secrétairerie d'État de l'Agriculture et du Travail. Service National de la Production Agricole et de l'Enseignement Rural.

Haiti Consulate in New York.
Haiti-*6900*.H5, 1935. *Handbook and Trade Directory of Haiti.*

Haiti: Summary of Biostatistics. Haiti-*5000*.H5, 1945.

Haitian Customs Receivership. Haiti-*6000*.R5.

The Haitian People. Haiti-*1900*.H5, 1941.

Hall, Charles E.
USA-*5120*.N5, 1920-32. *Negroes in the United States, 1920-32.*

Hall, Marguerite Franklyn.
NonG-*7110*.P5, 1942. *Public Health Statistics.*

Hance, Wendell D.
NonG-*6100*.S3. "Adequacy of Estimates Available for Computing Net Capital Formation."

Handbook and Trade Directory of Haiti. Haiti-*6900*.H5, 1935.

Handbook for Employees of the Bureau of the Census. USA-*3030*.H5, 1940.

Handbook for the Use of the Delegates of the Third Inter-American Conference on Agriculture. Int-*6200*.I4, 1945.2.

"Handbook: List of Members of the American Economic Association, 1946." USA-*6000*.A5, Sept. '46.

Handbook of International Organisations. Int-*8200*.H5, 1938.

Handbook of International Organizations in the Americas. Int-*1930*.H4, 1945.

Handbook of Labor Statistics, 1941. USA-*9000*.H5, 1941.

Handbook of Latin American Population Data. Int-*5100*.P8, 1945.

andbook of Latin American Studies. Int-*H1200*.H4.

Handbook of Mathematical Statistics. NonG-*3140*.H5, 1924.

Handbook of Scientific and Technical Societies and Institutions of the United States and Canada. Int-*1930*.H5, 1942.

Handbook on Death Registration and Certification. Can-*5250*.H2, 1937.

HANDBOOKS OF COUNTRIES AND PLACES—*see* 1950.

HANDICAPPED PERSONS—*see* 7170; 9951.

Hanna, Frank.
USA-*3000*.G7, 1946. *Government Statistics for Business Use.*

Hannay, Annie M.
Int-*6200*.M2, 1944. *Methods and Problems of International Statistics with Special Reference to Food and Agriculture: A List of References.*

Hansen, Morris H.
Int-*3040*.E5, Dec. '44. "A New Sample of the Population."
USA-*3040*.J5, Sept. '43. "On Some Census Aids to Sampling."
USA-*3040*.J5, Mar. '42. "Relative Efficiencies of Various Sampling Units in Population Inquiries."
USA-*3040*.J5, Dec. '40, Stephan. "The Sampling Procedure of the 1940 Population Census."
NonG-*3150*.S3. "Sample Surveys in Census Work."

Hanson, Alice Cable.
USA-*9700*.I5, 1942. *Income and Spending and Saving of City Families in Wartime.*
USA-*9700*.S5, 1942. *Spending and Saving of the Nation's Families in Wartime.*

Hanson, Earl Parker.
Int-*1950*.N5, 1943. *The New World Guides to the Latin American Republics.*
Int-*4000*.M4, 1945. *Index to Map of Hispanic America, 1:1,000.000.*

HARBORS—*see* 6615.

Harris, Seymour E.
Int-*6000*.E7, 1944. *Economic Problems of Latin America.*

Harrison, Frederick W.
Int-*3040*.E5, Sept. '45, Ely. "Cambios en la Clasificación de Estadísticas de Exportación de los Estados Unidos."

Hart, Albert Gailord.
USA-*6000*.S5, 1945. *The Social Framework of the American Economy.*

Hartkemeier, Harry Pelle.
NonG-*3110*.P5, 1942. *Principles of Punch-Card Machine Operation.*

Holmes, Irvin—cont.

and Livestock Statistics and Agricultural Census Work in Mexico.

USA-*2210*.C5, 1940.3. *Sixteenth Census of the United States: 1940: Agriculture: Value of Farm Products by Color and Tenure of Farm Operator.*

Holt, William Stull.

USA-*3030*.B5, 1929. *The Bureau of the Census: Its History, Activities and Organization.*

Homberger, Ludwig M.

Int-*6610*.T5, 1942. *Transportation in Latin America.*

HOME ECONOMICS—*see* 9700.

HOMICIDIOS—*véase* 5212; 8300.

Honduras. Comisión Nacional de Alimentación y Agricultura—*see - véase* Honduras. Secretaría de Agricultura, Fomento y Obras Públicas. Comisión Nacional de Alimentación y Agricultura.

——. Dirección General de Estadística—*see - véase* Honduras. Secretaría de Gobernación, Justicia, Sanidad y Beneficencia. Dirección General de Estadística.

——. Dirección General de Sanidad—*see - véase* Honduras. Secretaría de Gobernación, Justicia, Sanidad y Beneficencia. Dirección General de Sanidad.

——. Leyes, Estatutos, Etc.

Hond-*2100*.C5, 1940.3. *Reglamento para el Levantamiento del Censo de Población de 1940.*

Hond-*3030*.R5, 1944. *Reglamento de la Estadística Nacional de la República de Honduras.*

——. Secretaría de Agricultura, Fomento y Obras Públicas.

Hond-*6000*.I5. *Informe del Secretario de Agricultura, Fomento y Obras Públicas.*

——. ——. Comisión Nacional de Alimentación y Agricultura.

Hond-*6200*.C5, 1945. Cuestionario de la Comisión Nacional de Alimentación y Agricultura.

Hond-*6200*.H5. *Honduras Agrícola.*

——. Secretaría de Gobernación, Justicia, Sanidad y Beneficencia.

Hond-*2100*.C5, 1945.3. *Reglamento del Censo de Población de 1945.*

Hond-*8000*.I5. *Informe de los Actos Realizados por el Poder Ejecutivo en los Ramos de Gobernación, Justicia, Sanidad y Beneficencia.*

——. ——. Dirección General de Estadística.

Hond-*2100*.C5, 1945.1. *Censo General de Población Levantado el 24 de Junio de 1945: Métodos y Procedimientos.*

Hond-*2100*.C5, 1945.2. *Instrucciones para la Preparación y Práctica del Censo de Población de 1945.*

Hond-*2100*.C5, 1940.1. *Censo de 1940: Métodos y Procedimientos.*

Hond-*2100*.C5, 1940.2. *Instrucciones para la Preparación y Práctica del Censo de 1940.*

Hond-*2200*.C5, 1940. *Resumen del Censo General de Población Levantado el 30 de Junio de 1940.*

Hond-*5200*.M5, 1937-38. *Movimiento de Población, Nacimientos, Defunciones y Matrimonios Habidos en la República durante el Año Económico de 1937 a 1938.*

——. ——. Dirección General de Sanidad.

Hond-*7101*.R5, 1925. *Reglamento Interior de la . . . .*

Hond-*7110*.B5. *Boletín Sanitario de la . . .*

——. Secretaría de Hacienda, Crédito Público y Comercio.

Hond-*6900*.A5, 1934. *Arancel de Aduanas, 1934.*

Hond-*8800*.I5. *Informe de Hacienda, Crédito Público y Comercio.*

——. ——. Dirección General de Estadística—*see - véase* Honduras. Secretaría de Gobernación, Justicia, Sanidad y Beneficencia. Dirección General de Estadística.

Honduras Agrícola. Hond-*6200*.H5.

Honduras: Summary of Biostatistics. Hond-*5000*.H5, 1944.

HOOF AND MOUTH DISEASE OF ANIMALS—*see* 6230.

Hooper, Ofelia.

Pan-*2210*.C5, 1943.1. *Censo Agro-Pecuario del Distrito de Penonomé, Diciembre 1943.*

Hope, E. C.

Can-*3040*.C5, Aug. '43. "Agriculture's Share of the National Income."

HORAS DE TRABAJO—*véase* 9200.

HORTICULTURA—*véase* 6220.

Hosea, Harold R.

USA-*9000*.M5, May '45. "Fact-Finding Activities of the Bureau of Labor Statistics."

HOSPICIOS—*véase* 9951.

HOSPITALES—*véase* 7130.

HOSPITALS—*see* 7130.

Hours and Earnings. USA-*9200*.H4.

Hours and Earnings in the United States, 1932-40. USA-*9200*.H5, 1932-40.

HOURS OF LABOR—*see* 9200.

HOUSING—*see* 9900; 2220.

"Housing and Construction." USA-*3000*.G7, 1946.

Ramo acerca de las Labores Desarrolladas. Ec-*7110*.I5.
"Informe que la Contraloría General de la República Rinde al Sr. Ministro de Gobierno y a las Honorables Cámaras sobre el Levantamiento del Censo Civil de 1938." Col-*1420*.A5, Aug. '39.
Informe, Serie D, Demografía. Arg-*5000*.P2.
Informe, Serie I, Industria. Arg-*6400*.E5.
Informe sobre la Industria Pecuaria Presentado por la Delegación de Nicaragua a la 3a Conferencia Interamericana de Agricultura, en Caracas, EE. UU. de Venezuela. Nic-*6230*.I5, 1945.
Informe sobre Labores Agrícolas Presentado por la Delegación de Nicaragua a la Tercera Conferencia Interamericana de Agricultura, en Caracas, Estados Unidos de Venezuela. Nic-*6200*.I5, 1945.
Informe y Balances del Banco Hipotecario de Nicaragua. Nic-*6720*.I5.
Informe y Datos Estadísticos Presentados por el Superintendente Bancario al Señor Ministro de Hacienda y Crédito Público. Col-*6720*.I7.
Informes sobre la Estabilización de Precios, el Control de la Inflación y Solicitudes de Abastecimientos Adicionales para Cuba en el Año de 1945. Cuba-*6010*.I5, 1945.
Informes sobre las Principales Estadísticas Mexicanas. Méx-*3030*.I5, 1941.
INGENIERIA, TECNOLOGIA DE:
GENERAL—*véase* 6000.
ESPECIFICA—*véase* las respectivas clasificaciones.
INGRESOS:
GENERAL—*véase* 6721; 9700.
DE FAMILIAS—*véase* 9700.
DISTRIBUCION DE—*véase* 6100; 9200; 9700.
NACIONALES—*véase* 6100.
POR OCUPACIONES—*véase* 9200.
PROFESIONALES—*véase* 9200.
SEGUN EMPLEOS—*véase* 6100; 9200; 9700.
"Inhabitants per Physician." Int-*7110*.I6, June '44.
INHERITANCE TAX—*see* 8800.
Iniciação à Estatística Econômica. NonG-*3330*.Silva Rodrigues 2, 1942.
INJURIES—*see* 7160; 7150; 9400.
INMIGRACION—*véase* 5300.
"Inmigrantes, según Nacionalidad, Edades y Sexo, Año de 1944." Nic-*1420*.B5, June 15, '45, b.
INMUEBLES Y BIENES RAICES:
GENERAL—*véase* 2220; 6721; 6100.
PUBLICAS—*véase* 6100.
Inquérito Continental sôbre Fomento e Coordenação de Indústrias. Bras-*6400*.I5, 1946.
Inquéritos Econômicos para a Defesa Nacional. Bras-*6000*.I5.
INQUESTS—*see* 5212.
INQUILINOS AGRICOLAS—*véase* 9600.
"Inquiry Concerning Statistics of Wool Stocks in the Principal Wool-Producing and Wool-Consuming Countries of the World." Int-*3040*.R5, 1939, 2-3, Godfrey.
INSANE ASYLUMS—*see* 7130.
INSCRIPCION, INSUFICIENCIA DE—*véase* 5220.
INSPECCION E INSTRUCCION DEL PERSONAL—*véase* 1700.
INSTITUCIONES:
CENSOS DE—*véase* 2260.
ESTADISTICAS DE—*véase* las respectivas clasificaciones con que se relacionan las instituciones.
Institut Interaméricain de Statistique—*see - véase* Inter American Statistical Institute.
Institut International d'Agriculture—*see - véase* International Institute of Agriculture.
Institut International d'Agriculture. Service de la Statistique Générale—*see - véase* International Institute of Agriculture. Bureau of General Statistics.
L'Institut International de Statistique—*see - véase* International Statistical Institute.
Institute of Educational Research.
USA-*3040*.P5. *Psychometrika.*
Institute of International Education.
Int-*7200*.A7. *Annual Report of the Director of the Institute of International Education.*
Institute of Mathematical Statistics.
NonG-*3040*.A3. *The Annals of Mathematical Statistics.*
INSTITUTIONS:
CENSUSES OF—*see* 2260.
STATISTICS OF—*see* the particular type of institution.
Instituto Alejandro E. Bunge de Investigaciones Económicas y Sociales.
Arg-*3040*.R5. *Revista de Economía Argentina.*
Arg-*3040*.R5, July '46, a. "Bosquejo de una Economía Argentina para 1955: Segunda Parte (continuación)—Estudios Especiales."
Arg-*3040*.R5, June '46, a. "Bosquejo de una Economía Argentina para 1955: Segunda Parte (continuación)—Estudios Especiales."
Arg-*3040*.R5, May '46, a. "Bosquejo de una Economía Argentina para 1955:

Instituto Alejandro E. Bunge . . . —cont.

Segunda Parte (continuación)—Estudios Especiales."

Arg-*3040*.R5, Apr. '46. "Bosquejo de una Economía Argentina para 1955: Segunda Parte—Estudios Especiales."

Arg-*3040*.R5, Mar. '46. "Bosquejo de una Economía Argentina para 1955: Primera Parte—Plan General (continuación)."

Arg-*3040*.R5, Feb. '46, a. "Bosquejo de una Economía Argentina para 1955: Primera Parte—Plan General."

Arg-*3040*.R5, July '45, a. "A Propósito del IV Censo General."

Arg-*3040*.R5, June '45, a. "A Propósito del Cuarto Censo General."

Arg-*3040*.R5, May '45. "La Interpretación de los Indices del Costo de la Vida de la Familia Obrera."

Arg-*3040*.R5, Apr. '45. "Algunas Sugerencias a Propósito del Proyectado IV Censo General."

Arg-*3320*.E5. *Estadística: Programa Analítico.*

Instituto Brasileiro de Geografia e Estatística—*see - véase* Brasil. Instituto Brasileiro de Geografia e Estatística.

Instituto Brasileiro de Geografia e Estatística. Bras-*3030*.I5, 1940.

Instituto de Investigaciones Estadísticas, México.

Méx-*8300*.T5, 1939. *Tendencia y Ritmo de la Criminalidad en México, D. F.*

Instituto de Salubridad y Enfermedades Tropicales, México.

Méx-*7110*.R5. *Revista del. . . .*

Méx-*7110*.R5, June '41. "Esperanza de Vida en Veinte Estados de la República Mexicana."

Instituto do Açucar e do Álcool, Brasil.

Bras-*6210*.A4. *Anuário Açucareiro.*

——. Secção de Estatística.

Bras-*6210*.B6. *Boletim do Instituto. . . .*

Instituto Finlay, Cuba. Sección de Publicaciones Científicas, Biblioteca y Museo.

Cuba-*7110.B5*. *Boletín Semanal . . . Epidemiológico Sanitario.*

Instituto Interamericano de Estadística—*see - véase* Inter American Statistical Institute.

El Instituto Interamericano de Estadística. Int-*3030*.I6, 1943.

O Instituto Inter-Americano de Estatística. Int-*3030*.I6, 1943.

Instituto Internacional de Agricultura—*see - véase* International Institute of Agriculture.

Instituto Internacional de Agricultura. Servicio de la Estadística General—*see - véase* International Institute of Agriculture. Bureau of General Statistics.

Instituto Mexicano del Seguro Social—*see - véase* México. Instituto Mexicano del Seguro Social.

Instituto Nacional de Alimentación, Uruguay.

Ur-*9800*.B5. *Boletín Mensual del Instituto Nacional de Alimentación.*

Instituto Nacional de Estadística de la Universidad de Colombia. Col-*3030*.I5, 1945.

Instituto Nacional de Estudos Pedagógicos, Brasil—*see - véase* Brasil. Ministério da Educação e Saúde. Instituto Nacional de Estudos Pedagógicos.

Instituto Nacional de la Nutrición, Argentina—*see - véase* Argentina. Ministerio del Interior. Dirección Nacional de Salud Pública. Instituto Nacional de la Nutrición.

Instituto Nacional del Café, Venezuela.

Ven-*6210*.R5. *Revista del Instituto Nacional del Café.*

Instituto Panamericano de Bibliografía y Documentación.

Méx-*1200*.B5. *Boletín Bibliográfico Mexicano.*

Instituto Peruano de Estadística.

Int-*3040*.E5, Dec. '44, Luna Vegas. "La Estadística Peruana en el Sexenio 1938-44."

Perú-*3040*.E5. *Estadística Peruana.*

Perú-*3040*.E5, Jan.'45, a. "El Inventario del Potencial Económico de la Nación."

Perú-*3040*.E5, Jan.'45, b. "Legislación Estadística."

INSTRUCCION:

DE PERSONAL—*véase* 1700.

ESTADISTICA (MATERIALES Y METODOS DE ENSEÑANZA PARA PERSONAL TECNICO)—*véase* 3300-3399.

MANUALES DE (EN GENERAL)—*véase* 1100; 1960.

MANUALES DE INSTRUCCION ESPECIFICA—*véase* las respectivas clasificaciones.

PEDAGOGICA—*véase* 7200.

VOCACIONAL—*véase* 7200.

Instrucciones a los Importadores de Artículos y Materiales en Tránsito o Procedentes de los Estados Unidos de Norte América o del Canadá. Guat-*3210*.I5, 1943.

Instrucciones al Personal del Censo de la República. ElSal-*2100*.C5, 1930.4.

Instrucciones Complementarias a las "Normas para la Organización y el Levantamiento del Censo Escolar de la Nación." Arg-*2130*.C5, 1943.1.

INSTRUCCIONES ESTADISTICAS (REGLAMENTOS)—*véase* 3100-3199.

Instrucciones Generales para la Ejecución de los Censos de Edificios y Población. Col-*2100*.C5, 1938.

Instrucciones Generales para los Corresponsales Estadísticos. Méx-*6220*.I5, 1942.

Instrucciones Generales para los Jefes de Empadronadores y Personal Auxiliar. Guat-*2130*.C5, 1946.3.

Instrucciones para la División del Territorio Comunal en Zonas de Empadronamiento: XI Censo de Población. Chile-*2100*.C5, 1940.2.

Instrucciones para la Perforación de Tarjetas del 6° Censo de Población de 1940. Méx-*2100*.C5, 1940.3.

Instrucciones para la Preparación y Práctica del Censo de 1940. Hond-*2100*.C5, 1940.2.

Instrucciones para la Preparación y Práctica del Censo de Población de 1945. Hond-*2100*.C5, 1945.2.

INSTRUCTION:
- MANUALS, GENERAL—*see* 1100.
- MANUALS IN SPECIFIC SUBJECTS—*see* the particular field.
- PEDAGOGICAL—*see* 7200.
- STATISTICAL (MATERIALS AND METHODS OF TEACHING TECHNICAL PERSONNEL)—*see* 3300-3399.
- VOCATIONAL—*see* 7200.

Instructions for Processing Data of the Sixteenth Decennial Census. USA-*2100*.C5, 1940.7.

Instructions: Preparation of Plans of Division by Enumeration Districts, 1940. USA-*2100*.C5, 1940.6.

Instructivo para la Elaboración de la Estadística Demográfica en las Oficinas Sanitarias de los Estados y Territorios Federales. Méx-*5000*.I5, 1942.

INSUFICIENCIA DE LA INSCRIPCION (EN REGISTROS DEL ESTADO CIVIL)—*véase* 5250.

INSURANCE:
- COMMERCIAL—*see* 6730.
- HEALTH—*see* 9952; 7190.
- SOCIAL—*see* 9952.

INTELIGENCIA, PRUEBAS DE—*véase* 7200.

INTELLIGENCE TESTING—*see* 7200.

INTEMPERANCE—*see* 7150.

INTEMPERANCIA—*véase* 7150.

Inter-American Affairs. Int-*1910*.I5.

Inter-American Bibliographical and Library Association.
- Int-*1200*.I6. *Inter-American Bibliographical Review.*

Inter-American Bibliographical Review. Int-*1200*.I6.

Inter-American Committee on Social Security.
- Int-*3230*.P6, 1946. *Proyecto de Informe sobre Standardización de las Estadísticas de Atención Médica.*
- Int-*3230*.P7, 1946. *Proyecto de Informe sobre Standardización de las Estadísticas de Morbilidad.*

Inter-American Conference of Experts on Copyright.
- Int-*7300*.I5, 1946.2, Eng. *Inter-American Conference of Experts on Copyright, Pan American Union, June 1-22, 1946: Proceedings.*
- Int-*7300*.I5, 1946.2, Sp. *Conferencia Interamericana de Expertos para la Protección de los Derechos de Autor, 1-22 de Junio de 1946: Actas y Documentos.*

Inter-American Conference of Experts on Copyright, Pan American Union, June 1-22, 1946: Proceedings. Int-*7300*.I5, 1946.2, Eng.

Inter-American Conference of Police and Judicial Authorities.
- Int-*8300*.I5, 1942. *Final Act of the Inter-American Conference of Police and Judicial Authorities, Buenos Aires, Argentina, May 27-June 9, 1942.*

Inter-American Conference of Systems of Economic and Financial Control.
- Int-*6000*.I45, 1942, Eng. *Final Act of Inter-American Conference of Systems of Economic and Financial Control, Washington, June 30-July 10, 1942.*
- Int-*6000*.I45, 1942, Port. *Ata Final da Conferência Interamericana sôbre Medidas de Contrôle Econômico-Financeiro, Washington, 30 de Junho a 10 de Julho de 1942.*
- Int-*6000*.I45, 1942, Sp. *Acta Final de la Conferencia Interamericana sobre Sistemas del Control Económico y Financiero, Wáshington, 30 de Junio a 10 de Julio de 1942.*

Inter-American Conference on Agriculture.
- Int-*6200*.I4, 1945.1, Eng. *Final Act of the Third Inter-American Conference on Agriculture.*
- Int-*6200*.I4, 1942.1, Eng. *Final Act of the Second Inter-American Conference on Agriculture.*
- Int-*6200*.I4, 1942.1, Sp. *Acta Final de la Segunda Conferencia Interamericana de Agricultura.*
- Int-*6200*.I4, 1930.1. *Final Act of the Inter-American Conference on Agriculture. Acta Final de la Conferencia Interamericana de Agricultura. Ata Final*

Inter-American Development Commission—cont.

the Conference of Commissions of Inter-American Development (Held in New York, May 9-18, 1944) by the Peruvian Commission of Inter-American Development.

RepDom-*6000*.R5, 1944. *Report Presented to the Conference of Commissions of Inter-American Development (Held in New York, May 9-18, 1944) by the Dominican Commission of Inter-American Development.*

Ven-*6000*.R3, 1944. *Reports Presented to the Conference of Commissions of Inter-American Development (Held in New York, May 9-18, 1944) by the Venezuelan Commission of Inter-American Development.*

Inter-American Economic and Social Council—*see - véase* Pan American Union. Inter-American Economic and Social Council.

Inter-American Economic and Social Council: Report on the Organization of the Council Approved by the Governing Board of the Pan American Union at the Session of August 29, 1945. Int-*8200*.I45, 1945, Eng.

Inter-American Maritime Conference.

Int-*6615*.I5, 1940, Eng. *Final Act of the Inter-American Maritime Conference, Washington, November 25-December 2, 1940.*

Int-*6615*.I5, 1940, Sp. *Acta Final de la Conferencia Marítima Interamericana, Unión Panamericana, Wáshington, del 25 de Noviembre al 2 de Diciembre de 1940.*

Inter-American Series. Int-*8200*.I48.

Inter American Statistical Institute.

Int-*2100*.M5, 1945. *Métodos de los Censos de Población de las Naciones Americanas.*

Int-*3000*.D5, 1944. *Directory of Statistical Personnel in the American Nations. Directorio del Personal Estadístico en las Naciones Americanas. Nominata do Pessoal Estatístico das Nações Americanas.*

Int-*3000*.I5. *Intercambio Estadístico.*

Int-*3000*.I5, May-June '45. "Source Materials: IASI Survey of Census Methods and Procedures: Nicaragua, Panama, Paraguay, Peru, Republica Dominicana, Venezuela."

Int-*3000*.I5, Apr.'45. "Source Materials: IASI Survey of Census Methods and Procedures: Chile, Colombia, Cuba, El Salvador, Guatemala, Haiti, Honduras, Mexico."

Int-*3000*.I5, Nov.'44-Mar.'45. "Source Materials: IASI Survey of Census Methods and Procedures: Argentina, Bolivia, Brazil, Canada."

Int-*3000*.I5, Aug.-Oct.'44. "Source Materials: IASI Survey of Census Methods and Procedures: United States: Sixteenth Census, 1940."

Int-*3000*.P5. *Project Series of the Inter American Statistical Institute. Serie de Proyectos del Instituto Interamericano de Estadística.*

Int-*3010*.B5. *Bibliography Series, IASI.*

Int-*3010*.B5, Cen. 2, May 5, '45. *Census Methods and Procedures.*

Int-*3010*.B5, Dem. 2, Feb. 7, '45. *Vital Statistics Registration Methods in the Western Hemisphere.*

Int-*3010*.B5, Econ. 4a, Apr. 2, '45. *Bibliography of Selected References on Agricultural Statistics in the Western Hemisphere*

Int-*3010*.B5, Econ. 6a, June 1,'45. *Organization of Agriculture Statistics.*

Int-*3010*.B5, Econ. 8a, Oct. 1,'45. *Foreign Trade Classification in the American Nations.*

Int-*3010*.B5, Econ. 10, Apr. 15,'46. *National Income Bibliography.*

Int-*3010*.B5, Gov. 2, Mar.22,'46. *Justice and Crime Statistics.*

Int-*3010*.B5, Lab. 1, Jan.30,'46. *Labor and Living Conditions in the American Nations.*

Int-*3010*.B5, Stat. 7, Aug.10,'45. *International Statistical Standards.*

Int-*3030*.I6, 1943. *The Inter American Statistical Institute. El Instituto Interamericano de Estadística. O Instituto Inter-Americano de Estatística.*

Int-*3040*.E5. *Estadística.*

Int-*3040*.E5, Sept.'46, Rice. "ISI, IASI, and UN."

Int-*3040*.E5, Mar.'44. "National Statistical Societies in the Western Hemisphere."

Int-*3210*.1945. *Convertibility Index for Foreign Trade Statistical Classifications of the American Nations: Basic Classification Scheme Showing Detailed Export and Import Commodity Description by Classes of the Minimum List of Commodities for International Trade Statistics.*

Int-*6100*.N5, 1945. *National Income Estimates of Latin America.*

Chile-*6100*.N5, 1934-42. *The National Income of Chile.*

The Inter American Statistical Institute. Int-*3030*.I6, 1943.

International Conference of Economic Services.
Int-*6000*.I7, 1919-30. *International Abstract of Economic Statistics, 1919-1930.*

International Conference on Economic Statistics, Geneva, 1928: Preparatory Documents. Int-*6000*.I75, 1928.A.

International Conference Relating to Economic Statistics.
Int-*6000*.I75, 1928.B. *Conférence Internationale Concernant les Statistiques Économiques (26 Novembre au 14 Décembre): 1, Convention Internationale Concernant les Statistiques Économiques; 2, Protocole de la Convention Internationale; 3, Acte Final de la Conférence. International Conference Relating to Economic Statistics (November 26th to December 14th, 1928): 1, International Convention Relating to Economic Statistics; 2, Protocol to the International Convention; 3, Final Act of the Conference.*
Int-*6000*.I75, 1928.C. *Proceedings of the International Conference Relating to Economic Statistics.*

International Conference Relating to Economic Statistics (November 26th to December 14th, 1928). Int-*6000*.I75, 1928.B.

The International Conferences of American States, 1889-1928. Int-*8200*.I6, 1889-1928 Eng.

The International Conferences of American States, First Supplement, 1933-1940. Int-*8200*.I6, 1933-1940, Eng.

International Convention Relating to Economic Statistics. Int-*6000*.I75, 1928.B.

International Court of Justice—*see* - *véase* United Nations. International Court of Justice.

International Crop Report. Int-*6220*.M5.

International Crop-Reporting Service. Int-*6220* I5, 1910.

International Currency Experience, Lessons of the Inter-War Period. Int-*6720*.I6, 1944.

International Economic Conference.
Int-*6200*.A5, 1927. *Agricultural Problems in Their International Aspect: Documentation for the International Economic Conference, Geneva, 1927.*
Int-*9200*.E7, 1931-41. *Estimates of the Working Population of Certain Countries in 1931 and 1941.*

"International Financial Statistics." USA-*6720*.F4, Mar. '46.

"International Fishery Statistics." Int-*3020*.B5, 1909, Rew.

International Geological Congress.
Int-*6300*.C8, 1935. *Copper Resources of the World.*

International Health Organisation. Int-*7101*.I5, 1945.

International Health Year-Book. Int-*7100*.I6.

International Institute of Agriculture.
Int-*2110*.W5, 1940.1. *Programme of the World Agricultural Census of 1940.*
Int-*2110*.W5, 1940.2. *The World Agricultural Census of 1940: Proposed Standard Form.*
Int-*2110*.W5, 1940.3. *World Agricultural Census of 1940: Questionnaires.*
Int-*2210*.W5, 1930. *The First World Agricultural Census (1930.)*
Int-*3020*.C5, 1919. *Conference on International Co-operation in Statistics, August 14th and 15th, 1919.*
Int-*6200*.A5, 1927. *Agricultural Problems in Their International Aspect: Documentation for the International Economic Conference, Geneva, 1927.*
Int-*6200*.D3, 1942. *Decennial Index of Publications Issued by the International Institute of Agriculture, 1930-1939.*
Int-*6200*.I64, 1931. *The International Institute of Agriculture.*
Int-*6200*.I65, 1924. *The International Institute of Agriculture.*
Int-*6200*.I8. *International Review of Agriculture.*
Int-*6200*.I9. *International Yearbook of Agricultural Statistics. Annuaire International de Statistique Agricole.*
Int-*6200*.L6, 1939. *List of Publications Issued by the International Institute of Agriculture.*
Int-*6200*.L7, 1946. *Publications Éditées par l'Institut International d'Agriculture: Supplément à la Liste de Janvier 1943.*
Int-*6200*.L7, 1943. *Liste des Publications Éditées par l'Institut International d'Agriculture.*
Int-*6200*.M4. *Monthly Bulletin of Agricultural Economics and Sociology.*
Int-*6200*.O6, 1910-13. *L'Organisation des Services de Statistique Agricole dans les Divers Pays.*
Int-*6200*.W4, 1940-1945. *The Work of the International Institute of Agriculture During the War (1940-1945).*
Int-*6200*.*W6*. *The World Agricultural Situation.*
Int-*6210*.A4, 1934-38. *Agricultural Commodities and Raw Materials: Production*

International Labour Office—cont.

Consumo de Alimentos y la Nutrición en las Américas.
Int-*9800*.F4, 1942. *Food Consumption and Dietary Surveys in the Americas.*
Int-*9800*.I5, 1942. *International Comparisons of Food Costs.*
Int-*9900*.M5, 1928. *Methods of Compiling Housing Statistics.*
——. Committee of Statistical Experts.
Int-*3040*.R5, 1942, 1-2, Huber. "Le Comité d'Experts Statisticiens du Bureau International du Travail (1933-1939)."
International Labour Review. Int-*9000*.I5.
"International List of Causes of Death Adopted by the Fifth International Conference for Revision, Paris, October 3rd-7th, 1938." Int-*7100*.B6, Dec. '38.
International Migrations. Int-*5300*.I6, 1931.
International Monetary Fund.
Int-*6720*.I8, 1944. *International Monetary Fund and International Bank for Reconstruction and Development.*
International Review of Agriculture. Int-*6200*.I8.
The International Standardisation of Labour Statistics. Int-*9000*.I7, 1943.
International Statistical Institute.
Int-*1420*.R5. *Revue de l'Institut International de Statistique: Supplément Mensuel.*
Int-*3000*.R4, 1934. *Répertoire International des Institutions Statistiques.*
Int-*3020*.B5. *Bulletin de l'Institut International de Statistique.*
Int-*3020*.B5, 1934. "Statuts de l'Institut International de Statistique."
Int-*3020*.B5, 1887-1938. "Tables des Matières Contenues dans les Tomes I à XXX."
Int-*3020*.C5, 1919. *Conference on International Co-operation in Statistics, August 14th and 15th, 1919.*
Int-*3030*.C5, 1884-1934. *50 Années de l'Institut International de Statistique.*
Int-*3040*.E5, Sept. '46, Rice. "ISI, IASI, and UN."
Int-*3040*.R5. *Revue de l'Institut International de Statistique.*
Int-*3230*.N8, 1938. *Nomenclatures Internationales des Causes de Décès, 1938.*
Int-*5000*.A5, 1929-36. *Aperçu de la Démographie des Divers Pays du Monde, 1929-1936.*
Int-*5100*.S5, 1928-34. *Statistique Internationale des Grandes Villes: Territoire et Population des Grandes Villes, 1928-1934.*
Int-*5310*.S5, 1929-1934. *Statistique Internationale des Grandes Villes: Statistique du Tourisme dans les Grandes Villes, 1929-1934.*
Int-*6000*.I7, 1931-36. *International Abstract of Economic Statistics, 1931-1936.*
Int-*6000*.R4, 1931-36. *Recueil International de Statistiques Économiques, 1931-1936.*
Int-*6000*.R4, 1919-30. *Recueil International de Statistiques Économiques, 1919-1930.*
Int-*6400*.S5, 1934. *Statistique Internationale des Grandes Villes: Statistique de l'Electricité, du Gaz et de l'Eau dans les Grandes Villes, 1934.*
Int-*9900*.S5, 1928-34. *Statistique Intertionale des Grandes Villes: Statistique du Logement dans les Grandes Villes, 1928-1934.*
Col-*1420*.A5, May 5, '42, a. "Normas para la Elaboración de los Cuadros Estadísticos."
"International Statistical Needs of Public Health." Can-*7110*.C5, June '46, Dunn.
International Statistical Standards. Int-*3010*.B5, Stat. 7, Aug. 10, '45.
"International Trade and Payments." USA-*3000*.G7.
International Trade and the National Income Multiplier. NonG-*6100*.I5, 1943.
International Trade in Wood. Int-*6900*.I7, 1944.
International Trade Organization of the United Nations.
Int-*8200*.I55, Nov. '46. "Suggested Charter for an International Trade Organization of the United Nations."
International Trade Statistics. Int-*6900*.S6.
International Union for the Scientific Investigation of Population Problems.
Int-*5100*.P4. *Population.*
"International Vital Statistics: Summary." USA-*5200*.V8, May 2, '40.
International Yearbook of Agricultural Statistics. Int-*6200*.I9.
International Yearbook of Forestry Statistics. Int-*6250*.I5.
Interpolación y Ajustamiento de la Curva Logística Generalizada. NonG-*3140*.I5, 1938.
"La Interpretación de los Indices del Costo de la Vida de la Familia Obrera." Arg-*3040*.R5, May '45.
Interstate Migrations among the Native White as Indicated by Differences between State of Birth and State of Residence. USA-*5300*.I6, 1870-1930.

Jardim, Germano G.—cont.

Bras-*3030*.A5, 1941. *A Administração Pública e a Estatística.*

Bras-*3030*.R8, 1944. *Rumos da Organização Estatística Brasileira. Trends of Statistical Organization in Brazil.*

Bras-*7200*.R5, May '45. "A Coleta da Estatística Educacional (III)."

Bras-*7200*.R5, Nov. '44. "A Coleta da Estatística Educacional."

Bras-*7200*.R5, Sept. '44. "A Coleta da Estatística Educacional."

Jenks, Elizabeth.

USA-*6100*.N4, 1941. *National Income and Its Composition, 1919-1938.*

Jensen, Adolph.

Int-*3020*.B5, 1925. "Report on the Representative Method in Statistics."

Int-*3020*.B5, 1925. "The Representative Method in Practice."

Jessen, Raymond J.

USA-*3040*.J5, Mar. '45. "The Master Sample of Agriculture: II, Design."

NonG-*6200*.S5, 1942. *Statistical Investigation of a Sample Survey for Obtaining Farm Facts.*

Jewish Population Studies. USA-*5120*.J5, 1943.

Jewish Social Studies, Publications. USA-*9600*.J5.

JEWS—*see* 5120; 9600.

Jobim, José.

Bras-*1900*.B4, 1943. *Brazil in the Making.*

Johns Hopkins University. School of Hygiene and Public Health.

USA-*3230*.M5, 1944. *Manual for Coding Causes of Illness According to a Diagnosis Code for Tabulating Morbidity Statistics.*

Johnson, Elizabeth A.

USA-*9000*.S3, 1943. *Selected List of the Publications of the Bureau of Labor Statistics.*

USA-*9000*.S3, 1940. *A Selected List of the Publications of the Bureau of Labor Statistics.*

Joint Committee on Latin American Studies.

Int-*1200*.H4. *Handbook of Latin American Studies.*

Jones, Cecil Knight.

Int-*1200*.B5, 1942. *A Bibliography of Latin American Bibliographies.*

Jones, Lloyd A.

NonG-*3130*.N5, 1938. *Notes on Statistical Mapping with Special Reference to the Mapping of Population Phenomena.*

Jones, Robert C.

Int-*9900*.L5, 1943. *Low-Cost Housing in Latin America.*

USA-*9600*.M3, 1945. *Mexican War Workers in the United States.*

Journal de la Société de Statistique de Paris. Fr-*3040*.J5.

Journal du Conseil Economique et Social. Int-*8200*.J3.

Journal of Educational Psychology. USA-*7200*.J5.

Journal of Farm Economics. USA-*6200*.J5.

Journal of the American Statistical Association. USA-*3040*.J5.

Journal of the Economic and Social Council. Int-*8200*.J3.

Journal of the Royal Statistical Society. GrBr-*3040*.J5.

Journals Dealing with the Natural, Physical and Mathematical Sciences Published in Latin America. Int-*7300*.J5, 1944.

Judicial Criminal Statistics. USA-*8310*.J5.

JUDIOS—*véase* 5120; 9600.

Julin, Armand.

Int-*3020*.B5, 1936. "Premier Rapport sur l'Organization des Services Statistiques."

NonG-*3330*, 1924? *Curso de Estadística General y Aplicada.*

Junta Nacional del Turismo, El Salvador.

ElSal-*1950*.G5. *La Guía Turística de El Salvador.*

JURISPRUDENCE, MEDICAL—*see* 7180.

JURISPRUDENCIA MEDICA—*véase* 7180.

JUSTICE—*see* 8300-8399.

Justice and Crime Statistics. Int-*3010*.B5, Gov. 2, Mar. 22, '46.

JUSTICIA—*véase* 8300-8399.

Juvenile Court Statistics. USA-*8320*.J3.

JUVENILE DELINQUENCY—*see* 8320.

Juvenile Delinquents, for the Year Ended September 30, 1945. Jeunes Délinquents pour l'Année Expirée le 30 Septembre 1945. Can-*8320*.J5, 1945.

Kafuri, Jorge F.

NonG-*3330*, 1934. *Lições de Estatística Matemática.*

Kelley, Truman L.

NonG-*3140*.H5, 1924. *Handbook of Mathematical Statistics.*

Kellogg, Lester S.

USA-*3000*.G7, 1946. *Government Statistics for Business Use.*

USA-*3040*.J5, June '39. "Some Problems in Teaching Elementary Statistics."

NonG-*3330*.Brumbaugh, 1941. *Business Statistics.*

Kuznets, Solomon.
NonG-*6100*.S3. "Some Problems in Measuring Per Capita Labor Income."

La Paz (Ciudad)—*see - véase* Bolivia. La Paz (Ciudad).

La Rue, S. de.
Int-*3000*.S5, 1940. "The Statistical Activities of Haiti."

LABOR:
DEMAND—*see* 9200.
MARKET—*see* 9200.
ORGANIZATIONS—*see* 9000.
PRODUCTIVITY—*see* 9100.
SUPPLY—*see* 9200.

"Labor." USA-*3000*.G7, 1946.

LABOR AND LIVING CONDITIONS—*see* 9000-9999.

Labor and Living Conditions in the American Nations. Int-*3010*.B5, Lab. 1, Jan. 30, '46.

Labor Conditions in Latin America. USA-*9000*.L4.

Labor in the United States: Basic Statistics for Social Security. USA-*9000*.L6, 1938.

The Labor Market. USA-*9200*.L5.

Labor Savings in American Industry, 1899-1939. USA-*9100*.L5, 1899-1939.

"Labor Statistics for a Full Employment Program." USA-*3040*.J5, June '45, a.

Labor Trends and Social Welfare in Latin America, 1941 and 1942. Int-*9000*.L6, 1941-42.

Labor Turnover. USA-*9200*.L6.

LABORATORIO, MEDICINA DE (MICROSCOPIA)—*véase* 7180.

LABORATORY MEDICINE (MICROSCOPY)—*see* 7180.

"Labores de la Dirección General de Estadística y Censos." Ec-*1420*.T5, May '45, Lopez Muñoz.

Labour Force Bulletin. Can-*9200*.L5.

Lairet, Félix.
Int-*7110*.B5, Apr. '46. "La Sanidad en Venezuela."

LAND:
RECLAMATION—*see* 6200.
REGISTRATION—*see* 6200.
SETTLEMENT—*see* 5300.
TENURE—*see* 6200.
USE—*see* 6200.

Landazuri, Irma S.
Perú-*9800*.S7, 1944. *Situación Alimenticia en los Departamentos del Norte.*

Lander, Luis.
Int-*7110*.B5, Apr. '46, Lairet. "La Sanidad en Venezuela."

Landis, Paul Henry.
NonG-*5100*.P65, 1943. *Population Problems, a Cultural Interpretation.*

Landsberg, Hans H.
USA-*6200*.A7, 1942. *American Agriculture, 1899-1939: A Study of Output, Employment and Productivity.*

Lang, Richard O.
Int-*3020*.B5, 1938, Rice. "The Organization of Statistical Services in the United States and Great Britain."

LANGUAGE—*see* 7320.

Lankford, Frances G.
NonG-*3110*.S5, 1943. *Statistics: Collecting, Organizing, and Interpreting Data.*

Larke, William J.
NonG-*6400*.S5, 1943. *Statistical Methods in Industry.*

Lary, Hal B.
USA-*6910*.U5, 1943. *The United States in the World Economy.*

Laso, Luis Eduardo.
Ec-*6720*.B5, Mar. '42. "Breves Consideraciones sobre la Economía Ecuatoriana."

Latham, Earl.
USA-*3040*.J5, Sept. '46. "One Statistical World."

Latin America. Int-*1900*.L4, 1943.

Latin America. Int-*1900*.L5, 1942.

Latin America as a Source of Strategic and Other Essential Materials. Int-*6900*.L4, 1941.

Latin America in the Future World. Int-*1900*.L7, 1945.

Latin American Periodicals Currently Received in the Library of Congress and in the Library of the Department of Agriculture. Int-*1200*.L3, 1944.

"Latin American Periodicals Dealing with Labor and Social Welfare." Int-*1200*.H4, 1942, Rohen y Gálvez.

"Latin-American Price Trends." USA-*6900*.F7, July 27, '46, Bernstein.

Latin American Series of the Library of Congress. Int-*1200*.L5.

"Latin America's Economy as World Conflict Ended: Part I: The Region Readjusts to 'Postwar Imperatives'." USA-*6900*.F7, July 6, '46.

"Latin America's Economy as World Conflict Ended: Part II: Mexico, Central America, and the Caribbean Republics." USA-*6900*.F7, July 13, '46.

"Latin America's Economy as World Conflict

LAWS, STATUTES, ETC.—cont.

Estadística Nacional de la República de Honduras.

Hond-*7101*.R5, 1925. *Reglamento Interior de la Dirección General de Sanidad.*

Méx-*3030*.L5, 1941. *Ley Federal de Estadística y Su Reglamento.*

Nic-*1420*.B5, June 15, '44, b. "Ley sobre Estadística de Producción Agrícola."

Nic-*1420*.B5, June 15, '44, c. "Primera Ley Republicana que Centraliza la Estadística."

Nic-*1900*.G5, June 12, '44. "Ley No. 21 de 12 Junio de 1944, Obligación de Todos los Agricultores de la República . . ."

Nic-*1900*.G5, Apr. 25, '44. "Ley No. 91 de 25 de Abril de 1944, Créase la Oficina de Estadística Distritorial."

Nic-*3030*.L5, 1941. *Ley Orgánica de la Estadística Nacional.*

Nic-*6900*.L5, 1942. *Ley Arancelaria de la República de Nicaragua y Sus Reformas hasta el 30 de Junio de 1942.*

Pan-*6900*.A5, 1938. *Arancel de Importación de la República de Panamá.*

Par-*1410*.A5, 1936, a. "Censo de Población."

Par-*1410*.A5, 1936, b. "Ley de Estadística."

Par-*2100*.C5, 1936.2. *Censo de Población, Decreto N°. 2120 de 18 de Junio de 1936.*

Par-*2100*.C5, 1936.3. *Decreto N°. 60.287 de 21 de Noviembre de 1935.*

Par-*2100*.C5, 1936.4. *Ley N°. 1.509 de 1° de Noviembre de 1935.*

Par-*3030*.D5, 1942. *Decreto Ley 11.126 que Reorganiza y Coordina los Servicios Estadísticos de la República.*

Par-*5250*.L3, 1940. *Leyes del Registro del Estado Civil y Decretos Reglamentarios.*

Par-*5250*.L5, 1944. *Registro del Estado Civil de las Personas: Libreta de Familia.*

Par-*6900*.C5, 1940. *Consolidación de los Adicionales ad-Valorem de Importación, Decreto N. 105, Vigente desde el 15 de Marzo de 1940; Lista de los Cambios Efectuados en los Derechos de la Tarifa y Arancel de Aduanas, Ley N. 667; Nueva Tarifa de Exportación, Decreto N. 383 Vigente desde el 25 de Marzo de 1940.*

Par-*6900*.L5, 1924. *Ley N. 667, de Tarifa y Arancel de Aduanas, Septiembre 27 de 1924.*

Perú-*1410*.E5, 1942, a. "Decreto Supremo de 20 de Agosto de 1943 que Ordena Formar el Inventario del Potencial Económico de la Nación."

Perú-*1410*.E5, 1942, b. "Decreto Supremo de 1° de Enero de 1944, por el que se Organiza el Servicio de Estadística Nacional y Crea el Consejo Superior de Estadística."

Perú-*1410*.E5, 1942, d. "Ley No. 7567, de 5 de Agosto de 1932, que Transforma en Nacional la Dirección de Estadística y Establece las Normas para el Funcionamiento del Servicio de Estadística Nacional."

Perú-*1410*.E5, 1942, f. "Resolución Suprema que Aprueba el Diagrama de Organización del Servicio de Estadística Nacional."

Perú-*2200*.C5, 1940.1. *Censo Nacional de Población y Ocupación, 1940.*

Perú-*3040*.E5, Jan. '45, b. "Legislación Estadística."

Perú-*4000*.D5, 1821-1946. *Demarcación Política del Perú: Recopilación de Leyes y Decretos (1821-1946).*

RepDom-*2100*.C5, 1935.2. *Censo de 1935: Ley del Congreso Nacional que Dispone las Condiciones en que Deben Desarrollarse las Operaciones del Censo.*

RepDom-*6900*.E5, 1920. *Executive Order No. 332: . . . Tariff Law on Imports and Exports . . . Effective on and after January 1, 1920.*

RepDom-*6900*.O5, 1920. "Orden Ejecutiva No. 332 (de Septiembre 25, 1919) que Promulga la Ley sobre Aranceles de Importación y Exportación que Estará en Vigor desde el Día 1° de Enero de 1920."

USA-*5250*.B5, 1942. *Birth Certificates, a Digest of the Laws and Regulations of the Various States.*

Ur-*3030*.E4, 1914. *Estadística General de la República, Ley de Reorganización.*

Ur-*6200*.L5, 1918. *Ley y Reglamentación de la Estadística Agrícola.*

Ven-*1900*.G5, Nov. 27, '44. "Ley de Estadística y de Censos Nacionales."

Ven-*6900*.L5, 1941. *Ley de Arancel de Aduanas.*

NonG-*3330*.Rodríguez, 1946. *Lecciones de Estadística.*

League of Nations.

Int-*1420*.M5. *Bulletin Mensuel de Statistique. Monthly Bulletin of Statistics.*

Int-*6000*.C4, 1943. *Catalogue of Selected Publications on Economic and Financial Subjects.*

Int-*6000*.I75, 1928.A. *International Con-*

Lista en Castellano e Inglés de los Artículos Importables desde los Estados Unidos de Norte América (con Números del "Schedule B" y Unidades). Chile-*3210*.L5, 1943.
Liste des Publications Éditées par l'Institut International d'Agriculture. Int-*6200*.L7, 1943.
Liste Minimum de Marchandises pour les Statistiques du Commerce International. Int-*3210*.L5, 1938.
LITERATURA—*véase* 7300.
LITERATURE—*see* 7300.
Live Stock and Animal Products Statistics. Can-*6230*.L5.
Lively, C. E.
USA-*5300*.R6, 1939. *Rural Migration in the United States.*
LIVESTOCK—see 6230.
LIVESTOCK CENSUS—*see* 2110; 2210; 6230.
LIVING:
CONDITIONS—*see* 9600.
COST—*see* 9500.
STANDARDS—*see* 9700.
"Living Costs in Brazil." USA-*6900*.F7, Aug. 3, '46, b.
LLANTAS—*véase* 6400.
LLUVIA (CAIDA DE)—*véase* 4200.
LOANS—*see* 6720.
LOCAL GOVERNMENT—*see* 8000-8999.
LOCURA—*véase* 7170.
Loi sur le Service de l'État Civil. Haiti-*5203*.L5, Dec. 22, '22.
LONGEVIDAD—*véase* 5220.
LONGEVITY—*see* 5220.
Lopez, Francisca.
Chile-*7100*.R7, Jan. '43, Moroder. "Mortalidad Infantil en Chile."
Lopez Muñoz, Luis.
Ec-*1420*.T5, May '45. "Labores de la Dirección General de Estadística y Censos."
Lorimer, Frank.
USA-*5100*.F5, 1940. *Foundations of American Population Policy.*
LOSSES, FINANCIAL—*see* 6721.
Lotka, Alfred J.
Int-*1300*.A5, 1940.3. "The Place of the Intrinsic Rate of Natural Increase in Population Analysis."
USA-*3040*.A5, Jan. '45, Dublin. "Trends in Longevity."
USA-*5212*.T5, 1911-35. *Twenty-Five Years of Health Progress.*
NonG-*5220*.L5, 1936. *Length of Life.*
Low-Cost Housing in Latin America. Int-*9900*.L5, 1943.
Loyo González, Gilberto.
Int-*1300*.A5, 1940.3. "Bases Mínimas para la Uniformidad de los Censos Nacionales de Población en el Continente Americano."
LUGAR DE NACIMIENTO—*véase* 5211.
LUMBER—*see* 6400.
Luna Olmedo, Agustín.
Méx-*6000*.T5, Oct.-Dec. '45. "Factores que Influyen en la Balanza de Pagos de México."
Luna Vegas, Ricardo.
Int-*2100*.M5, 1945. *Métodos de los Censos de Población de las Naciones Americanas.*
Int-*3040*.E5, Dec. '44. "La Estadística Peruana en el Sexenio 1938-44."
Int-*5000*.M3, 1946. *Papers Presented by Special Guests of the Population Association of America at Its Meetings October 25-26, 1946.*
Perú-*3040*.E5, July '46. "Apuntes sobre la Estadística en los Estados Unidos."
Perú-*4000*.D5, 1821-1946. *Demarcación Política del Perú: Recopilación de Leyes y Decretos (1821-1946).*
LUNACY—*see* 7170.
LUNATIC ASYLUMS—*see* 7130.
Lunazzi, José M.
Arg-*7200*.D5, 1940. *Deserción Escolar y Analfabetismo.*
Lundberg, George A.
NonG-*9000*.S5, 1942. *Social Research.*
Lutz, Friedrich A.
USA-*6721*.C5, 1914-43. *Corporate Cash Balances, 1941-1943.*

Macaulay, Frederick R.
USA-*6100*.I4, 1921-22. *Income in the United States: Its Amount and Distribution.*
MacGregor, Donald.
Can-*6100*.N4, 1926-36. *National Income, Canada, 1926-36.*
Machine Methods of Accounting. NonG-*3110*.M3.
"Machine Tabulation." *Int-3040*.E5, Mar. '46, Mahaney.
Machlup, Fritz.
NonG-*6100*.I5, 1943. *International Trade and the National Income Multiplier.*
MacLean, M. G.
Can-*5120*.A5, 1943. *The American-Born in Canada, a Statistical Interpretation.*
MADERA:
ASERRADA—*véase* 6400.
EN TRONCO—*véase* 6250.
INDUSTRIA DE LA—*véase* 6250; 6400.
Maderas de Exportación. RepDom-*6900*.M5, 1926-37.

MARKETING—cont.
In specific subject fields, *see* the particular field.
"Marketing and Sampling Uses of Population and Housing Data." USA-*3040*.J5, Mar. '43, Eckler.
Marquez, Javier.
Int-*6721*.I5, 1945. *Inversiones Internacionales en América Latina.*
MARRIAGE AND MARRIAGE RATES—*see* 5231.
"Marriage Rates." Int-*5000*.P5, July '43, c.
Marsh, Leonard Charles.
Can-*9950*.R5, 1943. *Report on Social Security for Canada.*
Marshall, Herbert.
Int-*3020*.B5, 1938. "The Canadian Census of Distribution."
Marshall, John T.
Can-*7110*.C5, Feb.'46, Cameron. "State of Health of the People of Canada in 1944."
Martin, E. M.
NonG-*6100*.S3. "The Correction of Wealth and Income Estimates for Price Changes."
Martín Díaz, Mariano.
Int-*1300*.A5, 1940.3, Martínez-Fortún. "Las Estadísticas Sanitarias en Cuba."
Martínez, Rafael H.
Int-*7110*.I4, 1943. *Informe Biodemográfico y Epidemiológico de las Américas. Biostatistical and Epidemiological Report on the Americas.*
Martínez-Fortún, Ortelio.
Int-*1300*.A5, 1940.3. "Las Estadísticas Sanitarias en Cuba."
Martins, Alberto.
Int-*1300*.A5, 1940.3. "A Sociedade Brasileira de Estatística."
MASAS, PSICOLOGIA DE LAS—*véase* 9000.
"The Master Sample of Agriculture: I, Development and Use." USA-*3040*.J5, Mar. '45, King.
"The Master Sample of Agriculture: II, Design." USA-*3040*.J5, Mar. '45, Jessen.
Masters, Ruth D.
Int-*1930*.H4, 1945. *Handbook of International Organizations in the Americas.*
MATEMATICA:
CIENCIA—*véase* 7300.
ESTADISTICA—*véase* 3140.
PROBABILIDAD—*véase* 3140.
MATERNAL ASSISTANCE—*see* 9951.
Mateus, George.
NonG-*3330*, 1926. *Principios Técnicos de Estadística.*
MATHEMATICAL:
PROBABILITY—*see* 3140.
SCIENCE—*see* 7300.
STATISTICS—*see* 3140.
Mathematical Association of America. NonG-*3140*.M5, 1927.
"Mathematical Operations with Punched Cards." USA-*3040*.J5, June '42, McPherson.
Mathematical Reviews. USA-*3040*.M5.
Mathematical Statistics. NonG-*3140*.M5, 1927.
Mather, K.
NonG-*3330*, 1943. *Statistical Analysis of Biology.*
MATRIMONIOS Y TASAS DE MATRIMONIOS—*véase* 5231.
Mayo Clinic.
USA-*3230*.M5, 1944. *Manual for Coding Causes of Illness According to a Diagnosis Code for Tabulating Morbidity Statistics.*
MAYORISTAS, COMERCIO Y PRECIOS—*véase* 6710.
Mazza, Sigfrido C.
NonG-*3140*.A8, 1946. *Algunos Problemas de Inferencia Estadística.*
NonG-*3150*.S8, 1942. *Sobre la Teoría de las Muestras.*
NonG-*3330*.1, 1942. *Apuntes de Estadística.*
NonG-*3330*.2, 1940. *Cuestiones de Matemáticas Referentes a la Estadística. Fascículo I.*
McClure, Wallace.
Int-*6000*.W8, 1933. *World Prosperity as Sought through the Economic Work of the League of Nations.*
McCoy, H. B.
USA-*3000*.G7, 1946. *Government Statistics for Business Use.*
McKinley, Earl Baldwin.
Int-*7150*.G5 1935. *A Geography of Diseases.*
McPeak, William.
USA-*3040*.J5, June '45, b. "Management of Field Work and Collection of Statistics: A Round Table."
McPherson, J. C.
USA-*3040*.J5, June '42. "Mathematical Operations with Punched Cards."
Meade, J. E.
NonG-*6100*.N4, 1944. *National Income and Expenditure.*
Means, G. C.
NonG-*6100*.S3. "Problems in Estimating National Income Arising from Production by Government."
MEASUREMENT OF GROWTH—*see* 7180.
MEASUREMENTS, ANTHROPOMETRIC—*see* 7310.
MEASURES AND WEIGHTS—*see* 3270.

Memoria de la Caja Nacional de Jubilaciones y Pensiones Civiles. Arg-*9950*.M5.

Memoria de la Compañía Administradora del Guano. Perú-*6220*.M5.

Memoria de la Compañía Salvadoreña de Café, S.A. ElSal-*6210*.M5.

Memoria de la Contaduría de la Nación. Arg-*8800*.M4.

Memoria de la Contraloría General de la República y Balance General de la Hacienda Pública. Chile-*8800*.M5.

Memoria de la Dirección General de Correos y Telecomunicaciones. Arg-*6620*.M5.

Memoria de la Dirección General de Estadística. Par-*1410*.A5.

Memoria de la Dirección General de Tránsito. Perú-*6610*.M5.

Memoria de la Sección Estadística del Ministerio de Salud Pública. Par-*7110*.M5.

Memoria de la Secretaría de Agricultura. Guat-*6200*.M5.

Memoria de la Secretaría de Agricultura y Fomento. Méx-*6200*.M5.

Memoria de la Secretaría de Educación Pública. Guat-*7200*.M5.

Memoria de la Secretaría de Gobernación. Méx-*8000*.M5.

Memoria de la Secretaría de Hacienda y Comercio. CR-*8800*.M5.

Memoria de la Secretaría de la Economía Nacional. Méx-*6000*.M5.

Memoria de la Secretaría de Salubridad Pública y Protección Social. CR-*7110*.M5.

Memoria de la Sociedad Nacional Agraria. Perú-*6200*.M5.

Memoria de la Superintendencia de Bancos y Estadística Bancaria de Seguros y Capitalización. Perú-*6720*.M4.

Memoria de Labores (Secretaría de Trabajo y Previsión). Méx-*9000*.M5.

Memoria de las Labores del Ejecutivo en el Ramo de Hacienda y Crédito Público. Guat-*1410*.M5.

Memoria de las Labores del Poder Ejecutivo en el Ramo de Agricultura. Guat-*6200*.M5.

Memoria de las Labores del Poder Ejecutivo en el Ramo de Educación Pública. Guat-*7200*.M5.

Memoria de las Labores del Poder Ejecutivo en el Ramo de Hacienda y Crédito Público. Guat-*1410*.M5.

Memoria de las Labores Realizadas por el Hospital General y Sus Dependencias. Guat-*7130*.M5.

Memoria de los Censos Generales de Población, Agrícola, Ganadero e Industrial de 1930. Méx-*2200*.C5, 1930.1.

Memoria de los Ferrocarriles en Explotación. Arg-*6614*.M5.

Memoria del Banco Agrícola del Perú. Perú-*6720*.M6.

Memoria del Banco Central de Reserva de *El Salvador.* ElSal-*6720*.M5.

Memoria del Banco Central de Reserva del Perú. Perú-*6720*.M7.

Memoria del Banco Industrial del Perú. Perú-*6720*.M8.

Memoria del Contralor de Exportaciones e Importaciones. UR-*6900*.M5.

Memoria del Departamento de Hacienda. Arg-*8800*.M6.

Memoria del Departamento de Salubridad y Asistencia Pública. Méx-*7110*.M5.

Memoria del Directorio del Banco de la República Oriental del Uruguay. Ur-*6720*.M5.

Memoria del Instituto Municipal de la Nutrición. Arg-*9800*.M5.

"Memoria del Instituto Nacional de Higiene." *Informe que el Señor Director General de Sanidad Presenta* . . . . Ec-*7110*.I5, 1941-42.

Memoria del Ministerio de Agricultura y Trabajo. Nic-*6200*.M5.

Memoria del Ministerio de Comunicaciones y Obras Públicas. Guat-*6000*.M5.

Memoria del Ministerio de Economía. Ec-*6000*.M5.

Memoria del Ministerio de Fomento. Ven-*6000*.M3.

Memoria del Ministerio de Fomento y Obras Públicas. Perú-*6000*.M5.

Memoria del Ministerio de Justicia e Instrucción Pública. Arg-*7200*.M5.

Memoria del Ministerio de la Economía Nacional. Col-*6000*.M5.

Memoria del Ministerio de Salubridad y Obras Públicas. Pan-*7110*.M5.

Memoria del Ministerio de Salud Pública. Ur-*7110*.M5.

Memoria del Ministerio del Trabajo y de Comunicaciones. Ven-*9000*.M5.

Memoria (del Ministro de Previsión Social). Ec-*9000*.I5.

Memoria del Ramo de Hacienda, Crédito Público y Comercio. Hond-*8800*.I5.

Memoria del Recaudador General de Aduanas y Alta Comisión. Nic-*6900*.M5.

Memoria del Servicio Nacional de Salubridad. Chile-*7110*.M5.

Memoria Oficial de la Comisión Nacional de Transportes. Cuba-*6610*.M5.

Memoria, Primera Reunión de Técnicos sobre Problemas de Banca Central del Continente Americano. Int-*6720*.R5, 1946.

Memoria que el Ministro de Educación Nacional Presenta al Congreso Nacional. Ven-*7200*.M5.

"Metodología en Estadísticas Agrícolas de las Américas: Parte III, El Servicio de Estadística Agrícola en los Estados Unidos; Parte IV, Conclusiones." Int-*3040*.E5, Sept. '46, Montoya.
"Metodología Estadística Aplicada a las Ciencias Sociales y Biología Social." Int-1300.A5, 1940.3, Roa Reyes.
Métodos de Encuestas sobre las Condiciones de Vida de las Familias. Int-*9700*.M7, 1942.
Métodos de los Censos de Población de las Naciones Americanas. Int-*2100*.M5, 1945.
METODOS DE LA MUESTRA:
EN LOS CENSOS—*véase* 2100-2199.
EN LA ESTADISTICA—*véase* 3150.
Métodos Estadísticos. NonG-*3330*.Moncetz, 1941.
Métodos Estadísticos Aplicados a la Economía y los Negocios. NonG-*3330*.Mills, 1940, Sp.
"Métodos para Determinar el Grado de Mortalidad Infantil." Int-*1300*.A5, 1940.3, Bunge, a.
METODOS, PROCEDIMIENTOS Y TECNICAS:
ESPECIFICOS—*véase* las respectivas clasificaciones.
DE CENSOS—*véase* 2100-2199.
DE ESTADISTICA—*véase* 3100-3199; *véase también* las respectivas clasificaciones.
"Métodos Usados en los Registros del Estado Civil de las Personas y Estadística Vital." Nic-*1420*.B5, May 15, '45.
Metropolitan Life Insurance Company.
USA-*5212*.T5, 1911-35. *Twenty-Five Years of Health Progress.*
USA-*6730*.S5. *Statistical Bulletin of the Metropolitan Life Insurance Company.*
The Mexican Economy. Méx-*6000*.M7, 1942.
Mexican War Workers in the United States. USA-*9600*.M3, 1945.
México. Banco de la República—*see - véase* Banco de la República, México.
——. Banco de México, S.A.—*see - véase* Banco de México, S.A.
——. Comisión de Coordinación Económica Nacional—*see - véase* México. Secretaría de la Economía Nacional. Comisión de Coordinación Económica Nacional.
——. Departamento de la Estadística Nacional—*see - véase* México. Secretaría de la Economía Nacional. Dirección General de Estadística.
——. Departamento de los Censos—*see - véase* México. Secretaría de la Economía Nacional. Dirección General de Estadística. Departamento de los Censos.
——. Departamento de Salubridad Pública—*see - véase* México. Secretaría de Salubridad y Asistencia Pública.
——. Dirección de Economía Rural—*see - véase* México. Secretaría de Agricultura y Fomento. Dirección de Economía Rural.
——. Dirección General de Comercio Exterior—*see - véase* México. Secretaría de Relaciones Exteriores. Dirección General de Comercio Exterior.
——. Dirección General de Economía Rural—*see - véase* México. Secretaría de Agricultura y Fomento. Dirección de Economía Rural.
——. Dirección General de Educación Higiénica—*see - véase* México. Secretaría de Salubridad y Asistencia Pública. Dirección General de Educación Higiénica.
——. Dirección General de Estadística—*see - véase* México. Secretaría de la Economía Nacional. Dirección General de Estadística.
——. Dirección General de Población Rural, Tierras Nacionales y Colonización—*see - véase* México. Secretaría de Agricultura y Fomento. Dirección General de Población Rural, Tierras Nacionales y Colonización.
——. Dirección Nacional de Estadística—*see - véase* México. Secretaría de la Economía Nacional. Dirección General de Estadística.
——. Instituto de Investigaciones Estadísticas—*see - véase* Instituto de Investigaciones Estadísticas, México.
——. Instituto de Salubridad y Enfermedades Tropicales—*see - véase* Instituto de Salubridad y Enfermedades Tropicales, México.
——. Instituto Mexicano del Seguro Social.
USA-*9950*.S4, July '46, Cohen. "The First Two Years of Social Insurance in Mexico." *Social Security Bulletin.*
——. ——. Oficina de Estadística.
Int-*3040*.E5, Sept. '45, Rodríguez y Rodríguez. "La Oficina de Estadística del Instituto Mexicano del Seguro Social."
——. Oficina de Demografía y Estadística Sanitaria—*see - véase* México. Secretaría de Salubridad y Asistencia Pública. Departamento de Estadística y Demografía.
——. Oficina de Estadística Social—*see - véase* México. Secretaría de la Economía Nacional. Dirección General de Estadística. Oficina de Estadística Social.
——. Oficina del Censo Industrial—*see - véase* México. Secretaría de la Economía Nacional. Dirección General de Estadística. Oficina del Censo Industrial.
——. Secretaría de Agricultura y Fomento.
Méx-*6200*.M5. *Memoria de la . . .*

México—cont.

——. ——. Dirección de Economía Rural.
Méx-*6200*.B5. *Boletín Mensual de la Dirección General de Economía Rural.*
Méx-*6200*.B5, Jan. '45. "Indices Económicos."
Méx-*6200*.B5, Jan. '43. "Historia y Evolución de las Estadísticas Agrícolas Continuas, durante el Período 1925-42."
Méx-*6200*.P5, 1945. *Programa e Instructivo para el Servicio de Estadística Agrícola en las Agencias Generales.*
Méx-*6220*.I5, 1942. *Instrucciones Generales para los Corresponsales Estadísticos.*

——. ——. Dirección General de Población Rural, Tierras Nacionales y Colonización.
Méx-*5300*.D5, 1923-41. *Datos Estadísticos del Departamento de Terrenos Nacionales.*

——. Secretaría de Gobernación.
Méx-*8000*.M5. *Memoria de la . . .*

——. Secretaría de Hacienda y Crédito Público.
Méx-*8800*.B5, 1943. *Bibliografía de la Secretaría de Hacienda y Crédito Público.*

——. ——. Departamento de Biblioteca y Archivos Económicos.
Méx-*6900*.B5. *Boletín de Aduanas.*

——. ——. Dirección Nacional de Estadística—*see - véase* México. Secretaría de la Economía Nacional. Dirección General de Estadística.

——. Secretaría de la Economía Nacional.
Méx-*1200*.P5, 1933-42. *Publicaciones Oficiales, 1933-1942.*
Méx-*6000*.M5. *Memoria de la . . .*

——. ——. Comisión de Coordinación Económica Nacional.
Méx-*6000*.P5. *Planificación Económica.*

——. ——. Dirección General de Estadística.
Méx-*1400*.M5, 1938. *México en Cifras.*
Méx-*1410*.A5. *Anuario Estadístico de los Estados Unidos Mexicanos.*
Méx-*1410*.C5. *Compendio Estadístico.*
Méx-*1420*.R5. *Revista de Estadística.*
Méx-*2100*.C5, 1940.1. *Censo de 1940: Métodos y Procedimientos.*
Méx-*2100*.C5, 1940.3. *Instrucciones para la Perforación de Tarjetas del 6° Censo de Población de 1940.*
Méx-*2110*.C5, 1940.1. *Segundo Censo Agrícola-Ganadero, Marzo de 1940: Métodos y Procedimientos.*
Méx-*2110*.C7, 1940.1. *Segundo Censo Ejidal de los Estados Unidos Mexicanos, 6 de Marzo de 1940: Métodos y Procedimientos.*
Méx-*2120*.C5, 1939.1. *Segundo Censo de Edificios de los Estados Unidos Mexicanos, 20 de Octubre de 1939: Métodos y Procedimientos.*
Méx-*2150*.C3, 1940.1. *Primer Censo Comercial, Marzo de 1940: Métodos y Procedimientos.*
Méx-*2150*.C5, 1945.1. *Los Censos Cuarto Industrial, Segundo Comercial y Segundo de Transportes, 1945: Métodos y Procedimientos.*
Méx-*2150*.C5, 1940.1. *Tercer Censo Industrial de los Estados Unidos Mexicanos, 1940: Métodos y Procedimientos.*
Méx-*2180*.C5, 1940.1. *Primer Censo de Transportes, Febrero y Marzo de 1940: Métodos y Procedimientos.*
Méx-*2200*.C5, 1940.1. *Estados Unidos Mexicanos, Sexto Censo de Población, 1940.*
Méx-*2200*.C5, 1940.2. *Sexto Censo General de Población de los Estados Unidos Mexicanos, 6 de Marzo de 1940. Población Municipal.*
Méx-*2200*.C5, 1930.1. *Memoria de los Censos Generales de Población, Agrícola, Ganadero e Industrial de 1930.*
Méx-*2200*.C5, 1930.2. *Quinto Censo de Población, 15 de Mayo de 1930.*
Méx-*2210*.C7, 1940. *Segundo Censo Ejidal de los Estados Unidos Mexicanos, 6 de Marzo de 1940.*
Méx-*2210*.C7, 1935. *Primer Censo Ejidal, 1935.*
Méx-*2220*.C5, 1939.1. *Segundo Censo de Edificios de los Estados Unidos Mexicanos: Resumen General.*
Méx-*2220*.C5, 1939.2. *Segundo Censo de Edificios, 20 de Octubre de 1939: Datos Definitivos.*
Méx-*2250*.C5, 1940. *Tercer Censo Industrial de los Estados Unidos Mexicanos, 1940.*
Méx-*2250*.C5, 1940.3. *Características Principales de la Industria: Censo de 1940.*
Méx-*2250*.C5, 1935. *Segundo Censo Industrial de 1935.*
Méx-*2250*.C5, 1930. *Primer Censo Industrial de 1930.*
Méx-*3010*.B5, 1942. *Bibliografía Mexicana de Estadística.*
Méx-*3030*.I5, 1941. *Informes sobre las Principales Estadísticas Mexicanas.*
Méx-*3030*.L5, 1941. *Ley Federal de Estadística y Su Reglamento.*
Méx-*3220*.T5, 1940. *Tercer Censo Industrial: Catálogo General de las Industrias de Transformación y Extractivas, 1940.*

México—cont.

Méx-*3230*.S5, 1931. *Sinonimias Populares Mexicanas de las Enfermedades.*
Méx-*3240*.N5, 1940. *Nomenclatura Nacional de Ocupaciones, 1940.*
Méx-*5212*.M5, 1922-39. *Mortalidad en México.*
Méx-*6900*.A5. *Anuario Estadístico del Comercio Exterior de los Estados Unidos Mexicanos.*
NonG-*3330*.Gini 1. *Apuntes de Estadística Metodológica.*

——. ——. ——. Departamento de los Censos.
Méx-*2100*.C5, 1940.2. *Decreto y Bases de Organización de los Siete Censos Nacionales y de los Padrones Estadísticos que Se Levantarán en 1939-40.*

——. ——. ——. Oficina de Estadística Social.
Méx-*5100*.P5, 1921-38. *Población del País Calculada al 30 de Junio de Cada Año del Período 1921-1938.*

——. ——. ——. Oficina del Censo Industrial.
Méx-*3220*.C5, 1945. *Catálogo de Industrias, 1945.*

——. Secretaría de Relaciones Exteriores. Dirección General de Comercio Exterior.
Méx-*6900*.R5. *Revista del Comercio Exterior.*

——. Secretaría de Salubridad y Asistencia Pública.
Méx-*7110*.B7. *Salubridad y Asistencia.*
Méx-*7110*.M5. *Memoria del Departamento de Salubridad y Asistencia Pública.*

——. ——. Departamento de Estadística y Demografía.
Méx-*5000*.I5, 1942. *Instructivo para la Elaboración de la Estadística Demográfica en las Oficinas Sanitarias de los Estados y Territorios Federales.*
Méx-*7110*.B6. *Boletín de la Oficina de Demografía y Estadística Sanitaria.*

——. ——. Dirección General de Educación Higiénica.
Méx-*7110*.B5, 1943. *Bibliografía que el Departamento de Salubridad Pública Presenta en la Feria del Libro y Exposición Nacional de Periodismo.*

——. Secretaría del Trabajo y Previsión Social.
Méx-*9000*.A5. *Anuario de Estadísticas del Trabajo.*
Méx-*9000*.M5. *Memoria de Labores.*
Méx-*9000*.T5. *Trabajo y Previsión Social.*
Méx-*9200*.E5, 1942. *Estadística de Salarios.*

——. Sociedad Agronómica de México—*see* - *véase* Sociedad Agronómica de México.
——. Sociedad Mexicana de Geografía y Estadística—*see* - *véase* Sociedad Mexicana de Geografía y Estadística.
México Agrario. Méx-*6200*.M8.
Mexico: Demographic Data. Méx-*5000*.M5, 1943.
México en Cifras. Méx-*1400*.M5, 1938.
Mexico: Reports Presented to the Conference of Comissions of Inter-American Development. Méx-*6000*.R4.
"Mexico's Balance of Payments." Int-*6000*. C7, July-Aug. '44.
Meyer, H. J.
USA-*5120*.J5, 1943. *Jewish Population Studies.*
Míchalup, Eric H.
Int-*3040*.E5, Dec. '45. "La Composición Familiar de Asalariados en el Distrito Federal de Venezuela."
MICROCOPIA—*véase* 1600.
MICROCOPY—*see* 1600.
MICROFILM—*see* 1600.
MICROSCOPIA (MEDICINA DE LABORATORIO)—*véase* 7180.
MICROSCOPY (LABORATORY MEDICINE)—*see* 7180.
MIDWIFERY—*see* 5211.
Migone, Raúl C.
Int-*1410*.A7. *Anuario Estadístico Interamericano. Inter-American Statistical Yearbook. Anuário Estatístico Interamericano. Annuaire Statistique Interaméricain.*
MIGRACION—*véase* 5300-5399.
MIGRATION—*see* 5300-5399.
1944 Arancel Aduanero. Chile-*6900*.A5, 1944.
Milbank Memorial Fund. USA-*5000*.M5.
Milbank Memorial Fund Quarterly. USA-*5000*.M5.
MILK AND MILK PRODUCTS—*see* 6230.
"The Millionth Map of Hispanic America." USA-*4000*.G5. Apr. '37, Platt.
Mills, Frederick Cecil.
NonG-*3330*. 1940, Eng. *Statistical Methods Applied to Economics and Business.*
NonG-*3330*. 1940, Sp. *Métodos Estadísticos Aplicados a la Economía y los Negocios.*
Mineral Production of Canada. Can-*6300*. A5.
"Mineral Statistics." *Government Statistics for Business Use.* USA-*3000*.G7, 1946.
MINERALES—*véase* 6300.
MINERALS—*see* 6300.
Minerals Yearbook. USA-*6300*.M5.

MINERIA—*véase* 6300.

Minería. Chile-*6300*.M5.

Minería e Industria. Chile-*6300*.M5.

Minería, Transportes y Comunicaciones. Bol-*6000*.M5.

La Minería y el Petróleo en el Ecuador: Anuario. Ec-*6300*.M5.

Minería y Riqueza Minera de México. Méx-*6300*.M5, 1944.

Minimum List of Commodities for International Trade Statistics. Int-*3210*.M5, 1938.

MINIMUM LIST OF COMMODITIES IN INTERNATIONAL TRADE—*see* 3210; 6900.

MINING—*see* 6300.

Mining and Manufacturing Industries in Argentina. Arg-*6400*.M7, 1945.

Mining and Manufacturing Industries in Bolivia. Bol-*6400*.M5, 1945.

Mining and Manufacturing Industries in Brazil. Bras-*6400*.M5, 1945.

Mining and Manufacturing Industries in Chile. Chile-*6400*.M5, 1945.

Mining and Manufacturing Industries in Colombia. Col-*6400*.M5, 1945.

Mining and Manufacturing Industries in Costa Rica. CR-*6400*.M5, 1945.

Mining and Manufacturing Industries in Ecuador. Ec-*6400*.M5, 1945.

Mining and Manufacturing Industries in Haiti. Haiti-*6400*.M5, 1945.

Mining and Manufacturing Industries in Mexico. Méx-*6400*.M3, 1946.

Mining and Manufacturing Industries in Nicaragua. Nic-*6400*.M5, 1945.

Mining and Manufacturing Industries in Panama. Pan-*6400*.M5, 1945.

Mining and Manufacturing Industries in Paraguay. Par-*6400*.M5, 1945.

Mining and Manufacturing Industries in Peru. Perú-*6400*.M5, 1945.

Mining and Manufacturing Industries in the American Republics. Int-*6400*.M5.

Mining and Manufacturing Industries in Uruguay. Ur-*6400*.M5, 1945.

Mining and Manufacturing Industries in Venezuela. Ven-*6400*.M5, 1945.

The Mining Industries, 1899-1939: A Study of Output, Employment and Productivity. USA-*6300*.M7, 1899-1939.

Minogue, Adelaide E.
- *The Repair and Preservation of Records.* NonG-*1600*.R5, 1943.

MINORIAS, RECOLONIZACION DE GRUPOS DE—*véase* 5300.

MINORISTA, COMERCIO—*véase* 6710.

MINORISTAS, PRECIOS:
- GENERAL—*véase* 9500.
- DE MERCANCIAS ESPECIFICAS—*véase* las respectivas clasificaciones.

Minutes of the Special Meeting of the Governing Board of the Pan American Union Held on August 29, 1945. Int-*8200*.M3, 1945.

Mitchell, Westley C.
- USA-*6100*.I4, 1921-22. *Income in the United States: Its Amount and Distribution.*
- NonG-*3120*.M5, 1938. *The Making and Using of Index Numbers.*
- NonG-*6000*.B6, 1928. *Business Cycles, The Problem and Its Setting.*

MOB PSYCHOLOGY—*see* 9000.

"Modelos para Registrar las Cuentas Internacionales." Ec-*1420*.T5, July-Dec. '45, c.

Modely, Rudolf.
- NonG-3260.H6, 1937. *How to Use Pictorial Statistics.*

Molina Serrano, Francisco.
- Arg-*5100*.E5, 1946. *Ensayo sobre la Evolución de la Población de las Américas (Estimación en el Lapso 1938-1973): II, Argentina.*
- Can-*5100*.E7, 1946. *Ensayo sobre la Evolución de la Población de las Américas: I, Canadá.*

Molinari, Alessandro.
- Int-*3020*.B5, 1938. "Deuxième Rapport sur les Statistiques de la Distribution."
- Int-*3020*.B5, 1938, a. "Statistique du Tourisme International."
- Int-*3040*.R5, 1936.2. "Observations et Propositions pour une Statistique Internationale du Tourisme."

Moment, Samuel.
- USA-*5120*.J5, 1943. *Jewish Population Studies.*

Moncetz, A. de.
- NonG-*3330*, 1941. *Métodos Estadísticos.*

MONEDA:

MONEY—*see* 6720.

Money and Banking, 1942-44. Int-*6720*.M6, 1942-44.

"Money, Credit, and Banking." USA-*3000*.G7, 1946.

Monitor de Sociedades Anónimas. Arg-*6700*.M7.

"Monografía de la República de Nicaragua." Nic-*1420*.B5, Apr.-June '46, a.

Monografías Estadísticas de la Dirección General de Estadística. RepDom-*3000*.M5.

Monografías Industriales del Banco de México, S.A. Méx-*6400*.M5.

Monograph, Republic of Haiti, Compiled 1932. Haiti-*1400*.M5, 1932.

Monroe, Day.
- USA-*9700*.F5F1, 1941. *Family Food Con-*

Monroe, Day—cont.
sumption and Dietary Levels, Five Regions, Farm Series.
USA-*9700*.F511, 1941. *Family Income and Expenditures, Five Regions, Part 2, Family Expenditures, Farm Series.*
USA-*9700*.F512, 1940. *Family Income and Expenditures, Five Regions, Part 2, Family Expenditures, Urban and Village Series.*

Montesino Samperio, José V.
NonG-*7300*.S5, 1946. *Sobre la Cuantificación del Estilo Literario.*
RepDom-*3320*.P5, 1940. *Programa de un Curso de Estadística Metodológica (Introducción a la Estadística Matemática) Seguido de un Curso de Matemáticas Generales Necesarias para Su Estudio.*

Monthly Bulletin. (National Bank of the Republic of Haiti. Fiscal Department). Haiti-*1420*.M5.

Monthly Bulletin of Agricultural Economics and Sociology. Int-*6200*.M4.

Monthly Bulletin of Agricultural Statistics. Can-*6200*.Q5.

Monthly Bulletin of Statistics (United Nations). Int-*1420*.M5.

Monthly Bulletin of the Fiscal Department (National Bank of the Republic of Haiti). Haiti-*1420*.M5.

Monthly Crop Report and Agricultural Statistics. Int-*6220*.M5.

Monthly Epidemiological Report. Int-*7110*.I6.

Monthly Index of Retail Sales. Can-*6710*.M5.

Monthly Index of Wholesale Sales. Can-*6710*.M6.

Monthly Labor Review. USA-*9000*.M5.

Monthly Report, Biostatistics, Epidemiology. Int-*7110*.I6.

Monthly Report of Employment, Executive Branch of the Federal Government. USA-*9200*.M7.

Monthly Review of Business Statistics. Can-*6000*.M5.

The Monthly Review of the Railroad Retirement Board. USA-*9952*.M5.

Monthly Summary of Foreign Commerce of the United States. USA-*6900*.M5.

Monthly Vital Statistics Bulletin. USA-*5210*.M5.

Montoya, Hernán.
Int-*3040*.E5, Sept. '46. "Metodología en Estadísticas Agrícolas de las Américas: Parte III, El Servicio de Estadística Agrícola en los Estados Unidos; Parte IV, Conclusiones."
Int-*3040*.E5, June '46. "Metodología en Estadísticas Agrícolas de las Américas: Parte II, Sistemas Estadísticos Agrícolas de las Américas: Parte II, Sistemas Estadísticos Agrícolas en Algunos Países Latinoamericanos."
Int-*3040*.E5, Mar. '46. "Metodología en Estadísticas Agrícolas de las Américas."
Col-*1420*.A5, Oct. 20, '43. "Plan General para el Levantamiento del Primer Censo Agropecuario."

Montoya Canal, Aníbal.
Col-*1420*.A5, Feb. '45. "Estadística de la Delincuencia en Colombia."

M. O. P. (Ministerio de Obras Públicas y Comunicaciones). Par-*8000*.M5.

Morais, O. Alexander de.
Bras-*3040*.R5, Oct.-Dec. '43. " 'Números Indices', Suas Aplicações na Estatística Brasileira e Sugestões Atinentes à Sua Padronização Continental."

MORBILIDAD:
GENERAL—*véase* 7150.
CLASIFICACION Y CODIGOS DE—*véase* 7150; 3230.
INDUSTRIAL—*véase* 9400; 7150.

MORBIDITY:
GENERAL—*see* 7150.
CLASSIFICATION AND CODES—*see* 7150; 3230.
INDUSTRIAL—*see* 9400; 5212.

Morbidity: Instructions for Coding and Punching. USA-*7150*.M6.

Moreno Pérez, Ignacio.
Col-*3230*.N5, 1942. *Nomenclatura de Morbilidad y Mortalidad.*

Moreno y García, Roberto.
Int-*7200*.A5, 1941. *Analfabetismo y Cultura Popular en América.*

Morgan, Katherine Lenore.
Int-*7300*.J5, 1944. *Journals Dealing with the Natural, Physical and Mathematical Sciences Published in Latin America.*
Chile-*7210*.U5, 1944. *The Universities of Chile.*
Col-*7210*.U5, 1945. *The Universities of Colombia.*

Morgenstern, Oskar.
NonG-*6000*.T6, 1944. *Theory of Games and Economic Behavior.*

Moriyama, Iwao M.
Int-*3040*.E5, Sept. '45. "A Method for Indexing Statistical Tabulations."
USA-*7160*.U5, 1943. *Uniform Definitions of Motor Vehicle Accidents.*

Moroder, Juan.
Chile-*7100*.R7, Jan. '43. "Mortalidad Infantil en Chile."

MORTALIDAD:
GENERAL—*véase* 5212.

Muller, Hans J.
Int-*3000*.S5, 1940. "Las Actividades Estadísticas de Panamá."
Müller, Traugott.
International Crop-Reporting Service. Int-*6220*.I5, 1910.
MULTIGRAFO—*véase* 1800.
MULTIGRAPHING—*see* 1800.
MULTILITHING—*see* 1800.
MULTILITOGRAFIA—*véase* 1800.
Multiple Human Births. NonG-*5211*.M5, 1940.
MUNICIONES—*véase* 6400.
MUNICIPAL FINANCE—*see* 8800.
MUNICIPAL GOVERNMENT—*see* 8000-8999.
Municipal Year Book. USA-*8000*.M5.
MUNICIPIOS:
ADMINISTRACION DE—*véase* 8000-8999.
FINANZAS DE—*véase* 8800.
MUNITIONS—*see* 6400.
MURDER—*see* 5212; 8300.
MUSEOS—*véase* 7200.
MUSEUMS—*see* 7200.
MUSIC—*see* 7300.
MUSICA—*véase* 7300.
Myrdal, Gunnar.
Population, a Problem for Democracy. NonG-*5100*.P5, 1940.

NACIMIENTOS:
GENERAL—*véase* 5211.
LUGARES DE—*véase* 5211.
REGISTRO DE—*véase* 5250.
TASAS DE NATALIDAD—*véase* 5211.
TASAS DE REPRODUCCION—*véase* 5211.
NACIONALES:
DEUDAS—*véase* 8800.
FINANZAS—*véase* 8800.
GOBIERNOS—*véase* 8000-8999.
PLANIFICACIONES—*véase* 8500.
PRESUPUESTOS—*véase* 8800.
RECURSOS—*véase* 6100.
RENTAS—*véase* 6100.
RIQUEZAS—*véase* 6100.
NACIONALIDAD—*véase* 5120.
NACIONES FAVORECIDAS—*véase* 6900
Narancio, Edmundo M.
Ur-*5100*.H5, 1939. *Historia y Análisis Estadístico de la Población del Uruguay.*
NATALIDAD:
CONTROL DE—*véase* 5211.
TASAS DE—*véase* 5211.
"Natality in Venezuela." Int-*5000*.M3, 1946.
Nathan, Robert R.
USA-*6100*.N5, 1929-37. *National Income in the United States, 1929-1937.* NonG-*6100*.S3. "Problems in Estimating National Income Arising from Production by Government."
NonG-*6100*.S3. "Some Problems Involved in Allocating Incomes by States."
NATIONAL:
BUDGET—*see* 8800.
DEBT—*see* 8800.
FINANCE—*see* 8800.
GOVERNMENT—*see* 8000-8999.
INCOME—*see* 6100.
PARK LANDS—*see* 4300; 6800.
PLANNING—*see* 8500.
ROSTERS OF SCIENTIFIC PERSONNEL FOR EMPLOYMENT PURPOSES—*see* 1700.
WEALTH—*see* 6100.
National Bank of the Republic of Haiti—*see* - *véase* Banque Nationale de la Republique d'Haïti.
National Bureau of Economic Research, United States.
Int-*5300*.I6, 1931. *International Migrations.*
USA-*6000*.A6. *Annual Report of the . . .*
USA-*6000*.O4. *Occasional Papers of the . . .*
USA-*6000*.P5. *Publications of the . . .*
USA-*6100*.I4, 1921-22. *Income in the United States: Its Amount and Distribution.*
USA-*6100*.N2, 1946. *National Income: A Summary of Findings.*
USA-*6100*.N3, 1919-35. *National Income and Capital Formation, 1919-1935.*
USA-*6100*.N4, 1941. *National Income and Its Composition, 1919-1938.*
USA-*6100*.N7, 1945. *National Product in Wartime.*
USA-*6200*.A7, 1942. *American Agriculture, 1899-1939: A Study of Output, Employment, and Productivity.*
USA-*6300*.M7, 1899-1939. *The Mining Industries, 1899-1939: A Study of Output, Employment, and Productivity.*
USA-*6400*.O6, 1940. *The Output of Manufacturing Industries, 1899-1937.*
USA-*6721*.C5, 1914-43. *Corporate Cash Balances, 1914-1943.*
USA-*8500*.A5, 1943. *Aims, Organization, and Publications of the . . .*
USA-*9200*.E5, 1899-1939. *Employment in Manufacturing, 1899-1939, an Analysis of Its Relation to the Volume of Production.*
NonG-*6000*.B6, 1928. *Business Cycles, The Problem and Its Setting.*

NATURAL RESOURCES—*see* 6200-6299; 6300.
"Naturaleza y Alcances de la Llamada Balanza Internacional de Pagos." Perú-*3040*.E5, July '46, Barreto.
"Naturaleza y Alcances de la Llamada Balanza Internacional de Pagos." Perú-*6720*.B5, July '45, Barreto.
NATURALIZACION—*véase* 5300.
NATURALIZATION—*see* 5300.
Navarro, Hubert E.
Int-*7110*.B5, Sept. '46. "La Sanidad en Bolivia."
NAVEGACION (FLUVIAL Y MARITIMA) —*véase* 6615.
NAVIGATION—*see* 6615.
"Need and Plan for a Statistical Program in Furthering Conservation of Inter-American Fisheries." Int-*1300*.A5, 1940.3, Fiedler.
NEGOCIOS, GENERAL—*véase* 6700-6799.
NEGRO—*see - véase* 5120; 9600.
Negro Population, 1790-1915. USA-*5120*.N4, 1790-1915.
Negroes in the United States, 1920-32. USA-*5120*.N5, 1920-32.
NEGROS—*see* 5120; 9600.
The Negro's Share. USA-*9600*.N5, 1943.
Nehgme Rodríguez, Elías.
Chile-*1900*.E5, 1943. *La Economía Nacional y el Problema de las Subsistencias en Chile.*
Neiswanger, William Addison.
NonG-*3330*, 1943. *Elementary Statistical Methods.*
Nelson, Ernesto.
Arg-*7200*.A4, 1939. *El Analfabetismo en la República Argentina.*
Nelson, R. W.
NonG-*6100*.S3. "Allocation of Benefits from Government Expenditures."
NEO-NATAL DEATHS—*see* 5212.
Ness, Norman T.
Int-*1900*.L7, 1945. *Latin America in the Future World.*
The Network of World Trade. Int-*6900*.N5 1942.
Neumann, John von.
NonG-*6000*.T6, 1944. *Theory of Games and Economic Behavior.*
NEUMATICOS—*véase* 6400.
NEUROLOGIA—*véase* 7180.
NEUROLOGY—*see* 7180.
"A New Sample of the Population." Int-*3040*.E5, Dec. '44, Hansen.
The New World Guides to the Latin American Republics. Int-*1950*.N5, 1943.
New York Tuberculosis and Health Association.
USA-*7150*.T5. *Tuberculosis Reference Statistical Handbook.*
Newman, Horatio Hackett.
NonG-*5211*.M5, 1940. *Multiple Human Births.*
Nicaragua. Administración de Aduanas y Alta Comisión—*see - véase* Nicaragua. Ministerio de Hacienda y Crédito Público. Administración de Aduanas y Alta Comisión.
——. Banco Hipotecario de Nicaragua—*see - véase* Banco Hipotecario de Nicaragua.
——. Banco Nacional de Nicaragua—*see - véase* Banco Nacional de Nicaragua.
——. Dirección General de Estadística—*see - véase* Nicaragua. Ministerio de Hacienda y Crédito Público. Dirección General de Estadística.
——. Dirección General de Sanidad—*see - véase* Nicaragua. Ministerio de Higiene y Beneficencia Pública.
——. Leyes, Estatutos, Etc.
Nic-*3030*.L5, 1941. *Ley Orgánica de la Estadística Nacional.*
——. Ministerio de Agricultura y Trabajo.
Nic-*6200*.I5, 1945. *Informe sobre Labores Agrícolas Presentado por la Delegación de Nicaragua a la Tercera Conferencia Interamericana de Agricultura, en Caracas, Estados Unidos de Venezuela.*
Nic-*6200*.M5. *Memoria del . . . .*
Nic-*6230*.I5, 1945. *Informe sobre la Industria Pecuaria Presentado por la Delegación de Nicaragua a la 3a Conferencia Interamericana de Agricultura, en Caracas, EE. UU. de Venezuela.*
——. Ministerio de Fomento y Obras Públicas.
Nic-*8000*.M5. *Memoria de Fomento y Obras Públicas.*
——. Ministerio de Gobernación—*see - véase* Nicaragua. Ministerio de Hacienda y Crédito Público.
——. Ministerio de Hacienda y Crédito Público.
Nic-*8800*.M5. *Memoria de Hacienda y Crédito Público.*
——. ——. Administración de Aduanas y Alta Comisión.
Nic-*6900*.M5. *Memoria del Recaudador General de Aduanas y Alta Comisión.*
——. ——. Dirección General de Estadística.
Nic-*1410*.A5. *Anuario Estadístico de la República de Nicaragua.*
Nic-*1410*.G5. *Gráficos Estadísticos de Nicaragua.*

NIÑO—cont.

CRECIMIENTO DEL—*véase* 7180.

Nivel de Vida de la Familia Obrera. Arg-*9700*.N5, 1945.

Nixon, J. W.

Int-*3040*.R5, 1938.3. "On the Statistics Available Concerning the Occupied Population of the World and Its Distribution."

Int-*3040*.R5, 1936.1. "Committee on Family Budgets: Preliminary Report."

Int-*3040*.R5, 1933.1. "Index-Numbers of Wages: A Survey."

Nomenclador General de Areasy Lugares Habitados de Venezuela. Ven-*4100*.N5, 1941.

Nomenclador de Guatemala. Guat-*1950*.N5, 1937.

Nomenclatura de las Enfermedades. Chile-*3230*.N5, 1929.

Nomenclatura de Morbilidad y Mortalidad. Col-*3230*.N5, 1942.

Nomenclatura Internacional Abreviada de Causas de Morte. Bras-*3040*.R5, Jan.-Mar. '42, Freitas Filho.

"Nomenclatura Internacional de las Causas de Muerte." Int-*7110*.B5, June-Aug. '39.

Nomenclatura Internacional Detalhada de Causas de Morte. Bras-*3040*.R5, Jan.-Mar. '42, Freitas Filho.

Nomenclatura Internacional Intermediaria de Causas de Morte. Bras-*3040*.R5, Jan.-Mar. '42, Freitas Filho.

Nomenclatura Mínima de Industrias. Int-*3240*.E5, 1938.

Nomenclatura Nacional de Ocupaciones, 1940. Méx-*3240*.N5, 1940.

"Nomenclatura para Hospitales Dedicados a Enfermedades Mentales." Int-*7110*.B5. Aug. '36.

NOMENCLATURAS:

LISTA DE—*véase* NOMENCLATURES.

ESTADISTICAS, PARA CLASIFICACION—*véase* 3200-3299; *véase también* las respectivas clasificaciones.

Nomenclaturas Internacionales de las Enfermedades (Causas de Defunción—Causas de Incapacidad del Trabajo) para Establecer las Estadísticas Nosológicas (Clasificación "Bertillon"). Int-*3230*.N5, 1929.

Nomenclature Communes des Marchandises. Int-*3210*.C4, 1913.3.

Nomenclature Minimum des Industries. Int-*3240*.S6, 1938.

NOMENCLATURES

Int-*3010*.B5, Stat. 7, Aug. 10, '45. *International Statistical Standards.*

Int-*3200*.D5, 1946. *Banking Statistics, Recommendations on Scope and Principles of Classifications; Report of the Sub-Committee on Banking Statistics.*

Int-*3200*.S7, 1938. *Statistiques Relatives à la Formation des Capitaux.*

Int-*3200*.S8. *Studies and Reports on Statistical Methods. Etudes et Rapports sur les Méthodes Statistiques.*

Int-*3200*.T5, 1938. *Timber Statistics.*

Int-*3205*.B5, 1945, Annexes. *Annex I, League of Nations Estimated Balance of International Payments for the Year ... (8-page form for country to fill); and Annex II, Explanatory Notes to Scheme of International Balance of Payments (11 pages).*

Int-*3210*.C4, 1913.1. *Convention Respecting the Compilation of International Commercial Statistics.*

Int-*3210*.C4, 1913.2. *Algunos Documentos Relativos a la Adopción de un Sistema de Estadística Internacional, Convención de Bruselas, 1913.*

Int-*3210*.C4, 1913.3. *Documents et Procès-Verbaux de la Conférence Internationale de Statistique Commerciale, Bruxelles, 1913.*

Int-*3210*.C6, 1945. *Convertibility Index for Foreign Trade Statistical Classifications of the American Nations: Basic Classification Scheme Showing Detailed Export and Import Commodity Description by Classes of the Minimum List of Commodities for International Trade Statistics.*

Int-*3210*.D4, 1937. *Draft Customs Nomenclature.*

Int-*3210*.D6, 1929. *Draft Framework for a Customs Tariff Nomenclature and Draft Allocation of Goods to the Various Chapters of the Framework with Explanatory Notes.*

Int-*3210*.L5, 1938. *Liste Minimum de Marchandises pour les Statistiques du Commerce International.*

Int-*3210*.M5, 1938. *Minimum List of Commodities for International Trade Statistics.*

Int-*3210*.U5, 1930. *Unification of Customs Nomenclature.*

Int-*3220*.S5, 1923. *Systems of Classification of Industries and Occupations.*

Int-*3230*.C7, 1944. *Classification of Terms and Comparability of Title through Five Revisions of the International List of Causes of Death.*

Int-*3230*.M5, 1939. *Manual of the International List of Causes of Death and Manual of Joint Causes of Death.*

NOMENCLATURES—cont.

Int-*3230*.N5, 1929. *Nomenclaturas Internacionales de las Enfermedades (Causas de Defunción—Causas de Incapacidad del Trabajo) para Establecer las Estadísticas Nosológicas (Clasificación "Bertillon.").*

Int-*3230*.N8, 1938. *Nomenclatures Internationales des Causes de Décès, 1938.*

Int-*3230*.P5, 1946. *Proposed Statistical Classification of Diseases, Injuries, and Causes of Death.*

Int-*3230*.P6, 1946. *Proyecto de Informe sobre Standardización de las Estadísticas de Atención Médica.*

Int-*3230*.P7, 1946. *Proyecto de Informe sobre Standardización de las Estadísticas de Morbilidad.*

Int-*3240*.E5, 1938. *Estadísticas de la Población Activa.*

Int-*3240*.S4, 1938. *Statistics of the Gainfully-Occupied Population.*

Int-*3240*.S6, 1938. *Statistiques de la Population Active.*

Int-*6000*.I75, 1928.A. *International Conference on Economic Statistics, Geneva, 1928: Preparatory Documents.*

Int-*7110*.B5, June-July '41. "Clave Diagnóstica para la Clasificación en Columna (Tabulación) de las Causas de Morbidad."

Int-*7110*.B5, June-Aug. '39. "Nomenclatura Internacional de las Causas de Muerte."

Int-*7110*.B5, Aug. '36. "Nomenclatura para Hospitales Dedicados a Enfermedades Mentales."

Int-*9000*.E5, 1943, a. "La Clasificación de Industrias y Profesiones en las Estadísticas del Trabajo."

Arg-*3200*.C5, 1937.A. *Clasificador de Bienes.*

Arg-*3200*.C5, 1937.B. *Indice General Alfabético por Materia de los Bienes Nacionales del Estado.*

Arg-*5000*.P2, 1943, No. 10. *Clasificación Estadística de las Causas de las Defunciones.*

Bras-*1420*.B5, Sept. '44. "Anexo a Resolução No. 243: Nomenclatura Brasileira para a Classificação de Indústrias."

Can-*3220*.C5, 1941. *Classification of Industries: Eighth Census of Canada, 1941.*

Can-*3240*.C5, 1941. *Classification of Occupations: Eighth Census of Canada, 1941.*

Can-*5250*.H2, 1937. *Handbook on Death Registration and Certification.*

Chile-*3210*.L5, 1943. *Lista en Castellano e Inglés de los Artículos Importables desde los Estados Unidos de Norte América (con Números del "Schedule B" y Unidades).*

Chile-*3230*.N5, 1929. *Nomenclatura de las Enfermedades.*

Col-*3030*.E5, 1945. *Estatuto Orgánico de la Estadística Nacional.*

Col-*3210*.C5, 1936. *Codificación Estadística del Comercio Exterior.*

Col-*3230*.N5, 1942. *Nomenclatura de Morbilidad y Mortalidad.*

Ec-*3210*.C5, 1945. *Clasificación del Arancel Mínimum de Aduanas del Ecuador según la Lista de Mercaderías para las Estadísticas del Comercio Internacional.*

ElSal-*3200*.C5, 1943. *Claves y Nomenclaturas Usadas por la Dirección General de Estadística.*

Haiti-*6900*.H5, 1935. *Handbook and Trade Directory of Haiti.*

Méx-*3220*.C5, 1945. *Catálogo de Industrias, 1945.*

Méx-*3220*.T5, 1940. *Tercer Censo Industrial: Catálogo General de las Industrias de Transformación y Extractivas, 1940.*

Méx-*3230*.S5, 1931. *Sinonimias Populares Mexicanas de las Enfermedades.*

Méx-*3240*.N5, 1940. *Nomenclatura Nacional de Ocupaciones, 1940.*

Perú-*2200*.C5, 1940.1. *Censo Nacional de Población y Ocupación, 1940.*

USA-*3210*.C5. *Customs House Guide.*

USA-*3210*.S2, 1940. *Schedule A, Statistical Classification of Imports into the United States, Arranged by Tariff Schedules and Tariff Paragraphs of Act of 1930.*

USA-*3210*.S3, 1943. *Schedule A, Statistical Classification of Imports into the United States.*

USA-*3210*.S4, 1945. *Schedule B, Statistical Classification of Domestic and Foreign Commodities Exported from the United States.*

USA-*3210*.S6, 1944. *Standard Commodity Classification: Vol. II: Alphabetic Index.*

USA-*3210*.S6, 1943. *Standard Commodity Classification: Volume I, Standard Classified List of Commodities.*

USA-*3220*.C5, 1939. *Sixteenth Census of the United States: 1940: Manufactures: 1939: Industry Classification for the Census of Manufactures: 1939.*

NOMENCLATURES—cont.
USA-*3220*.S5, 1942. *Standard Industrial Classification Manual.*
USA-*3230*.C5, 1940. *Classification of Causes of Stillbirth.*
USA-*3240*.C5, 1940.1. *Sixteenth Census of the United States: Classified Index of Occupations.*
USA-*3240*.C5, 1940.2. *Sixteenth Census of the United States: Alphabetical Index of Occupations and Industries.*
USA-*5250*.P5, 1943. *Physician's Handbook on Birth and Death Registration.*
USA-*7110*.A4, June '36, Berkson. "A System of Codification of Medical Diagnoses for Application to Punch Cards, with a Plan for Operation."
USA-*7150*.M4, 1940. *Manual of Procedure for the Mechanical System of Reporting Morbidity, Treatment-Progress and Control of the Venereal Diseases.*
USA-*7160*.U5, 1943. *Uniform Definitions of Motor Vehicle Accidents.*
USA-*8300*.T5, 1930-39. *Ten Years of Uniform Crime Reporting, 1930-1939.*
Ven-*4100*.N5, 1941. *Nomenclador General de Areas y Lugares Habitados de Venezuela.*
NOMENCLATURES FOR STATISTICAL CLASSIFICATION—*see* 3200-3299; *see also* the particular subject field.
Nomenclatures Internationales des Causas de Décès, 1938. Int-*3230*.N8, 1938.
Nominata do Pessoal Estatístico das Nações Americanas. Int-*3000*.D5, 1944.
NORMAS—*véase* ESTANDARDS.
Normas de Apresentação da Estatística Brasileira. Bras-*3200*.N5, 1939.
"Normas para la Elaboración de los Cuadros Estadísticos." Col-*1420*.A5, May 5, '42, a.
Normas para la Organización y Levantamiento del Censo Escolar de la Nación en la Capital Federal. Arg-*2130*.C5, 1943.2.
Normas para la Organización y Levantamiento del Censo Escolar de la Nación en las Provincias. Arg-*2130*.C5, 1943.3.
Normas para la Organización y Levantamiento del Censo Escolar de la Nación en los Territorios. Arg-*2130*.C5, 1943.4.
North Carolina State College of Agriculture and Engineering. Department of Experimental Statistics.
NonG-*3150*.T8, 1942. *The Theory of Sampling.*
Notas para una Encuesta Continental sobre Productividad Obrera. Int-*9100*.N5, 1946.
"Notas sôbre as Estatísticas da Produção de Café na América." Bras-*6200*.
Note on Balance of Payments Statistics: Annex A. Int-*3205*.B5, 1945, Annex A.
NOTES, CIRCULATION OF—*see* 6720.
Notes on Statistical Mapping with Special Reference to the Mapping of Population Phenomena. NonG-*3130*.N5, 1938.
Una Nueva Argentina. Arg-*1900*.N5, 1940.
La Nueva Economía. Perú-*6000*.N8.
Nueva Tarifa de Exportación, Decreto No. 383 Vigente desde el 25 de Marzo de 1940. Par-*6900*.C5, 1940.
"Número Total de Empleados Públicos." Chile-*1420*.E5, Nov. '44.
NUMEROS INDICES:
GENERAL—*véase* 6010.
ESPECIFICOS—*véase* las respectivas clasificaciones.
DE ACCIONES Y BONOS—*véase* 6721.
DE LA CONSTRUCCION—*véase* 6500.
DE LA PRODUCCION—*véase* 6020; *véase también* las respectivas clasificaciones.
DE PRECIOS—*véase* 6010; *véase también* las respectivas clasificaciones.
DE SALARIOS—*véase* 9200.
DE TITULOS Y VALORES—*véase* 6721.
DEL COMERCIO (EN GENERAL)—*véase* 6700.
DEL COMERCIO AL POR MAYOR Y AL POR MENOR—*véase* 6710.
DEL COSTO DE LA VIDA—*véase* 9500.
DEL EMPLEO—*vease* 9200.
METODOLOGIA SOBRE LOS—*véase* 3120.
Números Indices. NonG-*3120*.N5, 1945.
Los Números Indices en Cuba. Cuba-*6010*.N5, 1938.
"'Números Indices', Suas Aplicações na Estatística Brasileira e Sugestões Atinentes à Sua Padronização Continental." Bras-*3040*.R5, Oct.-Dec. '43, Morais.
NURSING—*see* 7140.
NUTRICION—*véase* 9800.
NUTRITION—*see* 9800.
Nyström, Bertil.
Int-*3040*.R5, 1938.2. "Commission on Building and Housing Statistics: Preliminary Report."
Int-*3040*.R5, 1936.1. "Observations on the Possibility of Improving the International Comparability of Building and Housing Statistics."

OBRAS PUBLICAS—*véase* 8000.
OBREROS:
DEMANDA DE—*véase* 9200.
INDEMNIZACION DE—*véase* 9950.

1919) que Promulga la Ley sobre Aranceles de Importación y Exportación que Estará en Vigor desde el Día 1° de Enero de 1920." RepDom-*6900*.05, 1920.
Ordenamiento Económico-Social. Arg-*9000*.05, 1945.
Oreamuno, J. Rafael.
Int-*6400*.I3, 1946. "Industrialization in Latin America: An Outline of Its Development."
L'Organisation des Services de Statistique Agricole dans les Divers Pays. Int-*6200*.06, 1910-13.
ORGANISMOS:
ESTADISTICOS (SERVICIOS OFICIALES) —*véase* 3030.
ESTADISTICOS, LISTAS DE—*véase* 3000.
ORGANIZACION:
DE SERVICIOS Y ORGANISMOS ESTADISTICOS—*véase* 3030.
ESQUEMAS DE—*véase* ORGANIZATION CHARTS; *véase tambien* las respectivas clasificaciones.
"Organización de la Estadística Económica en el Uruguay." *Proceedings of the Eighth American Scientific Congress.* Int-*1300*.A5, 1940.3, Basabe Castellanos.
Organización de las Naciones Unidas para la Agricultura y la Alimentación—*see* - *véase* Food and Agriculture Organization of the United Nations; *see also* - *véase también* United Nations Interim Commission on Food and Agriculture.
*Organización de las Secciones de la Dirección General de Estadística. Cuba-3030.*O5, 1945?
"La Organización de los Servicios de Estadística Vital y Sanitaria en el Gobierno Federal de los Estados Unidos." *Boletín de la Oficina Sanitaria Panamericana.* Int-*7110*.B5, May '45, Collins.
"Organización de Servicios de Estadística: Resolución No. 64 de Enero de 1946." Col-*3030*.05, 1946.
"Organización del Servicio de Estadística Nacional en el Perú: Decreto y Esquema." Int-*3040*.E5, June '44, a.
Organización Sindical. Arg-*9000*.07, 1941.
ORGANIZACIONES:
DE OBREROS (incluyendo CONFEDERACIONES, GREMIOS, SINDICATOS Y UNIONES)—*véase* 9000.
ESTADISTICAS INTERNACIONALES —*véase* 3030.
INTERNACIONALES, MATERIAS DESCRIPTIVAS—*véase* 8200; *véase también* las respectivas clasificaciones.
ORGANIZATION:
CHARTS—*see* the particular subject field.
OF STATISTICAL SERVICE AGENCIES—*see* 3030.
Organization and activities of the statistical offices of the central government in Brazil; series of articles. *Revista Brasileira de Estatística.* Bras-*3040*.R5, Apr.-June '43, a.
Organization and activities of the statistical offices of the individual states; series of articles. *Revista Brasileira de Estatística.* Bras-*3040*.R5, Apr.-June '43, b.
Organization and Supervision of the Tabulating Department. NonG-*3110*.05, 1936.
ORGANIZATION CHARTS.
Int-*3040*.E5, Sept. '45, Rodríguez y Rodríguez. "La Oficina de Estadística del Instituto Mexicano del Seguro Social."
Int-*3040*.E5, June '44, a. "Organización del Servicio de Estadística Nacional en el Perú: Decreto y Esquema."
Int-*8200*.I55, Sept. '45. "The United Nations Charter."
Bras-*3030*.I5, 1940. *Instituto Brasileiro de Geografia e Estatística.*
Col-*2100*.C5, 1938. *Decretos y Bases de Organización para la Ejecución de los Censos, 1938.*
Col-*3030*.E5, 1945. *Estatuto Orgánico de la Estadística Nacional.*
Cuba-*3030*.05, 1945? *Organización de las Secciones de la Dirección General de Estadística.*
Pan-*9950*.B5. *Boletín de Previsión Social.*
USA-*9950*.S4, July '46, Cohen. "The First Two Years of Social Insurance in Mexico."
Organization of Agriculture Statistics. Int-*3010*.B5, Econ. 6a, June 1, '45.
"The Organization of Statistical Services in the United States and Great Britain." Int-*3020*.B5, 1938, Rice.
"Orientación, Organización y Resultados de un Curso Oficial de Estadística por Correspondencia. "Int-*3040*.E5, Mar. '43, Guthardt.
ORO—*véase* 7200.
ORPHAN ASYLUMS—*see* 9951.
ORPHANS—*see* 5110; 9951.
Ortiz, C., Luis B.
Col-*1420*.A5, Feb. '45. "El Censo Industrial de Colombia de 1945."
Col-*1420*.A5, July '39. "Ensayo sobre las Condiciones de la Vida Rural en el Municipio de Moniquirá-Boyacá."

Pan American Sanitary Bureau—Cont.

Sanidad, Abril 22-29, 1944. Fifth Pan American Conference of National Directors of Health, April 22-29, 1944.

Int-*7110*.P6. 1942. *Public Health in the Americas.*

Int-*7110*.P8. *Publications of the Pan American Sanitary Bureau. Publicaciones de la Oficina Sanitaria Panamericana.*

Int-*7110*.S5, 1942. *La Sanidad en las Américas.*

Int-*9950*.H5, 1942. *Housing and Hospital Projects of Latin American Social Security Systems.*

RepDom-*5200*.C5, 1946. *Commentary Explanations and Sample Copies of the Forms and Manuals Used in Vital Records and Vital Statistics Activities in the Dominican Republic, November 1946.*

RepDom-*7100*.C5, 1946. *Commentary Explanations and Sample Copies of the Forms Used in Medical and Public Health Statistics Activities (Not Including Vital Statistics) in the Dominican Republic, November 1946.*

Pan American Sanitary Conference.

Int-*9800*.E5, 1942. *Encuestas sobre el Consumo de Alimentos y la Nutrición en las Américas.*

Pan American Union.

Int-*1200*.B3. *Bibliographic Series of the . . .*

Int-*1300*.C8. *Congress and Conference Series.*

Int-*1300*.S5. *Serie de Congressos e Conferências.*

Int-*1300*.S6. *Serie sobre Congresos y Conferencias.*

Int-*1950*.A5. *American Nation Series.*

Int-*5310*.I4, 1941.1 Eng. *Final Act of the Second Inter-American Travel Congress.*

Int-*5310*.I4, 1941.1 Sp. *Acta Final del Segundo Congreso Interamericano de Turismo.*

Int-*6000*.C7. *Commercial Pan America.*

Int-*6000*.C7, July-Aug. '44. "Mexico's Balance of Payments."

Int-*6000*.I45, 1942, Eng. *Final Act of Inter-American Conference of Systems of Economic and Financial Control, Washington, June 30-July 10, 1942.*

Int-*6000*.I45, 1942, Port. *Ata Final da Conferência Interamericana sôbre Medidas de Contrôle Econômico-Financeiro, Washington, 30 de Junho a 10 de Julho de 1942.*

Int-*6000*.I45, 1942, Sp. *Acta Final de la Conferencia Interamericana sobre Sistemas de Control Económico y Financiero, Wáshington, 30 de Junio a 10 de Julio de 1942.*

Int-*6000*.P4. *Panamérica Comercial.*

Int-*6000*.P4, Jan.-Feb. '45. "El Comercio Internacional de Chile, 1904-1944."

Int-*6200*.I4, 1945.1, Eng. *Final Act of the Third Inter-American Conference on Agriculture.*

Int-*6200*.I4, 1945.2. *Handbook for the Use of the Delegates of the Third Inter-American Conference on Agriculture.*

Int-*6200*.I4, 1942.1, Eng. *Final Act of the Second Inter-American Conference on Agriculture.*

Int-*6200*.I4, 1942.1, Sp. *Acta Final de la Segunda Conferencia Interamericana de Agricultura.*

Int-*6210*.P5, 1945, Eng. *Fourth Pan American Conference, Mexico City, September 3-14, 1945.*

Int-*6615*.I5, 1940, Eng. *Final Act of the Inter-American Maritime Conference, Washington, November 25-December 2, 1940.*

Int-*6615*.I5, 1940, Sp. *Acta Final de la Conferencia Marítima Interamericana, Unión Panamericana, Wáshington, del 25 de Noviembre al 2 de Diciembre de 1940.*

Int-*6900*.C7. *Commodities of Commerce Series.*

Int-*6900*.F8. *Foreign Trade Series.*

Int-*7200*.C6, 1943.1, Eng. *Final Act of the First Conference of Ministers and Directors of Education of the American Republics, Held in Panama, September 27 to October 4, 1943.*

Int-*7200*.C6, 1943.1, Sp. *Acta Final de la Primera Conferencia de Ministros y Directores de Educación de las Repúblicas Americanas Celebrada en Panamá del 27 de Septiembre al 4 de Octubre de 1943.*

Int-*7300*.I5, 1946.2, Eng. *Inter-American Conference of Experts on Copyright, Pan American Union, June 1-22, 1946: Proceedings.*

Int-*7300*.I5, 1946.2, Sp. *Conferencia Interamericana de Expertos para la Protección de los Derechos de Autor, 1-22 de Junio de 1946: Actas y Documentos.*

Int-*8200*.B4. *Boletim da União Panamericana.*

Int-*8200*.B5. *Boletín de la Unión Panamericana.*

Int-*8200*.B7. *Bulletin of the . . .*

Int-*8200*.I26. *Informe Anual del Director General de la Unión Panamericana.*

Int-*8200*.I3, 1945.2, Eng. *Inter-American*

Panamá—cont.

Pecuario del Distrito de Penonomé, Diciembre 1943.

Pan-*6000*.E5, 1943. *Examen Económico de la República de Panamá para el Año de 1943.*

——. Departamento de Salubridad—*see - véase* Panamá. Ministerio de Salubridad y Obras Públicas. Departamento de Salubridad.

——. Dirección de Estadística y Censo—*see - véase* Panamá. Contraloría General de la República. Dirección de Estadística y Censo.

——. Dirección General de Estadística—*see - véase* Panamá. Contraloría General de la República. Dirección de Estadística y Censo.

——. División de Bio-Estadística y Educación Sanitaria—*see - véase* Panamá. Ministerio de Salubridad y Obras Públicas. Departamento de Salubridad. División de Bio-Estadística y Educación Sanitaria.

——. Ministerio de Agricultura y Comercio.

Pan-*6000*.R5. *Revista de Agricultura y Comercio.*

Pan-*6000*.R5, Apr.'39. "Análisis de los Resultados del Primer Censo Oficial de Empresas Industriales y Comerciales Radicadas en Panamá, Ejecutado en el Año de 1938."

——. ——. Sección de Economía Agrícola.

Pan-*2110*, 1945. *Censo Agro-Pecuario de las Provincias de Herrera y Los Santos, Diciembre 1945: Métodos y Procedimientos.*

Pan-*2110*.C5, 1945.1. *Censo Agro-Pecuario de las Provincias de Herrera y Los Santos, Diciembre 1945: Instrucciones a los Empadronadores.*

Pan-*2110*.C5, 1943.1. *Censo Agro-Pecuario, Diciembre de 1943, Distrito de Penonomé: Métodos y Procedimientos.*

Pan-*2210*.C5, 1943.1. *Censo Agro-Pecuario del Distrito de Penonomé, Diciembre 1943.*

Pan-*2210*.C5, 1942. *Censo Agro-Pecuario, 1942.*

——. Ministerio de Educación. Departamento de Estadística y Archivos.

Pan-*7210*.E5. *Estadística Cultural.*

——. Ministerio de Higiene, Beneficencia y Fomento—*see - véase* Panamá. Ministerio de Salubridad y Obras Públicas.

——. Ministerio de Salubridad y Obras Públicas.

Pan-*7110*.M5. *Memoria del . . .*

——. ——. Departamento de Salubridad. División de Bio-Estadística y Educación Sanitaria.

Pan-*7110*.B5. *Bio-Estadística—Informes Especiales.*

Pan-*7110*.E5. *Estadísticas Epidemiológicas.*

——. ——. ——. Sección de Salubridad.

Pan-*7110*.H5. *Higiene y Sanidad.*

——. Ministerio de Trabajo, Previsión Social y Salud Pública.

Pan-*9950*.B5. *Boletín de Previsión Social.*

——. Oficina del Censo—*see - véase* Panamá. Contraloría General de la República. Dirección de Estadística y Censo.

——. Sección de Economía Agrícola—*see - véase* Panamá. Ministerio de Agricultura y Comercio. Sección de Economía Agrícola.

——. Universidad Interamericana—*see - véase* Universidad Interamericana, Panamá.

Panama: Summary of Biostatistics. Pan-*5000*. P5, 1945.

Panamérica Comercial. Int-*6000*.P4.

Pando Gutierrez, Jorge.

Int-*3000*.S5, 1940. "Las Actividades Estadísticas de Bolivia."

"Panorama Demográfico de la Población de Colombia." Int-*1300*.A5, 1940.3, Guthardt.

Papers Presented by Special Guests of the Population Association of America at Its Meetings October 25-26, 1946.

Paradiso, Louis.

USA-*3000*.G7, 1946. *Government Statistics for Business Use.*

Paraguay. Banco del Paraguay—*see-véase* Banco del Paraguay. Contaduría General y Dirección del Tesoro—*see-véase* Paraguay. Ministerio de Hacienda. Contaduría General y Dirección del Tesoro

——. Dirección General de Estadística—*see-véase* Paraguay. Ministerio de Hacienda. Dirección General de Estadística.

——. Dirección General de Impuestos Internos—*see-véase* Paraguay. Ministerio de Hacienda. Dirección General de Impuestos Internos.

——. Leyes, Estatutos, Etc.

Par-*2100*.C5, 1936.2. *Censo de Población, Decreto N°. 2120 de 18 de Junio de 1936.*

Par-*2100*.C5, 1936.3. *Decreto N°. 60.287 de 21 de Noviembre de 1935.*

Par-*2100*.C5, 1936.4. *Ley N° 1.509 de 1° de Noviembre de 1935.*

Par-*3030*.D5, 1942. *Decreto Ley 11.126 que Reorganiza y Coordina los Servicios Estadísticos de la República.*

Par-*5250*.L3, 1940. *Leyes del Registro*

PERIODICALS—cont.

PERIODICALS—cont.

Can-*6710*.M5. *Monthly Index of Retail Sales.*

Can-*6710*.M6. *Monthly Index of Wholesale Sales.*

Can-*6720*.C5. *Commercial Failures in Canada.*

Can-*6900*.T6. *Trade of Canada: Exports of Canadian Produce.*

Can-*6900*.T7. *Trade of Canada: Imports Entered for Consumption.*

Can-*7110*.C5. *The Canadian Journal of Public Health.*

Can-*9200*.E5. *The Employment Situation. . .Together With Payrolls.*

Can-*9200*.M5. *Man Hours and Hourly Earnings.*

Can-*9500*.C6, 1913-42. *Cost of Living Index Numbers for Canada, 1913-1942.*

Can-*9952*.S5. *Statistical Report on the Operation of the Unemployment Insurance Act.*

Chile-*1420*.E5. *Estadística Chilena.*

Chile-*3040*.E5. *Economía.*

Chile-*6000*.E6. *Economía y Finanzas.*

Chile-*6200*.B5. *Boletín de la Sección Estadística de la Caja de Colonización Agrícola.*

Chile-*6720*.B5. *Boletín Mensual del Banco Central de Chile.*

Chile-*6900*.B5. *Boletín Oficial de la Superintendencia de Aduanas.*

Chile-*7100*.R7. *Revista Chilena de Higiene y Medicina Preventiva.*

Chile-*7110*.P5. *Previsión Social.*

Col-*1420*.A5. *Anales de Economía y Estadística.*

Col-*1900*.C5. *Colombia.*

Col-*3040*.E5. *Estadística* (Sociedad Colombiana de Estudios Estadísticos).

Col-*6000*.M7. *El Mes Financiero y Económico.*

Col-*6200*.T5. *Tierras y Aguas.*

Col-*6210*.B5. *Boletín de Estadística de la Federación Nacional de Cafeteros.*

Col-*6614*.R5. *Revista del Consejo Administrativo de los Ferrocarriles Nacionales.*

Col-*6720*.B5. *Boletín de Información Mensual* (Banco de la República).

Col-*6720*.B7. *Boletín de la Superintendencia Bancaria.*

Col-*6720*.R5. *Revista del Banco de la República.*

Col-*6900*.C5. *Comercio Exterior de Colombia.*

Col-*8800*.I3. *Información Fiscal de Colombia.*

Col-*8800*.R5. *Revista de Hacienda.*

CR-*6720*.B6. *Boletín Estadístico* (Banco Nacional de Costa Rica.)

CR-*6720*.R5. *Revista del Banco Nacional de Costa Rica.*

CR-*6720*.R6. *Revista del Banco Nacional de Seguros.*

Cuba-*1420*.B5. *Boletín Mensual de Estadísticas* (Dirección General de Estadística).

Cuba-*6000*.C5. *Cuba Económica y Financiera.*

Cuba-*6010*.B5. *Boletín de la Oficina de Números Indices.*

Cuba-*6200*.B5. *Boletín Agrícola para el Campesino Cubano.*

Cuba-*6210*.B5. *Boletín del Café.*

Cuba-*6900*.I5. *Importación y Exportación de la República de Cuba.*

Cuba-*7110*.B5. *Boletín Semanal . . . Ep - demiológico Sanitario.*

Cuba-*7110*.S5. *Salubridad y Asistencia Social.*

Cuba-*8800*.B5. *Boletín Oficial del Ministerio de Hacienda.*

Ec-*1420*.E5. *Estadística y Censos.*

Ec-*1420*.T5. *El Trimestre Estadístico del Ecuador.*

Ec-*6000*.I5. *Información Económica.*

Ec-*6000*.I5A. *Datos Estadísticos* (Dirección General de Estadística y Censos).

Ec-*6720*.B3. *Boletín de la Superintendencia de Bancos del Ecuador.*

Ec-*6720*.B4. *Boletín del Banco Hipotecario del Ecuador.*

Ec-*6720*.B5. *Banco Central del Ecuador, Boletín.*

Ec-*6900*.A4. *Aduanas del Ecuador.*

Ec-*8800*.B5. *Boletín del Ministerio del Tesoro.*

Ec-*8800*.R5. *Revista de Hacienda.*

ElSal-*1420*.B5. *Boletín Estadístico de la Dirección General de Estadística.*

ElSal-*6210*.C5. *El Café.*

ElSal-*6210*.C6. *El Café de El Salvador.*

ElSal-*6600*.B5. *Boletín* (Ministerio del Interior).

ElSal-*6720*.R5. *Revista Mensual del Banco Central de Reserva de El Salvador.*

Guat-*1420*.B5. *Boletín de la Dirección General de Estadística.*

Guat-*6000*.R5. *Revista de Economía.*

Guat-*6200*.R4. *Revista Agrícola.*

Guat-*6210*.R5. *Revista Cafetalera de Guatemala.*

Guat-*6620*.G5. *Gaceta de Comunicaciones.*

Guat-*7110*.R5. *Revista de Sanidad.*

Haiti-*1420*.B5. *Bulletin Mensuel du Dé-*

partement Fiscal (Banque Nationale de la République d'Haïti).
Haiti-*1420*.M5. *Monthly Bulletin of the Fiscal Department* (National Bank of the Republic of Haiti).
Haiti-*1900*.C5. *Cahiers d'Haïti.*
Haiti-*7110*.B5. *Bulletin du Service National d'Hygiène et d'Assistance Publique.*
Hond-*6200*.H5. *Honduras Agrícola.*
Hond-*7110*.B5. *Boletín Sanitario de la Dirección General de Sanidad.*
Méx-*1200*.B5. *Boletín Bibliográfico Mexicano.*
Méx-*1420*.R5. *Revista de Estadística (Dirección General de Estadística).*
Méx-*3040*.B5. *Boletín de la Sociedad Mexicana de Geografía y Estadística.*
Méx-*6000*.I5. *Investigación Económica.*
Méx-*6000*.P5. *Planificación Económica.*
Méx-*6000*.R5. *Revista de Economía.*
Méx-*6000*.T5. *El Trimestre Económico.*
Méx-*6200*.B5. *Boletín Mensual de la Dirección General de Economía Rural.*
Méx-*6200*.M8. *México Agrario.*
Méx-*6720*.R5. *Revista del Banco de la República.*
Méx-*6900*.B5. *Boletín de Aduanas.*
Méx-*6900*.R5. *Revista del Comercio Exterior.*
Méx-*7110*.B6. *Boletín de la Oficina de Demografía y Estadística Sanitaria.*
Méx-*7110*.B7. *Salubridad y Asistencia.*
Méx-*7110*.R5. *Revista del Instituto de Salubridad y Enfermedades Tropicales.*
Méx-*9000*.R5. *Revista Mexicana de Sociología.*
Méx-*9000*.T5. *Trabajo y Previsión Social.*
Nic-*1420*.B5. *Boletín Mensual de Estadística* (Dirección General de Estadística).
Nic-*1900*.G5. *La Gaceta.*
Nic-*6000*.E5. *Economía y Finanzas.*
Nic-*6720*.R5. *Revista del Banco Nacional de Nicaragua*
Nic-*7110*.B5. *Boletín Sanitario.*
Pan-*1420*.E5. *Estadística Panameña.*
Pan-*6000*.R5. *Revista de Agricultura y Comercio.*
Pan-*7110*.B5. *Bio-Estadística—Informes Especiales.*
Pan-*7110*.E5. *Estadísticas Epidemiológicas.*
Pan-*7110*.H5. *Higiene y Sanidad.*
Pan-*9950*.B5. *Boletín de Previsión Social.*
Par-*6000*.B5. *Boletín del Ministerio de Agricultura, Comercio e Industrias.*
Par-*6000*.P5. *Paraguay Industrial y Comercial.*
Par-*6200*.R5. *Revista de Agricultura y Ganadería.*
Par-*7110*.B5. *Boletín del Ministerio de Salud Pública y Previsión Social.*
Par-*7200*.R5. *Revista de Educación.*
Par-*8000*.M5. *M. O. P.* (Ministerio de Obras Públicas y Comunicaciones).
Par-*8800*.B5. *Boletín del Tesoro.*
Perú-*1420*.B4. *Boletín de la Dirección Nacional de Estadística.*
Perú-*3040*.E5. *Estadística Peruana.*
Perú-*6000*.G5. *La Gaceta Económica y Financiera.*
Perú-*6000*.I5. *Informaciones Comerciales, Económicas y Financieras del Perú.*
Perú-*6000*.N8. *La Nueva Economía.*
Perú-*6000*.R5. *Revista de Economía y Finanzas.*
Perú-*6000*.R7. *Revista de Hacienda.*
Perú-*6200*.B5. *Boletín de la Dirección de Agricultura.*
Perú-*6200*.V5. *La Vida Agrícola.*
Perú-*6210*.A5. *Algodón.*
Perú-*6220*.B4. *Aguas e Irrigación.*
Perú-*6220*.B6. *Boletín de la Compañía Administradora del Guano.*
Perú-*6300*.B7. *Boletín Oficial de la Dirección de Minas y Petróleo.*
Perú-*6610*.B5. *Boletín de la Dirección de Caminos y Ferrocarriles.*
Perú-*6611*.B5. *Boletín de la Dirección de Aeronáutica.*
Perú-*6720*.B5. *Boletín Mensual del Banco Central de Reserva del Perú.*
Perú-*6900*.B5. *Boletín de Aduanas.*
Perú-*7110*.B6. *Boletín del Departamento de Bioestadística.*
Perú-*7160*.B5. *Boletín del Cuerpo de Investigación y Vigilancia.*
Perú-*9000*.I5. *Informaciones Sociales.*
RepDom-*6720*.E5. *Estadística Bancaria.*
RepDom-*6900*.E7. *Exportación de la República Dominicana.*
RepDom-*6900*.I5. *Importación de la República Dominicana.*
USA-*3040*.A5. *The Annals of the American Academy of Political and Social Science.*
USA-*3040*.B5. *Biometrics Bulletin.*
USA-*3040*.J5. *Journal of the American Statistical Association.*
USA-*3040*.M5. *Mathematical Reviews.*
USA-*3040*.P5. *Psychometrika.*
USA-*3040*.R5. *The Review of Economic Statistics.*
USA-*4000*.G5. *Geographical Review.*

PERIODICALS—cont.

USA-*5000*.M5. *Milbank Memorial Fund Quarterly.*
USA-*5200*.V8. *Vital Statistics—Special Reports.*
USA-*5210*.M5. *Monthly Vital Statistics Bulletin.*
USA-*5211*.H5. *Human Fertility.*
USA-*5212*.C5. *Current Mortality Analysis.*
USA-*5212*.W5. *Weekly Mortality Index.*
USA-*6000*.A5. *American Economic Review.*
USA-*6000*.Q5. *Quarterly Journal of Economics.*
USA-*6000*.S7. *Survey of Current Business.*
USA-*6020*.S5. *Statistics of War Production.*
USA-*6200*.A45. *Agricultural Prices.*
USA-*6200*.C6. *Crops and Markets.*
USA-*6200*.F4. *The Farm Income Situation.*
USA-*6200*.J5. *Journal of Farm Economics.*
USA-*6220*.C7. *Crop Production.*
USA-*6400*.I3. *Industrial Corporation Reports.*
USA-*6500*.C3. *Construction.*
USA-*6700*.D4. *Domestic Commerce.*
USA-*6720*.F2. *Farm Credit Quarterly.*
USA-*6720*.F4. *Federal Reserve Bulletin.*
USA-*6720*.R5. *Report of the Reconstruction Finance Corporation.*
USA-*6721*.S5. *Statistical Bulletin* (Securities and Exchange Commission).
USA-*6730*.S5. *Statistical Bulletin of the Metropolitan Life Insurance Company.*
USA-*6900*.F7. *Foreign Commerce Weekly.*
USA-*6900*.M5. *Monthly Summary of Foreign Commerce of the United States.*
USA-*7110*.A4. *American Journal of Public Health and the Nation's Health.*
USA-*7110*.P5. *Public Health Reports.*
USA-*7150*.V5. *Venereal Disease Information.*
USA-*7200*.J5. *Journal of Educational Psychology.*
USA-*7300*.A5. *American Journal of Mathematics.*
USA-*7300*.Q5. *The Quartery Review of Biology.*
USA-*8200*.D5. *The Department of State Bulletin.*
USA-*8600*.P5. *Public Opinion Quarterly.*
USA-*8800*.S6. *Statement of the Public Debt.*
USA-*8800*.T5. *Treasury Bulletin.*
USA-*9000*.A4. *American Journal of Sociology.*
USA-*9000*.A6. *American Sociological Review.*
USA-*9000*.M5. *Monthly Labor Review.*
USA-*9000*.S5. *Social Forces.*
USA-*9200*.E4. *Employment and Pay Rolls.*
USA-*9200*.F5. *Farm Labor.*
USA-*9200*.H4. *Hours and Earnings.*
USA-*9200*.L5. *The Labor Market.*
USA-*9200*.L6. *Labor Turnover.*
USA-*9200*.M7. *Monthly Report of Employment, Executive Branch of the Federal Government.*
USA-*9300*.C5. *Conference Board Economic Record.*
USA-*9500*.C4. *Changes in Cost of Living.*
USA-*9900*.F5. *Federal Home Loan Bank Review.*
USA-*9950*.E5. *Employment Security Activities.*
USA-*9950*.S4. *Social Security Bulletin.*
USA-*9952*.M5. *The Monthly Review of the Railroad Retirement Board.*
USA PR-*1420*.P5. *Puerto Rico Monthly Statistical Report.*
USA PR-*7110*.P5. *The Puerto Rico Journal of Public Health and Tropical Medicine.*
Ur-*1420*.B5. *Boletín Estadístico de la República Oriental del Uruguay.*
Ur-*3040*.R5. *Revista de la Facultad de Ciencias Económicas y de Administración.*
Ur-*6000*.B5. *Boletín de Hacienda.*
Ur-*6000*.B6. *Boletín de Hacienda, Suplemento.*
Ur-*6000*.B7. *Boletín de la Dirección de Estadística Económica.*
Ur-*6000*.S5. *Suplemento Estadístico de la Revista Económica.*
Ur-*6200*.M5. *Mercados del Mundo.*
Ur-*6720*.R5. *Revista del Banco de la República Oriental del Uruguay.*
Ur-*6900*.B5. *Boletín Aduanero.*
Ur-*9500*.C5. *Costo de la Vida Obrera en Montevideo.*
Ur-*9800*.B5. *Boletín Mensual del Instituto Nacional de Alimentación.*
Ven-*1420*.B5. *Boletín Mensual de Estadística* (Dirección General de Estadística y de Censos).
Ven-*1900*.G5. *Gaceta Oficial de los Estados Unidos de Venezuela.*
Ven-*6000*.R5. *Revista de Fomento.*
Ven-*6210*.R5. *Revista del Instituto Nacional del Café.*
Ven-*6230*.B5. *Boletín Mensual de la Dirección de Ganadería.*
Ven-*6720*.B5. *Boletín del Banco Central de Venezuela.*

Perú—cont.

véase Perú. Ministerio de Salud Pública y Asistencia Social. Dirección General de Salubridad.

——. Dirección General de Tráfico y Rodaje —*see - véase* Perú. Ministerio de Gobierno y Policía. Dirección General de Tránsito.

——. Dirección General de Tránsito—*see - véase* Perú. Ministerio de Gobierno y Policía. Dirección General de Tránsito.

——. Dirección Nacional de Estadística—*see - véase* Perú. Ministerio de Hacienda y Comercio. Dirección Nacional de Estadística.

——. Instituto Peruano de Estadística—*see - véase* Instituto Peruano de Estadística.

——. Ministerio de Agricultura.

Perú-*6200*.S5, 1945, Sp. *Programa Estadístico para el Ministerio de Agricultura del Perú.*

——. ——. Dirección General de Agricultura.

Perú-*6200*.B5. *Boletín de la Dirección de Agricultura.*

Perú-*6210*.E5. *Estadística de la Producción del Algodón en el Perú.*

Perú-*6210*.E6. *Estadística de la Producción de Arroz.*

Perú-*6210*.E7. *Estadística de la Producción de Caña de Azúcar y Azúcar de Caña en el Perú.*

Perú-*6210*.E8. *Estadística de la Producción del Trigo en el Perú.*

——. ——. Servicio Cooperativo Inter-Americano de Producción de Alimentos.

Perú-*6230*.E5, 1944. *Estudio de la Situación de la Industria Lechera en el Departamento de Lambayeque.*

Perú-*9800*.F5, 1943. *The Food Situation in Southern Peru.*

Perú-*9800*.S4, 1945. *La Situación Alimenticia en el Perú, 1943-1944.*

Perú-*9800*.S5, 1944. *Situación Alimenticia en la Ciudad de Iquitos.*

Perú-*9800*.S7, 1944. *Situación Alimenticia en los Departamentos del Norte.*

——. Ministerio de Fomento y Obras Públicas.

Perú-*6000*.M5. *Memoria del . . .*

——. ——. Cuerpo de Ingenieros de Minas.

Perú-*6300*.B5. *Boletín del Cuerpo . . .*

——. ——. Dirección de Aguas e Irrigación.

Perú-*6220*.B4. *Aguas e Irrigación.*

——. ——. Dirección de Caminos y Ferrocarriles.

Perú-*6610*.B5. *Boletín de la Dirección . . .*

——. ——. Dirección de Comunicaciones y Meteorología de Aeronáutica. Departamento de Meteorología.

Perú-*4200*.B4. *Boletín Anual Meteorológico.*

——. ——. Dirección de Industrias.

Perú-*6400*.B5. *Boletín de la Dirección de Industrias del Ministerio . . .*

Perú-*6400*.B5, Aug. '46. "Estadística Industrial."

——. ——. Dirección de Minas y Petróleo.

Perú-*6300*.A5. *Anuario de la Industria Minera en el Perú.*

Perú-*6300*.B7. *Boletín Oficial de la Dirección . . .*

——. ——. Dirección General de Aeronáutica.

Perú-*6611*.B5. *Boletín de la Dirección de Aeronáutica.*

——. ——. Dirección General del Ramo.

Perú-*6400*.E5, 1942. *Estadística de los Servicios Eléctricos del Perú.*

——. Ministerio de Gobierno y Policía. Dirección General de Tránsito.

Perú-*6610*.M5. *Memoria de la Dirección . . .*

Perú-*7160*.B5. *Boletín del Cuerpo de Investigación y Vigilancia.*

——. Ministerio de Hacienda y Comercio.

Perú-*6000*.R7. *Revista de Hacienda.*

——. ——. Dirección Nacional de Estadística.

Int-*3040*.E5, Dec. '44, Luna Vegas. "La Estadística Peruana en el Sexenio 1938-44."

Int-*3040*.E5, June '44, a. "Organización del Servicio de Estadística Nacional en el Perú: Decreto y Esquema."

Perú-*1410*.E5. *Anuario Estadístico del Perú.*

Perú-*1410*.E5, 1942, a. "Decreto Supremo de 20 de Agosto de 1943 que Ordena Formar el Inventario del Potencial Económico de la Nación."

Perú-*1410*.E5, 1942, b. "Decreto Supremo de 1° de Enero de 1944, por el que se Organizà el Servicio de Estadística Nacional y Crea el Consejo Superior de Estadística."

Perú-*1410*.E5, 1942, c. "Encuesta sobre Presupuestos Familiares Obreros, en 1940, Correspondiente a Lima (Cercado)."

Perú-*1410*.E5, 1942, d. "Ley No. 7567, de 5 de Agosto de 1932, que Transforma en Nacional la Dirección de Estadística y Establece las Normas para el Funcionamiento del Servicio de Estadística Nacional."

Perú-*1410*.E5, 1942, e. "Producción de Energía."

Perú-*1410*.E5, 1942, f. "Resolución Su-

America." *Handbook of Latin American Studies, 1937.*
Phelps, Elizabeth.
USA-*9700*.F5I2, 1940. *Family Income and Expenditures, Five Regions, Part 2, Family Expenditures, Urban and Village Series.*
Phipard, Esther F.
USA-*9700*.F5F1, 1941. *Family Food Consumption and Dietary Levels, Five Regions, Farm Series.*
USA-*9800*.D5, 1939. *Diets of Families of Employed Wage Earners and Clerical Workers in Cities.*
PHYSICAL EXAMINATION PROGRAMS —*see* 7110.
PHYSICIANS—*see* 7100.
Physician's Handbook and Birth and Death Registration. USA-*5250*.P5, 1943.
PHYSICS—*see* 7300.
PHYSIOLOGY—*see* 7180.
Picón, Antonio H.
NonG-*5100*.T5, 1943. *Teoría Matemática del Desarrollo Numérico de la Población.*
Pineda de Castro, Alvaro.
Col-*1420*.A5, Mar. '45. "La Investigación del Costo de la Vida y las Condiciones Social-Económicas de los Obreros del Ferrocarril."
Pino, Ernesto.
NonG-*3120*.N5, 1945. *Números Indices.*
PIPE LINES:
CONSTRUCTION—*see* 6500.
OPERATION—*see* 6613.
"The Place of the Intrinsic Rate of Natural Increase in Population Analysis." Int-*1300*.A5, 1940.3, Lotka.
Plan for the Coordination of Vital Records and Vital Statistics in the United States. USA-*5200*.P5, 1944.
"Plan General para el Levantamiento del Primer Censo Agropecuario." Col-*1420*.A5, Oct. 20, '43, Montoya.
Plan para el Levantamiento del Censo de Población de la Ciudad de La Paz, 1942. Bol-*2100*.P5, 1942.
PLANIFICACION:
ESPECIFICA—*véase* las respectivas clasificaciones.
NACIONAL, REGIONAL Y CIVICA —*véase* 8500.
Planificación Económica. Méx-*6000*.P5.
Planned Parenthood Federation of America, Inc.
USA-*5211*.H5 *Human Fertility.*
PLANNING:
In specific subject fields, *see* the particular field.
NATIONAL, REGIONAL, AND CIVIC —*see* 8500.
Planning Pamphlets of the National Planning Association. USA-*8500*. P5.
Plano de Apuração do Censo Demográfico. Bras-*2100*.R5, 1940.6.
PLATA—*véase* 6720.
Platt, Raye R.
USA-*4000*.G5, Apr. '37. "The Millionth Map of Hispanic America."
Plenty of People. NonG-*5000*.P3, 1944.
POBLACION—*véase* 5100-5199.
POBLACION, MOVIMIENTO MIGRATORIO DE LA—*véase* 5300.
"La Poblacion Argentina, Su Desarrollo, Sus Características y Sus Tendencias." Int-*1300*.A5, 1940.3, Bunge, b.
"La Población Chilena: Primera Parte, Las Características Demográficas." Chile-*3040*.E5, July '44, Levine Bawden, b.
"La Población Chilena: Segunda Parte, El Trabajo de la Población Chilena." Chile-*3040*.E5, Apr. '45, Levine Bawden.
Población de Colombia por Ramas de Actividad Económica (según Censo del 5 de Julio de 1938). Col-*5110*.P5, 1946.
La Población de El Salvador. ElSal-*5100*.P5, 1942.
La Población de la República Argentina. Arg-*5100*.P5.
Población de la República Censada el 13 de Mayo de 1935, Clasificada por Grupos de Edades, Zona y Sexo. RepDom-*2200*.C5, 1935.2.
Población de la República de Costa Rica según el Censo General de Población Levantado al 11 de Mayo de 1927. CR-*2200*.P5, 1927.
Población de la República de El Salvador. ElSal-*2200*.C5, 1930.
"Población de la República de Nicaragua, Calculada hasta el 31 de Diciembre de 1944." Nic-*1420*.B5, Sept.-Oct. '45.
"Población de la República del Ecuador, al 1° de Enero de 1944." Ec-*5100*.P5, 1944.
Población de la República Dominicana, Distribuida por Nacionalidades: Cifras del Censo Nacional de 1935. RepDom-*2200*.C5, 1935.1.
Población de la República Dominicana, según las Sucesivas Modificaciones Territoriales a Partir del 13 de Mayo de 1935, Día en que Se Levantó el Segundo Censo Nacional de Población, Hasta el 1 de Enero de 1946. RepDom-*2200*.C5, 1935-46.
Población de la República, según el Censo Levantado el 24 de Diciembre de 1920, Clasificada por Sexo y Grupos de Edades. RepDom-*2200*.C5, 1920.
Población de Nicaragua. Nic-*2200*.C5, 1940.2.

POSTWAR PLANNING—*see* 8500.
POULTRY—*see* 6230.
Powell, Nora P.
USA-*3040*.A5, Jan. '45, Dunn. "Demographic Status of South America."
POZOS PETROLIFEROS—*véase* 6300.
Practical Applications of the Punched Card Method in College and Universities. NonG-*3110*.P4 1935.
Practical Business Statistics. NonG-*3330*. Croxton 3, 1934.
"Precio Medio de la Habitación Ofrecida en la Ciudad de Buenos Aires (1943-1945)." Arg-*3040*.R7, Jan.-June '46, a.
"Precio Medio de la Habitación Ofrecida en la Ciudad de Buenos Aires, 1943-44." Arg-*3040*.R7, Oct.-Dec. '44.
PRECIOS:
ESPECIFICOS—*véase* las respectivas clasificaciones.
AL POR MAYOR—*véase* 6710.
AL POR MENOR (Y COSTO DE LA VIDA)—*véase* 9500.
CONTROL DE—*véase* 6010.
INDICES GENERALES DE—*véase* 6010.
Precios. Arg-*6010*.P5, 1940.
"Precios de Exportación en la República Argentina, 1863-1933." Int-*3040*.E5, Sept. '45, Domínguez.
PRECIPITATION—*see* 4200.
PREDIOS—*véase* 6200.
Preliminary Annual Report, Vital Statistics of Canada. Can-*5200*.P4.
Preliminary Bibliography of Paraguay. Par-*1200*.S5, 1943.
Preliminary Cumulative List of Documents in Unrestricted Series Issued by the General Assembly, Security Council, Atomic Energy Commission, Economic and Social Council and Its Commissions, International Health Conference, to 31 July 1946. Int-*8200*.P45, 1946.
Preliminary Report on Births, Deaths and Marriages. Can-*5200*.P5.
"Premier Rapport sur l'Organisation des Services Statistiques." Int-*3020*.B5, 1936, Julin.
The Preparation of Statistical Tables: A Handbook. NonG-*3260*.P5.
PRESENTACION:
GRAFICA—*véase* 3260.
TABULAR—*véase* 3260.
PRESERVACION DE DOCUMENTOS—*véase* 1600.
PRESERVATION OF RECORDS—*see* 1600.
PRESIDIOS—*véase* 8330.
PRESION BAROMETRICA—*véase* 4200.
PRESTAMOS—*véase* 6720.
PRESUMPTIVE DEATH PRACTICES—*see* 5250.
Presupuesto General de Egresos e Ingresos. Nic-*8800*.P5.
Presupuesto General del Estado. Ur-*8800*.P5.
PRESUPUESTOS:
DE LA FAMILIA—*véase* 9700.
NACIONALES—*véase* 8800.
"Presupuestos de Ingresos y Gastos Públicos de Venezuela en los Años Económicos de 1935-1936 a 1942-1943." Ven-*8800*.R5, June 1944, c.
PREVENTIVE MEDICINE—*see* 7110.
Previsión Social. Chile-*7110*.P5.
PRICE:
In specific subject fields, *see* the particular field.
CONTROL—*see* 6010.
INDEXES, GENERAL—*see* 6010.
RETAIL, AND COST OF LIVING—*see* 9500.
WHOLESALE—*see* 6710.
The Price Control and Subsidy Program in Canada. Can-*6010*.P3, 1943.
Price Trends and Price Control in the American Republics, 1939-1945. Int-*6010*.P5, 1939-45.
"Prices." USA-*3000*.G7, 1946.
Prices and Price Indexes. Can-*6010*.P5.
Prices and Price Indexes. Can-*6010*.P6.
PRIMARY EDUCATIONAL INSTITUTIONS—*see* 7210.
Primer Censo Cafetalero Nacional, 1943. RepDom-*2210*.C5, 1943.1.
Primer Censo Comercial de Nicaragua, 1940: Instructivo. Nic-*2150*.C3, 1940.
Primer Censo Comercial, Marzo de 1940: Métodos y Procedimientos. Méx-*2150*.C3, 1940.1.
Primer Censo de Edificios: 18 de Enero de 1940: Cédula. Nic-*2120*.C5, 1940.
Primer Censo de la República Argentina, Verificado en los Días 15, 16 y 17 de Setiembre de 1869. Arg-*2200*.C5, 1869.
Primer Censo de Transportes, Febrero y Marzo de 1940: Métodos y Procedimientos. Méx-*2180*.C5, 1940.1.
Primer Censo Ejidal, 1935. Méx-*2210*.C7, 1935.
Primer Censo General de Población, 1940: Instructivo para Empadronadores; Interpretación de la Boleta de Población. Nic-*2100*.C5, 1940.2.
Primer Censo Industrial de 1930. Méx-*2250*.C5, 1930.
Primer Censo Industrial de Colombia, 1945: Boleta General. Col-*2150*.C5, 1945. c.
Primer Censo Industrial de Colombia, 1945: Cartilla de Instrucciones. Col-*2150*.C5, 1945. b.
Primer Censo Industrial de Nicaragua, 1940:

PRODUCTION—cont.
CONTROLS—*see* 6400.
INDUSTRIAL—*see* 6400.
MINING—*see* 6300.
Production: Wartime Achievements and the Reconversion Outlook. USA-*6020*.P5, 1945.
"Producto Nacional e Ingreso Nacional." Méx-*6000*.T5, Oct.-Dec. '45, Ortiz Mena.
PROFESIONALES:
CLASIFICACIONES—*véase* 3240; 9000.
DEFUNCIONES—*véase* 5212; 9400.
ENFERMEDADES—*véase* 9400.
INGRESOS—*véase* 9200.
NOMENCLATURAS—*véase* 3240; 9200.
RIESGOS—*véase* 9400.
PROFESSIONAL STANDARDS FOR STATISTICIANS—*see* 3250.
PROFIT SHARING, ECONOMIC—*see* 6000.
PROFITS, FINANCIAL—*see* 6721.
Programa de Biometría. Arg-*3320*.P3, 1943.
Programa de Estadística. Arg-*3320*.P5, 1944.
Programa de Estadística Social y Metodología Estadística. Cuba-*3320*.P5, 1944.
Programa de Estatística Geral e Aplicada. Bras-*3320*.P5, 1944.
"Programa de Estudios del Curso General de Estadística de la Contraloría General de la República." Col-*1420*.A5, May 5, '42, b.
Programa de Matemática Actuarial. Arg-*3320*.P6, 1944.
Programa de Matemática Financiera y Actuarial. Arg-*3320*.P7, 1944.
Programa de Matemáticas. Arg-*3320*.P8, 1944.
Programa de un Curso de Estadística Metodológica (Introducción a la Estadística Matemática) Seguido de un Curso de Matemáticas Generales Necesarias para Su Estudio. Rep Dom-*3320*.P5, 1940.
Programa e Instructivo para el Servicio de Estadística Agrícola en las Agencias Generales. Méx-*6200*.P5, 1945.
Programa Estadístico para el Ministerio de Agricultura del Perú. Perú-*6200*.S5, 1945, Sp.
Programa Estadístico y Económico para el Ministerio de Agricultura de la República Argentina. Arg-*6200*.P5, 1923.
Programme of the World Agricultural Census of 1940. Int-*2110*.W5, 1940.1.
"Progress in the Collection of Annual Agricultural Statistics and of Crop Reports during Growth in the Dominion of Canada." Int-*3040*.R5, 1936.4, Godfrey.
PROHIJAMIENTOS—*véase* 5250.
Project Series of the Inter American Statistical Institute. Int-*3000*.P5.
Projet d'un Recensement du Monde. Int-*2000*.P5, 1881.
PRONOSTICOS:
ECONOMICOS—*véase* 6000.
METEOROLOGICOS (Y DEL TIEMPO)—*véase* 4200.
Prontuario—Geográfico, Comercial, Estadístico, y Servicios Administrativos de El Salvador. ElSal-*1900*.P5, 1939.
PROPAGANDA CENSAL—*véase* 2100-2199.
PROPERTY AND REAL ESTATE—*see* 6721.
PROPERTY, REAL, AGRICULTURAL—*see* 6200.
PROPERTY, REAL, PUBLIC HOLDINGS—*see* 6100.
PROPIEDADES:
AGRICOLAS—*véase* 6200.
FINANCIERAS—*véase* 6721.
INMUEBLES (BIENES RAICES EN GENERAL)—*véase* 6721; 2220; 6100.
PUBLICAS—*véase* 6100.
Proposed Statistical Classification of Diseases, Injuries, and Causes of Death. Int-*3230*.P5, 1946.
"A Propósito del IV Censo General." Arg-*3040*.R5, July '45, a.
"A Propósito del Cuarto Censo General." Arg-*3040*.R5, June '45, a.
PRORRATEO—*véase* 8000.
"Protección a la Familia Argentina." Arg-*3040*.R5, Feb. '43, Casiello.
Protección Social. Bol-*9950*.P5.
Protocol to the International Convention (Relating to Economic Statistics). Int-*6000*.I75, 1928.B.
Proyecto de Informe sobre Standardización de las Estadísticas de Atención Médica. Int-*3230*.P6, 1946.
Proyecto de Informe sobre Standardización de las Estadísticas de Morbilidad. Int-*3230*.P7, 1946.
"Prueba de Registro de Nacimientos en Paraguay." Int-*3040*.E5, Mar. '46, Roig Ocampos.
PRUEBAS:
DE INTELIGENCIA—*véase* 7200.
DE PERSONALIDAD—*véase* 7200.
ESTADISTICAS—*véase* 3140.
PSICOLOGIA:
GENERAL—*véase* 7300.
PEDAGOGICA—*véase* 7200.
SOCIAL—*véase* 9000.
PSYCHOLOGY:
GENERAL—*see* 7300.
EDUCATIONAL—*see* 7200.
SOCIAL—*see* 9000.
Psychometrika. USA-*3040*.P5.
Ptoukha, Michel.
Int-*3040*.R5, 1934.2. "Méthodes pour Calculer les Taux de Mortalité Infantile selon les Mois de l'Année."

QUESTIONNAIRES—*see* 3100.
QUICKSILVER—*see* 6300.
QUIEBRAS—*véase* 6720.
QUIMICA:
GENERAL—*véase* 7300.
PRODUCCION—*véase* 6400.
Quinta Conferencia Panamericana de Directores Nacionales de Sanidad, Abril 22-29, 1944. Int-*7110*.P4, 1944.
Quinto Censo de Población, 15 de Mayo de 1930. Méx-*2200*.C5, 1930.2.
Quinto Censo General de Población, Levantado el 7 de Abril de 1940. Guat-*2200*.C5, 1940.

RACE—*see* 5120.
RACIAL PROBLEMS—*see* 9600.
RADIOS—*see-véase* 6622.
RADIOS, FABRICACION DE—*véase* 6400.
RADIOS, PRODUCTION OF—*see* 6400.
RAILROADS—*see* 6614.
Railway Statistical Terms. NonG-*6614*.R5, 1941.
Rainboth, Edith Dyer.
USA-*9700*.F511, 1941. *Family Income and Expenditures, Five Regions, Part 2, Family Expenditures, Farm Series.*
USA-*9700*.F512, 1940. *Family Income and Expenditures, Five Regions, Part 2, Family Expenditures, Urban and Village Series.*
RAINFALL—*see* 4200.
Ramos, C.
Cuba-*3030*.R5, 1945. *Reorganización de la Dirección General de Estadística de la República de Cuba.*
Random Sampling Distributions. NonG-*3140*. R5, 1943.
Random Sampling Numbers. NonG-*3150*.R5, 1927.
Rapport Annuel (Garde d'Haïti). Haiti-*8300*.R5.
Rapport Annuel de l'Administration Générale des Contributions. Haiti-*8800*.R3.
Rapport Annuel du Conseiller Financier—Receveur Général. Haiti-*6000*.R5.
Rapport Annuel du Département Fiscal (Banque Nationale de la République d'Haïti). Haiti-*6000*.R5.
Rapport Annuel du Directeur Général (Service National d'Hygiène et d'Assistance Publique). Haiti-*7110*.R5.
Rapport Annuel du Représentant Fiscal (Banque Nationale de la République d'Haïti). Haiti-*6000*.R5.
Rapport Annuel du Service National de la Production Agricole et de l'Enseignement Rural. Haiti-*6200*.R5.
Rapport Annuel du Sous-Secrétaire d'État et Ingénieur en Chêf. Haiti-*8000*.R5.
Rapport Annuel sur la Statistique de la Criminalité. Can-*8300*.A5.
Rapport Annuel sur les Institutions pour Maladies Mentales. Can-*7130*.A4.
Rapport Annuel sur les Institutions pour Tuberculeux. Can-*7130*.A6.
Rapport au Conseil sur les Travaux de la . . . Session (Société des Nations. Comité d'Experts Statisticiens). Int-*3000*.R6.
"Rapport de la Commission pour la Définition de la 'Population Rurale'." Int-*3040*.R5, 1938.2, Bunle.
Rapport Epidémiologique. Int-*7110*.R4.
Rapport Préliminaire Annual, Statistiques Vitales du Canada. Can-*5200*.P4.
Rapport Préliminaire sur les Naissances, les Décès et les Mariages. Can-*5200*.P5.
Rapport sur le . . . Exercice (Bureau International pour la Publication des Tarifs Douaniers). Int-*6900*.R4.
"Rapport sur les Travaux Préparatoires à la Vème Révision Décennale de la Nomenclature Internationale des Causes de Décès." Int-*3040*.R5, 1938.1, Huber.
RATES OF EXCHANGE—*see* 6720.
Ratio of Children to Women, 1920. USA-*5211*. R5, 1920.
Raymond, Natalie.
Int-*4000*.M4, 1945. *Index to Map of Hispanic America, 1:1,000.000.*
RAYON—*see-véase* 6400.
RAZA:
GENERAL (ASPECTO DE LA POBLACION)—*véase* 5120.
PROBLEMAS SOCIALES DE—*véase* 9600.
REAL ESTATE AND REAL PROPERTY INVESTMENTS AND HOLDINGS:
GENERAL—*see* 6721; 2220; 6100.
PUBLIC—*see* 6100.
La Realidad Médico-Social Chilena. Chile-*7100*.R5, 1939.
Rebuelto, Emilio.
Int-*1300*.A5, 1940.3. "La Sociedad Argentina Estadística."
Recenseamento de 1940: Metodos e Processos. Bras-*2100*.R5, 1940.1.
Recenseamento do Brasil Realizado em 1° de Setembro de 1920. Bras-*2200*.R4, 1920.
Recenseamento Geral da República em 1.° de Setembro de 1940: Esquema Geral da Operação (organization chart). Bras-*3030*.I5, 1940.
Recenseamento Geral de 1940: Finalidades do Censo Agrícola. Bras-*2110*.R5, 1940.
Recenseamento Geral do Brasil (1.° de Setembro

Reglamento del XI Censo de Población. Chile-*2100*.C5, 1940.4.
Reglamento del Censo de Población de 1945. Hond-*2100*.C5, 1945.3.
Reglamento Interior de la Dirección General de Sanidad. Hond-*7101*.R5, 1925.
Reglamento para el Censo General de Población del Año 1940. Guat-*2100*.C5, 1940.2.
Reglamento para el Levantamiento del Censo de Población de 1940. Hond-*2100*.C5, 1940.3.
REGLAMENTOS:
ESPECIFICOS—*véase* las respectivas clasificaciones.
DE CENSOS (incluyendo LEYES)—*véase* 2100-2199.
DE ORGANISMOS ESTADISTICOS (incluyendo ESTATUTOS)—*véase* 3030.
GUBERNAMENTALES (EN GENERAL)—*véase* 8400.
REHABILITACION DE LA TIERRA—*véase* 6200.
Relación Anual de la Sección de Estadística Vital. Ven-*5200*.A5.
RELACIONES:
CULTURALES, PROGRAMAS INTERNACIONALES DE—*véase* 8200.
INTERNACIONALES (EN GENERAL)—*véase* 8200-8299.
"Relative Efficiencies of Various Sampling Units in Population Inquiries." USA-*3040*.J5, Mar. '42, Hansen.
Relatório Anual do Serviço de Estatística da Previdência e Trabalho. Bras-*9950*.R5.
Relatório do Banco do Brasil. Bras-*6720*.R4.
Relatório do Departamento Administrativo do Serviço Público. Bras-*8000*.R4.
Relatório do Instituto Brasileiro de Geografia e Estatística. Bras-*3030*.R4.
Relevé Epidémiologique Hebdomadaire. Int-*7110*.R6.
RELIEF—*see* 9951.
RELIGION—*see* 7400.
Religious Bodies, 1936. USA-*2260*.R5, 1936.
Renda Nacional do Brasil: Colecção de Estudos Diversos. Bras-*6100*.R5, 1944.
Renooij, D. C.
Int-*3010*.E5, 1934. *The Economist's Handbook.*
RENTA NACIONAL—*véase* 6100.
"Renta Nacional." Arg-*3040*.R5, Feb. '46.
Renta Nacional, 1940-1945. Chile-*6100*.R5, 1940-45.
La Renta Nacional de la República Argentina. Arg-*6100*.R4, 1935-45.
Renta Nacional de la República de Panamá: National Income, 1940. Pan-*6100*.R5, 1940.
"La Renta Nacional del Perú. Perú-*6720*.B5, Aug. '45, Barreto.
"Renta Nacional, Su Significado y Medición." Int-*3040*.E5, Dec. '45, Domínguez.
"Renta Nacional, Su Significado y Medición, Parte II." Int-*3040*.E5, Mar. '46, Domínguez.
"Renta Nacional, Su Significado y Medición, Parte III." Int-*3040*.E5, June '46, Domínguez.
RENTS AND VALUES:
GENERAL—*see* 6721; 2220.
HOUSING—*see* 9900; 9500.
Reorganización de la Dirección General de Estadística de la República de Cuba. Cuba-*3030*.R5, 1945.
Reorganización del Servicio de Estadística en Bolivia. Bol-*3030*.R5, 1946.
The Repair and Preservation of Records. NonG-*1600*.R5, 1943.
Répertoire International des Institutions Statistiques. Int-*3000*.R4, 1934.
REPLANTACION FORESTAL—*véase* 6250.
Report, October 9, 1942-April 13, 1943, from Advisory Committee on Government Questionnaires. USA-*3110*.R5, 1942-43.
Report of the Anglo-American Caribbean Commission to the Governments of the United States and Great Britain. Int-*8200*.R3.
"Report of the Commission for the Study of Statistics of Consumption and of Carryover of Commodities." Int-*3040*.R5, 1933.3, Craig.
"Report of the Committee on Form and Content of Annual Vital Statistics Reports." Can-*7110*.C5, Sept. '46.
Report of the Committee on the Standardization of Mining Statistics. USA-*3200*.R5, 1918.
Report of the First Session of the Conference, Held at the City of Quebec, Canada, October 16 to November 1, 1945 (Food and Agriculture Organization of the United Nations). Int-*6200*.C5, 1945.2.
Report of the Grain Trade of Canada. Can-*6210*.R5.
Report of the National Health Survey. Can-*7110*.R5, 1945.
Report of the National Research Council. USA-*8500*.R3.
Report of the President's Committee on the Cost of Living. USA-*9500*.R3, 1945.
Report of the Reconstruction Finance Corporation. USA-*6720*.R5.
"Report of the Statistical Commission to the Economic and Social Council." Int-*8200*.J3, May 31, '46.
Report of the Sub-Committee on Banking Sta-

República Dominicana—cont.

RepDom-*2100*.C6, 1945. *Censo Especial Urbano de Ciudad Trujillo, 1945.*

RepDom-*2100*.C7, 1944. *Censo Especial Urbano San Cristóbal, 1944: Instrucciones a los Enumeradores.*

RepDom-*2110*.C5, 1940.1. *Censo Agro-Pecuario de 1940: Métodos y Procedimientos.*

RepDom-*2200*.C5, 1935-46. *Población de la República Dominicana, según las Sucesivas Modificaciones Territoriales a Partir del 13 de Mayo de 1935, Día en que Se Levantó el Segundo Censo Nacional de Población, Hasta el 1 de Enero de 1946.*

RepDom-*2200*.C5, 1935.2. *Población de la República Censada el 13 de Mayo de 1935, Clasificada por Grupos de Edades, Zona y Sexo.*

RepDom-*3000*.M5. *Monografías Estadísticas de la Dirección General de Estadística.*

RepDom-*3320*.P5, 1940. *Programa de un Curso de Estadística Metodológica (Introducción a la Estadística Matemática) Seguido de un Curso de Matemáticas Generales Necesarias para su Estudio.*

RepDom-*6400*.R5, 1938. *Resumen General de la Producción de la Pequeña Industria Rural, Año 1938.*

RepDom-*6900*.E3, 1941. *Estudio Estadístico de Algunos Aspectos del Comercio Exterior de la República Dominicana, 1920-1939.*

RepDom-*6900*.M5, 1926-37. *Maderas de Exportación.*

RepDom-*6900*.R5, 1937. *Resumen General de la Producción de Algunos Artículos de Consumo y Exportación de la República, durante el Año 1937.*

——. ——. Sección de Gobierno y Administración.

RepDom-*6600*.M5. *Movimiento Postal, Telefónico y Radiotelegráfico. Carreteras de la República Dominicana.*

RepDom-*8800*.F5. *Finanzas Municipales.*

——. ——. Sección de Producción y Economía.

RepDom-*6230*.S5. *Sacrificio de Ganado.*

RepDom-*6720*.E5. *Estadística Bancaria.*

RepDom-*6721*.R5. *Registro Público.*

RepDom-*6730*.E5. *Estadística de los Negocios de Seguros.*

RepDom-*6900*.E7. *Exportación de la República Dominicana.*

RepDom-*6900*.I5. *Importación de la República Dominicana.*

RepDom-*6900*.I5, 1944. *Importación de la República Dominicana, 1944.*

——. ——. Sección del Censo.

RepDom-*2100*.C5, 1935.1. *Censo de 1935: Métodos y Procedimientos.*

RepDom-*2100*.C5, 1935.3. *Censo de 1935: Instrucciones a los Enumeradores del Censo de Población.*

RepDom-*2100*.C5, 1935.4. *Censo de 1935: Instrucciones Generales sobre las Funciones de un Enumerador.*

RepDom-*2200*.C5, 1935.1. *Población de la República Dominicana, Distribuída por Nacionalidades: Cifras del Censo Nacional de 1935.*

RepDom-*2200*.C5, 1935.3. *Censo de Población Urbana y Rural, por Provincias y Comunas.*

——. ——. Sección Demográfica.

RepDom-*5200*.C5, 1946. *Commentary Explanations and Sample Copies of the Forms and Manuals Used in Vital Records and Vital Statistics Activities in the Dominican Republic, November 1946.*

——. ——. Sección Sociográfica.

RepDom-*5212*.S5. *Suicidios y Tentativas de Suicidios Verificados en la República.*

RepDom-*7100*.C5, 1946. *Commentary Explanations and Sample Copies of the Forms Used in Medical and Public Health Statistics Activities (Not Including Vital Statistics) in the Dominican Republic, November 1946.*

RepDom-*7110*.E5, *Estadística Sanitaria.*

RepDom-*7160*.A5. *Accidentes de Tránsito Terrestre por Circulación de Vehículos.*

——. Leyes, Estatutos, Etc.

RepDom-*2100*.C5, 1935.2. *Censo de 1935: Ley del Congreso Nacional que Dispone las Condiciones en que Deben Desarrollarse las Operaciones del Censo.*

RepDom-*6900*.E5, 1920. *Executive Order No. 332: . . .Tariff Law on Imports and Exports . . . Effective on and after January 1, 1920.*

RepDom-*6900*.O5, 1920. "Orden Ejecutiva No. 332 (de Septiembre 25, 1919) que Promulga la Ley sobre Aranceles de Importación y Exportación que Estará en Vigor desde el Día 1° de Enero de 1920."

——. Oficina Central del Estado Civil—*see-véase* República Dominicana. Departamento de Justicia. Oficina Central del Estado Civil.

——. Santo Domingo (Dist.) Consejo Administrativo.

RETAIL PRICES—cont.

Of particular commodities, *see* 6000, the particular subject field.

RETAIL TRADE—*see* 6710.

Reunión de Técnicos sobre Problemas de Banca Central del Continente Americano. Int-*6720*.R5, 1946.

Reuss, Carl F.
USA-*9000*.A6, June '43. "An Appraisal of the 1936 Religious Census."

The Review of Economic Statistics. USA-*3040*.R5.

"A Review of Recent Statistical Developments in Sampling and Sampling Surveys." GrBrit-*3040*.J5, Part 1, '46, Yates.

A Review of Statistics of Oil Pipe Lines, 1921-1941. USA-*6613*.R5, 1921-41.

Revista Agrícola. Guat-*6200*.R4.

Revista Brasileira de Estatística. Bras-*3040*.R5.

Revista Brasileira de Estudos Pedagógicos. Bras-*7200*.R5.

Revista Brasileira de Geografia. Bras-*4000*.R5.

Revista Cafetalera de Guatemala. Guat-*6210*.R5.

Revista Chilena de Higiene y Medicina Preventiva. Chile-*7100*.R7.

Revista de Agricultura y Comercio. Pan-*6000*.R5.

Revista de Agricultura y Ganadería. Par-*6200*.R5.

Revista de Ciencias Económicas. Arg-*3040*.R3.

Revista de Economía. Guat-*6000*.R5.

Revista de Economía. Méx-*6000*.R5.

Revista de Economía y Estadística. Arg-*3040*.R7.

Revista de Economía y Estadística. Méx-*1420*.R5.

Revista de Economía y Finanzas. Perú-*6000*.R5.

Revista de Educación. Par-*7200*.R5.

Revista de Estadística. Méx-*1420*.R5.

Revista de Fomento. Ven-*6000*.R5.

Revista de Hacienda. Col-*8800*.R5.

Revista de Hacienda. Ec-*8800*.R5.

Revista de Hacienda. Perú-*6000*.R7.

Revista de Hacienda. Ven-*8800*.R5.

Revista de Imigração. Bras-*5300*.R5.

Revista de la Facultad de Ciencias Económicas. Perú-*3040*.R5.

Revista de la Facultad de Ciencias Económicas, Comerciales y Políticas. Arg-*3040*.R8.

Revista de la Facultad de Ciencias Económicas y de Administración. Ur-*3040*.R5.

Revista de la Facultad de Ciencias Físicomatemáticas. Arg-*3040*.R9.

Revista de la Sociedad Colombiana de Estudios Estadísticos. Col-*3040*.E5.

Revista de la Unión Industrial Argentina. Arg-*6400*.R7.

Revista de Sanidad. Guat-*7110*.R5.

Revista de Sanidad y Asistencia Social. Ven-*7110*.R5.

Revista del Banco de la Nación Argentina. Arg-*6720*.R4.

Revista del Banco de la República. Col-*6720*.R5.

Revista del Banco de la República. Méx-*6720*.R5.

Revista del Banco de la República Oriental del Uruguay. Ur-*6720*.R5.

Revista del Banco Nacional de Costa Rica. CR-*6720*.R5.

Revista del Banco Nacional de Nicaragua. Nic-*6720*.R5.

Revista del Banco Nacional de Seguros. CR-*6720*.R6.

Revista del Comercio Exterior. Méx-*6900*.R5.

Revista del Consejo Administrativo de los Ferrocarriles Nacionales. Col-*6614*.R5.

Revista del Instituto de Salubridad y Enfermedades Tropicales. Méx-*7110*.R5.

Revista del Instituto Nacional del Café. Ven-*6210*.R5.

Revista del Ministerio de Obras Públicas. Par-*8000*.M5.

Revista del Trabajo. Méx-*9000*.T5.

Revista do Departamento Nacional do Café. Bras-*6210*.R5.

Revista Económica. Arg-*6720*.R6.

Revista Internacional del Trabajo. Int-*9000*.R6.

Revista Mensual (Dirección General de Estadística). Bol-*1420*.R5.

Revista Mensual del Banco Central de Reserva de El Salvador. ElSal-*6720*.R5.

Revista Mexicana de Sociología. Méx-*9000*.R5.

REVISTAS DE ESTADISTICA—*véase* 3040.

Revistas de Medicina y Sanidad de la América. Int-*7100*.R5, 1940.

Revue de l'Institut International de Statistique. Int-*3040*.R5.

Revue de l'Institut International de Statistique: Supplement Mensuel. Int-*1420*.R5.

Rew, R. H.
Int-*3020*.B5, 1923. "Statistics of the Fisheries."
Int-*3020*.B5, 1909. "International Fishery Statistics."
Int-*3020*.B5, 1909, a. "Statistique Internationale des Pêcheries."

Ribeiro Campos, Carlos Augusto.
Bras-*1400*.A7, 1941. *Atlas Estatístico do Brasil.*

Ribeiro Costa, Joaquim.
Bras-*6020*.E5, 1939. *A Estatística da Produção no Estado de Minas Gerais.*

1939: Datos Definitivos. Méx-*2220*.C5, 1939.2.
Segundo Censo de Edificios de los Estados Unidos Mexicanos, 20 de Octubre de 1939: Métodos y Procedimientos. Méx-*2120*.C5, 1939.1.
Segundo Censo de Edificios de los Estados Unidos Mexicanos: Resumen General. Méx-*2220*.C5, 1939.1.
Segundo Censo de la República Argentina, Mayo 10 de 1895. Arg-*2200*.C5, 1895.
Segundo Censo Ejidal de los Estados Unidos Mexicanos, 6 de Marzo de 1940. Méx-*2210*.C7, 1940.
Segundo Censo Ejidal de los Estados Unidos Mexicanos, 6 de Marzo de 1940: Métodos y Procedimientos. Méx-*2110*.C7, 1940.1.
Segundo Censo Industrial de 1935. Méx-*2250*.C5, 1935.
Segura, Pedro.
NonG-*3330*.Boddington, 1942. *Estadística y Su Aplicación al Comercio.*
SEGURIDAD SOCIAL—*véase* 9950-9959.
La Seguridad Social en Chile. Chile-*9950*.S5, 1942.
SEGURO:
COMERCIAL (EN GENERAL)—*véase* 6730.
CONTRA EL DESEMPLEO—*véase* 9952.
CONTRA ENFERMEDADES—*véase* 7190; 6730; 9952.
DE ENTIERRO—*véase* 6730; 9952.
DE LA VEJEZ—*véase* 6730; 9952.
DE SALUD—*véase* 7190; 9952.
DE VIDA—*véase* 6730.
SOCIAL—*véase* 9952.
Selected Bibliography in English on Latin American Resources, Economic Development, and Trade. Int-*6000*.S4, 1943.
Selected Bibliography on Medical Economics. USA-*7100*.S5, 1944.
A Selected List of Latin American Periodicals Containing Laws and Legal Information. Int-*8400*.S5, 1942.
Selected List of the Publications of the Bureau of Labor Statistics. USA-*9000*.S3, 1943.
SELECTIVE SERVICE—*see* 8100.
Un Semestre de Labor en los Sanatorios. Chile-*7130*.S6, 1946.
SENSORY DEFECTS—*see* 7170.
SENTIDOS, DEFECTOS DE LOS—*véase* 7170.
Septième Recensement du Canada, 1931. Can-*2200*.C5, 1931.2.
Séptimo Censo Nacional de Población, 1941: Instrucciones y Modelos. Ven-*2100*.C5, 1941.2.
Séptimo Censo Nacional de Población, Levantado el 7 de Diciembre de 1941. Ven-*2200*.C5, 1941.
Serial Publications of Foreign Governments, 1815-1931. Int-*1200*.S6, 1815-1931.
Serie B, Estadísticas y Censos, del Departamento Nacional del Trabajo. Arg-*9000*.S5.
Serie de Congressos e Conferências. Int-*1300*.S5.
Serie de Estudios Monográficos. Guat-*3000*.S5.
Serie de Proyectos del Instituto Interamericano de Estadística. Int-*3000*.P5.
Serie sobre Congresos y Conferencias. Int-*1300*.S6.
SERIES.
Int-*1200*.B3. *Bibliographic Series of the Pan American Union.*
Int-*1200*.E5. *Executive Agreement Series* (U. S. Department of State).
Int-*1200*.L5. *Latin American Series of the Library of Congress.*
Int-*1300*.C8. *Congress and Conference Series* (Pan American Union).
Int-*1300*.S5. *Serie de Congressos e Conferências* (União Panamericana).
Int-*1300*.S6. *Serie sobre Congresos y Conferencias* (Unión Panamericana).
Int-*1900*.S3. *Series of Descriptive Booklets on the American Nations.*
Int-*1950*.A5. *American Nation Series* (Pan American Union).
Int-*3000*.P5. *Project Series of the Inter-American Statistical Institute. Serie de Proyectos del Instituto Interamericano de Estadística.*
Int-*3010*.B5. *Bibliography Series, IASI.*
Int-*3200*.S8. *Studies and Reports on Statistical Methods* (League of Nations. Committee of Statistical Experts). *Etudes et Rapports sur les Méthodes Statistiques* (Société des Nations. Comité d'Experts Statisticiens).
Int-*5000*.S8. *Summary of Biostatistics Series* (Bureau of the Census).
Int-*5200*.S5. *Statistical Handbook Series: The Official Vital Statistics of . . . 1925- .*
Int-*5300*.S5. *Studies and Reports, Series O, Migration* (International Labour Office).
Int-*6000*.E2. *Economic Controls and Commercial Policy in the American Republics* (U. S. Tariff Commission).
Int-*6000*.I3. *Informaciones Económicas del Banco de México.*
Int-*6000*.S6. *Studies and Reports, Series*

SERIES—cont.
of the Bureau of Foreign and Domestic Commerce.
USA-*8210*.C5. *Conference Series of the United States Department of State.*
USA-*8500*.P3. *Pamphlets of the Brookings Institution.*
USA-*8500*.P5. *Planning Pamphlets of the National Planning Association.*
USA-*8500*.T5. *Technical Papers of the National Resources Planning Board.*
USA-*9000*.B4. *Bulletins of the Bureau of Labor Statistics.*
USA-*9000*.B6. *Bulletins of the Social Science Research Council.*
USA-*9000*.L4. *Labor Conditions in Latin America.*
USA-*9600*.B5. *Bulletins of the Women's Bureau.*
USA-*9600*.J5. *Jewish Social Studies, Publications.*
USA-*9700*.C7, 1935-36. *Consumer Purchases Study.*
USA-*9950*.B5. *Bureau Memorandums of the Bureau of Research and Statistics of the Social Security Board.*
USA-*9951*.P5. *Publications of the Children's Bureau.*
UR-*3000*.C5. *Cuadernos del Instituto de Estadística.*
WI-*6000*.D5. *Development and Welfare in the West Indies, Bulletins.*
SERIES ESTADISTICAS:
CRONOLOGICAS (EN GENERAL)—*véase* las respectivas clasificaciones.
DE LA ACTUALIDAD (EN GENERAL)—*véase* 1420.
DE LA ACTUALIDAD, SOBRE TOPICOS ESPECIALES DE INVESTIGACION—*véase* las respectivas clasificaciones.
ECONOMICAS—*véase* 6010; *véase también* las respectivas clasificaciones.
SERVICIO:
CIVIL (EMPLEADOS PUBLICOS)—*véase* 8000.
DE CORREOS—*véase* 8000.
DE POLICIA—*véase* 8300.
MILITAR—*véase* 8100.
SOCIAL—*véase* 9950-9959.
Servicio Cooperativo Inter-Americano de Producción de Alimentos—*see* - *véase* Perú. Ministerio de Agricultura. Servicio Cooperativo Inter-Americano de Producción de Alimentos.
"El Servicio de Bioestadística en México." Int-*7110*.B5, May '44, Granillo.
SERVICIOS:
AGRICOLAS (incluyendo ENSEÑANZA AGRICOLA—*véase* 6200.
POSTALES—*véase* 8000.
Seventh Census of 1931: Administrative Report of the Dominion Statistician. Can-*2100*.C5, 1931.
Seventh Census of Canada, 1931. Can-*2200*.C5, 1931.2.
SEWAGE AND SEWAGE DISPOSAL—*see* 7110.
SEX—*see* 5110.
SEX DETERMINATION—*see* 7300.
SEXO—*véase* 5110.
SEXO, DETERMINACION DE—*véase* 7300.
Sexto Censo de Población, 1940. Méx-*2200*.C5, 1940.1.
Sexto Censo General de Población de los Estados Unidos Mexicanos, 6 de Marzo de 1940: Población Municipal. Méx-*2200*.C5, 1940.2.
Sexto Censo Nacional, 1936. Ven-*2200*.C5, 1936.2.
SHARECROPPERS—*see* 9600.
Shattuck, George Cheever.
Guat-*7110*.M5, 1938. *A Medical Survey of the Republic of Guatemala.*
Shelby, Charmion.
Int-*1200*.L3, 1944. *Latin American Periodicals Currently Received in the Library of Congress and in the Library of the Department of Agriculture.*
Shelly Hernández, R. de.
Int-*5000*.M3, 1946. *Papers Presented by Special Guests of the Population Association of America at Its Meetings October 25-26, 1946.*
NonG-*3330*, 1941. *La Estadística Aplicada a las Ciencias Biológicas.*
Shewhart, Walter Andrew.
NonG-*3160*.E5, 1931. *Economic Control of Quality of Manufactured Product.*
NonG-*3160*.S5, 1939. *Statistical Method from the Viewpoint of Quality Control.*
SHIPBUILDING—*see* 6400.
SHIPPING—*see* 6615.
Shiskin, Julius.
USA-*6400*.O6, 1940. *The Output of Manufacturing Industries, 1899-1937.*
"Short Methods of Constructing Abridged Life Tables." USA-*6730*.R5, June '43, Greville.
Shoup, Carl.
NonG-*6100*.S3. "The Distinction Between 'Net' and 'Gross' in Income Taxation."
Sibley, Eldridge.
USA-*7110*.A4, Feb. '44. "Problems in Population Estimation."
SICKNESS—*see* 7150.

Irrigation of Agricultural Lands. USA-*2210.*C5, 1940.6.

Sixteenth Census of the United States: 1940: Manufactures, 1939. USA-*2250.*C7, 1939.1.

Sixteenth Census of the United States: 1940: Manufactures, 1939: Industry Classifications for the Census of Manufactures, 1939. USA-*3220.*C5, 1939.

Sixteenth Census of the United States: 1940: Methods and Procedures, USA-*2100.*C5, 1940.1.

Sixteenth Census of the United States: 1940: Methods and Procedures: Forms and Instructions Used in Taking the Census of Puerto Rico. USA PR-*2100.*C5, 1940.1.

Sixteenth Census of the United States: 1940: Mineral Industries, 1939. USA-*2270.*C5, 1939.

Sixteenth Census of the United States: 1940: Population. USA-*2200.*C5, 1940.1.

Sixteenth Census of the United States: 1940: Population and Housing: Families: Characteristics of Rural-Farm Families. USA-*2200.*C5, 1940.7b.

Sixteenth Census of the United States: 1940: Population and Housing: Families: General Characteristics. USA-*2200.*C5, 1940.7a.

Sixteenth Census of the United States: 1940: Population and Housing: Families: Income and Rent. USA-*2200.*C5, 1940.7e.

Sixteenth Census of the United States: 1940: Population and Housing: Families: Tenure and Rent. USA-*2200.*C5, 1940.7g.

Sixteenth Census of the United States: 1940: Population: Characteristics of Persons Not in the Labor Force 14 Years Old or Over. USA-*2200.*C5, 1940.2.

Sixteenth Census of the United States: 1940: Population: Characteristics of the Nonwhite Population by Race. USA-*2200.*C5, 1940.3.

Sixteenth Census of the United States: 1940: Population: Comparative Occupation Statistics for the United States, 1870 to 1940. USA-*2200.*C5, 1940.4.

Sixteenth Census of the United States: 1940: Population: Differential Fertility, 1940 and 1910: Fertility for States and Large Cities. USA-*2200.*C5, 1940.5a.

Sixteenth Census of the United States: 1940: Population: Differential Fertility, 1940 and 1910: Standardized Fertility Rates and Reproduction Rates. USA-*2200.*C5, 1940.5b.

Sixteenth Census of the United States: 1940: Population: Differential Fertility, 1940 and 1910: Women by Number of Children Ever Born. USA-*2200.*C5, 1940.5d.

Sixteenth Census of the United States: 1940: Population: Differential Fertility, 1940 and 1910: Women by Number of Children under 5 Years Old. USA-*2200.*C5, 1940.5c.

Sixteenth Census of the United States: 1940: Population: Education, Occupation and Household Relationship of Males 18 to 44 Years Old. USA-*2200.*C5, 1940.6.

Sixteenth Census of the United States: 1940: Population: Estimates of Labor Force, Employment and Unemployment in the United States, 1940 and 1930. USA-*2200.*C5, 1940.15.

Sixteenth Census of the United States: 1940: Population: Families: Employment Status. USA-*2200.*C5, 1940.7c.

Sixteenth Census of the United States: 1940: Population: Families: Family Wage or Salary Income in 1939. USA-*2200.*C5, 1940.7d.

Sixteenth Census of the United States: 1940: Population: Families: Size of Family and Age of Head. USA-*2200.*C5, 1940.7f.

Sixteenth Census of the United States, 1940: Population: Families: Types of Families. USA-*2200.*C5, 1940.7h.

Sixteenth Census of the United States: 1940: Population: Internal Migration 1935 to 1940. USA-*2200.*C5, 1940.8.

Sixteenth Census of the United States: 1940: Population: The Labor Force (Sample Statistics): Employment and Family Characteristics of Women. USA-*2200.*C5, 1940.10a.

Sixteenth Census of the United States: 1940: Population: The Labor Force (Sample Statistics): Employment and Personal Characteristics. USA-*2200.*C5, 1940.10b.

Sixteenth Census of the United States: 1940: Population: The Labor Force (Sample Statistics): Industrial Characteristics. USA-*2200.*C5, 1940.10c.

Sixteenth Census of the United States: 1940: Population: The Labor Force (Sample Statistics): Occupational Characteristics. USA-*2200.*C5, 1940.10d.

Sixteenth Census of the United States, 1940: Population: The Labor Force (Sample Statistics): Usual Occupation. USA-*2200.*C5, 1940.10e.

Sixteenth Census of the United States: 1940: Population: The Labor Force (Sample Statistics): Wage or Salary Income in 1939. USA-*2200.*C5, 1940.10f.

Sixteenth Census of the United States: 1940: Population: Nativity and Parentage of the White Population: Country of Origin of Foreign Stock. USA-*2200.*C5, 1940.11b.

Sixteenth Census of the United States: 1940: Population: Nativity and Parentage of the White Population: General Characteristics, USA-*2200.*C5, 1940.11a,

The Social Framework of the American Economy. USA-*6000*.S5, 1945.
Social Research. NonG-*9000*.S5, 1942.
Social Science Research Council, United States.
Int-*1200*.H4. *Handbook of Latin American Studies.*
USA-*5300*.R4, 1942. *A Research Memorandum on Internal Migration Resulting from the War Effort.*
USA-*9000*.B6. *Bulletins of the Social Science Research Council.*
NonG-*1800*.M5, 1936. *Manual on Methods of Reproducing Research Materials.*
——. Committee on Social Security.
USA-*9000*.L6, 1938. *Labor in the United States: Basic Statistics for Social Security.*
Social Security Bulletin. USA-*9950*.S4.
Social Security Bulletin. USA-*9950*.S6.
Social Security Yearbook. USA-*9950*.S6.
Sociedad Agronómica de México.
Méx-*6200*.M8. *México Agrario.*
"La Sociedad Argentina Estadística." Int-*1300*.A5, 1940.3, Rebuelto.
Sociedad Científica Argentina.
Arg-*3040*.A5. *Anales de la Sociedad Científica Argentina.*
Sociedad Colombiana de Estudios Estadísticos.
Col-*3040*.E5. *Estadística.*
Sociedad de Beneficencia Pública de Lima.
Int-*3230*.N5, 1929. *Nomenclaturas Internacionales de las Enfermedades (Causas de Defunción—Causas de Incapacidad del Trabajo) para Establecer las Estadísticas Nosológicas (Clasificación "Bertillon").*
——. Sección de Estadística.
Perú-*7130*.D5. *Demografía y Estadística de Hospitales y de Otros Establecimientos.*
Sociedad Mexicana de Geografía y Estadística.
Méx-*3040*.B5. *Boletín de la . . .*
"La Sociedad Mexicana de Geografía y Estadística." Int-*1300*.A5, 1940.3, Benavides.
Sociedad Nacional Agraria, Perú.
Perú-*6200*.M5. *Memoria de la . . .*
Sociedad Rural Argentina.
Arg-*6200*.A4. *Anales de la . . .*
Arg-*6200*.A5, 1928. *Anuario de la Sociedad Rural Argentina: Estadísticas Económicas y Agrarias.*
"A Sociedade Brasileira de Estatística." Int-*1300*.A5, 1940.3, Martins.
SOCIEDADES:
CIENTIFICAS—*véase* 7300.
DE CREDITO—*véase* 6720.
ESTADISTICAS—*véase* 3030.
"Sociedades Nacionales de Estadística en el Hemisferio Occidental." Int-*3040*.E5, Mar. '44.
Société de Statistique de Paris.
Fr-*3040*.J5. *Journal de la Société de Statistique de Paris.*
Société des Nations—*see - véase* League of Nations.
——. Comité d'Experts Statisticiens—*see-véase* League of Nations. Committee of Statistical Experts.
SOCIOLOGIA—*véase* 9900.
SOCIOLOGY—*see* 9900.
"Some Problems in Measuring per Capita Labor Income." NonG-*6100*.S3.
"Some Problems in Teaching Elementary Statistics." USA-*3040*.J5, June '39, Kellogg.
"Some Problems Involved in Allocating Incomes by States." NonG-*6100*.S3.
"Some Problems of Statistical Education and Teaching." USA-*3040*.J5, June '44, Kimelman.
Some Technical Aspects of Foreign Trade Statistics with Special Reference to Valuation. Int-*6900*.S3, 1946.
"Sondages Démographiques dans les Pays Sans État Civil Régulier." Int-*3040*.R5, 1937.2, Bunle.
SORDOS—*véase* 7170.
Soule, George.
Int-*1900*.L7, 1945. *Latin America in the Future World.*
A Source List of Selected Labor Statistics. USA-*9000*.S7, 1944.
"Source Materials: IASI Survey of Census Methods and Procedures: Argentina, Bolivia, Brazil, Canada." Int-*3000*.I5, Nov. '44-Mar. '45.
"Source Materials: IASI Survey of Census Methods and Procedures: Chile, Colombia, Cuba, El Salvador, Guatemala, Haiti, Honduras, Mexico." Int-*3000*.I5, Apr. '45.
"Source Materials: IASI Survey of Census Methods and Procedures: Nicaragua, Panama, Paraguay, Peru, Republica Dominicana, Venezuela." Int-*3000*.I5, May-June '45.
"Source Materials: IASI Survey of Census Methods and Procedures: United States: Sixteenth Census, 1940." Int-*3000*.I5, Aug.-Oct. '44.
"Sources of Current Economic Information of Latin America." Int-*1200*.H4, 1937, Phelps.
"Sources of Information on Social and Labor

Among Communities." NonG-*6100*.S3.

The Statistical Study of Literary Vocabulary. NonG-*7320*.S5, 1944.

Statistical Tables for Biological, Agricultural, and Medical Research. NonG-*3260*.S7, 1938.

Statistical Tables: Their Structure and Use. NonG-*3160*.S4, 1936.

Statistical Training in Canada: Methods and Materials. Can-*3300*, 1944-46.

Statistical Training in the American Nations: Methods and Materials. Int-*3300*, 1944-46.

Statistical Training in the United States: Methods and Materials. USA-*3300*, 1944-46.

Statistical Year Book of the League of Nations. Int-*1410*.A4.

Statistics. NonG-*3330*. Tippett, 1943.

Statistics and Their Application to Commerce. NonG-*3330*. Boddington, 1939.

Statistics: Collecting, Organizing, and Interpreting Data. NonG-*3110*.S5, 1943.

Statistics for Sociologists. NonG-*3330*. Hagood, 1941.

Statistics in Guatemala (A Brief Survey). Guat-*3030*.S5, 1945.

Statistics in Psychology and Education. NonG-*3330*.Garrett, 1937.

"Statistics in the World of Tomorrow." Int-*3040*.E5, June '46, Dunn.

Statistics of Electric Utilities in the United States, Classes A and B Privately Owned Companies. USA-*6400*.S5.

Statistics of Income. USA-*8800*.S7.

Statistics of Migration. Int-*5300*.S3, 1932.

Statistics of Railways in the United States. USA-*6614*.S6.

Statistics of Railways in the United States for the Year Ended December 31, 1943. USA-*6614*.S6, 1943.

Statistics of Steam Railways of Canada. Can-*6614*.S5.

Statistics of the Communications Industry in the United States. USA-*6620*.S5.

"Statistics of the Fisheries." Int-*3020*.B5, 1923, Rew.

Statistics of the Gainfully-Occupied Population. Int-*3240*.S4, 1938.

Statistics of War Production. USA-*6020*.S5.

Statistics Relating to Capital Formation. Int-*3200*.S3, 1938.

Statistique de Bétail et des Produits Animaux. Can-*6230*.L5.

Statistique de l'Electricité, du Gaz et de l'Eau dans les Grandes Villes, 1934. Int-*6400*.S5, 1934.

"Statistique de la Production Agricole." Int-*3020*.B5, 1923, Ricci.

"Statistique de la Radiodiffusion." Int-*3040*. R5, 1936.2, Idenburg.

"La Statistique des États de Culture." Int-*3020*.B5, 1913, March.

Statistique des Pêcheries. Can-*6240*.F5.

Statistique du Commerce Exterieur au Haïti: Méthodes et Procédes. Haiti-*6900*.

Statistique du Logement dans les Grandes Villes, 1928-1934. Int-*9900*.S5, 1928-34.

Statistique du Tourisme dans les Grandes Villes, 1929-1934. Int-*5310*.S5, 1929-1934.

"La Statistique Forestière Internationale." Int-*3040*.R5, 1936.2, Dore.

"La Statistique Forestière Internationale et ses Problèmes." Int-*3040*.R5, 1933.3, Dore.

"Statistique Internationale des États des Cultures." Int-*3020*.B5, 1911, Ricci.

"La Statistique Internationale des Forces Motrices." Int-*3040*.R5, 1933.2, Huber.

"La Statistique Internationale des Forces Motrices." Int-*3040*.R5, 1936.1, Huber.

Statistique Internationale des Grandes Villes: Statistique de l'Electricité, du Gaz et de l'Eau dans les Grandes Villes, 1934. Int-*6400*.S5, 1934.

Statistique Internationale des Grandes Villes: Statistique du Logement dans les Grandes Villes, 1928-1934. Int-*9900*.S5, 1928-34.

Statistique Internationale des Grandes Villes: Statistique du Tourisme dans les Grandes Villes, 1929-1934. Int-*5310*.S5, 1929-1934.

Statistique Internationale des Grandes Villes: Territoire et Population des Grandes Villes, 1928-1934. Int-*5100*.S5, 1928-34.

"Statistique Internationale des Pêcheries." Int-*3020*.B5, 1909, Rew, a.

"Statistique Internationale des Superficies et des Productions Agricoles." Int-*3020*.B5, 1913, Ricci.

Statistiques de la Population Active. Int-*3240*.S6, 1938.

Statistiques de l'Habitation. Int-*3200*.S5, 1939.

Statistiques des Canaux. Can-*6615*.C5.

Statistiques des Télégráphes. Can-*6622*.T5.

Statistiques du Bois. Int-*3200*.S6, 1938.

Statistiques du Commerce International. Int-*6900*.S6.

Statistiques du Logement. Can-*6500*.H5.

"Statistiques du Tourisme International." Int-*3020*.B5, 1938, Molinari, a.

Statistiques du Téléphone. Can-*6622*.T6.

Statistiques Relatives à la Formation des Capitaux. Int-*3200*.S7, 1938.

Statistiques Vitales. Can-*5200*.V5.

"Les Statistiques Vitales." Haiti-*1900*.C5, Dec. '43, Young.

Statistiques Vitales et Enregistrement en Haïti: Méthodes et Procédés. Haiti-*5200*.

Statute, Basic Principles, Social Leadership of

SUBSIDIES, AGRICULTURAL—*see* 6200.
SUBVENCIONES AGRICOLAS—*véase* 6200.
SUGAR—*see* 6210.
"Suggested Charter for an International Trade Organization of the United Nations." *International Conciliation.* Int-*8200*.I55, Nov. '46.
Suggested Standard Forms for Mine Statistics with Especial Reference to Accidents. USA-*3200*.S5, 1916.
Suggestions Concerning the Brazilian Policy of Social Assistance. Bras-*3030*.S7, 1940.
SUICIDES—*see* 5212.
SUICIDIOS—*véase* 5212.
Suicidios y Tentativas de Suicidos Verificados en la República. RepDom-*5212*.S5.
Suma de Productos Formados por Combinaciones con Números Dados. NonG-*3140*.S7, 1940.
Summary of Biostatistics Series. Int-*5000*.S8.
Summary of Foreign Trade of the United States. USA-*6900*.S5.
Summary of Monthly Railway Traffic Reports. Can-*6614*.S8.
Summary Report of National Bureau of Standards Research on Preservation of Records. NonG-*1600*.S5, 1937.
Summary Report of Research at the National Bureau of Standards on the Stability and Preservation of Records on Photographic Film. NonG-*1600*.S7, 1939.
SUPERFICIE DE HACIENDAS Y GRANJAS Y OTRAS TIERRAS CULTIVADAS—*véase* 6200.
Suplemento de la División Político-Territorial de la República. Ven-*4100*.D5, 1945.
Suplemento de la Guía Industrial y Comercial de Venezuela Correspondiente al Año 1943. Ven-*6000*.G5, 1943.A.
Suplemento Estadístico de la Revista Económica. Arg-*6000*.S8.
Suplemento Estadístico de la Revista Económica. Ur-*6000*.S5.
Supplement-A to Bibliography of the Latest Available Source Material on Foreign Trade Statistics. Int-*6900*.B5, 1944.2.
Supplement to American Republics Bibliography; A List of some of the Current Latin American Periodicals Relating to Agriculture, Commerce, Economics, Finances, Industry, Production in the Library of the Pan American Union. Int-*6000*.A5, 1943.B.
SUPPLY AND DEMAND—*see* 6000.
SUPREME COURT—*see* 8300.
SURGERY—*see* 7180.
Survey of American Listed Corporations: Data on Profits and Operations. USA-*6720*.S5.
Survey of Current Business. USA-*6000*.S7.
Survey of Economic Theory on Technological Change and Employment. NonG-*6000*.S7, 1940.
Survey of Production in Canada. Can-*6020*.S5.
Survey of Social Security Statistics. USA-*9950*.S8, 1944.
Surveying and Mapping. NonG-*4000*.S5.
SURVEYS OF HEALTH—*see* 7110.
"A System of Codification of Medical Diagnoses for Application to Punch Cards, with a Plan for Operation." USA-*7110*.A4, June '36, Berkson.
Systems of Classification of Industries and Occupations. Int-*3220*.S5, 1923.
Szturm de Sztrem, Edouard.
Int-*3020*.B5, 1936. "Observations au Sujet du Calcul des Indices des Prix et du 'Quantum' des Importations et Exportations."

"Tabla de Mortalidad de Colombia." Col-*1420*.A5, Mar. 20, '43, Rodríguez.
Tabla de Mortalidad de la República Argentina. Arg-*5220*.T5, 1936.
"Tabla de Mortalidad y Supervivencia para Colombia." Col-*1420*.A5, July '43, Mortara.
TABLAS DE VIDA—*véase* 5220.
Tablas de Vida de la Ciudad de la Habana. Cuba-*5220*.T5, 1920.
"Tablas de Vida de los Habitantes de los Estados Unidos Mexicanos." Méx-*7110*.R5, Bustamante.
"Tables des Matières Contenues dans les Tomes I à XXX." Int-*3020*.B5, 1887-1938.
Tables et Graphiques Statistiques de l'Enseignement Urbain, 1942-1943. Haiti-*7200*.T5, 1942-43.
Tables of Random Sampling Numbers. NonG-*3150*.T4, 1939.
Tables Statistiques de la Fréquentation des Ecoles de l'Enseignement Rural, 1939-1940 et 1940-1941. Haiti-*7200*.T6, 1942.
"Tábua de Mortalidade e de Sobrevivência para a Cidade de Lima (1933-35)." Bras-*3040*.R5, Oct.-Dec. '43, Mortara 2.
"Tábua de Mortalidade e de Sobrevivência para a Colômbia (1939-41)." Bras-*3040*.R5, July-Sept. '43, Mortara 1.
"Tábua de Mortalidade e de Sobrevivência para o México (1929-35)." Bras-*3040*.R5, Oct.-Dec. '43, Mortara 2.
Tábuas Brasileiras de Mortalidade e Sobrevivência. Bras-*5220*.T3, 1946.
Tábuas de Mortalidade e Sobrevivência Brasileiras: Distrito Federal e Município de São Paulo. Bras-*5220*.T52, 1946.

"The Trend Toward an Older Population." USA-*3040*.A5, Jan. '45, Durand.
"Trends, Determinants, and Control in Human Fertility." USA-*3040*.A5, Jan. '45, Whelpton.
Trends in American Progress. USA-*1400*.T5, 1946.
"Trends in Longevity." USA-*3040*.A5, Jan. '45, Dublin.
Trends of Statistical Organization in Brazil. Bras-*3030*.R8, 1944.
Triangulation de la République d'Haïti. Haiti-*6610*.T5, 1946?
TRIBUNAL SUPREMO—*véase* 8300.
TRIGO—*véase* 6210.
El Trimestre Económico. Méx-*6000*.T5.
El Trimestre Estadístico del Ecuador. Ec-*1420*.T5.
Trucco, Sixto E.
NonG-*3330*, 1944. *Análisis Estadístico Aplicado a los Trabajos de Investigación en Agricultura y Biología.*
TRUCK FARMING—*see* 6220.
TRUCKS—*see* 6612.
TRUEQUES (COMERCIALES)—*véase* 6900.
Truesdell, Leon E.
USA-*2220*.C5, 1940.2. *Sixteenth Census of the United States: 1940: Housing: Analytical Maps ...; Block Statistics.*
USA-*3040*.J5, June '45, a. "The Federal Government's Statistical Program for Reconversion and Postwar Adjustment: A Round Table."
USA-*5120*.C5, 1850-1930. *The Canadian-Born in the United States.*
Trujillo y la Estadística. RepDom-*3030*.T5, 1945.
Tuberculosis Reference Statistical Handbook. USA-*7150*.T5.
TUBERIAS:
CONSTRUCCION DE—*véase* 6500.
OPERACIONES DE—*véase* 6613.
Tucker, Rufus S.
USA-*3040*.J5, June '45, a. "The Federal Government's Statistical Program for Reconversion and Postwar Adjustment: A Round Table."
TULLIDOS:
GENERAL—*véase* 7170.
INSTITUCIONES PARA—*véase* 7130.
TUNGSTEN—*see* 6300.
TUNGSTENO—*véase* 6300.
TURISMO, ESTADISTICAS DE—*véase* 5310.
Turosienski, Severin K.
Cuba-*7200*.E5, 1943. *Education in Cuba.*
The T. V. A.: Lessons for International Application. Int-*6400*.T5, 1944.
Twentieth Report to Congress on Lend-Lease Operations, for the Period Ended June 30, 1945. USA-*8800*.R5, 1945.
Twenty-Five Years of Health Progress. USA-*5212*.T5, 1911-35.
Twins and Super-Twins. NonG-*5211*.M5, 1940.

Ugarte, Luis Rose.
Perú-*9800*.S5, 1944. *Situación Alimenticia en la Ciudad de Iquitos.*
Perú-*9800*.S7, 1944. *Situación Alimenticia en los Departamentos del Norte.*
Ullman, Morris B.
Int-*3040*.E5, Sept. '46. "Registros, Formas y Otra Documentación de una Investigación Estadística."
Ulmer, Melville J.
USA-*6010*.W5, 1939-41. *Wartime Prices: Part I, August 1939 to Pearl Harbor.*
Ulrich, Carolyn F.
Int-*1200*.U5, 1943. *Ulrich's Periodicals Directory. Directorio de Publicaciones Periódicas Ulrich.*
Ulrich's Periodicals Directory. Directorio de Publicaciones Periódicas Ulrich. Int-*1200*.U5, 1943.
Ultimas Disposiciones del Poder Ejecutivo Nacional sobre Estadística y Censos. Arg-*3030*.U5, 1944.
Ultimas Disposiciones Oficiales en Relación con la Estadística. Guat-*3030*.U5, 1946.
"XI Censo General de Población." Chile-*1420*.E5, Dec. '41.
UNDER-REGISTRATION—*see* 5220.
UNEMPLOYMENT—*see* 9200; 2200.
UNEMPLOYMENT INSURANCE—*see* 9952.
UNEMPLOYMENT RELIEF—*see* 9951.
União Panamericana—*see* - *véase* Pan American Union.
Unification of Customs Nomenclature. Int-*3210*.U5, 1930.
Uniform Crime Reports for the United States and Its Possessions. USA-*8300*.U5.
Uniform Definitions of Motor Vehicle Accidents. USA-*7160*.U5, 1943.
"L'Uniformité dans la Statistique des Accidents de la Circulation." Int-*3040*.R5, 1936.2, Zanten.
Unión Industrial Argentina.
Arg-*6400*.R7. *Revista de la Unión Industrial Argentina.*
Union Panaméricaine—*see* - *véase* Pan American Union.
Unión Panamericana—*see* - *véase* Pan American Union.
UNIONES OBRERAS—*véase* 9000.

UNIONS, CREDIT—*see* 6720.
UNIONS, LABOR—*see* 9000.
United Nations.
Int-*1420*.M5. *Bulletin Mensuel de Statistique. Monthly Bulletin of Statistics.*
Int-*3040*.E5, Sept. '46, Rice. "ISI, IASI, and UN."
Int-*8200*.I55, Sept. '45. "The United Nations Charter."
Int-*8200*.P45, 1946. *Preliminary Cumulative List of Documents in Unrestricted Series Issued by the General Assembly, Security Council, Atomic Energy Commission, Economic and Social Council and Its Commissions, International Health Conference, to 31 July 1946.*
Int-*8210*.C6, 1946. *Charter of the United Nations and Statute of the International Court of Justice. Charte des Nations Unies et Statut de la Cour Internationale de Justice.*
——. Committee on Contributions.
Int-*6100*.N3, 1946. *National Income and Related Subjects for the United Nations.*
——. Economic and Social Council.
Int-*6100*.W5, 1946. *Working Paper on the Status of National Income Statistics.*
Int-*8200*.J3. *Journal of the Economic ...*
——. ——. Statistical Commission.
Int-*3040*.E3, July '46, Rice. "The United Nations Statistical Commission."
Int-*6100*.W5, 1946. *Working Paper on the Status of National Income Statistics.*
Int-*8200*.J3, May 31, '46. "Report of the Statistical Commission to the Economic and Social Council."
——. International Court of Justice.
Int-*8210*.C6, 1946. *Charter of the United Nations and Statute of the International Court of Justice. Charte des Nations Unies et Statut de la Cour Internationale de Justice.*
——. Statistical Commission—*see-véase* United Nations. Economic and Social Council. Statistical Commission.
"The United Nations Charter." Int-*8200*.I55, Sept. '45.
United Nations Conference on Food and Agriculture—*see-véase* Conference of the Food and Agriculture Organization.
United Nations Food and Agriculture Organization—*see* Food and Agriculture Organization of the United Nations.
United Nations Information Organization. Reference Division.
Int-*7101*.I5, 1945. *International Health Organisation.*
United Nations Monetary and Financial Conference, Bretton Woods, New Hampshire, July 1 to July 22, 1944: Final Act and Related Documents. Int-*6720*.U6, 1944.
United Nations Relief and Rehabilitation Administration.
Int-*7110*.E4. *Epidemiological Information Bulletin.*
"The United Nations Statistical Commission." Int-*3040*.E3, July '46, Rice.
United States. Agricultural Marketing Service—*see-véase* United States. Department of Agriculture. Bureau of Agricultural Economics. Agricultural Marketing Service.
——. Agricultural Research Administration—*see-véase* United States. Department of Agriculture. Agricultural Research Administration.
——. Agriculture Department—*see-véase* United States. Department of Agriculture.
——. Archives of the United States—*see-véase* United States. National Archives.
——. Army Air Forces—*see-véase* United States. War Department. Army Air Forces.
——. Army Service Forces—*see-véase* United States. War Department. Army Service Forces.
——. Board of Economic Warfare—*see-véase* United States. Executive Office of the President. Office for Emergency Management. Foreign Economic Administration. Board of Economic Warfare.
——. Board of Governors of the Federal Reserve System—*see-véase* United States. Federal Reserve System. Board of Governors of the Federal Reserve System.
——. Bureau of Agricultural Economics—*see-véase* United States. Department of Agriculture. Bureau of Agricultural Economics.
——. Bureau of Foreign and Domestic Commerce—*see-véase* United States. Department of Commerce. Bureau of Foreign and Domestic Commerce.
——. Bureau of Home Economics—*see-véase* United States. Department of Agriculture. Agricultural Research Administration. Bureau of Human Nutrition and Home Economics.
——. Bureau of Human Nutrition and Home Economics—*see-véase* United States. Department of Agriculture. Agricultural Research Administration. Bureau of Human Nutrition and Home Economics.
——. Bureau of Internal Revenue—*see-véase* United States. Department of the Treasury. Bureau of Internal Revenue.
——. Bureau of Labor Statistics—*see-véase* United States. Department of Labor. Bureau of Labor Statistics.
——. Bureau of Mines—*see-véase* United

United States—cont.

States. Department of the Interior. Bureau of Mines.

——. Bureau of Prisons—*see-véase* United States. Department of Justice. Bureau of Prisons.

——. Bureau of Standards—*see-véase* United States. Department of Commerce. National Bureau of Standards.

——. Bureau of the Budget—*see-véase* United States. Executive Office of the President. Bureau of the Budget.

——. Bureau of the Census—*see-véase* United States. Department of Commerce. Bureau of the Census.

——. Bureau of the Comptroller of the Currency—*see-véase* United States. Department of the Treasury. Bureau of the Comptroller of the Currency.

——. Bureau of Transport Economics and Statistics—*see-véase* United States. Interstate Commerce Commission. Bureau of Transport Economics and Statistics.

——. Caribbean Fishery Mission.

Int-*6240*.F5, 1943. *The Fisheries and Fishery Resources of the Caribbean Area.*

——. Census Bureau—*see-véase* United States. Department of Commerce. Bureau of the Census.

——. Census Library Project.

Int-*5000*.G5, 1943. *General Censuses and Vital Statistics in the Americas.*

——. Census of Partial Employment, Unemployment, and Occupations.

USA-*2200*.C7, 1937. *Census of Unemployment, 1937: Final Report on Total and Partial Unemployment.*

USA-*2200*.C7, 1937.A. *Census of Unemployment, 1937: Final Report on Total and Partial Unemployment for Geographic Divisions.*

——. Central Statistical Board.

Int-*3000*.S5, 1940. "Statistical Activities of the United States."

USA-*1400*.F5, 1938. *The Federal Chart Book.*

USA-*3030*.A6. *Annual Report of the . . .*

——. Chief of Engineers—*see-véase* United States. War Department. Office of the Chief of Engineers.

——. Children's Bureau—*see-véase* United States. Department of Labor. Children's Bureau.

——. Civil Aeronautics Administration—*see-véase* United States. Department of Commerce. Civil Aeronautics Administration.

——. Civil Aeronautics Board—*see-véase* United States. Department of Commerce. Civil Aeronautics Board.

——. Civil Service Commission.

USA-*9200*.M7. *Monthly Report of Employment, Executive Branch of the Federal Government.*

——. Commerce Department—*see-véase* United States. Department of Commerce.

——. Committee for Congested Production Areas—*see - véase* United States. Executive Office of the President. Committee for Congested Production Areas.

——. Committee on Joint Causes of Death. Subcommittee on Classification of Diseases, Injuries, and Causes of Death.

Int-*3230*.P5, 1946. *Proposed Statistical Classification of Diseases, Injuries, and Causes of Death.*

——. Committee on the Cost of Living—*see - véase* United States. President's Committee on the Cost of Living.

——. Committee on the Cost of Medical Care—*see - véase* Committee on the Cost of Medical Care.

——. Comptroller of the Currency—*see - véase* United States. Department of the Treasury. Bureau of the Comptroller of the Currency.

——. Coordinator of Inter-American Affairs—*see - véase* United States. Executive Office of the President. Office for Emergency Management. Office of Inter-American Affairs.

——. Council on Vital Records and Vital Statistics.

USA-*5200*.P5, 1944. *Plan for the Coordination of Vital Records and Vital Statistics in the United States.*

——. Crop Reporting Board—*see - véase* United States. Department of Agriculture. Bureau of Agricultural Economics. Crop Reporting Board.

——. Departamento de Comercio. Oficina del Censo—*see - véase* United States. Department of Commerce. Bureau of the Census.

——. Department of Agriculture.

Int-*9700*.S5, 1935. *Studies of Family Living in the United States and Other Countries.*

USA-*6200*.A5. *Agricultural Statistics.*

USA-*6200*.C6. *Crops and Markets.*

USA-*6200*.F3, 1939. *The Farm-Housing Survey.*

USA-*6200*.Y5. *Yearbook of the . . .*

USA-*9700*.F5F1, 1941. *Family Food Consumption and Dietary Levels, Five Regions, Farm Series.*

USA-*9700*.F5I1, 1941. *Family Income*

United States—cont.

Measures with Factors for Conversion to United States Units.

——. Department of Commerce.

USA-*6700*.D4. *Domestic Commerce.*

——. ——. American Republics Division *see-véase* United States. Department of Commerce. Office of International Trade. American Republics Division.

——. ——. Bureau of Foreign and Domestic Commerce—*see also-véase también* United States. Department of Commerce. Office of International Trade.

Int-*1200*.I4, 1943. *Index of Latin American Publications.*

Int-*1950*.C5, 1939. *Commercial Travelers' Guide to Latin America.*

Int-*6000*.A5, 1943.B. *Supplement to American Republics Bibliography; A List of Some of the Current Latin American Periodicals Relating to Agriculture, Commerce, Economics, Finances, Industry, Production in the Library of the Pan American Union.*

Int-*6000*.S4, 1943. *Selected Bibliography in English on Latin American Resources, Economic Development, and Trade.*

Can-*6000*.C5, 1945. *Canada.*

Méx-*6400*.I5, 1943. *Industrial Development in Mexico.*

Perú-*6000*.P5, 1945. *Peru.*

RepDom-*6000*.D5, 1946. *Dominican Republic.*

USA-*6000*.S7. *Survey of Current Business.*

USA-*6100*.N5, 1929-37. *National Income in the United States, 1929-1937.*

USA-*6400*.04, 1929-39. *Output of Manufactured Commodities, 1929-1939.*

USA-*6700*.D6. *Domestic Commerce Series of the Bureau of Foreign and Domestic Commerce.*

USA-*6900*.E6, 1938. *Export and Import Practice.*

USA-*6900*.F7. *Foreign Commerce Weekly.*

USA-*6900*.F8. *Foreign Commerce Yearbook.*

USA-*6910*.B5. *Balance of International Payments of the United States.*

USA-*6910*.U5, 1943. *The United States in the World Economy.*

Ven-*6000*.V5, 1946. *Venezuela.*

——. ——. ——. Division of International Economy.

Chile-*6000*.C3, 1945. *Chile.*

——. ——. ——. ——. International Economics and Statistics Unit.

Int-*6900*.B5, 1944.1. *Bibliography of the Latest Available Source Material on Foreign Trade Statistics.*

Int-*6900*.B5, 1944.2. *Supplement-A to Bibliography of the Latest Available Source Material on Foreign Trade Statistics.*

Int-*6900*.B5, 1943.1. *Bibliography of the Latest Available Source Material on Foreign Trade Statistics Consulted by the International Economics and Statistics Unit.*

Int-*6900*.B5, 1943.2. *List of Publications References in Field of Foreign Trade Statistics for the American Republics.*

Int-*6900*.F4, 1944-45. *Foreign Commerce Series under the Fellowship Program of 1944-45.*

USA-*6900*.S5. *Summary of Foreign Trade of the United States.*

——. ——. Bureau of the Census.

Int-*3040*.E5, June '44, b. "Final Reports of the Sixteenth Census of the United States, 1940."

Int-*3230*.C7, 1944. *Classification of Terms and Comparability of Titles through Five Revisions of the International List of Causes of Death.*

Int-*3230*.M5, 1939. *Manual of the International List of Causes of Death and Manual of Joint Causes of Death.*

Int-*5000*.G5, 1943. *General Censuses and Vital Statistics in the Americas.*

Int-*6100*.N3, 1946. *National Income and Related Subjects for the United Nations.*

Arg-*5000*.A5, 1945. *Argentina: Summary of Biostatistics.*

Bol-*5000*.B5, 1945. *Bolivia: Summary of Biostatistics.*

Bras-*5000*.B5, 1945. *Brazil: Summary of Biostatistics.*

Chile-*5000*.C5, 1943. *Chile: Demographic Data.*

Col-*5000*.C5, 1944. *Colombia: Summary of Biostatistics.*

CR-*5000*.C5, 1944. *Costa Rica: Summary of Biostatistics.*

Cuba-*5000*.C5, 1945. *Cuba: Summary of Biostatistics.*

Ec-*5000*.E5, 1944. *Ecuador: Summary of Biostatistics.*

ELSal-*5000*.E5, 1944. *El Salvador: Summary of Biostatistics.*

Guat-*5000*.G5, 1944. *Guatemala: Summary of Biostatistics.*

Haiti-*2100*.S5, 1942. *Sampling Plans for a Census of Haiti.*

Haiti-*5000*.H5, 1945. *Haiti: Summary of Biostatistics.*

Hond-*5000*.H5, 1944. *Honduras: Summary of Biostatistics.*

United States—cont.

Special Report on Institutional Population, 14 Years Old and Over.

USA-*2200*.C5, 1940.10a. *Sixteenth Census of the United States: 1940: Population: The Labor Force (Sample Statistics): Employment and Family Characteristics of Women.*

USA-*2200*.C5, 1940, 10b. *Sixteenth Census of the United States: 1940: Population: The Labor Force (Sample Statistics): Employment and Personal Characteristics.*

USA-*2200*.C5, 1940.10c. *Sixteenth Census of the United States: 1940: Population: The Labor Force (Sample Statistics): Industrial Characteristics.*

USA-*2200*.C5, 1940.10d. *Sixteenth Census of the United States: 1940: Population: The Labor Force (Sample Statistics): Occupational Characteristics.*

USA-*2200*.C5, 1940.10e. *Sixteenth Census of the United States: 1940: Population: The Labor Force (Sample Statistics): Usual Occupation.*

USA-*2200*.C5, 1940.10f. *Sixteenth Census of the United States: 1940: Population: The Labor Force (Sample Statistics): Wage or Salary Income in 1939.*

USA-*2200*.C5, 1940.11a. *Sixteenth Census of the United States: 1940: Population: Nativity and Parentage of the White Population.*

USA-*2200*.C5, 1940.11b. *Sixteenth Census of the United States: 1940: Population: Nativity and Parentage of the White Population: Country of Origin of Foreign Stock.*

USA-*2200*.C5, 1940.11c. *Sixteenth Census of the United States: 1940: Population: Nativity and Parentage of the White Population: Mother Tongue.*

USA-*2200*.C5, 1940.12. *Sixteenth Census of the United States: 1940: Areas of the United States, 1940.*

USA-*2200*.C5, 1940.13. *Sixteenth Census of the United States: 1940: Population: Unincorporated Communities.*

USA-*2200*.C5, 1940.14. *Sixteenth Census of the United States: 1940: Population: State of Birth of the Native Population.*

USA-*2200*.C5, 1940.15. *Sixteenth Census of the United States: 1940: Population: Estimates of Labor Force, Employment and Unemployment in the United States, 1940 and 1930.*

USA-*2200*.C5, 1930. *Fifteenth Census of the United States: 1930: Abstract of the Fifteenth Census of the United States.*

USA-*2200*.C5, 1930.2. *Fifteenth Census of the United States: 1930: The Indian Population of the United States and Alaska.*

USA-*2200*.C8, 1910. *Indian Population in the United States and Alaska: 1910.*

USA-*2210*.C5, 1940.1. *Sixteenth Census of the United States: 1940: Agriculture.*

USA-*2210*.C5, 1940.2. *Sixteenth Census of the United States: 1940: Agriculture: Cross-Line Acreage.*

USA-*2210*.C5, 1940.3. *Sixteenth Census of the United States: 1940: Agriculture: Value of Farm Products by Color and Tenure of Farm Operator.*

USA-*2210*.C5, 1940.4. *Sixteenth Census of the United States: 1940: Drainage of Agricultural Land.*

USA-*2210*.C5, 1940.6. *Sixteenth Census of the United States: 1940: Irrigation of Agricultural Lands.*

USA-*2210*.C5, 1940.15. *Sixteenth Census of the United States: 1940: Agriculture: Abandoned or Idle Farms.*

USA-*2210*.C5, 1940.17. *Sixteenth Census of the United States: 1940: Agriculture: Cows Milked and Dairy Products.*

USA-*2210*.C5, 1940.25. *Sixteenth Census of the United States: 1940: Agriculture: Special Cotton Report.*

USA-*2210*.C5, 1940.27. *Sixteenth Census of the United States: 1940: Agriculture: Special Poultry Report.*

USA-*2220*.C5, 1940.1. *Sixteenth Census of the United States: 1940: Housing.*

USA-*2220*.C5, 1940.3. *Sixteenth Census of the United States: 1940: Housing: Characteristics by Type of Structure.*

USA-*2250*.C3, 1902-30. *Census of Electrical Industries: Telephones and Telegraphs.*

USA-*2250*.C5, 1940. *Sixteenth Census of the United States: 1940: Census of Business, 1939.*

USA-*2250*.C7, 1939.1. *Sixteenth Census of the United States: 1940: Manufactures, 1939.*

USA-*2260*.R5, 1936. *Religious Bodies, 1936.*

USA-*2270*.C5, 1939. *Sixteenth Census of the United States: 1940: Mineral Industries, 1939.*

USA-*3030*.H5, 1940. *Handbook for Employees of the Census.*

USA-*3110*.L5, 1945. *List of Schedules Used by the Bureau of the Census for Collecting Data: 1945.*

USA-*3210*.S3, 1943. *Schedule A, Statistical*

United States—cont.

letin No. 3: Occupations and Other Characteristics by Age.

USA PR-*2210*.C5, 1940. *Censo Décimosexto de los Estados Unidos: 1940: Agricultura. Sixteenth Census of the United States: 1940: Agriculture. Puerto Rico.*

USA PR-*2220*.C5, 1940. *Censo Décimosexto de los Estados Unidos: 1940: Puerto Rico: Viviendas. Sixteenth Census of the United States: 1940: Puerto Rico: Housing.*

USA PR-*5110*.S5, 1940. *Staff Report to the Interdepartmental Committee on Puerto Rico.*

Ven-*5000*.V5, 1944. *Venezuela: Summary of Biostatistics.*

NonG-*3150*.S3. "Sample Surveys in Census Work."

NonG-*5200*.V6, 1943. *Un Vocabulario Bioestadístico. A Vital Statistics Vocabulary.*

——. ——. ——. Division of Vital Statistics—*see - véase* United States. Department of Commerce. Bureau of the Census. Vital Statistics Division.

——. ——. ——. Vital Statistics Division.

Int-*5000*.S8. *Summary of Biostatistics Series.*

USA-*3230*.M5, 1944. *Manual for Coding Causes of Illness According to a Diagnosis Code for Tabulating Morbidity Statistics.*

——. ——. Civil Aeronautics Administration.

USA-*6611*.S5, 1944. *Statistical Handbook of Civil Aviation.*

——. ——. Civil Aeronautics Board.

USA-*6611*.A5. *Annual Airline Statistics: Domestic Carriers.*

——. ——. National Bureau of Standards.

NonG-*1600*.S5, 1937. *Summary Report of National Bureau of Standards Research on Preservation of Records.*

NonG-*1600*.S7, 1939. *Summary Report of Research at the National Bureau of Standards on the Stability and Preservation of Records on Photographic Film.*

——. ——. National Income Division.

Int-*6100*.W5, 1946. *Working Paper on the Status of National Income Statistics.*

——. ——. Office of International Trade—*see also - véase también* United States. Department of Commerce. Bureau of Foreign and Domestic Commerce.

——. ——. ——. American Republics Division.

USA-*6900*.F7, Aug. 3, '46, a. "Latin America's Economy as World Conflict Ended: Part IV: South American West-Coast Republics."

USA-*6900*.F7, July 27, '46. "Latin America's Economy as World Conflict Ended: Part III: South American North Coast and East Coast Republics."

USA-*6900*.F7, July 13, '46. "Latin America's Economy as World Conflict Ended: Part II: Mexico, Central America, and the Caribbean Republics."

USA-*6900*.F7, July 6, '46. "Latin America's Economy as World Conflict Ended: Part I: The Region Readjusts to 'Postwar Imperatives'."

——. Department of Justice. Bureau of Prisons.

USA-*8300*.F5. *Federal Prisons.*

——. ——. Federal Bureau of Investigation.

USA-*8300*.T5, 1930-39. *Ten Years of Uniform Crime Reporting, 1930-1939.*

USA-*8300*.U5. *Uniform Crime Reports for the United States and Its Possessions.*

——. ——. Immigration and Naturalization Service.

USA-*5120*.R5, 1943. *Report on Alien Registration.*

——. Department of Labor. Bureau of Labor Statistics.

Int-*6010*.P5, 1939-45. *Price Trends and Price Control in the American Republics, 1939-1945.*

USA-*6010*.W5, 1939-41. *Wartime Prices: Part I, August 1939 to Pearl Harbor.*

USA-*6500*.C3. *Construction.*

USA-*9000*.B4. *Bulletins of the Bureau . . .*

USA-*9000*.H5, 1941. *Handbook of Labor Statistics, 1941.*

USA-*9000*.L4. *Labor Conditions in Latin America.*

USA-*9000*.M5. *Monthly Labor Review.*

USA-*9000*.M5, May 1946. "Employment Situation in Latin America."

USA-*9000*.M5, May '45, Hosea. "Fact-Finding Activities of the Bureau of Labor Statistics."

USA-*9000*.S3, 1943. *Selected List of the Publications of the Bureau . . .*

USA-*9000*.S3, 1940. *A Selected List of the Publications of the Bureau . . .*

USA-*9200*.H5, 1932-40. *Hours and Earnings in the United States, 1932-40.*

USA-*9200*.I6, 1922. *Industrial Unemployment.*

USA-*9500*.C4. *Changes in Cost of Living.*

USA-*9700*.C7, 1935-36. *Consumer Purchases Study, 1935-36.*

United States—cont.

——. Division of Vital Statistics—*see - véase* United States. Department of Commerce. Bureau of the Census. Vital Statistics Division.

——. Employment Service—*see - véase* United States. Department of Labor. United States Employment Service.

——. Executive Office of the President. Bureau of the Budget.

Int-*6100*.N3, 1946. *National Income and Related Subjects for the United Nations.*

USA-*3210*.S6, 1944. *Standard Commodity Classification: Vol. II: Alphabetic Index.*

USA-*3210*.S6, 1943. *Standard Commodity Classification: Volume I: Standard Classified List of Commodities.*

USA-*3220*.S5, 1942. *Standard Industrial Classification Manual.*

USA-*8800*.B7. *The Budget of the United States Government.*

——. ——. ——. Division of Statistical Standards.

Int-*6100*.W5, 1946. *Working Paper on the Status of National Income Statistics.*

USA-*3030*.C5, 1946. *Collection of Economic Statistics in the United States.*

USA-*3030*.F5. *Federal Statistical Directory.*

USA-*3040*.S5. *Statistical Reporter.*

USA-*3110*.R5, 1942-43. *Report, October 9, 1942-April 13, 1943, from Advisory Committee on Government Questionnaires.*

USA-*3200*.T5. *Technical Papers of the Division of Statistical Standards.*

——. ——. Committee for Congested Production Areas.

USA-*2200*.C3, 1944. *Census of Congested Production Areas: 1944.*

USA-*5110*.F5, 1944. *Final Report of the Committee . . .*

——. ——. Office for Emergency Management. Foreign Economic Administration. Board of Economic Warfare.

Méx-*6000*.M7, 1942. *The Mexican Economy.*

——. ——. ——. Office of Economic Stabilization.

USA-*9500*.R3, 1945. *Report of the President's Committee on the Cost of Living.*

——. ——. ——. Office of Inter-American Affairs.

Int-*1400*.B5, 1945. *Basic Data on the Other American Republics.*

Int-*1950*.N5, 1943. *The New World Guides to the Latin American Republics.*

Int-*5000*.S8. *Summary of Biostatistics Series.*

Int-*5100*.P8, 1945. *Populations of the Other American Republics by Major Civil Divisions and by Cities of 5,000 or More Inhabitants.*

Int-*6240*.F5, 1943. *The Fisheries and Fishery Resources of the Caribbean Area.*

Arg-*5000*.A5, 1945. *Argentina: Summary of Biostatistics.*

Bol-*5000*.B5, 1945. *Bolivia: Summary of Biostatistics.*

Bras-*5000*.B5, 1945. *Brazil: Summary of Biostatistics.*

Chile-*5000*.C5, 1943. *Chile: Demographic Data.*

CR-*5000*.C5, 1944. *Costa Rica: Summary of Biostatistics.*

Col-*5000*.C5, 1944. *Colombia: Summary of Biostatistics.*

Cuba-*5000*.C5, 1945. *Cuba: Summary of Biostatistics.*

Ec-*5000*.E5, 1944. *Ecuador: Summary of Biostatistics.*

ElSal-*5000*.E5, 1944. *El Salvador: Summary of Biostatistics.*

Guat-*5000*.G5, 1944. *Guatemala: Summary of Biostatistics.*

Haiti-*5000*.H5, 1945. *Haiti: Summary of Biostatistics.*

Hond-*5000*.H5, 1944. *Honduras: Summary of Biostatistics.*

Méx-*5000*.M5, 1943. *Mexico: Demographic Data.*

Nic-*5000*.N5, 1945. *Nicaragua: Summary of Biostatistics.*

Pan-*5000*.P5, 1945. *Panama: Summary of Biostatistics.*

Par-*5000*.P5, 1944. *Paraguay: Summary of Biostatistics.*

Perú-*5000*.P5, 1944. *Peru: Summary of Biostatistics.*

Perú-*6200*.S5, 1945, Eng. *Statistical Program for the Ministry of Agriculture of Peru.*

Perú-*6200*.S5, 1945, Sp. *Programa Estadístico para el Ministerio de Agricultura del Perú.*

RepDom-*5000*.D5, 1945. *Dominican Republic: Summary of Biostatistics.*

Ur-*5000*.U5, 1944. *Uruguay: Summary of Biostatistics.*

Ven-*5000*.V5, 1944. *Venezuela: Summary of Biostatistics.*

——. ——. ——. ——. Research Division.

Par-*1200*.S5, 1943. *Strategic Index of the Americas: Preliminary Bibliography of Paraguay.*

——. ——. ——. Office of the Coordinator

United States—cont.

tion—*see - véase* United States. Federal Security Agency. Office of Education.

——. ——. United States Public Health Service—*see - véase* United States. Federal Security Agency. Public Health Service.

——. Federal Trade Commission.

USA-*6400*.I3. *Industrial Corporation Reports.*

——. Federal Works Agency. Work Projects Administration.

USA-*5300*.R6, 1939. *Rural Migration in the United States.*

USA-*9700*.F5C, 1935-36. *Family Income in Chicago, 1935-36.*

USA-*9700*.F5F1, 1941. *Family Food Consumption and Dietary Levels, Five Regions, Farm Series.*

USA-*9700*.F5I1, 1941. *Family Income and Expenditures, Five Regions, Part 2, Family Expenditures, Farm Series.*

USA-*9700*.F5I2, 1940. *Family Income and Expenditures, Five Regions, Part 2, Family Expenditures, Urban and Village Series.*

NonG-*6200*.S5, 1942. *Statistical Investigation of a Sample Survey for Obtaining Farm Facts.*

——. ——. ——. National Research Project on Reemployment Opportunities and Recent Changes in Industrial Techniques.

USA-*6000*.I4, 1939. *Industrial Change and Employment Opportunity—A Selected Bibliography.*

USA-*6000*.R5. *Reports of the National Research Project on Reemployment Opportunities and Recent Changes in Industrial Techniques.*

NonG-*6000*.S7, 1940. *Survey of Economic Theory on Technological Change and Employment.*

——. ——. Work Projects Administration, New York City.

USA-*2220*.C5, 1940.2. *Sixteenth Census of the United States: 1940: Housing: Analytical Maps ...; Block Statistics.*

——. Fish and Wildlife Service—*see - véase* United States. Department of the Interior. Fish and Wildlife Service.

——. Foreign Economic Administration—*see - véase* United States. Executive Office of the President. Office of Emergency Management. Foreign Economic Administration.

——. Graduate School of the Department of Agriculture—*see - véase* United States. Department of Agriculture. Graduate School.

——. Immigration and Naturalization Service—*see - véase* United States. Department of Justice. Immigration and Naturalization Service.

——. Interior Department—*see - véase* United States. Department of the Interior.

——. Interstate Commerce Commission. Bureau of Statistics—*see - véase* United States. Interstate Commerce Commission. Bureau of Transport Economics and Statistics.

——. ——. Bureau of Transport Economics and Statistics.

Int-*6614*.I5, 1939. *International Comparisons of Railway Ton-Mile Revenues.*

USA-*6610*.S5. *Statements of the Bureau of Transport Economics and Statistics.*

USA-*6613*.R5, 1942. *A Review of Statistics of Oil Pipe Lines, 1921-1941.*

USA-*6614*.S6. *Statistics of Railways in the United States.*

USA-*6614*.S6, 1943. *Fifty-Seventh Annual Report on the Statistics of Railways in the United States for the Year Ended December 31, 1943.*

NonG-*6614*.R5, 1941. *Railway Statistical Terms.*

——. Justice Department—*see - véase* United States. Department of Justice.

——. Labor Department—*see - véase* United States. Department of Labor.

——. Laws, Statutes, Etc.

USA-*3210*.C5, 1943, a. "United States Custom Tariff, Act of 1930."

——. Library of Congress.

Arg-*1200*.G5, 1945. *A Guide to the Official Publications of the Other American Republics: I, Argentina.*

Bol-*1200*.G5, 1945. *A Guide to the Official Publications of the Other American Republics: II, Bolivia.*

Cuba-*1200*.G5, 1945. *A Guide to the Official Publications of the Other American Republics: VII, Cuba.*

——. ——. Census Library Project—*see - véase* United States. Census Library Project.

——. ——. Division of Bibliography.

USA-*5300*.I4, 1943. *Immigration in the United States: A Selected List of Recent References.*

——. ——. Hispanic Foundation.

Int-*1200*.B5, 1942. *A Bibliography of Latin American Bibliographies.*

Int-*1200*.L3, 1944. *Latin American Periodicals Currently Received in the Library of Congress and in the Library of the Department of Agriculture.*

United States—cont.

——. Office of War Mobilization and Reconversion.

USA-*8100*.R5. *Report to the President, the Senate and the House of Representatives by the Director of War Mobilization and Reconversion.*

——. President.

USA-*8800*.R5, 1945. *Twentieth Report to Congress on Lend-Lease Operations, for the Period Ended June 30, 1945.*

——. President's Committee on the Cost of Living.

USA-*9500*.R3, 1945. *Report of the President's Committee on the Cost of Living.*

——. Public Health Service—*see - véase* United States. Federal Security Agency. Public Health Service.

——. Railroad Retirement Board.

USA-*9952*.A5. *Annual Report of the Railroad Retirement Board.*

USA-*9952*.M5. *The Monthly Review of the Railroad Retirement Board.*

——. Reconstruction Finance Corporation—*see - véase* United States. Federal Loan Agency. Reconstruction Finance Corporation.

——. Securities and Exchange Commission.

USA-*6720*.S5. *Survey of American Listed Corporations: Data on Profits and Operations.*

USA-*6721*.A5. *Annual Report of the Securities and Exchange Commission.*

USA-*6721*.S5. *Statistical Bulletin.*

——. Social Science Research Council—*see - véase* Social Science Research Council, United States.

——. Social Security Board—*see - véase* United States. Federal Security Agency. Social Security Board.

——. State Department—*see - véase* United States. Department of State.

——. Tariff Commission.

Int-*6000*.E2. *Economic Controls and Commercial Policy in the American Republics.*

Int-*6200*.A45. *Agricultural, Pastoral, and Forest Industries in the American Republics.*

Int-*6400*.M5. *Mining and Manufacturing Industries in the American Republics.*

Int-*6900*.C3, 1943. *Comercio Exterior de la América Latina.*

Int-*6900*.C5, 1943. *Commercial Policies and Trade Relations of European Possessions in the Caribbean Area.*

Int-*6900*.F6, 1942. *The Foreign Trade of Latin America.*

Int-*6900*.L4, 1941. *Latin America as a Source of Strategic and Other Essential Materials.*

Int-*6900*.R5. *Recent Developments in the Foreign Trade of the American Republics.*

Int-*6900*.R6. *Reports of the United States . . .*

Arg-*6000*.E6, 1945. *Economic Controls and Commercial Policy in Argentina.*

Arg-*6400*.M7, 1945. *Mining and Manufacturing Industries in Argentina.*

Bol-*6000*.E5, 1945. *Economic Controls and Commercial Policy in Bolivia.*

Bol-*6400*.M5, 1945. *Mining and Manufacturing Industries in Bolivia.*

Bras-*6000*.E3, 1945. *Economic Controls and Commercial Policy in Brazil.*

Bras-*6200*.A2, 1946. *Agricultural, Pastoral, and Forest Industries in Brazil.*

Bras-*6400*.M5, 1945. *Mining and Manufacturing Industries in Brazil.*

Chile-*6000*.E65, 1945. *Economic Controls and Commercial Policy in Chile.*

Chile-*6200*.A7, 1945. *Agricultural, Pastoral, and Forest Industries in Chile.*

Chile-*6400*.M5, 1945. *Mining and Manufacturing Industries in Chile.*

Chile-*6900*.R5, 1945. *Recent Developments in the Foreign Trade of Chile.*

Col-*6000*.E5, 1945. *Economic Controls and Commercial Policy in Colombia.*

Col-*6200*.A4, 1945. *Agricultural, Pastoral, and Forest Industries in Colombia.*

Col-*6400*.M5, 1945. *Mining and Manufacturing Industries in Colombia.*

CR-*6000*.E5, 1945. *Economic Controls and Commercial Policy in Costa Rica.*

CR-*6400*.M5, 1945. *Mining and Manufacturing Industries in Costa Rica.*

Cuba-*6000*.E5, 1946. *Economic Controls and Commercial Policy in Cuba.*

Ec-*6000*.E5, 1945. *Economic Controls and Commercial Policy in Ecuador.*

Ec-*6400*.M5, 1945. *Mining and Manufacturing Industries in Ecuador.*

Haiti-*6000*.E5, 1946. *Economic Controls and Commercial Policy in Haiti.*

Haiti-*6400*.M5, 1945. *Mining and Manufacturing Industries in Haiti.*

Méx-*6000*.E4, 1946. *Economic Controls and Commercial Policy in Mexico.*

Méx-*6400*.M3, 1946. *Mining and Manufacturing Industries in Mexico.*

Nic-*6400*.M5, 1945. *Mining and Manufacturing Industries in Nicaragua.*

Pan-*6400*.M5, 1945. *Mining and Manufacturing Industries in Panama.*

Par-*6000*.E5, 1945. *Economic Control and Commercial Policy in Paraguay.*

Universidad de la Habana—cont.

——. Facultad de Derecho y de Ciencias Sociales.
NonG-*3330*.Roa Reyes, 1929. *Curso de Estadística.*

Universidad de la República, Uruguay. Facultad de Ciencias Económicas y de Administración.
Ur-*3040*.R5. *Revista de la Facultad de Ciencias Económicas y de Administración.*

——. ——. Instituto de Estadística.
Ur-*3000*.C5. *Cuadernos del Instituto de Estadística.*
NonG-*3120*.E5, 1943. *Elementos Matemáticos para el Análisis de las Series de Tiempo.*
NonG-*3140*.A8, 1946. *Algunos Problemas de Inferencia Estadística.*
NonG-*3150*.S8, 1942. *Sobre la Teoría de las Muestras.*
NonG-*5100*.T5, 1943. *Teoría Matemática del Desarrollo Numérico de la Población.*

——. Facultad de Derecho y Ciencias Sociales de Montevideo. Biblioteca de Publicaciones Oficiales.
Ur-*5100*.H5, 1939. *Historia y Análisis Estadístico de la Población del Uruguay.*

Universidad de Puerto Rico. Consejo Superior de Enseñanza.
USA PR-*7200*.A5, 1945. *El Analfabetismo en Puerto Rico . . .*

Universidad Interamericana, Panamá. Instituto de Investigaciones Sociales y Económicas.
Int-*3040*.B5. *Boletín del Instituto de Investigaciones Sociales y Económicas. Bulletin of the Institute of Social and Economic Research. Bulletin de l'Institut des Recherches Sociales et Economiques. Boletim do Instituto de Investigações Sociais e Econômicas.*
Int-*3040*.B5, Feb. '45, a. "Directorio de Especialistas en Estudios Sociales y Económicos en Latinoamérica, Parte I."
Int-*3040*.B5, Feb. '45, b. "Directorio de Instituciones de Estudios e Investigaciones Sociales y Económicos en Latinoamérica," Parte I.

Universidad Mayor de San Marcos, Perú. Facultad de Ciencias Económicas.
Perú-*3040*.R5. *Revista de la Facultad de Ciencias Económicas.*

Universidad Nacional Autónoma de México. Facultad de Derecho y Ciencias Sociales. Escuela Nacional de Economía.
Méx-*3320*.C5, 1945. *Cursos de Invierno de 1945.*
Méx-*6000*.I5. *Investigación Económica.*
NonG-*3330*.Gavett, 1941. *Principios de Metodología Estadística.*

——. Instituto de Investigaciones Sociales.
Méx-*9000*.R5. *Revista Mexicana de Sociología.*

Universidad Nacional de Colombia. Instituto de Ciencias Económicas. Instituto Nacional de Estadística.
Col-*3030*.I5, 1945. *Instituto Nacional de Estadística de la Universidad de Colombia.*

Universidad Nacional de Córdoba. Escuela de Ciencias Económicas.
Arg-*3040*.R7. *Revista de Economía y Estadística.*
Arg-*3040*.R7, Oct.-Dec. '44. "Precio Medio de la Habitación Ofrecida en la Ciudad de Buenos Aires, 1943-44."

——. ——. Instituto de Estadística.
Arg-*3040*.R7, Jan.-June '46, a. "Precio Medio de la Habitación Ofrecida en la Ciudad de Buenos Aires (1943-1945)."

Universidad Nacional de La Plata. Facultad de Ciencias Físicomatemáticas.
Arg-*3040*.R9. *Revista de la Facultad . . .*

——. Facultad de Humanidades y Ciencias de la Educación.
Arg-*7200*.D5, 1940. *Deserción Escolar y Analfabetismo.*

Universidad Nacional del Litoral. Facultad de Ciencias Económicas, Comerciales y Políticas.
Arg-*3040*.R8. *Revista de la Facultad . . .*

——. ——. Instituto de Estadística.
NonG-*3100*.T5, 1935. *Teoría de la Correlación.*
NonG-*3330*.Dieulefait, 1938-40. *Elementos de Estadística Metodológica.*
NonG-*3330*.Mazza 2, 1940. *Cuestiones de Matemáticas Referentes a la Estadística. Fascículo I.*

Universidade do Brasil. Faculdade Nacional de Filosofia.
Bras-*3320*.P5, 1944. *Programa de Estatística Geral e Aplicada.*

UNIVERSIDADES—*véase* 7210.

UNIVERSITIES—*see* 7210.

The Universities of Chile. Chile-*7210*.U5, 1944.

The Universities of Colombia. Col-*7210*.U5, 1945.

University of London. University College. Department of Applied Statistics.
NonG-*3150*.R5, 1927. *Random Sampling Numbers.*
NonG-*3150*.T4, 1939. *Tables of Random Sampling Numbers.*

Uruguay—cont.

——. Ministerio de Industrias y Trabajo. Dirección de Estadística Económica.
Ur-*2250*.C5, 1936. *Censo Industrial de 1936.*
Ur-*6000*.B7. *Boletín de la Dirección de Estadística Económica.*

——. ——. Dirección General de Asuntos Económicos.
Ur-*9500*.C5. *Costo de la Vida Obrera en Montevideo.*

——. ——. Oficina de Estadística Agrícola.
Ur-*6200*.L5, 1918. *Ley y Reglamentación de la Estadística Agrícola.*

——. Ministerio de Instrucción Pública y Previsión Social. Dirección General del Registro del Estado Civil—*see* - *véase* Uruguay. Ministerio de Instrucción Pública y Previsión Social. Instituto de Jubilaciones y Pensiones. Dirección General del Registro del Estado Civil.

——. ——. Instituto de Jubilaciones y Pensiones. Dirección General del Registro del Estado Civil.
Ur-*5200*.M5. *El Movimiento del Estado Civil y la Mortalidad de la República Oriental del Uruguay.*

——. Ministerio de Salud Pública.
Ur-*7110*.M5. *Memoria del ...*
NonG-*5200*.M5, 1942. *Manual de Estadística Vital.*

——. Oficina de Estadística Agrícola—*see* - *véase* Uruguay. Ministerio de Industrias y Trabajo. Oficina de Estadística Agrícola.

——. Sección de Economía y Estadística Agraria—*see* - *véase* Uruguay. Ministerio de Ganadería y Agricultura. Dirección de Agronomía. Sección de Economía y Estadística Agraria.

——. Universidad de la República—*see* - *véase* Universidad de la República, Uruguay.

Uruguay: Summary of Biostatistics. Ur-*5000*.U5, 1944.

Urzúa Acevedo, Manfredo.
Chile-*5250*.M5, 1945. *Manual Práctico del Oficial Civil.*

"The Use of Income Tax Data in the National Resources Committee Estimate of the Distribution of Income by Size." NonG-*6100*.S3.

Usines Électriques Centrales du Canada. Can-*6400*.C6.

UTILIDADES:
FINANCIERAS (incluyendo VALORES Y TITULOS)—*véase* 6721.
PUBLICAS (INDUSTRIALES)—*véase* 6400.

UTILITIES, PUBLIC—*see* 6400.

Utilização de Medidas Estatísticas em Biologia, Medicina e Saúde Pública. NonG-*3330*. Jansen de Mello, 1944.

VACCINATION—*see* 7110.

VACUNACION—*véase* 7110.

"Valor por Kilogramo de las Exportaciones en los Años Comprendidos de 1935 a 1944." Ven-*8800*.R5, Mar. '46, b.

VALORES:
GENERAL (incluyendo ARRENDAMIENTOS Y TITULOS FINANCIEROS)—*véase* 6721; 2220.
DE HABITACIONES Y VIVIENDAS—*véase* 9900; 9500; 2220.

VALUES AND RENTS:
GENERAL—*see* 6721; 2220.
HOUSING—*see* 9900; 9500.

Van Deraa, Patricia.
USA-*3000*.G7, 1946. *Government Statistics for Business Use.*

Vandellós, José A.
Int-*3000*.S5, 1940. "Las Actividades Estadísticas de Venezuela."
NonG-*3330*.Gini 2, 1935. *Curso de Estadística.*

VAQUERIAS—*véase* 6230.

Vega Lamar.
Cuba-*5220*.T5, 1920. *Tablas de Vida de la Ciudad de la Habana.*

Veinte Años de Legislación Social. Chile-*9000*.V5, 1945.

Velarde B., César A.
Perú-*3040*.E5, July '46. "Estadística Económica Militar."
NonG-*3330*, 1945. *Curso de Estadística.*

A Velocidade de Circulação dos Estoques de Produtos Almentícios nos Mercados do Distrito Federal e de São Paulo, em 1944. Bras-*9800*.V5, 1945.

Venereal Disease Information. USA-*7150*.V5.

Venezuela. Banco Central de Venezuela—*see* - *véase* Banco Central de Venezuela.

——. Biblioteca Nacional.
Ven-*1200*.A5. *Anuario Bibliográfico Venezolano.*

——. Contraloría General de la Nación.
Ven-*8800*.I5. *Informe de la ...*

——. Dirección de Ganadería—*see* - *véase* Venezuela. Ministerio de Agricultura y Cría. Dirección de Ganadería.

——. Dirección de Salubridad Pública—*see* - *véase* Venezuela. Ministerio de Sanidad y Asistencia Social. Dirección de Salubridad Pública.

——. Dirección General de Estadística y de Censos—*see* - *véase* Venezuela. Ministerio de Fomento. Dirección General de Estadística y de Censos.

——. División de Epidemiología y Estadística Vital—*see* - *véase* Venezuela. Ministerio de Sanidad y Asistencia Social. Dirección

ployment Statistics in the Planning of a Full-Employment Program."
Woodbury, Robert Morse.
Int-*9700*.M4, 1941. *Méthodes d'Enquête sur les Conditions de Vie des Familles.*
Int-*9700*.M6, 1942. *Methods of Family Living Studies.*
Int-*9700*.M7, 1942. *Métodos de Encuestas sobre las Condiciones de Vida de las Familias.*
Int-*9800*.E5, 1942. *Encuestas sobre el Consumo de Alimentos y la Nutrición en las Américas.*
Int-*9800*.F4, 1942. *Food Consumption and Dietary Surveys in the Americas.*
Woodring, Wendell P.
Haiti-*4000*.G5, 1924. *Geology of the Republic of Haiti.*
"Work Accidents in Chile." USA-*9000*.M5, June '46, Cárcamo C.
The Work of the International Institute of Agriculture During the War (1940-1945). Int-*6200*.W4, 1940-1945.
Workbook in Applied General Statistics. NonG-*3330*.Croxton 2, 1941.
WORKING CONDITIONS—*see* 9100.
Working Paper on the Status of National Income Statistics. Int-*6100*.W5, 1946.
WORKMEN'S COMPENSATION—*see* 9950.
The World Agricultural Census of 1940: Proposed Standard Form. Int-*2110*.W5, 1940.2.
World Agricultural Census of 1940: Questionnaires. Int-*2110*.W5, 1940.3.
The World Agricultural Situation. Int-*6200*.W6.
The World Almanac and Book of Facts. Int-*1910*.W5.
World Economic Development. Int-*6000*.W4, 1944.
World Economic Survey. Int-*6000*.W5.
World Food Situation, 1946. Int-*9800*.W4, 1946.
World Grain Review and Outlook, 1945. Int-*6210*.W6, 1945.
World Price Movement. Can-*6010*.P5.
World Production and Prices. Int-*6000*.W7.
World Prosperity as Sought Through the Economic Work of the League of Nations. Int-*6000*.W8, 1933.
World Statistics of Aliens. Int-*5120*.W5, 1936.
World Trade in Agricultural Products. Int-*6900*.W4, 1943.
World Trade in Agricultural Products. Int-*6900*.W5, 1940.
Woscoboinik, Santiago.
Int-*1300*.A4, 1946. "Memorándum de las Actividades del IASI en el Campo de Estadística de Comercio Exterior."
Woytinsky, Wladimir S.
USA-*9000*.L6, 1938. *Labor in the United States: Basic Statistics for Social Security.*
USA-*9950*.S4, Mar. '46. "Postwar Economic Perspectives: IV, Aftermath of the War."
USA-*9950*.S4, Feb. '46. "Postwar Economic Perspectives: III, Prewar Experience—Production and Consumption."
USA-*9950*.S4, Jan. '46. "Postwar Economic Perspectives: II, Prewar Experience—The Labor Force and Employment."
USA-*9950*.S4, Dec. '45. "Postwar Economic Perspectives: I, Experience After World War I."
Wright, John Kirtland.
NonG-*3130*.N5, 1938. *Notes on Statistical Mapping with Special Reference to the Mapping of Population Phenomena.*
Wueller, P. H.
NonG-*6100*.S3. "Income and the Measurement of the Relative Capacities of the State."
Wythe, George.
Int-*6400*.I5, 1945. *Industry in Latin America.*

Yates, F.
GrBrit-*3040*.J5, Part 1, '46. "A Review of Recent Statistical Developments in Sampling and Sampling Surveys."
NonG-*3260*.S7, 1938. *Statistical Tables for Biological, Agricultural and Medical Research.*
Year Book of Labor Statistics. Int-*9000*.A5.
Yearbook of the Department of Agriculture. USA-*6200*.Y5.
YEARBOOKS:
LIST OF—*see* ANNUALS.
GENERAL DATA COMPILATIONS—*see* 1410.
GENERAL REFERENCE BOOKS—*see* 1910.
Yerushalmy, Jacob.
USA-*3040*.A5, Jan. '45. "Infant and Maternal Mortality in the Modern World."
Young, Allyn A.
NonG-*3140*.H5, 1924. *Handbook of Mathematical Statistics.*
Young, Chester W.
Haiti-*1900*.C5, Dec. '43. "Les Statistiques Vitales."
RepDom-*5200*.C5, 1946. *Commentary Explanations and Sample Copies of the*

SUBJECT INDEX—INDICE DE MATERIAS

NOTE

A publication is listed under the most specific subject which it covers, i.e., the finest subdivision in the Subject Index. For example, a book which is about telephones and telegraphs is listed under "ECONOMICS, Transport and communication, Communication, Telephone and telegraph"; one which covers not only telephones and telegraphs but also other phases of transport and communication is listed under "ECONOMICS, Transport and communication, General"; one which covers the entire economic life of a country, including in all probability information on telephones and telegraphs in the country, is listed under "ECONOMICS, General, General." This means that the reader wishing to find *all* the information about a specific subject such as telephones and telegraphs must look:

(a) First under the smallest applicable subheading for publications which are about that subject only;

(b) Then under the larger general heading above it for publications which cover that subject and also closely related subjects;

(c) Then under the main heading for books which cover that subject and also many other generally related subjects;

(d) In addition, the user must look under the largest "general" section, "GENERAL," for compendiums and general reference books which cover all fields of statistics. The user is particularly directed to 1400-1499, "GENERAL, Data compilations" for statistical yearbooks and general periodical statistical publications.

NOTA

Una publicación se clasifica de acuerdo con la materia más específica de que trata, es decir, según la subdivisión menor del Indice de Materias. Por ejemplo, un libro acerca de teléfonos y telégrafos se incluye bajo "ECONOMIA, Transporte y comunicaciones, Comunicaciones, Teléfonos y telégrafos"; otro que trate, no sólo de teléfonos y telégrafos, sino también de otros aspectos del transporte y las comunicaciones, aparecerá bajo "ECONOMIA, Transporte y comunicaciones, General"; uno que cubra completamente la vida económica de un país, que con toda probabilidad incluye informaciones sobre los teléfonos y telégrafos del país, se clasificará bajo "ECONOMIA, General, General." Esto significa que el lector que desee encontrar *toda* la información acerca de una materia específica tal como teléfonos y telégrafos, deberá buscar:

(a) Primero, bajo la menor de las subdivisiones, referente a las publicaciones sobre dicha materia, solamente;

(b) A continuación, bajo el grupo general más amplio que siga, para las publicaciones que cubran aquel tema y materias estrechamente relacionadas con él;

(c) Luego, bajo el grupo principal, referente a los libros acerca de esa materia y muchas otras generalmente relacionadas con el tema;

(d) Por último, el interesado debará buscar en la sección "general" más amplia, es decir, en la sección "GENERAL," que incluye los compendios y libros de referencia general, los cuales cubren todos los campos estadísticos. En particular, el consultante deberá examinar el grupo 1400-1499, "GENERAL, Compilaciones de datos," para anuarios estadísticos y publicaciones periódicas de estadística general

SUBJECT INDEX
INDICE DE MATERIAS

1000-1999 GENERAL

1100 Administrative controls.
Control administrativo.

Canada.
Manual of Instructions: Balance Sheets, Revenues and Expenditures and Other Accounting Statements of Municipal Corporations. Can-*1100*.M2, 1942.

Non-Geographic.
Accounting Control. NonG-*3110*.A5, 1936.
The Design of Tabulating Cards. NonG-*3110*.D4, 1936.
The Electric Accounting Machine Card. NonG-3110.E5, 1936.
Machine Methods of Accounting. NonG-*3110*.M3.
Organization and Supervision of the Tabulating Department. NonG-*3110*.05, 1936.

1200 Bibliography.
Bibliografía.

International.
Bibliographic Series of the Pan American Union. Int-*1200*.B3.
A Bibliography of Latin American Bibliographies. Int-*1200*. B5, 1942.
Executive Agreement Series. Int-*1200*.E5.
Handbook of Latin American Studies. Int-*1200*.H4.
Index of Latin American Publications. Int-*1200*.I4, 1943.
Inter-American Bibliographical Review. Int-*1200*.I6.
Latin American Periodicals Currently Received in the Library of Congress and in the Library of the Department of Agriculture. Int-*1200*.L3, 1944.
Latin American Series of the Library of Congress. Int-*1200*.L5.
The Pan American Book Shelf. Int-*1200*.P2.
Publications Issued by the League of Nations. Int-*1200*.P5.
Selected Bibliography in English on Latin American Resources, Economic Development, and Trade. Int-*6000*.S4, 1943.
A Selected List of Latin American Periodicals Containing Laws and Legal Information. Int-*8400*.S5, 1942.
Serial Publications of Foreign Governments, 1815-1931. Int-*1200*.S6, 1815-1931.
Ulrich's Periodicals Directory. Directorio de Publicaciones Periódicas Ulrich. Int-*1200*.U5, 1943.

Argentina.
Boletín Bibliográfico Argentino. Arg-*1200*.B6.
A Guide to the Official Publications of the Other American Republics: I, Argentina. Arg-*1200*.G5, 1945.

Bolivia.
A Guide to the Official Publications of the Other American Republics: II, Bolivia. Bol-*1200*.G5, 1945.

Canada.
"Publications of the Dominion Bureau of Statistics." Can-*1410*.C7, 1945, a.

Chile.
"Existencia de Anuarios, Censos y Otras Publicaciones Editadas por la Dirección General de Estadística." Chile -*1420*.E5, June '44, a.

Cuba.
A Guide to the Official Publications of the Other American Republics: VII, Cuba. Cuba-*1200*.G5, 1945.

Haiti.
Bibliographie Générale et Méthodique d'Haïti. Haiti-*1200*.B5, 1941.

México.
Bibliografía Mexicana de Estadística. Méx-*3010*.B5, 1942.
Boletín Bibliográfico Mexicano. Méx-*1200*.B5.
Publicaciones Oficiales, 1933-1942. Méx-*1200*.P5, 1933-42.

1000-1999 GENERAL—cont.

1200 **Bibliography—cont.**

Paraguay.

Strategic Index of the Americas: Preliminary Bibliography of Paraguay. Par-*1200*.S5, 1943.

Venezuela.

Anuario Bibliográfico Venezolano. Ven-*1200*.A5.

1300 **Congresses and conferences.**

Congresos y conferencias.

International.

Acta de Sesión No. 19 Celebrada el 1° de Agosto de 1946 (Aprobada el 5 de Septiembre de 1946). Int-*1300*.A4, 1946.

Congress and Conference Series. Int-*1300*.C8.

Proceedings of the Eighth American Scientific Congress. Int-*1300*.A5, 1940.3.

Serie de Congressos e Conferências. Int-*1300*.S5.

Serie sobre Congresos y Conferencias. Int-*1300*.S6.

1400-1499 **Data compilations.**

Compilaciones de datos.

1400 ——. **General.**

International.

Basic Data on the Other American Republics. Int-*1400*.B5, 1945.

Commercial Travelers' Guide to Latin America. Int-*1950*.C5, 1939.

Brasil.

Atlas Estatístico do Brasil. Statistic Atlas of Brazil. Atlas Statistique du Brésil. Bras-*1400*.A7, 1941.

Canada.

Bulletins No. F. Can-*1400*.B5.

Chile.

Chile a Través de Sus Zonas Geográfico-Económicas. Chile-*6000*.C34, 1944.

Colombia.

Colombia en Cifras. Col-*1400*.C5, 1944.

Síntesis Estadística de Colombia. Col-*1400*.S5.

Costa Rica.

Resúmenes Estadísticos, Años 1883 a 1910. CR-*1400*.R5, 1883-1910.

Cuba.

Condiciones Económicas y Sociales de la República de Cuba. Cuba-*1400*.C5, 1944.

Directorio Oficial de Exportación, Importación, Producción y Turismo, 1941. Cuba-*6700*.D5, 1941.

Ecuador.

"Ecuador, a Statistical Abstract." Int-*6000*.C7, Nov.-Dec. '44, Zier.

Ecuador en Cifras, 1938 a 1942. Ec-*1400*.E5, 1938-42.

Haiti.

Geology of the Republic of Haiti. Haiti-*4000*.G5, 1924.

Monograph, Republic of Haiti, Compiled 1932. Haiti-*1400*.M5, 1932.

México.

México en Cifras. Méx-*1400*.M5, 1938.

Nicaragua.

"Monografía de la República de Nicaragua." Nic-*1420*.B5, Apr.-June '46, a.

Panamá.

Examen Económico de la República de Panamá para el Año de 1943. Pan-*6000*.E5, 1943.

Paraguay.

Paraguay, Datos y Cifras Estadísticas. Par-*1400*.P5, 1939.

United States.

The Federal Chart Book. USA-*1400*.F5, 1938.

Fifteenth Census of the United States, 1930: Abstract of the Fifteenth Census of the United States. USA-*2200*.C5, 1930.

Trends in American Progress. USA-*1400*.T5, 1946.

1000-1999 GENERAL—cont.

1410 ——. Annuals.
Anuarios.

1000-1999 GENERAL—cont.

1410 ——. **Annuals—cont.**

Paraguay.
Anuario del Ministerio del Economía, 1938-39. Par-*6000*.A5, 1938-39.
Anuario Estadístico de la República del Paraguay. Par-*1410*.A5.
Boletín Semestral de la Dirección General de Estadística. Par-*1410*.B5.

Perú.
Anuario Estadístico del Perú. Perú-*1410*.E5.
Extracto Estadístico y Censo Electoral de la República. Perú-*1410*.E5, 1933.
Memoria que Presenta al Congreso Nacional el Ministro de Salud Pública y Asistencia Social. Perú-*7110*.M5.
Peru—Year Book of Foreign Trade Exporters, Importers, and Merchants. Perú-*6900*.P5.

República Dominicana.
Anuario Estadístico de la República Dominicana. RepDom-*1410*.A5.

United States.
Statistical Abstract of the United States. USA-*1410*.S5.
——. Puerto Rico.
Annual Report of the Governor of Puerto Rico. USA PR-*1410*.A5.

Uruguay.
Anuario Estadístico de la República Oriental del Uruguay. Ur-*1410*.A5.
Síntesis Estadística de la República Oriental del Uruguay. Ur-*1410*.S5.

Venezuela.
Anuario Estadístico de Venezuela. Ven-*1410*.A5.

1420 ——. **Periodicals.**
Publicaciones periódicas.

International.
Bulletin Mensuel de Statistique. Monthly Bulletin of Statistics. Int-*1420*.M5.
International Labour Review. Int-*9000*.I5.
Revista Internacional del Trabajo. Int-*9000*.R6.
Revue de l'Institut International de Statistique. Int-*3040*.R5.
Revue de l'Institut International de Statistique: Supplement Mensuel. Int-*1420*.R5.

Argentina.
Boletín Estadístico. Arg-*6000*.S8.
Boletín Estadístico del Ministerio de Agricultura. Arg-*6200*.B5.
Business Conditions in Argentina. Arg-*6000*.B8.
Economía y Finanzas de Mendoza. Arg-*6000*.E5.
Veritas. Arg-*6000*.V5.

Bolivia.
Boletín del Banco Central de Bolivia. Bol-*6720*.B5.
Revista Mensual. Bol-*1420*.R5.

Brasil.
Boletim do Ministério do Trabalho, Indústria e Comércio. Bras-*1420*.B5.
Boletim Estatístico do Instituto Brasileiro de Geografia e Estatística. Bras-*1420*.B7
Boletim Mensal do Serviço Federal de Bio-Estatística. Bras-*5200*.B5.
Dados e Índices Econômico-Financeiros Relativos ao Distrito Federal e á Cidade de São Paulo. Bras-*6010*.D5.
Estatísticas Econômicas. Bras-*6000*.E6.
Resumo da Bio-Estatística de Cidades Brasileiras. Bras-*5200*.R5.

Canada.
Monthly Index of Retail Sales. Can-*6710*.M5.
Monthly Index of Wholesale Sales. Can-*6710*.M6.
Monthly Review of Business Statistics. Can-*6000*.M5.
Trade of Canada: Exports of Canadian Produce. Can-*6900*.T6.
Trade of Canada: Imports Entered for Consumption. Can-*6900*.T7.

1000-1999 GENERAL—cont.

1420 ——. **Periodicals—cont.**

Chile.
Estadística Chilena. Chile-*1420*.E5.

Colombia.
Anales de Economía y Estadística. Col-*1420*.A5.
Colombia. Col-*1900*.C5.
Revista del Banco de la República. Col-*6720*.R5.

Cuba.
Boletín de la Oficina de Números Indices. Cuba-*6010*.B5.
Boletín Mensual de Estadísticas. Cuba-*1420*.B5.

Ecuador.
Datos Estadísticos. Ec-*6000*.I5A.
Estadística y Censos. Ec-*1420*.E5.
El Trimestre Estadístico del Ecuador. Ec-*1420*.T5.

El Salvador.
Boletín Estadístico de la Dirección General de Estadística. ElSal-*1420*.B5.
Revista Mensual del Banco Central de Reserva de El Salvador. ElSal-*6720*.R5.

Guatemala.
Boletín de la Dirección General de Estadística. Guat-*1420*.B5.

Haiti.
Bulletin Mensuel du Département Fiscal. Haiti-*1420*.B5.
Monthly Bulletin of the Fiscal Department. Haiti-*1420*.M5.

México.
Boletín Mensual de la Dirección General de Economía Rural. Méx-*6200*.B5.
Revista de Estadística. Méx-*1420*.R5.
Revista del Comercio Exterior. Méx-*6900*.R5.

Nicaragua.
Boletín Mensual de Estadística. Nic-*1420*.B5.
Economía y Finanzas. Nic-*6000*.E5.

Panamá.
Estadística Panameña. Pan-*1420*.E5.
Revista de Agricultura y Comercio. Pan-*6000*.R5.

Perú.
Algodón. Perú-*6210*.A5.
Boletín de la Dirección Nacional de Estadística. Perú-*1420*.B4.
Boletín Mensual del Banco Central de Reserva del Perú. Perú-*6720*.B5.
Informaciones Comerciales, Económicas y Financieras del Perú. Perú-*6000*.I5.

United States.
Federal Reserve Bulletin. USA-*6720*.F4.
Monthly Labor Review. USA-*9000*.M5.
Survey of Current Business. USA-*6000*.S7.
——. Puerto Rico.
Puerto Rico Monthly Statistical Report. USA PR-*1420*.P5.

Uruguay.
Boletín Aduanero. Ur-*6900*.B5.
Boletín de Hacienda, Suplemento. Ur-*6000*.B6.
Boletín de la Dirección de Estadística Económica. Ur-*6000*.B7.
Boletín Estadístico de la República Oriental del Uruguay. Ur-*1420*.B5.
Suplemento Estadístico de la Revista Económica. Ur-*6000*.S5.

Venezuela.
Boletín del Banco Central de Venezuela. Ven-*6720*.B5.
Boletín Mensual de Estadística. Ven-*1420*.B5.

1000-1999 GENERAL—cont.

1500 Filing and cataloging.
Archivo y catalogación.

Non-Geographic.
"Distribución Abecedaria de Apellidos." Chile-*1420*.E5, Mar. '46.
"A Method for Indexing Statistical Tabulations." Int-*3040*.E5, Sept. '45, Moriyama.

1600 Microfilm and preservation of records.
Microfilm y conservación de los datos.

Non-Geographic.
"Registros, Formas y Otra Documentación de una Investigación Estadística." Int-*3040*.E5, Sept. '46, Ullman.
The Repair and Preservation of Records. NonG-*1600*.R5, 1943.
Summary Report of National Bureau of Standards Research on Preservation of Records. NonG-*1600*.S5, 1937.
Summary Report of Research at the National Bureau of Standards on the Stability and Preservation of Records on Photographic Film. NonG-*1600*.S7, 1939.

1800 Publication processes.
Métodos de publicación.

Non-Geographic.
Manual on Methods of Reproducing Research Materials. NonG-*1800*.M5, 1936.

1900-1999 Reference books, general.
Libros de referencia, general.
1900 ——. General.

International.
Bulletin of the Pan American Union. Int-*8200*.B7.
The Caribbean Islands and the War. Int-*1900*.C5, 1943.
Good Neighbors, Argentina, Brazil and Chile and Seventeen Other Countries. Int-*1900*.G5, 1941.
Latin America. Int-*1900*.L4, 1943.
Latin America. Int-*1900*.L5, 1942.
Latin America in the Future World. Int-*1900*.L7, 1945.
The South American Handbook. Int-*1900*.S5.

Argentina.
Informaciones Argentinas. Arg-*1900*.I5.
Una Nueva Argentina. Arg-*1900*.N5, 1940.

Brasil.
Brazil in the Making. Bras-*1900*.B4, 1943.
Brazilian Culture. Bras-*2200*.R4, 1940, Eng.
Recenseamento Geral do Brasil (1°. de Setembro de 1940). Bras-*2200*.R4, 1940.

Chile.
La Economía Nacional y el Problema de las Subsistencias en Chile. Chile-*1900*.E5, 1943.
Chile: An Economy in Transition. Chile-*6000*.C35, 1945.

Colombia.
Colombia. Col-*1900*.C5.

El Salvador.
Prontuario—Geográfico, Comercial, Estadístico, y Servicios Administrativos de El Salvador. ElSal-*1900*.P5, 1939.

Haiti.
Cahiers d'Haïti. Haiti-*1900*.C5.
The Haitian People. Haiti-*1900*.H5, 1941.

Nicaragua.
La Gaceta. Nic-*1900*.G5.

1000-1999 GENERAL—cont.

1900 ——. **General—cont.**

United States.
Dun's Review. USA-*6000*.D5.
Venezuela.
Gaceta Oficial de los Estados Unidos de Venezuela. Ven-*1900*.G5.

1910 ——. **Annuals and almanacs.**
Anuarios y almanaques.

International.
Inter-American Affairs. Int-*1910*.I5.
The Pan American Yearbook. Int-*1410*.P5.
Political Handbook of the World. Int-*1910*.P5.
The Statesman's Year-Book. Int-*1910*.S5.
The West Indies Year Book, Including Also the Bermudas, the Bahamas, British Guiana and British Honduras, Etc. Int-*1410*.W5.
The World Almanac and Book of Facts. Int-*1910*.W5.
United States.
The American Year Book. USA-*1910*.A5.

1930 ——. **Directories of individuals and organizations.**
Directorios de personas y organizaciones.

International.
"Directorio de Especialistas en Estudios Sociales y Económicos en Latinoamérica, Parte I." Int-*3040*.B5, Feb. '45, a.
"Directorio de Instituciones de Estudios e Investigaciones Sociales y Económicos en Latinoamérica, Parte I." Int-*3040*.B5, Feb. '45, b.
Handbook of International Organizations in the Americas. Int-*1930*.H4, 1945.
Handbook of Scientific and Technical Societies and Institutions of the United States and Canada. Int-*1930*.H5, 1942.
International Agencies in which the United States Participates. Int-*8200*.I49, 1946.
"Who's who in Latin American Trade." Int-*1410*.P5, 1945.
Venezuela.
Guía Industrial y Comercial de Venezuela, 1943. Ven-*6000*.G5, 1943.
Suplemento de la Guía Industrial y Comercial de Venezuela Correspondiente al Año 1943. Ven-*6000*.G5, 1943.A.

1940 ——. **Encyclopedias.**
Enciclopedias.

Non-Geographic.
Encyclopaedia of the Social Sciences. NonG-*1940*.E5, 1937.

1950 ——. **Guide books and handbooks of countries and places.**
Guías y manuales de países y localidades.

International.
American Nation Series. Int-*1950*.A5.
Commercial Travelers' Guide to Latin America. Int-*1950*.C5, 1939.
The New World Guides to the Latin American Republics. Int-*1950*.N5, 1943.
Brasil.
Brazil. Bras-*6000*.B65, 1946.
El Salvador.
La Guía Turística de El Salvador. ElSal-*1950*.G5.
Guatemala.
Guía Kilométrica de las 23 Rutas Nacionales de la República de Guatemala. Guat-*1950*.G5, 1942.
Nomenclator de Guatemala. Guat-*1950*.N5, 1937.

1000-1999 GENERAL—cont.

1950 ——. Guide books and handbooks of countries and places—cont.

Haiti.
Monograph, Republic of Haiti, Compiled 1932. Haiti-*1400*.M5, 1932.

Nicaragua.
Nicaragua. Nic-*1950*.N5.

República Dominicana.
República Dominicana, Album Estadístico Gráfico. RepDom-*1950*.R5, 1944.

Venezuela.
Indicador Nacional de Vías de Comunicación de Venezuela. Ven-*6620*.I5, 1944.
Venezuela, 1945. Ven-*1950*.V5, 1945.

2000-2999 CENSUS
CENSOS

2000 General.

International.
"Bases Mínimas para la Uniformidad de los Censos Nacionales de Población en el Continente Americano." Int-*1300*.A5, 1940.3, Loyo González.
General Censuses and Vital Statistics in the Americas. Int-*5000*.G5, 1943.
Projet d'un Recensement du Monde. Int-*2000*.P5, 1881.

Argentina.
"Algunas Sugerencias a Propósito del Proyectado IV Censo General." Arg-*3040*.R5, Apr. '45.
Argentine Fourth General Census Postponed. Arg-*2100*.C5, 1945.
Boletín Censal y Estadístico. Arg-*3040*.B5.
"A Propósito del IV Censo General." Arg-*3040*.R5, July '45, a.
"A Propósito del Cuarto Censo General." Arg-*3040*.R5, June '45, a.
"Sobre Censo General: Observaciones y Sugerencias." Arg-*3040*.R8, Jan.-Apr. '45, Santone.
"Sobre los Servicios Estadísticos y Censales en la República Argentina." Arg.-*3040*.R8, Jan.-Apr. '45, Cortés.

Canada.
Seventh Census of 1931: Administrative Report of the Dominion Statistician. Can-*2100*.C5, 1931.

Ecuador.
Se Puede Levantar el Censo? Ec-*2000*.S5, 1939.

United States.
"An Appraisal of the 1936 Religious Census." USA-*9000*.A6, June '43, Reuss.
"On Some Census Aids to Sampling." USA-*3040*.J5, Sept. '43, Hansen.
"The Sampling Procedure of the 1940 Population Census." USA-*3040*.J5, Dec. '40, Stephan.

Non-Geographic.
"On Sample Inspection in the Processing of Census Returns." USA-*3040*.J5, Sept. '41, Deming.
"Sample Surveys in Census Work." NonG-*3150*.S3.
"Sobre Censo General: Observaciones y Sugerencias." Arg-*3040*.R8, Jan.-Apr. '45, Santone.

2000-2999 CENSUS—cont.

2000-2999 CENSUS—cont.

2100 ——. **General and population—cont.**

Chile.

Censo de 1940: Métodos y Procedimientos. Chile-*2100*.C5, 1940.1.
"Los Censos de Población en Chile." Int-*1300*.A5, 1940.3, Vergara.
Instrucciones para la División del Territorio Comunal en Zonas de Empadronamiento: XI Censo de Población. Chile-*2100*.C5, 1940.2.
Libreta de Empadronador: XI Censo de Población. Chile-*2100*.C5, 1940.3.
Reglamento del XI Censo de Población. Chile-*2100*.C5, 1940.4.

Colombia.

"Censo Continental: Una Sugerencia del Instituto Interamericano de Estadística." Col-*2100*.C5, 1950.
Decretos y Bases de Organización para la Ejecución de los Censos, 1938. Col-*2100*.C5, 1938.
"Esquema del Plan de Organización Puesto al Servicio del Levantamiento de los Censos de Edificios y Población Llevados a Cabo en Colombia en 1938." Int-*1300*.A5, 1940.3, Suárez Rivadeneira.
Estatuto Orgánico de la Estadística Nacional. Col-*3030*.E5, 1945.
"Informe que la Contraloría General de la República Rinde la Sr. Ministro de Gobierno y a las Honorables Cámaras sobre el Levantamiento del Censo Civil de 1938." Col-*1420*.A5, Aug. '39.

Cuba.

Censo de 1943. Cuba-*2100*.C5, 1943.2.
Censo de 1943: Métodos y Procedimientos. Cuba-*2100*.C5, 1943.1.

Ecuador.

"El Censo Continental de las Américas y el Primer Censo Demográfico Nacional." Ec-*1420*.T5, Apr.-June '45, Procaccia.

El Salvador.

Adiciones al Reglamento del Censo de Población. ElSal-*2100*.C5, 1930.2.
Censo de 1930: Métodos y Procedimientos. ElSal-*2100*.C5, 1930.1.
Censo de Población de 1930. ElSal-*2100*.C5, 1930.3.
Instrucciones al Personal del Censo de la República. ElSal-*2100*.C5, 1930.4.

Guatemala.

Censo de 1940: Métodos y Procedimientos. Guat-*2100*.C5, 1940.1.
Reglamento para el Censo General de Población del Año 1940. Guat-*2100*.C5, 1940.2.

Haiti.

Recensement Général del Communes de la République d'Haïti, Septembre 1918—Août 1919: Méthodes et Procédés et Résultats. Haiti-*2100*.R5, 1918-19.
Sampling Plans for a Census of Haiti. Haiti-*2100*.S5, 1942.

Honduras.

Censo de 1940: Métodos y Procedimientos. Hond-*2100*.C5, 1940.1.
Censo General de Población Levantado el 24 de Junio de 1945: Métodos y Procedimientos. Hond-*2100*.C5, 1945.1.
Instrucciones para la Preparación y Práctica del Censo de 1940. Hond-*2100*.C5, 1940.2.
Instrucciones para la Preparación y Práctica del Censo de Población de 1945. Hond-*2100*.C5, 1945.2.
Reglamento del Censo de Población de 1945. Hond-*2100*.C5, 1945.3.
Reglamento para el Levantamiento del Censo de Población de 1940. Hond-*2100*.C5, 1940.3.

México.

Censo de 1940: Métodos y Procedimientos. Méx-*2100*.C5, 1940.1.
Decreto y Bases de Organización de los Siete Censos Nacionales y de los Padrones Estadísticos que se Levantarán en 1939-40. Méx-*2100*.C5, 1940.2.
Instrucciones para la Perforación de Tarjetas del 6° Censo de Población de 1940. Méx-*2100*.C5, 1940.3.

2000-2999 CENSUS—cont.

2100 ——. General and population—cont.

Nicaragua.
Censo de 1940: Métodos y Procedimientos. Nic-*2100*.C5, 1940.1.
Primer Censo General de Población, 1940: Instructivo para Empadronadores; Interpretación de la Boleta de Población. Nic-*2100*.C5, 1940.2.

Panamá.
Censo de 1940: Métodos y Procedimientos. Pan-*2100*.C5, 1940.1.
Cuarto Censo Decenal de Población 1940: Instrucciones a los Empadronadores e Inspectores. Pan-*2100*.C5, 1940.2.

Paraguay.
"Censo de Población." Par-*1410*.A5, 1937, a.
"Censo de Población." Par-*1410*.A5, 1936, a.
Censo de 1936: Métodos y Procedimientos. Par-*2100*.C5, 1936.1.
Censo de Población, Decreto N°. 2120 de 18 de Junio de 1936. Par-*2100*.C5, 1936.2.
Decreto N°. 60.287 de 21 de Noviembre de 1935. Par-*2100*.C5, 1936.3.
Ley N°. 1.509 de 1° de Noviembre de 1935. Par-*2100*.C5, 1936.4.

Perú.
Censo Nacional de Población y Ocupación, 1940. Perú-*2200*.C5, 1940.1.
"Census of Peru, 1940." USA-*4000*.G5, Jan. '42, Arca Parró.
"La Estadística Peruana en el Sexenio 1938-44." Int-*3040*.E5, Dec. '44, Luna Vegas.
"Problemas y Soluciones para el Censo Demográfico Peruano de 1940." Int-*1300*.A5, 1940.3, Arca Parró, b.

República Dominicana.
Censo de 1935: Métodos y Procedimientos. RepDom-*2100*.C5, 1935.1.
Censo de 1935: Instrucciones a los Enumeradores del Censo de Población. RepDom-*2100*.C5, 1935.3.
Censo de 1935: Instrucciones Generales sobre las Funciones de un Enumerador. RepDom-*2100*.C5, 1935.4.
Censo Especial Urbano San Cristóbal, 1944: Instrucciones a los Enumeradores. RepDom-*2100*.C7, 1944.
Censo de 1935: Ley del Congreso Nacional que Dispone las Condiciones en que Deben Desarrollarse las Operaciones del Censo. RepDom-*2100*.C5, 1935.2.
Censo Especial Urbano de Ciudad Trujillo, 1945. RepDom-*2100*.C6, 1945.
Población de la República Dominicana, según las Sucesivas Modificaciones Territoriales a Partir del 13 de Mayo de 1935, Día en que Se Levantó el Segundo Censo Nacional de Población, Hasta el 1 de Enero de 1946. RepDom-*2200*.C5, 1935-46.

United States.
Handbook for Employees of the Bureau of the Census. USA-*3030*.H5, 1940.
Instructions for Processing Data of the Sixteenth Decennial Census. USA-*2100*.C5, 1940.7.
Instructions: Preparation of Plans of Division by Enumeration Districts, 1940. USA-*2100*.C5, 1940.6.
List of Schedules Used by the Bureau of the Census for Collecting Data, 1945. USA-*3110*.L5, 1945.
Sixteenth Census of the United States: 1940: Enumerator's Record Book. USA-*2100*.C5, 1940.4.
Sixteenth Census of the United States: 1940: Methods and Procedures. USA-*2100*.C5, 1940.1.
Sixteenth Decennial Census of the United States: Instructions to Enumerators: Population and Agriculture, 1940. USA-*2100*.C5, 1940.5.
"Source Materials: IASI Survey of Census Methods and Procedures: United States: Sixteenth Census, 1940." Int-*3000*.I5, Aug.-Oct. '44.

2000-2999 CENSUS—cont.

2100 ——. **General and population—cont.**

——. Puerto Rico.

Censo Décimosexto de los Estados Unidos: 1940: Instrucciones a los Enumeradores: Puerto Rico. USAPR-*2100*.C5, 1940.2.

Sixteenth Census of the United States: 1940: Methods and Procedures; Forms and Instructions Used in Taking the Census of Puerto Rico. USA-PR-*2100*.C5, 1940.1.

Venezuela.

Censo de 1941: Métodos y Procedimientos. Ven-*2100*.C5, 1941.1.

"Ley de Estadística y de Censos Nacionales." Ven-*1900*.G5, Nov. 27, '44.

Séptimo Censo Nacional de Población, 1941: Instrucciones y Modelos. Ven-*2100*.C5, 1941.2.

Séptimo Censo Nacional de Población, Levantado el 7 de Diciembre de 1941. Ven-*2200*.C5, 1941.

Other Countries.

West Indies.

Eighth Census of Jamaica, 1943: Administrative Report. WI-*2200*.E5, 1943.1.

2110 ——. **Agriculture.**

Agricultura.

International.

"Cuestiones Elementales en los Censos Agrícolas Aplicables en los Países de Latino-América." Int-*1300*.A5, 1940.3, Alarcón Mendizábal.

"Estudio sobre Estadística Agrícola." Col-*1420*.A5, July '39, Ricaurte Montoya.

The First World Agricultural Census (1930). Int-*2210*.W5, 1930.

Programme of the World Agricultural Census of 1940. Int-*2110*.W5, 1940.1.

"Le Recensement Agricole Mondial de 1940." Int-*3020*.B5, 1938, Dore.

The World Agricultural Census of 1940: Proposed Standard Form. Int-*2110*.W5, 1940.2.

World Agricultural Census of 1940: Questionnaires. Int-*2110*.W5, 1940.3.

Argentina.

Ley No. 12.343, Diciembre 29 de 1936. Arg-*2110*.L5, 1936.

Primer Censo Nacional de Floricultura, Ley 12.343, Años 1938-39. Arg-*2210*.C7, 1938-39.

Bolivia.

Censo Experimental Agropecuario, Departamento de Tarija, 1944-1945. Bol-*2210*.C5, 1944-45, Tarija.

Colombia.

"El Censo Cafetero 1939-40." Col-*6210*.B5, Aug. '39, b.

Estatuto Orgánico de la Estadística Nacional. Col-*3030*.E5, 1945.

"Plan General para el Levantamiento del Primer Censo Agropecuario." Col-*1420*.A5, Oct. 20, '43, Montoya.

El Salvador.

Primer Censo Nacional del Café, 1938-39. ElSal-*2210*.C5, 1944.

México.

Report on the Crop and Livestock Statistics and Agricultural Census Work in Mexico. Méx-*6200*.R8, 1946.

Segundo Censo Agrícola-Ganadero, Marzo de 1940: Métodos y Procedimientos. Méx-*2110*.C5, 1940.1.

Segundo Censo Ejidal de los Estados Unidos Mexicanos, 6 de Marzo de 1940: Métodos y Procedimientos. Méx-*2110*.C7, 1940.1.

Panamá.

Censo Agro-Pecuario de las Provincias de Herrera y Los Santos, Diciembre 1945: Instrucciones a los Empadronadores. Pan-*2110*.C5, 1945.1.

Censo Agro-Pecuario de las Provincias de Herrera y Los Santos, Diciembre 1945: Métodos y Procedimientos. Pan-*2110*, 1945.

2000-2999 CENSUS—cont.

2110 ——. **Agriculture—cont.**

Censo Agro-Pecuario, Diciembre de 1943, Distrito de Penonomé: Métodos y Procedimientos. Pan-*2110*.C5, 1943.1.

Paraguay.
Censo Familiar: Cédula. Par-*2110*.C5, 1944.5.

República Dominicana.
Censo Agro-Pecuario de 1940: Métodos y Procedimientos. RepDom-*2110*.C5, 1940.1.

United States.
Agriculture Handbook. USA-*2110*.C5, 1940.2.
Censo de Agricultura de los Estados Unidos. USA-*2110*.C5, 1940.1.

Uruguay.
Censo Ganadero 1943: Ley de Enero 27 de 1943. Ur-*2110*.C5, 1943.

2120 ——. **Buildings, real estate, investments.**
Edificios, bienes raíces, inversiones.

México.
Segundo Censo de Edificios de los Estados Unidos Mexicanos, 20 de Octubre de 1939: Métodos y Procedimientos, Méx-*2120*.C5, 1939.1.

Nicaragua.
Primer Censo de Edificios: 18 de Enero de 1940: Cédula. Nic-*2120*.C5, 1940.

United States.
Sixteenth Decennial Census of the United States: Instructions to Enumerators, Housing, 1940. USA-*2120*.C5, 1940.1.

2130 ——. **Education.**
Educación.

Argentina.
"El Censo Escolar Nacional de 1943." Int-*3040*.E5, Sept. '46, Coghlan.
Censo Escolar de la Nación, Ley No. 12.723: Cuestionario de Convivencia. Arg-*2130*.C5, 1943.8.
Censo Escolar de la Nación, Ley No. 12.723: Cuestionario Familiar. Arg-*2130*.C5, 1943.6.
Censo Escolar de la Nación, Ley No. 12.723: Cuestionario Individual. Arg-*2130*.C5, 1943.7.
Censo Escolar de la Nacón, Ley No. 12.723: Instrucciones para los Censistas. Arg-*2130*.C5, 1943.5.
Instrucciones Complementarias a las "Normas para la Organización y el Levantamiento del Censo Escolar de la Nación." Arg-*2130*.C5, 1943.1.
Normas para la Organización y Levantamiento del Censo Escolar de la Nación en la Capital Federal. Arg-*2130*.C5, 1943.2.
Normas para la Organización y Levantamiento del Censo Escolar de la Nación en las Provincias. Arg-*2130*.C5, 1943.3.
Normas para la Organización y Levantamiento del Censo Escolar de la Nación en los Territorios. Arg-*2130*.C5, 1943.4.

Guatemala.
Acuerdo Disponiendo el Levantamiento del Censo Escolar y Reglamento General del Mismo. Guat-*2130*.C5, 1946.2.
Censo Escolar de 1946: Métodos y Procedimientos. Guat-*2130*.C5, 1946.1.
Instrucciones Generales para los Jefes de Empadronadores y Personal Auxiliar. Guat-*2130*.C5, 1946.3.

2150 ——. **Industry, manufacturing, business.**
Industria, manufactura, negocios en general.

International.
"Conceptos Básicos de los Censos Industriales Recomendables para los Países Latino-Americanos." Int-*1300*.A5, 1940.3, Zertuche.

2000-2999 CENSUS—cont.

2150 ——. **Industry, manufacturing, business—cont.**

Canada.
"The Canadian Census of Distribution." Int-*3020*.B5, 1938, Marshall.

Chile.
El Censo Económico General de 1943. Chile-*2150*.C7, 1943.
Censo Industrial y Comercial, Año 1937. Chile-*2250*.C5, 1937.

Colombia.
Censo Industrial de Colombia, 1945: Métodos y Procedimientos. Col-*2150*.C5 1945, a.
"El Censo Industrial de Colombia de 1945." Col-*1420*.A5, Feb. '45, Ortiz C.
Estatuto Orgánico de la Estadística Nacional. Col-*3030*.E5, 1945.
Primer Censo Industrial de Colombia, 1945: Boleta General. Col-*2150*.C5, 1945, c.
Primer Censo Industrial de Colombia, 1945: Cartilla de Instrucciones. Col-*2150*.C5, 1945, b.

Guatemala.
"Censo Industrial." Guat-*1420*.B5, Nov. '46.

México.
"El Censo Comercial en México." Int-*1300*.A5, 1940.3, Villafuerte Flores.
Los Censos Cuarto Industrial, Segundo Comercial y Segundo de Transportes, 1945: Métodos y Procedimientos. Méx-*2150*.C5, 1945.1.
Primer Censo Comercial, Marzo de 1940: Métodos y Procedimientos. Méx-*2150*.C3, 1940.1.
Tercer Censo Industrial de los Estados Unidos Mexicanos, 1940: Métodos y Procedimientos. Méx-*2150*.C5, 1940.1.

Nicaragua.
Primer Censo Comercial de Nicaragua, 1940: Instructivo. Nic-*2150*.C3, 1940.
Primer Censo Industrial de Nicaragua, 1940: Instructivo para Proporcionar los Datos a los Empadronadores. Nic-*2150*.C5, 1940.

United States.
Sixteenth Census of the United States: 1940: Instructions to Enumerators for Business and Manufactures, 1939. USA-*2150*.C5, 1939.1.
Sixteenth Census of the United States: 1940: Manufactures, 1939: Industry Classification for the Census of Manufactures, 1939. USA-*3220*.C5, 1939.

2180 ——. **Transport and communication.**
Transportes y comunicaciones.

México.
Primer Censo de Transportes, Febrero y Marzo de 1940: Métodos y Procedimientos. Méx-*2180*.C5, 1940.1.

2200-2299 **Results.**
Resultados.

2200 ——. **General and population.**
General y de población.

Argentina.
Cuarto Censo General, 1936: Población 22-x-1936. Arg-*2200*.C4, 1936.
Primer Censo de la República Argentina, Verificado en los Días 15, 16 y 17 de Setiembre de 1869. Arg-*2200*.C5, 1869.
Segundo Censo de la República Argentina, Mayo 10 de 1895. Arg-*2200*.C5, 1895.
Tercer Censo Nacional, Levantado el 1 de Junio de 1914. Arg-*2200*.C5, 1914.

Bolivia.
Censo Demográfico de la Ciudad de La Paz, 15 de Octubre de 1942. Bol-*2200*.C4, 1942.

2000-2999 CENSUS—cont.

2200 ——. General and population—cont.

Censo General de la Población de la República de Bolivia según el Empadronamiento de lo de Septiembre de 1900. Bol-*2200*.C5, 1900.

Brasil.

Análises de Resultados do Censo Demográfico. Bras-*2200*.A5, 1940.

Brazilian Population: Preliminary Results of the 1940 Census. Bras-*2200*.B5, 1940.

Recenseamento do Brasil Realizado em 1° de Setembro de 1920. Bras-*2200*.R4, 1920.

Recenseamento Geral de 1940: Resultados Preliminares. Bras-*2200*.R6, 1940.

Recenseamento Geral do Brasil (1.° de Setembro de 1940). Bras-*2200*.R4, 1940.

Recenseamento Geral do Brasil, 1.° de Setembro de 1940: Sinopse do Censo Demográfico. Bras-*2200*.R7, 1940.

Recenseamento Geral do Brasil, Realizado em 1.° de Setembro de 1940: Sinopse Preliminar dos Resultados Demográficos segundo as Unidades da Federação e os Municípios. Bras-*2200*.R5, 1940.

Canada.

Census of Indians in Canada, 1944. Recensement des Indiens du Canada, 1944. Can-*2200*.C6, 1944.

Census of Indians in Canada, 1939. Recensement des Indiens du Canada, 1939. Can-*2200*.C6, 1939.

Eighth Census of Canada, 1941. Huitième Recensement du Canada, 1941. Can-*2200*.C5, 1941.2.

Eighth Census of Canada, 1941: Population. Can-*2200*.C5, 1941.2a.

Seventh Census of Canada, 1931. Septième Recensement du Canada, 1931. Can-*2200*.C5, 1931.2.

Chile.

Resultados del X Censo de la Población Efectuado el 27 de Noviembre de 1930 y Estadísticas Comparativas con Censos Anteriores. Chile-*2200*.C5, 1930.

"XI Censo General de Población." Chile-*1420*.E5, Dec. '41.

Colombia.

Censo General de Población, 5 de Julio de 1938. Col-*2200*.C5, 1938

Cuba.

Censo de 1943. Cuba-*2200*.C5, 1943.2.

Censo de 1931. Cuba-*2200*.C5, 1931-38.

Costa Rica.

Censo de Personas sin Trabajo, 1932. CR-*2200*.C5, 1932.

Población de la República de Costa Rica según el Censo General de Población Levantado al 11 de Mayo de 1927. CR-*2200*.P5, 1927.

El Salvador.

Población de la República de El Salvador. ElSal-*2200*.C5, 1930.

Gautemala.

Censo de la Población de la República, Levantado el 28 de Agosto de 1921. Guat-*2200*.C5, 1921.

Quinto Censo General de Población, Levantado el 7 de Abril de 1940. Guat-*2200*.C5, 1940.

Haiti.

"Résultats du Recensement Général des Communes de la République d'Haïti (Septembre 1918-Août 1919)." Haiti-*1900*.M5, Sept. 13, '19.

Honduras.

Resumen del Censo General de Población Levantado el 30 de Junio de 1940. Hond-*2200*.C5, 1940.

México.

Estados Unidos Mexicanos, Sexto Censo de Población, 1940. Méx-*2200*.C5, 1940.1.

Memoria de los Censos Generales de Población, Agrícola, Ganadero e Industrial de 1930. Méx-*2200*.C5, 1930.1.

2000-2999 CENSUS—cont.

2200 ——. General and population—cont.

Quinto Censo de Población, 15 de Mayo de 1930. Méx-*2200*.C5, 1930.2.
Sexto Censo General de Población de los Estados Unidos Mexicanos, 6 de Marzo de 1940: Población Municipal. Méx-*2200*.C5, 1940.2.

Nicaragua.
"Actividades de la Población de Nicaragua." Nic-*1420*.E5, Apr. 15, '45, a.
Censo General de 1920. Nic-*2200*.C5, 1920.
"Cifras Preliminares de Población por Provincias y Capitales de Provincias, según el Censo de 1940." Nic-*1420*.B5, July-Dec. '40.
"Población de la República de Nicàragua, Calculada hasta el 31 de Diciembre de 1944." Nic-*1420*.B5, Sept.-Oct. '45.
Población de Nicaragua. Nic-*2200*.C5, 1940.2.
Población de Nicaragua, por Sexos, Urbana y Rural (Censo de 1940). Nic-*2200*.C5, 1940.3.
"Población Economicamente Inactiva (Censo del 23 de Mayo de 1940)." Nic-*1420*.B5, April 15, '45, b.
Población según Edades y Sexo (Censo 1940). Nic-*2200*.C5, 1940.4.

Panamá.
Censo de Población, 1940. Pan-*2200*.C5, 1940.1.
Censo de Población, 1940. Pan-*2200*.C5, 1940.2.
Censo de Población, 1940: Informe Preliminar. Pan-*2200*.C5, 1940.3.

Perú.
Censo Nacional de 1940: Resultados Generales. Perú-*2200*.C5, 1940.2.
Censo Nacional de Población y Ocupación, 1940. Perú-*2200*.C5, 1940.1.
Estado de la Instrucción en el Perú según el Censo Nacional de 1940. Perú-*2200*.C5, 1940.4.

República Dominicana.
Anuario Estadístico de la República Dominicana. RepDom-*1410*.A5.
Censo de Población Urbana y Rural, por Provincias y Comunas. RepDom-*2200*.C5, 1935.3.
Población de la República Censada el 13 de Mayo de 1935, Clasificada por Grupos de Edades, Zona y Sexo. RepDom-*2200*.C5, 1935.2.
Población de la República Dominicana, Distribuida por Nacionalidades: Cifras del Censo Nacional de 1935. RepDom-*2200*.C5, 1935.1.
Población de la República Dominicana, según las Sucesivas Modificaciones Territoriales a Partir del 13 de Mayo de 1935, Día en que Se Levantó el Segundo Censo Nacional de Población, hasta el 1 de Enero de 1946. RepDom-*2200*.C5, 1935-46.
Población de la República, según el Censo Levantado el 24 de Diciembre de 1920, Clasificada por Sexo y Grupos de Edades. RepDom-*2200*.C5, 1920.

United States.
Census of Congested Production Areas: 1944. USA-*2200*.C3, 1944.
Census of Unemployment, 1937: Final Report on Total and Partial Unemployment. USA-*2200*.C7, 1937.
Census of Unemployment, 1937: Final Report on Total and Partial Unemployment for Geographic Divisions. USA-*2200*.C7, 1937.A.
Fifteenth Census of the United States: 1930: Abstract of the Fifteenth Census of the United States. USA-*2200*.C5, 1930.
Fifteenth Census of the United States: 1930: The Indian Population of the United States and Alaska. USA-*2200*.C5, 1930.2.
"Final Reports of the Sixteenth Census of the United States, 1940." Int-*3040*.E5, June '44, b.
Indian Population in the United States and Alaska, 1910. USA-*2200*.C8, 1910.
Sixteenth Census of the United States: 1940: Areas of the United States, 1940. USA-*2200*.C5, 1940.12.
Sixteenth Census of the United States: 1940: Population. USA-*2200*.C5, 1940.1.

2000-2999 CENSUS—cont.

2200—General and population—cont.

2000-2999 CENSUS—cont.

2200 ——. **General and population—cont.**

Sixteenth Census of the United States: 1940: Population: Nativity and Parentage of the White Population: Country of Origin of Foreign Stock. USA-*2200*.C5, 1940.11b.

Sixteenth Census of the United States: 1940: Population: Nativity and Parentage of the White Population: General Characteristics. USA-*2200*.C5, 1940.11a.

Sixteenth Census of the United States: 1940: Population: Nativity and Parentage of the White Population: Mother Tongue. USA-*2200*.C5, 1940.11c.

Sixteenth Census of the United States: 1940: Population: Special Report on Institutional Population, 14 Years Old and Over. USA-*2200*.C5, 1940.9.

Sixteenth Census of the United States: 1940: Population: State of Birth of the Native Population. USA-*2200*.C5, 1940.14.

Sixteenth Census of the United States: 1940: Population: Unincorporated Communities. USA-*2200*.C5, 1940.13.

——. Alaska.

Sixteenth Census of the United States: 1940: Alaska: Population: Characteristics of the Population (with Limited Data on Housing). USA Alsk-*2200*.C5, 1940.

——. Puerto Rico.

Censo Décimosexto de los Estados Unidos: 1940: Puerto Rico: Población: Boletín Núm. 2: Características de la Población. Sixteenth Census of the United States: 1940: Puerto Rico: Population: Bulletin No. 2: Characteristics of the Population. USA PR-*2200*.C5, 1940.2.

Sixteenth Census of the United States: 1940: Puerto Rico: Población: Boletín No. 1: Número de los Habitantes. Puerto Rico: Population: Bulletin No. 1: Number of Inhabitants. USA PR-*2200*.C5, 1940.1.

Sixteenth Census of the United States: 1940: Puerto Rico: Población: Boletín No. 3: Ocupaciones y Otras Características por Edad. Puerto Rico: Population: Bulletin No. 3: Occupations and Other Characteristics by Age. USA PR-*2200*.C5, 1940.3.

Uruguay.

"Censo General de la República en 1908." Ur-*1410*.A5, 1908.

Venezuela.

Séptimo Censo Nacional de Población, Levantado el 7 de Diciembre de 1941. Ven-*2200*.C5, 1941.

Sexto Censo Nacional, 1936. Ven-*2200*.C5, 1936.2.

Other Countries

West Indies.

Eighth Census of Jamaica and Its Dependencies, 1943: Population, Housing and Agriculture. WI-*2200*.E5, 1943.

Eighth Census of Jamaica, 1943: Administrative Report. WI-*2200*.E5, 1943.1.

Eighth Census of Jamaica, 1943: Introduction to Population and Housing. WI-*2200*.E5, 1943.2.

2210 ——. **Agriculture.**
Agricultura.

International.

The First World Agricultural Census (1930). Int-*2210*.W5, 1930.

Argentina.

Censo Algodonero de la República Argentina, Año 1935-36. Arg-*2210*.C4, 1935-36.

Censo Ganadero Nacional, Existencia al 1.° de Julio de 1930. Arg-*2210*.C5, 1930.

Censo Nacional Agropecuario, Año 1937. Arg-*2210*.C6, 1937.

Primer Censo Nacional de Floricultura, Ley 12.343, Años 1938-39. Arg-*2210*.C7, 1938-39.

2000-2999 CENSUS—cont.

2210 ——. **Agriculture—cont.**

Bolivia.

Censo Experimental Agropecuario del Departamento de Chuquisaca, 1942. Bol-*2210*.C5, 1942, Chuquisaca.

Brasil.

Recenseamento do Brasil Realizado em 1° de Setembro de 1920. Bras-*2200*.R4, 1920.

Canada.

Eighth Census of Canada, 1941: Census of Agriculture: Nova Scotia. Huitième Recensement du Canada, 1941: Recensement Agricole: Nouvelle-Ecosse. Can-*2210*.C5, 1941.2a.

Chile.

Agricultura 1935-36 Censo. Chile-*2210*.A5, 1935-36.

Costa Rica.

Resúmenes Estadísticos, Años 1883 a 1910. CR-*1400*.R5, 1883-1910.

Cuba.

"El Censo Ganadero de 1945." Cuba-*1420*.B5, Apr. '45, Queralt.

El Salvador.

Primer Censo Nacional del Café, 1938-39. ElSal-*2210*.C5, 1944.

México.

Primer Censo Ejidal, 1935. Méx-*2210*.C7, 1935.

Segundo Censo Ejidal de los Estados Unidos Mexicanos, 6 de Marzo de 1940. Méx-*2210*.C7, 1940.

Panamá.

Censo Agro-Pecuario, 1942. Pan-*2210*.C5, 1942.

Censo Agro-Pecuario del Distrito de Penonomé, Diciembre 1943. Pan-*2210*.C5, 1943.1.

República Dominicana.

Anuario Estadístico de la República Dominicana. RepDom-*1410*.A5.

Primer Censo Cafetalero Nacional, 1943. RepDom-*2210*.C5, 1943.1.

United States.

Sixteenth Census of the United States: 1940: Agriculture. USA-*2210*.C5, 1940.1.

Sixteenth Census of the United States: 1940: Agriculture: Abandoned or Idle Farms. USA-*2210*.C5, 1940.15.

Sixteenth Census of the United States: 1940: Agriculture: Cows Milked and Dairy Products. USA-*2210*.C5, 1940.17.

Sixteenth Census of the United States: 1940: Agriculture: Cross-Line Acreage. USA-*2210*.C5, 1940.2.

Sixteenth Census of the United States: 1940: Agriculture: Special Cotton Report. USA-*2210*.C5, 1940.25.

Sixteenth Census of the United States: 1940: Agriculture: Special Poultry Report. USA-*2210*.C5, 1940.27.

Sixteenth Census of the United States: 1940: Agriculture: Value of Farm Products by Color and Tenure of Farm Operator. USA-*2210*.C5, 1940.3.

Sixteenth Census of the United States: 1940: Drainage of Agricultural Lands. USA-*2210*.C5, 1940.4.

Sixteenth Census of the United States: 1940: Irrigation of Agricultural Lands. USA-*2210*.C5, 1940.6.

——. Puerto Rico.

Censo Décimosexto de los Estados Unidos: 1940: Agricultura. Sixteenth Census of the United States: 1940: Agriculture. USA PR-*2210*.C5, 1940.

Uruguay.

Censo Agropecuario, Año 1937. Ur-*2210*.C5, 1937.

Venezuela.

Censo Agrícola y Pecuario, 1937. Ven-*2210*.C5, 1937.

2000-2999 CENSUS—cont.

2220 ——. **Buildings, real estate, investments.**
Edificios, bienes raíces, inversiones.

Brasil.
Recenseamento do Brasil Realizado em 1.° de Setembro de 1920. Bras-*2200*.R4, 1920.

Colombia.
Primer Censo Nacional de Edificios, Efectuado el 20 de Abril de 1938. Col-*2220*.C5, 1938.

México.
Segundo Censo de Edificios, 20 de Octubre de 1939: Datos Definitivos. Méx-*2220*.C5, 1939.2.
Segundo Censo de Edificios de los Estados Unidos Mexicanos: Resumen General. Méx-*2220*.C5, 1939.1.

Nicaragua.
Censo de Edificios, 18 de Enero de 1940: Resumen por Departamentos y Municipios. Nic-*2220*.C5, 1940.
Nicaragua. Nic-*1950*.N5.

República Dominicana.
Anuario Estadístico de la República Dominicana. RepDom-*1410*.A5.

United States.
Sixteenth Census of the United States: 1940: Housing. USA-*2220*.C5, 1940.1.
Sixteenth Census of the United States: 1940: Housing: Analytical Maps . . .; Block Statistics. USA-*2220*.C5, 1940.2.
Sixteenth Census of the United States: 1940: Housing: Characteristics by Type of Structure. USA-*2220*.C5, 1940.3.

——. Puerto Rico.
Censo Décimosexto de los Estados Unidos: 1940: Puerto Rico: Viviendas. Sixteenth Census of the United States: 1940: Puerto Rico: Housing. USA PR-*2220*.C5, 1940.

2230 ——. **Education.**
Educación.

Argentina.
La Distribución por Zonas de la Población Argentina y Su Relación con los Hechos Culturales, Económicos y Sociales. Arg-*2230*.C5, 1943.2.

Chile.
Censo de Educación, Año 1933. Chile-*2230*.C5, 1933.

Costa Rica.
Alfabetismo y Analfabetismo en Costa Rica según el Censo General de Población, 11 de Mayo de 1927. CR-*2230*.A5, 1927.

Guatemala.
"Censo Escolar." Guat-*1420*.B5, May '46, a.
Censo Escolar, 1946. Guat-*2230*.C5, 1946.
"Resultados del Censo Escolar Levantado el 23 de Enero de 1946: Municipio de Guatemala." Guat-*1420*.B5, July '46, b.

2240 ——. **Electoral.**
Electorales.

Perú.
Extracto Estadístico y Censo Electoral de la República. Perú-*1410*.E5, 1933.

2250 ——. **Industry, manufacturing, business.**
Industria, manufactura, negocios en general.

Argentina.
Estadística Industrial. Arg-*2250*.E5.

2000-2999 CENSUS—cont.

2250 ——. **Industry, manufacturing, business—cont.**

Brasil.
"Censo Sindical do Brasil em 1939." Bras-*1420*.B5, Dec. '42.
Recenseamento do Brasil Realizado em 1.° de Setembro de 1920. Bras-*2200*.R4, 1920.

Chile.
Censo Industrial y Comercial, Año 1937. Chile-*2250*.C5, 1937.

Costa Rica.
Censo Comercial, 1915. CR-*2250*.C5, 1915.
Resúmenes Estadísticos, Años 1883 a 1910. CR-*1400*.R5, 1883-1910.

México.
Características Principales de la Industria: Censo de 1940. Méx-*2250*.C5, 1940.3.
Primer Censo Industrial de 1930. Méx-*2250*.C5, 1930.
Segundo Censo Industrial de 1935. Méx-*2250*.C5, 1935.
Tercer Censo Industrial de los Estados Unidos Mexicanos, 1940. Méx-*2250*.C5, 1940.

Panamá.
"Análisis de los Resultados del Primer Censo Oficial de Empresas Industriales y Comerciales Radicadas en Panamá, Ejecutado en el Año de 1938." Pan-*6000*.R5, Apr. '39.

Perú.
"Estadística Industrial." Perú-*6400*.B5, Aug. '46.

United States.
Census of Electrical Industries: Telephones and Telegraphs. USA-*2250*.C3, 1902-30.
Sixteenth Census of the United States: 1940: Census of Business, 1939. USA-*2250*.C5, 1940.
Sixteenth Census of the United States: 1940: Manufactures, 1939. USA-*2250*.C7, 1939.1.

Uruguay.
Censo Industrial de 1936. Ur-*2250*.C5, 1936.

Venezuela.
Censos Industrial, Comercial y Empresas que Prestan Servicios, 1936. Ven-*2250*.C5, 1936.

2260 ——. **Institutional.**
Instituciones.

Brasil.
"Censo Sindical Carioca em 1941." Bras-*1420*.B5, Aug. '43, Marinho de Andrade.

United States.
Religious Bodies, 1936. USA-*2260*.R5, 1936.

2270 ——. **Mining.**
Minería.

United States.
Sixteenth Census of the United States: 1940: Mineral Industries, 1939. USA-*2270*.C5, 1939.

3000-3999 STATISTICAL SCIENCE
CIENCIA ESTADISTICA

3000-3099 General.

3000 ——. General.

International.

Directory of Statistical Personnel in the American Nations. Directorio del Personal Estadístico en las Naciones Americanas. Nominata do Pessoal Estatístico das Nações Americanas. Int-*3000*.D5, 1944.

"La Estadística y el Mundo de Mañana." Col-*1420*.A5, Nov.-Dec. '45, Dunn.

"A Estatística no Mundo de Amanhã." Bras-*3040*.R5, Jan.-Mar. '46, Dunn.

Intercambio Estadístico. Int-*3000*.I5.

Project Series of the Inter American Statistical Institute. Serie de Proyectos del Instituto Interamericano de Estadística. Int-*3000*.P5.

Répertoire International des Institutions Statistiques. Int-*3000*.R4, 1934.

Report to the Council on the Work of the . . . Session. Rapport au Conseil sur les Travaux de la . . . Session. Int-*3000*.R6.

Statistical Activities of the American Nations, 1940. Int-*3000*.S5, 1940.

"Statistics in the World of Tomorrow."—"La Estadística en el Mundo de Mañana." Int-*3040*.E5, June '46, Dunn.

Brasil.

Anexo aos Questionários F 1, 2 e 3 do I.A.S.I. para Estabelecimento de uma Futura Definição das Exigências Mínimas para Admissão do Pessoal Estatístico. Bras-*3300*.A5, 1945.

Ecuador.

Cuadernos de Estadística. Ec-*3000*.C5.

El Salvador.

"Desarrollo Histórico de la Estadística en El Salvador." Int-*1300*.A5, 1940.3, Brannon.

Guatemala.

Serie de Estudios Monográficos. Guat-*3000*.S5.

Statistics in Guatemala (A Brief Survey). Guat-*3030*.S5, 1945.

Nicaragua.

Estudios Especiales de la Dirección General de Estadística. Nic-*3000*.E5.

República Dominicana.

Monografías Estadísticas de la Dirección General de Estadística. RepDom-*3000*.M5.

United States.

"The Federal Government's Statistical Program for Reconversion and Postwar Adjustment: A Round Table." USA-*3040*.J5, June '45, a.

Federal Statistical Directory. USA-*3030*.F5.

Government Statistics for Business Use. USA-*3000*.G7, 1946.

Opportunities for statisticians; a series of articles on statistical work in the federal government, in state and local governments, in teaching and research, and in industry. *Journal of the American Statistical Association.* USA-*3040*.J5, Mar. '45.

"Problems of Integrating Federal Statistics: A Round Table." USA-*3040*.J5, June '45, c.

Uruguay.

Cuadernos del Instituto de Estadística. Ur-*3000*.C5.

3005 ——. Dictionaries and glossaries of statistical terms.
Diccionarios y glosarios de términos estadísticos.

Non-Geographic.

Statistical Dictionary of Terms and Symbols. NonG-*3005*.S4, 1939.

Un Vocabulario Bioestadístico. A Vital Statistics Vocabulary. NonG-*5200*.V6, 1943.

"Vocabulário Brasileiro de Estatística." Bras-*3040*.R5, Apr.-June '44, Silva Rodrigues.

3000-3999 STATISTICAL SCIENCE—cont.

3010 ——. **Bibliography.**
Bibliografía.

International.
Bibliography Series, IASI. Int-*3010*.B5.
The Economic Literature of Latin America. Int-*6000*.E6, 1936.
The Economist's Handbook. Int-*3010*.E5, 1934.
Revue de l'Institut International de Statistique. Int-*3040*.R5.
Statistical Activities of the American Nations, 1940. Int-*3000*.S5, 1940.
"Tables des Matières Contenues dans les Tomes I à XXX." Int-*3020*.B5, 1887-1938.
Brasil.
Anexo aos Questionários F 1, 2 e 3 do I.A.S.I. para Estabelecimento de uma Futura Definição das Exigências Mínimas para Admissão do Pessoal Estatístico. Bras-*3300*.A5, 1945.
Canada.
"Publications of the Dominion Bureau of Statistics." Can-*1410*.C7, 1945, a.
Chile.
"Existencia de Anuarios, Censos y Otras Publicaciones Editadas por la Dirección General de Estadística." Chile-*1420*.E5, June '44, a.
México.
Bibliografía Mexicana de Estadística. Méx-*3010*.B5, 1942.
United States.
Government Statistics for Business Use. USA-*3000*.G7, 1946
Other Countries.
Great Britain.
Guide to Current Official Statistics of the United Kingdom. GrBr-*3010*.G5.
Non-Geographic.
The Second Yearbook of Research and Statistical Methodology. NonG-*3010*.S5, 1941.

3020 ——. **Congresses and conferences.**
Congresos y conferencias.

International.
Bulletin de l'Institut International de Statistique. Int-*3020*.B5.
Conference on International Co-operation in Statistics, August 14th and 15th, 1919. Int-*3020*.C5, 1919.
Proceedings of the Eighth American Scientific Congress. Int-*1300*.A5, 1940.3.
Brasil.
Convenção Nacional de Estatística. Bras-*3020*.C5, 1936.
Resolução Especial e Conjunta do Conselho Nacional de Estatística, Conselho Nacional de Geografia e Comissão Censitária Nacional, no. IX Aniversário do Instituto em 29 de Maio de 1945. Bras-*3020*.R4, 1945.

3030 ——. **Organization of statistical services, agencies, and societies—descriptive matter and laws.**
Organización de servicios estadísticos, organismos y sociedades—material descriptivo y leyes.

International.
50 Années de l'Institut International de Statistique. Int-*3030*.C5, 1884-1934.
"Le Comité d'Experts Statisticians de la Société des Nations (1931-1939)." Int-*3040*.R5, 1939, 2-3, Huber.
"Le Comité d'Experts Statisticiens du Bureau International du Travail (1933-1939)." Int-*3040*.R5, 1942, 1-2, Huber.
The Inter American Statistical Institute. El Instituto Interamericano de Estadística. O Instituto Inter-Americano de Estatística. Int-*3030*.I6, 1943.
"ISI, IASI, and UN." Int-*3040*.E5, Sept. '46, Rice.

3000-3999 STATISTICAL SCIENCE—cont.

3030 ——. **Organization of statistical sciences, agencies, and societies—descriptive matter and laws—cont.**

"National Statistical Societies in the Western Hemisphere."—"Sociedades Nacionales de Estadística en el Hemisferio Occidental." Int-*3040*.E5, Mar. '44.
"One Statistical World." USA-*3040*.J5, Sept. '46, Latham.
"Premier Rapport sur l'Organization des Services Statistiques." Int-*3020*.B5, 1936, Julin.
"Report of the Statistical Commission to the Economic and Social Council."—"Rapport de la Commission de Statistique au Conseil Economique et Social." Int-*8200*.J3, May 31, '46.
"Status de l'Institut International de Statistique." Int-*3020*.B5, 1934.
"The United Nations Statistical Commission." Int-*3040*.E3, July '46, Rice.

Argentina.
Ley de Estadística y Decreto Reglamentario de la Misma, Ley 3180. Arg-*3030*.L5, 1894.
Programa Estadístico y Económico para el Ministerio de Agricultura de la República Argentina. Arg-*6200*.P5, 1923.
"Sobre los Servicios Estadísticos y Censales en la República Argentina." Arg-*3040*.R8, Jan.-Apr. '45, Cortés.
"La Sociedad Argentina Estadística." Int-*1300*.A5, 1940.3, Rebuelto.
Ultimas Disposiciones del Poder Ejecutivo Nacional sobre Estadística y Censos. Arg-*3030*.U5, 1944.

Bolivia.
Reorganización del Servicio de Estadística en Bolivia. Bol-*3030*.R5, 1946.

Brasil.
A Administração Pública e a Estatística. Bras-*3030*.A5, 1941.
Convênios Nacionais de Estatística Municipal. Bras-*3030*.C5, 1944.
Instituto Brasileiro de Geografia e Estatística. Bras-*3030*.I5, 1940.
Organization and activities of the statistical offices of the central government in Brazil; series of articles. *Revista Brasileira de Estatística.* Bras-*3040*.R5, Apr.-June, '43, a.
Organization and activities of the statistical offices of the individual states; series of articles. *Revista Brasileira de Estatística.* Bras-*3040*.R5, Apr.-June '43, b.
Relatório do Instituto Brasileiro de Geografia e Estatística. Bras-*3030*.R4.
Resoluções da Junta Executiva Central do Conselho Nacional de Estatística. Bras-*3030*.R6.
Resoluções da Junta Executiva do Conselho Nacional de Estatística, no Ano de 1945. Bras-*3030*.R6, 1945, No. 204-234.
Rumos da Organização Estatística Brasileira. Trends of Statistical Organization in Brazil. Bras-*3030*.R8, 1944.
"A Sociedade Brasileira de Estatística." Int-*1300*.A5, 1940.3, Martins.
Statute, Basic Principles, Social Leadership of the Brazilian Statistical System. Bras-*3030*.S7, 1940.

Canada.
The Dominion Bureau of Statistics. Can-*3030*.D5, 1935.

Chile.
Dirección General de Estadística: Ley Orgánica. Chile-*3030*.L5, 1932.

Colombia.
"Censo Continental: Una Sugerencia del Instituto Interamericano de Estadística." Col-*2100*.C5, 1950.
"Decreto Número 75 de 1939 (Enero 14)." Col-*3030*.E5, 1945, No. 75.
La Estadística Nacional. Col-*3030*.E3, 1938.
Estatuto Orgánico de la Estadística Nacional. Col-*3030*.E5, 1945.
Instituto Nacional de Estadística de la Universidad de Colombia. Col-*3030*.I5, 1945.
Lecciones de Estadística. NonG-*3330*.Rodríguez, 1946.
"Ley 82 de 1935 (Diciembre 23)." Col-*3030*.E5, 1945, No. 82.

3000-3999 STATISTICAL SCIENCE—cont.

3030 ——. **Organization of statistical sciences, agencies, and societies—descriptive matter and laws—cont.**

"Organización de Servicios de Estadística: Resolución No. 64 de Enero de 1946." Col-*3030*.O5, 1946.

Cuba.

Consejo Nacional de Estadística. Cuba-*3030*.C5, 1945.

Gaceta Oficial, Edición Extraordinaria No. 115, de 28 de Junio de 1935, Creando la Dirección General de Estadística. Cuba-*3030*.G5, 1935.

Organización de las Secciones de la Dirección General de Estadística. Cuba-*3030*.O5, 1945?

Reorganización de la Dirección General de Estadística de la República de Cuba. Cuba-*3030*.R5, 1945.

Ecuador.

"La Estadística Nacional Ecuatoriana." Col-*1420*.A5, Mar.-Apr. '46, Abrisqueta 2.

"Labores de la Dirección General de Estadística y Censos." Ec-*1420*.T5, May '45, Lopez Muñoz.

"Ley de Estadística (de 9 de Agosto de 1944), Decreto No. 760." Ec-*1420*.T5 May '45.

El Salvador.

La Investigación Estadística, Origen, Desarrollo y Estado Actual Comparada con la de Varios Países Americanos y del Caribe. ElSal-*3030*.I5, 1935.

Ley Orgánica del Servicio Estadístico. ElSal-*3030*.L5, 1940.

Guatemala.

Acuerda: . . . Dictar las Disposiciones del Caso a Efecto de que la Dirección General de Estadística Llene Su Cometido. Guat-*3030*.A5, 1945.

Acuerdo Disponiendo el Levantamiento del Censo Escolar y Reglamento General del Mismo. Guat-*2130*.C5, 1946.2.

Acuerdo Gubernativo de 2 de Mayo de 1,945 por el Cual Fué Creada la Oficina Permanente del Censo. Guat-*3030*.A6, 1945.

"La Estadística en Guatemala: Una Breve Reseña." Guat-*6000*.R5, May-Aug. '46, Corcoran.

Ley de Estadística: Decreto Número 1820. Guat-*3030*.L5, 1936.

Statistics in Guatemala (A Brief Survey). Guat-*3030*.S5, 1945.

Ultimas Disposiciones Oficiales en Relación con la Estadística. Guat-*3030*.U5, 1946.

Honduras.

Reglamento de la Estadística Nacional de la República de Honduras. Hond-*3030*.R5, 1944.

México.

Informes sobre las Principales Estadísticas Mexicanas. Méx-*3030*.I5, 1941.

Ley Federal de Estadística y Su Reglamento. Méx-*3030*.L5, 1941.

"La Oficina de Estadística del Instituto Mexicano del Seguro Social." Int-*3040*.E5, Sept. '45, Rodriguez y Rodriguez.

"El Sistema de Coordinación en los Servicios Estadísticos Mexicanos." Int-*1300*.A5, 1940.3, González H.

"La Sociedad Mexicana de Geografía y Estadística." Int-*1300*.A5, 1940.3, Benavides.

Nicaragua.

"Ley No. 91 de 25 de Abril de 1944, Créase la Oficina de Estadística Distritorial." Nic-*1900*.G5, Apr. 25, '44.

"Ley Orgánica de la Estadística Nacional." Nic-*1420*.B5, July 15, '44.

Ley Orgánica de la Estadística Nacional. Nic-*3030*.L5, 1941.

"Primera Ley Republicana que Centraliza la Estadística." Nic-*1420*.B5, June 15, '44, c.

3000-3999 STATISTICAL SCIENCE—cont.

3030 ——. Organization of statistical services, agencies, and societies—descriptive matter and laws—cont.

Paraguay.

Decreto Ley 11.126 que Reorganiza y Coordina los Servicios Estadísticos de la República. Par-*3030*.D5, 1942.

"Ley de Estadística." Par-*1410*.A5, 1936, b.

Perú.

Censo Nacional de Población y Ocupación, 1940. Perú-*2200*.C5, 1940.1.

"Decreto Supremo de 1° de Enero de 1944, por el que se Organiza el Servicio de Estadística Nacional y Crea el Consejo Superior de Estadística." Perú-*1410*.E5, 1942, b.

"La Estadística Peruana en el Sexenio 1938-44." Int-*3040*.E5, Dec. '44, Luna Vegas.

"Legislación Estadística." Perú-*3040*.E5, Jan. '45, b.

"Ley No. 7567, dé 5 de Agosto de 1932, que Transforma en Nacional la Dirección de Estadística y Establece las Normas para el Funcionamiento del Servicio de Estadística Nacional." Perú-*1410*.E5, 1942, d.

"Organización del Servicio de Estadística Nacional en el Perú: Decreto y Esquema." Int-*3040*.E5, June '44, a.

Programa Estadístico para el Ministerio de Agricultura del Perú. Perú-*6200*.S5, 1945, Sp.

"Resolución Suprema que Aprueba el Diagrama de Organización del Servicio de Estadística Nacional." Perú-*1410*.E5, 1942, f.

Statistical Program for the Ministry of Agriculture of Peru. Perú-*6200*.S5, 1945, Eng.

República Dominicana.

Cuadros de Materias y Programas Oficiales de la Comisión Nacional del Servicio Civil. RepDom-*3030*.C5, 1944?

Trujillo y la Estadística. RepDom-*3030*.T5, 1945.

United States.

Aims, Organization, and Publications of the National Bureau of Economic Research. USA-*8500*.A5, 1943.

"The American Statistical Association." Int-*1300*.A5, 1940.3, Stephan.

Annual Report of the Central Statistical Board. USA-*3030*.A6.

"Apuntes sobre la Estadística en los Estados Unidos." Perú-*3040*.E5, July '46, Luna Vegas.

The Bureau of the Census: Its History, Activities and Organization. USA-*3030*.B5, 1929.

Collection of Economic Statistics in the United States. USA-*3030*.C5, 1946.

"Fact-Finding Activities of the Bureau of Labor Statistics." USA-*9000*.M5, May '45, Hosea.

Federal Statistical Directory. USA-*3030*.F5.

First Annual Report of the Central Statistical Board, Year Ended December 31, 1934. USA-*3030*.A6, 1934.

Fourth Annual Report of the Central Statistical Board, July 1, 1937, to June 30, 1938. USA-*3030*.A6, 1937-38.

Government Statistics for Business Use. USA-*3000*.G7, 1946.

Handbook for Employees of the Bureau of the Census. USA-*3030*.H5, 1940.

"The Organization of Statistical Services in the United States and Great Britain." Int-*3020*.B5, 1938, Rice.

Report, October 9, 1942-April 13, 1943, from Advisory Committee on Government Questionnaires. USA-*3110*.R5, 1942-43.

Uruguay.

Estadística General de la República, Ley de Reorganización. Ur-*3030*.E4, 1914.

"Organización de la Estadística Económica en el Uruguay." Int-*1300*.A5, 1940.3, Basabe Castellanos.

3000-3999 STATISTICAL SCIENCE—cont.

3040 ——. Periodicals—cont.

Uruguay.
Revista de la Facultad de Ciencias Económicas y de Administración. Ur-*3040*.R5.

Other Countries.
Great Britain.
Journal of the Royal Statistical Society. GrBr-*3040*.J5.
Biometrika. GrBr-*3040*.B5.
Non-Geographic.
The Annals of Mathematical Statistics. NonG-*3040*.A3.

3100-3199 Methods and techniques.
Métodos y técnicas.

3100 ——. General.

Colombia.
Estatuto Orgánico de la Estadística Nacional. Col-*3030*.E5, 1945.
"Factores de Examen de la Coyuntura Económica Colombiana." *Anales de Economía y Estadística.* Col-*1420*.A5, Apr. '45, Abrisqueta.

Perú.
"Determinación Estadística Matemática de la Población Electoral del Perú en Junio de 1945." Perú-*3040*.E5, Nov. '45, García Frías.

Non-Geographic.
"A Method for Indexing Statistical Tabulations." Int-*3040*.E5, Sept. '45, Moriyama.
"Metodología Estadística Aplicada a las Ciencias Sociales y Biología Social." Int-*1300*.A5, 1940.3, Roa Reyes.
"Registros, Formas y Otra Documentación de una Investigación Estadística." Int-*3040*.E5, Sept. '46, Ullman.
Social Research. NonG-*9000*.S5, 1942.
"Statistical Method in Forestry." USA-*3040*.B5, June '45, Schumacher.
"Statistical Methods for Delineation of Regions Applied to Data on Agriculture and Population." USA-*9000*.S5, Mar. '43, Hagood.
Statistical Methods in Industry. NonG-*6400*.S5, 1943.
Teoría de la Correlación. NonG-*3100*.T5, 1935.

3105 ——. Analysis of data.
Análisis de datos.

Non-Geographic.
An Introduction to Statistical Analysis. NonG-*3105*.I5, 1944.
Statistical Adjustment of Data. NonG-*3105*.S4, 1943.
Statistics: Collecting, Organizing, and Interpreting Data. NonG-*3110*.S5, 1943.

3110 ——. Data collection and treatment.
Recopilación y elaboración de datos.

Brasil.
"A Coleta da Estatística Educacional." Bras-*7200*.R5, Nov. '44, Jardim.
"A Coleta da Estatística Educacional." Bras-*7200*.R5, Sept. '44, Jardim.
"A Coleta da Estatística Educacional (III)." Bras-*7200*.R5, May '45, Jardim.

México.
Programa e Instructivo para el Servicio de Estadística Agrícola en las Agencias Generales. Méx-*6200*.P5, 1945.

United States.
List of Schedules Used by the Bureau of the Census for Collecting Data, 1945. USA-*3110*.L5, 1945.
"Management of Field Work and Collection of Statistics: A Round Table." USA-*3040*.J5, June '45, b.

3000-3999 STATISTICAL SCIENCE—cont.

3110 ——. Data collection and treatment—cont.

Manual of Procedure for the Mechanical System of Reporting Morbidity—Treatment-Progress and Control of the Venereal Diseases. USA-*7150*.M4, 1940.
"Methods Used in Processing Data from the Physical Examination Reports of the Selective Service System." USA-*3040*.J5, June '44, Edwards.
Morbidity: Instructions for Coding and Punching. USA-*7150*.M6.
Procedure for Selecting, Analyzing, Assembling and Perforating into Tabulating Cards the Economic and Financial Data Reported on Income Tax Returns. USA-*3110*.P5, 1933.
Questionnaire Manual. USA-*3110*.Q5, 1943.
Report, October 9, 1942-April 13, 1943, from Advisory Committee on Government Questionnaires. USA-*3110*.R5, 1942-43.

Non-Geographic.

Accounting Control. NonG-*3110*.A5, 1936.
"Adaptation of New Geometric Code to Multiple Punching in Mechanical Tabulation." USA-*3040*.J5, Sept. '32, Dunn.
"Application of Punched Card Methods to Hospital Statistics." USA-*3040*.J5, Mar. '35, Dunn.
A.S.T.M. Manual on Presentation of Data. NonG-*3160*.A5, 1945.
Bibliography: Use of IBM Electric Accounting Machines and Test Scoring Machine in Mathematics, Statistics and Scientific Research. NonG-*3110*.B5, 1944.
Consumer and Opinion Research. NonG-*3110*.C5, 1943.
The Design of Tabulating Cards. NonG-*3110*.D4, 1936.
The Electric Accounting Machine Card. NonG-*3110*.E5, 1936.
"Errors in Card Punching." USA-*3040*.J5, Dec. '42, Deming.
How to Use Crank-Driven Calculators. NonG-*3110*.H5, 1942.
Machine Methods of Accounting. NonG-*3110*.M3.
"Machine Tabulation." Int-*3040*.E5, Mar. '46, Mahaney.
Manual of Electric Punched Card Accounting Machines. NonG-*3110*.M5, 1940.
"Mathematical Operations with Punched Cards." USA-*3040*.J5, June '42, McPherson.
"On Sample Inspection in the Processing of Census Returns." USA-*3040*.J5, Sept. '41, Deming.
Organization and Supervision of the Tabulating Department. NonG-*3110*.O5, 1936.
Practical Applications of the Punched Card Method in Colleges and Universities. NonG-*3110*.P4, 1935.
Principles of Operation of the Alphabetical Accounting Machine. NonG-*3110*.P6, 1943.
Principles of Punch-Card Machine Operation. NonG-*3110*.P5, 1942.
"A Punch Card Designed to Contain Written Data and Coding." USA-*3040*.J5, Dec. '41, Berkson.
Statistics: Collecting, Organizing, and Interpreting Data. NonG-*3110*.S5, 1943.
"A System of Codification of Medical Diagnoses for Application to Punch Cards, with a Plan for Operation." USA-*7110*.A4, June '36, Berkson.
"Tabulating Machinery for Indexing and General Tabulation in the Vital Statistics Office." USA-*7110*.A4, July '36, Baehne.
Terminología, Inglés-Español Español-Inglés, Contiene las Voces y Expresiones mas Usuales con Respecto a las Máquinas Eléctricas de Contabilidad "International," Sus Funciones y Su Uso. NonG-*3110*.T5, 1942.

3000-3999 STATISTICAL SCIENCE—cont.

3120 —— **Index numbers.**
Números índices.

International.
" 'Números Índices', Suas Aplicações na Estatística Brasileira e Sugestões Atinentes à Sua Padronização Continental." Bras-*3040*.R5, Oct-Dec. '43, Morais.

México.
"Indices Generales de la Producción Agrícola." Méx-*6200*.B5, Mar. '45.

Non-Geographic.
Business Cycles, the Problem and Its Setting. NonG-*6000*.B6, 1928.
Economic Barometers. Int-*6010*.E5, 1924.
Elementos Matemáticos para el Análisis de las Series de Tiempo. NonG-*3120*.E5, 1943.
The Making and Using of Index Numbers. NonG-*3120*.M5, 1938.
Methods of Compiling Cost-of-Living Index Numbers. Int-*9500*.M5, 1925.
Números Indices. NonG-*3120*.N5, 1945.
"Observations au Sujet du Calcul des Indices des Prix et du 'Quantum' des Importations et Exportations." Int-*3020*.B5, 1936, Szturm de Sztrem.
Time-Series Charts. NonG-*3260*.T5, 1938.

3130 ——. **Mapping.**
Cartografía.

Non-Geographic.
Notes on Statistical Mapping with Special Reference to the Mapping of Population Phenomena. NonG-*3130*.N5, 1938.

3140 ——. **Mathematical statistics.**
Estadística matemática.

Non-Geographic.
The Advanced Theory of Statistics. NonG-*3140*.A6, 1943.
Algunos Problemas de Inferencia Estadística. NonG-*3140*.A8, 1946.
The Annals of Mathematical Statistics. NonG-*3040*.A3.
Elementos de Estadística Matemática. NonG-*3140*.E5, 1936.
Estadística Matemática. NonG-*3140*.E7, 1939.
Interpolación y Ajustamiento de la Curva Logística Generalizada. NonG-*3140*.I5, 1938.
Mathematical Statistics. NonG-*3140*.M5, 1927.
"Un Método Práctico para Suavizar a Mano las Curvas Estadísticas." Int-*1300*.A5, 1940.3, García-Mata.
Random Sampling Distributions. NonG-*3140*.R5, 1943.
"Sobre las Ecuaciones Diferenciales Ordinarias a Coeficientes Constantes y el Cálculo Operacional." Arg-*3040*.A5, Apr. '45, Dieulefait.
Statistical Deflation in the Analysis of Economic Series. NonG-*3140*.S5, 1941.
Suma de Productos Formados por Combinaciones con Números Dados. NonG-*3140*.S7, 1940.
Tests of Significance, What They Mean and How to Use Them. NonG-*3140*.T5, 1939.

3150 ——. **Sampling.**
Método de muestras.

Haiti.
Sampling Plans for a Census of Haiti. Haiti-*2100*.S5, 1942.

Perú.
"Aplicabilidad en la Agricultura Peruana del Método Estadístico de los 'Ejemplares-Tipo' (1)." Perú-*3040*.E5, July '46, Rose Jibaja.

3000-3999 STATISTICAL SCIENCE—cont.

3150 ——. **Sampling—cont.**

United States.

"A New Sample of the Population." Int-*3040*.E5, Dec. '44, Hansen.

"On Some Census Aids to Sampling." USA-*3040*.J5, Sept. '43, Hansen.

Procedures for Current Mortality Sample, 1945. USA-*5212*.P5, Feb. '45.

"The Sampling Procedure of the 1940 Population Census." USA-*3040*.J5, Dec. '40, Stephan.

Non-Geographic.

"Marketing and Sampling Uses of Population and Housing Data." USA-*3040*.J5, Mar. '43, Eckler.

"The Master Sample of Agriculture: I, Development and Use." USA-*3040*.J5, Mar. '45, King.

"The Master Sample of Agriculture: II, Design." USA-*3040*.J5, Mar. '45, Jessen.

"On Errors in Surveys." USA-*9000*.A6, Aug. '44, Deming.

"On Sample Inspection in the Processing of Census Returns." USA-*3040*.J5, Sept. '41, Deming.

"On Training in Sampling." USA-*3040*.J5, Sept. '45, Deming.

"El Problema del Muestreo en Estadística." Int-*1300*.A5, 1904.3, Toscano.

Random Sampling Numbers. NonG-*3150*.R5, 1927.

"Recent Development in Sampling for Agricultural Statistics." USA-*3040*.J5, Mar. '42, Snedecor.

"Relative Efficiencies of Various Sampling Units in Population Inquiries." USA-*3040*.J5, Mar. '42, Hansen.

"Report on the Representative Method in Statistics." Int-*3020*.B5, 1925, Jensen, a.

"The Representative Method in Practice. Int-*3020*.B5, 1925, Jensen, b.

"A Review of Recent Statistical Developments in Sampling and Sampling Surveys." GrBr-*3040*.J5, Part 1, '46, Yates.

"Sample Surveys in Census Work." NonG-*3150*.S3.

"The Sampling Approach to Economic Data." Can-*3040*.C5, Aug. '45, Keyfitz.

Sampling Inspection Tables: Single and Double Sampling. NonG-*3150*.S5, 1944.

Sampling Methods in Forestry and Range Management. NonG-*6250*.S5, 1942.

Sampling Statistics and Applications. NonG-*3150*.S7, 1945.

Sobre la Teoría de las Muestras. NonG-*3150*.S8, 1942.

"Some Problems of Statistical Education and Teaching." USA-*3040*.J5, June '44, Kimelman.

Tables of Random Sampling Numbers. NonG-*3150*.T4, 1939.

The Theory of Sampling. NonG-*3150*.T8, 1942.

3160 ——. **Quality control.**
Control de calidad.

Non-Geographic.

American War Standards: Guide for Quality Control and Control Chart Method of Analyzing Data. NonG-*3160*.G5, 1941.

Control Chart Method of Controlling Quality during Production. NonG-*3160*.C5, 1942.

Economic Control of Quality of Manufactured Product. NonG-*3160*.E5, 1931.

Statistical Method from the Viewpoint of Quality Control. NonG-*3160*.S5, 1939.

Statistical Tables; Their Structure and Use. NonG-*3160*.S4, 1936.

3000-3999 STATISTICAL SCIENCE—cont.

3200- Standards.
3299 Estandards.

3200 ——. General.

International.

Banking Statistics, Recommendations on Scope and Principles of Classification; Report of the Sub-Committee on Banking Statistics. Int-*3200*.D5, 1946.
Foreign Trade Classification in the American Nations. Int-*3010*.B5, Econ. 8a, Oct. 1, '45.
Housing Statistics. Int-*3200*.H5, 1939.
Indices de la Production Industrielle. Int-*3200*.I3, 1939.
Indices of Industrial Production. Int-*3200*.I5, 1939.
International Statistical Standards. Int-*3010*.B5, Stat. 7, Aug. 10, '45.
"Nomenclatura para Hospitales Dedicados a Enfermedades Mentàles." Int-*7110*.B5, Aug. '36.
Report of the Sub-Committee on National Income Statistics. Int-*6100*.R5, 1946.
Report to the Council on the Work of the Fifth Session. Int-*3000*.R6, 1936, Eng.
Report to the Council on the Work of the . . . Session. Rapport au Conseil sur les Travaux de la . . . Session. Int-*3000*.R6.
Reports Presented to the Conference of Commissions of Inter-American Development (Held in New York on May 9-18, 1944) by the Brazilian Commission of Inter-American Development. Bras-*6000*.R5, 1944.
Statistics Relating to Capital Formation. Int-*3200*.S3, 1938.
Statistiques de l'Habitation. Int-*3200*.S5, 1939.
Statistiques du Bois. Int-*3200*.S6, 1938.
Statistiques Relatives à la Formation des Capitaux. Int-*3200*.S7, 1938.
Studies and Reports on Statistical Methods. Etudes et Rapports sur les Méthodes Statistiques. Int-*3200*.S8.
Timber Statistics. Int-*3200*.T5, 1938.

Argentina.

Clasificador de Bienes. Arg-*3200*.C5, 1937.A.
Indice General Alfabético por Materia de los Bienes Nacionales del Estado. Arg-*3200*.C5, 1937.B.

Brasil.

Normas de Apresentação da Estatística Brasileira. Bras-*3200*.N5, 1939.

Colombia.

"Acuerdo de la Oficina Panamericana del Café en Nueva York, Acerca de la Reglamentación de la Estadística Cafetera Mundial." Col-*6210*.B5, Aug. '39, a.

El Salvador.

Claves y Nomenclaturas Usadas por la Dirección General de Estadística. ElSal-*3200*.C5, 1943.

United States.

Report of the Committee on the Standardization of Mining Statistics. USA-*3200*.R5, 1918.
Suggested Standard Forms for Mine Statistics with Especial Reference to Accidents. USA-*3200*.S5, 1916.
Technical Papers of the Division of Statistical Standards. USA-*3200*.T5.

3205 ——. Balance of payments.
Balanza de pagos.

International.

Annex I, League of Nations Estimated Balance of International Payments for the Year . . . (8-page form for country to fill); and Annex II, Explanatory Notes to Scheme of International Balance of Payments (11 pages). Int-*3205*.B5, 1945, Annexes.
Balance of Payments Statistics: Note by the Secretariat. Int-*3205*.B5, 1945.

3000-3999 STATISTICAL SCIENCE—cont.

3205 ——. **Balance of payments—cont.**

"Modelos para Registrar las Cuentas Internacionales." Ec-*1420*.T5, July-Dec. '45, c.

Note on Balance of Payments Statistics: Annex A. Int-*3205*.B5, 1945, Annex A

3210 ——. **Foreign trade classification.**
Clasificación de comercio exterior.

International.

Algunos Documentos Relativos a la Adopción de un Sistema de Estadística Internacional, Convención de Bruselas, 1913. Int-*3210*.C4, 1913.2.

Conférence Internationale Concernant les Statistiques Économiques (26 Novembre au 14 Décembre): . . . Int-*6000*.I75, 1928.B.

Convention Respecting the Compilation of International Commercial Statistics. Int-*3210*.C4, 1913.1.

Convertibility Index for Foreign Trade Statistical Classifications of the American Nations: Basic Classification Scheme Showing Detailed Export and Import Commodity Description by Classes of the Minimum List of Commodities for International Trade Statistics. Int-*3210*.C6, 1945.

Documents et Proces-Verbaux de la Conférence Internationale de Statistique Commerciale, Bruxelles, 1913. Int-*3210*.C4, 1913.3.

Draft Customs Nomenclature. Int-*3210*.D4, 1937.

Draft Framework for a Customs Tariff Nomenclature and Draft Allocation of Goods to the Various Chapters of the Framework with Explanatory Notes. Int-*3210*.D6, 1929.

International Conference on Economic Statistics, Geneva, 1928: Preparatory Documents. Int-*6000*-I75, 1928.A.

Liste Minimum de Marchandises pour les Statistiques du Commerce International. Int-*3210*.L5, 1938.

Minimum List of Commodities for International Trade Statistics. Int-*3210*.M5, 1938.

Proceedings of the International Conference Relating to Economic Statistics. Int-*6000*.I75, 1929.C.

Report. Int-*3000*.R6, 1931, Eng.

Report to the Council on the Work of the Eighth Session. Int-*3000*.R6, 1939, Eng.

Report to the Council on the Work of the Fourth Session. Int-*3000*.R6, 1935, Eng.

Report to the Council on the Work of the Second Session. Int-*3000*.R6, 1933, Eng.

Report to the Council on the Work of the Third Session. Int-*3000*.R6, 1934, Eng.

Some Technical Aspects of Foreign Trade Statistics with Special Reference to Valuation. Int-*6900*.S2, 1946.

Unification of Customs Nomenclature. Int-*3210*.U5, 1930.

Chile.

Lista en Castellano e Inglés de los Artículos Importables desde los Estados Unidos de Norte América (con Números del "Schedule B" y Unidades). Chile-*3210*.L5, 1943.

Colombia.

"La Clasificación Estadística del Comercio Exterior de Colombia." Col-*1420*.A5, Sept.-Oct. '45, Arévalo Correal.

Codificación Estadística del Comercio Exterior. Col-*3210*.C5, 1936.

Ecuador.

Clasificación del Arancel de Aduanas del Ecuador según la Lista Mínimum de Mercaderías para las Estadísticas del Comercio International. Ec-*3210*.C5, 1945.

Guatemala.

Instrucciones a los Importadores de Artículos y Materiales en Tránsito o Procedentes de los Estados Unidos de Norte América o del Canada. Guat-*3210*.I5, 1943.

3000-3999 STATISTICAL SCIENCE—cont.

3210 ——. **Foreign trade classification—cont.**

United States.

"Cambios en la Clasificación de Estadísticas de Exportación de los Estados Unidos." Int-*3040*.E5, Sept. '45, Ely.

Custom House Guide. USA-*3210*.C5.

Schedule A, Statistical Classification of Imports into the United States. USA-*3210*.S3, 1943.

Schedule A, Statistical Classification of Imports into the United States, Arranged by Tariff Schedules and Tariff Paragraphs of Act of 1930. USA-*3210*.S2, 1940.

Schedule B, Statistical Classification of Domestic and Foreign Commodities Exported from the United States. USA-*3210*.S4, 1945.

Standard Commodity Classification: Volume I: Standard Classified List of Commodities. USA-*3210*.S6, 1943.

Standard Commodity Classification: Vol. II: Alphabetic Index. USA-*3210*.S6, 1944.

Non-Geographic.

"Standard Commodity Classification." USA-*3040*.J5, Mar. '44, Kolesnikoff.

3220 ——. **Industrial classification.**
Clasificación industrial.

International.

Estadística de la Población Activa. Int-*3240*.E5, 1938.

Report to the Council on the Work of the Seventh Session. Int-*3000*.R6, 1938, Eng.

Report to the Council on the Work of the Sixth Session. Int-*3000*.R6, 1937, Eng.

Statistics of the Gainfully-Occupied Population. Int-*3240*.S4, 1938.

Statistiques de la Population Active. Int-*3240*.S6, 1938.

Systems of Classification of Industries and Occupations. Int-*3220*.S5, 1923.

Brasil.

"Anexo a Resolução No. 243: Nomenclatura Brasileira para a Classificação de Industrias." Bras-*1420*.B5, Sept. '44.

Canada.

Classification of Industries: Eighth Census of Canada, 1941. Can-*3220*.C5, 1941.

Colombia.

Estatuto Orgánico de la Estadística Nacional. Col-*3030*.E5, 1945.

México.

Catálogo de Industrias, 1945. Méx-*3220*.C5, 1945.

Tercer Censo Industrial: Catálogo General de las Industrias de Transformación y Extractivas, 1940. Méx-*3220*.T5, 1940.

United States.

Sixteenth Census of the United States: Alphabetical Index of Occupations and Industries. USA-*3240*.C5, 1940.2.

Sixteenth Census of the United States: 1940: Manufactures, 1939: Industry Classifications for the Census of Manufactures, 1939.

Standard Industrial Classification Manual. USA-*3220*.S5, 1942.

Non-Geographic.

"La Clasificación de Industrias y Profesiones en las Estadísticas del Trabajo." Int-*9000*.E5, 1943, a.

3230 ——. **Mortality and morbidity classifications.**
Clasificación de mortalidad y morbilidad.

International.

Clasificación Estadística de las Causas de las Defunciones. Arg-*5000*.P2, 1943, No. 10.

Classification of Terms and Comparability of Titles through Five Revisions of the International List of Causes of Death. Int–*3230*.C7, 1944.

"Clave Diagnóstica para la Clasificación en Columna (Tabulación) de las Causas de Morbidad." Int-*7110*.B5, June-July '41.

"O Clínico e a Bio-Estatística." Bras-*3040*.R5, Jan.-Mar. '42, Freitas Filho.

3000-3999 STATISTICAL SCIENCE—cont.

3000-3999 STATISTICAL SCIENCE—cont.

3240 ——. **Occupational classifications—cont.**

Sixteenth Census of the United States: Classified Index of Occupations. USA-*3240*.C5, 1940.1.

Non-Geographic.

"La Clasificación de Industrias y Profesiones en las Estadísticas del Trabajo." Int-*9000*.E5, 1943, a.

3250 ——. **Personnel.**
Personal.

Brasil.

Anexo aos Questionários F 1, 2 e 3 do I.A.S.I. para Establecimento de uma Futura Definição das Exigências Mínimas para Admissão de Pessoal Estatístico. Bras-*3300*.A5, 1945.

República Dominicana.

Cuadros de Materias y Programas Oficiales de la Comisión Nacional del Servicio Civil. RepDom-*3030*.C5, 1944?

3260 ——. **Tabular and graphic presentation.**
Presentación gráfica y tabular.

International.

"Normas para la Elaboración de los Cuadros Estadísticos." Col-*1420*.A5, May 5, '42, a.

Non-Geographic.

American Standard Engineering and Scientific Graphs for Publications. NonG-*3260*.A5, 1943.

Graphic Presentation. NonG-*3260*.G5, 1939.

Graphs: How to Make and Use Them. NonG-*3260*.G7, 1940.

Handbook of Statistical Monographs, Tables, and Formulas. NonG-*3260*.H4, 1932.

How to Use Pictorial Statistics. NonG-*3260*.H6, 1937.

The Preparation of Statistical Tables: A Handbook. NonG-*3260*.P5.

Standards of Presentation. NonG-*3260*.S3, 1944.

Statistical Control Manual. NonG-*3330*.ATC, 1945.

Statistical Tables for Biological, Agricultural and Medical Research. NonG-*3260*.S7, 1938.

Time-Series Charts. NonG-*3260*.T5, 1938.

3270 ——. **Weights and measures.**
Pesos y Medidas.

United States.

Foreign Weights and Measures with Factors for Conversion to United States Units. USA-*3270*.F5, 1944.

3300-3999 **Training materials and methods.**
Materiales y métodos de enseñanza.

3300 ——. **General.**

International.

Statistical Training in the American Nations: Methods and Materials. Int-*3300*, 1944-46.

Argentina.

Enseñanza Estadística en Argentina: Métodos y Materiales. Arg-*3300*, 1944-46.

Bolivia.

Enseñanza Estadística en Bolivia: Métodos y Materiales. Bol-*3300*, 1944-46.

Brasil.

Anexo aos Questionários F 1, 2 e 3 do I.A.S.I. para Establecimento de uma Futura Definição das Exigências Mínimas para Admissão do Pessoal Estatístico. Bras-*3300*.A5, 1945.

3000-3999 STATISTICAL SCIENCE—cont.

3300 ——. **General—cont.**

Aperfeiçoamento Estatístico no Brasil: Métodos e Recursos. Bras-*3300*, 1944-46.
"Curso de Estatística: Capítulo II, Séries Estatísticas, Análise das Séries Econômicas e Teoria da População." Bras-*3040*.R5, Apr.-June '46, Dieulefait.

Canada.
Statistical Training in Canada: Methods and Materials. Can-*3300*, 1944-46.

Chile.
Enseñanza Estadística en Chile: Métodos y Materiales. Chile-*3300*, 1944-46.

Colombia.
Enseñanza Estadística en Colombia: Métodos y Materiales. Col-*3300*, 1944-46.

Cuba.
Enseñanza Estadística en Cuba: Métodos y Materiales. Cuba-*3300*, 1944-46.

Ecuador.
Enseñanza Estadística en Ecuador: Métodos y Materiales. Ec-*3300*, 1944-46.

Guatemala.
Enseñanza Estadística en Guatemala: Métodos y Materiales. Guat-*3300*, 1944-46.

Haiti.
Enseignement Statistique en Haïti: Méthodes et Materiaux. Haiti-*3300*, 1944-46.

Honduras.
Enseñanza Estadística en Honduras: Métodos y Materiales. Hond-*3300*, 1944-46.

México.
"Enseñanza de la Estadística en México." Int-*3040*.E5, Mar. '44, Alanís Patiño. Méx-*3300*, 1944-46.
Enseñanza Estadística en México: Métodos y Materiales. Méx-*3300*, 1944-46.

Paraguay.
Enseñanza Estadística en Paraguay: Métodos y Materiales. Par-*3300*, 1944-46.

Perú.
Enseñanza Estadística en Perú: Métodos y Materiales. Perú-*3300*, 1944-46.

República Dominicana.
Enseñanza Estadística en la República Dominicana: Métodos y Materiales. RepDom-*3300*, 1944-46.

United States.
"An Outline of Training Activities in the U. S. Bureau of the Census." USA-*3300*.O5, 1945.
Statistical Training in the United States: Methods and Materials. USA-*3300*, 1944-46.

Uruguay.
Enseñanza Estadística en Uruguay: Métodos y Materiales. Ur-*3300*, 1944-46.

Venezuela.
Enseñanza Estadística en Venezuela. Métodos y Materiales. Ven-*3300*, 1944-46.

Non-Geographic.
"Fundamental Problem in the Case Method of Instruction in Statistics." USA-*3040*.J5, Mar. '28, Blackett.
"On the Training of Statisticians." USA-*3040*.J5, June '45, Silva Rodrigues.
"Curso de Estatística: Capítulo I, Introdução, Valores Indicativos de Distribuições Estatísticas, Elementos de Interpolação e de Ajustamento pelo Processo dos Mínimos Quadrados." Bras-*3040*.R5, Jan.-Mar. '44, Dieulefait.
"Requirements for Statisticians and Their Training." USA-*3040*.J5, Dec. '26.
"Some Problems in Teaching Elementary Statistics." USA-*3040*.J5, June '39, Kellogg.

3320 ——. **Outlines and course of study.**
Programas y cursos de estudio.

Argentina.
Estadística: Programa Analítico. Arg-*3320*.E5.
Programa de Biometría. Arg-*3320*.P3, 1943.

3000-3999 STATISTICAL SCIENCE—cont.

3320 ——. **Outlines and course of study—cont.**

Programa de Estadística. Arg-*3320*.P5, 1944.
Programa de Matemática Actuarial. Arg-*3320*.P6, 1944.
Programa de Matemática Financiera y Actuarial. Arg-*3320*.P7, 1944.
Programa de Matemáticas. Arg-*3320*.P8, 1944.

Brasil.
Programa de Estatística Geral e Aplicada. Bras-*3320*.P5, 1944.

Colombia.
"El Curso de Estadística de la Contraloría General." Col-*1420*.A5, Feb. 5, '42.
"Un Curso Oficial de Estadística." Col-*1420*.A5, Aug. 5, '42, Guthardt.
"Orientación, Organización y Resultados de un Curso Oficial de Estadística por Correspondencia." Int-*3040*.E5, Mar. '43, Guthardt.
"Programa de Estudios del Curso General de Estadística de la Contraloría General de la República." Col-*1420*.A5, May 5, '42, b.

Cuba.
Programa de Estadística Social y Metodología Estadística. Cuba-*3320*.P5, 1944.

Ecuador.
"Curso Rápido de Estadística Metodológica." Ec-*6000*.I5, Oct. '44, a.

México.
Cursos de Invierno de 1945. Méx-*3320*.C5, 1945.

República Dominicana.
Programa de un Curso de Estadística Metodológica (Introducción a la Estadística Matemática) Seguido de un Curso de Matemáticas Generales Necesarias para Su Estudio. RepDom-*3320*.P5, 1940.

3330 ——. **Textbooks.**
Textos de enseñanza.

Non-Geographic.
Análisis Estadístico Aplicado a los Trabajos de Investigación en Agricultura y Biología. NonG-*3330*.Trucco, 1944.
Applied General Statistics. NonG-*3330*.Croxton 1, 1943.
Apuntes de Estadística. NonG-*3330*.Mazza 1, 1942.
Biología Social. NonG-*5000*.B5.
Business Cycles and Forecasting. NonG-*6000*.B5, 1940.
Business Statistics. NonG-*3330*.Brumbaugh, 1941.
Cartilla de Estadística. NonG-*3330*.Colombia, 1937-38.
Compendio de la Obra Intitulada: Curso Elemental de Estadística Administrativa. NonG-*3330*.Bertillon, 1906.
Correlation Analysis. NonG-*3330*.Treloar 1, 1943.
Cuestiones de Matemáticas Referentes a la Estadística. Fascículo I. NonG-*3330*.Mazza 2, 1940.
Curso de Estadística. NonG-*3330*.Gini 2, 1935.
Curso de Estadística. NonG-*3330*.Roa Reyes, 1929.
Curso de Estadística. NonG-*3330*.Velarde B., 1945.
Curso de Estadística Expositiva o Aplicada. NonG-*3330*.Kürbs.
Curso de Estadística General y Aplicada. NonG-*3330*.Julin, 1924?
Elementary Statistical Methods. NonG-*3330*.Neiswanger, 1943.
Elementary Statistics and Applications. NonG-*3330*.Smith 1, 1944.
Elementos de Estadística Metodológica. NonG-*3330*.Dieulefait, 1938-40.
Elementos de Estatística Geral. NonG-*3330*.Silva Rodrigues 1, 1941.
Elementos de Método Estadístico. NonG-*3330*.García Pérez, 1945.
Elements of Statistical Reasoning. NonG-*3330*.Treloar 2, 1939.
Elements of Statistics. NonG-*3330*.Bowley, 1937.
Estadística. NonG-*3330*.Schott, 1934.
La Estadística Aplicada a las Ciencias Biológicas. NonG-*3330*.Shelly Hernández 1941.

3000-3999 STATISTICAL SCIENCE—cont.

3330 ——. **Textbooks—cont.**

4000–4999 GEOGRAPHY
GEOGRAFIA

4000 General.

International.
Index to Map of Hispanic America, 1:1,000,000. Int-*4000*.M4, 1945.
"The Map of Hispanic America on the Scale of 1:1,000,000." USA-*4000*.G5, Jan. '46.
Map of the Americas, 1:5,000,000. Int-*4000*.M5, 1942.
"The Millionth Map of Hispanic America." USA-*4000*.G5, Apr. '37, Platt.

Argentina.
Segundo Censo de la República Argentina, Mayo 10 de 1895. Arg-*2200*.C5, 1895.

Brasil.
Boletim Geográfico do Conselho Nacional de Geografia. Bras-*4000*.B5.
Revista Brasileira de Geografia. Bras-*4000*.R5.

Chile.
Sinopsis Geográfico-Estadística de la República de Chile, 1933. Chile-*1400*.S5, 1933.

Colombia.
Boletín Gráfico. Col-*6000*.B5.
Mapa Físico de Colombia. Col-*4000*.M5, 1946.

El Salvador.
Diccionario Geográfico de la República de El Salvador. ElSal-*4000*.D5, 1945.

Haiti.
Geology of the Republic of Haiti. Haiti-*4000*.G5, 1924.

México.
Boletín de la Sociedad Mexicana de Geografía y Estadística. Méx-*3040*.B5.

Perú.
Demarcación Política del Perú: Recopilación de Leyes y Decretos (1821-1946). Perú-*4000*.D5, 1821-1946.

United States.
Geographical Review. USA-*4000*.G5.

Non-Geographic.
Aerial Photographs and Their Applications. NonG-*4000*.A5, 1943.
Surveying and Mapping. NonG-*4000*.S5.

4100 Area and geographical characteristics.
Area y características geográficas.

Brasil.
Divisão Territorial do Brasil. Bras-*4100*.D6, 1945.

Chile.
Chile a Través de Sus Zonas Geográfico-Económicas. Chile-*6000*.C34, 1944.

Perú.
"El Medio Geográfico y la Población del Perú." Perú-*3040*.E5, Nov. '45, Arca Parró.
"El Medio Geográfico y la Población del Perú." Perú-*3040*.E5, Apr. '45, Arca Parró.

United States.
Annual Report of the Chief of Engineers, U. S. Army, Part I. USA-*6615*.A5.I.
Sixteenth Census of the United States: 1940: Areas of the United States, 1940. USA-*2200*.C5, 1940.12.

Venezuela.
División Político-Territorial de la República. Ven-*4100*.D5, 1944.
Nomenclador General de Areas y Lugares Habitados de Venezuela. Ven-*4100*.N5, 1941.
Suplemento de la División Político-Territorial de la República. Ven-*4100*.D5, 1945.

4000-4999 GEOGRAPHY—cont.

4200 Climate and weather.
Clima y condiciones del tiempo.

Argentina.
Registro de Observaciones Pluviométricas. Arg-*4200*.R5, 1943?
Colombia.
Anuario Meteorológico. Col-*4200*.A5.
Meteorología. Col-*4200*.M5, 1945.
Haiti.
Bulletin Annuel de l'Observatoire Météorologique du Petit Séminaire Collège St. Martial. Haiti-*4200*.B5.
Perú.
Boletín Anual Meteorológico. Perú-*4200*.B4.
Non-Geographic.
"Sobre la Interpolación y Reducción de Series Climatológicas." Arg-*3040*.A5, Oct. '45, Schneider.

5000-5999 DEMOGRAPHY
DEMOGRAFIA

5000 General.

International.
Aperçu de la Démographie des Divers Pays du Monde, 1929-1936. Int-*5000*.A5, 1929-36.
Books on Population, 1939-Mid-1944. Int-*5000*.B5, 1939-44.
Congrès International de la Population, Paris, 1937. Int-*5000*.C5, 1937.
"Demografía de las Repúblicas Americanas." Int-*7110*.B5, Jan. '40.
"Demographic Status of South America." USA-*3040*.A5, Jan. '45, Dunn.
The Economics of 1960. Int-*6000*.E9, *1960.*
Final Act of the First Inter American Demographic Congress. Int-*5000*.I5, 1943.1, Eng.
General Censuses and Vital Statistics in the Americas. Int-*5000*.G5, 1943.
Papers Presented by Special Guests of the Population Association of America at Its Meetings October 25-26, 1946. Int-*5000*.M3, 1946.
Population Index. Int-*5000*.P5.
Summary of Biostatistics Series. Int-*5000*.S8.
Argentina.
Argentina: Summary of Biostatistics. Arg-*5000*.A5, 1945.
Buenos Aires, Puerto del Río de la Plata, Capital de la Argentina; Estudio Crítico de Su Población, 1536-1936. Arg-*5000*.B5, 1536-1936.
El Comercio Exterior Argentino. Arg-*6900*.C5.
Estadística Demográfica. Arg-*5000*.E5.
Informe, Serie D, Demografía. Arg-*5000*.P2.
El Movimiento Demográfico en los Territorios Nacionales de la República Argentina en los Años 1933 a 1941. Arg-*5000*.P2, 1942, No. 9.
Bolivia.
Bolivia: Summary of Biostatistics. Bol-*5000*.B5, 1945.
Demografía. Bol-*5000*.D5.
Brasil.
Brasil: Summary of Biostatistics. Bras-*5000*.B5, 1945.
Estudios Brasileiros de Demografia. Bras-*5000*.E5.
Chile.
Chile: Demographic Data. Chile-*5000*.C5, 1943.
Demografía y Asistencia Social. Chile-*5000*.D5.
"La Población Chilena: Primera Parte, Las Características Demográficas." Chile-*3040*.E5, July '44, Levine Bawden, b.
"Población y Natalidad." Chile-*7100*.R7, Mar. '44, Cabello González.
La Realidad Medico-Social Chilena. Chile-*7100*.R5, 1939.

5000-5999 DEMOGRAPHY—cont.

5000 ——. General—cont.

Colombia.
Colombia: Summary of Biostatistics. Col-*5000*.C5, 1944.
Estadísticas Demográficas y Nosológica, Año de 1936. Col-*7150*.E5, 1936.
Costa Rica.
Apéndice al Informe de la Dirección General de Estadística. CR-*5000*.A5.
Costa Rica: Summary of Biostatistics. CR-*5000*.C5, 1944.
Cuba.
Cuba: Summary of Biostatistics. Cuba-*5000*.C5, 1945.
Ecuador.
Ecuador: Summary of Biostatistics. Ec-*5000*.E5, 1944.
El Salvador.
El Salvador: Summary of Biostatistics. ElSal-*5000*.E5, 1944.
Guatemala.
Guatemala: Summary of Biostatistics. Guat-*5000*.G5, 1944.
Revista de Sanidad. Guat-*7110*.R5.
Haiti.
Haiti: Summary of Biostatistics. Haiti-*5000*.H5, 1945.
Recensement et Démographie. Haiti-*5000*.R5, 1944.
Honduras.
Honduras: Summary of Biostatistics. Hond-*5000*.H5, 1944.
México.
Instructivo para la Elaboración de la Estadística Demográfica en las Oficinas Sanitarias de los Estados y Territorios Federales. Méx-*5000*.I5, 1942.
México: Demographic Data. Méx-*5000*.M5, 1943.
Nicaragua.
Indices Demográficos de Nicaragua, 1933-1942. Nic-*5000*.I5, 1933.
Nicaragua: Summary of Biostatistics. Nic-*5000*.N5, 1945.

Panamá.
Panama: Summary of Biostatistics. Pan-*5000*.P5, 1945.
Paraguay.
Paraguay: Summary of Biostatistics. Par-*5000*.P5, 1944.
Perú.
Demografía y Estadística de Hospitales y de Otros Establecimientos. Perú-*7130*.D5.
Peru: Summary of Biostatistics. Perú-*5000*.P5, 1944.
República Dominicana.
Dominican Republic: Summary of Biostatistics. RepDom-*5000*.D5, 1945.
United States.
Biometrics Bulletin. USA-*3040*.B5.
"Demographic Factors in Labor Force Growth." USA-*9000*.A6, Aug. '46, Wolfbein.
Milbank Memorial Fund Quarterly. USA-*5000*.M5.
The Natural History of Population. USA-*5000*.N5, 1939.
Studies in American Demography. USA-*5000*.S5, 1940.
Uruguay.
Uruguay: Summary of Biostatistics. Ur-*5000*.U5, 1944.
Venezuela.
Venezuela: Summary of Biostatistics. Ven-*5000*.V5, 1944.
Other Countries.
Great Britain.
Annals of Eugenics. GrBr-*7300*.A5.
Non-Geographic.
Biología Social. NonG-*5000*.B5.
Estudos Brasileiros de Demografia. Bras-*5000*.E5.
Human Biology. NonG-*5000*.H5.

5000-5999 DEMOGRAPHY—cont.

5000 ——. **General—cont.**

Introduction to Medical Biometry and Statistics. NonG-*7180*.I5, 1940.
"Metodología Estadística Aplicada a las Ciencias Sociales y Biología Social." Int-*1300*.A5, 1940.3, Roa Reyes.
Plenty of People. NonG-*5000*.P3, 1944.
Public Health Statistics. NonG-*7110*.P5, 1942.
"Sondages Démographiques dans les Pays Sans État Civil Régulier." Int-*3040*.R5, 1937.2, Bunle.
Studies in Human Biology. NonG-*5000*.S6, 1924.

5100-5199 **Population.**
Población.

5100 ——. **General.**

International.
"Bases Mínimas para la Uniformidad de los Censos Nacionales de Población en el Continente Americano." Int-*1300*.A5, 1940.3, Loyo Gonzáles.
Censo de las Américas de 1950: Antecedentes y Justificación del Proyecto. Int-*2000*.C5, 1950.1b.
"Comentarios y Sugestiones Respecto a las Bases Mínimas para el Censo de las Américas de 1950." Ec-*1420*.T5, July-Dec. '45, a.
"A Comparative Examination of Certain Aspects of the Population of the New World." NonG-*5000*.H5, Sept. '40, Pearl.
"The Development of Population Predictions in Europe and the Americas." Int-*3040*.E5, Sept. '44, Taeuber.
The Economic Pattern of World Population. Int-*5100*.E5, 1943.
"Las Estimaciones sobre la Población Futura de América." Arg-*3040*.R5, Aug. '45, Taeuber.
"Future Populations." Int-*5000*.P5, Jan. '44.
Latin America. Int-*1900*.L5, 1942.
Métodos de los Censos de Población de las Naciones Americanas. Int-*2100*.M5, 1945.
"Población y Salubridad Pública en el Nuevo Mundo." Int-*1300*.A5, 1940.3, León.
"Las Poblaciones y la Nutrición en el Nuevo Mundo." Int-*1300*.A5, 1940.3, Mardones Restat.
Population. Int-*5100*.P4.
"Population Problems in Central and Caribbean America." USA-*3040*.A5, Jan. '45, León.
Populations of the Other American Republics by Major Civil Divisions and by Cities of 5,000 or More Inhabitants. Int-*5100*.P8, 1945.
"Source Materials: IASI Survey of Census Methods and Procedures: Argentina, Bolivia, Brazil, Canada." Int-*3000*.I5, Nov. '44-Mar. '45.
"Source Materials: IASI Survey of Census Methods and Procedures: Chile, Colombia, Cuba, El Salvador, Guatemala, Haiti, Honduras, Mexico." Int-*3000*.I5, Apr. '45.
"Source Materials: IASI Survey of Census Methods and Procedures: Nicaragua, Panama, Paraguay, Peru, Republica Dominicana, Venezuela." Int-*3000*.I5, May-June '45.
Statistique Internationale des Grandes Villes: Territoire et Population des Grandes Villes, 1928-1934. Int-*5100*.S5, 1928-34.

Argentina.
Cuarto Censo General, 1936: Población 22-x-1936. Arg-*2200*.C4, 1936.
Dinámica Demográfica 1910-1940: Apéndice 1936-1945. Arg-*5000*.D5, 1945.
Ensayo sobre la Evolución de la Población de las Américas (Estimación en el Lapso 1938-1973): II, Argentina. Arg-*5100*.E5, 1946.

5000-5999 DEMOGRAPHY—cont.

5100 ——. General—cont.

"La Población Argentina, Su Desarrollo, Sus Características y Sus Tendencias.' Int-*1300*.A5, 1940.3, Bunge, b.
La Población de la República Argentina. Arg-*5100*.P5.
Primer Censo de la República Argentina, Verificado en los Días 15, 16 y 17 de Septiembre de 1869. Arg-*2200*.C5, 1869.
Segundo Censo de la República Argentina, Mayo 10 de 1895. Arg-*2200*.C5, 1895.
Tercer Censo Nacional, Levantado el 1 de Junio de 1914. Arg-*2200*.C5, 1914.

Bolivia.
Censo de 1900: Métodos y Procedimientos. Bol-*2100*.C4, 1900.
Censo Demográfico de la Ciudad de La Paz, 15 de Octubre de 1942. Bol-*2200*.C4, 1942.
Censo General de la Población de la República de Bolivia según el Empadronamiento de 1° de Septiembre de 1900. Bol-*2200*.C5, 1900.
Plan para el Levantamiento del Censo de Población de la Ciudad de La Paz, 1942. Bol-*2100*.P5, 1942.

Brasil.
Análises de Resultados do Censo Demográfico. Bras-*2200*.A5, 1940.
Brazilian Population: Preliminary Results of the 1940 Census. Bras-*2200*.B5, 1940.
Plano de Apuração do Censo Demográfico. Bras-*2100*.R5, 1940.6.
Quadros Estatísticos. Bras-*6000*.Q5.
Recenseamento de 1940: Métodos e Processos. Bras-*2100*.R5, 1940.1.
Recenseamento de Brasil Realizado em 1° de Setembro de 1920. Bras-*2200*.R4, 1920.
Recenseamento Geral de 1940: Resultados Preliminares. Bras-*2200*.R6, 1940.
Recenseamento Geral do Brasil (1° de Setembro de 1940). Bras-*2200*.R4, 1940.
Recenseamento Geral do Brasil, 1.° de Setembro de 1940: Sinopse do Censo Demográfico. Bras-*2200*.R7, 1940.
Recenseamento Geral do Brasil, Realizado em 1.° de Setembro de 1940: Sinopse Preliminar dos Resultados Demográficos segundo as Unidades da Federação e os Municípios. Bras-*2200*.R5, 1940.

Canada.
Eighth Census of Canada, 1941. Huitième Recensement du Canada, 1941. Can-*2200*.C5, 1941.2.
Eighth Census of Canada, 1941: Methods and Procedures. Can-*2100*.C5, 1941.1.
Eighth Census of Canada, 1941: Population. Can-*2200*.C5, 1941.2a.
Ensayo sobre la Evolución de la Población de las Américas: I, Canadá. Can-*5100*.E7, 1946.
The Future Population of Canada. Can-*5100*.F5, 1946.
Seventh Census of Canada, 1931. Septième Recensement du Canada, 1931. Can-*2200*.C5, 1931.2.

Chile.
Censo de 1940: Métodos y Procedimientos. Chile-*2100*.C5, 1940.1.
"Los Censos de Población en Chile." Int-*1300*.A5, 1940.3, Vergara.
Instrucciones para la División del Territorio Comunal en Zonas de Empadronamiento: XI Censo de Población. Chile-*2100*.C5, 1940.2.
Libreta de Empadronador: XI Censo de Población. Chile-*2100*.C5, 1940.3.
Reglamento del XI Censo de Población. Chile-*2100*.C5, 1940.4.
Resultados del X Censo de la Población Efectuado el 27 de Noviembre de 1930 y Estadísticas Comparativas con Censos Anteriores. Chile-*2200*.C5, 1930.
"XI Censo General de Población." Chile-*1420*.E5, Dec. '41.

Colombia.
Censo General de Población, 5 de Julio de 1938. Col-*2200*.C5, 1938.
"Esquema del Plan de Organización Puesto al Servicio del Levantamiento de los Censos de Edificios y Población Llevados a Cabo en Colombia en 1938." Int-*1300*.A5, 1940.3, Suárez Rivadeneira.

5000-5999 DEMOGRAPHY—cont.

5000-5999 DEMOGRAPHY—cont.

5100 ——. **General—cont.**

México.

Censo de 1940: Métodos y Procedimientos. Méx-*2100*.C5, 1940.1.

Decreto y Bases de Organización de los Siete Censos Nacionales y de los Padrones Estadísticos que se Levantarán en 1939-40. Méx-*2100*.C5, 1940.2.

Estados Unidos Mexicanos, Sexto Censo de Población, 1940. Méx-*2200*.C5, 1940.1.

Memoria de los Censos Generales de Población, Agrícola, Ganadero e Industrial de 1930. Méx-*2200*.C5, 1930.1.

Población del País Calculada al 30 de Junio de Cada Año del Período 1921-1938. Méx-*5100*.P5, 1921-38.

"Problemas de Población en la América del Caribe y Central." Méx-*3040*.B5, Jan.-Feb. '46, León.

Quinto Censo de Población, 15 de Mayo de 1930. Méx-*2200*.C5, 1930.2.

Sexto Censo General de Población de los Estados Unidos Mexicanos, 6 de Marzo de 1940. Población Municipal. Méx-*2200*.C5, 1940.2.

Nicaragua.

Censo de 1940: Métodos y Procedimientos. Nic-*2100*.C5, 1940.1.

Censo General de 1920. Nic-*2200*.C5, 1920.

"Cifras Preliminares de Población por Provincias y Capitales de Provincias, según el Censo de 1940." Nic-*1420*.B5, July-Dec. '40.

"Población de la República de Nicaragua, Calculada hasta el 31 de Diciembre de 1944." Nic-*1420*.B5, Sept.-Oct. '45.

Población de Nicaragua. Nic-*2200*.C5, 1940.2.

Población de Nicaragua, por Sexos, Urbana y Rural (Censo de 1940). Nic-*2200*.C5, 1940.3.

Población según Edades y Sexo (Censo 1940). Nic-*2200*.C5, 1940.4.

Primer Censo General de Población, 1940: Instructivo para Empadronadores; Interpretación de la Boleta de Población. Nic-*2100*.C5, 1940.2.

"Proceso Evolutivo de la Población de Nicaragua durante los Ultimos 25 Años." Nic-*1420*.B5, Apr.-June '46.

Panamá.

Censo de 1940: Métodos y Procedimientos. Pan-*2100*.C5, 1940.1.

Censo de Población, 1940. Pan-*2200*.C5, 1940.1.

Censo de Población, 1940. Pan-*2200*.C5, 1940.2.

Censo de Población, 1940: Informe Preliminar. Pan-*2200*.C5, 1940.3.

Cuarto Censo Decenal de Población, 1940: Instrucciones a los Empadronadores e Inspectores. Pan-*2100*.C5, 1940.2.

Paraguay.

Censo de 1936: Métodos y Procedimientos. Par-*2100*.C5, 1936.1.

Censo de Población, Decreto N°. 2120 de 18 de Junio de 1936. Par-*2100*.C5, 1936.2.

Decreto N°. 60.287 de 21 de Noviembre de 1935. Par-*2100*.C5, 1936.3.

Ley N°. 1.509 de 1° de Noviembre de 1935. Par-*2100*.C5, 1936.4.

Perú.

Censo Nacional de 1940: Resultados Generales. Perú-*2200*.C5, 1940.2.

Censo Nacional de Población y Ocupación, 1940. Perú-*2200*.C5, 1940.1.

"Determinación Estadística Matemática de la Población Electoral del Perú en Junio de 1945." *Estadística Peruana.* Perú-*3040*.E5, Nov. '45, García Frías.

"Problemas y Soluciones para el Censo Demográfico Peruano de 1940." *Proceedings of the Eighth American Scientific Congress.* Int-*1300*.A5, 1940.3, Arca Parró, b.

República Dominicana.

Anuario Estadístico de la República Dominicana. RepDom-*1410*.A5.

Censo de 1935: Instrucciones a los Enumeradores del Censo de Población. RepDom-*2100*.C5, 1935.3.

5000-5999 DEMOGRAPHY—cont.

5100 ——. General—cont.

Censo de 1935: Instrucciones Generales sobre las Funciones de un Enumerador. RepDom-*2100*.C5, 1935.4.

Censo de 1935: Ley del Congreso Nacional que Dispone las Condiciones en que Deben Desarrollarse las Operaciones del Censo. RepDom-*2100*.C5, 1935.2.

Censo de 1935: Métodos y Procedimientos. RepDom-*2100*.C5, 1935.1.

Censo de Población Urbana y Rural, por Provincias y Comunas. RepDom-*2200*.C5, 1935.3.

Población de la República Censada el 13 de Mayo de 1935, Clasificada por Grupos de Edades, Zona y Sexo. RepDom-*2200*.C5, 1935.2.

Población de la República Dominicana, Distribuida por Nacionalidades: Cifras del Censo Nacional de 1935. RepDom-*2200*.C5, 1935.1.

Población de la República Dominicana, según las Sucesivas Modificaciones Territoriales a Partir del 13 de Mayo de 1935, Día en que Se Levantó el Segundo Censo Nacional de Población, Hasta el 1 de Enero de 1946. RepDom-*2200*.C5, 1935-46.

Población de la República, según el Censo Levantado el 24 de Diciembre de 1920, Clasificada por Sexo y Grupos de Edades. RepDom-*2200*.C5, 1920.

United States.

"The American People: Studies in Population." USA-*3040*.A5, Nov. '36.

Census of Congested Production Areas: 1944. USA-*2200*.C3, 1944.

"The Changing Population of the United States." USA-*3040*.A5, Jan. '45, Hauser.

Estimates of Future Population of the United States, 1940-2000. USA-*5100*.E6, 1940-2000.

Foundations of American Population Policy. USA-*5100*.F5, 1940.

Membership List of the Population Association of America. USA-*5100*.M5, 1943.

"A New Sample of the Population." Int-*3040*.E5, Dec. '44, Hansen.

"The Sampling Procedure of the 1940 Population Census." USA-*3040*.J5, Dec. '40, Stephan.

Sixteenth Census of the United States: 1940: Methods and Procedures. USA-*2100*.C5, 1940.1.

Sixteenth Census of the United States: 1940: Population. USA-*2200*.C5, 1940.1.

Sixteenth Census of the United States: 1940: Population: Unincorporated Communities. USA-*2200*.C5, 1940.13.

Sixteenth Decennial Census of the United States: Instructions to Enumerators: Population and Agriculture, 1940. USA-*2100*.C5, 1940.5.

"Source Materials: IASI Survey of Census Methods and Procedures: United States: Sixteenth Census, 1940." Int-*3000*.I5, Aug.-Oct. '44.

——. Alaska.

Sixteenth Census of the United States: 1940: Alaska: Population: Characteristics of the Population (with Limited Data on Housing). USA Alsk-*2200*.C5, 1940.

——. Puerto Rico.

Sixteenth Census of the United States: 1940: Methods and Procedures; Forms and Instructions Used in Taking the Census of Puerto Rico. USA PR-*2100*.C5, 1940.1.

Sixteenth Census of the United States: 1940: Puerto Rico, Población, Boletín No. 1: Número de los Habitantes. Puerto Rico, Population, Bulletin No. 1: Number of Inhabitants. USA PR-*2200*.C5, 1940.1.

Uruguay.

"Censo General de la República en 1908." Ur-*1410*.A5, 1908.

Historia y Análisis Estadístico de la Población del Uruguay. Ur-*5100*.H5, 1939.

Venezuela.

Censo de 1941: Métodos y Procedimientos. Ven-*2100*.C5, 1941.1.

Séptimo Censo Nacional de Población, 1941: Instrucciones y Modelos. Ven-*2100*.C5, 1941.2.

5000-5999 DEMOGRAPHY—cont.

5100 ——. General—cont.

Séptimo Censo Nacional de Población, Levantado el 7 de Diciembre de 1941. Ven-*2200*.C5, 1941.
Sexto Censo Nacional, 1936. Ven-*2200*.C5, 1936.2.

Other Countries.

West Indies.

Eighth Census of Jamaica and Its Dependencies, 1943: Population, Housing and Agriculture. WI-*2200*.5, 1943.
Eighth Census of Jamaica, 1943: Introduction to Population and Housing. WI-*2200*.E5, 1943.2.

Non-Geographic.

Algunos Problemas de Inferencia Estadística. NonG-*3140*.A8, 1946.
"The Development of Population Predictions in Europe and the Americas." Int-*3040*.E5, Sept. '44, Taeuber.
"Marketing and Sampling Uses of Population and Housing Data." USA-*3040*.J5, Mar. '43, Eckler.
Notes on Statistical Mapping with Special Reference to the Mapping of Population Phenomena. NonG-*3130*.N5, 1938.
People, the Quantity and Quality of Population. NonG-*5100*.P2, 1939.
Population, a Problem for Democracy. NonG-*5100*.P5, 1940.
"Population and Per Capita Income." USA-*3040*.A5, Jan. '45, Spengler.
Population Problems. NonG-*5100*.P63, 1942.
Population Problems, a Cultural Interpretation. NonG-*5100*.P65, 1943.
"Problems in Population Estimation." USA-*7110*.A4, Feb. '44, Sibley.
"Relative Efficiencies of Various Sampling Units in Population Inquiries." USA-*3040*.J5, Mar. '42, Hansen.
Teoría Matemática de la Población. NonG-*5100*.T4, 1937.
Teoría Matemática del Desarrollo Numérico de la Población. NonG-*5100*.T5, 1943.

5110 ——. Characteristics.
Características.

International.

Analfabetismo y Cultura Popular en América. Int-*7200*.A5, 1941.
Estadísticas de la Población Activa. Int-*3240*.E5, 1938.
Estimates of the Working Population of Certain Countries in 1931 and 1941. Int-*9200*.E7, 1931-41.
International Migrations. Int-*5300*.I6, 1931.
National Income and Related Subjects for the United Nations. Int-*6100*.N3, 1946.
"On the Statistics Available Concerning the Occupied Population of the World and Its Distribution." Int-*3040*.R5, 1938.3, Nixon.
"A População Escolar e a Taxa de Analfabetos nas Estatísticas Educacionais Americanas." Int-*1300*.A5, 1940.3, Bergström Lourenço.
"Population Growth." Int-*5000*.P5, July '43.f.
Report to the Council on the Work of the Sixth Session. Int-*3000*.R6, 1937, Eng.
Statistics of the Gainfully-Occupied Population. Int-*3240*.S4, 1938.
Statistiques de la Population Active. Int-*3240*.S6, 1938.
"The Trend Toward an Older Population." USA-*3040*.A5, Jan. '45, Durand.

Argentina.

El Analfabetismo en la República Argentina. Arg-*7200*.A4, 1939.
El Analfabetismo y las Funciones del Consejo Nacional de Educación. Arg-*7200*.A6, 1938.
"Distribución de la Población Argentina por Regiones Geográficas." Arg-*3040*.R5, June '45, Daus.
La Distribución por Zonas de la Población Argentina y Su Relación con los Hechos Culturales, Económicos y Sociales. Arg-*2230*.C5, 1943.2.

5000-5999 DEMOGRAPHY—cont.

5000-5999 DEMOGRAPHY—cont.

5110 ——. Characteristics—cont.

Sixteenth Census of the United States: 1940: Population: The Labor Force (Sample Statistics): Occupational Characteristics. USA-*2200*.C5, 1940.10d.
Sixteenth Census of the United States: 1940: Population: The Labor Force (Sample Statistics): Usual Occupation. USA-*2200*.C5, 1940.10e.

——. Puerto Rico.
El Analfabetismo en Puerto Rico . . . USA PR-*7200*.A5, 1945.
Censo Décimosexto de los Estados Unidos: 1940: Puerto Rico, Población, Boletín Núm. 2: Características de la Población. Sixteenth Census of the United States: 1940: Puerto Rico, Population, Bulletin No. 2: Characteristics of the Population. USA PR-*2200*.C5, 1940.2.
"Population Growth in Puerto Rico and Its Relation to Time Changes in Vital Statistics." NonG-*5000*.H5, Dec. '45, Janer.
Sixteenth Census of the United States: 1940: Puerto Rico, Población, Boletín No. 3: Ocupaciones y Otras Características por Edad. Puerto Rico, Population, Bulletin No. 3: Occupations and Other Characteristics by Age. USA PR-*2200*.C5, 1940.3.
Staff Report to the Interdepartmental Committee on Puerto Rico. USA PR-*5110*.S5, 1940.

Uruguay.
El Analfabetismo. Ur-*7200*.A5, 1940.

Venezuela.
"La Composición Familiar de Asalariados en el Distrito Federal de Venezuela." Int-*3040*.E5, Dec. '45, Michalup.

Non-Geographic.
The Economics of a Declining Population. NonG-*5110*.E5, 1939.
"The Place of the Intrinsic Rate of Natural Increase in Population Analysis." Int-*1300*.A5, 1940.3, Lotka.
"La Population Rurale." Int-*3040*.R5, 1937.4, Bunle.
"Rapport de la Commission pour la Définition de la 'Population Rurale'." Int-*3040*.R5, 1938.2, Bunle.

5120 ——. Race and nationality.
Raza y nacionalidad.

International.
World Statistics of Aliens. Int-*5120*.W5, 1936.

Canada.
The American-Born in Canada, a Statistical Interpretation. Can-*5120*.A5, 1943.
Census of Indians in Canada, 1944. Recensement des Indiens du Canada, 1944. Can-*2200*.C6, 1944.
Census of Indians in Canada, 1939. Recensement des Indiens du Canada, 1939. Can-*2200*.C6, 1939.
"Racial Origins and Nativity of the Canadian People." Can-*2200*.C5, 1931.3d.

Nicaragua.
"Información y Estudio sobre los Aspectos Estadísticos de la Población Extranjera en Nicaragua—1937-1942." Nic-*1420*.B5, June 15, '44, a.
"Población Indígena en Nicaragua." Nic-*6000*.E5, Apr. 15, '44.

United States.
The Canadian-Born in the United States. USA-*5120*.C5, 1850-1930.
Fifteenth Census of the United States: 1930: The Indian Population of the United States and Alaska. USA-*2200*.C5, 1930.2.
Indian Population in the United States and Alaska, 1910. USA-*2200*.C8, 1910.
Jewish Population Studies. USA-*5120*.J5, 1943.
Negro Population, 1790-1915. USA-*5120*.N4, 1790-1915.
Negroes in the United States, 1920-32. USA-*5120*.N5, 1920-32.
Report on Alien Registration. USA-*5120*.R5, 1943.

5000-5999 DEMOGRAPHY—cont.

5120 ——. **Race and nationality—cont.**

Sixteenth Census of the United States: 1940: Population: Nativity and Parentage of the White Population: Country of Origin of Foreign Stock. USA-*2200*.C5, 1940.11b.

5200 **Vital statistics and vital records.**
5299 **Estadística vital y registros de estadística vital.**

5200 ——. **General.**
5209

5200 ——. ——. **General.**

International.

Informe Biodemográfico y Epidemiológico de las Américas. Biostatistical and Epidemiological Report on the Americas. Int-*7110*.I4, 1943.
Informe Epidemiológico Mensual. Monthly Epidemiological Report. Int-*7110*.I6.
"International Vital Statistics: Summary." USA-*5200*.V8, May 2, '40.
Statistical Handbook Series: The Official Vital Statistics of . . . Int-*5200*.S5.
Statistique Internationale des Grandes Villes: Territoire et Population des Grandes Villes, 1928-1934. Int-*5100*.S5, 1928-34.
Vital Statistics Registration Methods in the Western Hemisphere. Int-*3010*.B5, Dem. 2, Feb. 7, '45.

Argentina.

Anuario Demográfico. Arg-*5200*.A5.
Estadística Vital y Registro en Argentina: Métodos y Procedimientos. Arg-*5200*.

Bolivia.

Estadística Vital y Registro en Bolivia: Métodos y Procedimientos. Bol-*5200*.
Síntesis de Informes y Estadísticas de Salubridad, Año 1942. Bol-*7110*.S5, 1942.

Brasil.

Anuário de Bioestatística. Bras-*5200*.A5.
Boletim Mensal do Serviço Federal de Bio-Estatística. Bras-*5200*.B5.
"O Clínico e a Bio-Estatística." Bras-*3040*.R5, Jan.-Mar. '42, Freitas Filho.
Estatística Vital e Registro Civil no Brasil: Normas e Métodos. Bras-*5200*.
Resumo da Bio-Estatística de Cidades Brasileiras. Bras-*5200*.R5.
Vital Statistics and Registration Methods. Bras-*5200*.V5, 1946.

Canada.

Preliminary Annual Report, Vital Statistics of Canada. Rapport Préliminaire Annuel, Statistiques Vitales du Canada. Can-*5200*.P4.
Preliminary Report on Births, Deaths and Marriages. Rapport Préliminaire sur les Naissances, les Décès et les Mariages. Can-*5200*.P5.
Registration of Births, Deaths and Marriages. Enregistrement des Naissances, Décès et Mariages. Can-*5200*.R5.
"Report of the Committee on Form and Content of Annual Vital Statistics Reports." Can-*7110*.C5, Sept. '46.
Vital Statistics. Statistiques Vitales. Can-*5200*.V5.
Vital Statistics and Registration in Canada: Methods and Procedures. Can-*5200*.

Chile.

"La Estadística Biodemográfica y Epidemiológica en Chile." Int-*7110*.B5, Feb. '45, Cabello González.
Estadística Vital y Registro en Chile: Métodos y Procedimientos. Chile-*5200*.

Colombia.

Estadística Vital y Registro en Colombia: Métodos y Procedimientos. Col-*5200*.

Costa Rica.

Aspectos Biodemográficos de la Población de Costa Rica. CR-*5200*.A5, 1940.
Salud. CR-*7110*.S5.

Cuba.

Estadística Vital y Registro en Cuba: Métodos y Procedimientos. Cuba-*5200*.
Movimiento de Población. Cuba-*5200*.M5.
Salubridad y Asistencia Social. Cuba-*7110*.S5.

5000-5999 DEMOGRAPHY—cont.

5200 ——. General—cont.

Ecuador.

Bioestadística en el Cosmos Ecuatoriano. Ec-*5200*.B5, 1943.

"Las Estadísticas del Movimiento Demográfico en el Ecuador, y Su Perfeccionamiento." Ec-*1420*.T5, May '45, Procaccia.

El Salvador.

Estadística Vital y Registro en El Salvador: Métodos y Procedimientos. ElSal-*5200*.

Guatemala.

Estadística Vital y Registro en Guatemala: Métodos y Procedimientos. Guat-*5200*.

Statistics in Guatemala (A Brief Survey). Guat-*3030*.S5, 1945.

Haiti.

"Les Statistiques Vitales." Haïti-*1900*.C5, Dec. '43, Young.

Statistiques Vitales et Enregistrement en Haïti: Méthodes et Procédés. Haiti-*5200*.

Honduras.

Estadística Vital y Registro en Honduras: Métodos y Procedimientos. Hond-*5200*.

Informe de los Actos Realizados por el Poder Ejecutivo en los Ramos de Gobernación, Justicia, Sanidad y Beneficencia. Hond-*8000*.I5.

Movimiento de Población, Nacimientos, Defunciones y Matrimonios Habidos en la República durante el Año Económico de 1937 a 1938. Hond-*5200*.M5, 1937-38.

México.

Estadística Vital y Registro en México: Métodos y Procedimientos. Méx-*5200*.

Nicaragua.

"Estadística Vital, Año 1944." Nic-*1420*.B5, July '45.

Estadística Vital y Registro en Nicaragua: Métodos y Procedimientos. Nic-*5200*.

Movimiento Registrado en los Hospitales de la República durante los Años de 1944 y 1943. Nic-*7130*.M5, 1943-44.

Panamá.

Estadística Vital y Registro en Panamá: Métodos y Procedimientos. Pan-*5200*.

Paraguay.

Estadística Vital y Registro en Paraguay: Métodos y Procedimientos. Par-*5200*.

Memoria de la Sección Estadística. Par-*7110*.M5.

Perú.

Estadística Vital y Registro en Perú: Métodos y Procedimientos. Perú-*5200*.

República Dominicana.

Commentary Explanations and Sample Copies of the Forms and Manuals Used in Vital Records and Vital Statistics Activities in the Dominican Republic, November 1946. RepDom-*5200*.C5, 1946.

United States.

Introduction to the Vital Statistics of the United States, 1900 to 1930. USA-*5200*.I5, 1900-30.

Plan for the Coordination of Vital Records and Vital Statistics in the United States. USA-*5200*.P5, 1944.

Statistical Bulletin of the Metropolitan Life Insurance Company. USA-*6730*.S5.

"Vital and Health Statistics in the Federal Government." USA-*7110*.A4, Mar. '44, Collins.

Vital Statistics and Registration in the United States: Methods and Procedures. USA-*5200*.

Vital Statistics Instruction Manual. USA-*5200*.V5, 1944-45.

Vital Statistics of the United States. USA-*5200*.V6.

Vital Statistics Rates in the United States, 1900-1940. USA-*5200*.V7, 1900-40.

Vital Statistics—Special Reports. USA-*5200*.V8.

Alaska.

Evaluation of Vital Statistics Work in Alaska. August 26 to October 6, 1944. USA Alsk-*5200*.E5, 1944, Morrison.

5000-5999 DEMOGRAPHY—cont.

5200 ——. General—cont.

Uruguay.

Estadística Vital y Registro en Uruguay: Métodos y Procedimientos. Ur-*5200*.

El Movimiento del Estado Civil y la Mortalidad de la República Oriental del Uruguay. Ur-*5200*.M5.

"Los Problemas de la Bioestadística en la República Oriental del Uruguay." Int-*1300*.A5, 1940.3, Schiaffino.

Venezuela.

Anuario de Epidemiología y Estadística Vital. Ven-*5200*.A5.

"La Composición Familiar de Asalariados en el Distrito Federal de Venezuela." Int-*3040*.E5, Dec. '45, Michalup.

Estadística Vital y Registro en Venezuela: Métodos y Procedimientos. Ven-*5200*.

Non-Geographic.

"Clasificación Análisis de los Datos de Estadística Vital." Int-*7110*.B5, Aug. '43, Linder.

Conceptos de Estadística Demográfica. NonG-*5200*.C5, 1944.

The Fundamental Principles of Mathematical Statistics. NonG-*3330*.Wolfenden, 1942.

Manual de Estadística Vital. NonG-*5200*.M5, 1942.

"Quelques Possibilités de Reconstruction du Mouvement d'une Population à l'Aide des Données des Recensements Démographiques." Int-*3040*.R5, 1941, 1-2, Mortara.

Vital Statistics. NonG-*5200*.V4, 1923.

Un Vocabulario Bioestadístico. A Vital Statistics Vocabulary. NonG-*5200*.V6, 1943.

5203 ——. ——. Organizations and agencies—descriptive matter and laws. Organizaciones y agencias—material descriptivo y leyes.

Haiti.

Loi sur le Service de l'Etat Civil. Haiti-*5203*.L5, Dec. 22, '22.

México.

"El Servicio de Bioestadística en México." Int-*7110*.B5, May '44, Granillo.

United States.

"La Organización de los Servicios de Estadística Vital y Sanitaria en el Gobierno Federal de los Estados Unidos. Int-*7110*.B5, May '45, Collins.

5210-5219 ——. Birth, death, stillbirth. Nacimiento, mortalidad, nacidos muertos.

5210 ——. ——. General.

Argentina.

"La Denatalidad en la Argentina." Arg-*3040*.R5, Dec. '45, Coghlan.

"La Denatalidad en la Argentina." Arg-*3040*.R5, Nov. '45, Coghlan.

"La Denatalidad en la Argentina." Arg-*3040*.R5, Oct. '45, Coghlan.

"La Fecundidad y la Natalidad en el Campo y en las Ciudades." Arg-*3040*.R5, Sept. '45, Correa Avila.

Honduras.

Movimiento de Población, Nacimientos, Defunciones y Matrimonios Habidos en la República durante el Año Económico de 1937 a 1938. Hond-*5200*.M5, 1937-38.

United States.

Monthly Vital Statistics Bulletin. USA-*5210*.M5.

Uruguay.

El Movimiento del Estado Civil y la Mortalidad de la República Oriental del Uruguay. Ur-*5200*.M5.

5000-5999 DEMOGRAPHY—cont.

5211 ——. ——. **Birth.**
Nacimientos.

International.
"Birth Rates." Int-*5000*.P5, July '43, a.
"Trends, Determinants, and Control in Human Fertility." USA-*3040*.A5, Jan. '45, Whelpton.

Brasil.
Análises de Resultados do Censo Demográfico. Bras-*2200*.A5, 1940.

Canada.
"Gross and Net Reproduction Rates, Canada and Provinces, 1920-1942." Int-*3040*.E5, Sept. '45.

United States.
Human Fertility. USA-*5211*.H5.
Ratio of Children to Women, 1920. USA-*5211*.R5, 1920.
Sixteenth Census of the United States: 1940: Population: Differential Fertility, 1940 and 1910: Fertility for States and Large Cities. USA-*2200*.C5, 1940.5a.
Sixteenth Census of the United States: 1940: Population: Differential Fertility, 1940 and 1910: Standardized Fertility Rates and Reproduction Rates. USA-*2200*.C5, 1940.5b.
Sixteenth Census of the United States: 1940: Population: Differential Fertility, 1940 and 1910: Women by Number of Children Ever Born. USA-*2200*.C5, 1940.5d.
Sixteenth Census of the United States: 1940: Population: Differential Fertility, 1940 and 1910: Women by Number of Children Under 5 Years Old. USA-*2200*.C5, 1940.5.c

Non-Geographic.
Multiple Human Births. NonG-*5211*.M5, 1940.

5212 ——. ——. **Death.**
Defunciones.

International.
Clasificación Estadística de las Causas de las Defunciones. Arg-*5000*.P2, 1940, No. 5.
Classification of Terms and Comparability of Titles through Five Revisions of the International List of Causes of Death. Int-*3230*.C7, 1944.
"O Clínico e a Bio-Estatística." Bras-*3040*.R5, Jan.-Mar. '42, Freitas Filho.
"Conference Internationale pour la Cinquième Revision Decennale des Nomenclatures Internationales de Causes de Décès, 1938." Int-*3040*.R5, 1938.3, Huber.
"Death Rates." Int-*5000*.P5, July '43, c.
"Infant and Maternal Mortality in the Modern World." USA-*3040*.A5, Jan. '45, Yerushalmy.
"Infant Mortality Rates." Int-*5000*.P5, July '43, d.
"International List of Causes of Death Adopted by the Fifth International Conference for Revision, Paris, October 3rd-7th, 1938." Int-*7100*.B6, Dec. '38.
Manual of the International List of Causes of Death and Manual of Joint Causes of Death. Int-*3230*.M5, 1939.
"Nomenclatura Internacional de las Causas de Muerte." Int-*7110*.B5, June-Aug. '39.
Nomenclatures Internationales des Causes de Décès, 1938. Int-*3230*.N8, 1938.
Proposed Statistical Classification of Diseases, Injuries, and Causes of Death. Int-*3230*.P5, 1946.
"Rapport sur les Travaux Préparatoires à la Vème Révision Décennale de la Nomenclature Internationale des Causes de Dècès." Int-*3040*.R5, 1938.1, Huber.

5000-5999 DEMOGRAPHY—cont.

5000-5999 DEMOGRAPHY—cont.

5213 ——. ——. **Stillbirth.**
Nacidos muertos.

Brasil.
"Mortinatalidade nas Capitais Brasileiras, 1932-1941." Bras-*7100*.A5, Dec. '43, Scorzelli.

United States.
Classification of Causes of Stillbirth. USA-*3230*.C5, 1940.

5220 ——. **Life tables.**
Tablas de vida.

International.
"Complete Expectation of Life at Various Ages in Selected Countries." Int-*5000*.P5, July '43.b.
"Estudos de Demografia Interamericana: I, Tábua de Mortalidade e de Sobrevivência para a Colômbia (1939-41); II, Ajustamento de Tábua de Sobrevivência para a Colômbia (1939-41), segundo a Fórmula de Gompertz-Makeham." Bras-*3040*.R5, July-Sept. '43, Mortara, 1.
"Estudos de Demografia Interamericana: III, Tábua de Mortalidade e de Sobrevivência para o México (1929-33); IV, Tábua de Mortalidade e de Sobrevivência para a Cidade de Lima (1933-35)." Bras-*3040*.R5, Oct.-Dec. '43, Mortara, 2.
"Trends in Longevity." USA-*3040*.A5, Jan. '45, Dublin.

Argentina.
Tabla de Mortalidad de la República Argentina.

Brasil.
Estudos sôbre a Mortalidade nas Grandes Cidades Brasileiras. Bras-*5212*.E5, 1945.
"Estudos sôbre a Mortalidade no Distrito Federal e no Município de São Paulo." Bras-*3040*.R5, Oct.-Dec. '45.
Tábuas Brasileiras de Mortalidade e Sobrevivência. Bras-*5220*.T3, 1946.
Tábuas de Mortalidade e Sobrevivência Brasileiras: Distrito Federal e Município de São Paulo. Bras-*5220*.T52, 1946.
Tábuas de Sobrevivência, Conforme a Mortalidade do Período 1939-41, para o Distrito Federal e o Município de São Paulo. Bras-*5220*.T54, 1939-41.
Tábuas de Sobrevivência, Conforme a Mortalidade do Período 1920-21, para o Distrito Federal e o Município de São Paulo. Bras-*5220*.T53, 1920-21.

Canada.
Canadian Abridged Life Tables, 1871, 1881, 1921, 1931. Can-*5220*.C5, 1871-1931.
Canadian Life Tables. Can-*5220*.C5.
"Canadian Life Tables, 1931." Can-*2200*.C5, 1931.3m.

Chile.
Anexo Estadístico a la Memoria del Servicio Nacional de Salubridad. Chile-*7110*.M55.
"Sobre el Cálculo de la Duración Media de la Vida en Chile." Chile-*7100*.R7, Jan. '42, Mardones Restat.

Colombia.
"Tabla de Mortalidad de Colombia." Col-*1420*.A5, Mar. 20, '43, Rodríguez.
"Tabla de Mortalidad y Supervivencia para Colombia." Col-*1420*.A5, July '43, Mortara.

Costa Rica.
La Conmutación de Pensiones Futuras como Reparación Civil en Caso de Muerte. CR-*9952*.C5, 1944.

Cuba.
Tablas de Vida de la Ciudad de la Habana. Cuba-*5220*.T5, 1920.

5000-5999 DEMOGRAPHY—cont.

5220 ——. **Life tables—cont.**

México.

"Esperanza de Vida en Veinte Estados de la República Mexicana." Méx-*7110*.R5, June '41.

"Tablas de Vida de los Habitantes de los Estados Unidos Mexicanos." Méx-*7110*.R5, Bustamante.

Perú.

"Ensayo de Tabla de Mortalidad para los Habitantes de Lima, Ciudad Capital." Perú-*3040*.E5, July '46, García Frías.

United States.

United States Abridged Life Tables, 1939. USA-*5220*.U5, 1939.

United States Abridged Life Tables, 1930-1939. USA-*5220*.U5, 1930-39.

United States Life Tables, 1890, 1901, 1910, and 1901-1910. USA-*5220*.U5, 1890-1910.

——. Puerto Rico.

"Population Growth in Puerto Rico and Its Relation to Time Changes in Vital Statistics." NonG-*5000*.H5, Dec. '45, Janer.

Non-Geographic.

"Actuarial Note: 'Census' Methods of Constructing Mortality Tables and Their Relation to 'Insurance' Methods." USA-*6730*.R5, Oct. '42, Greville.

"Calculation of Life Tables for Population of Countries and Regions." Int-*3040*.E5, Mar. '46, Keyfitz.

"The Calculation of Life Tables for Populations Lacking Reliable Birth and Death Statistics, with Application to Brazil." Int-*1300*.A5, 1940.3, Mortara.

Length of Life. NonG-*5220*.L5, 1936.

"Short Methods of Constructing Abridged Life Tables." USA-*6730*.R5, June '43, Greville.

"As Tábuas de Sobrevivência e Suas Aplicações na Demografia." Bras-*3040*.R5, Jan.-Mar. '44 and July-Sept. '44, Mortara.

5230-5239 ——. **Marriage and divorce.**
Matrimonios y divorcios.

5230 ——. ——. **General.**

Honduras.

Movimiento de Población, Nacimientos, Defunciones y Matrimonios Habidos en la República durante el Año Económico de 1937 a 1938. Hond-*5200*.M5, 1937-38.

Uruguay.

El Movimiento del Estado Civil y la Mortalidad de la República Oriental del Uruguay. Ur-*5200*.M5.

5231 ——. ——. **Marriage.**
Matrimonios.

International.

"Marriage Rates." Int-*5000*.P5, July '43, e.

5250 ——. **Registration and registration practices.**
Registro y prácticas de registro.

International.

Vital Statistics Registration Methods in the Western Hemisphere. Int-*3010*.B5, Dem. 2, Feb. 7, '45.

Argentina.

Estadística Vital y Registro en Argentina: Métodos y Procedimientos. Arg-*5200*.

5000-5999 DEMOGRAPHY—cont.

5250 ——. Registration and registration practices—cont.

Bolivia.
Estadística Vital y Registro en Bolivia: Métodos y Procedimientos. Bol-*5200.*
Registro Civil de Bolivia. Bol-*5250.*R5, 1945.
Registro Civil de Bolivia, 1941. Bol-*5250.*R5, 1941.

Brasil.
Estatística Vital e Registro Civil no Brasil: Normas e Métodos. Bras-*5200.*
Vital Statistics and Registration Methods. Bras-*5200.*V5, 1946.

Canada.
Handbook on Death Registration and Certification. Can-*5250.*H2, 1937.
National Registration, August 1940. Can-*5250.*N2, 1940.A.
National Registration Regulations, 1940. Can-*5250.*N2, 1940.C.
"Report of the Committee on Form and Content of Annual Vital Statistics Reports." Can-*7110.*C5, Sept. '46.
Report on Completeness of Birth Registration in Canada. Can-*5250.*R5, 1943.
Specialized Occupations, National Registration, 1940. Occupations Spécialisées, l'Enregistrement National, 1940. Can-*5250.*N2, 1940.B.
Vital Statistics and Registration in Canada: Methods and Procedures. Can-*5200.*
The Vital Statistics Instruction Manual for Canada. Manuel d'Instructions sur les Statistiques Vitales. Can-*5240.*V5, 1942.

Chile.
Estadística Vital y Registro en Chile: Métodos y Procedimientos. Chile-*5200.*
Manual Práctico del Oficial Civil. Chile-*5250.*M5, 1945.

Colombia.
Estadística Vital y Registro en Colombia: Métodos y Procedimientos. Col-*5200.*
Registro del Estado Civil de las Personas. Col-*5250.*R5, 1940.
Registro del Estado Civil: Decreto 540 de 1934 y Disposiciones Vigentes sobre la Materia. Col-*5250.*R7, 1934.

Cuba.
Estadística Vital y Registro en Cuba: Métodos y Procedimientos. Cuba-*5200.*

Ecuador.
Ley de Registro Civil 25–29 de Octubre de 1.900 y Reglamento a la Misma 19 de Julio de 1.927. Ec-*5250.*L5, 1927.

El Salvador.
Estadística Vital y Registro en El Salvador: Métodos y Procedimientos. ElSal-*5200.*

Guatemala.
Estadística Vital y Registro en Guatemala: Métodos y Procedimientos. Guat-*5200.*

Haiti.
Loi sur le Service de l'État Civil. Haiti-*5203.*L5, Dec. 22, '22.

Honduras.
Estadística Vital y Registro en Honduras: Métodos y Procedimientos. Hond-*5200.*

México.
Estadística Vital y Registro en México: Métodos y Procedimientos. Méx-*5200.*

Nicaragua.
Estadística Vital y Registro en Nicaragua: Métodos y Procedimientos. Nic-*5200.*
"Métodos Usados en los Registros del Estado Civil de las Personas y Estadística Vital." Nic-*1420.*B5, May 15, '45.
"Registro Nacional de Identidad y del Estado Civil." Nic-*1420.*B5, June 15, '44, Castro Silva.

Panamá.
Estadística Vital y Registro en Panamá: Métodos y Procedimientos. Pan-*5200.*

Paraguay.
Estadística Vital y Registro en Paraguay: Métodos y Procedimientos. Par-*5200.*
Leyes del Registro del Estado Civil y Decretos Reglamentarios. Par-*5250.*L3, 1940.

5000-5999 DEMOGRAPHY—cont.

5250 ——. **Registration and registration practices—cont.**

"Prueba de Registro de Nacimientos en Paraguay." Int-*3040*.E5, Mar. '46, Roig Ocampos.
Registro del Estado Civil de las Personas: Libreta de Familia. Par-*5250*.L5, 1944.

Perú.
"La Estadística Peruana en el Sexenio 1938-44." Int-*3040*.E5, Dec. '44, Luna Vegas.

República Dominicana.
Commentary Explanations and Sample Copies of the Forms and Manuals Used in Vital Records and Vital Statistics Activities in the Dominican Republic, November 1946. RepDom-*5200*.C5, 1946.

United States.
Birth Certificates, a Digest of the Laws and Regulations of the Various States. USA-*5250*.B5, 1942.
"Breves Apuntes sobre los Registros de Nacimientos y Defunciones en los Estados Unidos." Int-*3040*.E5, Sept. '45, Uriarte.
"A National Test for Completeness of Birth Registration in the United States of America." Int-*3040*.R5, 1942, 1-2, Dunn.
Physician's Handbook on Birth and Death Registration. USA-*5250*.P5, 1943.
Vital Statistics and Registration in the United States: Methods and Procedures. USA-*5200*.

Uruguay.
Estadística Vital y Registro en Uruguay: Métodos y Procedimientos. Ur-*5200*.

Venezuela.
Estadística Vital y Registro en Venezuela: Métodos y Procedimientos. Ven-*5200*.

Non-Geographic.
"Perfeccionamiento del Registro Civil como Fuente de Información Estadística." Int-*1300*.A5, 1940.3, Parra Gómez.
"Problems in Population Estimation." USA-*7110*.A4, Feb. '44, Sibley.

5300-5399 **Migration.**
Movimiento migratorio.

5300 ——. **General.**

International.
International Migrations. Int-*5300*.I6, 1931.
Report to the Council on the Work of the Fifth Session. Int-*3000*.R6, 1936, Eng.
Statistics of Migration. Int-*5300*.S3, 1932.
Studies and Reports, Series O, Migration. Int-*5300*.S5.
World Statistics of Aliens. Int-*5120*.W5, 1936.

Brasil.
Estatística Geral Migratoria. Bras-*5300*.E5.
Quadros Estatísticos. Bras-*6000*.Q5.
Revista de Imigração e Colonização. Bras-*5300*.R5.

Chile.
Boletín de la Sección Estadística de la Caja de Colonización Agrícola. Chile-*6200*.B5.

México.
Datos Estadísticos del Departamento de Terrenos Nacionales. Méx-*5300*.D5, 1923-41.

Nicaragua.
"Balanza de Comercio, Año 1944." Nic-*1420*.B5, June 15, '45, a.
"Inmigrantes, Según Nacionalidad, Edades y Sexo, Año de 1944." Nic-*1420*.B5, June 15, '45, b.
"Movimiento Migratorio, Año 1944." Nic-*1420*.B5, June 15, '45, c.
"Movimiento Migratorio, Año 1943." Nic-*1420*.B5, Sept. 15, '44.
"Movimiento Migratorio Registrado en la República Según Procedencia y Destino, Año 1944." Nic-*1420*.B5, June 15, '45, d.

5000-5999 DEMOGRAPHY—cont.

5300 ——. General—cont.

República Dominicana.

Refugee Settlement in the Dominican Republic. RepDom-*5300*.R5, 1942.

United States.

Immigration in the United States: A Selected List of Recent References. USA-*5300*.I4, 1943.

Interstate Migrations among the Native White as Indicated by Differences between State of Birth and State of Residence. USA-*5300*.I6, 1870-1930.

A Research Memorandum on Internal Migration Resulting from the War Effort. USA-*5300*.R4, 1942.

Rural Migration in the United States. USA-*5300*.R6, 1939.

Sixteenth Census of the United States: 1940: Population: Internal Migration 1935 to 1940. USA-*2200*.C5, 1940.B.

5310 ——. Tourism.
Turismo.

International.

Acta Final del Segundo Congreso Interamericano de Turismo. Int-*5310*.I4, 1941.1, Sp.

Final Act of the Second Inter-American Travel Congress. Int-*5310*.I4, 1941.1, Eng.

"Observations et Propositions pour une Statistique Internationale du Tourisme." Int-*3040*.R5, 1936.2, Molinari.

Statistique Internationale des Grandes Villes: Statistique du Tourisme dans les Grandes Villes, 1929-1934. Int-*5310*.S5, 1929-1934.

Guatemala.

Statistics in Guatemala (A Brief Survey). Guat-*3030*.S5, 1945.

México.

Intercambio Turístico Continental con la República Mexicana. Méx-*5310*.I5, 1942.

Non-Geographic.

"Méthode-Frontière dans la Statistique du Tourisme." Int-*3020*.B5, 1938, Smutný.

6000-6999 ECONOMICS
ECONOMIA

6000-6099 General.

6000 ——. General.

International.

Acta Final de la Conferencia Interamericana sobre Sistemas de Control Económico y Financiero, Wáshington, 30 de Junio a 10 de Julio de 1942. Int-*6000*.I45, 1942, Sp.

Ata Final de Conferência Interamericana sôbre Medidas de Contrôle Econômico Financeiro, Washington, 30 de Junho a 10 de Julho de 1942. Int-*6000*.I45, 1942, Port.

Boletín del Instituto de Investigaciones Sociales y Económicas. Bulletin of the Institute of Social and Economic Research. Bulletin de l'Institut des Recherches Sociales et Economiques. Boletim do Instituto de Investigações Sociais e Econômicas. Int-*3040*.B5.

The Caribbean Islands and the War. Int-*1900*.C5, 1943.

Catalogue of Selected Publications on Economic and Financial Subjects. Int-*6000*.C4, 1943.

Commercial Pan America. Int-*6000*.C7.

Conférence Internationale Concernant les Statistiques Économiques (26 Novembre au 14 Décembre): 1, Convention Internationale Concernant les Statistiques

6000-6999 ECONOMICS—cont.

6000-6999 ECONOMICS—cont.

6000 ——. General—cont.

Argentina.

Argentina Económica e Industrial. Arg-*6000*.A6, 1944.
Boletín Estadístico. Arg-*6000*.S8.
Boletín Oficial de la Bolsa de Comercio de Buenos Aires. Arg-*6700*.B4.
Boletín Oficial de la Bolsa de Comercio Rosario. Arg-*6700*.B5.
"Bosquejo de una Economía Argentina para 1955: Primera Parte—Plan General." Arg-*3040*.R5, Feb. '46, a.
"Bosquejo de una Economía Argentina para 1955: Segunda Parte—Estudios Especiales." Arg-*3040*.R5, Apr. '46.
"Bosquejo de una Economía Argentina para 1955: Primera Parte—Plan General (continuación)." Arg-*3040*.R5, Mar. '46.
"Bosquejo de una Economía Argentina para 1955: Segunda Parte (continuación)—Estudios Especiales." Arg-*3040*.R5, July '46, a.
"Bosquejo de una Economía Argentina para 1955: Segunda Parte (continuación)—Estudios Especiales." Arg-*3040*.R5, June '46, a.
"Bosquejo de una Economía Argentina para 1955: Segunda Parte (continuación)—Estudios Especiales." Arg-*3040*.R5, May '46, a.
Business Conditions in Argentina. Arg-*6000*.B8.
"Economía Nacional de la Argentina." Int-*6000*.P4, July-Aug. '46.
Economía y Finanzas de Mendoza. Arg-*6000*.E5.
Economic Controls and Commercial Policy in Argentina. Arg-*6000*.E6, 1945.
Indices Básicos de la Economía Nacional. Arg-*6000*.I2, 1946.
Informaciones sobre las Actividades Económicas en Nuestro País: Informe. Arg-*6000*.I4.
Memoria del Departamento de Hacienda. Arg-*8800*.M6.
"National Economy of Argentina." Int-*6000*.C7, July-Aug. '46.
Ordenamiento Económico-Social. Arg-*9000*.O5, 1945.
Precios. Arg-*6010*.P5, 1940.
Reports Presented to the Conference of Commissions of Inter-American Development (Held in New York, May 9-18, 1944) by the Argentine Commission of Inter-American Development. Arg-*6000*.R5, 1945.
Revista de Ciencias Económicas. Arg-*3040*.R3.
Revista de Economía Argentina. Arg-*3040*.R5.
Revista de Economía y Estadística. Arg-*3040*.R7.
Revista de la Facultad de Ciencias Económicas, Comerciales y Políticas. Arg-*3040*.R8.
Revista Económica. Arg-*6720*.R6.
Situation in Argentina. Arg-*6000*.S4.
Veritas. Arg-*6000*.V5.

Bolivia.

Economic Controls and Commercial Policy in Bolivia. Bol-*6000*.E5, 1945.
Minería, Transportes y Comunicaciones. Bol-*6000*.M5.
Reports Presented to the Conference of Commissions of Inter-American Development (Held in New York, May 9-18, 1944) by the Bolivian Commission of Inter-American Development. Bol-*6000*.R5, 1944.

Brasil.

Boletim do Ministério do Trabalho, Indústria e Comércio. Bras-*1420*.B5.
Boletim Estatístico. Bras-*6720*.B5.
Brazil. Bras-*6000*.B65.
Brazil in the Making. Bras-*1900*.B4, 1943.
"Censo Sindical do Brasil em 1939." Bras-*1420*.B5, Dec. '42.
Economic Controls and Commercial Policy in Brazil. Bras-*6000*.E3, 1945.
Estatísticas Econômicas. Bras-*6000*.E6.
Informações Relativas ao Distrito Federal, em Agôsto de 1944. Bras-*6000*.I3. Aug. '44.

6000-6999 ECONOMICS—cont.

6000 ——. **General—cont.**

Informações Relativas ao Distrito Federal, em Setembro de 1944. Bras-*6000*.I3, Sept. '44.
Inquérito Continental sôbre Fomento e Coordenação de Indústrias. Bras-*6400*.I5, 1946.
Inquéritos Econômicos para a Defesa Nacional. Bras-*6000*.I5
Quadros Estatísticos. Bras-*6000*.Q5.
Recenseamento do Brasil Realizado em 1° de Setembro de 1920. Bras-*2200*.R4, 1920.
Recenseamento Geral do Brasil (1.° de Setembro de 1940). Bras-*2200*.R4, 1940.
Relatório do Banco do Brasil. Bras-*6720*.R4.
Reports Presented to the Conference of Commissions of Inter-American Development (Held in New York on May 9-18, 1944) by the Brazilian Commission of Inter-American Development. Bras-*6000*.R5, 1944.
Resumen de la Memoria del Banco do Brasil. Bras-*6720*.R6.
Situação e Movimento dos Estabelecimentos Comerciais e Industriais em Abril de 1944. Bras-*6000*.S7, Apr. '44.
Situação e Movimento dos Estabelecimentos Comerciais e Industriais em Março de 1944. Bras-*6000*.S7, Mar. '44.

Canada.
"A Bibliography of Current Publications on Canadian Economics." *The Canadian Journal of Economics and Political Science.* Can-*3040*.C5, Aug. '45.
Business Summary of the Bank of Montreal. Can-*6000*.B5.
Canada. Can-*6000*.C5, 1945.
Canadian Journal of Economics and Political Science. Can-*3040*.C5.
Monthly Review of Business Statistics. Can-*6000*.M5.

Chile.
El Censo Económico General de 1943. Chile-*2150*.C7, 1943.
Chile. Chile-*6000*.C3, 1945.
Chile a Través de Sus Zonas Geográfico-Económicas. Chile-*6000*.C34, 1944.
Chile: An Economy in Transition. Chile-*6000*.C35, 1945.
Cinco Años de Labor, 1939-1943. Chile-*6000*.C4, 1939-43.
Economía. Chile-*3040*.E5.
La Economía Nacional y el Problema de las Subsistencias en Chile. Chile-*1900*.E5, 1943.
Economía y Finanzas. Chile-*6000*.E6.
Economic Controls and Commercial Policy in Chile. Chile-*6000*.E65, 1945.
Memoria Anual Presentada a la Superintendencia de Bancos. Chile-*6720*.M5.
"Observaciones Preliminares sobre la Disponibilidad de Bienes de Consumo y la Creación de Bienes de Capital." Chile-*3040*.E5, Aug. '45, Levine Bawden.
Reports Presented to the Conference of Commissions of Inter-American Development (Held in New York on May 9-18, 1944) by the Chilean Commission of Inter-American Development. Chile-*6000*.R5, 1944.

Colombia.
Anales de Economía y Estadística. Col-*1420*.A5.
Annual Report Presented by the Manager to the Board of Directors. Col-*6720*.A5.
Boletín Gráfico. Col-*6000*.B5.
Comercio e Industrias. Col-*6000*.C5.
"Economía Nacional de Colombia." Int-*6000*.P4, Sept.-Oct. '46.
Economic Controls and Commercial Policy in Colombia. Col-*6000*.E5, 1945.
"La Estadística Aplicada a la Economía y a los Negocios." Col-*1420*.A5, Sept.-Oct. '45, Abrisqueta.
"Factores de Examen de la Coyuntura Económica Colombiana." Col-*1420*.A5, Apr. '45, Abrisqueta.
Memoria del Ministerio de la Economía Nacional. Col-*6000*.M5.
El Mes Financiero y Económico. Col-*6000*.M7.

6000-6999 ECONOMICS—cont.

6000 ——. General—cont.

Report Presented to the Conference of Commissions of Inter-American Development (Held in New York, May 9-18, 1944) by the Colombian Commission of Inter-American Development. Col-*6000*.R5, 1944.
Revista del Banco de la República. Col-*6720*.R5.

Costa Rica.

Economic Controls and Commercial Policy in Costa Rica. CR-*6000*.E5, 1945.
Memoria de la Secretaría de Hacienda y Comercio. CR-*8800*.M5.
Reports Presented to the Conference of Commissions of Inter-American Development (Held in New York, May 9-18, 1944), by the Costa Rican, Salvadorean, Guatemalan, Honduran and Nicaraguan Commissions of Inter-American Development. Int-*6000*.R7, 1944.
Revista del Banco Nacional de Costa Rica. CR-*6720*.R5.

Cuba.

Condiciones Económicas y Sociales de la República de Cuba. Cuba-*1400*.C5, 1944.
Cuba Económica y Financiera. Cuba-*6000*.C5.
Economic Controls and Commercial Policy in Cuba. Cuba-*6000*.E5, 1946.
Publicaciones de la Revista Trabajo. Cuba-*9000*.P5.

Ecuador.

"Algunos Datos sobre la Evolución Financiera y Económica en el Ecaudor." Ec-*6720*.B5, Apr.-May '43, Riofrio Villagomez.
Banco Central del Ecuador, Boletín. Ec-*6720*.B5.
Boletín del Banco Hipotecario del Ecuador. Ec-*6720*.B4.
"Breves Consideraciones sobre la Economía Ecuatoriana." Ec-*6720*.B5, Mar. '42, Laso.
Datos Estadísticos. Ec-*6000*.I5A.
Economic Controls and Commercial Policy in Ecuador. Ec-*6000*.E5, 1945.
Información Económica. Ec-*6000*.I5.
Informe que el Ministro de Hacienda y Crédito Público Presenta a la Nación. Ec-*8800*.I5.
Memoria del Ministerio de Economía. Ec-*6000*.M5.
Report Presented to the Conference of Commissions of Inter-American Development by the Ecuadorian Commission of Inter-American Development. Ec-*6000*.R5, 1944.
Revista de Hacienda. Ec-*8800*.R5.

El Salvador.

Memoria de Economía. ElSal-*6000*.M5.
Reports Presented to the Conference of Commissions of Inter-American Development (Held in New York, May 9-18, 1944), by the Costa Rican, Salvadorean, Guatemalan, Honduran and Nicaraguan Commissions of Inter-American Development. Int-*6000*.R7, 1944.

Guatemala.

Industrial Report on the Republic of Guatemala. Guat-*6400*.I5, 1946.
Informe Presentado por la Delegación de Guatemala ante la Conferencia de Comisiones Nacionales de Fomento Interamericano sobre la Estructura y Condiciones Económicas de Su País. Guat-*6000*.I5, 1944.
Memoria del Ministerio de Comunicaciones y Obras Públicas. Guat-*6000*.M5.
Reports Presented to the Conference of Commissions of Inter-American Development (Held in New York, May 9-18, 1944), by the Costa Rican, Salvadorean, Guatemalan, Honduran and Nicaraguan Commissions of Inter-American Development. Int-*6000*.R7, 1944.
Revista de Economía. Guat-*6000*.R5.

Haiti.

Annual Report of the Fiscal Department. Haiti-*6000*.A3.
Aspects de l'Économie et des Finances d'Haïti. Haiti-*6000*.A5, 1944.

6000-6999 ECONOMICS—cont.

6000 ——. General—cont.

Memoria del Ministerio de Fomento y Obras Públicas. Perú-*6000*.M5.
La Nueva Economía. Perú-*6000*.N8.
Peru. Perú-*6000*.P5, 1945.
Reports Presented to the Conference of Commissions of Inter-American Development (Held in New York, May 9-18, 1944) by the Peruvian Commission of Inter-American Development. Perú-*6000*.R3, 1944.
Revista de Economía y Finanzas. Perú-*6000*.R5.
Revista de Hacienda. Perú-*6000*.R7.
Revista de la Facultad de Ciencias Económicas. Perú-*3040*.R5.

República Dominicana.
Dominican Republic. RepDom-*6000*.D5, 1946.
Economic Controls and Commercial Policy in the Dominican Republic. RepDom-*6000*.E5, 1946.
Report Presented to the Conference of Commissions of Inter-American Development (Held in New York, May 9-18, 1944) by the Dominican Commission of Inter-American Development. RepDom-*6000*.R5, 1944.

United States.
American Economic Review. USA-*6000*.A5.
Annual Report of the National Bureau of Economic Research. USA-*6000*.A6.
Collection of Economic Statistics in the United States. USA-*3030*.C5, 1946.
Conference Board Economic Record. USA-*9300*.C5.
Dun's Review. USA-*6000*.D5.
"Handbook: List of Members of the American Economic Association, 1946." USA-*6000*.A5, Sept. '46.
Industrial Change and Employment Opportunity—A Selected Bibliography. USA-*6000*.I4, 1939.
Occasional Papers of the National Bureau of Economic Research. USA-*6000*.O4.
"Postwar Economic Perspectives: I, Experience after World War I." USA-*9950*.S4, Dec. '45, Woytinsky.
"Postwar Economic Perspectives: II, Prewar Experience—The Labor Force and Employment." USA-*9950*.S4, Jan. '46, Woytinsky.
"Postwar Economic Perspectives: III, Prewar Experience—Production and Consumption." USA-*9950*.S4, Feb. '46, Woytinsky.
"Postwar Economic Perspectives: IV, Aftermath of the War." USA-*9950*.S4, Mar. '46, Woytinsky.
Publications of the National Bureau of Economic Research. USA-*6000*.P5.
Quarterly Journal of Economics. USA-*6000*.Q5.
Reports of the National Research Project on Reemployment Opportunities and Recent Changes in Industrial Techniques. USA-*6000*.R5.
The Review of Economic Statistics. USA-*3040*.R5.
"Síntesis Estadística de los Problemas Económicos de los Estados Unidos de América en la Guerra y Postguerra." Perú-*3040*.E5, July '46, Kürbs.
Sixteenth Census of the United States: 1940: Census of Business, 1939. USA-*2250*.C5, 1940.
Sixteenth Census of the United States: 1940: Instructions to Enumerators for Business and Manufactures, 1939. USA-*2150*.C5, 1939.1.
The Social Framework of the American Economy. USA-*6000*.S5, 1945.
Standard Industrial Classification Manual. USA-*3220*.S5, 1942.
Survey of Current Business. USA-*6000*.S7.
Trends in American Progress. USA-*1400*.T5, 1946.
The United States in the World Economy. USA-*6910*.U5, 1943.

Uruguay.
Boletín de Hacienda. Ur-*6000*.B5.
Boletín de Hacienda, Suplemento. Ur-*6000*.B6.
Boletín de la Dirección de Estadística Económica. Ur-*6000*.B7.

6000-6999 ECONOMICS—cont.

6000 ——. **General—cont.**

Economic Controls and Commercial Policy in Uruguay. Ur-*6000*.E5, 1945.
"Organización de la Estadística Económica en el Uruguay." Int-*1300*.A5, 1940.3, Basabe Castellanos.
Revista de la Facultad de Ciencias Económicas y de Administración. Ur-*3040*.R5.
Revista del Banco de la República Oriental del Uruguay. Ur-*6720*.R5.
Suplemento Estadística de la Revista Económica. Ur-*6000*.S5.

Venezuela.
Banco Central de Venezuela: Memoria. Ven-*6720*.B3.
Economic Controls and Commercial Policy in Venezuela. Ven-*6000*.E3, 1945.
Guía Industrial y Comercial de Venezuela, 1943. Ven-*6000*.G5, 1943.
Memoria del Ministerio de Fomento. Ven-*6000*.M3.
Memoria y Cuenta del Ministerio de Obras Públicas. Ven-*6000*.M5.
Registro de Fomento. Ven-*6000*.R5.
Reports Presented to the Conference of Commissions of Inter-American Development (Held in New York, May 9-18, 1944) by the Venezuelan Commission of Inter-American Development. Ven-*6000*.R3, 1944.
Suplemento de la Guía Industrial y Comercial de Venezuela Correspondiente al Año 1943. Ven-*6000*.G5, 1943.A.
Venezuela. Ven-*6000*.V5, 1946.

Other Countries.
West Indies.
Development and Welfare in the West Indies, Bulletins. WI-*6000*.D5.

Non-Geographic.
Business Cycles and Forecasting. NonG-*6000*.B5, 1940.
Business Cycles, The Problem and Its Setting. NonG-*6000*.B6, 1928.
Commodity Flow and Capital Formation. NonG-*6000*.C4, 1938.
"El Concepto de Industria Nacional y la Protección del Estado." Int-*1300*.A5, 1940.3, Simón.
The Conditions of Economic Progress. NonG-*6000*.C6, 1940.
International Trade and the National Income Multiplier. NonG-*6100*.I5, 1943.
Métodos Estadísticos Aplicados a la Economía y los Negocios. NonG-*3330*.Mills, 1940, Sp.
Rudimentary Mathematics for Economists and Statisticians. NonG-*3330*.Crum, 1946.
Statistical Methods Applied to Economics and Business. NonG-*3330*.Mills, 1940, Eng.
Survey of Economic Theory on Technological Change and Employment. NonG-*6000*.S7, 1940.
Theory of Games and Economic Behavior. NonG-*6000*.T6, 1944.

6010 ——. **Prices and index numbers.**
Precios y números índices.

International.
Economic Barometers. Int-*6010*.E5, 1924.
Price Trends and Price Control in the American Republics, 1939-1945. Int-*6010*.P5, 1939-45.
World Production and Prices. Int-*6000*.W7.

Argentina.
"Precios de Exportación en la República Argentina, 1863-1943." Int-*3040*.E5, Sept. '45, Dominguez.

Brasil.
Dados e Indices Econômico-Financeiros Relativos ao Distrito Federal e á Cidade de São Paulo. Bras-*6010*.D5.

6000-6999 ECONOMICS—cont.

6010 ——. **Prices and index numbers—cont.**

Canada.
The Price Control and Subsidy Program in Canada. Can-*6010*.P3, 1943.
Prices and Price Indexes. Can-*6010*.P5.
Prices and Price Indexes. Can-*6010*.P6.

Colombia.
"El Indice de Precios del Comercio en General." Col-*1420*.A5, May-June '46, Abrisqueta.

Cuba.
Boletín de la Oficina de Números Indices. Cuba-*6010*.B5.
Informes sobre la Estabilización de Precios, el Control de la Inflación y Solicitudes de Abastecimientos Adicionales para Cuba en el Año de 1945. Cuba-*6010*.I5, 1945.
Los Números Indices en Cuba. Cuba-*6010*.N5, 1938.

México.
"Indices Económicos." Méx-*6200*.B5, Jan. '45.
"Indices Generales de la Producción Agrícola." Méx-*3040*.B5, Mar.-Apr. '45, Fernández y Fernández.

Paraguay.
Paraguay, a Study of Price Control, Cost-of-Living and Rationing. Par-*6010*.P5, 1943.

United States.
Wartime Prices: Part I, August 1939 to Pearl Harbor. USA-*6010*.W5, 1939-41.

Non-Geographic.
"Observations au Sujet du Calcul des Indices des Prix et du 'Quantum' des Importations et Exportations." Int-*3020*.B5, 1936, Szturm de Sztrem.

6020 ——. **Production, consumption, stocks.**
Producción, consumo, existencias.

International.
Boletín Informativo del Consejo Interamericano de Comercio y Producción. Int-*6700*.B5.
Circulares Documentales de la Comisión Ejecutiva del Consejo Interamericano de Comercio y Producción. Int-*6700*.C4.
Consejo Interamericano de Comercio y Producción: Recopilación de sus Acuerdos y Recomendaciones. Int-*6700*.C6, 1945.
Indices de la Production Industrielle. Int-*3200*.I3, 1939.
Indices of Industrial Production. Int-*3200*.I5, 1939.
Policies Adopted by the Second Plenary Meeting of the Permanent Council of American Associations of Commerce and Production. Int-*6700*.I6, 1944.2.
"Report of the Commission for the Study of Statistics of Consumption and of Carryover of Commodities." Int-*3040*.R5, 1933.3, Craig.
Resolutions of the First Plenary Meeting of the Permanent Council of American Associations of Commerce and Production. Int-*6700*.I6, 1942.1.
World Production and Prices. Int-*6000*.W7.

Argentina.
"Area Sembrada y Volumen Físico de la Producción: Años 1900 a 1944." Arg-*3040*.R5, June '45, b.

Bolivia.
Estadística de Consumo de Artículos de Procedencia Nacional en al Cuidad de Tarija, 1944. Bol-*6020*.E5, 1944.

Brasil.
A Estatística da Produção no Estado de Minas Gerais. Bras-*6020*.E5, 1939.

Canada.
Survey of Production in Canada. Can-*6020*.S5.

Chile.
"Indices de Producción Física." Chile-*3040*.E5, July '44, Levine Bawden, a.

6000-6999 ECONOMICS—cont.

6020 **——. Production, consumption, stocks—cont.**

Colombia.
"Estadísticas de la Producción." Col-*1420*.A5, Feb. '45, Abrisqueta.

Perú.
"Decreto Supremo de 20 de Agosto de 1943 que Ordena Formar el Inventario del Potencial Económico de la Nación." Perú-*1410*.E5, 1942, a.

United States.
National Product in Wartime. USA-*6100*.N7, 1945.
"Postwar Economic Perspectives: III, Prewar Experience—Production and Consumption." USA-*9950*.S4, Feb. '46, Woytinsky.
Production: Wartime Achievements and the Reconversion Outlook. USA-*6020*.P5, 1945.
Report to the President, the Senate and the House of Representatives by the Director of War Mobilization and Reconversion. USA-*8100*.R5.
Statistics of War Production. USA-*6020*.S5

Non-Geographic.
"The Statistical Measurement of Changes in National Productive Equipment." Int-*3040*.R5, 1938.2, Flux.

6100 **National income and national wealth.**
Renta nacional y riqueza nacional.

International.
National Income and Related Subjects for the United Nations. Int-*6100*.N3, 1946.
National Income Bibliography. Int-*3010*.B5, Econ. 10, Apr. 15, '46.
National Income Estimates of Latin America. Int-*6100*.N5, 1945.
Report of the Sub-Committee on National Income Statistics. Int-*6100*.R5, 1946.
Working Paper on the Status of National Income Statistics. Int-*6100*.W5, 1946.
World Economic Survey. Int-*6000*.W5.

Argentina.
Clasificador de Bienes. Arg-*3200*.C5, 1937.A.
"Estimación de la Renta Nacional Correspondiente al Año 1941." Arg-*3040*.R5, Dec. '44, a.
Indice General Alfabético por Materia de los Bienes Nacionales del Estado. Arg-*3200*.C5, 1937.B.
"La Población Ocupada y las Rentas Individuales en la Argentina." Arg-*3040*.R5, June '45, Di Tella.
"Renta Nacional." Arg-*3040*.R5, Feb. '46.
La Renta Nacional de la República Argentina. Arg-*6100*.R4, 1935-45.

Brasil.
"A Estimativa de Renda Geral do Brasil." Bras-*1420*.B5, June '44, Costa Miranda.
Renda Nacional do Brasil: Colecção de Estudos Diverses. Bras-*6100*.R5, 1944.

Canada.
"Agriculture's Share of the National Income." Can-*3040*.C5, Aug. '43, Hope.
National Income, Canada, 1926-36. Can-*6100*.N4, 1926-36.
National Income of Canada, 1919-1938. Can-*6100*.N5, 1919-38.
"Recent Studies on National Income." Can-*3040*.C5, May '45.
"Recent Studies on National Income." Can-*3040*.C5, Feb. '45.

Chile.
Determinación de la Entrada Nacional de Chile. Chile-*6100*.D5, 1935
The National Income of Chile. Chile-*6100*.N5, 1934-42.
Renta Nacional, 1940-1945. Chile-*6100*.R5, 1940-45.

Cuba.
Directorio Oficial de Exportación, Importación, Producción y Turismo, 1941. Cuba-*6700*.D5, 1941.

6000-6999 ECONOMICS—cont.

6100 ——. National income and national wealth—cont.

Ecuador.

"Algunos Datos sobre la Evolución Financiera y Económica en el Ecuador." Ec-*6720*.B5, Apr.-May '43, Riofrio Villagomez.

"Breves Consideraciones sobre la Economía Ecuatoriana." Ec-*6720*.B5, Mar. '42, Laso.

México.

"Producto Nacional e Ingreso Nacional." Méx-*6000*.T5, Oct.-Dec. '45, Ortiz Mena.

Panamá.

Renta Nacional de la República de Panamá: National Income, 1940. Pan-*6100*.R5, 1940.

Perú.

"El Inventario del Potencial Económico de la Nación." Perú-*3040*.E5, Jan. '45, a.

"La Renta Nacional del Perú." Perú-*6720*.B5, Aug. '45, Barreto.

United States.

"Gross National Product Projections: I, The Background and Relation to Current Issues." USA-*6000*.D5, March '45, George.

"Gross National Product Projections for Full Employment: II, Contrasting Estimates—Range and Reasons." USA-*6000*.D5, May '45, George.

"Gross National Product Projections for Full Employment: III, Measuring the Labor Force in 1950." USA-*6000*.D5, June '45, George.

Income in the United States: Its Amount and Distribution. USA-*6100*.I4, 1921-22.

The Income Structure of the United States. USA-*6100*.I6, 1938.

National Income: A Summary of Findings. USA-*6100*.N2, 1946.

National Income and Capital Formation, 1919-1935. USA-*6100*.N3, 1919-35.

National Income and Its Composition, 1919-1938. USA-*6100*.N4, 1941.

National Income in the United States, 1929-1937. USA-*6100*.N5, 1929-37.

National Product in Wartime. USA-*6100*.N7, 1945.

Our National Resources: Facts and Problems. USA-*6100*.O5, 1940.

Patterns of Resource Use. USA-*6100*.P5, 1939.

Wealth and Income of the People of the United States. USA-*6100*.W5, 1850-1900.

Venezuela.

"Exposición de Motivos del Proyecto de Ley de Presupuesto General de Rentas y Gastos Públicos para el Año Económico 1944-1945." Ven-*8800*.R5, June 1944, a.

"Movimiento General de las Rentas Nacionales por Años Económicos, desde el 1° de Julio de 1908 hasta el 30 de Junio de 1943." Ven-*8800*.R5, June 1944, b.

"Presupuestos de Ingresos y Gastos Públicos de Venezuela en los Años Económicos de 1935-1936 a 1942-1943." Ven-*8800*.R5, June 1944, c.

"Resumen del Presupuesto General de Rentas y Gastos Públicos que Regirá en el Año Económico de 1° de Julio de 1944 a 30 de Junio de 1945." Ven-*8800*.R5, June 1944, d.

Other Countries.

West Indies.

The National Income of Barbados, 1942. WI-*6100*.N3, 1942.

The National Income of Jamaica, 1942. WI-*6100*.N5, 1942.

The National Income of St. Vincent, 1942. WI-*6100*.N7, 1942.

Non-Geographic.

"La Balanza de Pagos y el Ingreso Nacional." Méx-*6000*.T5, Oct.-Dec. '46, Ortiz Mena.

International Trade and the National Income Multiplier. NonG-*6100*.I5, 1943.

National Income and Outlay. NonG-*6100*.N6, 1937.

6000-6999 ECONOMICS—cont.

6100 ——. **National income and national wealth—cont.**

"Renta Nacional, Su Significado y Medición." Int-*3040*.E5, Dec. '45, Domínguez.
"Renta Nacional, Su Significado y Medición, Parte II." Int-*3040*.E5, Mar. '46, Domínguez.
"Renta Nacional, Su Significado y Medición, Parte III." Int-*3040*.E5, June '46, Domínguez.
Studies in Income and Wealth. NonG-*6100*.S3.
Studies in the National Income, 1924-1938. NonG-*6100*.S5, 1942.

6200-6299 **Agriculture, livestock, forestry, fishery.**
Agricultura, ganadería, silvicultura, pesca.

6200 ——. **General.**

International.

Acta Final de la Segunda Conferencia Interamericana de Agricultura. Int-*6200*.I4, 1942.1 Sp.
Agricultural, Pastoral, and Forest Industries in the American Republics. Int *6200*.A45.
Agricultural Problems in Their International Aspect: Documentation for the International Economic Conference, Geneva, 1927. Int-*6200*.A5, 1927.
Agricultural Production and Trade by Countries. Int-*6200*.A6, 1945.
American Agricultural Series. Int-*6200*.A8.
Les Bases Théoriques de la Statistique Agricole Internationale. Int-*6200*.B5, 1914.
Bibliography of Selected References on Agricultural Statistics in the Western Hemisphere. Int-*3010*.B5, Econ. 4a, Apr. 2, '45.
"Cuestiones Elementales en los Censos Agrícolas Aplicables en los Países de Latino-América." Int-*1300*.A5, 1940.3, Alarcón Mendizábal.
Decennial Index of Publications Issued by the International Institute of Agriculture, 1930-1939. Int-*6200*.D3, 1942.
Documents Relating to the Food and Agriculture Organization of the United Nations, 1st August-14th December, 1944. Int-*6200*.D6, 1944.
"Estudio sobre Estadística Agrícola." Col-*1420*.A5, July '39, Ricaurte Montoya
Final Act of the Inter-American Conference on Agriculture. Acta Final de la Conferencia Interamericana de Agricultura. Acta Final da Conferência Interamericana de Agricultura. Acte Final de la Conférence Interaméricaine sur l'Agriculture. Int-*6200*.I4, 1930.1.
Final Act of the Second Inter-American Conference on Agriculture. Int-*6200*.I4, 1942.1, Eng.
Final Act of the Third Inter-American Conference on Agriculture. Int-*6200*.I4, 1945.1, Eng.
Final Act of the United Nations Conference on Food and Agriculture, Hot Springs, Virginia, United States of America, 18th May-3rd June, 1943. Int-*6200*.C5, 1943.
The First World Agricultural Census (1930). Int-*2210*.W5, 1930.
Les Grands Produits Agricoles—Compendium International de Statistiques, 1924-1938. Int-*6210*.G5, 1924-38.
Handbook for the Use of the Delegates of the Third Inter-American Conference on Agriculture. Int-*6200*.I4, 1945.2.
The International Institute of Agriculture. Int-*6200*.I64, 1931.
The International Institute of Agriculture. Int-*6200*.I65, 1924.
International Review of Agriculture. Int-*6200*.I8.
International Yearbook of Agricultural Statistics. Annuaire International de Statistique Agricole. Int-*6200*.I9.
List of Publications Issued by the International Institute of Agriculture. Int-*6200*.L6, 1939.
Liste des Publications Editées par l'Institut International d'Agriculture. Int-*6200*.L7, 1943.

6000-6999 ECONOMICS—cont.

6200 ——. General—cont.

Methods and Problems of International Statistics with Special Reference to Food and Agriculture: A List of References. Int-*6200*.M2, 1944.
"Metodología en Estadísticas Agrícolas de las Américas." Int-*3040*.E5, Mar. '46, Montoya.
"Metodología en Estadísticas Agrícolas de las Américas: Parte II, Sistemas Estadísticos Agrícolas en Algunos Países Latinoamericanos." Int-*3040*.E5, June '46, Montoya.
"Metodología en Estadísticas Agrícolas de las Américas: Parte III, El Servicio de Estadística Agrícola en los Estados Unidos; Parte IV, Conclusiones." Int-*3040*.E5, Sept. '46, Montoya.
Monthly Bulletin of Agricultural Economics and Sociology. Int-*6200*.M4.
L'Organisation des Services de Statistique Agricole dans les Divers Pays. Int-*6200*.O6, 1910-13.
Organization of Agriculture Statistics. Int-*3010*.B5, Econ. 6a, June 1, '45.
Programme of the World Agricultural Census of 1940. Int-*2110*.W5, 1940.1A.
Publications Editées par l'Institut International d'Agriculture: Supplément à la Liste de Janvier 1943. Int-*6200*.L7, 1946.
"Le Recensement Agricole Mondial de 1940." Int-*3020*.B5, 1938, Dore.
Report of the First Session of the Conference, Held at the City of Quebec, Canada, October 16 to November 1, 1945. Int-*6200*.C5, 1945.2.
Standing Advisory Committee on Statistics: First Report to the Director-General. Int-*6200*.S5, 1946.
"Statistique Internationale des Superficies et des Productions Agricoles." Int-*3020*.B5, 1913, Ricci.
The Work of the International Institute of Agriculture During the War (1940-1945). Int-*6200*.W4, 1940-1945.
The World Agricultural Census of 1940: Proposed Standard Form. Int-*2110*.W5, 1940.2.
World Agricultural Census of 1940: Questionnaires. Int-*2110*.W5, 1940.3.
The World Agricultural Situation. Int-*6200*.W6.
World Trade in Agricultural Products. Int-*6900*.W5, 1940.

Argentina.

"La Agrícola Argentina, País de Población Urbana." Arg-*3040*.R5, Oct. '45, Coghlan, b.
Almanaque del Ministerio de Agricultura. Arg-*6200*.A3.
Anales de la Sociedad Rural Argentina. Arg-*6200*.A4.
Anuario de la Sociedad Rural Argentina: Estadísticas Económicas y Agrarias. Arg-*6200*.A5, 1928.
Boletín Estadístico del Ministerio de Agricultura. Arg-*6200*.B5.
Censo Nacional Agropecuario, Año 1937. Arg-*2210*.C6, 1937.
Estadísticas Agrícolas en Argentina: Métodos y Procedimientos. Arg-*6200*.
Ley No. 12.343, Diciembre 29 de 1936. Arg-*2110*.L5, 1936.
Memoria. Arg-*6200*.M5.
Programa Estadístico y Económico para el Ministerio de Agricultura de la República Argentina. Arg-*6200*.P5, 1923.
Reports Presented to the Conference of Commissions of Inter-American Development (Held in New York, May 9-18, 1944) by the Argentine Commission of Inter-American Development. Arg-*6000*.R5, 1944.
Tercer Censo Nacional, Levantado el 1 de Junio de 1914. Arg-*2200*.C5, 1914.

Bolivia.

Censo Experimental Agropecuario del Departamento de Chuquisaca, 1942. Bol-*2210*.C5, 1942, Chuquisaca.
Censo Experimental Agropecuario, Departamento de Tarija, 1944-1945. Bol-*2210*.C5, 1944-45, Tarija.

6000-6999 **ECONOMICS—cont.**

6200 ——. **General—cont.**

Brasil.
Agricultural, Pastoral, and Forest Industries in Brazil. Bras-*6200*.A2, 1946.
Estatísticas Agrícolas no Brasil: Métodos e Processos. Bras-*6200*.
Mensário de Estatística da Produção. Bras-*6200*.M5.
Recenseamento do Brasil Realizado em 1° de Setembro de 1920. Bras-*2200*.R4, 1920.
Recenseamento Geral de 1940: Finalidades do Censo Agrícola. Bras-*2110*.R5, 1940.

Canada.
"Agriculture's Share of the National Income." Can-*3040*.C5, Aug. '43, Hope.
Canadian Farm Family Living Costs. Can-*9500*.C4.
Eighth Census of Canada, 1941: Census of Agriculture: Nova Scotia. Huitième Recensement du Canada, 1941: Recensement Agricole: Nouvelle-Ecosse, Can-*2210*.C5, 1941.2a.
Quarterly Bulletin of Agricultural Statistics. Can-*6200*.Q5.
"Progress in the Collection of Annual Agricultural Statistics and of Crop Reports during Growth in the Dominion of Canada." Int-*3040*.R5, 1936.4, Godfrey.
Seventh Census of Canada, 1931. Septième Recensement du Canada, 1931. Can-*2200*.C5, 1931.2.

Chile.
Agricultura 1935-36 Censo. Chile-*2210*.A5, 1935-36
Agricultura e Industrias Agropecuarias. Chile-*6200*.A5.
Agricultural, Pastoral, and Forest Industries in Chile. Chile-*6200*.A7, 1945.
Boletín de la Sección Estadística de la Caja de Colonización Agrícola. Chile-*6200*.B5.
Estadísticas Agrícolas en Chile: Métodos y Procedimientos. Chile-*6200*.
Memoria de la Caja de Crédito Agrario. Chile-*6200*.M5.

Colombia.
Agricultural, Pastoral, and Forest Industries in Colombia. Col-*6200*.A4, 1945.
Estadísticas Agrícolas en Colombia: Métodos y Procedimientos. Col-*6200*.
Estatuto Orgánico de la Estadística Nacional. Col-*3030*.E5, 1945.
"Plan General para el Levantamiento del Primer Censo Agropecuario." Col-*1420*.A5, Oct. 20, '43, Montoya.
Tierras y Aguas. Col-*6200*.T5.

Costa Rica.
Resúmenes Estadísticos, Años 1883 a 1910. CR-*1400*.R5, 1883-1910.

Cuba.
Boletín Agrícola para el Campesino Cubano. Cuba-*6200*.B5.

El Salvador.
Estadística Forestal y Agraria, Año 1943. ElSal-*6200*.E5, 1943.

Guatemala.
Memoria de las Labores del Poder Ejecutivo en el Ramo de Agricultura. Guat-*6200*.M5.
Revista Agrícola. Guat-*6200*.R4.
Statistics in Guatemala (A Brief Survey). Guat-*3030*.S5, 1945.

Haiti.
Bulletin Mensuel du Département Fiscal. Haiti-*1420*.B5.
Bulletins du Service National de la Production Agricole et de l'Enseignement Rural. Haiti-*6200*.B5.
Monthly Bulletin of the Fiscal Department. Haiti-*1420*.M5.
Rapport Annuel du Service National de la Production Agricole et de l'Enseignement Rural. Haiti-*6200*.R5.

6000-6999 ECONOMICS—cont.

6200 ——. General—cont.

Honduras.

Cuestionario de la Comisión Nacional de Alimentación y Agricultura. Hond-*6200*.C5, 1945.

Honduras Agrícola. Hond-*6200*.H5.

México.

Boletín Mensual de la Dirección General de Economía Rural. Méx-*6200*.B5.

"El Desarrollo y Estado Actual de las Estadísticas Agrícolas en México." Méx-*6200*.B5, Nov. '46, González H.

Estadísticas Agrícolas en México: Métodos y Procedimientos. Méx-*6200*.

"Historia y Evolución de las Estadísticas Agrícolas Continuas, durante el Período 1925-42." Méx-*6200*.B5, Jan. '43.

"Indices Económicos." Méx-*6200*.B5, Jan. '45.

"Indices Generales de la Producción Agrícola." Méx-*3040*.B5, Mar.-Apr. '45, Fernández y Fernández.

Memoria de la Secretaría de Agricultura y Fomento. Méx-*6200*.M5.

Memoria de los Censos Generales de Población, Agrícola, Ganadero e Industrial de 1930. Méx-*2200*.C5, 1930.1.

México Agrario. Méx-*6200*.M8.

"Observaciones sobre Algunas Estadísticas Agrícolas." Méx-*6000*.T5, Jan.-Mar. '46, Alanís Patiño.

Primer Censo Ejidal, 1935. Méx-*2210*.C7, 1935.

Programa e Instructivo para el Servicio de Estadística Agrícola en las Agencias Generales. Méx-*6200*.P5, 1945.

Report on the Crop and Livestock Statistics and Agricultural Census Work in Mexico. Méx-*6200*.R8, 1946.

Segundo Censo Agrícola-Ganadero, Marzo de 1940: Métodos y Procedimientos. Méx-*2110*.C5, 1940.1.

Segundo Censo Ejidal de los Estados Unidos Mexicanos, 6 de Marzo de 1940. Méx-*2210*.C7, 1940.

Segundo Censo Ejidal de los Estados Unidos Mexicanos, 6 de Marzo de 1940: Métodos y Procedimientos. Méx-*2110*.C7, 1940.1.

Nicaragua.

Informe sobre Labores Agrícolas Presentado por la Delegación de Nicaragua a la Tercera Conferencia Interamericana de Agricultura, en Caracas, Estados Unidos de Venezuela. Nic-*6200*.I5, 1945.

"Ley No. 21 de 12 de Junio de 1944, Obligación de Todos los Agricultores de la República . . ." Nic-*1900*.G5, June 12, '44.

Memoria del Ministerio de Agricultura y Trabajo. Nic-*6200*.M5.

Panamá.

Censo Agro-Pecuario, 1942. Pan-*2210*.C5, 1942.

Censo Agro-Pecuario de las Provincias de Herrera y Los Santos, Diciembre 1945: Instrucciones a los Empadronadores. Pan-*2110*.C5, 1945.1.

Censo Agro-Pecuario del Distrito de Penonomé, Diciembre 1943. Pan-*2210*.C5, 1943.1.

Censo Agro-Pecuario, Diciembre de 1943, Distrito de Penonomé: Métodos y Procedimientos. Pan-*2110*.C5, 1943.1.

Paraguay.

Revista de Agricultura y Ganadería. Par-*6200*.R5.

Technical Mission to Paraguay: Reports on the Forests, the Forest Industries, and Trade in Forest Products, Hides and Leather Industries, Edible Vegetable Oils Industries of Paraguay. Par-*6200*.T5, 1945.

Perú.

"Aplicabilidad en la Agricultura Peruana del Método Estadístico de los 'Ejemplares-Tipo'." Perú-*3040*.E5, July '46, Rose Jibaja.

Boletín de la Dirección de Agricultura. Perú-*6200*.B5.

6000-6999 ECONOMICS—cont.

6200 ——. **General—cont.**

"Decreto Supremo de 20 de Agosto de 1943 que Ordena Formar el Inventario del Potencial Económico de la Nación." Perú-*1410*.E5, 1942, a.
"Estadística de las Superficies." Perú-*3040*.E5, Nov. '45, Ricci.
Estadísticas Agrícolas en Perú: Métodos y Procedimientos. Perú-*6200*.
"El Inventario del Potencial Económico de la Nación." Perú-*3040*.E5, Jan. '45, a.
Memoria de la Sociedad Nacional Agraria. Perú-*6200*.M5.
Programa Estadístico para el Ministerio de Agricultura del Perú. Perú-*6200*.S5, 1945, Sp.
Reports Presented to the Conference of Commissions of Inter-American Development (Held in New York, May 9-18, 1944) by the Peruvian Commission of Inter-American Development. Perú-*6000*.R3, 1944.
Statistical Program for the Ministry of Agriculture of Peru. Perú-*6200*.S5, 1945, Eng.
La Vida Agrícola. Perú-*6200*.V5.

República Dominicana.
Anuario Estadístico de la República Dominicana. RepDom-*1410*.A5.
Censo Agro-Pecuario de 1940: Métodos y Procedimientos. RepDom-*2110*.C5, 1910.1.

United States.
Agricultural Finance Review. USA-*6200*.A4.
Agricultural Prices. USA-*6200*.A45.
Agricultural Statistics. USA-*6200*.A5.
Agricultural Statistics in the United States: Methods and Procedures. USA-6200.
Agriculture Handbook. USA-*2110*.C5, 1940.2.
American Agriculture, 1899-1939: A Study of Output, Employment and Productivity. USA-*6200*.A7, 1942.
Analysis of Specified Farm Characteristics for Farms Classified by Total Value of Products. USA-*6200*.A8, 1943.
Annual Report of the Farm Credit Administration. USA-*6720*.A7.
Censo de Agricultura de los Estados Unidos. USA-*2110*.C5, 1940.1.
Crops and Markets. USA-*6200*.C6.
Eleventh Annual Report of the Farm Credit Administration, 1943-44. USA-*6200*.A9, 1943-44.
Farm Credit Quarterly. USA-*6720*.F2.
The Farm Income Situation. USA-*6200*.F4.
Farm Labor. USA-*9200*.F5.
The Farm Real Estate Situation. USA-*6200*.F7.
Federal Reserve Bulletin. USA-*6720*.F4.
The Impact of the War on the Financial Structure of Agriculture. USA-*6200*.I4, 1944.
Journal of Farm Economics. USA-*6200*.J5.
Research Bulletin. USA-*6200*.R5.
Sixteenth Census of the United States: 1940: Agriculture. USA-*2210*.C5, 1940.1.
Sixteenth Census of the United States: 1940: Agriculture: Abandoned or Idle Farms. USA-*2210*.C5, 1940.15.
Sixteenth Census of the United States: 1940: Agriculture: Cows Milked and Dairy Products. USA-*2210*.C5, 1940.17.
Sixteenth Census of the United States: 1940: Agriculture: Cross-Line Acreage. USA-*2210*.C5, 1940.2.
Sixteenth Census of the United States: 1940: Agriculture: Special Cotton Report. USA-*2210*.C5, 1940.25.
Sixteenth Census of the United States: 1940: Agriculture: Special Poultry Report. USA-*2210*.C5, 1940.27.

6000-6999 ECONOMICS—cont.

6200 ——. **General—cont.**

Sixteenth Census of the United States: 1940: Agriculture: Value of Farm Products by Color and Tenure of Farm Operator. USA-*2210*.C5, 1940.3.
Sixteenth Census of the United States: 1940: Drainage of Agricultural Lands. USA-*2210*.C5, 1940.4.
Sixteenth Census of the United States: 1940: Irrigation of Agricultural Lands. USA-*2210*.C5, 1940.6.
Sixteenth Decennial Census of the United States: Instructions to Enumerators: Population and Agriculture, 1940. USA-*2100*.C5, 1940.5.
The Wages of Farm and Factory Laborers. USA-*9200*.W7, 1945.
Yearbook of the Department of Agriculture. USA-*6200*.Y5.

——. Puerto Rico.
Censo Décimosexto de los Estados Unidos: 1940: Agricultura. Sixteenth Census of the United States: 1940: Agriculture. USA PR-*2210*.C5, 1940.

Uruguay.
Anuario de Estadística Agrícola. Ur-*6200*.A5.
Boletín Mensual del Instituto Nacional de Alimentación. Ur-*9800*.B5
Censo Agropecuario, Año 1937. Ur-*2210*.C5, 1937.
Estadística Agrícola. Ur-*6200*.E4, 1943.
Estadísticas Agrícolas en Uruguay: Métodos y Procedimientos. Ur-*6200*.
Ley y Reglamentación de la Estadística Agrícola. Ur-*6200*.L5, 1918.
Mercados del Mundo. Ur-*6200*.M5.

Venezuela.
Agricultural, Pastoral, and Forest Industries in Venezuela. Ven-*6200*.A5, 1945.
Censos Agrícola y Pecuario, 1937. Ven-*2210*.C5, 1937.
Memoria y Cuenta del Ministerio de Agricultura y Cría. Ven-*6200*.M5.

Other Countries.
West Indies.
Eighth Census of Jamaica and Its Dependencies, 1943: Population, Housing and Agriculture. WI-*2200*.E5, 1943.

Non-Geographic.
"The Master Sample of Agriculture: I, Development and Use." USA-*3040*.J5, Mar. '45, King.
"The Master Sample of Agriculture: II, Design." USA-*3040*.J5, Mar. '45, Jessen.
"Recent Development in Sampling for Agricultural Statistics." USA-*3040*.J5, Mar. '42, Snedecor.
Statistical Methods Applied to Experiments in Agriculture and Biology. NonG-*3330*.Snedecor, 1946.
"Statistique de la Production Agricole." Int-*3020*.B5, 1923, Ricci.
The Theory of Sampling. NonG-*3150*.T8, 1942.

6210 ——. **Commodities.**
Mercancías.

International.
Agricultural Commodities and Raw Materials: Production and Consumption in the Different Parts of the World, 1934-1938. Int-*6210*.A4, 1934-38.
Coffee Statistics. Int-*6210*.C4.
Fourth Pan American Coffee Conference, Mexico City, September 3-14, 1945. Int-*6210*.P5, 1945, Eng.
Les Grands Produits Agricoles—Compendium International de Statistiques, 1924-1938. Int-*6210*.G5, 1924-38.
"Inquiry Concerning Statistics of the Wool Stoçks in the Principal Wool-Producing and Wool-Consuming Countries of the World." Int-*3040*.R5, 1939, 2-3, Godfrey.
Wheat Studies. Int-*6210*.W5.
World Food Situation, 1946. Int-*9800*.W4, 1946.

6000-6999 ECONOMICS—cont.

6210 ——. **Commodities—cont.**

World Grain Review and Outlook, 1945. Int-*6210*.W6, 1945.
World Trade in Agricultural Products. Int-*6900*.W4, 1943.

Argentina.
"Area Sembrada y Volumen Físico de la Producción: Años 1900 a 1944." Arg-*3040*.R5, June '45, b.
Boletín Frutas y Hortalizas. Arg-*6210*.B3.
Boletín Informativo. Arg-*6210*.B7.
Boletín Informativo de la Comisión Nacional de Granos y Elevadores. Arg-6210.B5.
Censo Algodonero de la República Argentina, Año 1935-36. Arg-*2210*.C4, 1935-36.
La Industria Azucarera. Arg-*6210*.I5.

Brasil.
Anuário Açucareiro. Bras-*6210*.A4.
Anuário Algodoeiro. Bras-*6210*.A5.
Anuário Estatístico do Café. Bras-*6210*.A6.
Boletim do Instituto do Açucar e do Álcool. Bras-*6210*.B6.
Brasil Açucareiro. Bras-*6210*.B8.
"The Elasticity of Substitution of Brazilian Coffee." Int-*1300*.A5, 1940.3, Kingston.
Revista do Departamento Nacional do Café. Bras-*6210*.R5.

Canada.
Report of the Grain Trade of Canada. Can-*6210*.R5.

Colombia.
"Acuerdo de la Oficina Panamericana del Café en Nueva York, acerca de la Reglamentación de la Estadística Cafetera Mundial." Col-*6210*.B5, Aug. '39, a.
Boletín de Estadística de la Federación Nacional de Cafeteros. Col-*6210*.B5.
"El Censo Cafetero 1939-40." Col-*6210*.B5, Aug. '39, b.
"Convenio Interamericano del Café, Noviembre 28 de 1940." Col-*6210*.B5, Dec. '40.

Cuba.
Anuario Azucarero de Cuba. Cuba Sugar Year Book. Cuba-*6210*.A5.
Boletín del Café. Cuba-*6210*.B5.
Compilación Estadística sobre Café. Cuba-*6210*.C5.
Zafra. Cuba-*6210*.Z5.

El Salvador.
El Café. ElSal-*6210*.C5.
El Café de El Salvador. ElSal-*6210*.C6.
Memoria de la Compañía Salvadoreña de Café. ElSal-*6210*.M5.
Primer Censo Nacional del Café, 1938-39. ElSal-*2210*.C5, 1940.

Guatemala.
Revista Cafetalera de Guatemala. Guat-*6210*.R5.

México.
"Indices Generales de la Producción Agrícola." Méx-*6200*.B5, Mar. '45.

Nicaragua.
"Ley sobre Estadística de Producción Agrícola." Nic-*1420*.B5, June 15, '44, b.

Paraguay.
Corporación Paraguaya de Alcoholes. Par-*6400*.C5.

Perú.
Algodón. Perú-*6210*.A5.
Estadística de la Producción de Algodón en el Perú Correspondiente al Año 1943. Perú-*6210*.E5, 1943.
Estadística de la Producción de Arroz. Perú-*6210*.E6.
Estadística de la Producción de Caña de Azúcar y Azúcar de Caña en el Perú. Perú-*6210*.E7.

6000-6999 ECONOMICS—cont.

6210 ——. Commodities—cont.

Estadística de la Producción del Algodón en el Perú. Perú-*6210*.E5.
Estadística de la Producción del Trigo en el Perú. Perú-*6210*.E8.
Memoria Anual de la Cámara Algodonera del Perú. Perú-*6210*.M5.
Memoria de la Sociedad Nacional Agraria. Perú-*6200*.M5.

República Dominicana.
Primer Censo Cafetalero Nacional, 1943. RepDom-*2210*.C5, 1943.1.

Uruguay.
Estadística de Cereales y Oleaginosos. Ur-*6210*.E5, 1944.
Estimaciones e Informes Finales de Area Sembrada y Producción: Maíz, Arroz, Cereales . . . 1943-45. Ur-*6210*.E6, 1943-45.

Venezuela.
Revista del Instituto Nacional del Café. Ven-*6210*.R5.

6220 ——. Agriculture.
Agricultura.

International.
International Crop-Reporting Service. Int-*6220*.I5, 1910.
Monthly Crop Report and Agricultural Statistics. Int-*6220*.M5.
"Statistique Internationale des États des Cultures." Int-*3020*.B5, 1911, Ricci.

Argentina.
Primer Censo Nacional de Floricultura, Ley 12.343, Años 1938-39. Arg-*2210*.C7, 1938-39.

México.
Instrucciones Generales para los Corresponsales Estadísticos. Méx-*6220*.I5, 1942.

Perú.
Aguas e Irrigación. Perú-*6220*.B4.
Boletín de la Compañía Administradora del Guano. Perú-*6220*.B6.
Memoria de la Compañía Administradora del Guano. Perú-*6220*.M5.

United States.
The Crop and Livestock Reporting Service of the United States. USA-*6220*.C5, 1933.
Crop Production. USA-*6220*.C7.

Non-Geographic.
"La Statistique des États de Culture." Int-*3020*.B5, 1913, March.

6230 ——. Livestock and small animal life.
Ganadería y animales pequeños.

International.
L'Aviculture dans le Monde. Int-*6230*.A5, 1933.

Argentina.
Anuario de Estadística Agro-Pecuaria. Arg-*6230*.A5.
Censo Ganadero Nacional, Existencia al 1.° de Julio de 1930. Arg-*2210*.C5, 1930.
Compras de Ganado Bovino, Ovino y Porcino Efectuadas por los Frigoríficos en las Estancias. Arg-*6230*.C5.

Bolivia.
Estadística Agropecuaria. Bol-*6230*.E5.

Canada.
Live Stock and Animal Products Statistics. Statistique de Bétail et des Produits Animaux. Can-*6230*.L5.

Cuba.
"El Censo Ganadero de 1945." Cuba-*1420*.B5, Apr. '45, Queralt.

Nicaragua.
Informe sobre la Industria Pecuaria Presentado por la Delegación de Nicaragua a la 3ª Conferencia Interamericana de Agricultura, en Caracas, EE. UU. de Venezuela. Nic-*6230*.I5, 1945.

6000-6999 ECONOMICS—cont.

6230 ——. **Livestock and small animal life—cont.**

Perú.
Estudio de la Situación de la Industria Lechera en el Departamento de Lambayeque. Perú-*6230*.E5, 1944.
República Dominicana.
Sacrificio de Ganado. RepDom-*6230*.S5.
Uruguay.
Censo Agropecuario, Año 1937. Ur-*2210*.C5, 1937.
Censo Ganadero 1943: Ley de Enero 27 de 1943. Ur-*2110*.C5, 1943.
Venezuela.
Boletín Mensual de la Dirección de Ganadería. Ven-*6230*.B5.

6240 ——. **Fishery.**
Pesca.

International.
The Fisheries and Fishery Resources of the Caribbean Area. Int-*6240*.F5, 1943.
"Need and Plan for a Statistical Program in Furthering Conservation of Inter-American Fisheries." Int-*1300*.A5, 1940.3, Fiedler.
Canada.
Fisheries Statistics of Canada. Statistique des Pêcheries. Can-*6240*.F5.
Chile.
"Pesca: Producción, Valor y Distribución de la Producción Pesquera." Chile-*1420*.E5, Mar.-Apr. '45.
United States.
Fishery Statistics of the United States. USA-*6240*.F4.
Statistical Digest Series of the Fish and Wildlife Service. USA-*6240*.S6.
Non-Geographic.
"International Fishery Statistics". Int-*3020*.B5, 1909, Rew.
"Statistics of the Fisheries". Int-*3020*.B5, 1923, Rew.
"Statistique Internationale des Pêcheries." Int-*3020*.B5, 1909, Rew, a.

6250 ——. **Forestry.**
Silvicultura.

International.
Actes du Ier Congrès International de Sylviculture. Int-*6250*.C5, 1926.
Forests and Forestry—Statistical and Other Information for Certain Countries. Int-*6250*.F5, 1925.
International Trade in Wood. Int-*6900*.I7, 1944.
International Yearbook of Forestry Statistics. Int-*6250*.I5.
"La Statistique Forestière Internationale." Int-*3040*.R5, 1936.2, Dore.
"La Statistique Forestière Internationale et ses Problèmes." Int-*3040*.R5. 1933.3, Dore.
Statistiques du Bois. Int-*3200*.S6, 1938.
Timber Statistics. Int-*3200*.T5, 1938.
Paraguay.
Technical Mission to Paraguay: Reports on the Forests, the Forest Industries, and Trade in Forests Products, Hides and Leather Industries, Edible Vegetable Oils Industries of Paraguay. Par-*6200*.T5, 1945.
Non-Geographic.
Sampling Methods in Forestry and Range Management. NonG-*6250*.S5. 1942.
"Statistical Method in Forestry." USA-*3040*.B5, June '45, Schumacher.

6300 **Mining and extractive industry.**
Minería e industrias extractivas.

International.
Copper Resources of the World. Int-*6300*.C8, 1935.
"Estadística de Metales y Minerales." Int-*3040*.E5, Sept. '46, Knox.

6000-6999 ECONOMICS—cont.

6300 ——. **Mining and extractive industry—cont.**

Foreign Minerals Quarterly. Int-*6300*.F5.
Mining and Manufacturing Industries in the American Republics. Int-*6400*.M5.
Report to the Council on the Work of the Fifth Session. Int-*3000*.R6, 1936, Eng.

Argentina.
Boletín de Informaciones Petroleras. Arg-*6300*.B5.
Estadística Minera de la Nación. Arg-*6300*.E5.
Estadísticas Industriales en Argentina: Métodos y Procedimientos. Arg-*6400*.
Mining and Manufacturing Industries in Argentina. Arg-*6400*.M7, 1945.

Bolivia.
Boletín de la Dirección General de Minas y Petróleo. Bol-*6300*.B4.
Carta Informativa. Bol-*6300*.C5.
Estadísticas Industriales en Bolivia: Métodos y Procedimientos. Bol-*6400*.
Industria. Bol-*6400*.I5.
Memoria Anual. Bol-*6300*.M5.
Minería, Transportes y Comunicaciones. Bol-*6000*.M5.
Mining and Manufacturing Industries in Bolivia. Bol-*6400*.M5, 1945.
Tasas e Impuestos sobre la Industria Minera en Bolivia. Bol-*6300*.T5, 1941.

Brasil.
Estatísticas Industriais no Brasil: Métodos e Processos. Bras-*6400*.
Mining and Manufacturing Industries in Brazil. Bras-*6400*.M5, 1945.

Canada.
Annual Report on the Mineral Production of Canada. Can-*6300*.A5.
Coal Statistics for Canada. Can-*6300*.C5.
Industrial Statistics in Canada: Methods and Procedures. Can-*6400*.

Chile.
Estadísticas Industriales en Chile: Métodos y Procedimientos. Chile-*6400*.
Minería. Chile-*6300*.M5.
Mining and Manufacturing Industries in Chile. Chile-*6400*.M5, 1945.

Colombia.
Mining and Manufacturing Industries in Colombia. Col-*6400*.M5, 1945.

Costa Rica.
Mining and Manufacturing Industries in Costa Rica. CR-*6400*.M5, 1945.

Cuba.
Boletín de Minas. Cuba-*6300*.B5.

Ecuador.
La Minería y el Petróleo en el Ecuador: Anuario. Ec-*6300*.M5.
Mining and Manufacturing Industries in Ecuador. Ec-*6400*.M5, 1945.

Guatemala.
Report on the Utilization of the Mineral Resources of the Republic of Guatemala. Guat-*6300*.R5, 1946.

Haiti.
Mining and Manufacturing Industries in Haiti. Haiti-*6400*.M5, 1945.

México.
Estadísticas Industriales en México: Métodos y Procedimientos. Méx-*6400*.
Minería y Riqueza Minera de México. Méx-*6300*.M5, 1944.
Mining and Manufacturing Industries in Mexico. Méx-*6400*.M3, 1946.
Tercer Censo Industrial: Catálogo General de las Industrias de Transformación y Extractivas, 1940. Méx-*3220*.T5, 1940.
"Victory Minerals of Mexico." Int-*6000*.C7, Dec. '45, Ysita.

Nicaragua.
Estadística Minera. Nic-*6300*.E5.
Mining and Manufacturing Industries in Nicaragua. Nic-*6400*.M5, 1945.

Panamá.
Mining and Manufacturing Industries in Panamá. Pan-*6400*.M5, 1945.

6000-6999 ECONOMICS—cont.

6300 Mining and extractive industry—cont.

Paraguay.
Estadísticas Industriales en Paraguay: Métodos y Procedimientos. Par-*6400*.
Mining and Manufacturing Industries in Paraguay. Par-*6400*.M5, 1945.

Perú.
Anuario de la Industria Minera en el Perú. Perú-*6300*.A5.
Boletín del Cuerpo de Ingenieros de Minas. Perú-*6300*,B5.
Boletín Oficial de la Dirección de Minas y Petróleo. Perú-*6300*.B7.
Mining and Manufacturing Industries in Peru. Perú-*6400*.M5, 1945.

United States.
Minerals Yearbook. USA-*6300*.M5.
The Mining Industries, 1899-1939: A Study of Output, Employment, and Productivity. USA-*6300*.M7, 1899-1939.
Report of the Committee on the Standardization of Mining Statistics. USA-*3200*.R5, 1918.
Sixteenth Census of the United States: 1940: Mineral Industries, 1939. USA-*2270*.C5, 1939.
Suggested Standard Forms for Mine Statistics with Especial Reference to Accidents. USA-*3200*.S5, 1916.
Technical Papers of the Bureau of Mines. USA-*6300*.T5.

Uruguay.
Estadísticas Industriales en Uruguay: Métodos y Procedimientos. Ur-*6400*.
Mining and Manufacturing Industries in Uruguay. Ur-*6400*.M5, 1945.

Venezuela.
Estadística del Petróleo. Ven-*6300*.E5.
Estadísticas Industriales en Venezuela: Métodos y Procedimientos. Ven-*6400*.
Memoria del Ministerio de Fomento. Ven-*6000*.M3.
Mining and Manufacturing Industries in Venezuela. Ven-*6400*.M5, 1945.

400 Manufacturing industry.
Industria manufacturera.

International.
"Conceptos Básicos de los Censos Industriales Recomendables para los Países Latino-Americanos." Int-*1300*.A5, 1940.3, Zertuche.
Estadísticas de la Población Activa. Int-*3240*.E5, 1938.
Indices de la Production Industrielle. Int-*3200*.I3, 1939.
Indices of Industrial Production. Int-*3200*.I5, 1939.
Industrialization and Foreign Trade. Int-*6000*.I2, 1945.
"Industrialization in Latin America: An Outline of Its Development." Int-*6400*.I3, 1946.
Industry in Latin America. Int-*6400*.I5, 1945.
Mining and Manufacturing Industries in the American Republics. Int-*6400*.M5.
Report to the Council on the Work of the Seventh Session. Int-*3000*.R6, 1938, Eng.
Report to the Council on the Work of the Sixth Session. Int-*3000*.R6, 1937, Eng.
Statistics of the Gainfully-Occupied Population. Int-*3240*.S4, 1938.
"La Statistique Internationale des Forces Motrices." Int-*3040*.R5, 1936.1, Huber.
Statistique Internationale des Grandes Villes: Statistique de l'Electricité, du Gaz et de l'Eau dans les Grandes Villes, 1934. Int-*6400*.S5, 1934.
Statistiques de la Population Active. Int-*3240*.S6, 1938.
Systems of Classification of Industries and Occupations. Int-*3220*.S5, 1923.
The T. V. A.: Lessons for International Application. Int-*6400*.T5, 1944.

Argentina.
La Actividad Industrial durante los Primeros . . . Meses de . . . , según los Números Indices Correspondientes a los Salarios Pagados, a la Ocupación de Obreros y a las Horas-Obrero de Trabajo en la Industria. Arg-*6400*.A3.

6000-6999 ECONOMICS—cont.

6400 Manufacturing industry—cont.

"Bosquejo de una Economía Argentina para 1955: Segunda Parte (continuación)—Estudios Especiales." Arg-*3040*.R5, July '46, a.
Estadística Industrial. Arg-*2250*.E5.
Estadísticas Industriales en Argentina: Métodos y Procedimientos. Arg-*6400*.
Informe, Serie I, Industria. Arg-*6400*.E5.
Mining and Manufacturing Industries in Argentina. Arg-*6400*.M7, 1945.
Reports Presented to the Conference of Commissions of Inter-American Development (Held in New York, May 9-18, 1944) by the Argentine Commission of Inter-American Development. Arg-*6000*.R5, 1944.
Revista de la Unión Industrial Argentina. Arg-*6400*.R7.
Tercer Censo Nacional, Levantado el 1 de Junio de 1914. Arg-*2200*.C5, 1914.

Bolivia.
Estadísticas Industriales en Bolivia: Métodos y Procedimientos. Bol-*6400*.
Industria. Bol-*6400*.I5.
Mining and Manufacturing Industries in Bolivia. Bol-*6400*.M5, 1945.
Síntesis de Estadística Industrial, 1941. Bol-*6400*.S5, 1941.

Brasil.
"Anexo a Resolução No. 243: Nomenclatura Brasileira para a Classificação de Indústrias." Bras-*1420*.B5, Sept. '44.
Estatísticas Industriais no Brasil: Métodos e Processos. Bras-*6400*.
Os Estoques de Matérias e Produtos Téxteis do Distrito Federal em 30 de Junho de 1943 e em 30 de Junho de 1944. Bras-*6400*.E7, 1944.
Inquérito Continental sôbre Fomento e Coordenação de Indústrias. Bras-*6400*.I5, 1946.
Mining and Manufacturing Industries in Brazil. Bras-*6400*.M5, 1945.
Recenseamento do Brasil Realizado em 1° de Setembro de 1920. Bras-*2200*.R4, 1920.
Reports Presented to the Conference of Commissions of Inter-American Development (Held in New York on May 9-18, 1944) by the Brazilian Commission of Inter-American Development. Bras-*6000*.R5, 1944.
Situação e Movimento dos Estabelecimentos Comerciais e Industriais em Abril de 1944. Bras-*6000*.S7, Apr. '44.
Situação e Movimento dos Estabelecimentos Comerciais e Industriais em Março de 1944. Bras-*6000*.S7, Mar. '44.

Canada.
Central Electric Stations in Canada. Can-*6400*.C6.
Classification of Industries: Eighth Census of Canada, 1941. Can-*3220*.C5, 1941.
Industrial Statistics in Canada: Methods and Procedures. Can-*6400*.
Iron and Steel and Their Products in Canada. Can-*6400*.I7.
Manufactures of the Non-Ferrous Metals in Canada, 1939-1943. Can-*6400*.M5, 1939-43.
The Manufacturing Industries of Canada. Can-*6400*.M6.
Survey of Production in Canada. Can-*6020*.S5.

Chile.
Censo Industrial y Comercial, Año 1937. Chile-*2250*.C5, 1937.
Estadísticas Industriales en Chile: Métodos y Procedimientos. Chile-*6400*.
Industrias. Chile-*6400*.I5.
Mining and Manufacturing Industries in Chile. Chile-*6400*.M5, 1945.
Reports Presented to the Conference of Commissions of Inter-American Development (Held in New York on May 9-18, 1944) by the Chilean Commission of Inter-American Development. Chile-*6000*.R5, 1944.

Colombia.
"El Censo Industrial de Colombia de 1945." Col-*1420*.A5, Feb. '45, Ortiz C.
Censo Industrial de Colombia, 1945: Métodos y Procedimientos. Col-*2150*.C5, 1945, a.
Estatuto Orgánico de la Estadística Nacional. Col-*3030*.E5, 1945.
Mining and Manufacturing Industries in Colombia. Col-*6400*.M5, 1945.

6000-6999 ECONOMICS—cont.

6400 Manufacturing industry—cont.

Primer Censo Industrial de Colombia, 1945: Boleta General. Col-*2150*.C5, 1945, c.
Primer Censo Industrial de Colombia, 1945: Cartilla de Instrucciones. Col-*2150*.C5, 1945, b.
Report Presented to the Conference of Commissions of Inter-American Development (Held in New York, May 9-18, 1944) by the Colombian Commission of Inter-American Development. Col-*6000*.R5, 1944.

Costa Rica.
Mining and Manufacturing Industries in Costa Rica. CR-*6400*.M5, 1945.
Resúmenes Estadísticos, Años 1883 a 1910. CR-*1400*.R5, 1883-1910.

Ecuador.
Mining and Manufacturing Industries in Ecuador. Ec-*6400*.M5, 1945.

Guatemala.
"Empresas Eléctricas de Servicio Público." Guat-*1420*.B5, July '46, a.
Industrial Report on the Republic of Guatemala. Guat-*6400*.I5, 1946.
Statistics in Guatemala (A Brief Survey). Guat-*3030*.S5, 1945.

Haiti.
Mining and Manufacturing Industries in Haiti. Haiti-*6400*.M5, 1945.

México.
Banco de México, S.A.: Vigésimacuarta Asamblea General Ordinaria de Accionistas. Méx-*6720*.B5, 1946.
Características Principales de la Industria: Censo de 1940. Méx-*2250*.C5, 1940.3.
Catálogo de Industrias, 1945. Méx-*3220*.C5, 1945.
Los Censos Cuarto Industrial, Segundo Comercial y Segundo de Transportes, 1945: Métodos y Procedimientos. Méx-*2150*.C5, 1945.1.
Estadísticas Industriales en México: Métodos y Procedimientos. Méx-*6400*.
"Growth of Industry and Aviation in Mexico, 1929-1942." Int-*6000*.C7, July-Aug. '44, Ysita.
Industrial Development in Mexico. Méx-*6400*.I5, 1943.
Memoria de la Secretaría de Agricultura y Fomento. Méx-*6200*.M5.
Memoria de los Censos Generales de Población, Agrícola, Ganadero e Industrial de 1930. Méx-*2200*.C5, 1930.1.
Mining and Manufacturing Industries in Mexico. Méx-*6400*.M3, 1946.
Monografías Industriales del Banco de México, S.A. Méx-*6400*.M5.
Primer Censo Industrial de 1930. Méx-*2250*.C5, 1930.
Reports Presented to the Conference of Commissions of Inter-American Development (Held in New York, May 9-18, 1944) by the Mexican Commission of Inter-American Development. Méx-*6000*.R4, 1944.
Segundo Censo Industrial de 1935. Méx-*2250*.C5, 1935.
Tercer Censo Industrial: Catálogo General de las Industrias de Transformación y Extractivas, 1940. Méx-*3220*.T5, 1940.
Tercer Censo Industrial de los Estados Unidos Mexicanos, 1940. Méx-*2250*.C5, 1940.
Tercer Censo Industrial de los Estados Unidos Mexicanos, 1940: Métodos y Procedimientos. Méx-*2150*.C5, 1940.1.

Nicaragua.
Mining and Manufacturing Industries in Nicaragua. Nic-*6400*.M5, 1945.
Primer Censo Industrial de Nicaragua, 1940: Instructivo para Proporcionar los Datos a los Empadronadores. Nic-*2150*.C5, 1940.

Panamá.
"Análisis de los Resultados del Primer Censo Oficial de Empresas Industriales y Comerciales Radicadas en Panamá, Ejecutado en el Año de 1938." Pan-*6000*.R5, Apr. '39.
Mining and Manufacturing Industries in Panama. Pan-*6400*.M5, 1945.

6000-6999 ECONOMICS—cont.

6400 Manufacturing industry—cont.

Paraguay.
Corporación Paraguaya de Alcoholes. Par-*6400*.C5.
Estadísticas Industriales en Paraguay: Métodos y Procedimientos. Par-*6400*.
Mining and Manufacturing Industries in Paraguay. Par-*6400*.M5, 1945.
Technical Mission to Paraguay: Reports on the Forests, the Forest Industries, and Trade in Forest Products, Hides and Leather Industries, Edible Vegetable Oils Industries of Paraguay. Par-*6200*.T5, 1945.

Perú.
Boletín de la Dirección de Industrias del Ministerio de Fomento y Obras Públicas. Perú-*6400*.B5.
"Cifras Estadísticas de la Industria Textil." Perú-*6400*.B5, Sept. '45, Arbulú Casanova.
Estadística de los Servicios Eléctricos del Perú. Perú-*6400*.E5, 1942.
"Estadística Industrial." Perú-*6400*.B5, Aug. '46.
Mining and Manufacturing Industries in Peru. Peru-*6400*.M5, 1945.
"Producción de Energía." Perú-*1410*.E5, 1942, e.

República Dominicana.
Resumen General de la Producción de la Pequeña Industria Rural, Año 1938. RepDom-*6400*.R5, 1938.

United States.
Census of Electrical Industries: Telephones and Telegraphs. USA-*2250*.C3, 1902-30.
Employment in Manufacturing, 1899-1939, an Analysis of Its Relation to the Volume of Production. USA-*9200*.E5, 1899-1939.
Industrial Corporation Reports. USA-*6400*.I3.
Industrial Corporation Reports for 1939: Summary by Industry Groups USA-*6400*.I3, 1939.
Output of Manufactured Commodities, 1929-1939. USA-*6400*.O4, 1929-39.
The Output of Manufacturing Industries, 1899-1937. USA-*6400*.06, 1940.
Sixteenth Census of the United States: 1940: Manufactures, 1939. USA-*2250*.C7, 1939.1.
Sixteenth Census of the United States: 1940: Manufactures, 1939: Industry Classifications for the Census of Manufactures, 1939. USA-*3320*.C5, 1939.
Sixteenth Census of the United States: Alphabetical Index of Occupations and Industries. USA-*3240*.C5, 1940.2.
Standard Industrial Classification Manual. USA-*3220*.S5, 1942.
Statistics of Electric Utilities in the United States, Class A and B Privately Owned Companies. USA-*6400*.S5.

Uruguay.
Censo Industrial de 1936. Ur-*2250*.C5, 1936.
Estadísticas Industriales en Uruguay: Métodos y Procedimientos. Ur-*6400*.
Mining and Manufacturing Industries in Uruguay. Ur-*6400*.M5, 1945.

Venezuela.
Censos Industrial, Comercial y Empresas que Prestan Servicios, 1936. Ven-*2250*.C5, 1936.
Estadísticas Industriales en Venezuela: Métodos y Procedimientos. Ven-*6400*.
Memoria del Ministerio de Fomento. Ven-*6000*.M3.
Mining and Manufacturing Industries in Venezuela. Ven-*6400*.M5, 1945.
Reports Presented to the Conference of Commissions of Inter-American Development (Held in New York, May 9-18, 1944) by the Venezuelan Commission of Inter-American Development. Ven-*6000*.R3, 1944.

Non-Geographic.
"La Clasificación de Industrias y Profesiones en las Estadísticas del Trabajo." Int-*9000*.E5, 1943, a.
"Estadística Económica Militar." Perú-*3040*.E5, July '46, Velarde B.

6000-6999 ECONOMICS—cont.

6400 **Manufacturing industry—cont.**

Statistical Methods in Industry. NonG-*6400*.S5, 1943.
"La Statistique Internationale des Forces Motrices." Int-*3040*.R5, 1933.2, Huber.

6500 **Building and other construction.**
Edificación y otras construcciones.

International.
"Commission on Building and Housing Statistics: Preliminary Report." Int-*3040*.R5, 1938.2, Nyström.
Housing and Hospital Projects of Latin American Social Security Systems. Int-*9950*.H5, 1942.
"Observations on the Possibility of Improving the International Comparability of Building and Housing Statistics." Int-*3040*.R5, 1936.1, Nyström.

Canada.
Housing Statistics. Statistiques du Logement. Can-*6500*.H5.
Report on the Construction Industry in Canada. Can-*6500*.R5.

United States.
Construction. USA-*6500*.C3.
The Construction Industry in the United States. USA-*6500*.C4, 1944.

6600-6699 **Transport and communication.**
Transportes y comunicaciones.

6600 ——. **General.**

Argentina.
Memoria. Arg-*6600*.M5.

El Salvador.
Boletín. ElSal-*6600*.B5.

Guatemala.
Vías de Comunicación . . . Guat-*6000*.V5, 1941.

República Dominicana.
Movimiento Postal, Telefónico y Radiotelegráfico. Carreteras de la República Dominicana. RepDom-*6600*.M5.

6610-6619 ——. **Transport.**
Transporte.

6610 ——. ——. **General.**

International.
Report on the Unification of Transport Statistics. Int-*6610*.R5, 1932.
Transportation in Latin America. Int-*6610*.T5, 1942.

Bolivia.
Minería, Transportes y Comunicaciones. Bol-*6000*.M5.

Brasil.
Reports Presented to the Conference of Commissions of Inter-American Development (Held in New York on May 9-18, 1944) by the Brazilian Commission of Inter-American Development. Bras-*6000*.R5, 1944.

Chile.
Comercio Interior y Comunicaciones. Chile-*6700*.C5.

Cuba.
Memoria Oficial de la Comisión Nacional de Transportes. Cuba-*6610*.M5.

El Salvador.
Boletín. ElSal-*6600*.B5.

Haiti.
Triangulation de la République d'Haïti. Haiti-*6610*.T5, 1946?

6000-6999 ECONOMICS—cont.

6610 ——. ——. **General—cont.**

México.
Los Censos Cuarto Industrial, Segundo Comercial y Segundo de Transportes, 1945: Métodos y Procedimientos. Méx-*2150*.C5, 1945.1.
Primer Censo de Transportes, Febrero y Marzo de 1940: Métodos y Procedimientos. Méx-*2180*.C5, 1940.1.

Perú.
Boletín de la Dirección de Caminos y Ferrocarriles. Perú-*6610*.B5.
Memoria de la Dirección General de Tránsito. Perú-*6610*.M5.

United States.
Fifty-Seventh Annual Report on the Statistics of Railways in the United States for the Year Ended December 31, 1943. USA-*6614*.S6, 1943.
Statements of the Bureau of Transport Economics and Statistics. USA-*6610*.S5.
Statistics of Railways in the United States. USA-*6614*.S6.
Transport Fact Book. USA-*6610*.T5.

6611 ——. ——. **Air.**
Aéreo.

International.
International Civil Aviation Conference, Chicago, Illinois, November 1 to December 7, 1944: Final Act and Related Documents. Int-*6611*.I5, 1944.

Canada.
Civil Aviation. Can-*6611*.C3.
Civil Aviation. Can-*6611*.C5.

México.
"Growth of Industry and Aviation in Mexico, 1929-1942." Int-*6000*.C7, July-Aug. '44, Ysita.

Perú.
Boletín de la Dirección de Aeronáutica. Perú-*6611*.B5.

United States.
Annual Airline Statistics: Domestic Carriers. USA-*6611*.A5.
Statistical Handbook of Civil Aviation. USA-*6611*.S5, 1944.

6612 ——. ——. **Highway.**
Carreteras.

Canada.
The Highway and the Motor Vehicle in Canada. Can-*6612*.H5.
Motor Carriers. Can-*6612*.M5.

Guatemala.
Vías de Comunicación . . . Guat-*6600*.V5, 1941.

6613 ——. ——. **Pipe line.**
Tuberías.

United States.
A Review of Statistics of Oil Pipe Lines, 1921-1941. USA-*6613*.R5, 1942.

6614 ——. ——. **Rail.**
Vías férreas.

International.
Boletín de la Asociación Permanente. Int-*6614*.B5.
International Comparisons of Railway Ton-Mile Revenues. Int-*6614*.I5, 1939.

Argentina.
Memoria de los Ferrocarriles en Explotación. Arg-*6614*.M5.

Brasil.
Estatísticas das Estradas de Ferro do Brasil. Bras-*6614*.E5.

6000-6999 ECONOMICS—cont.

6614 ——. ——. **Rail—cont.**

Canada.
Electric Railways of Canada. Can-*6614*.E5.
Statistics of Steam Railways of Canada. Can-*6614*.S5.
Summary of Monthly Railway Traffic Reports. Can-*6614*.S8.

Colombia.
Revista del Consejo Administrativo de los Ferrocarriles Nacionales. Col-*6614*.R5.

United States.
Statistics of Railways in the United States. USA-*6614*.S6.

Non-Geographic.
Railways Statistical Terms. NonG-*6614*.R5, 1941.

6615 ——. ——. **Water.**
Agua.

International.
Acta Final de la Conferencia Marítima Interamericana, Unión Panamericana, Wáshington, del 25 de Noviembre al 2 de Diciembre de 1940. Int-*6615*.I5, 1940, Sp.
Final Act of the Inter-American Maritime Conference, Washington, November 25-December 2, 1940. Int-*6615*.I5, 1940, Eng.

Argentina.
Anuario del Movimiento de los Puertos de la República Argentina. Arg-*6615*.A5.

Brasil.
Movimento Marítimo e Fluvial do Brasil. Bras-*6615*.M6.

Canada.
Canal Statistics. Can-*6615*.C5.

Guatemala.
Vías de Comunicación . . . Guat-*6600*.V5, 1941.

United States.
Annual Report of the Chief of Engineers, U. S. Army, Part I. USA-*6615*.A5.I.
Annual Report of the Chief of Engineers; Part 2, Commercial Statistics, Water-Borne Commerce of the United States. USA-*6615*.A5, II.
Foreign Commerce and Navigation of the United States. USA-*6900*.F4.

Venezuela.
Estadística Mercantil y Marítima. Ven-*6900*.E5.

6620-6629 ——. **Communication.**
Comunicaciones.

6620 ——. ——. **General.**

Argentina.
Memoria de la Dirección General de Correos y Telecomunicaciones. Arg-*6620*.M5.

Bolivia.
Minería, Transportes y Comunicaciones. Bol-*6000*.M5.

Chile.
Comercio Interior y Comunicaciones. Chile-*6700*.C5.

Guatemala.
Gaceta de Comunicaciones. Guat-*6620*.G5.

United States.
Statistics of the Communications Industry in the United States. USA-*6620*.S5.

Venezuela.
Indicador Nacional de Vías de Comunicación de Venezuela. Ven-*6620*.I5, 1944.
Memoria del Ministerio del Trabajo y de Comunicaciones. Ven-*9000*.M5.

6000-6999 ECONOMICS—cont.

6622 ——. ——. **Telephone and telegraph.**
Teléfono y telégrafo.

International.
"Directives pour l'Elaboration des Statistiques de la Radiodiffusion dans les Divers Pays." Int-*3040*.R5, 1938.2, Idenburg.
"Statistique de la Radiodiffusion." Int-*3040*.R5, 1936.2, Idenburg.

Canada.
Telegraph Statistics. Statistiques des Télégraphes. Can-*6622*.T5.
Telephone Statistics. Statistiques du Téléphone. Can-*6622*.T6.

6700-6799 **Commerce.**
Comercio.

6700 ——. **General.**

International.
Boletín Informativo del Consejo Interamericano de Comercio y Producción. Int-*6700*.B5.
Bulletin de Statistique Commerciale. Int-*6700*.B7.
Circulares Documentales de la Comisión Ejecutiva del Consejo Interamericano de Comercio y Producción. Int-*6700*.C4.
Consejo Interamericano de Comercio y Producción: Recopilación de Sus Acuerdos y Recomendaciones. Int-*6700*.C6, 1945.
Policies Adopted by the Second Plenary Meeting of the Permanent Council of American Associations of Commerce and Production. Int-*6700*.I6, 1944.2.
Resolutions of the First Plenary Meeting of the Permanent Council of American Associations of Commerce and Production. Int-*6700*.I6, 1942.1.

Argentina.
Boletín de la Cámara de Comercio Argentino-Paraguaya. Arg-*6700*.B2.
Boletín Oficial de la Bolsa de Comercio de Buenos Aires. Arg-*6700*.B4.
Boletín Oficial de la Bolsa de Comercio de Rosario. Arg-*6700*.B5.
Memoria de la Bolsa de Comercio de Buenos Aires. Arg-*6700*.M5.
Monitor de Sociedades Anónimas. Arg-*6700*.M7.
Tercer Censo Nacional, Levantado el 1 de Junio de 1914. Arg-*2200*.C5, 1914.

Brasil.
Situação e Movimento dos Estabelecimentos Comerciais e Industriais em Abril de 1944. Bras-*6000*.S7, Apr. '44.
Situação e Movimento dos Estabelecimentos Comerciais e Industriais em Março de 1944. Bras-*6000*.S7, Mar. '44.

Canada.
Seventh Census of Canada, 1931. Septième Recensement du Canada, 1931. Can-*2200*.C5, 1931.2.

Chile.
Censo Industrial y Commercial, Año 1937. Chile-*2250*.C5, 1937.
Comercio Interior y Comunicaciones. Chile-*6700*.C5.

Costa Rica.
Censo Comercial, 1915. CR-*2250*.C5, 1915.
Resúmenes Estadísticos, Año 1883 a 1910 CR-*1400*.R5, 1883-1910.

Cuba.
Directorio Oficial de Exportación, Importación, Producción y Turismo, 1941. Cuba-*6700*.D5, 1941.

Guatemala.
Statistics in Guatemala (A Brief Survey). Guat-*3030*.S5, 1945.

México.
"El Censo Comercial en México." Int-*1300*.A5, 1940.3, Villafuerte Flores.
Los Censos Cuarto Industrial, Segundo Comercial y Segundo de Transportes, 1945: Métodos y Procedimientos. Méx-*2150*.C5, 1945.1.

6000-6999 ECONOMICS—cont.

6700 ——. **General—cont.**

Primer Censo Comercial, Marzo de 1940: Métodos y Procedimientos. Méx-*2150*.C3, 1940.1.

Nicaragua.

Primer Censo Comercial de Nicaragua, 1940: Instructivo. Nic-*2150*.C3, 1940.

Panamá.

"Análisis de los Resultados del Primer Censo Oficial de Empresas Industriales y Comerciales Radicadas en Panamá." Pan-*6000*.R5, Apr. '39.

United States.

Domestic Commerce Series of the Bureau of Foreign and Domestic Commerce. USA-*6700*.D6.

Survey of Current Business. USA-*6000*.S7.

Uruguay.

Estadística de la Cámara Nacional de Comercio. Ur-*6700*.E5.

Venezuela.

Censo Industrial, Comercial y Empresas que Prestan Servicios, 1936. Ven-*2250*.C5, 1936.

6710 ——. **Wholesale and retail trade.**
Comercio mayorista y minorista.

International.

Studies and Reports, Series H, Cooperation. Int-*6710*.S4.

Brasil.

"Pela Livre Circulação de Mercadorias." Bras-*3040*.R5, July-Sept. '45, Carvalho.

Canada.

"The Canadian Census of Distribution." Int-*3020*.B5, 1938, Marshall.

Census of Merchandising and Service Establishments. Can-*6710*.C5, 1929-41.

Monthly Index of Retail Sales. Can-*6710*.M5.

Monthly Index of Wholesale Sales. Can-*6710*.M6.

Non-Geographic.

Consumer and Opinion Research. NonG-*3110*.C5, 1943.

"Deuxième Rapport sur les Statistiques de la Distribution." Int-*3020*.B5, 1938, Molinari.

"Marketing and Sampling Uses of Population and Housing Data." USA-*3040*.J5, Mar. '43, Eckler.

6720-6729 ——. **Banking and finance.**
Banca y finanzas.

6720 ——. ——. **General.**

International.

Banking Statistics, Recommendations on Scope and Principles of Classification; Report of the Sub-Committee on Banking Statistics. Int-*3200*.D5, 1946.

International Currency Experience, Lessons of the Inter-War Period. Int-*6720*.I6, 1944.

"International Financial Statistics." USA-*6720*.F4, Mar. '46.

International Monetary Fund and International Bank for Reconstruction and Development. Int-*6720*.I8, 1944.

Memoria, Primera Reunión de Técnicos sobre Problemas de Banca Central del Continente Americano. Int-*6720*.R5, 1946.

Money and Banking, 1942-44. Int-*6720*.M6, 1942-44.

United Nations Monetary and Financial Conference, Bretton Woods, New Hampshire, July 1 to July 22, 1944: Final Act and Related Documents. Int-*6720*.U6, 1944.

6000-6999 ECONOMICS—cont.

6720 ——. ——. General—cont.

Argentina.
Annual Report. Arg-*6720*.A5.
Annual Report of the National Mortgage Bank. Arg-*6720*.A7.
El Banco de la Nación Argentina en Su Cincuentenario, 1891-1941. Arg-*6720*.B5, 1891-1941.
Revista Económica. Arg-*6720*.R6.
Revista del Banco de la Nación Argentina. Arg-*6720*.R4.
Situation in Argentina. Arg-*6000*.S4.

Bolivia.
Boletín del Banco Central de Bolivia. Bol-*6720*.B5.
Carta Informativa. Bol-*6300*.C5.
Estadística Bancaria. Bol-*6720*.E4.
Finanzas. Bol-*6720*.E6.
Memoria Anual. Bol-*6300*.M5.
Memoria Anual. Bol-*6720*.M5.

Brasil.
Boletim Estatístico. Statistical Bulletin. Bras-*6720*.B5.
Movimento Bancário do Brasil. Bras-*6720*.M4.
Movimento Bancário do Brasil. Bras-*6720*.M5.
Relatório do Banco do Brasil. Bras-*6720*.R4.
Resumen de la Memoria del Banco do Brasil. Bras-*6720*.R6.

Canada.
Annual Report of Cheques Cashed Against Individual Accounts and Equation of Exchange. Can-*6720*.A5.
Commercial Failures in Canada. Can-*6720*.C5.

Chile.
Boletín Mensual del Banco Central de Chile. Chile-*6720*.B5.
Estadística Bancaria. Chile-*6720*.E5.
Finanzas, Bancos y Cajas Sociales. Chile-*6720*.F5.
Memoria Anual Presentada a la Superintendencia de Bancos. Chile-*6720*.M5.
Memoria de la Caja de Crédito Agrario. Chile-*6200*.M5.

Colombia.
Annual Report Presented by the Manager to the Board of Directors. Col-*6720*.A5.
Boletín de Información Mensual. Col-*6720*.B5.
Boletín de la Superintendencia Bancaria. Col-*6720*.B7.
Cuadros Estadísticos, 1928-1939. Col-*6720*.C8, 1928-39.
Informe Financiero del Contralor General. Col-*8800*.I5.
Informe y Datos Estadísticos Presentados por el Superintendente Bancario al Señor Ministro de Hacienda y Crédito Público. Col-*6720*.I7.
"Reglamentación sobre el Cálculo de los Medios de Pagos en Circulación y de las Exigibilidades Monetarias en Colombia." Col-*1420*.A5, Nov. '45, Abrisqueta.
Revista del Banco de la República. Col-*6720*.R5.

Costa Rica.
Revista del Banco Nacional de Costa Rica. CR-*6720*.R5.
Boletín Estadístico. CR-*6720*.B6.
"Memoria Anual de 1944, del Banco Nacional de Seguros." CR-*6720*.R6, Dec. '45.
Revista del Banco Nacional de Seguros. CR-*6720*.R6.

Ecuador.
"Algunas Instituciones Económicas Ecuatorianas: El Banco Central del Ecuador y Su Oficina de Investigaciones Económicas." Col-*1420*.A5, Mar.-Apr. '46, Abrisqueta 1.
Banco Central del Ecuador, Boletín. Ec-*6720*.B5.
Boletín de la Superintendencia de Bancos del Ecuador. Ec-*6720*.B3.
Boletín del Banco Hipotecario del Ecuador. Ec-*6720*.B4.

6000-6999 ECONOMICS—cont.

6720 ——. ——. **General—cont.**

Estadísticas y Otras Informaciones para el Fondo Monetario Internacional. Ec-*6720*.E5, 1946.
Informe del Superintendente de Bancos al Señor Ministro de Economía. Ec-*6720*.I5.
"Memoria Anual que el Consejo de Administración del Banco Central del Ecuador Presenta a la Asamblea Ordinaria de Accionistas por el Ejercicio Económico de 1945." Ec-*6720*.B5, Feb.-Mar. '46.

El Salvador.
Memoria del Banco Central de Reserva de El Salvador. ElSal-*6720*.M5.
Revista Mensual del Banco Central de Reserva de El Salvador. ElSal-*6720*.R5.

Haiti.
Annual Report of the Fiscal Department. Haiti-*6000*.A3.
Annual Report of the Fiscal Department for the Fiscal Year October 1943-September 1944. Haiti-*6000*.A3, 1943-44.
Aspects de l'Economie et des Finances d'Haïti. Haiti-*6000*.A5, 1944.
Bulletin Mensuel du Département Fiscal. Haiti-*1420*.B5.
Monthly Bulletin of the Fiscal Department. Haiti-*1420*.M5.
Rapport Annuel du Département Fiscal. Haiti-*6000*.R5.

México.
Banco de México, S.A.: Asamblea General Ordinaria de Accionistas. Méx-*6720*.B5.
Banco de México, S.A.: Vigésimacuarta Asamblea General Ordinaria de Accionistas. Méx-*6720*.B5, 1946.
"Factores que Influyen en la Balanza de Pagos de México." Méx-*6000*.T5, Oct-Dec. '45, Luna Olmedo.
Revista del Banco de la República. Méx-*6720*.R5.

Nicaragua.
Deuda Pública—Rentas—Bancos—Comercio Exterior, 1943. Nic-*8800*.D7, 1943.
Fondo de Nivelación de Cambios, 1941-1944. Nic-*6720*.F5, 1941-44.
Informe y Balances del Banco Hipotecario de Nicaragua. Nic-*6720*.I5.
Revista del Banco Nacional de Nicaragua. Nic-*6720*.R5.
Situación Monetaria, Año 1945. Nic-*6720*.S5, 1945.

Paraguay.
Memoria. Par-*6720*.M5.

Perú.
Boletín Mensual del Banco Central de Reserva del Perú. Perú-*6720*.B5.
Memoria de la Superintendencia de Bancos y Estadística Bancaria de Seguros y Capitalización. Perú-*6720*.M4.
Memoria del Banco Agrícola del Perú. Perú-*6720*.M6.
Memoria del Banco Central de Reserva del Perú. Perú-*6720*.M7.
Memoria del Banco Industrial del Perú. Perú-*6720*.M8.
Situación de las Empresas Bancarias del Perú. Perú-*6720*.S5.

República Dominicana.
Estadística Bancaria. RepDom-*6720*.E5.

United States.
Agricultural Finance Review. USA-*6200*.A4.
Annual Report of the Board of Governors of the Federal Reserve System. USA-*6720*.A3.
Annual Report of the Comptroller of the Currency. USA-*6720*.A5.
Annual Report of the Farm Credit Administration. USA-*6720*.A7.
Banking and Monetary Statistics. USA-*6720*.B5, 1943.
Chart Book I, Federal Reserve Charts on Bank Credit Money Rates, and Business. USA-*6720*.C5, 1945.
Eleventh Annual Report of the Farm Credit Administration, 1943-44. USA-*6200*.A9, 1943-44.

6720 ——. ——. General—cont.

Farm Credit Quarterly. USA-*6720*.F2.
Federal Home Loan Bank Review. USA-*9900*.F5.
Federal Reserve Bulletin. USA-*6720*.F4.
Report of the Reconstruction Finance Corporation. USA-*6720*.R5.

Uruguay.
Memoria del Directorio del Banco de la República Oriental del Uruguay. Ur-*6720*.M5.
Revista del Banco de la República Oriental del Uruguay. Ur-*6720*.R5.
Suplemento Estadístico de la Revista Económica. Ur-*6000*.S5.

Venezuela.
Banco Central de Venezuela: Memoria. Ven-*6720*.B3.
Boletín del Banco Central de Venezuela. Ven-*6720*.B5.
Revista de Hacienda. Ven-*8800*.R5.

6721 ——. ——. Securities, investments, holdings.
Valores, inversiones, títulos.

International.
Inversiones Internacionales en América Latina. Int-*6721*.I5, 1945.
Statistics Relating to Capital Formation. Int-*3200*.S3, 1938.
Statistiques Relatives à la Formation des Capitaux. Int-*3200*.S7, 1938.

Canada.
Prices and Price Indexes. Can-*6010*.P6.

República Dominicana.
Estadística de los Negocios de Seguros. RepDom-*6730*.E5.
Registro Público. RepDom-*6721*.R5.

United States.
Annual Report of the Securities and Exchange Commission. USA-*6721*.A5.
Census of Foreign-Owned Assets in the United States. USA-*6721*.C3, 1941.
Corporate Cash Balances, 1914-1943. USA-*6721*.C5, 1914-43.
"A National Survey of Liquid Assets." USA-*6720*.F4, June '46, b.
Statistical Bulletin. USA-*6721*.S5.
Survey of American Listed Corporations: Data on Profits and Operations. USA-*6720*.S5.

6730 ——. Insurance, commercial.
Seguro comercial.

República Dominicana.
Estadística de los Negocios de Seguros. RepDom-*6730*.E5.

United States.
Annual Report of the Federal Deposit Insurance Corporation. USA-*6730*.A5.
The Record of the American Institute of Actuaries. USA-*6730*.R5.
Statistical Bulletin of the Metropolitan Life Insurance Company. USA-*6730*.S5.

6900-6999 Foreign trade.
Comercio exterior.

6900 ——. General.

International.
Addendum to Annex III of "The Network of World Trade." Int-*6900*.N5, 1942.A.
Agricultural Production and Trade by Countries. Int-*6200*.A6, 1945.
Algunos Documentos Relativos a la Adopción de un Sistema de Estadística Internacional, Convención de Bruselas, 1913. Int-*3210*.C4, 1913.2.
Annuaire Statistique du Commerce des Armes et des Munitions. Statistical Year-Book of the Trade in Arms and Ammunition. Int-*6900*.A5.
Anteproyecto de Carta para la Organización Internacional de Comercio de las Naciones Unidas. Int-*6900*.A6, 1946.

6000-6999 ECONOMICS—cont.

6000-6999 ECONOMICS—cont.

6900 ——. General—cont.

National Income and Related Subjects for the United Nations. Int-*6100*.N3, 1946.
The Network of World Trade. Int-*6900*.N5, 1942.
Proceedings of the International Conference Relating to Economic Statistics. Int-*6000*.I75, 1928 C.
Rapport sur le . . . Exercice. Int-*6900*.R4.
Recent Developments in the Foreign Trade of the American Republics. Int-*6900*.R5.
Report. Int-*3000*.R6, 1931, Eng.
Report to the Council on the Work of the Eighth Session. Int-*3000*.R6, 1939, Eng.
Report to the Council on the Work of the Fourth Session. Int-*3000*.R6, 1935, Eng.
Report to the Council on the Work of the Second Session. Int-*3000*.R6, 1933, Eng.
Report to the Council on the Work of the Third Session. Int-*3000*.R6, 1934, Eng.
Reports of the United States Tariff Commission. Int-*6900*.R6.
Review of World Trade. Int-*6900*.R8.
Some Technical Aspects of Foreign Trade Statistics with Special Reference to Valuation. Int-*6900*.S2, 1946.
Statistiques du Commerce International. International Trade Statistics. Int-*6900*.S6.
"Suggested Charter for an International Trade Organization of the United Nations." Int-*8200*.I55, Nov. '46.
Supplement-A to Bibliography of the Latest Available Source Material on Foreign Trade Statistics. Int-*6900*.B5, 1944.2.
Unification of Customs Nomenclature. Int-*3210*.U5, 1930.
World Prosperity as Sought through the Economic Work of the League of Nations. Int-*6000*.W8, 1933.
World Trade in Agricultural Products. Int-*6900*.W4, 1943.
World Trade in Agricultural Products. Int-*6900*.W5, 1940.

Argentina.
Anuario del Comercio Exterior de la República Argentina. Arg-*6900*.A5.
Boletín de la Dirección General de Aduanas. Arg-*6900*.B5.
El Comercio Exterior Argentino. Arg-*6900*.C5.
El Comercio Exterior Argentino. Arg-*6900*.C6.
El Comercio Exterior Argentino en los Primeros Semestres de 1945 y 1944. Arg-*6900*.C5, Jan.-June '45.
El Comercio Exterior Argentino en . . . y Su Comparación con el Mismo Período del Año Anterior: Boletín Mensual. Arg-*6900*.C8.
Economic Controls and Commercial Policy in Argentina. Arg-*6000*.E6, 1945.
Estadísticas del Comercio Exterior en Argentina: Métodos y Procedimientos. Arg-*6900*.
Las Exportaciones en los . . . Comparadas con las del Mismo Período del Año Anterior. Arg-*6900*.E5.
"Precios de Exportación en la República Argentina, 1863-1943." Int-*3040*.E5, Sept. '45, Domínguez.
Tarifa de Avalúos y Arancel de Importación. Arg-*6900*.T5, 1939.

Bolivia.
Arancel Aduanero de Importaciones y Vocabulario para Su Aplicación. Bol-*6900*.A5, 1940.
Comercio Exterior. Bol-*6900*.C5.
Economic Controls and Commercial Policy in Bolivia. Bol-*6000*.E5, 1945.
Memoria Anual. Bol-*6720*.M5.

Brasil.
Boletim Estatístico da Diretoria das Rendas Aduaneiras. Bras-*6900*.B6.
Brazilian Customs Tariff. Tarifa das Alfândegas do Brasil. Bras-*6900*.B6, 1944.
Comércio Exterior do Brasil. Bras-*6900*.C3.
Comércio Exterior do Brasil, Intercambio por Países. Bras-*6900*.C4.
Comércio Exterior do Brasil: Resumo Mensal. Bras-*6900*.C5.

6000-6999 ECONOMICS—cont.

6000-6999 ECONOMICS—cont.

6900 ——. **General—cont.**

Economic Controls and Commercial Policy in Ecuador. Ec-*6000*.E5, 1945.
Ecuador. Ec-*6900*.E3, 1942.
Estadísticas de las Exportaciones del Ecuador, 1940-1942. Ec-*6900*.E5, 1940-42.
Estadísticas del Comercio Exterior en Ecuador: Métodos y Procedimientos. Ec-*6900*.
Ley Arancelaria de Aduanas, 1943. Ec-*6900*.L5, 1943.

El Salvador.
Estadísticas del Comercio Exterior en El Salvador: Métodos y Procedimientos. ElSal-*6900*.
Intercambio Comercial de El Salvador en el Año 1943. ElSal-*6900*.I5, 1943.
Tarifa de Aforos. ElSal-*6900*.T5, 1941.

Guatemala.
Cuadros de la Dirección General de Aduanas. Guat-*6900*.C5.
Decreto Gubernativo Número 1765, Diciembre 7, 1935: Arancel de Aduanas de la República de Guatemala. Guat-*6900*.E5, 1935.
Estadísticas del Comercio Exterior en Guatemala: Métodos y Procedimientos. Guat-*6900*.
Instrucciones a los Importadores de Artículos y Materiales en Tránsito o Procedentes de los Estados Unidos de Norte América o del Canadá. Guat-*3210*.I5, 1943.
Statistics in Guatemala (A Brief Survey). Guat-*3030*.S5, 1945.

Haiti.
Economic Controls and Commercial Policy in Haiti. Haiti-*6000*.E5, 1946.
Handbook and Trade Directory of Haiti. Haiti-*6900*.H5, 1935.
Quelques Données Statistiques sur le Commerce d'Haïti. Haiti-*6900*.Q5, 1942.
Statistique du Commerce Exterieur au Haïti: Méthodes et Procédés. Haiti-*6900*.

Honduras.
Arancel de Aduanas, 1934. Hond-*6900*.A5, 1934.
Informe de Hacienda, Crédito Público y Comercio. Hond-*8800*.I5.

México.
Anuario Estadístico del Comercio Exterior de los Estados Unidos Mexicanos. Méx-*6900*.A5.
Banco de México, S.A.: Vigésimacuarta Asamblea General Ordinaria de Accionistas. Méx-*6720*.B5, 1946.
Boletín de Aduanas. Méx-*6900*.B5.
Economic Controls and Commercial Policy in Mexico. Méx-*6000*.E4, 1946.
Estadísticas del Comercio Exterior en México: Métodos y Procedimientos. Méx-*6900*.
"Factores que Influyen en la Balanza de Pagos de México." Méx-*6000*.T5, Oct.-Dec. '45, Luna Olmedo.
Reports Presented to the Conference of Commissions of Inter-American Development (Held in New York, May 9-18, 1944) by the Mexican Commission of Inter-American Development. Méx-*6000*.R4, 1944.
Revista del Comercio Exterior. Méx-*6900*.R5.
Tarifa del Impuesto de Exportación. Méx-*6900*.T6.
Tarifas de los Derechos de Importación y Exportación de los Estados Unidos Mexicanos. Méx-*6900*.T5, 1937.

Nicaragua.
Deuda Pública—Rentas—Bancos—Comercio Exterior, 1943. Nic-*8800*.D7, 1943.
Ley Arancelaria de 1918 de la República de Nicaragua y Sus Reformas hasta el 30 de Junio de 1942. Nic-*6900*.L5, 1942.
Memoria del Recaudador General de Aduanas y Alta Comisión. Nic-*6900*.M5.

Panamá.
Arancel de Importación de la República de Panamá. Pan-*6900*.A5, 1938.
Estadísticas del Comercio Exterior en Panamá: Métodos y Procedimientos. Pan-*6900*.

6000-6999 ECONOMICS—cont.

6000-6999 ECONOMICS—cont.

6900 ——. General—cont.

Monthly Summary of Foreign Commerce of the United States. USA-*6900*.M5.
Schedule A, Statistical Classification of Imports into the United States. USA-*3210*.S3, 1943.
Schedule A, Statistical Classification of Imports into the United States, Arranged by Tariff Schedules and Tariff Paragraphs of Act of 1930. USA-*3210*.S2, 1940.
Schedule B, Statistical Classification of Domestic and Foreign Commodities Exported from the United States. USA-*3210*.S4, 1945.
Standard Commodity Classification: Volume I: Standard Classified List of Commodities. USA-*3210*.S6, 1943.
Standard Commodity Classification: Vol. II: Alphabetic Index. USA-*3210*.S6, 1944.
Summary of Foreign Trade of the United States. USA-*6900*.S5.
"United States Custom Tariff, Act of 1930." USA-*3210*.C5, 1943, a.

Uruguay.
Boletín Aduanero. Ur-*6900*.B5.
Economic Controls and Commercial Policy in Uruguay Ur-*6000*.E5, 1945.
Estadísticas del Comercio Exterior en Uruguay: Métodos y Procedimientos. Ur-*6900*.
Memoria del Contralor de Exportaciones e Importaciones. Ur-*6900*.M5.
Situación y Perspectivas para la Organización del Intercambio Comercial Interamericano. Ur-*6900*.S5, 1942.
Tarifa de Aforos de Importación. Ur-*6900*.T5, 1940-42.

Venezuela.
Análisis del Comercio Exterior. Ven-*6900*.A5, 1913-36.
Economic Controls and Commercial Policy in Venezuela. Ven-*6000*.E3, 1945.
Estadística Mercantil y Marítima. Ven-*6900*.E5.
Estadísticas del Comercio Exterior en Venezuela: Métodos y Procedimientos. Ven-*6900*.
Ley de Arancel de Aduanas. Ven-*6900*.L5, 1941.
Recent Developments in the Foreign Trade of Venezuela. Ven-*6900*.R5, 1945.
Venezuela. Ven-*6000*.V5, 1946.

Non-Geographic.
"Estimaciones Estadísticas de los Pagos Internacionales." Int-*1300*.A5, 1940.3, Ruíz Carranza.
International Trade and the National Income Multiplier. NonG-*6100*.I5, 1943.
Some Technical Aspects of Foreign Trade Statistics with Special Reference to Valuation. Int-*6900*.S2, 1946.
"Standard Commodity Classification." USA-*3040*.J5, Mar. '44, Kolesnikoff.

6910 ——. International balance of payment.
Balanza de pagos internacionales.

International.
Annex I, League of Nations Estimated Balance of International Payments for the Year . . . (8-page form for country to fill); and Annex II, Explanatory Notes to Scheme of International Balance of Payments (11 pages). Int-*3205*.B5, 1945, Annexes.
Balance of Payments. Int-*6910*.B5.
Balance of Payments, 1938. Int-*6910*.B5, 1938.
Balance of Payments Statistics: Note by the Secretariat. Int-*3205*.B5, 1945.
Memorandum on International Trade and Balances of Payments. Int-*6900*.M7.
"Modelos para Registrar las Cuentas Internacionales." Ec-*1420*.T5, July-Dec. '45, c.
National Income and Related Subjects for the United Nations. Int-*6100*.N3, 1946.
Note on Balance of Payments Statistics: Annex A. Int-*3205*.B5, 1945, Annex A.

Argentina.
Annual Report. Arg-*6720*.A5.

6000-6999 ECONOMICS—cont.

6910 **——. International balance of payment—cont.**

Bolivia.
Balanza Internacional de Pagos de Bolivia. Bol-*6910*.B5, 1938-40.
Memoria Anual. Bol-*6720*.M5.

Canada.
The Canadian Balance of International Payments. Can-*6910*.C5.

Chile.
Balanza de Pagos de Chile. Chile-*6910*.B5.

Guatemala.
"Balanza Comercial, 1935 a 1945." Guat-*1420*.B5, May '46, b.

México.
"Mexico's Balance of Payments." Int-*6000*.C7, July-Aug. '44.

Perú.
"La Balanza Internacional de Pagos del Perú, Años 1939-1943." Perú-*6720*.B5, Feb. '45, Barreto.
"Naturaleza y Alcances de la Llamada Balanza Internacional de Pagos." Perú-*3040*.E5, July '46, Barreto.
"Naturaleza y Alcances de la Llamada Balanza Internacional de Pagos." Perú-*6720*.B5, July '45, Barreto.
"Resumen de la Balanza de Operaciones Internacionales del Perú." Perú-*6720*.B5, July '45.

United States.
Balance of International Payments of the United States. USA-*6910*.B5.
The United States in the World Economy. USA-*6910*.U5, 1943.

——. Puerto Rico.
The Balance of External Payments of Puerto Rico, 1941-42 and 1942-43. USA PR-*6910*.B5, 1941-43.

Uruguay.
"Balance de Pagos del Uruguay: Estimación del Balance de Pagos del Uruguay Correspondiente al Año 1944." Ur-*6720*.R5, July '45.
"Balances de Pagos del Uruguay, Años 1940-1944." Ur-*6720*.R5, Jan. '46.

Non-Geographic.
"La Balanza de Pagos y el Ingreso Nacional." Méx-*6000*.T5, Oct.-Dec. '46, Ortiz Mena.

7000-7999 HEALTH AND CULTURAL STATISTICS ESTADISTICAS DE SALUBRIDAD Y CULTURA

7000 **General.**

International.
Proceedings of the Eighth American Scientific Congress. Int-*1300*.A5, 1940.3.

Brasil.
Brazilian Culture. Bras-*2200*.R4, 1940, Eng.
Educação e Saúde. Bras-*7000*.E5, 1942.
Recenseamento Geral do Brasil (1.° de Setembro de 1940). Bras-*2200*.R4, 1940.

Colombia.
"Diez Años de Educación y Cultura en Colombia." Col-*1420*.A5, Feb. 20, '42, Bermúdez V.

El Salvador.
Informe de las Labores Desarrolladas por el Poder Ejecutivo en los Ramos de Cultura y Asistencia Social. ElSal-*7000*.I5.
Memoria de Cultura Popular. ElSal-*7000*.M5.

7000-7999 HEALTH AND CULTURAL STATISTICS—cont.

7100-7199 Health.
Salubridad.

7100-7109 ——. General.

7100 ——. ——. General.

International.
Bulletin of the Health Organisation of the League of Nations. Int-*7100*.B6.
Classified List of Medical and Public Health Journals of Latin America. Int-*7100*.C5, 1943.
"Inhabitants per Physician." Int-*7110*.I6, June '44.
International Health Year-Book. Int-*7100*.I6.
Revistas de Medicina y Sanidad de la América. Int-*7100*.R5, 1940.

Brasil.
Arquivos de Higiene. Bras-*7100*.A5.

Chile.
"Indices Sanitarios para las Principales Ciudades de Chile." Int-*7110*.B5, May '45.
La Realidad Médico-Social Chilena. Chile-*7100*.R5, 1939.
Revista Chilena de Higiene y Medicina Preventiva. Chile-*7100*.R7.
Sinopsis Estadística de los Servicios de Beneficencia y Asistencia Social de Chile. Chile-*7110*.S5.

República Dominicana.
Commentary Explanations and Sample Copies of the Forms Used in Medical and Public Health Statistics Activities (Not Including Vital Statistics) in the Dominican Republic, November 1946. RepDom-*7100*.C5, 1946.

Venezuela.
Anuario de Epidemiología y Estadística Vital. Ven-*5200*.A5.

United States.
Milbank Memorial Fund Quarterly. USA-*5000*.M5.
Selected Bibliography on Medical Economics. USA-*7100*.S5, 1944.
Statistical Bulletin of the Metropolitan Life Insurance Company. USA-*6730*.S5.
"Vital and Health Statistics in the Federal Government." USA-*7110*.A4, Mar. '44, Collins.

Non-Geographic.
Utilização de Medidas Estatísticas em Biologia, Medicina e Saúde Pública. NonG-*3330*.Jansen de Mello, 1944.

7101 ——. ——. Organizations.
Organizaciones.

International.
International Health Organization. Int-*7101*.I5, 1945.

Brasil.
O Departamento Nacional de Saúde em 1943. Bras-*7101*.D5, 1943.

Honduras.
Reglamento Interior de la Dirección General de Sanidad. Hond-*7101*.R5, 1925.

República Dominicana.
Commentary Explanations and Sample Copies of the Forms Used in Medical and Public Health Statistics Activities (Not Including Vital Statistics) in the Dominican Republic, November 1946. RepDom-*7100*.C5, 1946.

United States.
Annual Report of the Surgeon General of the U. S. Public Health Service. USA-*7110*.A7.
"La Organización de los Servicios de Estadística Vital y Sanitaria en el Gobierno Federal de los Estados Unidos." Int-*7110*.B5, May '45, Collins.

7000-7999 HEALTH AND CULTURAL STATISTICS—cont.

7110 ——. Public health activities.
Actividades de la salud pública.

7000-7999 HEALTH AND CULTURAL STATISTICS—cont.

7110 ——. Public health activities—cont.

Sinopsis Estadística de los Servicios de Beneficencia y Asistencia Social de Chile. Chile-*7110*.S5.

Costa Rica.
Memoria de la Secretaría de Salubridad Pública y Protección Social. CR-*7110*.M5.
Salud. CR-*7110*.S5.

Cuba.
Boletín Semanal . . . Epidemiológico Sanitario. Cuba-*7110*.B5.
"Las Estadísticas Sanitarias en Cuba." Int-*1300*.A5, 1940.3, Martínez-Fortún.
Salubridad y Asistencia Social. Cuba-*7110*.S5.

Ecuador.
Documentos Anexos al Informe que el Ministro de Previsión Social y Trabajo Presenta a la Nación. Ec-*9000*.D5.
Informe a la Nación (del Ministro de Previsión Social y Trabajo). Ec-*9000*.I5.
Informe que el Señor Director General de Sanidad Presenta ante el Señor Ministro del Ramo acerca de las Labores Desarrolladas. Ec-*7110*.I5.

El Salvador.
Memoria de Asistencia Social y Beneficencia. ElSal-*9951*.M5.

Guatemala.
Boletín Sanitario de Guatemala. Guat-*7110*.B5.
A Medical Survey of the Republic of Guatemala. Guat-*7110*.M5, 1938.
Revista de Sanidad. Guat-*7110*.R5.
"La Sanidad en Guatemala." Int-*7110*.B5, Jan. '46, Herrera

Haiti.
Bulletin du Service National d'Hygiène et d'Assistance Publique. Haiti-*7110*.B5.
Rapport Annuel du Directeur Général. Haiti-*7110*.R5.

Honduras.
Boletín Sanitario de la Dirección General de Sanidad. Hond-*7110*.B5.
Informe de los Actos Realizados por el Poder Ejecutivo en los Ramos de Gobernación, Justicia, Sanidad y Beneficencia. Hond-*8000*.I5.

México.
Bibliografía que el Departamento de Salubridad Pública Presenta en la Feria del Libro y Exposición Nacional de Periodismo. Méx-*7110*.B5, 1943.
Boletín de la Oficina de Demografía y Estadística Sanitaria. Méx-*7110*.B6.
Memoria del Departamento de Salubridad y Asistencia Pública. Méx-*7110*.M5.
Revista del Instituto de Salubridad y Enfermedades Tropicales. Méx-*7110*.R5.
Salubridad y Asistencia. Méx-*7110*.B7.
"La Sanidad en México." Int-*7110*.B5, Mar. '46, Avila Camacho.
"El Servicio de Bioestadística en México." Int-*7110*.B5, May '44, Granillo.

Nicaragua.
Boletín Sanitario. Nic-*7110*.B5.
Informe de la Dirección General de Sanidad. Nic-*7110*.I5.
"La Sanidad en Nicaragua." Int-*7110*.B5, July '46, a.

Panamá.
Bio-Estadística—Informes Especiales. Pan-*7110*.B5.
Estadísticas Epidemiológicas. Pan-*7110*.E5.
Higiene y Sanidad. Pan-*7110*.H5.
Memoria del Ministerio de Salubridad y Obras Públicas. Pan-*7110*.M5.

Paraguay.
Boletín del Ministerio de Salud Pública y Previsión Social. Par-*7110*.B5.
Memoria de la Sección Estadística. Par-*7110*.M5.

Perú.
Boletín de la Dirección General de Salubridad. Perú-*7110*.B4.
Boletín del Departamento de Bioestadística. Perú-*7110*.B6.
Memoria que Presenta al Congreso Nacional el Ministro de Salud Pública y Asistencia Social. Perú-*7110*.M5.
"La Sanidad en el Perú." Int-*7110*.B5, Feb. '46, Carvallo.

7000-7999 HEALTH AND CULTURAL STATISTICS—cont.

7130 ——. Hospitals—cont.

United States.

Patients in Mental Institutions. USA-*7130*.P5.

Sixteenth Census of the United States: 1940: Population: Special Report on Institutional Population, 14 Years Old and Over. USA-*2200*.C5, 1940.9.

Non-Geographic.

"Application of Punched Card Methods to Hospital Statistics." USA-*3040*.J5, Mar. '35, Dunn.

7150 ——. Morbidity.
Morbilidad.

International.

"Clave Diagnóstica para la Clasificación en Columna (Tabulación) de las Causas de Morbidad." Int-*7110*.B5, June-July '41.

A Geography of Diseases. Int-*7150*.G5, 1935.

"Glossary of Communicable Diseases." Int-*7100*.B6, 1943-44, Biraud.

"Nomenclatura para Hospitales Dedicados a Enfermedades Mentales." Int-*7110*.B5, Aug. '36.

Nomenclaturas Internacionales de las Enfermedades (Causas de Defunción—Causas de Incapacidad del Trabajo) para Establecer las Estadísticas Nosológicas (Clasificación "Bertillon"). Int-*3230*.N5, 1929.

Proposed Statistical Classification of Diseases, Injuries, and Causes of Death. Int-*3230*.P5, 1946.

Proyecto de Informe sobre Standardización de las Estadísticas de Atención Médica. Int-*3230*.P6, 1946.

Proyecto de Informe sobre Standardización de las Estadísticas de Morbilidad. Int-*3230*.P7, 1946.

Canada.

Annual Report of Tuberculosis Institutions. Rapport Annuel sur les Institutions pour Tuberculeux. Can-*7130*.A6.

Chile.

"Estudios sobre Mortalidad por Tuberculosis, Gripe y Sífilis." Chile-*1420*.E5, Apr. '46.

Nomenclatura de las Enfermedades. Chile-*3230*.N5, 1929.

Colombia.

Estadísticas Demográficas y Nosológica, Año de 1936. Col-*7150*.E5, 1936.

Nomenclatura de Morbilidad y Mortalidad. Col-*3230*.N5, 1942.

México.

Sinonimias Populares Mexicanas de las Enfermedades. Méx-*3230*.S5, 1931.

Perú.

"Las Enfermedades Infecto-Contagiosas, Casos por Departamentos en 1945 y en la República en 1939-1945." Perú-*3040*.E5, July '46, Téllez Sarzola.

United States.

"The Incidence of Illness and the Volume of Medical Services among 9,000 Canvassed Families." USA-*7110*.P5, 1933-43.

Manual for Coding Causes of Illness According to a Diagnosis Code for Tabulating Morbidity Statistics. USA-*3230*.M5, 1944.

Manual of Procedure for the Mechanical System of Reporting Morbidity, Treatment-Progress and Control of the Venereal Diseases. USA-*7150*.M4, 1940.

Morbidity: Instructions for Coding and Punching. USA-*7150*.M6.

Tuberculosis Reference Statistical Handbook. USA-*7150*.T5.

Venereal Disease Information. USA-*7150*.V5.

Non-Geographic.

"Glossary of Communicable Diseases." Int-*7100*.B6, 1943-44, Biraud.

Problems of Aging, Biological and Medical Aspects. NonG-*7150*.P5, 1939.

"A System of Codification of Medical Diagnoses for Application to Punch Cards, with a Plan for Operation." USA-*7110*.A4, June '36, Berkson.

7000-7999 HEALTH AND CULTURAL STATISTICS—cont.

7160 ——. **Accidents, except industrial accidents.**
Accidentes, excepto los accidentes industriales.

International.
Methods of Compiling Statistics of Railway Accidents. Int-*7160*.M5, 1929.
"L'Uniformité dans la Statistique des Accidents de la Circulation." Int-*3040*.R5, 1936.2, Zanten.

Perú.
Boletín del Cuerpo de Investigación y Vigilancia. Perú-*7160*.B5.

República Dominicana.
Accidentes de Tránsito Terrestre por Circulación de Vehículos. RepDom-*7160*.A5.

United States.
Accident Facts. USA-*7160*.A5.
Uniform Definitions of Motor Vehicle Accidents. USA-*7160*.U5, 1943.

Non-Geographic.
"Les Bases de Comparaison du Nombre des Accidents de la Circulation Routière." Int-*3040*.R5, 1938.2, Zanten.

7170 ——. **Defects and deformities.**
Defectos y deformidades.

International.
"Nomenclatura para Hospitales Dedicados a Enfermedades Mentales." Int-*7110*.B5, Aug. '36.

United States.
The Blind and Deaf-Mutes in the United States. USA-*7170*.B5.

7180 ——. **Medical science.**
Ciencia médica.

Chile.
Revista Chilena de Higiene y Medicina Preventiva. Chile-*7100*.R7.

United States. Puerto Rico.
The Puerto Rico Journal of Public Health and Tropical Medicine. USA PR-*7110*.P5.

Non-Geographic.
Introduction to Medical Biometry and Statistics. NonG-*7180*.I5, 1940.
Principles of Medical Statistics. NonG-*7180*.P5, 1939.

7190 ——. **Group medicine and cost of medical care.**
Medicina social y costo de la atención médica.

International.
Proyecto de Informe sobre Standardización de las Estadísticas de Atención Médica. Int-*3230*.P6, 1946.

United States.
Medical Care and Costs in Relation to Family Income. USA-*7190*.M5, 1943.

7200-7299 **Education.**
Educación.

7200 ——. **General.**

International.
Acta Final de la Primera Conferencia de Ministros y Directores de Educación de las Repúblicas Americanas Celebrada en Panamá del 27 de Septiembre al 4 de Octubre de 1943. Int-*7200*.C6, 1943.1, Sp.
Analfabetismo y Cultura Popular en América. Int-*7200*.A5, 1941.
Annual Report of the Director of the Institute of International Education. Int-*7200*.A7.
"Bases para el Mejoramiento de las Estadísticas Escolares en los Países de América." Int-*1300*.A5, 1940.3, Alanís Patiño, a.

7000-7999 HEALTH AND CULTURAL STATISTICS—cont.

7200 ——. General—cont.

"Desarrollo de Normas Comunes en la Estadística Educacional Americana." Int-*1300*.A5, 1940.3, Sálas S.

Educational Trends in Latin America. Int-*7200*.E6, 1942.

Final Act of the First Conference of Ministers and Directors of Education of the American Republics, Held in Panama, September 27 to October 4, 1943. Int-*7200*.C6, 1943.1, Eng.

"A População Escolar e a Taxa de Analfabetos nas Estatísticas Educacionais Americanas." Int-*1300*.A5, 1940.3, Bergström Lourenço.

Argentina.

El Analfabetismo en la Argentina. Arg-*2230*.C5, 1943.1.

"El Analfabetismo en la Argentina." Arg-*3040*.R5, June '45, Correa Avila.

El Analfabetismo en la República Argentina. Arg-*7200*.A4, 1939.

El Analfabetismo y las Funciones del Consejo Nacional de Educación. Arg-*7200*.A6, 1938.

Censo Escolar de la Nación, Ley No. 12.723: Cuestionario de Convivencia. Arg-*2130*.C5, 1943.8.

Censo Escolar de la Nación, Ley No. 12.723: Cuestionario Familiar. Arg-*2130*.C5, 1943.6.

Censo Escolar de la Nación, Ley No. 12.723: Cuestionario Individual. Arg-*2130*.C5, 1943.7.

Censo Escolar de la Nación, Ley No. 12.723: Instrucciones para los Censistas. Arg-*2130*.C5, 1943.5.

"El Censo Escolar Nacional de 1943." Int-*3040*.E5, Sept. '46, Coghlan.

Deserción Escolar y Analfabetismo. Arg-*7200*.D5, 1940.

La Distribución por Zonas de la Población Argentina y Su Relación con los Hechos Culturales, Económicos y Sociales. Arg-*2230*.C5, 1943.2.

Instrucciones Complementarias a las "Normas para la Organización y el Levantamiento del Censo Escolar de la Nación." Arg-*2130*.C5, 1943.1.

Memoria del Ministerio de Justicia e Instrucción Pública. Arg-*7200*.M5.

Normas para la Organización y Levantamiento del Censo Escolar de la Nación en la Capital Federal. Arg-*2130*.C5, 1943.2.

Normas para la Organización y Levantamiento del Censo Escolar de la Nación en las Provincias. Arg-*2130*.C5, 1943.3.

Normas para la Organización y Levantamiento del Censo Escolar de la Nación en los Territorios. Arg-*2130*.C5, 1943.4.

Recopilación Estadística. Arg-*7200*.R5.

Tercer Censo Nacional, Levantado el 1 de Junio de 1914. Arg-*2200*.C5, 1914.

Bolivia.

El Estado Actual de la Educación en Bolivia. Bol-*7200*.E5, 1943.

Brasil.

Análises de Resultados do Censo Demográfico. Bras-*2200*.A5, 1940.

Boletim do Instituto Nacional de Estudos Pedagógicos. Bras-*7200*.B5.

"A Coleta da Estatística Educacional." Bras-*7200*.R5, Nov. '44, Jardim.

"A Coleta da Estatística Educacional." Bras-*7200*.R5, Sept. '44, Jardim.

"A Coleta da Estatística Educacional (III)." Bras-*7200*.R5, May '45, Jardim.

"I Conferência de Ministros e Diretores de Educação das Repúblicas Americanas." Bras-*7200*.R5, Aug. '44.

O Ensino no Brasil no Quinquênio 1936-1940. Bras-*7200*.E3, 1942.

O Ensino no Brasil no Quinquênio 1932-1936. Bras-*7200*.E4, 1939.

Estatística Intellectual do Brasil, 1929. Bras-*7200*.E6, 1929.

Outline of Education in Brazil. Bras-*7200*.O5, 1946.

Revista Brasileira de Estudos Pedagógicos. Bras-*7200*.R5.

Canada.

Biennial Survey of Education in Canada. Can-*7200*.B5.

Chile.

Censo de Educación, Año 1933. Chile-*2230*.C5, 1933.

7000-7999 HEALTH AND CULTURAL STATISTICS—cont.

7200 ——. General—cont.

Education in Chile. Chile-*7200*.E5, 1945.
Política, Administración, Justicia y Educación. Chile-*8000*.P5.
"XI Censo General de Población." Chile-*1420*.E5, Dec. '41.
Colombia.
"Analfabetismo Urbano y Rural." Col-*1420*.A5, May 5, '42, Rodríguez.
"Diez Años de Educación y Cultura en Colombia." Col-*1420*.A5, Feb. 20, '42, Bermúdez V.
Education in Colombia. Col-*7200*.E5, 1946.
Costa Rica.
Alfabetismo y Analfabetismo en Costa Rica según el Censo General de Población, 11 de Mayo de 1927. CR-*2230*.A5, 1927.
Education in Costa Rica. CR-*7200*.E5, 1946.
Cuba.
Education in Cuba. Cuba-*7200*.E5, 1943.
Ecuador.
Estadística y Escalafón. Ec-*7200*.E5, 1942.
"Síntesis de Estadística Educacional." Ec-*1420*.T5, Apr.-June '45, Romero.
El Salvador.
Informe de las Labores Desarrolladas por el Poder Ejecutivo en los Ramos de Cultura y Asistencia Social. ElSal-*7000*.I5.
Memoria de Cultura Popular. ElSal-*7000*.M5.
Guatemala.
Acuerdo Disponiendo el Levantamiento del Censo Escolar y Reglamento General del Mismo. Guat-*2130*.C5, 1946.2.
"Censo Escolar." Guat-*1420*.B5, May '46, a.
Censo Escolar de 1946: Métodos y Procedimientos. Guat-*2130*.C5, 1946.1.
La Edad de los Escolares y Su Valor como Determinante en el Fenómeno Educativo. Guat-*7200*.E3, 1946.
Estudio sobre la Enseñanza Parvularia. Guat-*7200*.E5, 1946.
Instrucciones Generales para los Jefes de Empadronadores y Personal Auxiliar. Guat-*2130*.C5, 1946.3.
Memoria de las Labores del Poder Ejecutivo en el Ramo de Educación Pública. Guat-*7200*.M5.
Publicaciones del Ministerio de Educación Pública. Guat-*7200*.P5.
"Resultados del Censo Escolar Levantado el 23 de Enero de 1946: Municipio de Guatemala." Guat-*1420*.B5, July '46, b.
Haiti.
Rapport Annuel du Service National de la Production Agricole et de l'Enseignement Rural. Haiti-*6200*.R5.
Les Résultats de la Troisième Année de Réforme de l'Enseignement Urbain, 1943-1944. Haiti-*7200*.R5, 1943-44.
Tables et Graphiques Statistiques de l'Enseignement Urbain, 1942-1943. Haiti-*7200*.T5, 1942-43.
Tables Statistiques de la Fréquentation des Écoles de l'Enseignement Rural, 1939-1940 et 1940-1941. Haiti-*7200*.T6, 1942.
Paraguay.
Revista de Educación. Par-*7200*.R5.
Perú.
Censo Nacional de Población y Ocupación, 1940. Perú-*2200*.C5, 1940.1.
Education in Peru. Perú-*7200*.E5, 1946.
Estado de la Instrucción en el Perú según el Censo Nacional de 1940. Perú-*2200*.C5, 1940.4.
United States.
Biennial Survey of Education. USA-*7200*.B3.
Journal of Educational Psychology. USA-*7200*.J5.
Research Bulletin of the National Education Association. USA-*7200*.R5.

7000-7999 HEALTH AND CULTURAL STATISTICS—cont.

7200 ——. General—cont.

Sixteenth Census of the United States: 1940: Population: Education, Occupation and Household Relationship of Males 18 to 44 Years Old. USA-*2200*.C5, 1940.6.
Special Salary Tabulations, 1944-45. USA-*9200*.S5.
——. Puerto Rico.
El Analfabetismo en Puerto Rico. USA PR-*7200*.A5, 1945.
Uruguay.
El Analfabetismo. Ur-*7200*.A5, 1940.
Venezuela.
Memoria que el Ministro de Educación Nacional Presenta al Congreso Nacional. Ven-*7200*.M5.

7210 ——. Educational institutions.
Instituciones educacionales.

International.
"Coordination Internationale des Statistiques Universitaires." Int-*3020*.B5, 1938, Castrilli.
Higher Education in Latin America. Int-*7210*.H5.
Argentina.
Recopilación Estadística, Años 1941-1942. Arg-*7200*.R5, 1941-42.
Brasil.
"O Ensino Primário Brasileiro no Decênio 1932-1941." Bras-*3040*.R5, Apr.-June '46, a.
Canada.
Educational Institutions in Canada. Can-*7210*.E5, 1944.
Chile.
The Universities of Chile. Chile-*7210*.U5, 1944.
Colombia.
"Decreto Número 75 de 1939 (Enero 14)." Col-*3030*.E5, 1945, No. 75.
The Universities of Colombia. Col-*7210*.U5, 1945.
Guatemala.
Cuatro Estudios Estadístico-Educativos de la Escuela Guatemalteca. Guat-*7210*.C5, 1946.
Estudio sobre el Costo de la Enseñanza Primaria. Guat-*7210*.E5, 1946.
Memoria de las Labores del Poder Ejecutivo en el Ramo de Educación Pública. Guat-*7200*.M5.
Haiti.
Tables Statistiques de la Fréquentation des Écoles de l'Enseignement Rural, 1939-1940 et 1940-1941. Haiti-*7200*.T6, 1942.
Panamá.
Estadística Cultural. Pan-*7210*.E5.

7300-7399 Arts, sciences, and public information services.
Artes, ciencias y servicios de información pública.

7300 ——. General.

International.
Conferencia Interamericana de Expertos para la Protección de los Derechos de Autor, 1-22 de Junio de 1946: Actas y Documentos. Int-*7300*.I5, 1946.2, Sp.
Handbook of Scientific and Technical Societies and Institutions of the United States and Canada. Int-*1930*.H5, 1942.
Inter-American Conference of Experts on Copyright, Pan American Union, June 1-22, 1946: Proceedings. Int-*7300*.I5, 1946.2, Eng.
Journals Dealing with the Natural, Physical and Mathematical Sciences Published in Latin America. Int-*7200*.J5, 1944.
Proceedings of the Eighth American Scientific Congress. Int-*1300*.A5, 1940.3.

8000-8999 GOVERNMENT AND POLITICAL STATISTICS—cont.

8000 General public administration and enterprise—cont.

Argentina.
Clasificador de Bienes. Arg-*3200*.C5, 1937.A.
Memoria. Arg-*6600*.M5.
Revista de la Facultad de Ciencias Económicas, Comerciales y Políticas. Arg-*3040*.R8.

Brasil.
"O Problema do Município no Brasil Atual." Bras-*3040*.R5, Oct.-Dec. '43. Teixeira de Freitas.
Relatório do Departamento Administrativo do Serviço Público. Bras-*8000*.R4.

Canada.
Canadian Journal of Economics and Political Science. Can-*3040*.C5.

Chile.
"Número Total de Empleados Públicos." Chile-*1420*.E5, Nov. '44.
Política, Administración, Justicia y Educación. Chile-*8000*.P5.

El Salvador.
Boletín. ElSal-*6600*.B5.

Guatemala.
Memoria del Ministerio de Comunicaciones y Obras Públicas. Guat-*6000*.M5.

Haiti.
Bulletin Mensuel du Département Fiscal. Haiti-*1420*.B5.
Monthly Bulletin of the Fiscal Department. Haiti-*1420*.M5.
Rapport Annuel du Sous-Secrétaire d'Etat et Ingénieur en Chêf. Haiti-*8000*.R5.

Honduras.
Informe de los Actos Realizados por el Poder Ejecutivo en los Ramos de Gobernación, Justicia, Sanidad y Beneficencia. Hond-*8000*.I5.
Informe del Secretario de Agricultura, Fomento y Obras Públicas. Hond-*6000*.I5.

México.
Memoria de la Secretaría de Gobernación. Méx-*8000*.M5.

Nicaragua.
Memoria de Fomento y Obras Públicas. Nic-*8000*.M5.

Paraguay.
M. O. P. (Ministerio de Obras Públicas y Comunicaciones). Par-*8000*.M5.

United States.
The Annals of the American Academy of Political and Social Science. USA-*3040*.A5.
Collection of Economic Statistics in the United States. USA-*3030*.C5, 1946.
Monthly Report of Employment, Executive Branch of the Federal Government. USA-*9200*.M7.
Municipal Year Book. USA-*8000*.M5.
Statistics of Income. USA-*8800*.S7.

Venezuela.
Memoria del Ministerio de Fomento. Ven-*6000*.M3.
Registro de Fomento. Ven-*6000*.R5.
Suplemento de la División Político-Territorial de la República. Ven-*4100*.D5, 1945.

Other Countries.
Great Britain.
Guide to Current Official Statistics of the United Kingdom. GrBr-*3010*.C5.

8100-8199 Defense and war.
Defensa y guerra.

8100 ——. General.

International.
Annuaire Statistique du Commerce des Armes et des Munitions. Statistical Year-Book of the Trade in Arms and Ammunition. Int-*6900*.A5.

8000-8999 GOVERNMENT AND POLITICAL STATISTICS—cont.

8100 ——. Defense and war—cont.

Brasil.
Inquéritos Econômicos para a Defesa Nacional. Bras-*6000*.I5.
Canada.
National Registration, August 1940. Can-*5250*.N2, 1940.A.
National Registration Regulations, 1940. Can-*5250*.N2, 1940.C.
Specialized Occupations, National Registration, 1940. Occupations Spécialisées, l'Enregistrement National, 1940. Can-*5250*.N2, 1940.B.
United States.
"Methods Used in Processing Data from the Physical Examination Reports of the Selective Service System." USA-*3040*.J5, June '44, Edwards.
Mexican War Workers in the United States. USA-*9600*.M3, 1945.
Production: Wartime Achievements and the Reconversion Outlook. USA-*6020*.P5, 1945.
Report to the President, the Senate and the House of Representatives by the Director of War Mobilization and Reconversion. USA-*8100*.R5.
Statistics of War Production. USA-*6020*.S5.
Non-Geographic.
"Estadística Económica Militar." *Estadística Peruana.* Perú-*3040*.E5, July '46, Velarde B.

8200-8299 International relations.
Relaciones internacionales.

8200 ——. General.

International.
Acta Final de la Conferencia de Comisiones de Fomento Interamericano, Nueva York, del 9 al 18 de Mayo de 1944. Int-*8200*.C6, 1944.1, Sp.
Boletim da União Panamericana. Int-*8200*.B4.
Boletín de la Unión Panamericana. Int-*8200*.B5.
Bulletin of the Pan American Union. Int-*8200*.B7.
Conference of Commissions of Inter-American Development, May 9 to 18, 1944: Final Act. Int-*8200*.C6, 1944.1, Eng.
Conferência Interamericana sôbre Problemas da Guerra e da Paz, México, D.F., México, 21 de Fevereiro a 8 de Março de 1945. Int-*8200*.I3, 1945.2, Port.
Conferencia Interamericana sobre Problemas de la Guerra y de la Paz, México, D.F., México, Febrero 21-Marzo 8, 1945. Int-*8200*.I3, 1945.2, Sp.
Consejo Interamericano Económico y Social: Informe sobre la Organización del Consejo Aprobado por el Consejo Directivo de la Unión Panamericana en la Sesión del 29 de Agosto de 1945. Int-*8200*.I45, 1945, Sp.
The Economic and Financial Organization of the League of Nations. Int-*8200*.E5, 1945.
Handbook of International Organisations. Int-*8200*.H5, 1938.
Handbook of International Organizations in the Americas. Int-*1930*.H4, 1945.
Informe Anual del Director General de la Unión Panamericana. Int-*8299*.I26.
Inter-American Affairs. Int-*1910*.I5.
Inter-American Conference on Problems of War and Peace, Mexico City, February 21-March 8, 1945. Int-*8200*.I3, 1945.2, Eng.
Inter-American Economic and Social Council: Report on the Organization of the Council Approved by the Governing Board of the Pan American Union at the Session of August 29, 1945. Int-*8200*.I45, 1945, Eng.
Inter-American Series. Int-*8200*.I48.
International Agencies in which the United States Participates. Int-*8200*.I49, 1946.
The International Administration of an International Secretariat. Int-*8200*.I5, 1945.
International Conciliation. Int-*8200*.I55.

8000-8999 GOVERNMENT AND POLITICAL STATISTICS—cont.

8200 ——. General—cont.

International Health Organisation. Int-*7101*.I5, 1945.
Journal of the Economic and Social Council. Journal du Conseil Economique et Social. Int-*8200*.J3.
Minutes of the Special Meeting of the Governing Board of the Pan American Union Held on August 29, 1945. Int-*8200*.M3, 1945.
Preliminary Cumulative List of Documents in Unrestricted Series Issued by the General Assembly, Security Council, Atomic Energy Commission, Economic and Social Council and Its Commissions, International Health Conference, to 31 July 1946. Int-*8200*.P45, 1946.
Report of the Anglo-American Caribbean Commission to the Governments of the United States and Great Britain. Int-*8200*.R3.
Report on the Work of the League. Int-*8200*.R7.
Report on the Work of the League During the War. Int-*8200*.R7, 1939-1945.

United States.
The Department of State Bulletin. USA-*8200*.D5.
Twentieth Report to Congress on Lend-Lease Operations, for the Period Ended June 30, 1945. USA-*8800*.R5, 1945.

8210 ——. Conventions and treaties.
Convenios y tratados.

International.
Charter of the United Nations and Statute of the International Court of Justice. Charte des Nations Unies et Statut de la Cour Internationale de Justice. Int-*8210*.C6, 1946.
Conferencias Internacionales Americanas, 1889-1936. Int-*8200*.I6, 1889-1936, Sp.
Executive Agreement Series. Int-*1200*.E5.
The International Conferences of American States, 1889-1928. Int-*8200*.I6, 1889-1928, Eng.
The International Conferences of American States First Supplement, 1933-1940. Int-*8200*.I6, 1933-1940, Eng.
"The United Nations Charter." Int-*8200*.I55, Sept. '45.

United States.
Conference Series of the United States Department of State. USA-*8210*.C5.

8300-8399 Justice, delinquency, and crime.
Justicia, delincuencia y crimen.

8300 ——. General.

International.
"Directives pour l'Elaboration des Statistiques Criminelles dans les Divers Pays." Int-*3040*.R5, 1936.2, Scäfer.
Final Act of the Inter-American Conference of Police and Judicial Authorities, Buenos Aires, Argentina, May 27-June 9, 1942. Int-*8300*.I5, 1942.
Justice and Crime Statistics. Int-*3010*.B5, Gov. 2, Mar. 22, '46.

Canada.
Annual Report of Statistics of Criminal and Other Offences. Rapport Annuel sur la Statistique de la Criminalité. Can-*8300*.A5.

Chile.
Política, Administración, Justicia y Educación. Chile-*8000*.P5.

Colombia.
"Estadística de la Delincuencia en Colombia." Col-*1420*.A5, Feb. '45, Montoya Canal.

Haiti.
Rapport Annuel. Haiti-*8300*.R5.

México.
Tendencia y Ritmo de la Criminalidad en México, D.F. Méx-*8300*.T5, 1939.

8000-8999 GOVERNMENT AND POLITICAL STATISTICS—cont.

8300 **——. General—cont.**

United States.
Federal Prisons. USA-*8300*.F5.
Judicial Criminal Statistics. USA-*8310*.J5.
Ten Years of Uniform Crime Reporting, 1930-1939. USA-*8300*.T5, 1930-39.
Uniform Crime Reports for the United States and Its Possessions. USA-*8300*.U5.

8320 **——. Juvenile delinquency.**
Delincuencia juvenil.

Canada.
Juvenile Delinquents, for the Year Ended September 30, 1945. Jeunes Délinquants pour l'Année Expirée le 30 Septembre 1945. Can-*8320*.J5, 1945.
United States.
Juvenile Court Statistics. USA-*8320*.J3.

8330 **——. Penal institutions.**
Instituciones penales.

United States.
Prisoners in State and Federal Prisons and Reformatories. USA-*8330*.P6.
Sixteenth Census of the United States: 1940: Population: Special Report on Institutional Population, 14 Years Old and Over. USA-*2200*.C5, 1940.9.

8400 **Law and legislation.**
Leyes y legislación.

International.
A Selected List of Latin American Periodicals Containing Laws and Legal Information. Int-*8400*.S5, 1942.

8500 **Planning and research—national, regional, and civic.**
Planificación e investigación—nacional, regional y cívica.

Argentina.
Ordenamiento Económico-Social. Arg-*9000*.O5, 1945.
México.
Planificación Económica. Méx-*6000*.P5.
United States.
Aims, Organization, and Publications of the National Bureau of Economic Research. USA-*8500*.A5, 1943.
Bureau of the Social Science Research Council. USA-*9000*.B6.
"The Federal Government's Statistical Program for Reconversion and Postwar Adjustment: A Round Table." USA-*3040*.J5, June '45, a.
Occasional Papers of the National Bureau of Economic Research. USA-*6000*.O4.
Pamphlets of the Brookings Institution. USA-*8500*.P3.
Planning Pamphlets of the National Planning Association. USA-*8500*.P5.
Publications of the National Bureau of Economic Research. USA-*6000*.P5.
Publications of the National Resources Planning Board, 1934-1943. USA. *8500*.P8, 1934-43.
Report of the National Research Council. USA-*8500*.R3.
Technical Papers of the National Resources Planning Board. USA-*8500*.T5.

8600 **Elections, political parties, public opinion polls.**
Elecciones, partidos políticos y encuestas de la opinión pública.

Perú.
Extracto Estadístico y Censo Electoral de la República. Perú-*1410*.E5, 1933.
United States.
Public Opinion Quarterly. USA-*8600*.P5.
Non-Geographic.
Consumer and Opinion Research. NonG-*3110*.C5, 1943.

8000-8999 GOVERNMENT AND POLITICAL STATISTICS—cont.

8800 Public finance and taxation—national, state, municipal, local.
Finanzas públicas, impuestos y contribuciones—nacionales, estatales, municipales, locales.

Argentina.
El Comercio Exterior Argentino. Arg-*6900*.C5.
Indice General Alfabético por Materia de los Bienes Nacionales del Estado. Arg-*3200*.C5, 1937.B.
Memoria de la Contaduría General de la Nación. Arg-*8800*.M4.
Memoria del Departamento de Hacienda. Arg-*8800*.M6.

Bolivia.
Boletín del Banco Central de Bolivia. Bol-*6720*.B5.
Finanzas. Bol-*6720*.E6.
Memoria Anual. Bol-*6720*.M5.
Tasas e Impuestos sobre la Industria Minera en Bolivia. Bol-*6300*.T5, 1941.

Canada.
Manual of Instructions: Balance Sheets, Revenues and Expenditures and Other Accounting Statements of Municipal Corporations. Can-*1100*.M2, 1942.

Chile.
Finanzas, Bancos y Cajas Sociales. Chile-*6720*.P5.
Memoria de la Contraloría General de la República y Balance General de la Hacienda Pública. Chile-*8800*.M5.

Colombia.
Estadística Fiscal y Administrativa. Col-*8800*.E5.
Información Fiscal de Colombia. Col-*8800*.I3.
Informe Financiero del Contralor General. Col-*8800*.I5.
"Reglamentación sobre el Cálculo de los Medios de Pagos en Circulación y de las Exigibilidades Monetarias en Colombia." Col-*1420*.A5, Nov. '45, Abrisqueta.
Revista de Hacienda. Col-*8800*.R5.

Costa Rica.
Memoria de la Secretaría de Hacienda y Comercio. CR-*8800*.M5.

Cuba.
Boletín Oficial del Ministerio de Hacienda. Cuba-*8800*.B5.

Ecuador.
Boletín del Ministerio del Tesoro. Ec-*8800*.B5.
Informe que el Ministro de Hacienda y Crédito Público Presenta a la Nación. Ec-*8800*.I5.
Revista de Hacienda. Ec-*8800*.R5.

Guatemala.
Análisis de los Sueldos Pagados por el Estado. Guat-*9200*.A5, 1946.
Memoria de las Labores del Poder Ejecutivo en el Ramo de Hacienda y Crédito Público. Guat-*1410*.M5.

Haiti.
Annual Report of the Fiscal Department. Haiti-*6000*.A3.
Aspects de l'Économie et des Finances d'Haïti. Haïti-*6000*.A5, 1944.
Rapport Annuel de l'Administration Générale des Contributions. Haiti-*8800*.R3.
Rapport Annuel du Département Fiscal. Haiti-*6000*.R5.

Honduras.
Informe de Hacienda, Crédito Público y Comercio. Hond-*8800*.I5.

México.
Bibliografía de la Secretaría de Hacienda y Crédito Público. Méx-*8800*.B5, 1943.

Nicaragua.
Deuda Pública—Rentas—Bancos—Comercio Exterior, 1943. Nic-*8800*.D7, 1943.
Economía y Finanzas. Nic-*6000*.E5.
Fondo de Nivelación de Cambios, 1941-1944. Nic-*6720*.F5, 1941-44.

8000-8999 GOVERNMENT AND POLITICAL STATISTICS—cont.

8800 ——. **Public finance and taxation—national, state, municipal, local—cont.**

Memoria de Hacienda y Crédito Público. Nic-*8800*.M5.
Movimiento Gráfico de los Presupuestos Nacionales, 1931/32 a 1943-44. Nic-*8800*.M7, 1931-44.
Presupuesto General de Egresos e Ingresos. Nic-*8800*.P5.
Situación Monetaria, Año 1945. Nic-*6720*.S5, 1945.

Panamá.
Informe del Contralor General de la República. Pan-*8800*.I5.

Paraguay.
Boletín de la Dirección General de Impuestos Internos. Par-*8800*.B3.
Boletín del Tesoro. Par-*8800*.B5.

República Dominicana.
Finanzas Municipales. RepDom-*8800*.F5.
Registro Público. RepDom-*6721*.R5.

United States.
Annual Report of the Secretary of the Treasury on the State of the Finances. USA-*8800*.A5.
The Budget of the United States Government. USA-*8800*.B7.
City Finances. USA-*8800*.C4.
County Finances. USA-*8800*.C6.
Financial Statistics of State and Local Government (Wealth, Public Debt, and Taxation). USA-*8800*.F5.
Report of the Reconstruction Finance Corporation. USA-*6720*.R5.
State Finances. USA-*8800*.S5.
Statement of the Public Debt. USA-*8800*.S6.
Treasury Bulletin. USA-*8800*.T5.
Twentieth Report to Congress on Lend-Lease Operations, for the Period Ended June 30, 1945. USA-*8800*.R5, 1945.

Uruguay.
Boletín de Hacienda. Ur-*6000*.B5.
Boletín de Hacienda, Suplemento. Ur-*6000*.B6.
Deuda Pública Nacional. Ur-*8800*.D5.
Presupuesto General del Estado. Ur-*8800*.P5.

Venezuela.
Banco Central de Venezuela: Memoria. Ven-*6720*.B3.
Boletín Informativo del Ministerio de Hacienda. Ven-*8800*.B5.
Cuenta General de Rentas y Gastos Públicos; Bienes Nacionales Inclusive Materias, y Cuenta del Departamento de Hacienda, Rentas y Gastos. Ven-*8800*.C5.
"Exposición de Motivos del Proyecto de Ley de Presupuesto General de Rentas y Gastos Públicos para el Año Económico 1944-1945." Ven-*8800*.R5. June 1944, a.
Informe de la Contraloría General de la Nación. Ven-*8800*.I5.
Memoria de Hacienda. Ven-*8800*.M5.
"Movimiento General de las Rentas Nacionales por Años Económicos, desde el 1° de Julio de 1908 hasta el 30 de Junio de 1943." Ven-*8800*.R5, June 1944, b.
"Presupuestos de Ingresos y Gastos Públicos de Venezuela en los Años Económicos de 1935-1936 a 1942-1943." Ven-*8800*.R5, June 1944, c.
"Resumen del Presupuesto General de Rentas y Gastos Públicos que Regirá en el Año Económico de 1° de Julio de 1944 a 30 de Junio de 1945." Ven-*8800*.R5, June 1944, d.
Revista de Hacienda. Ven-*8800*.R5.

9000-9999 LABOR AND LIVING CONDITIONS
CONDICIONES DEL TRABAJO Y DE VIDA

9000 General.

International.

Annuaire des Statistiques du Travail. Anuario de Estadísticas del Trabajo. Year Book of Labour Statistics. Int-*9000*.A5.

Bibliography on Labor and Social Welfare in Latin America. Int-*9000*.B5, 1944.

Boletín del Instituto de Investigaciones Sociales y Económicas. Bulletin of the Institute of Social and Economic Research. Bulletin de l'Institut des Recherches Sociales et Économiques. Boletim do Instituto de Investigações Sociais e Econômicas. Int-*3040*.B5.

The Caribbean Islands and the War. Int-*1900*.C5, 1943.

"Le Comité d'Experts Statisticiens du Bureau International du Travail (1933-1939)." Int-*3040*.R5, 1942, 1-2, Huber.

Cuadernos Sociales del Consejo. Int-*9000*.C8.

La Estandardización Internacional de las Estadísticas del Trabajo. Int-*9000*.E5, 1943.

Estudios y Documentos, Serie N, Estadísticas. Int-*9000*.E7.

Index to Publications and Articles on Latin America Issued by the United States Bureau of Labor Statistics, 1902-43. Int-*9000*.I2, 1945.

International Labour Review. Int-*9000*.I5.

The International Standardisation of Labour Statistics. Int-*9000*.I7, 1943.

Journal of the Economic and Social Council. Journal du Conseil Économique et Social. Int-*8200*.J3.

Labor and Living Conditions in the American Nations. Int-*3010*.B5, Lab. 1, Jan. 30, '46.

Labor Conditions in Latin America. USA-*9000*.L4.

Labor Trends and Social Welfare in Latin America, 1941 and 1942. Int-*9000*.L6, 1941-42.

"Latin American Periodicals Dealing with Labor and Social Welfare." Int-*1200*.H4, 1942, Rohen y Gálvez.

Proceedings of the Eighth American Scientific Congress. Int-*1300*.A5, 1940.3.

Revista International del Trabajo. Int-*9000*.R6.

"Sources of Information on Social and Labor Problems in Latin America." Int-*1200*.I6, Summer 1941, Owen.

Studies and Reports, Series N, Statistics. Int-*9000*.S5.

Argentina.

Boletín Informativo del Departamento Nacional del Trabajo. Arg-*9000*.B5.

Condiciones de Vida de la Familia Obrera. Arg-*9000*.C5, 1943.

Investigaciones Sociales. Arg-*9000*.I5.

Ordenamiento Económico-Social. Arg-*9000*.O5, 1945.

Organización Sindical. Arg-*9000*.O7, 1941.

Serie B, Estadísticas y Censos, del Departamento Nacional del Trabajo. Arg-*9000*.S5.

Bolivia.

Boletín del Despacho de Previsión Social. Bol-*9000*.B5.

Memorándum: Elementos Generales sobre las Condiciones de Trabajo, de Legislación y de Otros Problemas Sociales de la República de Bolivia. Bol-*9000*.M5, 1943.

El Problema Social en Bolivia—Condiciones de Vida y de Trabajo. Bol-*9000*.P5, 1941.

Brasil.

Boletim do Ministério do Trabalho, Indústria e Comércio. Bras-*1420*.B5.

"Censo Sindical Carioca em 1941." Bras-*1420*.B5, Aug. '43, Marinho de Andrade.

9000-9999 LABOR AND LIVING CONDITIONS—cont.

9000-9999 LABOR AND LIVING CONDITIONS—cont.

9000 General—cont.

"Population and Per Capita Income." USA-*3040*.A5, Jan. '45, Spengler.
"Representative Books in Sociology since 1938." USA-*9000*.A6, Dec. '44.
Social Forces. USA-*9000*.S5.
Social Research. NonG-*9000*.S5, 1942.
Statistics for Sociologists. NonG-*3330*.Hagood 1941
Studies and Reports, Series N, Statistics. Int-*9000*.S5.
Teoría y Métodos de Estadística del Trabajo. NonG-*9000*.T5, 1942.

9100 Labor conditions, general.
Condiciones del trabajo, general.

International.
Notas para una Encuesta Continental sobre Productividad Obrera. Int-*9100*.N5, 1946.

9200 Employment and wages.
Ocupación y salarios.

International.
Anteproyecto de Carta para la Organización Internacional de Comercio de las Naciones Unidas. Int-*6900*.A6, 1946.
Employment of United States Citizens in Latin America. Int-*9200*.E4, 1945.
Estadísticas de la Población Activa. Int-*3240*.E5, 1938.
"Employment Situation in Latin America." USA-*9000*.M5, May 1946.
Estimates of the Working Population of Certain Countries in 1931 and 1941. Int-*9200*.E7, 1931-41.
"Index-Numbers of Wages: A Survey." *Revue de l'Institut International de Statistique.* Int-*3040*.R5, 1933.1, Nixon.
Methods of Compiling Statistics of Unemployment. Int-*9200*.M5, 1922.
Methods of Statistics of Unemployment. Int-*9200*.M5, 1925.
Methods of Statistics of Wages and Hours of Labour. Int-*9200*.M7, 1923.
National Income and Related Subjects for the United Nations. Int-*6100*.N3, 1946.
"On the Statistics Available Concerning the Occupied Population of the World and Its Distribution." Int-*3040*.R5, 1938.3, Nixon.
Report to the Council on the Work of the Sixth Session. Int-*3000*.R6, 1937, Eng.
Statistics of the Gainfully-Occupied Population. Int-*3240*.S4, 1938.
Statistique de la Population Active. Int-*3240*.S6, 1938.
Studies and Reports, Series C, Employment and Unemployment. Int-*9200*.S7.
Studies and Reports, Series I, Employment of Women and Children. Int-*9200*.S8.
"Suggested Charter for an International Trade Organization of the United Nations." Int-*8200*.I55, Nov. '46.
Systems of Classification of Industries and Occupations. Int-*3220*.S5, 1923.
The War and Women's Employment. Int-*9200*.W5, 1946.

Argentina.
La Actividad Industrial durante los Primeros . . . Meses de . . . , según los Números Indices Correspondientes a los Salarios Pagados, a la Ocupación de Obreros y a las Horas-Obrero de Trabajo en la Industria. Arg-*6400*.A3.
Adaptación de los Salarios a las Fluctuaciones del Costo de la Vida. Arg-*9200*.A5, 1943.
La Desocupación en la Argentina, 1940. Arg-*9200*.D5, 1940.
Estadísticas de las Huelgas, 1940. Arg-*9300*.E5, 1940.
"La Población Ocupada y las Rentas Individuales en la Argentina." Arg-*3040*.R5, June '45, Di Tella.

Brasil.
Salário Mínimo. Bras-*9200*.S5, 1940.

Canada.
Classification of Occupations: Eighth Census of Canada, 1941. Can-*3240*.C5, 1941.

9000-9999 LABOR AND LIVING CONDITIONS—cont.

9200 **Employment and wages—cont.**

The Employment Situation . . . Together With Payrolls. Can-*9200*.E5.
Labour Force Bulletin. Can-*9200*.L5.
Man Hours and Hourly Earnings. Can-*9200*.M5.
Occupational Trends in Canada, 1891-1931. Can-*9200*.C5, 1891-1931.
Specialized Occupations, National Registration, 1940. Occupations Spécialisées, l'Enregistrement National, 1940. Can-*5250*.N2, 1940.B.

Chile.
"La Población Chilena: Segunda Parte, El Trabajo de la Población Chilena." Chile-*3040*.E5, Apr. '45, Levine Bawden.

Colombia.
Población de Colombia por Ramas de Actividad Económica (según Censo del 5 de Julio de 1938). Col-*5110*.P5, 1946.

Costa Rica.
Censo de Personas sin Trabajo, 1932. CR-*2200*.C5, 1932.

Guatemala.
Análisis de los Sueldos Pagados por el Estado. Guat-*9200*.A5, 1946.

México.
Estadística de Salarios. Méx-*9200*.E5, 1942.
Nomenclatura Nacional de Ocupaciones, 1940. Méx-*3240*.N5, 1940.

Nicaragua.
"Actividades de la Población de Nicaragua." Nic-*1420*.B5, Apr. 15, '45, a.
"Población Economicamente Inactiva (Censo del 23 de Mayo de 1940)." Nic-*1420*.B5, April 15, '45, B.

Perú.
Censo Nacional de Población y Ocupación, 1940. Perú-*2200*.C5, 1940.1.

United States.
Census of Unemployment, 1937: Final Report on Total and Partial Unemployment. USA-*2200*.C7, 1937.
Census of Unemployment, 1937: Final Report on Total and Partial Unemployment for Geographic Divisions. USA-*2200*.C7, 1937.A.
Changes in Distribution of Manufacturing Wage Earners, 1899-1939. USA-*9200*.C5, 1899-1939.
Death Rates by Occupation. USA-*5212*.D5, 1934.
"Demographic Factors in Labor Force Growth." USA-*9000*.A6, Aug. '46, Wolfbein.
Employment and Pay Rolls. USA-*9200*.E4.
Employment in Manufacturing, 1899-1939, an Analysis of Its Relation to the Volume of Production. USA-*9200*.E5, 1899-1939.
Farm Labor. USA-*9200*.F5.
"Gross National Product Projections for Full Employment: III, Measuring Labor Force in 1950." USA-*6000*.D5, June '45, George.
Hours and Earnings. USA-*9200*.H4.
Hours and Earnings in the United States, 1932-40. USA-*9200*.H5, 1932-40.
Income from Independent Professional Practice. USA-*9200*.14, 1945.
Industrial Change and Employment Opportunity—A Selected Bibliography. USA-*6000*.I4, 1939.
Industrial Unemployment. USA-*9200*.I6, 1922.
The Labor Market. USA-*9200*.L5.
Labor Turnover. USA-*9200*.L6.
Monthly Report of Employment, Executive Branch of the Federal Government. USA-*9200*.M7.
Reports of the National Research Project on Reemployment Opportunities and Recent Changes in Industrial Techniques. USA-*6000*.R5.
Sixteenth Census of the United States: Alphabetical Index of Occupations and Industries. USA-*3240*.C5, 1940.2.

9000-9999 LABOR AND LIVING CONDITIONS—cont.

9200 Employment and wages—cont.

Sixteenth Census of the United States: Classified Index of Occupations. USA-*3240*.C5, 1940.1.
Sixteenth Census of the United States: 1940: Population: Comparative Occupation Statistics for the United States, 1870 to 1940. USA-*2200*.C5, 1940.4.
Sixteenth Census of the United States: 1940: Population: Education, Occupation and Household Relationship of Males 18 to 44 Years Old. USA-*2200*.C5, 1940.6.
Sixteenth Census of the United States: 1940: Population: Estimates of Labor Force, Employment and Unemployment in the United States, 1940 and 1930. USA-*2200*.C5, *1940*.15.
Sixteenth Census of the United States: 1940: Population: The Labor Force (Sample Statistics): Employment and Family Characteristics of Women. USA-*2200*.C5, 1940.10a.
Sixteenth Census of the United States: 1940: Population: The Labor Force (Sample Statistics): Employment and Personal Characteristics. USA-*2200*.C5, 1940.10b.
Sixteenth Census of the United States: 1940: Population: The Labor Force (Sample Statistics): Industrial Characteristics. USA-*2200*.C5, 1940.10c.
Sixteenth Census of the United States: 1940: Population: The Labor Force (Sample Statistics): Occupational Characteristics. USA-*2200*.C5, 1940.10d.
Sixteenth Census of the United States: 1940: Population: The Labor Force (Sample Statistics): Usual Occupation. USA-*2200*.C5, 1940.10e.
Sixteenth Census of the United States: 1940: Population: The Labor Force (Sample Statistics): Wage or Salary Income in 1939. USA-*2200*.C5, 1940.10f.
Special Salary Tabulations, 1944-45. USA-*9200*.S5.
The Wages of Farm and Factory Laborers. USA-*9200*.W7, 1945.
Women in Factories. USA-*9200*.W9.

——. Puerto Rico.
Sixteenth Census of the United States: 1940: Puerto Rico, Población, Boletín No. 3: Ocupaciones y Otras Características por Edad. Puerto Rico, Population, Bulletin No. 3: Occupations and Other Characteristics by Age. USA PR-*2200*.C5, 1940.3.

Non-Geographic.
Survey of Economic Theory on Technological Change and Employment. NonG-*6000*.S7, 1940.
"La Clasificación de Industrias y Profesiones en las Estadísticas del Trabajo." Int-*9000*.E5, 1943, a.

9400 Industrial accidents and occupational disease.
Accidentes industriales y enfermedades provenientes de la ocupación.

International.
Industrial Accident Statistics. Int-*9400*.I3, 1938.
Methods of Statistics of Industrial Accidents. Int-*9400*.M5, 1923.
Statistical Methods for Measuring Occupational Morbidity and Mortality. Int-*9400*.S5, 1930.

Chile.
"Work Accidents in Chile." USA-*9000*.M5, June '46, Cárcamo C.

United States.
Accident Facts. USA-*7160*.A5.

9500 Cost of living.
Costo de la vida.

International.
A Contribution to the Study of International Comparisons of Costs of Living. Int-*9500*.C4, 1932.
El Costo de la Vida Obrera en América. Int-*9500*.C8, 1943.

9000-9999 LABOR AND LIVING CONDITIONS—cont.

9000-9999 LABOR AND LIVING CONDITIONS—cont.

9500 Cost of living—cont.

Uruguay.
Costo de la Vida Obrera en Montevideo. Ur-*9500*.C5.
Venezuela.
Investigación sobre el Costo de la Vida en Caracas, los Presupuestos Familiares, 1939. Ven-*9500*.I5, 1939.
Memoria del Ministerio del Trabajo y de Comunicaciones. Ven-*9000*.M5.

9600 Living conditions, general.
Condiciones de la vida, general.

International.
Consumo y Subconsumo. Int-*9600*.C5, 1943.
Argentina.
La Vivienda Popular. Arg-*2230*.C5, 1943.3.
Chile.
Encuesta Continental sobre el Consumo de Productos de Alimentación y Vestido y sobre la Vivienda Popular: Respuestatipo Referente a la República de Chile. Chile-*9600*.E5, 1943.
Colombia.
"Las Condiciones Económico-Sociales y el Costo de la Vida de la Clase Media en Bogotá." Col-*1420*.A5, Oct. '46.
"Ensayo sobre las Condiciones de la Vida Rural en el Municipio de Moniquirá-Boyacá." Col-*1420*.A5, July '39, Ortiz C.
"La Investigación del Costo de la Vida y las Condiciones Social-Económicas de los Obreros del Ferrocarril." Col-*1420*.A5, Mar. '45, Pineda de Castro.
México.
"Las Estadísticas sobre los Presupuestos Familiares en los Grupos Indígenas." Méx-*9600*.E5.
United States.
Bulletins of the Women's Bureau. USA-*9600*.B5.
The Canadian-Born in the United States. USA-*5120*.C5, 1850-1930.
Jewish Population Studies. USA-*5120*.J5, 1943.
Jewish Social Studies, Publications. USA-*9600*.J5.
Mexican War Workers in the United States. USA-*9600*.M3, 1945.
The Negro's Share. USA-*9600*.N5, 1943.
Women in Factories. USA-*9200*.W9.
Sixteenth Census of the United States; 1940: Population: The Labor Force (Sample Statistics): Employment and Family Characteristics of Women. USA-*2200*.C5, 1940.10a.
Non-Geographic.
Problems of Aging, Biological and Medical Aspects. NonG-*7150*.P5, 1939.

9700 Family living.
Condiciones de la vida familiar.

International.
"Committee on Family Budgets: Preliminary Report." Int-*3040*.R5, 1936.1, Nixon.
"Comparaciones Internacionales de los Modos de Vivir." Int-*1300*.A5, 1940.3, Alanís Patiño, b.
Encuestas sobre el Consumo de Alimentos y la Nutrición en las Américas. Int-*9800*.E5, 1942.
Food Consumption and Dietary Surveys in the Americas. Int-*9800*.F4, 1942.
Methods of Conducting Family Budget Inquiries. Int-*9700*.M5, 1926.
Methods of Family Living Studies. Int-*9700*.M6, 1942.
Méthodes d'Enquête sur les Conditions de Vie des Familles. Int-*9700*.M4, 1941.
Métodos de Encuestas sobre las Condiciones de Vida de las Familias. Int-*9700*.M7, 1942.

9000-9999 LABOR AND LIVING CONDITIONS—cont.

9700 Family living—cont.

Studies of Family Living in the United States and Other Countries. Int-*9700*.S5, 1935.

Argentina.

Nivel de Vida de la Familia Obrera. Arg-*9700*.N5, 1945.

"La Población Ocupada y las Rentas Individuales en la Argentina." Arg-*3040*.R5, June '45, Di Tella.

Colombia.

"Las Condiciones Económico-Sociales y el Costo de la Vida de la Clase Obrera en la Ciudad de Honda." Col-*1420*.A5, June '46.

México.

"Las Estadísticas sobre los Presupuestos Familiares en los Grupos Indígenas." Méx-*9600*.E5.

Perú.

"Encuesta sobre Presupuestos Familiares Obreros, en 1940, Correspondiente a Lima (Cercado)." Perú-*1410*.E5, 1942, c.

"Encuesta sobre Presupuestos Familiares Obreros, en 1940, Correspondiente a Lima (Cercado)." Perú-*1410*.E5, 1940, a.

"Encuesta sobre Presupuestos Familiares Obreros Realizada en la Ciudad de Lima, en 1940." Perú-*3040*.R5, 1944, Palacios.

United States.

Consumer Expenditures in the United States, Estimates for 1935-36. USA-*9700*.C5, 1935-36.

Consumer Incomes in the United States, Their Distribution in 1935-36. USA-*9700*.C6, 1935-36.

Family Expenditures in the United States, Statistical Tables and Appendixes. USA-*9700*.F5E9, 1935-36.

Family Income and Expenditures, Five Regions, Part 2, Family Expenditures, Farm Series. USA-*9700*.F5I1, 1941.

Family Income and Expenditures, Five Regions, Part 2, Family Expenditures, Urban and Village Series. USA-*9700*.F5I2, 1940.

Family Income in Chicago, 1935-36. USA-*9700*.F5C, 1935-36.

The Farm Income Situation. USA-*6200*.F4.

"The Incidence of Illness and the Volume of Medical Services among 9,000 Canvassed Families." USA-*7110*.P5, 1933-43.

Income and Spending and Saving of City Families in Wartime. USA-*9700*.I5, 1942.

Medical Care and Costs in Relation to Family Income. USA-*7190*.M5, 1943.

"Postwar Economic Perspectives: I, Experience after World War I." USA-*9950*.S4, Dec. '45, Woytinsky.

"Postwar Economic Perspectives: II, Prewar Experience—The Labor Force and Employment." USA-*9950*.S4, Jan. '46, Woytinsky.

"Postwar Economic Perspectives: III, Prewar Experience—Production and Consumption." USA-*9950*.S4, Feb. '46, Woytinsky.

"Postwar Economic Perspectives: IV, Aftermath of the War" USA-*9950*.S4, Mar. '46, Woytinsky.

Printed Publications for the Research Worker. USA-*9700*.P5, 1944.

Rural Family Spending and Saving in Wartime. USA-*9700*.R4, 1943.

Rural Level of Living Indexes for Counties of the United States, 1940. USA-*9700*.R6, 1940.

Sixteenth Census of the United States: 1940: Population: Families: Employment Status. USA-*2200*.C5, 1940.7c.

Sixteenth Census of the United States: 1940: Population: Families: Family Wage or Salary Income in 1939. USA-*2200*.C5, 1940.7d.

9000-9999 LABOR AND LIVING CONDITIONS—cont.

9700 Family living—cont.

Sixteenth Census of the United States: 1940: Population and Housing: Families: General Characteristics. USA-*2200*.C5, 1940.7a.
Sixteenth Census of the United States: 1940: Population and Housing: Families: Income and Rent. USA-*2200*.C5, 1940.7e.
Sixteenth Census of the United States: 1940: Population: Families: Size of Family and Age of Head. USA-*2200*.C5, 1940.7f.
Sixteenth Census of the United States: 1940: Population and Housing: Families: Tenure and Rent. USA-*2200*.C5, 1940.7g.
Sixteenth Census of the United States: 1940: Population: Families: Types of Families. USA-*2200*.C5, 1940.7h.
Sixteenth Census of the United States: 1940: Population: The Labor Force (Sample Statistics): Wage or Salary Income in 1939. USA-*2200*.C5, 1940.10f.
Spending and Saving of the Nation's Families in Wartime. USA-*9700*.S5, 1942.

Venezuela.
"Resumen de la Primera Parte de la Encuesta sobre las Condiciones de Vida de las Familias de Caracas." Ven-*8800*.R5, Mar. '46.

Non-Geographic.
"What Level of Living Indexes Measure." USA-*9000*.A6, Feb. '44, Hagood.

9800 Food consumption and nutrition.
Consumo de alimentos y nutrición.

International.
Documents Relating to the Food and Agriculture Organization of the United Nations, 1st August-14th December, 1944. Int-*6200*.D6, 1944.
Encuestas sobre el Consumo de Alimentos, y la Nutrición en las Américas. Int-*9800*.E5, 1942.
Final Act of the United Nations Conference on Food and Agriculture, Hot Springs, Virginia, United States of America, 18th May-3rd June, 1943. Int-*6200*.C5, 1943.
Food Consumption and Dietary Surveys in the Americas. Int-*9800*.F4, 1942.
International Comparisons of Food Costs. Int-*9800*.I5, 1942.
Methods and Problems of International Statistics with Special Reference to Food and Agriculture: A List of Reference. Int-*6200*.M2, 1944.
"Las Poblaciones y la Nutrición en el Nuevo Mundo." Int-*1300*.A5, 1940.3, Mardones Restat.
Report of the First Session of the Conference, Held at the City of Quebec, Canada, October 16 to November 1, 1945. Int-*6200*.C5, 1945.2.
El Subconsumo de Alimentos en América del Sur. Int-*9800*.S5, 1942.
World Food Situation. Int-*9800*.W4, 1946.

Argentina.
La Alimentación de la Familia en Buenos Aires. Arg-*9800*.A5.
El Costo de la Alimentación. Arg-*9800*.C5.
Memoria Anual del Instituto Nacional de la Nutrición. Arg-*9800*.M5.

Brasil.
"Sôbre o Consumo de Alguns Gêneros Alimentícios." Bras-*3040*.R5, Jan.-Mar. '46, Gualberto.
A Velocidade de Circulação dos Estoques de Produtos Alimentícios nos Mercados do Distrito Federal e de São Paulo, em 1944. Bras-*9800*.V5, 1945.

Chile.
"XI Censo General de Población." Chile-*1420*.E5, Dec. '41.

Ecuador.
Indices del Costo de la Vivienda Popular en Quito, de 1938 a 1944. Ec-*9800*.I5, 1945.

Perú.
The Food Situation in Southern Peru. Perú-*9800*.F5, 1943.
La Situación Alimenticia en el Perú, 1943-1944. Perú-*9800*.S4, 1945.

9000-9999 LABOR AND LIVING CONDITIONS—cont.

9800 **Food consumption and nutrition—cont.**

Situación Alimenticia en la Ciudad de Iquitos. Perú-*9800*.S5, 1944.
Situación Alimenticia en los Departamentos del Norte. Perú-*9800*.S7, 1944.

United States.
Diets of Families of Employed Wage Earners and Clerical Workers in Cities. USA-*9800*.D5, 1939.
Sixteenth Census of the United States: 1940: Population and Housing: Families: Characteristics of Rural-Farm Families. USA-*2200*.C5, 1940.7b.
Wartime Food Purchases. USA-*9800*.W5, 1945.

Uruguay.
Boletín Mensual del Instituto Nacional de Alimentación. Ur-*9800*.B5.

Non-Geographic.
"Técnicas y Procedimientos para las Encuestas o Censos Dietéticos." Int-*7110*.B5, June '42.

9900 **Housing conditions.**
Condiciones de la vivienda.

International.
"Commission on Building and Housing Statistics: Preliminary Report." Int-*3040*.R5, 1938.2, Nyström.
Housing and Hospital Projects of Latin American Social Security Systems. Int-*9950*.H5, 1942.
Housing Statistics. Int-*3200*.H5, 1939.
Low-Cost Housing in Latin America. Int-*9900*.L5, 1943.
Methods of Compiling Housing Statistics. Int-*9900*.M5, 1928.
"Observations on the Possibility of Improving the International Comparability of Building and Housing Statistics." Int-*3040*.R5, 1936.1, Nyström.
Report to the Council on the Work on the Seventh Session. Int-*3000*.R6, 1938, Eng.
Statistique Internationale des Grandes Villes: Statistique du Logement dans les Grandes Villes, 1928-1934. Int-*9900*.S5, 1928-34.
Statistiques de l'Habitation. Int-*3200*.S5, 1939.

Argentina.
"Precio Medio de la Habitación Ofrecida en la Ciudad de Buenos Aires (1943-1945)." Arg-*3040*.R7, Jan.-June '46, a.

Brasil.
Recenseamento do Brasil Realizado em 1° de Setembro de 1920. Bras-*2200*.R4, 1920.

Colombia.
Censo General de Población, 5 de Julio de 1938. Col-*2200*.C5, 1938.
"Esquema del Plan de Organización Puesto al Servicio del Levantamiento de los Censos de Edificios y Población Llevados a Cabo en Colombia en 1938." Int-*1300*.A5, 1940.3, Suárez Rivadeneira.
Primer Censo Nacional de Edificios, Efectuado el 20 de Abril de 1938. Col-*2220*.C5, 1938.
"El Problema de los Arrendamientos de Viviendas en Bogotá." Col-*1420*.A5, July-Aug. '45, Abrisqueta.

Ecuador.
"Indices del Costo de la Vivienda Popular en Quito de 1938 a 1945." Ec-*1420*.T5, July-Dec. '45, b.

México.
Segundo Censo de Edificios, 20 de Octubre de 1939: Datos Definitivos. Méx-*2220*.C5, 1939.2.
Segundo Censo de Edificios de los Estados Unidos Mexicanos, 20 de Octubre de 1939: Métodos y Procedimientos. Méx-*2120*.C5, 1939.1.
Segundo Censo de Edificios de los Estados Unidos Mexicanos: Resumen General. Méx-*2220*.C5, 1939.1.

9000-9999 LABOR AND LIVING CONDITIONS—cont.

9900 Housing conditions—cont.

Nicaragua.
Censo de Edificios, 18 de Enero de 1940: Resumen por Departamentos y Municipios. Nic-*2220*.C5, 1940.
Primer Censo de Edificios: 18 de Enero de 1940: Cédula. Nic-*2120*.C5, 1940.

Perú.
Censo Nacional de Población y Ocupación, 1940. Perú-*2200*.C5, 1940.1.

República Dominicana.
Anuario Estadístico de la República Dominicana. RepDom-*1410*.A5.

United States.
Annual Report, U. S. National Housing Agency. USA-*9900*.A5.
The Farm-Housing Survey. USA-*6200*.F3, 1939.
Federal Home Loan Bank Review. USA-*9900*.F5.
Sixteenth Census of the United States: 1940: Housing. USA-*2220*.C5, 1940.1.
Sixteenth Census of the United States, 1940: Housing: Analytical Maps . . . ; Block Statistics. USA-*2220*.C5, 1940.2.
Sixteenth Census of the United States: 1940: Housing: Characteristics by Type of Structure. USA-*2220*.C5, 1940.3.
Sixteenth Decennial Census of the United States: Instructions to Enumerators, Housing, 1940. USA-*2120*.C5, 1940.1.
War Housing in the United States. USA-*9900*.W5, 1945.

——. Alaska.
Sixteenth Census of the United States: 1940: Alaska: Population: Characteristics of the Population (with Limited Data on Housing). USA Alsk-*2200*.C5, 1940.

——. Puerto Rico.
Censo Décimosexto de los Estados Unidos: 1940: Puerto Rico: Viviendas. Sixteenth Census of the United States: 1940: Puerto Rico: Housing. USA PR-*2220*.C5, 1940.

Other Countries.
West Indies.
Eighth Census of Jamaica and Its Dependencies, 1943: Population, Housing and Agriculture. WI-*2200*.E5, 1943.
Eighth Census of Jamaica, 1943: Introduction to Population and Housing. WI-*2200*.E5, 1943.2.

Non-Geographic.
"Marketing and Sampling Uses of Population and Housing Data." USA-*3040*.J5, Mar. '43, Eckler.

9950-9959 Social security.
Seguridad social.

9950 ——. General.

International.
Housing and Hospital Projects of Latin American Social Security Systems. Int-*9950*.H5, 1942.

Bolivia.
Memoria de la Caja de Seguro y Ahorro Obrero. Bol-*9950*.M5.
Protección Social. Bol-*9950*.P5.

Brasil.
Relatório Anual do Serviço de Estatística da Previdência e Trabalho. Bras-*9950*.R5.

Canada.
Report on Social Security for Canada. Can-*9950*.R5, 1943.

Chile.
Estadísticas de la Caja de Seguro Obligatorio. Chile-*9950*.E5.
Finanzas, Bancos y Cajas Sociales. Chile-*6720*.F5.

9000-9999 LABOR AND LIVING CONDITIONS—cont.

9950 General—cont.

Previsión Social. Chile-*7110*.P5.
La Seguridad Social en Chile. Chile-*9950*.S5. 1942.

Costa Rica.
Memoria de la Secretaría de Salubridad Pública y Protección Social. CR-*7110*.M5.

Cuba.
Salubridad y Asistencia Social. Cuba-*7110*.S5.

Ecuador.
Documentos Anexos al Informe que el Ministro de Previsión Social y Trabajo Presenta a la Nación. Ec-*9000*.D5.
Informe a la Nación (del Ministro de Previsión Social y Trabajo). Ec-*9000*.I5.

Honduras.
Informe de los Actos Realizados por el Poder Ejecutivo en los Ramos de Gobernación, Justicia, Sanidad y Beneficencia. Hond-*8000*.I5.

México.
"La Oficina de Estadística del Instituto Mexicano del Seguro Social." Int-*3040*.E5, Sept. '45, Rodríguez y Rodríguez.

Panamá.
Boletín de Previsión Social. Pan-*9950*.B5.

Perú.
Informaciones Sociales. Perú-*9000*.I5.
Memoria Anual de la Caja Nacional de Seguro Social. Perú-*9950*.M5.

United States.
Bureau Memorandums of the Bureau of Research and Statistics of the Social Security Board. USA-*9950*.B5.
Employment Security Activities. USA-*9950*.E5.
Social Security Bulletin. USA-*9950*.S4.
Social Security Yearbook. USA-*9950*.S6.
Survey of Social Security Statistics. USA-*9950*.S8, 1944.

Venezuela.
Memoria del Ministerio del Trabajo y de Comunicaciones. Ven-*9000*.M5.

Non-Geographic.
"La Statistique de Morbidité á la Sécurité Sociale." Fr-*3040*.J5, Sept.-Oct. '46, Marx.

9951 ——. Assistance.
Asistencia.

International.
Acta Final del Octavo Congreso Panamericano del Niño, Wáshington del 2 al 9 de Mayo de 1942. Int-*9951*.P5, 1942, Sp.
Ata Final do Oitavo Congresso Panamericano da Criança, Washington, 2-9 de Maio de 1942. Int-*9951*.P5, 1942, Port.
Final Act of the Eighth Pan American Child Congress, Washington, D.C., May 2-9, 1942. Int-*9951*.P5, 1942, Eng.

Argentina.
"Protección a la Familia Argentina." Arg-*3040*.R5, Feb. '43, Casiello.

Chile.
Demografía y Asistencia Social. Chile-*5000*.D5.

El Salvador.
Informe de las Labores Desarrolladas por el Poder Ejecutivo en los Ramos de Cultura y Asistencia Social. ElSal-*7000*.I5.
Memoria de Asistencia Social y Beneficencia. ElSal-*9951*.M5.

Haiti.
Bulletin du Service National d'Hygiène et d'Assistance Publique. Haiti-*7110*.B5.

9000-9999 LABOR AND LIVING CONDITIONS—cont.

9951 ——. Assistance—cont.

United States.
Publications of the Children's Bureau. USA-*9951*.P5.
Sixteenth Census of the United States: 1940: Population: Special Report on Institutional Population, 14 Years Old and Over. USA-*2200*.C5, 1940.9.
Venezuela.
Memoria y Cuenta del Ministerio de Sanidad y Asistencia Social. Ven-*7110*.M5.
Revista de Sanidad y Asistencia Social. Ven-*7110*.R5.

9952 ——. Insurance.
Seguros.

Argentina.
Conceptos y Principios sobre Jubilaciones. Arg-*9952*.C5, 1941.
Canada.
Annual Report on Current Benefit Years Under the Unemployment Insurance Act. Can-*9952*.A5.
Statistical Report on the Operation of the Unemployment Insurance Act. Can-*9952*.S5.
Costa Rica.
La Conmutación de Pensiones Futuras como Reparación Civil en Caso de Muerte. CR-*9952*.C5, 1944.
United States.
"The First Two Years of Social Insurance in Mexico." USA-*9950*.S4, July '46, Cohen.
Annual Report of the Railroad Retirement Board. USA-*9952*.A5.
The Monthly Review of the Railroad Retirement Board. USA-*9952*.M5.

APPENDIX 1

SUBJECT CLASSIFICATION SCHEME FOR STATISTICAL MATERIALS

(Note: Many economies have resulted in the work of the Inter American Statistical Institute through the use of a single over-all subject classification scheme in the administration of the Permanent Office. The classification which follows is now used in the Statistical Source Files and for all bibliography purposes, in the Biographical Files of Statistical Personnel, the quarterly Journal (in the bibliography section), the correspondence files, and the account books.

It is the hope of the Institute that a subject classification for statistical materials can be established on a world-wide basis, since its general use would greatly facilitate the efficient handling and filing of materials, information, and data exchanged by international organizations and national statistical agencies. The Institute expects to continue using the present classification until or unless something of a corresponding nature emerges as a world standard.)

OUTLINE OF PRIMARY GROUPS (9)

1000–1999 GENERAL.

2000–2999 CENSUS.

3000–3999 STATISTICAL SCIENCE.

4000–4999 GEOGRAPHY.

5000–5999 DEMOGRAPHY.

6000–6999 ECONOMICS.

7000–7999 HEALTH AND CULTURAL STATISTICS.

8000–8999 GOVERNMENT AND POLITICAL STATISTICS.

9000–9999 LABOR AND LIVING CONDITIONS.

OUTLINE OF SECONDARY GROUPS (60)

1000–1999 GENERAL.

- 1000 General.
- 1100 Administrative controls.
- 1200 Bibliography.
- 1300 Congresses and conferences.
- 1400 Data compilations.
- 1500 Filing and cataloging.
- 1600 Microfilm and preservation of records.
- 1700 Personnel.
- 1800 Publication processes.
- 1900 Reference books, general.

2000–2999 CENSUS.

- 2000 General.
- 2100 Methods and procedures.
- 2200 Results.

3000–3999 STATISTICAL SCIENCE.

- 3000 General.
- 3100 Methods and techniques.
- 3200 Standards.
- 3300 Training materials and methods.

4000–4999 GEOGRAPHY.

- 4000 General.
- 4100 Area and geographical characteristics.
- 4200 Climate and weather.
- 4300 Natural resources, general surveys.

5000–5999 DEMOGRAPHY.

- 5000 General.
- 5100 Population.
- 5200 Vital statistics and vital records.
- 5300 Migration.

6000–6999 ECONOMICS.

- 6000 General.
- 6100 National income and national wealth.
- 6200 Agriculture, livestock, forestry, fishery.
- 6300 Mining and extractive industry.

ECONOMICS—Cont.

- 6400 Manufacturing industry.
- 6500 Building and other construction.
- 6600 Transport and communication.
- 6700 Commerce (except foreign trade).
- 6800 Personal and recreational services.
- 6900 Foreign trade.

7000–7999 HEALTH AND CULTURAL STATISTICS.

- 7000 General.
- 7100 Health.
- 7200 Education.
- 7300 Arts, sciences, and public information services.
- 7400 Religion.

8000–8999 GOVERNMENT AND POLITICAL STATISTICS.

- 8000 General public administration and enterprise.
- 8100 Defense and war.
- 8200 International relations.
- 8300 Justice, delinquency, and crime.
- 8400 Law and legislation.
- 8500 Planning and research—national, regional, and civic.
- 8600 Elections, political parties, and public opinion polls.
- 8700 Political systems.
- 8800 Public finance and taxation—national, state, municipal, local.

9000–9999 LABOR AND LIVING CONDITIONS.

- 9000 General.
- 9100 Labor conditions, general.
- 9200 Employment and wages.
- 9300 Labor relations.
- 9400 Industrial accidents and occupational disease.
- 9500 Cost of living.
- 9600 Living conditions, general.
- 9700 Family living.
- 9800 Food consumption and nutrition.
- 9900 Housing conditions.
- 9950 Social security.

COMPLETE OUTLINE

(9 primary, 60 secondary, 115 tertiary, and 24 quartic headings)

1000–1999 GENERAL

(Includes: Materials of general reference, or pertaining to more than one major subject field.)

1000 General.
1100 Administrative controls. (Includes: Accounting, budget, cost analysis, flow charts, inter-office procedures, inventory control, organization charts, production control.)
1200 Bibliography. (General in scope.)
1300 Congresses and conferences. (General in scope.)
Data compilations. (General in character and essentially statistical. Excludes censuses. For data compilations pertaining to specific subject matter, see those fields.)
1400 ——. General. (1-time statistical data compilations.)
1410 ——. Annuals. (Includes: All general national statistical yearbooks.)
1420 ——. Periodicals. (Containing the more important current statistical series.)
1500 Filing and cataloging. (Includes: Instruction manuals and classification schemes for filing and cataloging.)
1600 Microfilm and preservation of records. (Includes: Microcopy.)
1700 Personnel. (Includes: Job classification, morale and psychology, personnel management, promotion, supervision and training, rosters of scientific and other personnel, biographical files.)
1800 Publication processes. (Includes: Hectographing, multilithing, etc. Excludes: Microfilm.)
Reference books, general.
1900 ——. General.
1910 ——. Annuals and almanacs.
1920 ——. Dictionaries and glossaries.
1930 ——. Directories of individuals and organizations.
1940 ——. Encyclopedias.
1950 ——. Guide books and handbooks of countries and places.
1960 ——. Style manuals, grammars, etc.

2000–2999 CENSUS

(A "pull-together" of all official census material of each country.
Each entry appears also under the specific subject field.)

2000 General.
Methods and procedures. (Includes: Forms, instructions, laws, schedules, sampling and other techniques as applied to censuses.)
2100 ——. General and population.
2110 ——. Agriculture.
2120 ——. Buildings, real estate, investments.
2130 ——. Education.
2140 ——. Electoral.
2150 ——. Industry, manufacturing, business.
2160 ——. Institutions.
2170 ——. Mining.
2180 ——. Transport and communication.
Results.
2200 ——. General and population. (Includes: Occupation and unemployment censuses.)
2210 ——. Agriculture. (Includes: Drainage and irrigation censuses, livestock, etc.)
2220 ——. Buildings, real estate, investments. (Includes: Censuses of housing, mortgages, and personal property.)
2230 ——. Education. (Includes: School census.)
2240 ——. Electoral.
2250 ——. Industry, manufacturing, business. (Includes: Commercial censuses.)
2260 ——. Institutional. (Includes: Religious censuses.)
2270 ——. Mining.
2280 ——. Transport and communication.

3000–3999 STATISTICAL SCIENCE

(Primarily for statistical methodology which may apply to any subject field.)

General.
3000 ——. General.
3005 ——. Dictionaries and glossaries of statistical terms.
3010 ——. Bibliography.
3020 ——. Congresses and conferences. (Primarily statistical.)
3030 ——. Organization of statistical services, agencies, and societies—descriptive matter and laws.
3040 ——. Periodicals. (Containing important articles on statistical methodology.)
Methods and techniques. (Excludes: Nomenclatures to standardize statistical classifications. See also 3300.)
3100 ——. General. (Includes: Computation, codes, definitions, equations, forms, graphs, instructions.)
3105 ——. Analysis of data.
Census—see 2100–2199.
3110 ——. Data collection and treatment. (Includes: Instruction manuals on the subject, punched card applications, etc.)
3120 ——. Index numbers.
3130 ——. Mapping.
3140 ——. Mathematical statistics. (Includes: Mathematical probability.)
3150 ——. Sampling.
3160 ——. Quality control.
Standards. (Measures to obtain uniformity, including nomenclatures or other lists and schemes to standardize statistical classifications. A "pull-together" of all material on statistical standards. Each entry appears also under the specific subject field.)
3200 ——. General. (Includes: Bibliographies on nomenclatures.)
3205 ——. Balance of payments.
3210 ——. Foreign trade classification. (Includes: Commodity classifications.)
3220 ——. Industrial classification.
3230 ——. Mortality and morbidity classifications.
3240 ——. Occupational classification (gainfully occupied population.)
3250 ——. Personnel. (For statistical posts.)
3260 ——. Tabular and graphic presentation.
3270 ——. Weights and measures.
Training materials and methods.
3300 ——. General.
3310 ——. Fellowships, scholarships, and travel grants.
3320 ——. Outlines and courses of study.
3330 ——. Textbooks.

4000–4999 GEOGRAPHY

4000 General. (Includes: Atlases and maps.)
4100 Area and geographical characteristics. (Includes: Political divisions.)
4200 Climate and weather. (Includes: Barometric pressure, meteorology, precipitation, temperature, tides, wind velocity.)
4300 Natural resources, general surveys. (Includes: Forest reserves, mineral reserves, water resources, wild life and game, national park lands, etc.)

5000–5999 DEMOGRAPHY

5000 General. (Includes: Biometry and biometrics, demographic organizations, human biology.)
Population.
5100 ——. General. (Includes: Population estimates, population policy.)

DEMOGRAPHY—cont.

5110 ——. Characteristics. (Includes: Family characteristics, age groups, center of population, density, employment and occupations, growth and trends, illiteracy, marital status, sex, urban and rural.)

5120 ——. Race and nationality. (Includes: Alien, foreign born, Indian, Jew, Negro, etc.)

Vital statistics and vital records.

——. General.

5200 ——. ——. General.

5201 ——. ——. Bibliography.

5202 ——. ——. Congresses and conferences.

5203 ——. ——. Organizations and agencies—descriptive matter and laws.

——. Birth, death, stillbirth.

5210 ——. ——. General.

5211 ——. ——. Birth. (Includes: Birth control, birth rates, birthplace, fertility, fecundity, midwifery, reproductive rate, delayed births, sterility.)

5212 ——. ——. Death. (Includes: Death rates, death classification, International List of Causes of Death, selected causes of death lists, particular causes of death.)

5213 ——. ——. Stillbirth.

5220 ——. Life tables. (Includes: Expectation of life, longevity.)

——. Marriage and divorce.

5230 ——. ——. General.

5231 ——. ——. Marriage. (Includes: Marriage rates.)

5232 ——. ——. Divorce. (Includes: Divorce rates.)

5240 ——. Methods and procedures. (Includes: Certificates, forms, codes, indexing procedures, instruction manuals, etc.)

5250 ——. Registration and registration practices. (Includes: Adoptions, certification, completeness of registration, correction of records, legal use of records, legitimation, illegitimacy, personal identity—including fingerprinting—presumptive death practices, underregistration, etc.)

Migration.

5300 ——. General. (Includes: Emigration, immigration, and naturalization; colonization, land settlement, and resettlement of minority groups.)

5310 ——. Tourism.

6000–6999 ECONOMICS

General

6000 ——. General. (Includes: Depressions, economic geography, economic organizations, forecasting, profit sharing, purchasing power, theories of economics.)

6010 ——. Prices and index numbers. (General in scope and covering several fields. Includes: General indexes of economic activity, inflation, price control. For retail prices and indexes in connection with cost of living, see 9500. For other prices and indexes in specific subject fields, see those fields. For index number methods and techniques, see 3120.)

6020 ——. Production, consumption, stocks. (General in scope and covering several fields.)

6100 National income and national wealth. (Includes: Over-all estimates of gross product.)

Agriculture, livestock, forestry, fishery.

6200 ——. General. (Includes: Agricultural organizations; education and extension service; farm and ranch characteristics; land reclamation, registration, tenure, use; subsidies; world food supply.)

6210 ——. Commodities. (Includes: Production, marketing, stocks.)

6220 ——. Agriculture. (Includes: Crop reporting; drainage and irrigation—except construction; fertilizer; gardening; truck farming.)

6230 ——. Livestock and small animal life. (Includes: Beekeeping, dairying and poultry production; diseases of animals and veterinary medicine; game reserves, game protection, hunting.)

6240 ——. Fishery. (Includes: Catch, fishery industry, fish resources.)

6250 ——. Forestry. (Includes: Reforestation, resources.)

6300 Mining and extractive industry. (Includes: Bauxite, coal, natural gas, oil wells, petroleum, salt, tin, tungsten.)

ECONOMICS—cont.

6400 Manufacturing industry. (Includes: General statistics of industries which overlap with other headings; public utility industries for electric light and power, gas and coke, heat, water and water power.)

6500 Building and other construction. (Includes: Heavy construction, highway construction and maintenance, irrigation and reclamation works.)

Transport and communication. (Includes: Existing networks, transport equipment, traffic.)

6600 ——. General.

——. Transport.

6610 ——. ——. General.

6611 ——. ——. Air.

6612 ——. ——. Highway. (Includes: Automobiles, buses, trucks.)

6613 ——. ——. Pipe line.

6614 ——. ——. Rail. (Includes: Railroads, streetcars.)

6615 ——. ——. Water. (Includes: Canals, rivers, harbor, shipping and navigation.)

——. Communication.

6620 ——. ——. General.

6621 ——. ——. Posts and post offices.

6622 ——. ——. Telephone and telegraph. (Includes: Cable, radio, television, teletype.)

Commerce (except foreign trade).

6700 ——. General. (Includes: Data overlapping with the group "Personal and recreational services.")

6710 ——. Wholesale and retail trade. (Includes: Co-operatives, marketing, trademarks.)

——. Banking and finance. (Excludes: Public finance.)

6720 ——. ——. General. (Includes: Bank deposits, bankruptcy, building and loan associations, bullion, circulation of money and notes, clearinghouses, coinage, credit, credit unions, exchange rates, federal reserves, gold and silver reserves, interest and discount rates, loans.)

6721 ——. ——. Securities, investments, holdings. (Includes: Prices and indexes of bonds, mortgages, royalties, stocks, etc.; property and real estate holdings and transactions; capital, earnings, profits, losses.)

6730 ——. Insurance, commercial. (Excludes: Social insurance. Includes: Actuarial science training, annuities, etc.)

6800 Personal and recreational services. (Includes: Provision of lodging, food, and drink; cleaning and maintenance of goods and premises; personal care services; domestic service; amusement, physical training, recreation, and sport undertakings, including city playgrounds and national parks. Note that data overlapping with the group "Commerce" are carried under that heading.)

Foreign trade.

6900 ——. General. (Includes: Cartels; classification schemes; exports and imports; special trade arrangements such as barter, duties, favored nations, free ports, licenses, tariffs.)

6910 ——. International balance of payments.

7000–7999 HEALTH AND CULTURAL STATISTICS

7000 General.

Health.

——. General.

7100 ——. ——. General. (Includes: Directives of physicians, health education.)

7101 ——. ——. Organizations. (Health departments and agencies—descriptive matter and laws.)

7110 ——. Public health activities. (Includes: Epidemiology, mental hygiene, physical examination programs, preventive medicine, public health surveys, sanitation, vaccination.)

7120 ——. Dentistry.

7130 ——. Hospitals. (Includes: Clinics, private group clinics, institutions for feebleminded and mental defectives, sanitariums, rest homes, hospital medical records.)

7140 ——. Nursing.

HEALTH . . . —cont.

7150 ——. Morbidity. (Includes: Diseases, codes of morbidity, intemperance, alcoholism, drug addiction.)

7160 ——. Accidents, except industrial accidents. (Includes: Definitions, prevention and safety measures, transportation accidents, accident morbidity and mortality.)

7170 ——. Defects and deformities. (Includes: Congenital, traumatic, and sensory defects; feebleminded persons; handicapped persons.)

7180 ——. Medical science. (Includes: Cults and history of medical science. Specify particular medical science, as: Anatomy, bacteriology, clinical medicine, dermatology, laboratory medicine (microscopy), neurology, obstetrics and gynecology, ophthalmology, otology, laryngology and rhinology, pathology, pediatrics, pharmacology, physical growth, physiology, surgery.)

7190 ——. Group medicine and cost of medical care. (Includes: Group health plans, medical care, prepayment plans.)

Education.

7200 ——. General. (Includes: Educational psychology, illiteracy, intelligence and personality testing, libraries and museums, teachers' training, vocational guidance.)

7210 ——. Educational institutions. (Includes: Primary, secondary, and university education.)

7220 ——. Courses and curricula.

7230 ——. Fellowships and scholarships.

Arts, sciences, and public information services.

7300 ——. General. (Includes: Architecture, astronomy, biology, botany, chemistry, eugenics, geology, history, literature, mathematics—except mathematical statistics, music, physics, pyschology, scientific societies in these sciences, zoology.)

7310 ——. Anthropology, anthropometry, archaeology. (Includes: Anthropometric measurements, cephalic index.)

7320 ——. Language. (Includes: Articles on orthography.)

7330 ——. Public information services. (Includes: The press.)

7400 Religion. (Includes: Churches and other religious organizations and activities.)

8000–8999 GOVERNMENT AND POLITICAL STATISTICS

8000 General public administration and enterprise. (For specific public enterprise activities, see the particular subject field. Includes: Apportionment, civil service, general plans of government reorganization.)

Defense and war.

8100 ——. General. (Includes: Civilian defense, lend lease, selective service.)

8110 ——. Armed forces. (Specify whether air, army, marines, navy. Includes: Military deaths.)

International relations. (Includes: International law and general international organizations which are not dedicated to a specific subject.)

8200 ——. General.

8210 ——. Conventions and treaties. (Conventions in sense of agreements, not congresses.)

8220 ——. Military government (foreign occupation).

Justice, delinquency, and crime.

8300 ——. General. (Includes: Capital punishment, police system, prostitution, Supreme Court.)

8310 ——. Court statistics. (Includes: Judicial statistics.)

8320 ——. Juvenile delinquency.

8330 ——. Penal institutions. (Includes: Jails, penitentiaries, prisons, reformatories.)

8400 Law and legislation. (General in scope. For laws and legislation pertaining to specific fields, see those fields.)

8500 Planning and research—national, regional, and civic. (General in scope. By public or private bodies. Includes: Postwar planning.)

8600 Elections, political parties, public opinion polls.

8700 Political systems. (Includes: Bolshevism, communism, democracy, fascism.)

8800 Public finance and taxation—national, state, municipal, local. (Includes: Income and expenditures, defense expenditures, budget and debt, and all forms of taxes.)

9000–9999 LABOR AND LIVING CONDITIONS

(Note that material in this group may crosscut the major subject groups DEMOGRAPHY, ECONOMICS, HEALTH . . . , GOVERNMENT . . .)

9000 General. (Includes: Labor organizations, sociology, social and mob psychology, social surveys.)

9100 Labor conditions, general. (Includes: Studies of labor productivity.)

9200 Employment and wages. (Includes: Hours of work; rate and amount of wages and salaries—including professionals, managers, and directors; family allowances; gainfully employed; income by professions; indexes of employment and wages; labor market; labor supply and demand; occupational nomenclature; unemployment.)

9300 Labor relations. (Includes: Labor unions, collective agreements and industrial disputes; conciliation and arbitration; labor inspection.)

9400 Industrial accidents and occupational disease. (Includes: Measures for prevention of accidents; industrial fatigue, hazards, and hygiene.)

9500 Cost of living. (Includes: Retail prices and indexes of the various elements entering the cost of living; inter-regional and international comparisons.)

9600 Living conditions, general. (Includes: General living conditions of economic, social, or racial groups.)

9700 Family living. (Includes: Composition of family, distribution of expenditures, level and standard of family living, individual and family income and wealth, nature and source of income.)

9800 Food consumption and nutrition.

9900 Housing conditions. (Includes: Size, tenure, and type; sanitary and other facilities; occupancy and overcrowding; rental scales; low cost housing programs; rural housing.)

Social security. (Includes: Social assistance both public and private.)

9950 ——. General. (Includes: Employment exchanges, social welfare methods, workmen's compensation.)

9951 ——. Assistance. (Includes: Almshouses; homes for the blind; charity; child welfare; dependency; disability; handicapped persons; maternal, old age, orphan, and poor relief; public relief; unemployment relief.)

9952 ——. Insurance. (Includes: All forms of social as distinguished from commercial insurance; pensions.)

9953 ——. Social security systems and programs.

ALPHABETICAL INDEX

A

Abortion, 5212
Accident:
- General, 7160; 9400
- Deaths, 5212; 7160; 9400
- Industrial, 9400
- Insurance, 6730; 9952
- Morbidity, 7160
- Prevention, 9400; 7160
- Traffic, 7160

Accounting, 1100
Acreage of farms and ranches, 6200
Actuarial science and training programs, 6730
Administration:
- Personnel, 1700
- Public, 8000–8999
- Statistical, 3000–3099

Administrative controls, 1100
Adoptions, 5250
Advertising, 6710
Age:
- Aging population, 5110
- Diseases related to, 7150
- Groups in population, 5110

Agencies, statistical service, 3030
Agriculture:
- General (including associations and co-operatives, teaching, etc.), 6200
- Censuses of, 2210; 2110; 6200

Air, conditions of, 4200
Air forces, 8110
Air lines, 6611
Airplanes, manufacture of, 6400
Alcoholism, 7150
Aliens:
- Number of, 5120
- Social problems of, 9600

Almanacs, 1910
Almshouses, 9951
Aluminum, 6400
Ammunition, 6400
Amusement and show business, 6800
Analysis:
- Cost, 1100
- Statistical, 3105

Anatomy, 7180
Animals, diseases of, 6230
Annuals:
- General statistical, 1410
- In specific subject fields, see the particular field.
- Non-statistical, 1910

Annuities, 6730
Anthropology, 7310
Anthropometry and anthropometric measurements, 7310
Apportionment, 8000
Archaeology, 7310
Architecture, 7300
Army and armed forces, 8110
Arts, 7300–7399
Assistance, social, 9951
Associations, see the particular subject field.
Astronomy, 7300
Atlases:
- General, 4000
- Statistical, 1400; 3130

Automobile, 6612
Automobiles, manufacture of, 6400
Aviation:
- General, 6612
- Commercial, 6611
- Military, 8110

B

Bacteriology, 7180
Balance of international payments, 6910; 3205
Bankruptcy, 6720
Banks and banking, 6720
Barometric pressure, 4200
Barter, 6900
Bauxite, 6300
Beekeeping, 6230
Bibliography:
- General, 1200
- In specific subject fields, see the particular field.
- Statistical, 3010

Biographical material, 1930; 1700
Biology:
- General, 7300
- Human, 5000

Biometry and biometrics, 5000
Birth:
- General, 5211
- Control, 5211
- Rates, 5211
- Registration, 5250

Birthplace, 5211
Blind, 7170
Blind, homes for, 9951
Bolshevism, 8700
Bond prices and indexes, 6721
Botany, 7300
Budget:
- Administrative, 1100
- Family, 9700
- National, 8800

Building activity and indexes, 6500
Building and loan associations, 6720
Buildings, 2220; 9900
Bullion, 6720

Burial insurance, 6730; 9952
Buses, 6612
Business:
General, 6700–6799; 6800
Indexes, 6700–6799
Management, 1100

C

Cable, 6622
Calculating machines, 3110
Calendar reform, 7300
Canals, 6615
Capital, 6721
Capital punishment, 8300
Car loadings, 6614
Cartels, 6900
Cartography, 4000; 3130
Cataloging, 1500
Catch, fish, 6240
Cause of death, 5212
Cement, 6400
CENSUS:
General, 2000–2999; see also the particular subject field.
Agricultural, 2110; 2210; 6200
Building, 2120; 2220; 6500
Commercial, 2150; 2250; 6700
Communications, 2180; 2280; 6620
Construction, 2120; 2220; 6500
Education, 2130; 2230; 7200
Electoral, 2140; 2240; 8600
Forms, 2100–2199
Housing, 2120; 2220; 9900
Industrial, 2150; 2250; 6400
Institutional, 2160; 2260
Laws, 2100–2199
Livestock, 2110; 2210; 6230
Maps, 2100–2199
Mining, 2170; 2270; 6300
Population, 2100; 2200; 5100
Publicity, 2100–2199
Religion, 2160; 2260; 7400
Schedules, 2100–2199
Transportation, 2180; 2280; 6610
Center of population, 5110
Central heating, 6400; 9900
Cephalic index, 7310
Characteristics:
Families, 5110; 9700
Farms, 6200
Population, 5110
Charity, 9951
Charts, organization, 1100; see also the particular subject field.
Chemicals, 6400
Chemistry, 7300
Child growth, 7180
Child welfare, 9951
Childbirth mortality, 5212
Churches, 7400
Circulation of money and notes, 6720
Citizenship, 5110
Civic planning, 8500
Civil register, 5250
Civil service, 8000
Civilian defense, 8100
Classification:
Bibliography, 1500
Commodity, 3210; 6900
Foreign trade, 3210; 6900
Industrial, 3220; 6400
Morbidity, 3230; 7150
Mortality, 3230; 5212
Occupational, 3240; 1700; 9000
Statistical (including nomenclatures), 3200-3299; see also the particular subject field.
Tariff, 3210; 6900
Clearinghouses, 6720
Climatology, 4200
Clinical medicine, 7180
Clinics, including private group clinics, 7130
Clothing standards, 9700
Coal, 6300
Codes, 3100
Codes of morbidity, 7150; 3230
Coffee, 6210
Coinage, 6720
Coke, 6400
Collection of data, 3110
Colonization, 5300
Commerce, domestic, 6700–6799
Commerce, foreign, 6900–6999
Commodities:
Agricultural, 6210
Foreign trade, 6900
Manufacturing, 6400
Mining, 6300
Commodity classification, 3210; 6900
Commodity prices:
General in scope, 6010
Specific, see the particular subject field.
Communication:
General, 6620–6629
Censuses of, 2180; 2280; 6620
Communism, 8700
Compilations of data, 1400–1499
Computations, statistical, 3100–3199
Congenital defects, 7170
Congresses and conferences:
General, 1300
International, 8200; see also the particular field.
In specific subject fields, see the particular field.
Construction:
General, 6500
Censuses of, 2120; 2220; 6500

Engineering technology:
General, 6000
In specific subject fields, see the particular field.
English grammars for office use, 1960
Enterprise, public:
General, 8000
In specific subject fields, see the particular field.
Enumeration:
Methods, census, 2100–2199
Results, census, 2200–2299
Epidemics, 7110
Epidemiology, 7110
Eugenics, 7300
Exchange rates, 6720
Expectation of life, 5220
Expenditures:
Family, 9700
For defense, 8800
Public, 8800
Exports, 6900
Extension and education, agricultural, 6200
Extractives and extractive industries, 6300

F

Family:
Characteristics and composition, 5110; 9700
Income and wealth, 9700
Living, 9700
Size, 5110
Farm characteristics, 6200
Farming, truck, 6220
Fascism, 8700
Favored nations, 6900
Fecundity, 5211
Federal reserves, 6720
Feeble-minded, 7170; 9951
Feeble-minded, institutions for, 7130
Fellowships available to statisticians, 3310
Fellowships, statistics of, 7230
Fertility, 5211
Fertilizer, 6220
Filing and filing systems, 1500
Finance:
General, 6720–6729
Family, 9700
Government, 8800
International, 6720–6729; 6910; 8200–8299
Municipal, 8800
National, 8800; 6100
Public, 8800
State, 8800
Fingerprinting, 5250
Fiscal statistics, 8800
Fish resources, 4300; 6240
Fishery and fishery industry, 6240
Flax, 6210
Flow charts, 1100
Food consumption and nutrition, 9800
Food supply of world, 6200
Forecasting, economic, 6000
Foreign occupation, 8220
Foreign trade and commerce, 6900–6999
Foreign trade classification, 3210; 6900
Foreign-born, number of, 5120
Foreign-born, social problems of, 9600
Forest reserves, 4300
Forest resources, 4300; 6250
Forestry, 6250
Forms:
In specific subject fields, see the particular field.
Census, 2100–2199
Statistical, 3100
Free ports, 6900
Frequency distribution, 3100–3199

G

Gainfully employed, 9200; 5110
Game:
Protection, 6230
Reserves and resources, 6230; 4300
Gardening, 6220
Gas (manufactured) and coke, 6400
Genetics, 7300
Geodesy, 4000
GEOGRAPHY, 4000–4999
Geography, economic, 6000; 4300
Geology, 7300
Glossaries:
General, 1920
In specific subject fields, see the particular field.
Gold reserves, 6720
GOVERNMENT AND POLITICAL STATISTICS, 8000–8999
Government, military, 8220
Grammars, 1960
Graphic presentation, 3260
Graphs, 3100
Group medicine, 7190
Growth:
Curves, 3100
Of children, 7180
Of population, 5110
Guano, 6220
Guide books, 1950
Gynecology, 7180

H

Handbooks of countries and places, 1950
Handicapped persons, 7170; 9951
Harbors, 6615
Health:
General, 7100–7199
Agencies, 7101
Education, 7100

Methods, procedures, and techniques:
In specific subject fields, see the particular field.
Census, 2100–2199
Statistical, 3100–3199; see also the particular subject field.
Microcopy, 1600
Microfilm, 1600
Microscopy (laboratory medicine), 7180
Midwifery, 5211
Migration, 5300–5399
Military:
Deaths, 8110; 5212
Government, 8220
Milk and milk products, 6230
Mineral reserves (resources), 4300; 6300
Minerals, 6300; 4300
Minimum list of commodities in international trade, 3210; 6900
Mining, 6300
Mob psychology, 9000
Money, 6720
Morale, 1700
Morbidity:
General, 7150
Classification and codes, 7150; 3230
Industrial, 9400; 7150
Mortality:
General, 5212
Classification and codes, 5212; 3230
Industrial, 9400; 5212
Mortgages, 6721
Motor cars, production, 6400
Multigraphing, 1800
Multilithing, 1800
Municipal finance, 8800
Municipal government, 8000–8999
Munitions, 6400
Murder, 5212; 8300
Museums, 7200
Music, 7300

N

National:
Budget, 8800
Debt, 8800
Finance, 8800
Government, 8000–8999
Income, 6100
Park lands, 4300; 6800
Planning, 8500
Resources, 6100; 4300
Rosters of scientific personnel for employment purposes, 1700
Wealth, 6100
Nationality, 5120
Natural gas, 6300
Natural resources, 4300; 6200–6299; 6300
Naturalization, 5300
Navigation, 6615
Navy, 8110
Negro, 5120; 9600
Neo-natal deaths, 5212
Neurology, 7180
Nomenclatures for statistical classification, 3200–3299; see also the particular subject field.
Notes, circulation of, 6720
Nursing, 7140
Nutrition, 9800

O

Obstetrics, 7180
Occupation of foreign territory, 8220
Occupational:
Classifications, 3240; 1700; 9000
Deaths, 5212; 9400
Disease, 9400
Distribution of population, 5110
Hazards, 9400
Income, 9200
Nomenclature, 9200; 3240
Statistics, 9200
Office procedures, 1100
Oil wells, 6300
Old-age assistance, 9951
Ophthalmology, 7180
Organization:
Charts, 1100; see also the particular subject field.
Of statistical service agencies, 3030
Orphan asylums, 9951
Orphans, 5110; 9951
Otology, 7180
Overcrowding, 9900
Ownership, financial, 6721

P

Parks, 6800; 4300
Pathology, 7180
Pediatrics, 7180
Penal institutions, 8330
Penitentiaries, 8330
Pensions and pension systems, 9952
Periodical publications with statistical series, 1420
Periodogram analysis, 3140
Personal identity, 5250
Personality and intelligence testing, 7200
Personnel:
General, 1700
Management, 1700
Standards for statisticians, 3250
Petroleum, 6300
Pharmacology, 7180
Physical examination programs, 7110
Physical growth, 7180
Physicians, 7100

U

V

W

Y

Z

APENDICE 2

ESQUEMA DE CLASIFICACION POR MATERIAS ESTADISTICAS

(Nota: Se ha obtenido una gran economía en los trabajos del Instituto Interamericano de Estadística con la aplicación de un solo esquema general de clasificación por materias en la administración de la Oficina Permanente. La clasificación que sigue se emplea actualmente en los Archivos de Fuentes Estadísticas y para toda otra clasificación de carácter bibliográfico, en los Archivos Biográficos del Personal Estadístico, la revista trimestral (en su sección bibliográfica), los archivos de correspondencia, y los libros de contabilidad.

Desea el Instituto que pueda establecerse una clasificación por materias estadísticas, sobre una amplia base mundial, puesto que la generalización de su empleo facilitaría grandemente el archivo y el manejo eficiente del material, informaciones y datos intercambiados por las organizaciones internacionales y los organismos estadísticos nacionales. El Instituto espera continuar empleando la presente clasificación hasta que un sistema de similar naturaleza surja como estandard mundial.)

GRUPOS DE PRIMER ORDEN (9)

1000–1999 GENERAL.

2000–2999 CENSOS.

3000–3999 CIENCIA ESTADISTICA.

4000–4999 GEOGRAFIA.

5000–5999 DEMOGRAFIA.

6000–6999 ECONOMIA.

7000–7999 ESTADISTICAS DE SALUBRIDAD Y CULTURA.

8000–8999 ESTADISTICAS GUBERNAMENTALES Y POLITICAS.

9000–9999 CONDICIONES DE TRABAJO Y DE VIDA.

GRUPOS DE SEGUNDO ORDEN (60)

1000–1999 GENERAL.
- 1000 General.
- 1100 Controles administrativos.
- 1200 Bibliografía.
- 1300 Congresos y conferencias.
- 1400 Compilaciones de datos.
- 1500 Archivos y catálogos.
- 1600 Microfilm y conservación de los datos.
- 1700 Personal.
- 1800 Métodos de publicación.
- 1900 Libros de referencia, general.

2000–2999 CENSOS.
- 2000 General.
- 2100 Métodos y procedimientos.
- 2200 Resultados.

3000–3999 CIENCIA ESTADISTICA.
- 3000 General.
- 3100 Métodos y técnicas.
- 3200 Estandards.
- 3300 Materiales y métodos de enseñanza.

4000–4999 GEOGRAFIA.
- 4000 General.
- 4100 Area y características geográficas.
- 4200 Clima y condiciones del tiempo.
- 4300 Recursos naturales, estudios generales.

5000–5999 DEMOGRAFIA.
- 5000 General.
- 5100 Población.
- 5200 Estadística vital y registros de estadística vital.
- 5300 Migración

6000–6999 ECONOMIA.
- 6000 General.
- 6100 Renta nacional y riqueza nacional.
- 6200 Agricultura, ganadería, silvicultura, pesca.
- 6300 Minería e industrias extractivas.
- 6400 Industria manufacturera.
- 6500 Edificación y otras construcciones.
- 6600 Transportes y comunicaciones.

ECONOMIA—cont.
- 6700 Comercio (excepto comercio exterior).
- 6800 Servicios personales y de recreaciones.
- 6900 Comercio exterior.

7000–7999 ESTADISTICAS DE SALUBRIDAD Y CULTURA.
- 7000 General.
- 7100 Salubridad.
- 7200 Educación.
- 7300 Artes, ciencias y servicios de información pública.
- 7400 Religión.

8000–8999 ESTADISTICAS GUBERNAMENTALES Y POLITICAS.
- 8000 Administración pública y empresas públicas, general.
- 8100 Defensa y guerra.
- 8200 Relaciones internacionales.
- 8300 Justicia, delincuencia y crimen.
- 8400 Leyes y legislación.
- 8500 Planificación e investigación—nacional, regional y local.
- 8600 Elecciones, partidos políticos y encuestas de la opinión pública.
- 8700 Sistemas políticos.
- 8800 Finanzas públicas, impuestos y contribuciones—nacionales, estatales, municipales y locales.

9000–9999 CONDICIONES DEL TRABAJO Y DE VIDA.
- 9000 General.
- 9100 Condiciones del trabajo, general.
- 9200 Ocupación y salarios.
- 9300 Relaciones del trabajo
- 9400 Accidentes industriales y enfermedades provenientes de la ocupación.
- 9500 Costo de la vida.
- 9600 Condiciones de la vida, general.
- 9700 Condiciones de la vida familiar.
- 9800 Consumo de alimentos y nutrición.
- 9900 Condiciones de la vivienda.
- 9950 Seguridad social.

ESQUEMA COMPLETO

(9 grupos de primer orden, 60 de segundo, 115 de tercero y 24 de cuarto)

1000–1999 GENERAL

(Incluye: Materiales de referencia general, o pertenecientes a más de una clasificación principal de materias.)

1000 General.
1100 Controles administrativos. (Incluye: Contabilidad, presupuesto, análisis de costos, esquemas de inter-relación administrativa, procedimientos internos de las oficinas, control de inventario, esquemas de organización, control de la producción.)
1200 Bibliografía. (De carácter general.)
1300 Congresos y conferencias. (De carácter general.)
Compilaciones de datos. (De carácter general y esencialmente estadísticos. Excluye censos. Para compilaciones de datos relativos a materias específicas, véase las respectivas clasificaciones.)
1400 ——. General. (Publicación única de datos estadísticos compilados.)
1410 ——. Anuarios. (Incluye: Todos los anuarios generales de estadísticas nacionales.)
1420 ——. Publicaciones periódicas. (Contiene las más importantes series estadísticas de actualidad.)
1500 Achivo y catalogación. (Incluye: Manuales de instrucciones y esquemas de clasificación para archivar y catalogar.)
1600 Microfilm y conservación de los datos. (Incluye: Microcopias.)
1700 Personal. (Incluye: Clasificación de ocupaciones, moral y psicología, administración del personal, ascenso, supervigilancia e instrucción, nóminas de personal técnico y otro personal, archivos biográficos.)
1800 Métodos de publicación. (Incluye: Hectografía, multilitografía, etc. Excluye: Microfilm.)
Libros de referencia, general.
1900 ——. General.
1910 ——. Anuarios y almanaques.
1920 ——. Diccionarios y glosarios.
1930 ——. Directorios de personas y organizaciones.
1940 ——. Enciclopedias.
1950 ——. Guías y manuales de países y localidades.
1960 ——. Manuales de redacción, gramáticas, etc.

2000–2999 CENSOS

(Un conjunto de todo el material oficial sobre los censos de cada país. Cada partida aparece también en su respectiva clasificación de materia específica.)

2000 General.
Métodos y procedimientos. (Incluye: Formularios, instrucciones, leyes, cédulas, sistemas de muestras y otras técnicas en lo relacionado con censos.)
2100 ——. General y población.
2110 ——. Agricultura.
2120 ——. Edificios, bienes raíces, inversiones.
2130 ——. Educación.
2140 ——. Electorales.
2150 ——. Industria, manufactura, negocios en general.
2160 ——. Instituciones.
2170 ——. Minería.
2180 ——. Transportes y comunicaciones.
Resultados.
2200 ——. General y de población. (Incluye: Censos de ocupación y desocupación.)
2210 ——. Agricultura. (Incluye: Censos de drenaje e irrigación, ganadería, etc.)
2220 ——. Edificios, bienes raíces, inversiones. (Incluye: Censos de viviendas, hipotecas y bienes muebles.)
2230 ——. Educación. (Incluye: Censos escolares.)

CENSOS—cont.
2240 ——. Electorales.
2250 ——. Industria, manufactura, negocios en general. (Incluye: Censos comerciales.)
2260 ——. Instituciones. (Incluye: Censos religiosos.)
2270 ——. Minería.
2280 ——. Transportes y comunicaciones.

3000–3999 CIENCIA ESTADISTICA

(Principalmente sobre la metodología estadística que puede aplicarse a cualquier clasificación.)

General.
3000 ——. General.
3005 ——. Diccionarios y glosarios de términos estadísticos.
3010 ——. Bibliografía.
3020 ——. Congresos y conferencias. (Principalmente estadísticos.)
3030 ——. Organización de servicios estadísticos, organismos y sociedades—material descriptivo y leyes.
3040 ——. Publicaciones periódicas. (Conteniendo artículos importantes sobre metodología estadística.)
Métodos y técnicas. (Excluye: Nomenclaturas para estandardizar clasificaciones estadísticas. Véase también 3300.)
3100 ——. General. (Incluye: Cómputos, códigos, definiciones, ecuaciones, formularios, gráficos, instrucciones.)
3105 ——. Análisis de datos.
Censos—véase 2100–2199.
3110 ——. Recopilación y elaboración de datos. (Incluye: Manuales de instrucción sobre la materia, aplicación de tarjetas perforadas, etc.)
3120 ——. Números índices.
3130 ——. Cartografía.
3140 ——. Estadística matemática. (Incluye: Probabilidad matemática.)
3150 ——. Método de muestras.
3160 ——. Control de calidad.
Estandards. (Medidas para obtener uniformidad, incluyendo nomenclaturas u otras listas y esquemas para estandardizar clasificaciones estadísticas. Un conjunto de todo el material sobre los estandards estadísticos. Cada partida aparece también en su respectiva clasificación de materia específica.)
3200 ——. General. (Incluye: Bibliografías sobre nomenclaturas.)
3205 ——. Balanza de pagos.
3210 ——. Clasificación de comercio exterior. (Incluye: Clasificaciones de mercancías.)
3220 ——. Clasificación industrial.
3230 ——. Clasificación de mortalidad y morbilidad.
3240 ——. Clasificación de ocupaciones (población activa).
3250 ——. Personal. (Para funciones estadísticas.)
3260 ——. Presentación gráfica y tabular.
3270 ——. Pesos y medidas.
Materiales y métodos de enseñanza.
3300 ——. General.
3310 ——. Becas y asignaciones de viaje.
3320 ——. Programas y cursos de estudio.
3330 ——. Textos.

4000–4999 GEOGRAFIA

4000 General. (Incluye: Atlas y mapas.)
4100 Area y características geográficas. (Incluye: Divisiones políticas.)
4200 Clima y condiciones del tiempo. (Incluye: Presión barométrica, meteorología, precipitación, temperatura, mareas, velocidad del viento.)
4300 Recursos naturales, estudios generales. (Incluye: Reservas forestales, reservas minerales, recursos hidrológicos, caza mayor y menor, tierras para parques nacionales, etc.)

5000–5999 DEMOGRAFIA

5000 General. (Incluye: Biometría y estudios biométricos, organizaciones demográficas, biología humana.)
Población.
5100 ——. General. (Incluye: Estimaciones de la población, política de población.)
5110 ——. Características. (Incluye: Características de la familia, grupos por edades, centros de población, densidad, empleo y ocupaciones, crecimiento y tendencias, analfabetismo, estado civil, sexo, urbanos y rurales.)
5120 ——. Raza y nacionalidad. (Incluye: Extranjeros residentes, los nacidos en el exterior, indios, judíos, negros, etc.)
Estadística vital y registros de estadística vital.
——. General.
5200 ——. ——. General.
5201 ——. ——. Bibliografía.
5202 ——. ——. Congresos y conferencias.
5203 ——. ——. Organizaciones y agencias—material descriptivo y leyes.
——. Nacimientos, mortalidad, nacidos muertos.
5210 ——. ——. General.
5211 ——. ——. Nacimientos. (Incluye: Control de la natalidad, tasas de natalidad, lugar de nacimiento, fertilidad, fecundidad, partería, tasa de reproducción, natalidad retrasada, esterilidad.)
5212 ——. ——. Defunciones. (Incluye: Tasas de mortalidad, clasificación de mortalidad, Lista Internacional de las Causas de Muerte, listas selectivas de causas de muerte, causas especiales de muertes.)
5213 ——. ——. Nacidos muertos.
5220 ——. Tablas de vida. (Incluye: Esperanza de vida, longevidad.)
——. Matrimonios y divorcios.
5230 ——. ——. General.
5231 ——. ——. Matrimonios. (Incluye: Tasas de matrimonios.)
5232 ——. ——. Divorcios. (Incluye: Tasas de divorcios.)
5240 ——. Métodos y procedimientos. (Incluye: Certificados, formularios, códigos, procedimientos de índices, manuales de instrucción, etc.)
5250 ——. Registro y prácticas de registro. (Incluye: Adopciones, certificación, registros completos, corrección de los datos registrados, uso legal de los datos, legitimación, ilegitimidad, identidad personal—incluye impresiones digitales—práctica sobre muertes presuntas, insuficiencia de las inscripciones, etc.)
Migración.
5300 ——. General. (Incluye: Emigración, immigración y naturalización; colonización, colonización dirigida y recolonización de grupos minoritarios.)
5310 ——. Turismo.

6000–6999 ECONOMIA

General.
6000 ——. General. (Incluye: Depresiones, geografía económica, organizaciones económicas, pronósticos, participación en los beneficios, poder adquisitivo, teorías económicas.)
6010 ——. Precios y números índices. (De carácter general y cubriendo varios campos de materias. Incluye: Indices generales de actividad económica, inflación, control de precios. Para precios al por menor e índices en relación con el costo de la vida, véase 9500. Para otros precios y números índices en determinados campos, véase las respectivas clasificaciones. Para métodos y técnicas de números índices, véase 3120.)
6020 ——. Producción, consumo, existencias. (De carácter general y cubriendo varios campos de materias.)
6100 Renta nacional y riqueza nacional. (Incluye: Estimaciones globales del producto bruto.)
Agricultura, ganadería, silvicultura, pesca.
6200 ——. General. (Incluye: Organizaciones agrícolas; escuelas y servicios educacionales en general; características de haciendas y granjas; rehabilitación de la tierra, registro o catastro, posesión o propiedad, utilización; subvenciones; abastecimiento mundial de alimentos.)

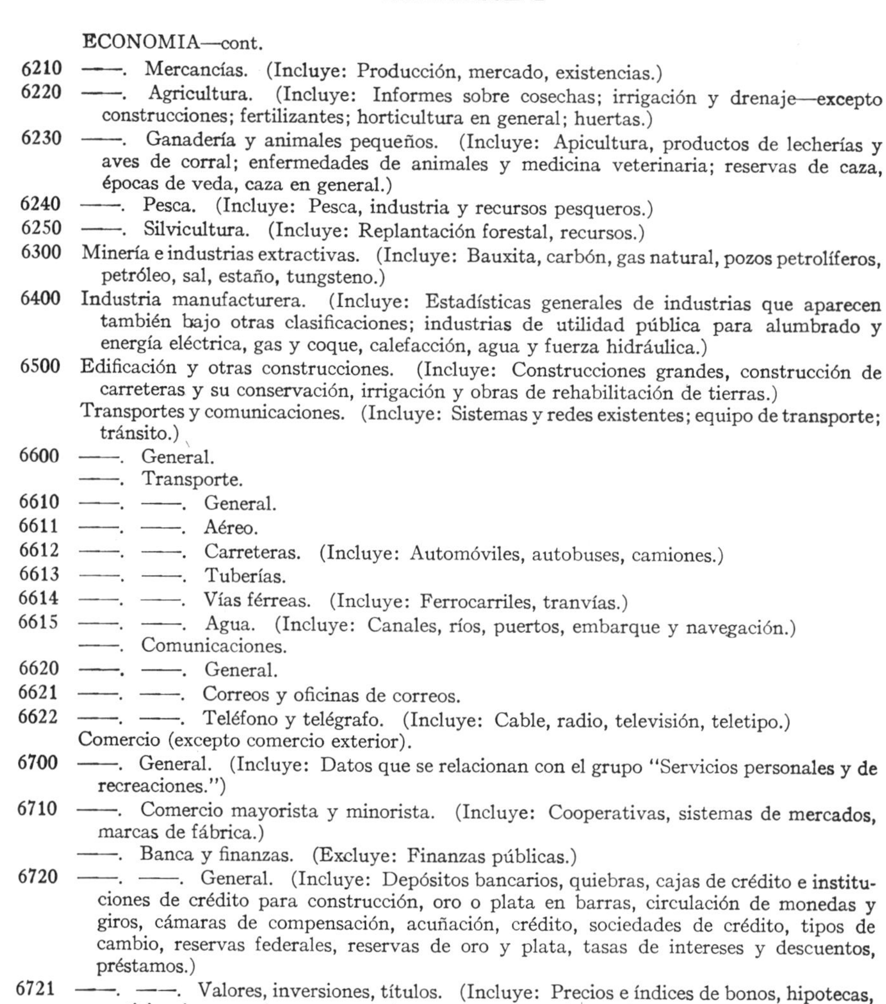

ECONOMIA—cont.

6210 ——. Mercancías. (Incluye: Producción, mercado, existencias.)

6220 ——. Agricultura. (Incluye: Informes sobre cosechas; irrigación y drenaje—excepto construcciones; fertilizantes; horticultura en general; huertas.)

6230 ——. Ganadería y animales pequeños. (Incluye: Apicultura, productos de lecherías y aves de corral; enfermedades de animales y medicina veterinaria; reservas de caza, épocas de veda, caza en general.)

6240 ——. Pesca. (Incluye: Pesca, industria y recursos pesqueros.)

6250 ——. Silvicultura. (Incluye: Replantación forestal, recursos.)

6300 Minería e industrias extractivas. (Incluye: Bauxita, carbón, gas natural, pozos petrolíferos, petróleo, sal, estaño, tungsteno.)

6400 Industria manufacturera. (Incluye: Estadísticas generales de industrias que aparecen también bajo otras clasificaciones; industrias de utilidad pública para alumbrado y energía eléctrica, gas y coque, calefacción, agua y fuerza hidráulica.)

6500 Edificación y otras construcciones. (Incluye: Construcciones grandes, construcción de carreteras y su conservación, irrigación y obras de rehabilitación de tierras.)

Transportes y comunicaciones. (Incluye: Sistemas y redes existentes; equipo de transporte; tránsito.)

6600 ——. General.

——. Transporte.

6610 ——. ——. General.

6611 ——. ——. Aéreo.

6612 ——. ——. Carreteras. (Incluye: Automóviles, autobuses, camiones.)

6613 ——. ——. Tuberías.

6614 ——. ——. Vías férreas. (Incluye: Ferrocarriles, tranvías.)

6615 ——. ——. Agua. (Incluye: Canales, ríos, puertos, embarque y navegación.)

——. Comunicaciones.

6620 ——. ——. General.

6621 ——. ——. Correos y oficinas de correos.

6622 ——. ——. Teléfono y telégrafo. (Incluye: Cable, radio, televisión, teletipo.)

Comercio (excepto comercio exterior).

6700 ——. General. (Incluye: Datos que se relacionan con el grupo "Servicios personales y de recreaciones.")

6710 ——. Comercio mayorista y minorista. (Incluye: Cooperativas, sistemas de mercados, marcas de fábrica.)

——. Banca y finanzas. (Excluye: Finanzas públicas.)

6720 ——. ——. General. (Incluye: Depósitos bancarios, quiebras, cajas de crédito e instituciones de crédito para construcción, oro o plata en barras, circulación de monedas y giros, cámaras de compensación, acuñación, crédito, sociedades de crédito, tipos de cambio, reservas federales, reservas de oro y plata, tasas de intereses y descuentos, préstamos.)

6721 ——. ——. Valores, inversiones, títulos. (Incluye: Precios e índices de bonos, hipotecas, participaciones, acciones, etc., títulos y transacciones de propiedades e inmuebles; capital, ingresos, ganancias, pérdidas.)

6730 ——. Seguro comercial. (Excluye: Seguro social. Incluye: Enseñanza de la ciencia actuarial, anualidades, etc.)

6800 Servicios personales y de recreaciones. (Incluye: Provisión de vivienda, alimentación y bedida; limpieza y manutención de los bienes y propiedades raíces; servicios personales; servicio doméstico; diversiones, ejercicio, recreación y deportes, incluyendo locales de recreo y parques nacionales. Nótese que los datos que se relacionan con el grupo "Comercio" se incluyen en ese grupo.)

Comercio exterior.

6900 ——. General. (Incluye: Carteles; esquemas de clasificación; exportación e importación; acuerdos comerciales especiales tales como cláusulas de trueque, derechos o impuestos, naciones más favorecidas, puertos libres, licencias, tarifas o aranceles.)

6910 ——. Balanza de pagos internacionales.

Economía
6210

7000–7999 ESTADISTICAS DE SALUBRIDAD Y CULTURA

7000 General.
Salubridad.
——. General.
7100 ——. ——. General. (Incluye: Directivas médicas, educación de higiene.)
7101 ——. ——. Organizaciones. (Departamentos y organismos de salubridad—material descriptivo y leyes.)
7110 ——. Actividades de la salud pública. (Incluye: Epidemiología, higiene mental, programas de exámenes físicos, medicina preventiva, investigaciones sobre salud pública, saneamiento, vacunación.)
7120 ——. Dentistería.
7130 ——. Hospitales. (Incluye: Clínicas, clínicas privadas, instituciones para enfermos mentales, sanatorios, asilos y refugios, registros médicos hospitalarios.)
7140 ——. Enfermería.
7150 ——. Morbilidad. (Incluye: Enfermedades, códigos de morbilidad, intemperancia, alcoholismo, adictos a drogas.)
7160 ——. Accidentes, excepto los accidentes industriales. (Incluye: Definiciones, prevención y medidas de seguridad, accidentes de transporte, morbilidad y mortalidad por accidentes.)
7170 ——. Defectos y deformidades. (Incluye: Congénitos, traumáticos y defectos de los sentidos; enfermos mentales; personas incapacitadas.)
7180 ——. Ciencia médica. (Incluye: Culto e historia de la ciencia médica. Especifíquese la ciencia médica en particular, tal como: Anatomía, bacteriología, medicina clínica, dermatología, medicina de laboratorio (microscopia), neurología, obstetricia y ginecología, oftalmología, otología, laringología y rinología, patología, pediatría, farmacología, desarrollo físico, fisiología, cirugía.)
7190 ——. Medicina social y costo de la atención médica. (Incluye: Planes de salud social, atención médica, sistemas de pagos de servicios médicos.)
Educación.
7200 ——. General. (Incluye: Psicología educacional, analfabetismo, pruebas de inteligencia y personalidad, bibliotecas y museos, pedagogía, instrucción vocacional.)
7210 ——. Instituciones educacionales. (Incluye: Educación primaria, secundaria y universitaria.)
7220 ——. Cursos y programas de estudios.
7230 ——. Becas.
Artes, ciencias y servicios de información pública.
7300 ——. General. (Incluye: Arquitectura, astronomía, biología, botánica, química, eugenesia, geología, historia, literatura, matemáticas—excepto estadística matemática, música, física, psicología, sociedades científicas en estas ciencias, zoología.)
7310 ——. Antropología, antropometría, arqueología. (Incluye: Medidas antropométricas, índices cefálicos.)
7320 ——. Idioma. (Incluye: Artículos sobre ortografía.)
7330 ——. Servicios de información pública. (Incluye: La prensa.)
7400 Religión. (Incluye: Iglesias y otras congregaciones y actividades religiosas.)

8000–8999 ESTADISTICAS GUBERNAMENTALES Y POLITICAS

8000 Administración pública y empresas públicas, general. (Para actividades de las distintas empresas públicas en determinados campos, véase las respectivas clasificaciones. Incluye: Empleados públicos, planes generales de reorganización gubernativa.)
Defensa y guerra.
8100 ——. General. (Incluye: Defensa civil, préstamo y arrendamiento, servicio militar.)
8110 ——. Fuerzas armadas. (Especifíquese si es aérea, ejército o marina. Incluye: Mortalidad militar.)
Relaciones internacionales. (Incluye: Derecho internacional y organismos generales internacionales que no se dedican a una materia específica.)
8200 ——. General.
8210 ——. Convenios y tratados.
8220 ——. Gobierno militar. (Ocupación de tierras extranjeras.)
Justicia, delincuencia y crimen.

ESTADISTICAS GUBERNAMENTALES Y POLITICAS—cont.

8300 ——. General. (Incluye: Pena capital, sistema policial, prostitución, Corte Suprema.)
8310 ——. Estadísticas de las cortes. (Incluye: Estadísticas judiciales.)
8320 ——. Delincuencia juvenil.
8330 ——. Instituciones penales. (Incluye: Cárceles, penitenciarías, prisiones, reformatorios.)
8400 Leyes y legislación. (De carácter general. Para leyes y legislación sobre materias específicas, véase las respectivas clasificaciones.)
8500 Planificación e investigación—nacional, regional y local. (De carácter general. Por instituciones públicas o privadas. Incluye: Planes para el período de la postguerra.)
8600 Elecciones, partidos políticos y encuestas de la opinión pública.
8700 Sistemas políticos. (Incluye: Bolchevismo, comunismo, democracia, fascismo.)
8800 Finanzas públicas, impuestos y contribuciones—nacionales, estatales, municipales, locales. (Incluye: Ingresos y egresos, gastos para defensa, presupuesto y deuda, y toda forma de impuestos y contribuciones.)

9000–9999 CONDICIONES DEL TRABAJO Y DE VIDA

(Nótese que el material de este grupo puede cruzarse con los grupos principales de materias DEMOGRAFIA, ECONOMIA, ESTADISTICAS DE SALUBRIDAD . . . , ESTADISTICAS GUBERNAMENTALES . . .)

9000 General. (Incluye: Organizaciones del trabajo, sociología, psicología social y de las masas, investigaciones sociales.)
9100 Condiciones del trabajo, general. (Incluye: Estudios sobre la productividad del trabajo.)
9200 Ocupación y salarios. (Incluye: Horario de trabajo, tasa y monto de salarios y sueldos—incluyendo profesionales, gerentes y directores; asignaciones a las familias; población activa; ingresos por profesiones; índices de ocupación y salarios; mercado del trabajo; oferta y demanda de trabajo; nomenclatura de ocupaciones; desocupación.)
9300 Relaciones del trabajo. (Incluye: Sindicatos y gremios obreros, convenios colectivos y conflictos industriales; conciliación y arbitraje; inspección del trabajo.)
9400 Accidentes industriales y enfermedades provenientes de la ocupación. (Incluye: Medidas para la prevención de accidentes; fatiga, riesgos e higiene.)
9500 Costo de la vida. (Incluye: Precios al por menor e índices de los varios elementos que integran el costo de la vida; comparaciones internacionales e inter-regionales.)
9600 Condiciones de la vida, general. (Incluye: Condiciones generales de grupos económicos, sociales o raciales.)
9700 Condiciones de la vida familiar. (Incluye: Composición de la familia, distribución de los gastos, nivel y estandard de vida de la familia, ingresos y renta individual y de la familia, naturaleza y fuente de los ingresos.)
9800 Consumo de alimentos y nutrición.
9900 Condiciones de la vivienda. (Incluye: Tamaño, propiedad y tipo; equipos sanitarios y otros servicios; ocupación de la vivienda y congestionamiento; escala de alquileres; programas de viviendas de bajo costo; viviendas rurales.)
Seguridad social. (Incluye: Asistencia social, tanto pública como privada.)
9950 ——. General. (Incluye: Bolsas del trabajo, métodos de asistencia social, indemnización para trabajadores.)
9951 ——. Asistencia. (Incluye: Hospicios; asilos para ciegos; obras de caridad; beneficencia para niños; instituciones para necesitados; incapacitados; personas impedidas; ayuda para la maternidad, para ancianos, para huérfanos y para pobres; asistencia pública; ayuda para los desocupados.)
9952 ——. Seguros. (Incluye: Toda forma de seguro social, excluyendo el comercial; pensiones.)
9953 ——. Sistemas y programas de seguridad social.

INDICE ALFABETICO

Ciegos:
General, 7170
Asilos para, 9951
Ciencia:
General (pura), 7300–7399
Específica, véase las respectivas clasificaciones.
Actuarial (de seguros, incluyendo cursos y programas de enseñanza), 6730
Aplicada, véase las respectivas clasificaciones.
Bibliográfica, 7200
Doméstica, 9700
Estadística, 3000–3999
Matemática, 7300
Médica, 7180
Política, 8000–8999
Social (sociología), 9000
Cierres y huelgas, 9300
Circulación de monedas y giros, 6720
Cirugía, 7180
Cirugía dental, 7120
Ciudadanía, 5110
Clasificación:
Aduanera, 3210; 6900
Bibliográfica, 1500
De comercio exterior, 3210; 6900
De mercancías, 3210; 6900
De morbilidad, 3230; 7150
De mortalidad, 3230; 5212
De ocupaciones, 3240; 1700n; 9000
Estadística (incluyendo nomeclaturas), 3200–3299; véase también las respectivas clasificaciones.
Industrial, 3220; 6400
Profesional, 3240; 1700; 9000
Climatología, 4200
Clínicas (incluyendo clínicas de grupos particulares), 7130
Cloacas y sistemas de cloacas, 7110
Cobre, 6300
Códigos, véase clasificación
Colonización, 5300
Comercio exterior, 6900–6999
Comercio interior, 6700–6799
Compañías comerciales:
General, 6710
Específicas, véase las respectivas clasificaciones.
Compendios estadísticos, 1400–1499
Cómputos estadísticos, 3100–3199
Comunicaciones:
General, 6620–6629
Censos de, 2180; 2280; 6620
Comunismo, 8700
Confederaciones de trabajo, 9000; 9300
Conferencias, véase Congresos y conferencias.
Conflictos del trabajo, 9300
Congestionamiento (en viviendas), 9900
Congresos y conferencias:
General, 1300
Específicos, véase las respectivas clasificaciones.
Internacionales, 8200; véase también las respectivas clasificaciones.
Conservación de registros, 1600
Construcción:
General, 6500
Censos de, 2120; 2220; 6500
Consumo:
General, 6020
De productos determinados, véase las respectivas clasificaciones.
Cooperativas de, 6710
De productos alimenticios (nutrición de familias), 9800
Por familias (no alimenticio), 9700
Contabilidad (en general, incluyendo contabilidad de costos), 1100
Contribuciones (en general y de todas clases, incluyendo impuestos nacionales, estatales, municipales y locales, sobre herencias, ventas, etc.), 8800
Convenios:
General (incluyendo convenios políticos internacionales), 8210
Comerciales, 6900
Cooperativas:
Agrícolas, 6200
De consumo, 6710
De crédito, 6720
Coque, 6400
Corte suprema, 8300
Cortes, estadísticas de, 8310
Corral, aves de, 6230
Corrección, casas de, 8330
Correos:
Casas de (incluyendo estafetas y oficinas), 6621
Servicios de (administración), 8000
Cosechas, informes sobre, 6220
Costo de la vida (incluyendo índices), 9500
Costos:
Análisis y contabilidad de, 1100
De la atención médica, 7190; 9700
De la producción, 6400
Crecimiento:
Curvas de, 3100
Del niño, 7180
De la población, 5110
Medición del, 7180
Crédito (en general, incluyendo cajas, cooperativas y sociedades de crédito), 6720
Criminalidad, 8300–8399
Crisis económicas, 6000
Cuerpo de aviación, 8110
Cuestionarios (estadísticos), 3100

J

L

M